CIRCUITS

McGRAW-HILL SERIES ON THE FUNDAMENTALS OF ELECTRONIC SCIENCE

Consulting Editor
James F. Gibbons, Stanford University

GRINICH AND JACKSON Introduction to Integrated Circuits
SIEGMAN Introduction to Lasers and Masers
TUTTLE Circuits

CIRCUITS

David F. Tuttle, Jr.
Professor of Electrical Engineering
Stanford University

McGraw-Hill Book Company

New York St. Louis San Francisco Auckland Bogotá
Düsseldorf Johannesburg London Madrid Mexico
Montreal New Delhi Panama Paris São Paulo
Singapore Sydney Tokyo Toronto

CIRCUITS

1234567890 KPKP 783210987

This book was set in Times New Roman. The editors were Peter D. Nalle and Madelaine Eichberg; the designer was Joseph Gillians; the production supervisor was Leroy A. Young. The drawings were done by J & R Services, Inc. Kingsport Press, Inc. was printer and binder.

Library of Congress Cataloging in Publication Data

Tuttle, David F
Circuits.

(McGraw-Hill series on the fundamentals of electronic science)
Includes index.
1. Electric circuits. 2. Electric networks.
I. Title.
TK454.T87 621.319'2 76-22750
ISBN 0-07-065591-X

CONTENTS

PREFACE

Books on electric circuits and circuit analysis are legion. Another such textbook must therefore begin with an apologia. That this book has novelties, that it uses simple, relevant, very informative, easily reproduced physical demonstrations of the soundness of circuit models as accurate representations of reality; that it makes good use of the digital computer; that it is based on the perspective that comes only from experience (of both teacher and student)—these make up an apologia of sorts. But only in the text itself can the justification for such a book be found. (Chapter 1 sketches out its rationale, describes what the book tries to do, its method of attack, and the demands it places on the reader.) In particular, the approach to generality is *gradual*: from simple, easily understood (special, perhaps, but useful and informative) cases, cases that require only a *minimum* of background, through more elaborate examples eventually to fulldress discussions, but only when the student is ready. Students who work their way through it will have built a sound foundation in a basic practical discipline of electrical engineering.

The whole derives from the teaching opportunities offered me at Stanford University over many years, including the wholesome benefits of sabbatical years spent at other institutions with different viewpoints. Innumerable discussions with colleagues and students have contributed; in particular, certain ideas stem from Professor L. A. Manning. Especially does the book depend on my students (see Sec. 1.01).

Finally to those who helped perhaps not so much in technical matters but infinitely much with patience and encouragement and support, the other members of my family, Becky, Jacqueline, and Robert, are due thanks immeasurable.

David F. Tuttle, Jr.

Chapter **1**

INTRODUCTION

Begin at the beginning
and go on till you come to the end:
Then stop.

—The King of Hearts

At the start of a voyage one needs some sort of chart, the proposed course being plotted thereon. This first chapter plots *what* it is we are to study in this book, *why* it is interesting and important to engineers and engineering students, and *how* we are going to study it. Moreover, it will reveal some of the most important results to be obtained later. These will of course be unsubstantiated, without proof, mere "say so" (though often with some physical or intuitive justification). The intent is to help readers by giving them a clear idea of what lies ahead, hopefully without robbing them of the joy of discovery, study, and even of making proofs. And browsers will surely know, in a few pages, whether or not this book is what they want to invest many of their hours in.

1.01 BACKGROUND

There can be no better advice than "begin at the beginning." But the real beginning was long ago, when we first learned to put numbers together, and even before that, when we first touched something and felt a response. Now as we start this study of electric circuits we have already come a long way from the beginning, and already have an impressive background.

As to numbers (mathematics) we suppose a reasonable knowledge of the calculus; as to experimentation we suppose a reasonable background in general physics and particularly in electricity, and that not purely empirical but developed with the aid of the calculus. We start, that is, with a year or two of college work behind us—and that is long after the beginning. But it is a convenient place to start, and a suitable one for most engineering students, and so a good one for this book. Ability to program and use a digital computer and access to one is important: the computer need not be large, nor does the language matter. But beginning in Chap. 4 its use is important. Finally, contemporary (or previous) study of electronics is important for understanding many of the hypotheses taken as starting points and whence come some of the models used.

The book has evolved from teaching a "circuits" course over a number of years, for students at Stanford University. They have generally been third-year electrical engineering students; but there have been students, undergraduate and graduate, from other disciplines, whose dissimilar reactions and questions have been very valuable. The debt I owe these students is incalculable and unstatable, for they are responsible for many of the questions asked and answered in the book, the manner and detail thereof. It is they who have really written most of it.

To read this book with profit, one must of course have some interest in the subject. And one must be willing to work: the problems, of various sorts, are numerous (see Sec. 1.12), and a reasonable number thereof *must* be "done"; the "demonstrations" have to be seen and felt, and that requires some work too, which can profitably be shared by professor and student. An engineering "bent" is at least desirable. The book really deals with applied mathematics but the emphasis is on the adjective there, and a certain liking for the physical approach will be very helpful. In any event the book aims to develop a physical feel for "what goes," the perspective of a good engineer.

1.02 METHOD

Electric circuits are fundamental to electrical engineering—that we here postulate. We leave demonstration of this fact to experience, to come in later years, and other studies; we cannot do it here. Exactly what "electric circuits" are we shall define in due course. Definitions of terms, after all, are often profitably postponed until one knows what one is talking about; as long as one has a rough idea of the meaning to work with, one may better wait until more ideas and vocabulary come. The impatient reader need only consult the index to locate any precise definition.

Electric circuits are therefore deserving of careful study. Such a study must cover a number of ideas that are sufficiently different one from another that the study can easily become very disjointed. But they are closely related, too, so the study ought to be a smooth, continuous, and fascinating story, and we shall try to make it so.

Circuit theory depends heavily on mathematics. It has sufficiently precise premises (the models with which it deals) to permit lengthy and complicated mathematical developments. And such developments are often of great engineering value. But they also have a Lorelei character of which we must beware. It is engineering that we are to study here, and we must resist temptations (they will come) to develop theory for elegance's sake alone when it has no immediate engineering application. That it may some day have some important application is of course possible; only a fool will say categorically that some particular abstract mathematical derivation can never have engineering relevance. But our concern here is not with tomorrow's research developments; it is with today's circuit theory and its applications. We shall therefore concentrate on those aspects of circuit analysis that have demonstrated their permanent utility and are in wide use. Certain more advanced viewpoints and techniques we shall briefly mention, for the sake of perspective, but then leave (usually with regret) for detailed study elsewhere (a phrase we'll often use). Their detailed development here would be confusing indeed and completely obscure the forest with their trees.

Because its history is a long one, and because it builds on models (of physical systems) that are mathematically tractable, circuit theory is today a very imposing edifice. It is logically strong, and we need not fear climbing to the upper levels. But we must not forget that our principal concern is with *engineering*. Because the foundations are *models*, and not actual engineering devices or systems, there may be discrepancies between a prediction made by circuit theory from the twenty-third floor, and actual performance of the engineering artifact on the ground. Such discrepancies are extremely important to the engineer. It will not do to ignore them or to postpone or delegate consideration of them. Nor is it difficult to monitor the distance between our circuit-theoretical devices, as we make them, and "reality," the actual performance of corresponding physical, engineering circuits. Simple concrete realizations for most of our work are not hard to build. The cathode-ray oscilloscope will display their performance simply and lucidly. Moreover, the displays can be photographed and some are presented here, so that even in the pages of this book one can continually evaluate the discrepancies. This we shall do, and so estimate the physical validity of our results.

Looking at the still photographs of cathode-ray-tube faces in this book is of course no substitute for three-dimensional viewing of actual apparatus. Especially does it *not* replace watching the screen as controls on the apparatus are moved to vary parameter values. Readers of this book, especially instructors of classes, should make every effort to assemble the demonstrations and see for themselves. It is even better to arrange things so that each member of the class can handle the apparatus, "twiddle the knobs," and really see for himself. Most of the "hardware" is simple and readily available. The instruments necessary are conventional, with few exceptions. And the descriptions given will minimize the time needed to set up and adjust the demonstrations. The important thing is to *do* them! And in doing them, modifications for improvements, variations to give better insight, and additional demonstrations will tumble head over heels to suggest themselves. To *see* along with the mathematical development the physical development of a waveshape as a circuit actually operates on a signal is an experience of tremendous educational value. The value comes from the *feeling* of the effect; "tremendous," trite though it be, is accurate. The photographs here do what they can; keeping your feet on the ground by using actual instruments and seeing for yourself is up to you.

We shall devise models of engineering systems, models simple enough to analyze. After each mathematical analysis we shall observe the actual behavior of simple realizations of the models. We can then decide on the value of the analysis. If it seems valid, we add it to our repertoire, and may use it in *synthesis*, in the design of engineering systems. When we come to the end of this book we shall have a reasonably large repertoire. This is not the real end, of course; but we shall then have added to the background with which we started a solid foundation in electrical circuits on which to build the next stages of our educational and engineering work.

Had the book 10 times its pages, it could not be complete; it could not discuss "circuits" in all detail, for too much is known about them. So we shall compromise, distill, select; the best of yesterday's work, organized and compressed, constitutes most of the book. But there is an attempt to show where circuits seem to be going tomorrow. It is incumbent on us to prepare ourselves for reading other books, to fill lacunae here that may some day become very interesting, and to scan and to study the flood of journals of today and tomorrow. In short, one objective of the book is to teach you "how to read."

We shall try to maintain standards of literary quality. The jargon of technology proliferates without restraint; much of it is nonsense, inaccurate, of little value. But some technical slang is colorful, descriptive, precise, and concise, and a valuable addition to our language. This we accept and use.

1.03 SUBJECT

We have some idea already of what electric circuits are. But some clarification is necessary here.

Communication (the transmission of information) and control (of energy) constitute most of what electrical engineering is about. And from a few moderately complicated systems, of either type, we could easily obtain examples enough to develop all the circuit theory we may need. These could be the backbone of our story, the skeleton of this book. To provide this, and to keep us close to engineering reality, we begin by looking at three such systems.

The first is a long-distance telephone connection, say between California and France, some 10,000 km apart. To connect a telephone in San Francisco with one in Marseille is today a routine matter; it takes only seconds to establish the connection (by machine switching almost entirely), and the quality thereof is excellent. But the complexity of the system involved is great enough to defy description here. Any analysis thereof is beyond us—and in fact probably never was nor ever will be made. The design and construction work had to be limited to subsystems, so made that when connected together (in unpredictable ways, according to demand), they would operate well as a large system. The elementary diagrams of Fig. 1.03-A suggest this structure. The blocks represent local apparatus; the connecting lines represent the transmission media between them.

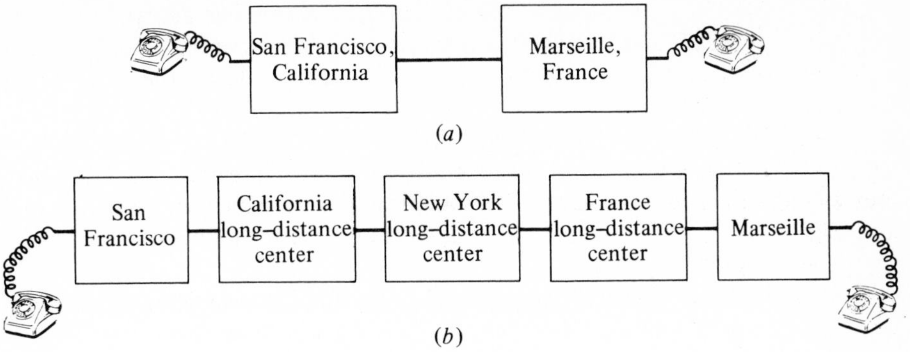

FIG. 1.03-A

We need not be expert communication engineers to know that switching centers of some sort must be provided, as in Fig. 1.03-A*b*, and that the transmission systems that connect them may be quite different one from another. From the telephone subscriber's instrument to the local central office the connection is probably a pair of insulated wires tightly packed with many others into a cable. From the local central office to the nearest long-distance switching center the connection is probably made by a "short-haul" multiplex system that uses wires in another type of cable and handles a large number of different conversations simultaneously. The transcontinental link will be a more complicated multiplex system that uses coaxial cable or microwave radio links. The transatlantic link may be either a submarine cable system or a radio system that uses an earth satellite as a relay station. Similar systems will complete the connection in France.

We need not remark further on the complexity of the complete system. But we can make a partial list of the principal kinds of subsystems that must make it up. There must be *amplifiers*, to compensate for the attenuation that signals experience as they travel over the transmission systems. There must be *modulators* and *demodulators*, samplers, coders and decoders, and devices for combining many signal channels at one end of a transmission system and separating them at the other. (The high cost of long-distance systems demands that they be efficiently and economically used, by serving many customers simultaneously.) Both cable and radio multiplex systems require transmitters and receivers at their terminals. Other subsystems provide the all-important power (energy) for transmission, to switch calls properly, and record enough information about calls made that toll charges can be calculated and collected. Sufficiently broken down into small blocks, every one of these subsystems can give us circuits (and devices) to study in wide variety.

For a second example, we consider the problem of precisely controlling the position of a massive load (a ship's rudder for example). Moving the load requires much power, but the command signal is weak. Not only must amplification be provided but some insurance that the load will actually obey the com-

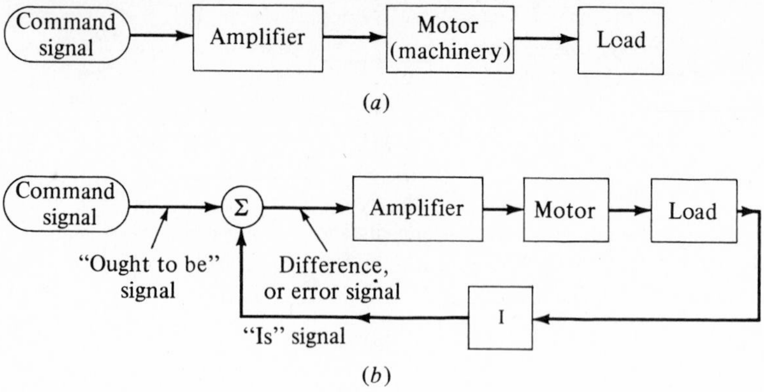

FIG. 1.03-B

mands. Directing a machine tool to cut and shape a steel block according to a previous pattern, steering a large ship, pointing a 60-ton antenna toward the satellite of the transatlantic radio link above (or making it follow a rapidly moving radar target to control air traffic, or a slowly moving stellar object if a telescope)—those illustrate such problems. Figure 1.03-B*a* shows elements of such a control system. The arrows indicate an important difference from the telephone connection: the signal transmission here is in one direction only. The command signal comes from a recorded pattern, from a prescribed course, from carefully selected spatial coordinates and observations, or from a hand-varied control—or perhaps it comes over a telephone line from elsewhere. It is amplified and applied to machinery that is appropriate for moving the load in translation or in rotation about various axes.

Accurate motion is not likely to result. The position of the (heavy) load is probably not going to be close enough to that commanded; error will exist. But long ago someone brilliantly conceived that if this *error* could be measured, *it* could be used to make the motion accurate. This is *feedback* control: to compare what *is* with what *ought to be*, and to use the difference for the actual command. We supply measuring apparatus (I in Fig. 1.03-B*b*) to convert the actual load position into an "is" signal that is consonant with the command signal. It is now the comparing device Σ that produces, in the error, the command that actually drives the machinery. If the system is properly designed and adjusted, the machinery will presumably operate to move the load until the error is reduced to negligible size. For as long as there is error, there will be torque (or force) to act on the load to reduce the error.

Feedback, be it carefully noted, raises new problems. For the closed loop is much like an Ozark hoopsnake once its tail is in its mouth: it can roll down a hill in a fashion quite different from its normal motion. Improperly designed or adjusted, a closed-loop system can circulate and amplify some random noise or disturbance into large, annoying, and even dangerous (usually oscillatory) motion of the load. So the design of the various blocks in the system is not

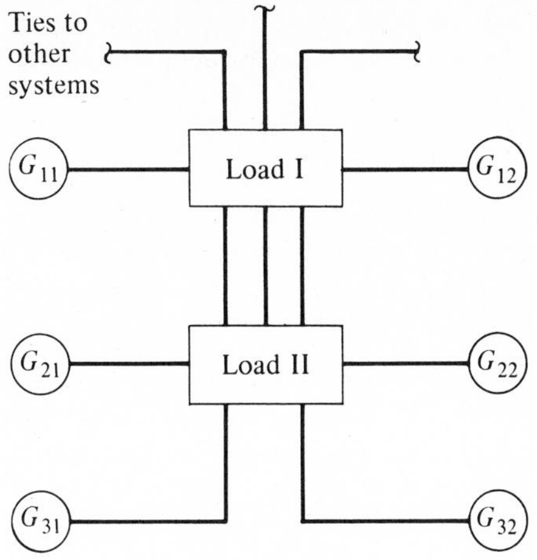

FIG. 1.03-C

necessarily easy; it poses problems relevant to circuit theory, fascinating problems that we shall eventually touch on (Sec. 6.25). Sometimes, incidentally, one may design a closed-loop system deliberately to oscillate or vibrate "on its own," without any command signal. Then the system is a *generator* of motion, or of electrical signals (an oscillator) if the machinery and load are replaced by appropriate apparatus.

Both examples assume the existence of adequate supplies of power, probably electrical, and are usually taken for granted. In a third example we shall sketch a system for providing such power. It starts with the burning of coal or fuel oil or gas, or in a nuclear reaction, or from the potential energy of falling water. The energy is used to drive turbines to turn electrical generators. It remains to transmit the electrical power to the user and to convert it to a form (voltage) appropriate for use: 6 kV alternating at 25 Hz, 115 V alternating at 60 Hz, or 24 V direct (constant). Figure 1.03-C shows the rudiments of part of a large power system, but gives no idea of its true complexity. Each generating plant G has energy sources, prime movers, and electrical generators. But these require auxiliary apparatus in profusion: fuel-oil pumps, lubricating-oil pumps, compressors for air to operate switches and protective devices, auxiliary power sources, measuring apparatus, relaying apparatus, communications apparatus, etc. Transmission lines connect distant generating plants to loads, to sections of cities for example, with customers whose demands range from a few lamps and home appliances to the elevator motors and air-conditioning apparatus in large buildings, to the machinery of industry, and to the many large motors of electrical railways. Other transmission lines connect (tie) geographically distant systems together so that power may be generated and distributed efficiently and reliably. The problems of designing and operating such a system: determining what distribution of power production between generators is most efficient, switching generators in and out to obtain such efficient operation, handling changing demands, guarding against instabilities, and the service interruptions they may cause (for it is a feedback system, and a very complicated one); these problems are tremendous.

But that is enough of examples. These three alone have in them, in the details of their blocks once opened up and subdivided into other blocks small enough for us to comprehend, electric circuits galore. They also give one at least a general idea of the rôle engineering plays in society, of the importance of systems built on circuits in the day-to-day "operation" of the world.

Circuit theory concerns itself not only with the purely electrical parts of such systems, but with the electromechanical and mechanical and other parts also; by analogy, the models by which they are described often have precisely the same mathematical forms. So simply to choose where to start is itself a bewildering problem.

1.04 A START

We have to go down to some simple block, simple enough for us to begin with. But it should not be so simple that it has no engineering significance. Somewhat arbitrarily we shall choose an *amplifier* box. Amplifiers are obviously prolific and important in the first two examples above, and are surely used in the communication and control apparatus necessary in the third. Ubiquitous, they must be important; some at least must be simple. An amplifier block (Fig. 1.04-A) receives signals and power; its output signals are stronger, amplified, and more useful. It need not be entirely electrical; hydraulic, pneumatic, and rotating electric machine amplifiers are useful in control systems. But we are interested here in amplifiers that are based on electronic devices. These are, broadly speaking, half electronic devices dependent on phenomena that occur in semiconductors (or in a vacuum) and half *circuits.* The circuits are ideal for us to study. They are necessary for utilizing the devices, they are vital to system operation, and they are prolific in engineering. Some simple and some complicated, they are exactly what we seek. Later we may look into other blocks. But the amplifier, in its circuit parts, is a good place to start.

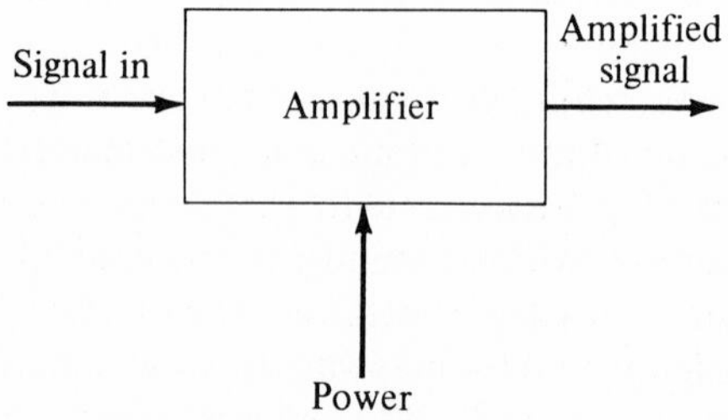

FIG. 1.04-A

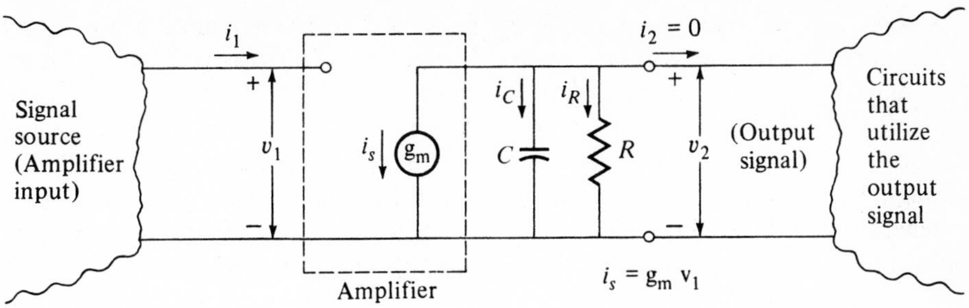

FIG. 1.05-A

1.05 A MODEL

Even a small amplifier that contains but one transistor is a complicated affair. But by concentrating on the most important behavior, we can devise a useful model to express that behavior. Such a model will be simple enough to analyze. And it will not be so unlike the real amplifier that its behavior will differ radically. Analysis of the simple model will be both practical and useful.

It is not our province to discuss the mechanism of transport of charge within semiconducting materials (or in vacua); nor is it for us to consider the design and construction of electron devices that are based thereon. We simply state a result of such studies in electronics. Many small (one-device) amplifiers can be satisfactorily modeled as in Fig. 1.05-A.

The transistor (or vacuum tube) is modeled by the dashed box. (Some part at least of the C and R elements comes also from the electron device.) The solid lines now represent wires that carry electric current. They connect the elements g_m, C, and R to make a *circuit* or *network*. One peculiar property of the transistor model is that it draws no current at its input terminals: $i_1 = 0$. A second property is that the (signal) voltage at the input, v_1, *controls* the generator of current labeled g_m: $i_s = g_m v_1$. It is from this generator that the energy of the amplified signal comes in the model, so that g_m is an *active*, not a *passive*, element of the actual network. It represents the electronic mechanism that controls the flow of energy from the power source of Fig. 1.04-A, absent as such in Fig. 1.05-A.

The currents and voltages in the model are not the actual ones of the amplifier. They represent the *dynamic* (time-varying, and most interesting and useful) parts thereof; the static parts have been subtracted out. Our i's and v's are *increments* only, about the quiescent values that are important in making the electron device operate but are uninteresting as signals; it is the *increments* that carry the signal information. We shall not discuss how the signal is introduced, combined with the static parts and later removed. Our interest is only in the dynamic parts, and that is all that the model displays.

The capacitor C and resistor R are passive elements of the network. Part of C represents a parasitic effect in the transistor (tube) that is ideally absent; this may be the major part or even all of C, or it may be insignificant. Similarly, R is in part due to the electron device, but is chiefly supplied by a resistor external thereto. The elements R and C constitute the *load* into which the transistor operates, across which the output signal, the voltage v_2, is developed. Their effect on the signal's size and form is something we are about to discuss. In this model the next part of the system utilizes the output signal, also without drawing any current ($i_2 = 0$).

We assume that linear laws govern the capacitor and resistor. This postulate implies that the currents and voltages remain *small*, for the mechanism of operation of the devices usually becomes nonlinear as signal size increases. Hence what "small" means is a matter for engineering judgment to decide; it depends on the situation. It is surprising how accurate these linear models can be, however!

For the capacitor, characterized by its capacitance

$$C = \frac{q}{v} \qquad \text{(farads, F)} \tag{1.05-1}$$

in which q (coulombs, C) measures its electrical charge and v its voltage, we have

$$q = Cv_2 \tag{1.05-2}$$

and

$$\frac{dq}{dt} = C\frac{dv_2}{dt} = i_C \qquad \text{(amperes, A).} \tag{1.05-3}$$

And for the resistor, by Ohm's law,†

$$v = Ri \tag{1.05-4}$$

or

$$i = Gv, \tag{1.05-5}$$

in which $v = v_2$ and $i = i_R$, its voltage and current. The resistance R is of course measured in ohms (Ω); its reciprocal, $G = 1/R$, is measured in mhos (or Siemens).

The linear nature of these laws implies again that the currents and voltages, as well as their rates of change, remain small. Note also the tacit assumption

† The history of the experimental and theoretical development of the basic laws, which we merely postulate, is an important part of circuit analysis but beyond our reach in this book; there are starting points for its study in Appendix B.

that effects can be localized (lumped), that spatial coordinates (x, y, z) are not relevant, and that no partial derivatives are necessary. We also assume that our element values remain constant as time goes on; that is, g_m (mhos), R, and C do not change. (Section 4.02 will discuss these postulates in more detail.)

Because the mathematical model of one of the network elements involves a rate of change (a derivative), we have to solve a differential equation if we are to find the output signal $v_2(t)$ that a given input signal $v_1(t)$ produces. The differential equation comes from Kirchhoff's law:

$$i_s + i_C + i_R = 0 \tag{1.05-6}$$

and is

$$g_m v_1 + C\frac{dv_2}{dt} + Gv_2 = 0. \tag{1.05-7}$$

We must of course know the value of v_2 at some starting instant of time and we must know the input signal $v_1(t)$ thereafter. (This initial value was established by the circuit's past history, and v_1 will determine its future history.)

There we have a "circuit" or network, and a problem in network analysis: to find $v_2(t)$ after the starting instant. It is a simple one, a practical and useful one, and a good place to start.

1.06 MODELS

The model of Sec. 1.05 has there two different forms. One is *physical*, a schematic diagram that describes the physical network. Redrawn to emphasize the network part over the electronic part, it is Fig. 1.06-A. The notational change in direction of current source (and of signal in the equation below) is merely for neatness. The other form is *mathematical*, a differential equation rewritten to fit Fig. 1.06-A. It is

$$\begin{aligned} E[v] = C\frac{dv}{dt} + Gv - i_F(t) = 0,& \\ v(0) = V_0,& \end{aligned} \tag{1.06-1}$$

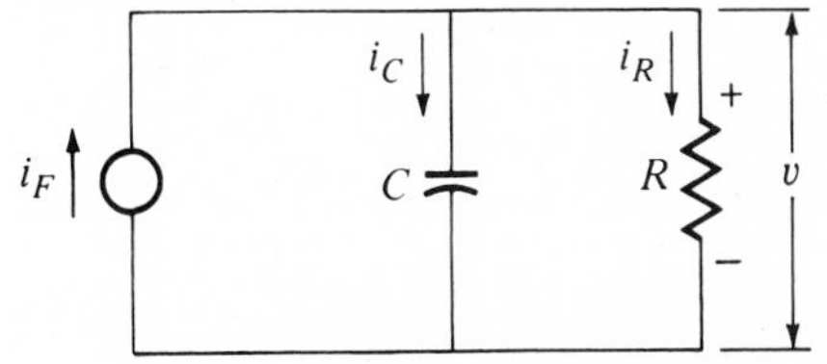

FIG. 1.06-A

in which E stands for equation, and the operational notation $E[v]$ emphasizes that it is an equation in which v is the unknown. The necessary starting or initial condition (we shall call $t = 0$ the starting time) is labeled V_0.

We shall set up our models generally in both physical and mathematical forms. They are equivalent, but it is often very helpful to have both. The physical form is close to the system that the model represents; the mathematical form is what we shall find convenient for the work of analysis.

The accuracy with which the model represents the system is a matter that requires discussion in every real problem. The modeling process inevitably introduces error: its aim is to set out those effects in the actual system that are important and to discard those that are not. An intractable (because complicated, because real) system is replaced by a tractable (because simplified) model. The simpler the model (and the easier its analysis), the less realistic it is; the more accurate and realistic the model, the more difficult and complicated and expensive is its analysis. Good engineering requires a sense of proportion, an ability to judge and to decide on a practical point of balance between these tendencies—an ability to balance the amount of information to be gained (complexity of model) against the cost thereof (complexity of analysis). Where the balance is depends, of course, on the particular job at hand. The essence of engineering judgment is this decision. It is unfortunate that a book can point this out again and again, but it can only *talk* about it. Practical experience alone can *give* it to you.

1.07 ANALYSIS AND SYNTHESIS

Once a model is decided upon, the work of analysis can begin. This is to determine what the output signal [$v(t)$ above] is when the input signal $i_F(t)$ and initial condition V_0 are known. Physically the problem is to find out how the model behaves; mathematically the problem is to solve the differential equation.

Because amplifiers are sluggish, a phenomenon found in the behavior of every physical device, we have to use derivatives and the infinitesimal calculus. Inertia is unwanted, it is parasitic and undesirable for good amplification, but it cannot be avoided. It can be reduced by using more carefully (and expensively) made electron devices; but its effect, modelled by the capacitor and the derivative, will always be present. (Sometimes we shall find that such "reactive" or energy-storing elements are deliberately used to obtain shaping effects desired for some reason.) Anyhow, we find derivatives in every really interesting analysis problem. Analysis is therefore in part, often in very large part, the solution of differential equations. And so we need the calculus.

A schematic diagram that is very convenient for discussing our work is Fig. 1.07-A. It calls the input signal to the network (circuit model) the *excitation*, $e(t)$; The output signal is the *response*, $r(t)$. The initial condition of the network, for example, the value of V_0 in (1.06-1) or some equivalent such as the amount of energy stored in the capacitor at $t = 0$, is essential and therefore

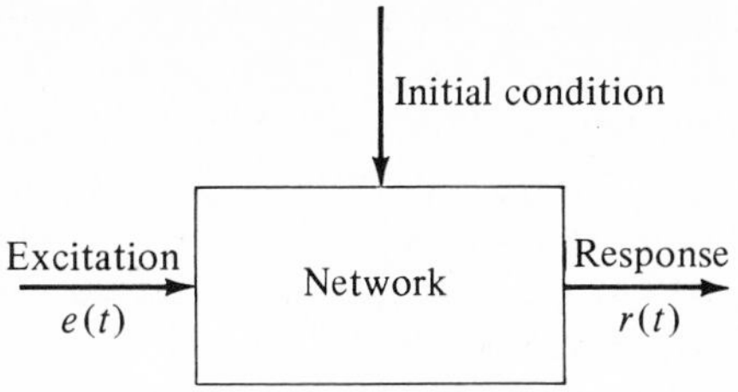

FIG. 1.07-A

shown as an additional "input" to the network or problem. (We shall find that initial conditions act, and are generally best thought of, as additional excitations.) *Analysis* is (the work of) finding the response, given the excitation and the network. It is a basic engineering task, and usually a straightforward one, and is sometimes not tremendously interesting. Much more interesting, because usually *not* straightforward but instead demanding of imagination and ideas, is *synthesis*, the design process. In it, the network is not given, but only the excitation and the response desired. Fascinating as synthesis may be, its practice requires solid background in analysis. Ability to analyze is a part of the foundation necessary for synthesis.

Analysis of extant systems is also a necessary part of the work of maintenance and operation thereof. It may, incidentally, equally well be the determination of input, given the response (and network)—for this is a basic problem in making and using measurements.

And so analysis is where we have to start. In this book we shall try to perform synthesis as often as we can, but it is perforce subordinate to analysis in the attention that it gets.

1.08 SIGNALS

In Secs. 1.05 and 1.06 we have a model of an amplifier, simple but useful. It is a very small amplifier, only one "stage"; a real amplifier may well be made up of several such stages, for one may not amplify enough. In a real amplifier there may be feedback connections in and around the stages, for that can be very advantageous. But in that little model we do have the essence of a real amplifier. It is a real electronic circuit (network), and it has the seeds of interesting and important phenomena that will grow and grow as we study it.

To analyze it is to find the response $v(t)$, implied in (1.06-1). But however well founded we may be in differential equations, we can make little progress in the analysis until we know the excitation $i(t)$. And so our first job is to study excitations. It would be inefficient indeed to analyze the model at length for some particular excitation only to find later that it was an uncommon and uninspiring one. At the other extreme, it would be impossible in one's lifetime to consider every possible excitation. We have somehow to attain some perspective on useful and important input signals. For which ones is it worthwhile to analyze systems? Signal theory is one of the many aspects of engineering for which our study of circuits prepares us. It has been developed to an awesome structure which may profitably be studied thoroughly after a good introduction to circuits. We can only touch on it here, but that we must do, to get a start.

It is the nature of signals to be unpredictable. If we *know* what a signal is, then it carries no information: it can tell us nothing. To analyze a network for a given input is therefore an academic exercise; the only really important signals are those yet to come, with information as yet unknown. But analyze we must, to have a basis for design. The paradox can be resolved, in a practical way, by recognizing that the input signals of interest, unknown though they may be, yet belong to some family, some category with describable characteristics. Important inputs to our amplifier are not completely arbitrary signals. The amplifier is presumably part of a telegraph or telephone or television communication system—or it may be a link in a control system where the commands are of limited sorts. Signal theory and communication theory are able to make precise statements, in fact, about the *statistical* nature of the signals to be expected in a given application, even though no one input signal can be exactly predicted. The types of signals that teletypewriters or digital-data-producing machines emit are not completely arbitrary. The pitch range of the human voice is limited, even if we consider every inhabitant (or visitor likely to come) in a large country. The notes played by musical instruments lie between known limits. Some sort of spectrum limitation of tones exists. Be more precise here we cannot, but later, when we discover the ideas of pure tones, overtones, and harmonics, and how signals are composed thereof, we may return to this.

What we *can* grasp now is the idea of *prototype* or test signals, the responses to which are worth calculating. Though a prototype itself never exists as a real input signal, the response thereto may tell us enough about the system (model) that we can confidently say we begin to understand it. We look now for useful prototype signals.

Such a signal should certainly be *simple;* it should also be informative in that it somehow typifies signals that may occur in the actual system. A happy property of networks is that the response to almost any signal, once found, can be used to find the response to almost any other signal. This is expressed in a very important theorem that we shall derive in due course. We limit ourselves therefore to very simple signals for prototypes.

Suppose the signal source in Fig. 1.05-A contains simply a battery, a telegraph key or switch, and a resistor. Then closing the switch (operating the key) at some instant t_0 on some time scale gives to the voltage v_1 the form shown in Fig. 1.08-A, in which $V_0 = E_0$ (volts). [Unless otherwise stated, *voltages* are measured in *volts* (!).] Mathematically, $v_1(t)$ is a *function* with the value zero for $t < t_0$, the value V_0 for $t > t_0$, but undefined at $t = t_0$. Physically it is a *wave,* visible to the eye on a cathode-ray tube that is suitably operated.

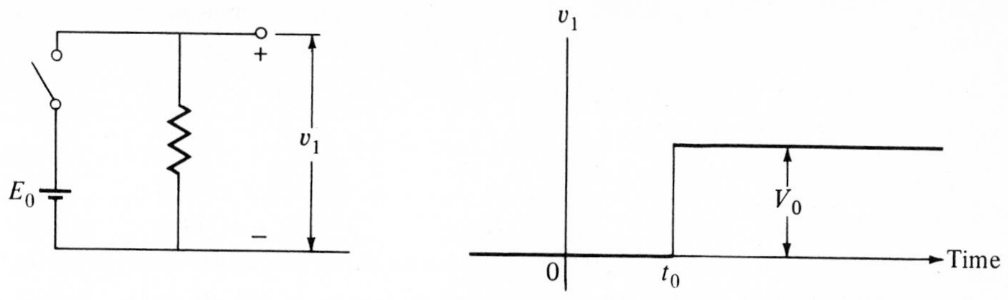

FIG. 1.08-A

The perspective on engineering and mathematics that we have set up as a goal requires that we actually *look* at such waves as often as we can. In a book, only photographs of the cathode-ray-tube screen can be given. It is up to the readers of this book, teachers or students, actually to assemble the apparatus for each such "demonstration" (it is actually quite simple) and examine the screen themselves, with care. The photographs are vicariously helpful, but not enough; the instructor and student who neglect to examine actual demonstrations are missing something extremely valuable. Figure 1.08-B suggests a demonstration setup; more detail of this, and others to come, is given in Appendix A.†

In Fig. 1.08-C this wave can be seen in two real manifestations. (The apparatus necessary for conveniently examining it is not exactly that of Fig. 1.08-B, but it is the wave that interests us here and not the mode of creating it.) In each of the two cases shown, part (*a*) is much like the *step* of Fig. 1.08-A. But magnification of the time scale changes the pattern: part (*b*) shows that the step is far from instantaneous. The upper "corner" is definitely rounded and not sharp; the lower one is also not square, though it may be more nearly so. Examine the wave with as many different magnifications as your apparatus permits.

The "moral" of this experiment is a basic one that we must not forget. Idealizations like the wave of Fig. 1.08-A do not exist in actual engineering artifacts. Corners that seem sharp when the sweep is slow are really rounded;

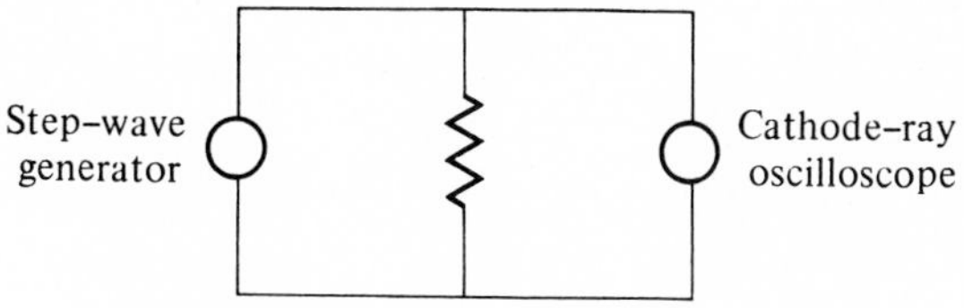

(See Appendix A for other data.)

FIG. 1.08-B

† To Professor J. F. Gibbons is due great credit for stimulating the use of demonstrations, not only in the classroom but in the book.

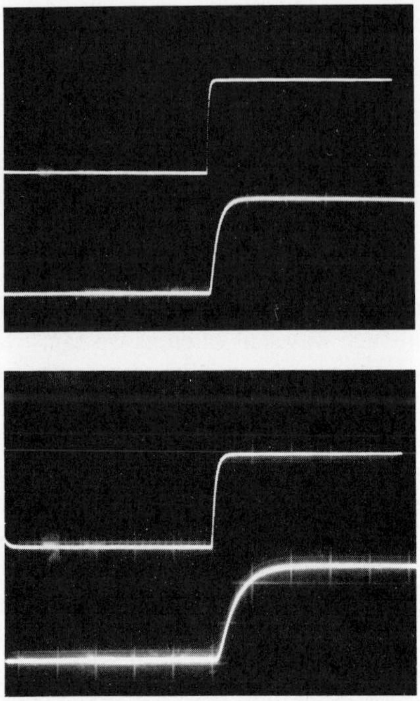

FIG. 1.08-C

a sufficiently fast sweep will show them up for what they are. Instead of no time at all, to rise from zero to V_0 requires time. The *rise time*, t_r in Fig. 1.08-D, has two important characteristics: it is not precisely definable, and its numerical magnitude varies greatly from one device to another. It cannot be defined because, with rounded corners, there is no definite time at which the rise is complete. We could arbitrarily choose to define t_r as the time taken to rise to say 90 or 95 percent of the final value, and we might arbitrarily measure its beginning from the 5 or 10 percent points; the word *arbitrarily* speaks for itself. The second characteristic is that the rise time, however defined, may in electronic apparatus be a few nanoseconds, microseconds, milliseconds, or even seconds.

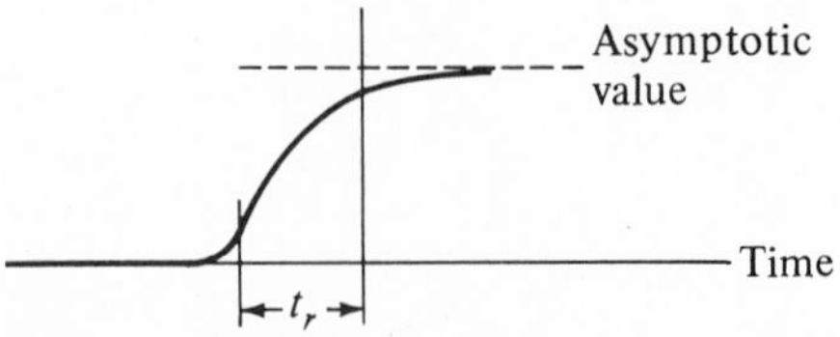

FIG. 1.08-D

Any particular application requires a relatively fast rise time, but what "relatively fast" means depends on the situation. Rapid rises are, of course, expensive; the design engineer makes the rise only as fast as need be.

Our first prototype signal, the step function of Fig. 1.08-A, is engineering fiction, mathematical reality. But our first look at a wave (Fig. 1.08-C*a*) showed substantially that step function: only magnification revealed the real rounding of corners. And such expanded time may not be important in engineering applications: we may often be able to ignore the rounding and perform our analysis with the idealized function, without appreciable error. This judgment or decision is an example of the use of engineering perspective, the practical wisdom one needs in engineering analysis. If the circuit under consideration is sluggish and responds slowly in comparison with the rise time, we may just as well use the idealized step. If the circuit is relatively rapid in responding, then it will be sensitive to the nonidealness, to the sluggishness, of the step's rise, and then we must consider the latter.

No mathematics can decide; only engineering experience and perspective can balance the gain in simplicity and speed of analysis (in using the step) against the loss in accuracy (because we did not use the rounded rise).

1.09 STEPS AND PULSES

Resolution of the problem in favor of the idealized step is common enough that we need a precise definition of that idealized unit step function (Fig. 1.09-A):

$$u(t) = \begin{cases} 0 & t < 0, \\ +1 & t > 0. \end{cases} \tag{1.09-1}$$

The function is not defined at $t = 0$ (that is not necessary); but we do add that it rises there infinitely rapidly, from the value 0 to the value $+1$, without any extraordinary behavior, so that its integral is

$$\int_{-T}^{t} u(x)\,dx = \begin{cases} 0 & t \leqq 0, \\ t & t \geqq 0, \end{cases} \tag{1.09-2}$$

in which T is positive but otherwise arbitrary.

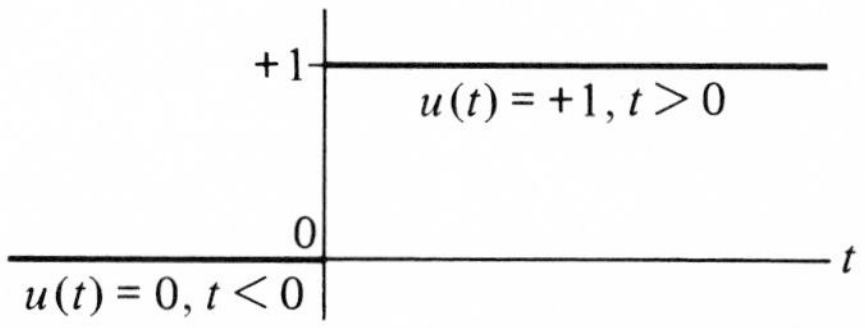

FIG. 1.09-A. The unit step function, $u(t)$.

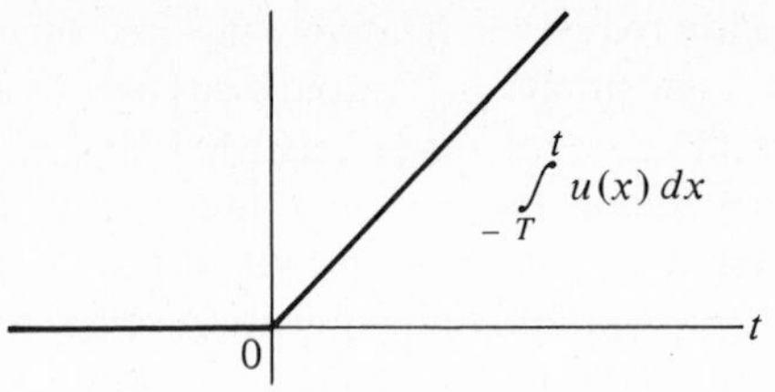

FIG. 1.09-B. A ramp function.

The variable of integration under the integral sign, here written x, is but a dummy variable that disappears when the (definite) integration is performed. One should not confuse x with t!

We call the function $u(t)$ the *unit* step function, and multiply it by whatever constant may be appropriate. In Sec. 1.08, for example, we need $V_0 u(t)$ to express the voltage v_1 (Fig. 1.08-A).

Note that $u(t)$ *has no dimensions*, by definition. The function obtained by integration [(1.09-2), Fig. 1.09-B], often called a *ramp function*, does have the dimensions of time, unless altered by multiplication by a suitable constant.

The function

$$\begin{aligned} v(t) &= V_0 \frac{t}{T_0} u(t) \\ &= \int_{-T}^{t} \frac{V_0}{T_0} u(x)\, dx \end{aligned} \tag{1.09-3}$$

has the shape of the ramp and the dimension of voltage. It resembles the beginning of the expanded "step" of Fig. 1.08-C*b*.

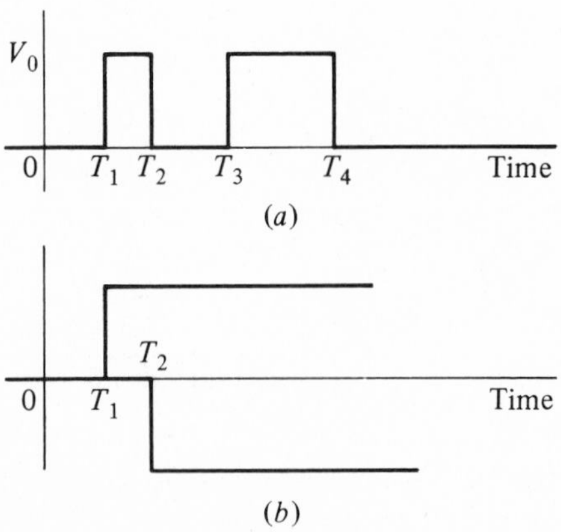

FIG. 1.09-C

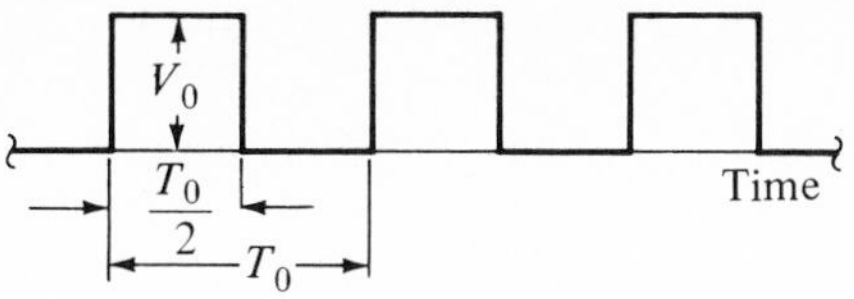

FIG. 1.09-D. A regular square wave.

Note the natural utility of the step function: multiplication by $u(t)$ obviates the need for a two-part definition as in (1.09-2); this is a handy mathematical (shorthand) notational device.

To transmit information, one may operate the telegraph key or switch of Fig. 1.08-A at various times to turn the voltage v_1 on and off and so produce *pulses*. The pulses in Fig. 1.09-C*a*, for example, represent the letter *A* in the international Morse telegraph code. The rises and falls are here idealized once more; in practice the corners are rounded and the rise and fall times are not zero. The vertical lines in Figs. 1.09-C and 1.09-D are to guide the eye only; they are not parts of the functions. But for circuits sufficiently sluggish in comparison, we may use the idealization without error. And we may conveniently express the pulses of Fig. 1.09-C*a* mathematically by using the unit step function. The representation is:

$$V_0[u(t - T_1) - u(t - T_2) + (u - T_3) - u(t - T_4)], \qquad (1.09\text{-}4)$$

since a step delayed by T_0 (seconds) is simply $u(t - T_0)$, according to its definition, (1.09-1). Figure 1.09-C*b* shows the two components that add to constitute the first pulse; their addition or "superposition" clearly makes the pulse.

The information-carrying ability of a train of pulses is obvious. They may, for example, be of different durations and heights in various combinations to represent samples of a complicated wave that carries information and so at least indicate that information. Or they may be identical in duration and size, and by their mere presence or absence at agreed-on times represent the digits 1 or 0 of a binary code. A sequence of these digits can represent any message one desires, and this is a very practical way to transmit it.

To obtain a very simple train of pulses for study, suppose they occur regularly in the fashion of Fig. 1.09-D. Since this symmetrical "square wave" is so regular, it can carry no information at all, for we know exactly what is coming. Even so it will be very useful. So also will small numbers of pulses. Single pulses especially we shall find to be very helpful.

1.10 THE SINGLE PULSE AND THE IMPULSE

One isolated pulse (Fig. 1.10-A*a*) is a simple function. Yet it ought to be useful in giving at least some indication of a circuit's behavior in general. Its *turning-on* step must excite the circuit into some sort of action, which we can calculate

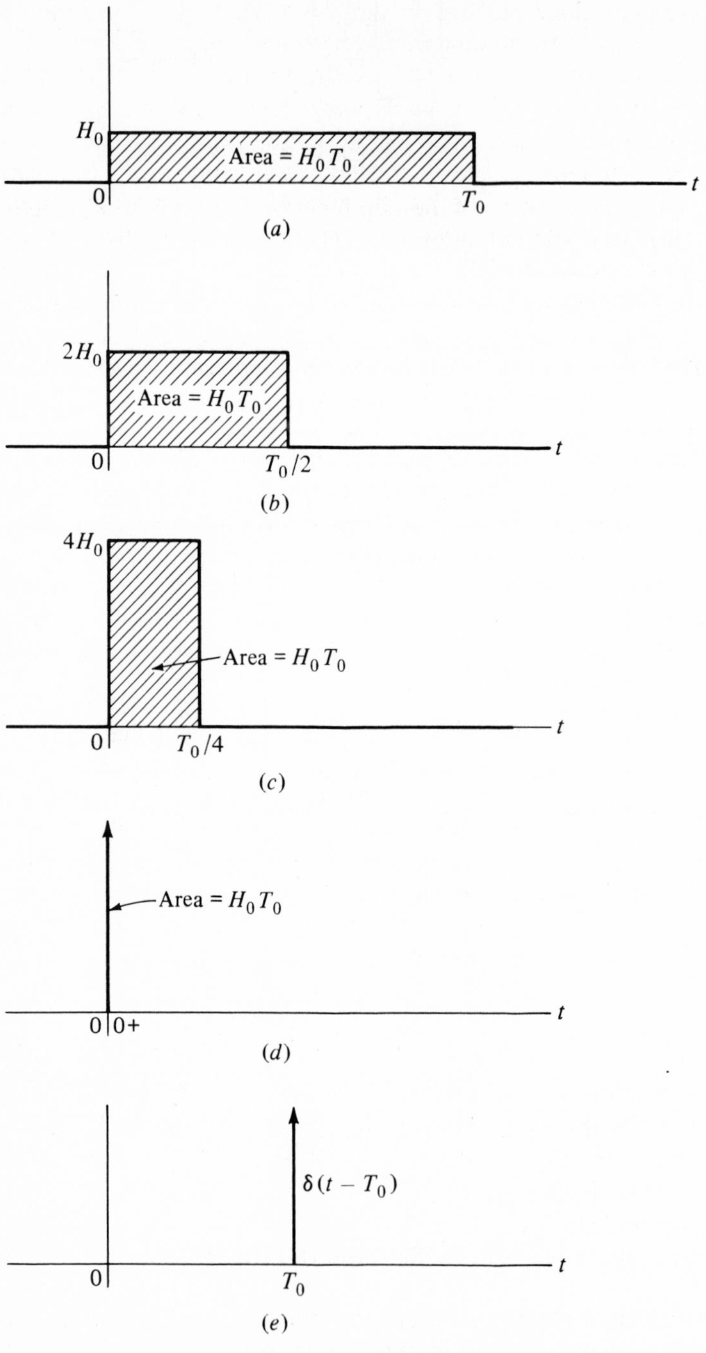

H_0
Area = H_0T_0
0
T_0
t
(a)
$2H_0$
Area = H_0T_0
0
$T_0/2$
t
(b)
$4H_0$
Area = H_0T_0
0
$T_0/4$
t
(c)
Area = H_0T_0
0
0+
t
(d)
$\delta(t - T_0)$
0
T_0
t
(e)

FIG. 1.10-A

or observe with profit. And its *turning-off* step must be similarly useful. Even one of these steps alone can lead to very profitable results. And this is why the function $u(t)$ is so often encountered in network studies.

The duration of the pulse may be rather long in comparison with the innate sluggishness most circuits have, as implied above in turning on and turning off. A brief pulse, on the other hand, may be interesting because it delivers some sort of "kick" or impulse but then disappears before much can happen. Yet its effect must last for a while just because of the slowness of response of the circuit. Such an excitation is, we shall find, about the simplest we can conceive of. And its effect, the circuit's response thereto, is extraordinarily useful to have. It is worth stopping here to make a precise definition of such a (short-lived) pulse.

Suppose we halve the duration of the single pulse but double its height (Fig. 1.10-A*b*). This makes it brief, but retains the area, which we feel somehow must measure the pulse's effect. And let us repeat this to obtain Fig. 1.10-A*c*. More repetitions will make the pulse too narrow in duration and too large in amplitude to plot. The schematic arrow of Fig. 1.10-A*d* is a convenient symbol for the very narrow pulse. It begins at $t = 0$ and ends at so small a value of t that we call it simply $0+$; its height is too great to plot.

The *mathematical* representation of a very short pulse is simplest in the limit, the extreme case where the duration is infinitesimal and the height increases without limit. Figure 1.10-A*d* is the best portrait we can give of the idealization. As for a name we may call it an *impulse* function because its effect is like that of a short blow delivered to a heavy object: the object acquires some momentum virtually instantaneously, though its motion is not immediately apparent. It is convenient to define a *unit* impulse function by setting the area equal to unity. For a symbol we use δ to remind us of its finder and other common name, the *Dirac delta function:*

$$\delta(t) = \begin{cases} 0 & t < 0, \\ 0 & t > 0, \end{cases} \qquad \int_{0-}^{0+} \delta(t)\, dt = 1. \tag{1.10-1}$$

It appears that the actual shape of the impulse during its brief life is of little importance; only the integrated value (unity, above) matters. (A more rigorous mathematical view requires the concept of *generalized function* or *distribution*, inappropriate to develop here.)

This Dirac delta function, or unit impulse function, we shall use again and again. Note that it is given the dimensions of inverse time, so that the area is dimensionless.

The important properties of this function are that it is zero except where its argument is zero, and that its integral from any point to the left thereof to any point to the right thereof is unity. An impulse of current occurring at time

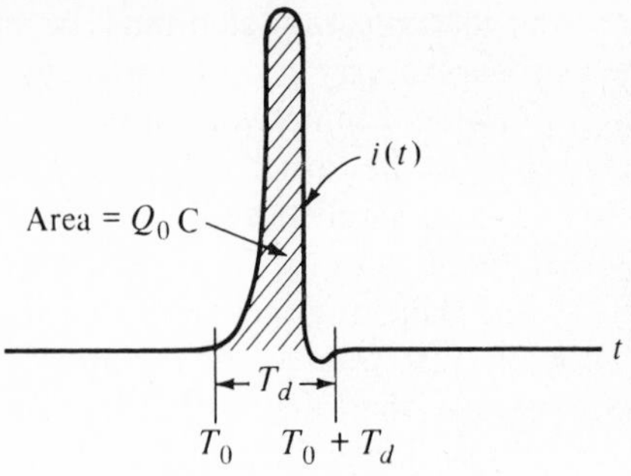

FIG. 1.10-B

T_0, of integrated value Q_0 (coulombs) can thus be represented by $Q_0\,\delta(t - T_0)$ (amperes). An impulse of voltage is similarly represented by $\Lambda_0\,\delta(t - T_0)$, the *size constant* Λ_0 being in weber turns or volt seconds (V·s) (see Fig. 1.10-A*e*).

Extraordinary mathematical simplicity attends the use of the delta function $\delta(t)$. It is therefore important to discuss when it can be used in practical engineering analysis. For the present we simply observe, again, that it is eminently reasonable to assume that:

> *If* the effective duration of a real pulse, for example T_d in Fig. 1.10-B, is short compared to the reaction time of a network,
>
> *Then* the pulse can be replaced in the analysis of the network by the mathematical delta function without appreciable error. (1.10-2)

Figure 1.10-C gives physical corroboration. In it we see the response of a simple network to

a An essentially square pulse, like that of Fig. 1.10-A,

b A pulse of the same shape but of half the duration and twice the amplitude,

c A pulse of the same shape as that of (*a*) but of one quarter the duration and four times the amplitude.

In each case the photograph shows both response and excitation, for easy comparison; horizontal and vertical scales are unchanged between the three photographs. Note that in (*a*) the pulse duration is not effectively short, and its replacement by a mathematical impulse would cause some error in the calculated response. But in (*b*) and (*c*) the pulse duration *is* effectively short and the responses to pulses (*b*) and (*c*) and to a mathematical impulse would, all three, be practically the same. Careful comparison of the responses shown, particularly those in (*b*) and (*c*), which are virtually identical, will corroborate this.

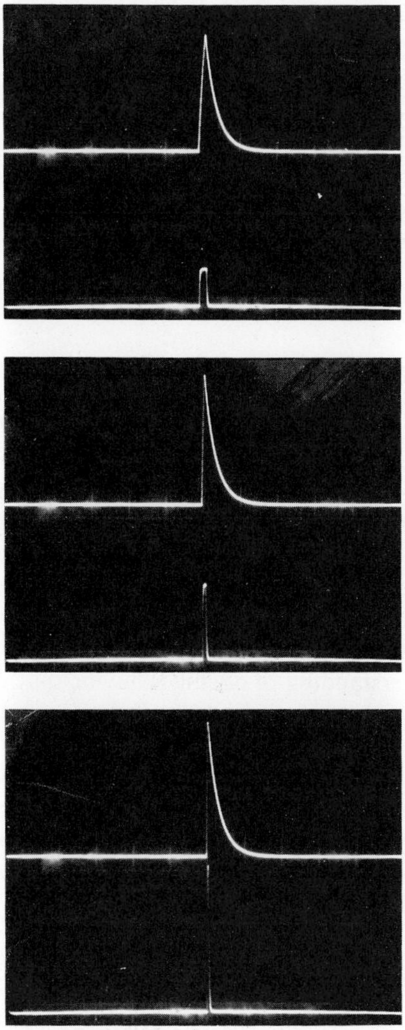

FIG. 1.10-C

To replace the "real" pulse of current $i(t)$ in Fig. 1.10-B—whose duration is short (compared to the response time of a network) and whose "effect' or area, is Q_0 (Coulombs), and which occurs at time T_0—by the idealized "unreal' function

$$i_{\text{substitute}}(t) = Q_0\,\delta(t - T_0) \tag{1.10-3}$$

is greatly to simplify the analysis of that network. What the actual error is, and more substantial proof of this thesis, are things we shall deal with later.

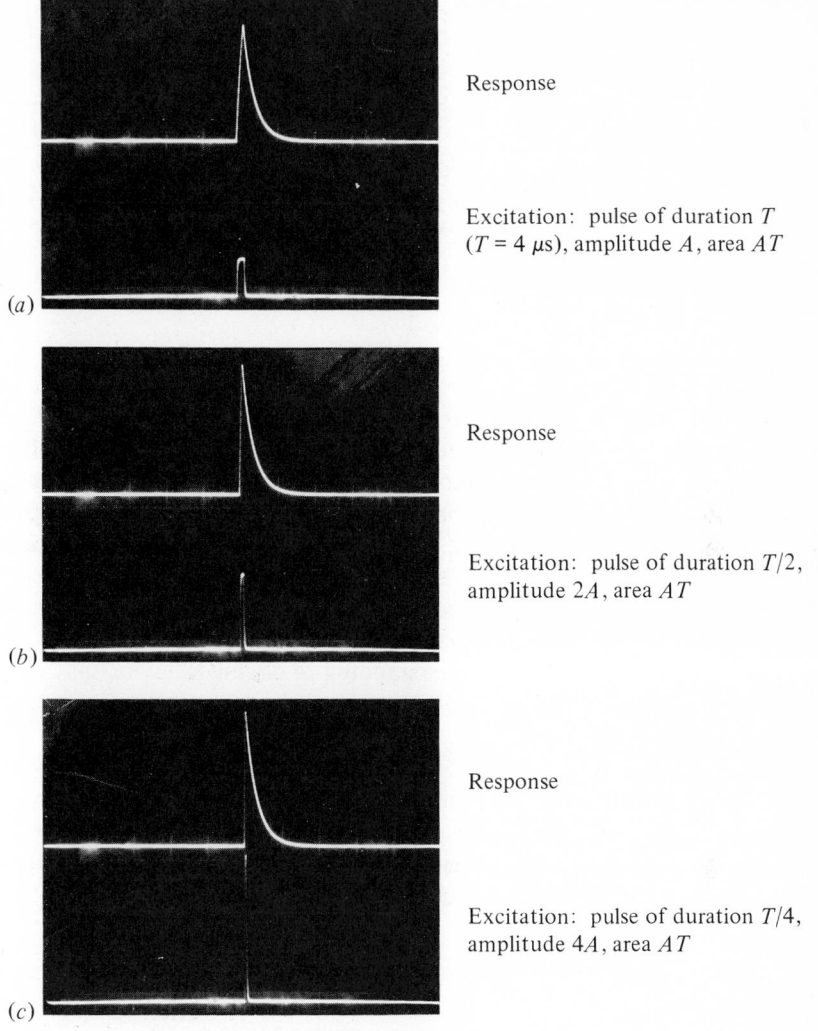

The word *impulse* we shall use in two different senses, distinguishing them when necessary by the adjectives *physical* and *mathematical.* A *physical impulse* is a current (or voltage or ...) wave whose appreciable values are concentrated in a relatively short time interval; its principal characteristics are the time of its occurrence and its strength or integrated value or area (T_0 and Q_0 in Fig. 1.10–B). A *mathematical impulse* is a Dirac delta function, such as $Q_0\ \delta(t - T_0)$.

The latter is often practically equivalent to the former. Use of the latter can often simplify our analysis. So the delta function can be useful, however novel it may be.

1.11 SUMMARY

Our first chapter has provided, if not a detailed chart, at least an introduction to guide our work. Electric circuits evidently constitute a vast field for study. In a sense it doesn't matter where one starts, provided one has a good guide at the beginning. But if one's time is limited, an efficient choice should be made by the guide. Here, that choice is first to study an amplifier component of a large system and its response to pulses of various sorts. [We shall then move on to more complicated networks, always keeping the network example as simple as is consistent with gaining the understanding of circuit behavior that we seek; eventually we shall even discuss network analysis in the completely general case (see Secs. 4.04 and 6.24).]

Those who wish to read more deeply about the ideas we develop will find suggestions in Appendix B. The list there is by no means exhaustive, but it suggests the more important original and classic works, good textbooks, and some contemporary work. It is particularly desirable to examine the more readable original works; good historical discussions of the development of circuit analysis speak for themselves.

1.12 EXERCISES, PROBLEMS, AND EPSTEAN'S LAW

We assume the reader has some desire or need to learn something about network analysis. Reading the text of this book could, in principle, satisfy it, were it not for Epstean's law.† Common sense and personal experience say that the exertion of working exercises and solving problems is essential for real understanding. For fixing ideas, for absorbing their meaning, and for mastering their use, *drill* on stated principles, and attacking of genuine *problems* that require their application and even modification, are both essential.

† "Man tends always to satisfy his needs and desires with the least possible exertion" (A. J. Nock, *Memoirs of a Superfluous Man*, Harper, New York, p. 133).

We mentioned this displeasing fact at the very beginning (in Sec. 1.01) but have since ignored it. Now we face it. There follow here (and at the end of every chapter) both

a *Exercises* that require only mental exertion or a small amount of paper and pencil work. Generally these only use and drill on stated principles.

b *Problems* that are just that. Of larger size, they generally require some thought and calculation, and genuine effort.

But you must not wait for each chapter's end; the exercises and problems are collected at the chapter ends merely for technical convenience. Go at them from the beginning of the chapter! Half the job is deciding, at every stage, which are relevant, and how many to "do." And do a reasonable number and selection you must, Epstean's law or no. There is no other way to learn.

A final word: these exercises and problems are but suggestions: construct others by extensions of your own, seek out others, be never content!

And don't forget the demonstrations!

1.13 EXERCISES

1-1. In Fig. 1.04-A suppose $A = k$, that is, that $e_{\text{out}} = ke_{\text{in}}$, k being a dimensionless constant. There is then in the amplifier no energy storage, no delay; it is truly ideal. Let e_{in} be the wave shown here (Fig. 1-1). (*a*) If $k = +10$, draw

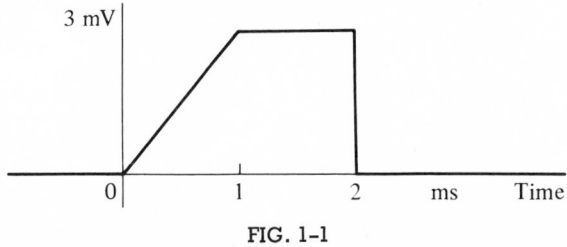

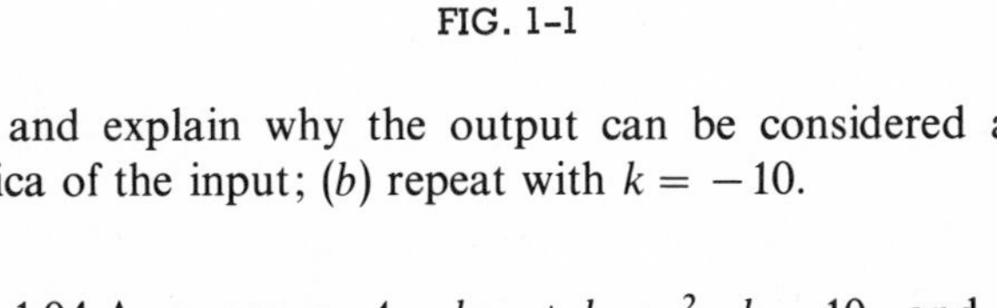

FIG. 1-1

the e_{out} wave and explain why the output can be considered an undistorted amplified replica of the input; (*b*) repeat with $k = -10$.

1-2. In Fig. 1.04-A, suppose $A = ke_{\text{in}} + k_2 e_{\text{in}}{}^2$, $k = 10$, and $k_2 = 10\ \text{V}^{-1}$. Sketch the e_{out} wave when e_{in} has the waveform of Exercise 1-1. Explain why this amplifier introduces distortion and roughly express the amount of distortion. Now change k_2 to $100\ \text{V}^{-1}$ and repeat.

1-3. In the model of Fig. 1.06-A, to find v given i_F is an *analysis* problem (Chap. 2). Here consider this far simpler problem: if v is to be some given wave, what i_F wave is necessary? For the v wave given here (Fig. 1-3) evaluate and sketch

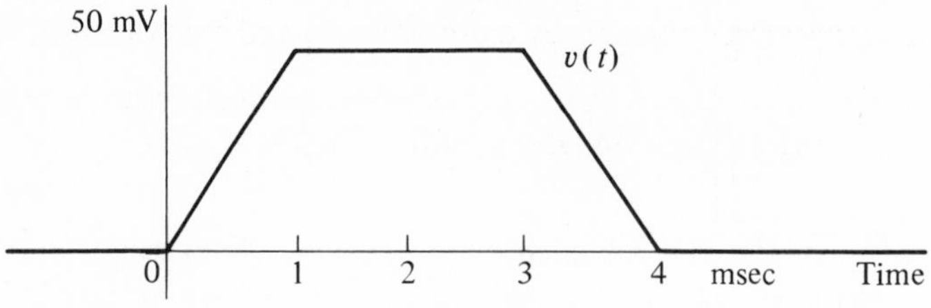

FIG. 1-3

i_C, i_G, and i_F versus time. Align the sketches vertically and compare the three currents. Do this (*a*) with $C = 0.5\ \mu F$ and $R = 1000\ \Omega$, and (*b*) with $C = 5000$ pF and $R = 1000\ \Omega$. (*c*) Describe and explain the differences carefully.

1-4. Consider the fitting (by trial, and approximately) of wave *b* in the second part of Fig. 1.08-C, with the form $v = V_0(1 - e^{-\alpha t})$. Why, with the data there, is V_0 not determinable? Assign it some convenient value (in centimeters). Then find the value of α within 10 percent.

1-5. Describe the wave of Fig. 1.09-C*a* in terms of unit step functions.

1-6. Sketch the waves described formally below.

(*a*) $V_0[u(t) + u(t-1) - 2u(t-3)]$,
(*b*) $V_0[3u(t-3) - 4u(t-4)]$,
(*c*) The sum of (*a*) and (*b*) above,
(*d*) $V_0[u(t) - 2u(t-1) + 2u(t-2) - 2u(t-3) + u(t-4)]$.

1-7. Describe formally the waves shown in Fig. 1-7, using powers of t and $u(t)$.

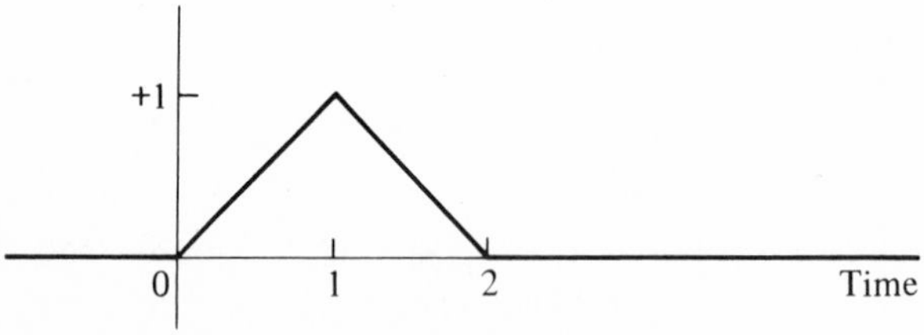

(*a*)

FIG. 1-7

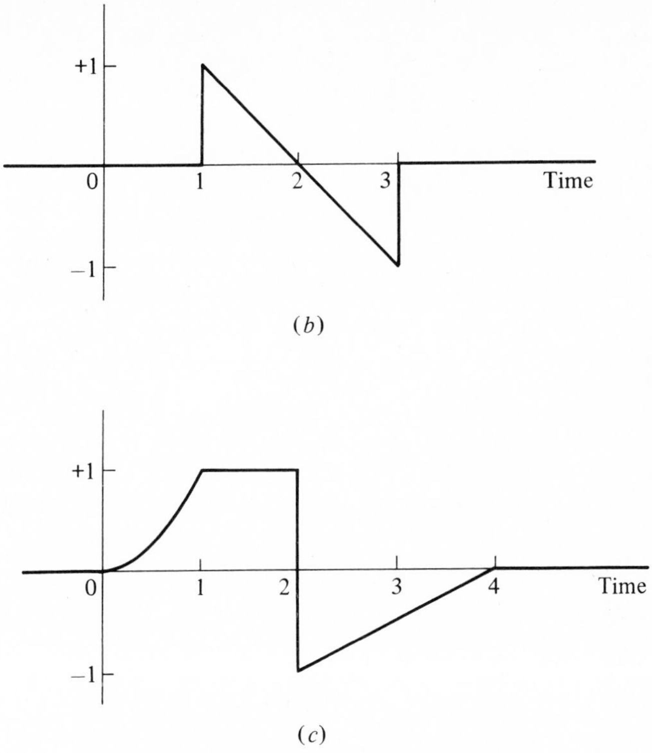

FIG. 1-7 (continued)

1.14 PROBLEMS

1-A. Which of the following functions, when multiplied by a suitable constant (depending, of course, on the parameters in the function), in the limit as the appropriate parameter approaches zero or infinity, becomes an impulse function in the sense of (1.10-1)? Explain, with sketches and calculations.

(*a*) $e^{-\alpha t}$, (*b*) $te^{-\alpha t}$,

(*c*) $t^2 e^{-\alpha t}$, (*d*) $e^{-\alpha t^2}$,

(*e*) $\dfrac{d}{dt}(\tan^{-1} \alpha t)$, (*f*) $\dfrac{\sin \alpha t}{t}$,

(*g*) $e^{-\alpha t} \cos \beta t$, (*h*) $[u(t) - u(t - T_0)]$.

1-B. Integration of the ramp function (1.09-2) produces a parabolic function. Discuss its dimensions and plot it.

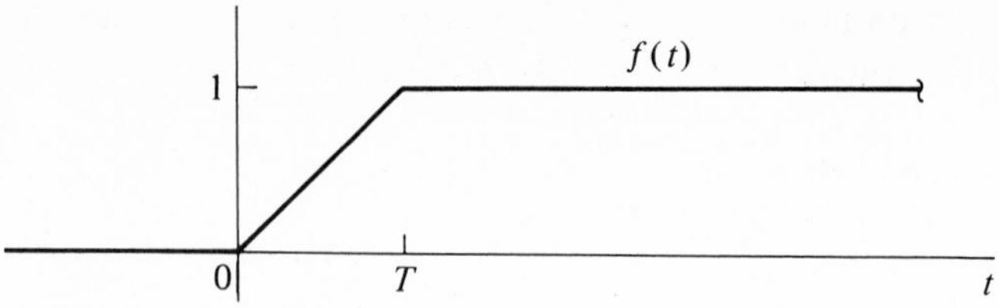

FIG. 1-B

Sketch the derivative of the function $f(t)$ shown in Fig. 1-B. In the limit as $T \to 0$, to what function does $f(t)$ tend? To what function does $f'(t)$ tend? What relation between the unit step and the unit impulse functions does this calculation illustrate?

1-C. What difficulty arises in discussing the function obtained by *differentiation* of the single pulse (Fig. 1.10-A*c*, for example). Consider Fig. 1.10-B and resolve the difficulty, at least in spirit. Sketch the derivative's behavior versus time.

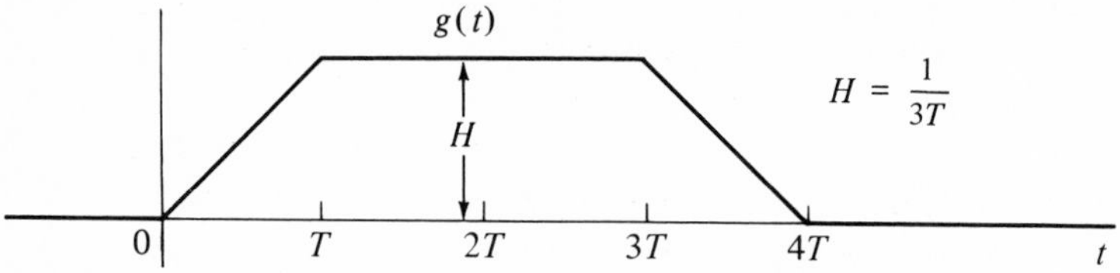

FIG. 1-C

In the limit as $T \to 0$, show that $g(t)$ in Fig. 1-C approaches the unit impulse function. Discuss the limiting behavior of $g'(t)$.

1-D. Multiply an arbitrary function $f(t)$ by $p(t)$, a narrow rectangular pulse of area 1, centered at $t = T$ (Fig. 1-D). Show that

$$\int_{-\infty}^{\infty} f(t)p(t)\, dt$$

is the average value of $f(t)$ over the interval $(T - L/2)$ to $(T + L/2)$. Show also that in the limit as $L \to 0$, if $f(t)$ is continuous at $t = T$,

$$\int_{-\infty}^{\infty} \delta(t) f(t)\, dt = f(T).$$

This is known as the *sifting* or *filtering* property of the impulse function.

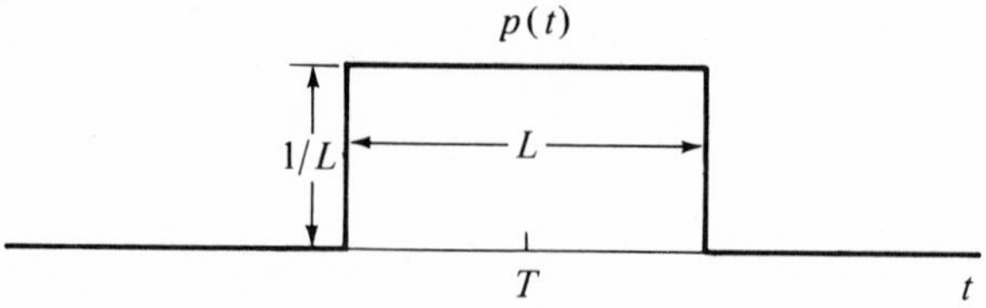

FIG. 1-D

If $f(t)$ is discontinuous at $t = T$, the value $f(T)$ must, of course, be specifically defined. What value for $f(T)$ results from calculating $\int_{-\infty}^{\infty} \delta(t) f(t)\, dt$ in the limit, as above?

What support can you adduce for the statement "The only reasonable value to assign $f(T)$ when $f(t)$ is discontinuous at $t = T$ is the average value of $f(T+)$ and $f(T-)$"?

Explain the filtering property of the impulse function in terms of

$$\int_{-\infty}^{\infty} f(x)\, \delta(t - x)\, dx = f(t).$$

1-E. Scaling and positioning of the impulse function (in time) sometimes require care.

Positioning: sketch, versus time, the three functions $\delta(t)$, $\delta(t - T)$, and $\delta(t + T)$. Now evaluate and sketch versus t, $\int_{-\infty}^{t} f(x)\, dx$ in which f is each of the three functions in turn.

Scaling: evaluate $\int_{-\infty}^{\infty} \delta(t/a)\, dt$, in which a is a positive, dimensionless constant, both formally (as by a change of variable) and graphically (by sketches and a limiting process).

1-F. Consider the "function" $[\delta(t)]^2$. In the limit as $T_0 \to 0$, what happens to the rectangular pulse of height $(1/T_0)^{-2}$ and duration T_0?

This illustrates the difficulty of treating $\delta(t)$ as a function in the ordinary sense. For our purposes $\delta(t)$ and its integral suffice. We leave $\delta'(t)$, $[\delta(t)]^2$, and the like for study elsewhere.

Chapter **2**

THE *RC* CIRCUIT

A pawn goes two squares in its first move, you know. So you'll go *very* quickly through the Third Square—by railway, I should think—and you'll find yourself in the Fourth Square in no time. Well, *that* square belongs to Tweedledum and Tweedledee

—The Red Queen

A very useful category of circuits is that made up entirely of resistors. We pass *very* quickly over it, however (such circuits are not dynamic), so quickly in fact that we begin our study of networks as such with a dynamic yet very simple example. It has two dynamically interacting elements (Tweedledum and Tweedledee if you will), and we shall study it in detail that may be painful, so simple is the model. But the essence of the results (which will therefore be easy to grasp) will still be valid and very useful to us in the most general case. And with this experience we shall find very complicated networks much easier to understand.

2.01 INTRODUCTION

Here we begin our actual study of circuit (network) analysis. Chiefly it will be a series of experiments, performed largely with paper and pencil and a slide rule or computer. Physical verification we shall make as often as we can, as planned. But by and large we build on the experimental work of Kirchhoff, Ohm, Faraday, Coulomb, and a host of others, whose laws we take, in effect, as postulates. When we seem to have discovered a general principle, we shall enunciate and demonstrate it as best we can. If often we have to leave proofs that are really general to later in the book, or even to subsequent studies, we should not be dismayed. To stop for complete investigation of every principle we come across would be inevitably to lose ourselves in the dense trees within and never to see the forest itself. What *is* important, for intellectual and engineering honesty, is to note carefully that which we have *not* proved or based on definitive knowledge.

Thus we make a list of agenda for future study. It will be a list far too long for one lifetime—but from it we can later select what is for us most important, most relevant, to develop as may then seem appropriate.

2.02 EXAMPLE ONE

To the simple amplifier stage model of Fig. 1.05-A let us apply an impulsive signal or excitation. Stripped (for circuit analysis purposes) of its nonessentials, the model becomes the simple circuit of Fig. 1.06-A and the differential equation of (1.06-1). Restated to take advantage of our discussion of signals, it is Fig. 2.02-A and the differential equation (2.02-1).

$$C\frac{dv}{dt} + Gv - Q_0\,\delta(t) = 0 \qquad v(0) = 0. \tag{2.02-1}$$

We here assume that there is no energy stored in the capacitor at the instant of excitation, which we call the origin of time, $t = 0$. Simple as it is, the differential equation is typical of all the differential equations we shall meet, of whatever order and complexity. It is appropriate to study it carefully because its simplicity is deceiving.

To begin with, the impulsive or short-lived nature of the excitation, together with the innate sluggishness of the circuit, suggest that there should be a simple way to analyze its behavior in at least the first few seconds of the response.

The $C\,dv/dt$ term represents sluggishness of response in the circuit. The derivative appears because the circuit is slow to react. It is because all physical systems have this property that derivatives appear in any reasonably accurate mathematical characterization. It means that in response to an ordinary (finite) excitation, the system does not immediately "jump"; it merely assumes a rate of change, i.e., *decides* to move. One must wait a little (integrate the differential equation) for the response to become noticeable because it starts very slowly.

An impulsive excitation, however, is extraordinary in that it is very strong during its brief life, strong enough perhaps even to overcome this inertia. Our

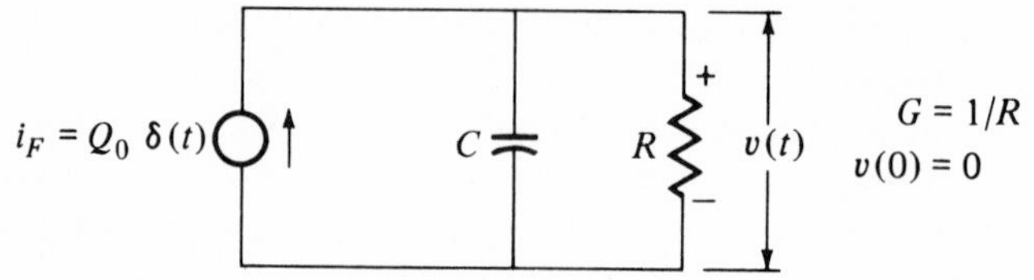

FIG. 2.02-A.

impulse actually represents a compression in time of a real pulse of appropriately short duration. Therefore we can perhaps evaluate the integrated effect, the condition of the circuit *after* the impulse is over and *before* any "motion" really begins, with little effort. At least this seems reasonable. Let us try.

We rewrite the equation as

$$C\frac{dv}{dt} + Gv = Q_0\,\delta(t) \tag{2.02-2}$$

or

$$C\,dv + Gv\,dt = Q_0\,\delta(t)\,dt. \tag{2.02-3}$$

We can integrate between $t = 0$ and $t = T_0$ and obtain [see (1.10-1)]

$$C\,[v(T_0) - v(0)] + G\int_0^{T_0} v\,dt = Q_0 \tag{2.02-4}$$

or

$$C\,v(T_0) = Q_0 - G\int_0^{T_0} v\,dt. \tag{2.02-5}$$

If we now let T_0 become small, reduce to our $0+$, then the integral on the right becomes negligible and we find

$$C\,v(0+) = Q_0 \tag{2.02-6}$$

or

$$v(0+) = \frac{Q_0}{C}. \tag{2.02-7}$$

The mathematical evaluation of the effect left by the impulse, before the system has had any time really to begin to stir, is not difficult, we conclude.

Physically and with hindsight prompted thereby, the result (2.02-7) is readily predicted. During the brief life of the impulse the large current flows essentially through the capacitor alone. The opposition offered there is virtually nil, since the initial condition is $v(0) = 0$. To flow through the resistor would require developing a voltage Ri and that would be a large voltage because i is large. To flow through the capacitor requires developing some voltage, true ...; but that is only the result of accumulating charge, and it rises from zero to only the (reasonable) value Q_0/C and is comparatively small. For a very short-lived pulse of current, one can say that nearly all the current flows through the capacitor and leaves its integrated value, the charge Q_0, thereon at the end of the pulse. If we represent the very short pulse by the impulse function $Q_0\,\delta(t)$, that is consider its duration infinitesimal, then *all* the current flows through C, and the voltage steps instantaneously from 0 to Q_0/C. In reality, for a relatively short-lived pulse, the same is virtually true, though a small amount of current does of course flow through R as voltage builds up. *Relatively* is here the important word. It requires comparison of the pulse duration, T_0 in Fig. 2.02-B, with the *wake-up time* T_c of the circuit. If $T_0 \ll T_c$ we can use the δ function without sensible error—and with much gain in mathematical simplicity. Exactly what error is committed, and exactly how "$\ll$" and "relatively" are nearly synonymous—these we shall discuss later. For the present we simply observe how reasonable it is to adopt the above point of view (see Figs. 2.02-B and 1.10-C).

The analysis above may seem painfully extended and detailed. It explains an extremely important point of view, however, and one we shall often adopt. The calculation of conditions at "$t = 0+$," just after a comparatively brief pulse of excitation, can often be rapidly made in this fashion, the results are often useful, and the method gives excellent physical understanding—and represents, to an engineer, mere common sense. For all these reasons we have made the analysis with elaborate care. Hereafter we shall simply say, in a situation like that of Fig. 2.02-A, that the short-lived current pulse flows entirely through C with the result $v(0+) = Q_0/C$.

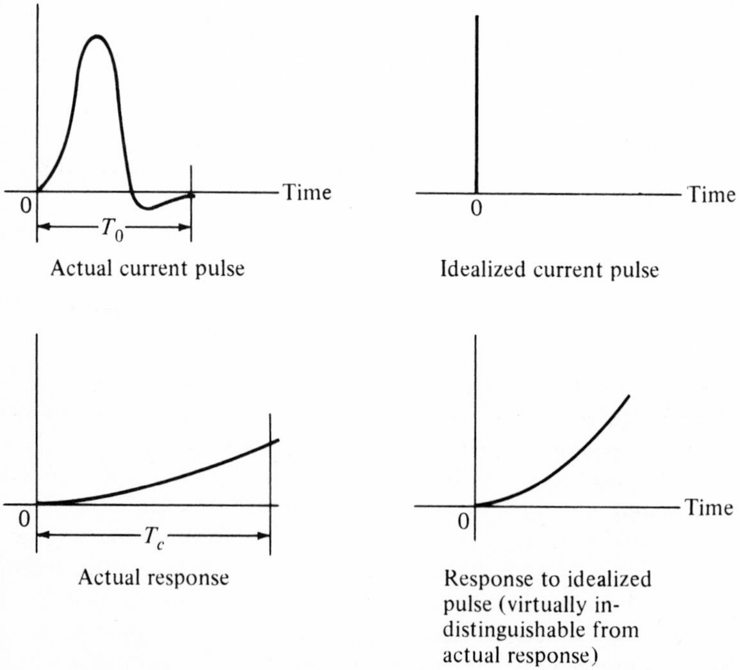

FIG. 2.02-B.

After the pulse has gone, the behavior is governed by

$$C\frac{dv}{dt} + Gv = 0, \qquad t > 0+,$$
$$v(0+) = V_0 = \frac{Q_0}{C}. \tag{2.02-8}$$

This differential equation *implicitly* expresses $v(t)$, the voltage wave, for times *after* $0+$. Since the charging of the capacitor is in comparison instantaneous, we may omit the + in $0+$ except in describing that charging process. The rest of the analysis problem is to extract $v(t)$ in *explicit* form from (2.02-8).

We note that from now on the analysis is the same for these two histories:

a The capacitor holds no energy at $t = 0$ but is charged by an impulsive current $Q_0\,\delta(t)$ between $t = 0$ and $t = 0+$;

b there is at $t = 0$ a charge Q_0 on the capacitor as a result of previous action, unknown to us, but no further excitation is applied. (2.02-9)

The two situations are equivalent in that, for $t > 0$ (or $t > 0+$), the behaviors are the same. It follows that we may, if we wish, replace an initially charged capacitor by the same capacitor uncharged *and* an impulsive current source of the proper strength, $Q_0\,\delta(t)$.

We observe that in a more complicated situation this may have to be qualified, if there is any possibility that the impulsive current can flow elsewhere—as might occur if there are other capacitors nearby. This replacement idea will be very useful later.

2.03 NATURAL BEHAVIOR AND NATURAL FREQUENCY

To analyze the circuit's behavior for $t > 0+$ is to solve the differential equation (2.02-8). Since there is now no excitation, we call the result the *force-free* (excitation-free) or *natural* behavior of the circuit.

So simple is the equation that one can immediately separate variables and integrate to obtain

$$v(t) = V_0 e^{-\alpha t}, \qquad \alpha = \frac{G}{C}. \tag{2.03-1}$$

The circuit's natural mode of behavior is exponential. In form, it is e^{pt}. Here p is a constant, characteristic of the circuit alone and quite independent of the

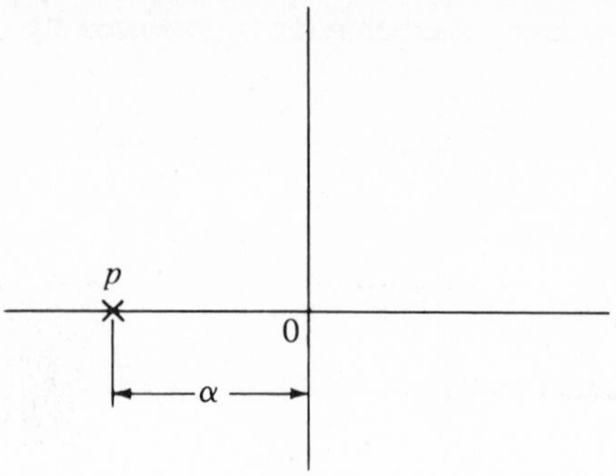

FIG. 2.03-A.

excitation; its value is $-\alpha = -1/(RC)$. Its dimensions are those of inverse time, since the argument of the exponential function must be dimensionless. It (and similar numbers p for networks in general) describes the circuit's most important characteristic, its natural behavior, and therefore deserves an appropriate, distinctive name. That most commonly used is *natural frequency*. The ordinary "oftenness" concept of the word "frequency" is here generalized in a way that will become more and more reasonable as we study more extensive networks. For the moment, we accept it as useful jargon, correct in its dimensions if nothing more.

The physical behavior of the circuit is that observed in Fig. 1.10-C [in which, incidentally, we find confirmation of (2.02-9)]; the "simple" circuit there was an *RC* one.

It will be convenient, and very informative in more complicated situations where there are several natural frequencies, to plot them. For the *RC* network there is but one; it is of course a real number, and negative (Fig. 2.03-A).

2.04 CALCULATION OF THE NATURAL BEHAVIOR

The circuit analysis of Fig. 2.02-A is, mathematically, to solve the differential equation (2.02-8). So simple is this equation that we can write out its solution, (2.03-1), immediately. In less simple cases, it will surely *not* be possible to write the solution merely by inspection. Yet (we shall find) there is really nothing more complicated in the analysis of a network with, say, 15 elements. That may require the solution of a differential equation of order 15, to be sure. But the solution is basically the same; and so we stop to see what the "basis" of the solution of (2.02-8) really is.

A differential equation is a rule for prediction. What (2.02-8) really expresses is

$$\frac{dv}{dt} = -\frac{G}{C}v \tag{2.04-1}$$

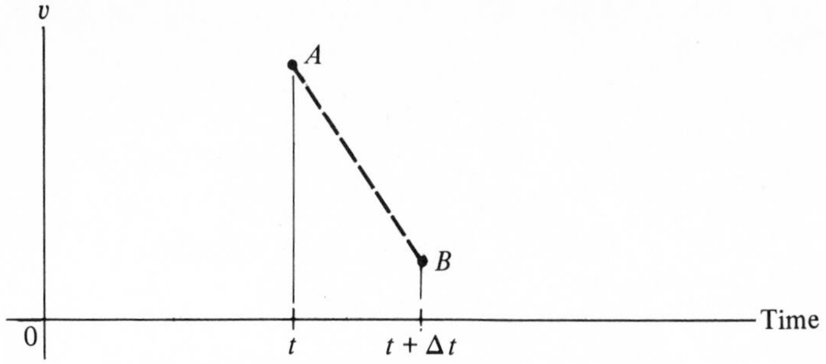

FIG. 2.04-A.

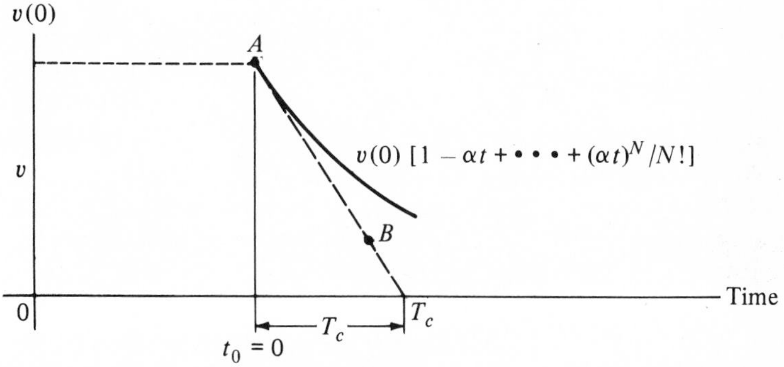

FIG. 2.04-B.

at any instant. If we know the value of v at any time t, we therefore know how fast it is changing at that instant. The value of v a short time Δt later is therefore approximately

$$v(t + \Delta t) = v(t) - \frac{G}{C}\, v(t)\, \Delta t. \tag{2.04-2}$$

Graphically (Fig. 2.04-A) we have extrapolated (linearly) from the present known value at A to an estimate of the later value at B. The actual behavior of the function $v(t)$ is, of course, more complicated; it is *not* simply linear, but curves in some way. It is the genius of the differential equation (2.04-1) that it contains (*implicitly*, of course) all the information one needs to calculate the "curvy" course of $v(t)$, starting from its initial value V_0.

One way of extracting this information and making it explicit is to try the experiment of expressing $v(t)$ as a polynomial in t,

$$v(t) = a_0 + a_1 t + a_2 t^2 + \cdots + a_N t^N. \tag{2.04-3}$$

The first two terms can express the initial value and the linear start; the higher powers of t seem just what one needs to express curvature. The solution of differential equations is in general a purely experimental (empirical) process; this particular (polynomial) technique is a well-known instrument for this purpose, often very successful. Here in particular it "works" beautifully.

We merely place (2.04-3) in (2.02-8) and obtain

$$\begin{aligned}\frac{dv}{dt} &= a_1 + 2a_2 t + 3a_3 t^2 + \cdots + Na_N t^{N-1} \\ &= -\alpha(a_0 + a_1 t + a_2 t^2 + \cdots + a_N t^N), \\ \alpha &= \frac{G}{C}.\end{aligned} \tag{2.04-4}$$

Because the highest powers of t differ on the two sides, this cannot be true. That bridge we shall cross later. Meanwhile we consider evaluation of at least the first few unknown coefficients, the a's, which is not at all difficult. Placing $t = 0$ gives, from (2.04-3) and (2.02-8),

$$v(0) = a_0 = V_0. \tag{2.04-5}$$

Placing $t = 0$ in (2.04-4) gives

$$a_1 = -\alpha a_0 = -\alpha V_0. \tag{2.04-6}$$

To find the value of a_2, we can differentiate (2.04-4),

$$\begin{aligned}\frac{d^2v}{dt^2} &= 2a_2 + (2)(3a_3 t) + \cdots \\ &= -\alpha(a_1 + 2a_2 t + \cdots),\end{aligned} \tag{2.04-7}$$

and then set $t = 0$ to obtain

$$a_2 = -\frac{\alpha a_1}{2} = \alpha^2 \frac{V_0}{2}. \tag{2.04-8}$$

Further differentiation and setting $t = 0$ gives

$$\begin{aligned}a_3 &= -\frac{\alpha a_2}{3} = -\frac{\alpha^2 V_0}{(2)(3)}, \\ a_4 &= -\frac{\alpha a_3}{4} = +\frac{\alpha^3 V_0}{4!}.\end{aligned} \tag{2.04-9}$$

Finally we obtain

$$v(t) = V_0\left(1 - \alpha t + \frac{\alpha^2 t^2}{2!} - \frac{\alpha^3 t^3}{3!} + \cdots + \frac{\alpha^N t^N}{N!}\right) \tag{2.04-10}$$

We observed above that there is difficulty with the last term or two. But in (2.04-10) we see that the value of a_j decreases rapidly as j increases, since $(j!)$ is a very rapidly increasing function of j. Hence we have in (2.04-10) at least a very good *approximate* solution of the differential equation.

At least this is true if the value of αt is not too large, if we don't to too far out in Fig. 2.04-B (p. 37) in which this is plotted. If αt is too large, we need to increase N. But then, for larger t, that value of N will still not suffice. And so we are led to a fundamental concept of the infinitesimal calculus: to let N increase without limit. Then we have

$$v(t) = V_0\left[1 - (\alpha t) + \frac{(\alpha t)^2}{2!} + \cdots + \frac{(\alpha t)^n}{n!} + \cdots\right]. \tag{2.04-11}$$

We assume that such infinite *series* can be studied and that this one at least has meaning, i.e., *converges*. The series in (2.04-11) is a MacLaurin series, an expansion of the function $v(t)$ about $t = 0$ in a series of powers of t. It can be shown to have meaning (to converge) for all values of (αt), and so it does indeed represent a function.

This function is well known, being the exponential function

$$f(x) = e^x = \exp(x) = 1 + x + \frac{x^2}{2!} + \cdots \tag{2.04-12}$$

In effect it was defined originally [see (2.02-8)] by

$$\frac{df}{dx} = f \tag{2.04-13}$$

which leads to (2.04-12) as we have seen. Its value for any value of x can be calculated from (2.04-12) by taking enough terms to obtain the accuracy we need. The result is expressed (for negative arguments) by the well-known curve of Fig. 2.04-C and the sample values given. Note therein these useful parameter values:

The x value at which the decay is 50 percent of the original value, the *half-life*: $\ln 2 = 0.693$; (2.04-14)

The decay factor in an x (or time) change of unity: $e^{-1} = 0.368$. (2.04-15)

x	Value of x in time constants	e^{-x}
0		1.00
0.25	$\frac{1}{4}$	0.779
0.50	$\frac{1}{2}$	0.607
0.693		0.500 ← "Half-life" point
1.000	1	0.368 ← Value at one time constant
1.386		0.250
1.500	$\frac{3}{2}$	0.223
2.00	2	0.135

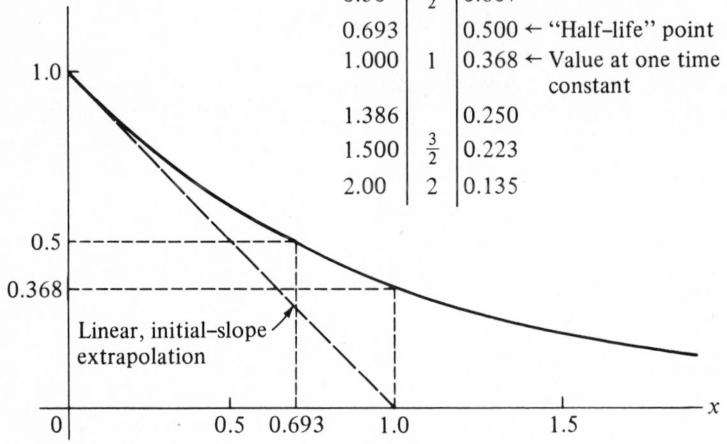

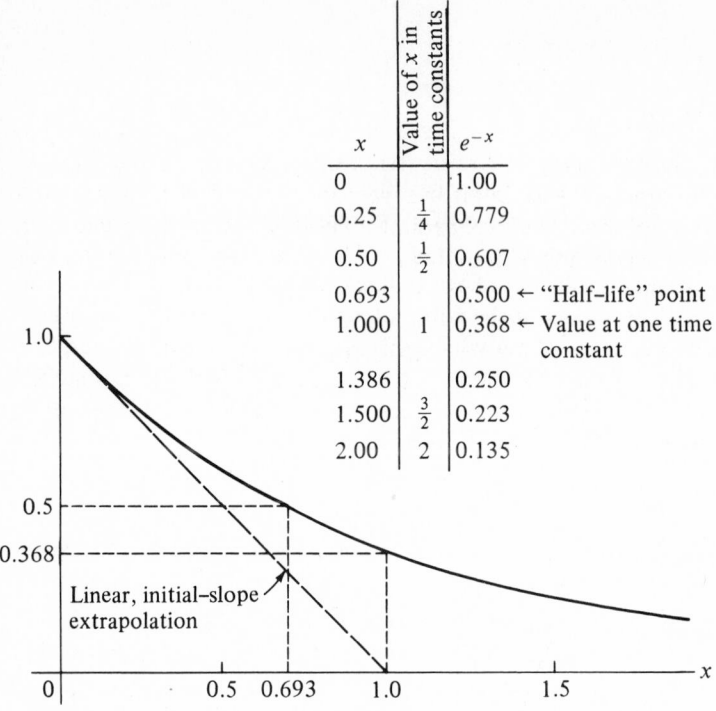

FIG. 2.04-C. The function $\exp(-x) = e^{-x}$.

One could go on and note the quarter-life value of x, the decay when the exponent has reached the value -2, etc. In our work the most useful parameter is the *time constant*, which corresponds to $x = 1$ (see Sec. 2.06). Note that the initial slope is -1, so that linear extrapolation therealong to the x axis leads to the x value of unity.

It is a function essential to circuit analysis and to the study of any other physical situation in which some quantity varies at a rate proportional to that quantity. It is *natural*, it *belongs* to such a situation.

We note that not only is the derivative of e^x equal to e^x itself but so in essence is its integral:

$$\frac{d(e^x)}{dx} = e^x, \tag{2.04-16}$$

$$\int e^x \, dx = e^x + \text{constant}. \tag{2.04-17}$$

We shall soon need to study this function for complex values of its argument, when it reveals an intimate connection with the trigonometric sine and cosine functions. It has many other interesting properties as well, and will be for us the most important of all functions.

2.05 THE *RC* CIRCUIT AGAIN

The simplicity of the circuit analysis problem of (2.02-8) and Fig. 2.02-A is deceptive. We have already extracted therefrom a wealth of information about the nature of impulses and impulsive response, about the nature of differential equations, and about the exponential function. All of these ideas we are going to apply to vastly more complicated circuits, and then our thorough understanding of this simple one will aid us immensely.

Let us now recapitulate, in as compact form as possible, the analysis of the problem of Sec. 2.02. Its formal mathematical statement is

$$C\frac{dv}{dt} + Gv = Q_0\,\delta(t),$$

$$v(0) = 0, \tag{2.05-1}$$

$$v(t) = ?, \qquad 0 < t.$$

Our first step is to observe that the behavior during the life of the impulse is simple: the (large) current flows through the capacitor C (Fig. 2.02-A) leaving it charged with Q_0 (coulombs) at $t = 0+$. Thereafter we have

$$C\frac{dv}{dt} + Gv = 0,$$

$$v(0+) = V_0 = \frac{Q_0}{C}, \tag{2.05-2}$$

$$v(t) = ?, \qquad 0+ < t.$$

The behavior is now the natural, force-free behavior, which is *exponential* since dv/dt is proportional to v. The independent variable is t (time), but the argument x of the exponential e^x can have no dimensions. And so we write $x = pt$, in which the (unknown) constant p has dimensions inverse to those of time. We use this in (2.05-2) as an experiment (of whose success we are now confident), with $v(t) = Ke^{pt}$, in which K is also unknown. The constant K is necessary, again for proper dimensions. The homogeneity of the differential equation allows, nay encourages, this. We find:

$$Cp(Ke^{pt}) + G(Ke^{pt}) = 0,$$

$$v(0+) = Ke^0 = V_0, \tag{2.05-3}$$

$$v(t) = Ke^{pt}, \qquad 0+ < t.$$

The first equation gives (since Ke^{pt} is surely not zero)

$$p = -\frac{G}{C} \tag{2.05-4}$$

for the "natural-frequency" constant. The second gives [since $e^{(0+)(-G/C)}$ is essentially unity]

$$K = V_0. \tag{2.05-5}$$

Hence we find

$$v(t) = V_0 e^{-\alpha t}, \qquad 0+ < t \tag{2.05-6}$$

for the solution, having written $\alpha = G/C = -p$ for brevity.

This sort of solution (analysis) will hold for all our problems. The only complications introduced by large numbers of circuit elements are those that go with increased differential-equation order, large numbers of initial conditions, large numbers of natural frequencies, and many simultaneous equations. But the basic technique—*try* the exponential function, and deal then with the resulting algebra—we shall find always to suffice. Its basic simplicity remains, however complicated the ancillary computations become.

The reasons for this are rooted in our basic postulates about the elements of the circuit: they are *lumped*, *linear*, and *constant*.

Before exploiting our mathematical (exponential) result, it behooves us to physically verify its correctness. Observe now Fig. 2.05-A whose photographs show cathode-ray-tube presentations† of excitation and response in an *RC* circuit, that of Fig. 2.02-A in essence. The excitation is a very narrow pulse, the same in each of the three cases, and essentially an impulse or delta function; its tip is visible below the response in each case. The responses are obviously very much decaying-exponential-like. We note with satisfaction the corroboration of our analysis: (*a*) in $0 < t < 0+$ there is rapid charging of the capacitor, steplike increase of the response v; (*b*) in $0+ < t$ there is essentially exponential decay. The agreement with the predictions of the model is good, though of course not perfect. We note also vertification of the relation $V_0 = v(0+) = Q_0/C$: Q_0 is the same for all three (the excitation pulse is the same for all), so that

a In comparing (*b*) to (*a*), since C doubles, V_0 should halve (it does).

b In comparing (*c*) to (*b*), since C doubles again, V_0 should again halve (it does).

† Suggestions for apparatus and control settings for this and all such "demonstrations" are given in Appendix A.

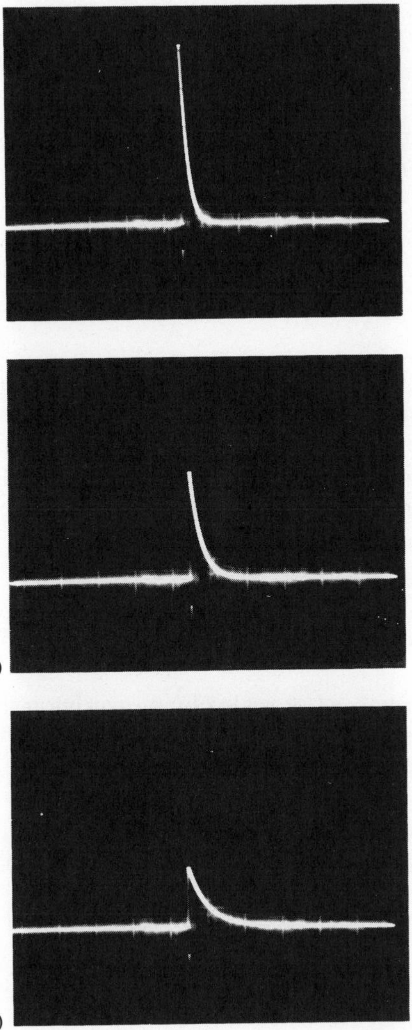

Response

$C = 0.001\ \mu\text{F}$
Time constant = 10 μs

Excitation

(*a*)

Response

$C = 0.002\ \mu\text{F}$
Time constant = 20 μs

Excitation

(*b*)

Response

$C = 0.004\ \mu\text{F}$
Time constant = 40 μs

Excitation

(*c*)

NOTE: $R = 10{,}000\ \Omega$ in all cases; both response and time scales are the same in all cases (the latter is 40 μs per division).

FIG. 2.05-A.

Finally we note, as far as we can tell in a reduced photograph, that the time constants are in the predicted ratios (1 : 2 : 4) and agree with the time scale [one division is one time constant of case (*c*)].

2.06 NORMALIZATION

The result (2.05-6) resembles the basic exponential function of (2.04-12). To use our reservoir of numbers (Fig. 2.04-C) we need only write (2.05-6) as

$$v(t) = V_0 e^{-\alpha t} = V_0 e^{-x}, \qquad x = \alpha t = \frac{t}{T}, \tag{2.06-1}$$

in which we replace $\alpha = G/C$ by T^{-1} in order to emphasize its dimensions. We recognize that $T = C/G$ is the *time constant* of the circuit, with dimension of time. The variable x is free of dimensions, yet measures time, but in units of T. We call it a *normalized* time variable. We can enter curve or table with values of x and immediately find values of v, apart from its multiplier V_0.

We can also normalize the response itself by writing

$$r_n(t) = \frac{v(t)}{V_0} = \frac{v(t)}{v(0+)} = e^{-t/T} \tag{2.06-2}$$

in which r_n is the response voltage normalized or scaled by measuring it in units of its initial value. The subscript n reminds us of the normalization. If we use it for time also, writing $t_n = t/T = (G/C)t$, we have the extraordinarily simple result

$$r_n(t) = e^{-t_n}. \tag{2.06-3}$$

Figure 2.04-C is now all we need to find numerical values therefor.

The simplicity we can gain from these normalizations we shall always find useful. Both *time* and *response* scales can in general conveniently be normalized.

2.07 NUMERICAL CALCULATION OF THE RESPONSE

One important use of normalization (scaling) is to simplify the numerical work that must be done. To plot v against t we need a series of numerical values. They should be close enough yet not too close; they may well be equally spaced, say at $t = 0, T, 2T, 3T, \ldots$. The initial value $v(0)$ we know; it represents the boundary condition from which we start. Then

$$v(T) = v(0)e^{-\alpha T} = Kv(0), \qquad K = e^{-\alpha T}. \tag{2.07-1}$$

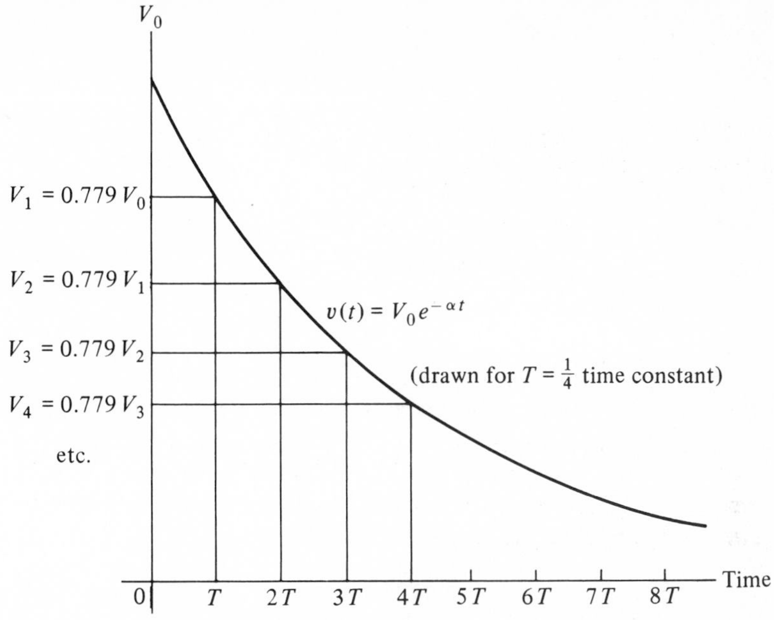

FIG. 2.07-A.

We calculate the value of K from (2.04-10) or look it up in a table that someone else has already calculated. To evaluate $v(2T)$ we do not *need* to calculate any further values of the exponential function, for we know that

$$e^{A+B} = e^A e^B \tag{2.07-2}$$

[or we can show it from (2.04-12)]. Hence,

$$\begin{aligned} v(0) &= v(0), \\ v(T) &= Kv(0), \\ v(2T) &= K^2v(0) = Kv(T), \\ v(3T) &= K^3v(0) = Kv(2T), \\ &\vdots \end{aligned} \tag{2.07-3}$$

To plot the curve of $v(t)$, Fig. 2.07-A, we need only one value of the exponential function,

$$e^{-\alpha T} = K = 1 - \alpha T + \frac{(\alpha T)^2}{2!} - \cdots. \tag{2.07-4}$$

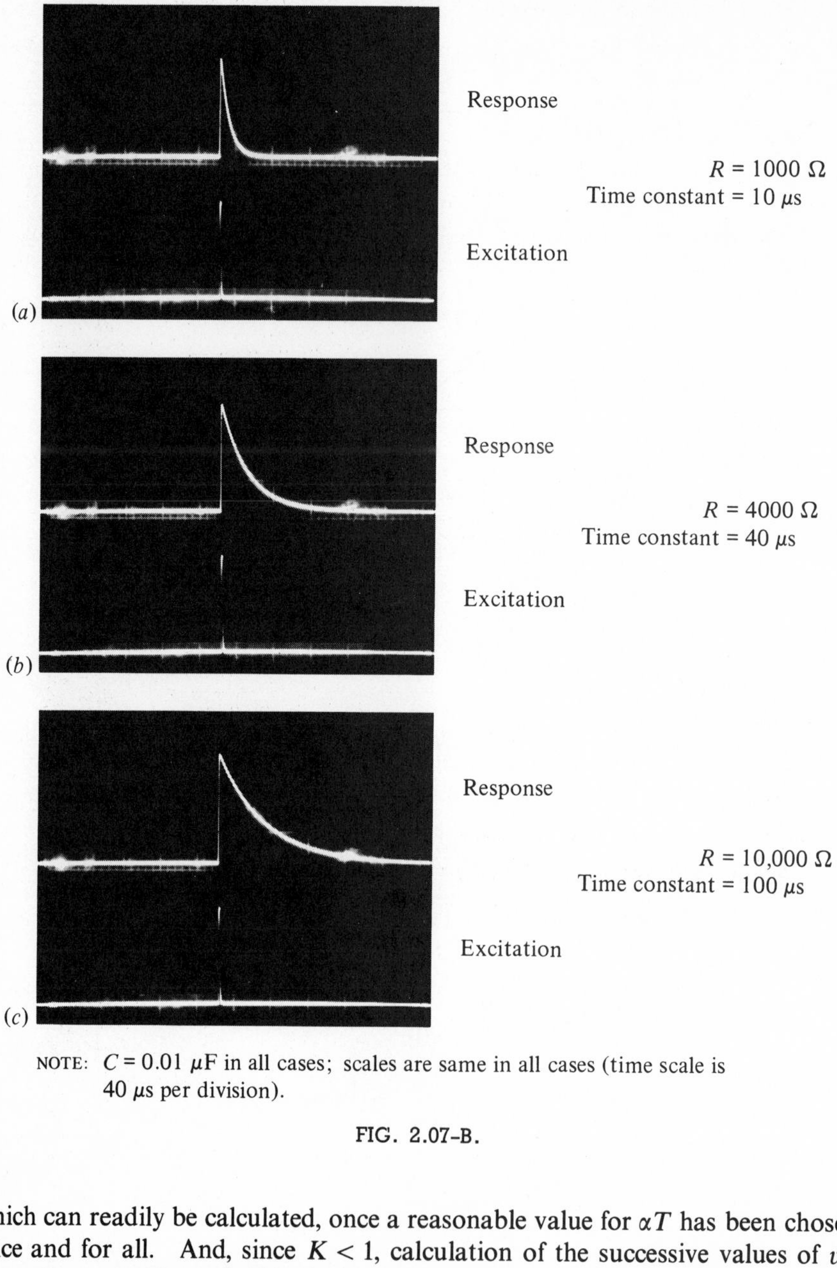

NOTE: $C = 0.01\ \mu$F in all cases; scales are same in all cases (time scale is 40 μs per division).

FIG. 2.07-B.

which can readily be calculated, once a reasonable value for αT has been chosen, once and for all. And, since $K < 1$, calculation of the successive values of $v(t)$ presents no problem of numerical accuracy.

The cathode-ray-tube photographs of Fig. 2.07-B once more corroborate our analysis. Here (in contrast to Fig. 2.05-A) the capacitance in the *RC* circuit is unchanged; the *resistance* value is altered, and hence the time constant, in the three parts. The initial value $V_0 = Q_0/C$ should therefore be the same for all three cases, since the impulse excitation is unchanged. Note in the photographs that V_0 does not change and that the time constant does change as expected (check the numerical values).

We shall later find that the response of even elaborate circuits can still usually be performed in essentially the same, simple way [as in (2.07-3)], though of course more calculations and a somewhat different interpretation of the symbols are needed. But there are many things to study before we can discuss that.

We shall consider the same *RC* circuit with different types of excitation, but it is important also to understand the distribution and variation of energy in the circuit. This we discuss first.

2.08 ENERGY

In electronic circuits the amount of energy stored in capacitors (and inductors) is often numerically very small. Stored in a 1000-pF capacitor charged to 1.2 V, for example, there are only

$$0.5\,(1000 \times 10^{-12})1.2^2 = 720 \times 10^{-12} \text{ joules.} \qquad (2.08\text{-}1)$$

This is not a "power" situation. Nevertheless it is interesting to note how the stored energy varies with time—and in later examples it will be far more important.

To begin with, we recall that energy flows into an element or circuit at the rate vi (watts), the instantaneous *power* p_i of Fig. 2.08-A. The energy that passes the dotted line into the capacitor in Fig. 2.08-B in the time interval t_a to t_b is therefore

$$W = \int_{t_a}^{t_b} vi\,dt = \frac{1}{C}\int_{t_a}^{t_b} q\,dq = C\int_{t_a}^{t_b} v\,dv. \qquad (2.08\text{-}2)$$

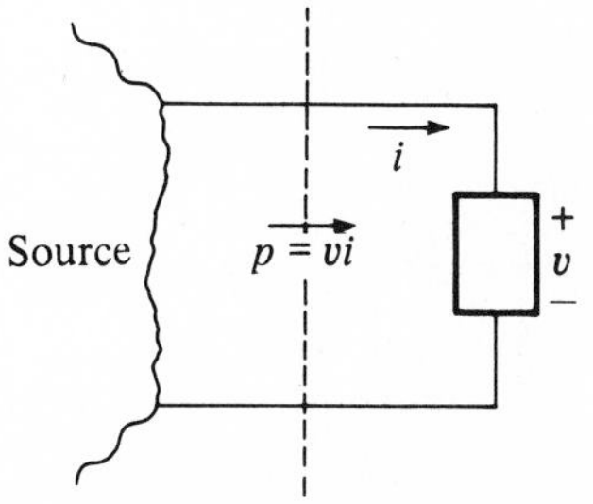

FIG. 2.08–A.

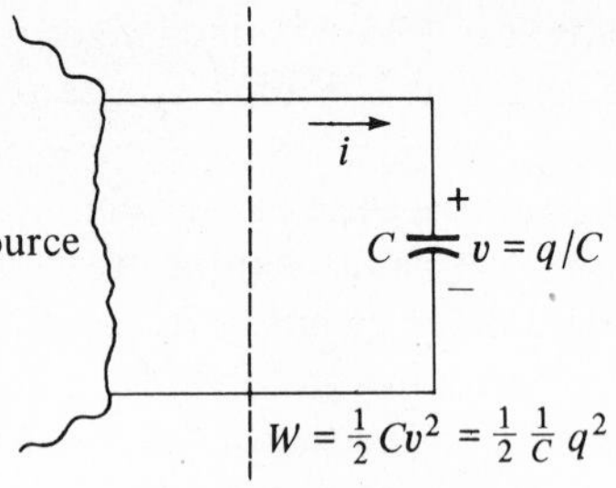

FIG. 2.08-B.

To calculate the energy stored at any instant t, we set $t_a = -\infty$ (or any other instant at which v or q or W is zero); then we have

$$W(t) = \tfrac{1}{2}Dq^2 = \tfrac{1}{2}Cv^2, \tag{2.08-3}$$

in which $D = 1/C$. This is the formula used in (2.08-1) above.

Since the stored energy varies as the square of the voltage, its behavior versus time is like that of the response voltage, except for a factor of 2 in the exponent. (It is always positive, even if v is negative.) Figure 2.08-C shows this. We note the relevant time intervals of Table 2.08-A.

We note in passing that one could base the analysis of the RC circuit on energy principles. Assuming that all the energy from the source flows into the elements R and C implies (in terms of rates of energy flow, in the notation of Fig. 1.06-A) that

$$v \times (i_F) = v \times (i_R) + v \times \left(C\frac{dv}{dt}\right) \tag{2.08-4}$$

from which (1.06-1) follows on division by v.

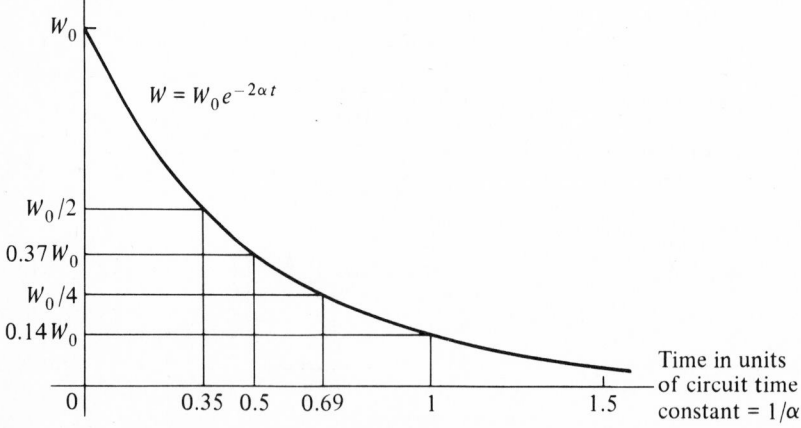

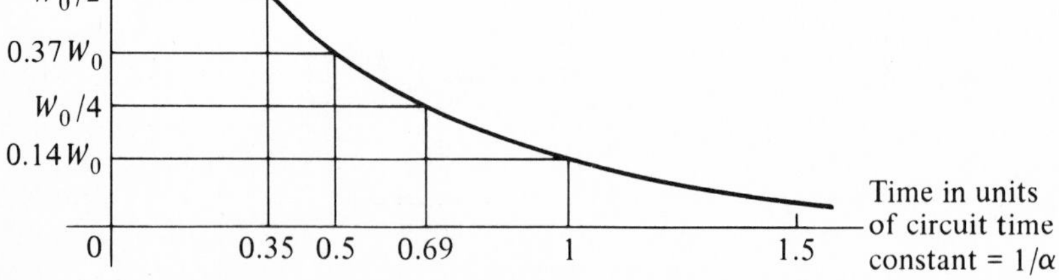

FIG. 2.08-C.

TABLE 2.08-A

Time Increment in Time Constants, *RC*	Decay Factors: Voltage Magnitude	Decay Factors: Stored Energy
0.35	$e^{-0.35} = 0.71$	$e^{-0.69} = 0.50$
0.50	$e^{-0.50} = 0.61$	$e^{-1.00} = 0.37$
0.69	$e^{-0.69} = 0.50$	$e^{-1.39} = 0.25$
1.00	$e^{-1.00} = 0.37$	$e^{-2.00} = 0.14$

2.09 STEP RESPONSE

In our discussion of excitation waves or signal types (Sec. 1.08-9), after the impulse came the step, its integral. Physically this may represent the opening or closing of a switch (Fig. 2.09-A). If the action takes place in a sufficiently short time, we can represent such an excitation by the step function of (1.09-1). And since switch closings and openings are common engineering events, both in control of energy (turning it on and off) and in communication (forming the characters of a Morse code or computer-data message), it is profitable for us to study the response of our simple *RC* circuit amplifier model thereto.

The interesting analysis problem is that of Fig. 2.09-A*a*, to which we add the possibility of initial energy storage (from previous history) to obtain

$$C\frac{dv}{dt} + Gv - I_0 u(t) = 0,$$
$$v(0) = V_0,$$
$$v(t) = ?, \qquad 0 < t. \tag{2.09-1}$$

The natural (exponential) behavior of the *RC* circuit must almost certainly enter into the solution. So also must the (constant) nature of the excitation. A

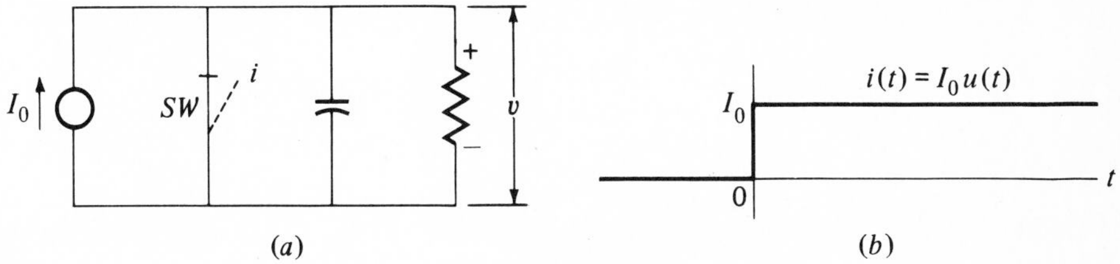

FIG. 2.09-A.

simple experiment will answer the questions "Is that true?" and "If so, how?" We write, for $t > 0$,

$$v(t) = K_1 e^{pt} + K_0, \tag{2.09-2}$$

allowing complete freedom to the three constants (K_1, p, and K_0), and substitute it in (2.09-1):

$$\begin{aligned} &CK_1 p e^{pt} + GK_1 e^{pt} + GK_0 - I_0 \\ &\qquad = (Cp + G)K_1 e^{pt} + GK_0 - I_0 = 0, \qquad t > 0 \\ &V_0 = K_1 + K_0 . \end{aligned} \tag{2.09-3}$$

Since the first equation must hold for all positive values of t, and only e^{pt} therein varies with t, the coefficient of e^{pt} must be zero. Hence

$$p = -\frac{G}{C} = -\alpha \tag{2.09-4}$$

as we could well have predicted. Also

$$K_0 = \frac{I_0}{G} = RI_0 \tag{2.09-5}$$

and

$$K_1 = V_0 - RI_0 \tag{2.09-6}$$

and so

$$v(t) = (V_0 - RI_0)e^{-\alpha t} + RI_0 \tag{2.09-7}$$

$$= V_0 e^{-\alpha t} + RI_0(1 - e^{-\alpha t}), \qquad t > 0. \tag{2.09-8}$$

Figure 2.09-B shows the behavior of the response voltage for various sorts of initial condition. It is, as expected, a combination of the circuit's natural behavior $e^{-\alpha t}$ and the excitation's form (constant). To name the parts accordingly is very convenient:

$$v(t) = \underbrace{(V_0 - RI_0)e^{-\alpha t}}_{v_N} + \underbrace{(RI_0)}_{v_F}, \qquad t > 0. \tag{2.09-9}$$

The subscript N denotes the natural-behavior part of the response; the subscript F indicates the forced (excitationlike) part. We shall find practically all our network responses, however complicated, to have these two parts.

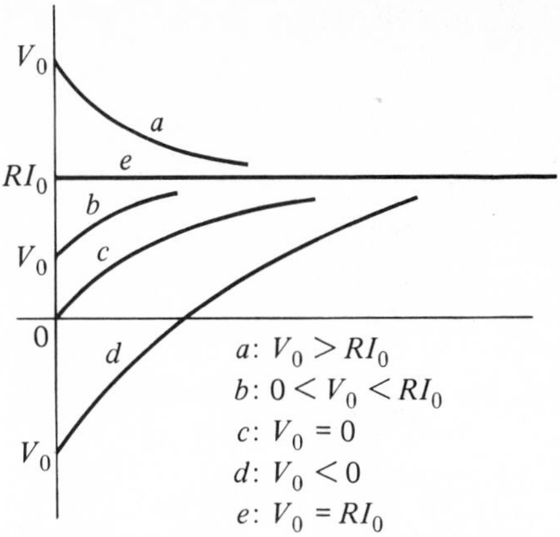

FIG. 2.09-B.

We also rewrite (2.09-9) as

$$v(t) = [v(0) - v(\infty)]e^{-\alpha t} + v(\infty) \tag{2.09-10}$$

which emphasizes the initial and final values of v. [Many practical circuits are in essence merely *RC* circuits whose time constant or α is easily calculated, as are $v(0)$ and $v(\infty)$, almost by inspection; for them (2.09-10) is very useful.]

We note that in any time interval equal to one time constant $[1/\alpha = C/G = RC$ (seconds)$]$ the natural part decays by 63 percent, for its multiplier changes by a factor of $e^{-1} = 0.37$. The response from "rest," i.e., the case $V_0 = 0$, is important enough to draw separately, in normalized form (Fig. 2.09-C), with some interesting numerical values.

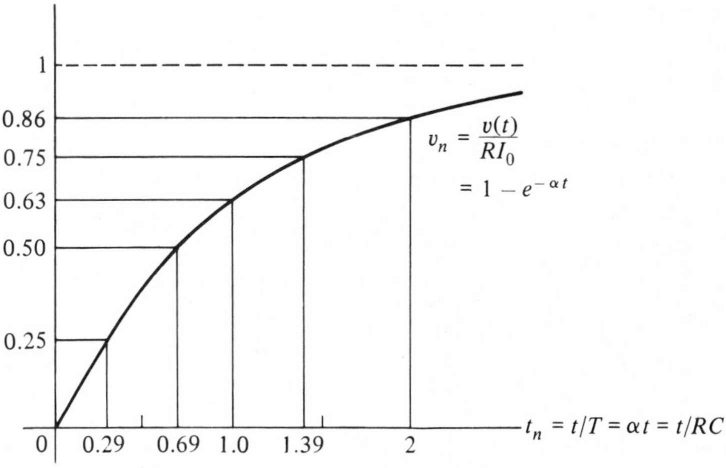

FIG. 2.09-C.

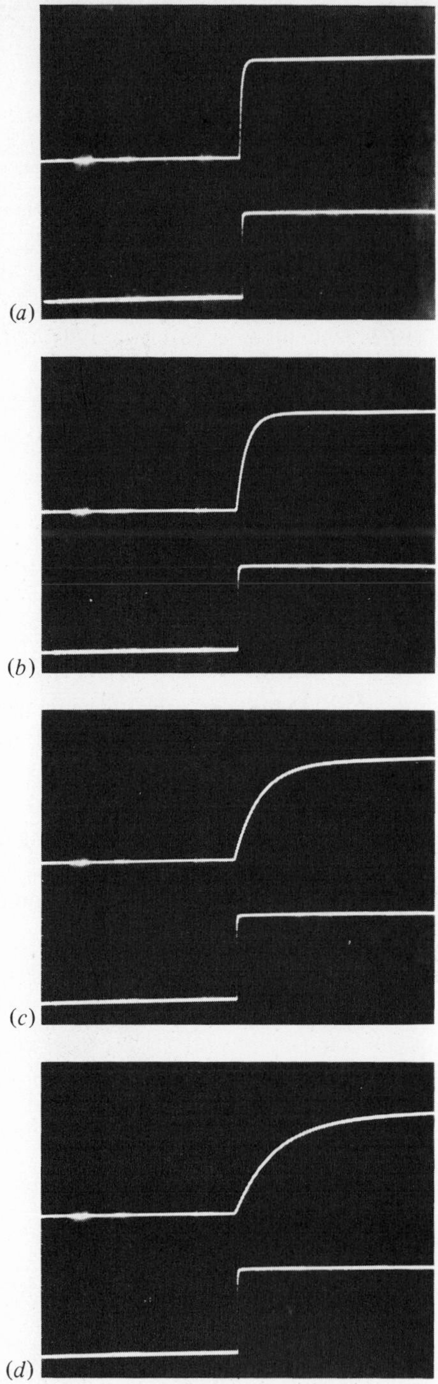

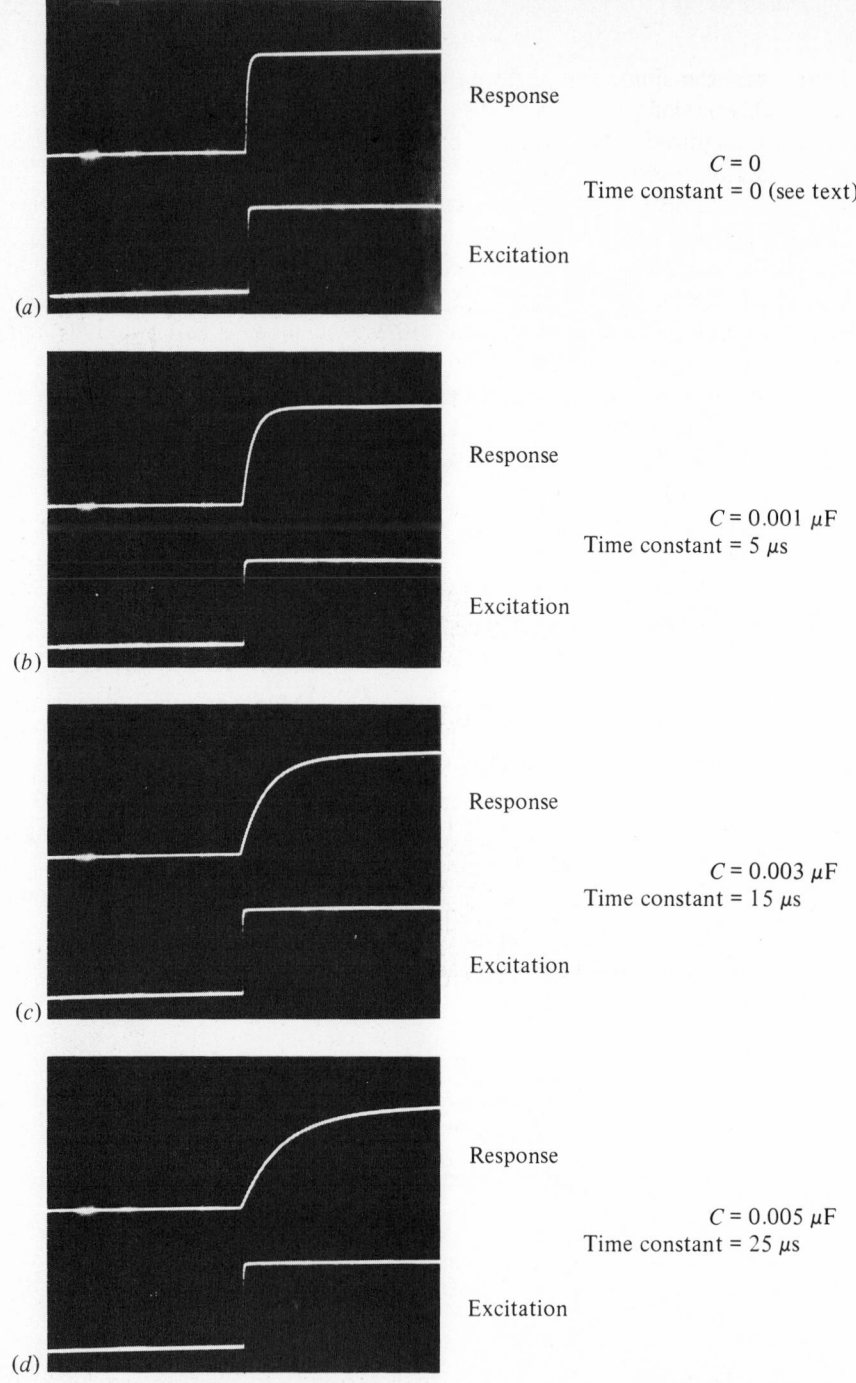

(a) Response / Excitation — $C = 0$, Time constant = 0 (see text)

(b) Response / Excitation — $C = 0.001\ \mu F$, Time constant = 5 μs

(c) Response / Excitation — $C = 0.003\ \mu F$, Time constant = 15 μs

(d) Response / Excitation — $C = 0.005\ \mu F$, Time constant = 25 μs

NOTE: $R = 5000\ \Omega$ in all parts; time scale is 20 μs/division in all parts; scales are the same in all parts.

FIG. 2.09-D.

Note again the simplicity of the result, once the importance of the natural (exponential) behavior and of the excitation waveshape, and the utility of normalization, are recognized. For physical corroboration, Fig. 2.09-D presents four actual *RC* circuit responses, from rest, to step excitations. In them R is unchanged; C is altered to give the four cases. In (*a*) $C = 0$, but some rounding is noticeable because *parasitic* capacitance (and inductance) effects in wiring cannot be completely eliminated. (A *part* of the rounding is also due, of course, to the nonsquare corner of the excitation step.) Physical corroboration of our results is therefore extremely important, lest we become so fascinated by our circuits that we forget that they are but models and we ignore the realities of the engineering world! A refined model could take account of these parasitic effects, but is not worth the digression here. In Fig. 2.09-D*a* the time constant is in theory zero; in practice the response corner is never rectangular.

In (*b*), (*c*), and (*d*) the capacitance C is given values in the ratios 1 : 3 : 5; R is constant, and so the time constants should have the ratios 1 : 3 : 5. The initial response slopes I_0/C should take inverse ratios, since I_0 is constant. Examination of the photographs of Fig. 2.09-D corroborates these. Again we find good (not perfect) corroboration of the results of our analysis.

2.10 STEP AND IMPULSE RESPONSE

An interesting observation is that the response from rest to a step, and the response from rest to an impulse, are very simply related to each other. We have

$$\text{Response to current step } I_0\, u(t) = RI_0(1 - e^{-\alpha t})u(t) \qquad (2.10\text{-}1)$$

and

$$\text{Response to current impulse } Q_0\, \delta(t) = \frac{Q_0}{C}(e^{-\alpha t})u(t) \qquad (2.10\text{-}2)$$

in which

$$\alpha = \frac{G}{C} = \frac{1}{RC} \qquad (\text{s}^{-1}) \qquad (2.10\text{-}3)$$

We observe that the step response is essentially the integral of the impulse response. Since the step function is itself the integral of the impulse (Secs. 1.09 and 1.10) it is not greatly surprising that the corresponding responses should be similarly related.

The relation is an important one and requires clarification. It is simplest first to define *step response* as the response to a *unit* step from rest; thus:

$$\begin{aligned} r_u(t) &= RI_0(1 - e^{-\alpha t})u(t) \qquad \text{when } I_0 = 1\text{ A} \qquad (\text{V}) \\ &= R(1 - e^{-\alpha t})u(t) \qquad (\text{V/A} = \Omega). \end{aligned} \tag{2.10-4}$$

We similarly define *impulse response* as the response to a *unit* impulse, from rest:

$$\begin{aligned} r_\delta(t) &= \frac{Q_0}{C} e^{-\alpha t}u(t) \qquad \text{when } Q_0 = 1\text{ C} \qquad (\text{V}) \\ &= \frac{1}{C} e^{-\alpha t}u(t) \qquad (\text{V/C} = \text{F}^{-1}). \end{aligned} \tag{2.10-5}$$

Then we have

$$\begin{aligned} \int_0^t r_\delta(x)\,dx &= \frac{e^{-\alpha t} - 1}{C(-\alpha)} = R(1 - e^{-\alpha t})u(t) \\ &= r_u(t) \qquad (s/F) = \Omega = \text{V/A}). \end{aligned} \tag{2.10-6}$$

[The use of x as the (dummy) variable of integration, instead of t, obviates confusion between this variable and the (quite different) upper limit of the integral.]

This relation we shall find to hold in general:

The step response is the integral of the impulse response (2.10-7)

or

The impulse response is the derivative of the step response. (2.10-8)

Note the need, for the relations to hold precisely, for defining the two responses to be calculated from an initial condition of *rest*, and for *unit* excitations, and for consequent *careful attention to dimensions.* From now on, the symbols $r_u(t)$ and $r_\delta(t)$ shall mean the responses so defined.

2.11 PULSE RESPONSE

The pulse signal (Secs. 1.09 and 1.10) is another important type of excitation. We can immediately write the response of the *RC* circuit thereto by expressing the pulse as the superposition of two steps (Sec. 1.09):

$$i(t) = I_0\, u(t) - I_0\, u(t - T_0) \tag{2.11-1}$$

as shown in Fig. 2.11-A.

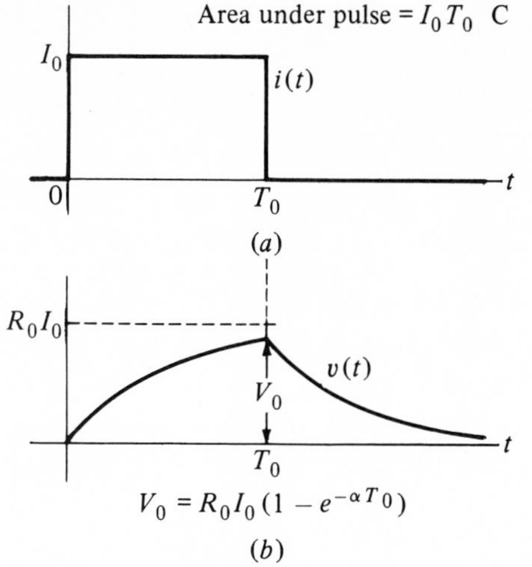

FIG. 2.11-A.

For now, by using superposition we find the response, from rest, to be

$$
\begin{aligned}
v(t) &= I_0 r_u(t) - I_0 r_u(t - T_0) \\
&= RI_0(1 - e^{-\alpha t}) && 0 < t < T_0 \\
&= RI_0(1 - e^{-\alpha T_0})e^{-\alpha(t - T_0)} && \\
&= RI_0(e^{\alpha T_0} - 1)e^{-\alpha t} && T_0 < t
\end{aligned}
\tag{2.11-2}
$$

as shown in Fig. 2.11-A*b*.

What strikes us most about the curves therein is the *distortion* introduced by the *RC* circuit. If this indeed represents an amplifier, and the voltage level RI_0 represents a gain in amplitude over that of the input (excitation) signal, the amplification is achieved only at the cost of waveshape distortion. This is normal, and one objective of amplifier design is to keep the distortion within tolerable limits. A more complicated (and more expensive) amplifier may produce an amplified wave that more closely follows the input pulse. Or we may devise a (nonlinear) regenerative device to improve the "pulse" of Fig. 2.11-A*b*. The subject is too complicated for discussion here.

The distortion is simply evidence of the inherently sluggish character of the amplifier, modeled by the physical capacitor C or the mathematical derivative dv/dt. The device cannot respond instantaneously. The time constant RC is a convenient measure of this slowness. If the apparatus that follows this amplifier in the complete system is not greatly inconvenienced by this sluggishness, fine; if it is, then ways of improving the response must be found.

The wave of Fig. 2.11-A*b* is still quite pulselike if the "duration" T_0 is much longer than the time constant RC. Figure 2.11-B, drawn to scale for $RC = 0.1T_0$, shows this. In fact, by decreasing the value of C, and so increasing the

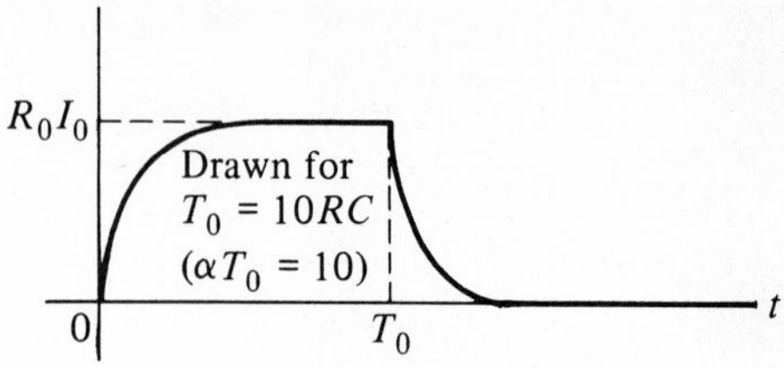

FIG. 2.11-B.

value of α, the response wave can be made arbitrarily like the rectangular excitation pulse. In the limit, since the capacitor disappears, then the response v, being proportional to the excitation current ($v = Ri$), is of course rectangular also.

Physical corroboration is found in Fig. 2.11-C, which shows four actual cases, presumably much like the theoretical cases of Figs. 2.11-A and 2.11-B (see also Fig. 1.10-A). The excitation pulse duration is 150 μs in each case; its amplitude is unchanged, as is the value of the resistor R (10,000 Ω). Hence the asymptote of the first piece of the response, the quantity RI_0, is the same for all four. The capacitance C is varied so that the time constants (80, 40, 20, 10 μs) are respectively about $\frac{1}{2}$, $\frac{1}{4}$, $\frac{1}{8}$, and $\frac{1}{16}$ of the pulse width. The agreement with the predictions of the model (circuit) is again excellent. (The scales are the same in all cases; the time scale is 40 μs per division.)

If T_0 is small, and the amplitude large, the response even resembles an impulse. Let us write

$$Q_0 = I_0 T_0 \qquad \text{(C)} \tag{2.11-3}$$

to represent the total charge delivered by the current excitation pulse (Fig. 2.11-A*a*). Then the response is

$$\begin{aligned} v(t) &= \frac{RQ_0}{T_0}(1 - e^{-\alpha t}), && 0 < t < T_0, \\ &= \frac{RQ_0}{T_0}(e^{\alpha T_0} - 1)e^{-\alpha t} && T_0 < t. \end{aligned} \tag{2.11-4}$$

Suppose we now make T_0 smaller and smaller, increasing I_0 to compensate, so that the total charge delivered, Q_0, remains constant. Then, in the limit, two things of interest occur. In the first place, the excitation becomes an impulse:

$$i(t) \to Q_0\,\delta(t). \tag{2.11-5}$$

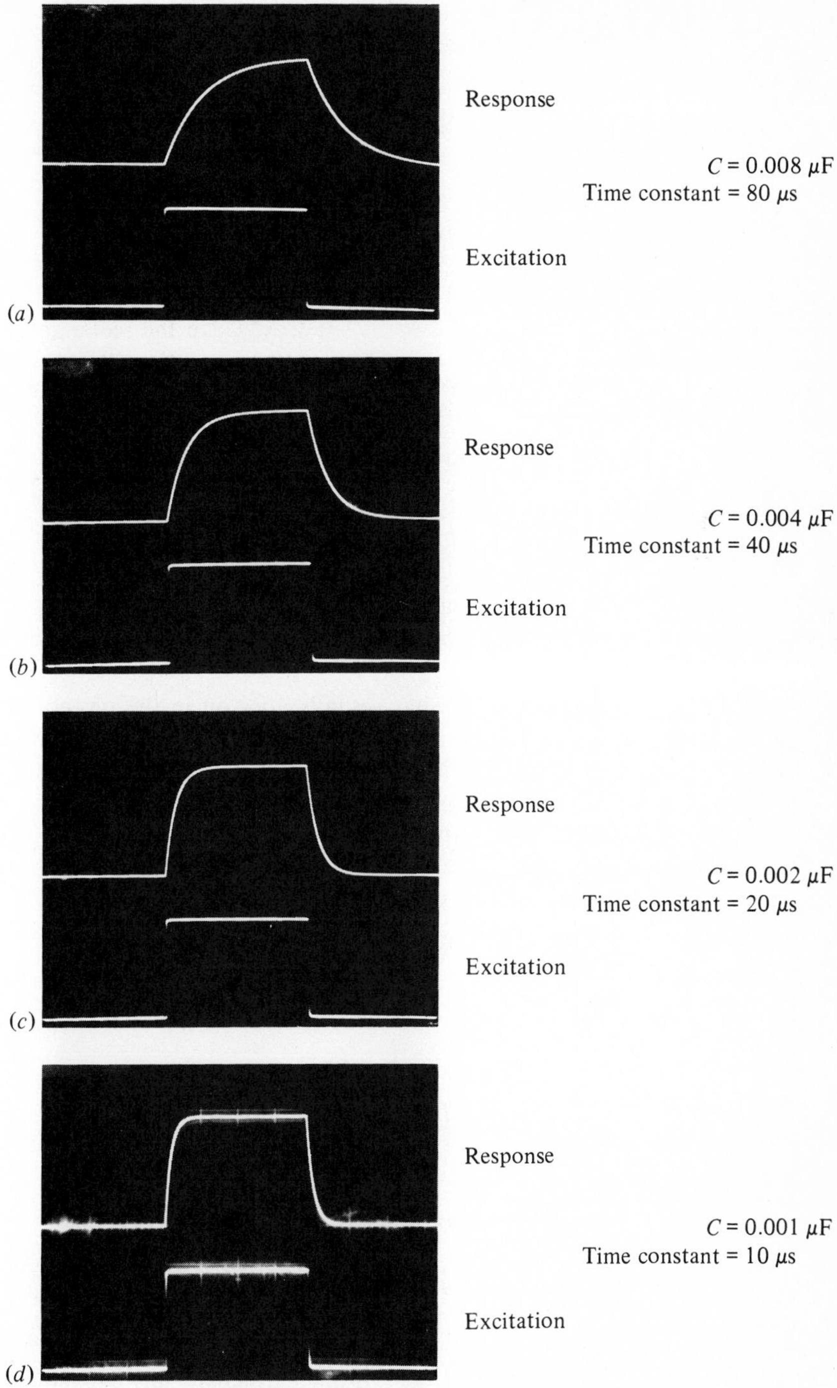
Response
$C = 0.008\ \mu F$
Time constant = 80 μs
Excitation
(a)
Response
$C = 0.004\ \mu F$
Time constant = 40 μs
Excitation
(b)
Response
$C = 0.002\ \mu F$
Time constant = 20 μs
Excitation
(c)
Response
$C = 0.001\ \mu F$
Time constant = 10 μs
Excitation
(d)

FIG. 2.11-C.

Secondly, the response becomes a simple exponential, as it should, for the first line of (2.11-4) is no longer of interest, and the second develops thus:

$$\begin{aligned} v(t) &= \frac{RQ_0}{T_0}\left(1 + \alpha T_0 + \frac{\alpha^2 T_0{}^2}{2!} + \cdots - 1\right)e^{-\alpha t} \\ &= RQ_0\left(\alpha + \frac{\alpha^2 T_0}{2!} + \cdots\right)e^{-\alpha t} \\ &= \frac{Q_0}{C}e^{-\alpha t}u(t) \qquad \text{in the limit as } T_0 \to 0. \end{aligned} \tag{2.11-6}$$

This merely checks our work in Sec. 2.05, for (2.11-6) is the same as (2.05-6). Finally, if we both decrease T_0 as above *and* decrease C, then $v(t)$ becomes an impulse:

$$v(t) \to \lim_{C \to 0}\left[\frac{RQ_0}{RC}e^{-t/(RC)}\right] = RQ_0\,\delta(t). \tag{2.11-7}$$

The area under the wave in the brackets above is RQ_0, and its limit is the impulse shown, as it should be. Note that here is another way of writing the impulse function: $\delta(t) = \lim_{\alpha \to \infty}(\alpha e^{(-\alpha t)})$.

Very helpful in understanding what actually goes on in the physical circuit during impulsive excitation is the set of curves of Fig. 2.11-D. These are drawn for $T_0 = 0.1RC$, for a pulse of duration quite short compared to the circuit's time constant. Then I_0 must be rather large if Q_0 is to have a respectable value. The response voltage has barely time to get started before the pulse is gone. It is virtually linear with time, since in

$$\begin{aligned} v(t) &= RI_0(1 - e^{-\alpha t}) = RI_0\left[1 - \left(1 - \alpha t + \frac{\alpha^2 t^2}{2!} - \cdots\right)\right] \\ &= RI_0\left(\alpha t - \frac{\alpha^2 t^2}{2!} + \cdots\right) \\ &= \frac{Q_0}{C}\left[\frac{t}{T_0} - \frac{T_0}{2RC}\left(\frac{t}{T_0}\right)^2 + \cdots\right], \end{aligned} \tag{2.11-8}$$

only the first term (αt or t/T_0) need be considered. The curves of resistor current ($i_R = Gv$) and capacitor current ($i_C = C\,dv/dt$) are added to complete the picture.

When T_0 is small enough to consider the current an impulse, then i_C clearly eclipses i_R: practically all the excitation current flows through C, very little flows through R, and the voltage rises very rapidly to Q_0/C. This is the physical behavior for a "real" impulse excitation. In the mathematically convenient (and physical equivalent if $T_0 \ll RC$) case, *all* the current flows through C and the

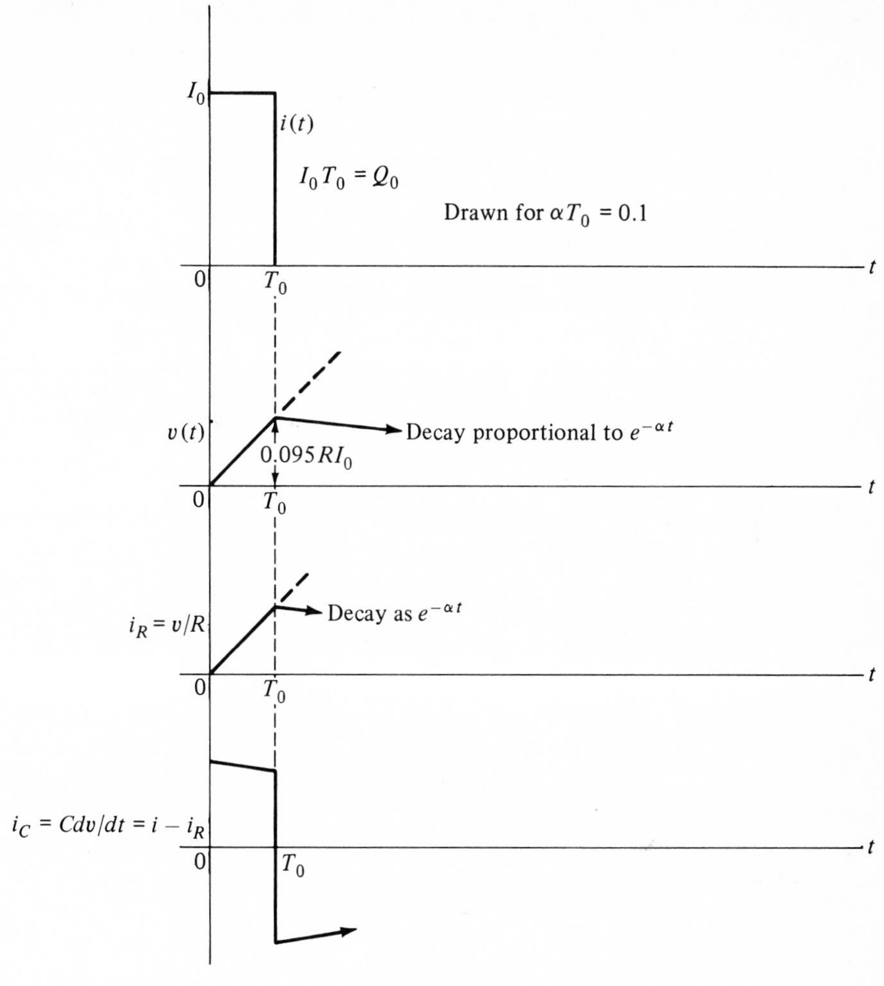

FIG. 2.11-D.

voltage steps immediately to Q_0/C. It is helpful, to retain perspective, to remember that what actually happens is the behavior shown in Fig. 2.11-D, T_0 being very short indeed.

2.12 MULTIPLE-PULSE RESPONSE

It is a simple matter now to write down the response to an excitation composed of many steps or pulses. Figure 2.12-A*a* illustrates such a *piecewise-constant* wave, composed of a series of steps (or pulses), something like an erratic

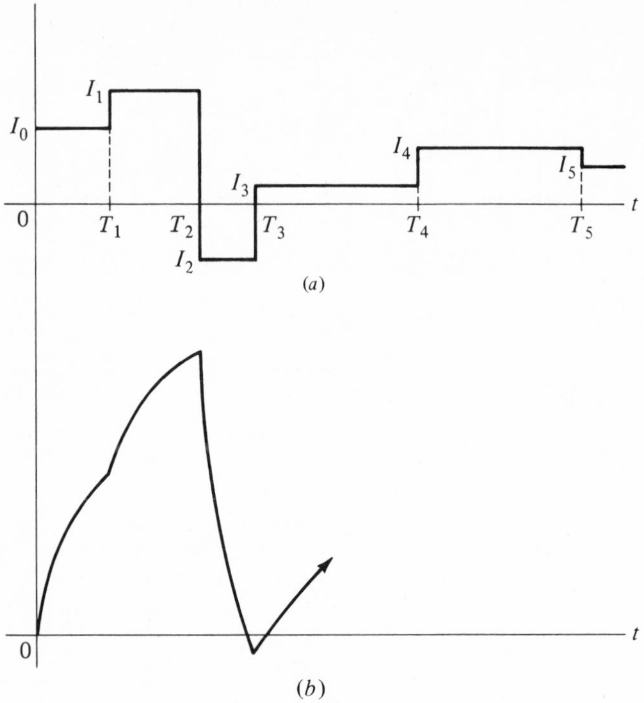

FIG. 2.12-A.

staircase. The response thereto, if $v(0) = V_0$, is

$$v(t) = V_0 e^{-\alpha t}u(t) + I_0 r_u(t) + (I_1 - I_0)r_u(t - T_1) + (I_2 - I_1)r_u(t - T_2) + (I_3 - I_2)r_u(t - T_3) + (I_4 - I_3)r_u(t - T_4), \tag{2.12-1}$$

drawn in Fig. 2.12-A*b*. Note how useful is the step-function notation $u(t)$, which is of course included in $r_u(t)$ of (2.10-4). It obviates the need for writing many cumbersome notes like "$0 < t < T_1$," "$T_1 < t < T_2$," etc., to express the piecewise nature of the response expressions (which of course it *is* essential to state).

The response to a *regular*, repetitive pulselike excitation is simpler and more tractable. And it is interesting and informative. Consider now the symmetrical, indefinitely recurring, square-wave excitation of Fig. 2.12-B. The response, starting from V_0, is

$$v(t) = [V_0 e^{-\alpha t}u(t)] + I_0[r_u(t) - r_u(t - T_0) + r_u(t - 2T_0) - \cdots]. \tag{2.12-2}$$

The character of this response wave is interesting. It obviously does *not* repeat, as the excitation does. *But* it does seem to *approach* a steady, repetitive state

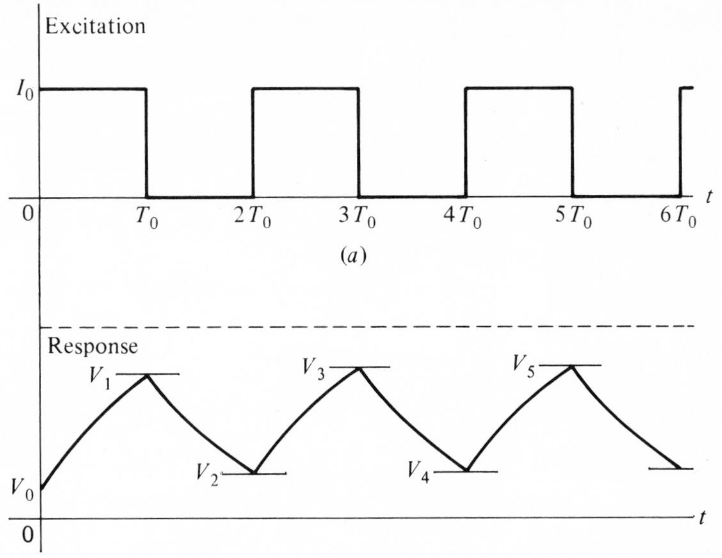

(b)

FIG. 2.12-B.

as time goes on. And it seems entirely reasonable physically that this *transient* (nonrepetitive) state should eventually lead to a *steady* (repetitive) state, because of the repetitive nature of the excitation.

We can investigate this mathematically by calculating and tabulating the values of v for a series of values of time, say, at multiples of T_0, the "change-over" times. If we can find a trend and show the existence of a limit of some sort, we shall have proved the point.

Given that $r_u(t) = R(1 - e^{-\alpha t})u(t)$, (2.10-4), we easily calculate the values given in Table 2.12-A, from (2.12-2). It is convenient first to define $K = e^{-\alpha T_0}$, which

TABLE 2.12-A

$\frac{t}{T_0}$	**Response Voltage** $(K = e^{-\alpha T_0})$
0	V_0
1	$V_0 K + RI_0(1 - K)$
2	$V_0 K^2 + RI_0(1 - K^2) - RI_0(1 - K)$
3	$V_0 K^3 + RI_0(1 - K^3) - RI_0(1 - K^2) + RI_0(1 - K)$
4	$V_0 K^4 + RI_0(1 - K^4) - RI_0(1 - K^3) + RI_0(1 - K^2) - RI_0(1 - K)$
⋮	⋮
n (odd)	$V_0 K^n + RI_0(1 - K + K^2 - K^3 + \cdots - K^n) = V_0 K^n + RI_0 \frac{1 - K^{n+1}}{1 + K}$
$n + 1$ (even)	$V_0 K^{n+1} + RI_0(K - K^2 + K^3 - \cdots + K^{n+1}) = V_0 K^{n+1} + RI_0 K \frac{1 - K^{n+1}}{1 + K}$

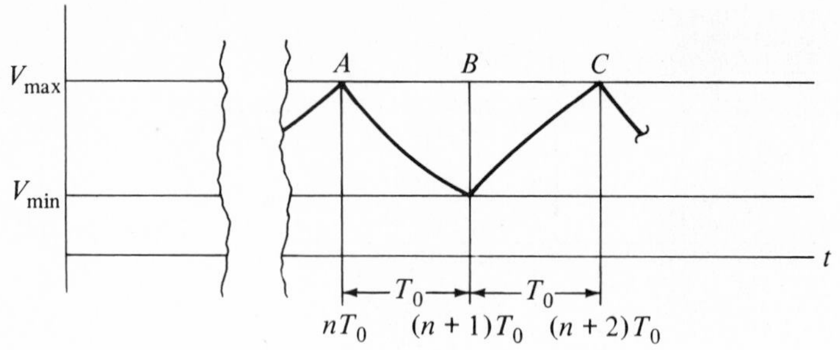

FIG. 2.12-C.

we know to be less than unity, for it appears many times. The first five lines of the table come from using (2.10-4) in (2.12-2) at the appropriate values of time. The n and $n+1$ lines are written by extension therefrom; the sums there are obtained from the conventional geometric series formula. We see that as t (or n) increases, the maximum and minimum values *do* approach definite limits:

$$V_{max} = \lim_{n\to\infty}\left(V_0 K^n + RI_0 \frac{1-K^{n+1}}{1+K}\right) = \frac{RI_0}{1+K}, \tag{2.12-3}$$

$$V_{min} = \lim_{n\to\infty}\left(V_0 K^{n+1} + RI_0 K \frac{1-K^{n+1}}{1+K}\right) = \frac{RI_0 K}{1+K}, \tag{2.12-4}$$

since $K = e^{-\alpha T_0} < 1$. Figure 2.12-C shows the ultimate steady-state behavior. Theoretically, it is never reached, but once $V_0 K^n$ becomes small enough, the maxima and minima differ negligibly from the values above, and the steady state has, in effect, set in.

An alternative approach is to *assume* that a steady state exists and to treat V_{max} and V_{min} in Fig. 2.12-C as unknowns. We can write two simultaneous equations to determine them:

$$\begin{aligned} V_{min} &= V_{max} K, \\ V_{max} &= V_{min} K + RI_0(1-K), \end{aligned} \tag{2.12-5}$$

in which $K = e^{-\alpha T_0}$. The first expresses the force-free (natural) behavior that obtains in the interval AB; the second comes from (2.09-8) which applies in the interval BC. Their solution gives (2.12-3) and (2.12-4), as it should.

The steady state of Fig. 2.12-C is readily observed with simple equipment, for physical corroboration. See Fig. 2.12-D in which the excitation is a symmetrical square wave, "on" for 0.5 ms and "off" for 0.5 ms. The time scale is 0.2 ms per division, the value of R in the RC circuit is the same (2000 Ω) for all cases, and C is varied to give several illustrative time constants. Figure 2.12-E shows a variation in which the pulse length is 200 μs and the pulse separation

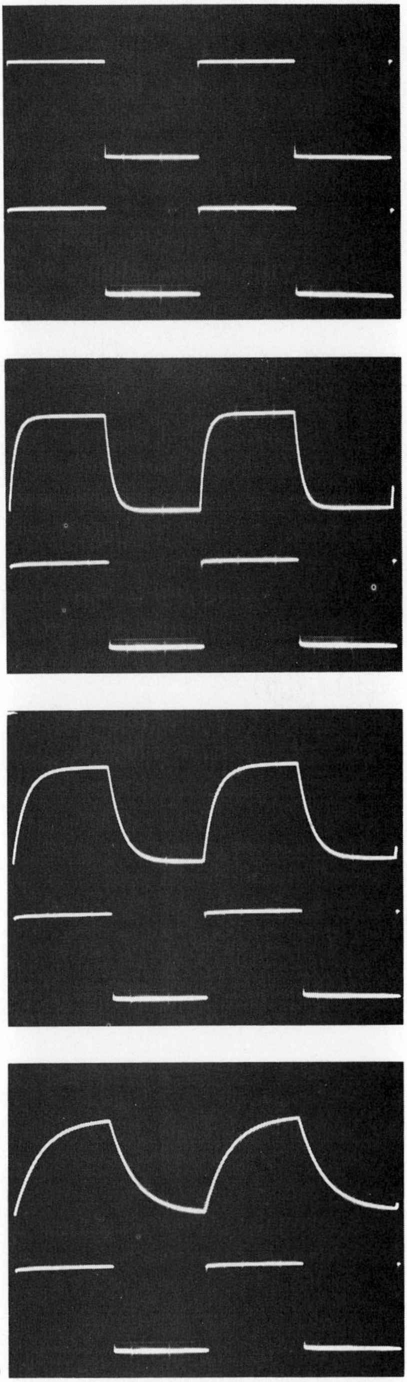

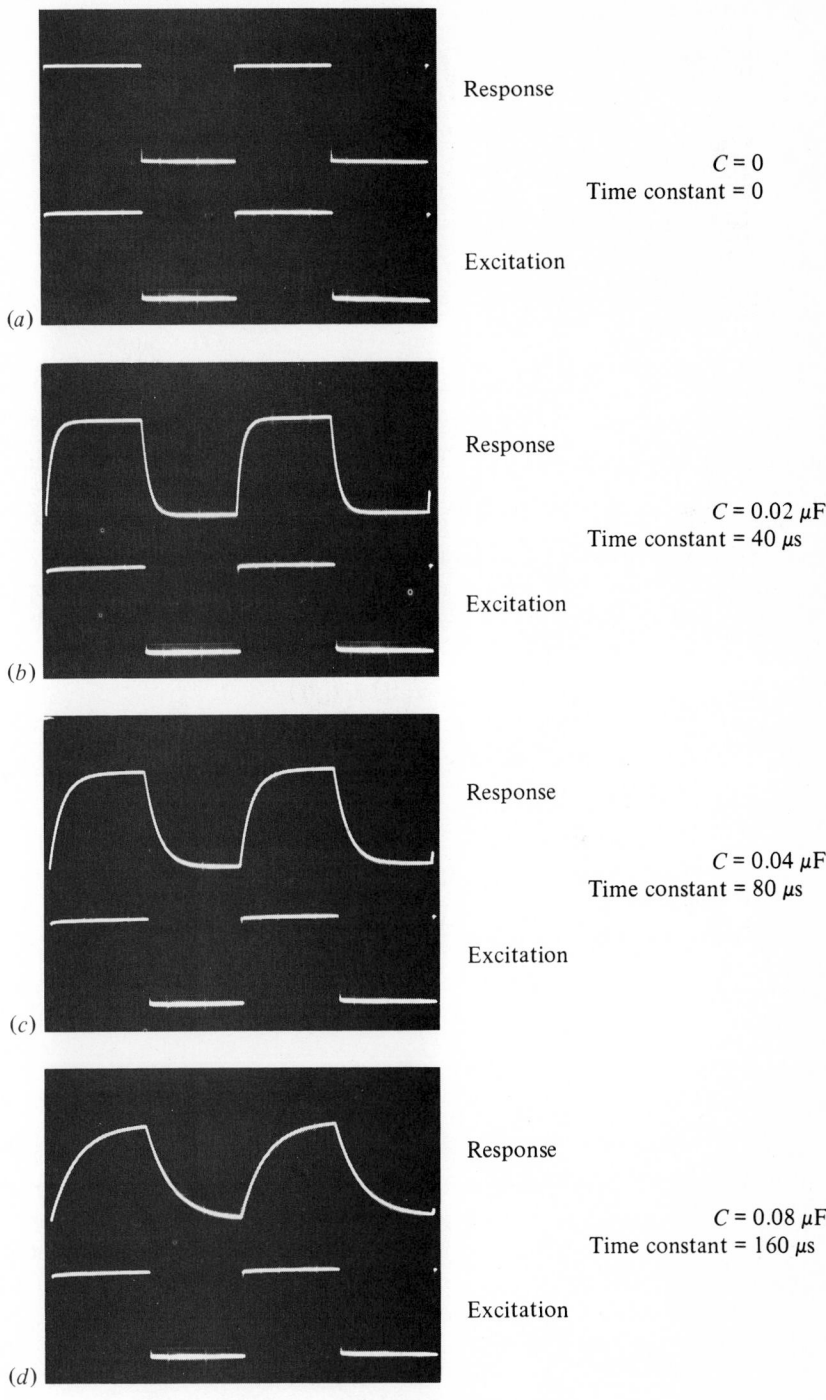

(*a*)
Response
$C = 0$
Time constant = 0
Excitation

(*b*)
Response
$C = 0.02\ \mu F$
Time constant = 40 μs
Excitation

(*c*)
Response
$C = 0.04\ \mu F$
Time constant = 80 μs
Excitation

(*d*)
Response
$C = 0.08\ \mu F$
Time constant = 160 μs
Excitation

FIG. 2.12-D.

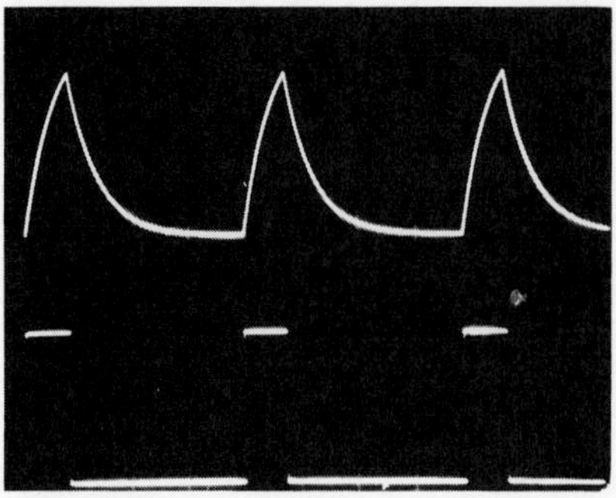

Response

$R = 2000\ \Omega$
$C = 0.08\ \mu$F
Time constant $= 160\ \mu$s

Excitation

Pulse width $= 0.2$ ms
Pulse separation $= 1.0$ ms
(1:5 ratio)

FIG. 2.12-E.

is 1.0 ms. In this (unsymmetrical) case R is again 2000 Ω, and C is 0.08 μF, so that the time constant is 160 μs. The agreement with the model's behavior is again excellent.

Since the response goes through a transient (nonpermanent) state into a steady (permanent) state, we naturally wonder whether the expression (2.12-2) can easily be divided into two such parts. *Both* terms there are nonrepetitive. But we can write $r_{ss}(t)$ for the steady-state wave of Fig. 2.12-C, and then write

$$v(t) = r_{tr}(t) + r_{ss}(t) \tag{2.12-6}$$

as a definition of the transient part, r_{tr}. Then

$$\begin{aligned} r_{tr}(t) &= v(t) - r_{ss}(t) \\ &= \{V_0 e^{-\alpha t} + I_0[r_u(t) - r_u(t - T_0) + r_u(t - 2T_0) - \cdots]\} \\ &\quad - \{[V_{\min} e^{-\alpha t} + I_0 r_u(t)][u(t) - u(t - T_0)] \\ &\quad + (V_{\max} e^{-\alpha(t - T_0)})[u(t - T_0) - u(t - 2T_0)] \\ &\quad + [V_{\min} e^{-\alpha(t - 2T_0)} + I_0 r_u(t - 2T_0)][u(t - 2T_0) - u(t - 3T_0)] \\ &\quad + \cdots\} \\ &= (V_0 - V_{\min})e^{-\alpha t}u(t). \end{aligned} \tag{2.12-7}$$

The result, such a simple expression, is both surprising and confirming of our suspicion. It should be carefully checked (Prob. 2-K); the checking is an excellent exercise in understanding what is happening, and in the interpretation of combinations of step functions. Figure 2.12-F depicts this resolution of the response into *transient* and *steady-state* components. It is drawn for the simple case:

$$V_0 = 0 \text{ (initial rest)} \qquad \text{and} \qquad \alpha T_0 = 1.$$

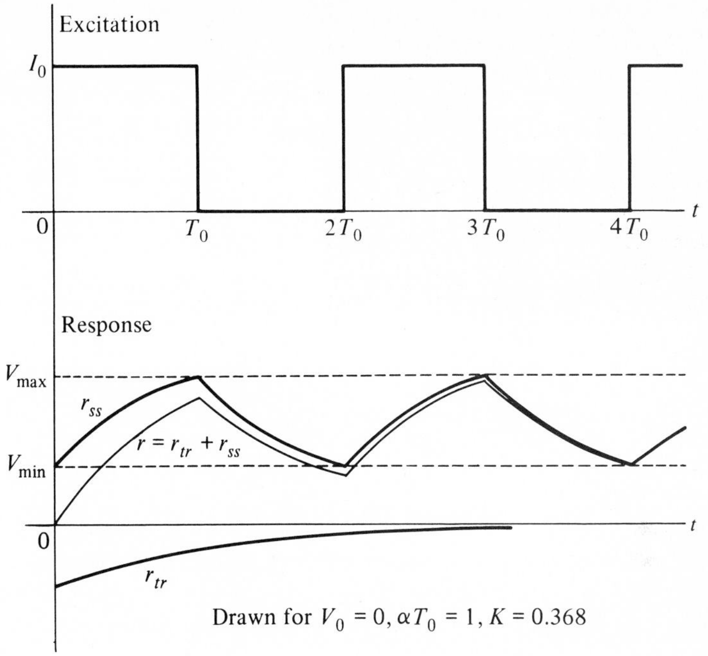

FIG. 2.12-F.

We note that if the circuit's previous history makes V_0 have the value $V_0 = V_{min} = RI_0[e^{+T_0/(RC)} + 1]^{-1}$, then there is *no* transient and the steady state obtains immediately. Common sense says "of course": for if conditions are just right, then there is no need for the transient, whose basic task is to bridge the gap between arbitrary initial condition and the steady-state condition.

It remains to ask whether we may properly call the transient part the *natural* part and the steady-state part the *forced* part. In this example there is no part of the response that has the same waveshape as the excitation. Nevertheless, what we have called r_{ss} is proportional in size (though not in shape) to the excitation, by virtue of the multiplier I_0 therein. The part we have called r_{tr} does not involve I_0, and does have the circuit's natural form of behavior. Consequently we (reasonably) write

$$
\begin{aligned}
v(t) &= \underbrace{(V_0 - V_{min})e^{-\alpha t}} + \underbrace{\text{the piecewise exponential wave of Fig. 2.12-C}} \\
&= \qquad r_{tr} \qquad + \qquad r_{ss} \qquad (2.12\text{-}8) \\
&= \qquad r_N \qquad + \qquad r_F .
\end{aligned}
$$

Under what conditions the forced part has the *same waveshape* as the excitation is, evidently, a question we cannot yet answer. In one example (Sec. 2.09) it does; in another (here) it does not. We hope to answer the question in due time. It is clearly related to the *distortion* of signals by the circuit.

2.13 EXISTENCE AND UNIQUENESS

We have modeled the practical engineering analysis problem of finding the response of an amplifier stage to a given excitation with the simple *RC* circuit of Fig. 1.06-A and the (equivalent) differential equation of (2.02-1). We assume that the physical amplifier does respond, and responds always in the same way, to a given impulsive, steplike, pulselike, etc., excitation. If the circuit model is accurate, it too should respond, and always in the same way, to any given excitation and initial condition: the solution of the differential equation should exist and be unique.

Regarded purely as a mathematical question, the existence and uniqueness of the solution(s) of differential equations provide much food for thought and argument. The sort of differential equation with which we deal (ordinary, linear, having only coefficients that are constant) is a very special and simple sort. That a solution exists in the present case is demonstrated by the success of the experiments of previous sections in which we actually found solutions. They obviously exist, and they *are* solutions; we have only to substitute them in the differential equations and the boundary conditions to see that. As for uniqueness, the power series attack of Sec. 2.04 in effect demonstrates that.

If physical reasoning is not enough, then, we can develop a mathematical proof of the existence and uniqueness of a solution to our analysis problem. And the same arguments will hold for all our network analysis discussions. (We observe in passing that if we were to admit nonlinear circuit elements, or elements that vary with time, it might be necessary to revise this proposition.)

2.14 EXPONENTIAL RESPONSE

Very interesting should be the response of the *RC* circuit to an excitation whose waveform is like the circuit's own natural behavior. And this is also extremely important, as we shall see again and again. Hence we calculate it next.

Let the excitation be

$$i_0(t) = I_0 e^{st} u(t) \tag{2.14-1}$$

in which s is some (known) constant with the dimension of $(\text{time})^{-1}$ or "frequency." Figure 2.14-A shows the behavior of such excitation for negative and positive values of s. Note that $s = 0$ brings us back to the *step*. If s is positive, the excitation cannot continue indefinitely; sooner or later it will become large enough to vitiate our postulate of small signals (small currents and voltages) and perhaps even damage the physical amplifier. But calculation of the response to the truncated or pulselike exponential excitation $i_1(t)$ *is* a reasonable problem,

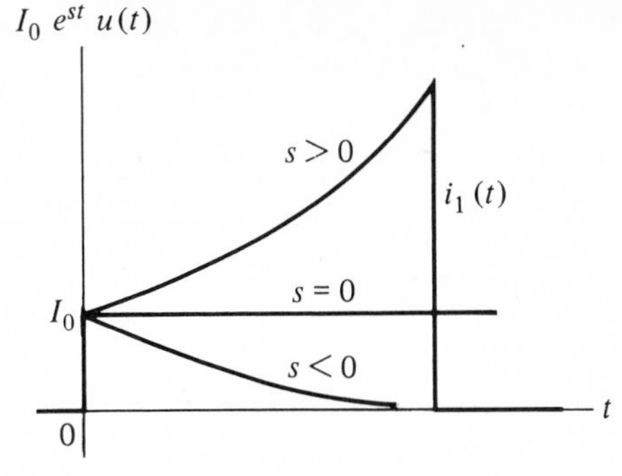

FIG. 2.14-A.

as is calculation of the response to a decaying exponential. The case $s = -\alpha$ should be particularly interesting, because the circuit may be somewhat sensitive to excitation that is like its own, natural-behavior waveform. For the moment we assume $s \neq -\alpha$.

A most reasonable form of response, suggested by our experience so far, is

$$v(t) = v_N + v_F = Ke^{pt} + V_F\, e^{st}, \qquad t > 0. \tag{2.14-2}$$

We have written a *natural* part of unknown coefficient K and "frequency" p [which we expect to be $-1/(RC)$], and a *forced* part of unknown coefficient V_F but shape identical to the excitation. A simple experiment will corroborate or disprove (2.14-2). We substitute it in (2.02-1) and obtain

$$[(Cp + G)K]e^{pt} + [(Cs + G)V_F - I_0]e^{st} = 0, \qquad t > 0, \tag{2.14-3}$$

and

$$K + V_F = V_0, \tag{2.14-4}$$

in which V_0 is the (known) boundary or initial value of the response voltage.

In (2.14-3) are two *different* functions of time, e^{pt} and e^{st}, as long as $s \neq p$. If (2.14-3) is to hold as t varies and e^{pt} and e^{st} go their separate ways, then each of the two must in effect vanish; there is no other way in which the equation can hold as t varies. The only way in which this can occur is for each bracketed expression to vanish:

$$\begin{aligned} Cp + G &= 0, \\ (Cs + G)V_F - I_0 &= 0. \end{aligned} \tag{2.14-5}$$

In other words we need

$$p = \frac{-G}{C} = \frac{-1}{RC} = -\alpha \tag{2.14-6}$$

and

$$V_F = \frac{I_0}{Cs + G}. \tag{2.14-7}$$

These are reasonable values for our previously unknown p and V_F. The former is once again the natural-behavior frequency constant. The latter evaluates the amplitude of the forced component in terms of known quantities. Finally the other unknown, K, is determined by (2.14-4) in combination with (2.14-7):

$$K = V_0 - V_F = V_0 - \frac{I_0}{Cs + G}. \tag{2.14-8}$$

So we arrive at

$$\begin{aligned} v(t) &= \left(V_0 - \frac{I_0}{Cs + G}\right)e^{-\alpha t} + \frac{I_0}{Cs + G}e^{st} \qquad (2.14\text{-}9) \\ &= V_0 e^{-\alpha t} + \frac{I_0/C}{s + \alpha}(e^{st} - e^{-\alpha t}). \end{aligned}$$

which satisfies the differential equation and the boundary (initial) condition and therefore is our solution. It could have been obtained by mathematically more elegant (less empirical) methods, perhaps, but the physical concept of forced and natural components of response is important to keep in mind. Figure 2.14-B*a* shows how the response (from rest) appears for negative values of s; positive values are illustrated by Fig. 2.14-B*b*, the excitation being cut off at T_0 and the response thereafter being of course

$$v(t) = v(T_0)e^{-\alpha(t - T_0)}, \qquad T_0 < t. \tag{2.14-10}$$

(For the case $s = 0$, see Sec. 2.09.)

For negative values of s not equal to $-\alpha$, the response has a pulselike character: it starts to rise to meet the force shape, and then the transient dies, and the forced part also, since the excitation does too. The natural part is transient in character of course; here the forced part is transient also, since the excitation itself is transitory.

One naturally wonders whether the response is in some way extraordinary when the excitation has the form of the natural behavior, when $s = -\alpha$. In the

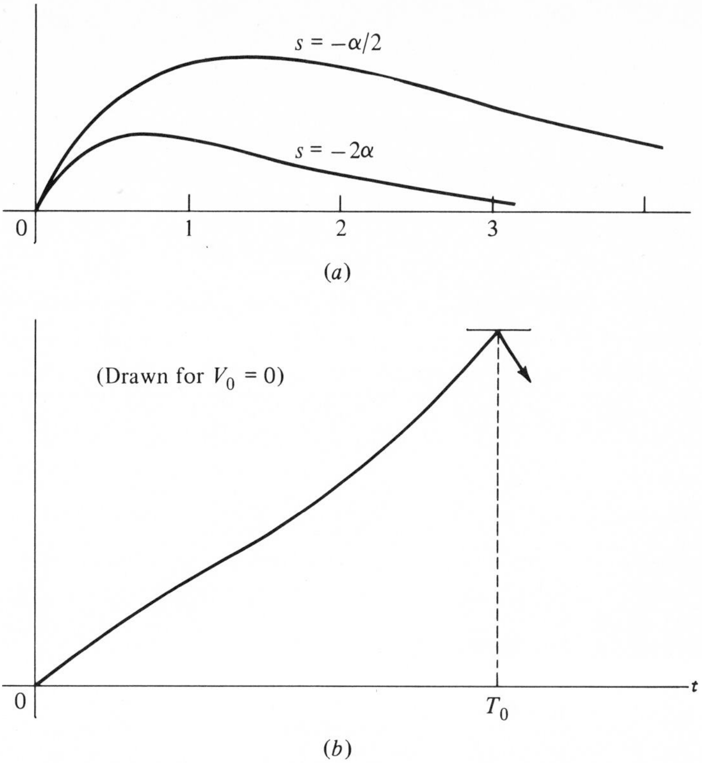

FIG. 2.14-B.

formula (2.14-9), the denominator factor $(s + \alpha)$ vanishes; one is tempted to say: the response *is* extraordinary, being very large indeed when the circuit is excited "at its own frequency." Such behavior can indeed occur in more complicated circuits when the phenomenon of resonance exists—but here it cannot. For the numerator factor $(e^{st} - e^{-\alpha t})$ *also* becomes zero, and we are left with the meaningless expression 0/0. But it is not altogether meaningless; rather it tells us that a limiting process is necessary for the analysis, even though Fig. 2.14-B*a* implies that the response to $e^{-\alpha t}$ is in no way extraordinary.

Factoring out $e^{-\alpha t}$ and using the series (2.04-12) for the exponential left inside the bracket gives us

$$
\begin{aligned}
v(t) &= V_0 e^{-\alpha t} + \frac{I_0/C}{s+\alpha} e^{-\alpha t}[e^{(s+\alpha)t} - 1] \\
&= V_0 e^{-\alpha t} + \frac{I_0/C}{s+\alpha} e^{-\alpha t}\left[1 + (s+\alpha)t + \frac{(s+\alpha)^2 t^2}{2!} + \cdots - 1\right] \\
&= V_0 e^{-\alpha t} + \frac{I_0/C}{s+\alpha} e^{-\alpha t}(s+\alpha)t\left[1 + \frac{(s+\alpha)t}{2} + \cdots\right] \qquad (2.14\text{-}11)
\end{aligned}
$$

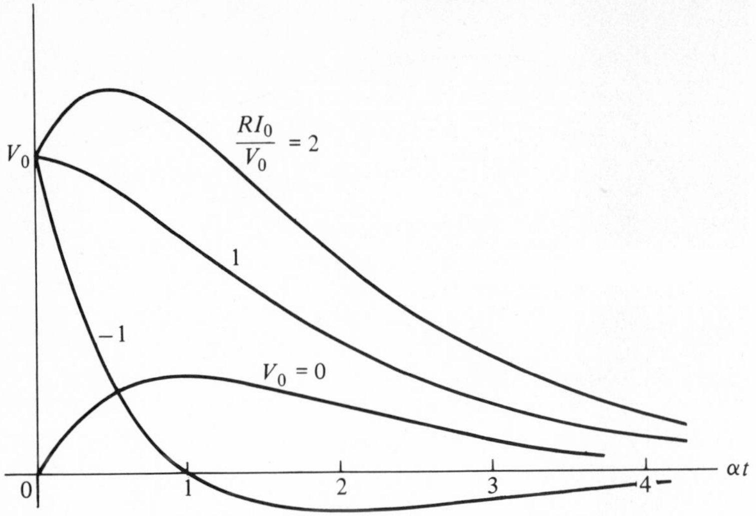

FIG. 2.14-C.

from which follows immediately, for $s = -\alpha$,

$$\begin{aligned} v(t) &= V_0 e^{-\alpha t} + \frac{I_0}{C} t e^{-\alpha t} \\ &= RI_0\left(\frac{V_0}{RI_0} + \alpha t\right) e^{-\alpha t}. \end{aligned} \tag{2.14-12}$$

This is plotted in Fig. 2.14-C for several values of the ratio of V_0 to RI_0.

There is nothing extraordinary about the response of the *RC* circuit to an excitation of the same form as its own, inherent natural behavior. The response curves of Fig. 2.14-C differ in no noticeable way from those of Fig. 2.14-B*a*. And Fig. 2.14-D taken for photographic-physical corroboration, confirms this: the second case there *is* that of excitation time constant equal to circuit time constant.

This figure (2.14-D) shows an *RC* circuit's actual response (from rest) to exponential excitation (which is also shown). The four different cases corroborate the analysis. The excitation is the same in all, as is *R*. Increases in *C* change the circuit's time constant, reducing the initial slope and the maximum value of the response, as (2.14-9) indicates (Prob. 2-5).

It is not always true that excitation at its own peculiar frequency produces no extraordinary response in a circuit. In Chap. 3 we shall discuss a very common example of natural-frequency excitation that produces very large response indeed. But that cannot happen unless at least one more energy-storing element is present.

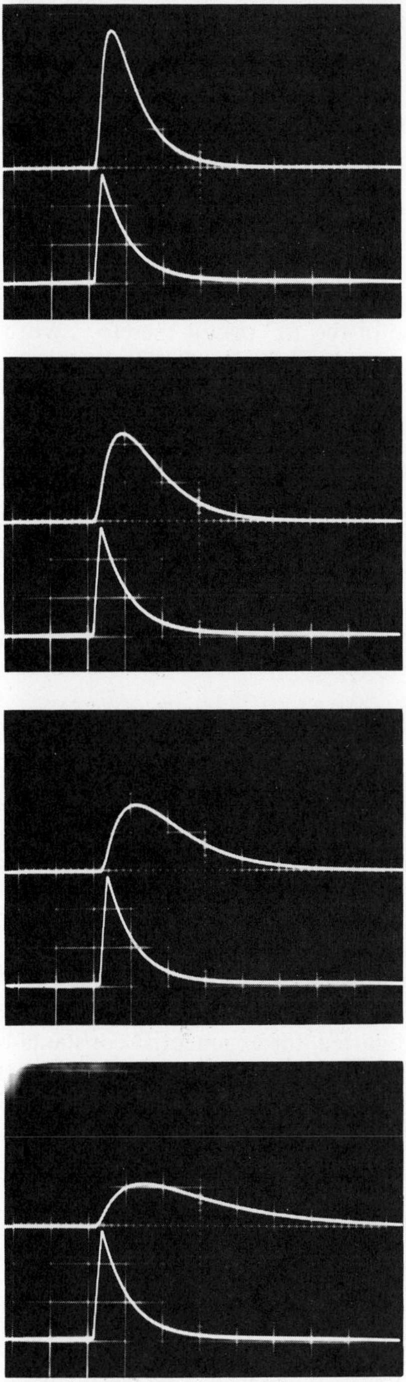

NOTES: In all cases: $R = 1000\ \Omega$, time scale is 40 μs per division, vertical scales are alike; time constant of excitation exponential is 30 μs ($\frac{3}{4}$ division)

FIG. 2.14-D.

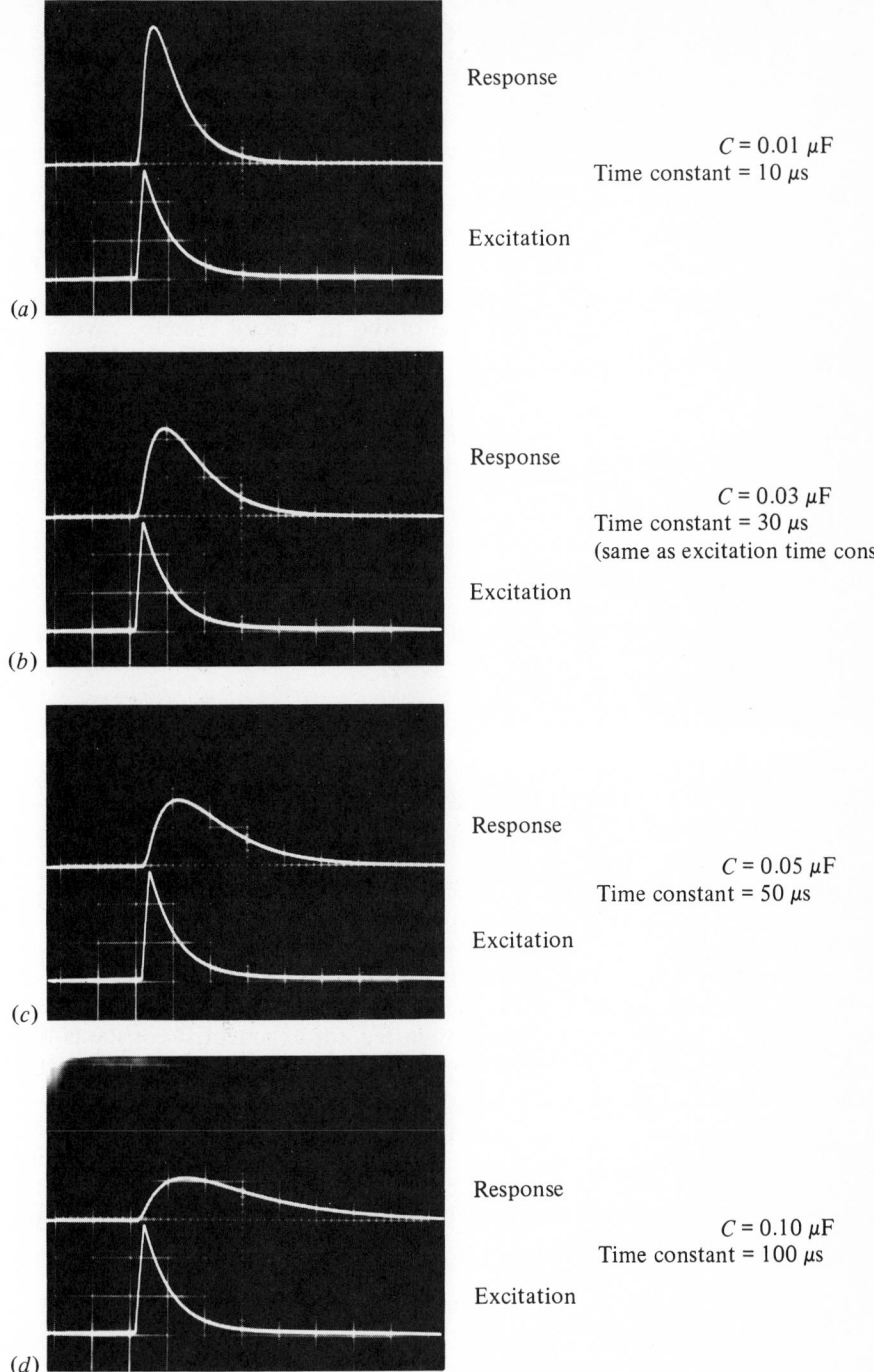

2.15 SINUSOIDAL EXCITATION: COMPLEX NUMBERS AND PHASORS

Even casual observation of electronic and electrical apparatus tells us that engineering makes great use of sinusoidal waves. Accepting that fact for the moment, and postponing discussion of why the sinusoid is so useful (and important), let us calculate the response of the *RC* circuit thereto. We write for the excitation (force)

$$i_F(t) = I_m \cos(\omega t + \psi) \tag{2.15-1}$$

as a general sinusoidal wave, in which, by definition:

$$\begin{aligned} I_m &= \textit{amplitude} \text{ or } \textit{peak (maximum) value}, \\ \omega &= \textit{angular frequency} \text{ or } \textit{pulsation}, \text{ rad/s}, \\ (\omega t + \psi) &= \textit{phase}, \\ \psi &= \textit{initial phase}. \end{aligned} \tag{2.15-2}$$

The term *frequency* of ordinary usage is, for this wave,

$$\frac{\omega}{2\pi} = f \qquad \text{(H)}. \tag{2.15-3}$$

Since both ω and f measure the same attribute of the wave (though in different units), we shall use the word *frequency* for both, as may be convenient. The more general use of the word already started, for exponential constants like $-\alpha$, is quite consistent, as we shall see, with a generalized definition of this, the ordinary garden variety of frequency. Figure 2.15-A shows the wave; for convenience three scales of time are shown there (all are useful). Also useful is

$$T = \frac{1}{f} = \frac{2\pi}{\omega} = \textit{period} \text{ of wave (in seconds)}. \tag{2.15-4}$$

One of the most striking advances in the history of mathematics was the discovery of a relation between the trigonometric sinusoidal functions and the exponential function. In Euler's famous formula, using j to represent the imaginary unit $\sqrt{-1}$,

$$e^{j\beta} = \cos\beta + j\sin\beta. \tag{2.15-5}$$

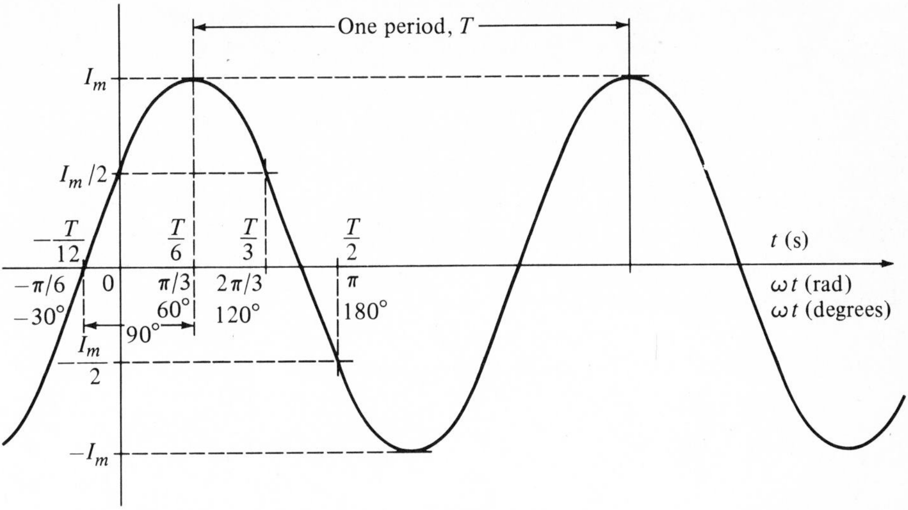

FIG. 2.15-A. A section of the sinusoidal wave (2.15-1) $i_F(t) = I_m \cos(\omega_0 t + \psi)$, drawn for $\psi = -\pi/3$ rad (that is, $-60°$).

The reverse point of view is

$$\cos \beta = \frac{e^{j\beta} + e^{-j\beta}}{2} \tag{2.15-6}$$

and

$$\sin \beta = \frac{e^{j\beta} - e^{-j\beta}}{2j}. \tag{2.15-7}$$

These relations are surely of interest here, because the natural behavior of the *RC* circuit is exponential (albeit with a real exponent) and the excitation is sinusoidal.

We shall be using complex numbers and functions from now on, primarily because sinusoids will be extremely important, and because it greatly simplifies matters to describe sinusoids with complex quantities. Note that the mathematician's symbol for $\sqrt{-1}$, that is, i, would lead to hopeless confusion with the symbol for electric current; common engineering usage is j for $\sqrt{-1}$.

Complex numbers are in principle not necessary in network analysis; as a practical matter, however, they are essential. We can immediately demonstrate their utility in the present analysis problem. The excitation is

$$i_F(t) = I_m \cos(\omega_0 t + \psi) = \frac{I_m}{2}(e^{j\psi}e^{j\omega t} + e^{-j\psi}e^{-j\omega t}). \tag{2.15-8}$$

We note that in (2.15-8) both negative and positive frequencies appear: $+\omega$ and $-\omega$ play equal roles. It is as though Euler recognized that frequency can equally well be considered positive or negative and so he used both, taking the *average* of the two [see (2.15-6) and (2.15-7)]. We shall consistently find that negative frequency and positive frequency are equally valid, and that using both simplifies matters.

In Sec. 2.14 we calculated the response to an exponential excitation, $e^{(st)}$, without restriction on s. Mathematically, we can use that result to calculate the response to the first of the two exponentials in (2.15-7), placing $s = +j\omega$. The same can be said for the second, with $s = -j\omega$. Physically, of course, they cannot be separately treated. So we apply (2.14-9) twice to obtain for the response here:

$$\begin{aligned} v(t) = &\left[\frac{(I_m/2)e^{j\psi}}{C(j\omega) + G} e^{j\omega t} + \frac{(I_m/2)e^{-j\psi}}{C(-j\omega) + G} e^{-j\omega t}\right] \\ &+ \left[V_0 - \frac{(I_m/2)e^{j\psi}}{C(j\omega) + G} - \frac{(I_m/2)e^{-j\psi}}{C(-j\omega) + G}\right] e^{-\alpha t}. \end{aligned} \tag{2.15-9}$$

The superposition can be completely justified by the linearity of the model. Nevertheless, one should carefully verify that (2.15-9) does indeed satisfy both the differential equation and the initial condition.

The form of (2.15-9) is unnecessarily complicated. We have used the transformation from (real) sinusoids to (complex) exponentials for convenience in obtaining the solution. But we have not taken the complementary step of transforming the (complex) solution back to the form of sinusoids. The function $v(t)$ must, of course, be real; the complex numbers in (2.15-9) must therefore be *conjugate* so that their imaginary parts cancel each other. We see by inspection that for each appearance of $+j$ there is a corresponding appearance of $-j$, so that the cancellation does occur. We might add the terms to verify this, but the calculation is tedious: it requires rationalization of the complex numbers (to obtain a common denominator). Instead we simply write

$$\begin{aligned} v(t) = &\left[\frac{(I_m/2)e^{j\psi}}{G + jC\omega_0} e^{j\omega_0 t} + \text{conjugate term}\right] \\ &+ \left[V_0 - \frac{(I_m/2)e^{j\psi}}{G + jC\omega_0} + \text{conjugate term}\right] e^{-\alpha t}. \end{aligned} \tag{2.15-10}$$

Certain operations with complex numbers that are useful in simplifying things may well be recapitulated here for review purposes. In brief, a complex number $\overline{N}$ can be written in rectangular or cartesian form, $(a + jb)$, or in polar form as $\rho e^{j\beta}$:

$$\overline{N} = a + jb = \rho e^{j\beta} = \rho\angle\beta. \tag{2.15-11}$$

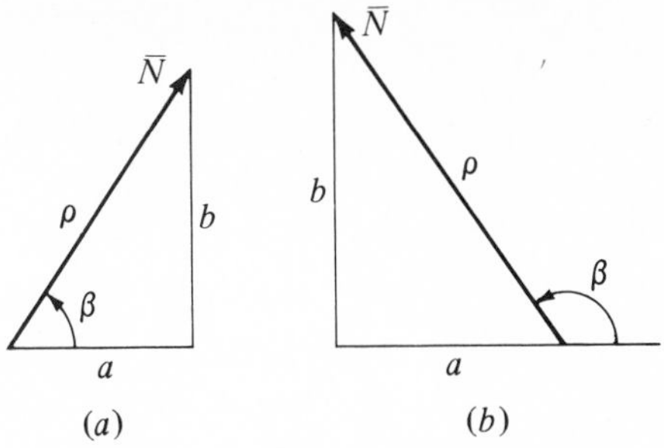

FIG. 2.15-B.

Customary nomenclature and useful symbolism and formulas are:

$$\begin{aligned} a &= \text{real part of } \overline{N} = \mathbf{RE}\,(\overline{N}) = \rho \cos \beta, \\ b &= \text{imaginary part of } \overline{N} = \mathbf{IM}(\overline{N}) = \rho \sin \beta, \\ \rho &= \text{magnitude or modulus of } \overline{N} = \sqrt{a^2 + b^2}, \\ \beta &= \text{angle or argument of } \overline{N} = \tan^{-1} \frac{b}{a}. \end{aligned} \tag{2.15-12}$$

Note that it is b (and not jb) that is called the imaginary part; a misnomer, but a standard convention.

Figure 2.15-B shows two examples: in (a) both a and b are positive and β is an angle in quadrant I; in (b) a is negative, b is positive, and β lies in quadrant II. In calculating the angle β from the arctan expression one must be careful to place the angle in the correct quadrant (the $\tan^{-1}$ function is multivalued). In Fig. 2.15-B, for example, β is approximately 125° (and *not* $-55°$). The notation $\underline{/\beta}$ is much more convenient than $e^{j\beta}$; it means the same thing and loses us nothing since $|e^{j\beta}| = 1$.

The rectangular form is convenient for addition and subtraction; the polar form is convenient for multiplication and division:

$$\overline{N}_1 + \overline{N}_2 = (a_1 + jb_1) + (a_2 + jb_2) = (a_1 + a_2) + j(b_1 + b_2), \tag{2.15-13}$$

$$\overline{N}_1 \overline{N}_2 = \rho_1 e^{j\beta_1} \rho_2 e^{j\beta_2} = \rho_1 \rho_2 \underline{/\beta_1 + \beta_2}. \tag{2.15-14}$$

In multiplication the angles add; if a logarithmic measure of magnitude is used (which we shall find convenient later), the magnitudes also add in that measure.

Note that the *conjugate* of a complex number has the same magnitude, but the negative angle;

$$\overline{N}^* = \text{conjugate of } \overline{N} = a - jb = \rho \underline{/-\beta}; \tag{2.15-15}$$

that the sum of a complex number and its conjugate is twice the real part of either;

$$\bar{N} + \bar{N}^* = (a + jb) + (a - jb) = 2a; \tag{2.15-16}$$

and the product of a complex number and its conjugate is

$$\bar{N}\bar{N}^* = \rho^2 \angle 0 = \rho^2. \tag{2.15-17}$$

Note also that the reciprocal of a complex number can conveniently be obtained by "rationalization":

$$\begin{aligned} \frac{1}{\bar{N}} &= \frac{1}{a + jb} = \frac{1}{a + jb}\,\frac{a - jb}{a - jb} = \frac{a - jb}{a^2 + b^2} \\ &= \frac{a}{a^2 + b^2} + j\,\frac{-b}{a^2 + b^2} \end{aligned} \tag{2.15-18}$$

if we want the rectangular form of the result; if we want the polar form of the reciprocal, then we write

$$\frac{1}{\bar{N}} = \frac{1}{\rho e^{j\beta}} = \frac{1}{\rho} e^{-j\beta} = \frac{1}{\rho} \angle -\beta. \tag{2.15-19}$$

We shall use complex numbers a great deal, so that a standard notation, particularly one that simply relates the complex number itself and its magnitude, is worth establishing. It is often convenient to use, as we have been doing, a superscript bar to denote that a number is *complex*, and to use the same letter without the bar to indicate its *magnitude*. Then (2.15-11) becomes

$$\bar{N} = N \angle \beta. \tag{2.15-20}$$

To return to our analysis problem: the expression (2.15-10) can be greatly simplified by using these ideas. We have

$$\begin{aligned} v(t) &= \left(\frac{(I_m/2)e^{j\psi}}{G + jC\omega} e^{j\omega t} + \text{conjugate}\right) + \left[V_0 - \frac{(I_m/2)e^{j\psi}}{G + jC\omega} + \text{conjugate}\right] e^{-\alpha t} \\ &= \left[2\mathbf{RE}\left[\frac{(I_m/2)e^{j\psi}}{G + jC\omega} e^{j\omega t}\right]\right] + \left\{V_0 - 2\mathbf{RE}\left[\frac{(I_m/2)e^{j\psi}}{G + jC\omega}\right]\right\} e^{-\alpha t} \\ &= \mathbf{RE}\left[\frac{I_m}{\sqrt{G^2 + C^2\omega^2}} e^{j(\psi - \theta + \omega t)}\right] + \left\{V_0 - \mathbf{RE}\left[\frac{I_m}{\sqrt{G^2 + C^2\omega^2}} e^{j(\psi - \theta)}\right]\right\} e^{-\alpha t} \\ &= \underbrace{\frac{I_m}{\sqrt{G^2 + C^2\omega^2}} \cos(\omega t + \phi)}_{v_F} + \underbrace{\left(V_0 - \frac{I_m}{\sqrt{G^2 + C^2\omega^2}} \cos\phi\right) e^{-\alpha t}}_{v_N} \\ &= \quad v_F \quad + \quad v_N \end{aligned} \tag{2.15-21}$$

in which

$$\theta = \tan^{-1}\frac{C\omega}{G} \qquad \text{(in quadrant I)} \tag{2.15-22}$$

and

$$\phi = \psi - \theta. \tag{2.15-23}$$

We find, as now seems usual, that the response has two parts:

a The forced part, here of the same waveshape as the excitation (though shifted in time) and steady state in character, (2.15-24)

b A natural part, again e^{pt} in form, $p = -\alpha = -1/(RC)$, and of amplitude sufficient to "absorb" the actual initial condition vis-à-vis the forced component, at $t = 0$. (2.15-25)

The manipulations in (2.15-21) and Euler's formula (2.15-5) also suggest one other convenient shorthand. Since

$$\cos\beta = \frac{e^{j\beta} + e^{-j\beta}}{2} = \mathbf{RE}(e^{j\beta}) \tag{2.15-26}$$

and since the exponential form is so useful to us, we shall often *represent* a sinusoidal wave by an associated $e^{j\beta}$. For example, we write (2.15-8) thus:

$$\begin{aligned} i_F(t) &= I_m \cos(\omega t + \psi) = \mathbf{RE}(I_m e^{j\psi} e^{j\omega t}) \\ &= \mathbf{RE}(\bar{I}_F e^{j\omega t}) \end{aligned} \tag{2.15-27}$$

in which the (constant) complex number $\bar{I}_F = I_m e^{j\psi}$ is called the *phasor representation* of the sinusoidal wave $i_F(t)$. (See Fig. 2.15-C.) Note that it contains all the information that the wave itself does, save the frequency ω. Its utility is immediately evident if we look at the phasor representation of the forced component of the response. From (2.15-21) we have

$$\begin{aligned} v_F(t) &= \frac{I_m}{\sqrt{G^2 + C^2\omega^2}} \cos(\omega t + \psi - \theta) \\ &= \mathbf{RE}\left(\frac{I_m e^{j\psi}}{G + jC\omega} e^{j\omega t}\right) \end{aligned} \tag{2.15-28}$$

so that the phasor representation of the sinusoid $v_F(t)$ is

$$\bar{V}_F = \frac{I_m e^{j\psi}}{G + jC\omega} = \frac{\bar{I}_F}{\bar{Y}} \tag{2.15-29}$$

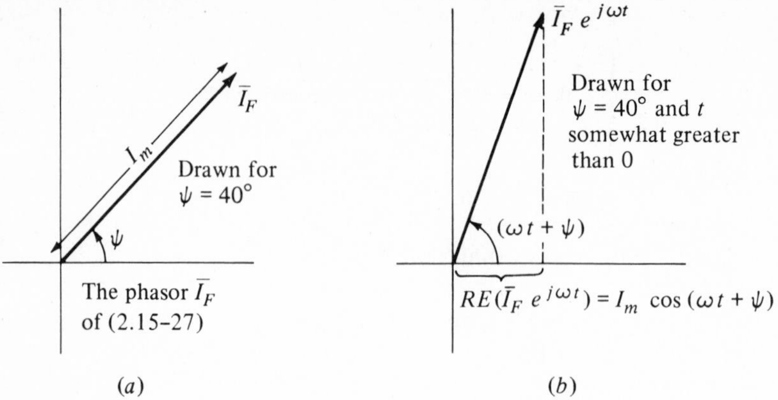

FIG. 2.15-C. Representation of a sinusoid by rotating and projecting its phasor.

in which we have written

$$\bar{Y} = G + jC\omega. \tag{2.15-30}$$

We can find the phasor representation of the forced (steady-state) component of the response simply by multiplying the phasor representation of the excitation by the complex number $\bar{Z} = 1/\bar{Y}$, which is a characteristic of the circuit. The relation (2.15-29) appears to be a generalization of Ohm's law to this more complicated situation; that this is true, and that it is not hard to find the factor by which one must multiply the phasor representation of the excitation to obtain the phasor representation of the forced component of the response, is something to which we shall later devote considerable attention.

These ideas of phasor representation of sinusoids, and certain complex numbers characteristic of circuits, are of general value. Here we simply note that they greatly simplify the calculation of the response of the RC circuit to the sinusoidal excitation (2.15-1). The actual calculations come down in fact simply to these, given the excitation and circuit parameters:

$$\begin{aligned}
\bar{Y} &= G + jC\omega = Y\angle\theta, \\
\bar{Z} &= \frac{1}{\bar{Y}}, \\
\bar{I}_F &= I_m\angle\psi, \\
\bar{V}_F &= \bar{I}_F \bar{Z}, \\
v(t) &= v_F + v_N, \\
v_F &= \mathbf{RE}(\bar{V}_F e^{j\omega t}) = \frac{I_m}{Y}\cos(\omega t + \psi - \theta), \\
v_N &= [V_0 - v_F(0)]e^{-\alpha t}.
\end{aligned} \tag{2.15-31}$$

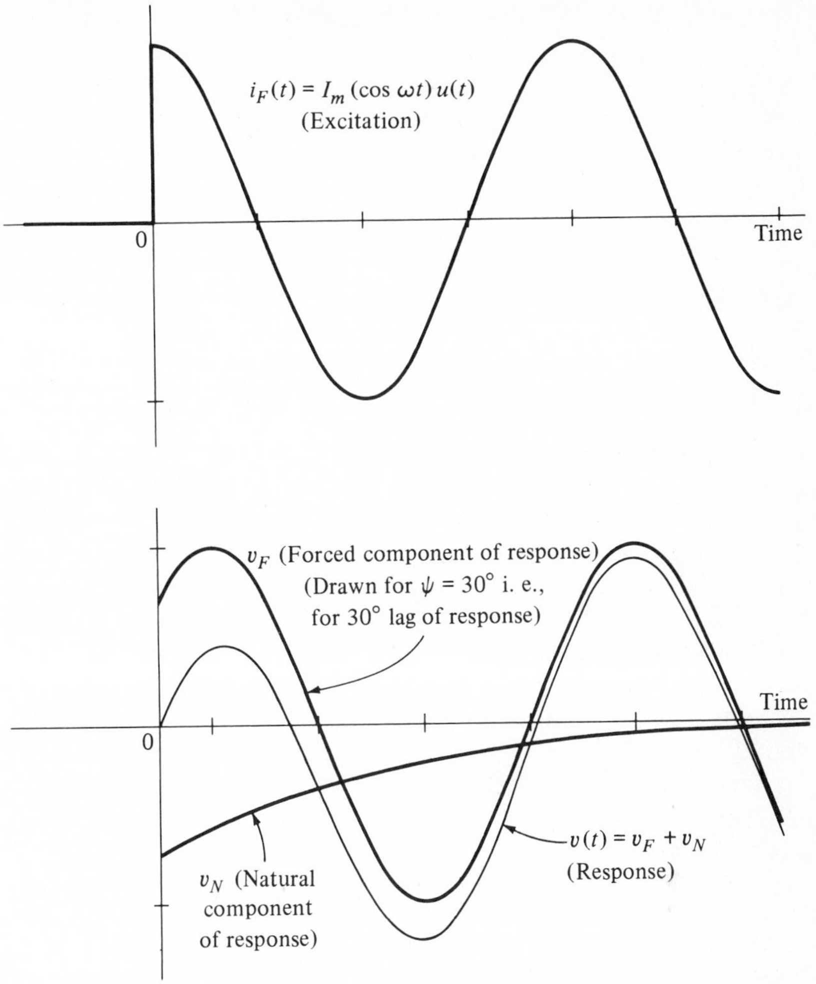

FIG. 2.15-D.

Figure 2.15-D shows a typical case. Note again how the natural part makes up the difference between the initial value V_0 and the steady-state part (projected back) at that instant, and then gradually fades away as the behavior settles into its steady state.

We also recapitulate here in compact form the (direct) transformation of a sinusoidal function of time into its phasor representation [see (2.15-7)]:

$$f(t) = F_m \cos(\omega t + \psi) \quad \rightarrow \quad \bar{F} = F_m \underline{/\psi} \qquad (2.15\text{-}32)$$

and the (inverse) transformation of a phasor into the sinusoidal function that it represents:

$$\bar{G} = G_m \underline{/\phi} \quad \rightarrow \quad g(t) = \mathbf{RE}(\bar{G}e^{j\omega t}) = G_m \cos(\omega t + \phi). \qquad (2.15\text{-}33)$$

The *graphical* interpretation (Fig. 2.15-C) is also very informative.

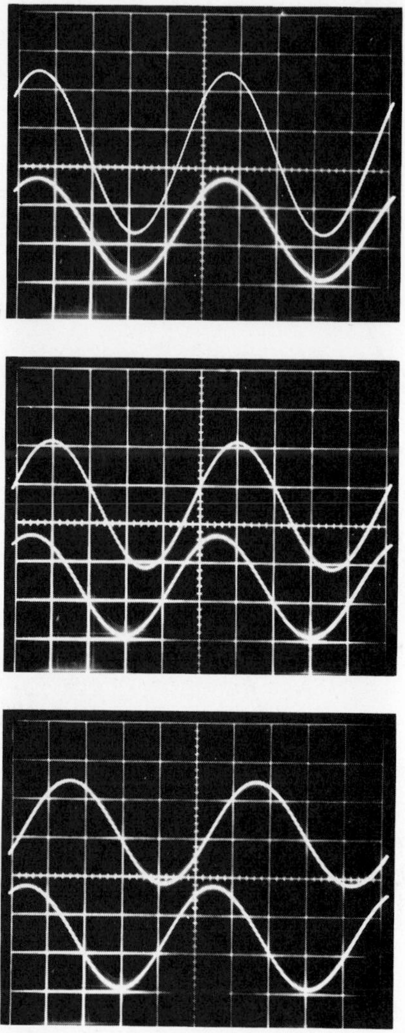

(a) v_F / Excitation

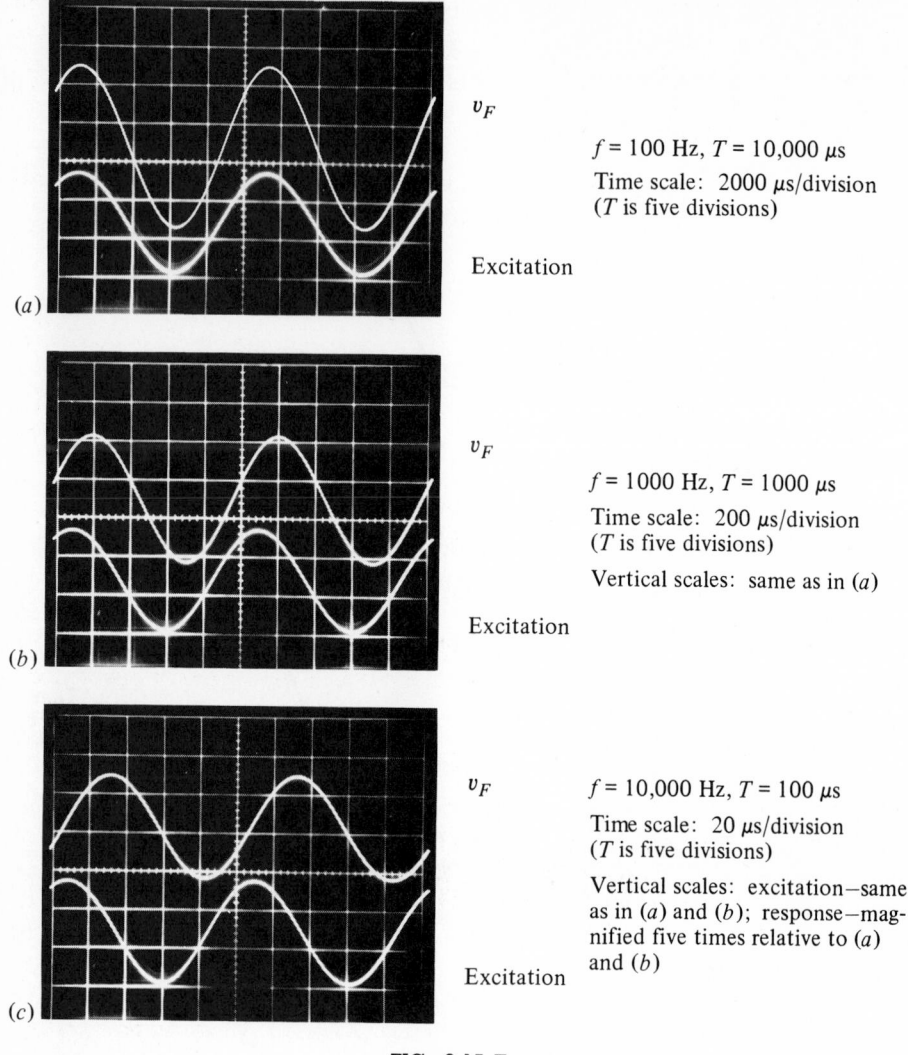

$f = 100$ Hz, $T = 10{,}000$ μs
Time scale: 2000 μs/division
(T is five divisions)

(b) v_F / Excitation

$f = 1000$ Hz, $T = 1000$ μs
Time scale: 200 μs/division
(T is five divisions)
Vertical scales: same as in (a)

(c) v_F / Excitation

$f = 10{,}000$ Hz, $T = 100$ μs
Time scale: 20 μs/division
(T is five divisions)
Vertical scales: excitation—same as in (a) and (b); response—magnified five times relative to (a) and (b)

FIG. 2.15-E.

It is refreshing now after so much mathematics to turn to physical corroboration of our results. (The mathematics is necessary, of course, and very useful.) We limit our observations chiefly to the *forced* part of the response, which is itself sinusoidal and therefore permanent or steady state in character. (The natural part, transient as it is, soon dies away, and is somewhat more difficult to photograph.) The three photographs in Fig. 2.15-E show sinusoidal excitation and the forced component of the RC circuit's response for three different frequencies of excitation (and response). The parameter values are: $R = 2000\ \Omega$ ($G = 500$ μmhos), $C = 0.08$ μF, and circuit time constant $= RC = 160$ μs.

To choose interesting frequency values we observe, from (2.15-31), that the important circuit function is $\overline{Y} = G + jC\omega$, which can be rewritten (to make the frequency stand out) as

$$\begin{aligned} \overline{Y} &= G(1 + jRC\omega) = G[1 + j(2\pi RC)f] = Y\angle\theta \\ &= 500 \times 10^{-6}\left(1 + \frac{jf}{1000}\right) \qquad (2.15\text{-}34) \\ &= G\sqrt{1 + \left(\frac{f}{1000}\right)^2}\,\underline{/\tan^{-1} f/1000}. \end{aligned}$$

Evidently interesting frequencies are

- **a** Very small values, for which there is no appreciable phase difference between excitation and (forced) response ($\theta \cong 0,\ Y \cong G$),
- **b** "Medium" values, $f \cong (2\pi RC)^{-1} = 1000$ Hz, for which the phase difference is near 45° and Y is near $1.4G$,
- **c** Large values, for which θ approaches 90° and Y approaches $jC\omega$.

For such values we should be able clearly to observe changes in lag and relative amplitude of v_F. Figure 2.15-E shows excitation and (forced component of) response for $(a)\ f = 100$, $(b)\ f = 1000$, and $(c)\ f = 10{,}000$ Hz.

In (a) there is virtually no displacement in time; peaks and valleys are practically in synchronism. (The vertical scales are arbitrarily chosen.)

In (b) the phase difference (the response *lags*, since t increases to the right) is about half a division or one-tenth of a period, reasonably near the 45° mentioned above. (See Fig. 2.15-D, where the lag is about 30°.) Note that the response amplitude is here reduced, in comparison with (a), to about three-fourths of its size in (a); since the scales are unchanged, the (reciprocal) 1.4 ratio is also about what it should be.

In (c) the lag has increased to nearly 90° (note the relative positions of peaks, valleys, and "zero crossings"); the response magnitude (note the 5 × magnification) is now "down" to about 10 or 15 percent of its value in (a).

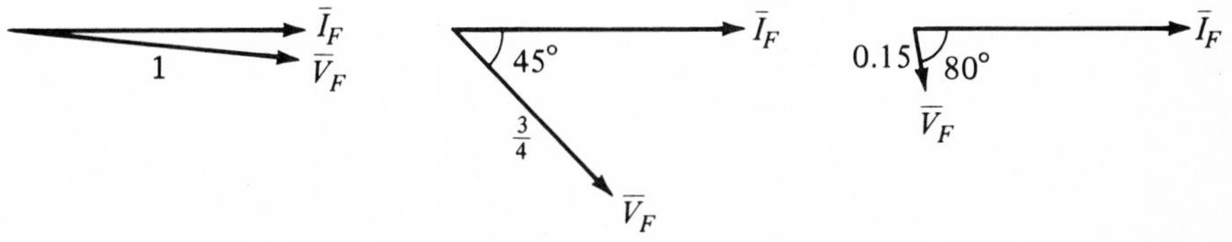

FIG. 2.15-F. Phasor diagrams of the response-excitation relations in Fig. 2.15-E. NOTE: The excitation phasors are marked $\overline{I}_F$; their lengths are the same. The response phasors are marked $\overline{V}_F$; their relative lengths are indicated on the diagrams. Since the dimensions of $\overline{I}_F$ and $\overline{V}_F$ differ, their lengths *cannot* be compared, regardless of the scales used; their angular relations, involving no dimensions, *are* correct.

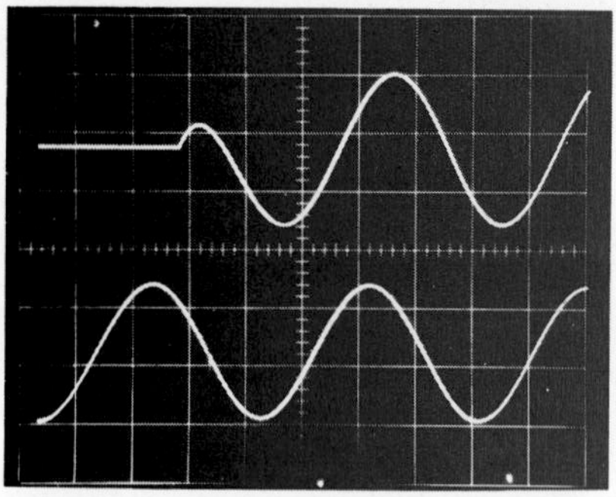

FIG. 2.15-G.

Phasor descriptions of the same three cases are drawn in Fig. 2.15-F (p. 81). Note the clarity of these phasor diagrams: simple though they are, they almost completely describe the forced (steady-state) responses of the *RC* circuit.

Physical observation of the transient state preceding the steady state here is, by the nature of the phenomenon (!), more difficult. Figure 2.15-G shows the total response and the excitation in one case. Compare it with the curves of Fig. 2.15-D and note the agreement.

We find again that the physical behavior of actual *RC* circuits is much as the mathematical analysis of the model predicts. The numerical agreement, in fact, is usually found to be excellent when accurate measurements are made.

2.16 SOME OTHER SIMPLE MODELS

Mathematically, the letter symbols used in the differential equation form of our model have no significance. It could just as well be written

$$A\frac{dy}{dt} + By - F(t) = 0, \qquad y(0) = Y_0. \tag{2.16-1}$$

The solution, if the excitation wave $F(t)$ is one of those we have studied, can merely be copied from our previous work, with careful changing of the letters. What makes this interesting is that *other* physical systems' models have the *same* differential equation.

For example, the *series RL* circuit of Fig. 2.16-A is governed by

$$L\frac{di}{dt} + Ri - e_0(t) = 0, \qquad i(0) = I_0. \tag{2.16-2}$$

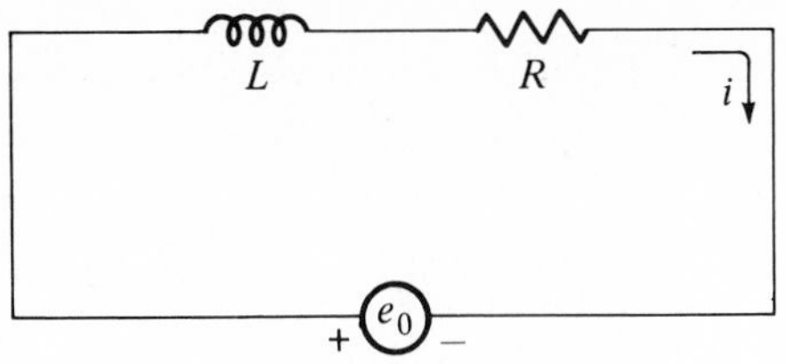

FIG. 2.16-A.

Another example is the mechanical-translation model of Fig. 2.16-B, for which

$$M\frac{dv}{dt} + Bv - F_0(t) = 0, \qquad v(0) = V_0. \tag{2.16-3}$$

Another is the mechanical-rotation model of Fig. 2.16-C, for which

$$J\frac{d\omega}{dt} + B\omega - T_0(t) = 0, \qquad \omega(0) = \Omega_0. \tag{2.16-4}$$

A simple heat-conduction model, an acoustic model, and various others can be found without difficulty, all with the same differential equation. None of these, if encountered, need additional analysis, if the corresponding *RC* circuit problem has been solved. For the mathematics is the same!

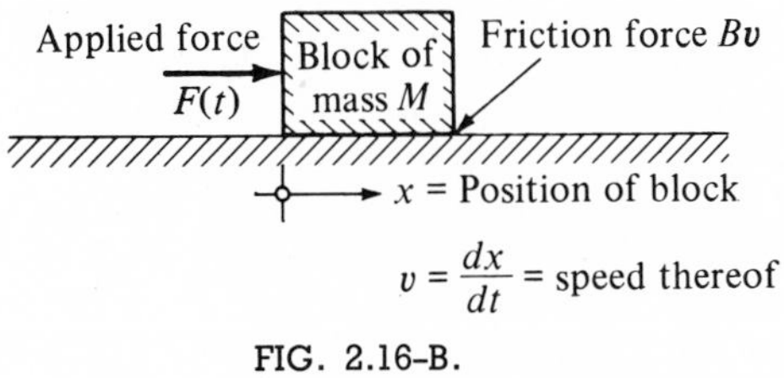

FIG. 2.16-B.

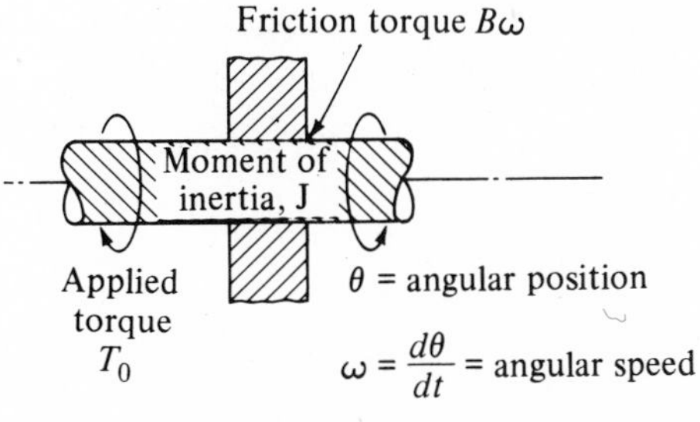

FIG. 2.16-C.

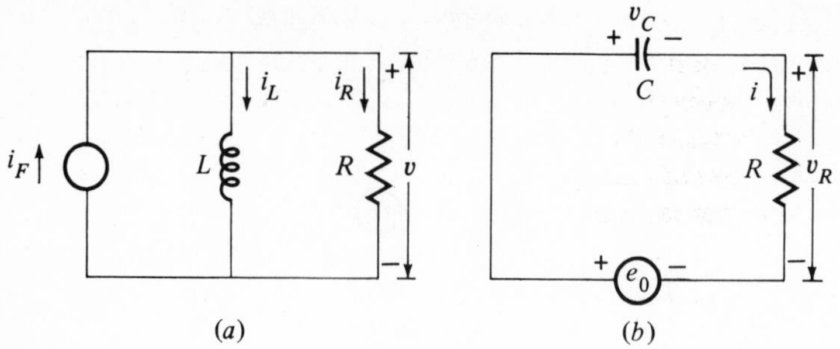

FIG. 2.16-D.

Two models with the same differential equations are called *analogs*, one of the other. The examples above, and the *RC* circuit, are all analogs of each other.

If two analogous models represent systems from the same physical category of system, they are also called *dual* to each other. Our (parallel) *RC* circuit and the series *RL* circuit of Fig. 2.16-A are duals, since both represent *electrical* systems. We note simply by comparison the correspondences between the two given in Table 2.16-A. Duality is useful as well as time-saving, and we shall return to it.

TABLE 2.16-A Quantities Dual (Corresponding) to Each Other in Two Circuits

Parallel *RC* Circuit	**Series *RL* Circuit**
C (capacitance)	L (inductance)
v (voltage)	i (current)
G (conductance)	R (resistance)
i (current)	v (voltage)
parallel connection.	series connection
$q = Cv$ (charge)	$\lambda = Li$ (flux linkages)

Two other simple electric circuit models are shown in Fig. 2.16-D. Their equations are, for positive time,

$$\Sigma\, i = \frac{1}{L}\int^{t} v\, dt + i_L(0) + Gv - i_0(t) = 0, \qquad (2.16\text{-}5)$$

and

$$\Sigma\, v = \frac{1}{C}\int_0^{t} i\, dt + v_C(0) + Ri - e_0(t) = 0. \qquad (2.16\text{-}6)$$

They are obviously duals. But their equations are not (strictly) differential: they are *integral* equations. Note that they incorporate the initial conditions directly; these are not separately stated. But there is no real difference between such integral equations and the previous differential equation. One method of solution is to differentiate and to restore the separate statement of the initial condition in appropriate form (by common sense). This gives us

$$G\frac{dv}{dt} + \Gamma v - i_0'(t) = 0, \qquad v(0) = Ri_R(0) = R[i_L(0) - i_o(0)], \tag{2.16-7}$$

and

$$R\frac{di}{dt} + Di - e_0'(t) = 0, \qquad i(0) = Gv_R(0) = G[v_C(0) - e_0(0)]. \tag{2.16-8}$$

They are now the same equation as that of the *RC* circuit and we need say no more about their solution. Since the equations as originally written are *not* the same as the *RC* differential equation, however, these two circuits are *not* dual thereto.

A more satisfying solution method is to proceed, experimentally, exactly as we did with the *RC* circuit, making use of our experience there. Consider, for example, the series *CR* circuit with exponential excitation: $e_0(t) = E_0 e^{st}$. We write, instinctively,

$$i(t) = Ke^{pt} + I_F e^{st}, \qquad 0 < t. \tag{2.16-9}$$

On substituting in the governing integral equation (2.16-6) we obtain

$$K\left(\frac{1}{Cp} + R\right)e^{pt} + \left[I_F\left(\frac{1}{Cs} + R\right) - E_0\right]e^{st} + \left[v_C(0) - \frac{K}{Cp} - \frac{I_F}{Cs}\right] = 0, \qquad 0 < t. \tag{2.16-10}$$

As before, we conclude that for this to hold it is necessary that

$$p = -\frac{1}{RC} = -\alpha,$$

$$I_F = \frac{E_0}{1/(Cs) + R}, \tag{2.16-11}$$

$$K = \left[v_C(0) + \frac{E_0}{1 + RCs}\right]Cp.$$

We find this is sufficient also, so that the response is

$$\begin{aligned} i(t) &= \left[Cpv_C(0) + \frac{GpE_0}{s+\alpha}\right]e^{pt} + \frac{E_0}{\dfrac{1}{Cs}+R}e^{st} \\ &= G\left[-v_C(0) + \frac{E_0}{RCs+1}\right]e^{-\alpha t} + \frac{E_0}{\dfrac{1}{Cs}+R}e^{st} \\ &= \qquad\qquad i_N \qquad\qquad + \quad i_F\,. \end{aligned} \tag{2.16-12}$$

The calculation of response to other types of excitation also proceeds essentially as before; we need not elaborate further on the series *CR* circuit, or its dual.

2.17 RECAPITULATION

The simple parallel *RC* circuit has been beaten to death, one may think. Our discussion thereof has indeed been lengthy. But an extraordinary number of ideas and concepts, and of mathematical and computational devices, have come out of it. Some of them are probably not novel; some are. All of them will be useful in the study of more complicated networks. That is why we have developed them in such detail for this simple, first case. We stop here to recapitulate the most important ones.

First, let us list our postulates again, and some definitions. We deal with *networks* that are interconnected sets of *elements* of the capacitor, inductor, resistor, and controlled-source types (see Secs. 1.05 and 1.06), excited by independent current and voltage sources. (The term *circuit* is a sort of friendly diminutive used for simple networks.) We postulate that these elements be *linear*, *lumped*, and *constant*.

Physically this means that for expediency we replace actual systems by simplified *models* composed of such simplified parts. As a result, the mathematical form of a model or network consists of ordinary differential equation(s) that are linear, with only constant coefficients. On these foundations rest all the structures we shall build.

We now restate briefly but cogently the principles we have found useful in the analysis of the parallel *RC* circuit (and of other simple circuits). We state them in a general form as though they were valid for all networks. That is here mere conjecture: they hold, we know, for our simple circuits; we do not know whether they hold for networks in general. But they do, and an objective of the rest of this book is to demonstrate that fact.

The response of a network may conveniently be divided into two parts:

a The natural part, exponential in form as a function of t, like Ke^{pt} in which the constant in the exponent, p, is the *natural frequency*, a characteristic of the network, and K is another constant, used to satisfy initial conditions.

b The forced part, of the same form as the excitation in some cases; it has at least roughly similar form in others, and is proportional in strength (multiplier) thereto.

Mathematicians refer to (b) as the *particular integral* and to (a) as *the complementary function.* The terms are not inappropriate, but they do not stress the physical meaning of the parts. Their physical rôles are worth restating:

> The *forced part* (particular integral) represents the network's reaction to the excitation, with such fidelity as it may have; it is always distorted somewhat (by design or as an inevitable nuisance, depending on the situation), if only by the addition of the natural part; the forced part is often the important part of the response, that which the network was designed to produce.
>
> The *natural part* complements the forced part: at the initial instant it brings the latter into harmony with the actual initial condition of the network, which is imposed by its history and is often quite different from the value of the forced part then; thereafter, at a rate determined by the network's property of innate sluggishness (its time constants), the natural part disappears, to leave the forced part dominant if it is capable of dominating the response. If the forced part is permanent (steady-state) in nature, it can do so; if the forced part is transitory also, then it becomes a matter of comparison of time constants.

As a device for analysis of the network (solution of the model differential equations) we can sometimes successfully write

$$\text{Response} = r(t) = r_N + r_F = Ke^{pt} + R_F\, e(t) \tag{2.17-1}$$

in which $e(t)$ is the excitation wave. Success is determined by substituting this in the differential equation; if we can determine values for the parameters K, p, and K_F so that $r(t)$ satisfies the differential equation (and boundary condition), we succeed. If we do not, we try decomposing the excitation into component parts, obtaining the individual responses thereto, and adding (superposing) them. *Superposition* is in general justified by the linear character of the model; attention to details is essential, however, for certain things (additive constants, for example) do not necessarily simply superpose. We have not yet had need for any other devices.

We note again that the *exponential function* appears to have a very special place in network analysis. The natural behavior is exponential, as e^{pt}. Common excitation functions (sinusoids, steps, even impulses) are exponential functions or combinations of exponential functions. Complex numbers simplify calculating the relations between excitation and response (natural and forced parts, both).

Recall once again that these statements have been demonstrated for the simple circuits of this chapter. It remains to show that they are true (and useful) for networks in general; that is our agenda.

As a reasonable step toward generalization, we turn now to a circuit with *two* elements that can store energy. Its analysis will be easier because we have the tools above; but its behavior may well be more complicated because of the second energy-storing element.

2.18 EXERCISES

In all the exercises of Sec. 2.18 use a parallel *RC* circuit and an initial state of rest, unless otherwise directed.

2-1. State, or find from basic electric relations, the dimensions of each of the quantities listed, in terms of these four basic dimensions:

$$M = \text{mass},$$
$$L = \text{length},$$
$$T = \text{time},$$
$$Q = \text{electric charge}.$$

Start from

$$\text{Voltage} = \text{force per unit electric charge} = \frac{[ML]}{[T^2Q]}$$

and find the dimensions of current, resistance, conductance, capacitance, inductance, frequency, flux linkages, power, and energy (work). What is the name of the standard SI unit for each quantity? Justify using the name of the *unit* instead of the *dimensions* in making dimensional checks.

2-2. Make two plots† of (2.06-1) versus x: (*a*) using an arithmetic scale for v; (*b*) using a logarithmic scale for v (show two decades from the initial value).

† By *sketch* is meant a rapidly made, freehand drawing, based on the minimum amount of calculation adequate for answering the question; a *plot*, on the other hand, is to be based on appropriate calculation and drawn carefully to scale.

Use an arithmetic scale for x, showing $0 < x < 4$. What advantage can you see in (*b*) as compared with (*a*)? These curves may be useful as templates in the exercises that follow.

2-3. Answer the following two questions (1) very quickly, making only a rough, mental calculation, and (2) finding the answers to three significant figures. Compare your results.

(*a*) The impulse response of an *RC* circuit decays 10 percent (to 90% of its initial value) in what fraction of a time constant?

(*b*) The step response of an *RC* circuit rises from 0 to 15 percent of its final value in what fraction of a time constant?

2-4. Explain carefully the validity of the statements below, which are often useful in finding the responses of circuits to step excitation.

(*a*) At $t = 0+$ an uncharged capacitor acts as a short circuit;
(*b*) At $t = 0+$ an uncharged inductor acts as an open circuit;
(*c*) As $t \to \infty$ a capacitor becomes an open circuit;
(*d*) As $t \to \infty$ an inductor becomes a short circuit.

Illustrate by the following example (Fig. 2-4).

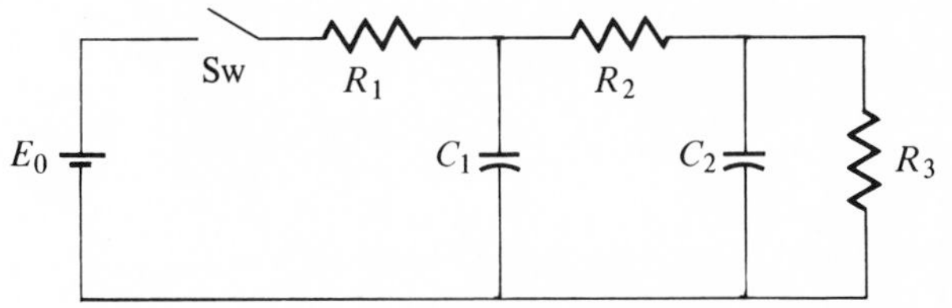

FIG. 2-4.

The capacitors are initially uncharged and the switch is closed at $t = 0$. What is the initial ($t = 0+$) current in each capacitor? What is the final ($t \to \infty$) current in each capacitor?

2-5. Mentally evaluate (by using one or two terms of the appropriate power series) each of the following; then evaluate them correctly to three significant figures. Compare and discuss the results:

$$e^{0.05} \qquad e^{-0.07} \qquad \sin 12^\circ \qquad \cos 86^\circ \qquad \cos 12^\circ \qquad \sin 86^\circ$$

$$\sqrt{1 + 0.08^2} \qquad 1 + j0.2 \qquad (1 - j0.08)^{-1}$$

(Convert the last two to polar form.)

2-6. Of the following (approximate, quick, mental) calculations, four are grossly incorrect. Find them, correct them, and try to explain how the error probably occurred.

$$1 + j10 = 10\underline{/6^\circ} \qquad 3 + j4 = 5\underline{/55^\circ}$$

$$(1 + j10)^{-1} = 0.1\underline{/-84^\circ} \qquad \sqrt{1^2 + 0.1^2} = 1.005$$

$$e^{0.2} = 0.8 \qquad e^{-0.2} = 0.8$$

$$\tan 30^\circ = 0.5 \qquad \tan 75^\circ = 0.25$$

$$\sin 22^\circ = 0.37 \qquad \cos 15^\circ = 0.25$$

2-7. Of each of the complex numbers given below calculate the other (rectangular or polar) form and draw the corresponding triangle [see (2.15-11) and Fig. 2.15-B].

$0 + j\,3.2$	$-1 + j\,0$
$1.3\underline{/49.3^\circ}$	$-1 - j\,1$
$0.6\underline{/-16.7^\circ}$	$-2.73 - j\,0.96$
$4.26\underline{/207.0^\circ}$	$1\underline{/60^\circ}$
$0 + j1$	$1\underline{/30^\circ}$
$1 + j1$	$1\underline{/120^\circ}$
$-1 + j1$	$1\underline{/-130^\circ}$

2-8. Sketch† two sinusoidal waves, 30° apart in phase. Explain the two words *lag* and *lead* in terms of the waves and of the time-function expressions and of their phasor representations. Sketch two waves that are "in phase" and two that are "out of phase," explaining your interpretations of these expressions. Does ambiguity exist?

2-9. Give the time-function expression of the sinusoidal waves whose phasor representations are (*a*) A, (*b*), B, (*c*) AB, (*d*) A/B, (*e*) $A + B$, and (*f*) $A - B$, in which $A = 6.2\underline{/37^\circ}$ and $B = 0.7\underline{/-56^\circ}$.

2-10. What is the phasor representation of each of the following sinusoids? Give both formal replies and drawings of the phasors, to scale. The waves to be considered are all given by $E_m \cos(\omega t + \phi)$ in which the numerical values are

† See footnote to Exercise 2-2.

E_m, V	ω, rad/s	ϕ, °
32	$2\pi(1000)$	4.6
64	$2\pi(1000)$	47.2
48	$2\pi(1000)$	136.2
105	$2\pi(3500)$	276.9

2-11. Give the phasor result of each of the following operations, and the time-function formula for the corresponding sinusoid, and sketch the four sinusoids versus time.

(*a*) $4\underline{/\ 16^\circ} - 2\underline{/\ -30^\circ} + 1\underline{/\ 160^\circ}$,

(*b*) $(4\underline{/\ 16^\circ})(2\underline{/\ -30^\circ})(1\underline{/\ 160^\circ})$

(*c*) $\dfrac{4\underline{/\ 16^\circ}}{2\underline{/\ -30^\circ}}$,

(*d*) $\dfrac{110\underline{/\ 72^\circ}}{32\underline{/\ 28^\circ}}$.

2-12. For each of the element-value pairs given in the table (for a parallel *RC* circuit) find the circuit time constant and the natural frequency. [If the time constant is given, verify or correct it.] Express the time constant in units (milliseconds, microseconds, nanoseconds, ...) such that the numerical value of the time constant is between 1 and 1000 inclusive (this is a convenient range for discussions and specifications). Similarly express the natural frequencies in units (mega-, kilo-, nano-, ..., nepers per second) such that the numerical values lie between 1 and 1000 inclusive.

Add a column below listing frequencies at which, in steady state response to sinusoidal excitation lags 45°. For what frequencies will the response lead?

R, (Ω)	*C*	**Time Constant**
100	0.01 μ*F*	
1*K*†	0.01 μ*F*	
10*K*	1000 *pF*	10 μs
1*K*	1000 *pF*	
200*K*	0.05 μ*F*	
1*K*	10 *pF*	10 ps
10*K*	0.01 μ*F*	
200	5 *pF*	
20*K*	4000 *pF*	80 ps
600	20 *pF*	
75	200 *pF*	

† *K* indicates thousands of ohms.

2-13. A 0.01-μF capacitor is charged to 10 mV. An impulsive charge of 1000 pC is then placed thereon at $t = 0$. At what time will the capacitor voltage return to 10 mV: (*a*) If a 50-Ω resistor is connected across the capacitor? (*b*) If no resistor is connected across it?

2-14. A one 1-nC charge is suddenly deposited on an uncharged 1000-pF capacitor in parallel with a 500-Ω resistor. What is the capacitor voltage immediately thereafter? How much time has elapsed when the capacitor voltage is one-half of its original value? One-fourth thereof? One-eighth? Change R to 2000 Ω and repeat.

2-15. In the circuit shown $R = 2000\ \Omega$, $C = 0.01\ \mu$F. What is the time constant? If $v(0) = 10$ V, $i_F(t) = 0.05u(t)$ A, what is $v(\infty)$? Sketch $v(t)$ versus t. (See Fig. 2-15.)

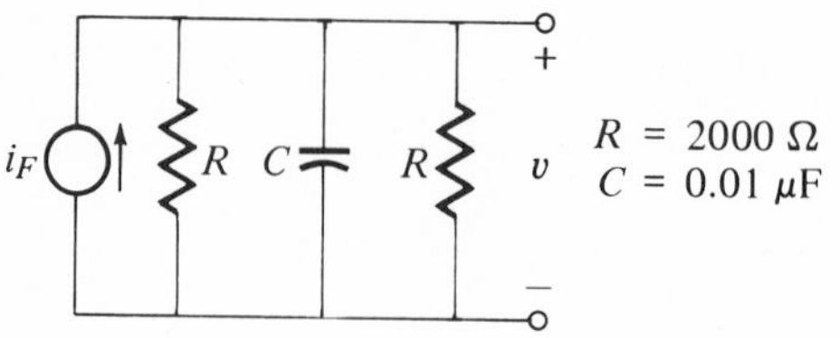

FIG. 2-15.

2-16. In Fig. 2-15 $v(0) = 0$, $i_F(t) = 50\ e^{-t/T_0}$ mA, and $T_0 = 20\ \mu$s. What is the final value of v? What is $v(t)$? Sketch $v(t)$ versus t. Locate and evaluate its maximum value.

2-17. In Fig. 2-15 $v(0) = -25$ V, $i_F = 0.01u(t)$ A. What is $v(\infty)$? Sketch v versus t. At what instant is $v = 0$?

2-18. The voltage of a parallel RC combination, in which $R = 3000\ \Omega$, responds to step excitation from rest by rising to 95 percent of its final value in 0.1 μs. How large is the capacitance?

2-19. In a parallel RC circuit $R = 3000\ \Omega$ and $C = 0.03\ \mu$F. The excitation is a step function of current. Let V_0 represent the ultimate, steady-state value of the voltage response. The initial value is given below for three different cases. In each case how long does it take to reach the value of interest? Answer (*a*) in time constants, (*b*) in seconds.

$v(0)$	Value of Interest
$0.5V_0$	$0.8V_0$
$1.0V_0$	$1.0V_0$
$2.0V_0$	$1.5V_0$

2-20. Find the ultimate, steady-state (repetitive) response of the *RC* circuit when excited by a train of alternately positive and negative impulses of equal strength, occurring at intervals of one time constant.

2-21. The step response of a parallel *RC* circuit is interrupted every *RC* (seconds) and brought back to zero by applying negative impulses of strength Q_0. That is, $i_F(t) = I_0 u(t) - Q_0(t - RC) - Q_0(t - 2RC) - Q_0(t - 3RC) - \dots$. What is the value of Q_0 in terms of I_0 and RC?

2-22. Consider the *RC* circuit with step excitation. Express the voltage of the capacitor, the energy stored in the electric field of the capacitor, and the rate at which energy is being stored therein, all as functions of time. When, and for what value of voltage, is the capacitor receiving energy at the maximum rate? State and solve the dual circuit problem.

2-23. In the circuit shown $i_F(t) = I_0 u(t)$ and $v(0) = 0$. What is $v(t)$ for $t > 0$? Give a sketch as well as a formula.

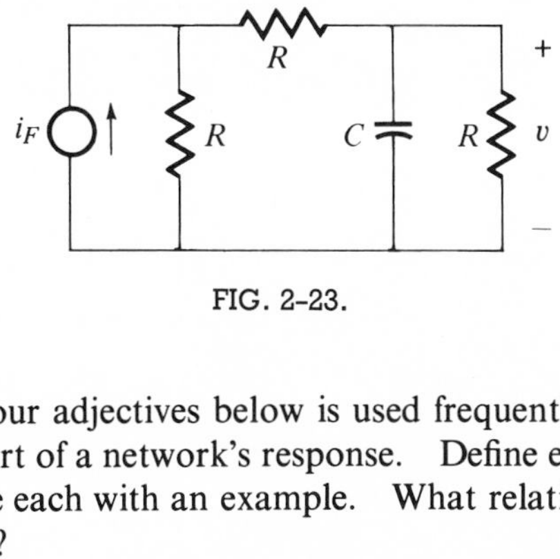

FIG. 2-23.

2-24. Each of the four adjectives below is used frequently in circuit analysis to characterize some part of a network's response. Define each one clearly in your own words; illustrate each with an example. What relation does each one have with the other three?

Forced, natural, steady-state, transient.

Why is it that in many circuit problems the parts of the response that bear these names are all exponential functions of time? Explain fully, with necessary qualifications and reservations. Then give four simple examples, one for each adjective. Can you give simple qualitative examples of nonexponential character of each?

2-25. Why cannot the response of the parallel *RC* circuit to the simple excitation $i_F = I_0(t/T_0)/(1 + t/T_0)$ be calculated by the methods of this chapter?

2-26. The response of the *RC* circuit to a step contains a transient component; the steady state is not immediately reached. What additional excitation would cause the steady state to be immediately reached?

2-27. Initially at rest, the circuit shown is excited impulsively: $i_F(t) = Q_0\,\delta(t)$. What is the initial ($t = 0+$) value of $v(t)$? The final value? The formula for $v(t)$, $t > 0$? Sketch $v(t)$ versus t.

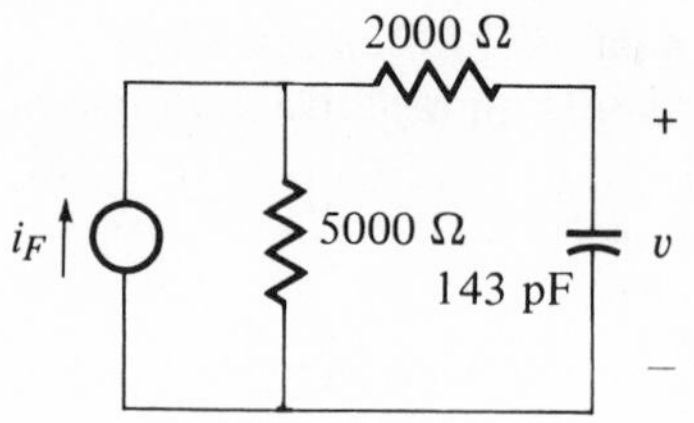

FIG. 2-27.

2-28. An RC circuit is excited by the wave $i_F(t) = I_F\,u(t) - I_F\,u(t - t_1) - Q_0\,\delta(t - t_2)$. For the response to be identically zero for $t > t_2$, what must be the value of Q_0 in terms of I_F, t_1, t_2, R, and C? Use sketches as appropriate.

2-29. Sketch the response from rest of the RC circuit to the several impulse trains defined below.

	Occurrence time, in Units of RC	Strength
(a)	0	1
	1	1
	2	1
	3	x (x is such that for $t > 3RC$, the response is identically zero)
(b)	0	1
	1	−1
	2	1
	3	−1
(c)	0	+1
	1	+1
	2	−1
	3	−1

2-30. A parallel RC circuit is excited by an impulse train. Each impulse is of strength Q_0, and their spacing is T_0. Tabulate the value of the response voltage at $t = 0-, 0+, T_0-, T_0+, 2T_0-, 2T_0+, \ldots, 10T_0+$. What is a simple expression for $v(t)$ within any one period of the steady state?

2-31. An impulse train, $Q_0\,\delta(t) + Q_0\,\delta(t - T_0) + Q_0\,\delta(t - 2T_0) + \ldots$ is applied to an RC circuit. What is the average value of the response voltage v in the steady state? Compare the exact result with the approximate result obtained by drawing a straight line mentally between a maximum point on the curve of $v(t)$ and the point just before the application of the next impulse. What is the percentage error in using the approximate result for $T_0 = RC$? $2RC$? $3RC$?

2-32. A parallel *RC* circuit is excited from rest by the succession of six impulses:

$$i_F(t) = Q_0[\delta(t) - \delta(t - T) + \delta(t - 2T) + \delta(t - 3T) - \delta(t - 4T) - \delta(t - 5T)].$$

Criticize carefully this statement: "Since the total charge supplied is zero, the response will be zero for $t > 5T$."

2-33. A 100-pF capacitor and a 50,000-Ω resistor are connected in parallel. At $t = 0$ a charge is suddenly deposited on the capacitor, such that the voltage jumps to 80 mV. When the voltage has dropped to 40 mV the charge is impulsively renewed, raising the voltage to 80 mV again. This action is repeated when next the voltage is 40 mV, and so on, again and again, world without end. What charge must be supplied at each renewal, and how often must it be supplied?

2-34. What excitation (in form of a train of impulses) will produce the steady-state response shown in Fig. 2-34?

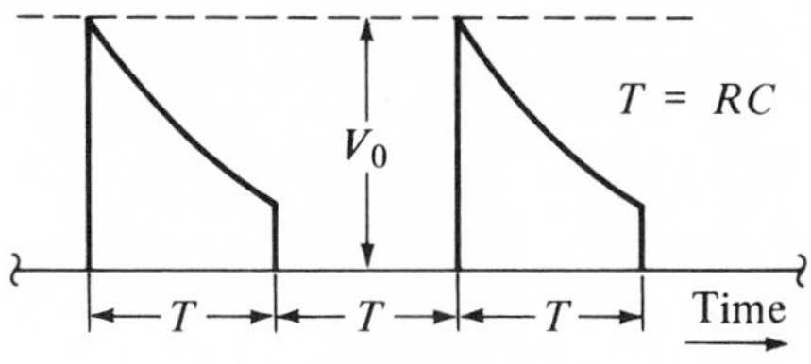

FIG. 2-34.

2-35. For the circuit shown, initially at rest, find the initial ($t = 0+$) and final values of $v(t)$, and the circuit time constant. Then find and sketch $v(t)$ by applying (2.09-10), first explaining why it is applicable.

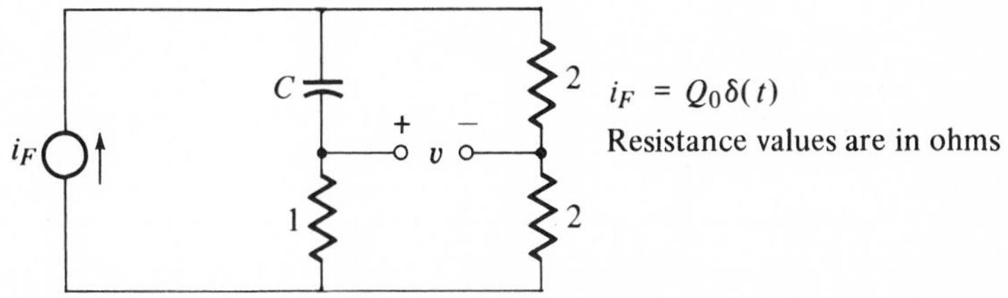

FIG. 2-35.

2-36. Repeat Exercise 2-35 for each of the two circuits shown in Fig. 2-36.

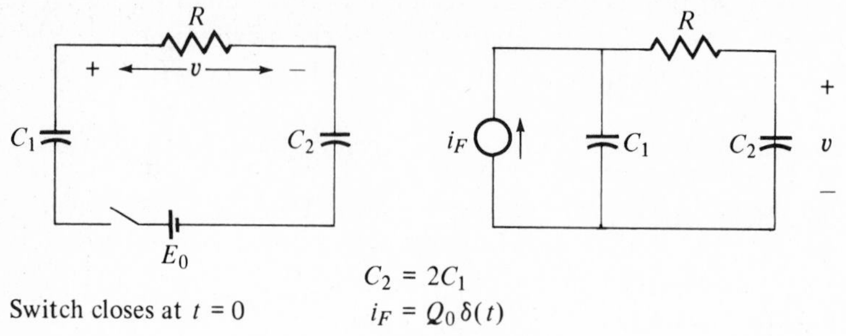

FIG. 2-36.

2-37. The circuit of Fig. 2-37*a* is composed of six identical 1000-Ω resistors and a 1000-μF capacitor. What is its time constant? Why, in the calculation, may one temporarily assume all resistors to have unit value? Of what help is this device? (Consult the second part of the figure if necessary.)·

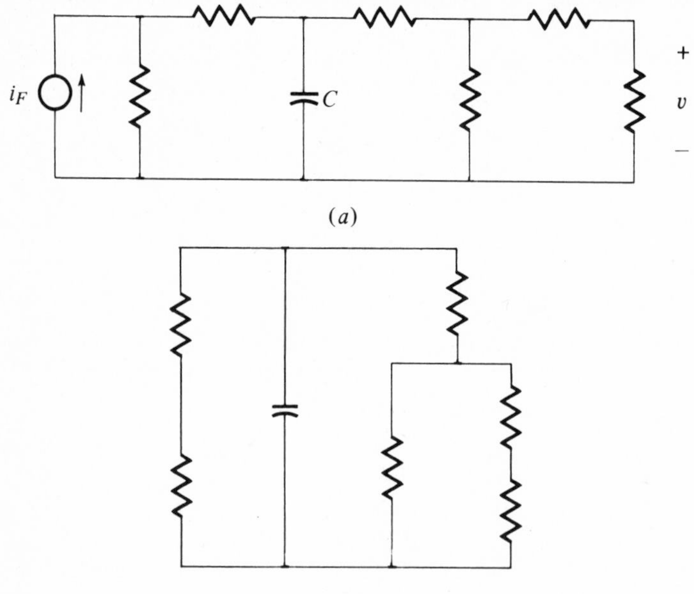

FIG. 2-37.

The initial voltage on C is 30 mV; $i_F(t) = 22u(t)$ mA. What is $v(0)$? $v(\infty)$? Plot $v(t)$, using and adapting (2.09-10), with explanations.

2-38. Calculate, for cases in Fig. 2.15-E, what phase and amplitude differences between excitation and response should be. As closely as you can, read the actual values from the photographs. Compare the two sets. Explain the discrepancies. To what extent can you similarly verify response-excitation amplitude ratios?

2-39. In Fig. 2.15-G, the time scale, for both curves, is 0.5 s/cm. The upper curve is the response of an *RC* circuit, from rest, to the sinusoidal excitation shown by the lower curve. What, approximately, is the amount of steady-state phase shift? The time constant of the circuit? Why can you not determine, even approximately, the values of *R* and of *C*?

2-40. An *RC* circuit is excited sinusoidally. Take the excitation phasor to have unit length at zero angle, for simplicity. Show that the (steady-state) response phasor is then $R/(1 + jRC\omega)$. Draw, to scale, response phasors for excitation frequencies such that the phasor angle is, successively, $0, -10°, -20°, \ldots, -80°, -90°$. Sketch in the locus of the tips of the phasors as the frequency varies. What geometrical figure does it seem to be? Verify your conjecture by analysis.

For what excitation frequencies will the response phasor have angle $-15°$? $-45°$? $-75°$? Answer by expressing the excitation *periods* as multiples of the circuit time constant.

2-41. In proportion to RI_F, the value for very-low-frequency excitation, how large is the steady-state response of an *RC* circuit to sinusoidal excitation of each of these periods: RC, $RC/10$, $10RC$, $20RC$? What, in each case, is the angle of lag (lead) of the response with respect to the excitation?

2.19 PROBLEMS

In all problems of Sec. 2.19, unless otherwise stated, the circuit is a parallel *RC* combination, initially at rest.

2-A. Apply the theorem of the mean (the mean-value theorem) to show that, in (2.02-5), $\int_0^{T_0} Gv\,dt = Gv(\eta)$, $0 < \eta < \varepsilon$ $(\varepsilon = T_0)$, and so show that its contribution is negligible if ε is small enough.

2-B. An uncharged capacitor *C* in parallel with a resistor *R* is charged by an impulse. The energy stored in the capacitor just after the charging is $W_C = \frac{1}{2}C(Q_0/C)^2$ (verify this). Why cannot this energy be computed by integrating $v_C i$, i being the source (excitation) current during the charging period? Check that energy is conserved thereafter by evaluating $\int_0^\infty Ri^2\,dt$, i being the resistor current.

2-C. A battery of internal resistance R_0 replaces the current source as excitation for the parallel *RC* circuit (Fig. 2-C*a*), the switch being closed at $t = 0$ when the capacitor is uncharged. Show that, as far as *C*, *G*, and *v* are concerned, the circuit of Fig. 2-C*b* is equivalent. (*Suggestion:* write the differential equation for *v* in terms of *G*, *C*, etc.; write also the equation for *i* in terms of E_0, R_0, and

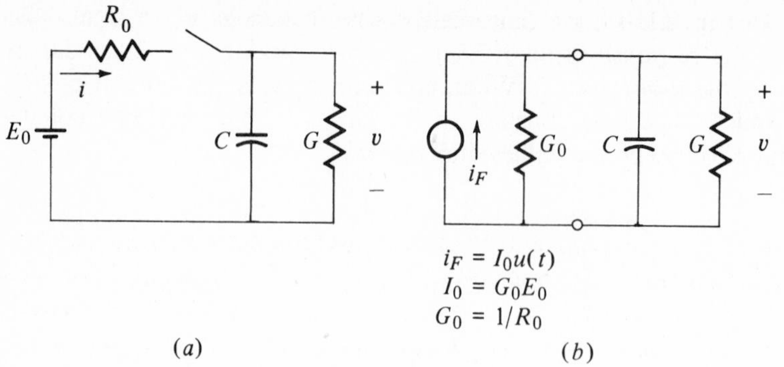

FIG. 2-C.

v.) What can you say about the *external* equivalence of the two sources, one of voltage, one of current? What is the time constant of the circuit? Find the voltage v as a function of time.

2-D. A narrow current pulse of duration T (seconds) and area unity (coulombs) produces a voltage in the RC circuit of $(R/T)(1 - e^{-\alpha T})e^{-\alpha(t-T)}$ for $t > T$. First verify this statement, then obtain the limit as $T \to 0$ and show that the result is the impulse response. Explain how this justifies neglecting the difference between narrow-pulse and impulse excitation. Carefully plot† the difference between the two, versus time, for several reasonably small values of $T/(RC)$; compare the absolute and fractional differences.

2-E. The capacitor C in Fig. 2-E is initially uncharged, $i_1 = Q_1\,\delta(t)$, $e_2(t) = E_2 u(t)$. The response v can be calculated as the sum of two components, v_1 calculated as though e_2 were zero and v_2 calculated as though i_1 were zero. Show, using the appropriate differential equation, that this statement is true, and find both $v_1(t)$ and $v_2(t)$ for $t > 0$. Why is the term *superposition* relevant to this manner of analysis? Show that, for the calculation of $v(t)$, a simple parallel RC circuit, excited by $i_F(t)$, is equivalent, in that its voltage will be the same as $v(t)$ here, provided $i_F(t)$, R, and C have therein certain values. What are those values? Show also that if E_2 and Q_1 are properly related, the response will step immediately (at $t = 0+$) to its steady-state value and remain there, without any transient component. What is that relation?

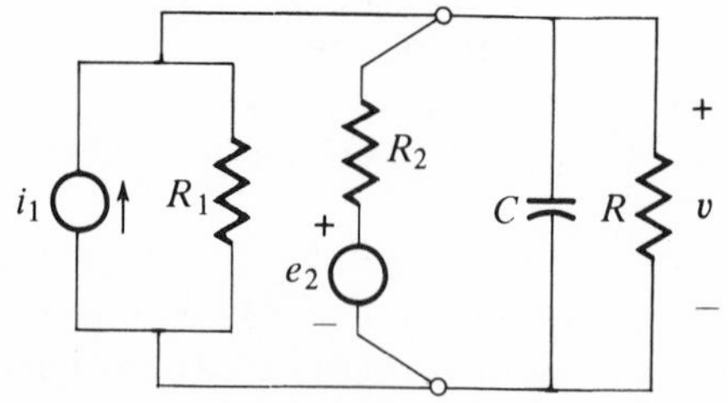

FIG. 2-E.

†See the footnote to Exercise 2-2.

2-F. The capacitor C of a parallel RC circuit initially at rest is charged

1 Impulsively, by $i_F(t) = Q_0\,\delta(t)$;

2 By a short pulse, $i_F(t) = \dfrac{Q_0}{T_0}[u(t) - u(t - T_0)]$.

(*a*) Let the capacitor voltage be represented by $r_a(t)$ in case (1), and by $r_b(t)$ in case 2. Show that, for $t > T_0$, the fractional difference between the two responses, $x = (r_b - r_a)/r_a$, is approximately equal to $T_0/(2RC)$.
(*b*) Explain how this result supports the common allegation that, for *short* pulse excitation, the *response* is virtually indistinguishable from the response to an impulse of the same total charge. Why is this equivalence of practical interest?
(*c*) If in case 2 we assume linear charging of the capacitor in $0 < t < T_0$ (i.e., we neglect the current in R), show that the fractional difference is approximately $T_0/(RC)$. Explain the change by a factor of 2. Does this alter your view of the approximate equivalence of part (*b*) of this problem?
(*d*) If T_0 is large enough compared with the circuit time constant, the difference between the two responses becomes large. Plot x (the fractional difference) versus the ratio $T_0/(RC)$ to show this.

2-G. In the circuit shown $R_1 = 2R$, $R_2 = 3R$, $R_3 = 4R$, $R = 6200\ \Omega$, $C = 300$ pF. What is the time constant with the switch closed? If $v(0) = 0$, and the switch is closed at $t = 0$, what is the ultimate value of v? Write the expression for $v(t)$ immediately from the two preceding results, using (2.09-10). Why is the analysis of this five-element circuit so simple?

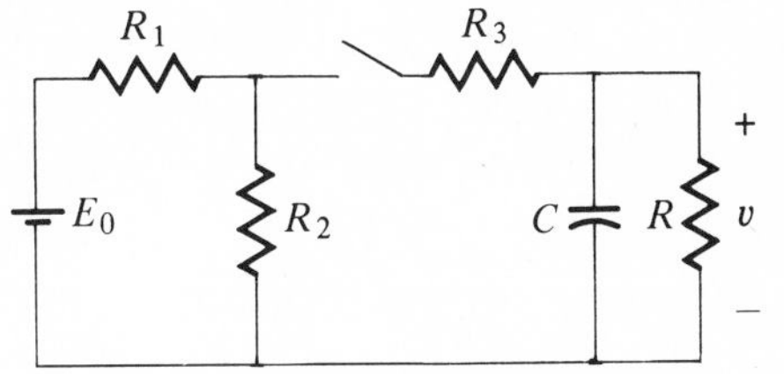

FIG. 2-G.

2-H. A switch and two uncharged capacitors are connected to a current source as shown (Fig. 2-H).

(*a*) A current impulse of strength Q_0 (coulombs) is applied to the first capacitor by i_F, S being open. Then S is closed. The problem is to find the voltages v_1 and v_2 just after the closing of the switch. Show that application of the principle of conservation of energy (no *energy* is lost or gained during the operation of the switch) leads to $v_2 = (Q_0/C_1)(1 + C_2/C_1)^{-1/2}$ but that application of the principle of conser-

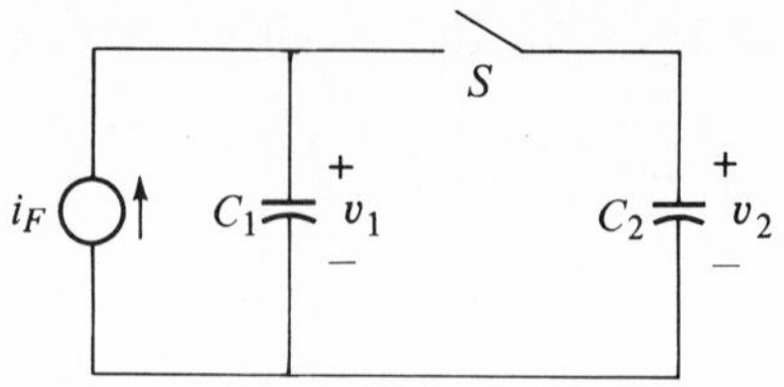

FIG. 2-H.

vation of charge (no charge is lost or gained during the operation of the switch) leads to $v_2 = (Q_0/C_1)(1 + C_2/C_1)^{-1}$, which is different. Can you determine by physical arguments at this point which is correct? If so, resolve the paradox thoroughly. If not so, proceed to (*b*).

(*b*) Suppose the circuit (model) changed by the addition of a resistor r in series with C_2, between C_2 and S. Again the two capacitors are initially uncharged and $i_F = Q_0\,\delta(t)$, S being closed after the impulse is over. Find $v_1(t)$ and $v_2(t)$ thereafter; sketch their behavior versus time. How do the curves change as r is made smaller and smaller? Now use these results to answer the following questions. (1) What is the basic cause of the paradox of (*a*)? (2) Why are *both* conservation principles (of charge, of energy) valid *if properly applied?*

(*c*) What is "the correct answer" to the original problem?

(*d*) Construct and explain the dual problem.

2-I. Sketch the current response of the series RC circuit to (*a*) step excitation, (*b*) impulse excitation, and (*c*) pulse excitation. Explain the physical action in each case. Make your sketch as accurate (include formal mathematical expressions therefor) as possible without carrying out calculations except in your head. Sketch the forced and natural parts separately in each case, as well as their sum.

2-J. Perform an analysis similar to that of Sec. 2.12 for excitation composed of a train of impulses rather than of pulses. The impulses occur regularly, but every other impulse is in strength only half that of the others:

$$Q_0[\delta(t) + \tfrac{1}{2}\,\delta(t - T_0) + \delta(t - 2T_0) + \tfrac{1}{2}\,\delta(t - 3T_0) + \delta(t - 4T_0) + \cdots].$$

2-K. Carefully verify the resolution of the square-wave response (Fig. 2.12-B*b*) into transient and steady-state components, (2.12-7) and (2.12-8).

2-L. Square-pulse signals can be formed easily in practice and are widely used in communication and control. They suffer distortion in passing through any real apparatus: any circuit or network, amplifier, or transmission line, or even a simple RC circuit. They can, however, be "regenerated" by nonlinear electronic devices that restore their rectangularity. There is of course some signal-change price to be paid for this, which we here study. Consider a signal composed of two rectangular pulses (Fig. 2-L). Suppose it passes through an

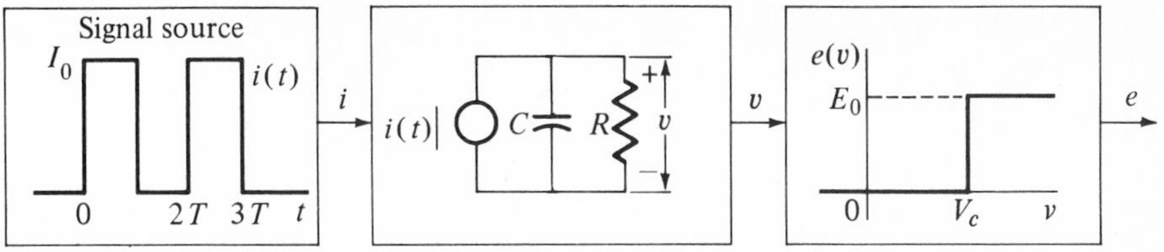

FIG. 2-L.

amplifier stage that can be modelled as an *RC* circuit, $RC = \frac{2}{3}T$. Suppose it then passes through a regenerator which we assume is simply an instantaneously acting on-off device triggered at the voltage V_c, $V_c = 0.6RI_0$. Plot the original two-pulse signal *i*, the amplifier output *v*, and the regenerated signal *e*. Are the widths of the rectangular pulses of the regenerated signal the same as those of the input signal? Does the spacing between pulses change? Explain. Do you recommend a change in the ratio of *T* to *RC*?

2-M. Determine the values of V_A and of V_B in the steady state in terms of $V_0 = RI_0$, T_1, and T_2 (see Fig. 2-M). Use an algebraic method [see (2.12-5)]. If $V_A = 2V_B$, what is T_1 in terms of the time constant *RC*? What is the value of the ratio V_A/V_B in terms of T_2 and $\alpha = (RC)^{-1}$? Sketch V_A/V_B versus αT_2.

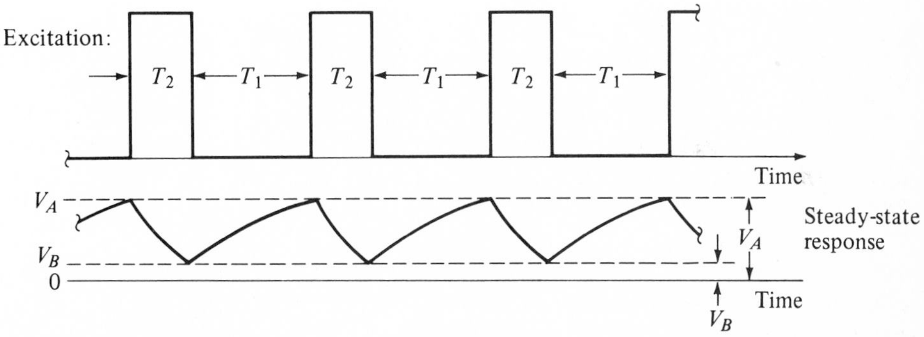

FIG. 2-M.

What is a simple approximate formula for V_A/V_B for small values of αT_2? Now let T_2 approach zero in such a way that $I_0 T_2 = Q_0$ remains constant. What are the limiting values of V_A and of V_B? Check against the impulse response calculated in the text.

2-N. The circuit of Fig. 2-N is a simplified model of a practical rectifier circuit whose purpose is to convert the alternating voltage *e* (a source of power) to a dc or constant voltage at *v*, across the load *R*.

The diode, a highly nonlinear electronic device, is idealized to become in effect a switch. Its characteristic (see figure) is as follows. When current tends to flow from left to right (in the $+i_D$ direction), $v_D = 0$ and i_D flows as determined by the circuit external to the diode (the diode state is a short circuit,

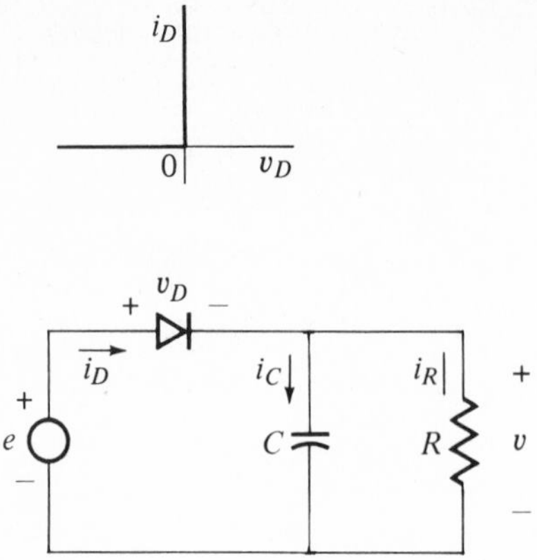

$e = E_m \cos \omega t$

$RC\omega = 4$

FIG. 2-N.

a closed switch); when current tends to flow from right to left (in the $-i_D$ direction), as determined by i_C, i_R, v, and e, the diode becomes an open circuit, the switch opens, $i_D = 0$, and v_D (negative) is determined by the external circuit.

With this idealization of the diode, the circuit is piecewise linear in time, changing back and forth between (*a*) an isolated *RC* circuit and (*b*) an *RC* circuit connected directly to the voltage source *e*. On proceeding in steps, in time, therefore, only linear circuit analysis is needed.

Calculate and plot the voltage v versus time, starting from $t = 0$ when $v = 0$, the switch is closed, until two cycles of the steady state have been displayed. The following questions may be useful: How does i_D behave initially? When does i_D become zero? What then happens in the diode? What then happens in the *RC* circuit? When does the diode state again change?

To what extent is the rectification objective met? (Compare the average value of v with the size of the *rectification error* or "ripple," $v - v_{av}$.) Briefly discuss the effect of varying the value of $RC\omega$. Should it be large or small?

2-0. A voltage source *e*, a resistor *R*, and a capacitor *C* are connected in series to form a closed loop. The (repetitive) voltage wave is $e(t) = E_m[u(t) - u(t - T) + u(t - 2T) - u(t - 3T) + \cdots]$. Find the steady-state waveform of the current and plot two cycles thereof. What is its average value? Contrast with square-wave response of *parallel RC* circuit and explain differences.

2-P. It is becoming increasingly common to transmit information of whatever sort (telephone conversations, music, TV, etc.) not in the form of a continuous wave that follows the actual physical signal, but in a modified form that consists merely of a series of numbers. This problem deals with one part of such a *digital* system in a very elementary way. The basic modus operandi of the system is this. The signal wave $e(t)$ is "sampled" (its value is read and noted) every T_s (seconds). This produces a set of values $e(t)$, $e(T_s)$, $e(2T_s)$, ... These values are converted into numbers that are transmitted in the form of corresponding sets of pulses, coded to represent those numbers (as in Prob. 2-T). (There are engineering advantages to having only pulses, or pulse groups, and not speech or the like itself, to transmit.) At the receiving end the pulses are manipulated to give back the original samples (with some error, of course, but very little). It then remains to reconstitute the signal wave from the samples. It is with this final stage of the system that this problem deals.

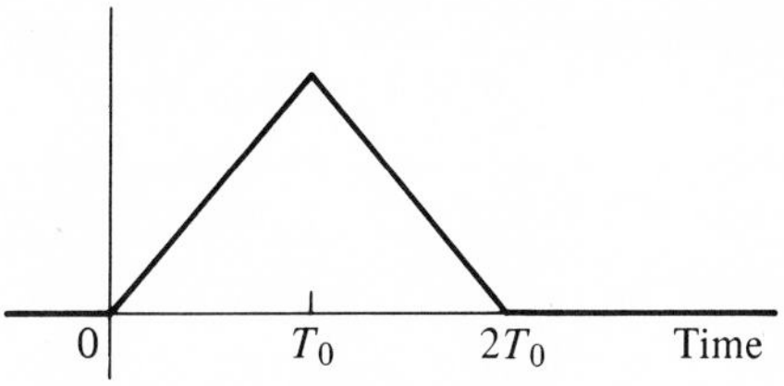

FIG. 2-P.

Take as a signal the simple triangular pulse (Fig. 2-P). Suppose it sampled every $0.2T_0$ (seconds); i.e., take $T_s = 0.2T_0$. Tabulate the values of the samples. Suppose these numbers transmitted 5000 km and delivered to the receiving apparatus without error, in spite of the coding, transmission, and decoding. To reconstitute the signal from these, its samples, apply to a simple parallel *RC* circuit a current excitation that is a train of impulses, each of strength (area) proportional to one sample, which it represents, and occurring at the appropriate sample time:

$$i_F(t) = K[e(0)\,\delta(t) + e(T_s)\,\delta(t - T_s) + e(2T_s)\,\delta(t - 2T_s) + \cdots].$$

Plot the response of the *RC* circuit to the impulse train, versus time, for (*a*) $RC = 0.2T_s$, (*b*) $RC = 5T_s$, and (*c*) $RC = T_s$. On the basis of your results, choose one of the alternatives in the statement below and justify your choice. "The *RC* circuit (is/is not) capable of reconstituting a signal from a sequence of samples of that signal, in acceptable form." What characteristics, broadly speaking, should a (more complicated) circuit have to be able to do a better job of reconstituting a wave from samples thereof?

2-Q. A parallel *RC* circuit is excited, from rest, by the ramp $i_F = I_0(t/T_0)u(t)$. Show that $v(t) = V_0 + V_1(t/T_0) + Ke^{-\alpha t}$ and evaluate the parameters therein in terms of R, C, T_0, I_0. Sketch the general behavior of $v(t)$.

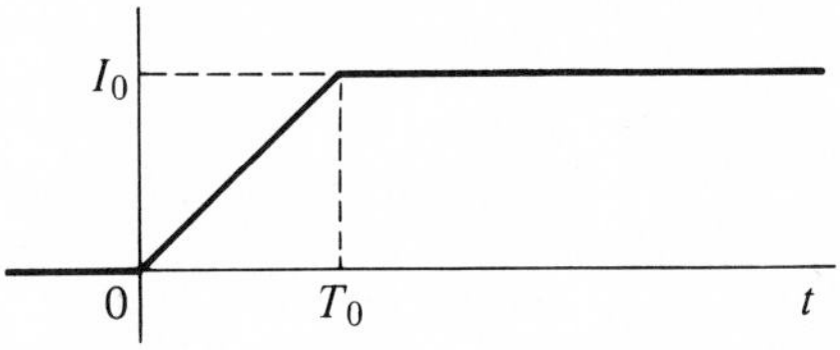

FIG. 2-Q.

Change the excitation to the sloping step of Fig. 2-Q. Calculate and plot the response wave versus time for $\alpha T_0 = 2, 1, 0.5$. Is it possible that there be no transient (natural) component in v after $t = T_0$?

2-R. Sketch, approximately and as rapidly as you can, one cycle of the ultimate, steady-state response of a parallel RC circuit to the repetitive "sawtooth" wave shown. Now formally and carefully evaluate this steady-state response. Plot

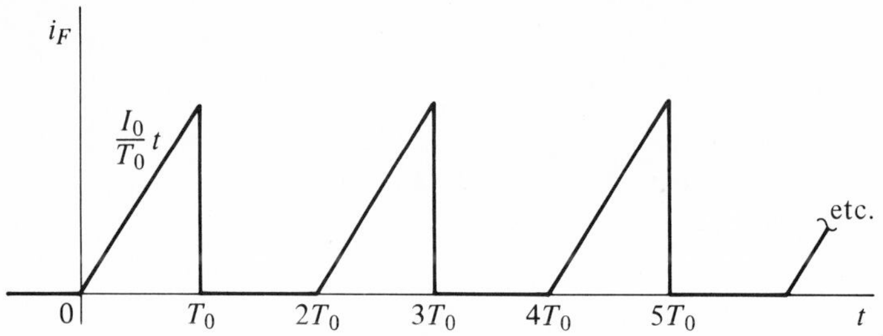

FIG. 2-R.

two cycles thereof, taking $T_0 = 0.693RC$. Briefly describe, in words, the distortion undergone by the excitation signal in passing through the RC circuit. Would the distortion be reduced by increasing the value of the time constant RC? By decreasing it? Explain carefully.

2-S. Consider the response (from rest) to exponential excitation, $i_F = I_0 e^{-\beta t}$, in which I_0 and β are (real) positive. Evaluate the initial slope of the response, $v'(0+)$. Explain why $v(t)$ is never negative thereafter, regardless of the value of β.

If β and I_0 and R are held constant, but C is varied, how do the initial slope and the time location of the maximum response and the value of the maximum response vary with C? Correlate with the response of Fig. 2.14-D.

Plot those three quantities versus $\beta RC = \beta/\alpha$ (only C varying) as a convenient parameter. Explain the behavior at $\beta RC = 1$. For what value of β will the maximum response occur at $t = 2RC$?

Now suppose that R alone varies. How will the corresponding set of response curves differ from those of Fig. 2.14-D?

2-T. Many electronic systems transmit signals that are merely successions of numbers—either ipso facto, as in computers, or as measured values of samples of speech or music or picture waves, taken often enough to retain the essential nature of the wave. It is very convenient (physically simple) to use the base 2, i.e. to express the numbers in binary form, because simple physical switches have *two* positions (open, closed), transistors can be made to operate with *two* states, etc. And the *presence or absence* of pulses at agreed-on instants of time (two simple, unambiguous, practical signal "states") can easily represent the two necessary digits or *bits* (0 and 1) so that a pulse train is a convenient binary representation of a number. For example the (decimal) number 25 in binary form becomes 11001 because $25 = 1 \times 1 + 0 \times 2 + 0 \times 4 + 1 \times 8 + 1 \times 16$. Note the convention (following that used in decimal notation) of writing the 16's bit first, then the 8's bit, ..., and the 1's bit last. The number 25 requires five bits or digits (0 or 1 each) because of its size: a five-bit notation can accomodate the 32 numbers from zero to $(2^5 - 1) = 31$, inclusive.

A (current) pulse train expression for 25_{10} or 11001_2 (the subscripts indicate the base used) is the signal

$$Q_0\,\delta(t) + 0\,\delta(t-T) + 0\,\delta(t-2T) + Q_0\,\delta(t-3T) + Q_0\,\delta(t-4T)$$

in which impulses of strength Q_0 represent the narrow pulses actually used, the 1's bit pulse (or no pulse) is sent first, the 2's bit (here a "no pulse") second, and the 16's bit (pulse or no pulse) last; T is the agreed-on pulse spacing (Fig. 2-T*a*).

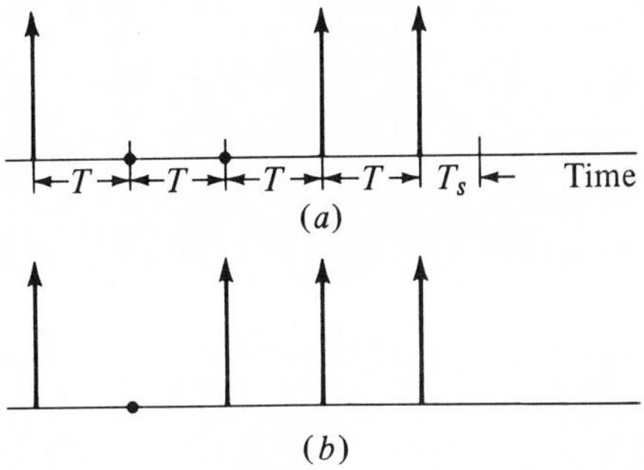

FIG. 2-T.

Similarly Fig. 2-T*b* shows a pulse-train representation of the number 29_{10}. This transmission sequence (1's bit first) is interesting because such a pulse train can be "decoded" [(re)converted to the number it represents in "analog" form, i.e., a physically measurable quantity proportional thereto, as opposed to "digital" form], by a simple *RC* circuit.

Show that if the time constant is properly chosen, then the parallel *RC* circuit's response to the pulse train excitation of Fig. 2-T*a*, measured at a fixed, agreed-on time T_s after the fifth (last) pulse, is proportional to the number 25. What must be the time constant, in units of T? What is the proportionality factor in volts per number represented (25)? Use the symbols Q_0, C, and $k = e^{-\alpha T_s}$.

Verify that the pulse train of Fig. 2-T*b* represents the number 29, and that the same proportionality factor will be correct for the *RC* circuit conversion thereof.

Explain why *any* five-bit binary number, in appropriate pulse train form, can be similarly decoded by such an *RC* circuit. (In principle this decoding method is satisfactory, but other, more practical, methods exist: consult the literature on pulse code modulation for example; see also Prob. 3-V.)

2-U. The excitation applied to an RC circuit initially at rest is

$$
\begin{aligned}
i_F(t) &= I_F e^{\alpha t} && 0 < t < T_1 \\
&= I_F e^{\alpha T_1} e^{-\alpha(t-T_1)} && T_1 < t < T_2 \\
&= 0 && T_2 < t.
\end{aligned}
$$

For simplicity, take $T_1 = T_2/2 = RC = \alpha^{-1}$. Compute and plot the response voltage for all positive time.

2-V. A single rectangular pulse signal passes through an amplifier composed of two simple RC stages and a pulse regenerator. (Fig. 2-V). Each RC stage is like that of Fig. 1.05-A. The pulse regenerator turns *on* when the voltage applied to it (rising) reaches E_c (volts); it turns *off* when the applied voltage

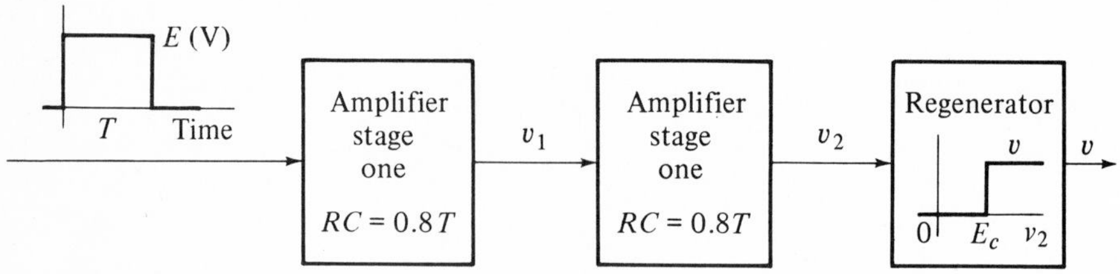

FIG. 2-V.

(falling) reaches E_c (volts). Its output v is therefore a rectangular pulse whose occurrence time and width depend on E_c and the characteristics of the two preceding stages. Assume no device in any way loads the preceding device. Plot the waves v_1 and v_2 versus t. Then plot a curve of the width of the output pulse versus its delay (lag), E_c being allowed to vary (compare leading edges of the input and output pulses). At what value must E_c be set to retain the input pulse width?

2-W. Find and plot the steady-state response of a parallel RC circuit to each of the three excitation waves shown in Fig. 2-W. Then find the natural and forced parts, for an initial condition of rest. For what initial condition will the natural part of the response be zero? Explain. Repeat for the *series* RC circuit, interpreting i_F as e.

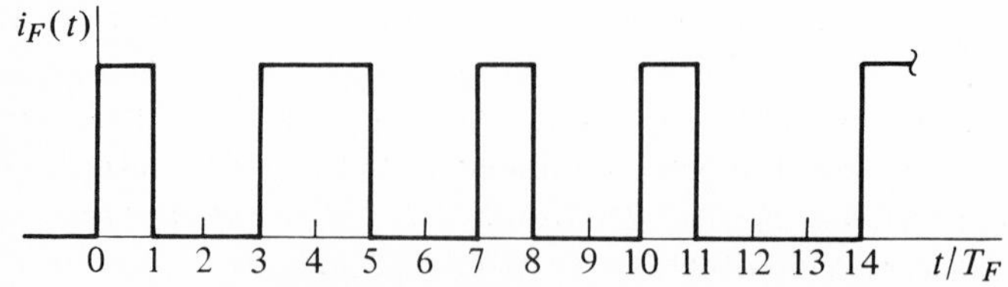

$T_F = RC$, $i_F(t + 14T_F) = i_F(t)$, $0 < t$

(a)

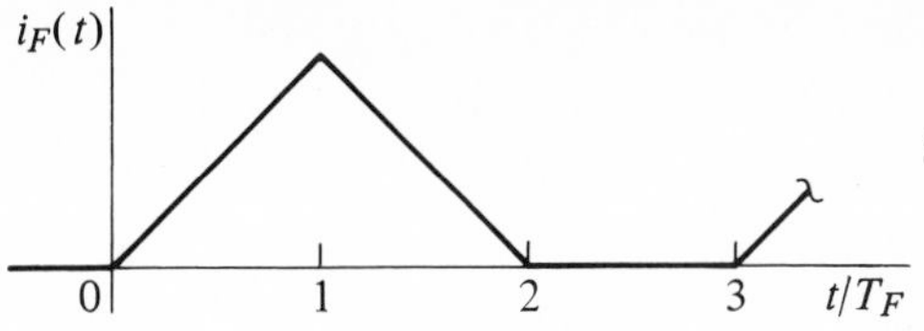

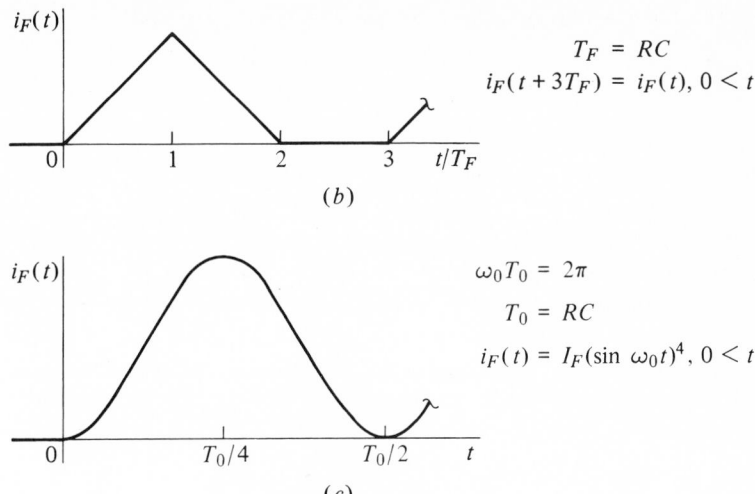

$$T_F = RC$$
$$i_F(t+3T_F) = i_F(t),\ 0 < t$$

(*b*)

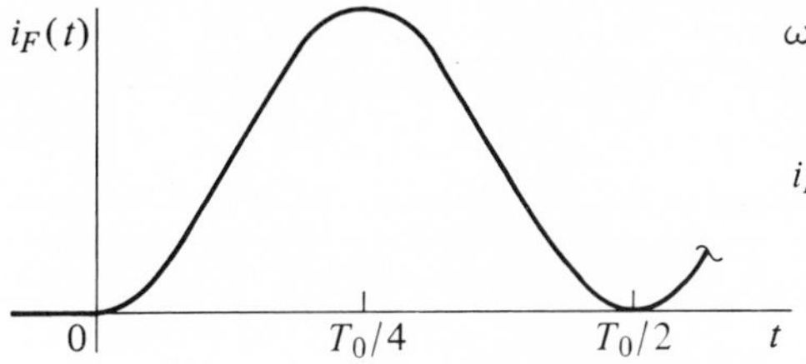

$$\omega_0 T_0 = 2\pi$$
$$T_0 = RC$$
$$i_F(t) = I_F(\sin \omega_0 t)^4,\ 0 < t$$

(*c*)

(Suggestion: how can $\sin^4 x$ be simply expressed in terms of a constant, $\cos 2x$, and $\cos 4x$?)

FIG. 2-W.

2-X. Excite a parallel *RC* circuit simusoidally, $i_F = I_m \cos \omega_0 t \ \ u(t)$, to be specific. Let ω_0 be variable, all other parameters being constant.

(*a*) Plot to scale the phasor representations of the forced components of the response for excitation-wave periods of 12, 6, 4, and 2 time constants (*RC*). Join their tips with a smooth curve and determine its geometric form. How far, both "before" and "after" is it sensible to extend your locus of tips?

(*b*) Let the excitation-wave period be 5*RC*. If the transient component of the response is to be zero, what initial condition is required?

(*c*) If in (*b*) the initial condition is $-RI_m$ (volts), what is the transient component of the response? Plot the forced, natural, and complete responses from $t = 0$ until the transient part is reduced to one-tenth of the forced part in magnitude.

2-Y. A series *RC* circuit is excited by a sinusoidal voltage source of frequency 400 Hz and (peak) amplitude 45 V. At this frequency, the steady-state response current leads the excitation by $\tan^{-1} 0.3$. The resistance is 5000 Ω. At $t = 0$, when the switch is closed, no energy is stored and the source voltage is passing through zero.

(*a*) Calculate and plot (with some care, so that important details are evident): (1) the forced component of the response current from $t = 0$ through two cycles of the excitation wave, (2) the total response current for the same period of time.

(*b*) What initial capacitor voltage would be necessary for your curves 1 and 2 to be identical?

(*c*) What, in terms of R, C, and ω in general, is the relation between the *power factor* of the circuit and its time constant? (Power factor is the cosine of the phase difference between excitation and steady-state response waves.)

2-Z. How much power, on the average, is delivered by the generator to the load in the circuit of Fig. 2-Z, R_L being equal to $10R_0$? If R_L is adjustable in resistance value, what value will maximize the power delivered to it? (This amount is known as the *available power* of the generator.) By how many decibels (see Appendix D) does the second power differ from the first one?

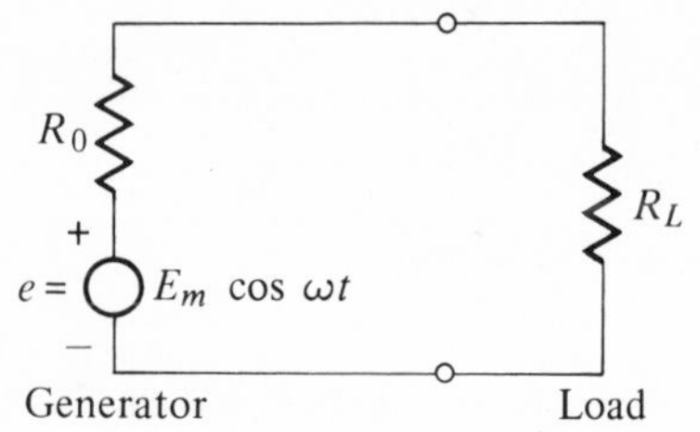

FIG. 2-Z.

One might reason, in an effort to improve the (steady-state) power delivery situation when $R_L = 10R_0$ and is fixed at that value, as follows. Since R_L is so large in comparison, why not put a capacitor C across R_L and thereby reduce it to something more like R_0? Either (*a*) show the fallacy of this and demonstrate that at no frequency ω will this help matters or (*b*) determine the appropriate value of C to make the power delivered to R_L when $\omega = \omega_0$ a maximum; how many decibels of improvement have you then achieved?

Chapter **3**

THE *RLC* CIRCUIT

I must have *two*, you know—to come and go.
One to come, and one to go.

—The White King

To add a second energy-storing element is the next step in complexity. The coming and going of energy between the two should give rise to interesting new phenomena. It does indeed.

3.01 INTRODUCTION: THE PARALLEL *RLC* CIRCUIT

Looking at only a few diagrams of electronic apparatus shows that a very common amplifier stage is constructed by adding to the circuit of Fig. 2.02-A an inductor in parallel with the capacitor and resistor. For the moment we model it by an idealized inductor L (Fig. 3.01-A). We should not forget, however, that real inductors dissipate energy, and we should not long delay in refining the model to account for that fact. But we shall find that in many applications the resulting changes are negligible.

This combination of R, L, and C elements in parallel is commonly referred to as a "tuned circuit" in the jargon of the trade. In analyzing it we shall find out why. The essential difference from the RC circuit is the addition of a second element that is capable of storing energy; it adds another derivative to the differential equation model, and so presumably makes the response more sluggish. Other novel, remarkable effects can occur too, as we soon shall see.

The mathematical form of the model is written as easily as was (2.02-1), by summing currents:

$$\begin{aligned}\sum i &= i_C + i_R + i_L - i_F\\ &= C\frac{dv}{dt} + Gv + \frac{1}{L}\int_0^t v\,dx + i_L(0) - i_F(t) \qquad (3.01\text{-}1)\\ &= 0,\\ v(0) &= V_0.\end{aligned}$$

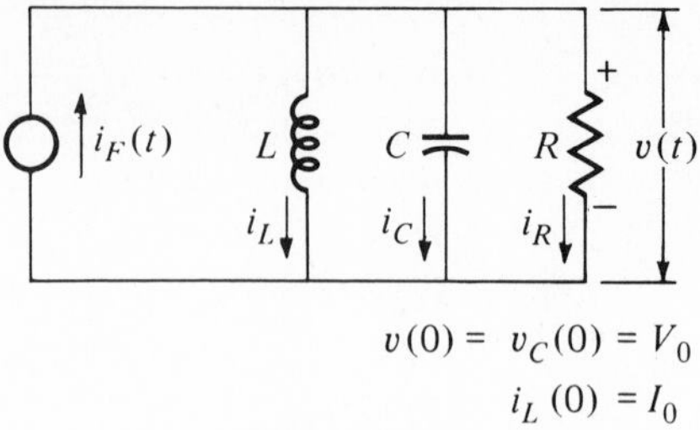

FIG. 3.01-A.

We note the addition of a second initial condition, necessary to take account of the fact that the inductor too stores energy; this one is incorporated automatically into the equation itself, since it is represented by the current in the inductor, initially $i_L(0)$. The equation is now an *integrodifferential* equation, since it contains both integral and derivative of the unknown, v. Its properties are not different from those of purely differential equations of our postulated category, however, as Sec. 2.16 suggested. The two initial conditions measure energy storage in the two elements that can store energy; they represent past history of those elements, and are therefore arbitrary. But is essential to know their values, as well as the excitation function $i_F(t)$. The additional necessary information is of course the values of the three elements: R, L, C. As a guide to study, we follow our discussion of the RC circuit as far as it is useful.

3.02 IMPULSE RESPONSE

If the excitation is impulsive, $i_F(t) = Q_0\,\delta(t)$, we can calculate the initial behavior exactly as we did in Sec. 2.02. We can integrate the equation over the brief life of the impulse:

$$C[v(0+) - v(0)] + G\int_0^{0+} v\,dt + \int_0^{0+} \frac{1}{L}\int_0^t v\,dx\,dt + \int_0^{0+} i_L(0)\,dt \qquad (3.02\text{-}1)$$
$$= Q_0\int_0^{0+} \delta(t)\,dt = Q_0$$

gives

$$v(0+) = v(0) + \frac{Q_0}{C}. \qquad (3.02\text{-}2)$$

Somewhat to our surprise, perhaps, the immediate action of the impulse is to charge the capacitor exactly as before; the inductor has no effect. On reflection we see that this is no real surprise at all. Inductors are very sluggish; current will flow in them only as a result of integrating the voltage across them, and that takes time! If the current were to change in the infinitesimal time interval $0-$ to $0+$, the voltage would have to be tremendous. And there can be no very large voltage because the capacitor current would then be unreasonable. The capacitor, in other words, controls the situation, and charges stepwise, exactly as before. In the inductor, there is no immediate change:

$$i_L(0+) = i_L(0). \tag{3.02-3}$$

Physical verification is given in Fig. 3.02-A. Here we see the response under discussion (to impulse excitation, from rest). In part (*a*) the upper trace is i_L, and the lower one is the capacitor current i_C. Note that in i_L there *is* no immediate change, and the response is slower than i_C, which starts (essentially) with an impulse that charges the capacitor stepwise. In part (*b*) the upper trace is v, which *steps* as it should; the lower trace is again i_C, with its impulsive start, repeated here for closer comparison. The corroboration is excellent.

We come then to the problem of determining the response without excitation, for $t > 0+$. There is of course energy stored in the capacitor now, due in part to the action of the impulse and in part to the energy already there at $t = 0$ from previous history. The inductor may also have energy stored in it, but this is not due to the action of the impulse. Our "initial conditions" are then

$$\begin{aligned} v(0+) &= v(0) + \frac{Q_0}{C} = V_0 + \frac{Q_0}{C}, \\ i_L(0+) &= i_L(0). \end{aligned} \tag{3.02-4}$$

And i_F, the excitation, is of course zero for $t > 0+$. The only action of the impulse excitation is to alter the initial condition; we have, as before, a natural-behavior problem to solve once the impulse has gone.

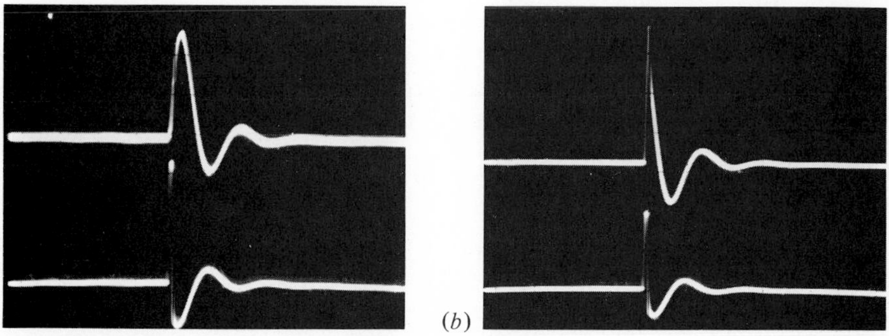

(*a*) (*b*)

FIG. 3.02-A.

In the photograph (Fig. 3.02-A) oscillations occur for $t > 0+$, which is no great surprise. But we proceed in orderly fashion, from the basic equation (3.01-1).

3.03 NATURAL BEHAVIOR

We try the same experiment as before. We write

$$v(t) = Ke^{pt} \tag{3.03-1}$$

and substitute in the integrodifferential equation (3.01-1) to obtain

$$\begin{aligned} \left(Cp + G + \frac{1}{Lp}\right)Ke^{pt} - \frac{K}{Lp} + i_L(0) = 0 \qquad 0+ < t, \\ K = v(0+) = V_0 + \frac{Q_0}{C}. \end{aligned} \tag{3.03-2}$$

It is obvious that the experiment fails. Even if we determine p so that

$$Cp + G + \frac{1}{Lp} = 0 \tag{3.03-3}$$

and the term in e^{pt} vanishes and so removes time from (3.03-2), we still have to make

$$-\frac{v(0+)}{Lp} + i_L(0) = 0. \tag{3.03-4}$$

This is impossible, except in very special cases, since both $v(0+) = V_0 + Q_0/C$ and $i_L(0)$ are entirely arbitrary and are not necessarily related in any way.

But there is a suggestion for a second experiment concealed in (3.03-3). For that gives a quadratic equation and therefore *two* values for p:

$$\begin{aligned} LCp^2 + GLp + 1 = 0, \\ p = \frac{-GL \pm \sqrt{G^2L^2 - 4LC}}{2LC} \\ = -\frac{G}{2C} \pm \sqrt{\frac{G^2}{4C^2} - \frac{1}{LC}}. \end{aligned} \tag{3.03-5}$$

(We leave for later discussion the very special case that results when the two roots are the same') Our first try, (3.03-1), provided for only *one* value of p. Evidently the circuit, having now *two* energy-storing elements, is characterized by *two* characteristic values, p_1 and p_2, rather than by one. With hindsight, as is usual, this seems eminently reasonable, if not obvious (!).

And so we try again, writing

$$v(t) = K_1 e^{p_1 t} + K_2 e^{p_2 t}. \tag{3.03-6}$$

Substitution in the integrodifferential equation now gives

$$\left(Cp_1 + G + \frac{1}{Lp_1}\right)K_1 e^{p_1 t} + \left(Cp_2 + G + \frac{1}{Lp_2}\right)K_2 e^{p_2 t} - \frac{K_1}{Lp_1} - \frac{K_2}{Lp_2} + i_L(0) = 0,$$

$$K_1 + K_2 = v(0+) = V_0 + \frac{Q_0}{C}. \tag{3.03-7}$$

Since we now have two constants K, the solution is straightforward. First, both $[Cp_1 + G + 1/(Lp_1)]$ and $[Cp_2 + G + 1/(Lp_2)]$ must vanish, since they multiply two different functions of time. The two values for p given by (3.03-5) are exactly the p_1 and p_2 we need for that:

$$\begin{aligned} p_1 &= -\frac{G}{2C} + \sqrt{\frac{G^2}{4C^2} - \frac{1}{LC}}, \\ p_2 &= -\frac{G}{2C} - \sqrt{\frac{G^2}{4C^2} - \frac{1}{LC}}. \end{aligned} \tag{3.03-8}$$

Then the K's must satisfy

$$\begin{aligned} &(a) \qquad K_1 + K_2 = v(0+), \\ &(b) \qquad \frac{K_1}{p_1} + \frac{K_2}{p_2} = Li_L(0). \end{aligned} \tag{3.03-9}$$

These are two simultaneous linear algebraic equations in two unknowns and readily solvable. They give

$$\begin{aligned} K_1 &= \frac{p_1 p_2}{p_1 - p_2}\left[\frac{v(0+)}{p_2} - Li_L(0)\right], \\ K_2 &= \frac{p_2 p_1}{p_2 - p_1}\left[\frac{v(0+)}{p_1} - Li_L(0)\right]. \end{aligned} \tag{3.03-10}$$

With these values for the p's and the K's, our experiment is a success and (3.03-6) is the (natural-behavior) response after the impulse has gone.

We note that the initial voltage to be used is $v(0+)$ with two components, as given by (3.02-2). And we see that it is not really the initial inductor current (the same at $t = 0+$ as at $t = 0$) but rather the product $[Li_L(0)]$ that we use. But since the law postulated for inductor behavior really uses flux linkages rather than current, this is no surprise. We could replace $Li_L(0)$ by, say, Λ_0 to emphasize this—but since initial data is more likely to be given in terms of current, we retain that. For brevity, however, we now write $I_0 = i_L(0) = i_L(0+)$.

An interesting and informative calculation is that of the *derivative* of the initial voltage. We find, using (3.03-6) and (3.03-10), after some algebra:

$$\begin{aligned}\left.\frac{dv}{dt}\right|_{t=0+} &= v'(0+) = K_1 p_1 + K_2 p_2 \\ &= \frac{p_1{}^2 p_2}{p_1 - p_2}\left[\frac{v(0+)}{p_2} - LI_0\right] + \frac{p_1 p_2{}^2}{p_1 - p_2}\left[-\frac{v(0+)}{p_1} + LI_0\right] \\ &= -\frac{G}{C}\, v(0+) - \frac{1}{C} I_0. \end{aligned} \qquad (3.03\text{-}11)$$

(The prime notation for the derivative is convenient, for its brevity and for indicating specific values of t.) With hindsight, we can see that we should have been able to write (3.03-11) by inspection and common sense. For dv/dt is essentially the capacitor current, which is equal to the negative of the resistor and inductor currents together. Hence

$$\begin{aligned} Cv'(0+) &= -Gv(0+) - i_L(0+) \\ &= -G\left(V_0 + \frac{Q_0}{C}\right) - I_0. \end{aligned} \qquad (3.03\text{-}12)$$

This not only provides a valuable check on the involved previous algebra; it also shows the value of physical common sense.

An additional valuable check is to verify dimensions (or units) in every term of complicated formulas such as (3.03-7). Simple checks like these will often point to errors of computation and can be extremely valuable. Good perspective requires that they be always ready and often used.

We observe an increase in complexity of algebra and calculations necessary. There is, however, no real change in nature of the response, nor of the natural behavior: the sum of two exponentials is only a natural extension of a single exponential to a more complicated (two-energy-storing-element) category.

We add now to our vocabulary the term *order* of a network. This is the number of energy-storing elements, or the degree of the *characteristic equation* (that determines the p's and hence the number of such p, or *natural frequencies*). In ordinary situations these two are the same. But there are occasional exceptions. So, for precision, we define:

> The *order* of a network is the degree of its characteristic (p-determining) equation, that is, the number of natural frequencies it has. (3.03-13)

In the present example, $N = 2$ is the order of the network; in the previous (RC) example, $N = 1$.

The order of a network is also the number of initial conditions (such as capacitor voltage and inductor current in the present example) that can be *independently* specified, and *must* be known for the solution. Usually (but not always) the number of energy-storing elements is equal to the order.

3.04 THE NATURAL FREQUENCIES

The characteristic (p-determining) equation for this *RLC* circuit is quadratic and has two roots, (3.03-5). In addition to increasing the amount of calculation required, the higher degree presents us with the interesting possibility that the p's may be not real but *complex*.

Before discussing this and other possibilities, the notation can profitably be simplified and normalized. We define

$$\alpha = \frac{G}{2C},$$

$$\omega_0 = \frac{1}{\sqrt{LC}}, \tag{3.04-1}$$

$$\beta = \sqrt{\left(\frac{G}{2C}\right)^2 - \frac{1}{LC}} = \sqrt{\alpha^2 - \omega_0{}^2}.$$

The dimensions of all of these are those of inverse time, or "frequency"; all are measured in (seconds)$^{-1}$. It is customary, for reasons that will develop soon, to use the unit "radians per second" for ω_0 and the unit "nepers per second" for α, to distinguish them, but neither "radian" nor "neper" contributes to the dimensions.

Then the p's are determined by this variant of the characteristic equation (3.03-3):

$$p^2 + 2\alpha p + \omega_0{}^2 = 0. \tag{3.04-2}$$

They are

$$\left.\begin{matrix} p_1 \\ p_2 \end{matrix}\right\} = -\alpha \pm \beta. \tag{3.04-3}$$

Normalization to the reference frequency ω_0 is convenient:

$$\left.\begin{matrix} p_{1n} \\ \\ p_{2n} \end{matrix}\right\} = \begin{matrix} \dfrac{p_1}{\omega_0} \\ \\ \dfrac{p_2}{\omega_0} \end{matrix} = -\frac{\alpha}{\omega_0} \pm \frac{\beta}{\omega_0} = -\zeta \pm \sqrt{\zeta^2 - 1} \tag{3.04-4}$$

in which

$$\zeta = \frac{\alpha}{\omega_0} = \frac{G}{2\sqrt{C/L}} \tag{3.04-5}$$

is dimensionless.

The various possibilities are three:

Case 1:

$$\begin{aligned} &\alpha > \omega_0, \\ &\zeta > 1, \qquad\qquad p_1 = -\alpha + \beta = -\alpha_1, \\ & \qquad\qquad\qquad p_2 = -\alpha - \beta = -\alpha_2. \\ &G > 2\sqrt{\frac{C}{L}}, \end{aligned} \tag{3.04-6}$$

The two values of p are real, negative, distinct.

Case 2:

$$\begin{aligned} &\alpha = \omega_0, \\ &\zeta = 1, \qquad\qquad p_1 = p_2 = -\alpha. \\ &G = 2\sqrt{\frac{C}{L}}, \end{aligned} \tag{3.04-7}$$

The two values of p are real, negative, identical.

Case 3:

$$\begin{aligned} &\alpha < \omega_0, \\ &\zeta < 1, \qquad\qquad \left.\begin{matrix} p_1 \\ p_2 \end{matrix}\right\} = -\alpha \pm j\omega_N, \\ &G < 2\sqrt{\frac{C}{L}}, \qquad\qquad = \omega_0(-\zeta \pm j\sqrt{1-\zeta^2}), \\ &\omega_N^2 \equiv \omega_0^2 - \alpha^2. \end{aligned} \tag{3.04-8}$$

The two values of p are complex, conjugates, have real parts that are negative.

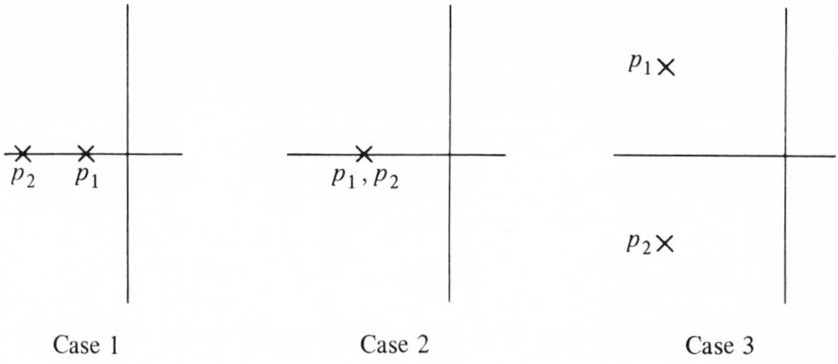

FIG. 3.04-A. The natural frequencies of the parallel *RLC* circuit in the (complex) plane of p.

Since one of the possibilities is complex, we must now use a two-dimensional plane to represent p; Fig. 3.04-A shows graphically how the three cases differ. We proceed to discuss them in turn.

3.05 THE SLUGGISH CASE (CASE 1)

The least interesting case is the first one. There $\alpha > \omega_0$ and the dissipation is relatively large, for $G > 2\sqrt{C/L}$; that is, the conductance G is larger than the reference conductance established by the capacitor and inductor: $2\sqrt{C/L}$. The impulse response, and the natural behavior, is

$$\begin{aligned} v(t) &= K_1 e^{p_1 t} + K_2 e^{p_2 t} \\ &= K_1 e^{-\alpha_1 t} + K_2 e^{-\alpha_2 t} \\ &= K_1 e^{-\alpha_1 t}\left[1 + \frac{K_2}{K_1} e^{-(\alpha_2 - \alpha_1)t}\right] \\ &= K_1 e^{-\alpha_1 t}(1 - K_0 e^{-2\beta t}) \end{aligned} \tag{3.05-1}$$

in which

$$K_0 = -\frac{K_2}{K_1} = \frac{Li_L(0) + v(0+)/\alpha_1}{Li_L(0) + v(0+)/\alpha_2}. \tag{3.05-2}$$

The nature of the response depends somewhat on the initial conditions. But it is still decaying-exponential in character. It can resemble the *RC* network's natural behavior (a single exponential), like curve (*a*) in Fig. 3.05-A, or its negative. This occurs if the initial conditions are such that $K_0 < 1$. If $K_0 = 1$ the behavior of curve (*b*) (or its negative) results. If $K_0 > 1$, then curves like (*c*) describe the natural behavior.

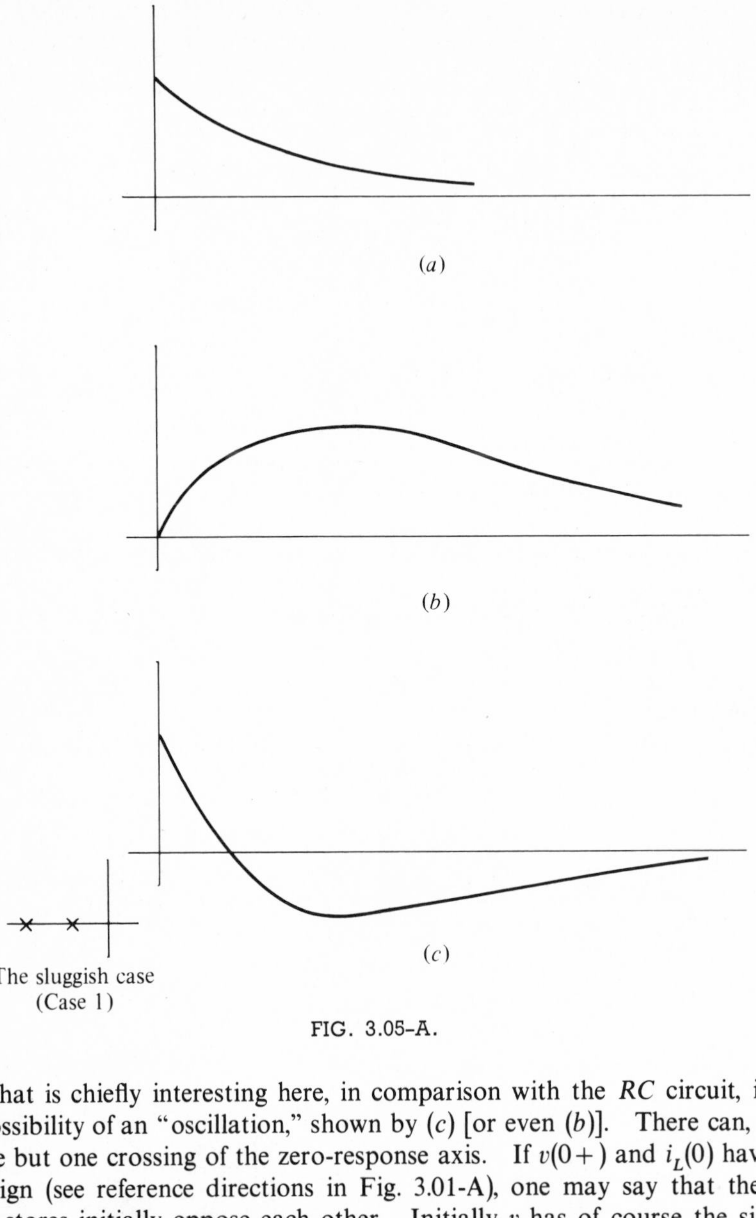

FIG. 3.05-A.

What is chiefly interesting here, in comparison with the *RC* circuit, is the new possibility of an "oscillation," shown by (*c*) [or even (*b*)]. There can, however, be but one crossing of the zero-response axis. If $v(0+)$ and $i_L(0)$ have the same sign (see reference directions in Fig. 3.01-A), one may say that the two energy stores initially oppose each other. Initially v has of course the sign of $v_C(0)$; but the inductor current acts initially to discharge the capacitor and eventually the energy stored in L will overcome and reverse the sign of v and charge C in the opposite sense. As time goes on, the sign of v will reverse: the behavior (*c*) in Fig. 3.05-A occurs. [If $v_C(0)$ and $i_L(0)$ have opposite signs, this may or may not occur.] A *second* crossing of the zero-voltage axis cannot occur because the dissipation of energy in the resistive element is too great (G

is too large, R is too small). But if the value of G is reduced enough (in comparison with $\sqrt{C/L}$) to make the natural frequencies complex, then the natural behavior is quite different.

We note that the behavior in case 1 is essentially the same as that of the RC circuit when excited by a decaying exponential (Sec. 2.14). And that the case of equal natural frequencies (case 2) is like that of the RC circuit excited at its own natural frequency (Sec. 2.14). This can be established simply by comparison of formulas, or by a limiting process applied to (3.05-1). There is nothing extraordinary about case 2, and no need to discuss it. In fact neither case 1 nor case 2 is of much practical importance.

3.06 THE RESONANT (HIGH-Q) CASE

It is otherwise with the third case. Here is one of the most interesting and useful of electronic circuits, a veritable workhorse of electrical engineering. To study it with care is our next task, and an important one.

The meaning of a *complex* natural frequency is both simple and fascinating. We note first that K_1 and K_2, (3.03-10), are now also complex. And we note that they are conjugates, as evidenced by the consistent permutation of subscripts in the p's between K_2 and K_1. From (3.03-10) and (3.04-8) we obtain in detail:

$$\begin{aligned} K_1 &= \frac{{\omega_0}^2}{j2\omega_N}\left[\frac{v(0+)}{-\alpha - j\omega_N} - Li_L(0)\right] \\ &= \frac{{\omega_0}^2}{2\omega_N}\left\{\frac{\omega_N v(0+)}{{\omega_N}^2 + \alpha^2} + j\left[Li_L(0) + \frac{\alpha v(0+)}{{\omega_N}^2+\alpha^2}\right]\right\}. \end{aligned} \tag{3.06-1}$$

For brevity we write (see Fig. 3.06-A)

$$\begin{aligned} K_1 &= K_{1r} + jK_{1i} = |K_1| \underline{/\phi_1}, \\ K_2 &= K_1^* = K_{1r} - jK_{1i} = |K_1| \underline{/-\phi_1}. \end{aligned} \tag{3.06-2}$$

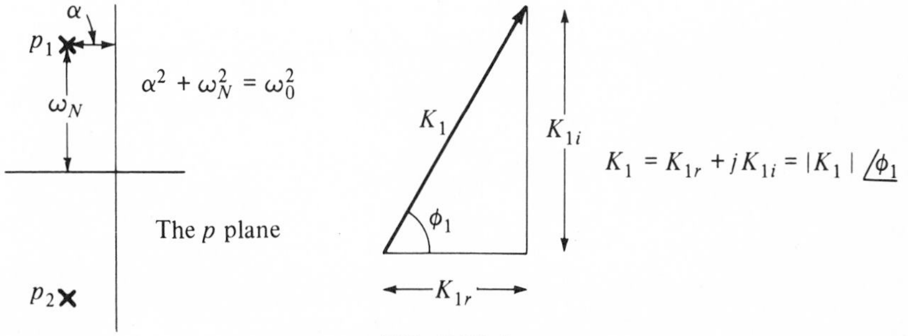

FIG. 3.06-A.

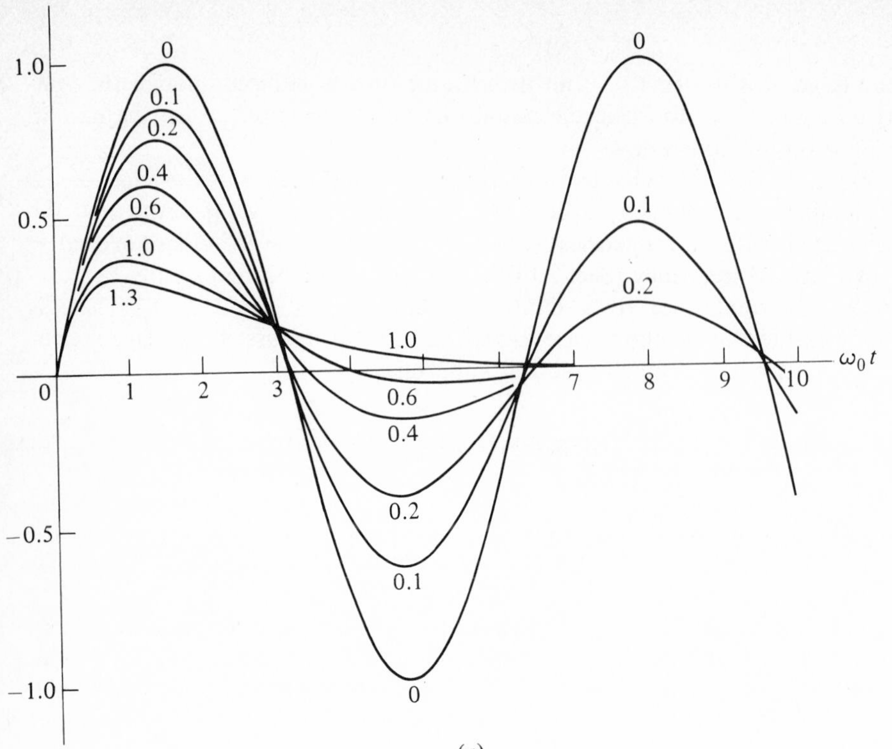

1.0
0.5
−0.5
−1.0
0
0.1
0.2
0.4
0.6
1.0
1.3
0
1
2
3
7
8
9
10
$\omega_0 t$

(a)

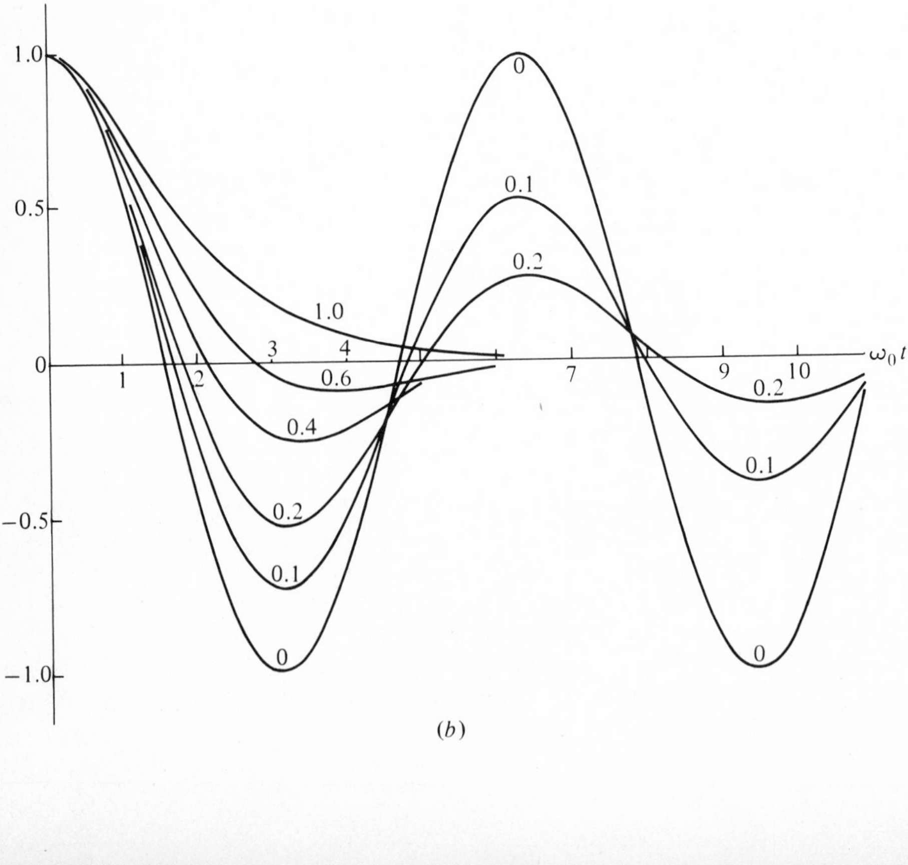

1.0
0.5
0
−0.5
−1.0
0
0.1
0.2
0.4
0.6
1.0
1
2
3
4
7
9
10
$\omega_0 t$

(b)

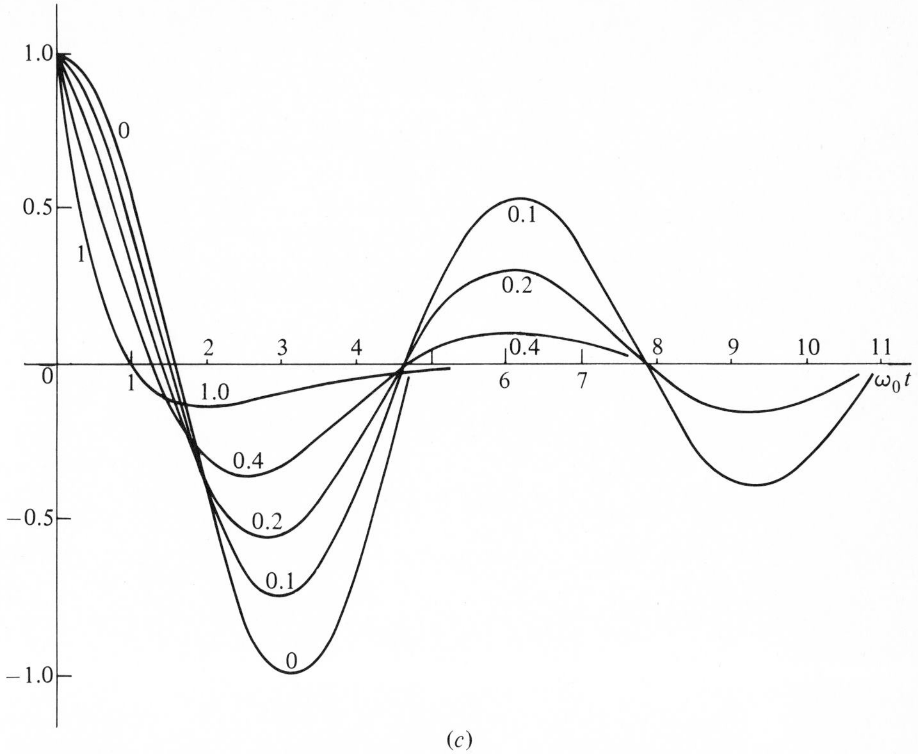

FIG. 3.06-B. The value of the parameter $\zeta = \alpha/\omega_0$ is given on each curve.

The response (3.03-6) can now be written

$$\begin{aligned} v(t) &= K_1 e^{p_1 t} + \text{conjugate} \\ &= 2\ \mathbf{RE}\ (K_1 e^{p_1 t}) \qquad\qquad (3.06\text{-}3) \\ &= 2|K_1|e^{-\alpha t}\cos(\omega_N t + \phi_1), \qquad 0+ < t. \end{aligned}$$

From this point on we shall omit the qualifier $(t > 0+)$; we should not forget, however, that our expressions are valid only *after* the charging.

The response is clearly a damped but truly oscillatory wave. If the situation is such that ω_N is small (α is only slightly smaller than ω_0), there is no outstanding change in the natural behavior from that of curve (*c*) [or curve (*b*)] in Fig. 3.05-A. But for less dissipation, there *is* a marked difference. But the infinite number of zero crossings that the oscillation gives is hidden by the amplitude decay unless ω_N is appreciably greater than α.

The curves of Fig. 3.06-B show this. In them L and C are constant; only G is varied. Alternatively, ω_0 is constant and only α (and hence $\omega_N = \sqrt{\omega_0^2 - \alpha^2}$)

is varied. Those of set (*a*) are all drawn for the initial condition of zero voltage and the same (negative) value of inductor current. As time increases the voltage first builds up, therefore, and then oscillates as shown. The initial slope, $v'(0+)$, is the same for all. For ζ (that is, α/ω_0) values only slightly less than unity, no oscillation is visible. For $\zeta = 0.6$ we begin to see oscillations, for $\zeta = 0.2$ many oscillations are clearly visible, and for $\zeta = 0$ (the limiting case of no energy dissipation, $G = 0$) the wave is that of simple harmonic motion, purely sinusoidal.

The curves of set (*b*) are all drawn for the initial condition of a certain voltage and zero initial slope thereof. The initial inductor current is adjusted in each case so that $v'(0+) = 0$.

Those of set (*c*) are all drawn for the initial condition of a certain voltage and zero initial inductor current. The initial slope, $v'(0+)$, changes accordingly.

They illustrate the point nicely: only for ω_N/α reasonably greater than unity do the oscillations become visible. The transition from *real* natural frequencies (case 1 in Fig. 3.04-A) to *complex* natural frequencies (case 3 in Fig. 3.04-A) is smooth and in no way remarkable. It is only when the dissipation becomes rather small, when α is much smaller than ω_0, that we begin to see a really oscillatory response.

The common character of α and of the ω's (ω_N and ω_0) is once more evidenced; dimensionally they are both $(\text{time})^{-1}$ or "frequency." Since α obviously relates to decay, it is again usually measured in nepers per second (neper, from the Latin form of Napier, whose connection with logarithmic or exponential decay requires no comment). Since the ω's relate to frequency in the usual sense, they are usually measured in radians per second (see Sec. 3.04).

The interesting cases are those for which ω_N is of the order of magnitude of ω_0; that is, when $\alpha \ll \omega_0$. Then the natural frequencies lie close to the imaginary axis, the oscillation decays slowly, and the phenomenon known as *resonance* appears. We here take resonance to mean that the principal action is a periodic exchange of energy between inductor and capacitor, with very little damping. We shall give a more usual, but equivalent, definition later (Sec. 3.14).

An informative way of expressing the amount of dissipation, one that in effect shows "how many oscillations are visible," is to use the pseudoperiod

$$T = \frac{2\pi}{\omega_N}. \tag{3.06-4}$$

This is twice the time interval between successive zero values of the response, as (3.06-3) shows. We rewrite (3.06-3) as

$$v(t) = 2|K_1|e^{-(\alpha/\omega_N)2\pi(t/T)} \cos\left(2\pi\frac{t}{T} + \phi_1\right). \tag{3.06-5}$$

Our previous measure of relative damping, $\zeta = \alpha/\omega_0$, does not appear in the exponent. But for lightly damped cases we have

$$\begin{aligned} \omega_N &= \sqrt{\omega_0^2 - \alpha^2} = \omega_0\sqrt{1-\zeta^2} \\ &= \omega_0\left(1 - \frac{\zeta^2}{2} - \frac{1}{8}\zeta^4 + \cdots\right). \end{aligned} \tag{3.06-6}$$

We do not ordinarily measure the decay of the amplitude with a great precision. Hence we can well neglect the difference between ω_N and ω_0 in what we shall call the resonant situation. For $\zeta = 0.05$, to write $\omega_N = \omega_0$ is to be in error by only 0.1 percent; many interesting practical cases involve values of ζ this small or even smaller. So we can rewrite (3.06-3) as

$$v(t) \cong 2|K_1|e^{-\zeta 2\pi(t/T)} \cos\left(2\pi\frac{t}{T} + \phi_1\right) \tag{3.06-7}$$

or

$$v(t) \cong 2|K_1|e^{-\zeta\omega_0 t} \cos(\omega_0 t + \phi_1) \tag{3.06-8}$$

with very little error. The damping factor "per cycle" is then $e^{-2\pi\zeta}$. For $\zeta = 0.05$ this is 0.73; for $\zeta = 0.01$ it is 0.94, a drop of only 6 percent.

Another common measure of the "quality" of the resonance or oscillation is essentially the reciprocal of ζ, a number which becomes very *large* when the damping is *small*. It is

$$Q = \frac{1}{2\zeta} = \frac{\omega_0}{2\alpha} = \frac{\sqrt{C/L}}{G}. \tag{3.06-9}$$

For $\zeta = 0.05$, $Q = 10$; for $\zeta = 0.01$, $Q = 50$. In terms of Q, the response oscillation is

$$\begin{aligned} v(t) &= 2|K_1|e^{-\omega_0 t/(2Q)} \cos(\omega_N t + \phi_1) \\ &\cong 2|K_1|e^{-\omega_0 t/(2Q)} \cos(\omega_0 t + \phi_1). \end{aligned} \tag{3.06-10}$$

In the second form the approximation consists in replacing ω_N by ω_0. We note in that connection that (3.06-6) is, in terms of Q,

$$\omega_N = \omega_0\sqrt{1 - \frac{1}{4Q^2}} = \omega_0\left(1 - \frac{1}{8Q^2} + \cdots\right). \tag{3.06-11}$$

The resonant (lightly damped) case is commonly referred to as the *high-Q* case. A Q value of 10 or 20 is, let us say, high Q. The term is, of course, relative; the value of Q is seldom calculated or measured to more than one significant figure. Even for $Q = 10$ we note that $\omega_N = \omega_0(1 - 0.00125 + \cdots)$, so that in the second form in (3.06-10) the error is merely 0.1 percent, and that in phase only.

We note the following additional formulas for Q, which has still other very interesting interpretations, yet to come:

$$Q = \frac{\omega_0}{2\alpha} = \frac{\sqrt{C/L}}{G} = \frac{R}{\sqrt{L/C}} = RC\omega_0 = \frac{R}{L\omega_0} = \frac{C\omega_0}{G}. \qquad (3.06\text{-}12)$$

The meaning of a *complex* natural frequency is now very clear. We have here "two-dimensional" natural behavior: *oscillation* (at frequency ω_N) and *damping* (at logarithmic rate α). Since a complex number is simply a two-dimensional number, it is "naturally" suited to describe such natural behavior. Complex natural frequency is a simple, natural extension of ordinary frequency.

3.07 NATURAL BEHAVIOR IN THE RESONANT CASE

Typical of the response to an impulse and of the natural behavior in the high-Q case is the curve of Fig. 3.07-A. It is calculated for the common initial conditions $v(0+) = V_0$, $i_L(0) = 0$, as would be given by impulse excitation from rest.

Figure 3.07-B shows us that our model is a good one. It displays (to two different time scales) the actual physical response to an impulse, from rest, of a *tuned circuit*, that is, a parallel *RLC* circuit of reasonably high Q. The value of Q is about 15, not tremendously high, but high enough to show many oscilla-

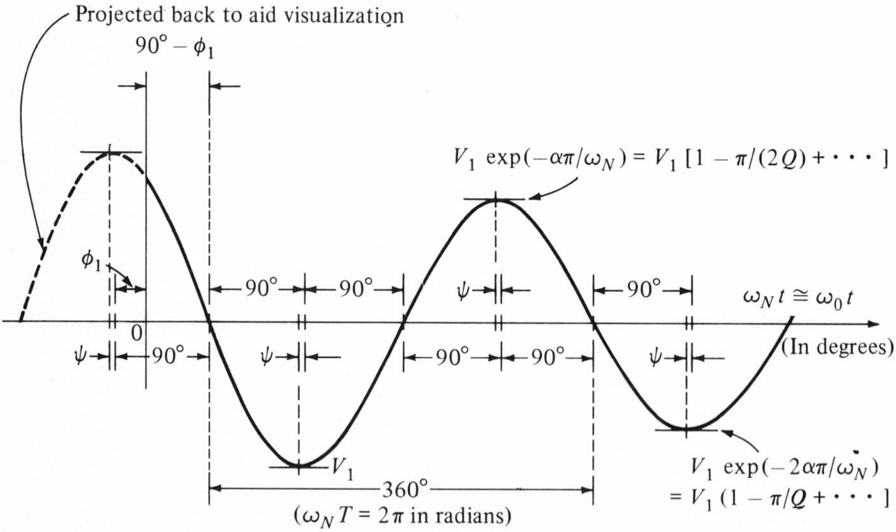

FIG. 3.07-A. The natural behavior of the high-Q circuit, $V_m e^{-\alpha t} \cos(\omega_N t + \phi_1)$, per (3.06-10) (drawn for $Q = 10$, $\phi_1 = 30°$).

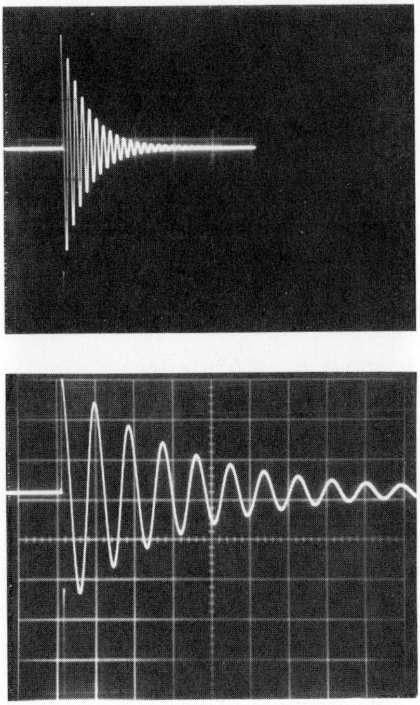

(*a*) Time scale: 100 μs/division

(*b*) Time scale: 20 μs/division

FIG. 3.07-B.

tions; the circuit is damped but not enough to alter the inherent oscillatory character of the natural behavior. In both photographs the exciting impulse can be seen. The first shows many oscillations and the final decay to zero; the second (magnified horizontally five times) shows more detail, enough to enable us to calculate the values of Q and of ω_N (see below).

The initial slope is negative (if V_0 is positive), for (see Fig. 3.01-A)

$$i_C = C\frac{dv}{dt} = -i_R - i_L = -GV_0. \tag{3.07-1}$$

at $t = 0+$. We note in passing, from physical common sense (see Fig. 3.01-A again) that for $t \geqq 0+$

$$i'_C = C\frac{d^2v}{dt^2} = Cv'' = -i'_R - i'_L = -Gv' - \frac{v}{L} \tag{3.07-2}$$

since $L\,di/dt = v$. Hence

$$v''(0+) = \frac{G^2}{C^2}V_0 - \frac{V_0}{LC}. \tag{3.07-3}$$

We could continue in this vein and calculate the $t = 0+$ values of higher derivatives also:

$$v'''(0+) = -\frac{1}{C}(i_R'' + i_L'')_0 = -\frac{1}{C}\left(Gv'' + \frac{v'}{L}\right)_0 \tag{3.07-4}$$
$$= -\frac{G^3}{C^3}V_0 + \frac{2G}{LC^2}V_0 .$$

Then we could write the power series form of the response,

$$v(t) = V_0\left[1 - \frac{G}{C}t + \left(\frac{G^2}{C^2} - \frac{1}{LC}\right)\frac{t^2}{2} - \left(\frac{G^3}{C^3} - \frac{2G}{LC^2}\right)\frac{t^3}{3!} + \cdots\right]. \tag{3.07-5}$$

We shall discuss this series form from time to time, for it is sometimes useful.

The curve of Fig. 3.07-A will serve for *any* initial condition. Different initial conditions mean simply that the $t = 0$ point is at some other point of the wave shown there: we need merely shift the origin of time horizontally to the appropriate point, erase the curve to the left of that point, and modify the vertical scale as necessary. On it are indicated various interesting features: where the extrema (maxima and minima) occur, the time intervals between them and the zeros, the damping rate, etc. We note that the extrema do not occur exactly at the points where the phase is a multiple of 180° (where the cosine function has its extrema), but slightly in advance thereof. The advance is

$$t_{\text{adv}} = \frac{\psi}{\omega_N} = \frac{1}{(2\pi f_0)(2Q)} = 0.08\,\frac{T_0}{Q} \quad \text{(s)}, \tag{3.07-6}$$

in which we have defined the *undamped natural frequency of oscillation* f_0 and *undamped natural period of oscillation* T_0:

$$f_0 = \frac{1}{2\pi\sqrt{LC}} = \frac{1}{T_0}. \tag{3.07-7}$$

Here "frequency" is used in the strict sense of "rate of occurrence." The symbols f_0 and T_0 refer to the natural behavior of the circuit with infinitely high Q, in the theoretical case of complete absence of dissipation (the simple harmonic oscillator with its famous simple harmonic motion). The *damped* natural frequency and period in this sense are given by [See (3.06-11)]

$$f_N = \frac{\omega_N}{2\pi} = f_0\left(1 - \frac{1}{8Q^2} + \cdots\right) = \frac{1}{T_N}. \tag{3.07-8}$$

We note that for high Q the advance, t_{adv}, is negligible, as is the difference between f_0 and f_N.

The outstanding feature is of course the *oscillatory* behavior, essentially at the frequency of undamped natural behavior, f_0 or ω_0. But damping does occur, at the rate of a π/Q fractional decrease "per cycle" (for high Q), since $e^{-\pi/Q} = 1 - \pi/Q + \cdots$ (Fig. 3.07-A). In one "period" the amplitude decreases by some $300/Q$ percent for practical purposes. The term *logarithmic decrement* is sometimes applied to the corresponding decrease in the exponent of $e^{-\alpha t}$, $2\pi\alpha/\omega_N \cong \pi/Q$.

We have described the high-Q case as one in which the energy surges back and forth between the two energy-storing elements, with very little loss in one cycle. For this reason the very descriptive term *tank circuit* is often used as its name in the jargon of the trade. For $Q = 15$, the amplitude of the voltage drops about 20 percent in one cycle, so that the energy stored in the capacitor drops some 40 percent from one of its peak values to the next. These numbers may seem somewhat large to be used in the approximate formulas, but the exact figures are not far away: $1 - e^{-\pi/15} = 0.19$, as opposed to 20 percent from $300/Q$, and $100(1 - e^{-2\pi/15}) = 0.34$, as opposed to 40 percent. For $Q = 30$ the agreement is excellent: the figures are 10 percent for the amplitude, from both calculations, and 20 percent (19 percent exactly) for the energy drop.

In Fig. 3.07-B*b* we observe that the decay in the first cycle is from 2.8 to 2.2 divisions. We can then calculate the circuit's Q from $\pi/Q = 0.6/2.8$ as $Q = (\pi)(28/6) = 15$, or from $300/Q = 60/2.8$ as $Q = 14$. Either value will do, as Q is not that precise a number. The circuit's natural or resonant frequency f_0 can also be calculated from the photograph. The period is slightly less than one division: 10 oscillations occur in 9 divisions so that $T_N = (9)(20/10) = 18\ \mu\text{s} = T_0$. Then $f_0 = 1/T_0 = 56$ kHz. Neglecting R and using the known value of C, 0.003 μf, then gives us the value of the inductance: $L = 1/(C\omega_0^2) = 2.8$ mH.

In general we have

$$\frac{\text{Energy stored in capacitor at one maximum thereof}}{\text{Energy stored in capacitor at previous maximum}}$$

$$= \frac{\frac{1}{2}C(V_1{}^2 e^{-\pi/Q})^2}{\frac{1}{2}CV_1{}^2} = e^{-2\pi/Q} = 1 - 2\pi/Q + \cdots \qquad (3.07\text{-}9)$$

and so the loss in energy stored in the capacitor between two successive maxima is

$$\frac{2\pi}{Q} \times \text{energy stored at the first of the maxima.} \qquad (3.07\text{-}10)$$

Hence (referring to the capacitor)

$$Q = \frac{2\pi\ (\text{energy stored})}{\text{energy lost per cycle}} \qquad (3.07\text{-}11)$$

which offers an alternative definition of Q, if the value thereof is high. At a maximum of the v wave, since $dv/dt = 0$, $i_C = 0$; and so $i_L = 0$ if we neglect i_G (which is reasonable). At such an instant, therefore, *all* the stored energy is in the capacitor, and (3.07-11) may be interpreted as referring to *all* the stored energy, that in both capacitor and inductor. In fact the sum of the two stored energies is substantially constant except for the slow decay factor $e^{-2\alpha t}$; we can therefore write (Exercise 3-44):

$$\begin{aligned} W &= \text{total stored energy} = W_{\text{capac.}} + W_{\text{induc.}} \\ &= W_0 e^{-2\alpha t} = W_0(1 - 2\alpha t + \cdots) \\ &= W_0\left(1 - \frac{\omega_0 t}{Q} + \cdots\right) \end{aligned} \qquad (3.07\text{-}12)$$

which again gives the energy formula for Q of (3.07-11), but now in terms of *all* the energy stored at any instant. We shall meet this form again [see (3.16-10)].

We add to our repertoire of formulas for the high-Q case the simple forms taken by the K's of (3.03-10):

$$\begin{aligned} K_1 &= \frac{\omega_0^2}{2j\omega_N}\left(\frac{V_0}{-\alpha - j\omega_N} - LI_0\right) \\ &= \frac{\omega_0^2}{2\omega_N}\left(\frac{V_0}{\omega_N - j\alpha} + jLI_0\right) \\ &= \frac{\omega_0}{2}\left[\frac{V_0}{\omega_N}\left(1 + \frac{j\alpha}{\omega_N} + \cdots\right) + jLI_0\right] \\ &= \frac{V_0}{2}\left(1 + \frac{j}{2Q} + \cdots\right) + j\frac{\omega_0 LI_0}{2} \\ &= \frac{V_0}{2} + j\left(\frac{L\omega_0 I_0}{2} + \frac{V_0}{4Q}\right) \\ &= K_2^*. \end{aligned} \qquad (3.07\text{-}13)$$

Thus (3.06-10) gives

$$\begin{aligned} v(t) &= 2|K_1|e^{-\alpha t}\cos(\omega_N t + \phi) \\ &= \sqrt{V_0^2 + (L\omega_0 I_0)^2}\, e^{-\omega_0 t/(2Q)} \cos\left(\omega_0 t + \tan^{-1}\frac{L\omega_0 I_0}{V_0}\right) \end{aligned} \qquad (3.07\text{-}14)$$

to an approximation usually sufficient (for high Q).

As the value of Q is raised, increasing the oscillatory quality of the circuit, the response becomes more nearly a sinusoid. This is physically demonstrated

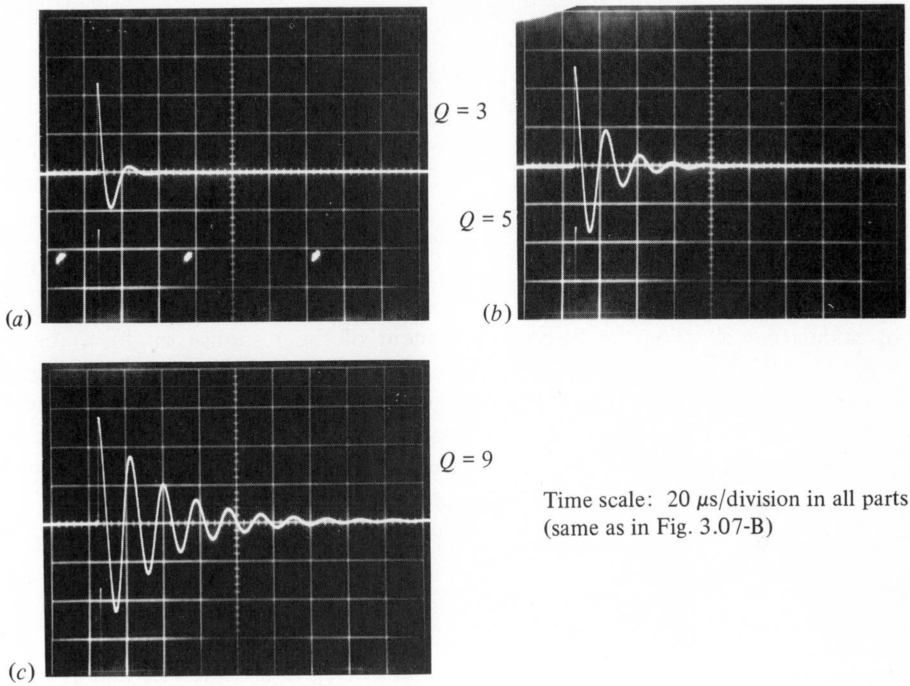

Time scale: 20 μs/division in all parts (same as in Fig. 3.07-B)

FIG. 3.07-C.

in the photographs of Fig. 3.07-C, for $Q = 3$, 5, and 9 in turn; one should mentally add Fig. 3.07-B ($Q = 15$) and a pure sinusoid to complete the set.

Such a response is useful, for example, as a generator of time markers. The zero crossings are equally spaced, at intervals $\pi\sqrt{LC} = 1/(2f_0) = T_0/2$. The wave can be amplified, limited (clipped), differentiated, or otherwise manipulated to produce markers, though some nonlinear apparatus will be necessary.

But the high-Q resonant circuit has many other uses, which justify our detailed study of it.

3.08 STEP RESPONSE OF THE HIGH-Q CIRCUIT

When excited by a step of current, when $i_F(t) = I_F u(t)$ in the situation of Fig. 3.08-A, the response is very little different. We discover this by performing a mathematical experiment (suggested by our success in Sec. 2.14): we substitute the "obvious" form

$$v(t) = v_N + v_F = K_1 e^{p_1 t} + K_2 e^{p_2 t} + V_F \tag{3.08-1}$$

into the differential equation (3.01-1), and we find (Exercise 3-40) that p_1 and p_2 must have the same values as before (the natural frequencies), and that $V_F = 0$.

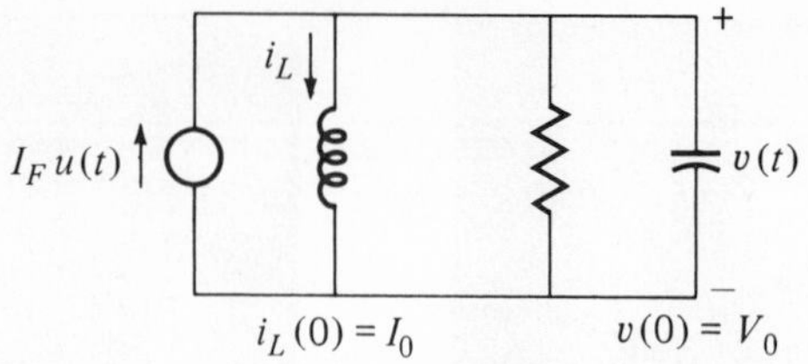

FIG. 3.08-A.

Physical common sense might well have told us this and saved us the labor of calculating it. For the forced component of the response ought to be a constant, like the excitation. But a constant voltage v is steady (not transient) in nature, and if it existed, it would eventually require a linear, forever-increasing current in the inductor, because $L\, di/dt = v$. There is, then, *no* forced (or steady-state) component of response, as the previous experiment has just told us. That experiment, being different, is good to have, of course.

The excitation constant I_F must appear *somewhere* in the response

$$v(t) = K_1 e^{p_1 t} + K_2 e^{p_2 t}. \tag{3.08-2}$$

The experiment with the differential equation also gave us (Exercise 3-40)

$$\frac{K_1}{Lp_1} + \frac{K_2}{Lp_2} = -I_F + I_0 \tag{3.08-3}$$

instead of (3.03-9*b*). In addition, of course,

$$K_1 + K_2 = V_0 \tag{3.08-4}$$

exactly like (3.03-9*a*). The physical common sense approach tells us that at $t = 0+$ (Fig. 3.08-A)

$$i_C = I_F - i_R - i_L = I_F - GV_0 - I_0 = C\frac{dv}{dt} \tag{3.08-5}$$

and so

$$\frac{dv}{dt} = K_1 p_1 + K_2 p_2 = \frac{1}{C}(-GV_0 - I_0 + I_F). \tag{3.08-6}$$

That (3.08-6) and (3.08-3) are completely equivalent is easily established (Exercise 3-40). We may then use any two of the equations (3.08-3), (3.08-4), or (3.08-6) to determine the K's in this case.

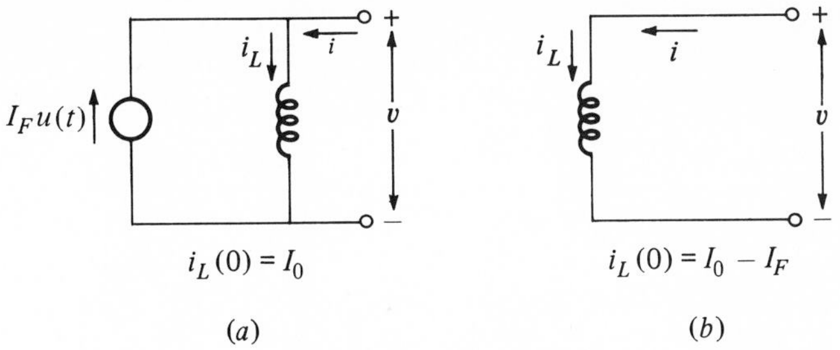

FIG. 3.08-B.

Comparison of (3.08-3) with (3.03-9) indicates that the response to the step excitation is exactly the same as the natural behavior (or impulse response) for initial conditions

$$\begin{aligned} v(0+) &= V_0, \\ i_L(0+) &= I_0 - I_F. \end{aligned} \tag{3.08-7}$$

It appears that initially charging the inductor with current $-I_F$ produces the same response as exciting the network with the current step $I_F u(t)$. The analysis above says in effect that the two little circuits in Fig. 3.08-B are *externally* equivalent. That this is so is easily established independently by writing the governing equations for the two, for $t \geqq 0+$:

$$\begin{aligned} &(a) \qquad i = -I_F + \frac{1}{L}\int_{0+}^{t} v\,dx + I_0, \\ &(b) \qquad i = \frac{1}{L}\int_{0+}^{t} v\,dx + (I_0 - I_F). \end{aligned} \tag{3.08-8}$$

That there is no forced (constant) component in $v(t)$ is again shown by this possibility of exchanging one for the other. The first two elements in Fig. 3.08-A can thus be replaced immediately, according to Fig. 3.08-B, to give Fig. 3.08-C, a natural-behavior situation with the initial conditions of (3.08-7).

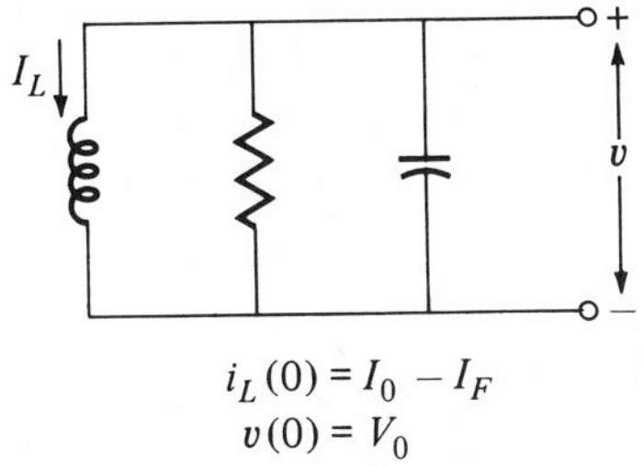

FIG. 3.08-C.

This is the second time that we have discovered an ability to replace a source by initial energy charge in an L or C element. (The first, observed in Sec. 2.02, was that an impulse current could be replaced by a charge on a capacitor, provided no other capacitive path was connected thereto.) The exchanges can, of course, equally well be made in the opposite sense, to replace initially charged elements by initially uncharged elements together with supplementary sources. We shall return to this, and to other possible exchanges, in Sec. 7.22.

We have found that the step response of our parallel RLC circuit is identical to its impulse response, provided the initial inductor current I_0 is modified to $(I_0 - I_F)$. There is no need then to discuss the behavior further. Figure 3.07-A describes it.

We do note that the *step response*, defined as the response from rest to unit step excitation, is

$$v_u(t) = K_1 e^{p_1 t} + K_2 e^{p_2 t},$$

$$K_1 + K_2 = 0, \qquad (3.08\text{-}9)$$

$$K_1 p_1 + K_2 p_2 = \frac{I_F}{C} = \frac{1}{C}.$$

The step response starts from a positive-going, zero-crossing point on the wave of Fig. 3.07-A [that is, $\phi_1 = -90°$ and $v(t) = V_m e^{-\alpha t} \sin \omega_N t$]. Figure 3.08-D shows step excitation and the actual step response of the tuned circuit used for

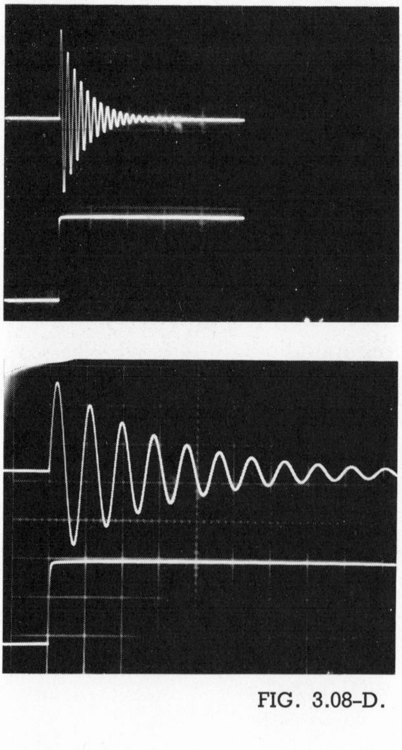

(*a*) Time scale: 100 μs/division

(*b*) Time scale: 20 μs/division

FIG. 3.08-D.

Fig. 3.07-B ($Q = 15, f_0 = 45$ kHz), in compressed and expanded views. Careful examination of the initial rise shows clearly that it is *not* fast and sudden as in the impulse response: the trace width (its comparative thickness) as well as its evident inclination make this clear. As (3.08-5) and (3.08-9) state,

$$v(0+) = 0,$$
$$v'(0+) = \frac{1}{C} \qquad \text{(a positive number)}, \tag{3.08-10}$$

which the photograph corroborates. Note that formal solution of the set of equations in (3.08-9) gives

$$v(t) = \frac{1}{C\omega_N} e^{-\alpha t} \sin \omega_N t \tag{3.08-11}$$

but that physical common sense enables one to write (3.08-11) almost immediately, once one knows that $v(t)$ must be a damped sinusoid (the Q is high) and recognizes (3.08-10) from the physics of the situation.

Degrading the Q (increasing the dissipation by decreasing the value of R in Fig. 3.08-A) makes the step response decay faster. Figure 3.08-E gives four such responses, for successively lower Q's.

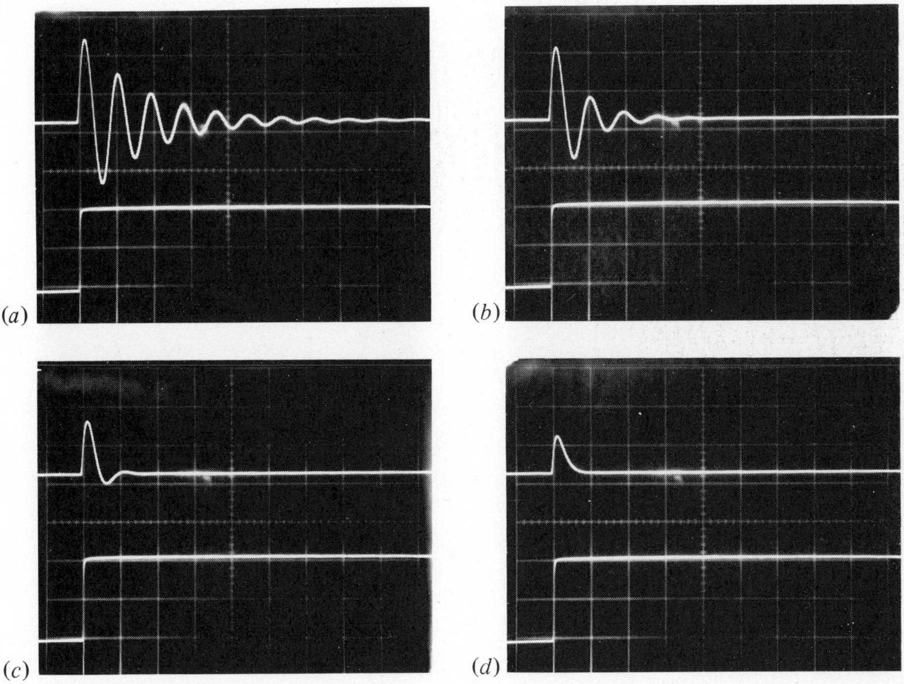

(*a*) (*b*) (*c*) (*d*)

Time scale: 20 μs/division in all four photographs

FIG. 3.08-E.

3.09 PULSE RESPONSE OF THE HIGH-Q CIRCUIT

Changing the excitation to a *pulse* (Fig. 1.10-A*a*) of current introduces no additional work of analysis. But it does show how versatile the tuned circuit is. To find the response to the current pulse

$$i_F(t) = I_F u(t) - I_F u(t - T_D) \tag{3.09-1}$$

we need merely practice superposition and use (3.08-11) twice:

$$v(t) = I_F v_u(t) - I_F v_u(t - T_D). \tag{3.09-2}$$

(This supposes an initial condition of rest, to be sure; the effect of a nonrest initial condition is simply to add another natural-behavior component to the response.)

The two components of the response are shown in Fig. 3.09-A. The first (*a*) is like that of Fig. 3.06-B*a* but drawn for $\alpha = 0.02$ ($Q = 25$), being calculated from (3.08-11). To obtain the second (*b*) we can imagine a template made from the first, turned over, and moved T_D (seconds) to the right. The versatility of the pulse response lies in the multifarious forms the sum of these two components can take as the "delay" T_D is varied. Curve (*c*) shows the response when the pulse duration T_D is equal to one period (T_N) of the tuned circuit's natural behavior: it is essentially one cycle of a sine wave.

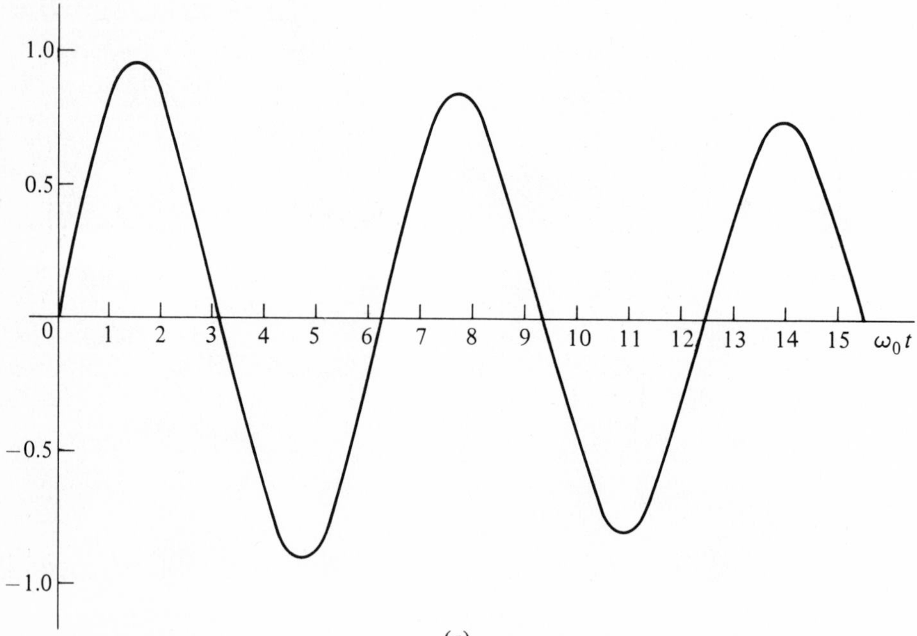

(*a*)

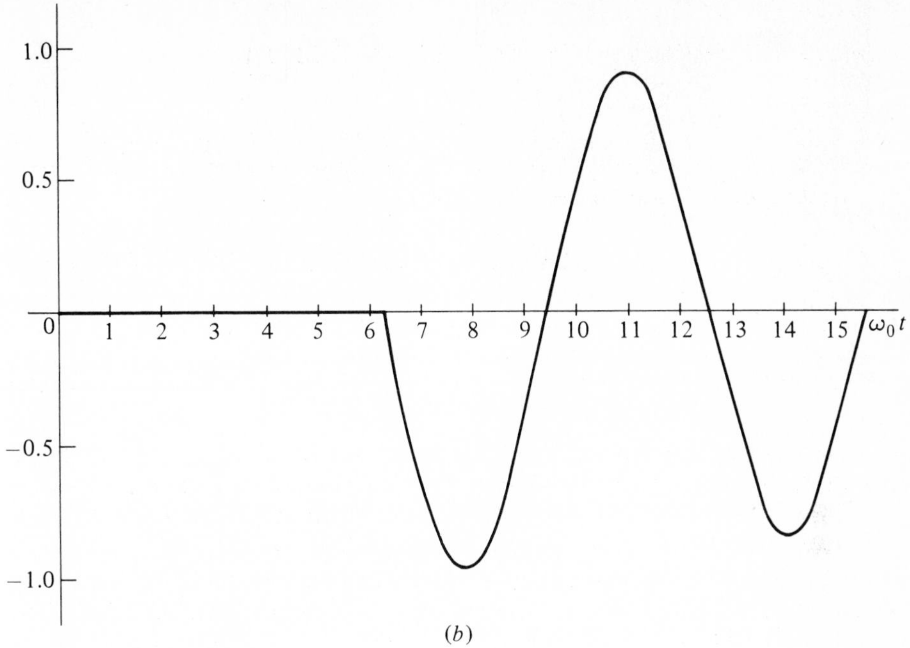

(*b*)

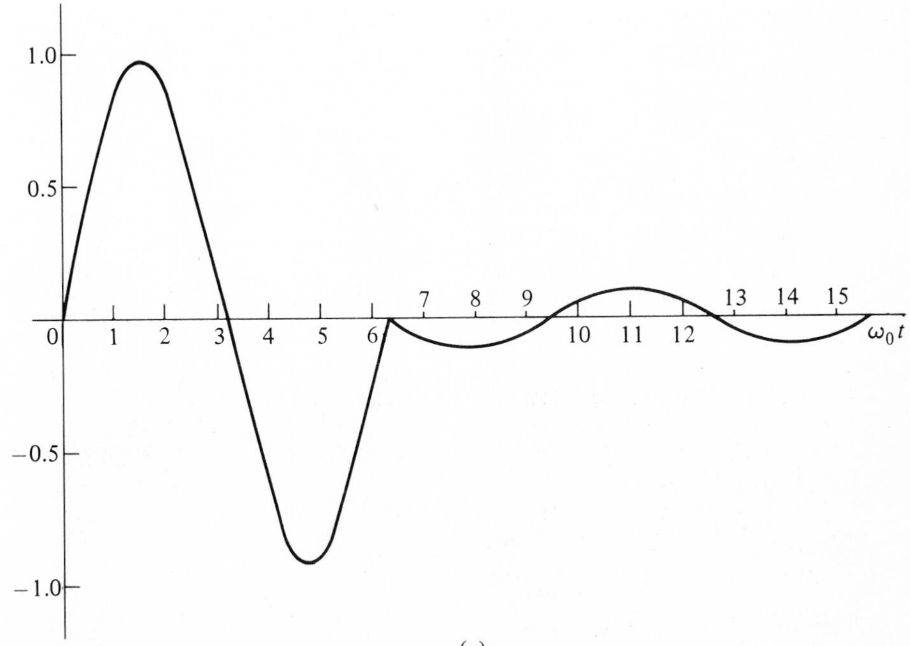

(*c*)

FIG. 3.09-A. (a) $r_u(t)$; $\alpha = 0.02$, $Q = 25$. (b) $-r_u(t - T_D)$; $T_D = 2\pi/\omega_N$. (c) $r_p(t) = r_u(t) - r_u(t - T_D)$.

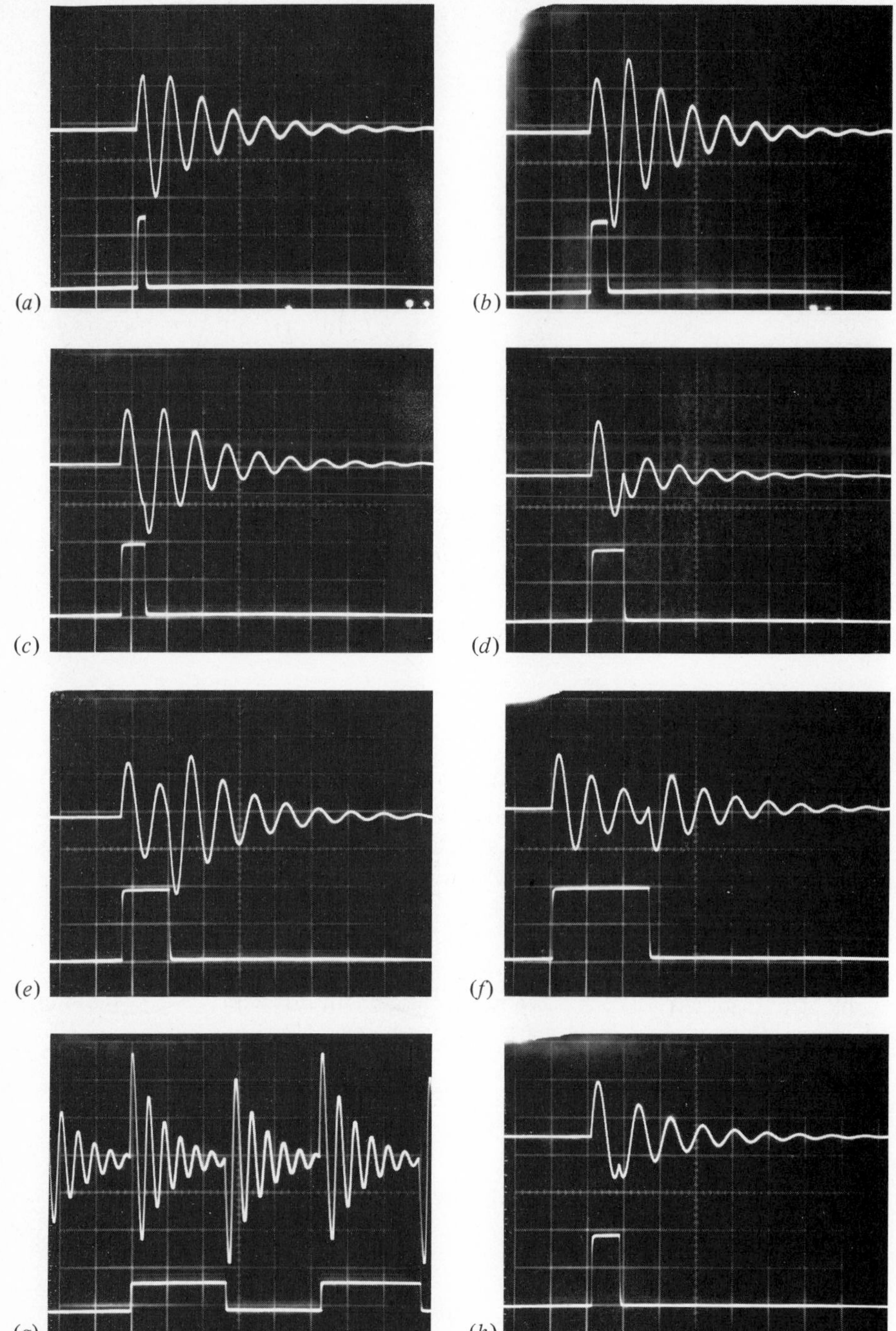

FIG. 3.09-B.

It is extremely useful, in studying a response as complicated as this one, to have a means of rapid visualization of the various possibilities. Since this is a very simple circuit, the actual response, seen on the cathode-ray tube, is eminently practical. (Later we shall find that additional "computational machinery" is necessary, and discuss it as appropriate.)

Various actual responses are shown in Fig. 3.09-B, together with the exciting pulses, for a lower-Q circuit ($Q = 8$). They provide a variety of interesting shapes, as well as physical corroboration of our theoretical results. The responses differ from each other only in that the exciting pulse width T_D is varied (see Table 3.09-A). All time scales are 20 μs per division. The combining of the

TABLE 3.09-A

Figure 3.09-B Part	Approximate Value of T_D, in Units of T_N
(*a*)	$\frac{1}{4}$
(*b*)	$\frac{1}{2}$
(*c*)	$\frac{3}{4}$
(*d*)	1
(*e*)	$1\frac{1}{2}$
(*f*)	3
(*g*)	6
(*h*)	1^-

two components is less striking here because the Q is much lower than in Fig. 3.09-A, but the tendencies are evident. (Compare Fig. 3.09-B*d* with Fig. 3.09-A.) Study of the curves for what they should be with high Q is very instructive! Response *h*, incidentally, is for a pulse width slightly less than T_N, adjusted to make the first two negative humps alike, an interesting sight.

Among the many uses one can imagine for the tuned circuit is the production of "bursts" of sinusoids, by, in effect, turning its natural behavior on and off by a pulse excitation. If the value of Q is high enough, adjusting T_D to a multiple of T_N can give essentially this sort of response (Fig. 3.09-A).

3.10 EXPONENTIAL RESPONSE; THE TRANSFER FUNCTION

Excitation that is exponential in form

$$i_F(t) = I_F e^{st} u(t) \tag{3.10-1}$$

should certainly produce a response of this form:

$$v(t) = v_N + v_F = K_1 e^{p_1 t} + K_2 e^{p_2 t} + V_F e^{st}. \tag{3.10-2}$$

All of our experience so far indicates that. Yet it must be verified, however reasonable it seem. Substituting in the differential equation gives, much as before,

$$\left(Cp_1 + G + \frac{1}{Lp_1}\right)K_1 e^{p_1 t} + \left(Cp_2 + G + \frac{1}{Lp_2}\right)K_2 e^{p_2 t}$$
$$+ \left[\left(Cs + G + \frac{1}{Ls}\right)V_F - I_F\right]e^{st} - \left(\frac{K_1}{Lp_1} + \frac{K_2}{Lp_2} + \frac{V_F}{Ls}\right) + I_0 = 0. \qquad (3.10\text{-}3)$$

The fact that the expression $Cx + G + 1/(Lx)$ occurs, albeit with differing x, in all three exponential-term coefficients suggests that it may have some general significance. It does indeed. Looking back at the RC circuit, we see now that it occurred there, too, as $Cp + G$ or $Cs + G$. Looking ahead is more difficult, but we shall often encounter this sort of function of frequency, be it natural frequency p or excitation frequency s. Accordingly we define the *admittance function*

$$Y(x) = Cx + G + \frac{1}{Lx}. \qquad (3.10\text{-}4)$$

The symbol Y and the name *admittance* are traditional; they come from other uses of the function which we shall discuss later; here we simply accept them as arbitrary notation. We note that Y is measured in mhos and that the dimensions of each term in (3.10-3) are those of current, the product of admittance (mhos) and voltage.

Rewriting (3.10-3) using (3.10-4) gives us

$$Y(p_1)K_1 e^{p_1 t} + Y(p_2)K_2 e^{p_2 t} + [Y(s)V_F - I_F]e^{st} - \left(\frac{K_1}{Lp_1} + \frac{K_2}{Lp_2} + \frac{V_F}{Ls}\right) + I_0 = 0. \qquad (3.10\text{-}5)$$

Because of the independent character of each of the three exponential functions, it is necessary that

$$Y(p_1) = 0, \qquad (3.10\text{-}6)$$

$$Y(p_2) = 0, \qquad (3.10\text{-}7)$$

$$V_F = \frac{I_F}{Y(s)}, \qquad (3.10\text{-}8)$$

and

$$\frac{K_1}{Lp_1} + \frac{K_2}{Lp_2} + \frac{V_F}{Ls} = I_0. \qquad (3.10\text{-}9)$$

The first two equations state that p_1 and p_2 must be the roots of the characteristic equation (3.03-3); that is, they must be the natural frequencies. This is no surprise. The third equation evaluates V_F for us in terms of known quantities. Finally, (3.10-9) gives us one equation for the two unknown K's; the necessary second equation is of course $v(0+) = V_0$, or (3.10-10):

$$K_1 + K_2 + V_F = 0. \tag{3.10-10}$$

An alternate to (3.10-9) is

$$K_1 p_1 + K_2 p_2 + V_F s = v'(0+) = \frac{1}{C}(-GV_0 - I_0 + I_F). \tag{3.10-11}$$

This can be written by inspection, using physical common sense. When the five unknown parameters in (3.10-2) are so determined, (3.10-2) is the response; the verification is complete.

A very interesting by-product of this analysis is the admittance function (3.10-4), so strongly suggested as useful by (3.10-3). We note again its two uses here:

a The natural frequencies p are determined by setting the admittance function $Y(p) = Cp + G + 1/(Lp)$ equal to zero, (3.10-12)

and

b The "amplitude" V_F of the forced component of the response is determined by dividing the amplitude of the (exponential) excitation by the admittance function $Y(s)$. (3.10-13)

We shall find this and corresponding functions of tremendous value in all network analysis and even in network synthesis.

One useful variant is the reciprocal of Y, that is, the function that *multiplies* the excitation to give the forced component. Since this measures the effect of, or the "transfer" from, excitation to (forced component of) response, it is called the *transfer function* (of s),

$$T(s) = \frac{1}{Cs + G + 1/(Ls)} = \frac{V_F}{I_F} = \frac{V_F e^{st}}{I_F e^{st}}. \tag{3.10-14}$$

It is here equal to the reciprocal of the admittance function; in more complicated cases, there may be differences between the two. But we here define, once and for all, the transfer function of a network.

We assume that, on excitation by an exponential function $E_F e^{st}$, a network response will contain natural and forced components, and that the forced component will be of the same form, say $R_F e^{st}$. Then, by definition,

$$\text{the transfer function of the network} = T(s) = \frac{\text{forced component of response}}{\text{excitation, in form exponential}}$$

$$= \frac{R_F e^{st}}{E_F e^{st}} = \frac{R_F}{E_F}. \tag{3.10-15}$$

That e^{st} cancels is due to the linear postulate and makes it possible to speak either of the ratio of the *amplitudes* of the two or of the things themselves. For our parallel *RLC* circuit, the transfer function is accordingly given by (3.10-14). [When $L \rightarrow \infty$ this becomes $(Cs + G)^{-1}$, the transfer function of the *RC* circuit; see (2.14-7).] Note that the natural component of the response is not considered in the definition, nor are the initial conditions. That *exponential* excitation should be so important and warrant such attention should by now seem eminently reasonable because of the inherent exponential nature of the natural behavior. But it remains to verify this in general terms, as we shall do, and to discuss how the transfer function is best calculated in its own right.

The nature of the response wave is evident from (3.10-2). There will be an exponential component like the excitation, and a natural component. Many variations are possible, but for the high-Q circuit, the lightly damped natural oscillatory part will predominate. When the natural frequencies are real, rather than complex, the response consists merely of three ordinary exponential terms.

By way of illustration and physical corroboration we examine here not exactly the response to an ordinary damped exponential excitation (which is not very important as an excitation anyway) but a more interesting response: that to an exponential plus constant, a sort of ramp excitation. The excitation (Fig. 3.10-A) is in form

$$i_F(t) = I_F(1 - e^{-\alpha_F t}). \tag{3.10-16}$$

The response, as we easily calculate on the basis of experience [see (3.10-2) and Sec. 3.08, using superposition], should have the form

$$v(t) = v_N + v_F = (K_1 e^{p_1 t} + K_2 e^{p_2 t}) + (V_{F0} + V_{F1} e^{-\alpha_F t}). \tag{3.10-17}$$

That

$$V_{F0} = 0,$$

$$V_{F1} = -\frac{I_F}{Y(-\alpha_F)} = -T(-\alpha_F)I_F, \tag{3.10-18}$$

p_1 and p_2 are the natural frequencies of the circuit,

K_1 and K_2 must be evaluated to suit the initial conditions (rest),

should be verified, but they seem almost obvious.

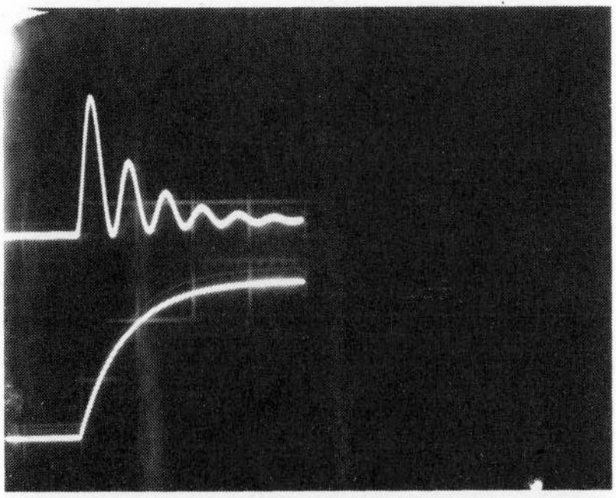

(*a*) Time scale: 100 μs/division

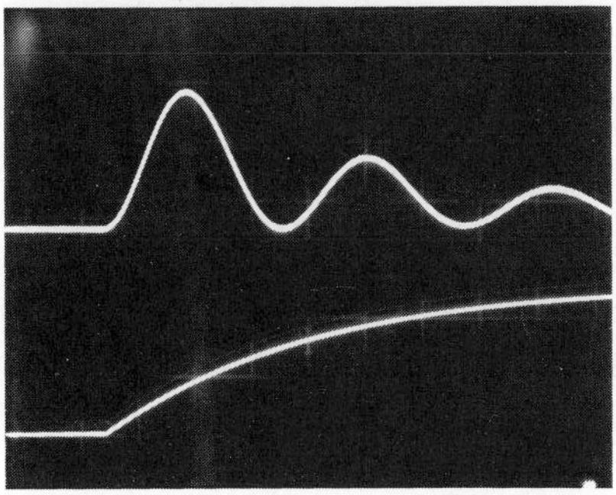

(*b*) Time scale: 20 μs/division

FIG. 3.10-A.

The actual physical response wave (Fig. 3.10-A) corroborates the analysis and is interesting in its own right, the usual damped natural oscillation being completely offset by the (forced) exponential component, so that there is no negative going. Note that in a way this *is* the response to exponential excitation, since the constant component of the excitation produces no corresponding response component. But because of that constant component of excitation, the initial slope of the response is zero (as is evident in the expanded response, Fig. 3.10-A*b*, if not in Fig. 3.10-A*a*). It would not be zero if the excitation were purely exponential.

The response to one (or more) specific, ordinary damped exponentials is in practice not important. But in and for the theory of network analysis, the concept that exponential excitation introduces, the *transfer function*, is vital. We shall find it increasingly significant, beginning with the response to sinusoidal excitation, which *is* very important.

3.11 SINUSOIDAL RESPONSE

Exciting the high-Q circuit with a sinusoidal current leads to complexity of behavior that is intensely interesting and practically very useful. The tuned circuit is indeed worthy of respect! The formal part of the analysis here requires no new concepts; we shall make full use of our tools to get through it quickly and efficiently. It is in the manifold nature of the results that we shall find a new dimension of complexity and fascination.

Suppose the excitation is sinusoidal, as in Sec. 2.15:

$$\begin{aligned} i_F(t) &= I_m \cos(\omega_F t + \psi) \\ &= \frac{I_m}{2} e^{j\psi} e^{j\omega_F t} + * \qquad (3.11\text{-}1) \\ &= \mathbf{RE}\,(\bar{I}_F\, e^{j\omega_F t}), \end{aligned}$$

in which we have emphasized the exponential nature of the sinusoid and used the phasor representation of $i_F(t)$,

$$\bar{I}_F = I_m\, e^{j\psi}. \qquad (3.11\text{-}2)$$

The subscript F (unnecessary in Sec. 2.15) has here been added to the symbol for the frequency of the exciting sinusoid because we now have to deal with three different ω's (ω_0, ω_N, ω_F), which should not be confused. The asterisk, as usual, denotes a term conjugate to that preceding it, in itself uninformative and unnecessary of explicit expression, but essential for completeness.

Instinctively we use the results of Sec. 3.10 to find the (formal) response to each of the two exponential terms, and invoke superposition to write the response

$$\begin{aligned} v(t) &= v_F + v_N \\ &= \left[\frac{I_m}{2} e^{j\psi} T(j\omega_F) e^{j\omega_F t} + *\right] + (K_1 e^{p_1 t} + K_2 e^{p_2 t}). \qquad (3.11\text{-}3) \end{aligned}$$

The transfer function (3.10-14) is evidently to be evaluated here with $s = j\omega_F$. It is but a slight extension of the work of Sec. 3.10 to verify that (3.11-3) *is* the response. We note that, since each is complex, the two exponentials must be considered simultaneously; but their conjugate nature obviates the need of writing both out in full: an asterisk will do for one of them.

The forced part must be evaluated first, before the K's of the natural part can be found. Permanent or steady-state in nature, it is

$$\begin{aligned} v_F(t) &= \mathbf{RE}\,[I_m e^{j\psi}\, T(j\omega_F) e^{j\omega_F t}] \\ &= \mathbf{RE}\,[\bar{I}_F\, T(j\omega_F) e^{j\omega_F t}] \qquad (3.11\text{-}4) \\ &= \mathbf{RE}\,(\bar{V}_F\, e^{j\omega_F t}), \end{aligned}$$

in which we have written $\bar{V}_F$ for the phasor representation of this sinusoid:

$$\bar{V}_F = \bar{I}_F\, T(j\omega_F) = \frac{\bar{I}_F}{\bar{Y}(j\omega_F)} = \bar{Z}(j\omega_F)\bar{I}_F. \qquad (3.11\text{-}5)$$

We note with interest that the response phasor is very simply calculated from the excitation phasor: we need merely multiply by the transfer function, evaluated for $s = j\omega_F$. The formula for $\bar{V}_F$, (3.11-5), is again (see Sec. 2.14) a generalization (to phasors) of Ohm's law. In it appear the admittance $\bar{Y}(j\omega_F)$, written with a bar here to emphasize its complex character, and its reciprocal, $\bar{Z} = 1/\bar{Y}$, called *impedance*. Both are but slight extensions of familiar impedance and admittance concepts, but we postpone their general discussion to a more appropriate moment; the important function here is the transfer function $T(s)$, evaluated for $s = j\omega_F$.

We note that $T(j\omega_F)$ is *complex*, since its argument is no longer real. To clarify this we write [see (3.10-14)]:

$$T(j\omega_F) = \frac{1}{G + j[C\omega_F - 1/(L\omega_F)]}$$
$$= \frac{1}{G + jB} = \frac{1}{\sqrt{G^2 + B^2}} \angle{-\theta}\,. \tag{3.11-6}$$

The symbols B and θ, used for brevity, are defined in Fig. 3.11-A. Note that the angle of T is the negative of the angle of $\bar{Y}$, and that the latter, θ, may lie in either quadrant I or quadrant IV, or be zero.

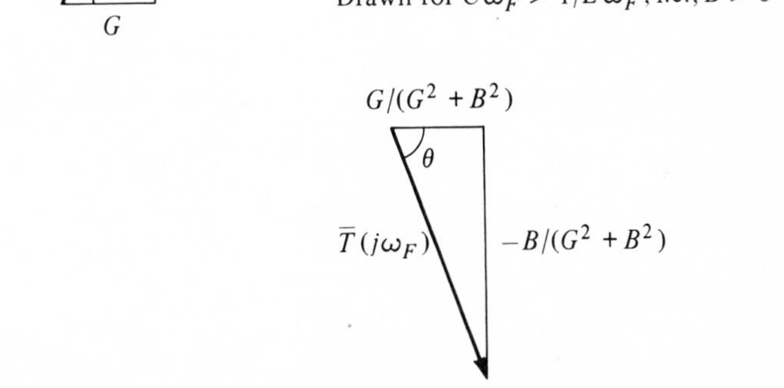

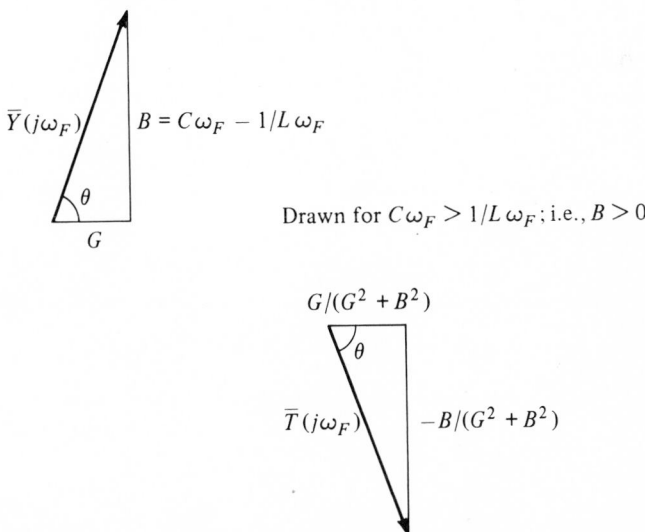

FIG. 3.11-A. The complex admittance $\bar{Y}(j\omega_F)$ and the transfer function $\bar{T}(j\omega_F) = 1/\bar{Y}(j\omega_F)$.

From (3.11-5) we now extract the following details about the forced (steady-state) component of the response voltage and its phasor representation:

$$\begin{aligned} \bar{V}_F &= I_m |T(j\omega_F)| \underline{/ \psi - \theta} \\ &= \frac{I_m}{\sqrt{G^2 + B^2}} \underline{/ \phi}, \qquad \phi = \psi - \theta, \\ &= \frac{I_m}{\sqrt{(G)^2 + [C\omega_F - 1/(L\omega_F)]^2}} \underline{/ \psi - \tan^{-1} \frac{C\omega_F - 1/(L\omega_F)}{G}} \end{aligned} \tag{3.11-7}$$

and

$$\begin{aligned} v_F(t) &= \frac{I_m}{\sqrt{G^2 + B^2}} \cos(\omega_F t + \phi) \\ &= V_F \cos(\omega_F t + \phi). \end{aligned} \tag{3.11-8}$$

It is, of course, a sinusoid of the same frequency as the excitation. What makes it interesting, and complicated, is the dependence of its amplitude (and initial phase) upon the frequency ω_F. This we shall soon discuss.

We turn now to the natural part of the response,

$$v_N = v - v_F = K_1 e^{p_1 t} + K_2 e^{p_2 t}. \tag{3.11-9}$$

Since the initial conditions are given in terms of capacitor voltage and inductor current, that is, $v(0+) = V_0$ and $i_L(0) = I_0$, the K's will be determined in part by the initial values of the forced part of the response also. The determining equations may be any two of the following three.

$$\frac{K_1}{Lp_1} + \frac{K_2}{Lp_2} = I_0 - \frac{I_m \sin \phi}{L\omega_F \sqrt{G^2 + B^2}}, \tag{3.11-10}$$

$$K_1 + K_2 = V_0 - \frac{I_m \cos \phi}{\sqrt{G^2 + B^2}}, \tag{3.11-11}$$

$$K_1 p_1 + K_2 p_2 = \frac{1}{C}(-GV_0 - I_0 + I_m \cos \psi) + \frac{I_m \omega_F \sin \phi}{\sqrt{G^2 + B^2}}. \tag{3.11-12}$$

The foregoing results are generally valid, whether the Q be high or low, whether the natural frequencies be complex or real. But our chief interest here is in the high-Q (resonant) case, in which the natural frequencies are complex, as are K_1 and K_2. Then the natural (transient) part of the response is

$$\begin{aligned} v_N(t) &= \bar{K}_1 e^{p_1 t} + * \\ &= 2Ke^{-\alpha t} \cos(\omega_N t + \phi_K) \end{aligned} \tag{3.11-13}$$

in which the bar is added above K_1 to emphasize that it is complex. Its magnitude K and angle ϕ_K are defined in Fig. 3.11-B. They can be evaluated indirectly from two of the equations above (3.11-10 to 3.11-12); perhaps the second and third provide the most direct route. We write

$$v_N(0+) = \bar{K}_1 + \bar{K}_2 = 2K_{1r} = V_0 - \frac{I_m \cos\phi}{\sqrt{G^2 + B^2}} = D, \qquad (3.11\text{-}14)$$

and

$$\begin{aligned} v'_N(0+) &= \bar{K}_1 p_1 + \bar{K}_2 p_2 = -2(K_{1r}\alpha + K_{1i}\omega_N) \\ &= \frac{1}{C}(-GV_0 - I_0 + I_m \cos\psi) + \frac{I_m \omega_F \sin\phi}{\sqrt{G^2 + B^2}} = E, \end{aligned} \qquad (3.11\text{-}15)$$

in which D and E represent real (initial-condition) quantities, known once the forced part has been calculated. Their solution gives

$$\begin{aligned} K_{1r} &= \frac{1}{2} v_N(0+) = \frac{D}{2}, \\ K_{1i} &= \frac{-[v'_N(0+) + \alpha v_N(0+)]}{2\omega_N} = \frac{-(\alpha D + E)}{2\omega_N}. \end{aligned} \qquad (3.11\text{-}16)$$

Since K_{1r} and K_{1i}, the real and imaginary parts of $\bar{K}_1$, can each have either sign, ϕ_K can lie in any of the four quadrants.

This natural behavior is still a lightly damped oscillatory wave, regardless of the initial conditions and the excitation. Its "amplitude" K and initial phase ϕ_K, that is, K_1 and K_2 in effect, can be evaluated only *after* the forced component has been calculated. They then take up whatever values are necessary to meet the initial conditions. Thereafter the natural part discreetly fades from the picture, eventually leaving the steady-state (forced) part dominant.

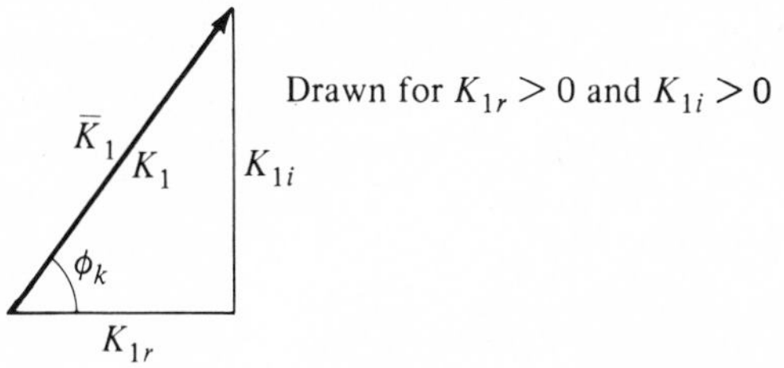

FIG. 3.11-B.

The manifold character of the response of the high-Q circuit to sinusoidal excitation is due to the additional parameter ω_F. When excited by one exponential wave (Sec. 3.10) there is only one exponent to deal with: s in (3.10-2), which must be real. A sinusoid is actually *two* exponentials, however, and an additional dimension of complexity enters. True, the two exponents are not really different, for one is the negative of the other. But they are *two*, and their effect appears in the *two* parameters ω_F and ψ. The result is a very wide range of possibilities, because the natural frequencies can also be complex.

The complicated formulas of this section give one the impression that many different sorts of behavior can occur in response to sinusoidal excitation. This is true. To obtain an adequate sampling we really need help, help in the form of a computational and display device that will relieve us of the drudgery of evaluating the formulas for many cases and plotting many points. This we consider in Sec. 3.12. Then, in Sec. 3.13, we shall return to the response of the high-Q tuned circuit.

3.12 COMPUTATION: DRUDGERY OR DELIGHT?

The response calculations we have encountered thus far have not been very demanding. There is no great difficulty in the numerical work of the simple arithmetic operations required, nor in the evaluation of the simple functions needed (exponentials, sinusoids), for which tables are readily available. Tedious it may be when done by hand, because we may need many points, for long sequences of values in time, from which to make the plots (or tables) that concisely present the real fruits of the analysis. The many numbers, necessary though their calculation and accumulation be, are only intermediate results: it is what they show when properly displayed, the insight then gained, that is the goal. "The purpose of computing is insight, not numbers."† It is drudgery to obtain the quantity of numbers needed; there can be delight in perceiving what they signify.

For the sinusoidally excited tuned circuit, the response-calculation problem augments significantly in length (number of calculations) if not in complexity. It is imperative now to consider means of reducing the drudgery (sure to intensify), lest it completely obscures the real goal. "It is unworthy of excellent men to lose hours like slaves in the labor of calculation."‡

Yet computation is of the essence of engineering: engineers need numbers to apply their knowledge to plan and predict (design), to avoid mistakes, and to construct safe and efficient artifacts. They cannot do without numbers, nor the computations that lead to them. The most intellectual analytical effort comes

† Richard W. Hamming.

‡ Attributed to Gottfried Wilhelm Leibnitz.

down for practical purposes, sooner or later, to computing a mass of numbers, tabulating or plotting them—and then extracting information therefrom. Computing is done not for numbers *qua* numbers, but for the insight to be gained by contemplation thereof. Yet the horde of numbers needed is tremendous and must be kept under control. We must keep the quantity to the minimum necessary and use efficient computing tools. In computing, the engineer needs perspective par excellence: there is more than one way to skin a cat. The engineer has available computational machinery of tremendous variety in capability and cost. Which piece of this machinery should be used in a given situation? Something simple and quick but limited in accuracy? Or a complicated and slower but very accurate method?

Some engineering computations should be made on the spot, mentally, quickly, to one or two significant figures only: that's all that's needed, and speed is of the essence. It is not appropriate to calculate more because it would take too long, or cost too much, or the data's accuracy cannot support such effort. An estimate is enough, or an estimate is all that can be justified. Even for simple mental calculation, incidentally, the infinitesimal calculus offers useful tools. One such is the power series, truncated to one or two terms; Secs. 3.06 and 3.07 had striking examples of the power of this tool, and more will come.

Some engineering computations require great precision; our network circuit models are usually amazingly accurate and can often justify six-significant-figure calculations, and we may *need* such accuracy, in both analysis and synthesis. Or the three significant figures of a few rapid slide-rule calculations may be more appropriate.

The history of computation is largely the development of tools useful in reducing the human drudgery required for lengthy calculations: logarithms, the slide rule, mechanical then electric digital calculators, and finally electronic digital computers of pocket size, giant size, and in-between. [Analog computers, ranging from the circuits themselves (!) to complicated electronic differential analyzers, are also useful at times.]

Sooner or later circuit-network-analysis computation requires the use of a large-scale, programmed digital computer. Once the relevant programs are written, it easily outclasses all other computation methods in speed and cost for what we want: the calculation of values of the response at a series of points in time, then a display thereof (plot or table), all repeated for a variety of parameter values. It becomes hopeless to do this by hand in any reasonable time with any reasonable effort. But the calculations are quickly and cheaply made with the mechanized (programmed) machine. The development of several such programs for our use will be discussed in Sec. 7.18; here we merely take one of them (RESSINET) and use it as a drudgery-relieving computational tool for our immediate computational chore.† We call on it (in Sec. 3.13) for a set of curves of the response of a tuned circuit to sinusoidal excitation, continuing Sec. 3.11.

†Drudgery is not absent in the *development* of such programs; many thanks go to my patient wife Becky for performing much of it—author.

3.13 SINUSOIDAL RESPONSE OF THE *RLC* CIRCUIT (Continued)

We now *use* such a calculating machine to observe the response voltage (3.11-3). The essential character of that response (in the high-Q case, with sinusoidal excitation) is, from (3.11-13) and (3.11-8),

$$v(t) = v_N + v_F = 2Ke^{-\alpha t} \cos (\omega_N t + \phi_K) + V_F \cos (\omega_F t + \phi). \quad (3.13\text{-}1)$$

It consists of an undamped sinusoid at the excitation frequency ω_F and a damped one at the "natural" frequency ω_N. Of the many possibilities, we look at only a few, the most important. The machine makes it easy to do this, cheaply and with little hand labor. It removes the drudgery of evaluating (3.13-1) hundreds of times, for a sequence of time values, and for various excitation parameters (ω_F, ϕ).

The curves in Figs. 3.13-A to 3.13-D are so "generated." The excitation amplitude I_F and the circuit parameters (most important: $Q = 15$) are the same for all the curves. [Since $Q = 15$, ω_N is the same as ω_0 for all practical purposes, being within one-twentieth of one percent, (3.06-11).] The excitation frequency is altered from one curve to another, and the scales differ.

For our first example, suppose the excitation frequency to be low compared to the natural circuit frequency. We choose $\omega_F = \omega_0/6$, initial conditions of rest, and initial phase (or starting point of the excitation wave) such that $v'(0)$ is zero as well as $v(0)$: $\phi = -90°$. (The excitation wave is initially zero, so the start is as slow as possible.) One may succinctly describe the response (Fig. 3.13-A) thus: during its brief life, the high-frequency transient rides on top of the slower-moving steady-state response.

Secondly, at the other extreme, suppose the excitation frequency is large, $\omega_F = 6\omega_0$. With the same initial condition and initial phase of excitation, we obtain Fig. 3.13-B. This time one may aptly say that the steady-state wave rides up and down on the transient, gradually settling down to a mean value of zero.

The case "in the middle," $\omega_F = \omega_N \cong \omega_0$, is rather interesting. It seems (Fig. 3.13-C) as though the forced component simply builds up slowly in amplitude to its final steady state. (The initial conditions and initial phase remain the same.) Especially interesting is the large amplitude of this steady-state response: it is many times larger than in the other two examples, though the circuit and the amplitude (not the frequency) of the excitation are the same for all three. (The scales differ in Figs. 3.13-A to 3.13-C.) The ratio of the amplitude of the steady-state response (keeping excitation amplitude and circuit parameters constant) when $\omega_F = \omega_0$ to its value when ω_F is small is about Q/ω_F, to its value when ω_F is large is about $Q\omega_F$ (see Exercise 3-41 and Sec. 3.14); for our cases here, both ratios are 90 : 1. It is here that the adjective *resonant* really applies: sinusoidal excitation at (or near) the natural frequency of a high-Q circuit produces a very large steady-state response; at frequencies not close

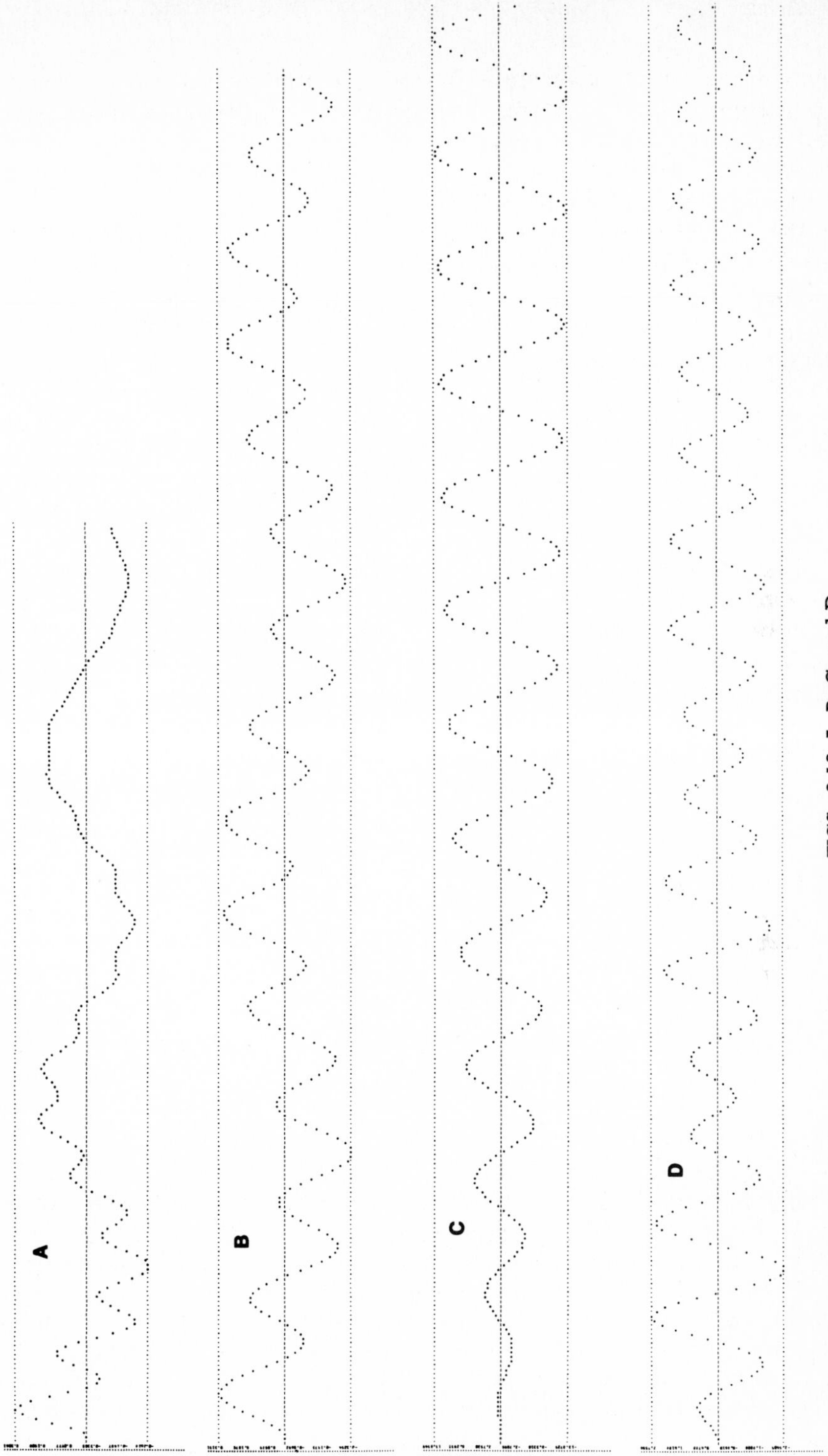

FIGS. 3.13–A, B, C, and D.

thereto, the response is comparatively weak. The phenomenon is well known, both in nature (sounds in caves, in seashells) and in human artifacts (organ pipes); we shall return to it in Sec. 3.14. The adjective *tuned* also applies here, but in a reverse fashion: it implies rather that ω_N (the circuit) is adjusted to be very close to a given ω_F (of the excitation).

Also very interesting are cases where ω_F is not equal to ω_N but is very close to it: $\omega_F = \omega_N(1 + \varepsilon)$, ε being much smaller than unity. We should expect a sort of resonance here. But, because of the slight difference in frequencies, the buildup of Fig. 3.13-C does not occur so smoothly, but pulsates at a small frequency, related to the difference, $\varepsilon\omega_N$, as in Fig. 3.13-D. The effect is something like a modulation of the steady-state wave at that small frequency (see Probs. 3-X and 7-P). It requires a rather high value of Q to be seen; otherwise the damping obscures it. The name given the phenomenon is *beats;* it again is not unfamiliar in nature: witness the sounds heard when tuning a violin, as the violin-string frequency approaches the reference frequency. (The value of ε in Fig. 3.13-D is 1/3.)

These four curves were not difficult to obtain: the computer offers a much less laborious way to get them than to use slide rule and compute and plot by hand. No scales are given in the drawings since everything was normalized in any event. The original plots have scales, of course, from which (together with the normalization factors) actual response values can be computed. (See Fig. 7.18-E for additional, similar response curves.)

The response of Fig. 3.13-C is particularly interesting because it is some 100 times as large as those of Figs. 3-13-A and 3.13-B. (The scales of the drawings are quite different.) We now need to discuss this phenomenon of resonance in more detail.

3.14 RESONANCE

Very important in engineering is the resonance phenomenon encountered in Fig. 3.13-C. Excitation of a high-Q circuit by a sinusoid at frequency ω_F equal or nearly equal to ω_0 produces a very large forced component of response. The effect may be disagreeable and even dangerous (especially in analogous mechanical situations). But it can be very useful to electrical engineers in its *selectivity* property: the high-Q tuned circuit is very responsive to (sinusoidal) excitation at frequencies in a narrow band near ω_0, and unresponsive at frequencies removed therefrom. There is no simpler way (in a radio receiver) to separate the signal of a desired station from those unwanted; "tuning in" a station is everyday usage. (And it has enriched the language by extension far from engineering!) We must investigate this carefully.

Our concern here is only with the forced component of response. And that is most compactly described by its phasor [see (3.11-5)]

$$\bar{V}_F = \bar{I}_F\, T(j\omega_F). \tag{3.14-1}$$

The normalized response is even more compact: it is by definition, or from (3.14-1), equal simply to the transfer function evaluated for $s = j\omega_F$. So it is to $T(j\omega_F)$ that we look for assistance in discussing resonance. From (3.11-6)

$$\begin{aligned} T(j\omega_F) &= \left[\frac{1}{Cs + G + 1/(Ls)}\right]_{s=j\omega_F} \\ &= \frac{1}{G + j[C\omega_F - 1/(L\omega_F)]} = \frac{1}{Y(j\omega_F)} \\ &= \frac{R}{1 + j[C\omega_0/G(\omega_F/\omega_0 - \omega_0/\omega_F)]} \\ &= \frac{R}{1 + j[(\sqrt{C/L}/G)(\omega_F/\omega_0 - \omega_0/\omega_F)]} \\ &= \frac{R}{1 + jQ(\omega_F/\omega_0 - \omega_0/\omega_F)}. \end{aligned} \tag{3.14-2}$$

The manipulations in (3.14-2) are intended to cast the transfer function into forms in which the most important features of the resonance are most clear.

It is evident, first of all, that $|T(j\omega_F)|$ is largest when $\omega_F = \omega_0$. This is the frequency of resonance, or the "tuning frequency." We note also that the angle of $T(j\omega_F)$ is then zero. The transfer function is equal to the real number R (ohms), for the L and C elements in the circuit (Fig. 3.01-A) are effectively absent, being busily occupied with energy exchange between themselves, as we shall soon see.

The variation of the magnitude of the response with ω_F is that shown in Fig. 3.14-A: there is a marked peak (resonance) at $\omega_F = \omega_0$ (unless Q is very small indeed). One might as well say "at $\omega_F = \omega_N$," since ω_N and ω_0 differ by

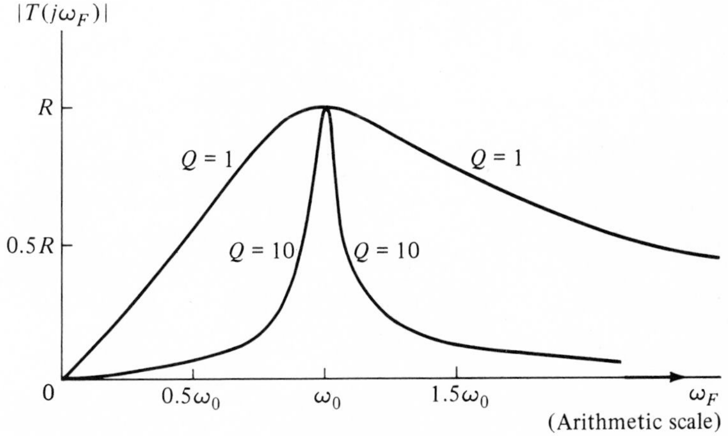

FIG. 3.14-A.

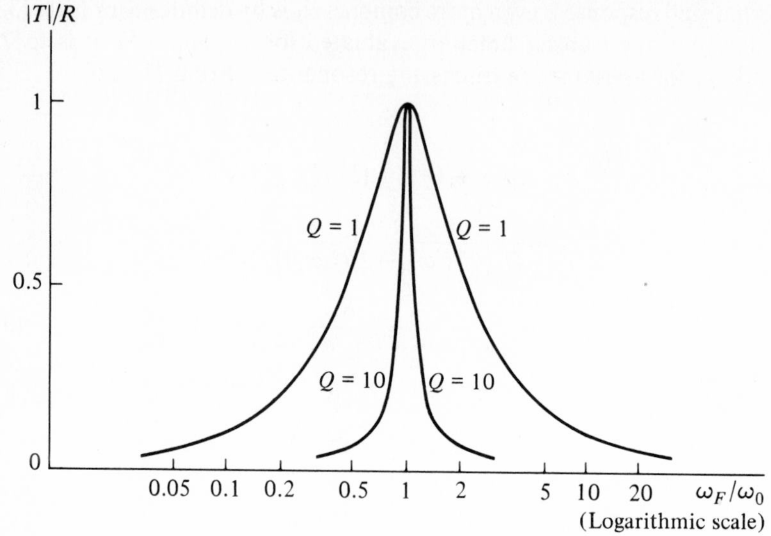

FIG. 3.14-B.

only $10/Q^2$ percent according to (3.07-8), a very small amount indeed, being, for example, merely 0.1 percent when Q is only 10. For $Q = 100$, an often not impracticable value, the curve drawn to this frequency scale would to all intents and purposes be a single vertical line, a sort of impulse function in frequency!

The third line in (3.14-2) states that $|T|$ has the same value at frequencies

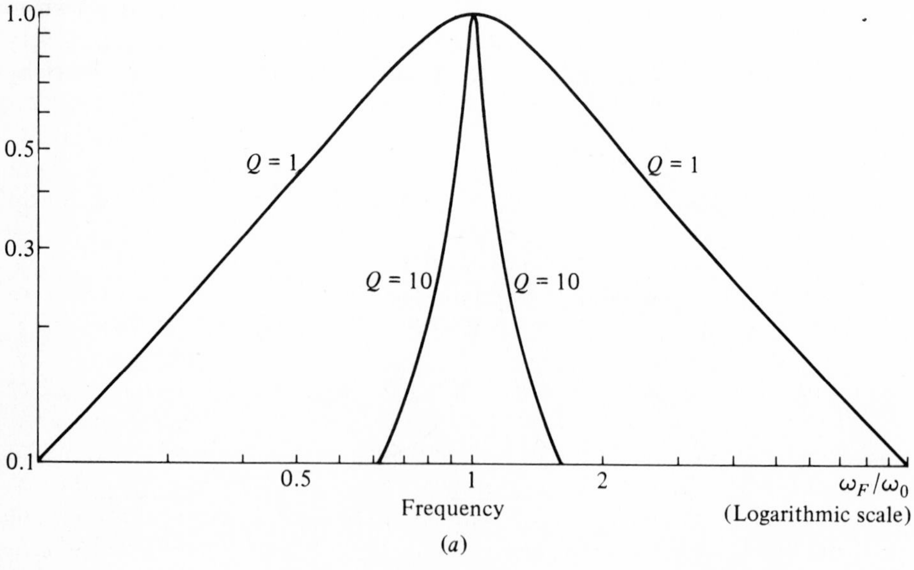

FIG. 3.14-C.

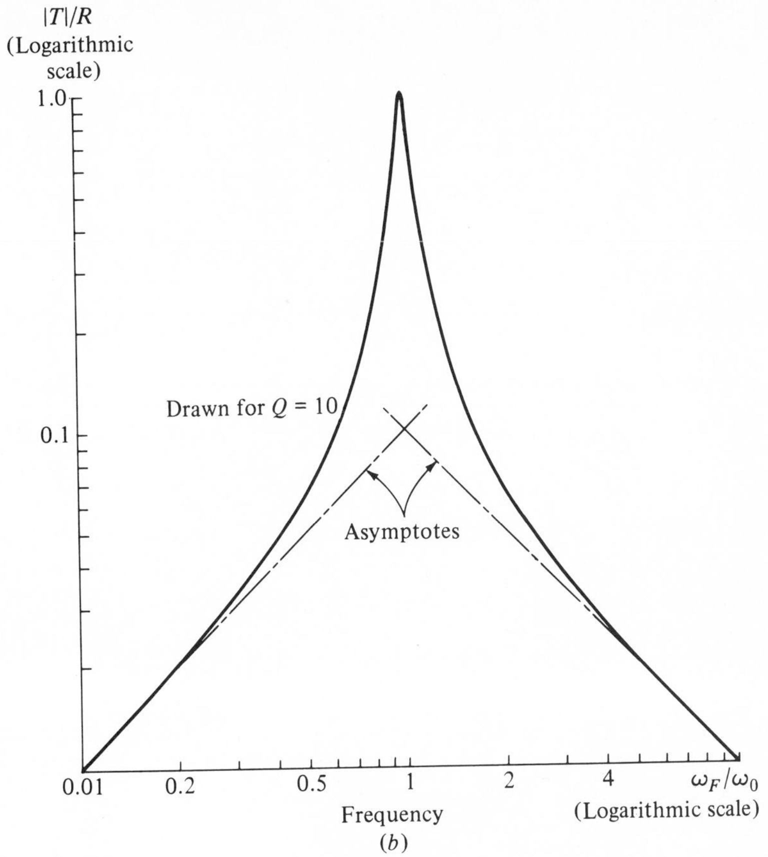

FIG. 3.14-C.

ω_{F1} and ω_{F2} such that $\omega_{F1}/\omega_0 = \omega_0/\omega_{F2}$, that is, such that ω_{F1} and ω_{F2} have ω_0 for their geometric mean: $\omega_0^2 = \omega_{F1}\omega_{F2}$. The use of a logarithmic frequency scale (Fig. 3.14-B) makes this evident, for then ω_{F1} and ω_{F2} are symmetrically located about ω_0. Note that even the (very low) $Q = 1$ curve is now symmetrical about the resonance point, in contrast to Fig. 3.14-A.

The curves become even simpler if we use a logarithmic scale for $|T|$ also, as in Fig. 3.14-C. One must of course normalize in both cases, since the argument of a logarithm can have no dimensions. In Figs. 3.14-B and 3.14-C horizontal distances are proportional to the logarithm (to any base) of ω_F/ω_0; in Fig. 3.14-C the ordinates are proportional to the logarithm (to any base) of $|T|/R$. We note the asymptotic simplicity of the curve in Fig. 3.14-C: it approaches straight lines at very low and very high frequencies, a subject to which we shall return. There is little difference between logarithmic and arithmetic (frequency) scale presentations near the resonance, for all reasonable values of Q. But we shall often appreciate the symmetry that results from using logarithmic scales (Appendix D).

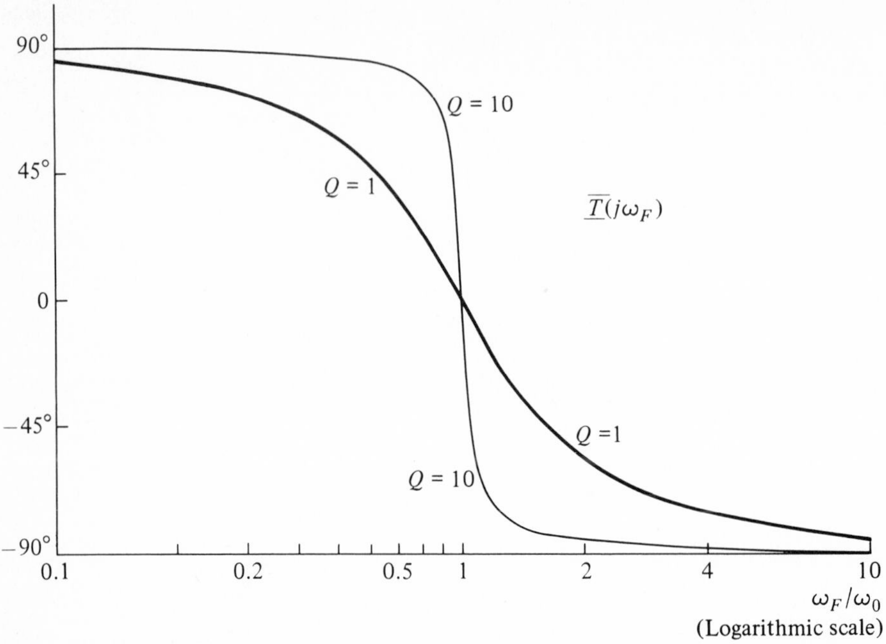

FIG. 3.14-D.

The angle of $T(j\omega_F)$ is usually less important. It is essentially $+90°$ at low frequencies, where the inductor predominates, and $-90°$ at high frequencies, where the capacitor is in control. At resonance, where the resistor alone determines T, the angle is zero. Figure 3.14-D shows the behavior. If the value of Q is high, most of the change occurs near ω_0, and the central part of the curve looks much the same on arithmetic and on logarithmic frequency scales. As a matter of fact, the same is true for the $|T|$ curve: the peak and most of the dropping off occur near ω_0, in a range of ω_F small enough that the choice of frequency scale matters little.

The sharpness (or "stiffness") of the selectivity curve (Fig. 3.14-A or 3.14-B or 3.14-C) should, according to the last line of (3.14-2), depend on Q alone. And so it does. Suppose we consider two values of ω_F, one below and one above resonance, at which $|T|$ has dropped to a fraction γ of its peak value, such as ω_a and ω_b in Fig. 3.14-E. They define, in an arbitrary way, a sort of "bandwidth," a range of excitation frequency centered on the resonance frequency, over which the response is "appreciable." The choice of γ is arbitrary: one could reasonably make it 0.5, for example. But calculations are somewhat simpler and easier to remember if we instead make it such that

$$Q\left(\frac{\omega_F}{\omega_0} - \frac{\omega_0}{\omega_F}\right) = -1 \text{ and } +1 \tag{3.14-3}$$

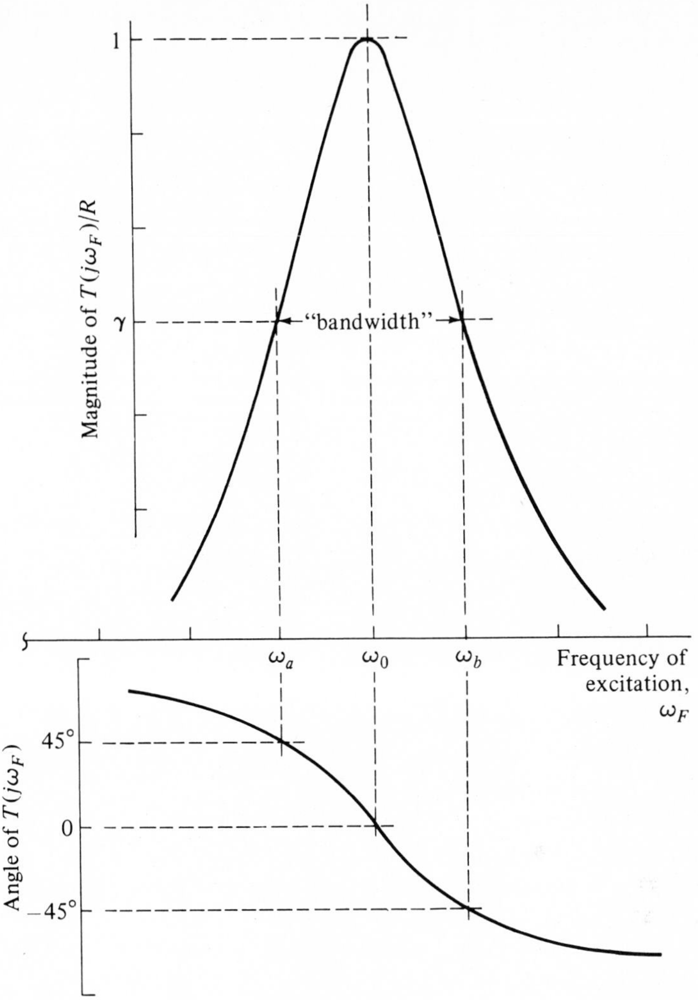

FIG. 3.14-E.

at $\omega_F = \omega_a$ and at $\omega_F = \omega_b$, respectively. For then

$$\frac{T}{R} = \begin{cases} \dfrac{1}{1 - j1} & \text{at } \omega_F = \omega_a \\[2ex] \dfrac{1}{1 + j1} & \text{at } \omega_F = \omega_b \end{cases} \tag{3.14-4}$$

and $\gamma = 1/\sqrt{2} = 0.7$. And for the values of these two frequencies we find

$$\omega_a = \omega_0\left[1 - \frac{1}{2Q}\right], \qquad \omega_b = \omega_0\left[1 + \frac{1}{2Q}\right] \tag{3.14-5}$$

to a good approximation. We see that

$$\frac{\text{bandwidth}}{\omega_0} = \frac{\omega_b - \omega_a}{\omega_0} = \frac{1}{Q} \tag{3.14-6}$$

and that

$$\begin{aligned} Q &= \frac{\omega_0}{\text{bandwidth}} = \frac{\omega_0}{\omega_b - \omega_a} \\ &= \frac{\text{frequency of resonance}}{\text{width of frequency band between 70\% points}} \\ &= \frac{1}{\text{fractional bandwidth}}. \end{aligned} \tag{3.14-7}$$

Q is thus a good measure of the selectivity of a tuned circuit. And (3.14-7) is in fact often taken as the definition of Q. We add it to our collection of (substantially) equivalent formulas for Q.

The angle of $T(j\omega_F)$ takes the value $+45°$ at $\omega_F = \omega_a$, $-45°$ at $\omega_F = \omega_b$, and is of course zero at $\omega_F = \omega_0$. Its curve is also shown in Fig. 3.14-E. Another interpretation of ω_a and ω_b will be given in Sec. 3.16 (Fig. 3.16-B).

An additional approximation, very useful, is to simplify the frequency scale in Fig. 3.14-E. We continue the manipulations of (3.14-2), using $\Delta = \omega_F - \omega_0$ to represent the (arithmetic) frequency departure from resonance:

$$\begin{aligned} \frac{T(j\omega_F)}{R} &= \frac{1}{1 + jQ(\omega_F/\omega_0 - \omega_0/\omega_F)} \\ &= \frac{1}{1 + jQ[(\omega_0 + \Delta)/\omega_0 - \omega_0/(\omega_0 + \Delta)]} \\ &= \frac{1}{1 + jQ[1 + \Delta/\omega_0 - (1 - \Delta/\omega_0 + \Delta^2/{\omega_0}^2 + \cdots)]} \\ &= \frac{1}{1 + j2Q[(\omega_F - \omega_0)/\omega_0 + \cdots]} = \frac{1}{1 + j[(\omega_F - \omega_0)/\alpha + \cdots]} \\ &\cong \frac{1}{1 + jx} \end{aligned} \tag{3.14-8}$$

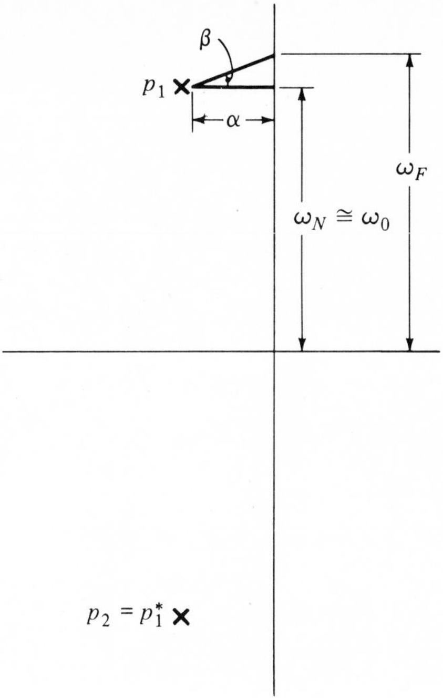

FIG. 3.14-F.

in which

$$x = \frac{2Q}{\omega_0}(\omega_F - \omega_0) = 2\frac{\omega_F - \omega_0}{\text{bandwidth}} \tag{3.14-9}$$

measures, in the simplest terms, the frequency of excitation as a (weighted) fractional departure from the resonant frequency. The graphical interpretation of x as the tangent of the angle β in Fig. 3.14-F is particularly simple.

Not only is the form $(1 + jx)^{-1}$ an extremely simple (though approximate) "universal" representation of the (complex) resonance curve—it is surprisingly accurate at all frequencies. Near ω_0 the terms represented by dots in (3.14-8) are indeed negligible; far away from ω_0 they are not, but there the curve has magnitude so small that the error is tolerable. For example, when $\omega_F = 10\omega_0$, the exact value of $T(j\omega_F)/R$ is $(1 + j9.9Q)^{-1}$ and the approximate value is $(1 + j18Q)^{-1}$; for $Q = 10$ these values are $0.010\underline{/-89.4^\circ}$ and $0.006\underline{/-89.7^\circ}$, respectively. The magnitudes are both so small that the apparent 50 or 100 percent error is of no consequence; nor is the 0.3° angle difference at all important—mental calculation (Sec. 3.12) suffices.

Those who work often with resonances may find useful a sturdy version of Fig. 3.14-E based on the $(1 + jx)^{-1}$ formula, with x as the abscissa.

Figures 3.14-E and 3.14-F are fascinating and useful tools of analysis. But they lure us into a can't-see-the-forest-for-the-trees situation. Their *meaning* we tend to forget: they represent transformations, removed from the actual physical phenomena. In the real (t) world are two sinusoidal waves, the excitation and the response, i_F and v. The transfer function, in its magnitude and angle as functions of ω_F, is a very useful device for analysis, but it is in a sense *unreal*. So we present Fig. 3.14-G from the real world. Its three photographs show the actual (time-domain) excitation and response waves in the sinusoidal steady state for a parallel RLC circuit of reasonably high Q (about 20), at the three frequencies marked in Fig. 3.14-E. The excitation *amplitude* is the same for all

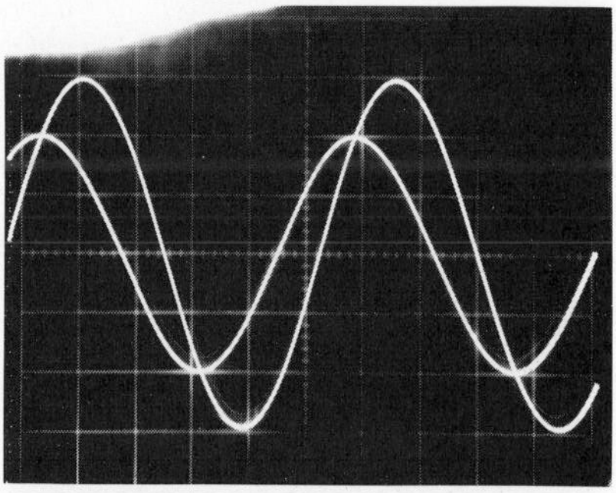

(a)

Frequency = 17.7 kHz
($\omega_F = \omega_a$ in Fig. 3.14-E)

Response leads excitation by 45°, has 0.7 amplitude [referred to (b) below]

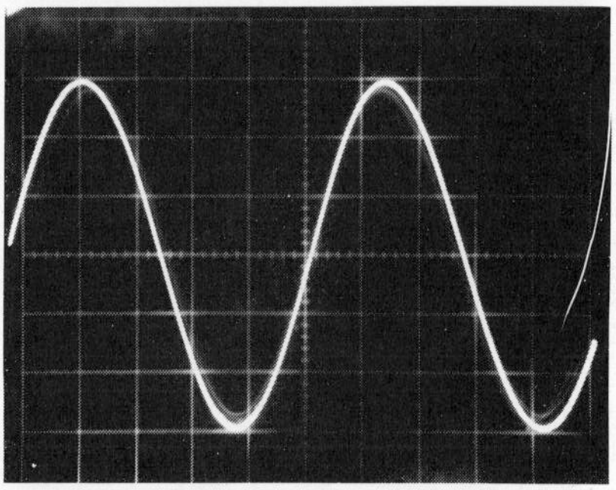

(b)

Frequency = 18.2 kHz
($\omega_F = \omega_0$ in Fig. 3.14-E)

Response is in phase with excitation, at resonant (maximum) amplitude; waves practically coincide

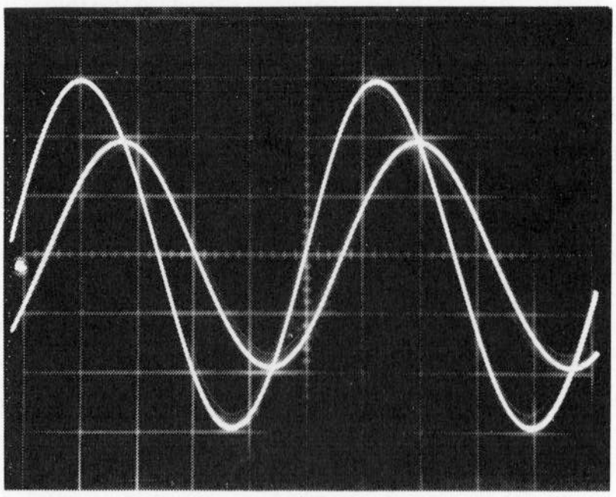

(c)

Frequency = 18.6 kHz
($\omega_F = \omega_b$ in Fig. 3.14-E)

Response lags excitation by 45°, has 0.7 amplitude [referred to (b) above]

FIG. 3.14-G.

three photographs; the excitation *frequency* is varied from one to another. The time scale is 10 μs per division for all three; and the amplitude scales are the same in all three, the gains being adjusted, once and for all, so that response and excitation waves coincide at the resonant condition, part (*b*). This is a real-world presentation of what Fig. 3.14-E shows in the unreal but very useful transfer function world. To retain perspective and understanding of the relations between these two worlds (of t, of ω_F) is essential for the engineer.

We observe that variation of the excitation frequency does actually produce in the physical response the magnitude and phase differences predicted by the transfer function.

Now we naturally should like to have direct verification (by photograph of actual physical behavior) of the resonance curve of Fig. 3.14-E (and Fig. 3.14-F). This curve represents the results of a series of measurements made by applying sinusoidal excitation at some particular frequency, waiting for the steady state to set in (the transient to decay), measuring the amplitude of the sinusoidal (forced, steady-state) response, plotting a point for that value of ω_F, and then changing the frequency and repeating the process until we have enough points to draw in a smooth curve connecting those points. We can hardly expect, therefore, to be able simply to display such a curve on a cathode-ray tube and photograph it. Yet it *is* possible to do just that, in effect. (See Appendix A.) The display is slowly painted on the cathode-ray tube, as the measurements are made, by a series of vertical lines of heights proportional to the steady-state response amplitudes for a fixed amplitude of excitation for various frequencies. These are held by the (storage) cathode-ray tube, and the photograph is made when the process is finished. The painting is done slowly enough (say 5 or 10 s for one sweep) that one can "follow" the process visually. Here we can only show the result, photographed when done, as in Fig. 3.14-H. One must mentally suppress the bars and look only at the curve outlined by their tips to obtain the likeness of Fig. 3.14-E; the result is essentially $|T|$ versus ω_F. (But the bars themselves also help to visualize the curve.) By this device we in effect *can* photograph the resonance curve of Fig. 3.14-E. Figure 3.14-H shows it (*a*) for a reasonably high value of Q (between 15 and 20), (*b*) for a lower value of Q (about 10), and (*c*) for a very low value, $Q = 5$. The circuit is unchanged in the three cases, except for altering the value of R. Note that once more our physical apparatus confirms the mathematical calculations made from the model: the sharpness of the resonance depends on the value of Q, as it should, and the height of the maximum upon the value of R; the resonance frequency is practically the same over this range of R.

We add Fig. 3.14-I, in which the inductor is removed so that the *RC* circuit has returned. It is interesting to consider this as a limiting case of the *RLC* circuit in which $Q = 0$ and the resonant frequency is zero.

The formulas and curves of this section, obtained mathematically from our model (3.01-1), describe the commonest, most widely used properties of the high-Q tuned circuit. But there are still other ways of describing resonance, and other properties which we cannot overlook.

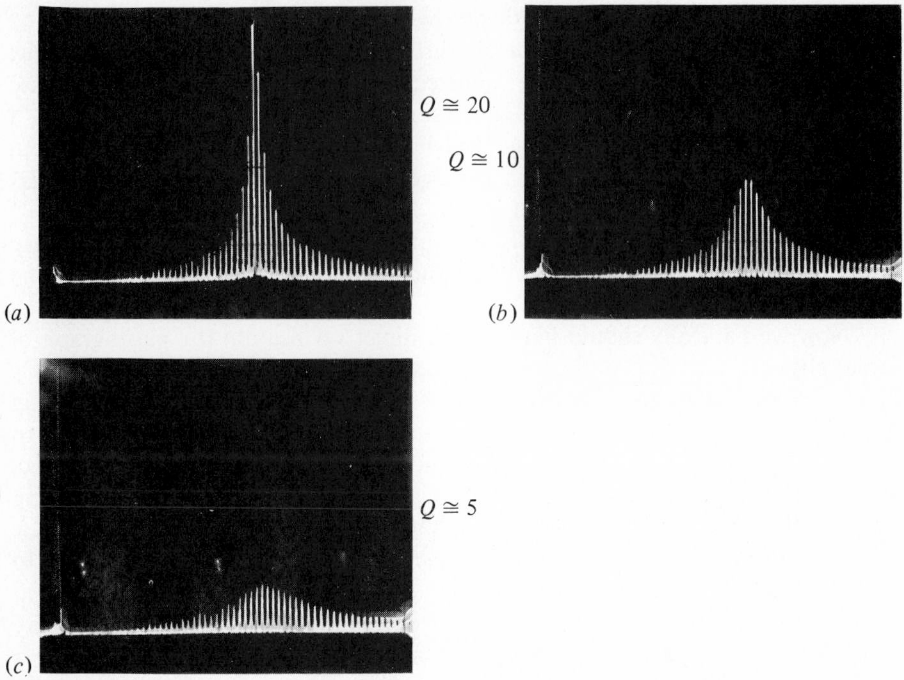

FIG. 3.14–H. $|T(j\omega_F)|$ versus ω_F (*RLC* tuned circuit) scales are the same for all three cases; frequency = 0 at left end, marked by (separated) bar; resonance frequency $\cong$20 kHz.

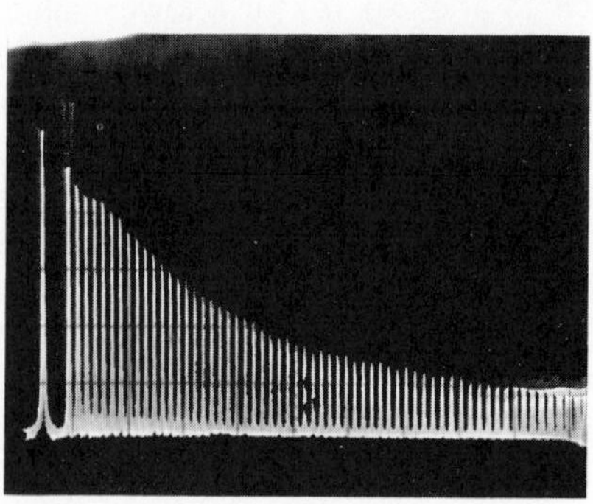

FIG. 3.14–I. $|T(j\omega_F)|$ versus ω_F for *RC* circuit ($Q = 0, f_0 = 0$); frequency scale is same as in Fig. 3.14–H, vertical scale is altered.

3.15 CURRENTS IN THE *RLC* CIRCUIT

The most striking feature of the high-Q resonance curve (Fig. 3.14-B) is the sharp peak concentrated at the resonant frequency. But there are two other significant regions: the low frequencies and the high frequencies, where the response is equally striking if totally different, very feeble as it is. It is very instructive to examine these three regions further, both mathematically and especially with more attention to physical action. And it is to the division of the excitation current between the three branches (R, L, and C) that we can most profitably turn.

Physical common sense tells us that the response at low frequencies should be small. For at the limiting value of *zero* frequency, the (idealized) inductor develops no voltage at all, since $v = L\,di/dt$. "At dc," to use a common and expressive bit of jargon, the inductor is a *short circuit*. Now passage of any direct current through the resistor would produce a nonzero voltage across both resistor and inductor, a ridiculous situation. And the capacitor cannot pass direct current. Hence *all* of i_F flows through the inductor when $\omega_F = 0$.

The more rapidly changing the current in an inductor, the greater the opposition voltage developed. In the sinusoidal steady state the current is perpetually changing, and at a rate proportional to the frequency. Hence the opposition increases as the frequency is raised. At very high frequencies the opposition is very large and the current is very small; "at infinity," to coin another bit of useful jargon, the inductor becomes an *open circuit*. (The inductor model is not valid at high frequencies, to be sure, but this concept is still useful as an approximate indicator of high-frequency behavior.)

We return now to the low end of the frequency spectrum. For nonzero but small values of ω_F, there can be no sudden dramatic change from the dc behavior: the bulk of the excitation current must still pass through the inductor (which offers only small opposition) and very little through R and C. And a simple calculation tells us approximately how much. If we neglect the current in R and in C, then

$$i_L = i_F = I_F \cos(\omega_F t + \psi), \tag{3.15-1}$$

and

$$v = L\frac{di_F}{dt} = -L\omega_F I_F \sin(\omega_F t + \psi). \tag{3.15-2}$$

We now proceed to lift ourselves by our bootstraps, to calculate the neglected i_R and i_C from (3.15-2). The results must be approximate, of course, but if they are indeed very much smaller than i_L, then the approximation will be justified.

We find

$$i_R = Gv = -GL\omega_F I_F \sin(\omega_F t + \psi) = \frac{1}{Q}\frac{\omega_F}{\omega_0} I_F \sin(\omega_F t + \psi), \quad (3.15\text{-}3)$$

and

$$i_C = C\frac{dv}{dt} = -CL\omega_F{}^2 I_F \cos(\omega_F t + \psi) = \left(\frac{\omega_F}{\omega_0}\right)^2 I_F \cos(\omega_F t + \psi). \quad (3.15\text{-}4)$$

The peak values are indeed small compared to that of i_F, provided we restrict ourselves to very small values of ω_F. Hence our physical intuition that at low frequencies the inductor controls the distribution of currents is correct. An alternative point of view, which we shall discuss later, is to say simply that the impedance of the inductor (Ls, with $s = j\omega_F$) is small enough compared to the impedances of the other branches [R, $1/(Cs)$] to ensure that virtually all of the entering current i_F pass through it. We note that the peak value of the response voltage at low frequencies is $L\omega_F I_F$, which is indeed small.

This sort of *asymptotic* calculation is useful both for providing information quickly and for checking the results of complicated, exact calculations. We should try to develop this skill and use it whenever we can. It will often point out absurd results (due to analysis error, for example, or to improper modelling) and help us catch errors in the bud.

To illustrate its use in checking, we develop the transfer function, our exact mechanism, in a Maclaurin series:

$$T(s) = \frac{1}{Cs + G + 1/(Ls)} = Ls(1 - GLs + \cdots). \quad (3.15\text{-}5)$$

When $s = j\omega_F$ and ω_F is small, this gives exactly what we wrote so glibly in (3.15-2). Careful calculation of the other two currents shows that the results above, (3.15-3) and (3.15-4), also correctly give the first significant terms of appropriate series expansions (Table 3.15-B, p. 167).

To summarize our low-frequency (asymptotic, as $\omega_F \to 0$) results, we construct Fig. 3.15-A*a*.

The behavior of the capacitor is dual to that of the inductor: instead of $v = L\,di/dt$, we have $i = C\,dv/dt$. Here the tables are turned: in the sinusoidal steady state the opposition voltage is *inversely* proportional to the frequency and becomes smaller and smaller as the frequency is raised. At infinity the capacitor is a *short circuit* and *all* of i_F flows through C. (At dc the capacitor is an open circuit, as we have already observed.)

For large values of ω_F the bulk of the excitation current flows through the

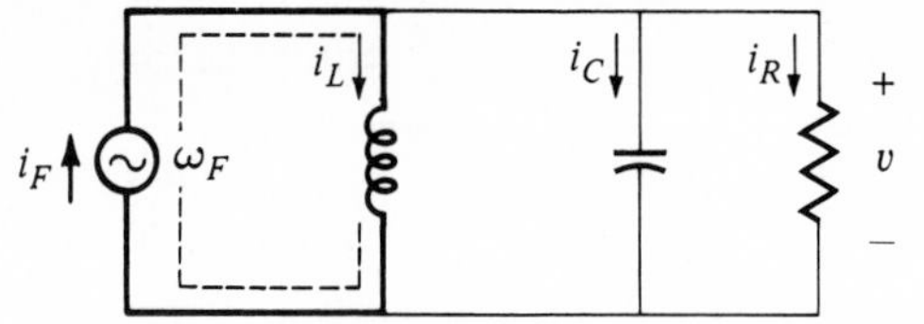

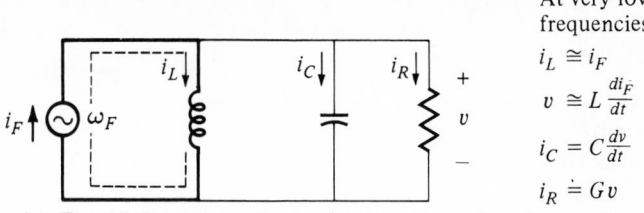

(*a*) The *RLC* circuit in the steady state at very low frequencies.

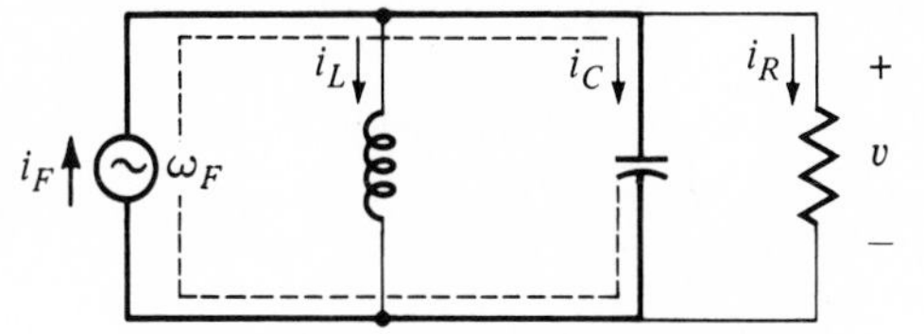

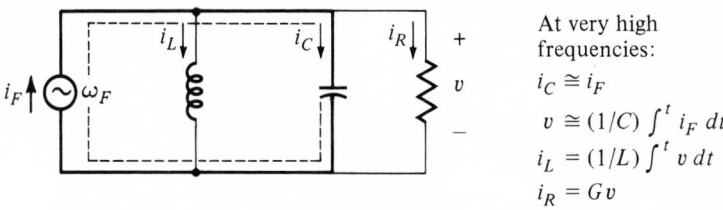

(*b*) The *RLC* circuit in the steady state at very high frequencies.

FIG. 3.15-A.

capacitor (which offers only small opposition) and very little through R and L. Approximate values of those two currents are easily calculated:

$$i_C = i_F = I_F \cos(\omega_F t + \psi), \tag{3.15-6}$$

$$v = \frac{1}{C} \int^t i_F \, dt = \frac{I_F}{C\omega_F} \sin(\omega_F t + \psi), \tag{3.15-7}$$

$$i_R = GV = \frac{G}{C\omega_F} I_F \sin(\omega_F t + \psi), \tag{3.15-8}$$

$$i_L = \frac{1}{L} \int^t v \, dt = -\frac{I_F}{LC\omega_F^{\,2}} \cos(\omega_F t + \psi). \tag{3.15-9}$$

We verify that i_R and i_L are indeed negligible at high frequencies. The (valid) intuitive point of view is that at high frequencies the capacitor controls the distribution of current. Alternatively, the impedance of the capacitor $[1/(Cs)$ with $s = j\omega_F]$ is comparatively small at high frequencies. Finally, we note that the peak value of the response voltage at high frequencies is $I_F/(C\omega_F)$, which *is* small.

Asymptotic high-frequency calculation from the transfer function requires developing $T(s)$ in a power series in $1/s$ instead of s:

$$T(s) = \frac{1}{Cs + G + 1/(Ls)} = \frac{1}{Cs}\left(1 - \frac{G}{C}\frac{1}{s} + \cdots\right). \tag{3.15-10}$$

Our physical calculation above, (3.15-7), can be used to check the first term here. And exact calculations with series in $1/s$ also confirm the results above for i_R and i_L (Table 3.15-B). Figure 3.15-A*b* summarizes graphically the high-frequency situation; it should be carefully compared with Fig. 3.15-A*a*.

Let us look now at the steady-state distribution of currents in the tuned circuit, at an *arbitrary* frequency ω_F, and without approximation. From (3.11-8) we have, in the steady state:

$$v(t) = v_F(t) = V_F \cos(\omega_F t + \phi) = V_F \cos A \tag{3.15-11}$$

in which we have written, for brevity and clarity,

$$A = \omega_F t + \phi = \omega_F t + (\psi - \theta). \tag{3.15-12}$$

And of course (see Fig. 3.11-A)

$$\begin{aligned} V_F &= I_F |\bar{T}(j\omega_F)| = \frac{I_F}{Y} = \frac{I_F}{|\bar{Y}(j\omega_F)|} \\ &= \frac{I_F}{|G + jB|} = \frac{I_F}{|G + j[C\omega_F - 1/(L\omega_F)]|}. \end{aligned} \tag{3.15-13}$$

We now redraw the circuit as in Fig. 3.15-B, placing the energy-*storage* ("tank") part of the circuit, the LC combination, somewhat apart from the energy *sink*, the resistor.

Now all currents are sinusoidal, of frequency ω_F, in form $\pm I_x \begin{smallmatrix}\sin\\ \cos\end{smallmatrix} A$. The appropriate subscript x is attached to the peak value I in each case in (3.15-14):

$$\begin{aligned} i_C &= C\frac{dv}{dt} = -C\omega_F V_F \sin A = -\frac{C\omega_F I_F}{Y}\sin A = -I_C \sin A, \\ i_L &= \frac{1}{L}\int^t v\,dx = \frac{V_F}{L\omega_F}\sin A = \frac{I_F}{L\omega_F Y}\sin A = I_L \sin A, \\ i_R &= Gv = GV_F \cos A = \frac{GI_F}{Y}\cos A = I_R \cos A, \end{aligned} \tag{3.15-14}$$

$$i_{LC} = i_L + i_C = (I_L - I_C)\sin A = \frac{I_F}{Y}(-B)\sin A = \frac{I_F}{Y}\left(-C\omega_F + \frac{1}{L\omega_F}\right)\sin A.$$

As a check, we compute the sum

$$\begin{aligned} i_R + i_{LC} &= \frac{I_F}{Y}(G\cos A - B\sin A) = I_F \cos(A + \theta) \\ &= I_F \cos(\omega_F t + \psi) = i_F(t), \end{aligned} \tag{3.15-15}$$

which checks very nicely.

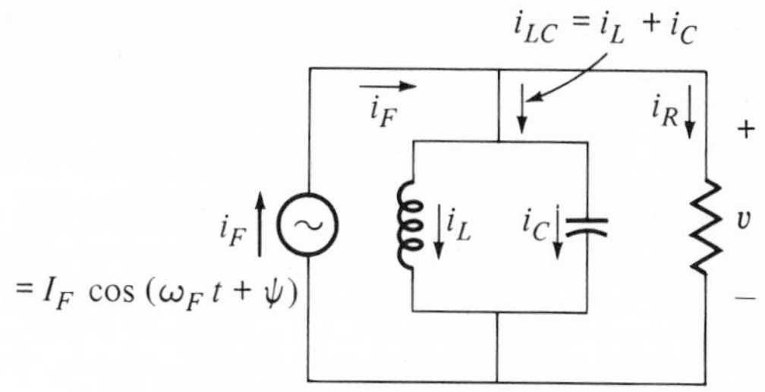

FIG. 3.15-B.

But the formulas for the various currents are not as lucid as are diagrams of the corresponding phasors. The purpose of *phasor diagrams* is to show graphically the phase relations and the magnitude relations between the various steady-state currents and voltage(s) by taking advantage of the simplicity and cogency of using complex numbers to represent sinusoids. Since each phasor is a (constant) complex number (see Sec. 2.15), it is very easily drawn.

We should of course place voltage and current phasors on different drawings, for their dimensions are not the same. Table 3.15-A gives the notation for the

TABLE 3.15-A

Sinusoidal Time Function	Phasor Representations: Original	Phasor Representations: Shifted
$i_F(t) = I_F \cos(A + \psi)$	$\bar{I}_F = I_F\underline{/\psi} = I_F\underline{/\phi + \theta}$	$I_F\underline{/\theta}$
$v(t) = v_F(t) = V_F \cos A$	$\bar{V}_F = V_F\underline{/\phi}$	$V_F\underline{/0}$
$i_C(t) = -I_C \sin A$	$\bar{I}_C = I_C\underline{/\phi + 90^\circ}$	$I_C\underline{/90^\circ}$
$i_L(t) = I_L \sin A$	$\bar{I}_L = I_L\underline{/\phi - 90^\circ}$	$I_L\underline{/-90^\circ}$
$i_R(t) = I_R \cos A$	$\bar{I}_R = I_R\underline{/\phi}$	$I_R\underline{/0}$

phasors of interest here. Since the angle ϕ appears in all of them, we can simplify the diagram somewhat by rotating it through an angle $-\phi$: this has the effect of adding $-\phi$ to all phasor angles. Or, since in the steady state the initial phase is of no significance (because the $t = 0$ instant is in the very distant past), we can simply take the $\bar{V}_F$ phasor (or any other one, for that matter) as reference for all angles, draw it horizontally, and call its angle zero. We then measure the angles of the others therefrom. And so we shall do from now on, as in Table 3.15-A and Fig. 3.15-C.

We note first that the current in the resistor is "in phase" with the voltage, and that the energy-storing-element currents are "90° out of phase"; that in the capacitor leads, that in the inductor lags. The angular position of the excitation-

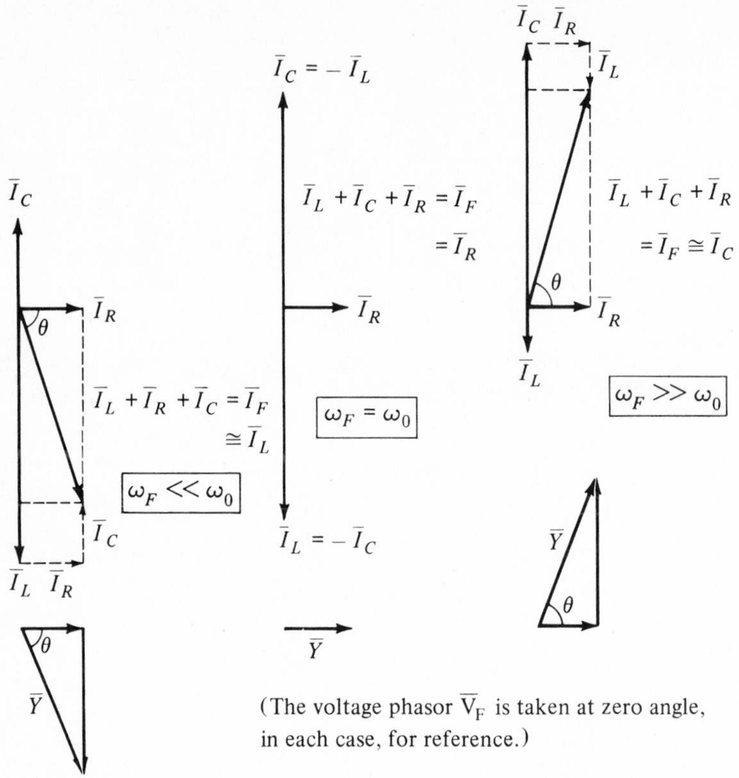

FIG. 3.15-C.

current phasor $\bar{I}_F$ varies, because we have chosen $\bar{V}_F$ as angular reference. One might prefer to take $\bar{I}_F$ at angle zero, since it is the (fixed) excitation, but the usage of Fig. 3.15-C is common and perhaps clearer.

Secondly we note that the character of these diagrams depends greatly on the frequency of excitation. In Fig. 3.15-C there are three such diagrams, drawn for an excitation frequency (*a*) low compared to the resonant frequency, (*b*) equal to the resonant frequency, (*c*) large compared thereto.

The mathematical variations of the three currents are summarized in Table 3.15-B, derived from (3.15-14). In the low-frequency and high-frequency asymptotic formulas only the first terms of the series are kept: the dots represent terms that are asymptotically negligible. We note that I_C is small at low frequencies, very large at resonance, and of course approaches I_F at high frequencies; that I_R is small at low and at high frequencies and equal to I_F at resonance; and that I_L is approximately equal to I_F at low frequencies, very large at resonance, and small at high frequencies. All of this agrees with Fig. 3.15-C; Fig. 3.15-D shows curves of these three current amplitudes versus frequency.

These mathematical manipulations are interesting and useful, but it is very important to understand the physical action at these three frequencies. An

TABLE 3.15-B Peak Values of (Steady-State) Currents in the Three Elements of the *RLC* Circuit

Element	Complete Formula	Low-Frequency Asymptotic Formula	Value at Resonance	High-Frequency Asymptotic Formula
Capacitor	$I_C = \dfrac{C\omega_F I_F}{\sqrt{G^2 + [C\omega_F - 1/(L\omega_F)]^2}}$	$I_F \dfrac{\omega_F{}^2}{\omega_0{}^2}(1 + \cdots)$	$I_F Q$	$I_F(1 + \cdots)$
Resistor	$I_R = \dfrac{GI_F}{\sqrt{G^2 + [C\omega_F - 1/(L\omega_F)]^2}}$	$I_F LG\omega_F(1 + \cdots)$	I_F	$I_F \dfrac{G}{C\omega_F}(1 + \cdots)$
Inductor	$I_L = \dfrac{[1/(L\omega_F)]I_F}{\sqrt{G^2 + [C\omega_F - 1/(L\omega_F)]^2}}$	$I_F(1 + \cdots)$	$I_F Q$	$I_F \dfrac{\omega_0{}^2}{\omega_F{}^2}(1 + \cdots)$

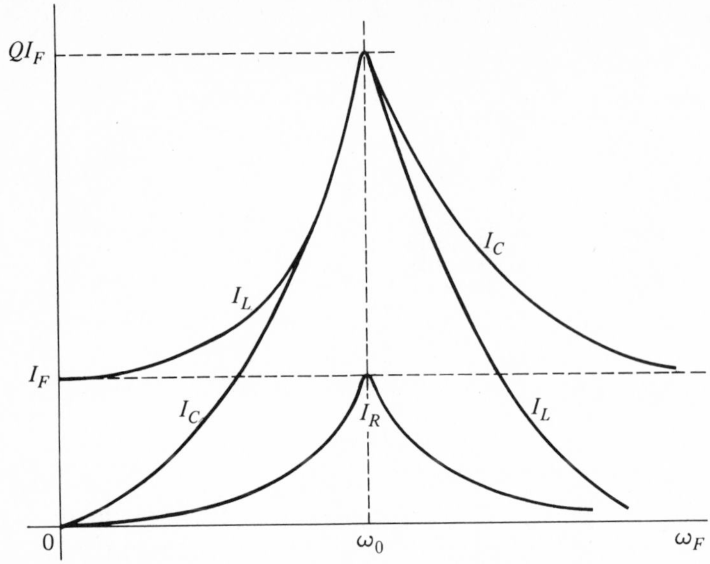

FIG. 3.15-D. Steady-state currents in the RLC circuit: peak amplitudes versus frequency ω_F (not to scale).

ability to "see what happens" by inspection of a circuit schematic diagram, even if only approximately, is very useful: it provides useful information quickly and cheaply; and it gives very useful checks. The excitation current is faced with a choice: it sees three paths and must decide how to divide itself between them. At very low and at very high frequencies the choice is simple, as we have seen. And at resonance, this is also true. For i_{LC} can be written [see (3.15-14)]

$$\begin{aligned} i_{LC} &= \frac{I_F}{Y}\left(-C\omega_F + \frac{1}{L\omega_F}\right)\sin A \\ &= I_F \frac{C}{\omega_F}\frac{\omega_0{}^2 - \omega_F{}^2}{\sqrt{G^2+B^2}}\sin A. \end{aligned} \tag{3.15-16}$$

When $\omega_F = \omega_0$, i_{LC} is clearly zero! The reason is that at resonance i_L and i_C become equal except for sign, and add to zero. Neither one is zero; in fact each is in peak value quite large, QI_F.

The steady-state resonance situation is this (Fig. 3.15-E): all the excitation current flows through the resistor, and the (idealized) L and C combination draws no current whatever from the source. But within the LC combination circulates a *very large* sinusoidal current of peak value QI_F. The L and the C elements are attuned to each other at this frequency: $L\omega_F = 1/(C\omega_F)$ and $B = 0$. We could take this as a good definition of resonance, in fact. The peak voltage is $RI_F = Q\sqrt{L/C}\,I_F$, which is also very large (as expected), for any reasonably high value of Q and common L/C ratio.

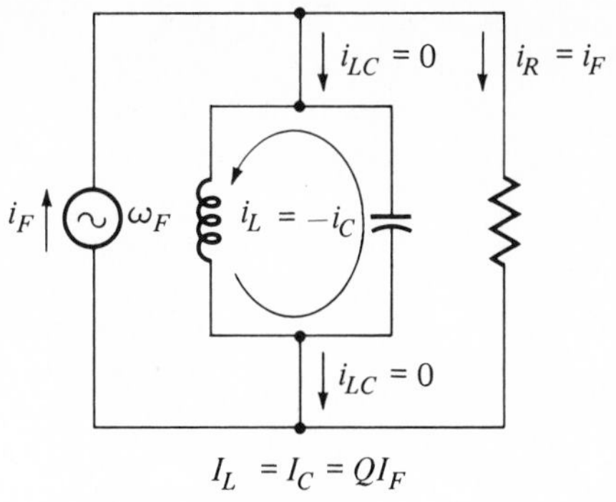

FIG. 3.15-E. The *RLC* circuit in the steady state at resonance $\omega_F = \omega_0$.

The inductor is idealized here, remember; its model dissipates no energy (has no reşistance), a physical impossibility. But this model makes the parallel *RLC* circuit easy to understand, and we shall later see (*mirabile dictu!*) that adding a small resistor to account for dissipation in the inductor makes very little difference in the properties of the resonant circuit.

The three extreme situations shown in Figs. 3.15-A and 3.15-E (low frequencies, resonance, high frequencies) characterize the important and useful properties of resonance. Also of interest are the corresponding energy situations.

3.16 ENERGY IN THE *RLC* CIRCUIT

In the transient state, energy flows from the (sinusoidal) source into the three passive elements of the circuit in a rather complicated fashion. Once the steady state is established, however, the flow of energy is simpler, particularly at resonance.

Let us first calculate the energy stored instantaneously in the L and C elements. Starting from (3.15-11), (3.15-13), and (3.15-14), we find

$$\begin{aligned} W_L &= \tfrac{1}{2}Li_L^{\,2} = \tfrac{1}{2}LI_L^{\,2}\sin^2 A = \frac{1}{2}\frac{I_F^{\,2}}{Y^2}\frac{1}{L\omega_F^{\,2}}\frac{1-\cos 2A}{2} \\ &= \tfrac{1}{4}I_F^{\,2}\frac{\sqrt{L/C}/\omega_F}{1/Q^2 + (\omega_F/\omega_0 - \omega_0/\omega_F)^2}(1-\cos 2A)\frac{\omega_0}{\omega_F} \end{aligned} \tag{3.16-1}$$

and

$$\begin{aligned} W_C &= \tfrac{1}{2}Cv^2 = \tfrac{1}{2}CV_F^{\,2}\cos^2 A = \frac{1}{2}\frac{I_F^{\,2}}{Y^2}C\frac{1+\cos 2A}{2} \\ &= \tfrac{1}{4}I_F^{\,2}\frac{\sqrt{L/C}/\omega_F}{1/Q^2 + (\omega_F/\omega_0 - \omega_0/\omega_F)^2}(1+\cos 2A)\frac{\omega_F}{\omega_0}, \end{aligned} \tag{3.16-2}$$

in which $A = \omega_F t + \phi$, as in Sec. 3.15. In each of these elements the stored energy oscillates, at double frequency, between zero and a peak value that depends on the excitation frequency as well as the size of the excitation. Curves for the two stored energies are similar, but of different peak values, and shifted in phase (Fig. 3.16-A*a*).

Especially interesting is the sum, the total energy stored instantaneously in the *LC* "tank." It is

$$W_{LC} = \frac{1}{4}\frac{I_F^2}{Y^2}\left[\left(C + \frac{1}{L\omega_F^2}\right) + \left(C - \frac{1}{L\omega_F^2}\right)\cos 2A\right]$$

$$= \tfrac{1}{4}I_F^2 \frac{\sqrt{L/C}/\omega_F}{1/Q^2 + (\omega_F/\omega_0 - \omega_0/\omega_F)^2}\left[\left(\frac{\omega_F}{\omega_0} + \frac{\omega_0}{\omega_F}\right) + \left(\frac{\omega_F}{\omega_0} - \frac{\omega_0}{\omega_F}\right)\cos 2A\right]. \qquad (3.16\text{-}3)$$

This also oscillates at double frequency (Fig. 3.16-A*b*), but the fluctuation is smaller, and the total *LC* stored energy is nearly constant. And at resonance it *is* constant, for then the W_L and W_C curves have the same peak values and add up to the constant

$$W_{LC} = \tfrac{1}{2}L(QI_F)^2 = \tfrac{1}{2}C(RI_F)^2$$
$$= \tfrac{1}{2}LI_L^2 = \tfrac{1}{2}CV_F^2.$$

The outstanding characteristic of the energy situation at resonance is exactly what Fig. 3.15-E predicts: the *LC* "tank" is in effect isolated. A large amount of energy oscillates therein, back and forth between inductor and capacitor. The resonant circuit could be used to accumulate this large amount of energy during the transient (charging) state; then, after switching to change the circuit,

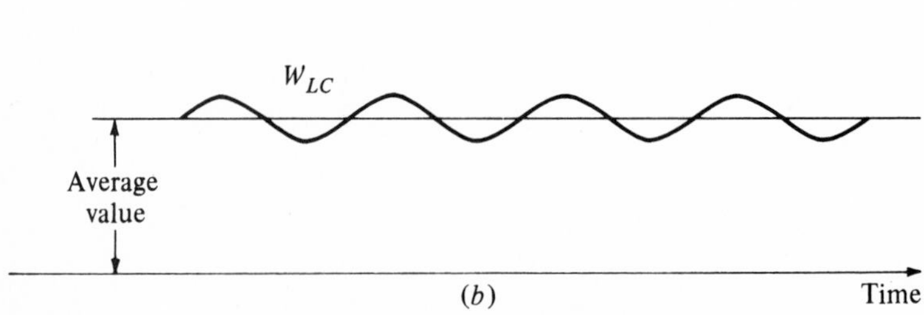

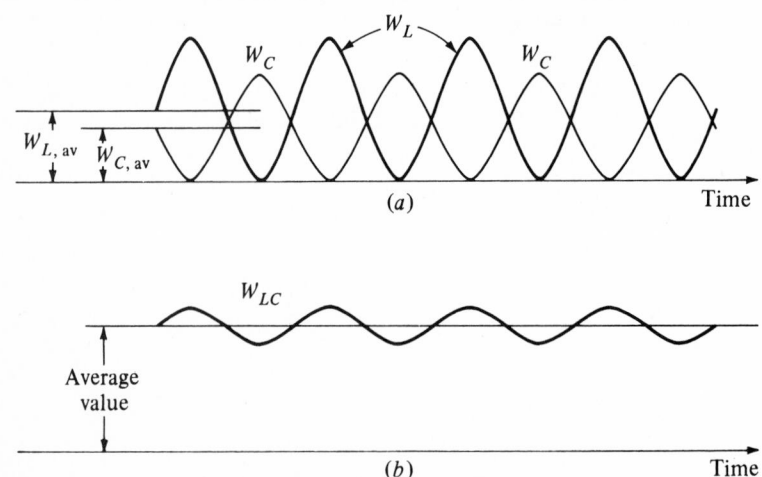

FIG. 3.16-A. Drawn for $\omega_F < \omega_0$.

the energy could be discharged to do work elsewhere. The analogy with a flywheel is apt.

Externally one is conscious at resonance only of the flow of energy into the resistor at the rate

$$\begin{aligned} vi_R &= Ri_F{}^2 = RI_F{}^2 \cos^2 (\omega_0 t + \psi) \\ &= \frac{RI_F{}^2}{2} [1 + \cos 2(\omega_0 t + \psi)] \end{aligned} \tag{3.16-4}$$

exactly as though the current source and the resistor were alone in the circuit. This energy goes of course into heat (or whatever form of energy sink the resistor models) and leaves the circuit at the average rate (power) $RI_F{}^2/2$ (watts).

Off resonance, when $\omega_F \neq \omega_0$, the energy flow into the resistor is of course reduced sharply. It is

$$vi_R = \frac{GI_F{}^2}{G^2 + B^2} \cos^2 A = \frac{\frac{1}{2}RI_F{}^2(1 + \cos 2A)}{1 + Q^2(\omega_F/\omega_0 - \omega_0/\omega_F)^2}. \tag{3.16-5}$$

Its average value falls off rapidly as the excitation frequency departs from the resonance frequency (Fig. 3.16-B).

The energy flow into the L and C elements is

$$\frac{-BI_F{}^2}{G^2 + B^2} \sin A \cos A = -\frac{1}{2} \frac{Q^2(\omega_F/\omega_0 - \omega_F/\omega_0)L\omega_0 I_F{}^2 \sin 2A}{1 + Q^2(\omega_F/\omega_0 - \omega_0/\omega_F)^2} \tag{3.16-6}$$

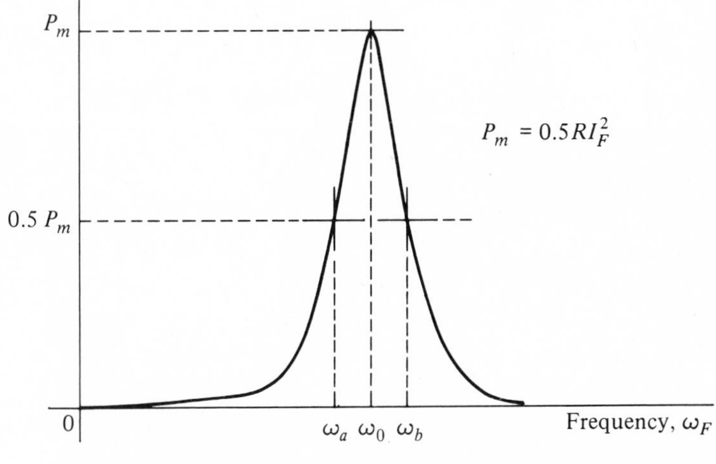

FIG. 3.16-B. Average power (rate of energy flow) into R.

which may be conveniently checked from (3.16-3). Its average value is of course zero; the energy stored in the LC combination is on the average constant in the steady state.

The total energy flow rate, from the source into the RLC combination, is

$$\begin{aligned}\text{power from source} &= \text{power into } R + \text{power into } L \text{ and } C\\ &= \frac{I_F^2}{G^2+B^2}\,(G\cos^2 A \quad - B\sin A\cos A)\\ &= \frac{I_F^2}{Y}\,(\cos\theta\cos^2 A - \sin\theta\sin A\cos A)\\ &= \frac{1}{2}\frac{I_F^2}{Y}\,[\cos\theta + \cos(2A+\theta)]\end{aligned} \tag{3.16-7}$$

of which the average value, $\frac{1}{2}I_F^2\cos\theta/Y$ (watts), is of course the same as that given by (3.16-5):

$$\frac{1}{2}\frac{I_F^2}{Y}\cos\theta = \tfrac{1}{2}I_F^2 R\,\frac{1}{1+Q^2(\omega_F/\omega_0-\omega_0/\omega_F)^2}. \tag{3.16-8}$$

This is the average value of the power delivered to the resistor in the steady state. A complex-number representation of the total power (3.16-7) can be convenient, as well as the use of the power factor ($\cos\theta$) and of rms values instead of peak values (Probs. 3-Z and 5-N).

To recapitulate the energy situation in the sinusoidal steady state: *At resonance* the flow of energy from the source is entirely into the resistor, at double frequency, with average value $\frac{1}{2}RI_F^2$ (watts); a large amount of energy is present in the tank elements, and remains therein, oscillating back and forth between the inductor and the capacitor. Off resonance the energy flow to the resistor drops sharply, and there is a flow between source and LC elements whose size depends on frequency.

The curve of Fig. 3.16-B is a resonance curve much like that of Figs. 3.14-A or 3.14-E. But this one is proportional to the *square* of those, since it represents *power*. The frequencies ω_a, ω_b of Fig. 3.14-E are therefore those at which the average power delivered to the resistor is *half* of its peak value. From this comes the name *half-power frequency*, often used for the 70 percent points of the resonance curve of Fig. 3.14-E.

The ubiquitous parameter Q, which measures the fractional distance between half-power frequencies by $Q = \omega_0/(\omega_b - \omega_a)$, has also a simple meaning in energy terms. The energy delivered to the resistor during one cycle of the excitation current is at resonance

$$\tfrac{1}{2}RI_F^2\,\frac{2\pi}{\omega_0} \quad \text{joules.} \tag{3.16-9}$$

The (total) energy stored in the tank (L, C) elements at resonance is $\frac{1}{2}CR^2I_F{}^2$. Hence the ratio of the energy stored in the tank to the energy lost in the resistor per cycle is simply $RC\omega_0/(2\pi)$, or $Q/(2\pi)$, by (3.06-12). Explicitly,

$$Q = 2\pi \frac{\text{energy stored in } L \text{ and } C}{\text{energy lost in } R \text{ per cycle}}, \tag{3.16-10}$$

the calculations being made in the sinusoidal steady state, at the resonance frequency. We found, in (3.07-11), that Q could be interpreted in exactly the same way in the transient (natural-behavior) state, provided that Q had a very large value. Under those conditions, the circuit behaves very much like the sinusoidal steady state presently under consideration, so it should be no surprise that (3.07-11) and (3.16-10) are nearly identical. Q is remarkable in its versatility!

3.17 THE COMPLEX-FREQUENCY PLANE: GEOMETRY VERSUS ALGEBRA

We have become rather well acquainted with the *RLC* circuit. We have in fact analyzed it in several different ways, and have thus acquired several different tools which will be useful later on. But there are others that we have not yet used. And it is efficient and pleasant and easy to apply some of them here, to learn about them in terms of a familiar example. Then they will be sharp and ready.

Such a tool is the *transfer function*, whose abilities we have not even begun to exploit. We recall its definition as the ratio of the forced component of the response to the excitation, the latter being exponential, e^{st}, in form. Since s is not fixed but is arbitrary, we look on s as a variable, and on that ratio as a *function* of s, $T(s)$. We recall that for sinusoidal excitation it is $T(j\omega_F)$ that multiplies the excitation phasor to give the response phasor. Let us examine this evaluation with $s = j\omega_F$ more carefully. We have

$$T(s) = \frac{1}{Cs + G + 1/(Ls)} = \frac{(1/C)s}{s^2 + 2\alpha s + \omega_0{}^2}. \tag{3.17-1}$$

We note again that it is a rational function of s (the ratio of two polynomials in s) with real coefficients, a fact traceable to our basic postulates. We note also that it is easy to write $T(s)$ in factored form. For we already know the factors of the denominator from our calculation of the natural frequencies [see (3.04-2) and (3.04-3)]. The numerator *is* factored. Hence we have

$$T(s) = \frac{1}{C} \frac{s - 0}{(s - p_1)(s - p_2)}. \tag{3.17-2}$$

The values of s that make $T(s)$ vanish are called its *zeros;* the values of s that make T infinite are called its *poles.*

From (3.17-2) we see that the poles of this transfer function are two in number, and that they are the natural frequencies p_1 and p_2. There is a very close connection therefore between s and p. Both represent generalized, complex frequency: p for the natural component of response, and s for the forced component (when the excitation is e^{st}). So their general characters should be the same, their dimensions must agree, and we shall from now on plot them in the same plane. This we shall call the *complex-frequency plane.* In it we denote real parts σ and imaginary parts ω, so that

$$s = \sigma + j\omega \tag{3.17-3}$$

is the explicit form of the complex-frequency variable.

The natural frequencies represent two particular points thereon, at which $T(s) \to \infty$. And this is eminently reasonable. For a natural mode of behavior, Ke^{pt} in form, can exist without excitation: that is its nature! But such behavior is an exponential function of time, and we might look on it as the forced component of the response to an exponential excitation $I_F e^{st}$ *when* $s = p$ *and* I_F *is zero.* Then K (which is *not* zero) would equal

$$\left[T(s)I_F\right]_{\substack{s=p,\\ I_F=0}} \tag{3.17-4}$$

which could only occur if $T(p) \to \infty$. But this is exactly what happens to $T(s)$ as s approaches a natural frequency [see (3.17-2), and also (3.10-6) to (3.10-8), where this has already been said]. To recapitulate, somewhat colloquially: For certain special values of s the "forced" response can exist without excitation; these are the values of s that make T infinite, the poles of T. But response extant without excitation is natural behavior:

the poles of the transfer function are the natural frequencies. (3.17-5)

We represent poles by $\times$ (Fig. 3.17-A).

As for *zeros*, it is clear that $T(0) = 0$, so that the origin (the point $s = 0$) is a zero of $T(s)$. Physical common sense confirms this, for the inductor is a short circuit at dc (at $s = 0$). We represent zeros by $\circ$ (Fig. 3.17-A). Now there are *two* energy-storing elements in the circuit, the order of the network is 2, and $T(s)$ has *two* poles. It would be consistent, and elegant, if $T(s)$ had two zeros also. So it has, in fact, for the capacitor also becomes a short circuit, and makes $T(s)$ zero, but at high frequencies. The value of s that corresponds we cannot plot, no matter what scale we use, for it is indefinitely large. We can speak of it as a (hypothetical) point, however: "the point at infinity," or, simply, ∞. So we say that infinity is a zero of $T(s)$, and represent it by a $\circ$ on the margin of Fig. 3.17-A, a sort of reminder.

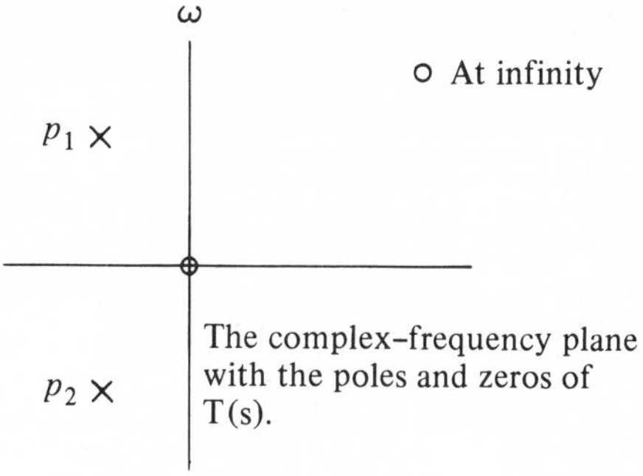

FIG. 3.17–A.

In general, equality between (a) the order of a network, defined by (3.03-13), (b) the number of zeros of $T(s)$, and (c) the number of poles of $T(s)$ commonly exists (as in this example, where all equal 2). There are exceptional cases where this may seem not to hold, but strictly speaking it always does, if we include the behavior at infinity.

In the complex-frequency plane we can plot the poles of $T(s)$ and its zeros. All of $T(s)$ is represented then, except a multiplier or scale factor, which is often of only secondary importance, being relative. The poles of $T(s)$ and the natural frequencies of the network are the same points in general. To find their values, a polynomial must be factored [a characteristic equation like (3.04-2) must be solved]. The zeros may be easier to calculate: frequently some are obvious by inspection (as here). The principal utility of the complex-frequency-plane diagram of $T(s)$ is that from it, once the zeros and poles are plotted, a wealth of useful information can be obtained.

The *natural behavior* and the natural part of a response are of course in form $\sum Ke^{pt}$, whose p are the × on the plane. Since the K have to do with initial conditions, one cannot expect to find their values from the complex-frequency plane. But the p, the natural frequencies, are the most important parameters the network has.

The phasor that represents the forced component of the response to sinusoidal excitation, on the other hand, *is* essentially visible. For it is proportional to the transfer function evaluated for $s = j\omega_F$, and that is

$$T(j\omega_F) = \frac{1}{C} \frac{j\omega_F - 0}{(j\omega_F - p_1)(j\omega_F - p_2)}. \tag{3.17-6}$$

The points corresponding to the natural frequencies p_1 and p_2 are labeled P_1 and P_2 in the complex-frequency plane of Fig. 3.17-B. The point that corresponds to $j\omega_F$ is P, shown for an arbitrary value of ω_F. The factors of the

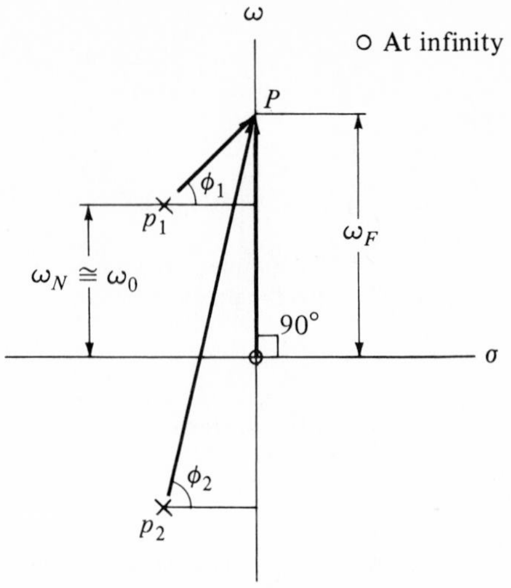

FIG. 3.17-B. (Not to scale.)

numerator and denominator of $T(j\omega_F)$ are the directed line segments (complex numbers) $\overline{OP}$, $\overline{P_1P}$, and $\overline{P_2P}$. Hence

$$\begin{aligned} T(j\omega_F) &= \frac{1}{C}\frac{\overline{OP}}{\overline{P_1P}\;\overline{P_2P}} \\ &= \frac{1}{C}\frac{|OP|}{|P_1P||P_2P|} \underline{/90^\circ - \phi_1 - \phi_2} \end{aligned} \tag{3.17-7}$$

so that the behavior both of magnitude (within a constant multiplier) and of angle of the response phasor can easily be visualized and in fact calculated from this diagram as ω_F varies. Note how magnitude and angle of line segments associated with zeros and of those associated with poles enter (in particular, the pole-associated angles are to be *subtracted*). The calculation of the magnitude requires measuring the three distances, and multiplication; the calculation of the angle requires measuring the three angles, and addition. Instruments have been developed for making these calculations on drawing boards, but they are of limited accuracy.

Far more important is the ease of *visualization* of the nature of the variation of the magnitude and of the angle of the response phasor, or, better, the transfer function, that the pole-zero diagram of $T(s)$ affords. At low frequencies the point P is near the origin, as in Fig. 3.17-C*a*: the angle of T is by inspection slightly less than 90° (note that here ϕ_1 is negative); the magnitude of T is very small; and the asymptotic values as $s \to 0$ are 90° and zero. At high frequencies, where the three distances (and the three angles) are approximately equal (see Fig. 3.17-C*b*, drawn to a different scale), the net angle is approximately $-90°$

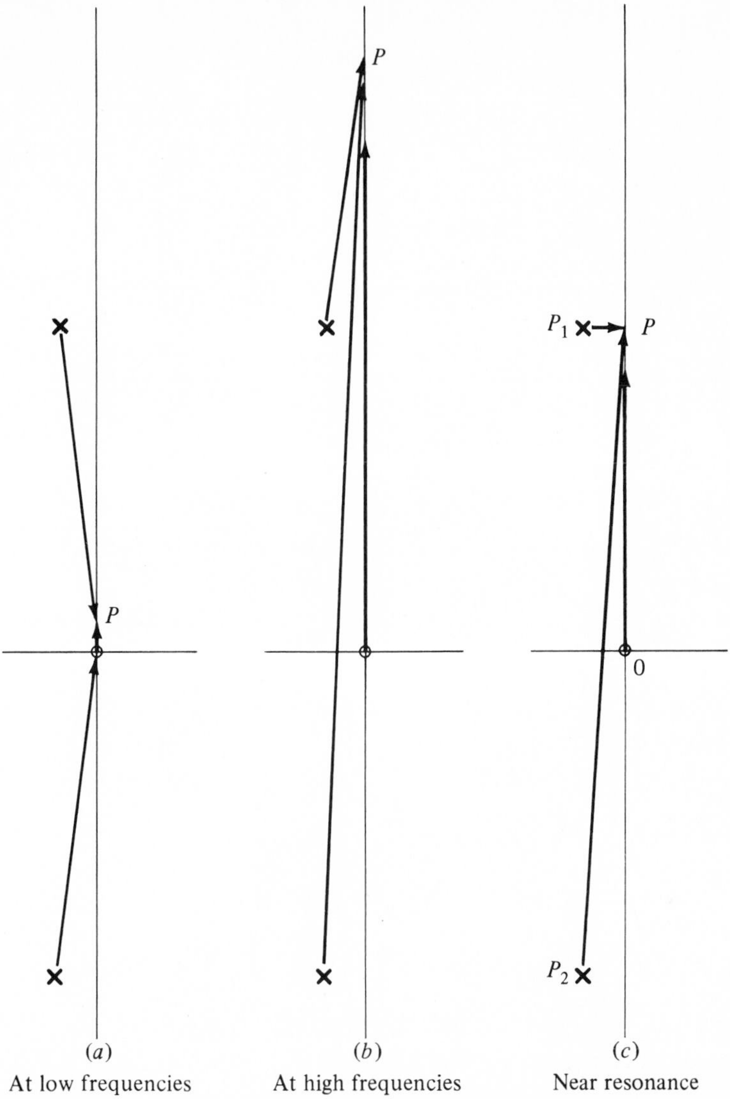

FIG. 3.17-C. (Drawn for a low value of Q, for clarity.)

and the magnitude is again very small. Near resonance (Fig. 3.17-C*c*) there appears at first glance to be no such simple indication of behavior, except that $|P_1 P|$ does go through a minimum. If that occurs more quickly than do the variations of $|P_2 P|$ and $|OP|$, which tend to neutralize each other anyway, then $|T|$ has a maximum thereabout, and the response curve "peaks."

A sharp-resonance situation, however, is different. For if Q is large and if Fig. 3.17-B is drawn to a scale reasonable for ω_N, then the point P_1 will appear to be on the ω axis, so small is α in comparison. If we expand the region near

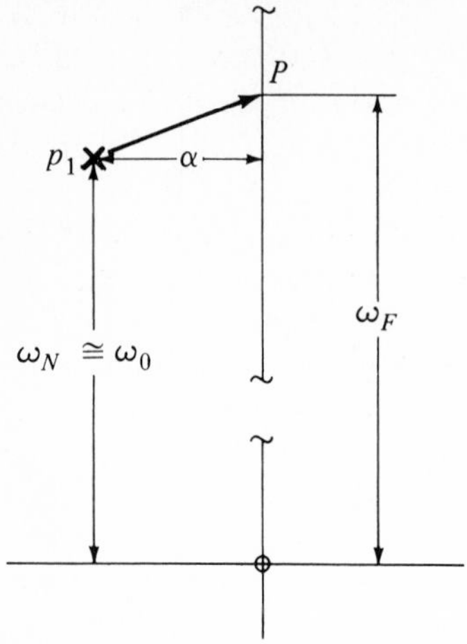

(a) The resonance region, expanded.

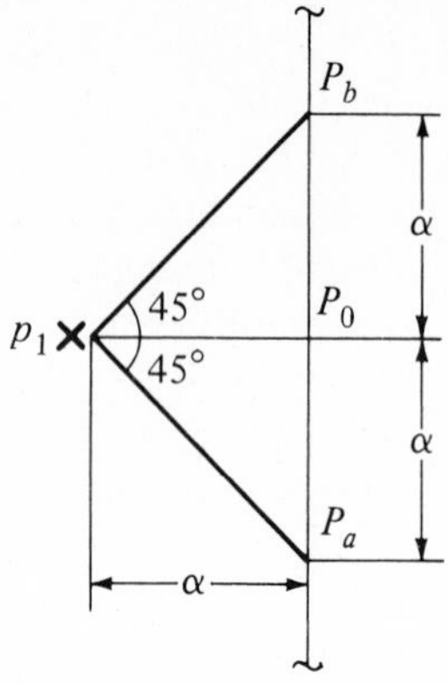

(b) The vicinity of p_1, greatly expanded.

FIG. 3.17-D.

the natural frequency P_1, as in Fig. 3.17-D*a*, then the origin and the point P_2 recede and lie well off the paper. For frequencies ω_F such that P is in the vicinity of P_1,

$$T(j\omega_F) = \left[\frac{1}{C}\frac{\overline{OP}}{\overline{P_2P}}\right]\frac{1}{\overline{P_1P}} = \frac{H}{\overline{P_1P}} \tag{3.17-8}$$

in which H is essentially a (real, positive) constant. The resonance effect is caused by the rapid variation of $\overline{P_1P}$ as P moves past P_1. A very sharp peak results, for, α being very small in comparison with ω_0, a large variation in $|P_1P|$ occurs in a very small range of ω_F. The maximum occurs for $\omega_F = \omega_N \cong \omega_0$. We can write, for P near P_1,

$$T(j\omega_F) = \frac{H}{\alpha + j(\omega_F - \omega_N)} = \frac{T_0}{1 + j(\omega_F - \omega_N)/\alpha} \tag{3.17-9}$$

in which

$$T_0 = \frac{H}{\alpha} = \text{maximum value of } |T| = \frac{1}{G} = R. \tag{3.17-10}$$

We recognize this (approximate, but for high Q very accurate) expression as the $(1 + jx)^{-1}$ approximate formula of (3.14-6).

A geometrically convenient characterization of the sharpness of the peak is the frequency separation of the points P_a and P_b, positions of P for which $\overline{P_1P}$ is at -45 and $+45°$ angles (Fig. 3.17-D*b*). This separation is 2α, and at both P_a and P_b, $|T| = T_0/\sqrt{2}$. We recognize that these are the 70 percent points of Fig. 3.14-E and the half-power points of Fig. 3.16-B, and that the reciprocal of the fractional bandwidth is of course $Q = \omega_0/(2\alpha)$. The pole-zero diagram in the complex-frequency plane thus gives us Fig. 3.14-E, and the universal resonance formulation of Sec. 3.14, *by mere inspection.*

It is this rapid insight that often makes the pole-zero diagram of a transfer function very useful indeed. The asymptotic behavior for P near the origin, and for P "near infinity," is immediately obvious, not only in this simple case, but in more complicated ones to come.

That a sharp resonance exists opposite a natural frequency which is very close to the ω axis is made crystal clear by the simple diagram of Fig. 3.17-D*b*. The "width" thereof is 2α; its "location" is ω_0.

With these pieces of information one can immediately sketch a curve of $|T|$ versus frequency; it will be only a sketch, but it may delineate the character of the steady-state response quite well enough. A sketch of the angle of $\bar{T}$ is easily added.

The salient characteristic of a high-Q circuit is that a natural frequency lies very close to the ω axis. There is a very large sinusoidal steady-state response when $\omega_F \cong \omega_0$. This is *not* excitation at the natural frequency, for s is not equal to p_1. The common expression "excitation at a natural frequency produces a large, resonant response" means $\omega_F \cong \omega_N$ and *not* $s = p_1$. Excitation *at* the natural frequency ($s = p_1$) would give a response that would die out. Resonant response comes from excitation by a permanent (not transient) wave whose s is *near* the natural frequency p_1: $s = j\omega_F \cong j\omega_0$.

True, if the natural frequency were *on* the ω axis, $p_1 = 0 + j\omega_0$, then the response for $\omega_F = \omega_0$ would be extraordinary! For then Q would be infinite and the response would be unlimited. In systems with high-Q but finite resonances, the response can be very large indeed, however, and perhaps useful on that account, perhaps troublesome.

In more complicated situations there may be a number of such resonances (well separated), to each of which the simple analysis [with appropriate evaluation of the constants in (3.17-9) for each] applies.

3.18 DISSIPATION IN THE INDUCTOR (AND CAPACITOR)

Resonance, that remarkable behavior of the high-Q circuit, can be very useful indeed. There is often an engineering need to separate a signal made up of sinusoidal components that are concentrated in a narrow-frequency band from other signals also present, but concentrated at different frequencies. It is interesting to speculate on how much radio broadcasting and wire-less point-to-

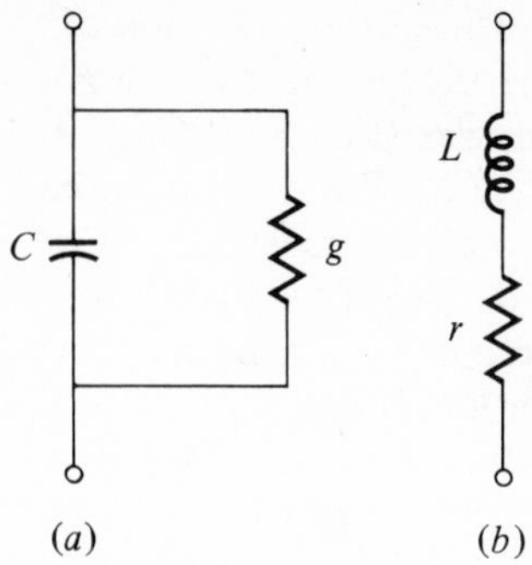

FIG. 3.18-A.

point communication would be possible without the tuned circuit ... and how different our civilization would be!

Our analysis has assumed that the inductor and the capacitor are non-dissipative elements. And this, of course, is not true. The capacitor dissipates some energy in its dielectric, and is more accurately represented by the two-element model of Fig. 3.18-A*a*. The inductor dissipates energy in its wire (and perhaps in its core); it also requires a two-element model (Fig. 3.18-A*b*) to take account of this. And it is extremely important to know how this realism affects the resonance and its selectivity (Q).

It is only reasonable now to discard the separate resistor (G in Fig. 3.01-A); the dissipation in the inductor, represented by r in Fig. 3.18-B, and that in the capacitor, represented by the conductance g, are almost certainly going to limit the value of Q enough, without assistance. (In those rare cases where one wishes deliberately to lower the Q, a separate resistor in parallel, taken account of in an increased value of g, will do the job.) We deal now then with the more accurate (in fact, very accurate) model of Fig. 3.18-B, in which the important question is: "To what extent must we now alter the results of our previous analysis, made with $r = 0$?"

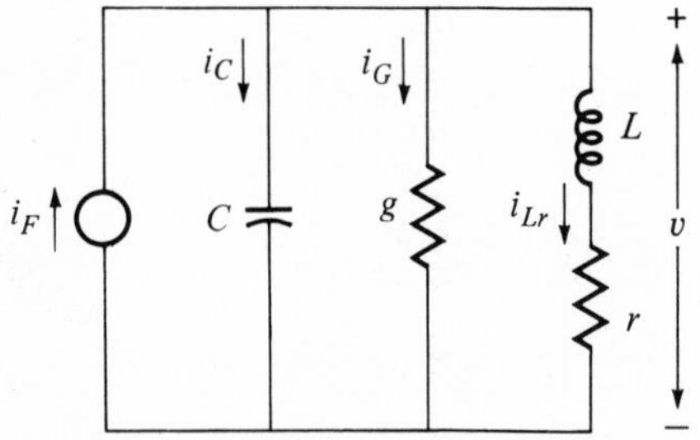

Initial conditions: $v(0) = V_0,\ i_{Lr}(0) = I_0$

FIG. 3.18-B.

The order of the circuit is still 2, and it is only reasonable to expect a similar differential equation. Summing currents as before [see (3.01-1)], we can write (Fig. 3.18-B)

$$\sum i = C\frac{dv}{dt} + gv + i_{Lr} - i_F(t) = 0, \qquad v(0) = V_0. \tag{3.18-1}$$

Here we are forced to write i_{Lr} as such; we cannot simply write $\int v\,dt$ as we did before. We need to get rid of i_{Lr}. More systematic and less clumsy methods of analysis will be discussed later; here we observe that

$$v = L\frac{di_{Lr}}{dt} + ri_{Lr}, \qquad i_{Lr}(0) = I_0. \tag{3.18-2}$$

Solving (3.18-1) for i_{Lr} and placing the result in (3.18-2) gives

$$L\left(\frac{di_F}{dt} - g\frac{dv}{dt} - C\frac{d^2v}{dt^2}\right) + r\left(i_F - gv - C\frac{dv}{dt}\right) - v = 0 \tag{3.18-3}$$

which is free of i_{Lr}. Rearranged in neater form it is

$$\frac{d^2v}{dt^2} + \left(\frac{g}{C} + \frac{r}{L}\right)\frac{dv}{dt} + \frac{1+rg}{LC}v - \frac{1}{LC}\left(L\frac{di_F}{dt} + ri_F\right) = 0, \qquad v(0) = V_0, \qquad i_{Lr}(0) = I_0. \tag{3.18-4}$$

The contrast with (3.01-1) is not as great as it may seem. If that is written as a differential equation rather than an integrodifferential equation, it is

$$\frac{d^2v}{dt^2} + \frac{G}{C}\frac{dv}{dt} + \frac{1}{LC}v - \frac{1}{C}\frac{di_F}{dt} = 0, \qquad v(0) = V_0, \qquad i_L(0) = I_0. \tag{3.18-5}$$

And (3.18-5) is quite close to (3.18-4): if we set $r = 0$ and $g = G$ in the latter, we do obtain the former, a good check. An additional check is to verify the dimensions of each term in (3.18-4); they should all be (voltage)(time)$^{-2}$, or, in units, volts per second per second.

We might laboriously proceed to find of this new model the impulse response, the step response, and so on. Since our present interest is in *resonance*, however, we move immediately to the sinusoidal steady-state response. We first calculate the transfer function. For an excitation $i_F = I_F e^{st}$, the forced component of the response should be $V_F e^{st}$. Assuming this, and substituting in (3.18-4), gives

$$\left[s^2 + \left(\frac{g}{C} + \frac{r}{L}\right)s + \frac{1 + rg}{LC}\right] V_F e^{st} = \frac{1}{LC}(Ls + r)I_F e^{st} \tag{3.18-6}$$

and

$$T(s) = \frac{V_F e^{st}}{I_F e^{st}} = \frac{(1/C)(s + r/L)}{s^2 + 2\alpha' s + \omega_0'^2} \tag{3.18-7}$$

in which

$$\begin{aligned} 2\alpha' &= \frac{g}{C} + \frac{r}{L}, \\ \omega_0'^2 &= \frac{1 + rg}{LC}. \end{aligned} \tag{3.18-8}$$

Since the phasor that represents the steady-state part of the response is the product of $T(j\omega_F)$ and the excitation phasor, we expect (3.18-7), when compared with the previous transfer function (3.17-1), to point out the effect of dissipation on resonance. The only differences that the comparison reveals are (*a*) addition of a constant term to the numerator and (*b*) changes in the coefficients in the denominator. We say that the dissipation is *small* if these differences have little effect on the resonance.

Probably the easiest way to see what "small" must therefore mean is to draw the pole-zero diagram of $T(s)$. "Small" must first mean that the new natural frequencies, the poles of $T(s)$, are *complex*, and hence that $\alpha' < \omega_0'$; then the natural frequencies are $-\alpha' \pm j\omega_N'$, $\omega_N'^2 = \omega_0'^2 - \alpha'^2$ (Fig. 3.18-C). Secondly it must mean that they lie very close to the ω axis: $\alpha' \ll \omega_N'$ (and so $\omega_N' \cong \omega_0'$). The symbol $\ll$ is of course imprecise; but the connotation is a ratio of say 0.05 or 0.02 or less, for we know from the analysis of Sec. 3.17 that a sharp resonance will then exist. The Q will in fact be

$$Q = \frac{\omega_N'}{2\alpha'} \cong \frac{\omega_0'}{2\alpha'} = \frac{\omega_0'}{(g/C + r/L)}. \tag{3.18-9}$$

So *small dissipation* comes down to: g and r are small enough that

$$\frac{g}{C\omega_0'} + \frac{r}{L\omega_0'} = \frac{1}{Q} \ll 1. \tag{3.18-10}$$

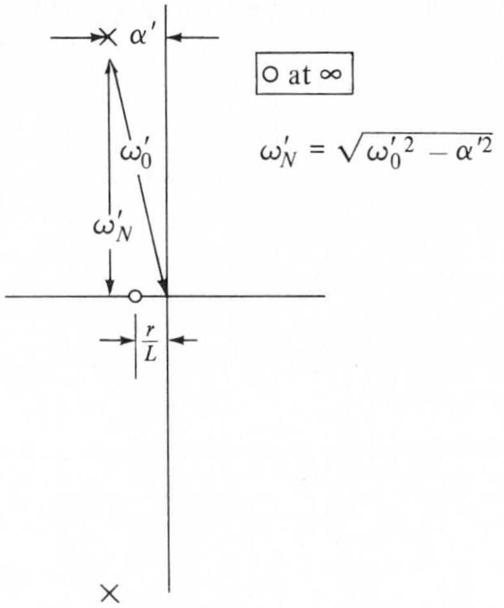

FIG. 3.18-C. (Not to scale.)

Then for all practical purposes there is no numerical difference between ω_N', ω_0', ω_0, and ω_N. The resonance still occurs at the resonance frequency $\omega_0 = 1/\sqrt{LC}$ rad/s, or $f_0 = 1/(2\pi\sqrt{LC})$ Hz. But the α to be used is now α', that is, $\frac{1}{2}(g/C + r/L)$ instead of $G/(2C)$. The dissipation of both elements affects the resonance, in the manner of (3.18-9). As far as resonance goes, Fig. 3.18-C is merely Fig. 3.17-A with that change (α replaced by α').

A figure of merit or quality factor Q for each of the individual energy-storing elements suggests itself in (3.18-10). We define them by

$$Q_C = \frac{C\omega_0}{g}, \qquad Q_L = \frac{L\omega_0}{r}, \tag{3.18-11}$$

and note that their reciprocals add to give the reciprocal of the circuit's Q:

$$Q = \left(\frac{1}{Q_C} + \frac{1}{Q_L}\right)^{-1}. \tag{3.18-12}$$

For a sharp resonance, both element Q's must of course be high. It often happens that $Q_C \gg Q_L$, say 2000 versus 100, so that with normal accuracy Q_C can be neglected in calculating Q:

$$Q \cong Q_L = \frac{L\omega_0}{r}. \tag{3.18-13}$$

Individual element Q's may be defined without reference to their use in a tuned circuit. Then we say that

$$Q_C = \frac{C\omega_F}{g}, \qquad Q_L = \frac{L\omega_F}{r}, \tag{3.18-14}$$

and note that these element Q's depend on frequency. Their values at the *resonant* frequency are the useful ones here.

The answer to the question: "*How does the dissipation in the inductor affect resonance?*" is what one would expect: *not at all*, provided that we take Q_L as the circuit Q. [If Q_C is low in comparison with Q_L, that is, if the effect of g (Figs. 3.18-A and 3.18-B) is appreciable, then it too must of course be considered.]

This discussion has suggested one other thing: how great the power of the transfer function is to describe a circuit's behavior in a very concise way. We shall later find many more instances of this, and shall soon turn to a careful study of transfer functions.

It would not do, of course, to turn away from a discussion of so practical a thing as the effect of dissipation in elements without observing the actual physical behavior of real tuned (resonant) circuits. But this is not really necessary and we present no further photographs or demonstrations of resonance here. The (obvious) reason is that *all* of the photographs of tuned circuit behavior, the demonstrations previously presented in this chapter, were actually made with real elements with dissipation; there is of course no other way. The highest Q observed (15 or 20) is essentially the Q of the inductor at the resonant frequency; the inductor provided virtually all the dissipation in the highest-Q examples. In those cases where the Q was reduced this was accomplished simply by using a resistor R of small enough resistance ($R = 1/g$) to degrade the Q per (3.18-9). Note carefully that the validity of those demonstrations at the time is supported by the discussion of this section. The discussion to follow, Sec. 3.19, needs the same comment.

There is, however, one qualification to be made, a footnote to the inductor dissipation discussion. Because $T(0)$ is no longer zero [the zero of $T(s)$ has moved to the left in Fig. 3.18-C], the forced response to excitations in which $s = 0$ or $|s|$ is very small will change somewhat. The forced (steady-state) response to the step excitation $I_F u(t)$ is now rI_F instead of zero. The forced (steady-state) response to very-low-frequency sinusoidal excitation is greater than it was when $r = 0$.

Physical verification is easily made by examining the behavior of $|T(j\omega_F)|$ near the origin. The demonstration of Fig. 3.14-H is therefore repeated in Fig. 3.18-D but with special attention to the low-frequency region. Part (a) is essentially a magnified version of Fig. 3.14-Ha, showing $|T(j\omega_F)|$ versus ω_F near

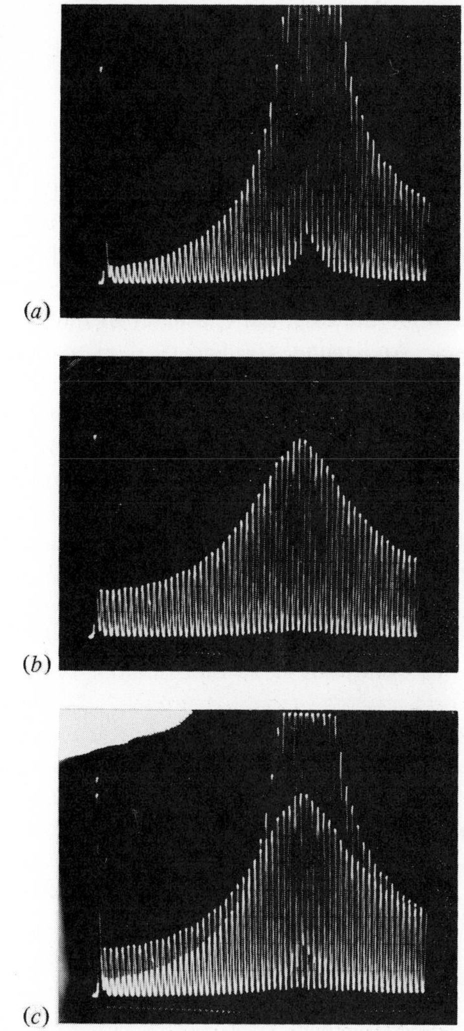

FIG. 3.18-D. The low-frequency effect of inductor dissipation.

the origin, for a fairly high-Q case, a case in which the inductor resistance is probably negligible. In part (*b*) the r of Fig. 3.18-B has been increased enough to move the zero well to the left of the origin and visibly to increase the low-frequency or dc transfer function values. (All three parts of Fig. 3.18-A have the same scales.) Part (*c*) superposes the two displays for convenient comparison at low frequencies. The predicted low-frequency effect of r is clearly verified. It is, however, generally unimportant.

3.19 RESPONSE TO OTHER EXCITATIONS, WITH DISSIPATION

It is reasonable to suppose that the impulse response, the step response, and so on, for the model of Fig. 3.18-B will in high-Q cases differ little from those of the model of Fig. 3.01-A provided we replace α by α'. Certainly the immediate response to an impulse (Sec. 3.02), to charge the capacitor and nothing more, is the same: $v(0+) = v(0) + Q_0/C$. True, $v'(0+)$ may be different (if $g = 0$), and in calculating the K's perhaps there will be slight changes. But it is the *natural frequencies* that determine the essential character of the natural behavior and of the impulse response. And using α' (instead of α) makes them correct. The smaller changes are not worth discussion here. (But they too are implicit in the transfer function, as we shall soon see.)

It may seem significant that the finite zero of $T(s)$ has moved away from the origin, to the point $-r/L$ (Fig. 3.18-C). But this (small) change affects $T(s)$ only when s is near the origin. It has little effect on the resonance, as we have seen. And the impulse response, one can show, is virtually unaffected by it. It *will* affect the step response for large values of time, for the forced part of that response is proportional to $T(0)$, whose value is now $r/(1 + rg) \cong r$, no longer zero. A step of I_F (amperes) results then in a forced response $[T(0)I_F e^{0 \times t}] = rI_F =$ constant, and this eventually is all that remains of the step response. But it is very small.

We leave for later discussion, by a more systematic (and perhaps simpler) method of analysis the calculation with $r \neq 0$ of other responses in the high-Q case and of the response in general in the low-Q case, and the discussion of other two-energy-storing-element (order 2) and higher-order circuits (Chaps. 6 and 7).

3.20 STATE-DESCRIBING VARIABLES

The state or condition of a circuit, at any instant, is described by the ensemble of values of the currents and voltages (or energies) of its various elements. It is the situation in which the network finds itself at that moment. We have been able to express the state of the RLC circuits by one variable, the voltage $v(t)$ in Figs. 3.01-A and 3.18-B. For from v we can immediately derive (should we wish to) the other (internal) element voltages and currents in those circuits. Hence $v(t)$, the solution of a second-order integrodifferential equation (3.01-1) or a second-order differential equation (3.18-5), or (3.18-4), gives (implicitly) a complete description of the state of the network.

The augmentation of the order of the differential equation from *1* for the RC circuit of Chap. 2 to *2* for the RLC circuits of this chapter is caused by the addition of a second energy-storing element, the inductor. The new element introduces its own derivative or integral into the analysis. Even so, should one prefer, he may limit analysis to first-order differential equations, provided he uses enough of them: the *RLC* circuits must have *two* such equations. Then the state description (by two variables) is more explicit.

In more complicated analyses, into which we do not go, such first-order differential equation formulation of the equations of analysis can be useful or even necessary (particularly when nonlinear elements appear). We shall now briefly discuss such an analysis for the *RLC* circuits, but shall leave further development of this method of analysis for study elsewhere.

We must first choose two dependent variables for the unknowns in our two first-order differential equations. They, when found, will describe the state of the network in a different, more explicit way than did $v(t)$ alone. Such variables are often called "*state-describing variables*" or simply "*state variables.*"

We are ineluctably drawn to the two energy-storing elements L and C for the choice: capacitor voltage (or charge, $q = Cv$) and inductor current (or flux linkages $\lambda = Li$) seem the "natural" choices. So, instead of (3.01-1) we look for a pair of equations of this form:

$$\begin{aligned} \frac{dv_c}{dt} &= \text{a function of } v_c,\ i_L, \text{ and } i_F \text{ in which no derivatives appear,} \\ \frac{di_L}{dt} &= \text{a similar function,} \end{aligned} \tag{3.20-1}$$

for the parallel *RLC* circuit of Fig. 3.01-A. From (3.01-1), retaining i_L and using the inductor relation $v = L\, di_L/dt$, we obtain

$$\begin{aligned} C\frac{dv_C}{dt} &= -Gv_C - i_L + i_F, \\ L\frac{di_L}{dt} &= v_C, \end{aligned} \tag{3.20-2}$$

or

$$\begin{aligned} \frac{dv_C}{dt} &= -\frac{G}{C}v_C - \frac{1}{C}i_L + \frac{1}{C}i_F, \\ \frac{di_L}{dt} &= \frac{1}{L}v_C. \end{aligned} \tag{3.20-3}$$

The simultaneous solution of this pair of first-order differential equations need not detain us here, for it proceeds essentially as did that of (3.01-1).

For the circuit of Fig. 3.18-B we have, from (3.18-1) and (3.18-2),

$$C\frac{dv_C}{dt} = -Gv_C - i_L + i_F,$$
$$L\frac{di_L}{dt} = v_C - ri_L, \tag{3.20-4}$$

in which i_L represents the inductor current i_{Lr}. We note here that for this second form of the tuned circuit the two-state-variable formulation (3.20-4) is quicker and simpler to obtain than was (3.18-4). As circuit complexity increases, similar comparisons are often apt. [But we have other methods to develop (Chap. 6) which must also be considered.]

The systematic simultaneous solution of the set of first-order differential equations of this state-variable method is in general best studied with the devices (determinants and matrices) of linear algebra. Accordingly we leave the multi-first-order differential equation or "*state-variable*" method for detailed study elsewhere. But there is one very interesting phenomenon associated with the state-variable approach that merits attention even here.

Instead of v_C and i_L we might choose as state-describing variables some linear combination thereof:

$$u_1 = a_{11}v_C + a_{12}i_L,$$
$$u_2 = a_{21}v_C + a_{22}i_L, \tag{3.20-5}$$

in which the *a*'s are some constants. The equations for the *u*'s would still be a pair of simultaneous first-order differential equations and be equally valid as (implicit) descriptions of the network's behavior. There is no limit to the possibilities: the state variables (the *u*'s) can be chosen in an infinite number of ways.

We look only at one particularly interesting such combination (for Fig. 3.01-A):

$$u_1 = v_C - p_1 L i_L,$$
$$u_2 = v_C - p_2 L i_L. \tag{3.20-6}$$

The differential equations are

$$\frac{du_1}{dt} = \frac{dv_C}{dt} - p_1 L\frac{di_L}{dt}$$
$$= -\frac{G}{C}v_C - \frac{1}{C}i_L + \frac{1}{C}i_F - p_1 L\frac{1}{L}v_C, \tag{3.20-7}$$
$$\frac{du_2}{dt} = -\frac{G}{C}v_C - \frac{1}{C}i_L + \frac{1}{C}i_F - p_2 L\frac{1}{L}v_C,$$

which, at first glance, appear complicated and not at all interesting. But they can be manipulated by using the relations between the natural frequencies and the elements. The characteristic polynomial [see (3.03-5) and (3.03-8)] gives us

$$\begin{aligned} p^2 + \frac{G}{C}p + \frac{1}{LC} &= (p - p_1)(p - p_2) \\ &= p^2 - (p_1 + p_2)p + (p_1 p_2) \end{aligned} \tag{3.20-8}$$

and so

$$\begin{aligned} p_1 + p_2 &= -\frac{G}{C}, \\ p_1 p_2 &= \frac{1}{LC}. \end{aligned} \tag{3.20-9}$$

which are relations well known in algebra [see (5.05-4)]. Using (3.20-9) we can transform (3.20-7) to

$$\begin{aligned} \frac{du_1}{dt} &= \left(-\frac{G}{C} - p_1\right)v_C - \frac{1}{C}i_L + \frac{1}{C}i_F \\ &= p_2 v_C - Lp_1 p_2 i_L + \frac{1}{C}i_F \\ &= p_2 u_1 + \frac{1}{C}i_F \end{aligned} \tag{3.20-10}$$

and

$$\frac{du_2}{dt} = p_1 u_2 + \frac{1}{C}i_F. \tag{3.20-11}$$

The remarkable, truly remarkable, feature of these two first-order differential equations is that in them the two state variables defined by (3.20-6) have been *separated:* each can be found, independently of the other, by solving a single first-order differential equation! It is as though we had transformed the parallel *RLC* circuit into two independent *RC* circuits, to each of which corresponds *one* of the natural frequencies (p_1, p_2) of the *RLC* circuit. Their solution requires only the methods of Chap. 2. For example, exponential excitation $i_F(t) = I_F e^{st}$ gives [see (2.14-9)]

$$u_1(t) = \left[u_1(0) - \frac{(1/C)I_F}{s - p_2}\right]e^{p_2 t} + \frac{(1/C)I_F}{s - p_2}e^{st} \tag{3.20-12}$$

and a similar result for $u_2(t)$. The forced and natural parts of the response have been separated; the transfer function "from i_F to u_1" is evident. From $u_1(t)$ and $u_2(t)$ we can of course easily obtain any of the currents and voltages in the network, solving (3.20-6).

The physical significance of the separated variables u_1 and u_2 of (3.20-6) (which are also known as *normal coordinates*) is unfortunately not easy to see. Generally neither one is even real, because only one of the natural frequencies appears in its equations. But the analysis that determines the particular choice of the *a*'s in (3.20-5) that will thus "uncouple" the two variables is well known as a salient part of the apparatus of linear algebra, the algebra of simultaneous equations. It profitably uses matrix notation and the properties of matrices (especially their inherent or *eigen* numbers) in the separation process. In that notation, remarkable for its brevity and clarity, the pair of equations (3.20-10) appears as

$$\begin{bmatrix} \dfrac{du_1}{dt} \\ \dfrac{du_2}{dt} \end{bmatrix} = \begin{bmatrix} p_2 & 0 \\ 0 & p_1 \end{bmatrix} \begin{bmatrix} u_1 \\ u_2 \end{bmatrix} + \begin{bmatrix} 1/C & 0 \\ 0 & 1/C \end{bmatrix} \begin{bmatrix} i_F \\ i_F \end{bmatrix}. \tag{3.20-13}$$

The general (separated-variable) form that corresponds to (3.20-13) can be obtained in the analysis of *all* linear system models (electrical, mechanical, or other) and is particularly useful in designing control systems. But this is general system theory which, regretfully, we must leave for further study elsewhere.

3.21 DUALS AND ANALOGS

We must not leave the current-driven parallel *RLC* circuit without mention of its dual, the voltage-driven series *RLC* circuit of Fig. 3.21-A. Its differential equation is

$$L\frac{di}{dt} + Ri + \frac{1}{C}\int_0^t i\,dx + v_C(0) - e_F(t) = 0, \tag{3.21-1}$$
$$i(0) = I_0.$$

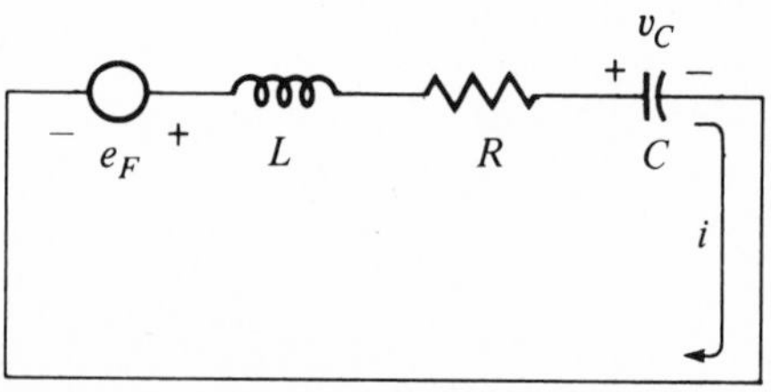

FIG. 3.21-A.

This is obviously the same as (3.01-1) for all mathematical purposes. The analysis of the series *RLC* circuit requires then no additional work: merely the changing (or reinterpretation) of symbols in all that we have done for the parallel *RLC* circuit, according to Table 3.21-A. (See Table 2.16-A.)

The dual of the second form of the parallel circuit (Fig. 3.18-B) is the network of Fig. 3.21-B, which can account for dissipation in both of the energy-storing elements. Again no additional analysis is necessary: we have already done the work and need merely change symbols (Table 3.21-B).

If to the mechanical models of Figs. 2.16-B and -C we now add springs (elastic, Hooke's law restraining elements), we obtain additional *RLC* circuit analogs which we have also, in effect, already analyzed.

TABLE 3.21-A Duals

Parallel *RLC* Circuit (Fig. 3.01-A)	**Series *RLC* Circuit (Fig. 3.21-A)**
C	L
v	i
G	R
L	C
i	v
Parallel connection	Series connection
Current excitation	Voltage excitation

TABLE 3.21-B Duals

Parallel *RLC* Circuit Modified for Dissipation (Fig. 3.18-B)	**Series *RLC* Circuit Modified for Dissipation (Fig. 3.21-B)**
C	L
g	r
L	C
r	g
Voltage	Current
Current	Voltage
Series	Parallel
Parallel	Series

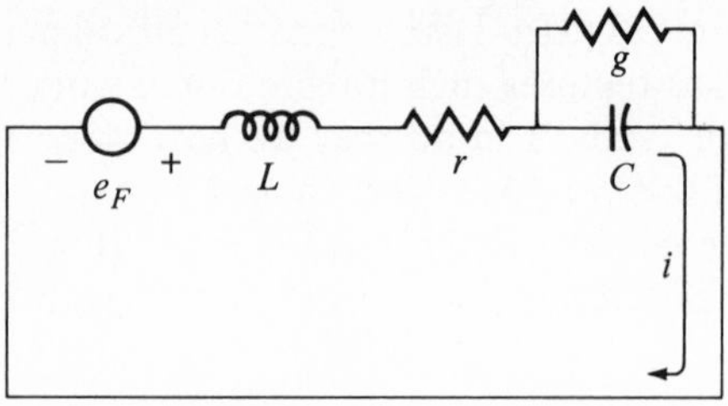

FIG. 3.21-B.

3.22 CONCLUSION

Our discussion of the *RLC* circuit has been very lengthy. But it has taught us a great deal: it has shown us a number of novel things in a simple circuit context, where ideas are easily grasped. We shall find it easy, with this experience, to apply these ideas to the analysis of complicated networks.

Let us briefly restate and summarize some of them.

First, the high-*Q* tuned circuit is characterized essentially by *two* numbers: ω_0, which shows *where* the resonance occurs, and *Q*, which shows *how sharp* it is. Figure 3.22-A shows an abbreviated model of the parallel circuit that has only the essentials: in the schematic diagram, the capacitor and inductor that resonate, and in the pole-zero diagram, the two natural frequencies. The presence of

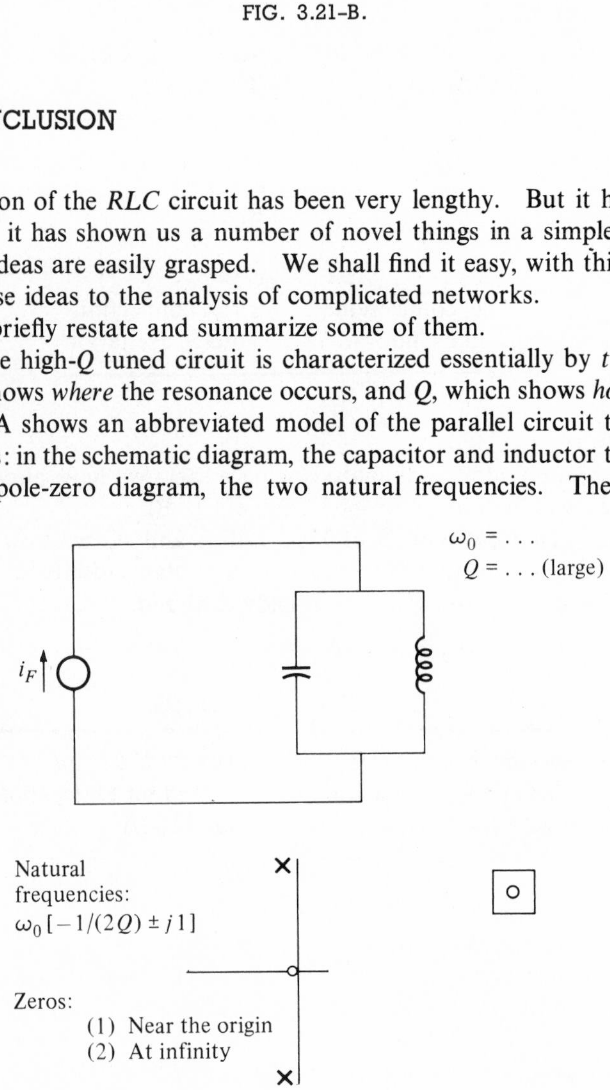

FIG. 3.22-A.

(small) energy dissipation in the elements is indicated by the value of Q attached to the schematic diagram, and by the horizontal coordinate of the natural frequencies in the pole-zero diagram. What we lack, given only these two numbers, are two other numbers: the peak value of $|T(j\omega_F)|$, and the value of $T(0)$. Both are in many analyses relatively unimportant, though technically necessary. From this crude diagram we can write, using only the two numbers,

$$T(s) = \frac{\text{constant}}{(s - p_1)(s - p_2)} \tag{3.22-1}$$

provided s is near one of the natural frequencies. From this follows

$$T(j\omega_F) = \frac{\text{constant}}{1 + j[2Q/\omega_0](\omega_F - \omega_0)} \tag{3.22-2}$$

provided ω_F is near ω_0. In (3.22-2) we have again all the important characteristics of (sharp) resonance behavior, obtained entirely if you will from the locations of the natural frequencies. These must of course be (*a*) *close to the* ω *axis* ($\alpha \ll \omega_0$) and (*b*) isolated (no other poles or zeros may lie nearby). In more complicated circuits, each natural frequency that has these two properties will cause a resonance describable by (3.22-2). More and more we see how important the natural frequencies of a circuit or network are!

Second, the natural behavior and the impulse response consist of exponential terms whose character is determined by the natural frequencies. We know this to be true for the *RC* and the *RLC* circuits that we have analyzed; we shall find it to be a very general network property, and shall become very familiar with the expression

$$\left.\begin{matrix} r_N \\ r_\delta \end{matrix}\right\} = \sum_{m=1}^{N} K_m e^{p_m t} u(t) \tag{3.22-3}$$

in which the p_m are the natural frequencies.

Third, the response to an excitation has two parts, readily distinguished and conveniently calculated separately (though not independently): (*a*) the forced part r_F and (*b*) the natural part r_N. Mathematicians prefer to call these the particular integral and the complementary function, respectively. Fortunately for us, this decomposition is possible for practically all the cases of interest in ordinary circuit analysis; it is, however, not *always* possible.

Fourth, to exponential excitation (which includes steps and sinusoids) the forced part of the response is readily calculated by using the *transfer function*.

Fifth, the K's in the natural part of the response are determined by the initial conditions (and the forced part).

The importance of the exponential function, which clearly has some intrinsic significance in network analysis, cannot be underestimated. We shall encounter it again and again, and slowly come to understand its role.

To exponential excitation, if we can find the transfer function, the calculation of the response is easy. But even in the *RC* and *RLC* cases we have not considered the response calculation for *arbitrary* excitation. This is an important item for our agenda. So is the determination of the differential equation(s) and the calculation of the transfer function in general cases. By the time we have thoroughly considered these things, we shall have read if not the whole story of network analysis, at least a very large part of it.

It is interesting to note, in passing, how the jargon of electrical engineering, of circuits in particular, is like argot anywhere. It develops itself to fit a need—the need for precise expression of a particular technical idea or description of a particular technical thing. It is often colorful, descriptive, concise, accurate, and mnemonic, though perhaps meaningless to the noninitiated. We have not hesitated to add to our vocabulary such useful expressions as tank circuit, tuned circuit, Q, high Q, and low Q. And we shall continue to adopt *useful* and *accurate* jargon.

Finally let us remind ourselves, once more, how deeply we are indebted to our fundamental postulates (Sec. 2.17): if the elements were not lumped, linear, and constant, how far would we have gotten with the *RLC* circuit? Nor should we forget the practical applications of the electric tuned circuit, consistent therewith: the tuned circuit is used not merely for its selectivity and discrimination in the sinusoidal steady state (as in radio receivers), but as a generator (by its "ringing") of oscillations and time marks. We cannot here go into these, or into its use as a model to explain and to help design mechanical shock absorbers, lasers, microwave cavities, and a host of other phenomena that are, in essence, simply resonant (tuned) circuits.

3.23 EXERCISES

3-1. Make a list of physical situations in which resonance can be disagreeable and dangerous. Which can reasonably be called analogous to the *RLC* tuned circuits of this chapter?

Make a list of pleasant and useful applications of the resonance phenomenon.

3-2. Consider a rather lightly damped tuned circuit. Sketch† several "cycles" of the natural behavior. Why is it appropriate to place the word *cycles* between inverted commas? Answer each of the following questions (*a*) with a mathematically exact formula and (*b*) with a simpler, but usually adequate, approximate formula based on the assumption that the damping is very small. But retain some indication of the damping's effect, such as one term of a series. [Use ω_0 rather than ω_N in (*a*).]

† See the footnote to Exercise 2-2.

What is the time interval (1) between two successive zero crossings? (2) From one zero crossing to the second one thereafter? (3) Between two successive extrema (e.g., from one maximum to the succeeding minimum)? (4) Between two successive maxima? (5) From a zero where the response is increasing to the immediately succeeding maximum?

In which cases does your answer vary as time goes on? In which does it remain the same?

In answering item 3 one may pass from the complex form ($K_1 e^{p_1 t}$ + conjugate) to the real-number-only form (exponential, cosine, sine) either before or after the needed differentiation. Which is simpler? Why?

3-3. Let V_0 be the initial value of the response of a tuned circuit (capacitor in parallel with dissipative inductor), and V_1 its value at the first succeeding maximum. The expression $(V_0 - V_1)/V_0 = Q/\pi$ is accurate for large values of Q. For what value of Q is it 10 percent in error?

3-4. Explain in your own words why the statements "a capacitor is initially a short circuit" and "an inductor is initially an open circuit" are legitimate statements. Trace the path of the charging current i_F through the network shown. Just after the impulse, what are the voltage across and current in each of the 10 elements?

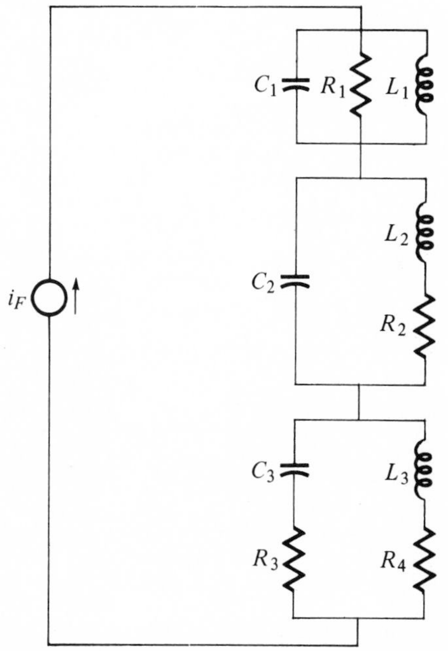

FIG. 3-4.

3-5. The natural behavior of a tuned circuit is recorded as an oscillogram. Subsequent examination shows three successive maxima to have the (relative) values 1.00, 0.90, 0.81. What is the Q of the circuit? In general, if the ratio of two successive maxima is r, what is the Q of the circuit in terms of r?

3-6. Let $Ke^{-\alpha t}\cos(\omega_N t + \phi)$ represent the (voltage) response of a parallel RLC circuit to impulse excitation, $i_F = Q_0\,\delta(t)$, when the initial conditions are $v(0) = 0$ and $i_L(0) = I_0$. Let $v(0+) = V_0$. Find the values of the three elements and of I_0 in terms of Q_0/V_0 and the response parameters K, α, ω_N, ϕ. Show that the circuit can be synthesized (all element values will be positive) for *any* prescribed set of response parameter values. What relation must hold among the response parameters if I_0 is to be zero? Perform the synthesis for the case: $\phi = -30°$, $K = 2$ V, $\alpha = \omega_N = 10^5\ \text{s}^{-1}$.

3-7. Listed below are the L and C element values of a number of tuned circuits, as well as the Q of the inductor as defined in (3.18-11). (No other dissipation need be considered.) For each tuned circuit find, to two significant figures, the resonant frequency in Hertz, the Q, the half-power frequencies. Calculate mentally as far as possible.

L	C	Q_L
1 mH	1 μF	100
1 mH	1 pF	100
1 μH	1 μF	50
1 μH	1000 pF	50
3.4 mH	3000 pF	25
250 mH	5 μF	30
42 mH	0.025 μF	10
0.5 H	4.7 μF	20

3-8. Verify the formulas shown in Fig. 3-8 for the element values of a tuned circuit in terms of its Q, resonant frequency ω_0, and impedance at resonance, Z_0. Note that the L and C formulas are the *same* for the two circuit forms.

3-9. A resonant circuit with a Q of 100 at 1 MHz, suitable for tuning from 600 KHz to 1500 KHz, is wanted for a radio receiver. (*a*) If the inductor is to be fixed, the capacitor variable, what range of C will be required? (*b*) If the capacitor is fixed, the inductor varied, what range of L will be needed?

3-10. Some of the entries in the table (Fig. 3-10) are incorrect. Find them, correct them, and explain how they probably occurred. Then fill in the blanks. All entries refer to the tuned circuit, in one or the other of the two forms (Fig. 3-8), in the sinusoidal steady state.

$$L = \frac{Z_0}{Q\omega_0}$$

$$C = \frac{Q}{Z_0\omega_0}$$

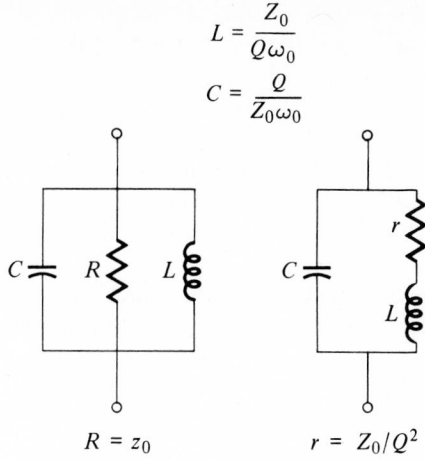

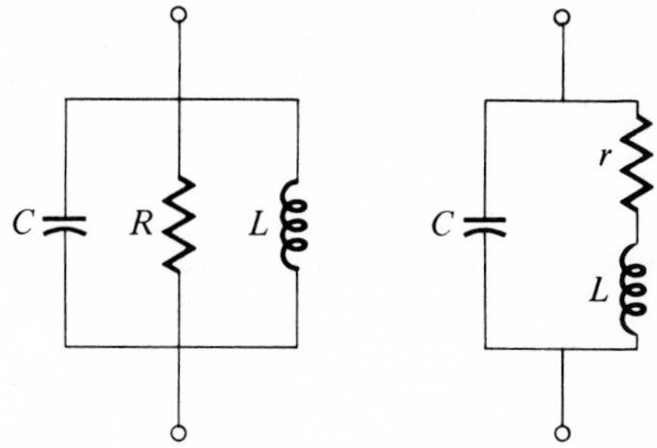

$R = z_0$ $\quad\quad$ $r = Z_0/Q^2$

FIG. 3-8.

Q	**Resonant Frequency**	**Impedance (Resistance) at Resonance,** Ω	R Ω	L	C, μF	r, Ω
50	1000 Hz	1000	1000	3.2 mH	8.0	0.40
30	60 kHz	5000	5000	0.44 mH	0.100	5.6
100	340 kHz	75	75			
8	440 Hz	90	90	25.6 mH	32.1	1.41
40	1.16 MHz	100	100	0.34 μH	0.055	0.063
10	1.5 MHz	300	300	3.18 μH	0.354	3.0
150	12 kHz	600	600			

FIG. 3-10.

3-11. The response of the parallel *RLC* circuit of Fig. 3.01-A from rest, to the excitation $i_F = I_0[u(t) - u(t - T_0)]$, is shown in the sketch (Fig. 3-11), which also

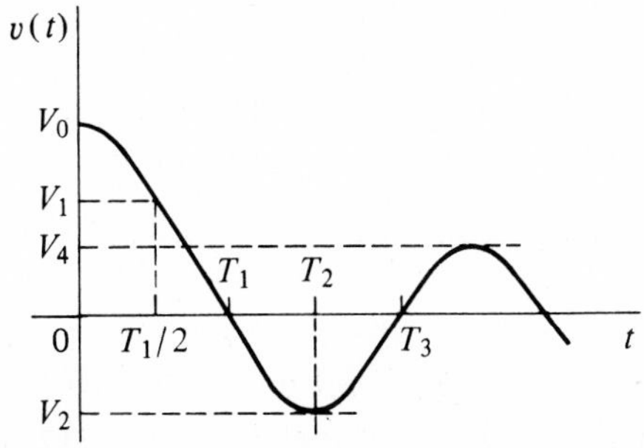

FIG. 3-11.

defines the symbols used below. Take $I_0 = 0.5$ mA, $T_0 = 1$ μs, $L = 50$ mH, $C = 0.02$ μF, $R = 67$ kΩ. Find the values (two significant figures are adequate) of V_0, T_1, V_1, T_2, V_2, T_3, V_4.

3-12. Refer to Sec. 3.02. Suppose that there *were* a steplike change in inductor current between $t = 0$ and $t = 0+$. What then would be the current in the resistor? In the capacitor? Explain in sufficient detail completely to justify (3.02-3).

3-13. Someone wants to generate a waveshape like that sketched (we don't know why). A parallel tuned circuit of the correct frequency ($\omega_0 \cong \omega_N = 6\pi/T$ rad/s) is handy; its Q is 50, which is good enough (the damping that occurs during each of the "on" parts of the wave is acceptable, provided the "signal" is *zero* where it is shown to be zero). You are to design a (current) excitation function made up solely of impulses and steps as may be appropriate; these are to occur only at the time points 0, T, $3T$, $3.5T$, $4T$, as may be necessary, and with whatever strengths (coulombs) and step sizes (amperes) may be necessary. The tuned circuit is at rest at $t = 0$. Thereafter it reacts to your excitation to produce a voltage waveshape. Fill values in the blocks of a table like that below. If there is a difference between "before" and "after" values at any time, subdivide the block and give both. Explain how you reach your results. (For "excitation" indicate whether step or impulse, and give numerical values, relative to some fixed value of your choosing.)

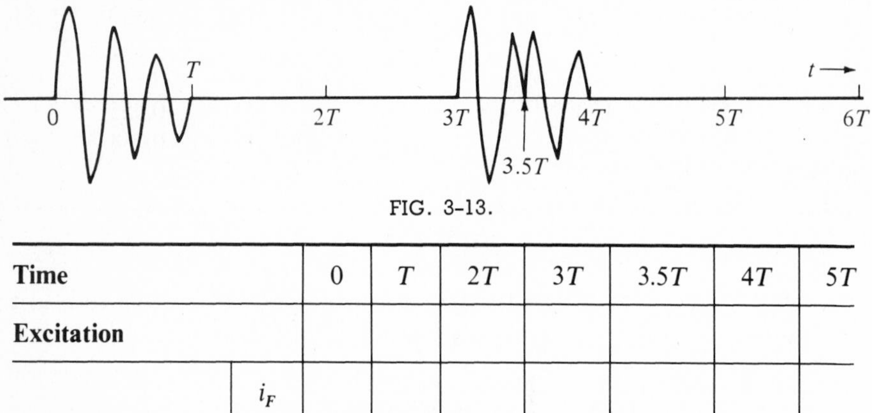

FIG. 3-13.

Time		0	T	$2T$	$3T$	$3.5T$	$4T$	$5T$
Excitation								
Conditions in the Circuit Model (Currents)	i_F							
	i_C							
	i_R							
	i_L							

3-14. In a simple radio receiver signal selection is to be accomplished by an ordinary tuned circuit. If the desired signal is at 1 MHz and the selectivity necessary is that at the next adjacent channels (990 and 1010 kHz) the response be down at least 20 dBV (see Appendix D), what Q is necessary? Find the L and C element values and the necessary Q of the inductor, assuming infinite Q for the capacitor. If the next adjacent channels are at 980 and 1020 kHz, how do your results change? How much does the response vary between frequencies 5 kHz above and below 1 MHz in each case? (The signal usually contains components in this range, which may need careful attention.)

3-15. Find the element values (C, L, Q_L) for tuned circuits that meet the following specifications (use the model of Fig. 3.18-B with $g = 0$). Which, if any, of these seem difficult to realize? Explain.

Resonant Frequency	**Half-power Frequencies**	Q	**Impedance (Resistance) at Resonance**
3000 Hz	2000 Hz 4000 Hz		10 kΩ
41 kHz	40 kHz 43 kHz		600 Ω
150 kHz	145 kHz 155 kHz		1000 Ω
1.25 MHz		50	75 Ω
22.0 MHz	21.5 MHz 22.5 MHz		50 Ω

3-16. In this variant of (3.06-10), $v(t) = V_m e^{-\alpha t} \cos(\omega_N t + \phi)$, evaluate V_m and ϕ in terms of arbitrary initial conditions $v(0+) = V_0$, $i_L(0+) = I_0$. Explain fully the comment in Sec. 3.07: "The curve of Fig. 3.07-A will serve for *any* initial conditions." Where thereon is the point "$t = 0$" for positive I_0 and positive V_0? For negative I_0 and negative V_0? For what $V_0 - I_0$ relationship will ϕ be zero?

3-17. Fill in the blanks. An accuracy of ± 1 percent is adequate; perform the calculations with the minimum amount of work consistent therewith, mentally if possible.

f_0	Q	f_N
1000 Hz	10	
50 MHz	5	
	50	60 kHz
2 MHz		3 MHz
25 kHz	2	

3-18. The sketch shows the natural frequency of a tuned circuit that lies in the second quadrant: $p_1 = -\alpha + j\omega_N$. Evaluate (*a*) the distance $AB = \omega_0 - \omega_N$, (*b*) the angle β, in terms of ω_0 and Q.

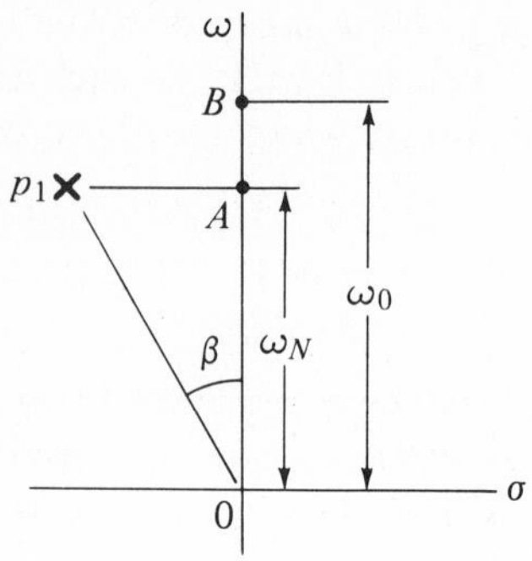

FIG. 3-18.

3-19. The figure shows (to scale) part of an oscillogram of the natural behavior of a mechanical shock-absorber mechanism. Assume that it is adequately modelled by a second-order system, in electrical form a parallel *RLC* circuit. The behavior shown is that of the analogous voltage. The actual time interval between successive zeros of the response on the oscillogram in 10 ms. What are the natural frequencies of the shock absorber? Plot with care the response, from rest, to a rectangular pulse of duration 20 ms. To what figure in the text is this related?

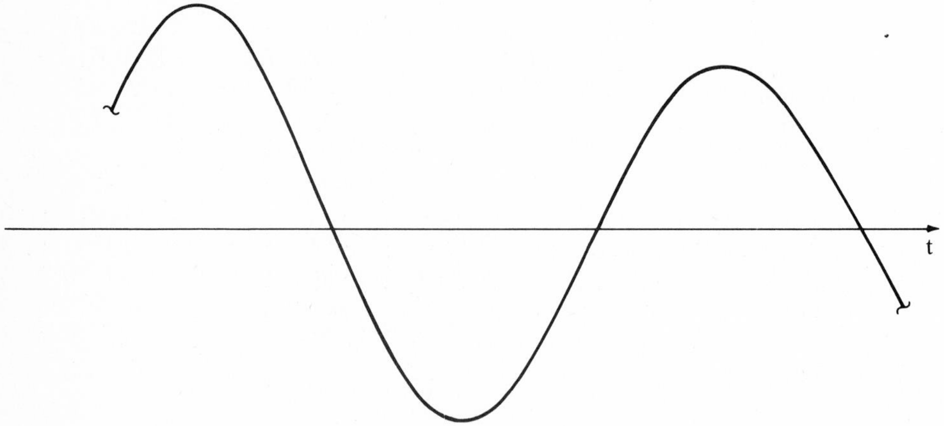

FIG. 3-19.

3-20. Plot the locus, in the plane of (complex) p, of the natural frequencies of the parallel *RLC* circuit as R varies from infinity to zero. Explain why you do not need values of L and C to do this. What is the geometrical figure that is the locus? Which part corresponds to very small dissipation of energy in the circuit? Repeat for the tuned circuit in the form of a capacitor in parallel with a dissipative inductor (replace R with r).

3-21. Draw to scale the second quadrant of the p plane, showing therein the natural frequency (p_1) of a tuned circuit of $Q = 10, 20, 30$. As ω_F (in $s = j\omega_F$) varies from $(\omega_0 - \alpha)$ to $(\omega_0 + \alpha)$, how much do the factors (s) and $(s - p_2)$ in the transfer function vary? What useful approximate formula(s) of the text are thereby verified?

3-22. The ratio (C/L) (or its reciprocal) is sometimes referred to as the *stiffness* parameter of the LC part of a tuned circuit. Consider sketches of the *susceptance B* (Fig. 3.11-A) versus ω_F for several values of (C/L), and explain why this term "stiffness" is appropriate. What is the physical situation when G is neglected? If G is not neglected, what is the significance of the ratio of G and the stiffness parameter?

3-23. Do the minimum amount of work, for only 5 percent accuracy is needed here! An ordinary tuned circuit (Fig. 3.01-A) has a Q of 40, resonates at $f_0 = 60$ kHz, has $R = 1000\ \Omega$, is excited by $i_F = 180 \sin(\omega_F t)\ \mu\text{A}$, $f_F = f_0$. Find the steady-state response voltage in both phasor and time-function forms for each of the cases below:

(*a*) Initial conditions: $V_0 = 0$, $I_0 = 0$;

(*b*) Initial conditions: $V_0 = 180$ mV, $I_0 = 0$;

(*c*) Initial conditions: $V_0 = 0$, $I_0 = -180\ \mu\text{A}$;

(*d*) Change f_F to 120 kHz and repeat (*a*), (*b*), (*c*);

(*e*) Change f_F to 30 kHz and repeat (*a*), (*b*), (*c*).

3-24. Sketch versus frequency $f = \omega_F/(2\pi)$ the peak amplitude V_F of the (sinusoidal) voltage developed across each of the tuned circuits described below, when excited by $i_F = I_F \cos \omega_F t$ in which ω_F (and only ω_F) is varied. What is the frequency range, in each case, over which V_F varies from its maximum value by no more than ten percent? Circuit descriptions: $f_0 = 10$ kHz, $Q = 10, 20, 50, 100$.

3-25. A mechanical linear system, modellable by a parallel RLC circuit, responds as shown on the following oscillogram to a sinusoidal excitation of frequency ω_F. Answer the following questions with reasonable accuracy. What is its resonant frequency, in units of ω_F? What are its half-power frequencies in units of ω_F? What is the maximum amplitude of its transfer function (for $s = j\omega$) compared to its value at ω_F? What is its Q?

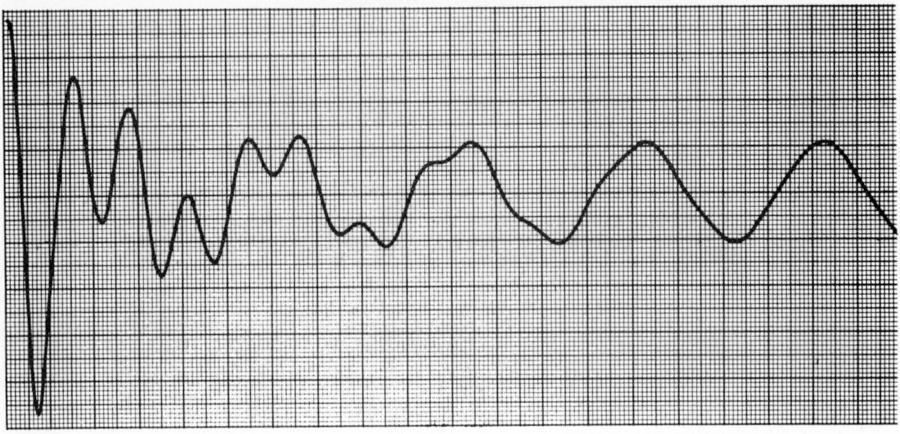

FIG. 3-25.

3-26. Consider an amplifier constructed by cascading three stages, each of the form shown (Fig. 3-26), so that the e of one stage is the v of the preceding stage. There is no loading of (reaction on) a stage by the following stage: each v is the same, whether the following stage is connected or not. The box Z is: in stage 1 an RC circuit of natural frequency $-\alpha + j0$; in stage 2 a tuned circuit of $Q = 20$, $\omega_0 = 10\alpha$; in stage 3 a tuned circuit of $Q = 40$, $\omega_0 = 20\alpha$, all in seconds^{-1}.

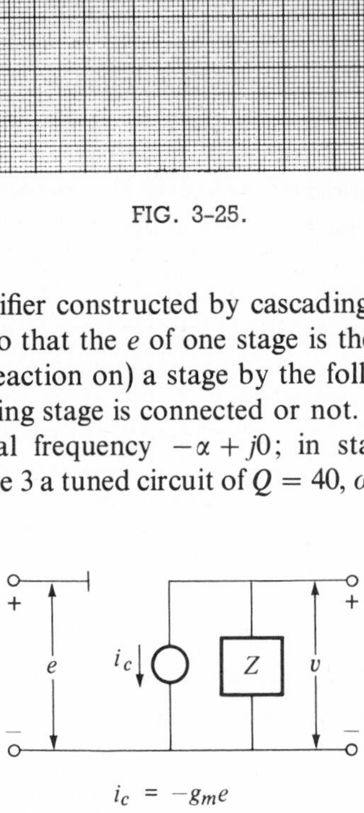

FIG. 3-26.

(*a*) Show, on a complex-frequency plane, the set of natural frequencies of the complete amplifier. For what frequencies is the transfer function infinite? zero?

(*b*) Write the *form* of the impulse response. Include all terms; use numerical values that can be obtained without calculation, or else letters.

(*c*) Sketch the magnitude of the sinusoidal-steady-state response versus the exciting frequency ω_F, the amplitude of the excitation being held constant.

(*d*) What is the ratio (within 5 percent) of the heights of the two resonant peaks in (*c*)?

3-27. A simple AM radio receiver is to select (tune to) stations with an ordinary tuned circuit of fixed inductor and variable capacitor. The signal from the "nearest" interfering station (which will be at a frequency 20 kHz or more different from that tuned to) must be down 6 dBV; that is, the quarter-power point on the resonance curve is to lie no further away from the frequency tuned to than 20 kHz. The tuning range to be covered is 600 kHz to 1500 kHz.

One surely wants to use a (fixed) inductor of the minimum possible Q. What is that minimum value of Q_L, at what frequency? Suppose the inductor used (with that Q at that frequency) has an inductance of 50 μH. What is the range of the (variable) capacitance C that must be provided? Sketch the resulting circuit Q versus frequency.

3-28. Suppose a tuned circuit models a power amplifier in which $i_F = 0.15 \cos(\omega_F t + \phi_F)$ (amperes), $R = 1000\ \Omega$, $Q = 70$, and the frequency of excitation is the resonance frequency ($\omega_F = \omega_0$). Find the steady-state peak voltage and the peak value of the steady-state current in each of the three elements. Sketch (versus time) over one cycle: the rate of dissipation of energy in the resistor, the energy stored in the capacitor, the energy stored in the inductor. How much energy is dissipated in the load (resistor) in one cycle?

3-29. A tuned circuit is excited by $i_F = I_F \cos(\omega_F t + \phi_F)$, when initially at rest. Is it possible to apply the excitation at such an instant that there is no transient component in the response (i.e., so to choose ϕ_F)? Reason physically as far as possible, and explain. If the excitation is applied at a positive peak ($\phi_F = 0$), can the initial conditions be chosen so that there will be no transient component in the response? Explain.

3-30. For unit (sinusoidal) excitation (1 A, 1 V, respectively), for each case given (Fig. 3-30): (1) find the steady-state response phasor; (2) find the steady-state

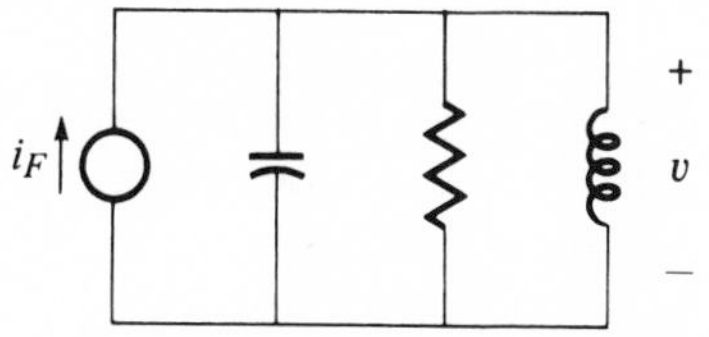

(a)

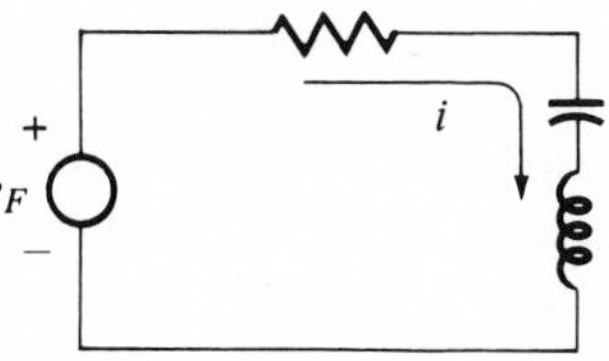

(b)

FIG. 3-30.

response time function; and (3) sketch both steady-state response and excitation waves.

Circuit	f_0, Hz	Q	f_F, Hz
b	60	1	60
b	30	1	60
b	120	1	60
a	2000	10	1000
a	2000	10	2000
a	2000	10	5000

3-31. Draw to scale, and explain, phasor diagrams for Fig. 3.14-G. To what extent is Fig. 3.15-C a valid "answer" for this exercise?

3-32. It is frequently stated that "the impedance of a tuned circuit at resonance is QX." Here $X = L\omega_0$ is the "reactance" of the inductor at resonance, equal also to $1/(C\omega_0)$, the (negative of the) "reactance" of the capacitor at resonance; Q is the circuit Q, defined generally by (3.18-12). Verify the statement for the (general) circuit of Fig. 3.18-B, taking the "impedance" to be the same as the transfer function from i_F to v, at resonance. Note that the statement applies equally well to the "ordinary" forms of tuned circuit in which $g = 0$ or $r = 0$.

3-33. In the sinusoidal steady state the (voltage) response of a tuned circuit resonant at frequency f_0 leads the (current) excitation by an angle θ. Find a simple (approximate but usually adequate) expression for the frequency of excitation in terms of the (high) Q of the circuit, θ, and f_0. How must it be altered if the Q is very low?

3-34. Explain the (straight-line) asymptotic behavior in Fig. 3.14-C. What are the slopes of the asymptotes if arbitrary logarithmic scales are used? Look up and explain the terms *octave, decade, decilog.*

3-35. (*a*) The step response of a parallel *RLC* circuit (from rest) is to pass through these four points: 0 at $t = 0$, V_1 at $t = T$, V_2 at $t = 2T$, 0 at $t = 3T$. Show that this is possible, if $V_2 < V_1$, by finding the element values of such a circuit in terms of T, V_1, V_2.

(*b*) Change the prescription at $t = 3T$ to V_3 (instead of zero). Find conditions (on the three prescribed values of response) necessary and sufficient for the synthesis to succeed.

3-36. Show that the response of a parallel *RLC* circuit to the (peculiar) excitation, a polynomial in t for positive time, $i_F = (a_0 + a_1 t + a_2 t^2)u(t)$, is strictly linear, $v = b_0 + b_1 t$, provided that certain initial conditions obtain. Evaluate b_0, b_1, $v(0)$, $i_L(0)$, in terms of a_0, a_1, a_2, and the element values. Why is there *no natural component* in the response?

3-37. In Fig. 3.07-C verify the Q values given. Does the value of f_N differ sensibly from that of f_0 (45 kHz) in any case?

3-38. Suppose a curve for $Q = 50$ is added to Fig. 3.14-A. What is the "width" of the sharp (resonant) portion on a scale where ω_0 corresponds to 10 cm? And if $Q = 100$? What would the widths be for Fig. 3.14-B, one logarithmic cycle covering 10 cm?

3-39. Under what conditions will the response (3.13-1) take the form $v(t) = (1 - e^{-\alpha t}) \cos(\omega_F t + \phi) \times$ constant, which Fig. 3.13-C so strongly suggests? What latitude exists in excitation frequency and initial phase, and in circuit initial conditions? To the case of Fig. 3.13-C does the formula above actually apply?

3-40. Substitute (3.08-1) into the appropriate differential equation and so verify that (*a*) p_1 and p_2 are the natural frequencies; (*b*) $V_F = 0$; (*c*) the K's must obey (3.08-3) and (3.08-4); (*d*) (3.08-5) and (3.08-3) are equivalent; (*e*) (3.08-1) *is* then the solution of the differential equation.

3-41. Verify the statements in the fifth paragraph of Sec. 3.13 about Q/ω_F and $Q\omega_F$ amplitude ratios.

3-42. In Sec. 3.15 it is stated that current and voltage phasors cannot properly be drawn on the same diagram. Explain fully. Nevertheless one very commonly finds phasor diagrams in which phasors with different dimensions do appear. Explain why the *angular* relations between all of them *can* be correct. This makes the diagrams very useful, even though a purist may argue about the dimensions of the moduli (phasor lengths).

3-43. A parallel *RLC* circuit has a Q of 50. Sketch the circuit's usual (voltage) impulse response versus time from $t = 0$ to the fourth zero crossing thereafter. Indicate every point where (*a*) the capacitor current is zero, (*b*) the resistor current is zero, (*c*) the inductor current is zero. Explain, in terms of the curve itself, how each is determined.

The same circuit is excited from rest by the sequence of impulses: (1) an impulse at $t = 0$, (2) a second impulse at $t = T_2$, the time of the first stationary point (point where the derivative of voltage is zero) after $t = 0$, of strength just sufficient to change the voltage immediately to zero, (3) a third impulse at time T_3, the next zero crossing of v after $t = T_2$, of strength one-half that of the second impulse. Fill in the blanks in the accompanying table with +, 0, −, according as the current involved is at that instant positive, zero, or negative. Note that this is to be done for "just before" and "just after" times in each case.

Time	Capacitor Current	Resistor Current	Inductor Current
0			
0^+			
T_2^-			
T_2^+			
T_3^-			
T_3^+			

Use the reference directions of Fig. 3.01-A. Explain succinctly but accurately how you determine each value.

Now prepare an accurate plot of the voltage versus time. At what instant of time, T_1, in units of ω_N^{-1}, within 1 percent, is the voltage next stationary after T_3? When is it next zero?

3-44. Show that in a high-Q resonant circuit the *total* stored energy (in capacitor *and* in inductor) is given approximately by (3.07-12). Discuss the departure of the exact formula therefrom in detail. Plot to scale, for $Q = 15$, the quantities W_C, W_L, $(W_C + W_L)$.

3.24 PROBLEMS

3-A. Show that the formula for each of the curves of Fig. 3.06-B may be written $v(t) = A \cos \omega_N t + B \sin \omega_N t$, $t > 0$, in which A and B are (real) constants. Evaluate each for each of the three cases in terms of the three quantities $V_0 = v(0+)$, $V_a = Ri_L(0+)$, $\zeta = \alpha/\omega_N$. Check all dimensions. Evaluate ϕ_1 in each case and compare with the angular position of p_1 in the (complex) p plane. Determine the value of $i_L(0+)$ in b both mathematically and physically. Repeat for $i_C(0+)$ in c.

3-B. What are the natural frequencies of the circuit shown (a) with the switch open, (b) with the switch closed? (e is a voltage source.) Calculate the transfer function from e to v with the switch closed. Where are its zeros? its poles? Sketch the magnitude of the transfer function versus ω_F, for $s = j\omega_F$. Why might it be called a "low-pass filter"? Find and sketch versus time the voltage v in response to a step voltage e. Now change the element values to 600 Ω, 600 Ω, $600/(2\pi \times 10^6)$ H, $(600 \times 2\pi \times 10^6)^{-1}$ F. Relabel your previous step-response plot to apply to this new network and explain why it is easy so to do.

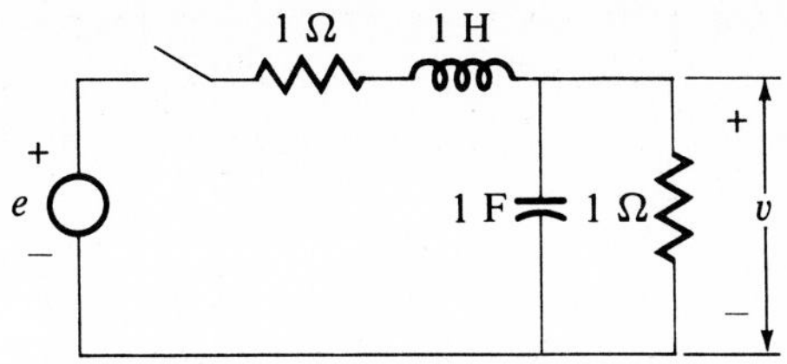

FIG. 3-B.

3-C. Refer to Fig. 3.07-A. Define the *envelope* of the damped oscillation wave as the "bounding" curves of the exponential multiplier: $\pm V_0 e^{-\alpha t}$. At what points does the damped oscillation touch its envelope? Are these the same as the extrema of $v(t)$? Explain.

3-D. The table gives observed values of the natural behavior of a parallel *RLC* circuit. What is its Q, within 10 percent? What are the (normalized) natural frequencies, within 0.002? How long (to two significant figures) does it take the circuit to lose half of its stored energy? How does this depend on the time at which you start the calculation? How long does it take the energy to drop to zero? Is your last reply realistic?

Suppose a sinusoidal excitation of frequency ω_F (in the same normalized units) applied to the circuit; allow the steady state to set in. In the following consider a wide range of variation of ω_F. Let V_F represent the amplitude of the

Time (normalized units)	Voltage (normalized units)	Time (normalized units)	Voltage (normalized units)	Time (normalized units)	Voltage (normalized units)
0.000	10.00	2.967	−9.28	5.934	8.35
0.175	9.81	3.142	−9.39	6.109	8.72
0.349	9.33	3.316	−9.22	6.283	8.82
0.524	8.57	3.491	−8.76	6.458	8.65
0.698	7.55	3.665	−8.05	6.632	8.23
0.873	6.32	3.840	−7.09	6.807	7.56
1.047	4.90	4.014	−5.93	6.981	6.66
1.222	3.34	4.189	−4.60	7.156	5.57
1.396	1.69	4.363	−3.13	7.330	4.32
1.571	0.00	4.538	−1.59	7.505	2.94
1.745	−1.68	4.712	0.00	7.679	1.49
1.920	−3.29	4.887	1.57	7.854	0.00
2.094	−4.79	5.061	3.09	8.029	−1.48
2.269	−6.14	5.236	4.50	8.203	−2.90
2.443	−7.30	5.411	5.77	8.378	−4.23
2.618	−8.22	5.585	6.85	8.552	−5.42
2.793	−8.89	5.760	7.72	8.727	−6.43

(forced) response. Determine the frequencies asked for within 0.002 normalized units. At what value(s) of ω_F will: (*a*) V_F be largest? (*b*) V_F be reduced to 80 percent of its maximum value? (*c*) the phasor $\bar{V}_F$ be 30° ahead of the excitation phasor? (*d*) the phasor $\bar{V}_F$ be 20° behind the excitation phasor?

Find the L and C element values (within 5 percent) on an $R = 1\ \Omega$, normalized frequency basis. Find their values for $R = 50\ \text{k}\Omega$ and a frequency scale such that $t = 1$ (normalized) corresponds to $t = 1\ \mu s$ (actual).

3-E. To force a pulse of current through the load resistor R (which represents some device such as a radar oscillator to be "pulsed"), a capacitor C is first *charged* from the voltage source E (the switch being open), then suddenly, quickly, *discharged* through R by closing the switch (Fig. 3-E, left). The lapse of time before opening the switch and repeating the process is enough that the capacitor is effectively discharged completely.

In the (complete) charging process how much energy is dissipated in R_0? In R? After closing the switch (when the capacitor is fully charged) how much energy is delivered (in the pulse) to R? (Note the relative time constants.)

Figure 3-E (right) shows a modified scheme in which a high-Q inductor replaces the charging resistor R_0. The switch is closed when i is zero, decreasing. How much energy is then dissipated in R? What advantage(s) does this "resonant charging" scheme have?

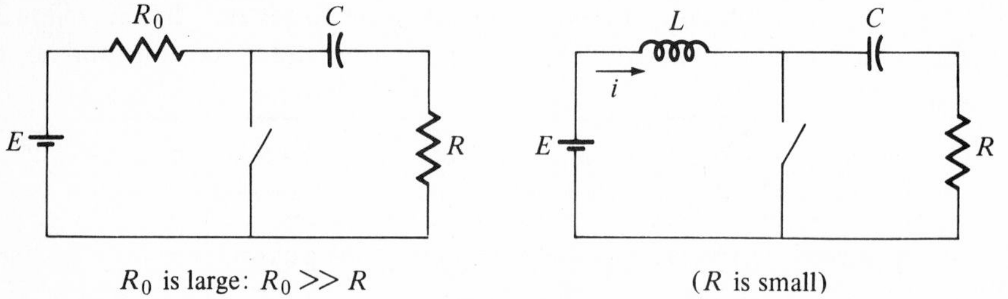

FIG. 3-E.

3-F. The photograph shows the response of the tuned circuit to a rectangular pulse of duration T (seconds). Each centimeter (large division) represents 40 μs.

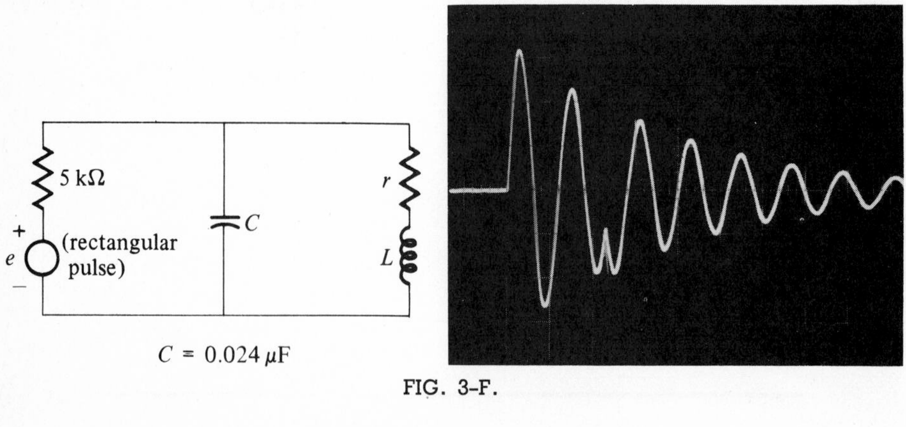

FIG. 3-F.

Find the values of T, Q, L, f_0, and ω_0, in each case stating the limits of accuracy obtainable. Plot the step response of the circuit versus time. Superpose a replica of the same, properly translated and scaled, and thus obtain a plot of the response to the rectangular pulse. Does it agree with the photograph? Explain.

3-G. The *series* connection of three elements (R, L, C), voltage excited, is analogous (dual) to the parallel *RLC* circuit discussed here at length in the text. Write the differential equation for such a series circuit and compare with (3.01-1). List, opposite to each other, the quantities dual to each other (see Sec. 2.16). Make a brief translation guide that will enable you to apply all the analysis of this chapter to series *RLC* circuit analysis.

Discuss the inclusion of (small) *inductor* dissipation in the analysis of the series *RLC* circuit. Why is no additional analysis necessary, and why does a small change in one element value suffice? Why does the dual of Fig. 3.18-B not apply? To what modification of the series circuit *does* that figure apply (dually)?

3-H. A sinusoidal *voltage* source is applied to a tuned circuit (Fig. 3-H). Show, from fundamental time-domain (differential-equation) relations that when the steady state has arrived, the phasor representations of the voltage $\bar{E}$ and the three currents ($\bar{I}$, $\bar{I}_C$, $\bar{I}_{RL}$) have these relations: $\bar{I}_C = jC\omega_F\bar{E}$, $\bar{I}_{RL} = \bar{E}/(R + jL\omega_F)$, $\bar{I} = \bar{I}_C + \bar{I}_{RL}$. Show that the point P, the tip of the phasor $\bar{I}$, may lie in either the first or fourth quadrant.

Let $x = \omega_F/\omega_0 = \omega_F\sqrt{LC}$ and Q = circuit $Q = \sqrt{L/C}/R$. What is the angle ϕ in terms of Q and x? What is the length of the phasor $\bar{I}$ in terms of Q and x?

Draw the phasor diagram to scale and locate the point P for $Q = 10$ and $x = 0, 0.01, 0.1, 1, 10$. Sketch the locus of P as ω_F varies from very low to very high values. Answer the two questions below (1) formally, in terms of Q, and (2) numerically for the case above.

(*a*) For what frequency is the magnitude of $\bar{I}$ a minimum?

(*b*) For what frequency is the angle ϕ equal to zero?

What arguments can you adduce for defining the resonant frequency of the circuit as the frequency found in (*a*)? As that found in (*b*)? What difference does it make?

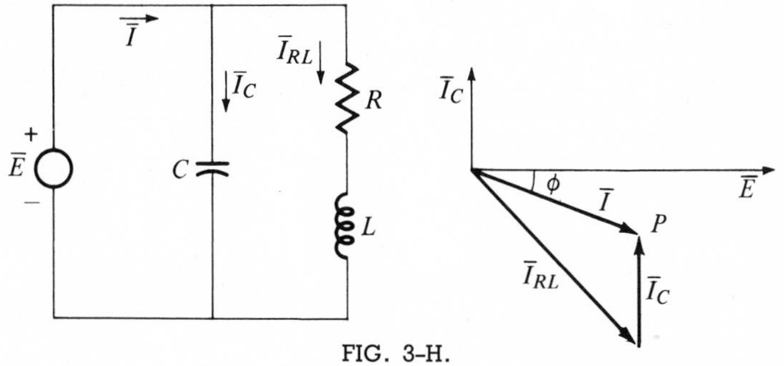

FIG. 3-H.

3-I. The circuit shown samples the signal $e(t)$ by transferring its value, when wanted, to the capacitor C_2, by briefly closing the switch, then opening it. The inductor L is adjusted so that one half cycle at the resonant frequency of the C_1-C_2-L loop is equal to the time the switch is closed. After the switch is reopened, the sample value (the voltage on C_2) is taken off to the right for "processing." When C_2 has been discharged, C_1 now being charged with a new voltage to be sampled, the sampling is repeated. For simplicity, let $e(t)$ be sinusoidal, at the frequency corresponding to a period of 125 μs. Close the switch for 1 μs. Find an appropriate value for L, and sketch the waveforms of i_1, i_L, and the voltages on C_1 and C_2. What is gained by the presence of L (in contrast to the switch alone)?

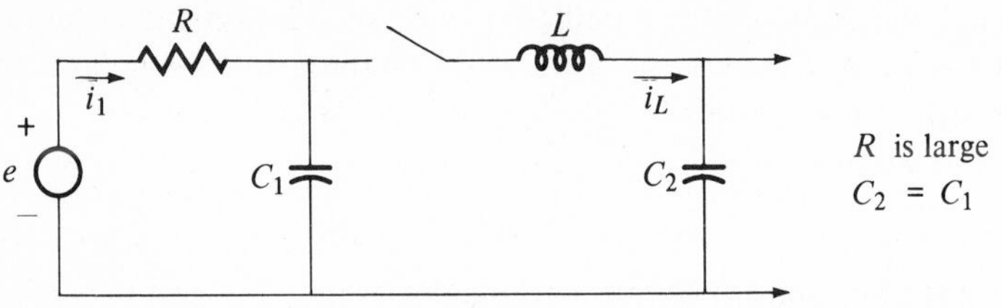

FIG. 3-I.

3-J. A tuned circuit (characterized by ω_0 and $Q = 10$) is excited sinusoidally, $i_F = I_F \cos \omega_F t$. Let $\bar{I}_F$ and $\bar{V}_F$ be the phasor representations of the excitation current and the steady-state response voltage. Let I_F be constant, and let ω_F vary. Plot, in a complex-number plane, the phasor $\bar{V}_F$ for $(\omega_F/\omega_0) = 0.1$, 0.5, 0.9, 0.95, 0.97, 1.00. Then rewrite (3.14-8), without approximation, as $T(j\omega_F)/R = (1 + j \tan \theta)^{-1}$. Where, in your plot, is the angle θ? How does θ vary with ω_F? Add to the plot phasors for (ω_F/ω_0) equal to the *reciprocals* of each of the values above. Why is very little additional computation necessary therefor? What is the geometrical figure that is the locus of the tips of the phasors? (See Exercise 2-40.)

3-K. In the circuit arrangement shown in Fig. 3-K, the two source currents are $i_{F1} = 10 \cos \omega_1 t$ mA, $i_{F2} = -2 \cos \omega_3 t$ mA, and $\omega_3 = 3\omega_1$. The half-power frequencies of the tuned circuit are $\omega_a = \omega_1$ and $\omega_b = 3\omega_1$.

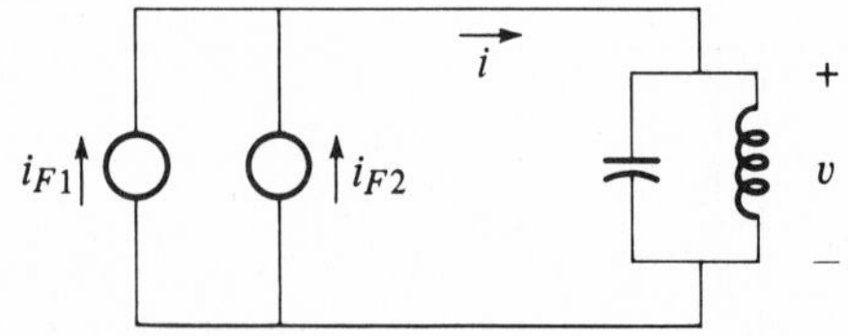

FIG. 3-K.

(*a*) Sketch the waves i_{F1}, i_{F2}, i, v for one full cycle of the steady state. Describe the nature and cause of the waveshape distortion (from wave i to wave v) caused by the tuned circuit.

(*b*) Find the Q of the tuned circuit and sketch its resonance curve.
(*c*) Change source frequency ω_1 to ω_0 (the resonant frequency of the tuned circuit), keeping $\omega_3 = 3\omega_1$, and repeat (*a*).
(*d*) Change source frequency ω_1 to $(\omega_0/3)$, keeping $\omega_3 = 3\omega_1$. Repeat (*a*).
(*e*) If a signal is the sum of several sinusoids of different frequencies, under what conditions will it be little distorted in passing through a tuned circuit? (Consider circuit Q and frequencies "in" the signal.)

3-L. Two very-high-Q tuned circuits are connected in series (Fig. 3-L). What is the transfer function (from i_F to v), completely neglecting dissipation of energy? (Express it in terms of s, C_1, C_3, ω_1, ω_3.) For what values of s is the transfer function zero? What is the physical meaning of these zeros? Locate the zeros with respect to the poles of the transfer function. Why do you suppose the second tuned circuit is characterized with subscript 3 instead of 2? Sketch $T(j\omega_F)$ versus ω_F.

Let $i_F = I_F \sin \omega_F t$; keep I_F constant and vary ω_F. Assume initial rest. What is the form of the natural part of v? Of the forced part? Sketch the peak value of the forced part versus ω_F. [*Note:* the properties found here, the mutual locations of zeros and poles and the form of the variation of $T(j\omega_F)$ with ω_F, are generally true of all two-terminal nondissipative networks.]

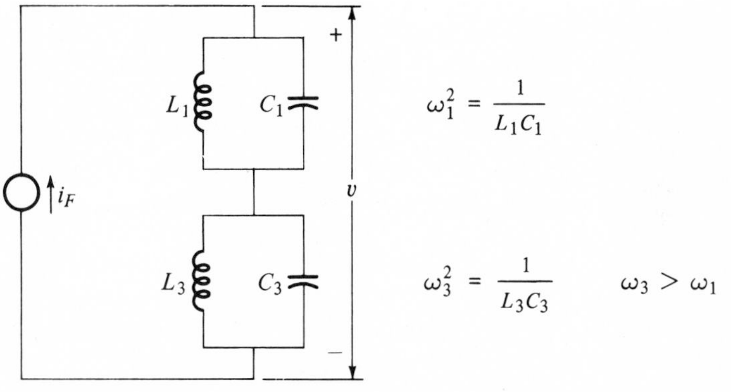

FIG. 3-L.

3-M. The voltage response, from rest, to impulse excitation $i_F = Q_0\,\delta(t)$, of the two-resistor form of the tuned circuit (Fig. 3.18-B) is to have the prescribed values 3, 2, 1, 0.4 (volts) at $t = 0$, T, $2T$, $3T$ (seconds), respectively.

(*a*) Find the requisite natural frequencies p_1 and p_2 and the coefficients K_1 and K_2 in the response form (3.03-6). Then find the four element values of the circuit in terms of Q_0 and T.
(*b*) Now change the prescribed value at $t = 3T$ from 0.4 to x, unknown. For what range of x can the prescription be met?

Suggestion: Let $\lambda_1 = e^{p_1 T}$ and $\lambda_2 = e^{p_2 T}$; then write the prescription thus:

$$\begin{aligned}
&1 \quad K_1 + K_2 = 3,\\
&2 \quad K_1\lambda_1 + K_2\lambda_2 = 2,\\
&3 \quad K_1\lambda_1^2 + K_2\lambda_2^2 = 1,\\
&4 \quad K_1\lambda_1^3 + K_2\lambda_2^3 = 0.4.
\end{aligned}$$

Now show, by multiplying (1) through by $(\lambda_1\lambda_2)$, (2) by $-(\lambda_1 + \lambda_2)$, and adding three equations, that

$$3(\lambda_1\lambda_2) - 2(\lambda_1 + \lambda_2) + 1 = 0,$$

which is one equation in the two unknowns $(\lambda_1\lambda_2)$ and $(\lambda_1 + \lambda_2)$. Obtain a second equation by multiplying (2) by $(\lambda_1\lambda_2)$, (3) by $-(\lambda_1 + \lambda_2)$ and then adding these two equations and (4). Solve the two equations for $(\lambda_1\lambda_2)$ and $(\lambda_1 + \lambda_2)$ and thus find a quadratic equation whose roots must be λ_1 and λ_2. Then find p_1 and p_2, K_1 and K_2. Then proceed to find the element values from the impulse response.

Note: a more elegant version of this technique for passing the sum of two (or more) exponentials through given points (interpolation by exponential sums) is given in Sec. 7.06.

3-N. An engineer is interested in a network whose response to impulse excitation is of the form shown in Fig. 3-N, made up of the three straight-line segments *ABCD*, *DE*, zero. He asks: "Can this (voltage) response wave be generated by an *RLC* circuit (with, if necessary, two resistors, as shown), initially at rest and impulse excited: $i_F(t) = Q_0\,\delta(t)$?"

(*a*) Why is the answer *no?* Reply in the simplest possible terms.

Faced with your reply the engineer says: "All right, I relax the stipulations to require only that the response pass through the *points* *A*, *B*, *C*, and *D*; let it curve as may be necessary in between them, and after $t = 3T$; but let the departure from the straight line *ABC* (in $0 < t < 2T$) be no more than necessary (no large oscillations, please)." As a flexibility bonus he also changes the stipulated value at $t = 3T$ to $-x$ and allows you to vary x as may be convenient: it need no longer be unity (but point *D* is to remain below the axis at a negative value).

(*b*) Design the circuit (find the element values) for the case $x = 0.5$ (which is possible). Check that the response does pass through *A*, *B*, and *C* as well as (the revised) *D*. Show, by a careful plot, how much it departs from the straight-line segments over the range $0 < t < 6T$.

(*c*) For what *range* of x can the circuit meet the requirements? At the extremes which element(s) disappear from the circuit?

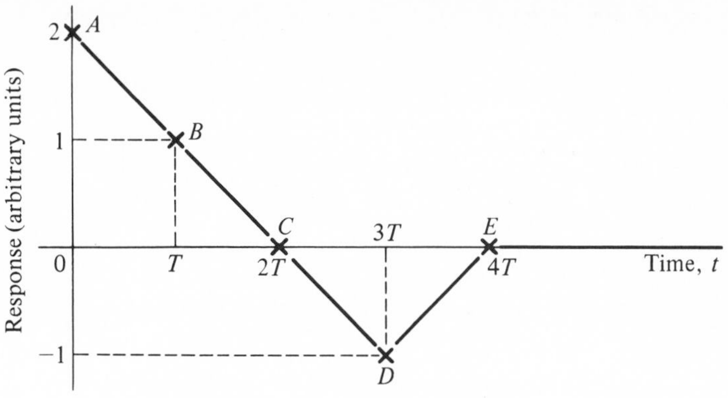

(*a*)

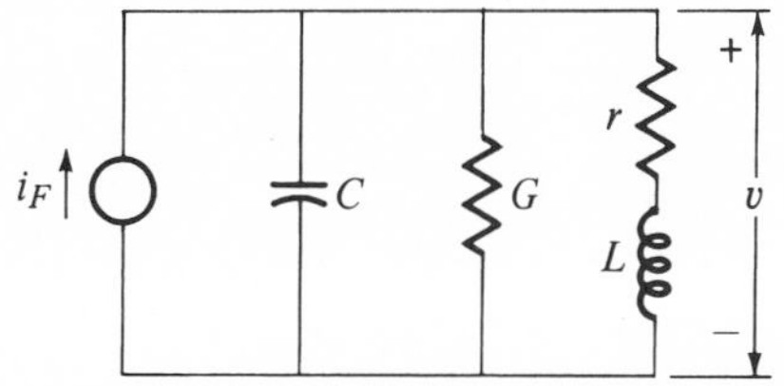

(*b*)

FIG. 3-N.

(*d*) Show that, if x lies outside the range determined in (*c*), a nonzero initial inductor current will enable you to meet the requirements. What must be the value of that initial current?

(*e*) Discuss the other solutions that become possible if the "no large departures from *ABC*" requirement is removed, and oscillations are permitted.

3-O. The *RC* circuit shown is a model of a system (such as an amplifier interstage network, or part of a mechanical control system) in which the resistor *R* represents a (fixed) load. The capacitor *C* represents an irreducible parasitic effect and i_F represents a command or excitation signal. The system is initially at rest; $i_F(t) = I_0 u(t)$. The response of interest, $i(t) = I_0(1 - e^{-\alpha t})$, is annoyingly sluggish in its rise to the final value, I_0 (Fig. 3-O*b*).

(*a*) Sketch $i(t)$ and explain concisely and forcefully why *increasing* the value of *C* (which is physically possible) will not improve matters at all. (*C* cannot be *de*creased.)

(*b*) Consider amelioration by adding an inductor in the circuit model at the point *X*, so that oscillations will be present in the response. Let

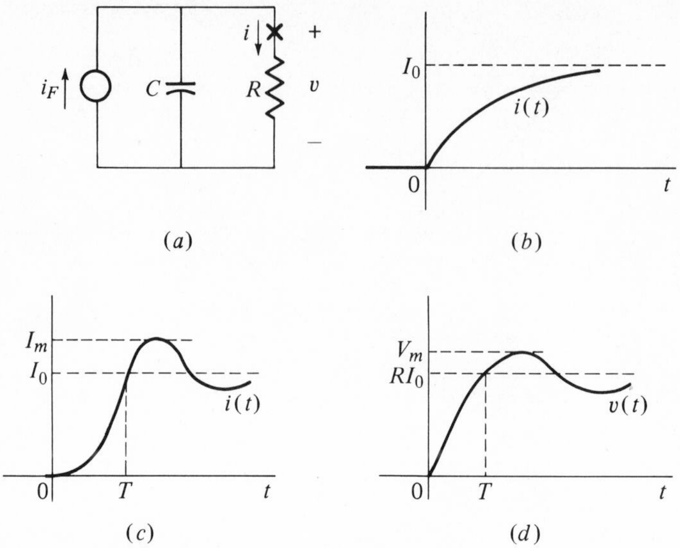

FIG. 3-O.

p_1 and p_2 be the new natural frequencies. What are their values in terms of R, C, L? What is the nature of the natural component of the response $i(t)$? Of its forced component? Show that now

$$i(t) = I_0[1 - (\sec \phi_i)e^{-\alpha t} \cos (\omega_N t + \phi_i]$$

and

$$v(t) = RI_0[1 - (\sec \phi_v)e^{-\alpha t} \cos (\omega_N t + \phi_v].$$

Evaluate ϕ_i and ϕ_v as simply as possible. [*Suggestion:* use the initial values of i, v and their derivatives; note that α here is different from α in (*a*).]

(*c*) Evaluate the improvement in response speed in terms of the price paid (the degradation in response smoothness) by plotting overshoot I_m versus response time T, as defined in Fig. 3-O*c*.

(*d*) Repeat (*c*) for the *voltage* response v (Fig. 3-O*d*).

(*e*) Which of the two responses, i and v, is "better" after the inductor is inserted? Explain.

3-P. This problem deals with the "head end" or initial stages of a simple radio receiver in which tuned circuits are put to work. Assume that the electromagnetic field produced by the radio transmitter acts on the antenna to produce a signal voltage v_0, that v_0 produces a current i_1, and that i_1 produces a voltage v_1. (See Fig. 3-P*a*.) Later stages in the receiver take the signal v_1, and amplify

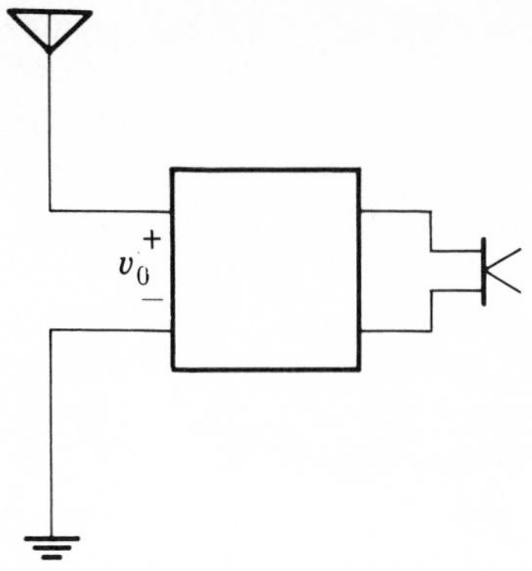

(*a*)

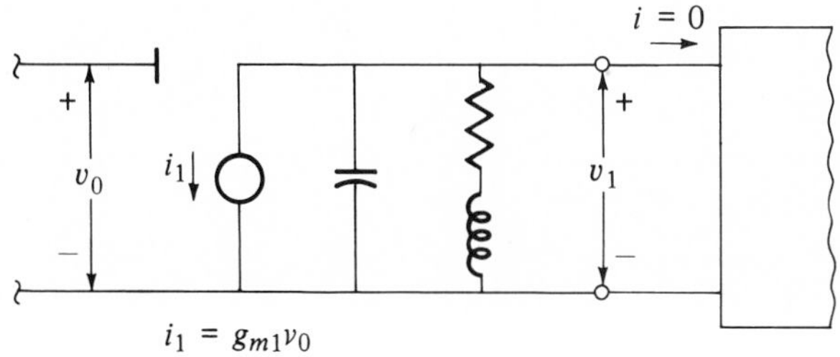

(*b*)

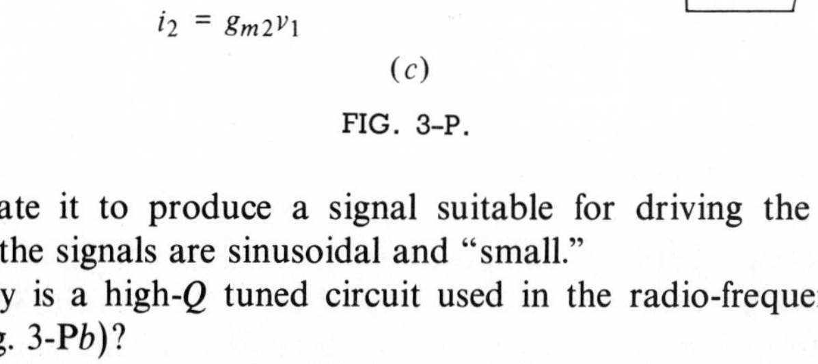

(*c*)

FIG. 3-P.

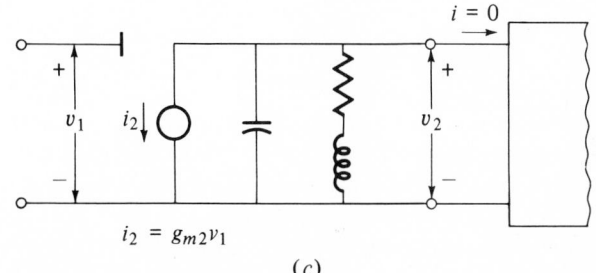

and manipulate it to produce a signal suitable for driving the loudspeaker. Assume that the signals are sinusoidal and "small."

(*a*) Why is a high-Q tuned circuit used in the radio-frequency amplifier (Fig. 3-P*b*)?

(*b*) To what frequency should the circuit be tuned to receive your favorite (AM, broadcast band) radio station? Call this number f_0 in the rest of the problem.

When the station is broadcasting, it emits signals that are very complicated. We assume here that they add to the "carrier" frequency f_0 additional "side" frequencies that extend 5 kHz on each side of f_0. To receive well, then, one would like the response, as a function of driving-signal frequency, (1) to be fairly flat in the band from $(f_0 - 5\text{ kHz})$ to $(f_0 + 5\text{ kHz})$. At the same time, other stations are assigned positions in the radio-frequency spectrum, some not far away from f_0. The nominal spacing of station frequencies is 10 kHz, by law, though one that close in frequency is unlikely to be close geographically. Still the receiver response ought to be poor at frequencies above $(f_0 + 5\text{ kHz})$ and below $(f_0 - 5\text{ kHz})$ because we want to reject other-station signals. Hence the response ought also (2) to drop off rapidly beyond the limits of the 10-kHz band of interest, for discrimination.

(*c*) Explain why a simple *RLC* circuit of the kind studied in this chapter cannot really do either of these jobs (1) and (2) well, but can do both in a "sort of" or approximate fashion.

(*d*) What Q should the amplifier tuned circuit have if the steady-state response is supposed to be down to half of its peak value at frequency $(f_0 + 5\text{ kHz})$? What then is its response at frequency $(f_0 - 5\text{ kHz})$?

(*e*) Plot the response of your tuned circuit, with the chosen Q, in dBV (Appendix D) over the frequency range $(f_0 - 10\text{ kHz})$ to $(f_0 + 10\text{ kHz})$. Use an arithmetic frequency scale. Criticize the response in terms of (1) and (2) above.

Suppose that we now consider making this radio-frequency amplifier out of *two* cascaded stages, each with its own tuned circuit. Consider "stagger tuning" them so that the two f_0 are not the same.

(*f*) Show that the decibel measure of the response of two cascaded stages (Fig. 3-P*c*) is the *sum* of the individual decibel measures of stage transmission.

(*g*) Let both stages have the same Q as before. Tune them, one to $(f_0 - F)$, the other to $(f_0 + F)$. Tabulate the overall response, in dBV, from $(f_0 - 10\text{ kHz})$ to $(f_0 + 10\text{ kHz})$ at intervals of at most 0.5 kHz, for each of these cases: $F = 0, 1, 2, 3, 4$ kHz. Now discuss, using sketches, the relative merits of the various tunings, with due regard for (1) and (2) Which seems the best? Why?

Note: Much work can be saved by judicious use of a template and dividers; alternatively, a simple computer program for this job is not hard to write.

(*h*) Find the L and C element values, and the requisite inductor Q, for your chosen amplifier stages.

3-Q. Consider an amplifier composed of four cascaded (but noninteracting) stages, each of which is essentially a tuned circuit. (See Prob. 3-P.) Plot the decibel measure of the overall transmission (transfer function magnitude) in the sinusoidal steady state versus frequency on scales that will clearly exhibit the amplifier's interesting properties. Plot also the amplifier's phase shift. Make the plots for two cases, the eight natural frequencies being (*a*) $-\alpha_1 \pm j(\omega_0 \pm \beta_1)$, $-\alpha_1 \pm j(\omega_0 \pm \alpha_1)$ and (*b*) $-\alpha_1 \pm j(\omega_0 \pm \beta_1)$, $-\beta_1 \pm j(\omega_0 \pm \alpha_1)$, in which $\alpha_1 = 0.003827\omega_0$, $\beta_1 = 0.009239\omega_0$. (Skillful use of a template *may* require less time than using a computer.)

Compare the two transmission curves in flatness at and near ω_0. (The positioning of the natural frequencies for response flatness or other desired shape is an important part of amplifier design, not necessarily a simple process.) Can you detect any corresponding tendencies in the phase shifts?

3-R. For a high-Q tuned-circuit amplifier stage (see Prob. 3-P) for frequencies near resonance,

$$|T(j\omega_F)|^2 = \frac{\text{constant}}{1 + 4Q^2[(\omega_F - \omega_0)/\omega_0]^2}.$$

(*a*) Verify this statement and explain why it is technically only approximately true, but practically is virtually correct.

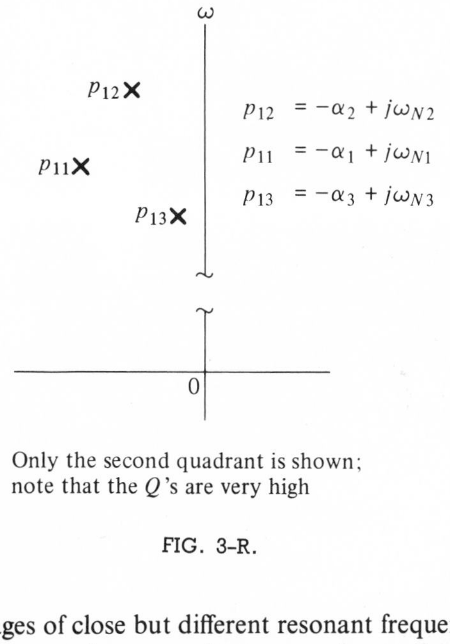

Only the second quadrant is shown;
note that the Q's are very high

FIG. 3-R.

When several stages of close but different resonant frequencies are cascaded, it is convenient to measure frequency from some (central) reference frequency ω_C according to $\Omega = (\omega_F - \omega_c)/\omega_c = (f_F - f_c)/f_c$.

(*b*) Show that then

$$|T|^2 = \frac{\text{constant}}{1 + a_0 + a_1\Omega + a_2\Omega^2}$$

and evaluate the constants a_0, a_1, a_2 in terms of f_0, f_c, and Q.

For the multistage amplifier, properly constructed, the individual $|T|^2$ expressions simply multiply to give the overall $|T|^2$.

(*c*) Show that, for a three-stage amplifier,

$$|T|^2 = \frac{\text{constant}}{1 + b_1\Omega + b_2\Omega^2 + \cdots + b_6\Omega^6}.$$

(*d*) If the transmission characteristic is to be maximally flat (as flat as possible at frequency ω_c, given the constraints implied), show that b_1, $b_2, \ldots, b_5$ must vanish.

(*e*) Show that b_1, b_3, and b_5 can be made zero simply by suitably positioning the natural frequencies p_{12} and p_{13} (see Fig. 3-R) with respect to each other and to p_{11}, taking p_{11} as given. (*Suggestion:* Consider symmetry.) Take $\omega_c = \omega_{N1}$.

(*f*) Show that, in addition, b_2 and b_4 can be made zero by placing p_{12} in a certain definite position. (All natural frequencies are now determined.)

(*g*) Take $\alpha_1 = 0.005\omega_{N1}$. Locate the remaining natural frequencies according to the results of (*e*) and (*f*). Plot the (logarithmic) transmission versus frequency from $0.98\omega_c$ to $1.02\omega_c$, choosing scales that clearly display the details. (Choose the constant to give zero dB at ω_c.)

(*h*) Superpose the amplifier transmission characteristic as it would be if the three stages were identical, all (second-quadrant) natural frequencies at p_{11}. Compare the two curves as to flatness and point out what has been changed by the "stagger-tuning" process above.

(*i*) Find the element values of the circuits of an amplifier that is "maximally flat stagger-tuned" at 3.6 MHz, with 1000-Ω resistors in the parallel *RLC* circuits. Use the *Q* value of (*g*) for the center stage.

3-S. A high-*Q* parallel *RLC* circuit is excited, from rest, by a sinusoidal current of the resonant frequency (that is, $\omega_F = \omega_N$). Neglect the difference between ω_N and ω_0.

(*a*) Show by a sketch the relation between the excitation current and the forced component of the response voltage.

(*b*) Explain why there *must* be a natural (transient) component in the response; no matter what the initial phase of the excitation, the natural part cannot be zero.

(*c*) Suppose the excitation is applied at a positive peak: $i_F = I_F \cos \omega_0 t$. Show that if an impulse excitation is *also* applied, the transient can be eliminated. What is the requisite strength of the impulse, Q_0?

(*d*) Find the natural component of the response of the circuit to excitation $I_F \cos \omega_0 t$ (alone), from rest, using previous results and a minimum of labor.

(*e*) Find a simple formula for the total response in (*d*) and sketch it from $t = 0$ into the steady state.

(*f*) Suppose the excitation applied at a *zero* of its wave, $i_F = I_F \sin \omega_0 t$. What additional excitation, very simple in form, applied also at $t = 0$, will remove the transient and cause the circuit to go immediately into the steady state?

(*g*) Find the natural component of the response of the circuit to excitation $I_F \sin \omega_0 t$ alone, from rest, again with the least possible work.

(*h*) Find a simple formula for the total response in (*g*) and sketch it from $t = 0$ into the steady state.

(*i*) If the circuit Q is 20, what are the relative sizes of the (peak) steady-state currents in the three elements?

(*j*) If the excitation frequency is *not* equal to the resonance frequency, there can be no simple formula like those of (*e*) and (*h*) above, for the total response. Why?

3-T. A high-Q tuned circuit, resonant at frequency f_0, is excited by a series of short current pulses recurring at the rate f_c pulses per second, $f_c = f_0$. Sketch several cycles of the steady-state response wave. Explain why it is almost a sinusoid and evaluate its peak value in terms of Q_0 (the charge in each pulse), C (the capacitance), and Q (the circuit Q). (See Fig. 3-T.)

The pulses may, for example, be the peaks (clipped off at a certain level) of a sinusoid, the output of a highly nonlinear (class C) amplifier, operated in

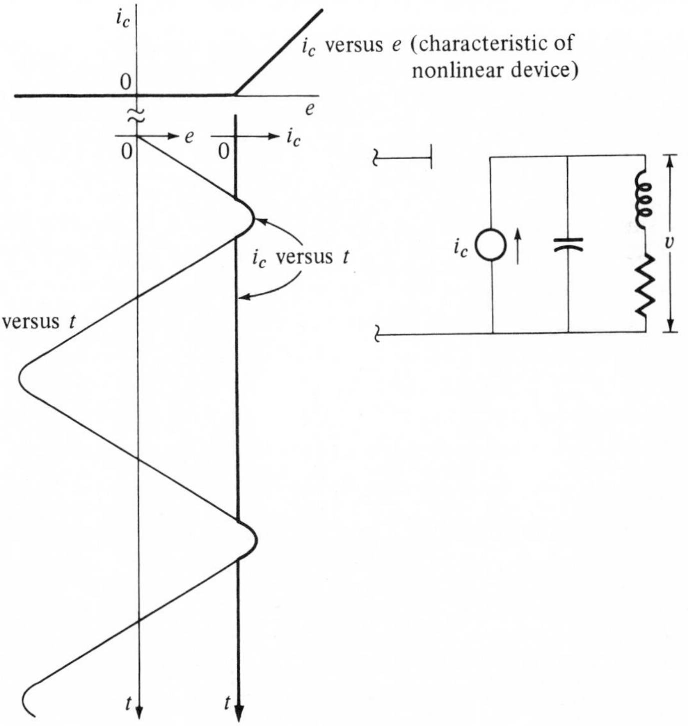

FIG. 3-T.

this fashion because the energy efficiency is then high (large amounts of power may be involved). Suppose the (slowly varying) signal information being amplified is contained in the amplitude of the wave that is clipped. Explain how the tuned-circuit voltage response is virtually a sinusoid whose amplitude is modulated (varied) according to the signal information.

Why may the combination aptly be called a *linear* class C amplifier in spite of the highly nonlinear character of the electronic part? Where is the (large) signal output power dissipated or where does it go?

3-U. The transfer function (3.10-14) evaluated for $s = j\omega_F$, for the sinusoidal steady state, is of course (*a*) a function of ω_F and (*b*) a complex number having two parts, for each value of ω_F. A complete graphical representation would require a three-dimensional "plot." More practical is a set of two-dimensional plots that give the same information. In the text, for example, the magnitude and the angle of $T(j\omega_F)$ are separately plotted against ω_F. Here we discuss alternatives.

Plot, for $Q = 15$, say, the *real* and *imaginary* parts of $T(j\omega_F)$ versus ω_F. Why do you suppose these are less often used than the magnitude and angle plots?

Suppress the variable ω_F and plot $T(j\omega_F)$ as a complex number in its own plane: x = real part of $T(j\omega_F)$, y = imaginary part thereof. Show that, for the circuit of Fig. 3.01-A, the *locus* of the tip of the complex number T is a circle. In what way does this depend on the value of Q? How is it altered by the introduction of dissipation in the inductor (Fig. 3.18-B)? *Suggestion:* Write $T(j\omega_F)$ in the forms (*a*) constant/$(1 + j \tan \theta)$ and (*b*) constant $\times$ $(1 + jL\omega_F/r)/(1 + j \tan \theta)$ in which the constants and the definition of θ are not necessarily exactly the same.

3-V. The single-RC-circuit decoding scheme of Prob. 2-T requires a certain precision of pulse and measurement timing. An improvement is to put "shelves" in the impulse response of the decoding network (Fig. 3-V*a*), for it then becomes less important that the pulses occur at exactly equal intervals and that the measurement be made at exactly a certain instant.

(*a*) Explain the meaning of the last sentence above in your own words.

(*b*) Show that the *RC* and *RLC* circuit combination of Fig. 3-V*b* can be designed to have just such a staircaselike impulse response. One procedure is:

1 Sketch the impulse response of an *RC* circuit alone.

2 Sketch the impulse response of an *RLC* circuit alone (assume a moderate value of Q, say 5 or 10).

3 On a third sketch, based on these two, show that the slope of the sum thereof can be made zero at time T_1 and repetitively thereafter (Fig. 3-V*a*), thus producing some semblance of shelves. Show also that the time T_1 must lie between certain specific limits the upper of which is less than T.

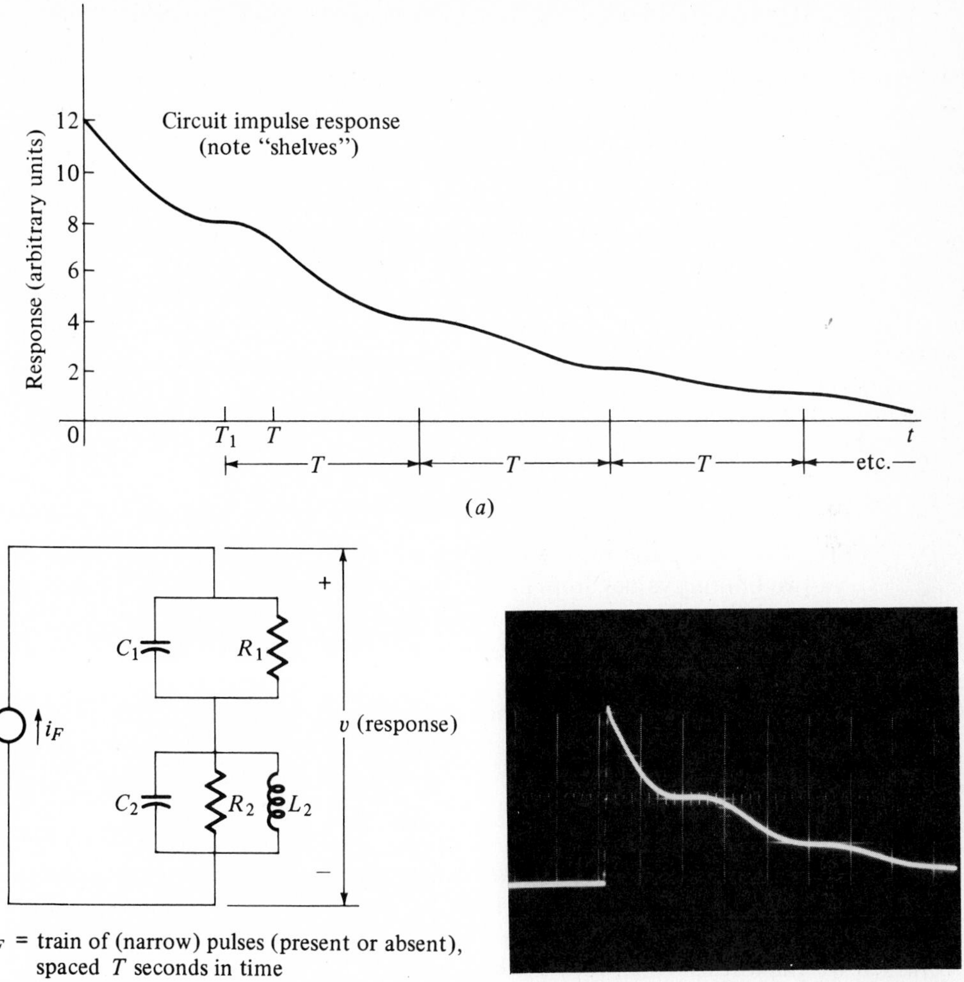

(*b*) (*c*)

FIG. 3-V.

(*c*) Explain why, as a matter of common sense, the following parameter relations must hold:

1 $\alpha_1 = \dfrac{1}{R_1C_1} = \alpha_2 = \dfrac{1}{2R_2C_2}$,

2 $\omega_N = \dfrac{2\pi}{T}$,

3 $e^{-\alpha T} = \frac{1}{2}$ in which $\alpha = \alpha_1 = \alpha_2$.

Plot the resulting natural frequencies to scale in the p plane. What is the Q of the tuned circuit?

(*d*) Since T_1 must be less than T, the incident pulse is not reduced to half its value at T_1, though it is halved in each succeeding shelf interval. Explain why this is of no consequence in the decoding process. [The measurement of the decoded value will be made at the shelf following the last pulse (or no-pulse) instant.] Explain again why the shelf *does* reduce the precision demanded in pulse spacings and in location of the measurement instant.

(*e*) Three parameter relations have been established in (*c*) so that in effect three of the five element values are now fixed. There remains essentially only one parameter (one parameter is the vertical scale, not important except in deciding on a practical impedance level). For this parameter, yet to be determined, use the symbol $k = C_1/C_2$. Show that the value of T_1 must lie between two well-defined numerical limits if a shelf (defined as a point of zero slope of the impulse response) is to occur. What are these limits?

(*f*) Plot the impulse response, from rest, for T_1 equal to the mean of the two limiting values, from $t = 0$ through at least the third shelf thereafter. Why is a logarithmic response scale informative?

(*g*) Show that if k is given a certain particular value within the allowed range of (*e*), the *second* derivative of the impulse response is *also* made zero at the shelf points, thus improving the flatness of the shelves. Plot the curve for this value of k and compare with the previous plot.

(*h*) Find the element values for the network for a pulse spacing of 25 μs and a 1000-Ω R_1.

The photograph (Fig. 3-V*c*) shows the actual narrow-pulse response of such a circuit combination. Does it perform properly? (For the genesis of this problem, see *BSTJ*, January 1948, pp. 36–38. Contemporary PCM systems use other, weighted-resistor-network methods instead.)

3-W. (*a*) Obtain the differential equation for the voltage v in Fig. 3-W. [*Suggestion:* Verify the equations $i_F = i_1 + i$, $i_1 = G_1(v + (1/C)\int_0^t i\,dx)$, $i = i_L + i_R$, $v = L\,di/dt$, $v = Ri_R$; substitute into the first of these, differentiate, and differentiate again.] Evaluate the natural frequencies. What is the circuit Q?

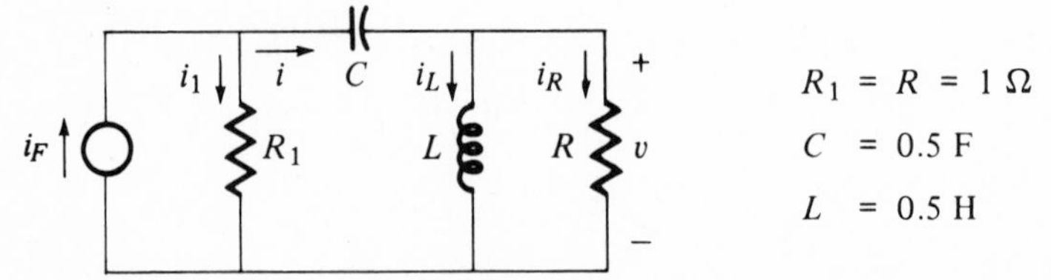

FIG. 3-W.

(*b*) Calculate and plot the response v (from rest) for (1) $i_F = Q_0\,\delta(t)$, (2) $i_F = I_0\,u(t)$, $i_F = I_0\,e^{-\beta t}u(t)$, $\beta = 1$ Np/s. [*Suggestion:* Determine what happens in the circuit in the interval $0 < t < 0+$; mark the $t = 0+$ conditions on the circuit diagram; evaluate $v(0+)$ and $v'(0+)$.]

By inserting a comparatively small resistor in series with the inductor, and observing the voltage across it, obtain a photograph of the inductor current to add to Fig. 3.08-D. Does this current start as a ramp (proportional to time)? Explain formally and use the photograph for corroboration.

3-X. Show that the response (3.13-1), in high-Q cases, when $\omega_F = \omega_N(1 + \varepsilon)$, ε being small, can be written $v = V_F[1 + e^{-\alpha t}\cos(xt + \phi_x)]\cos(yt + \phi_y)$ with excellent accuracy. Here x represents a very small ("beat") frequency and y is essentially the excitation frequency. What are x and y in terms of ω_N and ε? What differences occur if the sign of ε is reversed? Show that the response can alternatively be written $v = 2V_F\cos(At + \phi_A)\cos(Bt + \phi_B)$ in which the frequency A is one-half the difference frequency $\varepsilon\omega_N$, and B is essentially the excitation frequency. Explain Fig. 3.13-D and the expression "beats" in terms of these two formulas.

3-Y. (*a*) Find the differential equation satisfied by v in Fig. 3-Y. [One method is: (1) Obtain the three equations $v = Ri_G$, $v = L\,di_L/dt + r_2 i_L$, $C\,dv/dt = r_1 C\,di_C/dt + i_C$; (2) calculate $r_1\,C$ times the derivative of the second one, and L times the derivative of the third, add these and the second and third equations themselves, replace $(i_C + i_L)$ by $(i_F - Gv)$, which is a sort of simultaneous-equation solution process; a much neater, simpler, clearer view is to come, in Secs. 5.11 and 6.12.]

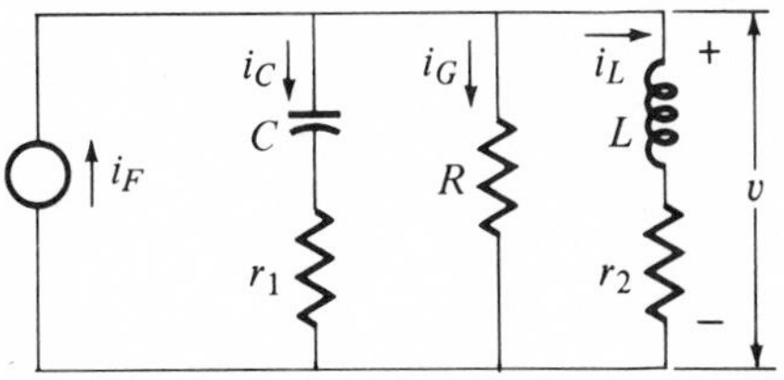

FIG. 3-Y.

(*b*) Find the (characteristic) equation, whose roots are the natural frequencies.

(*c*) Assume these are complex and evaluate ω_0 and α in the $p^2 + 2\alpha p + \omega_0{}^2 = 0$ form of the characteristic equation. Evaluate the circuit Q.

(*d*) Starting from rest, apply impulse excitation, $i_F = Q_0\,\delta(t)$. Trace the physical path of the charging current i_F and evaluate $v(0+)$.

(*e*) By considering the derivative of the current conservation equation (Kirchhoff's current law) $i_C + i_L + i_G = i_F$, and previous equations, evaluate $v'(0+)$.

(f) Obtain the resonant frequency, the Q, and the impulse response for the case $L = 200$ mH, $C = 5000$ pF, $r_1 = 0.9\ \Omega$, $r_2 = 90\ \Omega$, $R = 100$ kΩ.

(g) Obtain the step response for the same set of element values.

3-Z. The utility of resonance is not limited to high frequencies, nor to radio communication. It is useful also in electric power supply at (for example) 60 Hz, albeit under another name.

Consider the supply of power to an industrial load. For precision, let the load be 500 kW supplied at 13 kV (rms), at 80 percent power factor. A simple model that adequately translates this jargon is that shown in Fig. 3-Z*a*, in which R represents the actual energy-consuming load; L represents unavoidable inductive effects therein (due, for example, to the iron used in motors that form part of the load). The steady-state current $i_L = i_{Lm} \cos(\omega_F t - \phi)$ is such that the average power dissipated in the load is $\frac{1}{2}I_{Lm}{}^2R = 500$ kW, and that $\cos\phi = 0.8$. The cosine of the phase difference between e and i_L is called the *power factor* because it expresses the reduction in actual power delivered from the nominal (what one might casually expect) $\frac{1}{2}E_m I_{Lm}$.

Because of the inductance, the utility company must supply a current I_L (rms value) that is larger than it need be if the objective is simply to deliver 500 kW. This is a nuisance and a costly one since the utility company's transmission losses depend primarily on the square of the current. Suppose now that a corrective capacitance C is added (Fig. 3-Z*b*), one that resonates with L at the supply frequency ω_F. Note how the current to be supplied changes: the utility company need now supply only the current I, which is less than I_L by the power factor of the load, $\cos\phi$. The same power is delivered to the load, but,

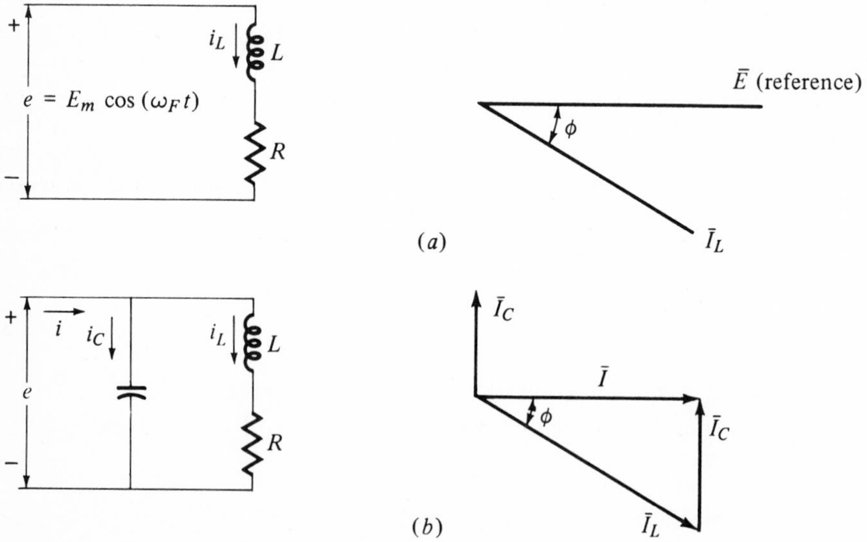

FIG. 3-Z.

because of the *power-factor correction* that has been made, the line current needed is smaller and therefore less costly. This is reflected in this simplified tariff:

Energy charge: 1.7 cents per kilowatthour;

Power-factor penalty: 0.1 percent of energy charge for each 0.01 that power factor is less than 0.90;

Power-factor credit: 0.1 percent of energy charge for each 0.01 that power factor exceeds 0.90.

A power factor of 0.9 is considered reasonable, and the energy charge is based thereon. (Various other factors affect the actual billing: there are demand charges, energy rates that vary with consumption, etc.; these we neglect.) A better power factor will reduce the net energy cost; a poorer one will increase it.

(*a*) Calculate the instantaneous power flow into the load in Fig. 3-*Za*, $p = ei_L$, and show that its average value (the value of interest) is $\frac{1}{2}E_m I_{Lm} \cos\phi$ (see Sec. 3.16). In rms notation, commonly used in power work ($E = E_m/\sqrt{2}$, $I_L = I_{Lm}/\sqrt{2}$, $I_C = I_{Cm}/\sqrt{2}$, $I = I_m/\sqrt{2}$), this is $EI_L \cos\phi$.

(*b*) Calculate the capacitor current in Fig. 3-*Zb* and call its peak (maximum) value I_{Cm}.

(*c*) Verify the phasor diagrams given (and note how simple they are!).

(*d*) Find the capacitance C necessary to fully correct the power factor in the case given. (This may, in practice, be supplied by a capacitor as such, or, especially for large installations, by a rotating machine whose terminal current can be made to lead the voltage by nearly 90°.) What is the Q of the "tuned circuit"?

(*e*) Suppose the load operates 600 hours a month. What is the net energy cost per month (1) without power-factor correction, (2) with full power-factor correction?

(*f*) What is the maximum economically feasible cost of the corrective capacitance per annum?

Chapter **4**

SUPERPOSITION INTEGRALS AND CONVOLUTION

Twas brillig, and the slithy toves
Did gyre and gimble in the wabe:
All mimsy were the borogoves,
And the mome raths outgrabe.

4.01 INTRODUCTION

We have thus far discussed only one circuit, really, that of Fig. 3.18-B. (Making L indefinitely large therein reduces it to the RC circuit of Chap. 2.) We have done that at great length, and in so doing we have introduced, in this simple case, many of the ideas we need for network analysis in general.

The question: "What should we do next?" has no unique answer. But it does seem expedient to now discuss excitation waves of arbitrary shape, for our excitation examples have been rather limited: only the impulse, the step, and the exponential. We shall try to do this more generally than just for the RLC circuit, or even for some third- or fourth-order network. And we shall try to develop a method for handling *any* reasonable excitation function.

We are going to take a big step now, for we divorce ourselves from the RC and RLC circuits to deal with *arbitrary* networks. But the study of those two circuits has been very profitable, because they are very important in practice and because they have given us some general ideas of what network responses are like.

We assume that somehow (by analysis, from experiment, from ideas for a network design) we know a network's response to some simple wave, but nothing more about the network. We then ask

> Can the response of a network to an *arbitrary* excitation wave be calculated from its response to a *simple* wave, such as those we have already discussed, without additional information? (4.01-1)

So to calculate would be elegant and practical: we could never complete the analyses for all possible excitation waves if taken one by one (and that not just because of laziness). It should be very efficient. And so we look into the matter.

We have observed that *superposition* of waves can sometimes be very convenient. We can from two step functions make a pulse function (Sec. 1.09). We can then calculate the response to a pulse from the response to a step (Sec. 2.11). So it is possible that we have here the beginnings of an answer to our question. At any rate it suggests two further questions:

> **a** Can an arbitrary excitation $e(t)$ be looked on as a sum of simple waves? If so, of what simple ones?
>
> **b** If so, can a circuit's responses to the individual waves in that sum be legitimately added to obtain the response to $e(t)$? (4.01-2)

We have work to do.

4.02 SUPERPOSITION

There seem to be two kinds of superposition. The *first* (see Sec. 1.09) is the reverse of the mathematical decomposition of a wave into an equivalent set of additive components: summing these components to regain the original wave or function. It requires no precautions except the obvious one: to make sure the work is properly done, to verify that the sum is correct. Such decomposition of a function is never unique; and if the number of components is infinite, this superposition may become a delicate matter.

The *second*, the sort practiced in Sec. 2.11, is a different matter. And it is sufficiently important in network theory to warrant careful discussion here. This superposition is of a network's responses to various excitations to obtain (presumably) the network's response to all the excitations applied together. And it certainly is not always valid. Figure 4.02-A is a schematic diagram of such a situation. The same network N, excited by three different excitations, gives the three responses shown. We ask: "If $e_3 = e_1 + e_2$, does $r_3 = r_1 + r_2$?"

Superposing r_1 and r_2 is not like the simple addition of e_1 and e_2 to form e_3. An operation has been performed (by the network) on e_1 and on e_2, the

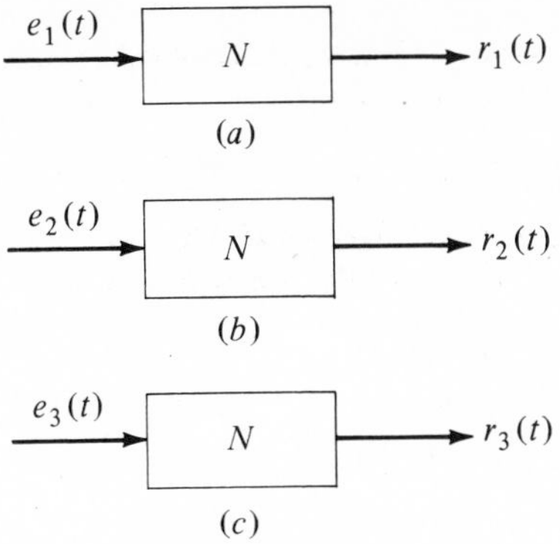

FIG. 4.02-A.

components of the decomposition of e_3, to obtain r_1 and r_2. And there is no reason that the same operation on $(e_1 + e_2)$ should give $(r_1 + r_2)$. A simple example is a network that consists of a slightly nonlinear resistive device such that the response r is related to the excitation e by

$$r = A_1 e + A_2 e^2. \tag{4.02-1}$$

[The nonlinearity lies in the e^2 term, not a (straight) line when plotted.] Then, in the notation of Fig. 4.02-A,

$$\begin{aligned} r_1 &= A_1 e_1 + A_2 e_1{}^2, \\ r_2 &= A_1 e_2 + A_2 e_2{}^2, \\ r_3 &= A_1(e_1 + e_2) + A_2(e_1 + e_2)^2 \\ &= r_1 + r_2 + 2A_2 e_1 e_2 . \end{aligned} \tag{4.02-2}$$

Superposition (addition) of r_1 and r_2 does *not* give r_3.

One says immediately: but such a network is *nonlinear* and so the result is no surprise; if it were *linear*, superposition would be valid. But even that statement requires careful interpretation. If the network response excitation characteristic were the linear relation

$$r = A_0 + A_1 e, \tag{4.02-3}$$

then the calculations for Fig. 4.02-A would be

$$\begin{aligned} r_1 &= A_0 + A_1 e_1, \\ r_2 &= A_0 + A_1 e_2, \\ r_3 &= A_0 + A_1(e_1 + e_2) \\ &= r_1 + r_2 - A_0 . \end{aligned} \tag{4.02-4}$$

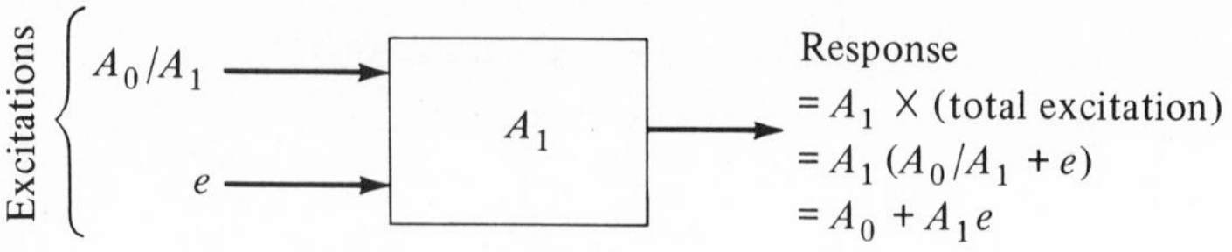

FIG. 4.02-B.

Superposition (addition) of r_1 and r_2 to give r_3 is again *not* valid.

Something more than linearity of characteristics in the ordinary sense is required to *ensure* that superposition be valid. To limit the discussion to essentials, let us require that a device with a linear characteristic like (4.02-3), with offset A_0, be modelled in our networks as a *source* (A_0/A_1) and a *network element* A_1 whose characteristic is linear *and* passes through the origin, $r = A_1 e$. We then have to consider not Fig. 4.02-A but Fig. 4.02-B for which the response *is* clearly the superposition of the responses to the various excitations: (A_0/A_1) and each of the (additive) components of a decomposition of e.

A simpler method is to require that we work only with *incremental* signals, measured from some convenient reference(s). This is how linear network theory is made applicable to many inherently nonlinear devices, such as transistors, or iron-core inductors. It is of course also necessary to limit the magnitudes of all incremental currents and voltages for the linear models to be valid.

The resistor, inductor, capacitor, and controlled-source elements used in previous chapters are all characterized by straight lines that pass through the origin (Fig. 4.02-C). The presence of derivatives and integrals in the laws of element behavior does not upset "linearity through the origin," *provided* the lower limits of integrals are properly treated. (They correspond to nonzero initial conditions and are properly interpreted as sources anyway; see Sec. 7.22 and Exercise 4-7.)

Any variation of an element's character with time could easily make superposition invalid, also. But if we limit our network models to the elements of Fig. 4.02-C, all characterized by the basic postulates "lumped," "linear," "constant," we may then safely postulate that superposition of the responses to individual excitations *will* give the response to the sum of those excitations. (The converses of these arguments are not relevant to our work here.)

To recapitulate, we consider only linear, "time-invariant" *networks*, to which superposition can be applied, and which, by definition, contain only elements that are

Lumped (i.e., of "network" type, with no external fields),

Linear [the element characteristic is a straight line through the origin (Fig. 4.02-C)], (4.02-5)

Constant (the element characteristic does not change with time).

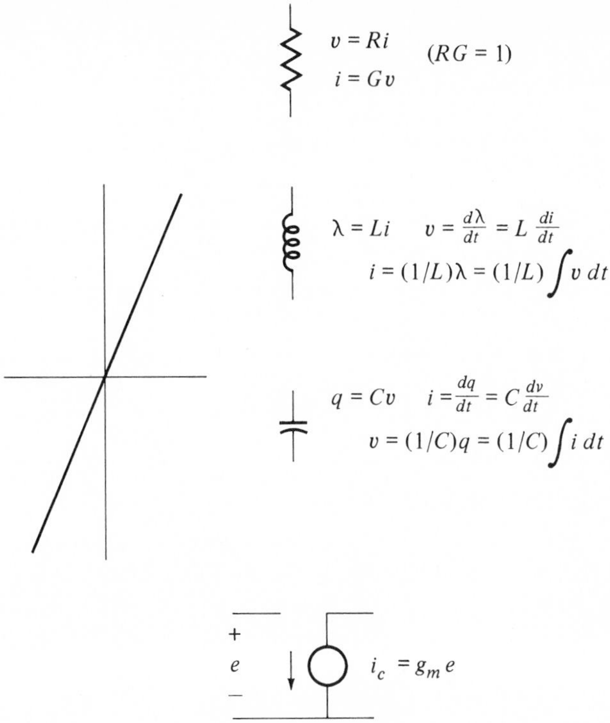

FIG. 4.02-C.

Nonzero initial conditions are to be treated as *sources* (see Sec. 7.22), as are characteristic offsets.

Some definitions of linearity use network behavior; we prefer to define linearity in terms of network *construction*, as above. [The analysis of networks that contain nonlinear or time-varying elements is essentially different, and not discussed here; elements or systems with fields (distributed, not lumped) require partial differential equations in their description.]

4.03 IDEAS

To the two questions of (4.01-2) we now have satisfactory answers. We proceed to the basic question, (4.01-1). The reply is almost certainly *yes*, we feel, and probably the "yes" is to be qualified by "and in various different ways."

Let us take the simplest excitation wave we know, and attempt to decompose an arbitrary excitation into a sum of such waves. That simplest excitation

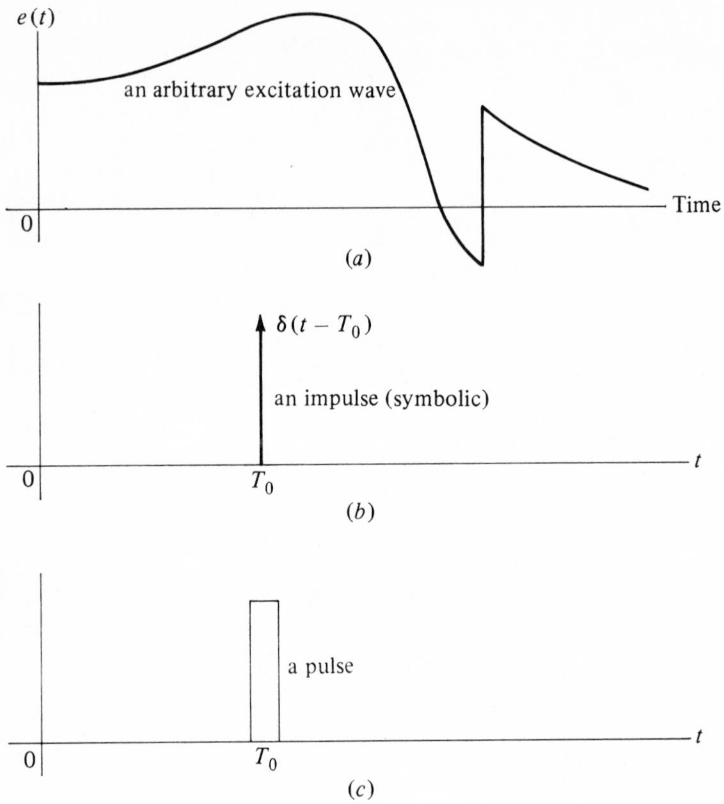

FIG. 4.03-A.

is the impulse, $\delta(t)$; for the response thereto is simply the natural behavior, adjusted to those initial conditions that the impulse creates at $t = 0+$. But it is hard to see how a wave like $e(t)$ in Fig. 4.03-A*a* could be decomposed into a sum of waves like that of Fig. 4.03-A*b*. And in fact it cannot be so treated.

But our interest is not in the excitation, it is in the response. So the "decomposition" of the excitation can be anything that permits us successfully to sum individual responses. Between excitation and response lies the network, a sort of buffer that may very well act to permit very crude manipulations on the excitation side without gross error on the response side. In fact we have already observed such a phenomenon: the response to a short pulse (Fig. 4.03-A*c*) may be almost indistinguishable from the response to the impulse (Fig. 4.03-A*b*) if the pulse is short enough and large enough. And the calculus makes use of step and pulse approximations to functions in studying the idea of the integral. Hence pulses can approximate arbitrary excitations if we take enough of them, properly constructed. With these ideas we have all we need.

We propose to

a Replace the excitation by a train of pulses,

b Replace the pulses by a train of impulses,

c Obtain the response to the impulse train by summing the appropriate impulse responses of the network, (4.03-1)

d Examine the difference between the result of (*c*) and the true response to the original excitation.

It remains to make the experiment, to determine if this be (1) valid, and (2) fruitful. It may answer (4.01-1) very nicely.

4.04 A SUPERPOSITION INTEGRAL

The superposition theorem that will result from our experiment is central to network theory. It were well then to be precise about our *postulates*. They are:

a A network N exists, at least as a model, and obeys the limitations of Sec. 4.02: its elements are lumped and constant, and are characterized by straight lines through the origin, and superposition is valid;

b Its impulse response $r_\delta(t)$ (see Sec. 2.10 and Fig. 4.04-A*a*) has been calculated (or measured, or somehow conceived) and is available;

c The initial condition is *rest* (no energy is stored in the network at $t = 0$, when the analysis is to begin); (4.04-1)

d The given excitation, $e(t)$ of Fig. 4.04-A*b*, is a reasonable wave: it may have steps (discontinuities) but it has no impulses, and it is differentiable (except at the discontinuities).

Postulate (*c*) is inserted here only for convenience; nonrest initial conditions are in effect additional excitations whose contributions to the response we can calculate separately (see Sec. 7.22). Similarly we rule out impulsive components in (*d*) because their contributions can be calculated separately, by using (*b*); differentiability will be convenient in Sec. 4.09. We seek to express $r(t)$, the response to $e(t)$, in terms of $r_\delta(t)$, without prying into the network. The reference situation of Fig. 4.04-A*a* gives us a prototype response; the actual situation of interest is that of Fig. 4.04-A*b*.

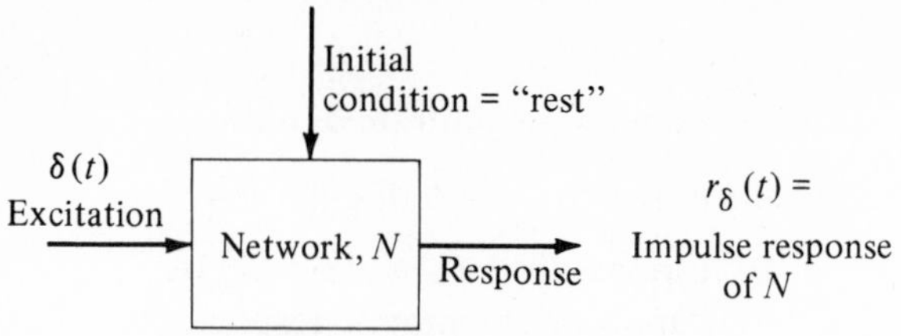

(*a*) The reference (prototype) situation.

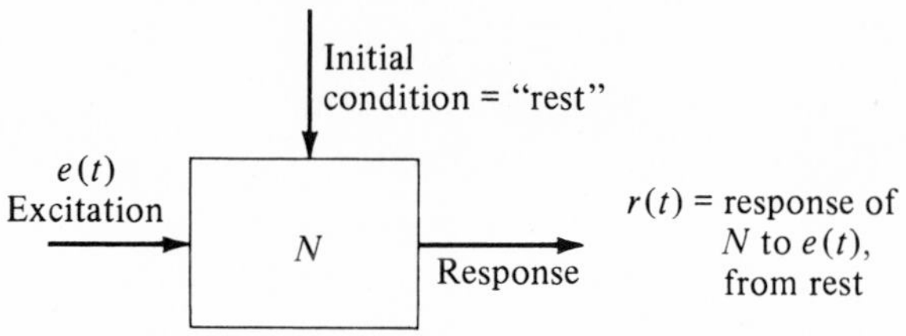

(*b*) The actual (operating) situation.

FIG. 4.04-A.

At any particular instant, the time at which we are calculating the response [the t in $r(t)$], the value of the response certainly depends on that part of the excitation wave that has occurred (is past), and not at all on the part that has not yet occurred (is future). We must therefore in effect stop the clock at t long enough to examine the history of the excitation's effect on the network from the initial instant (the start) up to time t (the present). To express this historical time we use a different symbol, x; confusion confounded would else result. Thus

> t represents "now," the present instant, at which we want the value of the response, $r(t)$, and
>
> x represents historical time, the running variable for the examination from the beginning (time zero) up to now (time t). (4.04-2)

The first step, according to (4.03-1), is to replace the excitation $e(t)$ by a set of pulses. To keep the error small, we must make the pulses short ... and we immediately encounter a familiar problem: What is "short?" Evading that question for the moment, we simply call the pulse length Δx. That is, we divide the past into say M equal intervals, so $\Delta x = t/M$. (If M is large enough, Δx is surely "small.") Figure 4.04-B*a* shows the (given) excitation wave, past, present, and future, and the division of the past time scale. Let the steps occur at the division points, at which

$$x = m\,\Delta x, \qquad m = 0, 1, 2, \ldots, M. \tag{4.04-3}$$

Let each pulse height take the value that e has at the beginning of the pulse. Then the pulses are defined by a staircaselike function that follows e closely (Fig. 4.04-B*b*). Figure 4.04-C*a* shows a small section, enlarged.

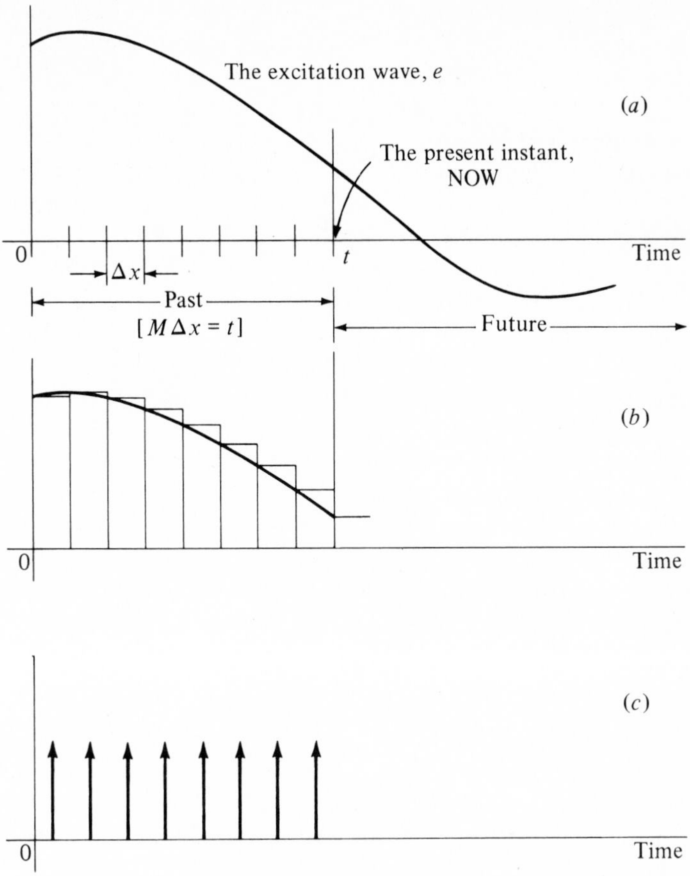

FIG. 4.04-B.

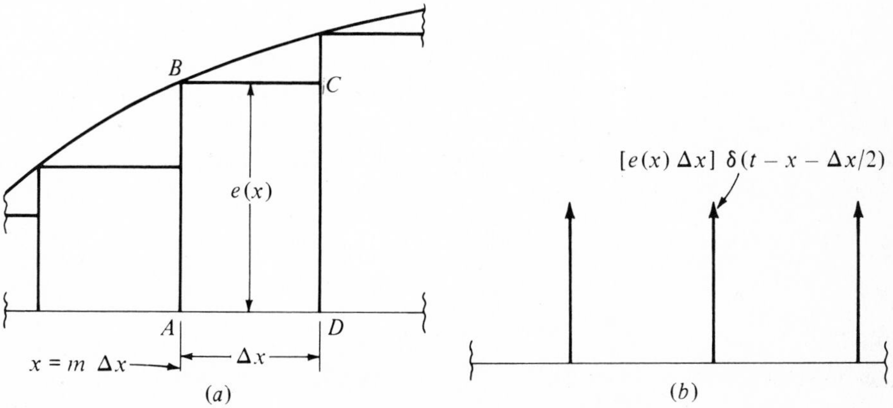

FIG. 4.04-C.

We now replace each pulse by an impulse, located at the center of the pulse and of strength appropriate for the impulse to have about the same effect as the pulse. Thus the typical pulse *ABCD*, of area $[e(x)\,\Delta x]$ gives us the impulse $[e(x)\,\Delta x]\,\delta(t - x - \Delta x/2)$ of Fig. 4.04-C*b*. The two adjacent impulses are also shown; they are of different strengths, but the diagram does not show strengths. The result is a train of M impulses; they occur at equal intervals from $x = \Delta x/2$ to $x = (M - \frac{1}{2})\,\Delta x$, but they have varying strengths, each proportional to the value of the excitation in the vicinity, Fig. 4.04-B*c*.

The staircase or pulse train looks very much like the excitation wave, except for a certain roughness (Fig. 4.04-B*b*). The impulse train (Fig. 4.04-B*c*) does not in the least resemble the excitation. But the *response* of the network to the impulse train will not differ greatly from the response of the network to the actual excitation e, if Δx is sufficiently small. We have seen before (Sec. 2.11) that as far as effect goes, a "δ" (impulse) can be replaced by a pulse (constant over a short time), provided the *areas* or integrated values are the same. And this is exactly (albeit reversed) our modus operandi here. "Short" evidently requires that Δx be much smaller than the reaction time (say the time constant or time constants of the network, or the reciprocals of the real parts of its natural frequencies, to be more precise). We do not know what this reaction time is, but whatever it is, we can make Δx small enough by letting it approach *zero*. The number of components will become infinite, true, but the summation of impulse responses will become an integral, much easier to deal with.

The response to the single impulse of Fig. 4.04-C*b* is $r_\delta(t - x - \Delta x/2)$. The response to the impulse *train* is easily obtained by superposing such components, each properly scaled and shifted:

$$r_{\text{imp. tr.}}(t) = \sum_{\substack{m=0 \\ (x=0)}}^{\substack{(x=t-\Delta x) \\ m=M-1}} \left\{[e(x)\,\Delta x] r_\delta\left(t - x - \frac{\Delta x}{2}\right)\right\}, \qquad x = m\,\Delta x. \tag{4.04-4}$$

Now we let Δx become infinitesimal, in the manner of the calculus writing it dx; in the limit it will surely be short enough (!) to make the pulse and impulse responses indistinguishable. And $\Delta x/2$ becomes negligible in the argument of r_δ and in the summation's upper limit. Then x becomes a continuous historical-time variable rather than the discrete variable it is in (4.04-4). The summation therein becomes an integral, and the impulse-train response is no longer distinguishable from the network's response to e:

$$r(t) = \int_0^t e(x) r_\delta(t - x)\,dx. \tag{4.04-5}$$

We note how easy it is to keep (4.04-5) in mind, how admirably mnemonic it is, for it is simply the superposition ($\int$) of the responses to all the impulse components:

$$r(t) = \text{sum}\{[e(x)\,dx] r_\delta(t - x)\}. \tag{4.04-6}$$

The sum must of course be *from* the beginning $(x = 0)$ *to* now $(x = t)$, and of integral type ($\int$ instead of Σ).

The conventional form (4.04-5) is called a *superposition integral* or *superposition theorem*, because it depends on adding component responses. If we can calculate a network's impulse response, and if the postulates (4.04-1) hold, then we can *in principle* calculate the response of the network to any excitation, from (4.04-5). To say "in principle" is important, because the integral may be difficult to evaluate. But (approximate) numerical evaluation is always possible, at the worst.

There are other superposition integrals (see Sec. 4.09), but this is the simplest. It can take various forms and has many uses.

4.05 OTHER FORMS OF THIS SUPERPOSITION INTEGRAL

The appearance of the superposition integral (4.04-5) can be altered by changing the limits and the arguments of excitation and response. Some of the different forms that result can be useful.

The upper limit can be t or anything greater than t, for $r_\delta(y) = 0$ when $y < 0$ (see Sec. 2.09). It is sometimes convenient to make this limit infinite, which gives

$$r(t) = \int_0^\infty e(x) r_\delta(t - x)\, dx. \tag{4.05-1}$$

The lower limit can be zero or anything less than zero, provided we understand, as is entirely reasonable (and as Fig. 4.04-B*a* implies) that $e(x) = 0$ when $x < 0$. For example, one often finds the rather pleasing, symmetrical form

$$r(t) = \int_{-\infty}^\infty e(x) r_\delta(t - x)\, dx. \tag{4.05-2}$$

And the arguments, surprisingly, can be interchanged. The substitution $y = t - x$ followed by manipulation of the integral gives

$$\begin{aligned} r(t) &= \int_0^t e(x) r_\delta(t - x)\, dx = -\int_t^0 e(t - y) r_\delta(y)\, dy \\ &= \int_0^t e(t - x) r_\delta(x)\, dx \\ &= \int_0^\infty e(t - x) r_\delta(x)\, dx \\ &= \int_{-\infty}^\infty e(t - x) r_\delta(x)\, dx. \end{aligned} \tag{4.05-3}$$

4.06 USE OF THE SUPERPOSITION INTEGRAL

Application of the integral to network-response calculation is in principle straightforward. But there are stumbling blocks; some of them appear in the following examples.

Example 1: Suppose the impulse response is a simple exponential: $r_\delta(t) = R_0 e^{-\alpha t}u(t)$. And let the excitation be like it: $e(t) = E_0 e^{-\alpha t}u(t)$. The two constants R_0 and E_0 are here generic, with appropriate units (or dimensions): E_0 is in excitation units, R_0 in response units per excitation unit-second. Then the response (from rest) is easily calculated:

$$
\begin{aligned}
r(t) &= \int_0^t [E_0 e^{-\alpha x}u(x)][R_0 e^{-\alpha(t-x)}u(t-x)]\,dx \\
&= E_0 R_0 e^{-\alpha t}\int_0^t dx = E_0 R_0 te^{-\alpha t}, \qquad t > 0.
\end{aligned}
\tag{4.06-1}
$$

The remarkable thing about this is that it calculates with great ease the response of the network to an excitation identical in form with its own natural behavior (see Fig. 4.06-A). Compare this with the laborious calculations of Sec. 2.14, and the discussion there of this situation, and Fig. 2.14-D.

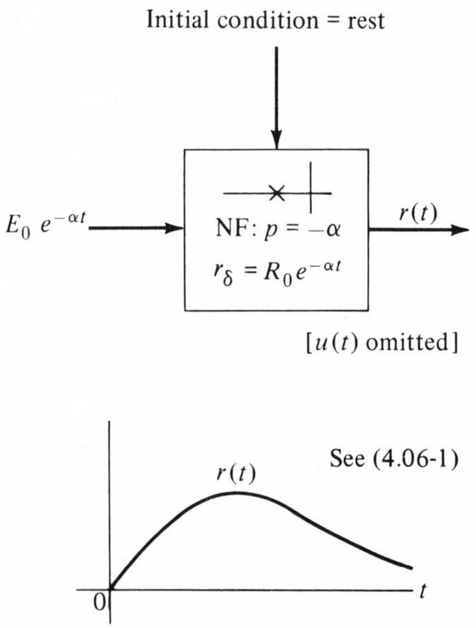

FIG. 4.06-A.

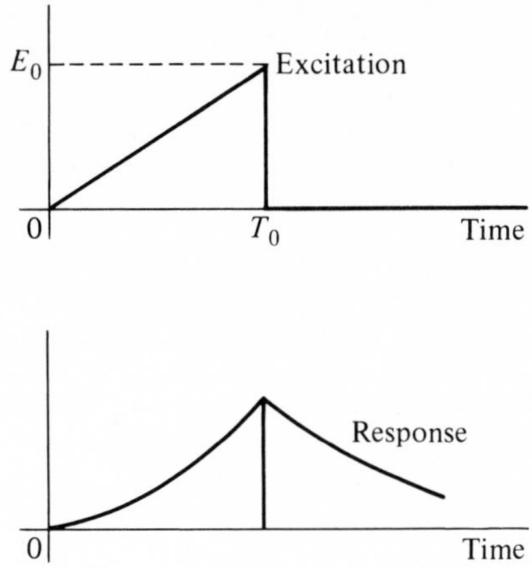

FIG. 4.06-B.

Example 2: For simplicity, keep the same impulse response, but let the excitation be the truncated ramp of Fig. 4.06-B:

$$e(t) = E_0 \frac{t}{T_0} [u(t) - u(t - T_0)]. \tag{4.06-2}$$

The response calculation begins thus:

$$\begin{aligned} r(t) &= \int_0^t \frac{E_0}{T_0} x[u(x) - u(x - T_0)]R_0 e^{-\alpha(t-x)} dx \\ &= \frac{E_0 R_0}{T_0} e^{-\alpha t} \int_0^t x e^{\alpha x}[u(x) - u(x - T_0)]\, dx. \end{aligned} \tag{4.06-3}$$

The piecewise nature of the excitation makes it convenient (if not essential, for clarity) to calculate the response in corresponding pieces. The two step functions in the brackets make this almost automatic:

$$\begin{aligned} 0 < t \leqq T_0: \quad r(t) &= \frac{E_0 R_0}{T_0} e^{-\alpha t} \int_0^t x e^{\alpha x}\, dx = \frac{K_0}{\alpha^2} (\alpha t - 1 + e^{-\alpha t}), \\ T_0 \leqq t: \quad r(t) &= \frac{E_0 R_0}{T_0} e^{-\alpha t} \int_0^{T_0} x e^{\alpha x}\, dx + 0 \\ &= \frac{K_0}{\alpha^2} (\alpha T_0 - 1 + e^{-\alpha T_0}) e^{-\alpha(t - T_0)}. \end{aligned} \tag{4.06-4}$$

At $t = T_0$ the two results of course agree (Fig. 4.06-B). And for $t > T_0$ the response is simply the natural behavior with appropriate "initial condition."

One could omit the step functions in (4.06-2) and instead write the response in two pieces, recognizing that the excitation no longer acts after time T_0. But their use in (4.06-2) makes the calculation more nearly foolproof.

It is characteristic of piecewise excitations to require piecewise response expressions. Each can of course be calculated individually by other methods, and the pieces can then be properly joined. But the superposition integral is particularly convenient for their calculation.

Example 3: Retaining the same simple impulse response, let us consider the simple excitation wave

$$e(t) = \frac{E_0}{1 + t/T_0} u(t). \tag{4.06-5}$$

This is not exponential, but it looks very much like one (Fig. 4.06-C). The response calculation begins

$$\begin{aligned} r(t) &= \int_0^t \frac{E_0}{1 + x/T_0} u(x) R_0 e^{-\alpha(t-x)}\, dx \\ &= E_0 R_0 e^{-\alpha t} \int_0^t \frac{e^{\alpha x}}{1 + x/T_0}\, dx. \end{aligned} \tag{4.06-6}$$

The formal evaluation of this integral is not easy: simple as it seems, it is not to be found in tables of elementary integrals. But approximate, practical evaluation, as by series, or numerical integration, *is* simple (and a "higher" function, the exponential integral, can be used—see Prob. 4-T). What is astonishing is that so simple an excitation wave, the response to which is surely not much different from the response to an exponential, should be so much more difficult to treat.

It tells us once more that the exponential function is peculiarly *natural* for our networks. (An interesting idea, should it really be necessary to find the

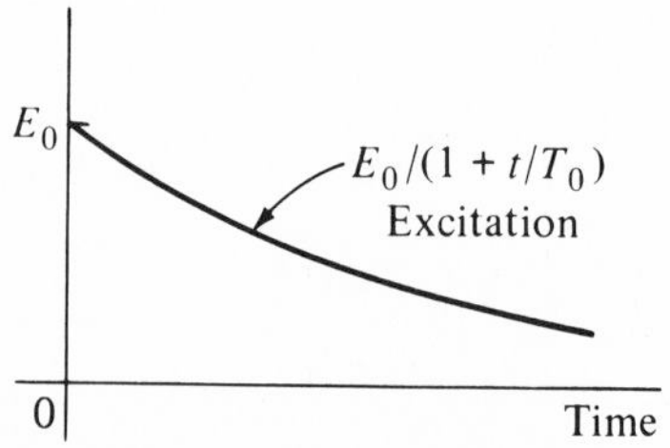

FIG. 4.06-C.

response to such nonexponential, nontractable waves, is to *approximate* the excitation by a sum of exponentials, after which one has clear sailing.)

Example 4: It is an interesting exercise if nothing more to calculate the response to an impulse generally by means of the superposition integral. If the excitation be $E_0\,\delta(t)$ the response is

$$r(t) = \int_0^t E_0\,\delta(x) r_\delta(t - x)\,dx = E_0\,r_\delta(t) \tag{4.06-7}$$

as of course it should be. One should be sure to understand the integral's evaluation in (4.06-7), based on the "sifting" or "filtering" action of the δ function. Note also that the dimensions of E_0 in this example must be those of E_0 of previous examples multiplied by the dimension of time.

Example 5: Another interesting example is to calculate the general response to a step, $e(t) = E_0\,u(t)$:

$$\begin{aligned} r_u(t) &= \int_0^t E_0\,u(t - x) r_\delta(x)\,dx \\ &= E_0 \int_0^t r_\delta(x)\,dx, \qquad 0 \leqq t. \end{aligned} \tag{4.06-8}$$

We note that it is slightly more convenient here to use the "other" form, from (4.05-3), so that the argument of r_δ is simply x. One is always at liberty, of course, to use *any* of the forms (see Sec. 4.05). We have seen before that the step response is sometimes equal to the integral of the impulse response (Sec. 2.10). Here we have evidence that it is generally true:

$$r_u(t) = \int_0^t r_\delta(x)\,dx. \tag{4.06-9}$$

By differentiating (4.06-9) with respect to t, with careful attention to detail in the process, we verify the complementary relation

$$r_\delta(t) = r_u'(t), \qquad t > 0. \tag{4.06-10}$$

Note that because r_u may be discontinuous at $t = 0$, (4.06-10) may not hold *at* $t = 0$.

We do not stop for additional examples here. The above are simple, yet illustrative, and more will occur. When natural frequencies are complex, and when there is a large number of them, the calculations become lengthy. But real complications occur only when the excitation, or perhaps the response (or both) are *not exponentials.*

4.07 CONVOLUTION

Our superposition integral has many applications; it appears in many fields, some very different indeed from network analysis. It may therefore carry other names, of which the commonest comes from a geometrical interpretation of the integral in (4.04-5). Figure 4.07-A, in its various parts, describes this view. Parts (*a*), (*b*), and (*c*) show the impulse response (*a*) as given, (*b*) with reversed argument, and (*c*) shifted to correspond to argument $(t - x)$. Note that it is zero for negative values of the argument, a property of importance if one is to extend the limits of integration as in Sec. 4.05. Part (*d*) shows the excitation, and part (*e*) the appropriate product, the integrand of the superposition integral, for a particular instant t:

$$e(x)r_\delta(t - x). \tag{4.07-1}$$

The value of the response at this instant t is of course the *area* crosshatched in part (*e*), according to (4.04-5):

$$r(t) = \int_0^t e(x)r_\delta(t - x)\,dx. \tag{4.07-2}$$

To obtain the response as a function of t, we must imagine t varying in Fig. 4.07-A from zero up to the largest value of interest. If t is very small, there is little overlap of the curves in parts (*c*) and (*d*) and the area in (*e*) is small: the response begins with small values, as one expects from impulse response and excitation curves of the types shown. For larger values of t, the accumulated area depends on the excursions of the two curves: if both are always positive, and do not decay too rapidly, the response will ultimately approach a nonzero steady-state value. If positive and negative values occur in one or both, the response may be complicated.

Actual evaluation of the response can in principle be made by area measurement on such a careful drawing, as by a planimeter. But the chief value of this pictorial approach is in its clarity, and in the insight and understanding it gives—and in the names it suggests.

In the graphical approach, we first lay out the impulse response (or the excitation) and then *fold* it over (reverse it). From this comes the name *folding theorem* (*faltungssatz* in German). The function is then translated, multiplied or "rolled" into the excitation (or the impulse response), and an area measured. From this comes the commoner name *convolution theorem*, or *operation* (from the Latin *con* + *volvo*). And occasionally the name *composition* is used. But *convolution* is the name most often met, sometimes in fields quite remote from circuit analysis. Hence (4.04-5) may be called a *convolution integral*. (Other convolution integrals exist.)

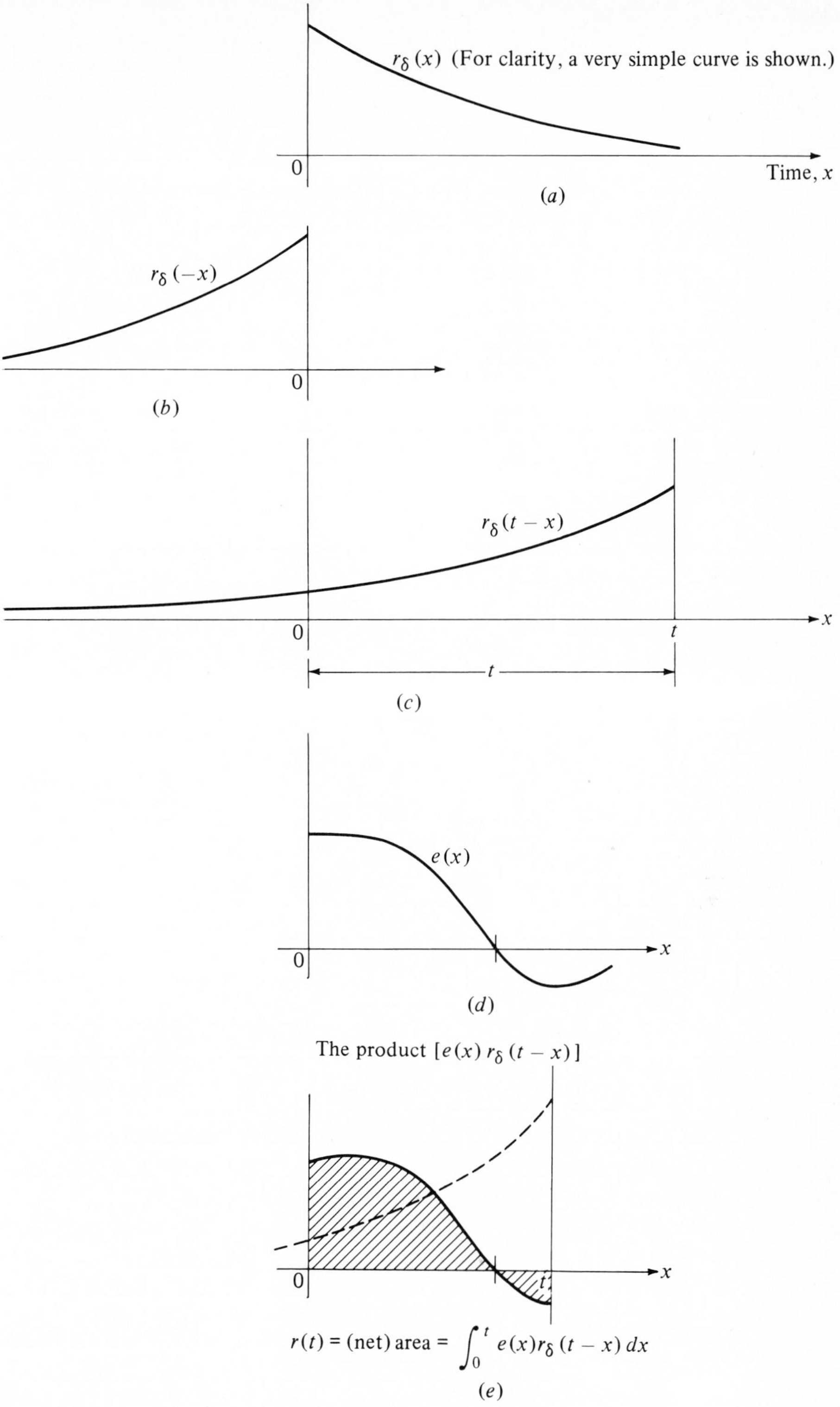

$r_\delta(x)$ (For clarity, a very simple curve is shown.)
0
Time, x
(a)
$r_\delta(-x)$
0
(b)
$r_\delta(t-x)$
0
t
x
t
(c)
$e(x)$
0
x
(d)
The product $[e(x)\, r_\delta(t-x)]$
0
t
x
$r(t) = \text{(net) area} = \int_0^t e(x) r_\delta(t-x)\, dx$
(e)

FIG. 4.07-A.

A common notation for the convolution of two arbitrary functions $f_1(t)$ and $f_2(t)$ is simply a star:

$$f_1 * f_2 = \int_0^t f_1(x) f_2(t-x)\,dx = \int_0^t f_1(t-x) f_2(x)\,dx = f_2 * f_1. \qquad (4.07\text{-}3)$$

The limits can be extended in both directions if f_1 and f_2 behave properly (Sec. 4.05).

In this notation, the response of a network whose impulse response is $r_\delta(t)$, from rest, to an excitation $e(t)$, is very simply expressed:

$$r(t) = e * r_\delta = r_\delta * e. \qquad (4.07\text{-}4)$$

The actual *evaluation* of the response is not necessarily simple. If only exponential functions are involved in both r_δ and e, the calculation is straightforward, though perhaps lengthy (see Sec. 4.06 and Chap. 5). If other functions are involved, one first examines tables of definite integrals; if the integral cannot be found, then some form of numerical integration must be used. Since all function evaluation comes down eventually to numerical evaluation (of some series, or continued fraction or the like) this is in no way a defeat. Numerical evaluation of the convolution integral may indeed be the best method.

4.08 NUMERICAL CONVOLUTION

The *finite-difference* or summation form of the response, (4.04-4), gives us, in retrospect, a practical evaluation procedure for approximate calculation of the response. We have, from Sec. 4.04 (see Fig. 4.08-A),

$$\begin{aligned} r(t) = \int_0^t e(x) r_\delta(t-x)\,dx &\cong \sum_{m=0}^{m=M-1} e(mx) r_\delta[(M-m)\,\Delta x]\,\Delta x \\ &= \sum_{n=0}^{n=M-1} e[(M-n)\,\Delta x] r_\delta(n\,\Delta x)\,\Delta x. \end{aligned} \qquad (4.08\text{-}1)$$

The sums are approximate, rather than exact, values of the response because they are finite sums, corresponding to noninfinitesimal increments Δx; in them the limit has not been taken. But because these *are* finite sums and not integrals, they are capable of simple numerical evaluation and can give useful results, approximate though they be.

The sums actually spell out a calculation procedure for us (see Table 4.08-A and Fig. 4.08-B):

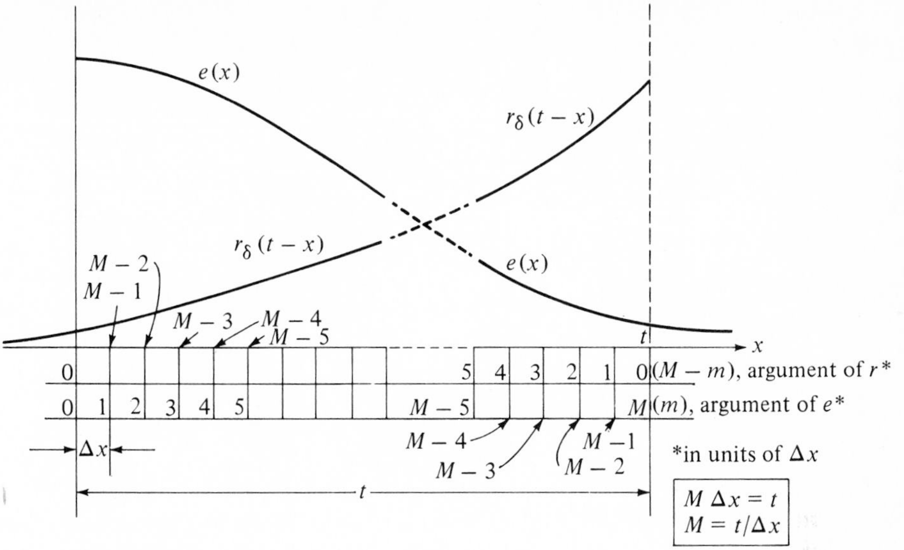

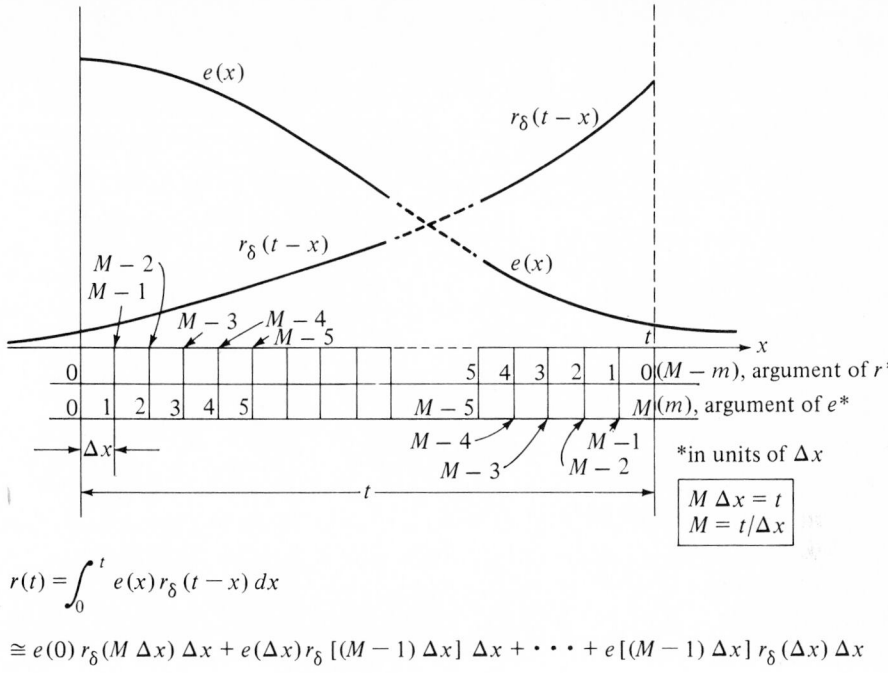

$$r(t) = \int_0^t e(x)\, r_\delta(t-x)\, dx$$

$$\cong e(0)\, r_\delta(M\,\Delta x)\,\Delta x + e(\Delta x)\, r_\delta[(M-1)\,\Delta x]\,\Delta x + \cdots + e[(M-1)\,\Delta x]\, r_\delta(\Delta x)\,\Delta x$$

$$\cong e(M\,\Delta x)\, r_\delta(0)\,\Delta x + e[(M-1)\,\Delta x]\, r_\delta(\Delta x)\Delta x + \cdots + e(\Delta x) r_\delta[(M-1)\,\Delta x]\,\Delta x$$

FIG. 4.08-A.

a Calculate e and r_δ for arguments 0, Δx, $2\Delta x$, ..., $N\,\Delta x$ (choose for N an integer large enough to accommodate all t of interest, $N = t_{max}/\Delta x$);

b Tabulate the values of e and of r_δ in parallel columns (1 and 2 in Table 4.08-A); then reverse or "turn upside down" one of them (either one); cut this column to form a strip that can be moved relatively to the other one (column 3);

c Choose a value of t, and hence of $M = t/\Delta x$ (an integer);

d Slide the movable column (strip) so that the initial values are separated by M entries (column 3); (4.08-2)

e On each level multiply the two entries (in 2 and 3) and write the product in the next (a new) column (4);

f Accumulate (sum) the products in this last column (4); this sum, multiplied by Δx, is the (approximate) value of $r(t)$;

g Choose a new value of t and repeat the process, starting again with step (c).

TABLE 4.08-A

	1	2	3	4	
x (Time Variable, in Units of T_0)	$e(x)$ (Excitation, in Units of E_0)	$r_\delta(x)$ (Impulse Response, in Units of $R_0 T_0$)	$e(t-x)$ (Excitation, Reversed and Shifted by t, in Units of E_0)	$e(t-x)r_\delta(x)$ (Products, in Units of $E_0 R_0 T_0$)	
.	0	0	.	0	
.	0	0	.	0	
.	0	0	.	0	
−0.3	0	0	0.372	0	
−0.2	0	0	0.410	0	
−0.1	0	0	0.452	0	
0	1.000	0.000	0.500	0.000	
0.1	0.990	0.082	0.552	0.045	
0.2	0.962	0.134	0.610	0.082	
0.3	0.917	0.165	0.671	0.111	
0.4	0.862	0.180	0.735	0.133	
0.5	0.800	0.184	0.800	0.147	Sum = 1.272 (rows 0 to 1.0)
0.6	0.735	0.181	0.862	0.157	
0.7	0.671	0.173	0.917	0.159	
0.8	0.610	0.162	0.962	0.156	
0.9	0.552	0.148	0.990	0.147	
1.0	0.500	0.135	1.000	0.135	
1.1	0.452	0.122	0	0	$r_{t=T_0} \cong 1.272 R_0 E_0 T_0 \Delta x$
1.2	0.410	0.109	0	0	$= 0.127 R_0 E_0 T_0^2$
1.3	0.372	0.097	0	0	
.	.	.	0	0	
.	.	.	0	0	
.	.	.	0	0	

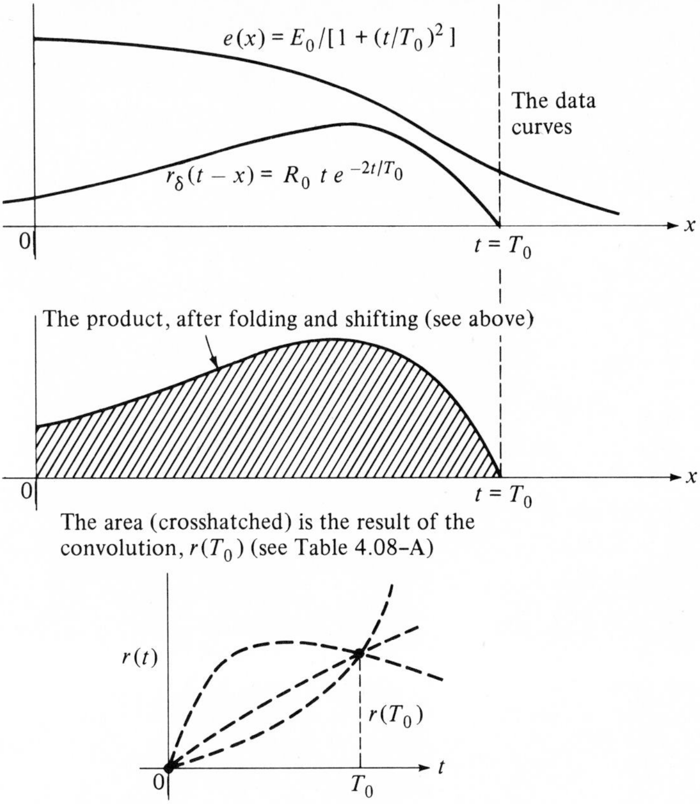

FIG. 4.08-B.

In the specific example of Table 4.08-A, the excitation is taken proportional to $[1 + (t/T_0)^2]^{-1}$ and the impulse response to (te^{-2t/T_0}). The granularity used corresponds to $\Delta x = 0.1T_0$, and the calculations are performed for $t = T_0$, so that $M = t/\Delta x = 10$. [To the approximate value obtained, 0.127, there corresponds the exact value 0.121, obtained by a more elaborate calculation (Table 4.08-B): the accuracy obtained here by this rough calculation is evidently about 5 percent.]

The result of the calculations with the movable strip in one position is a single point on the curve of $r(t)$, say $r(T_0)$ in Fig. 4.08-B. To determine the course of the curve, to see for example whether $r(t)$ follows one of the dashed curves, many additional points must be calculated, starting each time with step (*c*). This "sliding paper strip" experiment can be carried out with considerable educational profit, because of the *insight* that actually doing it gives one.

In practice, of course, the calculation is done with a very small value of Δx, M is a large number, and a computer does the work. The elements of a simple digital computer program to accomplish this are described below. (The programming language used is FORTRAN, but this is only for specificity; we shall discuss computation and programming in detail in Secs. 7.17 and 7.18.)

TABLE 4.08-B Results RE(T) of the Convolution of RD(T) and E(T)

RESULTS RE(T) CF THE CCNVOLUTION OF RD(T) AND E(T)

TIME	EXCITATION	RD	RE
......	....	....	....
0.CCC000	1.000000	0.000000	0.000000
0.2C0000	C.961538	0.134064	0.015277
C.4CC000	0.862069	0.179732	0.046335
0.6CC000	C.735294	0.180717	0.078405
0.8CC000	0.609756	0.161517	0.104046
1.CCC000	C.5C00C0	0.135335	0.120650
1.2CC000	0.409836	0.108862	0.128449
1.4CCCC0	C.337838	0.085134	0.129024
1.6CC000	0.280899	0.065219	0.124358
1.8C0000	0.235849	0.049183	0.116313
2.CC0C00	C.200000	0.036631	0.106413
2.2CCCC0	0.171233	0.027010	0.095792
2.4CCCC1	0.147929	0.019751	0.085225
2.6CCC00	0.128866	0.014343	0.075207
2.8CCC00	0.113122	0.010354	0.066009
3.CCC000	C.100000	0.007436	0.057755
3.2CCCC1	0.088968	0.005317	0.050469
3.4CCCC1	0.079618	0.003787	0.044113
3.6CCCC0	0.071633	0.002688	0.038613
3.8CCCC0	0.064767	0.001902	0.033881
4.CC0001	C.058824	0.001342	0.029823
4.2C0001	0.053648	0.000944	0.026348
4.400001	C.049116	0.000663	0.023373
4.6CCCC0	0.045126	0.000465	0.020824
4.8CCCC1	0.041597	0.000325	0.018636
5.CCC001	0.038462	0.000227	0.016752

Time is in units of T_0, excitation in units of E_0, impulse response (RD) units of $R_0 T_0$ and response obtained by convolution (RE) is in units of $E_0 R_0 T_0^2$; $\Delta x = 0.1$. Note the value at $t = T_0$: 0.1207 (to be compared with 0,127 in Table 4.08-A); more refined computation does not change these figures.

a Determine the largest argument (time value) needed, t_{max}, the increment to be used, Δx = DX, and the corresponding integer L = $t_{max}/\Delta x$;

b Calculate values of the impulse response and excitation, and save these in arrays:

$$\begin{aligned} \text{RD(I)} &= r_\delta(k\,\Delta x) \\ \text{E(I)} &= e(k\,\Delta x) \end{aligned} \quad \begin{cases} k = 0, 1, 2, \ldots, L \\ I = 1, 2, 3, \ldots, (L+1); \end{cases} \tag{4.08-3}$$

c Carry out the summation for each of a series of values of time t equal to M units Δx: $M = t/\Delta x$:

```
      DO 200 M=1,L
      SUM = 0.0
      DO 100 IJ=1,M
      IM = M+2-IJ
100   SUM = SUM + E(IJ)*RD(IM)*DX
200   R(M) = SUM
```

(4.08-3) con.

d The values in the array elements R(M) are the (approximate) values of the response at times $t = M\,\Delta x$, that is, at $t = \Delta x$, $2\Delta x$, $3\Delta x$, ..., $L\,\Delta x = t_{max}$. [To this one may add: $r(0) = 0.0$.]

Since what we really wish, however, is the value of the integral itself, we can improve on this by using numerical integration formulas. For example, Simpson's (parabolic approximation) rule applies weights $\frac{1}{3}, \frac{4}{3}, \frac{2}{3}, \frac{4}{3}, \frac{2}{3}, \frac{4}{3}, \ldots, \frac{4}{3}, \frac{2}{3}, \frac{4}{3}, \frac{1}{3}$ to the (integrand) terms in the sum. Then the summation statement should be modified to incorporate the weights, W(IJ), from an array in which W(1) = $\frac{1}{3}$, W(2) = $\frac{4}{3}$, W(3) = $\frac{2}{3}$, ..., W(M − 1) = $\frac{2}{3}$, W(M) = $\frac{4}{3}$, W(M + 1) = $\frac{1}{3}$. Note that M must now be an *even* integer, so that the response is calculated only at $t = 0, 2\,\Delta x, 4\,\Delta x, \ldots, L\,\Delta x$; so also must L be even. Finally the DO statements must be changed accordingly. The revised program element in (c) above is then

```
      DO 200 M=2,L,2
      SUM = 0.0
      DO 100 IJ=1,M+1
      IM = M + 2 - IJ
100   SUM = SUM + E(IJ)*RD(IM)*DX*W(IJ)
200   R(M) = SUM
```

(4.08-4)

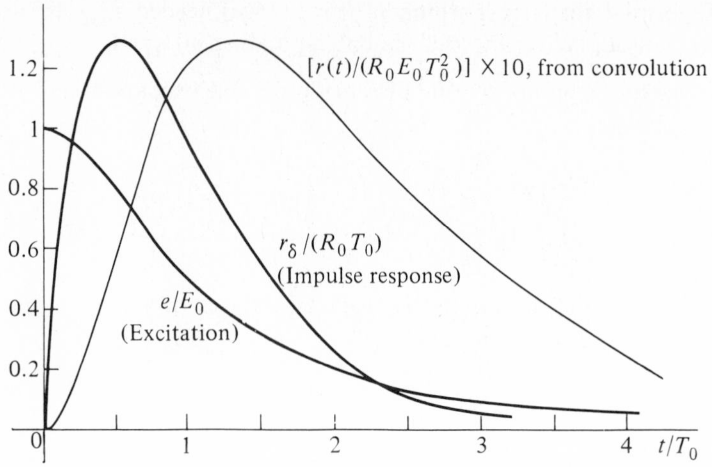

FIG. 4.08-C.

Further refinement is seldom necessary, and we leave the subject of numerical convolution at that. By way of illustration, the example of Fig. 4.08-B and Table 4.08-A, studied with the aid of a computer program based on the element above, leads to the results of Table 4.08-B and Fig. 4.08-C. (A more elaborate convolution program is described in Sec. 7.18.)

4.09 CONVOLUTION AND SUPERPOSITION INTEGRALS

Convolution provides a convenient, practical way to calculate responses in cases where one must resort to numerical methods, such as those that involve a great many exponential terms, or nonexponential or otherwise peculiar excitations. If the excitation is given only as a sketch (a proposal or an idea, for example), or by a set of measured or computed points, numerical convolution is appropriate. Sometimes the impulse response of the network (model) is known only by measurement, for example as an oscillogram, such as the photograph of Fig. 4.09-A (though probably on a larger scale, from which points can be read with adequate accuracy). Then numerical convolution again is indicated.

A subject of great interest to which we can only allude in passing is the part convolution has played in the historical development of computing machinery. Long an important integral, often difficult to evaluate except numerically, convolution has stimulated the development of such interesting devices as the cinema integraph (see GB in Appendix B). This device also utilized moving strips, but of photographic film on which the functions were outlined; a beam of light was passed through the films to effect multiplication, and then on into an integrating sphere.

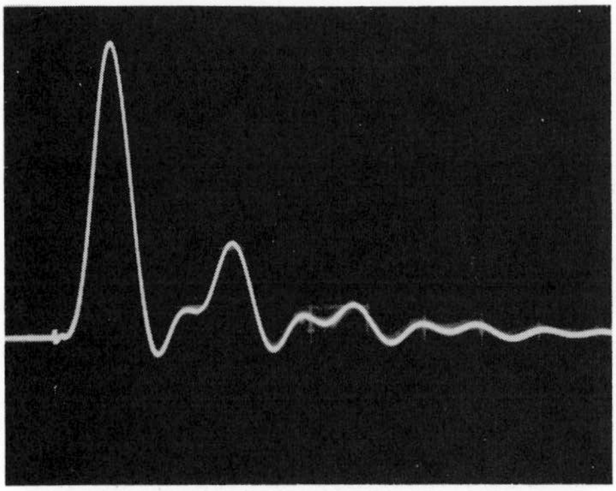

FIG. 4.09-A.

Another interesting path down which we can only glance leads to the use of other prototype functions. Instead of using the impulse response as a base, one could, for example, use the step response, $r_u(t)$. By an argument similar to that of Sec. 4.05 we arrive then at this expression for the response, from rest, to an excitation $e(t)$:

$$\begin{aligned} r(t) &= e(0+)r_u(t) + \int_{0+}^{t} e'(t-x)r_u(x)\,dx \\ &= e(0+)r_u(t) + \int_{0+}^{t} e'(x)r_u(t-x)\,dx. \end{aligned} \tag{4.09-1}$$

This is the famous Heaviside form of superposition integral, known also as the Duhamel integral. It can be derived by using Fig. 4.04-C*a*, thinking in terms of steps rather than rectangles—or it can be derived from our first superposition integral (Sec. 4.05) by integration by parts. It presents an alternative computation procedure, useful in cases where the step response is for some reason more convenient to use than is the impulse response, and where the derivative of the excitation is readily obtainable.

Particular attention must be given to the initial instant, if $e(0+) \neq 0$, for then a step occurs in the excitation and its derivative requires very careful consideration. The first, separate, term is a consequence of such a step. It could be omitted by changing the lower limit on the integral from $0+$ to 0 and carefully evaluating the effect of the step by means of an impulse function in the derivative of e. The theory of generalized functions, or distributions, has been developed to treat such (and far more complicated) unusual situations.

In a similar way we can develop a superposition integral that utilizes the response to an exponential as prototype. And the set of superposition integrals can be further extended. The ordinary convolution integral, however, is the superposition integral of most value. It is also found in many other fields of analysis.

4.10 CONCLUSION

Our comparatively brief investigation of the problem posed in (4.01-2) has produced important results. We can now find the response of a network to *any* excitation. We need know nothing about the network except the impulse response. It can be a simple *RC* or *RLC* circuit, or it can be a very complicated network whose impulse response has been calculated or measured for us, but of whose inner details we know nothing. Or the network may not even exist: What *would* be its response to ... *if* a network with such and such an impulse response existed? Convolution can tell us.

Convolution is an operation one may expect to encounter elsewhere: its utility is not limited to network analysis.

4.11 EXERCISES

4-1. If a network's impulse response is written $r_\delta(t) = R_0 f(t)$, in which $f(t)$ has no dimensions, give the dimensions of R_0 by filling in the table. In each case, reply with one simple dimension or unit name, as in the first line.

Excitation	Response	Dimensions or Units of R_0
Current	Voltage	farad^{-1}
Current	Current	
Voltage	Voltage	
Voltage	Current	

4-2. If Examples 1, 2, and 3 of Sec. 4.06 refer to the *RC* circuit of Chap. 2, what are the dimensions of E_0 and of R_0? If they apply to the *RLC* circuits of Chap. 3, what then are the dimensions of E_0 and of R_0? What are the dimensions of E_0 in Examples 4 and 5?

4-3. Show graphically, by using Fig. 4.07-A (with perhaps some redrawing) that it *makes no difference which curve of the two is folded* (or convolved). Correlate this with the mathematical form of the superposition integral.

4-4. Apply the mean-value theorem of the calculus to show that the area under the excitation wave of Fig. 4.04-C*a* between *A* and *D* is *exactly* $[e(x)\,\Delta x]$ provided x in $e(x)$ is redefined. Then explain why the superposition integral is unaffected by such a change in its development.

4-5. Explain how the situation shown in the figure relates to (4.02-3) and (4.02-4). How can the individual responses to two excitations e_1 and e_2 be used to obtain the response to $(e_1 + e_2)$?

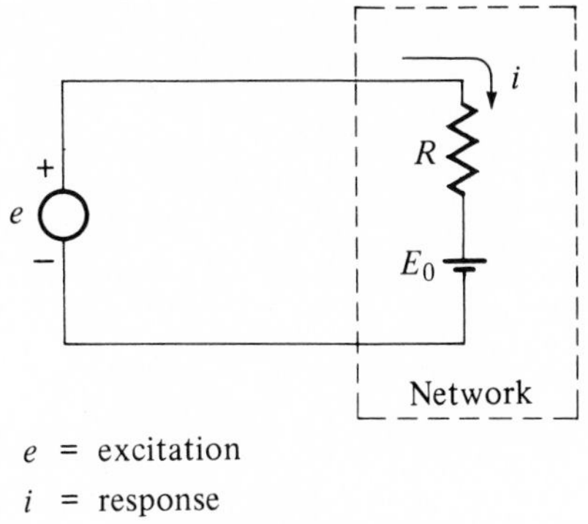

FIG. 4-5.

4-6. The initial conditions of the circuit shown are not zero. Explain carefully how the response to an excitation $i_F = i_{F1} + i_{F2} + \cdots$ can be calculated from the responses to i_{F1}, to i_{F2}, ..., individually obtained (from rest).

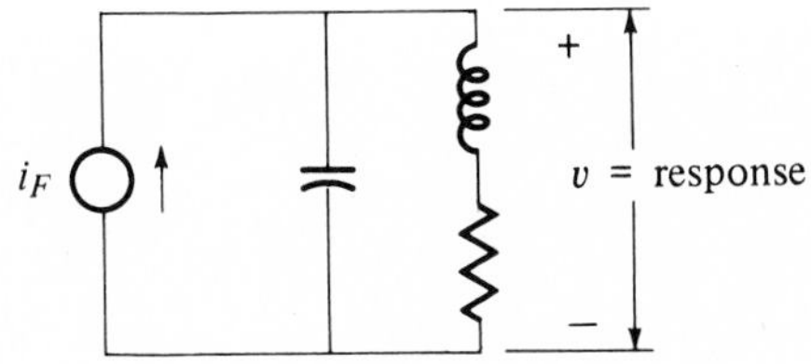

FIG. 4-6.

4-7. Discuss thoroughly the two sentences in the seventh paragraph of Sec. 4.02 that begin "The presence of derivatives and ..." and "They correspond to nonzero...." Correlate with the "source exchange" discussions of Sec. 3.08 (Fig. 3.08-B in particular) and of Sec. 2.02. Show why the superpositions practiced in Sec. 2.12 are valid.

4-8. If the response $r(t)$ is known formally (and not merely by a curve or numerical data) and the network is also known, it is not difficult to calculate the excitation that gave rise to $r(t)$. Explain carefully, at least for the *RC* and *RLC* circuits. Why, if the response is known only numerically, is the process of finding the excitation more difficult (see Prob. 4-W)? Illustrate by finding the excitation of an *RC* circuit that causes the response $v(t) = V_0 e^{-\alpha t}(1 - \cos \omega_N t)$ in which $\alpha = 1/(RC)$. Find also the excitation of an *RLC* circuit that will produce the same response, assuming that in the (parallel) *RLC* circuit $G/C = 2\alpha$ and $\omega_0^2 = \omega_N^2 + \alpha^2$.

4-9. Excitation may be given, as by measurement, only as a sequence of sample values, for example those of the figure. What assumptions must be made if the response of a given network to such excitation is to be calculated?

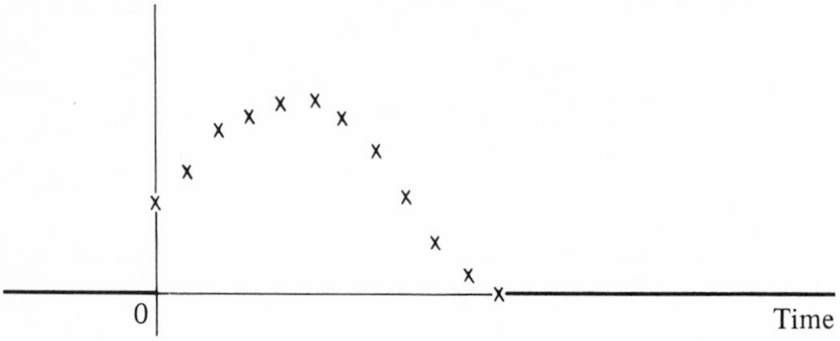

FIG. 4-9.

4-10. An *RC* circuit is excited by a wave whose initial value is zero, which takes the value I_0 at $t = T_0, 2T_0, 3T_0$, and is zero at and after $t = 4T_0$. (T_0 is the time constant of the *RC* circuit.) Sketch, as best you can, the response wave. State the assumptions you make, and why you make them. Explain carefully what limits the accuracy with which the response can be calculated.

4-11. Under what conditions in Fig. 4.07-A will the response $r(t)$ step immediately (at $t = 0+$) to some nonzero value? Why does this "not ordinarily occur"?

Under what conditions will the ultimate response $r(\infty)$ be a positive constant? Zero? A negative constant?

4-12. A hypothetical network is characterized by the impulse response $r_\delta(t) = E_0(1 - t/T)[u(t) - u(t - T)]$. Sketch its response to a rectangular pulse of duration T. Verify your result by performing the convolution.

4-13. An *RC* circuit excited (from rest) by a current wave that is $t^n e^{-\alpha t}$ in shape generates what voltage? [n is a positive integer, $\alpha = 1/(RC)$.]

4-14. A high-Q *RLC* circuit excited by the wave $te^{-\alpha t}$ produces what voltage? [$\alpha = G/(2C)$.]

4-15. Can the arguments be interchanged in the integral of (4.09-1), as they can in the integral of (4.05-2)?

4-16. Explain why the slight difference in the sums of (4.08-1) and (4.08-4) is of no consequence. Show also that any of these four forms can be used without appreciable error:

(a) The two in (4.08-1);

(b) $\sum_{0}^{M-1} e(m\,\Delta x) r_\delta[(M-1-m)\,\Delta x]\,\Delta x,$

(c) $\sum_{0}^{M-1} e[(M-1-m)\,\Delta x] r_\delta(m\,\Delta x)\,\Delta x.$

4-17. Show, entirely by means of sketches (like those of Sec. 4.07) why the response of a network whose impulse response is $R_0 e^{-\alpha t} u(t)$, to a sinusoidal excitation, goes through a transient state and then settles down to a sinusoidal steady state.

4-18. Explain how impulsive components in the excitation are to be handled in performing convolution. Take, for illustration, the RC circuit excited by $i_F = I_F u(t) + Q_0\,\delta(t-T) - Q_0\,\delta(t-2T)$, $T = RC$.

4-19. Suppose the impulse response of a network has the character shown: it is a "delayed," narrow, smooth "pulse." Suppose also that the excitation varies only slowly with time. What features of the response wave can you then predict?

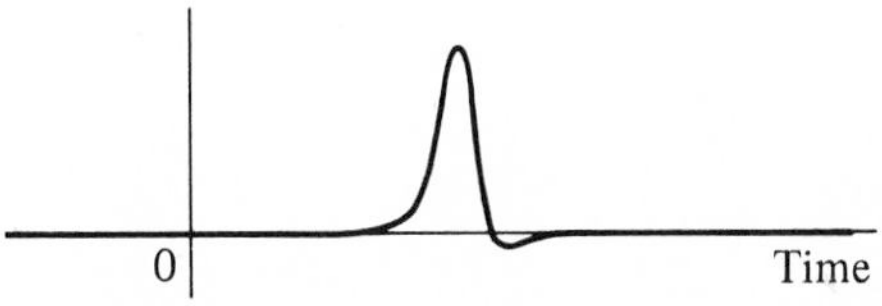

FIG. 4-19.

4-20. Show, by appropriate sketches based on the convolution integral, that a network's response to a rectangular pulse tends to be altered when the pulse excitation becomes smooth rather than square (see Fig. 4-20) thus: (*a*) for small values of time, the response decreases; (*b*) for larger values of time, it increases.

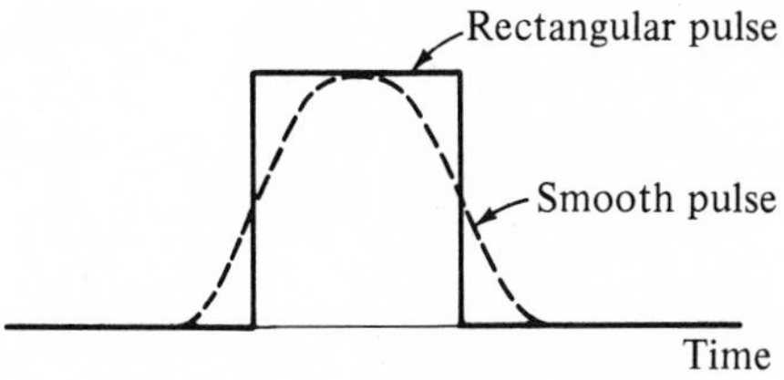

FIG. 4-20.

4-21. For each of the networks described by the impulse responses sketched, state the initial and final values of the step response within 15 percent.

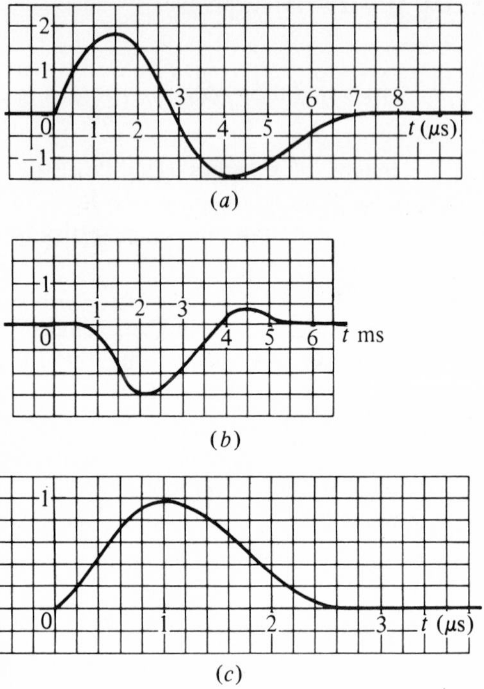

FIG. 4-21.

4-22. Derive (4.06-10) by differentiation of (4.06-9). Explain carefully the necessary continuity conditions. Under what conditions will (4.06-10) be valid *at* $t = 0$? Find a simple example of such a condition. Find also one where it is *not* so valid, in the *RLC* networks of Chap. 3. Explain the physical action in each case.

4-23. The step response of a network is proportional to $(e^{-\alpha t} - e^{-\beta t})$, in which $\beta > \alpha$. When excited by a wave proportional to $e^{-\alpha t}$, what is the response? (Give formula, and sketch for $\beta = 2\alpha$.)

4-24. An *RC* circuit at rest is excited by $i_F = I_F(t/T_0)^n u(t)$. Sketch excitation and response waves for $n = 1, 0.5, 0, -0.5$. Can you do the same for $n = -1$? Explain. (Take $T_0 = RC$.)

4-25. Suppose the impulse response of a network is a rectangular pulse of height R_0 and duration T_0. Ordinary networks cannot possibly produce such sharp-cornered responses, though complicated networks can be made to behave approximately so. Ideal, nondissipative transmission lines (to which the convolution principle applies) *can* have such behavior. Find the response of this network to an excitation identical in form to its impulse response. A (carefully drawn) sketch is adequate answer.

Suppose the impulse response of a transmission-line network is $\delta(t) - \delta(t - 2T)$. Sketch its step response. (This behavior is the basis of pulse-forming networks for radar use.)

4.12 PROBLEMS

4-A. Suppose $r = R_0 e^{-\alpha t}$ and $e = E_0 e^{-\alpha t}$ in the situation of Sec. 4.07. For $t = 1/\alpha$ plot the curves to scale, one folded and one not, and their product. By careful measurement of the area determine $r(t)$ and compare it with the correct value.

4-B. An ordinary RC circuit responds, it is found by observation, to ramp excitation $i_F = I_F t/T_F$ by increasing from zero at $t = 0$ to $V_0 = 1.5$ V at $t = T_F$. Given that the resistor is 1 kΩ, $I_F = 3$ mA, $T_F = 22$ μs, find (within 10 percent) the time constant and the capacitance. Another RC circuit, with $R = 5000$ Ω, reaches 7.5 V in the same time (22 μs) when similarly excited. What is its time constant? Its capacitance?

4-C. A network whose impulse response is simply $e^{-\alpha t}$ is excited by the wave $e^{\alpha(t-T)}[u(t) - u(t - T)]$. Sketch the form of the response wave you expect. Where is its value greatest? Why? Now perform a careful analysis and plot the response for comparison with your preliminary answers above.

4-D. Compare in detail the responses of an RC circuit of time constant T to the two (symmetrical) excitation waves shown in Fig. 4-D.

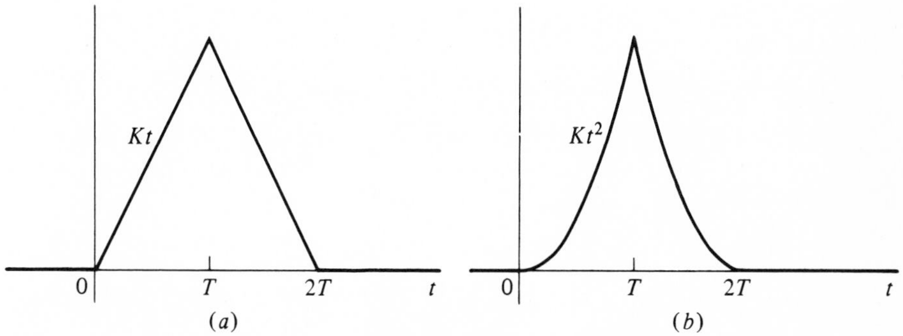

FIG. 4-D.

4-E. An RC circuit at rest is excited by the peculiar pulse shown in Fig. 4-E. First sketch the response as you expect it to be and then find its response formally and carefully plot it. Compare.

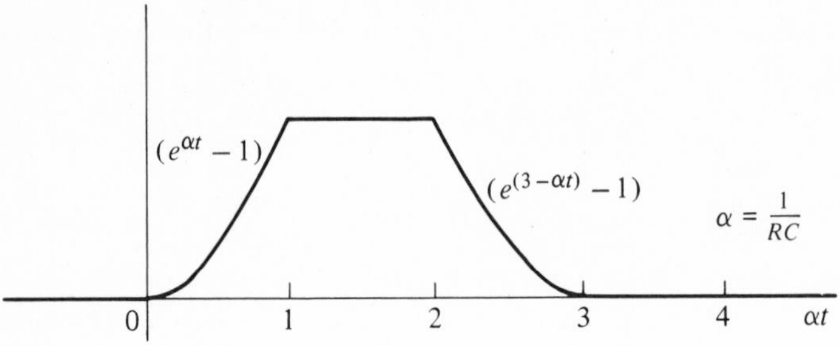

FIG. 4-E.

4-F. Someone wants to know the shape of the response of a parallel RC circuit excited (from rest) by a current "pulse" of semicircular shape:

$$i_F = I_0 \sqrt{z(2-z)}[u(z) - u(z-2)] \quad \text{in which} \quad z = \alpha t = \frac{t}{RC}.$$

(a) Solely by mental visualization of the convolution integrand and mental calculations, sketch the shape of the response versus time and determine its value to one significant figure, in units of RI_0, at $z = 0, 1, 2, 3$.

(b) Calculate the value of the response at each of those four points by subdividing the time scale into sections of length 0.5 and using the trapezoidal rule for integration: $\int_a^b f(x)\,dx = (b-a)[f(a) + f(b)]/2$. Compare these results with your first estimates, in (*a*).

(c) Repeat, using subdivisions of length 0.25.

(d) Calculate the response at such additional times as you deem necessary, and make a plot from which response values can be read within 5 percent.

4-G. Find the response of a tuned circuit of moderately high Q to excitation identical to its own impulse response (the circuit is excited at its own natural frequency). Does the factor t appear, as it does for the RC circuit similarly excited [see (4.06-1)]? Is the response bounded or does it increase indefinitely because of the resonant excitation? Explain fully. {*Hint:* $\cos A \cos B = \frac{1}{2}[\cos(A+B) + \cos(A-B)]$.}

4-H. An RC circuit ($R = 1000\ \Omega$) is excited from rest by a ramp function of current that varies linearly with time from zero at $t = 0$ to 1 mA at $t = 1\ \mu$s. The response is 1.5 V at $t = 5\ \mu$s. Find the value of the time constant and of the capacitance within 2 percent.

4-I. An RC circuit is excited by the three current pulses shown. For each case (a) sketch the general form of the response without making any calculations (intermediate sketches may be used) and (b) determine as far as possible without calculations whether the maximum response occurs before or after $t = T$. Explain your thinking. Then proceed to formal analysis and make a careful plot of the response. Discuss.

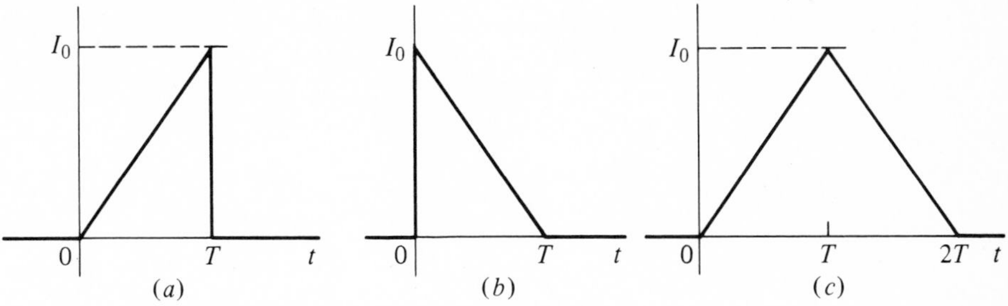

FIG. 4-I.

4-J. The step response of a network is $R_0(1 - e^{-\alpha t})u(t)$. Find its response, from rest, to these excitations: (a) $E_0 e^{-\alpha t}u(t)$; (b) $K(\alpha t)(1 - \alpha t)[u(t) - u(t - 1/\alpha)]$. In each case perform the analysis by using (1) superposition integral (4.04-5) and (2) superposition integral (4.09-1). Compare the work.

4-K. The (measured) step response of a network is given in the figure. Plot, as accurately as you can, (a) its impulse response and (b) its response (from rest) to ramp excitation. In each case state clearly how accurate your answer is.

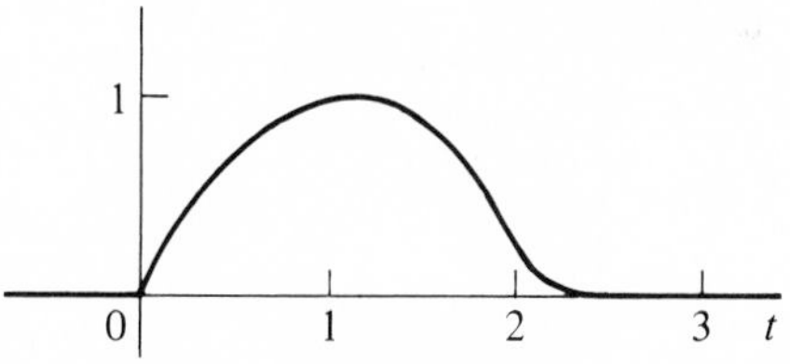

FIG. 4-K.

4-L. Consider a cascade (tandem) connection of identical RC circuits with amplifying units g_m (Fig. 4-La). Find the responses requested below, using the notation $\omega_c = g_m/C$ for simplicity, and $\alpha = 1/(RC)$.

(a) The excitation $e_1 = \Lambda_0\,\delta(t)$, a voltage impulse of strength Λ_0 Weber-turns or volt-seconds, so that the first response $r_1 = -\Lambda_0\,\omega_c e^{-\alpha t}$. Find r_2, $r_3, r_4, \ldots, r_n$ both literally and pictorially (sketches will do).

(b) Change e_1 to a *step*. Sketch the responses $r_1, r_2, \ldots, r_n$, using previous results and doing the minimum work necessary.

(c) Change e_1 to the triangular wave e_1 shown. Plot r_3 versus time. Compare the result with the original triangle and briefly describe the distortion thereof, and its "delay" or lateness of appearance.

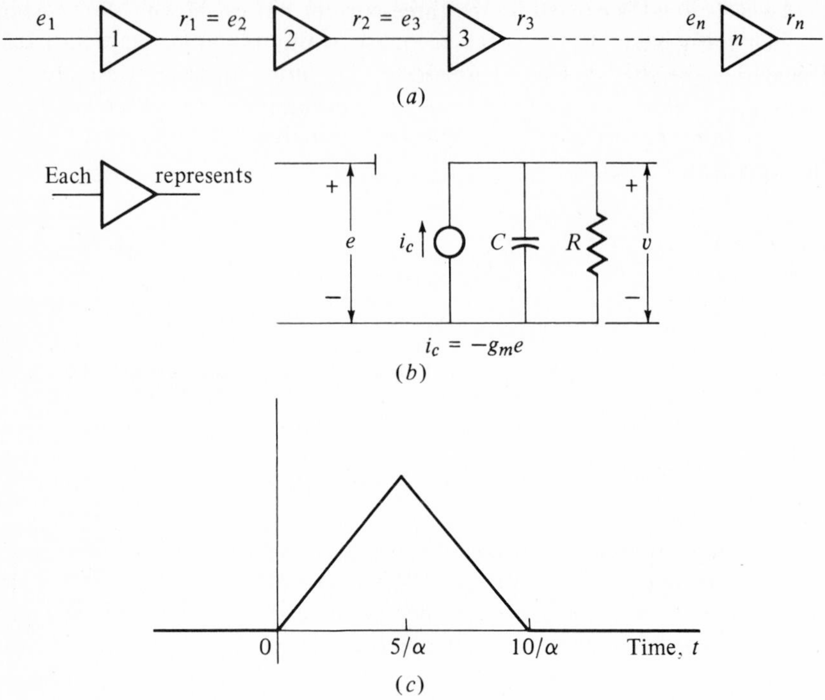

FIG. 4-L.

4-M. Since the unit step is the integral of the unit impulse, it is not unreasonable that a network's response (from rest) to unit step excitation is the integral of its impulse response [(4.06-9)]. Verify now, from the convolution integral, that its response to the unit *ramp* function, $[tu(t)]$, the integral of the unit step function, is the integral of its step response. (*Suggestion:* Integrate by parts.) Proceed to show that its response to the integral of the ramp, the unit parabola $[0.5t^2u(t)]$, is the integral of the ramp response. Is, in general, the response to $e_1(t) = \int_0^t e(x)\,dx$ equal to the integral of the network's response to $e(t)$? [*Suggestion:* If integration by parts no longer helps (why?), try interchanging the order of integration, first justifying this step.]

4-N. The figure gives an engineer's sketch of the expected impulse response of a receiver under design. Assume zero value for $t > 2.5$ and $t < 0$. The engineer wants to know what the response of such a receiver would be to (*a*) excitation identical in shape to the impulse response and (*b*) excitation like the impulse response reversed in time. Plot and compare the two responses. Explain any striking features in terms of convolution sketches. The system the engineer is working on is intended for echo ranging: a pulse is sent out, reflected by an object, and returned at time T, which is equivalent to excitation of the receiver by $e(t - T)$ instead of $e(t)$. For such a system, intended to measure T (i.e., the distance to the object), which pulse (excitation) wave is preferable? Why?

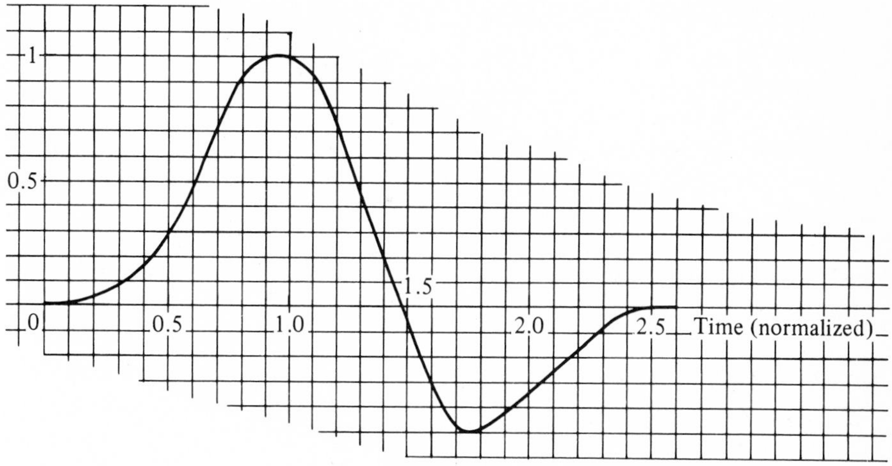

FIG. 4-N.

4-O. The response of an RC circuit to excitation of "gaussian" nature, $e^{(-t^2/T_0{}^2)u(t)}$, is needed. Compare the labor of numerical evaluation of the convolution integral with that involved in transforming the integral to the form $\int e^{-u^2}\,du$ and consulting a table of the integral of the normal error probability (distribution) function.

4-P. Consider the problem of finding the response, to given excitation $e(t)$, of a network whose impulse response is that shown in Fig. 4.09-A. The following possibilities exist (if not more):

1 Find the differential equation corresponding to Fig. 4.09-A; solve it with excitation $e(t)$;

2 Fit the wave of Fig. 4.09-A with an exponential sum, $\sum_{j=1}^{N} K_j e^{p_j t}$, and convolve this sum with $e(t)$ analytically;

3 Convolve the wave of Fig. 4.09-A numerically with $e(t)$, directly.

(a) Compare these generally as to relative difficulty and ease of execution.
(b) Find the response of the network to (1) step excitation, (2) excitation $e^{-\alpha t}$, taking the time constant α^{-1} to be 1 cm (one division) in Fig. 4.09-A.

4-Q. Let a network's impulse response take the form $r_\delta(t) = \sum_{j=1}^{N} K_j e^{p_j t}$ in which $p_1, p_2, \ldots, p_N$ are its natural frequencies. Let it be excited at its own natural frequencies in this sense: excitation $= \sum_{j=1}^{N} C_j e^{p_j t}$, in which the C's are constants. What *types* of terms will appear in the response? Illustrate for the *RLC* circuit of moderately high Q; give your answer in purely real (exponential, sine, cosine) terms format.

4-R. Derive (4.09-1) in the same general way that Sec. 4.04 derives (4.04-5). State all assumptions clearly and describe your procedure fully. (*Suggestion:* Fig. 4.04-B*b* may be looked on as a succession of steps just as well as a series of blocks.) Also obtain (4.09-1) from (4.04-5) by integration by parts. Also write (4.09-1) in a single-term form by eliminating its first term by incorporating an impulse function into $e'(t)$; explain fully.

4-S. The text gives two superposition integrals, one based on a network's impulse response as the prototype, (4.04-5), and one based on its step response, (4.09-1). Develop a superposition integral based on the network's response to exponential excitation $e^{-\alpha t}u(t)$. Contrast the three integrals in simplicity and probable ease of use. Why do you suppose your new (third) superposition integral is rarely encountered in the literature?

4-T. Show that the convolution integral (4.06-6) can be evaluated by series methods. (*Suggestion:* Let $y = 1 + x/T_0$.) Construct a table of values of the response (*a*) by using a series, (*b*) by numerical integration, and (*c*) by consulting a table of the *exponential integral function.*

4-U. The power series is a mathematical tool sometimes valuable in network analysis, particularly in discussing ideal or limiting cases. Consider a hypothetical network whose impulse response is $r_\delta(t) = K[(\sin \omega_c t)/t]u(t)$. (See Sec. 8.09 for remarks about such a network, if interested.) Sketch impulse and step response versus time. Show that

$$r_u(t) = K\omega_c t\left[1 - \frac{(\omega_c t)^2}{3 \times 3!} + \frac{(\omega_c t)^4}{5 \times 5!} - \cdots\right].$$

Answer the following questions (1) by considering your sketches only *or* (2) by calculating with the power series, whichever requires less work.

(a) At what time does the step response reach its first maximum (first overshoot)?

(b) At what time does it reach its first minimum thereafter (first undershoot)?

(c) Evaluate (in terms of K) the first maximum of r_u, within 5 percent.

4-V. Complete the writing of the digital computer program outlined in Sec. 4.08. Then (*a*) verify its correctness by calculating $r(t)$ in the case $e(t) = E_0 e^{-t/T_0}u(t)$, $r_\delta(t) = R_0 e^{-t/T_0}u(t)$, and comparing your result with the known response $r(t) = R_0 E_0(t/T_0)e^{-t/T_0}u(t)$ [see Secs. 2.14 and (4.06-1)]; (*b*) apply it to the example of Table 4.08-A, using it to obtain enough points to plot an accurate curve of $r(t)$. What is the error in the value of $r(T_0)$ obtained in Table 4.08-A (0.127)?

Change the excitation to $E_0/[1 + (t/T_0)^n]$ and repeat for $n = 1, 3, 4$. Compare the response curves and explain their differences.

4-W. The process of "reverse" or "inverse" convolution in which the response is known, as is *either* the excitation or the impulse response (but not the other), in principle requires the solution of an integral equation, for the unknown appears under the integral sign in (4.05-1). In practice the numerical treatment corresponding to that of Sec. 4.08 is essentially the same. Suppose for example that the response $r(t)$ is known (observed and measured), as is the impulse response (the network is given), and the excitation $e(t)$ is unknown. Show that in the simplified notation

Time	Excitation Value	Impulse Response Value
0	$e(0)$	$r_\delta(0)$
Δ	$e(1)$	$r_\delta(1)$
2Δ	$e(2)$	$r_\delta(2)$
.....	.....	.....

(a) If the trapezoidal rule is used for numerical integration, then

$$e(1) = \frac{(2/\Delta)r(1) - e(0)r_\delta(1)}{r_\delta(0)},$$

$$e(2) = \frac{(2/\Delta)r(2) - e(0)r_\delta(2) - 2e(1)r_\delta(1)}{r_\delta(0)},$$

$$e(3) = \frac{(2/\Delta)r(3) - e(0)r_\delta(3) - 2e(1)r_\delta(2) - 2e(2)r_\delta(1)}{r_\delta(0)},$$

. . . .

(b) If Simpson's rule is used, then

$$e(2) = \frac{(3/\Delta)r(2) - e(0)r_\delta(2) - 4e(1)r_\delta(1)}{r_\delta(0)},$$

$$e(4) = \frac{(3/\Delta)r(4) - e(0)r_\delta(4) - 4e(1)r_\delta(3) - 2e(2)r_\delta(2) - 4e(3)r_\delta(1)}{r_\delta(0)},$$

. . . .

Why is $e(0)$ not obtainable in (a)? Why are both $e(0)$ and $e(1)$ unobtainable in (b)? How much difference does this ignorance make in the result, after the first few points on the e curve have been found? What remedy can you suggest, if necessary to improve matters? How must the formula be modified if $r_\delta(0) = 0$? How does the process go if it is $r_\delta(t)$ that is unknown and e is known? Find as best you can the excitation (in volts) from the data below.

Time, μs	Impulse Response, μs^{-1}	Response, V
0	1.00	0.00
1	0.78	0.02
2	0.55	0.08
3	0.28	0.22
4	0.00	0.49
5	−0.15	0.33
6	−0.19	0.01
7	−0.20	0.04
8	−0.21	0.18
9	−0.18	0.22
10	−0.14	0.01
11	−0.07	0.07
12	0.00	0.17
13	0.06	0.06
14	0.08	−0.11
15	0.09	−0.13
16	0.09	−0.09
17	0.07	−0.05
18	0.03	−0.03
19	0.01	−0.02
20	0.00	−0.01

By using power series one may obtain the initial value of e. Show that $e(0) = r'(0)/r_\delta(0)$, $e'(0) = 2[r''(0) - r'_\delta(0)e(0)]/r_\delta(0)$, ... These may be used to start the process, if it seems worthwhile (and possible).

Chapter 5

IMPULSE RESPONSES AND TRANSFER FUNCTIONS

The Lion and the Unicorn were fighting for the
crown:
The Lion beat the Unicorn all round the town.
Some gave them white bread, some gave them
brown:
Some gave them plum-cake and drummed them
out of town.

Many paths, many problems, now open before us. Should we go back and continue our analysis of successively more complicated networks? Move on from *RLC* to something with four or five elements? Should we consider a wider variety of excitation waves for those circuits we now understand? What about problems posed by initial conditions in general? Should we look further into convolution and superposition integrals?

5.01 INTRODUCTION

We have so far "done" two circuits in great detail. We have also developed a *general* procedure (convolution) for the calculation of response, a procedure that is applicable not only to those two circuits, but to any model whose impulse response is known.

For each of the two circuits, we have calculated the impulse response, the step response, the exponential response, the sinusoidal response. (But we recognize that all these are but examples of one, the exponential excitation

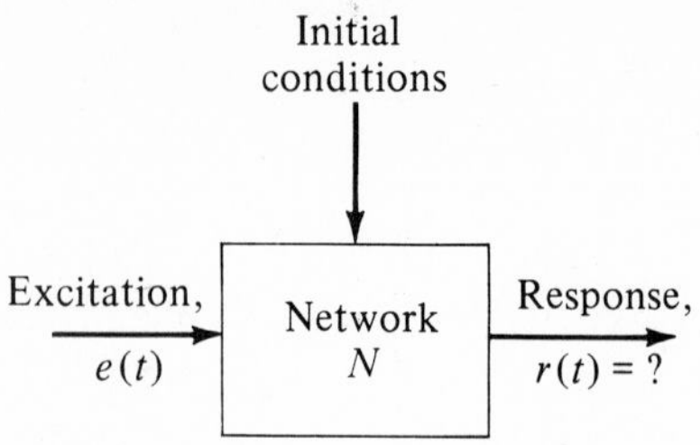

FIG. 5.01-A. The general problem of network analysis.

case. The impulse case requires taking a limit, of course; in the others, s in e^{st} simply takes the proper value.)

To calculate the response to an *arbitrary* excitation we have the general superposition (convolution) theorem.

Arbitrary initial conditions therein are also tractable: we need merely superpose their effects, looked on as due to separate sources that produce the given initial conditions from rest.

Of the general case (Fig. 5.01-A) we have therefore completed the analysis if the network N is an RC or RLC circuit. To generalize so that we can analyze *any* network for *any* initial conditions, for *any* excitation, is a long process. But that is our goal.

In a sense all networks, however complicated, are but combinations or "multiples" of these two basic circuits. Expanding and generalizing the contents of the box N can only increase the number and multiplicity of the natural frequencies; it cannot introduce any new kinds thereof, there are only real and complex varieties. It can, to be sure, immensely complicate the calculations, if only by sheer volume. We must therefore find out how to systematize, order, and normalize them.

The two circuits on which we have spent so much time are virtually "basic building blocks." We therefore carefully tabulate in one place the essentials of what we have learned about them. Figure 5.01-B does this. It recapitulates what we have learned in Chaps. 2 and 3. Note that for the RLC circuit $T(s)$ has also been presented here in a novel form, its expansion as a sum of two simple fractions, in order to parallel the form for the RC circuit. It is remarkable that the coefficients therein, K_1 and K_2, are the same as those in $r_\delta(t)$.

5.02 THE LESSONS OF *RC* AND *RLC*

When the properties of these two circuits are neatly arranged, as in Fig. 5.01-B, something "reaches out and hits us." We had not realized, perhaps, how closely the transfer function and the impulse function are related. The transfer function has to do with the forced part of the response to exponential excitation; the im-

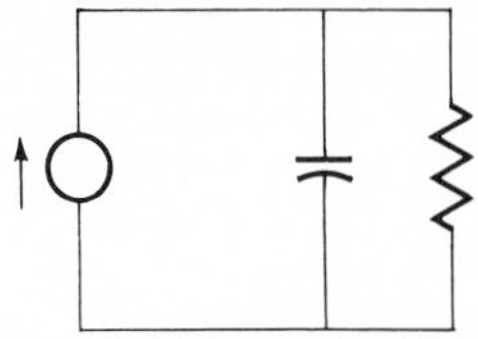

The RC Circuit

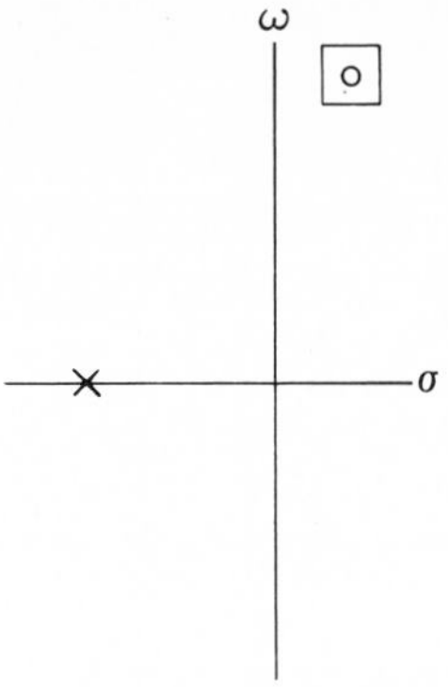

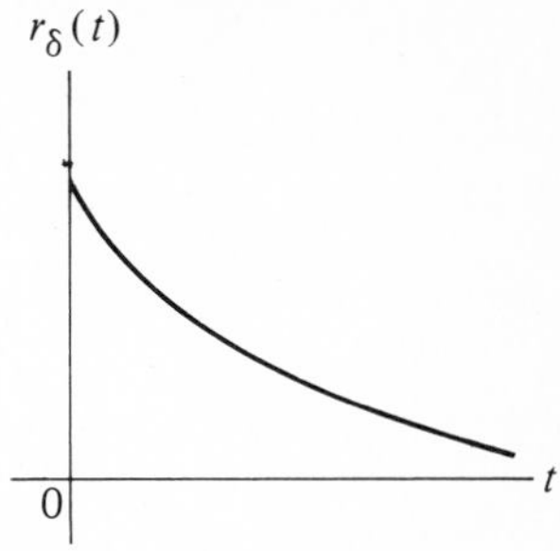

p = natural frequency
$= -\alpha + j0 = -1/(RC)$
K = constant coefficient
$= v(0+)$

$$T(s) = \frac{1/C}{s+\alpha} = \frac{K}{s-p}$$

$$r_\delta(t) = (1/C)\, e^{-\alpha t} = Ke^{pt}$$

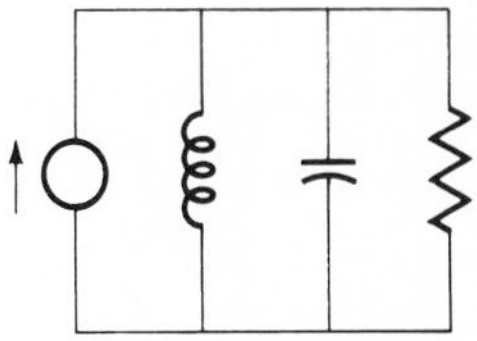

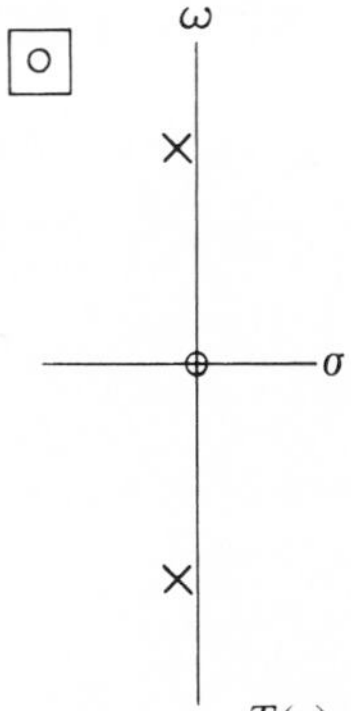

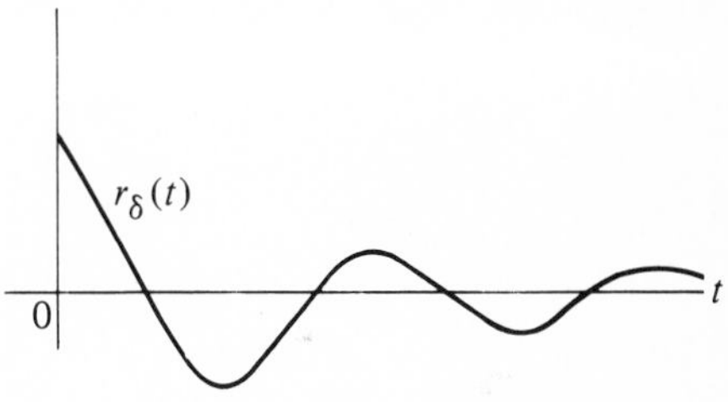

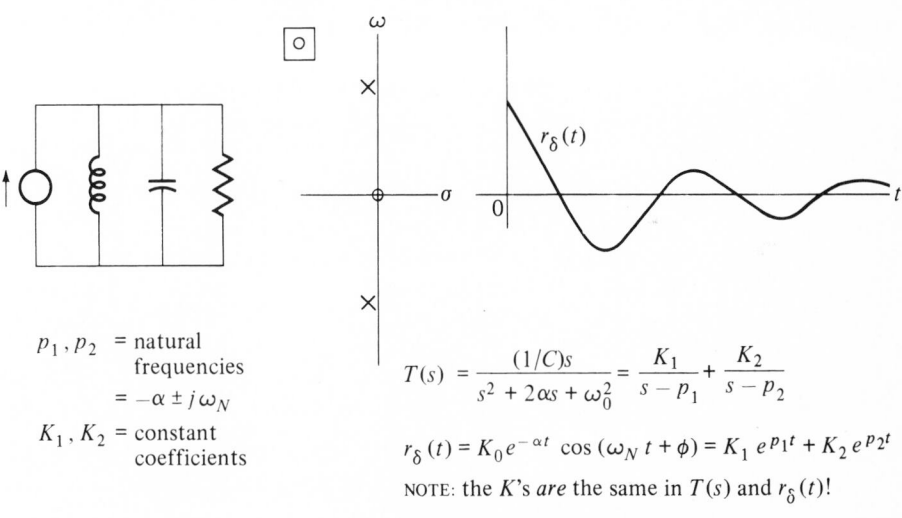

p_1, p_2 = natural frequencies
$= -\alpha \pm j\omega_N$
K_1, K_2 = constant coefficients

$$T(s) = \frac{(1/C)s}{s^2 + 2\alpha s + \omega_0^2} = \frac{K_1}{s-p_1} + \frac{K_2}{s-p_2}$$

$$r_\delta(t) = K_0 e^{-\alpha t} \cos(\omega_N t + \phi) = K_1\, e^{p_1 t} + K_2\, e^{p_2 t}$$

NOTE: the K's *are* the same in $T(s)$ and $r_\delta(t)$!

FIG. 5.01-B.

pulse response is what happens, in the natural-behavior manner, when the excitation is purely impulsive. Yet there they are: coefficients K and natural frequencies p in each, the *same* K's and p's, pairwise, in the two functions, one of s, one of t! Two particular cases prove nothing, of course, about the general case. But this is unlikely to be a coincidence; at the very least, it suggests something interesting that should be looked into.

And so, basing our decision on these two cases, we propose the study of the *general* validity of the following propositions:

a The impulse response is a sum of exponentials, in form

$$r_\delta(t) = \sum_{j=1}^{N} K_j e^{p_j t}; \tag{5.02-1}$$

b The transfer function is a sum made up of the *same* K's and p's, and of course s:

$$T(s) = \sum_{j=1}^{N} \frac{K_j}{s - p_j}. \tag{5.02-2}$$

We know this to be true for $N = 1$ and for $N = 2$ in our two examples, the RC and the RLC circuits. To what extent is it *generally* true? In the next two sections we shall attempt to answer this in different ways.

5.03 A FORMAL MODEL

Let us on the strength of the observations of Sec. 5.02 establish a class of networks, those for which the impulse response *is* a sum of exponential terms in time, those for which (5.02-1) holds. We have two examples already; there are many more. In fact, nearly all networks of practical interest belong to this class. A formal proof that having such an impulse response is merely a consequence of our postulates can wait; here we simply postulate or assume that $r_\delta(t) = \sum_1^N K_j e^{p_j t}$. There are N natural frequencies, and we also postulate that all N are distinct, no two coincide. Such networks as may exist with other, different sorts of impulse response we set aside for now; we limit our attention to those with the property (5.02-1).

To such networks (circuits) Fig. 5.03-A then applies; it presents our hypothesis (5.02-1) in a diagrammatic fashion. In a purely formal way we can then model the network as in Fig. 5.03-B. There each box is a (hypothetical, not necessarily realizable) small network with one natural frequency. The figure is a one-line (not two-wire), purely schematic sort of picture; hence in it a box *can* have but one natural frequency, even if that be complex (one of the other boxes then has the conjugate natural frequency). In it the boxes do not interact in any way; the diagram merely states pictorially, in detail, what the hypothesis (5.02-1) is.

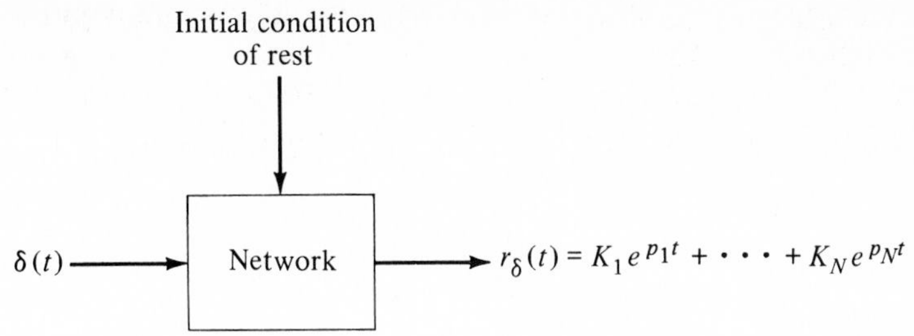

FIG. 5.03-A.

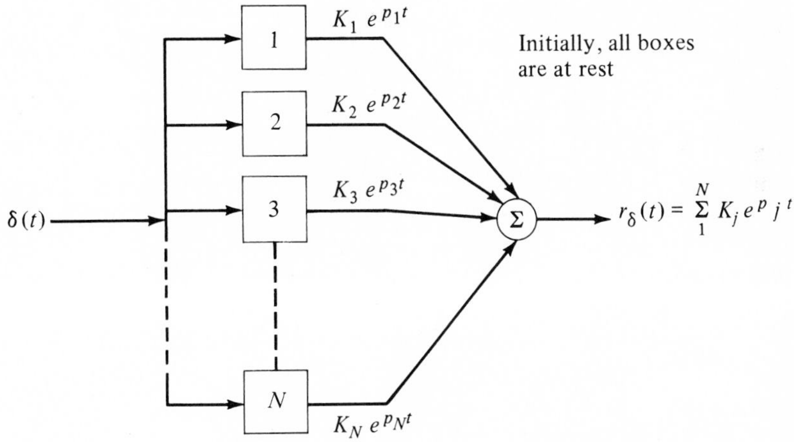

FIG. 5.03-B.

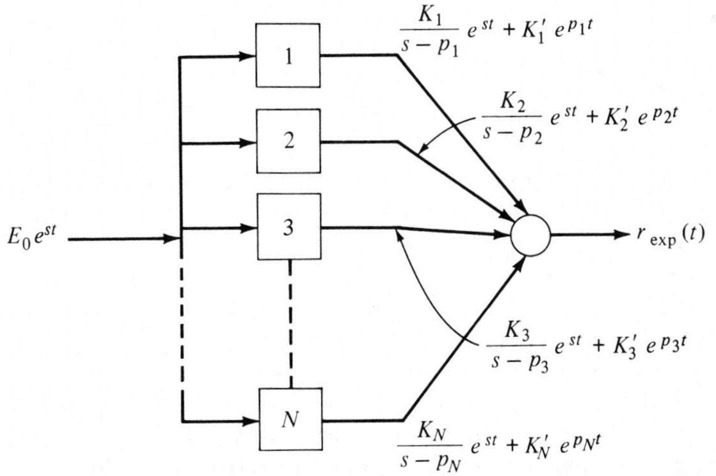

FIG. 5.03-C. For initial conditions of rest, $K_1' = -K_1/(s-p_1)$, $K_2' = -K_2/(s-p_2)$, $K_N = -K_N/(s-p_N) \cdots$.

To that model let us now apply, from rest, instead of a unit impulse, a typical exponential excitation $E_0 e^{st}$, as in Fig. 5.03-C. To calculate the response from the diagram is perfectly straightforward. From Fig. 5.01-B we assume, as seems only reasonable, that each box gives a contributing term $\{[K/(s-p)]e^{st} + K'e^{pt}\}$, and hence that

$$r_{exp}(t) = \underbrace{\sum_{j=1}^{N} \frac{K_j}{s-p_j} E_0 e^{st}}_{r_F(t)} + \underbrace{\sum_{j=1}^{N} K'_j e^{p_j t}}_{r_N(t)}. \tag{5.03-1}$$

The first sum, apart from the excitation factor $E_0 e^{st}$, is by definition the transfer function [(3.10-15)].

Hence,

$$T(s) = \frac{K_1}{s-p_1} + \frac{K_2}{s-p_2} + \cdots + \frac{K_N}{s-p_N}, \tag{5.03-2}$$

i.e., (5.02-2) follows from (5.02-1). Tautological, true, but the relations are very important and worth emphasis and repetition.

The second sum in (5.03-1) represents the natural part of the response. If the initial conditions for each box are "rest," $r(0) = 0$, then the K'_j must each satisfy this simple relation to the corresponding K_j:

$$K'_j = \frac{-K_j}{s-p_j} E_0 \qquad (j = 1, 2, 3, \ldots, N). \tag{5.03-3}$$

We shall later return to (5.03-3), which seems to relate the coefficients in the *natural* part of the response to the forced part of the response (and therefore also to the transfer function) in a simple and very interesting way (see Sec. 7.03). For the present let us continue to study the relation of the *transfer function* to the impulse response. This is important enough to justify a second, different method of deriving (5.02-2) from (5.02-1).

5.04 A MATHEMATICAL DERIVATION

Let us start again from (5.02-1) as a definition of a (perhaps special) class of networks, i.e., from Fig. 5.03-A. And again let us imagine an ordinary exponential excitation applied from rest (Fig. 5.04-A). But let us this time use the superposition or convolution integral to evaluate the response:

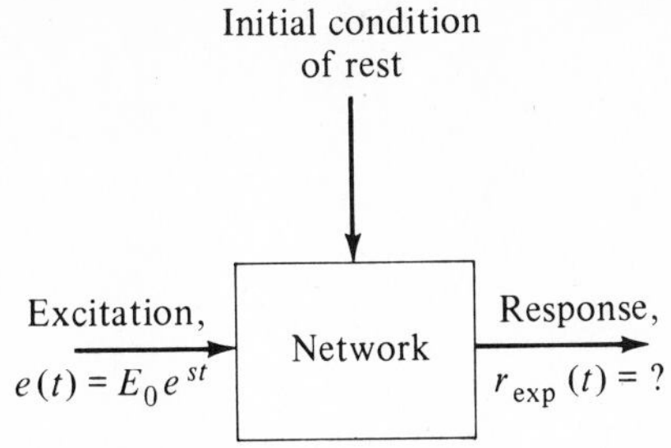

FIG. 5.04-A.

$$
\begin{aligned}
r_{exp}(t) &= e(t) * r_\delta(t) \\
&= E_0 e^{st} * \sum_{j=1}^{N} K_j e^{p_j t} \\
&= \int_0^t E_0 e^{s(t-x)} \sum_{j=1}^{N} K_j e^{p_j x}\, dx \\
&= E_0 e^{st} \int_0^t \sum_1^N K_j e^{(p_j - s)x}\, dx \\
&= E_0 e^{st} \sum_1^N K_j \frac{e^{(p_j - s)t} - 1}{p_j - s} \\
&= E_0 \sum_1^N \frac{K_j e^{p_j t}}{p_j - s} - E_0 e^{st} \sum_1^N \frac{K_j}{p_j - s} \\
&= \underbrace{\sum_1^N \frac{K_j}{s - p_j} E_0 e^{st}}_{r_F} + \underbrace{E_0 \sum_1^N K'_j e^{p_j t}}_{r_N} \\
&= T(s) E_0 e^{st} + r_N(t).
\end{aligned}
\tag{5.04-1}
$$

These results are identical with (5.03-1), (5.03-2), and (5.03-3)—no great surprise, but a comforting corroboration.

5.05 THE TRANSFER FUNCTION; POLES AND ZEROS

We can now state with assurance what is at one and the same time a very important proposition of network theory and a very useful, practical tool of network analysis:

If:

$$r_\delta(t) = \sum_{j=1}^{N} K_j e^{p_j t}, \tag{5.05-1}$$

Then:

$$T(s) = \sum_{j=1}^{N} \frac{K_j}{s - p_j}. \tag{5.05-2}$$

The *postulate*, the *if*, which to be sure does somewhat restrict the networks to which (5.05-1) and (5.05-2) apply, is, once more: There are N (separate, distinct) natural frequencies and the impulse response consists simply of N exponential terms in which *time* is the variable. As a *consequence* the transfer function is another sum, of terms in s, closely related to the first sum. And this implies additional properties of the transfer function that are worth examining here.

On collecting terms over a common denominator, we have

$$\begin{aligned} T(s) &= \sum_{j=1}^{N} \frac{K_j}{s - p_j} = \frac{\text{a polynomial in } s \text{ of degree } (N-1) \text{ at most}}{(s - p_1)(s - p_2)\cdots(s - p_N)} \\ &= \frac{A_0 s^{N-1} + A_1 s^{N-2} + \cdots + A_{N-2} s + A_{N-1}}{s^N + B_1 s^{N-1} + B_2 s^{N-2} + \cdots + B_{N-1} s + B_N} \\ &= H \frac{s^M + \cdots}{s^N + \cdots} = H \frac{Q_M(s)}{Q_N(s)}. \end{aligned} \tag{5.05-3}$$

The system of subscripts used here is chosen for simplicity in later developments, where the cumbersomeness will disappear.

The numerator polynomial's degree, M, is at most $(N - 1)$ because of the form assumed for the impulse response. Its actual value depends on the values of the coefficients K_j: whether they are positive or negative, whether cancellations occur in collecting terms. Any of the A's may be zero; that depends on how the K's combine. (The multiplier H is equal to the first nonzero A.) These A's are of course real numbers; any of them may be positive or negative (or zero).

The denominator polynomial's degree, N, is the number of terms in the impulse response, the number of natural frequencies. And these are of course the zeros of the denominator polynomial Q_N. Note that in this polynomial the coefficients B_j are the symmetric functions of the negatives of the natural frequencies:

$$B_1 = -p_1 - p_2 - \cdots - p_N = \sum_{1}^{N} (-p_j)$$

$$B_2 = (-p_1)(-p_2) + (-p_1)(-p_3) + \cdots = \text{sum of all possible products of the } (-p_j) \text{ taken two at a time}$$

$$\vdots$$

$$B_N = (-p_1)(-p_2)\cdots(-p_N) = \text{product of all the } (-p_j). \tag{5.05-4}$$

We shall shortly illustrate these points with convincing real examples.

The form of the transfer function displayed in (5.05-3) has certain noteworthy properties which we repeat and summarize:

> **a** $T(s)$ is a rational function of s (a quotient of two polynomials) with real coefficients;
>
> **b** The denominator, a polynomial of degree N, has N zeros which are the natural frequencies, the poles of T; (5.05-5)
>
> **c** The degree of the numerator polynomial, M, is at most $(N - 1)$; cancellation of terms may reduce it below $(N - 1)$, even to zero.

These properties, simple and clear though they are, constitute most of the important properties of transfer functions in general. They lead in turn to other properties which we now first extract formally and then exemplify (Sec. 5.06).

We add to (*b*) in (5.05-5): the natural frequencies have in general *negative* real parts, simply because dissipation of energy in resistive elements must eventually cause the natural behavior (the impulse response) to disappear. A model idealized by removing dissipative elements could conceivably have one or more natural frequencies with *zero* real part; sometimes it is reasonable to consider such an "infinite-Q" model for certain purposes. But no natural frequency can have a *positive* real part unless the network contain (internally) a source of energy. Such a network we shall call *active* as opposed to *passive* networks which contain no sources of energy. (The excitation or signal source is by definition external.) An active network may (but need not) have (some) positive-real-part natural frequencies.

For the present we limit our attention to passive, dissipative networks, whose natural frequencies (poles of the transfer function) obey the restriction

$$\left.\begin{array}{c} \mathbf{RE}(p_j) < 0 \qquad j = 1, 2, 3, \ldots, N \\ \textit{or} \\ \text{The } p_j \text{ lie in the left half of the } s \text{ plane.} \end{array}\right\} \tag{5.05-6}$$

[Even if a network is *active*, it must obey (5.05-6) if it is to be *stable*, by definition of the term.] Observe that if (5.05-6) holds, then it follows that all the coefficients B_j are (real and) positive, by (5.05-4). (The converse is not necessarily true.) Such a polynomial, one whose zeros lie only in the left half plane, we call a *Hurwitz* polynomial. The denominator Q_N in (5.05-3) is such a polynomial, by our postulate above.

As for (*c*) in (5.05-5): the numerator is a polynomial Q_M whose degree can be anything from zero to $(N - 1)$. *Its* zeros form a totally different class of points in the s plane: these are values of s for which there is no *forced* component in the response to an exponential excitation. (There remains, of course, the natural part, r_N, usually *not* zero.) The network is opaque to $(e^{s_0 t})$ if s_0 is equal to one of the zeros of $T(s)$. There is, one may say, infinite attenuation ("infinite loss") for such signals. The physical cause may be in effect an open circuit due to a resonance, or, more subtly, a cancellation effect between two or more components in the output.

The *zeros* of the transfer function, in other words, represent values of s for which the forced component of the response to exponential excitation e^{st} is zero: the network is unable or unwilling to transmit e^{st} signals when s is a zero of the transfer function, a point of infinite loss. These zeros may lie anywhere in the s plane, for they have nothing to do with natural behavior, decay or nondecay, resistance or energy dissipation, or the like. Since M is less than N in (5.05-3), the value of T is also zero for infinitely large values of s. We say that T has $(N - M)$ zeros "at infinity" because it behaves essentially as $s^{-(N-M)}$ for large s. Remote though these zeros are, they are still zeros in the same sense; we indicate them on the margin of a pole-zero diagram in a box, as in Fig. 5.01-B, or Fig. 5.07-A*c*. We note that the *total* number of zeros of $T(s)$, M finite ones plus $(N - M)$ at infinity, is also N, a property of rational functions:

> The number of zeros of a rational function is equal to the number of its poles, if those at infinity are counted. (5.05-7)

The *poles* of the transfer function, on the other hand, represent values of s for which the opposite occurs: exponential terms $e^{p_j t}$, in which the p_j are the poles of the transfer function, appear in the response without being forced. The same sort of colorful, useful mnemonic jargon may be used for them, the natural frequencies in (*b*): these are the points of "infinite gain," since such terms appear "without excitation." The poles may NOT lie in the *right* half of the plane, nor even on the imaginary axis (except in very idealized models) because, lacking drive behind them, the corresponding terms must decay if there is any resistance (energy-dissipation modeling) element present. And dissipation there is in realistic models. Models with infinite Q, simplifying the analysis as they do, represent legitimate objects of study under certain circumstances: those where the energy dissipation is not important. For example, take the resonant *RLC* circuit, examined when s is not *at* or near a natural frequency; the gain in simplicity of analysis, at the cost of the loss of only a little accuracy, may be very attractive. Poles of $T(s)$ may then lie on the ω axis.

The foregoing discussion of the properties of transfer functions is rather dry and formal; logically valid, it is yet a "bald and unconvincing narrative" badly in need of "artistic verisimilitude." More convincing perhaps, and certainly refreshing and illuminating, are the demonstrative illustrative examples that follow in later sections. [We should never forget, incidentally, that the transfer function concept depends on our *linearity* postulate (Sec. 4.02); for analyzing nonlinear networks it is of limited utility.]

5.06 EXAMPLES

It behooves us now to illustrate by concrete physical examples, and so to verify, some at least of the general properties of transfer functions stated in Sec. 5.05. It is all well and good to propose mathematically, for example, that the K's in the impulse response $r_\delta(t) = \sum K_j e^{p_j t}$ may be so disposed, positive and negative, that the transfer function's numerator polynomial (5.05-3) may have any degree less than N. But without concrete evidence in the form of network (circuit) examples, that has a hollow, theoretical sound. We need assurance that such networks do exist; and it will be very interesting to see some physical structures that correspond to certain novel locations of the zeros of $T(s)$.

We shall encounter many more examples in later parts of this book; here we limit ourselves to a few, chosen simply to make the points; they are no more complicated than they need to be to give clarifying examples of particularly interesting zeros of transfer functions. In fact we need nothing more than our two simple basic circuits, the RC circuit (Chap. 2) and the RLC circuit (Chap. 3), properly combined.

5.07 CASCADE CONNECTIONS

So long as the tandem (cascade) connection of two networks does not alter the transfer function of either one by loading it, then the overall transfer function is the product of the two individual transfer functions. Cascade connection evidently merely multiplies transfer functions. In Fig. 5.07-A, the electronic g_m devices (see Sec. 1.05) ensure that T_1, the transfer function from v_1 to v_2, is unaffected by the connection of circuit 2 to circuit 1: no current flows in the connection between them, $i_{12} = 0$. Isolation (buffer) action is provided by the g_{m2} element. Similarly T_2 and T_3 represent the transfer functions from v_2 to v_3, and from v_3 to v_4, both before *and* after the interconnection. The transfer function from v_1 to v_4 is then simply the product of the three individual transfer functions, for

$$T = \frac{v_4}{v_1} = \frac{v_2}{v_1}\frac{v_3}{v_2}\frac{v_4}{v_3} = T_1 T_2 T_3 = T \tag{5.07-1}$$

in which the usual conditions for defining transfer functions, (3.10-15), of course are understood to obtain.

For the particular cascade connection of Fig. 5.07-A*b*, symbolized in one-line-diagram form in Fig. 5.07-A*a*,

$$T = \frac{-(1/C_1)}{s+\alpha_1}\,\frac{-(1/C_2)(s + r_2/L_2)}{s^2 + 2\alpha_2 s + \omega_2{}^2}\,\frac{-(1/C_3)}{s+\alpha_3} \tag{5.07-2}$$

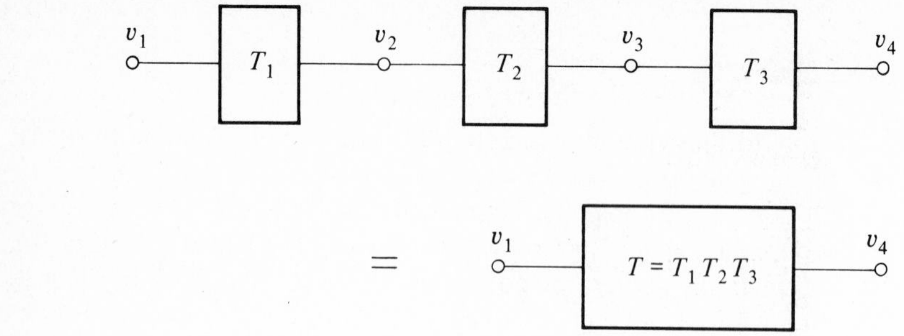

(*a*) One-line diagrams

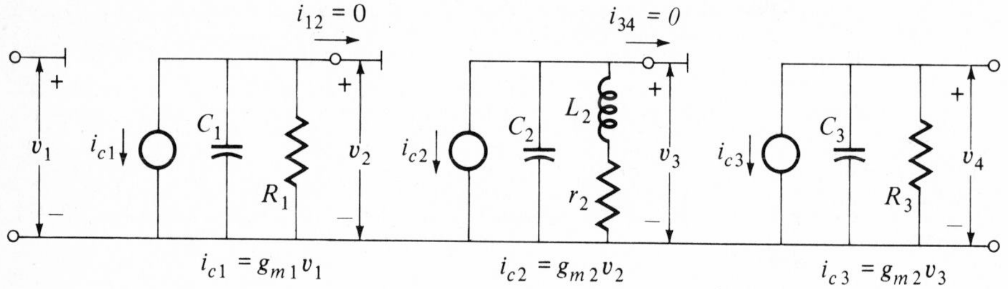

(*b*) Schematic two-wire diagram

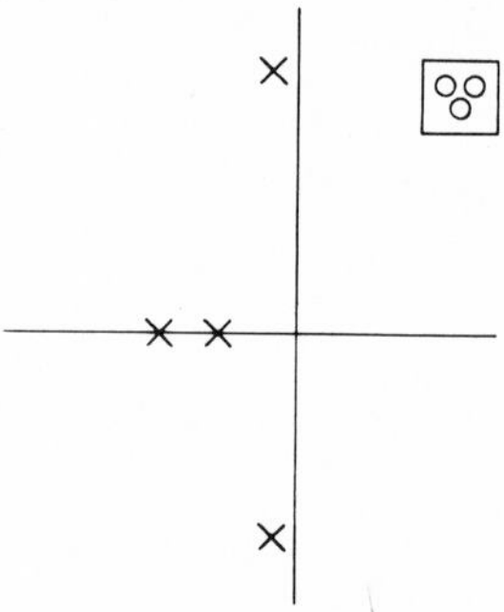

(*c*) Pole-zero diagram

FIG. 5.07-A.

whose pole-zero diagram is perhaps (it depends somewhat on element values) that of Fig. 5.07-A*c*.

Cascade (tandem) connection produces more complicated transfer functions merely by accumulating the poles and zeros of the component transfer functions; it does not alter (move) the poles and zeros. There are evidently limits on the novelties (as to zero positions) that cascade connection can introduce: no zero or pole can appear in the overall transfer function other than those in the individual, component transfer functions.

5.08 PARALLEL CONNECTIONS

It is otherwise with parallel connections. This method of interconnection of networks *can* drastically alter zeros.

The poles, being natural frequencies, will not change, on connecting networks in parallel, unless there is some essential change in the circulatory paths for currents through the network elements. This we avoid by again ensuring that there be no loading (interaction), so that individual transfer functions remain unaltered.

It is the possibility of combining outputs of individual networks by *addition* or *subtraction* (as opposed to multiplication) that can alter the zeros. We make the parallel connection, again (as in Sec. 5.07) without drawing currents that could upset individual transfer functions, as the purely schematic (one-line) diagrams of Fig. 5.08-A indicate. The network N, excited by $e(t)$ and giving the response $r(t)$, is made up of two networks, N_1 and N_2, connected in parallel. We suppose the output from the paralleled network combination, $r(t)$, to be measured by an infinite-impedance (draw-no-current) device, an ideal voltmeter or amplifier, for example. The circle labeled $\sum$ then adds (or subtracts) the individual responses to form $r = r_1 \pm r_2$.

For a specific example, consider Fig. 5.08-B. The two boxes Z_1 and Z_2 represent component circuits that may be of either of our two fundamental types, RC or RLC, for the moment; later we shall perhaps discuss other "boxes."

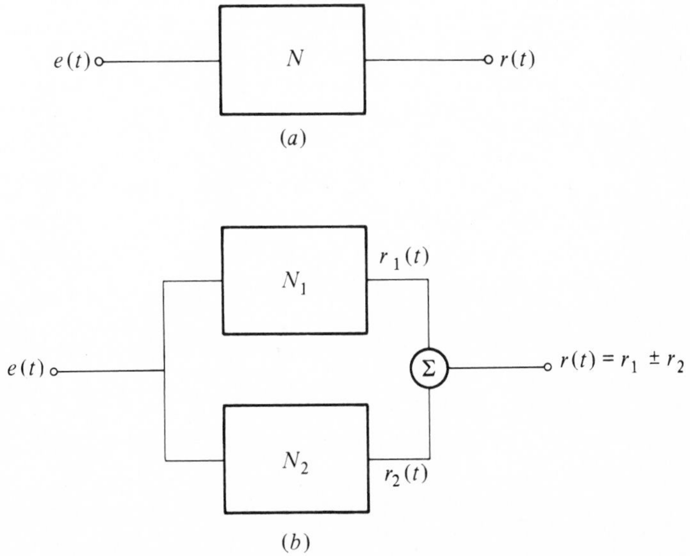

FIG. 5.08-A.

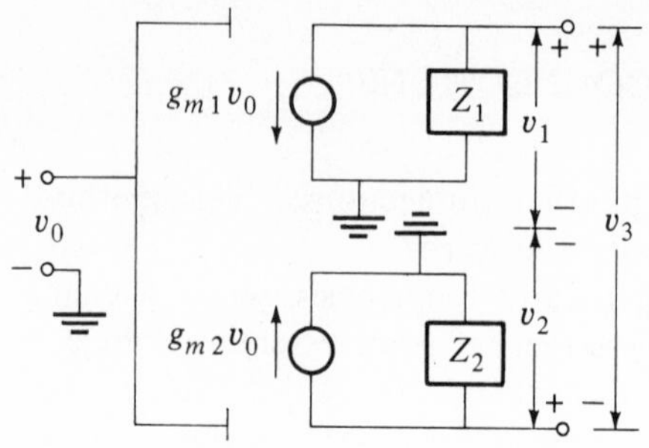

FIG. 5.08-B.

The connection (sometimes called "push-pull") clearly makes the output voltage v_3 the *difference* of the output voltages of the component networks:

$$v_3 = v_1 - v_2 . \tag{5.08-1}$$

It is this *subtraction* that can move the zeros; multiplication or cascade connection cannot. There are two components in v_3; they arrive at the output having come from the input by two different (parallel and different) paths.

Since no current is drawn from the devices, the individual transfer functions are still valid for the individual networks. The overall transfer function is therefore the *difference* of the component transfer functions:

$$T = \frac{v_3}{v_0} = \frac{v_1 - v_2}{v_0} = T_1 - T_2 . \tag{5.08-2}$$

The polarity reversal caused by the g_m devices is of no importance here: our interest is in the zeros of $T(s)$ and the effect of providing *two* paths therefor, We now therefore relegate the sign to the left side. If for example both circuits are of the *RC* type,

$$-T(s) = \frac{g_{m1}/C_1}{s+\alpha_1} - \frac{g_{m2}/C_2}{s+\alpha_2} = \frac{(g_{m1}/C_1 - g_{m2}/C_2)s + (g_{m1}\alpha_2/C_1 - g_{m2}\alpha_1/C_2)}{(s+\alpha_1)(s+\alpha_2)}$$

$$= \frac{H(s-s_0)}{(s+\alpha_1)(s+\alpha_2)} \tag{5.08-3}$$

in which

$$H = \frac{g_{m1}}{C_1} - \frac{g_{m2}}{C_2},$$

$$s_0 = \frac{g_{m2}\alpha_1/C_2 - g_{m1}\alpha_2/C_1}{g_{m1}/C_1 - g_{m2}/C_2} = \frac{g_{m1}G_1 - g_{m2}G_2}{g_{m1}C_2 - g_{m2}C_1}. \tag{5.08-4}$$

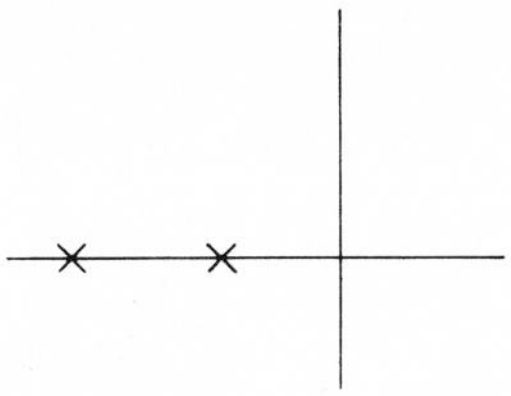

(a) The poles of the transfer function of the parallel combination of two RC circuits

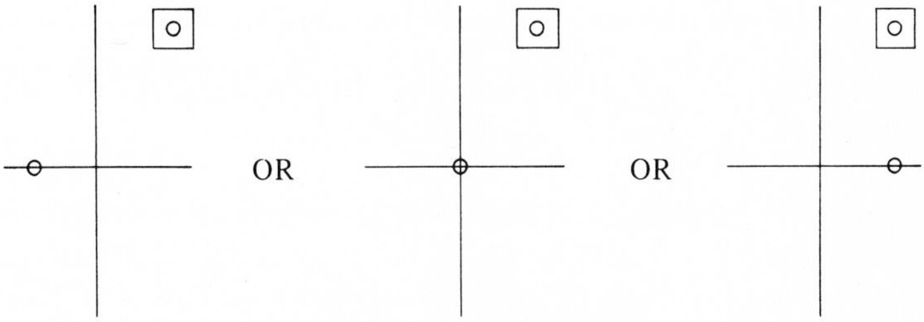

(b) The zeros thereof

FIG. 5.08-C.

The two natural frequencies of the composite network are simply those of the individual circuits. But one of the zeros of the overall transfer function, s_0, can now be placed anywhere on the real axis (Fig. 5.08-C) by adjustment of the values of the component elements! (The other zero remains at infinity.) It requires only that (g_{m1}/C_1) be greater than (g_{m2}/C_2) to place it in the right half plane, a distinct novelty.

If one circuit is RC and the other RLC, we find

$$\begin{aligned}
-T(s) &= \frac{g_{m1}/C_1}{s+\alpha_1} - \frac{(g_{m2}/C_2)s}{s^2 + 2\alpha_2 s + \omega_2{}^2} \\
&= \frac{(g_{m1}/C_1 - g_{m2}/C_2)s^2 + 2(g_{m1}\alpha_2/C_1 - g_{m2}\alpha_1/C_2)s + g_{m1}\omega_2{}^2/C_1}{(s+\alpha_1)(s^2 + 2\alpha_2 s + \omega_2{}^2)} \\
&= H\,\frac{s^2 + 2\alpha_0 s + \omega_0{}^2}{(s+\alpha_1)(s^2 + 2\alpha_2 s + \omega_2{}^2)},
\end{aligned} \tag{5.08-5}$$

in which

$$\begin{aligned}
H &= \frac{g_{m1}}{C_1} - \frac{g_{m2}}{C_2}, \\
\alpha_0 &= \frac{g_{m1}}{C_1}\alpha_2 - \frac{g_{m2}}{C_2}\alpha_1, \\
\omega_0{}^2 &= \frac{g_{m1}}{C_1}\omega_2{}^2.
\end{aligned} \tag{5.08-6}$$

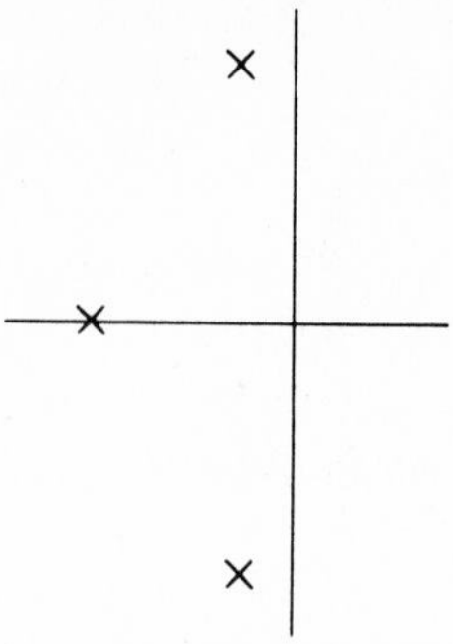

(*a*) The poles of the transfer function of the parallel combination of an *RC* and an *RLC* circuit

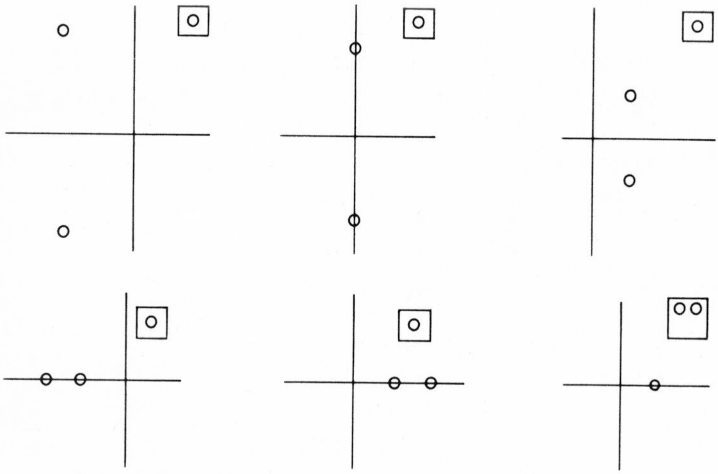

(*b*) Some of the possible zero locations thereof

FIG. 5.08-D.

The natural frequencies are again simply the union of those of the component networks. One zero remains at infinity. But the remaining two zeros can be anywhere in the plane (the origin excepted), provided only that they be conjugates if they are complex (Fig. 5.08-D).

If the *RLC* circuit contains a resistor r associated in series with the inductor, then there may be one or two zeros of $T(s)$ at the origin.

If both circuits are *RLC*, then there are four zeros in $T(s)$, of which one is at infinity; the remaining three can be placed anywhere (subject to the usual restriction on conjugate pairs if complex) by adjusting the element values. (One zero may have to be at, or near, the origin.)

5.09 THE ZEROS OF $T(s)$

We conclude that by using nothing more complicated than two of our basic circuits, we can place zeros of the transfer function anywhere in the s plane. The key to success here is being able to *subtract* components in the output, and to this end the component networks are connected in parallel, in order that two (different) transmission paths be provided through the composite network. The proposition of Sec. 5.05, that the "points of infinite loss" can lie anywhere in the plane, is thus verified. *Zeros of transfer functions are not necessarily restricted as to location.* We shall see many other examples of this in more complicated networks. Where they lie for a given network was presumably determined by its synthesizer, to meet some objective (such as filtering out unwanted signals with certain s values).

Note again that there has been no change in the *natural frequencies* because of the interconnection of the networks: these simply combine in union to form a larger set. Their real parts are still negative, because the networks are essentially passive and dissipative and their values (point locations) are unchanged. [If changes are made in the paths through which currents can circulate, as, for example, by adding feedback, then the natural frequencies may change in location and even move into the right half plane, with attendant problems of instability, as we shall later see (Sec. 6.25).]

The number of *zeros* is unchanged in the parallel connection of Sec. 5.08; it is still equal to that of the union of the component networks, but the zeros have *moved*. And they can be anywhere. The reason, once more, is that the output signal has two components, which can combine positively or negatively: cancellation is possible—and that anywhere, for any value of s. But the natural frequencies, where the gain (not the loss) becomes infinite, remain the *same*.

Networks whose transfer functions have one or more zeros in the right half plane do form a special class; they are, as one might suspect, "different." (There is actually a jargon name for them—"non-minimum-phase"—which we may investigate elsewhere.)

5.10 BRIDGES AND LATTICES

The "difference" has nothing to do with the presence of an active element however. The g_m of Sec. 5.08 are used merely for convenience, for simplicity of construction of the example. To emphasize that fact, we look now at another, purely passive, type of network in which the output has more than one component. Again because of cancellation of components of the output, some zeros of $T(s)$ can appear in the right half plane.

The network of Fig. 5-10A*a* exemplifies a class of networks known as *bridges.* They are important exactly because their outputs do have two components which can be made to cancel at various desirable values of s; this makes them useful in the construction of measuring apparatus, for example. One of the elements may be an unknown and we can adjust some other element(s) until the cancellation occurs at some convenient, observable value of s, such as $j\omega$ (i.e., in the sinusoidal steady state), for example. Then we can calculate (in effect measure) the value of the unknown.

The excitation is the voltage e; the response is measured (without drawing current, in this example) as the voltage v. Since current flows in both branches, in R_1 and R_2, and in R_4 and R_3, the output voltage v is the *difference* between the voltage developed across R_2 and that developed across R_3:

$$v = -R_3 i_{34} + R_2 i_{12}. \tag{5.10-1}$$

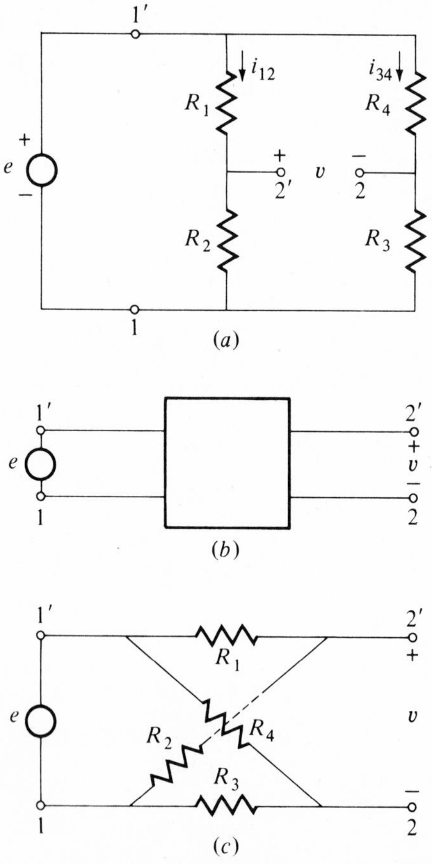

FIG. 5.10-A.

Again we find the fundamental subtraction property: the response contains two components, each of which has arrived at the output by its own (different) route.

Redrawn to fit the conventional input-output-box diagram of Fig. 5.10-A*b*, the network takes the form of Fig. 5.10-A*c*. Here an "X" or *lattice* structure is formed by the crossing or bridging over of leads, whence the name "bridge" network. Its most interesting property is that zero transmission occurs when the "arms" of the bridge are properly adjusted, or the bridge is *balanced*. Application of Ohm's law to determine the currents gives, from (5.10-1),

$$\begin{aligned} v &= R_2 \frac{e}{R_1 + R_2} - R_3 \frac{e}{R_3 + R_4} \\ &= \frac{R_2 R_4 - R_1 R_3}{(R_1 + R_2)(R_3 + R_4)} e. \end{aligned} \tag{5.10-2}$$

If $R_2 R_4 = R_1 R_3$, clearly the output voltage (by virtue of cancellation between its components) is zero.

Used as a measuring instrument, this is the Wheatstone bridge in its primitive form: in it an "unknown" resistor to be measured is made one arm; the other three arms are known resistors, at least one of which is adjustable. The latter is varied until a detector at v reads zero. Then the unknown has in principle been measured, according to (5.10-3). Actual bridges and actual operation are usually not quite that simple, but the principle is just that.

The balance condition

$$R_1 R_3 = R_2 R_4 \tag{5.10-3}$$

is simply that the "cross-resistor products" (Fig. 5.10-A*a*) be equal, or that the "near and far products" (Fig. 5.10-A*c*) be equal; both views are handy mnemonics. In Fig. 5.10-B the network is drawn in an opened, more conventional "bridge" form to which the first mnemonic is easily applied. The second

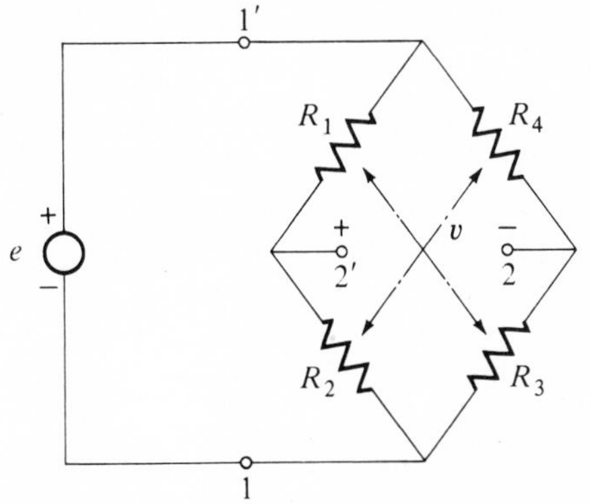

FIG. 5.10-B.

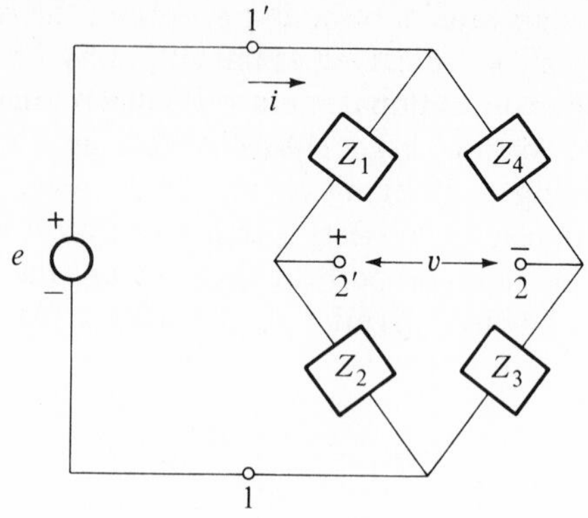

FIG. 5.10-C.

applies to Fig. 5.10-A*c*, looking first at the center crossing arms and then at the outer ones.

The transfer function of this network is simply a constant [see (5.10-2)], independent of *s*, since no energy-storing elements are present, and no derivatives or integrals appear in the equations. To obtain a transfer function with some zeros, we replace one or more of the resistors by (combinations of) elements that include energy-storing capability. The result is Fig. 5.10-C in which each *Z* box represents some sort of simple combination of elements. Now the transfer function can have zeros in the right half plane, strictly passive though the network be.

Before giving examples, we consider, separately, the calculations that are involved when such "boxes" appear within a network. The principles are of great general importance and utility and therefore worth the digression (Sec. 5.11).

5.11 IMPEDANCES AND ADMITTANCES

For example, suppose a capacitor C added in parallel with R_4 (Fig. 5.11-A). Calculation of the transfer function from e to v now requires more sophistication. By definition [(3.10-15)]

$$T(s) = \frac{\text{forced component of response}}{\text{excitation, } e\text{, in form exponential}} = \frac{v_F}{E_0 e^{st}} \tag{5.11-1}$$

in which v_F is the forced component of v, and e is exponential in form, $e = E_0 e^{st}$. The relation (5.10-1) is still valid, as is $i_{12} = e/(R_1 + R_2)$. But the other current, i_{34}, because of the energy-storing element C, now contains both forced and natural components, and we cannot immediately and simply divide e by something analogous to $(R_3 + R_4)$ to obtain i_{34}. In principle a differential equation must be solved. What we can do, instead, however, is to draw on our experience and reason as follows, in self-explanatory symbols and simple steps. The subscript F designates the forced component of a current or voltage, the subscript N its natural component (see Fig. 5.11-A for definitions of the variables).

$$\begin{aligned}
i_{34} &= i_{34F} + i_{34N},\\
i_{34} &= C\frac{dv_{RC}}{dt} + G_4 v_{RC},\\
v_{RC} &= v_{RCF} + v_{RCN} = V_{RC}e^{st} + v_{RCN},\\
i_{34F} + i_{34N} &= \left(C\frac{dv_{RCF}}{dt} + G_4 v_{RCF}\right) + \left(C\frac{dv_{RCN}}{dt} + G_4 v_{RCN}\right),\\
i_{34F} &= (Cs + G_4)v_{RCF},\\
v_{RCF} &= \frac{i_{34F}}{Cs + G_4},\\
e = E_0 e^{st} &= v_{R3} + v_{RC} = v_{R3F} + v_{RCF},\\
v_{R3} &= R_3 i_{34} = R_3 i_{34F} + R_3 i_{34N},\\
E_0 e^{st} &= R_3 i_{34F} + \frac{i_{34F}}{Cs + G_4},\\
i_{34F} &= \frac{E_0 e^{st}}{R_3 + \dfrac{1}{Cs + G_4}},
\end{aligned} \tag{5.11-2}$$

$$v_F = -v_{R3F} + v_{R2F} = -R_3 i_{34F} + \frac{R_2}{R_1 + R_2}E_0 e^{st},$$

$$\begin{aligned}
T(s) = \frac{v_F}{e} &= \frac{R_2}{R_1 + R_2} - \frac{R_3}{R_3 + 1/(Cs + G_4)}\\
&= \frac{R_2}{R_1 + R_2} - \frac{R_3 Cs + R_3 G_4}{1 + R_3 Cs + R_3 G_4}\\
&= \frac{-R_1 R_3 Cs + R_2 - R_1 R_3 G_4}{(R_1 + R_2)(R_3 Cs + R_3 G_4 + 1)}\\
&= \frac{-R_1 R_3 C(s - R_2/R_1 R_3 C + G_4/C)}{(R_1 + R_2)(R_3 Cs + R_3 G_4 + 1)}
\end{aligned} \tag{5.11-3}$$

which does indeed have a zero in the right half plane (and a pole in the left half plane). We shall return to this idea in Sec. 5.13.

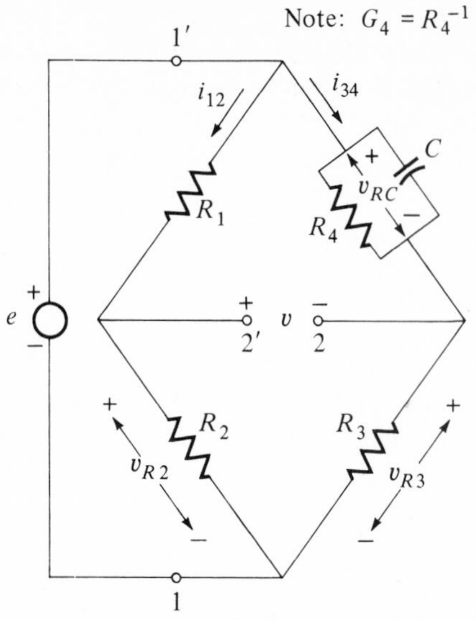

FIG. 5.11-A.

At this point it is imperative that we stop and collect our thoughts, to be sure that we understand what we have done, to realize how simple it really is, once stripped of unnecessaries, to recognize that it is nothing new but that we have been here before, and to see what the real rule, the heart of the procedure is, for future use.

In the first place we *have* been here before, the transfer function from i to v for the parallel RC combination is simply that of Chap. 2, the *impedance*

$$T(s) = \left.\frac{v_{RC}}{i_{34}}\right|_{\ldots} = Z(s) = \frac{1}{Cs + G_4}. \tag{5.11-4}$$

[The three dots, used for brevity, refer to the definition condition (5.11-1).] Alternatively we can say that the transfer function from v to i here is the *admittance*

$$T(s) = \left.\frac{i_{34}}{v_{RC}}\right|_{\ldots} = Y(s) = Cs + G_4 = \frac{1}{Z(s)}. \tag{5.11-5}$$

The principle involved is extremely simple, even though technically a differential equation must be solved: assume exponential excitation and that consequently all currents and voltages have both natural and forced components. These vary with time in different ways and can therefore be sorted into groups, each of which, as a whole, must vanish or balance out. (True, cases where natural frequencies "p" and s are equal may be troublesome; we avoid them for the present, noting only that one can always handle them by properly taking limits as one approaches the other.) We need therefore write out in detail *only the forced parts*, and from *them* we can deduce the transfer function of interest. This is what the eleven lines of equations in (5.11-2) do, resulting in (5.11-3).

We note now a pair of simple but general rules that will be extremely useful. The first relates to elements in parallel (Fig. 5.11-B). Since

$$T(s) = \frac{i}{v}\bigg|_{\ldots} = \frac{\sum_{j=1}^{N} i_j}{v} = \sum_{j=1}^{N} \frac{i_j}{v} = \sum_{1}^{N} Y_j = Y(s) = \frac{1}{Z(s)}, \tag{5.11-6}$$

the transfer function from voltage to current (the "admittance") of a number of elements in parallel is simply the sum of those of the individual elements; the transfer function from current to voltage is its reciprocal, the "impedance." (5.11-7)

The first three cases in Fig. 5.11-C are familiar examples, from Chaps. 2 and 3 (q.v.); the fourth provides a new, more complicated, but readily comprehended example. The fifth gives a generic symbol for such combinations. Again we see that Ohm's law, properly generalized to include the variable s, is the basis of (5.11-6).

The second general rule relates to elements connected in series (Fig. 5.11-D). Since

$$T(s) = \frac{v}{i}\bigg|_{\ldots} = \frac{\sum_{j=1}^{N} v_j}{i} = \sum_{j=1}^{N} \frac{v_j}{i} = \sum_{1}^{N} Z_j = Z(s) = \frac{1}{Y(s)}, \tag{5.11-8}$$

the transfer function from current to voltage (the "impedance") of a number of elements in series is simply the sum of those of the individual elements; the transfer function from voltage to current is its reciprocal, the "admittance." (5.11-9)

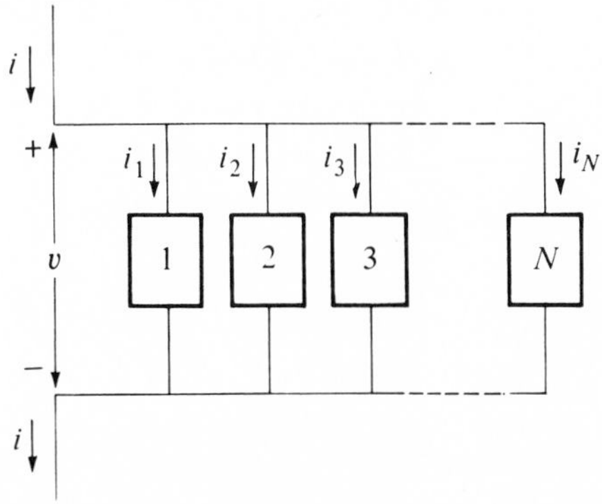

FIG. 5.11-B.

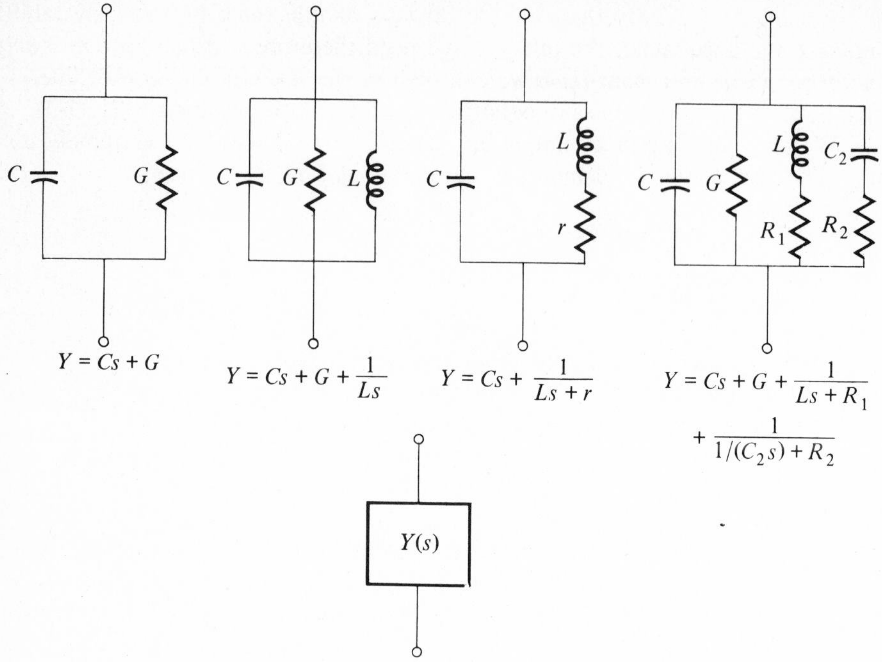

FIG. 5.11-C.

Figure 5.11-E shows simple examples, analogous to those of Fig. 5.11-C.

We observe that the term "element" or the box symbol, the fifth (last) example in Figs. 5.11-C and 5.11-E, can actually be a collection of individual *R*, *L*, *C* elements, connected in series and parallel combinations. One need only apply the two rules repeatedly, alternating from one to the other. The fourth examples in Figs. 5.11-C and 5.11-E illustrate this.

In this way, the *Z* and *Y* transfer functions of many complicated networks that have only two external terminals can easily (if tediously, lengthily) be calculated. Networks for which this is not true, and networks with more than two external terminals, we postpone consideration of.

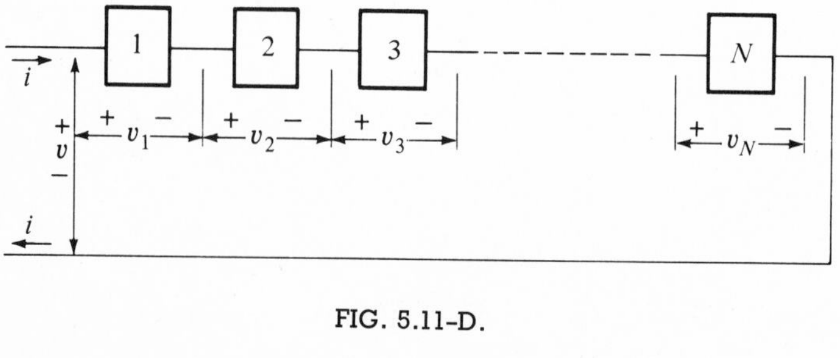

FIG. 5.11-D.

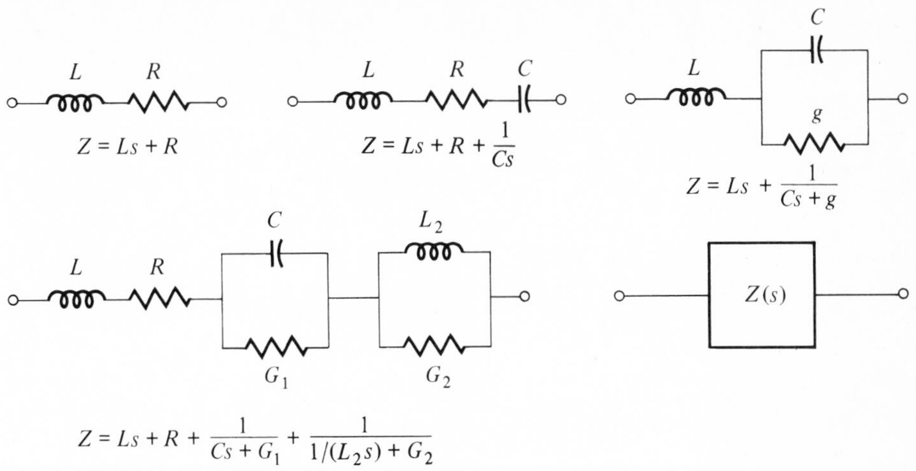

FIG. 5.11-E.

The terms *impedance* and *admittance*, and the portmanteau term *immittance*† that covers both (and is much more euphonious than the equally possible "adpedance") are also used in more general senses for such networks (with more than two external terminals). When they are used in connection with *one* pair of terminals, as here, the modifier *driving-point* is often prefixed, for precision, because the response is then measured *at the excitation* (*driving*) *point*. In the sinusoidal steady state ($s = j\omega$) the real and imginary parts of a driving-point impedance are called respectively *resistance* and *reactance*; the real and imaginary parts of a driving-point admittance are called respectively *conductance* and *susceptance*. Note that they are all functions of frequency ω.

Distinction, in using these expressions, between (*a*) *mathematical* functions (of s) and (*b*) *physical* elements and the interconnection thereof soon becomes hopeless—and pointless. We shall not often discriminate between Z and Y as mathematical symbols and as physical "boxes."

It is amazing, and very helpful in analysis, that Ohm's law, the familiar "$v = Ri$," *can* be used to obtain Z and Y transfer *functions* with so little generalization effort. The series (add-impedance) and parallel (add-admittance) rules still apply: we have only to use Ls for inductor impedance and Cs for capacitor admittance. It is very informative to stop to reflect here on the reason that the differential equations that must be solved can in effect be ignored or, better, replaced by such simple algebra. (The linearity and the constancy of elements that we have postulated are responsible.)

What we might call "Ohm's-law box manipulation" will soon be a familiar tool. Its primary use is to calculate $Z(s)$ and $Y(s)$ expressions that are, if you will, equivalent single "boxes" that can replace series and parallel combinations of elements and more complicated immittances, according to (5.11-7) and (5.11-9).

† Coined by H. W. Bode; see BO in Appendix B.

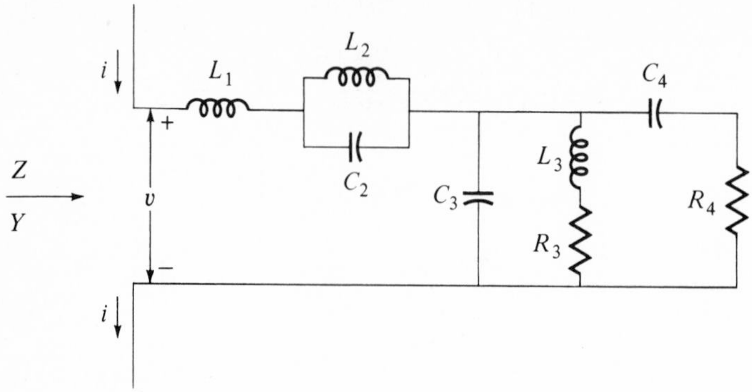

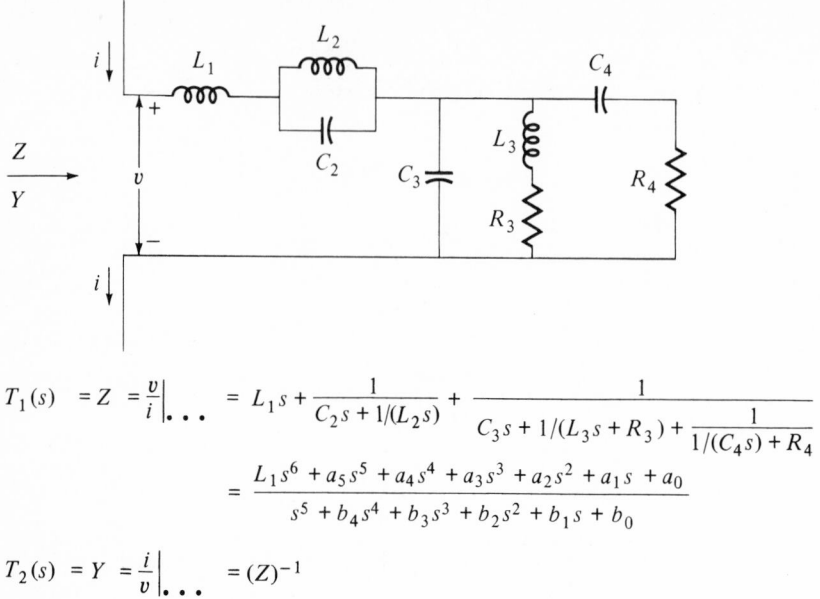

$$T_1(s) = Z = \frac{v}{i}\Big|_{\ldots} = L_1 s + \frac{1}{C_2 s + 1/(L_2 s)} + \frac{1}{C_3 s + 1/(L_3 s + R_3) + \dfrac{1}{1/(C_4 s) + R_4}}$$

$$= \frac{L_1 s^6 + a_5 s^5 + a_4 s^4 + a_3 s^3 + a_2 s^2 + a_1 s + a_0}{s^5 + b_4 s^4 + b_3 s^3 + b_2 s^2 + b_1 s + b_0}$$

$$T_2(s) = Y = \frac{i}{v}\Big|_{\ldots} = (Z)^{-1}$$

FIG. 5.11-F.

Figure 5.11-F gives a rather complicated driving-point-impedance example which should be carefully studied until understood. Note in particular how the "input impedance," as the transfer function T_1 is also called, can be written in its "expanded" form by inspection. The subsequent work of condensing it into a rational function of s with real coefficients [see (5.05-5)] is straightforward, if tedious algebra; once done, the a and b coefficients will be expressed in terms of the eight element values.

Figure 5.11-G gives another example, to which the same remarks apply. There is a striking resemblance between these last two examples, which requires investigation (see Sec. 5.12).

Two very useful general applications enable us to formulate, once and for all, the laws of voltage and current division between two immittances connected in series or in parallel, respectively, in the form of handy rules. Since in Fig. 5.11-H, $v = (Z_1 + Z_2)i$, $v_1 = Z_1 i$, and $v_2 = Z_2 i$, according to Ohm's law, the current i cancels in the ratios and we have

> The division of a voltage between two series-connected impedances is in the ratio of the relevant ("here") impedance to the sum of the two impedances (Fig. 5.11-H). (5.11-10)

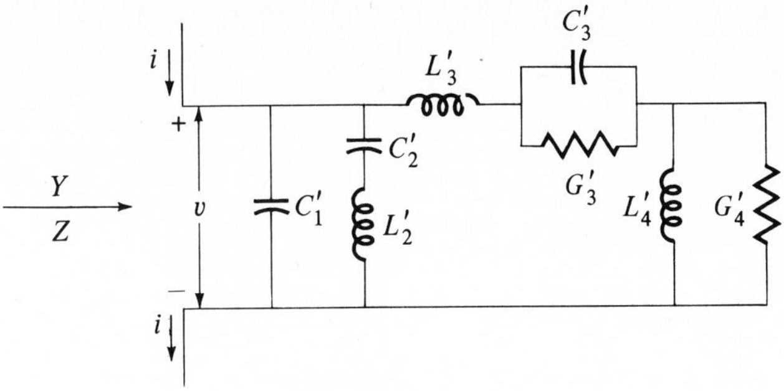

$$Y(s) = \frac{i}{v}\bigg|_{\cdots} = C'_1 s + \frac{1}{L'_2 s + 1/(C'_2 s)} + \frac{1}{L'_3 s + \dfrac{1}{C'_3 s + G'_3 s} + \dfrac{1}{1/(L'_4 s) + G'_4}}$$

$$Z(s) = [Y]^{-1} = \frac{C'_1 s^6 + a'_5 s^5 + \cdots a'_0}{s^5 + b'_4 + \cdots + b'_0}$$

FIG. 5.11-G.

This is the *voltage-divider rule*, often useful. For example, the voltage v in Fig. 5.10-C (the difference of the voltages across Z_4 and Z_1) is

$$v = \frac{Z_4}{Z_3 + Z_4} e - \frac{Z_1}{Z_1 + Z_2} e \tag{5.11-11}$$

which immediately gives us the transfer function from e to v. In Fig. 5.11-F the voltage across the capacitor C_3 is v times the fraction (Z_3/Z) in which

$$Z_3 = \left(C_3 s + \frac{1}{L_3 s + R_3} + \frac{1}{1/(C_4 s) + R_4}\right)^{-1}. \tag{5.11-12}$$

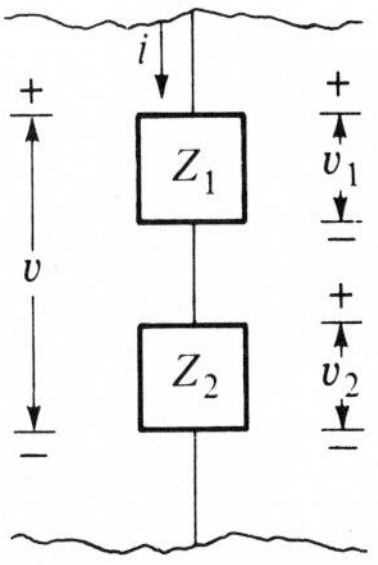

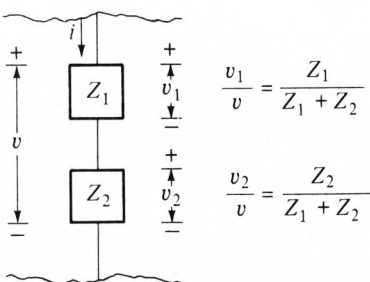

$$\frac{v_1}{v} = \frac{Z_1}{Z_1 + Z_2}$$

$$\frac{v_2}{v} = \frac{Z_2}{Z_1 + Z_2}$$

FIG. 5.11-H. The voltage divider.

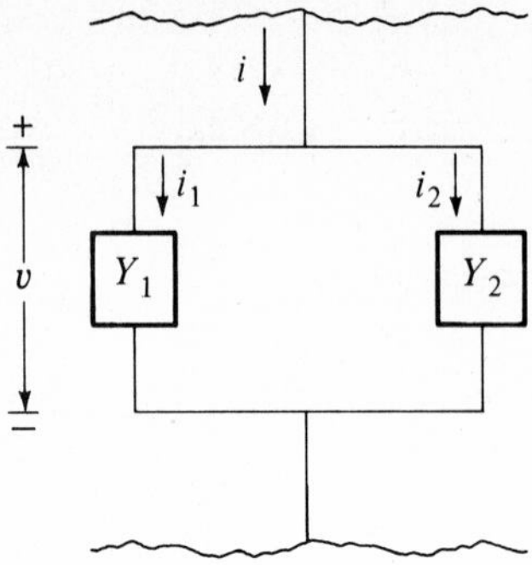

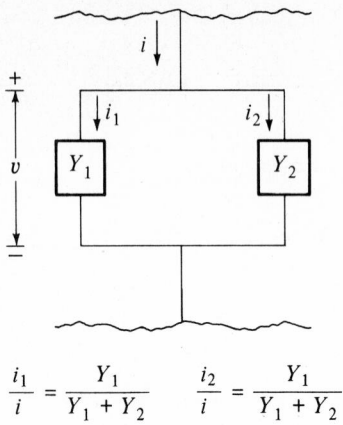

$$\frac{i_1}{i} = \frac{Y_1}{Y_1 + Y_2} \qquad \frac{i_2}{i} = \frac{Y_1}{Y_1 + Y_2}$$

FIG. 5.11-I. The current divider.

A similar division, but of *current*, is shown in Fig. 5.11-I. Since $i = (Y_1 + Y_2)v$, $i_1 = Y_1 v$, and $i_2 = Y_2 v$, the voltage cancels in the ratios shown:

> The division of a current between two parallel-connected admittances is in the ratio of the relevant ("near") admittance to the sum of the two admittances (Fig. 5.11-I). (5.11-13)

This is the *current-divider rule*, equally useful with (5.11-10). For example, in Fig. 5.10-C the current i divides so that the current in Z_1 and Z_2 is

$$\frac{1/(Z_1 + Z_2)}{1/(Z_1 + Z_2) + 1/(Z_3 + Z_4)}\, i. \tag{5.11-14}$$

For combinations of more than two immittances, one may repeatedly apply the rule(s). They will be found rather useful from time to time.

5.12 DUALITY

Certain parallelisms in Sec. 5.11 are very striking, as between these pairs:

(5.11-6)	and	(5.11-8),
(5.11-7)	and	(5.11-9),
Fig. 5.11-B	and	Fig. 5.11-D,
Fig. 5.11-C	and	Fig. 5.11-E,
Fig. 5.11-F	and	Fig. 5.11-G,
Fig. 5.11-H	and	Fig. 5.11-I.

(5.12-1)

These parallelisms are examples of the phenomenon of *analogy* in its special form of *duality* (see Secs. 2.16, 3.21): two models have identical mathematical descriptions, and their differential equations are identical (except for differing symbols, which is mathematically trivial). Graphically, in their schematic diagrams, there is also parallelism, but of a different sort: series connections in one versus parallel connections in the other, inductance versus capacitance, resistance versus conductance.

To sum up in tabular form, the analogy pairs we note are

i (current)	$\leftrightarrow$	v (voltage)
Y (admittance)	$\leftrightarrow$	Z (impedance)
series connection	$\leftrightarrow$	parallel connection
L	$\leftrightarrow$	C
R	$\leftrightarrow$	G
$V = ZI$	$\leftrightarrow$	$I = YV$

(5.12-2)

But s, we note, plays the *same* role in both members.

The usual name for this analogy is *duality*. It refers to those special analogies in which *both* members are of the *same* physical form; here both are *electrical*. (A *mechanical* system model could be analogous to an *electrical* model, but "duality" is then not applicable.)

In analysis, an understanding of duality is useful because it points out that "one-half" of all analysis problems are essentially the same as "the other half"; the analysis of a circuit is unnecessary if the dual problem has already been solved! We note, for example, that when the analysis (calculation of the transfer function Z) of the network in Fig. 5.11-F has been made, we have also the analysis (calculation of the transfer function Y) of the network in Fig. 5.11-G. One need only replace, in the expression for $T_1(s) = Z(s)$ in Fig. 5.11-F, L_1 by C_1', L_2 by C_2', C_2 by L_2', ..., R_3 by G_3', C_4 by L_4', and R_4 by G_4'.

In synthesis, we now expect that a design once made is probably in effect also a second (dual) design, whose synthesis should be essentially effortless.

We shall keep duality in mind as we go along, and use it whenever it will give us insight or save us work.

5.13 LATTICES AND BRIDGES

Our concern (in Secs. 5.02 to 5.04) with those networks whose impulse response is simply a sum of exponential terms has discovered for us a number of unexpected things:

The transfer function is then a rational function of s (with real coefficients) in which the degree of the numerator is less than the degree of the denominator; (5.13-1)

Its poles are perforce in the left half plane (at least if the network is passive, exception made of nondissipative models); (5.13-2)

The zeros, in contrast, may lie anywhere. (5.13-3)

To be sure, we have found right-half-plane zeros of $T(s)$ only in networks with a certain amount of sophistication: the paralleled g_m networks of Fig. 5.08-A. Now, equipped as we are with the tools of Sec. 5.11, we can look for more examples. Our particular aim is to continue the argument of Sec. 5.10, to show that right-half-plane zeros of $T(s)$ can exist even if the network is purely passive, and that the active elements have nothing to do with this property.

We choose the *symmetrical* lattice (bridge), unloaded, for simplicity. It is a very important network form in its own right, as well as excellent for making our point here. It is the general bridge of Fig. 5.10-C specialized in that Z_1 and Z_3 are now identical, called Z_a, and Z_2 and Z_4 are now identical, called Z_b (Fig. 5.13-A*a*). The symmetry is in that the e and v terminals, the excitation and detector positions, may be interchanged without altering the transfer function. "Unloaded" means that no impedance (load) is connected at the output, between terminals 2 and 2′.

Redrawn in lattice form, Fig. 5.13-A*b*, the network displays its symmetry even more clearly; the dotted lines represent the other two "boxes," which it is not necessary to draw, identical with those shown as they are. The schematic two-port form of Fig. 5.13-A*c* may be reversed (the "2" terminals taken as the input or excitation point, and the "1" terminals as output or response point),

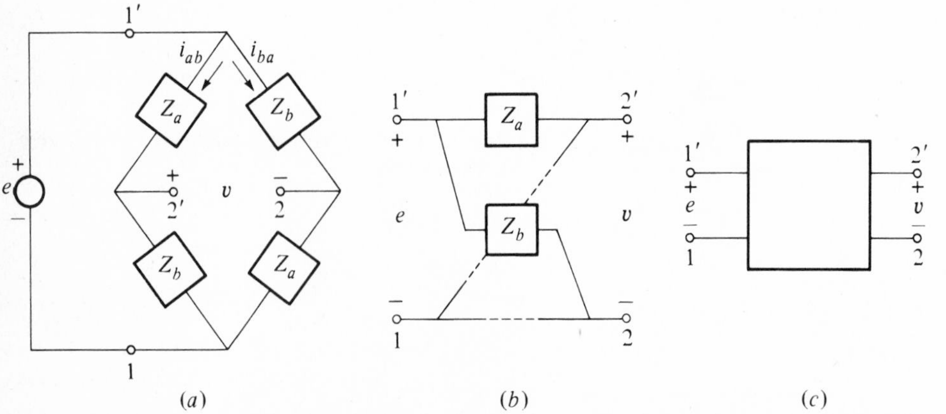

FIG. 5.13-A.

again without change in transfer function, because of the symmetry. We shall meet this *two-terminal-pair* or *two-port-network* format more and more often, since most networks are designed for *transmission* purposes, which it clearly suggests (see Sec. 5.08).

Calculation of the transfer function of the network of Fig. 5.13-A is now no more complicated than the corresponding analysis of the resistance network of Fig. 5.10-A or Fig. 5.10-C*a*. We proceed exactly as in Sec. 5.10, using $Z(s)$ instead of R. (The actual calculation *is* that simple, though we should not forget the logical necessity of justifying it, the work of Sec. 5.11.)

We devise symbols for the coefficients of the forced components, those that vary as e^{st} when $e = Ee^{st}$:

$$
\begin{aligned}
&v_F = Ve^{st}, \\
&i_{abF} = I_{ab}e^{st}, \qquad i_{baF} = I_{ba}e^{st}.
\end{aligned} \tag{5.13-4}
$$

Then we make the calculation as in Sec. 5.10, paying no attention to the natural components (which must also balance out, but as a separate group, and may therefore be ignored). We write only the coefficients, since e^{st} is everywhere a common factor that can be canceled out:

$$
\begin{aligned}
V &= -Z_aI_{ba} + Z_bI_{ab} \\
&= \frac{-Z_aE}{Z_b + Z_a} + \frac{Z_bE}{Z_a + Z_b},
\end{aligned} \tag{5.13-5}
$$

$$
T(s) = \frac{v}{e}\bigg|_{\dots} = \frac{V}{E} = \frac{Z_b - Z_a}{Z_b + Z_a}. \tag{5.13-6}
$$

This transfer function, remarkable in its simplicity, is very instructive. Its numerator indicates when the bridge is balanced:

> The zeros of $T(s)$ are those values of s for which $Z_b(s) = Z_a(s)$. (5.13-7)

Its denominator indicates what the natural frequencies are:

> The poles of $T(s)$ are those values of s for which $Z_b(s) = -Z_a(s)$. (5.13-8)

The bridge balance condition should, by extension of (5.10-3), be $Z_aZ_a = Z_bZ_b$, which seems to give not only (5.13-7) but also (5.13-8)! The paradox is due to the special symmetry of the bridge, and may be resolved by approaching the situation as a limit of the general case: a factor $(Z_b + Z_a)$, present in both numerator and denominator, may be canceled, leaving (5.13-7) and (5.13-8) vindicated and correct [see (6.23-15)].

The natural frequencies (infinite-gain points) are the poles of $T(s)$, the zeros of $(Z_b + Z_a)$, as of course they should be: currents can flow without excitation when the impedance they encounter is zero, as it is (Fig. 5.13-A) for either path, completed through the generator e whose impedance is zero. (Its voltage remains e, regardless of the current through it!) It is a general principle (see Sec. 6.03) and an often useful one, that the natural frequencies are the roots of the equation

$$\text{The sum of the impedances around a closed loop (carefully and accurately computed)} = 0. \tag{5.13-9}$$

If Z_a and Z_b both represent purely passive networks, so also does the sum $(Z_b + Z_a)$, in effect the impedance of their series connection. Hence the natural frequencies lie in the left half of the s plane. The usual exception, for idealized, nondissipative models, allows them also on the imaginary axis.

Infinite loss $[T(s) = 0]$ comes about by a *cancellation* process, a very sophisticated process, though here exemplified by a simple action: $Z_b = Z_a$ and the component voltages are equal and cancel! This can be made to occur for any desired value(s) of s if the impedances Z_a and Z_b are properly designed; hence right-half-plane *zeros* are as physically reasonable as are left-half-plane or "$j\omega$" ones. We note with interest that in effect this balance can be thought of as coming about by connecting a normal "positive" impedance in series with a "negative" one (made up of negative elements if you like):

$$\underbrace{[Z_b]}_{\substack{\text{Ordinary}\\\text{network of}\\\text{positive}\\\text{elements}\\\text{(passive)}}} \quad \underbrace{+}_{\substack{\text{In}\\\text{series}\\\text{with}}} \quad \underbrace{[-Z_a]}_{\substack{\text{Hypothetical}\\\text{network in}\\\text{which all}\\\text{elements are}\\\text{negative}}} = 0. \tag{5.13-10}$$

There is now no doubt about the validity of (5.13-1), (5.13-2), and (5.13-3), whether the network contains active elements or not. But a few examples will clinch the point. These examples have interest, and are important in their own right, also.

The network of Fig. 5.11-A suggests a simple symmetrical network in which (Fig. 5.13-A) $Z_a = R_a$, $Z_b = 1/(C_b s)$. For it,

$$T(s) = \frac{Z_b - Z_a}{Z_b + Z_a} = \frac{1/(C_b s) - R_a}{1/(C_b s) + R_a} = \frac{1 - R_a C_b s}{1 + R_a C_b s}. \tag{5.13-11}$$

We note that it has a right-half-plane zero, passive though the network is; we note also a symmetry between the pole and the zero, in their locations in the two halves of the plane, with respect to the origin. The significance and utility of this we shall discuss later.

Figure 5.13-B tabulates this as the second of a number of lattice (bridge) networks, with their transfer functions. The first is of course the prototype, the

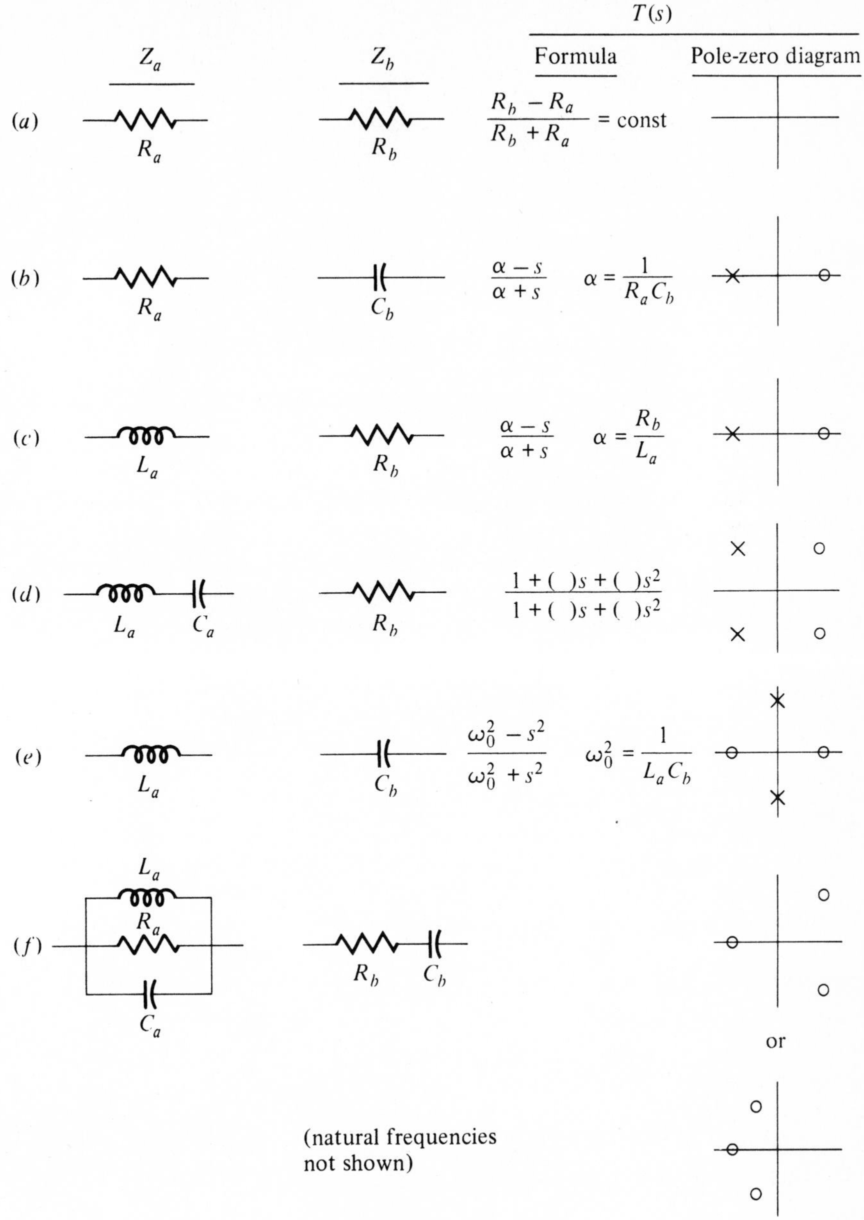
$T(s)$
Z_a
Z_b
Formula
Pole-zero diagram
(a)
R_a
R_b
$\frac{R_b - R_a}{R_b + R_a}$ = const
(b)
R_a
C_b
$\frac{\alpha - s}{\alpha + s}$
$\alpha = \frac{1}{R_a C_b}$
(c)
L_a
R_b
$\frac{\alpha - s}{\alpha + s}$
$\alpha = \frac{R_b}{L_a}$
(d)
L_a
C_a
R_b
$\frac{1 + (\ )s + (\ )s^2}{1 + (\ )s + (\ )s^2}$
(e)
L_a
C_b
$\frac{\omega_0^2 - s^2}{\omega_0^2 + s^2}$
$\omega_0^2 = \frac{1}{L_a C_b}$
(f)
L_a
R_a
C_a
R_b
C_b
or
(natural frequencies not shown)

FIG. 5.13-B.

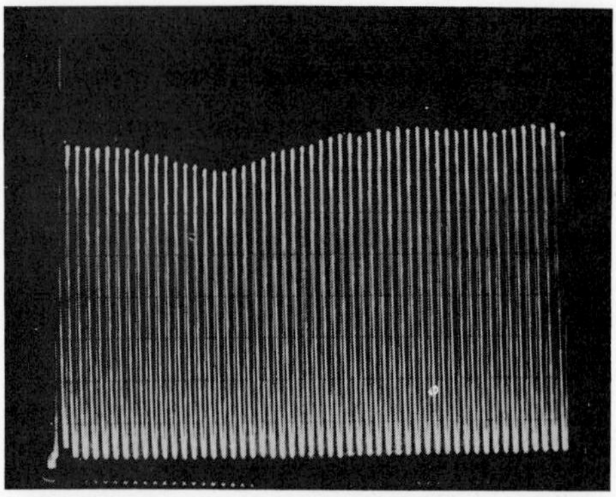

(g)

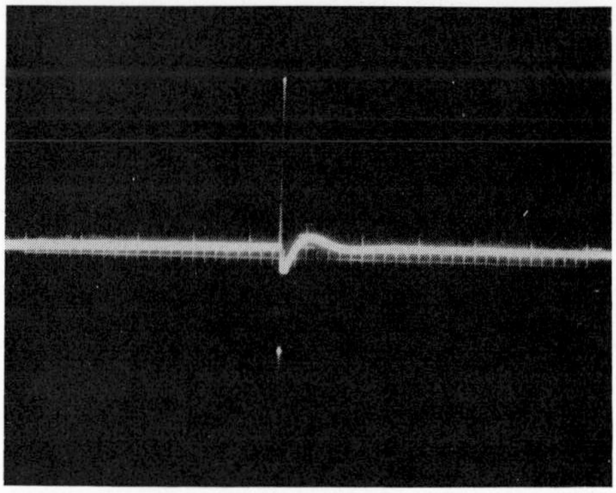

(h)

FIG. 5.13-B *(Cont.)*

Wheatstone bridge. The others give ample additional evidence that the zeros of a transfer function can indeed lie in the right half plane for strictly passive networks. (It seems, however, that the bridge's property of composing the output signal of two components that have arrived by different paths is important to if not essential for such zeros.)

One additional observation should be recorded here. Some of the transfer functions in Fig. 5.13-B have poles and zeros symmetrically located about the origin: the infinite-loss points are the negatives of the infinite-gain points. In general form such symmetry is expressible thus:

$$\begin{aligned} T(s) &= K\frac{(s-z_1)(s-z_2)\cdots(s-z_N)}{(s-p_1)(s-p_2)\cdots(s-p_N)} \\ &= K\frac{(s+p_1)(s+p_2)\cdots(s+p_N)}{(s-p_1)(s-p_2)\cdots(s-p_N)} \end{aligned} \tag{5.13-12}$$

in which $p_1, p_2, \ldots, p_N$ are the natural frequencies (in the left half plane). As a consequence of this pole-zero symmetry we find, in calculating the forced component of the response to sinusoidal excitation (setting $s = j\omega$), that

$$T(j\omega) = K' \frac{(p_1 + j\omega)(p_2 + j\omega)\cdots(p_N + j\omega)}{(p_1 - j\omega)(p_2 - j\omega)\cdots(p_N - j\omega)} \tag{5.13-13}$$

in which the constant $K' = (-1)^N K$. The noteworthy property is that $|T(j\omega)|$ is constant with frequency, that the magnitude of the (forced-component) response is the same at all frequencies for given excitation magnitude: "all frequencies are transmitted alike" in magnitude. Figure 5.13-B*g* gives us physical corroboration of this interesting behavior: $|T|$ there is essentially constant with frequency. For the network used in this demonstration (essentially a lattice bridge, though in a different physical format), the pole-zero diagram is that of Fig. 5.13-B*d*. Because of the symmetry, $|T|$ should be constant with frequency; and the actual behavior is practically that.

"All frequencies" do *not* however "receive the same treatment" in *angle* (phase shift); that is, the angle of $T(j\omega)$ does vary with frequency, and in a more or less complicated fashion. The impulse response of the network (Fig. 5.13-B*h*) clearly shows that the network distorts input signals: in this case a distorting oscillation follows the impulse in the output. Presumably this distortion is due entirely to the phase characteristic since $|T|$ is essentially "correct" (see Sec. 8.07). Such "all-pass" networks may be useful in manipulating a system's phase-shift characteristic (as a function of frequency) without altering the magnitude characteristic.

In network *synthesis*, the symmetrical lattice form is very useful—and not merely in the special case just mentioned. Because of its simplicity we have [see (5.13-6) and Fig. 5.13-A]:

$$T(s) = \frac{Z_b(s) - Z_a(s)}{Z_b(s) + Z_a(s)} = \frac{Z_b/Z_a - 1}{Z_b/Z_a + 1} \tag{5.13-14}$$

and so

$$\frac{Z_b(s)}{Z_a(s)} = \frac{1 + T(s)}{1 - T(s)}. \tag{5.13-15}$$

Hence the function (Z_b/Z_a) is immediately obtainable from a desired (to-be-synthesized) transfer function. Whether Z_b and Z_a can be individually extracted therefrom, and under what circumstances they will be physically realizable with (stable, passive) networks, is another matter.

The category of bridge (lattice) networks is inexhaustible. Some are particularly useful in constructing measurement apparatus, usually in the more general (nonsymmetrical) form of Fig. 5.10-C*b*. We can use known (standard) impedances in all arms but one, in which we place an unknown (to be measured).

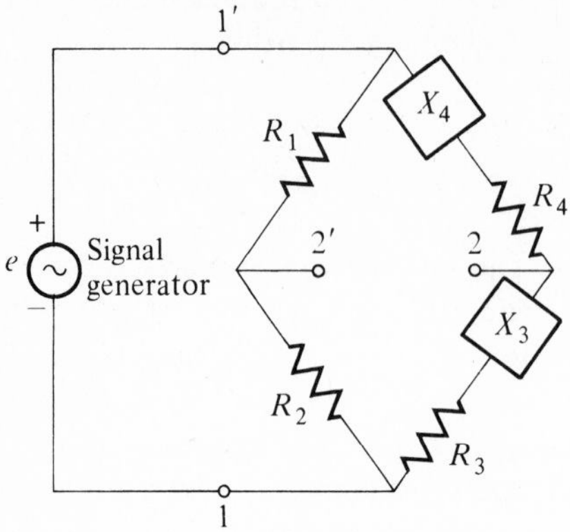

Results of balance condition (at $s = j\omega$)

$$R_3 = \frac{R_2}{R_1} R_4$$

and

$$X_3 = \frac{R_2}{R_1} X_4$$

↑ Unknowns, measured as ↑ Knowns (calibrated, standards)

(*a*) The "resistance-ratio" bridge

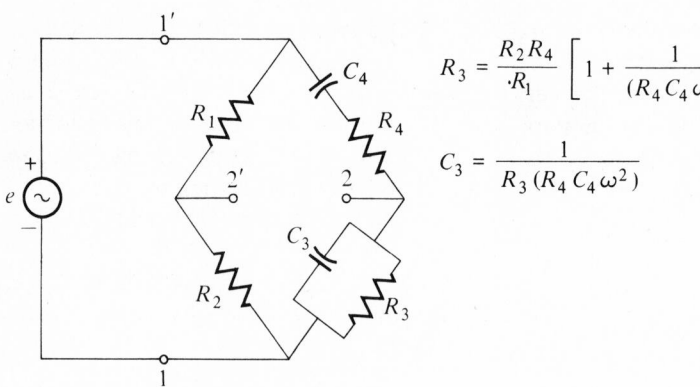

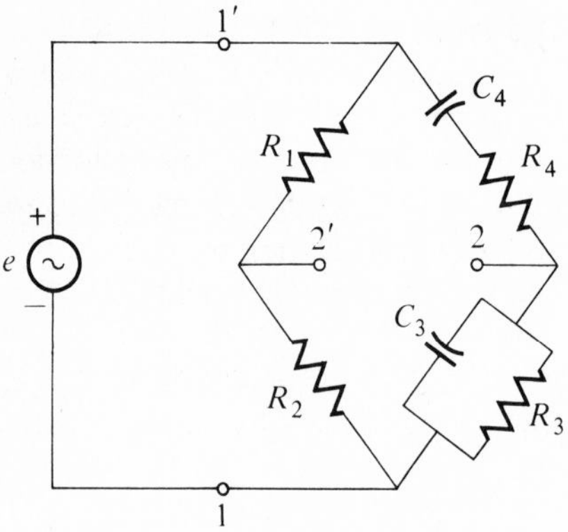

$$R_3 = \frac{R_2 R_4}{R_1}\left[1 + \frac{1}{(R_4 C_4 \omega^2)}\right]$$

$$C_3 = \frac{1}{R_3 (R_4 C_4 \omega^2)}$$

(*b*) The Wien bridge

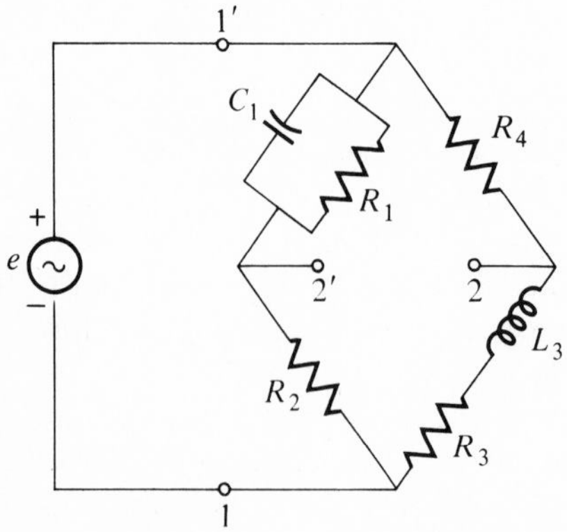

$$R_3 = \frac{R_2}{R_1} R_4$$

$$L_3 = R_2 R_4 C_1$$

(*c*) The Maxwell bridge

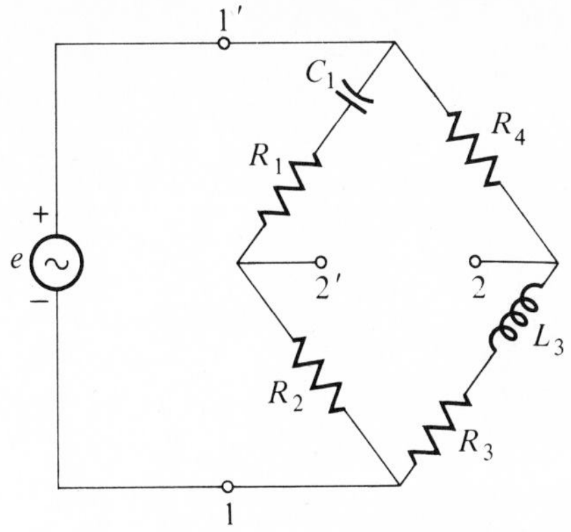

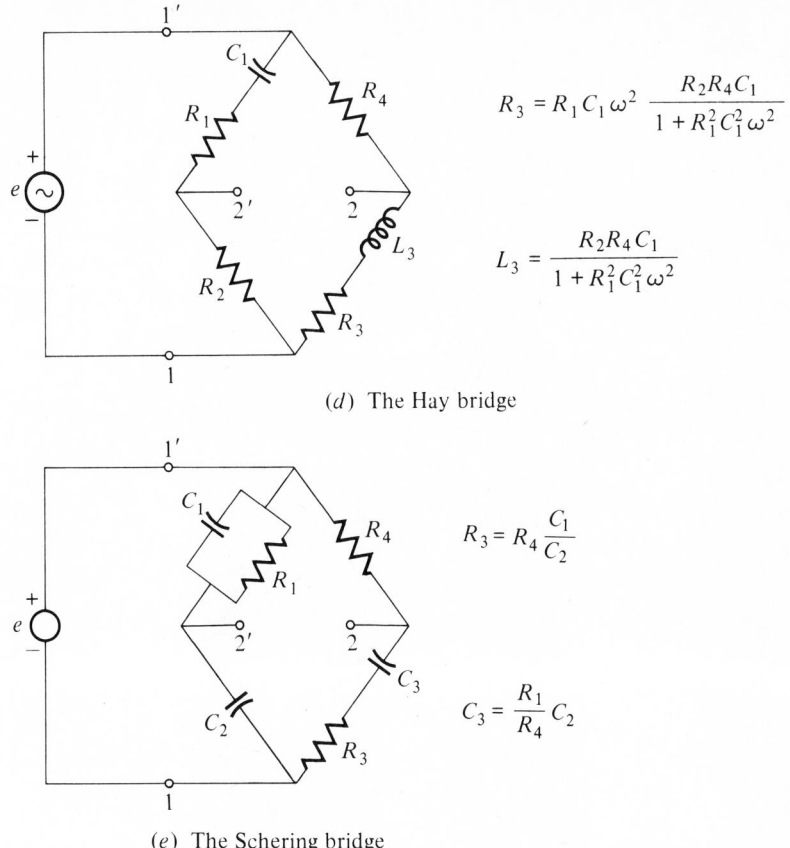

$$R_3 = R_1 C_1 \omega^2 \frac{R_2 R_4 C_1}{1 + R_1^2 C_1^2 \omega^2}$$

$$L_3 = \frac{R_2 R_4 C_1}{1 + R_1^2 C_1^2 \omega^2}$$

(*d*) The Hay bridge

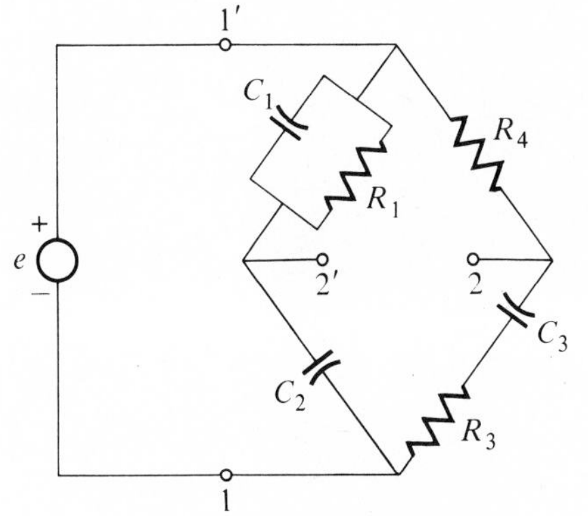

$$R_3 = R_4 \frac{C_1}{C_2}$$

$$C_3 = \frac{R_1}{R_4} C_2$$

(*e*) The Schering bridge

FIG. 5.13-C.

We use a sinusoidal signal generator of frequency ω; we make some of the standard elements variable, and adjust them until zero voltage (as measured by a sensitive, high-impedance detector) appears at the 2-2′ terminals. Since the forced component is "permanent" and not transient, we can do this leisurely, in a very practical way, in the (sinusoidal) steady state. *Two* adjustments are necessary, of course, since the sinusoidal waves must balance in both amplitude and phase. In our symbolic or transformed notation, Z_a and Z_b must be equal in both magnitude and angle (both real and imaginary parts must vanish).

Many variations, many types of measuring bridges exist. In particular, the networks of Fig. 5.13-C are useful in making such measurements.

The action of bridge balance is physically verified in Fig. 5.13-D. There the sinusoidal excitation $e(t)$ applied to a Maxwell bridge (Fig. 5.13-C*c*) appears as the lower wave in each of the three parts of the figure. The upper wave is, in each case, the "detector" voltage between terminals 2 and 2′ of Fig. 5.13-C*c*. The three situations differ in the value of C_1 in the network; all else remains

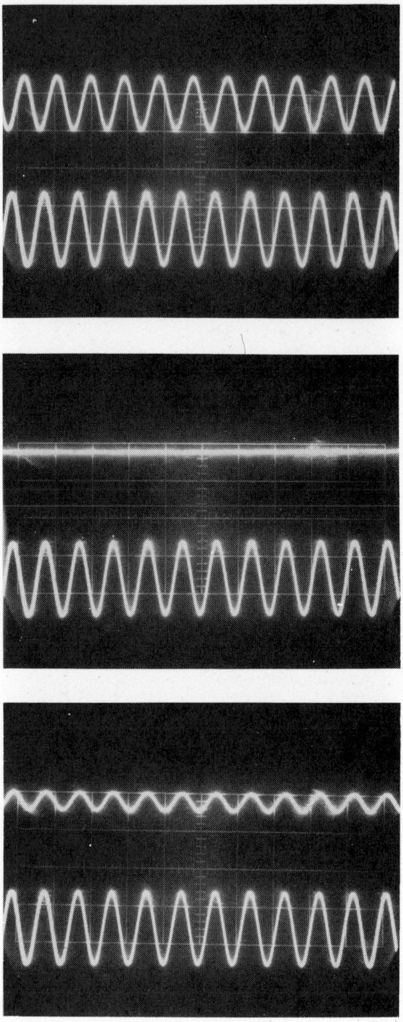

(a) (b) (c)

FIG. 5.13-D.

the same, including the sensitivity (scale) settings. The bridge has been carefully balanced at the excitation frequency; then C_1 is altered. In the photographs, its capacitance varies from a value appreciably below the critical (balance) value, in (*a*), through the critical (balance) value, in (*b*), to a value appreciably greater than the balance value, in (*c*). As C_1 passes through the critical value, the detector voltage goes from an appreciable value (*a*) through essentially zero at balance (*b*) to another appreciable value (*c*). (A much more vivid demonstration is to watch the "movie" action as C_1 is varied before your eyes; this is of course the way a bridge is actually used.) But the three photographs, as a sequence, show that bridge balance action is a very real thing.

Finally, we note that for the general, unsymmetrical bridge of Fig. 5.10-C, the balance condition is

$$Z_1(s)Z_3(s) = Z_2(s)Z_4(s). \tag{5.13-16}$$

This is simply an extension of (5.10-3) by the methods and analysis of Sec. 5.11, from the purely resistive ("dc") condition $(R_1R_3 = R_2R_4)$ to a more general algebraic equation in s. Its roots are the values of s for which bridge balance occurs, the zeros of the transfer function. And presumably $(Z_1Z_3 - Z_2Z_4)$ is the numerator of the transfer function from e to v even in the completely general case in which source and detector immittances are added. Calculation of the transfer function in that case presents problems we shall face in Sec. 6.23.

5.14 RECAPITULATION

It is remarkable that impulse response, in the (real) domain of time, and transfer function, in the (artificial) domain of s, should term by term be this simply related:

$$Ke^{pt} \sim \frac{K}{s-p}. \tag{5.14-1}$$

Suggested first by the results of our experiments with two simple circuits (*RC*, Chap. 2, and *RLC*, Chap. 3), it seems now to hold more generally. With minor alterations necessary for generalization, it does indeed hold for all "networks." To verify that statement is one of our most important tasks.

So important is the relation that we repeat here what we have actually demonstrated so far:

a *If* $$r_\delta(t) = \sum_{j=1}^{N} K_j e^{p_j t}, \tag{5.14-2}$$

b *Then* $$T(s) = \sum_{j=1}^{N} \frac{K_j}{s-p_j}. \tag{5.14-3}$$

The numbers p_j need not be real, but if they are not, they must occur in conjugate pairs. The same is true of the corresponding K_j. And all of the p_j are separate and distinct, one from another. (This last restriction need not hold, we shall find, and is not really important.) To (*b*) we may add the following consequences [see (5.03-3)]:

$$\textbf{b (con.) } \textit{And} \qquad T(s) = \frac{\text{a polynomial in } s \text{ of degree } (N-1) \text{ or less}}{(s-p_1)(s-p_2)(\quad)\cdots(s-p_N)}$$
$$= \frac{A_0 s^{N-1} + A_1 s + \cdots + A_{N-1}}{s^N + B_1 s^{N-1} + \cdots + B_N} \tag{5.14-4}$$
$$= K_0 \frac{s^M + \cdots}{s^N + \cdots} = K_0 \frac{Q_M(s)}{Q_N(s)}$$
$$= \text{a rational function of } s \text{ with real coefficients.}$$

We recall that $M < N$; M can have any of the values $(N-1)$, $(N-2)$, ..., 3, 2, 1, even zero.

The zeros of the denominator $Q_N(s)$, the poles of $T(s)$, are the natural frequencies or infinite-gain points. For passive networks they generally lie in the left half of the s plane. (Idealization by removing dissipation (resistance) elements may place some or all of them on the ω axis.)

The zeros of the numerator $Q_M(s)$ are the zeros of $T(s)$ or the infinite-loss points. They may lie anywhere, without restriction. True, the particular *structure* of a network seems to have a great deal to do with the locations of its infinite-loss points: the discussion of bridge networks (Sec. 5.13) suggests this, and the discussion of ladder networks (Chap. 6) will confirm it.

The principal outstanding unanswered question raised by our previous discussions is

> When *does* (5.14-2) hold? When may we remove the *if* and simply state that the transfer functions *does* have all the properties listed above? (5.14-5)

Beyond the *RC* and *RLC* circuits, and the bridges and lattices of this chapter, lie a host of other networks for which (5.14-2) holds. In studying them, the idea of duality (Sec. 5.12) will be helpful, and the ability to calculate Z and Y transfer functions (Sec. 5.11) will be extraordinarily useful. [The synonymity of (5.14-2) and (5.14-3) will be found to be central to understanding network analysis.]

The concepts of this chapter, summarized in (5.14-2) and (5.14-3), may seem far removed from physical reality. Abstract mathematics and intellectual exercise, models—that they are. Yet they do, as models, represent, describe, and characterize real networks. They can, nay *must*, be used in network analysis. Here they may seem remote but their ties to reality will soon be more evident.

The photograph in Fig. 4.09-A, for example, hardly seems to show a sum of exponentials. But it *is* the impulse response of a network, that of Fig. 5.14-A.

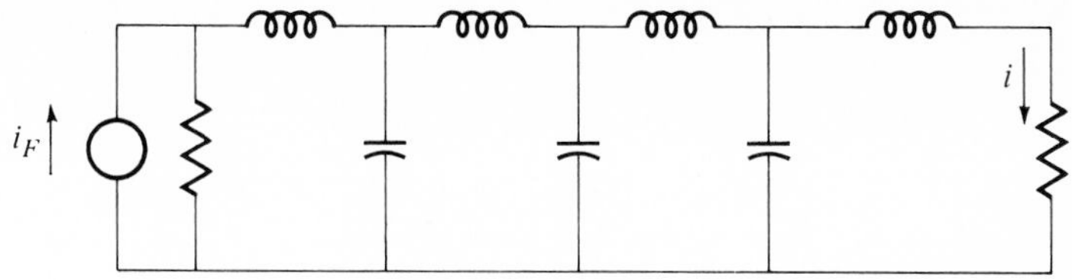

Excitation: i_F
Response: i

FIG. 5.14-A.

(We shall verify the details of this example later—see Prob. 7-A.) When N is large, the exponential sum can take on almost any shape; here $N = 7$ and it is not unreasonable that $r_\delta(t)$ be quite unlike the impulse response of a simple RC or RLC circuit! When N is large, the mathematics of this chapter becomes essential to the analysis; we shall try not to stray too far from physical reality, but the mathematics *is* essential! And so a perennial problem arises for us—to maintain a balance between the two, a good engineering perspective.

5.15 EXERCISES

5-1. In Fig. 5.11-F why can one say, merely by inspection, that the coefficient a_6 is equal to L_1? Why can one similarly say that $a_0 = R_3 b_0$? Are there any other "physically obvious simple relations"? Repeat, in corresponding fashion, for Fig. 5.11-G.

5-2. It can be shown that, by proper choice of element values, the impedances $Z_2(s)$, $Z_3(s)$, and $Z_4(s)$ in Fig. 5-2 can be made identical to $Z_1(s)$ (the networks are then equivalent as to driving-point immittances). Assuming this, evaluate by inspection the element values C_2, L_2, L_3, and C_4. Can you evaluate similarly (by physical inspection) any other elements?

5-3. What does the ammeter read in Fig. 5-3 (p. 306) if at the center the two perpendicularly crossing wires are (*a*) connected? (*b*) not connected? Exploit symmetry.

5-4. What is the voltage v in the steady state in each case (Fig. 5-4, p. 307)?.

5-5. Discuss the zeros of the transfer function from v_0 to v_3 in Fig. 5.08-B when v_3 is made (by an adding device) to be the *sum* of v_1 and v_2. Consider each of the three compositions for Z_1 and Z_2 described in Sec. 5.08.

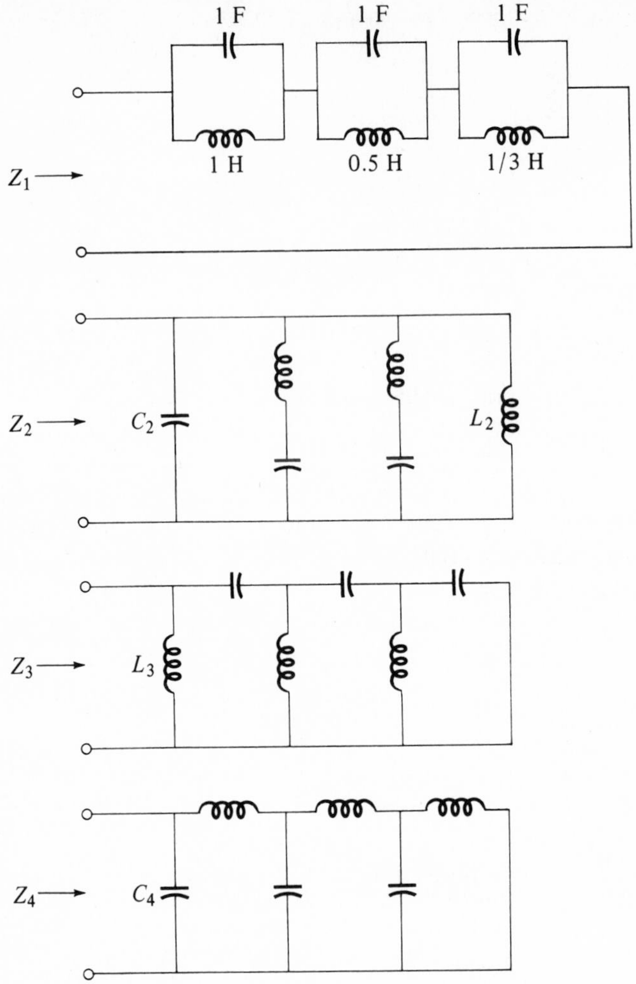

FIG. 5-2.

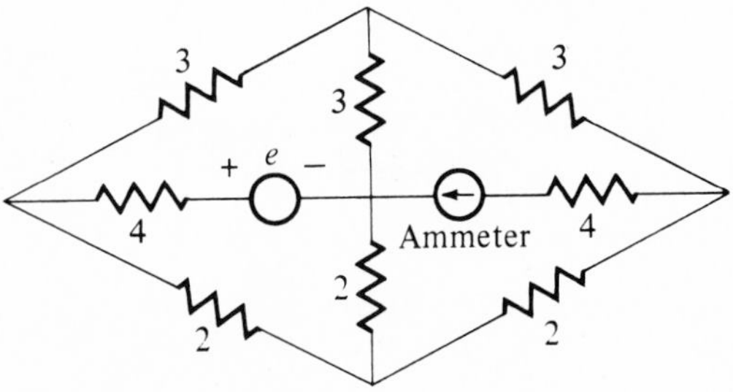

$e = E_m \cos \omega_F t$

(All resistance values are in ohms)

FIG. 5-3.

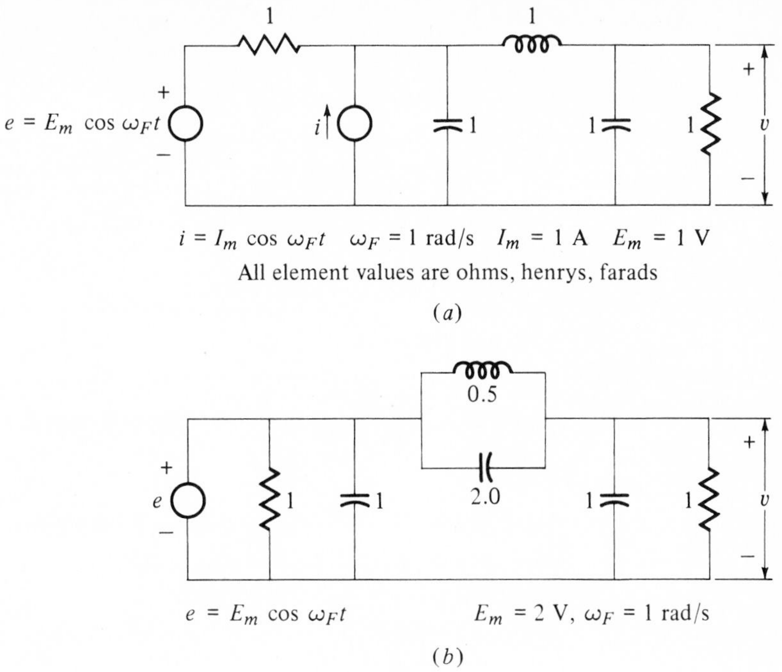

FIG. 5-4.

5-6. Form the polynomial whose zeros are $(-1 + j0)$, $(-1 + j1)$, $(-1 - j1)$, $(1 + j1)$, $(1 - j1)$, with unity leading coefficient. Why are its coefficients *real?* Does this property depend in general on the distribution of a polynomial's zeros between the left and right halves of the plane?

Factor the polynomial $(as^2 + bs + c)$ in which a, b, c are all real, and $(4ac) > b^2$. Why are the (complex) zeros in a conjugate pair?

Show that if the zeros of a polynomial are all real or, if complex, occur in conjugate pairs, the coefficients of the polynomial are all real. Is the converse true? (Do real coefficients imply that all the zeros are either real or, if complex, occur in conjugate pairs?)

5-7. Show that for a polynomial $Q = a_0 + a_1 s + \dots + a_N s^N$ to be Hurwitz, all the (real) coefficients must have the same sign and that (a) for $N = 1$ and $N = 2$ this is necessary and sufficient, (b) for $N = 3$ it is necessary but not sufficient. (*Suggestion:* Consider the coefficients of polynomials whose zeros have the two sorts of locations shown in Fig. 5-7 on page 308; see Prob. 5-L.)

5-8. Show by means of carefully drawn phasor diagrams how the forced component of the response to sinusoidal excitation can be zero for the parallel connection of two *RLC* circuits (as in Fig. 5.08-B) but not for their cascade connection (Sec. 5.07). What are the conditions on the element values for this zero of the overall transfer function to occur at a particular frequency (at $s = j\omega_F$)?

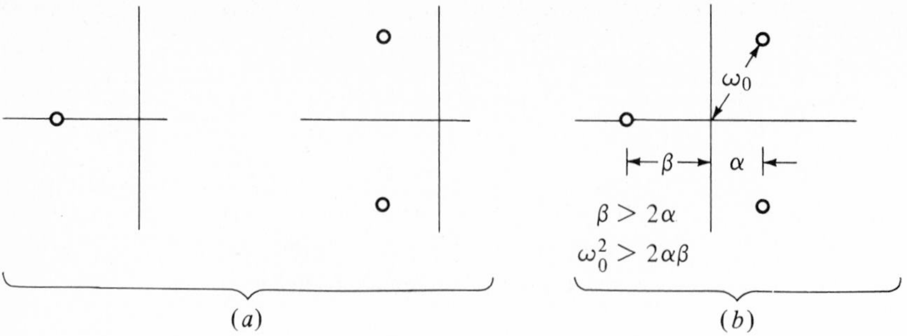

FIG. 5.7.

5-9. A voltage excitation of exponential form, $e = E_0 e^{s_0 t}u(t)$, is applied to terminals 1 and 1′ in the network of Fig. 5.10-C when no energy is stored in the network. Each arm is a function of s, of which it is known only that the bridge balances at $s = s_0$. If a resistor of moderate size is connected to terminals 2 and 2′, will any current flow therein at the instant $t = 0+$? At subsequent instants? Explain.

5-10. A bridge network (Fig. 5.10-C) is excited by a voltage source of internal impedance R_G; a detector impedance R_D is connected between terminals 2 and 2′. Which of the specific bridges, constructed according to the table below, can, by proper adjustment of the element values specified, be made to balance in the sinusoidal steady state? What are their frequencies of balance in terms of the literal element values?

	Z_1	Z_4	Z_2	Z_3
a	R_A	R_B	R_C	R_D
b	R_A and C_A in parallel	R_B	L_C and R_C in series	R_D
c	L_A	L_B	C_C	C_D
d	L_A	C_B	C_C	L_D
e	C_A	R_B	R_C	C_D
f	C_A	C_B	R_C	R_D
g	R_A	R_B	R_C	R_D, L_D, and C_D in series
h	R_A and C_A in series	R_B	R_C	R_D and L_D in series

5-11. Two arms of a bridge are resistors—one is an inductor, one a capacitor. Where, in the s plane, are the balance frequencies for each of the possible bridge formations? Which (if any) can be made to balance for $s = j\omega$ ("at real frequencies")? Explain any interesting anomalies.

5-12. Consider the network of Fig. 5.13-B*f*. Discuss fully the possible locations of the zeros of $T(s)$ if all five elements are positive. Verify that the two situations shown are possible. Can all three zeros lie in the right half plane? Can only one lie in the right half plane? What other possibilities exist?

5-13. Explain carefully the statement of Sec. 5.13 that detector and excitation may be exchanged in Fig. 5.13-A without affecting the transfer function. Is this true of Fig. 5.10-C? The symmetrical lattice is said to be its own dual or is "self dual." Explain this statement.

5-14. Write out the condition for balance at $s = 0 + j\omega_F$ of each of the bridges in Fig. 5.13-C and verify the formulas given there for the "unknowns" being measured.

5-15. A Maxwell bridge (Fig. 5.13-C*c*) balances at frequency 20 kHz with these settings: $R_2 = 900\ \Omega$, $R_4 = 200\ \Omega$, $C_1 = 0.014\ \mu\text{F}$, $R_1 = 30\ \text{k}\Omega$. Find the values of L_3 and R_3 of the (unknown) *RF* choke coil represented thereby. What is the coil's Q at 20 kHz? Correlate with Fig. 5.13-D, taken with such a network; there $C_1 = 0.003\ \mu\text{F}$ in (*a*), 0.014 μF in (*b*), and 0.023 μF in (*c*), the other elements being as above.

5-16. Show that the symmetrical lattice (bridge) network of Sec. 5.13 can be replaced by a simpler (and often cheaper) network, the *semilattice* or *half-lattice* network (Fig. 5-16). Verify that the transfer function of the simpler network is essentially the same as (5.13-6). Assume that controlled-source devices (such as transformers or semiconductor active devices) equivalent to the box shown in broken lines can be obtained; see also Fig. 5.08-B.

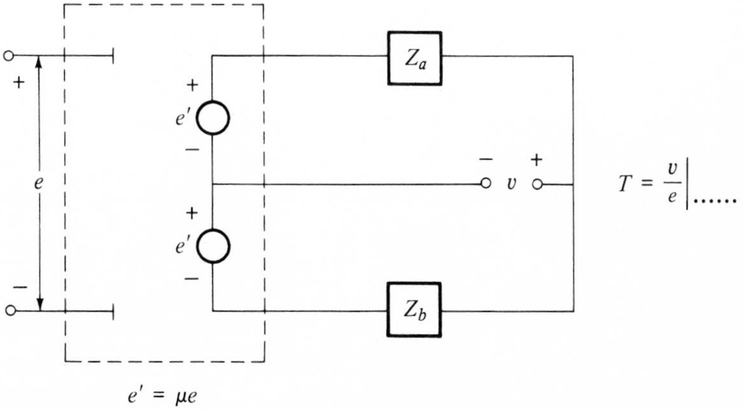

FIG. 5-16.

5.16 PROBLEMS

5-A. Show that the two coefficients K_1 and K_2 in the algebraic partial-fraction expansion of $T(s)$ for the *RLC* circuit in Fig. 5.01-B are *exactly* those in the expression for $r_\delta(t)$ directly under it. (Recall how from the conditions at $t = 0+$, one calculates the values of K_1 and K_2 in the impulse response.)

5-B. Show that network form 2 is potentially *equivalent* to network form 1 as to driving-point immittances, i.e., that $Z_2(s)$ can be made identical to $Z_1(s)$ by proper choice of element values, all of which will be positive and hence realizable. What must be the element values of network 2? (Many such equivalences exist, for network synthesis is not unique.)

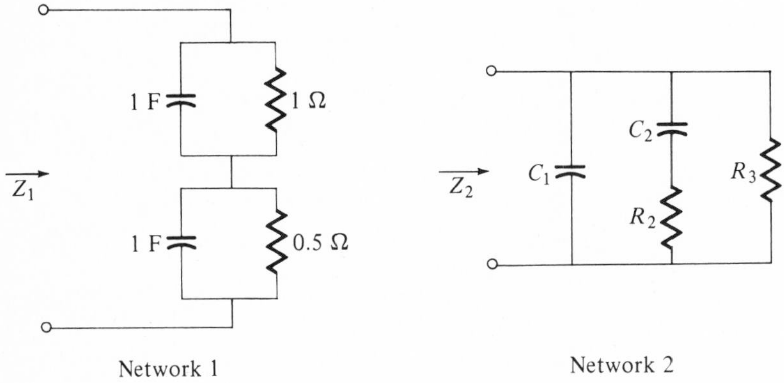

FIG. 5-B.

5-C. By applying the voltage-divider relation and the rules for combining impedances (Sec. 5.11) find the transfer function from v_F to v in terms of Z_1, Z_2, Z_3, Z_4 (each of which is, presumably, a rational function of s). What is the equation that determines the natural frequencies? Describe the dual situation in detail. (Fortunately there are simpler methods for calculating such transfer functions, as Chap. 6 will show you.)

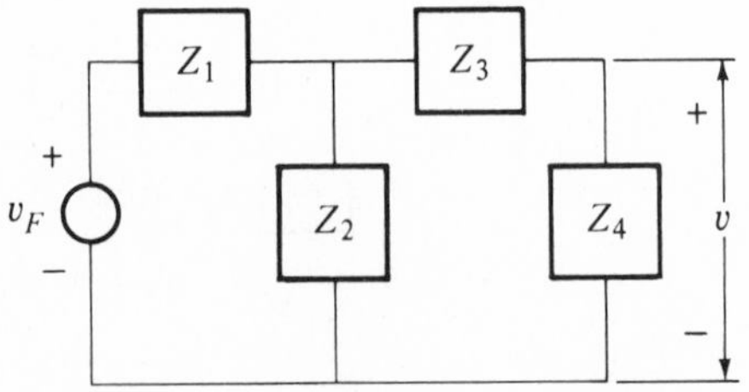

FIG. 5-C.

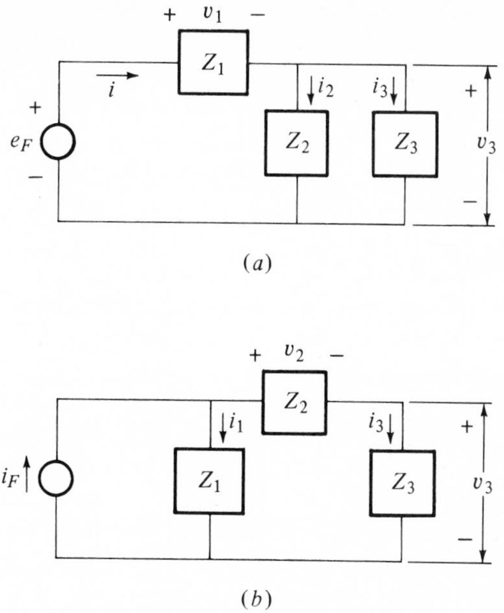

FIG. 5-D.

5-D. (*a*) Find the transfer function from e_F to i_3 in Fig. 5-D, using voltage- and current-divider principles, in terms of Z_1, Z_2, Z_3 and their reciprocals, Y_1, Y_2, Y_3. Choose between impedance and admittance as most efficient. Now, in the sinusoidal steady state, suppose the impedances have the values given below. Draw, to scale, phasor diagrams that show the three currents and the three voltages and their relations. Do this in two ways: (1) calculate I_3 from E_F by way of the transfer function; (2) assume $I_3 = 1\angle 0$ (arbitrarily) and proceed to calculate V_3 therefrom, then I_2, I, V_1, and E_F in sequence. Since E_F will almost certainly be incorrect, rescale (by simple multiplication by a real number) the phasors just computed and thereby obtain the correct result. Why are both methods legitimate? Which is the easier? [*Data:* (α) $Z_3 = 80 + j60$, $Z_2 = 80 - j60$, $Z_1 = 10 + j30\ \Omega$, $E = 100$ V; (β) change Z_2 to $80 + j60$, retain the other values.]

(*b*) Find the transfer function from i_F to i_3 in terms of the three Z symbols and their reciprocals, choosing between admittance and impedance as is most convenient. In the sinusoidal steady state suppose the impedances have the values given below. Calculate and draw the complete phasor diagrams, as in (*a*). [*Data:* (α) $Z_3 = 100 + j50$, $Z_2 = 10 + j20$, $Z_3 = 100 - j50\ \Omega$, $E = 150$ V; (β) change Z_1 to $100 + j50\ \Omega$ and repeat.]

5-E. A network with two accessible terminals is composed of the series combination of a number of parallel RC combinations. Let $Z(s)$ be the driving-point impedance so defined. Show that the zeros and poles of $Z(s)$ are all on the negative half of the real axis, that zeros and poles alternate therealong, that nearest the origin (and possibly *at* the origin) is a pole and all of them are simple (of first order). [*Suggestion:* Sketch Z versus σ when $s = \sigma$ (s is real).] It can be shown that all two-terminal networks composed exclusively of resistors and capacitors present impedances with this set of properties. What must then be true of the *admittances* of *RC* driving-point-admittance functions of such *RC* networks?

5-F. A circuit is composed of a sequence of (different) infinite-Q tuned circuits connected in series, as shown in Fig. 5-F. Show that $Z(s)$, the driving-point impedance at the two terminals, is the quotient of an *odd* polynomial and an *even* polynomial and is therefore an *odd* function of s. What is the real part of $Z(j\omega)$? Explain physically why it has that value. Sketch the imaginary part of $Z(j\omega)$ versus ω. Where are the zeros and poles of $Z(s)$? What relation do you find between zeros and poles? Explain in terms of the individual circuits why the slope of the reactance [the imaginary part of $Z(j\omega)$], $dX/d\omega$, is always positive. (These properties can be shown to hold generally for networks composed only of inductors and capacitors—networks in which there is no resistance.)

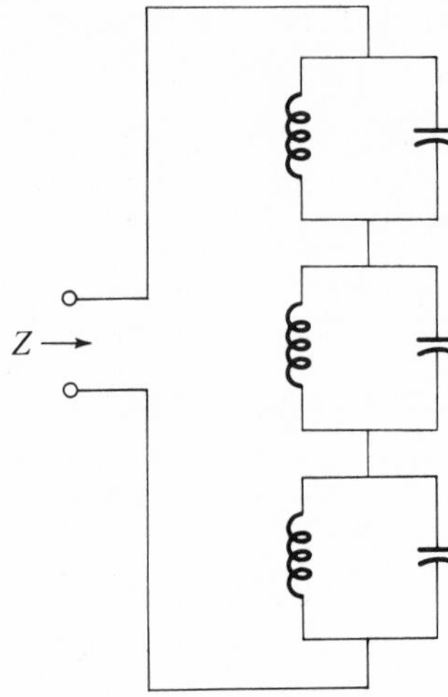

FIG. 5-F.

5-G. Show that the transfer function of the general bridge network of Fig. 5.10-C is

$$T(s) = \frac{Z_2 Z_4 - Z_1 Z_3}{(Z_1 + Z_2)(Z_3 + Z_4)}.$$

Now let $Z_1 = Z_a$, $Z_2 = Z_b$, $Z_3 = Z_a + z$, $Z_4 = Z_3 + z$. Express $T(s)$ in terms of Z_a, Z_b, and z. Then let z approach zero and carefully resolve the paradox described in Sec. 5.13.

5-H. Consider a symmetrical bridge (lattice) like that of Fig. 5.13-A. Let the two types of arms be

1 A parallel tuned circuit of high Q that consists of a capacitor C_a and a (real) inductance coil modelable as an ideal inductance L_a in series with a (small) resistor r_a,

2 A series tuned circuit of high Q that consists of a capacitor C_b and a (real) inductor identical to that used in arm 1.

The "resonant" frequency of arm 1, defined by $\omega_{0a} = (L_a C_a)^{-1/2}$ is one-half that of arm 2, defined by $\omega_{0b} = (L_b C_b)^{-1/2}$; that is, $\omega_{0b} = 2\omega_{0a}$. The Q of impedance 1, defined as $\omega_{0a}/(2\alpha_a) = \omega_{0a}/(r_a/L_a)$, is 100. The excitation is provided by a current source in parallel with a resistor R_G, connected to the bridge at terminals 1 and 1′, replacing the e of Fig. 5.13-A. The detector, connected to terminals 2 and 2′, is of very high impedance and draws essentially no current.

(*a*) Explain briefly, with the aid of a small equation or two and a sketch of the paths in the network, but chiefly in words, what conditions determine the natural frequencies. What paths are taken by which natural-mode currents?

(*b*) For an ideal source (R_G infinite), where are the natural frequencies? Plot them, to scale. (*Note:* There are *eight* energy-storing elements in the network.)

(*c*) Let the source resistance R_G vary downward from very large values to very small values. Sketch the loci that describe the motions of the natural frequencies from their positions in (*b*) as R_G decreases. Where are they when $R_G = 0$?

(*d*) Change the excitation source from a current generator to a voltage generator, as in Fig. 5.13-A, but with an internal (series) resistor R_G. Repeat (*b*) for the new ideal situation ($R_G = 0$).

(*e*) Repeat (*c*) for the new situation, letting R_G *rise* to infinity.

(*f*) Explain the differences between the loci of (*e*) and those of (*c*).

5-I. Design symmetrical lattice networks whose transfer functions (from e to v in Fig. 5.13-A) are those given. In each case find at least one such network in complete detail.

(*a*) $\dfrac{s^2 - 2\alpha s + \omega_0^2}{s^2 + 2\alpha s + \omega_0^2}$, ($\alpha$, ω_0 are real and positive),

(*b*) $\dfrac{(s-1)(s^2 - 0.2s + 2)}{(s+1)(s^2 + 0.2s + 2)}$, (*c*) $\dfrac{\alpha}{\alpha + s}$, (*d*) $\dfrac{s}{s + \alpha}$,

(*e*) $\dfrac{s^2 + A}{s^2 + B}$, (1) $0 < A < B$, (2) $A < 0 < B$, $B > |A|$,

(*f*) $\dfrac{3s + 2}{2s^2 + 3s + 2}$.

5-J. Network synthesis often leads to the problems of realizing (finding a circuit for) a given driving-point immittance function. Suppose, for example, that we have to realize

$$Z(s) = Z_0 \frac{s^2 + as + \omega_1{}^2}{s^2 + bs + \omega_2{}^2}$$

in which the five parameters have been determined elsewhere. The realization process is not necessarily easy; not any set of five parameter values will do: negative element values, for example, are of no interest here. For simplicity, suppose further that $\omega_1 = \omega_2$. Show that if Z_0, a, b, are all positive, and $a = b$, then realization is possible in one or the other of the two circuit forms shown. What are the element values in terms of the given parameters? What happens if $a = b$? (*Suggestion:* Consideration, both mathematically and physically, of the behavior at very low frequencies and at very high frequencies may speed the calculations.)

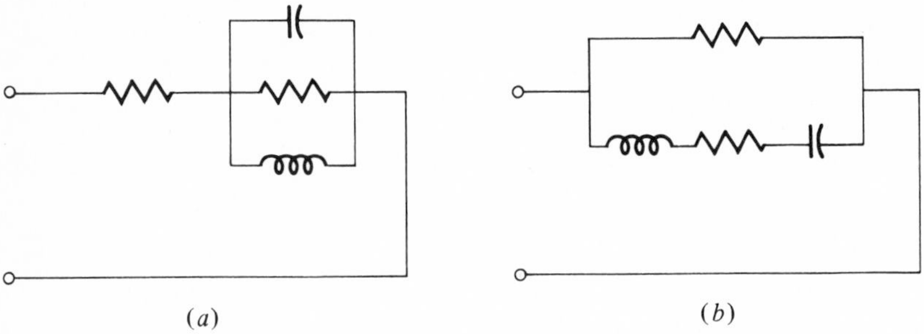

FIG. 5-J.

5-K. Find the driving-point impedance Z as a rational function of s. It represents the transfer function from what excitation to what response? Suppose $Z(s)$ has the (special) form

$$Z(s) = Z_0 \frac{s^2 + as + \omega_0{}^2}{s^2 + bs + \omega_0{}^2};$$

express the element values in terms of Z_0, a, b, ω_0.

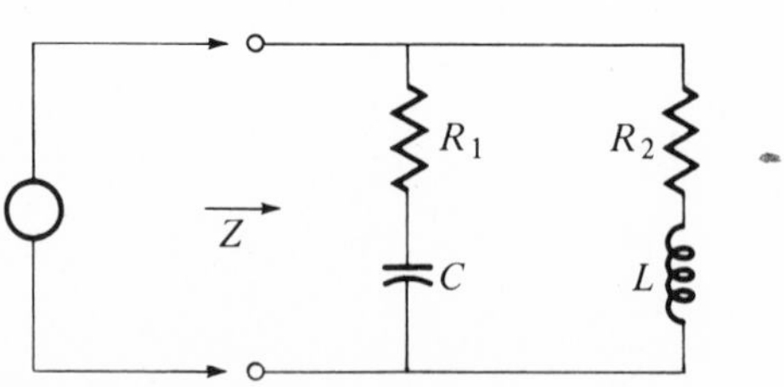

FIG. 5-K.

What restrictions must be obeyed by prescribed values of these four parameters if positive element values are to result therefrom? Find the element values, that is, synthesize the network, for the two cases:

Z_0, Ω	a, (rad/s)	b, (rad/s)
600	1.3×10^6	10^6
75	0.8×10^6	0.2×10^6

5-L. Engineering analysis and design studies frequently require testing a given system model or circuit for *stability*. The problem is to determine whether all of its natural frequencies lie in the left half plane (then, and only then, is it stable). The actual test commonly comes down to determining whether a given (the characteristic) polynomial is or is not *Hurwitz*. For polynomials of degree 3 or higher this is not obvious from the coefficients (but all coefficients *must have the same sign* for a polynomial to be Hurwitz, see Exercise 5-7). A "Hurwitz" or "Hurwitz-Routh" polynomial test that is often useful is closely related to a simple circuit immittance calculation, which we proceed to examine.

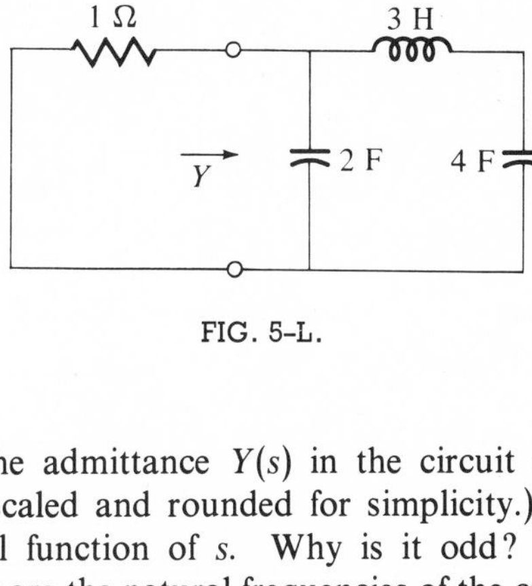

FIG. 5-L.

(a) Calculate the admittance $Y(s)$ in the circuit shown. (The numbers have been scaled and rounded for simplicity.) Note that $Y(s)$ is an *odd* rational function of s. Why is it odd? What is the equation whose roots are the natural frequencies of the complete circuit? What specific polynomial have you here shown to be Hurwitz?

(b) To test a given polynomial for Hurwitz character, you may reverse the process of (a) by first forming the quotient of the even and odd parts of the given polynomial, placing the part of higher degree in the numerator. Call the resulting odd rational function of s, $Y(s)$. Carry out the division (it has only one step), invert the remaining fraction, and repeat. Continue this divide-invert process to its end. Symbolically, the calculation is, given a polynomial $Q = a_N s^N + a_{N-1}s^{N-1} + \cdots + a_1 s + a_0 = A + sB$, in which A and B are even polynomials, A is the sum of the even-power terms in Q, sB is the sum of the odd-power terms,

$$Y = \frac{A}{sB} \qquad or \qquad \frac{sB}{A} = C_1 s + \cfrac{1}{L_2 s + \cfrac{1}{C_3 s + \cdots}}$$

The process (called *continued-fraction* expansion of Y) should determine N constants. (If it does not so work out, Q is not Hurwitz.) Show, by constructing a circuit and reasoning therefrom, that if N constants *are* determined, and all of them are positive, then Q *is* a Hurwitz polynomial. This demonstrates the sufficiency part of an argument for the test's validity. The necessity part, that if Q is Hurwitz then the continued-fraction development must proceed as above and give N positive constants, is more difficult, and we merely assume it.

The following example is compressed by using synthetic division. Verify all its steps and note how little calculation is really necessary.

Polynomial to be tested:

$$Q = s^3 + 5s^2 + 3s + 2$$

Divide-invert calculation:

$$\begin{array}{llllll}
\underline{5 \quad 2}| & 1 & 3 & |\tfrac{1}{5}(+) & & \\
 & 1 & \tfrac{2}{5} & & & \\
\hline
 & & \tfrac{13}{5}| & 5 & 2 & |\tfrac{25}{13}(+) \\
 & & & 5 & & \\
\hline
 & & & & 2| & \tfrac{13}{5} \quad |\tfrac{13}{10}(+) \\
 & & & & & \tfrac{13}{5}
\end{array}$$

$N = 3$ and three positive quotients result; Q *is* Hurwitz. [The formal result, not necessary, but perhaps helpful to understanding, is:

$$\frac{s^3 + 3s}{5s^2 + 2} = \tfrac{1}{5}s + \frac{1}{\tfrac{25}{13}s + 1/\tfrac{13}{10}s}.\Bigg]$$

(*c*) Assuming the test to be valid (we have *not* completely demonstrated that it is), test the following polynomials for Hurwitz character:

1 $s^3 + 3s^2 + 2s + 1$,

2 $s^3 + 2s^2 + 3s + 1$,

3 $4s^4 + 3s^3 + 2s^2 + s + 5$,

4 $s^4 + 1.8 \times 10^6 s^3 + 3.1 \times 10^{12} s^2 + 0.9 \times 10^{18} s + 4.6 \times 10^{24}$ (*Suggestion:* Let $s = 10^6 p$),

5 $4s^5 + 5s^4 + 6s^3 + 7s^2 + 8s + 9$.

(*d*) Demonstrate that for $N = 1$ and $N = 2$, it is necessary and sufficient, for Q to be Hurwitz, that its coefficients all be of the same sign. For the polynomial $Q = s^3 + as^2 + bs + c$, what must be true of the coefficients for Q to be Hurwitz?

5-M. Many routine day-to-day calculations on electric power systems involve only the sinusoidal steady state; the immittances used are (complex) numerical constants, simple and very useful in combination with phasors. To calculate the current drawn by a given load, for example (Fig. 5-M*a*), we use $\bar{I} = \bar{E}/\bar{Z}$ or $\bar{I} = \bar{Y}\bar{E}$ (see Secs. 2.15, 3.11) in which

$\bar{Z}$	$=$	R	$+j$	X	$= Z\angle\theta$
(Complex impedance)		(Resistance)		(Reactance)	

and

$\bar{Y}$	$=$	G	$+j$	B	$= Y\angle -\theta.$ (mhos)
(Complex admittance)		(Conductance)		(Susceptance)	

Note that $\bar{Z}$ is the *complex* impedance but Z is its *magnitude* (a real number), etc. Here $e(t) = E_m \cos \omega_F t$ is represented by phasor $\bar{E}$ and $i(t) = I_m \cos(\omega_F t - \theta)$ is represented by phasor $\bar{I}$ in which it is convenient to take $I = I_m/\sqrt{2}$ and $E = E_m/\sqrt{2}$ for simplicity; see below. (I and E are the *root-mean-square* or simply *rms* current and voltage.) Taking $\bar{E}$ as reference (setting the angle of $\bar{E}$ equal to zero) and assuming an inductive load (the common case) places the current phasor $\bar{I}$ as shown, at an angle $-\theta$ (Fig. 5-M*c*)- $\bar{I} = I\angle -\theta$. Its value is easily calculated by complex-number arithmetic.

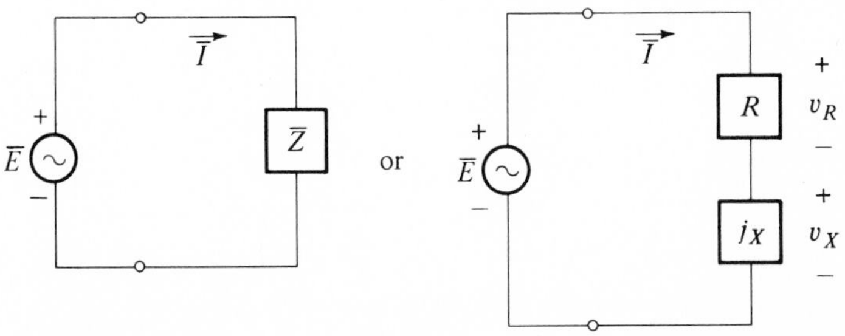

FIG. 5-M(a).

Show that if $\bar{V}_R$ is the (rms) phasor that represents $v_R(t)$ and $\bar{V}_X$ is the (rms) phasor that represents $v_X(t)$, then $\bar{E} = \bar{V}_R + \bar{V}_X$. Establish this rigorously in the time domain which is, after all, the source of our knowledge. Multiply $\bar{I}$ by R in the diagram (Fig. 5-M*c*) and complete the voltage triangle with $\bar{V}_X$. What is the angle between $\bar{V}_R$ and $\bar{V}_X$? Why? Now multiply all legs of the voltage triangle by I. Show that in the triangle so formed (rotated so that the RI leg is horizontal for neatness, Fig. 5-Md), $P + jQ = \overline{E}\overline{I}^*$ in which

$$P = V_R I = \text{average power consumed by the load} = \mathbf{RE}(\overline{E}\overline{I}^*).$$

Explain why the use of rms values simplifies the notation. Show further that

EI = triangle hypotenuse = apparent power taken by the load, the volt-ampere product at its terminals, the "volt-amperes."

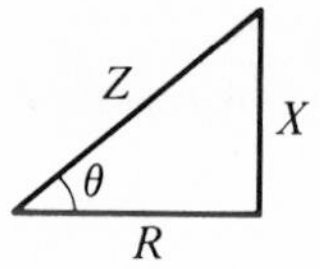

The impedance triangle
$\overline{Z} = R + jX = Z\underline{/\theta}$

FIG. 5-M(b).

Why is it appropriate to call EI the "apparent" power? Show further that

Q = maximum rate of flow of energy into (and out of) storage in the reactance (inductance) element, called *reactive power* for nomenclatural consistency and mnemonic reasons, though it is *not* power in the usual sense of the word.

$$= V_X I = \left.\frac{dW_X}{dt}\right|_{\max} = \mathbf{IM}(EI^*).$$

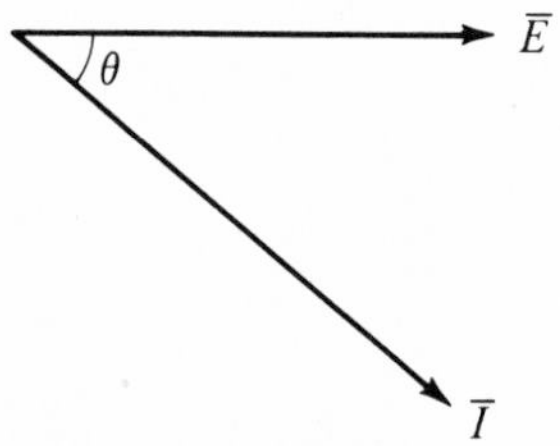

FIG. 5-M(c).

Why is it mnemonically useful to adopt the term "reactive power"? Show also that

$$P = EI \cos\theta \qquad \text{and} \qquad Q = EI \sin\theta.$$

The factor by which the apparent power (the volt-amperes) EI is multiplied to give the real (dissipated, used in work) power consumption, $\cos\theta$, is called the *power factor* of the load.

Loads are commonly expressed in terms of P and the power factor, that is in terms of their power demands and power factors (and voltage, of course). Show that the equivalent (model) R and X elements are then given by

$$R = \frac{E^2 \cos^2\theta}{P} \quad \text{and} \quad X = \frac{E^2 \cos\theta \sin\theta}{P}.$$

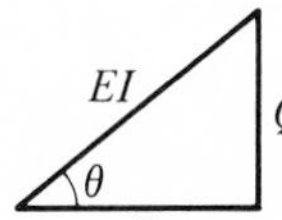

The (complex)
power-voltampere triangle:
$P + jQ = \bar{E}\bar{I}^* = EI\angle\theta$

FIG. 5-M(d).

5-N. Apply the ideas and principles of Prob. 5-M to calculate the current drawn by a load center maintained locally at voltage V and connected through a reactance X (which may represent the principal component of the impedance of a transmission line) to a source (generator) $\bar{E} = E\angle\alpha$, Fig. 5-N. Then use your results to calculate $P + jQ$ as seen (a) by the generator $\bar{E}$ and (b) at the load. Show formally that the difference between the two P's is zero and explain from common sense why that must be so. Calculate the difference between the two Q's and show that it is equal to the "I^2X loss" in the line: "The Q loss in the line is I^2X." Explain why this must be considered in power-system design and operation, and why Q is of importance even though it does *not* represent actual power. (See Prob. 3-Z.) (*Suggestion:* The law of cosines may be useful.)

What is the maximum power transferable from the generator $\bar{E}$ to the load center in terms of E, X, V, and the angle between the phasors $\bar{E}$ and $\bar{V}$? Explain this result in physical terms as well as you can.

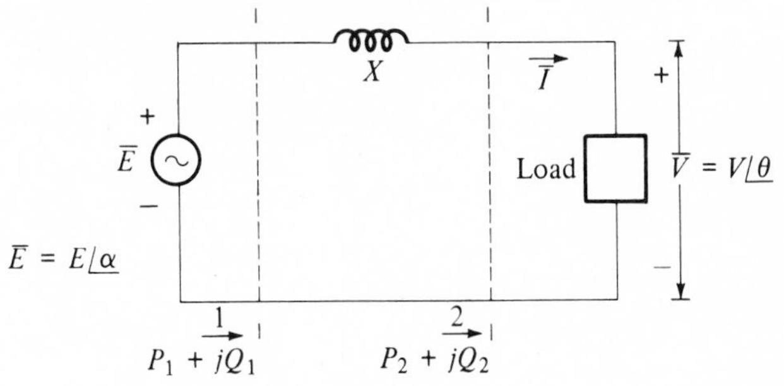

FIG. 5-N.

Chapter **6**

NETWORKS AND TRANSFER FUNCTIONS

A pawn goes two squares in its first move, you know. So you'll go very quickly through the Third Square—by railway, I should think—and you'll find yourself in the Fourth Square in no time. Well, *that* square belongs to Tweedledum and Tweedledee—the Fifth is mostly water—the Sixth belongs to Humpty Dumpty—

—The Red Queen

"... When *I* use a word," Humpty Dumpty said, in rather a scornful tone, "it means just what I choose it to mean—neither more nor less." ... "I meant by 'impenetrability' that we've had enough of that subject, and it would be just as well if you'd mention what you mean to do next, as I suppose you don't mean to stop here all the rest of your life...."

Neither the simple *RC* and *RLC* networks, nor bridges of arbitrary complexity, exhaust the category of networks that are most interesting, those that are useful and important, and those for which the properties of Sec. 5.14 hold. To the largest subcategory of all therein we now direct our attention.

6.01 LADDER NETWORKS

This largest subcategory of important network types is the *ladder* (Fig. 6.01-A). It is very widely used, it has great practical importance, and for it the impulse response–transfer function relation of Chap. 5 holds. And it is remarkably simple to analyze: to obtain the transfer function, we shall see, is a straightforward, soon-to-become-routine calculation.

In the figure each box represents an *immittance*, that is an *im*pedance $[Z(s)]$, or an ad*mittance* $[Y(s)]$. Bode's graphic word "immittance" (Sec. 5.11) is very useful, for we shall find that $Z(s)$ and $Y(s)$ have the same properties, and should therefore be described generically by a single word.

The boxes form the rungs and sides of the ladder structure. Since typically the currents i_a and i_b between rungs (Fig. 6.01-A*a*) should be identical, it makes no difference whether the Z_a and Z_b boxes are located separately in the two sides (Fig. 6.01-A*a*) or combined in one side as $Z_c = Z_a + Z_b$ (Fig. 6.01-A*b*). This

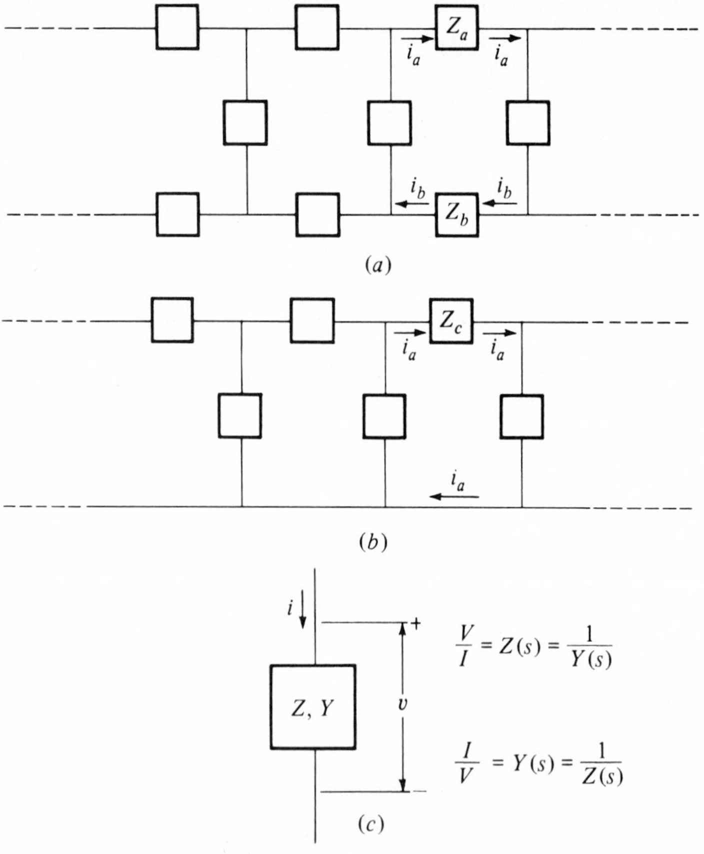

FIG. 6.01-A.

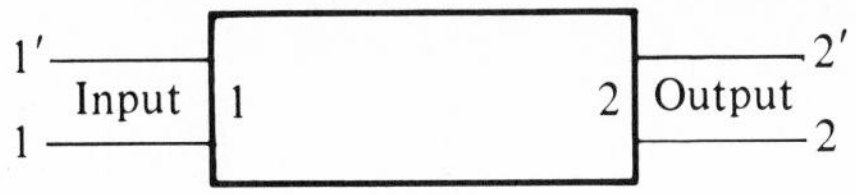

FIG. 6.01-B.

implies that our interest is limited to the *external* behavior of the network, to its transmission from input to output (Fig. 6.01-B). And it implies that there is no other, parallel path from input to output, no other network connected in parallel with the ladder proper (the properties could then be quite different!).

For precision, we include these matters in our formal definition of a ladder network:

> A network of the two-port variety (Fig. 6.01-B) is a ladder network if, on examining the internal structure, we find *only* a network of the Fig. 6.01-A type(s). (6.01-1)

Then $i_a = i_b$ as above. Another signal path, from input to output, by feedback or feedforward transmission of some sort, would obviously vitiate the ladder idea.

The network of Fig. 6.01-A*b* has the great practical advantage that it can be solidly grounded throughout on one side; the network is then called "unbalanced" (with respect to ground): Fig. 6.01-C*a*. That of Fig. 6.01-A*a* is useful,

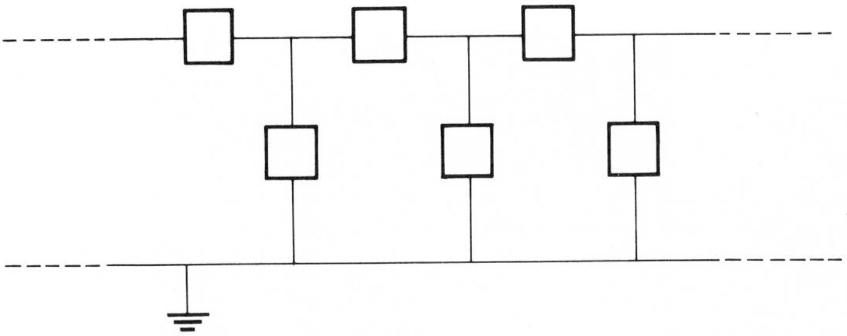

(*a*) Section of a one-side-grounded or "unbalanced" ladder network

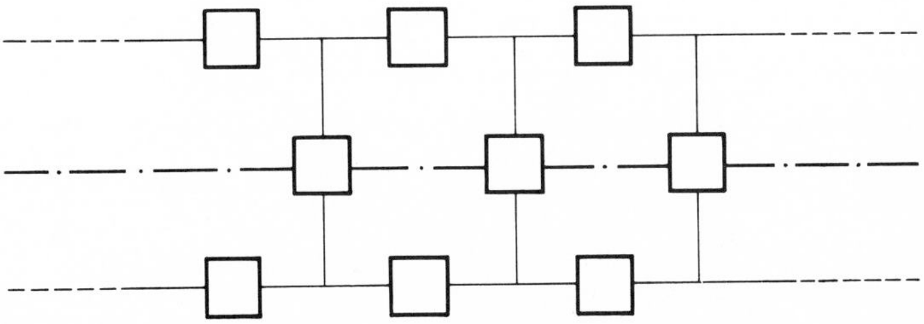

(*b*) Section of a "balanced" (to ground) ladder network: the center line indicates ground potential, the sides being symmetric with respect thereto

FIG. 6.01-C.

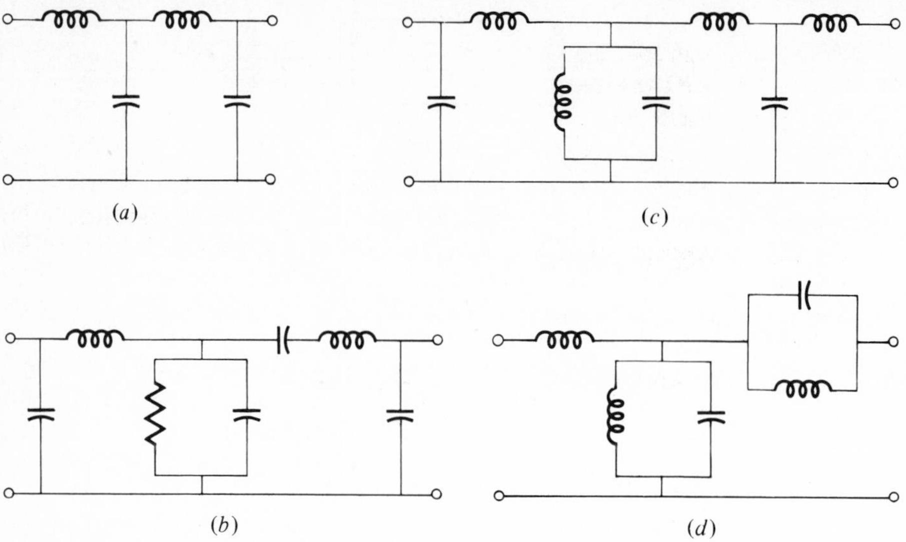

FIG. 6.01-D.

when symmetrically constructed ($Z_a = Z_b$), in balanced-to-ground system situations, where a ground potential plane runs down the middle: Fig. 6.01-C*b*.

The figures show only typical sections or pieces of ladder networks. They may terminate, at the two ends, with either a "series" or a "shunt" box. Figure 6.01-D gives several examples that illustrate the various possible ending combinations: series-shunt, shunt-shunt, shunt-series, series-series. Here each box has but one or two elements in it, rather simple cases. Each box could be complicated, as, for example, those in Figs. 5.11-F and 5.11-G. What describes each box is its "local transfer function," its impedance function $Z(s)$ or admittance function $Y(s)$.

Note again that the definition precludes any bypassing, or parallel connected networks (as in Sec. 5.08). In a genuine ladder there is only one route from input to output.

For simplicity we postulate for the present that there be no active (g_m) elements in the boxes, so that they are strictly *passive*. The transfer function of each box, or the relation between (forced) current and voltage therefor, with e^{st} excitation, is described in Fig. 6.01-A*c*.

We shall deal largely, in the next sections, with passive ladder networks. Their simplicity and ease of analysis, and their *very* common occurrence in real-life engineering, make this reasonable. And in them we can easily *see* and *feel* many important physical properties of networks. But our results in this chapter are generally valid for all networks; we concentrate on ladders only for convenience and simplicity.

In Sec. 6.23 we shall again discuss the bridge, and in Sec. 6.24 we shall briefly discuss completely general schemes, suitable for the analysis of any network. Section 6.25 discusses some (nonladder) active networks that are particularly interesting because of feedback therein.

6.02 THE TRANSFER FUNCTION OF A LADDER NETWORK

It is the one-route-only nature of a ladder network that makes rather easy the calculation of its transfer function. Our postulate of linearity presumably makes the *ratio* of the forced component of response to (exponential) excitation independent of the amount of excitation. To calculate the transfer function we can therefore *either*

a As before, calculate forward from a given exponential excitation to the forced component of the corresponding response, and divide the (forced) response by the excitation,

or (6.02-1)

b Assume a forced component of response in exponential form, calculate back to find the (exponential) excitation required therefor, and divide the (forced) response by the excitation.

For ladder networks procedure (*b*) is much the simpler of the two.

For example, consider the network of Fig. 6.02-A. Suppose the excitation voltage to be $e(t) = E_0 e^{st}$. Then all currents and voltages will presumably contain forced components that vary as e^{st}, and natural components that vary as $\sum K_j e^{p_j t}$, the p_j being the natural frequencies and the K_j various coefficients suitable to the various currents and voltages. We can both verify this statement (entirely reasonable, on the basis of our experience) *and* calculate the transfer function by answering the question: "Suppose the response voltage *is*

$$v(t) = V_F e^{st} + \sum_j K_j e^{p_j t}; \tag{6.02-2}$$

what then must $e(t)$ be?"

In this first analysis of a ladder we write out the calculations that answer the question in careful detail; once completed, it will show us that much of that detail is unnecessary and may be omitted in future ladder-network analyses.

The calculation (we omit the subscript F in V_F as unnecessary and cumbersome) is:

$$\textit{If} \qquad v(t) = \underbrace{Ve^{st}}_{v_F} + \underbrace{\sum_{j=1}^{N} K_j e^{p_j t}}_{v_N} \tag{6.02-3}$$

$$\textit{Then} \qquad i_3(t) = G_3 v = G_3 V e^{st} + G_3 \sum K_j e^{p_j t},$$

$$v_m(t) = v + L_3 \frac{di_3}{dt} = Ve^{st} + L_3 G_3 s V e^{st} + \cdots,\dagger$$

$$i_2(t) = C_2 \frac{dv_m}{dt} = (C_2 s + L_3 G_3 C_2 s^2) V e^{st} + \cdots,$$

$$i_1 = i_2 + i_3 = (G_3 + C_2 s + L_3 G_3 C_2 s^2) V e^{st} + \cdots, \tag{6.02-4}$$

$$\begin{aligned} e(t) &= v_m + L_1 \frac{di_1}{dt} + R_1 i_1 \\ &= [(1 + L_3 G_3 s) + (L_1 G_3 s + L_1 C_2 s^2 + L_1 C_2 L_3 G_3 s^3) \\ &\quad + (R_1 G_3 + R_1 C_2 s + R_1 L_3 G_3 C_2 s^2)] V e^{st} + \cdots \\ &= [(1 + R_1 G_3) + (L_3 G_3 + R_1 C_2 + L_1 G_3) s \\ &\quad + (R_1 C_2 L_3 G_3 + L_1 C_2) s^2 + (L_1 C_2 L_3 G_3) s^3] V e^{st} + (\cdots). \end{aligned}$$

Now what we want is purely exponential excitation, e^{st} in form. We therefore require that the dots within the last parentheses in (6.02-4) add up to zero. This will be true if all the (different) K_j and (same) p_j therein are properly determined: if the initial conditions are properly met and the natural frequencies properly evaluated. We assume this done. [Then the natural part of $v(t)$ in (6.02-3), if calculated out, will be correctly evaluated.]

Then, to continue, we have

$$\begin{aligned} T(s)\Big|_{e \text{ to } v} &= \frac{v(t)}{e(t)}\Big|_{\ldots} = \frac{Ve^{st}}{[(1 + R_1 G_3) + \cdots] V e^{st}} \\ &= \frac{1}{\left[\begin{aligned}&(1 + R_1 G_3) + (L_3 G_3 + R_1 C_2 + L_1 G_3) s \\ &\quad + (R_1 C_2 L_3 G_3 + L_1 C_2) s^2 + (L_1 C_2 L_3 G_3) s^3\end{aligned}\right]}. \end{aligned} \tag{6.02-5}$$

We have both answered the question following (6.02-2) and verified the format used for the response of the network, $v(t)$. Most important is that we have calculated the ladder network's transfer function in a simple, straightforward fashion.

† The non-e^{st} terms (the natural part) are becoming cumbersome and we abbreviate: the dots that represent them are both vivid and adequate (but not, of course, identical in meaning from one line to the next).

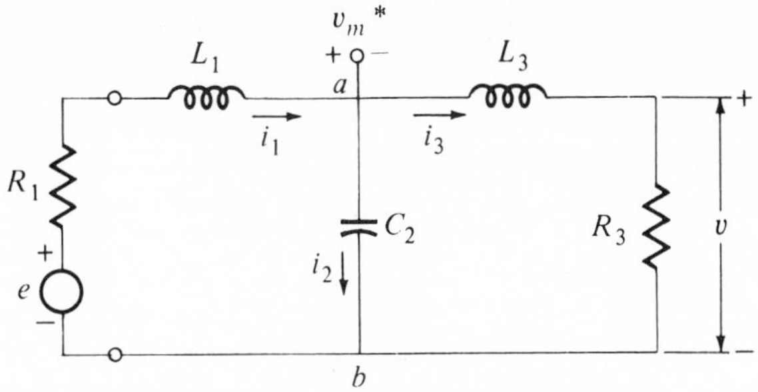

$*v_m$ is the potential of point a with respect to point b, "the bottom," often grounded

(a)

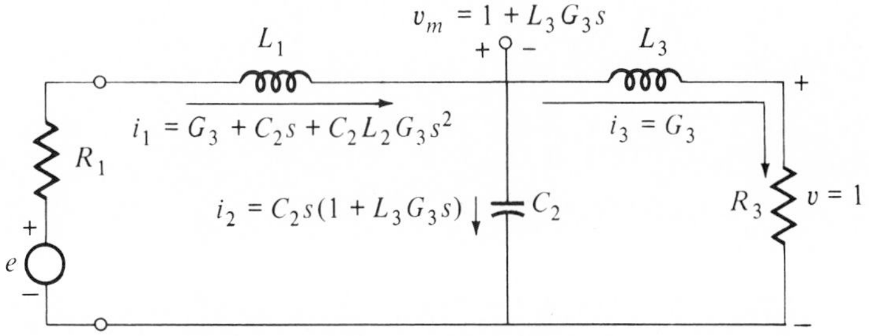

$$e = (1 + L_3 G_3 s) + (R_1 + L_1 s)(G_3 + C_2 s + C_2 L_3 G_3 s^2)$$
$$= (1 + R_1 G_3) + (L_3 G_3 + R_1 C_2 + L_1 G_3)s + (R_1 C_2 L_3 G_3 + L_1 C_2)s^2 + (L_1 C_2 L_3 G_3)s^2$$

$$T(s)\Big|_{e \text{ to } v} = \frac{v}{e}\Big| \ldots$$
$$= \frac{1}{[(1 + R_1 G_3) + (L_3 G_3 + R_1 C_2 + L_1 G_3)s + (R_1 C_2 L_3 G_3 + L_1 C_2)s^2 + (L_1 C_2 L_3 G_3)s^3]}$$

(b)

FIG. 6.02-A.

In retrospect we see now that much of the calculation is indeed unnecessary. In fact Fig. 6.02-A*b* shows the complete calculation, all that is necessary. We may conveniently let $v = 1$, omitting the natural part (which will take care of itself) entirely, omitting e^{st}, which will eventually cancel anyway, and taking $V_F = 1$ for convenience. Then the current in R_3 is simply $i_3 = G_3$, the voltage v_m is simply $(1 + L_3 G_3 s)$, e^{st} being omitted, ..., and so on, as in the calculations made directly (completely) on the diagram itself.

Why *is* the calculation so easy? Why is no consideration of differential equations as such necessary? It is the ladder structure that makes it so! Working back from response location to source, we encounter no forks, no

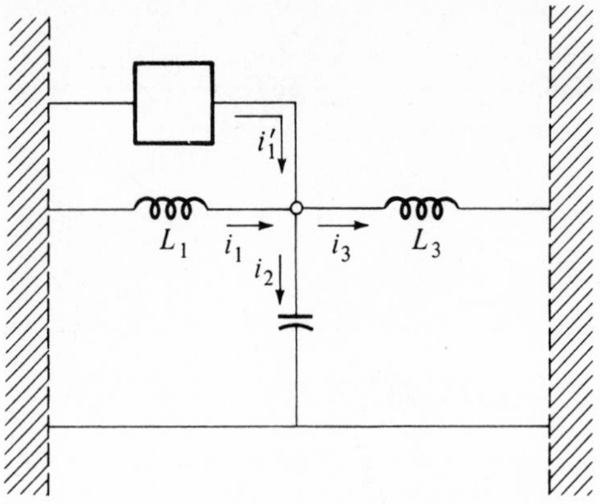

FIG. 6.02-B.

branches "upstream"; there is no difficulty in determining "earlier" currents in terms of "later" currents, as there would be, for example, if there were another "box" joined to L_1 (Fig. 6.02-B): then the determination of i_1 and i_1' from i_2 and i_3 would present problems. But then the network is no longer a pure ladder; and how to deal with such networks we postpone consideration of (to Sec. 6.23). In a genuine ladder, the calculation of $T(s)$ by working back from an assumed

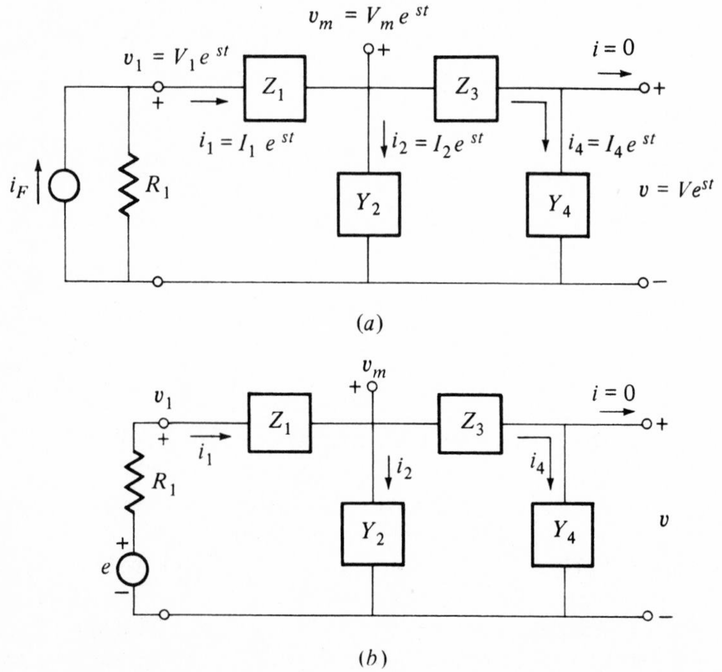

FIG. 6.02-C.

(say unit-value) response presents no difficulty other than the algebra of perhaps individually complicated Z and Y functions (Figs. 6.01-A and 6.01-B). Linearity makes the assumed value of response irrelevant; it is only the ratio of response to excitation that counts. As for the differential equations, their special character, and the fact that we want only forced components, enables us to abbreviate them as in Fig. 6.02-A*b*—for that is all that it really does.

In the calculations above, simple though the network is, the algebraic symbols accumulate rapidly and profusely. The complexity of the expressions is unavoidable, but it is less when numbers rather than literal element values are used, in realistic network analysis. An intermediate stage of complexity appears in the example of Fig. 6.02-C, where $Z(s)$ and $Y(s)$ are used for the (arbitrarily complicated) impedance and admittance immittance expressions of each rung or side piece of the ladder. In every case $Z_j Y_j = 1$, and Z or Y is used as is most convenient at the moment. We also find it convenient to use $G = 1/R$ (with subscripts) from time to time. Omitting the natural parts, writing only the forced parts, and omitting e^{st} and writing the coefficients thereof only, reduces the calculation to its essentials (Fig. 6.02-C*a*):

Let:

$$V = 1;$$

Then

$$\begin{aligned}
I_4 &= Y_4 V = Y_4,\\
V_m &= V + Z_3 I_4 = 1 + Z_3 Y_4,\\
I_2 &= Y_2 V_m = Y_2 + Y_2 Z_3 Y_4,\\
I_1 &= I_4 + I_2 = Y_2 + Y_4 + Y_2 Z_3 Y_4,\\
V_1 &= V_m + Z_1 I_1 = 1 + Z_3 Y_4 + Z_1 Y_2 + Z_1 Y_4 + Z_1 Y_2 Z_3 Y_4,\\
I_F &= I_1 + G_1 V_1 \qquad (6.02\text{-}6)\\
&= Y_2 + Y_4 + Y_2 Z_3 Y_4 + G_1 + G_1 Z_3 Y_4 + G_1 Z_1 Y_2 + G_1 Z_1 Y_4\\
&\quad + G_1 Z_1 Y_2 Z_3 Y_4,\\
T &= \frac{v}{i_F}\bigg|_{\ldots}\\
&= \frac{1}{G_1 + Y_2 + Y_4 + G_1 Z_3 Y_4 + G_1 Z_1 Y_2 + G_1 Z_1 Y_4 + Y_2 Z_3 Y_4 + G_1 Z_1 Y_2 Z_3 Y_4}.
\end{aligned}$$

In Fig. 6.02-C*b* the excitation (signal generator) composed of the current source $i_F(t)$ and an associated resistor R_1 is replaced by a voltage source $e(t)$

and a series resistor. The analysis here, we note, is identical up through the calculation of V_1. Then

$$E = V_1 + R_1 I_1 = 1 + Z_3 Y_4 + Z_1 Y_2 + Z_1 Y_4 + Z_1 Y_2 Z_3 Y_4 + R_1 Y_2 + R_1 Y_4 + R_1 Y_2 Z_3 Y_4,$$

$$T = \frac{v}{e}\bigg|_{\ldots} = \frac{1}{1 + R_1 Y_2 + R_1 Y_4 + Z_3 Y_4 + Z_1 Y_2 + Z_1 Y_4 + R_1 Y_2 Z_3 Y_4 + Z_1 Y_2 Z_3 Y_4}. \tag{6.02-7}$$

The only difference between the two transfer functions, (6.02-6) and (6.02-7), is a factor R_1; if e and i_F are related by

$$e(t) = R_1 i_F(t) \tag{6.02-8}$$

in fact, one transfer function can be derived immediately from the other.

This suggests an *equivalence*, as far as results to the right of the terminals to which the (either) generator is attached, between the two signal generators: (*a*) i_F in parallel with R_1 and (*b*) e in series with R_1. This is a special, perhaps already familiar, instance of a general theorem of "equivalent sources" due to Helmholtz. [The current type (*a*) is sometimes called the "Norton form," and the voltage type (*b*) the "Thévenin form."]

There are two other transfer functions of interest, those from the two sources to the (output) *current* i_4, which we can write immediately from the calculations above,

$$T = \frac{i_4}{i_F}\bigg|_{\ldots} = \frac{Y_4}{G_1 + Y_2 + Y_4 + \cdots + G_1 Z_1 Y_2 Z_3 Y_4}, \tag{6.02-9}$$

and

$$T = \frac{i_4}{e}\bigg|_{\ldots} = \frac{Y_4}{1 + R_1 Y_2 + R_1 Y_4 + \cdots + Z_1 Y_2 Z_3 Y_4}. \tag{6.02-10}$$

One can easily imagine the complexity of a transfer function such as (6.02-10) when written out in complete detail for a case in which each "box" contains two or three elements, and every Z and Y function has second- or third-degree numerator and denominator! Nevertheless, this is what must be done in actual network analysis. Use of actual numerical element values will make specific-case transfer functions more tractable; usually the numbers can even be made of the order of magnitude of unity by normalization (which removes awkward powers of 10 and other factors as we shall see).

Our experience above leads to these observations:

a A ladder-network transfer function can, in principle, easily be calculated by working backward from response observation point to excitation point; (6.02-11)

b There are four different transfer functions that depend on the nature of the response and of the excitation (current or voltage), and are closely related to each other; (6.02-12)

c If the two types of signal source (excitation) considered have the same internal impedance (R_1 in Fig. 6.02-C), then all four transfer functions have the same denominator, within a constant multiplier; for the example of Fig. 6.02-C it is (6.02-13)

$$\begin{aligned} Q(s) = 1 &+ R_1 Y_2(s) + R_1 Y_4(s) + Z_3(s) Y_4(s) + Z_1(s) Y_2(s) \\ &+ Z_1(s) Y_4(s) + R_1 Y_2(s) Z_3(s) Y_4(s) + Z_1(s) Y_2(s) Z_3(s) Y_4(s). \end{aligned} \tag{6.02-14}$$

A true current generator has infinite impedance (zero admittance), since the current through it (i_F in Fig. 6.02-C, for example) cannot be changed by voltage applied across it; a true voltage generator has infinite admittance (zero impedance), since the voltage across it is not altered by the flow of current through it. Hence one may say that, in a sense, R_1 in Fig. 6.02-C should really be considered part of the network and not of (either) excitation. Then in (6.02-13), the source being considered "pure," (*c*) becomes

c′ All four transfer functions have the same denominator (within a constant multiplier). (6.02-15)

This denominator is an important characteristic of the network, for it determines the natural frequencies; those of the network of Fig. 6.02-C are the zeros of $Q(s)$ of (6.02-14).

For emphasis and convenient reference we restate the source-equivalence (Helmholtz) theorem discovered above in more general terms, replacing R_1 by an arbitrary impedance Z_0, according to Sec. 5.11. The two source (excitation) forms of Fig. 6.02-D are completely equivalent, *externally*, and may be freely interchanged in networks without affecting conditions external thereto, since $v = e_0 - Z_0 i$ [which is true in part (*a*)] and $i = i_0 - Y_0 v$ [which is true in (*b*)] are the *same* relation. Note the duality between the two.

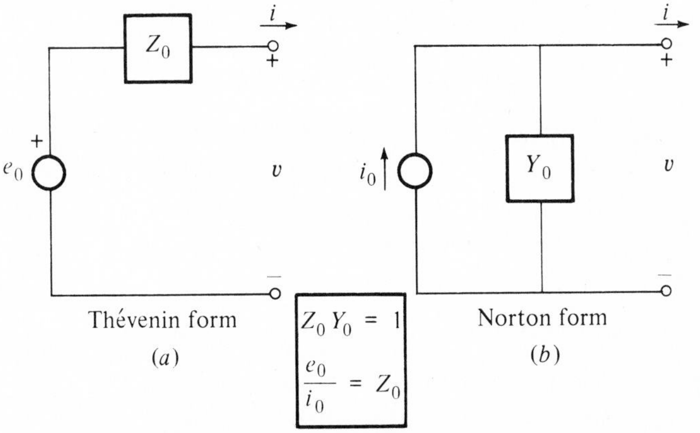

FIG. 6.02-D. Source Equivalences.

6.03 NATURAL MODES AND NATURAL FREQUENCIES

The natural behavior of a network, the current-voltage behavior that can exist in the absence of excitation, caused only by initial stored-energy conditions, consists of a sum of exponential terms, each Ke^{pt} in form, in which

$$\begin{aligned} &e^{pt} \text{ represents a } \textit{natural mode} \text{ of behavior,} \\ &p \text{ represents a } \textit{natural frequency,} \\ &K \text{ is a constant, to be evaluated to fit the initial conditions.} \end{aligned} \tag{6.03-1}$$

Here, on the basis of our experience so far, this *seems* eminently reasonable; a completely general *proof* will come later. (We note that if excitation *is* present, the K's in the natural part of the response depend on the initial condition of the network *and* on the initial value of the excitation.)

Natural-mode currents flow in the absence of excitation (other than initial energy storage). They do so because they encounter no opposition; they "see" zero impedance. Phrased more elegantly, the total impedance around any closed circuit (or loop) is zero if calculated at a natural frequency:

$$\sum_{\substack{\text{Around a} \\ \text{closed} \\ \text{circuit}}} Z(s) = 0. \tag{6.03-2}$$

The roots of this equation are the natural frequencies: $p_1, p_2, \ldots, p_N$. t does not matter what loop is chosen for the calculation; the same equation 5.03-2) results, because the p's are characteristic of the network as a whole, not of any particular path through it. One must be careful, however, to include *all* the impedances encountered in going around the loop. (The principle is valid for any network, ladder or not, passive or active.)

For example, in Fig. 6.02-A, traversing the first loop (R_1, L_1, and the combination of C_2 in parallel with the remaining elements) gives

$$Z(s) = R_1 + L_1 s + \frac{1}{C_2 s + 1/(L_3 s + R_3)} \tag{6.03-3}$$

whose zeros are the natural frequencies. Alternatively, traversing the second loop gives

$$Z(s) = R_3 + L_3 s + \frac{1}{C_2 s + 1/(L_1 s + R_1)} \tag{6.03-4}$$

whose zeros are also the (same) natural frequencies.

In more general terms, the natural frequencies of the network of Fig. 6.02-C are the roots of

$$R_1 + Z_1(s) + \frac{1}{Y_2(s) + 1/[Z_3(s) + 1/Y_4(s)]} = 0. \qquad (6.03\text{-}5)$$

Such expressions rapidly become very complicated when literal element-value symbols are used. In any actual analysis one must of course use numerical element values, with some reduction in complexity of expressions.

The natural frequencies may equally well be determined from a dual point of view. We need only "dualize" the paragraph that contains (6.03-2), word for word:

Natural-mode voltages exist in the absence of excitation (other than initial energy storage). They do so because they encounter no opposition; they "see" zero admittance. Phrased more elegantly, the total admittance between any two nodes (element terminals) is zero if calculated at a natural frequency:

$$\sum_{\substack{\text{Between}\\\text{any two}\\\text{nodes}}} Y(s) = 0. \qquad (6.03\text{-}6)$$

The roots of this equation are the natural frequencies: $p_1, p_2, \ldots, p_N$. It does not matter what node pair is chosen for the calculation; the same equation (6.03-6) results, because the p's are characteristic of the network as a whole, not of any particular path through it. One must be careful, however, to include *all* the admittances attached to the nodes used.

Finally we recall (Sec. 5.05) that the natural frequencies are the poles of the transfer functions (they correspond to response without excitation), which of course gives us the same characteristic equation to determine the natural frequencies.

Figure 6.03-A resumes all this in compact form.

The network of Fig. 6.03-B has simple, integral numerical element values for illustration (which may however be not unrealistic, because of normalization, soon to be explained). Its characteristic equation, calculated from the first loop marked, is

$$\begin{aligned} 0 &= 1 + \frac{2}{s} + \cfrac{1}{\cfrac{1}{2s} + \cfrac{1}{1/(4s + \frac{1}{5} + 1/2s) + 1/(3s + 1)}} \\ &= \frac{240s^5 + 712s^4 + 312s^3 + 240s^2 + 49s + 10}{240s^5 + 92s^4 + 104s^3 + 22s^2 + 5s} \\ &= 0. \end{aligned} \qquad (6.03\text{-}7)$$

The current view

There is no opposition to the flow of current, a *short circuit* exists when s equals a natural frequency, which encourages natural-mode current flow.

$$\sum_{\text{loop}} Z(s) = F(s) = 0.$$

The voltage view

There is no opposition to the existence of voltage, an *open circuit* exists when s equals a natural frequency, which encourages natural-mode voltage existence:

$$\sum_{\substack{\text{node}\\ \text{pair}}} Y(s) = G(s) = 0.$$

Any transfer function of the network, $T(s) = N(s)/D(s)$ in form, may also be used. The roots of the equation

$$F(s) = 0, \qquad G(s) = 0, \qquad D(s) = 0,$$

are the same and are the natural frequencies of the network.

In any given network the natural frequencies are the natural frequencies, and the same characteristic (natural-frequency-determining) equation will be obtained, no matter what loop, what node pair, what transfer function be used.

WARNING: in some cases, numerator and denominator of the rational function encountered (F or G or T) may have common (identical) factors; these must NOT be canceled if the above statements are to hold literally. (This matter will be discussed later.)

FIG. 6.03-A. Determination of the natural frequencies of a network.

The natural frequencies are accordingly the zeros of the characteristic polynomial

$$240s^5 + 712s^4 + 312s^3 + 240s^2 + 49s + 10. \qquad (6.03\text{-}8)$$

Going around either of the other loops marked, or the two other possible (unmarked) loops, gives the same equation. And if the excitation is a current source in parallel with R_1, the same natural frequencies of course result.

Finding the natural frequencies calls for factoring the polynomial (6.03-8). This is typical of the numerical problems of network analysis. It can only be done numerically, by some sort of successive-approximation procedure. Except

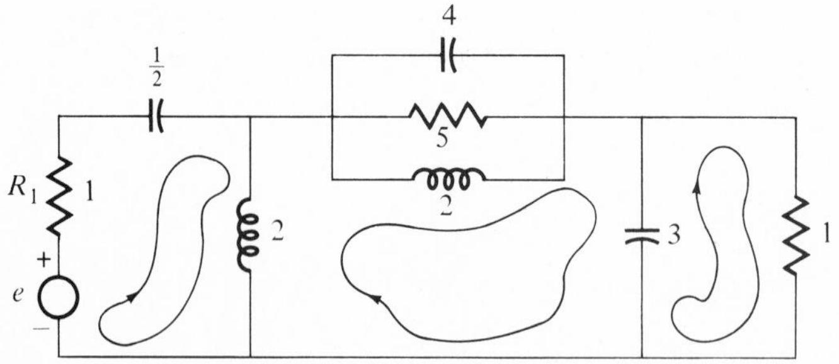

Element values are ohms, henrys, farads (normalized)

FIG. 6.03-B.

```
FACTORING (BY THE SUBROUTINE POLRTT) OF THE POLYNOMIAL OF DEGREE  5, WHOSE COEFFICIENTS
(STARTING WITH     THE CONSTANT TERM) ARE

     10.000000     49.000000     240.000000     312 000000     712.000000     240.000000

GIVES THESE ZEROS:


  -0.121774 + J   0.212098

  -0.121774 + J  -0.212098

  -0.059420 + J   0.513762

  -0.059420 + J  -0.513762

  -2.604278 + J   0.000000



TO ESTIMATE THE PRECISION ATTAINED, NOTE THAT THE   SUM OF THE ZEROS IS      -2.966665

                                          FOR COMPARISON, 712.0/240.0 =       2.966666

AND THE PRODUCT OF THE ZEROS IS      -0.041667

            AGAINST 10.0/240.0 =      0.041667



CORE USAGE          OBJECT CODE=     5824 BYTES,ARRAY AREA=        496 BYTES,TOTAL AREA AVAILABLE=  106584  BYTES

COMPILE TIME=     0.54 SEC,EXECUTION TIME=      0.10 SEC,   WATFIV - VERSION 1 LEVEL 1 JANUARY  1970      DATE=  74/044
```

FIG. 6.03-C. $ Data. Factoring polynomial (6.03-8) by subroutine POLRITT. Zeros are natural frequencies.

in trivial cases, one must use a (digital) computer, for which finding the zeros of a polynomial is a routine task. Figure 6.03-C gives the results for the characteristic polynomial above—and some indication of the numerical accuracy attained, some hint of the problems of numerical analysis, and some hint of a computer's speed.

A dual approach is to calculate the admittance between two terminals, say those of the 3-farad capacitor, and set it equal to zero:

$$0 = 1 + 3s + \cfrac{1}{\cfrac{1}{4s + \frac{1}{5} + 1/2s} + \cfrac{1}{1/2s + 1/(2/s + 1)}}$$

$$= \frac{240s^5 + 712s^4 + 312s^3 + 240s^2 + 49s + 10}{80s^4 + 184s^3 + 28s^2 + 40s}. \qquad (6.03\text{-}9)$$

This equation has of course the same roots as does (6.03-8).

6.04 FACTORING POLYNOMIALS

The literature of this particular kind of calculation is enormous. It is inappropriate to develop the matter in detail here, for every computer installation has a library of programs for factoring polynomials (but see Appendix C).

Methods vary, but generally they (*a*) pick a trial value of *s*, either arbitrarily or making use of whatever information about the polynomial is available, and

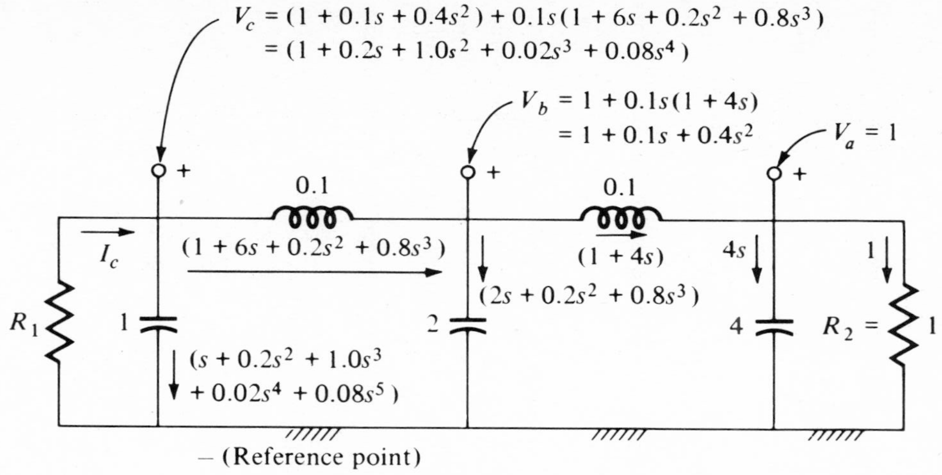

[Element values are (normalized) ohms, henrys, farads]

$$I_c = 1 + 7s + 0.4s^2 + 1.8s^3 + 0.02s^4 + 0.08s^5$$

$$Z_c = \frac{V_c}{I_c} = \frac{1 + 0.2s + 1.0s^2 + 0.02s^3 + 0.08s^4}{1 + 7s + 0.4s^2 + 1.8s^3 + 0.02s^4 + 0.08s^5}$$

Characteristic equation: $R_1 + Z_c(s) = 0$:

$$Q(s) = (1 + R_1) + (7R_1 + 0.2)s + (0.4R_1 + 1.0)s^2 + (1.8R_1 + 0.02)s^3 + (0.02R_1 + 0.08)s^4 + 0.08R_1 s^5 = 0$$

FIG. 6.04–A.

(*b*) improve upon that trial value. The successive improvements can be made by (1) calculating the derivative of the polynomial at that point and using this to estimate an appropriate increment in *s*, or (2) "boxing a zero in" by reducing the size of a square inside which a zero occurs. Method 1, Newton's method, is usually very efficient; it becomes impractical, however, when zeros lie close together so that the derivative (by which one must divide) becomes small. Method 2, which utilizes d'Alembert's principle, does not require such division and will find close or coincident zeros without difficulty, though perhaps somewhat more slowly. Many variations and refinements of these and other techniques have been developed.

In most cases, network natural frequencies do not lie very close to one another. Coincidence, or near coincidence of zeros of a characteristic polynomial, represents either poor (inefficient) design or (occasionally) some very special situation. Some form of Newton's method (adapted of course to deal with complex zeros) is therefore most frequently used.

Appendix C gives programs for one form of each method. Figure 6.03-C is one illustration of the results of their use; Fig. 6.04-B is another. The latter explores the variation of the natural frequencies of the network of Fig. 6.04-A as the source impedance R_1 therein varies in value. The calculation of the natural-frequency-determining (characteristic) equation, and of the transfer function from source to load, is made on Fig. 6.04-A itself; Fig. 6.04-B gives typical computer results, finding the natural frequencies for a sequence of values of R_1. The loci of the natural frequencies are given in Fig. 6.04-C, in which only the second quadrant is shown; the two conjugate natural frequencies in quadrant III are omitted. Several values of R_1 are marked on the three loci,

```
    R1                NATURAL FREQUENCIES
 .........          ........................

 0.100000          -0.443152  + J   1.061072
                   -0.443152  + J  -1.061072
                   -8.944170  + J   0.000000
                   -0.209764  + J   3.403285
                   -0.209764  + J  -3.403285

    R1                NATURAL FREQUENCIES
 .........          ........................

 0.200000          -2.206196  + J   0.000000
                   -1.108625  + J   1.222881
                   -1.108625  + J  -1.222881
                   -0.413276  + J   3.508109
                   -0.413276  + J  -3.508109

    R1                NATURAL FREQUENCIES
 .........          ........................

 0.300000          -0.743325  + J   0.000000
                   -0.897833  + J   2.077745
                   -0.897833  + J  -2.077745
                   -0.522172  + J   3.735131
                   -0.522172  + J  -3.735131

    R1                NATURAL FREQUENCIES
 .........          ........................

 1.000000          -0.287938  + J   0.000000
                   -0.254102  + J   2.232863
                   -0.254102  + J  -2.232863
                   -0.226929  + J   4.140119
                   -0.226929  + J  -4.140119


CORE USAGE      OBJECT CODE=   5216 BYTES,ARRAY AREA=    276 BYTES,TOTAL AREA AVAILABLE=  106624  BYTES

COMPILE TIME=    0.52 SEC,EXECUTION TIME=    0.46 SEC,  WATFIV - VERSION 1 LEVEL 1 JANUARY 1970   DATE= 74/048
```

FIG. 6.04-B. $ Data. Calculation, using subroutine POLRITT, of the natural frequencies of the network of Fig. 6.04-A, for various values of R_1.

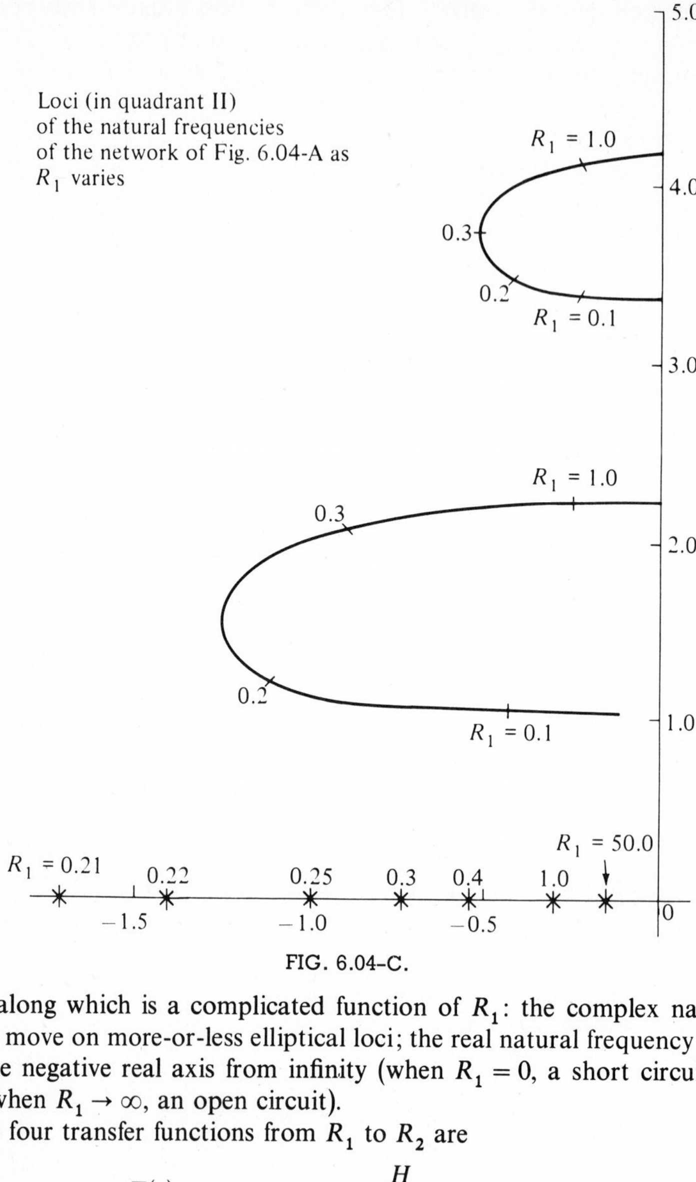

FIG. 6.04-C.

motion along which is a complicated function of R_1: the complex natural frequencies move on more-or-less elliptical loci; the real natural frequency moves in along the negative real axis from infinity (when $R_1 = 0$, a short circuit) to the origin (when $R_1 \to \infty$, an open circuit).

The four transfer functions from R_1 to R_2 are

$$T(s) = \frac{H}{[Q(s) = (1 + R_1) + \cdots + 0.08R_1 s^5]} \tag{6.04-1}$$

in which the constant H has the value unity if the excitation is a voltage source in series with R_1, and the value R_1 if the excitation is a current source in parallel with R_1. [Since R_2 has the value 1.0, dimensions have been lost in (6.04-1)].

Finally, Fig. 6.04-D illustrates the difficulties encountered when zeros coincide, or nearly coincide. In the first two of the five examples there, the poly-

```
......................................................................................................................

EXAMPLE 1:

THE POLYNOMIAL (OF DEGREE  5) HAS THESE COEFFICIENTS:

     1.99996     7.99992     12.99994     10.99998     5.00000     1.00000

THE ZEROS FOUND ARE:
                    BY POLRITT                      BY DALBER                       CORRECT ZEROS
              ........................      ........................      ........................
      1       -0.998722  + J   0.000000     -1.000000  + J -0.000000     -1.000000  + J  1.000000       1
      2       -1.000000  + J   1.000000     -0.995525  + J -0.000000     -1.000000  + J  1.000000       2
      3       -1.000000  + J  -1.000000     -1.004475  + J -0.000000     -1.000000  + J  0.000000       3
      4       -0.996408  + J   0.000000     -1.000000  + J  1.000000     -1.004472  + J  0.000000       4
      5       -1.004969  + J   0.000000     -1.000000  + J -1.000000     -0.995528  + J  0.000000       5

......................................................................................................................

EXAMPLE 2:

THE POLYNOMIAL (OF DEGREE  5) HAS THESE COEFFICIENTS:

     2.00004     8.00008     13.00006     11.00002     5.00000     1.00000

THE ZEROS FOUND ARE:
                    BY POLRITT                      BY DALBER                       CORRECT ZEROS
              ........................      ........................      ........................
      1       -0.991226  + J  0.000000      -1.000000  + J -0.000000     -1.000000  + J  1.000000       1
      2       -1.004385  + J  0.008649      -1.000000  + J  0.004475     -1.000000  + J -1.000000       2
      3       -1.004385  + J -0.008649      -1.000000  + J -0.004475     -1.000000  + J  0.000000       3
      4       -1.000001  + J  1.000002      -1.000000  + J  1.000000     -1.000000  + J  0.004472       4
      5       -1.000001  + J -1.000002      -1.000000  + J -1.000000     -1.000000  + J -0.004472       5

......................................................................................................................

EXAMPLE 3:

THE POLYNOMIAL (OF DEGREE  5) HAS THESE COEFFICIENTS:

     2.00002     8.00004     13.00003     11.00001     5.00000     1.00000

THE ZEROS FOUND ARE:
                    BY POLRITT                      BY DALBER                       CORRECT ZEROS
              ........................      ........................      ........................
      1       -1.000000  + J  0.999995      -1.000000  + J -0.000000     -1.000000  + J  1.000000       1
      2       -1.000000  + J -0.999995      -1.000000  + J  0.003088     -1.000000  + J -1.000000       2
      3       -1.009022  + J  0.000000      -1.000000  + J -0.003088     -1.000000  + J  0.000000       3
      4       -0.995489  + J  0.009053      -1.000000  + J  1.000000     -1.000000  + J  0.003162       4
      5       -0.995489  + J -0.009053      -1.000000  + J -1.000000     -1.000000  + J -0.003162       5

......................................................................................................................

EXAMPLE 4:

THE POLYNOMIAL (OF DEGREE  5) HAS THESE COEFFICIENTS:

     1.99998     7.99996     12.99997     10.99999     5.00000     1.00000

THE ZEROS FOUND ARE:

******** ERROR RETURN FROM SUBROUTINE POLRTT (DOES NOT CONVERGE):  N2 =  3, IN =11, ICT =  25       ****************************
         ZEROS FOUND,  IF ANY:          1   -1.000000   + J        0.999999
                                        2   -1.000000   + J       -0.999999
                                        3************   + J**************
                                        4************   + J**************
                                        5************   + J**************
                                        6************   + J**************

                    BY POLRTT                       BY DALBER                       CORRECT ZEROS
              ........................      ........................      ........................
      1              -1.000000  + J  0.999999   -1.000000  + J -0.000000     -1.000000  + J  1.000000       1
      2              -1.000000  + J -0.999999   -0.996511  + J -0.000000     -1.000000  + J -1.000000       2
      3  99999870000.000000  + J**********      -1.003088  + J -0.000000     -1.003162  + J  0.000000       3
      4  99999870000.000000  + J**********      -1.000000  + J  1.000000     -0.996838  + J  0.000000       4
      5  99999870000.000000  + J**********      -1.000000  + J -1.000000     -1.000000  + J  0.000000       5

......................................................................................................................

EXAMPLE 5:

THE POLYNOMIAL (OF DEGREE  5) HAS THESE COEFFICIENTS:

     1.00000     5.00000     10.00000     10.00000     5.00000     1.00000

THE ZEROS FOUND ARE:

******** ERROR RETURN FROM SUBROUTINE POLRTT (DOES NOT CONVERGE):  N2 =  1, IN =11, ICT =  25       ****************************
         ZEROS FOUND,  IF ANY:          1************   + J**************
                                        2************   + J**************
                                        3************   + J**************
                                        4************   + J**************
                                        5************   + J**************
                                        6************   + J**************

                    BY POLRTT                       BY DALBER                       CORRECT ZEROS
              ........................      ........................      ........................
      1  99999870000.000000  + J**********      -1.000000  + J -0.000000     -1.000000  + J  0.000000       1
      2  99999870000.000000  + J**********      -1.000000  + J -0.000000     -1.000000  + J  0.000000       2
      3  99999870000.000000  + J**********      -1.000000  + J -0.000000     -1.000000  + J  0.000000       3
      4  99999870000.000000  + J**********      -1.000000  + J -0.000000     -1.000000  + J  0.000000       4
      5  99999870000.000000  + J**********      -1.000000  + J -0.000000     -1.000000  + J  0.000000       5

......................................................................................................................

CORE USAGE       OBJECT CODE=   12720 BYTES,ARRAY AREA=     520 BYTES,TOTAL AREA AVAILABLE=  181088  BYTES

COMPILE TIME=      1.52 SEC,EXECUTION TIME=      7.27 SEC,  WATFIV - VERSION 1 LEVEL 1 JANUARY  1970    DATE=  74/075
```

FIG. 6.04-D. Comparison of zeros found by POLRITT and DALBER for polynomial with nearly coincident zeros.

nomial's three clustered zeros, though close to each other, are sufficiently far apart that both methods "work," though Newton's method (employed by subroutine POLRTT) is appreciably less accurate. In example 3 the cluster is tighter and the same is true. In example 4 the Newton's-method subroutine returns mostly nonsense; d'Alembert's performance is good. In the fifth example, where all five zeros are coincident, only d'Alembert's performs satisfactorily. [In fairness one should note that some of d'Alembert's success is due to its use of double precision (POLRTT is a single-precision subroutine), but the principal cause of the differences is the *method* used: Newton's by POLRTT and d'Alembert's by DALBER.]

6.05 DIMENSIONAL CHECKS

It is phenomenally easy to make mistakes: *errare humanum est* is as valid in the calculation of transfer functions as it is in any other work. We should therefore know and use such *checks* as may be appropriate. There are several practical ones, particularly those based on *dimensions*, which we now discuss, and those based on *asymptotic behavior* (see Sec. 6.06).

Dimensions are usually easy to verify, if literal symbols are used, and they provide useful checks on the results of involved calculations. These checks are not complete, of course, but they will "catch" most errors. It is axiomatic that our formulas have dimensional homogeneity, that all terms on the same "level" have consistent (the same) dimensions. For example, transfer functions of the type voltage-to-voltage, as (6.02-7), and those of the type current-to-current, as (6.02-9), must have null dimensions. When reduced to ordinary rational function form (the quotient of two polynomials), the dimensions of the two polynomials will not be null as a rule, but they must be the *same*. And within each polynomial, all terms must have the same dimensions. This makes dimensional checking easy.

For example, in the denominator of (6.02-6), all terms are dimensionally *admittances*, evident by inspection. A calculation error that produced $G_1 Y_1 Y_2$ would be glaringly evident in the light of its dimensions. The terms in the denominator of (6.02-7) are consistently free of dimensions, another check. Both polynomials in (6.02-9) are consistently admittances. In (6.02-10) the numerator is admittance, and the denominator terms are all free of dimensions—exactly as they should be.

As a practical matter, we can speak of the *units* of an expression equally well. That is, "having units of ohms" and "having dimensions of impedance" are equivalent, for we hereby agree to use only the *ohm* as a unit for measuring impedances. Similarly, to have consistent units of mhos (or seconds, or volts, or amperes) constitutes a dimensional check. The *RC* circuit impulse response $v(t) = (Q_0/C)\,e^{-\alpha t}$ is expressed in volts on the left, and in coulombs per farad or volts on the right; the two are consistent, as they must be in an equation.

Note that αt is free of dimensions, in $e^{-\alpha t}$, as must be every argument of the exponential and similar functions.

Again, the current response

$$i(t) = GV_0 e^{-\alpha t} \cos(\omega t + \theta) + K_1 e^{p_1 t} \tag{6.05-1}$$

is presumably in amperes on the left, and in mho-volts = amperes on the right; the units of K_1 must also be amperes. The quantities α, ω, p_1 are all in seconds^{-1} so that αt, ωt, and $p_1 t$ have null dimensions, as they must, being arguments of the exponential function, real or complex as the case may be. Note that the angle θ has null dimensions, though we speak of it (and of ωt) as measured in *radians* (or perhaps degrees). We may speak of real exponents such as αt and the real part of $p_1 t$ as measured in *nepers*, merely to distinguish them from angles; both are dimension-free.

When numerical values are used, as in (6.03-7) for example, dimensional checking is impossible. Whenever the dimensions of terms are known, however, a quick dimensional check is well worth the effort. It should become habitual to look at the dimensions of every term we write.

Note that the input immittances of ladder networks naturally take a continued-fraction form when first written, each level of which is impedance *or* admittance. Examples are (6.03-3), (6.03-4,), (6.03-5); in (6.03-7) and (6.03-9), however, no dimensions are evident, since numerical values are used. The same is true of Fig. 6.04-A.

Table 6.05-A lists some often-met combinations whose dimensions should be instinctively recognized.

TABLE 6.05-A Some Useful Dimensions

Type of Term	Dimensions	Units
R	resistance	ohm
G	conductance	mho
L	inductance	henry = ohm-sec
λ	flux linkage	weber-turn = volt-second = ohm-coulomb
C	capacitance	farad = mho-second
q	charge	coulomb
Ls	impedance	ohm
Cs	admittance	mho
L/C	impedance^2	ohm^2
$RC = C/G$	time	second
L/R	time	second
LC	time^2	second^2

TABLE 6.05-A. (*Con.*)

Some Common Terms Whose Dimensions Are Null				
RG	YZ	RCs	LGs	LCs^2
ωt	αt	pt	st	

6.06 INITIAL BEHAVIOR OF IMPULSE RESPONSE: THE PHYSICAL VIEW

Complicated though response functions may be, it is often quite easy to evaluate their principal components at the ends of the range of the independent variable. Useful checks result. More important, however, is the exercise of physical insight that is needed to make such calculations, an essential attribute of a good engineer. It is tempting to take refuge in the delightful mathematics of a model and escape the grim facts of physical behavior. It can be dangerous so to do, for the physical view may reveal absurdities not noticeable in the mathematics, due to mistakes in calculation (or perhaps to inadequacies in the model). In any event, since it is the physical behavior that is ultimately of interest, one does well to keep it continually in mind.

We have become well acquainted with two domains, those of time, t, and of complex frequency, s and p. Asymptotic calculations can be made in each domain. Both require physical insight if made on the model itself. Both require some mathematical sophistication if made on the transfer function. It is in the matching up of the two that we obtain checks: the two are of course by no means independent.

Let us begin by reviewing what happens in the simple RC circuit of Chap. 2 during the brief lifetime of an impulsive excitation. Conditions in the network at $t = 0+$, just after the impulse is over, are different from those at $t = 0$, before the excitation is applied, very brief (infinitesimal even) though this time difference be. The network's behavior in the interval $0 < t < 0+$ can of course be calculated mathematically by integrating the appropriate differential equation (as in Secs. 2.02 and 3.02). It can equally well be determined by examination of the physical action in the network, from physical understanding of "what goes on." And because the latter is closer to reality it is not to be lost sight of! We assume a very brief excitation life, we neglect all but the principal effects, and we reach the same result (if no mistakes are made in either calculation) as does the mathematics in the limiting case of zero impulse duration. In other words, we obtain a check on our work. Figure 6.06-A sums this simple example up. First, (*a*) shows the model, (*b*) an excitation of short (but not infinitesimal) duration, and (*c*) the response (from initial rest) thereto. Then (*d*) and (*e*) show the same excitation and response when T is infinitesimal, together with the corresponding (limiting-case) mathematics.

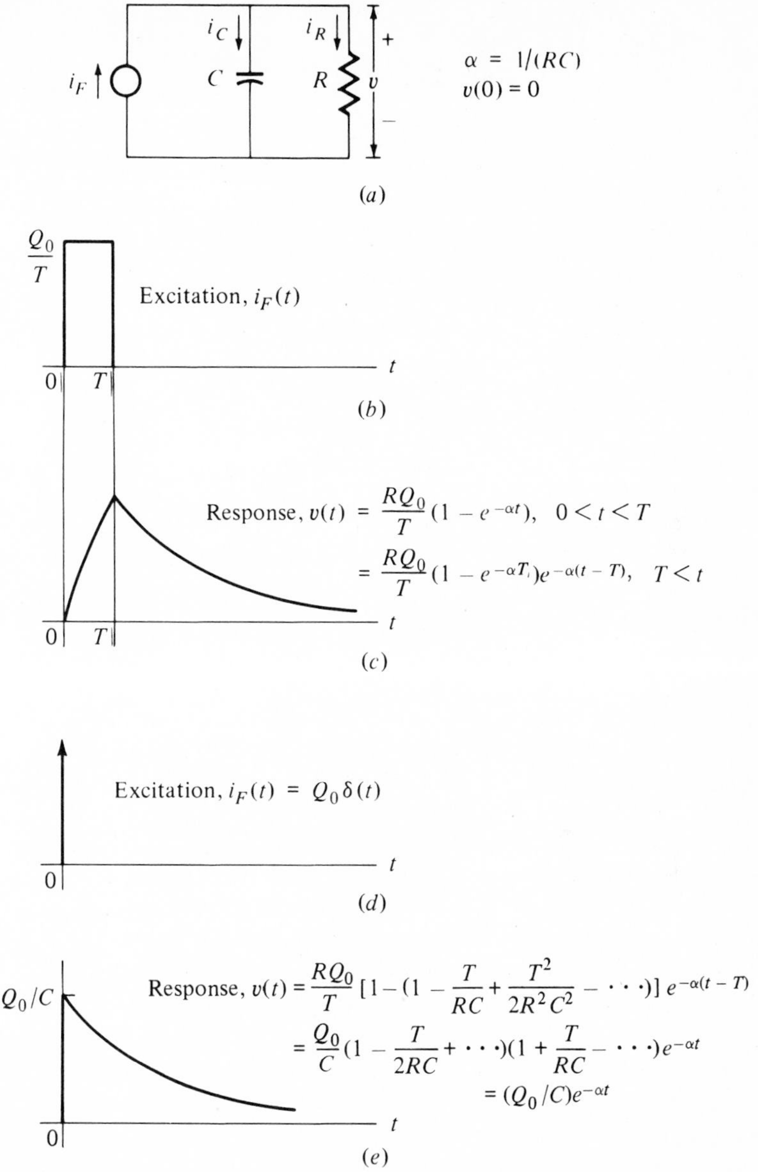

FIG. 6.06-A.

So long as the pulse duration T is very small compared to the network's time constant (for example, $T = 0.001\ RC$) we may regard it as simply an impulse of strength (area) Q_0 (not necessarily small). Even though what actually happens is, if sufficiently magnified, what (a) and (c) show, still the action during its life is essentially the following. No current flows in the resistor ($i_R = 0$) because its impedance is tremendous compared to that of the capacitor (the capacitor voltage is zero and a resistor current would require a nonzero Ri_R voltage). In effect the impedance of the capacitor is zero. Hence all the excitation current flows through the capacitor and (suddenly) deposits its Q_0 (coulombs) thereon. The result is the voltage $v(0+) = Q_0/C$. Thereafter events occur at normal speed, measured if you like in circuit-time-constant (RC) units; the usual exponential natural behavior (decay) takes place. Actual physical verification can be found in Figs. 2.05-A and 2.07-B (q.v.).

We conclude that the physical view of the RC circuit behavior is simple, realistic, and useful. The check is that the mathematical analysis also gave $v(0+) = (Q_0/C)e^{-\alpha t}]_{t=0} = Q_0/C$. We have examined this simple case in painful detail because the ideas involved remain the same in complicated cases and will be used again and again.

With very little additional effort, even more information can be gotten from the physical analysis. We note that (Fig. 6.06-A)

$$i_R(0+) = Gv(0+) = \frac{Q_0}{RC} = -i_C(0+) = -Cv'(0+) \tag{6.06-1}$$

so that

$$v'(0+) = -\frac{Q_0}{RC^2}. \tag{6.06-2}$$

We have obtained the value of the initial rate of change of the response voltage, again by physical examination of the model. We go no further into derivatives here but note the ability of the physical, asymptotic observation to give the 0+ values of both the response and its derivative. (Differentiation of $v = (Q_0/C)e^{-\alpha t}$ and setting $t = 0+$ gives the same result.)

Physical analysis of the initial behavior of the RLC circuit of Chap. 3 in similar circumstances is slightly more complicated. In return, somewhat more information is obtained—as one might expect, since there are now two energy-storing elements in the network. Figure 6.06-B shows two models of the tuned

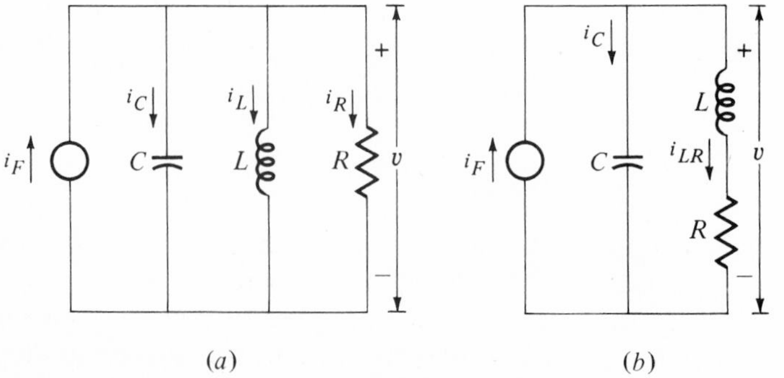

FIG. 6.06-B.

(resonant) circuit. We assume initial rest, $v(0)$ and $i_L(0)$ both zero, and excitation by the same current pulse (Fig. 6.06-A). During the life of the (short) pulse, the action is exactly the same, for an inductor offers even more opposition to current flow: since $v = L\,di/dt$, not only is the initial inductor current zero; so also is its rate of change, because $v(0) = 0$. (To cause current to flow in a resistor requires only a finite voltage; to cause current suddenly to flow in an inductor requires infinite voltage, of which there is here no possible source.) The capacitor offers zero impedance, all of the impulsive current flows through it, and there results $v(0+) = Q_0/C$ in both models. The four graphs of Fig. 6.06-A apply here also; Figs. 3.07-B and 3.07-C (q.v.) give direct physical corroboration.

Physical analysis of the initial ($t = 0+$) currents here resembles (6.06-1) and (6.06-2), but with additions. For model (*a*) of Fig. 6.06-B,

$$\begin{aligned} i_R(0+) &= Gv(0+) = \frac{Q_0}{RC}, \\ i_L(0+) &= 0, \\ i_C(0+) &= -i_R(0+) = -\frac{Q_0}{RC}, \end{aligned} \tag{6.06-3}$$

and so

$$v'(0+) = -\frac{Q_0}{RC^2}. \tag{6.06-4}$$

Because we have here a second energy-storing element, we can also easily obtain initial rates of change of currents, and hence the initial value of the second derivative of the voltage:

$$\begin{aligned} i_R'(0+) &= Gv'(0+) = -\frac{Q_0}{R^2C^2}, \\ i_L'(0+) &= \frac{1}{L}\,v(0+) = \frac{Q_0}{LC}, \\ i_C'(0+) &= -i_R'(0+) - i_L'(0+) = \frac{Q_0}{C}\left(\frac{1}{R^2C} - \frac{1}{L}\right), \end{aligned} \tag{6.06-5}$$

and so

$$v''(0+) = \frac{Q_0}{C^2}\left(\frac{1}{R^2C} - \frac{1}{L}\right). \tag{6.06-6}$$

These equations provide an excellent opportunity to practice the checking of dimensions. We note the units of $Q_0/(RC)$: coulombs per second (Fig. 6.05-A), or amperes, as they should be, since this quantity is supposed to be a current in (6.06-3). Similarly, we verify the dimensions of the other quantities above:

$$\frac{Q_0}{RC^2}: \text{amperes per farad} = \text{amperes per mho-second} = \text{volts per second},$$

$$\frac{Q_0}{R^2C^2}: \text{volts per ohm-second} = \text{amperes per second},$$

$$\frac{Q_0}{LC}: \text{coulombs per second}^2 = \text{amperes per second}, \tag{6.06-7}$$

$$\frac{1}{R^2C}: 1/\text{ohm}^2\text{-mho-second} = 1/\text{henry},$$

$$\frac{Q_0}{R^2C^3}: \text{amperes per second-mho-second} = \text{volts per second}^2.$$

For model (b) we have

$$\begin{aligned} i_{RL}(0+) &= 0, \\ i_C(0+) &= 0, \qquad \text{and} \qquad v'(0+) = 0; \\ v(0+) &= Ri_{RL}(0+) + Li'_{RL}(0+), \\ i'_{RL}(0+) &= \frac{Q_0}{LC}, \\ i'_C(0+) &= -\frac{Q_0}{LC}, \end{aligned} \tag{6.06-8}$$

and

$$v''(0+) = -\frac{Q_0}{LC^2}. \tag{6.06-9}$$

The dimensions of these terms have been checked, except for the last one: the units of $Q_0/(LC^2)$ are amperes per second-mho-second = volts per second^2, as they should be.

Note that physical examination of model (b) gives, almost by inspection, the initial (0+) values of v, of v', and of v'', of three quantities instead of two, because of the second energy-storing element. To continue to higher derivatives is possible, but involves calculations extended enough to amount more to mathematical analysis than to physical checking.

Now we use these results, obtained physically, for checking the formal analysis. In Chap. 3 we obtained the results below, for initial rest conditions and $Q_0\,\delta(t)$ excitation.

For model (a),

$$
\begin{aligned}
v(t) &= K_1 e^{p_1 t} + K_2 e^{p_2 t} \\
&= \frac{Q_0}{C} \frac{p_1}{p_1 - p_2} e^{p_1 t} + \frac{Q_0}{C} \frac{(-p_2)}{p_1 - p_2} e^{p_2 t}
\end{aligned}
\tag{6.06-10}
$$

for which

$$
p^2 + \frac{G}{C} p + \frac{1}{LC} = 0 \tag{6.06-11}
$$

is the characteristic equation, so that

$$
\begin{aligned}
p_1 + p_2 &= -\frac{G}{C}, \\
p_1 p_2 &= \frac{1}{LC}.
\end{aligned}
\tag{6.06-12}
$$

For model (b),

$$
\begin{aligned}
v(t) &= K_1 e^{p_1 t} + K_2 e^{p_2 t} \\
&= \frac{Q_0}{C} \frac{(-p_2)}{p_1 - p_2} e^{p_1 t} + \frac{Q_0}{C} \frac{p_1}{p_1 - p_2} e^{p_2 t}
\end{aligned}
\tag{6.06-13}
$$

for which

$$
p^2 + \frac{R}{L} p + \frac{1}{LC} = 0 \tag{6.06-14}
$$

and

$$
\begin{aligned}
p_1 + p_2 &= -\frac{R}{L}, \\
p_1 p_2 &= \frac{1}{LC}.
\end{aligned}
\tag{6.06-15}
$$

Evaluation of the initial conditions from these formal results gives

$$
v(0+) = \frac{Q_0}{C} \tag{6.06-16}
$$

for both models, good checks,

$$v'(0+) = \frac{Q_0}{C}\frac{p_1{}^2 - p_2{}^2}{p_1 - p_2} = \frac{Q_0}{C}(p_1 + p_2) = \frac{Q_0}{C}\left(-\frac{G}{C}\right) \qquad (6.06\text{-}17)$$

for model (a), again a good check,

$$v'(0+) = \frac{Q_0}{C}\frac{(-p_1p_2) + (p_1p_2)}{p_1 - p_2} = 0 \qquad (6.06\text{-}18)$$

for model (b) (check),

$$\begin{aligned} v''(0+) &= \frac{Q_0}{C}\frac{p_1{}^3 - p_2{}^3}{p_1 - p_2} = \frac{Q_0}{C}(p_1{}^2 + p_1p_2 + p_2{}^2) \\ &= \frac{Q_0}{C}\left(\frac{G^2}{C^2} - \frac{1}{LC}\right) \end{aligned} \qquad (6.06\text{-}19)$$

for model (a), and

$$v''(0+) = \frac{Q_0}{C}\frac{(-p_1{}^2p_2 + p_1p_2{}^2)}{p_1 - p_2} = \frac{Q_0}{C}(-p_1p_2) = \frac{Q_0}{C}\left(-\frac{1}{LC}\right) \qquad (6.06\text{-}20)$$

for model (b), both checking. We have here three checks, some of which were made in Chap. 3. But there are four quantities to be checked: p_1, p_2, K_1, and K_2; the check is not complete. But it is a rare series of errors indeed that would not be caught by the first two checks alone, let alone by three. Note that the first alone is *not* very good because it does not really involve the p's at all. But the first two together (easy to make) do constitute a good check.

We seldom apply physical checking beyond the first derivative of the response; but its use in obtaining initial values of the response and its first derivative is important.

A semantic difficulty can occur in situations where, because of impulsive excitation, a network's state changes suddenly. When confusion can arise, let us agree to define *initial condition* and *initial value* thus:

> *Initial condition* refers to the state of a network at $t = 0$;
>
> *Initial value* refers to its state at $t = 0+$. (6.06-21)

There need not be any difference between, for example, $r_\delta(0)$ and $r_\delta(0+)$; if there is, however, $r_\delta(0)$ is the initial condition, and $r_\delta(0+)$ is the initial value of the response.

6.07 INITIAL BEHAVIOR OF IMPULSE RESPONSE: THE TRANSFER-FUNCTION VIEW

To have and to be able to use physical insight is an essential attribute of a good engineer, be the engineer a "circuits" engineer or some other sort. In Sec. 6.06 we discussed an application of such physical insight: the determination, from the network schematic model, of conditions at $t = 0+$, just after the application of impulse excitation. Now let us use these results for checking purposes.

A transfer function has been obtained, let us say, by formal calculations, for example like those of Secs. 6.01 to 6.03. If the network is complicated, it would be comforting to have a good check on the resulting $T(s)$. And this we can obtain by calculating the $t = 0+$ conditions *from the transfer function*, and then comparing these with those obtained *physically* (Sec. 6.06). To do this we need to recognize that the values of $r_\delta(0+)$, $r'_\delta(0+)$, ..., which are certainly implicit in $T(s)$, are in fact practically *explicit* there.

Suppose we reverse the point of view of (5.05-1) and (5.14-2),

$$\sum_{j=1}^{N} \frac{K_j}{s - p_j} = T(s),$$

to read

$$T(s) = \sum_{j=1}^{N} \frac{K_j}{s - p_j};$$

that is, (5.05-2) and (5.14-3). We are here performing a development of $T(s)$, a rational function of s, a *fraction*, into a sum of very simple "partial" fractions. To do this requires, from algebra, the theory and techniques of *partial-fraction expansion* (which we shall develop in Chap. 7; we do not need them here). On the continuing assumption that $T(s)$ has one or more zeros at infinity and that all the natural frequencies are distinct, and taking $Q_0 = 1$, we then have

$$T(s) = \frac{A_0 s^{N-1} + A_1 s^{N-2} + \cdots}{s^N + B_1 s^{N-1} + B_2 s^{N-2} + \cdots} \tag{6.07-1}$$

$$= \frac{K_1}{s - p_1} + \frac{K_2}{s - p_2} + \cdots = \sum_{j=1}^{N} \frac{K_j}{s - p_j}, \tag{6.07-2}$$

$$r_\delta(t) = K_1 e^{p_1 t} + K_2 e^{p_2 t} + \cdots = \sum_{j=1}^{N} K_j e^{p_j t}, \qquad 0+ < t, \tag{6.07-3}$$

$$r_\delta(0+) = K_1 + K_2 + \cdots = \sum_{j=1}^{N} K_j. \tag{6.07-4}$$

We can relate $\sum K_j$ to the A's and B's in $T(s)$ by a *deus ex machina:* multiply T by s and take the limit for large s:

$$\begin{aligned} \lim_{s\to\infty} [sT(s)] &= A_0 && \text{from (6.07-1)} \\ &= \sum_{j=1}^{N} K_j && \text{from (6.07-2)} \end{aligned} \tag{6.07-5}$$

so that

$$r_\delta(0+) = A_0 = [sT(s)]_\infty \tag{6.07-6}$$

in which the last symbol represents of course the *limit* as $s \to \infty$. If the physical calculation of $r_\delta(0+)$ from the network itself gives the same result as does the formal evaluation of $(sT)_\infty$ from $T(s)$, then we have a check on the calculation of the transfer function.

If the partial-fraction expansion of the transfer function (6.07-2) has been made, we also compare the sum of the K_j therein with this same number; agreement here gives a check on the partial-fraction expansion.

To illustrate, we look again at the familiar RC circuit (Sec. 6.06). With $Q_0 = 1$ we have

$$T(s) = \frac{1/C}{s + 1/(RC)}, \qquad N = 1, \qquad A_0 = \frac{1}{C};$$

$$\text{Physical calculation from the network gave: } r_\delta(0+) = \frac{1}{C}; \tag{6.07-7}$$

$$\text{The sum of the } K\text{'s in the partial-fraction expansion of } T(s) \text{ is} \qquad \frac{1}{C}.$$

The checks are obvious. For the familiar RLC circuit (Sec. 6.06) we have, with $Q_0 = 1$,

$$\begin{aligned} &\text{For model } (a){:} && T(s) = \frac{(1/C)s}{s^2 + (G/C)s + 1/(LC)}, \\ &\text{For model } (b){:} && T(s) = \frac{(1/C)(s + R/L)}{s^2 + (R/L)s + 1/(LC)}. \end{aligned} \tag{6.07-8}$$

For both models we have

$$\begin{aligned} &N = 2, && A_0 = \frac{1}{C}, \\ &\text{Physical calculation gives:} && r_\delta(0+) = \frac{1}{C}, \\ &\text{From the partial-fraction expansions of } T{:} && \sum K_j = \frac{1}{C}. \end{aligned} \tag{6.07-9}$$

Again, no comment about the checks is necessary.

For a more complicated and more interesting example, consider the network of Fig. 6.07-A*a*. Calculation of the transfer function from i_F to v_3 (a somewhat tedious calculation which we omit) gives

$$T(s) = \frac{(1/C_3)s^3}{s^4 + B_1 s^3 + B_2 s^2 + B_3 s + B_4},$$

in which

$$N = 4, \qquad A_0 = \frac{1}{C_3}, \qquad A_1 = 0, \qquad A_2 = 0, \qquad A_3 = 0,$$
$$B_1 = \frac{G_1}{C_3} + \frac{G_1}{C_2} + \frac{G_3}{C_3}. \tag{6.07-10}$$

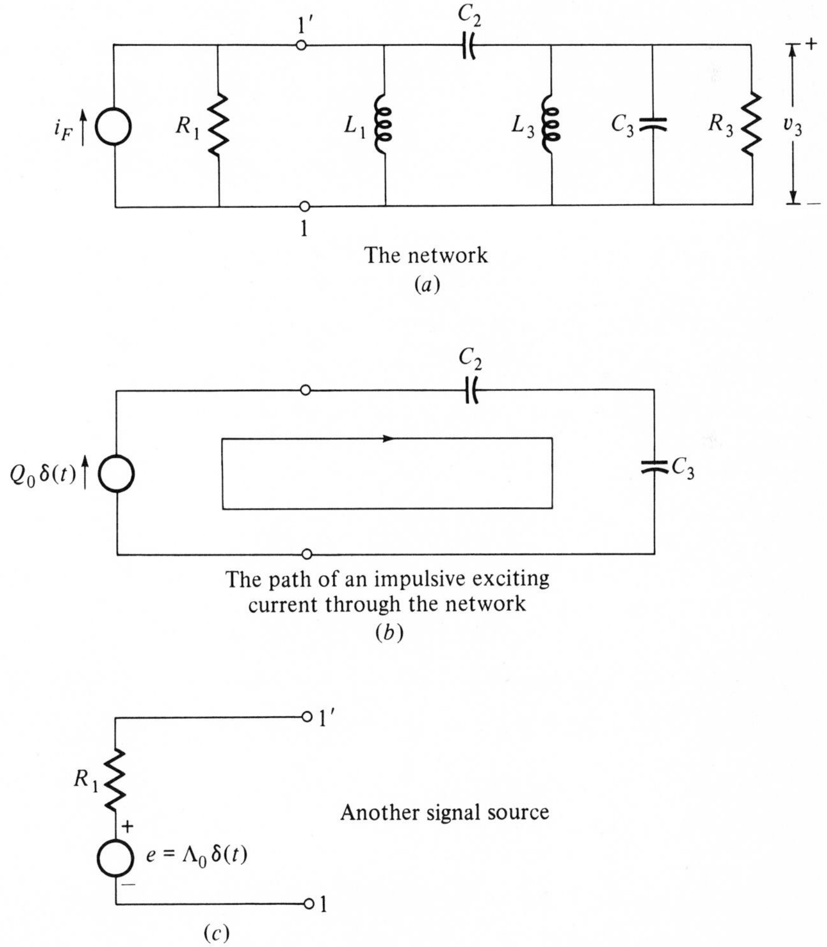

The network
(*a*)

The path of an impulsive exciting current through the network
(*b*)

Another signal source
(*c*)

FIG. 6.07-A.

(The values of B_2, B_3, and B_4 we do not need here.) Note that the units of $T(s)$ in (6.07-10) are ohms, as they should be, and that the units of B_1 are seconds^{-1} so that the second term in the denominator is dimensionally consistent with the first. Physical consideration of the network shows, since the impedances of the resistors and the inductors are substantially infinite in comparison with the opposition offered by the capacitors, that the impulsive exciting current flows essentially along the path shown in Fig. 6.07-A*b*: through capacitors alone (and so will it be generally, whenever a purely capacitive path exists).

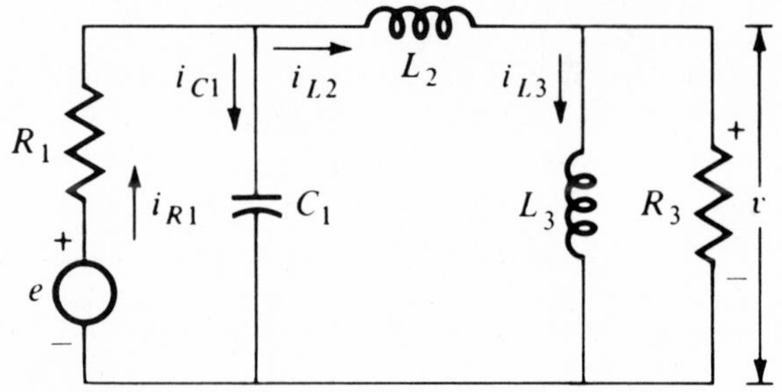

$e(t) = \Lambda_0 \delta(t)$ Initial condition: rest

$$N = 3$$

$$A_0 = 0$$

$$T(s) = \frac{A_1 s}{s^3 + B_1 s^2 + B_2 s + B_3} \qquad A_1 = \frac{R_3}{R_1 C_1 L_2}$$

(e to v)

$$B_1 = \frac{1}{R_1 C_1} + \frac{R_3}{L_2} + \frac{R_3}{L_3} \qquad A_2 = 0$$

Physical view: in $0 < t < 0+$: $\quad i_{R1} = i_{C1} = \dfrac{\Lambda_0}{R_1}\delta(t),$

$$i_{L2} = 0,$$

Leaving these conditions at $t = 0+$: $\quad v_{C1} = \Lambda_0/(R_1 C_1),$

$$i_{L2} = i_{L3} = 0,$$

$$v = 0.$$

From the transfer function: $[sT]_\infty = A_0 = 0$ so that $v(0+) = 0$.

In the partial-fraction expansion of $T(s)$, $\sum K = 0$.

FIG. 6.07-B.

If $Q_0\,\delta(t)$ flows through a capacitor, the initial situation is to all intents and purposes again as it was in the *RC* and *RLC* circuits; hence both capacitors (initially uncharged) are left with charges equal to Q_0 at $t = 0+$, and $v_3(0+) = Q_0/C_3$. From the transfer function (6.07-10) we find

$$(sT)_\infty = A_0 = \frac{1}{C_3} \tag{6.07-11}$$

so that the transfer function also gives $v_3(0+) = Q_0/C_3$. If the partial-fraction expansion of $T(s)$ is made, then $(K_1 + K_2 + K_3 + K_4)$ will equal (Q_0/C_3) also, if no mistakes are made.

If the impulsive voltage generator of Fig. 6.07-A*c* replaces the current generator, then an impulsive current flows, equal to e/R_1, for the remaining, capacitive, elements in the current path (that of Fig. 6.07-A*b* again) offer no impedance in comparison. The same checks apply.

Two additional examples are shown in Figs. 6.07-B and 6.07-C. The bare essentials of the necessary formulas are given therein; one should of course verify them carefully. Note especially the physical reasons for the absence of current flow in the inductors and in R_3 (and C_3) during the life of the impulse.

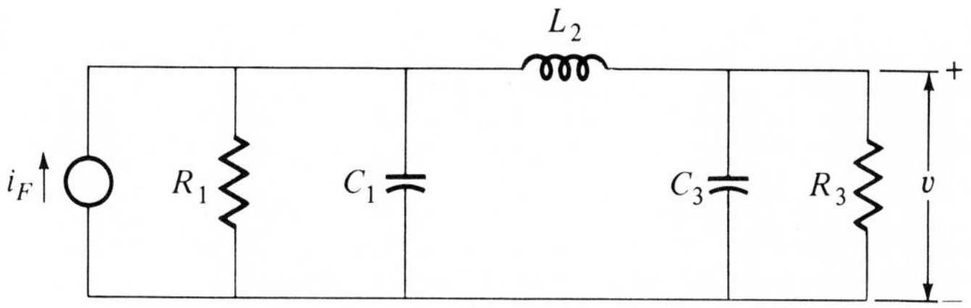

$i_F = Q_0\,\delta(t)$ Initial condition: rest

$$T(s) \;(i_F \text{ to } v) = \frac{A_2}{s^3 + B_1 s^2 + B_2 s + B_3}$$

$$B_1 = \frac{1}{R_1 C_1} + \frac{1}{R_3 C_3}$$

$N = 3$

$A_0 = 0$

$A_1 = 0$

$$A_2 = \frac{1}{C_1 L_2 C_3}$$

Physical view: in $0 < t < 0+$: i_F flows through C_1 alone, all other currents are zero; this leaves these conditions at $t = 0+$:

$$v_{C1} = Q_0/C_1, \quad i_{R1} = G_1 Q_0/C_1,$$

$$i_{L2} = i_{C3} = i_{R3} = 0, \quad v = 0.$$

From the transfer function: $[sT]_\infty = A_0 = 0$ so that $v(0+) = 0$.

In the partial-fraction expansion of $T(s)$, $\Sigma K = 0$.

FIG. 6.07-C.

6.08 INITIAL AND FINAL BEHAVIOR OF STEP RESPONSE

It has been interesting and profitable to examine what happens within a network during the life of an impulse excitation. Is it now equally informative to examine the initial response to *step* excitation?

Let us consider the step response (from rest) of the same examples. Physical consideration of the RC network (Fig. 6.06-A) shows that when $i_F(t) = I_0 u(t)$, then $v(0+) = v(0)$. There is no causative agent for any sudden change in the capacitor voltage, there is no source of the necessary impulsive current. The excitation current flows initially through the capacitor alone and we have $i_R(0+) = 0$, $i_C(0+) = I_0$, and $v_C'(0+) = +I_0/C$. The formulas (Sec. 2.09) give the same result. [The formal connection with $T(s)$ we leave to Sec. 6.09.] We note that the capacitor current *does* step, that more dramatic changes do occur at this level (one level down, one might say), as might be expected from the impulsive character of the derivative of the step.

In both models of the tuned circuit (Fig. 6.06-B) we find, from physical considerations, essentially the same behavior: there is no initial change in the capacitor voltage, the capacitor current steps immediately to I_0, and $v'(0+) = +I_0/C$. The $0+$ currents in the other elements (i_L, i_R, i_{LR}) are still zero. And the formulas (Sec. 3.08) of course corroborate these results.

In the network of Fig. 6.07-A, initially at rest, we find only slightly different behavior, readily understood with our experience. All of a step excitation current $i_F = I_0 u(t)$ flows initially through the purely capacitive path ($C_2 - C_3$), where there is no opposition; initially the other four elements carry zero current and all six element voltages are zero. If the excitation is a voltage step, $e = E_0 u(t)$, then the initial current is again entirely in the two capacitors, of value E_0/R_1 (amperes). Note that the *derivatives* of the capacitor voltages step suddenly from zero to I_0/C_2, I_0/C_3, $E_0/(R_1C_2)$, $E_0/(R_1C_3)$, respectively.

The network of Fig. 6.07-B presents no difficulty: $e(t) = E_0 u(t)$ produces these initial conditions:

$$i_{C1}(0+) = \frac{E_0}{R_1},$$

$$i_{L2}, i_{L3}, v = 0 \qquad \text{(both at } t = 0 \text{ and } t = 0+\text{)},$$

$$v_{C1}'(0+) = \frac{E_0}{R_1 C_1}.$$

Similar behavior is found in Fig. 6.07-C: if $i_F = I_0 u(t)$, then

$$i_{C1}(0+) = I_0, \qquad v_{C1}'(0+) = \frac{I_0}{C_1},$$

All other element currents remain zero at $t = 0+$, and of course $v(0+) = 0$.

Somewhat more interesting is the *ultimate* behavior of the step response, as $t \to \infty$. Here at the other end of the range of the independent variable t the analysis is extremely simple. The final response to impulsive excitation must of course be zero, if all the natural frequencies lie in the left half plane, as they do in these examples, for $\lim_{t \to \infty} (Ke^{pt}) = 0$ if $\mathbf{RE}(p) < 0$. All currents and voltages decay with the passage of time as the energy initially introduced is gradually consumed in the resistors. (Special idealized networks without resistance are sometimes useful; we may discuss them elsewhere.)

But the final response to *step* excitation will not be zero if there is a forced component in the response. Formally this depends on the value of $T(0)$ because (Sec. 3.10):

$$\text{Exponential excitation } E_0 e^{st} \to r_F = E_0 T(s) e^{st},$$

step excitation is the same with $s = 0$; and

$$\therefore r_{uF} = T(0),$$

so that

$$r_u(\infty) = T(0).$$

The physical view is equally simple: after the transients have died away (we again assume all natural frequencies to lie in the left half plane), the inductors are in effect short circuits, the capacitors are open circuits ($s = 0$), the resistors alone remain "active" and determine what happens. In a very useful jargon we have "dc" conditions and accompanying easy calculations. The results must agree with $T(0)$, and so provide one more check. Table 6.08-A gives both for

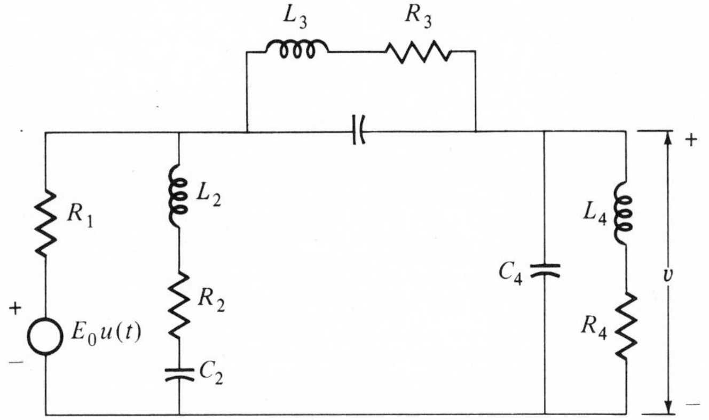

FIG. 6.08-A.

the same set of examples and one new one (Fig. 6.08-A). It is particularly informative to note the different paths taken by (*a*) impulse excitation current in $0 < t < 0+$, and (*b*) step excitation current as $t \to \infty$ (under dc conditions).

The *final* value of a response does not play as sophisticated a rôle as does the *initial* value, but it can be extremely important in engineering. *Impulse* excitation can cause complicated behavior within a network in $0 < t < 0+$; it usually produces no final response. *Step* excitation usually produces no particularly startling initial behavior, but can lead to nonzero, useful final response. So can sinusoidal excitation—in fact any persistent, periodic excitation, for example, a combination of sinusoids of commensurate periods, can ultimately produce an interesting steady state, one in which the response repeats indefinitely, the transient components being gone.

TABLE 6.08-A

Network	**Physical View of Condition As $t \to \infty$**		**Formal View**	
Initially at rest, excitation $= I_0 u(t)$ or $E_0 u(t)$	Current flows in these elements *only*	$v(\infty)$	$T(0)$	$v(\infty)$
RC (Fig. 6.06-A)	R	RI_0	R	RI_0
RLC model (*a*) (Fig. 6.06-B*a*)	L	0	0	0
RLC model (*b*) (Fig. 6.06-B*b*)	L, R	RI_0	R	RI_0
Fig. 6.07-A	L_1	0	0	0
Fig. 6.07-B	R_1, L_2, L_3	0	0	0
Fig. 6.07-C	R_1, R_2	$\dfrac{I_0}{G_1 + G_3}$	$\dfrac{1}{G_1 + G_3}$	$\dfrac{I_0}{G_1 + G_3}$
Fig. 6.08-A	R_1, R_3, L_3 L_4, R_4	$\dfrac{R_4 E_0}{R_4 + R_3 + R_1}$	$\dfrac{R_4}{R_4 + R_3 + R_1}$	$\dfrac{R_4 E_0}{R_4 + R_3 + R_1}$

6.09 INITIAL VALUES OF DERIVATIVES OF IMPULSE RESPONSE

To consider the initial value of step response, as well as of impulse response, and to consider the final values of both (Sec. 6.08), was logical—and one could go on to the examination of the initial and ultimate values of the responses to many other excitations. But we have lost sight of our objectives. We became better acquainted with the nature and value of transfer functions in Chap. 5. In Sec. 6.02 we found their calculation for ladder networks surprisingly easy in principle; the actual calculation may be lengthy, but it is eminently do-able. This is one reason that we use ladder networks for our transfer function examples; another is that the physical behavior of ladder networks is readily seen and comprehended. Our general results, however, do not depend on that; they apply to any transfer function of any network.

When the network is of any real complexity, the transfer function becomes large and full of numbers. It contains $2N$ parameters, even in the special form we have been assuming ($T \to 0$ as $s \to \infty$); there is risk of error, the calculations need checking. We have found one check in the examination of dimensions. This is helpful for small, literal parameter analyses, but it is of no help for large, necessarily numerical parameter problems. We need more. We have found another in the application of physical insight and common sense to obtain the $t = 0+$ conditions when a network is impulse excited. For we can also obtain the value of $r_\delta(0+)$ from $T(s)$; it is $(sT)_\infty$. Figure 6.09-A recapitulates the derivation of that formula in an interesting and (we shall find) suggestive, useful way. There the tilde ($\sim$) symbolizes the close connection that exists between the two functions on its left and right: one in the s domain, $T(s)$, and the other in

$$T(s) = \frac{A_0 s^{N-1} + A_1 s^{N-2} + \cdots}{s^N + B_1 s^{N-1} + \cdots} = \frac{K_1}{s - p_1} + \frac{K_2}{s - y_2} + \cdots$$

$$\sim r_\delta(t) = K_1 e^{p_1 t} + K_2 e^{p_2 t} + \cdots \quad 0+ < t$$

$$sT = \frac{A_0 s^N + A_1 s^{N-1} + \cdots}{s^N + B_1 s^{N-1} + \cdots} = \frac{K_1 s}{s - p_1} + \frac{K_2 s}{s - p_2} + \cdots$$

$$= A_0 + \frac{(A_1 - A_0 B_1)s^{N-1} + (A_2 - A_0 B_2)s^{N-2} + \cdots}{s^N + B_1 s^{N-1} + \cdots}$$

$$= [K_1 + K_2 + \cdots] + \frac{K_1 p_1}{s - p_1} + \frac{K_2 p_2}{s - p_2} + \cdots$$

RESULT I: $\boxed{[sT]_\infty = A_0 = \sum_1^N K_j = r_\delta(0+)}$

FIG. 6.09-A.

the t domain, $r_\delta(t)$. The multiplication of T by s produces a new function, sT, which is not a proper fraction. We have performed the division and written sT as the sum of A_0 (the value of sT at infinity) and a (new) proper fraction. The partial-fraction expansion of the latter is intimately related to the partial-fraction expansion of $T(s)$ itself: the poles are the same, and the coefficients K become simply Kp; there is little new. One can see this either from the general partial-fraction expansion formula [see Secs. 6.07 and 7.12] or from the simple partial-fraction expansion of the typical term:

$$\frac{Ks}{s-p} = K + \frac{Kp}{s-p}. \tag{6.09-1}$$

The result that is so useful in checking is repeated in the box at the bottom of Fig. 6.09-A.

Since there are $2N$ parameters in $T(s)$ we need $2N$ independent checks in principle: one is not enough. That would of course amount to a complete, separate, independent calculation of $T(s)$, and represent a ridiculous extreme. But using only *one* check is also an extreme, and a ridiculous one also—it's not enough! A simple, homely example will make the point: Consider the simple *RLC* circuit [model (*b*), Fig. 6.06-B] with impulse excitation, taking Q_0, L, C to have unit values and $R = 0.1$, on some normalized scale. The transfer function is of order 2 so that *four* checks are needed in principle to verify the four parameters A_0, A_1, B_1, and B_2 therein. We have specifically

$$T(s) = \frac{s+0.1}{s^2+0.1s+1}, \qquad p_1 = -0.05 + j1.00. \tag{6.09-2}$$

Mental calculation suffices to find p_1: the real part is $-0.1/2 = -0.05$, and the imaginary part is $\sqrt{1^2 - 0.05^2} = 1(1 - 0.0025/2 + \cdots) = 1.00$ to two decimal places. The corresponding K is

$$K_1 = \frac{+0.05 + j1.00}{0.00 + j2.00} = 0.50\angle 3.0^\circ \tag{6.09-3}$$

in which, in addition to (6.06-13) we have again used truncated series for quick mental calculation (see Fig. 6.09-B). A certain vocabulary of simple series expansions (to two or three terms) will expedite calculations like these, common in engineering, where three or two significant figures suffice. Note also the use of "60" for $57.3 = (\pi/2)/180$ in converting radians to degrees. The resulting value, so quickly obtained (3.0°), should actually be 2.86°; but the mental calculation

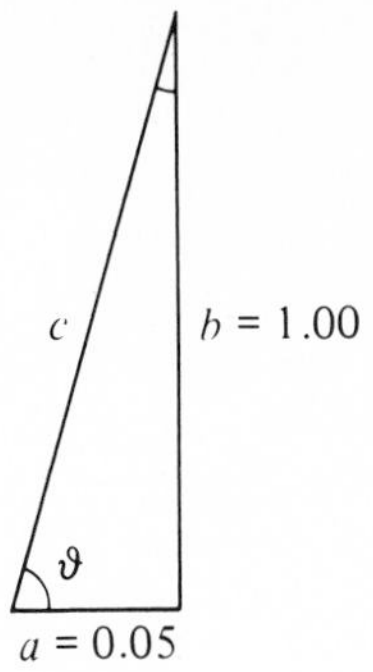

$$c = \sqrt{a^2 + b^2} = b\sqrt{1 + a^2/b^2} = b(1 + \tfrac{1}{2}(a^2/b^2) + \cdots)$$
$$= 1.00\,(1 + 0.5 \times 0.0025/1.00 + \cdots) = 1.00$$
$$\vartheta = \tan^{-1}(1.00 - 0.05) = 90^\circ - \tan^{-1}(0.05/1.00)$$
$$= 90^\circ - 60\,(x - \tfrac{1}{3}x^3 + \cdots)_{x = 0.05}$$
$$= 90^\circ - 3.0^\circ$$
$$\vartheta - 90^\circ = -3.0^\circ$$

FIG. 6.09-B.

gives almost two significant figures and its speed offsets the slight loss of accuracy. To continue the example, we calculate

$$\begin{aligned} r_\delta(t) &= 2\,\mathbf{RE}(K_1 e^{p_1 t}) \\ &= 1.00 e^{-0.05t} \cos\,[1.00t \text{ (radians)} + 3.0^\circ]. \end{aligned} \tag{6.09-4}$$

We now apply our first check:

From physical observation (Fig. 6.06-B*b*):	$r_\delta(0+) = \dfrac{Q_0}{C} = 1.00,$	
From $T(s)$, by Fig. 6.09-A:	$r_\delta(0+) = A_0 = 1.00,$	(6.09-5)
From (6.09-4):	$r_\delta(0+) \quad = 1.00,$	

[since $\cos 3^\circ = 1 - (\frac{3}{60})^2/2! + \cdots = 1.00$]. Here all is well, the three numbers agree. To show that this one check is not enough, we look at $r'(0+)$. From physical observation (Fig. 6.06-B*b*), $r'(0+) = 0$. But from (6.09-4) we find

$$\begin{aligned} r'_\delta(0+) &= 1.00e^{-0.05t}[-1.00 \sin\,(1.00t + 3.0^\circ) - 0.05 \cos\,(1.00t + 3.0^\circ)]_{t=0} \\ &= 1.00[-1.00 \times 0.05 - 0.05 \times 1.00] = -0.10 \end{aligned} \tag{6.09-6}$$

(since $\sin 3° = \frac{3}{60} - (\frac{3}{60})^3/3! + \cdots = 0.05$). An error has certainly been made, since the two values do not agree. The sharp-eyed reader noticed an error in the sign of the 3° angle in (6.09-3). This is a common sort of error, easy to make, and illustrative of why one has to make checks—and enough checks! In (6.09-3) we should of course have written $K_1 = 0.5\angle -3.0°$. Note that this does not affect the calculation of $r_\delta(0+)$ in (6.09-5) at all, since $\cos(-3°) = \cos(+3°)$, corroboration of the need for a second check. The last line in (6.09-6), K_1 being corrected, becomes $1.00[+0.05 - 0.05] = 0.00$, and all is now well in the second check.

In principle we need still more (two more checks, since $N = 2$ and $2N = 4$), but experience shows that two checks are usually enough—they will detect most errors, exactly as in this simple example. So we shall be content with formalizing our second check: the calculation of the value of $r'_\delta(0+)$ in three ways, paralleling (6.09-5). Physical considerations give $r'_\delta(0+)$ without great difficulty, we have found (see the examples in Sec. 6.08). Evaluation from $r_\delta(t)$ presents no difficulty. It remains to develop the calculation of $r'_\delta(0+)$ from $T(s)$.

But this calculation is already performed in Fig. 6.09-A; there is no need to look further. If we move the constant A_0 to the left side of the expression for sT developed in Fig. 6.09-A, we obtain the new function $T_1(s) = (sT - A_0)$ in Fig. 6.09-C, whose partial-fraction expansion has coefficients Kp. Its "mate" in the time domain, we can therefore say immediately, is $\sum (Kp)e^{pt}$. But this is $r'_\delta(t)$! And this suggests the ersatz network situation of Fig. 6.09-D. The hypothetical network of transfer function $T_1 = (sT - A_0)$ has an impulse response equal to the derivative of the impulse response of our network. It should be no surprise that T_1 is essentially the product of T and s, because to differentiate e^{st} we merely multiply by s.

To T_1 we now apply the same idea as before: multiply by s and take the limit for larger and larger s. This gives the $t = 0+$ value of $r'_\delta(t)$ and result II

$$T_1(s) = [sT - A_0] = \frac{A_{01}s^{N-1} + A_{11}s^{N-2} + \cdots}{s^N + B_1 s^{N-1} + \cdots}$$

$$= \frac{K_1 p_1}{s - p_1} + \frac{K_2 p_2}{s - p_2} + \cdots$$

$$\sim r'_\delta(t) = K_1 p_1 e^{p_1 t} + K_2 p_2 e^{p_2 t} + \cdots \qquad 0+ < t$$

$$sT_1 = A_{01} + \frac{(A_{11} - A_{01})s^{N-1} + \cdots}{s^N + B_1 s^{N-1} + \cdots}$$

$$= [K_1 p_1 + K_2 p_2 + \cdots] + \frac{K_1 {p_1}^2}{s - p_1} + \frac{K_2 {p_2}^2}{s - p_2} + \cdots$$

$$\text{RESULT II:} \quad \boxed{[sT_1]_\infty = A_{01} = A_1 - A_0 B_1 = \sum_1^N K_j p_j = r'_\delta(0+)}$$

FIG. 6.09-C.

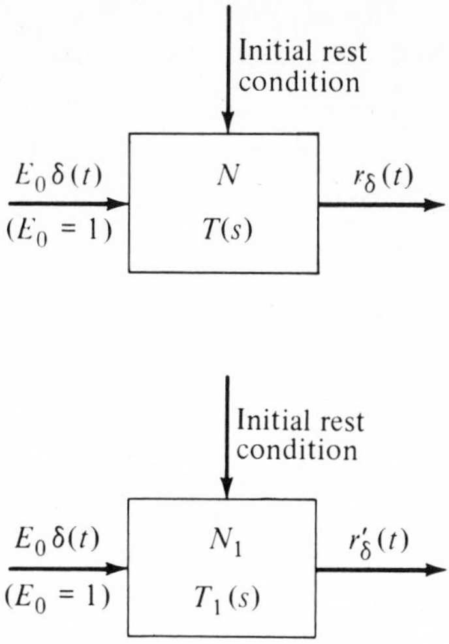

$$T_1 = [sT - r_\delta(0+)]$$
$$= (sT - A_0)$$

N_1 = hypothetical network with above transfer function

FIG. 6.09-D.

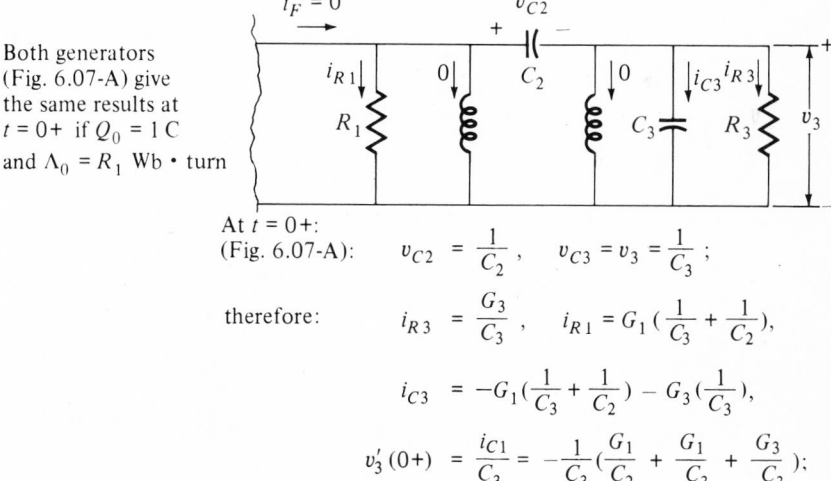

Both generators (Fig. 6.07-A) give the same results at $t = 0+$ if $Q_0 = 1$ C and $\Lambda_0 = R_1$ Wb • turn

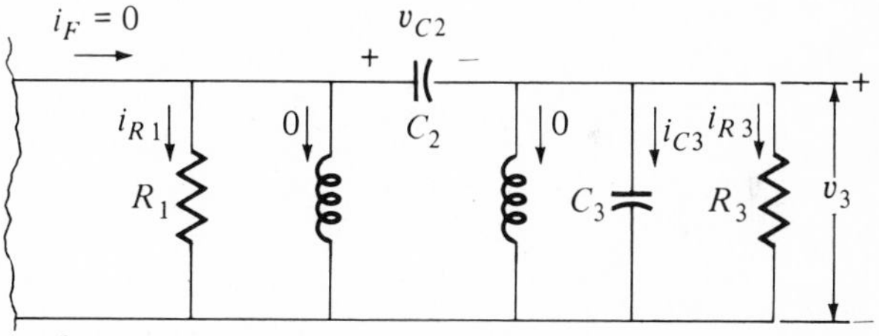

At $t = 0+$:
(Fig. 6.07-A): $\quad v_{C2} = \dfrac{1}{C_2}, \quad v_{C3} = v_3 = \dfrac{1}{C_3}$;

therefore: $\quad i_{R3} = \dfrac{G_3}{C_3}, \quad i_{R1} = G_1\left(\dfrac{1}{C_3} + \dfrac{1}{C_2}\right),$

$$i_{C3} = -G_1\left(\frac{1}{C_3} + \frac{1}{C_2}\right) - G_3\left(\frac{1}{C_3}\right),$$

$$v_3'(0+) = \frac{i_{C1}}{C_3} = -\frac{1}{C_3}\left(\frac{G_1}{C_2} + \frac{G_1}{C_3} + \frac{G_3}{C_3}\right);$$

from the transfer function (6.07-10),

$$v_3'(0+) = A_1 - A_0B_1 = 0 - \frac{1}{C_3}\left(\frac{G_1}{C_3} + \frac{G_1}{C_2} + \frac{G_3}{C_3}\right).$$

FIG. 6.09-E.

in Fig. 6.09-C, exactly what we need. The calculation of the value of $r'_\delta(0+)$ from $T(s)$ is almost as simple as the calculation of $r_\delta(0+)$:

$$r'_\delta(0+) = (sT_1)_\infty = [s(sT - A_0)]_\infty = A_1 - A_0 B_1. \qquad (6.09\text{-}7)$$

Applied to the example above, it gives $r'_\delta(0+) = 0.1 - 1 \times 0.1 = 0.0$; the check is now satisfactory.

Application of this second check to our basic RC and RLC networks is simple but illuminating. For the RC circuit (Fig. 6.06-A) we have $r'_\delta(0+) = -1/(RC^2)$ when $Q_0 = 1$ by physical reasoning, (6.06-2); from the transfer function $T = (1/C)s/[s + (1/(RC)]$ we obtain $r'_\delta(0+) = 0 - (1/C)[1/(RC)] = -1/(RC)^2$, a good check. For the RLC circuit (Fig. 6.06-B), we have, for model (a): physical reasoning (6.06-4) gives $r'_\delta(0+) = -Q_0/(RC^2)$, and the transfer function,

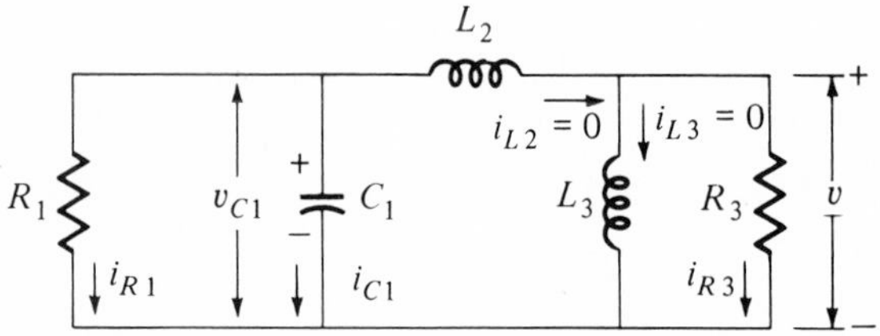

Physically from Fig. 6.07-B, at $t = 0+$: $\quad v_{C1} = \dfrac{\Lambda_0}{R_1 C_1},$

$$v = R_3 i_{R3} = L_3 i'_{L3} = 0,$$

$$i_{R1} = G_1 \frac{\Lambda_0}{R_1 C_1} = C_1 v'_{C1};$$

$$v_{C1} = L_3 i'_{L3} + L_2 i'_{L2} = 0 + L_2 i'_{L2},$$

$$i'_{L2} = \frac{1}{L_2}\frac{\Lambda_0}{R_1 C_1};$$

$$i_{L2} = i_{L3} + i_{R3};$$

$$i'_{L2} = i'_{L3} + i'_{R3},$$

$$\frac{1}{L_2}\frac{\Lambda_0}{R_1 C_1} = 0 + G_3 v',$$

$$v'(0+) = \frac{\Lambda_0}{R_1 C_1 L_2 G_3}$$

(units check: V·s/s·s = V/s)

From the transfer function (Fig. 6.07-B): $\quad v'(0+) = (A_1 - A_0 B_1)\Lambda_0 = \dfrac{R_3 \Lambda_0}{R_1 C_1 L_2} - 0.$

FIG. 6.09-F.

$T = (1/C)s/[s^2 + (G/C)s + 1/(LC)]$ gives $r'_\delta(0+) = 0 - (1/C)(G/C) = -1/(RC^2)$. For model (*b*), the physical result (6.06-5) is $r'_\delta(0+) = 0$; from the transfer function $T = (1/C)(s + R/L)/[s^2 + (R/L)s + 1/(LC)]$ comes $r'_\delta(0+) = R/(CL) - (1/C)(R/L) = 0$.

For further, more elaborate illustrations, let us first return to the network of Fig. 6.07-A. The voltages at $t = 0+$ obtained by physical consideration in Sec. 6.07 are shown in Fig. 6.09-E (p. 361). Now the resistor currents must be those shown, by Ohm's law; there can be no currents in the inductors (for there has been no impulsive voltage), so that the current in C_3 must be that shown, from which follows the value of $v'(0+)$. From the transfer function comes the same value; when the partial-fraction expansion of $T(s)$ is made, it can be checked by comparing $\sum Kp$ obtained therefrom with this value.

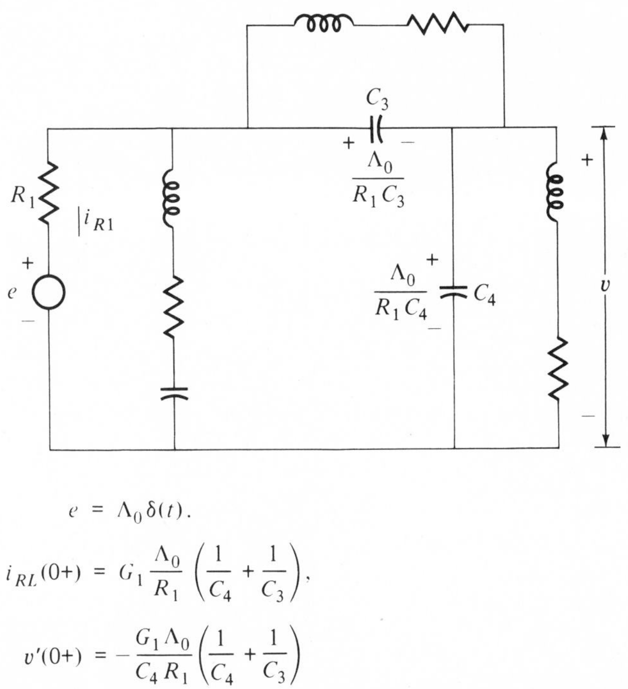

$$e = \Lambda_0 \delta(t).$$

$$i_{RL}(0+) = G_1 \frac{\Lambda_0}{R_1}\left(\frac{1}{C_4} + \frac{1}{C_3}\right),$$

$$v'(0+) = -\frac{G_1 \Lambda_0}{C_4 R_1}\left(\frac{1}{C_4} + \frac{1}{C_3}\right)$$

$$T\Big|_{e \text{ to } v} = \frac{A_0 s^5 + A_1 s^4 + \cdots + A_5}{s^6 + B_1 s^5 + \cdots + B_6}.$$

$$A_0 = \frac{1}{R_1 C_4}, \quad A_1 = \frac{1}{R_1 C_4}\left(\frac{R_4}{L_4} + \frac{R_3}{L_4} + \frac{R_2}{L_2}\right).$$

$$B_1 = \frac{1}{R_1 C_4}\left(1 + \frac{C_4}{C_3} + R_1 R_2 \frac{C_4}{L_2} + R_1 R_4 \frac{C_4}{L_4} + R_1 R_3 \frac{C_4}{L_3}\right)$$

FIG. 6.09-G.

The network of Fig. 6.07-B is shown again in Fig. 6.09-F (p. 362), with the $t = 0+$ conditions already found. To obtain, physically, the initial value of v', we calculate i_{R1}, relate v_{C1} to the inductor voltages, and so find i'_{L2} and then v' as shown. The calculation of $v'(0+)$ from the transfer function checks nicely.

In the network of Fig. 6.07-C we have already found by physical arguments that the current in C_3 is zero at $t = 0+$. Hence $v'(0+) = (1/C_3)i'_{C3}(0+)$ is also zero. From the transfer function we obtain $v'(0+) = A_1 - A_0 B_1 = 0$, a good check.

The network of Fig. 6.08-A is reproduced in Fig. 6.09-G (p. 363). *Impulse* excitation causes impulsive current to flow through R_1, C_3, and C_4 and leaves the voltages shown, at $t = 0+$. It is then easy to calculate the current i_{R_1}, which flows through C_4 and C_3 (the inductors can carry no current, initially), and the initial rate of change of v. The calculation of $T(s)$ is laborious, but with perseverance one obtains the results given, whose dimensions check, and agreement with the "physical" results.

These examples suffice to show how this second check (Fig. 6.09-C) complements the first one (Fig. 6.09-A). Theoretically not enough, still the *two* checks will in practice detect most errors.

6.10 INITIAL-VALUE CHECKS

In retrospect, we see that in formulating our two initial-value ($t = 0+$) checks we have in fact obtained something more. Figure 6.10-A shows the relations between the quantities that we have been discussing, and indicates the uses to which they can be put. If first we obtain the values of $r_\delta(0+)$ and $r'_\delta(0+)$ by physical insight (from examination of the network schematic diagram), we can then ask whether

a $A_0 = r_\delta(0+)$? And $A_1 - A_0 B_1 = r'_\delta(0+)$?
[If *yes*, we have an excellent check on the calculation of $T(s)$];

b $\sum K = A_0$? And $\sum Kp = A_1 - A_0 B_1$?
[If *yes*, we have an excellent check on the calculation of the partial-expansion of $T(s)$, an unexpected benefit];

c $\sum K = r_\delta(0+)$? And $\sum Kp = r'_\delta(0+)$?

If (c) is answered *yes*, we have an excellent check on the overall analysis, whether made in the t domain by writing and solving differential equations (which we presumably could do if we wanted to), or in the s domain by calculating and using the transfer function (which we can do, at least for ladder networks).

Our prolonged discussion of initial values [even with our restriction that the network be such that $T(\infty) = 0$ and all natural frequencies be distinct] is justified just for the three practical checking procedures it has given us. But it

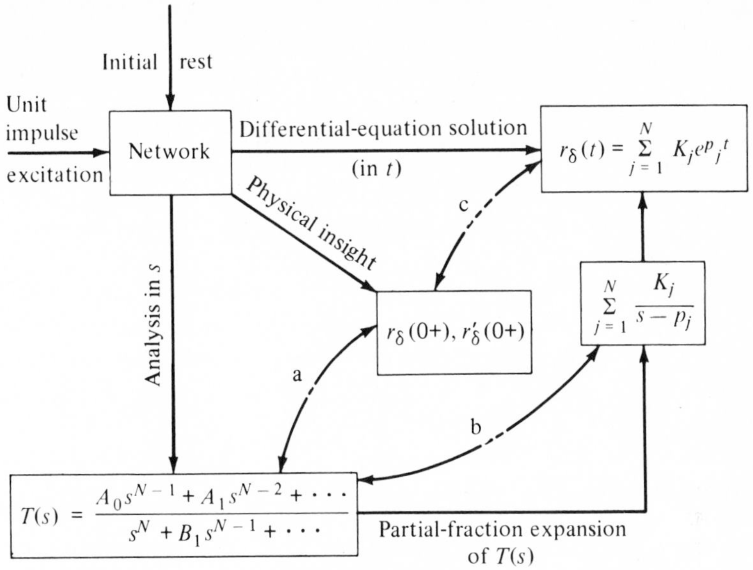

FIG. 6.10-A.

has shown us much more: it has sharpened our insight about physical behavior in the charging interval $(0 < t < 0+)$ and it has revealed some previously unknown aspects of transfer functions. We shall develop greater and greater respect for the transfer function's abilities as we go on.

6.11 THE INITIAL-VALUE TABLEAU

It is tempting to explore the possibility of similarly calculating the initial values of the second and higher derivatives of the impulse response, of $r''(0+)$ and so on. The spirit of discovery urges us on, with a certain excitement. But there is some danger in pushing too far down side roads, in leaving the main highway of analysis. Yet there is also the chance of discovering something new and useful, as we have already seen. We compromise here by merely indicating the results, without development or proof. That can be worked out when and if it's appropriate.

"It turns out that" one *can* go on indefinitely, finding the relations of $r_\delta^{(n)}(0+)$ to the coefficients of $T(s)$, for $n = 2, 3, 4, 5, \ldots$. But they become more complicated. The "physical" calculations from the schematic become more and more the mathematical manipulation of equations, rather clumsy and lengthy; in effect they constitute a somewhat different method of analysis. But the formal tie between the initial values of the derivatives, and the coefficients of $T(s)$, is worth stating, for this can be done in a rather neat form. The construction of

Fig. 6.11-A. in a spirit of imaginative extrapolation-exploration is this: we write our result I (Fig. 6.09-A) as the first line of the "first form" of the tableau; we write result II (Fig. 6.09-C) as the second line thereof, choosing to write r instead of A_0. Vertical extrapolation suggests the third line, and then the fourth, and so on. That these are correct must, of course, be shown. But this is how research often proceeds: idea first, verification second. Note how the initial $(0+)$ values of all the derivatives can easily be calculated, recursively, from these equations. (Their validity will be established in Sec. 6.12.)

The second form of the tableau, in which all the initial values and B's have been moved to the left sides, suggests a differential equation, an extremely important idea to which we shall return. The third form expresses each initial value in terms of the coefficients of the transfer function alone. It is more clumsy, but here too a rule of formation suggests itself, if not so stridently. (If our postulate that all the natural frequencies be distinct is removed, these relations remain valid, if properly interpreted.)

Another path that opens itself up is to extend the ersatz network concept of Fig. 6.09-D. Continuing it downward, to $r_\delta''(0+)$, ..., gives the lower part of Fig. 6.11-B. It can also be "continued" upward, as shown there. Network N_{-1} and its transfer function T_{-1} can be verified by going downward from it, or

$$T(s) = \frac{A_0 s^{N-1} + A_1 s^{N-2} + \cdots}{s^N + B_1 s^{N-1} + \cdots} \sim r_\delta(t)$$

First form:

$$\left\{\begin{aligned} r &= A_0, \\ r' &= A_1 - B_1 r, \\ r'' &= A_2 - B_1 r' - B_2 r, \\ r''' &= A_3 - B_1 r'' - B_2 r' - B_3 r, \\ &\vdots \end{aligned}\right.$$

Second form:

$$\left.\begin{aligned} r &= A_0, \\ r' + B_1 r &= A_1, \\ r'' + B_1 r' + B_2 r &= A_2, \\ r''' + B_1 r'' + B_2 r' + B_3 r &= A_3, \\ &\vdots \end{aligned}\right\}$$

Third form:

$$\left.\begin{aligned} r &= A_0, & &= \sum K, \\ r' &= A_1 - A_0 B_1 & &= \sum Kp, \\ r'' &= A_2 - A_1 B_1 + A_0(B_1{}^2 - B_2) & &= \sum Kp^2, \\ r''' &= A_3 - A_2 B_1 + A_1(B_1{}^2 - B_2) - A_0(B_1{}^3 - B_1 B_2 - B_2 B_1 + B_3) & &= \sum Kp^3, \\ &\vdots \end{aligned}\right\}$$

NOTE: in these equations $(0+)$, and subscripts, are omitted for brevity and clarity; that is, r'' means $r_\delta''(0+)$, and $\sum Kp^2$ means $\sum_1^N K_j p_j{}^2$, and so on. Also: A_N, A_{N+1}, ..., and B_{N+1}, B_{N+2}, $\cdots = 0$.

FIG. 6.11-A. The initial-value tableau.

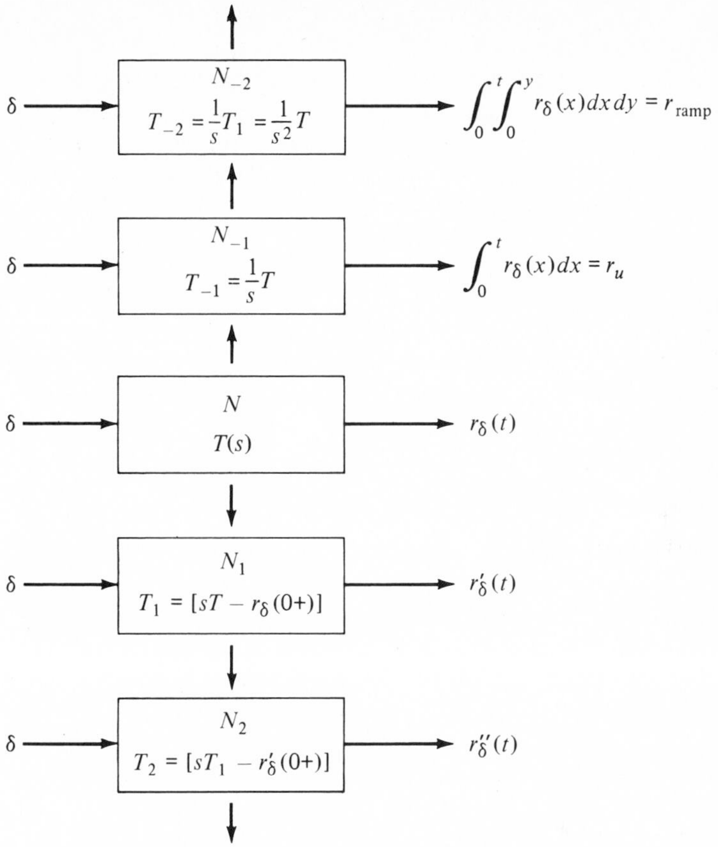

FIG. 6.11-B.

constructed by applying the relation between step and impulse responses (4.06-9) and using convolution to determine T_{-1}. Then network N_{-2} and its T_{-2} can be similarly obtained from N_{-1}.

The excitations in Fig. 6.11-B are all unit impulses; the initial condition is rest. In those cases where the action in $0 < t < 0+$ is problematic, the responses are at least valid for $t > 0+$. The "r_{ramp}" referred to at N_{-2} is the response to the integral of the unit step, to the unit ramp function (1.09-3).

6.12 THE DIFFERENTIAL EQUATION OF A TWO-PORT NETWORK

A fascinating by-product of the manipulations in Fig. 6.11-A is the suggestion of a series of differential equations, related to the coefficients in $T(s)$. To be sure, the equations of the second form there are not really differential equations, but merely algebraic relations between initial values and coefficients. Yet they *must* imply somehow that the transfer function from e to r (Fig. 6.12-A) is not really different from the differential equation that implicitly expresses $r(t)$ as a function of $e(t)$ and the initial condition of the network. That differential equation must be of order N, because N initial voltage (or current) conditions are independently determined by the initial energy stores in the N energy-storing elements assumed in the network. It must be *linear*, by our postulate of linear elements. Its coefficients must be *constant*, by hypothesis. It therefore is

$$r^{(N)}(t) + C_1 r^{(N-1)}(t) + \cdots + C_{N-1} r'(t) + C_N r(t) = \text{some function of } e(t) \qquad (6.12\text{-}1)$$

in which the coefficients C are constants. [The initial condition of the network must also be given in detail, of course, if (6.12-1) is to be solved.] This equation is not in the initial-value tableau. But there is a close suggestion of it there, the line in the second form that reads

$$r^{(N)} + B_1 r^{(N-1)} + \cdots + B_{N-1} r' + B_N r = 0, \qquad (6.12\text{-}2)$$

even though these r are but constants (initial values), and the excitation is absent.

For example, in the network of Fig. 6.12-B, the response voltage v must be determined by the differential equation

$$v''' + C_1 v'' + C_2 v' + C_3 v = \text{some function of } i_F \qquad (6.12\text{-}3)$$

together with initial-condition information. One is tempted, by (6.12-2), to identify the C's of (6.12-3) with the B's of the denominator of the transfer function from i_F to v. And that is, in fact, correct—but it needs both justification and interpretation.

The basic definition of the transfer function, (3.10-15), is

$$T(s) = \frac{r}{e}\bigg|_{\cdots} = \frac{\text{forced part of response}}{\text{excitation, of exponential form}} = \frac{R(s)e^{st}}{E_0 e^{st}} = \frac{R(s)}{E_0} \qquad (6.12\text{-}4)$$

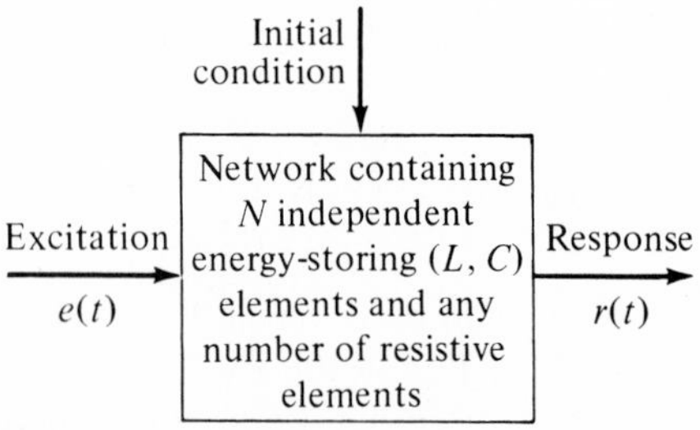

FIG. 6.12-A.

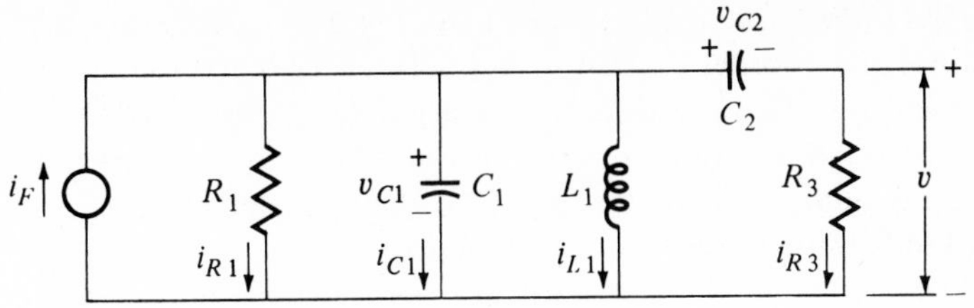

$$T(s) = \frac{A_0 s^2 + A_1 s + A_2}{s^3 + B_1 s^2 + B_2 s + B_3}.$$

$$A_0 = \frac{1}{C_1} \quad B_1 = \frac{G_3}{C_1} + \frac{G_3}{C_2} + \frac{G_1}{C_1}$$

$$A_1 = 0, \quad B_2 = \frac{G_1 G_3}{C_1 C_2} + \frac{1}{L_1 C_1},$$

$$A_2 = 0, \quad B_3 = \frac{G_3}{L_1 C_1 C_2}.$$

FIG. 6-12-B.

in which we have written $r_F(t) = R(s)e^{st}$. Now, for the networks we are considering,

$$T(s) = \frac{A_0 s^{N-1} + A_1 s^{N-2} + \cdots}{s^N + B_1 s^{N-1} + \cdots} = \frac{r}{e}\bigg|_{\ldots}. \tag{6.12-5}$$

We can therefore write

$$(s^N + B_1 s^{N-1} + \cdots)r = (A_0 s^{N-1} + A_1 s^{N-2} + \cdots)e \tag{6.12-6}$$

in which

$$r = r_F = Re^{st}, \qquad \text{and} \qquad e = \text{exponential} = E_0 e^{st}. \tag{6.12-7}$$

The basic property of the exponential is that its derivative is s times itself (Sec. 2.04), and its second derivative is s^2 times itself, etc. Applying this "backwards," with (6.12-7), to (6.12-6) gives the differential equation

$$r_F^{(N)} + B_1 r_F^{(N-1)} + \cdots = A_0 e^{(N-1)} + A_1 e^{(N-2)} + \cdots \tag{6.12-8}$$

Note that in (6.12-8) and in (6.12-9) below, e refers to the excitation function and *not* to the base of natural logarithms. From this we are led to write

$$r^{(N)} + B_1 r^{(N-1)} + \cdots + B_{N-1} r' + B_N r$$
$$= A_0 e^{(N-1)} + A_1 e^{(N-2)} + \cdots + A_{N-2} e' + A_{N-1} e \tag{6.12-9}$$

as the (single) differential equation, which, together with a statement of the initial condition, determines the response r in terms of the excitation e. The idea is crystal clear, and it *is* valid, though the argument above needs bolstering. If one were to write all the individual differential equations for the innards of the network, and eliminate all the internal voltages and currents, leaving only the response r, the result would presumably be (6.12-9).

Let us be content with verifying it for the case of impulse excitation with $N = 3$. The differential equation (6.12-9) is then the fourth line of Table 6.12-A (marked with arrows). We let $e = E_0\, \delta(t)$ and then in the equation find ourselves face to face with e' and e'', "functions" which are singular indeed. The impulse δ is bad enough, with its infinity; its derivatives raise troublesome questions indeed. If the equation holds, these singularities must be balanced by similar singularities in r, r', r'', r''' on the other side. We might struggle here with the consequent behavior of the response in $0 < t < 0+$—but an easier way is to try to remove the singularities and alter the equation into a less troublesome form. Since integrating an impulse removes the infinity and gives a more reasonable function, the step $u(t)$, let us try to integrate the differential equation. For brevity and clarity we use the notation shown for the integral from 0 to t; since we start from rest, the lower limits of the integration produce nothing; only the parts from the upper limits, functions of t, remain. In this new equation, one line up (i.e., the third line) in Table 6.12-A, we still have the singular e'. So we integrate again. Now we have only e; one more integration gives the top line, with no singular functions at all except the step, $\int e$. This is a purely integral equation, in place of the original differential equation, but equivalent thereto, which one may proceed to solve. Its right side, since $e(t) = E_0\, \delta(t)$, is easily written out. With $E_0 = 1$ for simplicity, it is

$$\left(A_0 + A_1 t + \frac{A_2 t^2}{2}\right) u(t). \tag{6.12-10}$$

Its left side must have the same form, to balance. But here we find $r(t)$ and its integrals, much more complicated [for $r(t)$ is of course a sum of exponentials]. A little thought shows that in the interval $0 < t < 0+$, r must step; it can have no singular functions at all except the step $\int e$. This is a purely integral equation, 6.12-C, exactly as in the simple RC circuit, our mentor (Sec. 2.05).

But the step alone accounts only for the initial behavior; for $t > 0+$ one needs more, for example,

$$r(t) = (R_0 + R_1 t + R_2 t^2 + \cdots)\, u(t). \tag{6.12-11}$$

The series must be infinite so it can balance the mere three terms on the right side; it is, of course, the MacLaurin series for $r(t)$. Since we already have an excellent method of determining $r(t)$ from the transfer function, we do not pursue this any further.

TABLE 6.12-A

$$
\begin{array}{lllllllll}
 & & & & \vdots & & & & \\
r & +B_1\int r & +B_2\iint r & +B_3\iiint r & = & A_0\int e & +A_1\iint e & +A_2\iiint e & \\
 & & & & \uparrow & & & & \\
r' & +B_1 r & +B_2\int r & +B_3\iint r & = & A_0 e & +A_1\int e & +A_2\iint e & \\
 & & & & \uparrow & & & & \\
r'' & +B_1 r' & +B_2 r & +B_3\int r & = & A_0 e' & +A_1 e & +A_2\int e & \\
 & & & & \uparrow & & & & \\
\rightarrow\ r''' & +B_1 r'' & +B_2 r' & +B_3 r & = & A_0 e'' & +A_1 e' & +A_2 e & \leftarrow \\
 & & & & \downarrow & & & & \\
r^{(iv)} & +B_1 r''' & +B_2 r'' & +B_3 r' & = & A_0 e''' & +A_1 e'' & +A_2 e' & \\
 & & & & \vdots & & & &
\end{array}
$$

Notation: $\int r \equiv \int_0^t r(x)\,dx, \qquad \iint r \equiv \int_0^t \int_0^y r(x)\,dx\,dy, \qquad$ etc.

At $t = 0$ (initial condition):		In $0 < t < 0+$ (excitation interval)	At $t = 0+$ (initial values):	
$r = 0,$	$e = 0,$	e is	$e = 0,$	$r = ? = A_0 E_0,$
$r' = 0,$	$e' = 0,$	impulsive,	$\int e = E_0,$	$\int r = 0,$
$r'' = 0,$	$e'' = 0,$	r steps	$\iint e = 0,$	$\iint r = 0,$
$\vdots$	$\vdots$		$\vdots$	$\vdots$

Hence the initial-value table, obtained from these, with $E_0 = 1$, is

$$
\begin{aligned}
r \qquad\qquad\quad &= A_0, \\
r' + B_1 r \qquad\quad &= A_1, \\
r'' + B_1 r' + B_2 r &= A_2, \\
&\ \vdots
\end{aligned}
$$

It is interesting to set $t = 0+$ in the first line of Table 6.12-A, and then in the second line, and then the third. The result is the small initial-value table at the bottom of Table 6.12-A. It is the same, of course, as the (second form in) Fig. 6.11-A. We observe that extension to the general case, of arbitrary N, is obvious. We have therefore established the validity of the initial-value tableau, Fig. 6.11-A.

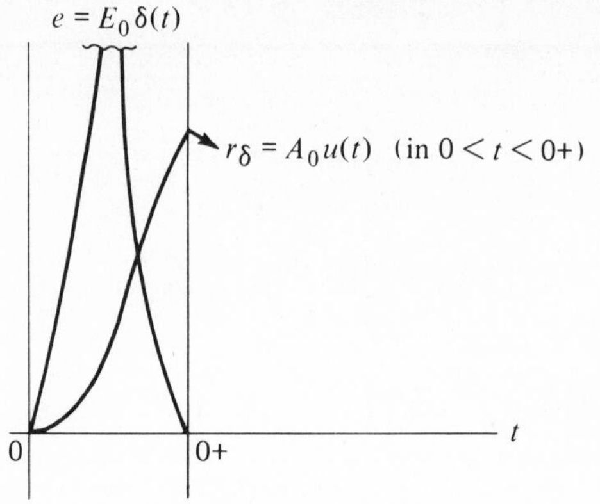

FIG. 6.12-C. Schematic behavior of excitation and reponse in $0<t<0+$.

We add to our knowledge the facts that

From the transfer function one can obtain the differential equation that (implicitly) expresses the response in terms of the excitation, valid for $0 < t$, not just for $0+ < t$, (6.12-12)

and that

From this differential equation one can obtain the initial ($t = 0+$) values of the impulse response and its derivatives, the initial-value tableau. (6.12-13)

The really important discovery of this section is (6.12-12) that

The transfer function *is* the differential equation, (6.12-14)

that there is no real difference between them. Equally important is the reason therefor: that we are dealing with a very special sort of problem, that our assumptions restrict us to an extraordinarily simple category of differential equation. Physically we postulate that the network elements all be *lu*mped, *li*near, *con*stant; the mathematical equivalent is that the differential equations be ordinary, linear, have only constant coefficients. From the physical formulation we coin the not uneuphonious acronym *lulicon* as a concise expression thereof, a handy tag and reminder.

Since differential equation and transfer function are "the same," if the calculation of $T(s)$ is easy (as for ladder networks), so also should the calculation of the overall differential equation be relatively easy. See Prob. 6-A for a simple example.

The "lumped" postulate is not really important; it is the linearity and the constancy of element values with time that matter. For they make the exponential function truly *eigen* (characteristic, innate) in that the network's natural behavior (the complementary function) is simply a sum of exponentials when the differential equation is (lu)licon:

$$r^{(n)} + B_1 r^{(N-1)} + \cdots + B_{N-1} r' + B_N = 0 \tag{6.12-15}$$

is solved by

$$r = Ke^{pt} \tag{6.12-16}$$

in which K is completely arbitrary, provided p is a root of the characteristic equation

$$p^N + B_1 p^{N-1} + \cdots + B_{N-1} p + B_N = 0. \tag{6.12-17}$$

Since there are N such roots, all equally eigen, the differential equation (because linear and therefore amenable to superposition) is also solved by

$$r = \sum_1^N K_j e^{p_j t} \tag{6.12-18}$$

in which the K's are still completely arbitrary. If certain initial conditions are to be met, then of course the K's are no longer arbitrary but must be determined to meet those conditions, as we well know. If excitation is present, superposition permits (and convolution in particular enables) us to calculate the response (from rest) to an arbitrary excitation in terms of (6.12-18), the K's being evaluated to make it the impulse response $r_\delta(t)$. (If the initial condition is not rest, the initial stores of energy produce additional components in the response, separately calculable by similar methods.)

It requires only simple counterexamples to show that the entire transfer function concept is destroyed when an element becomes nonlinear, or varies with time.

We choose to work with the transfer function, rather than with the differential equations, fundamental though they are, because $T(s)$ is simpler, being merely algebraic. Its use reduces the analysis to its essentials: the calculation of $T(s)$ from the network, factoring its denominator, the characteristic polynomial, and evaluating the coefficients K.

We have discovered here (independently) that from $T(s)$ one should be able (by going to and solving the differential equation) to calculate the response to any excitation. Convolution, we already knew, makes possible the calculation of r (for any e) from T in a different way, requiring integration but not differential equation solution as such. A natural question here is: "Can we perhaps calculate r (for any e) by using $T(s)$ and algebra alone?" Verification of a suspicion that we *can* often do so, and details of the (important) method will be discussed in Chap. 7.

Now let us look again at the example of Fig. 6.12-B. Calculation of the transfer function, a routine matter, gives the results shown. The unit of A_0 is farads^{-1} = (mho-seconds)$^{-1}$ which is correct, since T is an impedance. We note also that all the B's check dimensionally. The differential equation [see (6.12-3)] is specifically

$$v''' + B_1 v'' + B_2 v' + B_3 v = i_F''. \tag{6.12-19}$$

For impulse excitation $Q_0\,\delta(t)$ from rest we find these initial values from $T(s)$:

$$v(0+) = Q_0 A_0 = \frac{Q_0}{C_1},$$

and (6.12-20)

$$v'(0+) = Q_0(A_1 - A_0 B_1) = -\frac{Q_0}{C_1}\left(\frac{G_3}{C_1} + \frac{G_3}{C_2} + \frac{G_1}{C_1}\right).$$

Physical observation tells us that the charging current flows entirely through C_1, leaving $v_{C1}(0+) = Q_0/C_1$, $v_{C2}(0+) = 0$, and hence $v(0+) = Q_0/C_1$. The initial currents are then:

$$\begin{aligned} i_{R1} &= \frac{G_1 Q_0}{C_1}, & i_{R3} &= \frac{G_3 Q_0}{C_1} = i_{C2} = C_2 v'_{C2}, \\ i_{L1} &= 0, & i_{C1} &= -i_{R1} - i_{L1} - i_{R3} = C_1 v'_{C1}. \end{aligned} \tag{6.12-21}$$

Furthermore, $v + v_{C2} = v_{C1}$, so that

$$v'(0+) = v'_{C1} - v'_{C2} = -\frac{Q_0}{C_1}\left(\frac{G_1}{C_1} + \frac{G_3}{C_1} + \frac{G_3}{C_2}\right). \tag{6.12-22}$$

The check is good.

To show the initial-value table, we take, for simplicity, all (normalized) element values to be unity and $Q_0 = 1$. Then the initial values are

$$\begin{aligned} r &= A_0 && = 1, \\ r' &= A_1 - B_1 r = 0 - 3(1) && = -3, \\ r'' &= A_2 - B_1 r' - B_2 r = 0 - 3(-3) - 2(1) && = +7, \\ r''' &= A_3 - B_1 r'' - B_2 r' - B_3 r = 0 - 3(7) - 2(-3) - 1(1) && = -16, \\ r^{iv} &= 0 - 3(-16) - 2(7) - 1(-3) + 0 = 48 - 14 + 3 && = 37, \\ r^{v} &= -3(37) - 2(-16) - 1(7) = -111 + 32 - 7 && = -86, \end{aligned} \tag{6.12-23}$$

....

We have here, incidentally, directly and without additional effort, the coefficients of the Maclaurin series for $r_\delta(t)$. That series is, from (6.12-23),

$$r_\delta(t) = 1 - 3t + \frac{7t^2}{2} - \frac{16t^3}{3!} + \frac{37t^4}{4!} - \frac{86t^5}{5!} + \cdots \qquad 0+ < t. \qquad (6.12\text{-}24)$$

This form of solution, technically equally valid with the sum-of-exponentials form, is practically useful only for very small values of t. But it is interesting, particularly for its close relation to the transfer function. And that connection will bear closer scrutiny.

6.13 THE POWER SERIES FORM OF $r_\delta(t)$

The initial-value table is in effect also the Maclaurin series form of the impulse response. This series must therefore be very closely related to the transfer function. Since the transfer function *is* the differential equation, and the only *general* method of solving differential equations is to obtain the solution in power series form, this is not at all surprising. Let us look more closely into the relation.

The Maclaurin series (valid for $t > 0+$) is:

$$r_\delta(t) = r_\delta(0+) + r'_\delta(0+)t + \frac{r''_\delta(0+)t^2}{2!} + \cdots \qquad (6.13\text{-}1)$$

The coefficients, the entries in the initial-value table, come from evaluating sT "at infinity." Surely this operation will be much more lucid if we can first somehow *expand* $T(s)$ "about infinity." And this is not at all dangerous or difficult: we need merely look for a development in powers of $(1/s)$ rather than s. Then evaluating $(sT)_\infty$ will be very little work indeed.

Very simply and straightforwardly, we need only let $p = 1/s$ and think in terms of p. (An alternative, very helpful view is to look on this as "turning the plane inside out" or mapping it on a sphere.) We first divide numerator and denominator of $T(s)$ by s^N and then make the substitution:

$$\begin{aligned} T(s) &= \frac{A_0 s^{N-1} + A_1 s^{N-2} + \cdots + A_{N-1}}{s^N + B_1 s^{N-1} + \cdots + B_N} \\ &= \frac{A_0(1/s) + A_1(1/s)^2 + \cdots + A_{N-1}(1/s)^N}{1 + B_1(1/s) + \cdots + B_N(1/s)^N} \\ &= \frac{A_0 p + A_1 p^2 + \cdots + A_{N-1}p^N}{1 + B_1 p + \cdots + B_N p^N}. \end{aligned} \qquad (6.13\text{-}2)$$

The value of T when $p = 0$ is of course zero; its Maclaurin series in p is

$$T = D_0 p + D_1 p^2 + D_2 p^3 + \cdots \tag{6.13-3}$$

in which, according to the usual development, or by ordinary (long) division, which is equivalent and perhaps simpler,

$$\begin{aligned} D_0 &= A_0, \\ D_1 &= A_1 - A_0 B_1, \\ &\cdots\cdots\cdots\cdots\cdots \end{aligned} \tag{6.13-4}$$

The coefficients seem to be the initial values. To determine whether this is indeed true, we return to s. We have

$$T(s) = D_0 \frac{1}{s} + D_1 \left(\frac{1}{s}\right)^2 + D_2 \left(\frac{1}{s}\right)^3 + \cdots. \tag{6.13-5}$$

Note that the expansion about infinity has been successful and very easily made.

We now recall that [Fig. 6.09-A, (6.07-6), Fig. 6.09-C, (6.09-7)]

$$\begin{aligned} r_\delta(0+) &= (sT)_\infty \\ r'_\delta(0+) &= \{s[sT - (sT)_\infty]\}_\infty = (sT_1)_\infty \\ &\cdots\cdots\cdots\cdots\cdots\cdots\cdots\cdots \end{aligned} \tag{6.13-6}$$

The developed-about-infinity form of the transfer function (6.13-5) tells us immediately and simply that

$$sT = D_0 + D_1 \left(\frac{1}{s}\right) + D_2 \left(\frac{1}{s}\right)^2 + \cdots \tag{6.13-7}$$

so that

$$r_\delta(0+) = (sT)_\infty = D_0, \tag{6.13-8}$$

that

$$T_1 = sT - D_0 = D_1 \left(\frac{1}{s}\right) + D_2 \left(\frac{1}{s}\right)^2 + \cdots \tag{6.13-9}$$

so that

$$r'_\delta(0+) = (sT_1)_\infty = D_1, \tag{6.13-10}$$

and so on. Clearly the coefficients in the expansion of the transfer function about infinity *are* indeed the initial values of the impulse response. For an illustrative example, we return to the network of Fig. 6.12-B with unit excitation and unit element values. We have

$$\begin{aligned} T(s) &= \frac{s^2}{s^3 + 3s^2 + 2s + 1} = \frac{p}{1 + 3p + 2p^2 + p^3} \\ &= D_0 p + D_1 p^3 + \cdots. \end{aligned} \tag{6.13-11}$$

The calculation of the D's by (long) division (writing only the coefficients) is:

$$\begin{array}{l|cccccccc} 1\;\;3\;\;2\;\;1] & 1 & & & & & & & \quad [1,\ -3,\ +7,\ -16,\ +37,\ -86,\ldots \\ & 1 & 3 & 2 & 1 & & & & \\ \hline & 0 & -3 & -2 & -1 & & & & \\ & & -3 & -9 & -6 & -3 & & & \\ \hline & & 0 & +7 & +5 & +3 & & & \\ & & & +7 & 21 & 14 & 7 & & \\ \hline & & & 0 & -16 & -11 & -7 & & \\ & & & & -16 & -48 & -32 & -16 & \\ \hline & & & & 0 & 37 & 25 & 16 & \\ & & & & & 37 & 111 & 74 & 37 \\ \hline & & & & & 0 & -86 & \ldots & \ldots \end{array} \tag{6.13-12}$$

so that

$$r_\delta(t) = 1 - 3t + \frac{7t^2}{2!} - \frac{16t^3}{3!} + \frac{37t^4}{4!} - \frac{86t^5}{5!} + \cdots \tag{6.13-13}$$

which should be compared with (6.12-24).

We conclude that the following remarkable relation exists between the expansion of the transfer function about infinity and the impulse response:

$$\begin{array}{ccccccc} T(s) = D_0\left(\dfrac{1}{s}\right) & + & D_1\left(\dfrac{1}{s}\right)^2 & + & D_2\left(\dfrac{1}{s}\right)^3 & + \cdots, \\ \updownarrow & & \updownarrow & & \updownarrow & \\ r_\delta(t) = D_0 & + & D_1 t & + & \dfrac{D_2 t^2}{2!} & + \cdots; \end{array} \tag{6.13-14}$$

i.e., the coefficients of the expansion of T, the D's, *are* the initial values of $r_\delta(t)$:

$$\begin{aligned} T(s) &= \sum_{j=0}^{\infty} D_j s^{-j-1}, \\ r_\delta(t) &= \sum_{j=0}^{\infty} D_j \frac{t^j}{j!}. \end{aligned} \tag{6.13-15}$$

There is evidently a close mathematical relation between the large-s (or high-frequency) behavior of a transfer function on the one hand, and the initial behavior of the corresponding impulse response on the other. Our commonsense study of the initial charging of capacitors and of initial currents gives the physical counterpart. The practical engineer expresses it by saying that it is the "high-frequency transmission" that influences the "transmission of steep wavefronts (such as narrow pulses, or steps)."

The expansions in (6.13-15) suggest visualizing the network as a number of component networks in "parallel," somewhat as in Fig. 5.03-B. The transfer function of each network would be proportional to s^{-n} and its impulse response to $t^{n-1}u(t)$. But we really have this already in the upper half of Fig. 6.11-B. It is worth putting this in tabular form, the specific transfer function–impulse response "pairs" in the upper part of Table 6.13-A. Note the nature of the pairs

TABLE 6.13-A

$T(s)$	$r_\delta(t)$
1	(t)
$\frac{1}{s}$	$u(t)$
$\frac{1}{s^2}$	$tu(t)$
$\frac{1}{s^3}$	$\frac{t^2}{2!}u(t)$
$\frac{1}{s^4}$	$\frac{t^3}{3!}u(t)$
$\vdots$	$\vdots$
$\frac{1}{s^n}$	$\frac{t^{n-1}}{(n-1)!}u(t)$

we have developed here: the transfer functions have (multiple) poles at the origin only, a very special category. On the time-domain side we find powers of t. (Multiple poles in a transfer function have much the same effect in general, as we shall see in Chap. 7.)

This discussion of power series in t and expansion of $T(s)$ about infinity suggest, in retrospect, a different general analysis method: to develop $T(s)$ in a series of powers of $(1/s)$, from which the series for $r_\delta(t)$ follows immediately. This is useful for small t, at least. Figure 6.13-A expresses and compares the "standard" method and this one. For arbitrary excitation we can then apply convolution, though again usefully only for small values of time.

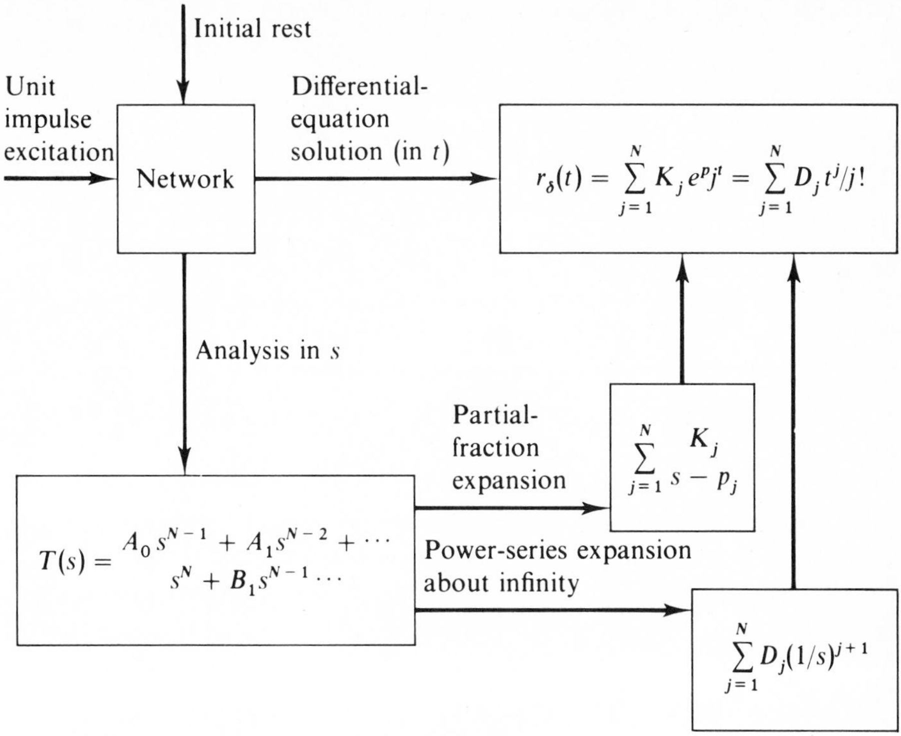

FIG. 6.13-A.

We note, in comparing the two methods, that one requires factoring a polynomial to obtain the natural frequencies, the other does not, and that calculating D's is probably less laborious than calculating K's. That is: the series method requires less work. But its utility is less because it is limited (by convergence problems in evaluating series) to small values of t. In special circumstances this may be satisfactory; then one takes the shorter route.

6.14 A PROBLEM IN SYNTHESIS

Many indeed are the fascinating side roads that open and tempt us as we go along. We began this chapter by studying transfer functions of ladder networks, which have led us to look at many related things. We cannot explore *all* the side roads, but the more important we must at least mark, hoping to return later. Here is one other interesting offshoot of the initial-value table.

On occasion we may know the impulse response but not the transfer function or network. For example, a special problem in *synthesis* (there are many, many others!) is this: the impulse response is prescribed for some system block (see Sec. 1.04) and we are asked to design a network with that $r_\delta(t)$. If the data is in exponential-sum form, we can immediately write down $T(s)$ in partial-fraction form and proceed to the realization of a network therefrom. This last is generally a difficult problem, which we set aside. More likely, the data is *not* in exponentials but is a curve, or merely a series of points. Or the data may be in the form of a nonexponential function. In either case we have to concoct an exponential sum that approximates the prescription, since networks' impulse responses are exponential—we have to be content with "almost" behavior.

One particular method, among the many that exist, is suggested by the initial-value table. In principle one should be able to determine the transfer function from the first $2N$ initial values [the first $2N$ D's in (6.13-15)], since there are $2N$ coefficients in $T(s)$, or N p's and N K's. Our initial-value checks (Sec. 6.10) are in effect steps in this direction. The implementation for synthesis is to

a Choose a tentative value of N, the order of the network (subject to various constraints, particularly to the use of sound judgement);

b List the first $2N$ initial values of the datum function, either given as such or calculated from the prescribed function;

c Determine the B's of a hypothetical transfer function therefrom; in the second form in the initial-value table, Fig. 6.11-A, the second N equations are N simultaneous linear algebraic equations to be solved therefor:

$$\left.\begin{aligned} r^{(N)} + B_1 r^{(N-1)} + \cdots + B_N r &= 0, \\ r^{(N+1)} + B_1 r^{(N)} + \cdots + B_N r' &= 0, \\ \cdots\cdots\cdots\cdots\cdots\cdots\cdots\cdots\cdots \\ r^{(2N-1)} + B_1 r^{(2N-2)} + \cdots + B_N r^{(N-1)} &= 0; \end{aligned}\right\} \qquad [6.14\text{-}1(a\text{–}g)]$$

d Determine its A's from the first N equations of the same form:

$$\begin{aligned} A_0 &= r, \\ A_1 &= B_1 r + r', \\ \cdots\cdots\cdots\cdots\cdots\cdots\cdots\cdots\cdots \\ A_{N-1} &= B_{N-1} r + B_{N-2} r' + \cdots + B_1 r^{(N-2)} + r^{(N-1)}; \end{aligned}$$

e Evaluate the result:

(1) Are the poles of T in an appropriate part of the plane? (They are, after all, the putative natural frequencies of the network, but they will not necessarily be in the left half plane.)

(2) Is the behavior of the impulse response satisfactory? [The first $2N$ terms of its Maclaurin series, having been matched to the prescription, must be, but what about terms beyond? Better, how does the behavior of $r_\delta(t)$ compare with the prescription for all t?]

f Repeat the process with different values of N as the results of (e) may indicate.

g Realize network(s) from the $T(s)$ decided upon (if one can be found).

This is a rather extensive subject, the above suggestion and outline of which are enough for now. Here is one brief example.

Suppose a network is to generate a logarithmic (instead of linear) sweep voltage for a cathode-ray tube, in response to a step trigger (excitation), as in Fig. 6.14-A. The corresponding impulse response is of course proportional to the derivative of the logarithmic voltage (Fig. 6.14-B). Our prescription is then (K_0 = some constant)

$$r_\delta(t) = \frac{K_0}{1 + t/T_0} = \frac{K_0}{T_0}\left[1 - \frac{t}{T_0} + \left(\frac{t}{T_0}\right)^2 - \left(\frac{t}{T_0}\right)^3 + \left(\frac{t}{T_0}\right)^4 - \cdots\right], \tag{6.14-2}$$

the Maclaurin series being very easily obtained by division. For simplicity, we normalize the response by measuring it in units of K_0/T_0, and measure time in units of T_0. These scales are practically important details but are not important for the present discussion; they are easily adjusted later. We proceed:

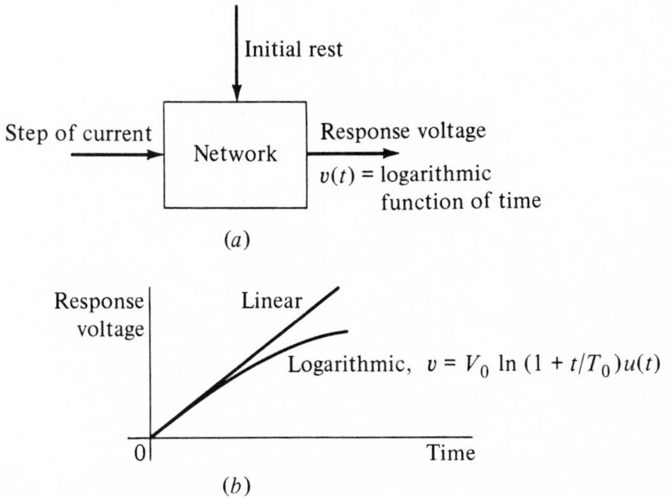

FIG. 6.14-A.

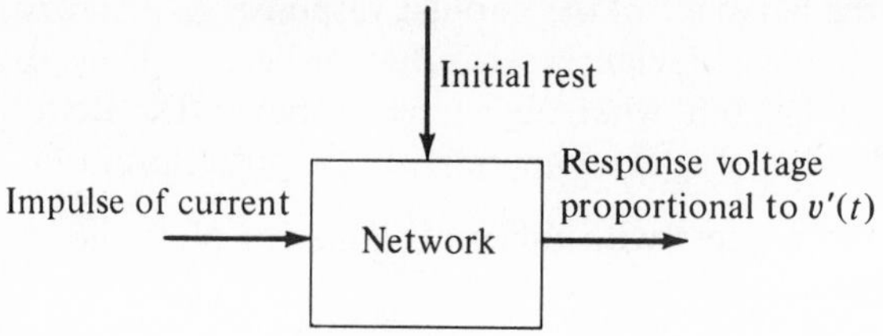

FIG. 6.14-B.

a Take $N = 2$ by way of (simple) experiment.

b The initial values are: $+1,\ -1,\ +2,\ -6,\ +24,\ -120,\ \ldots$. (We need only four, but have listed two more for comparisons later on.)

c The equations for the B's are

$$2 - B_1 + B_2 = 0,$$
$$-6 + 2B_1 - B_2 = 0,$$

whose solution is easy and gives:

$$B_1 = 4, \qquad B_2 = 2.$$

d The equations for the A's are

$$A_0 = 1,$$
$$A_1 = 4 \times 1 - 1 = 3.$$

e Thus

$$\begin{aligned} T = \frac{s+3}{s^2+4s+2} &= \frac{K_1}{s-p_1} + \frac{K_2}{s-p_2} \\ &= \frac{1}{s} - \left(\frac{1}{s}\right)^2 + 2\left(\frac{1}{s}\right)^3 - 6\left(\frac{1}{s}\right)^4 \\ &\quad + 20\left(\frac{1}{s}\right)^5 - 72\left(\frac{1}{s}\right)^6 + 120\left(\frac{1}{s}\right)^7 + \cdots, \end{aligned} \tag{6.14-3}$$

in which

$$\begin{aligned} p_1 &= -0.586, \qquad & p_2 &= -3.414, \\ K_1 &= 0.854, & K_2 &= 0.146, \end{aligned} \tag{6.14-4}$$

so that

$$r_\delta(t) = 0.85e^{-0.59t} + 0.15e^{-3.14t} \tag{6.14-5}$$

and

$$r_u(t) = \int_0^t r(x)\,dx = -1.46e^{-0.59t} - 0.04e^{-3.41t} + 1.50. \tag{6.14-6}$$

In the expansion of T about infinity, the first four terms do agree of course with the prescription; the fifth term and those beyond do not, though the discrepancies are not tremendous. Figure 6.14-C shows the impulse response obtained and the prescription, Fig. 6.14-D the step response obtained and the desired (logarithmic) response. The approximation is excellent for small t, as one would expect, becoming poorer as t increases. For larger N presumably the close agreement would extend to larger t.

This example is extremely simple: all goes well, the natural frequencies are in the left half plane, real, and simple. We can easily find a network realization by associating the partial-fraction expansion of T as an impedance with the series combination of two RC circuits (Fig. 6.14-E*a*). But synthesis is not unique either in the manner of approximating a prescription *or* in the manner of realizing a given transfer function. For example, the transfer function may be inverted to form an admittance and then expanded as in Fig. 6.14-E*b*, which yields a second realization. And there are more; synthesis is a large subject, into which we go no further here.

For larger values of N one encounters numerical calculation accuracy problems, even in this simple example. For more complicated problems the natural frequencies found may be complex, may lie on the ω axis or in the right half plane, and generally be troublesome. The sailing is not always smooth. But the procedure outlined in (6.14-1) can be useful, and we can always return and

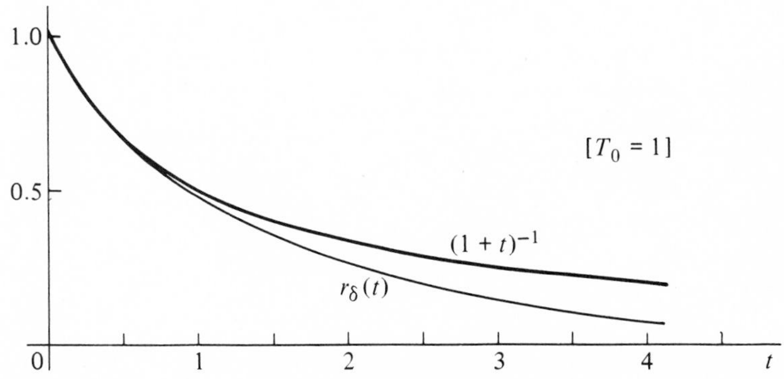

FIG. 6.14-C.

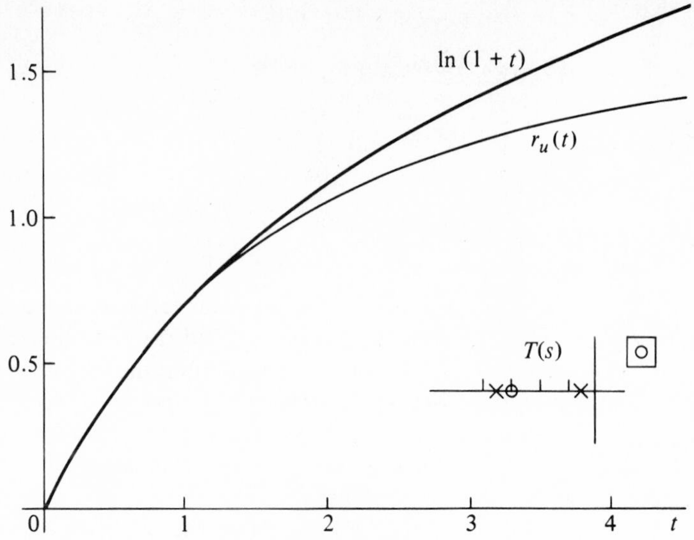

FIG. 6.14-D.

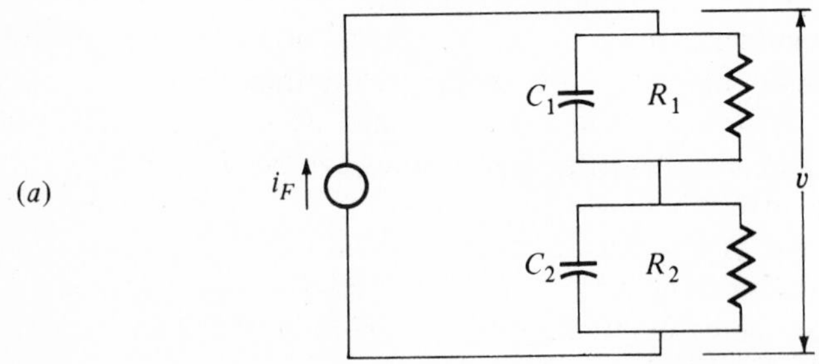

$$T = Z = \frac{0.854}{s + 0.586} + \frac{0.146}{s + 3.414} = \frac{1}{C_1 s + G_1} + \frac{1}{C_2 s + G_2}$$

$$C_1 = \frac{1}{0.854} = 1.17, \quad C_2 = 6.86, \quad R_1 = 1.46, \quad R_2 = 0.0428 \text{ (norm. units).}$$

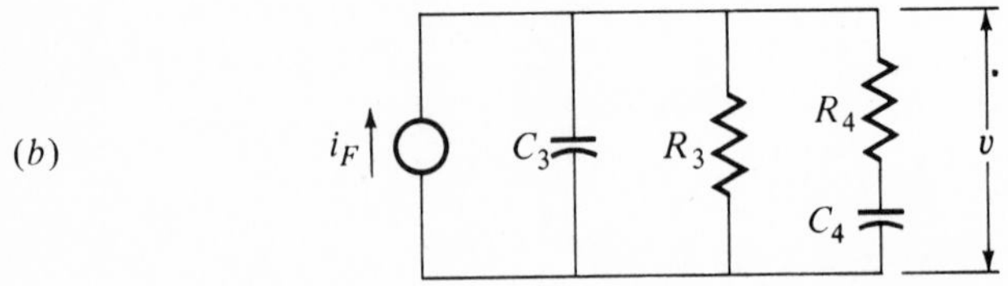

$$\frac{1}{T} = Y = s + \frac{2}{3} + \frac{s}{3s + 9} = C_3 s + G_3 + \frac{1}{R_4 + 1/(C_4 s)},$$

$$C_3 = 1, \quad C_4 = \tfrac{1}{9}, \quad R_3 = 1.5, \quad R_4 = 3 \text{ (norm. units).}$$

FIG. 6.14-E.

develop it further. (See *PR* in Appendix B for reading suggestions; see also Sec. 7.06.)

The same situation (impulse response known, network unknown) may also occur in *maintenance* work, less glamorous but equally important. Something "goes bad," we suspect a certain network in a system, we measure (for example) its impulse response, which we find is indeed not what it should be. We ask then: "What transfer function, or network, could give such a response?" The answer may indicate that an element in the network has changed value, "opened up," or "shorted out," or some such thing. The data here (the measured impulse response) are usually a series of points in time, or a curve, and *not* in the form of initial values. We postpone the discussion of this case. (See Sec. 7.06.)

6.15 ASYMPTOTIC BEHAVIOR IN s

"Many indeed are the fascinating side roads," but we must remember the main road: ladder-network analysis, transfer function properties, checks. We have found it possible (and advisable) to independently check $T(s)$. We have found initial-value checking both feasible and useful: $r_\delta(0+)$ and $r'_\delta(0+)$ can usually be easily evaluated both physically (by commonsense examination of the network) and analytically (from the coefficients of T). Sometimes (Sec. 6.08) we can profitably go also to the other end of the time scale, checking $T(0)$ against the network's physical "dc behavior." What we have really done is to examine behavior at the two ends of the time scale, approached from the interior: $t = 0+$ means $t \to 0$ from above, really, and $t \to \infty$ means an approach from below. We treat the end "pieces" separately if they are troublesome; for example the region $0 < t < 0+$ can be complicated if behavior is impulsive.

It seems eminently reasonable to attempt the same sort of thing in the domain of s: to examine and correlate the analytical behavior of $T(s)$ and the physical behavior of the network at zero and infinity in s. Is this possible? profitable?

Analytically, the two evaluations are easy and we already know how to make them. For s at or near zero (the origin) we expand $T(s)$ in a (Maclaurin) series of powers of s: the leading (first nonzero) terms give the asymptotic behavior near $s = 0$ and the behavior *at* $s = 0$. For s at or near infinity, that is, for very large values of s, we expand $T(s)$ in a series of powers of $(1/s)$ and use the leading terms in the same way.

Physically, the two evaluations are sometimes easy but sometimes require thought. Placing $s = 0$ means considering constant (step) excitation. This is the "dc," zero-frequency case in which capacitors are open circuits, inductors are short circuits, and both kinds of energy-storing elements literally remove themselves from the picture. We need only calculate what happens in the remaining, purely resistive network. Letting s go to infinity is the "high-frequency" case: the capacitors are short circuits and the inductors are open circuits, and we are again reduced to a resistive network.

It appears that we have the material for another set of checks on a given $T(s)$, based on physical examination of dc and high-frequency behavior. We should not expect totally different procedures, though, because we have found that some connection exists between high-frequency behavior in s and initial behavior in t. But there is much to learn, and we shall receive some unexpected dividends: more insight into the nature of transfer functions and additional calculation methods. Examples are the best teachers here, so we begin by examining our previous ones for $s = 0$ and $s = \infty$.

Consider the network of Fig. 6.02-A. At low frequencies the physical behavior is obvious: the generator "sees" R_3 directly; the network is transparent and we have merely the $R_3 R_1$ voltage divider:

$$T(0) = \frac{R_3}{R_3 + R_1} = \frac{1}{1 + R_1 G_3}. \tag{6.15-1}$$

The transfer function (Fig. 6.02-A) has neither zero nor pole at the origin, and the leading term of the Maclaurin series is equally obvious:

$$T(0) = \frac{1}{1 + R_1 G_3}, \tag{6.15-2}$$

and we have a check.

At high frequencies the physical behavior is still obvious: no signal comes through because all three energy-storing elements act to prevent it, either by opening the connection between source and load, or by short-circuiting it. Analytically, so also does $T(\infty) = 0$, a check. It is not a satisfying check, however, since no values of elements or coefficients are actually used in it: it is a consequence of the particular network configuration and independent of element values. But it is not difficult to obtain the first *nonzero* term, both physically and analytically, and this *does* give a good check. Note the difference between (*a*) setting s *equal to* infinity, which does not add much to our knowledge, and (*b*) considering the *approach* to infinity, the asymptotic, large-s behavior.

Analytically we have, writing only leading terms,

$$\begin{aligned} T(s) &= \frac{1}{L_1 C_2 L_3 G_3 s^3 + \cdots} = \frac{A_2}{s^3 + \cdots} \\ &= \frac{A_2}{s^3(1 + \cdots)} = \frac{A_2}{s^3} + \cdots, \end{aligned} \tag{6.15-3}$$

$$A_2 = \frac{1}{L_1 C_2 L_3 G_3},$$

obtained by formal expansion in series, by division, or (since only one term is wanted) by mere inspection. The leading term is the $(1/s)^3$ term, indicating not only that $T(\infty) = 0$ but also that the behavior near infinity is as $1/s^3$, that is, that $T(s)$ has three zeros at infinity: three times over, T goes to zero at infinity. Since there are three energy-storing elements, all acting to prevent high-frequency transmission, this is not surprising.

Let us look more closely at the physical behavior of the network for exponential excitation, $e = E_0 e^{st}$, when s is large. The forced component of i_1 (Fig. 6.02-A) is determined by the transfer function from e to i_1, which is simply the reciprocal of the impedance of the series combination of R_1, L_1, and the remaining group of three elements [see (6.03-3)]:

$$\begin{aligned} T\bigg|_{e \text{ to } i_1} &= \frac{1}{R_1 + L_1 s + 1/[C_2 s + 1/(L_3 s + R_3)]} \\ &= \frac{1}{L_1 s + \cdots} = \frac{1}{L_1 s} + \cdots . \end{aligned} \tag{6.15-4}$$

In the second line of (6.15-4) only the most important (leading) term is kept in both forms. The physical interpretation is simply that the impedance of L_1 is comparatively very large and completely determines the current, to one-term accuracy in the series form. At the v_m node (junction) this current has to divide between the two paths open to it. But the impedance of C_2 is so much smaller that (to one-term accuracy) all of i_1 goes through C_2. The leading term of the forced component of the voltage v_m is thus

$$E_0 e^{st} \frac{1}{L_1 s} \frac{1}{C_2 s}. \tag{6.15-5}$$

As for i_3, since L_3 eclipses R_3 in impedance, the forced component of i_3, to one term, is $v_m/(L_3 s)$. Finally, since $v = R_3 i_3$,

$$T = \frac{v}{e}\bigg|_{\ldots} = \frac{R_3 i_3}{e}\bigg|_{\ldots} = \frac{R_3 v_m}{L_3 s e}\bigg|_{\ldots} = \frac{R_3}{L_3 s L_1 s C_2 s} = \frac{1}{L_1 C_2 L_3 G_3 s^3}, \tag{6.15-6}$$

in which the dots remind us of the definition of T, (3.10-15), and only leading terms have been written. This result is the same as the analytical result A_2/s^3, (6.15-3), so that we have a good check.

The physical view makes it clear that the three energy-storing elements *do* act, each one separately, to block high-frequency transmission and to produce a zero in T in infinity: the three zeros of T at ∞ are due thereto. Asymptotic (high-frequency) calculations like these are often useful, are not difficult, and should become second nature to make.

Since the transfer function has three zeros at infinity, the high-frequency

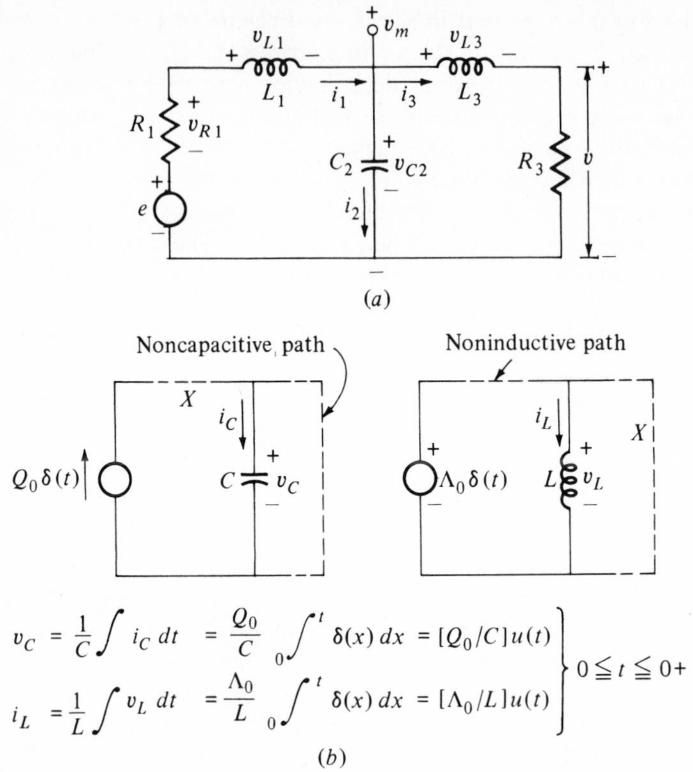

FIG. 6.15-A.

transmission is very poor. It may therefore be profitable and informative to look closely at the initial-value check for this network, because of the connection we have sensed between high-frequency transmission and initial values of impulse response. With impulse excitation $e = \Lambda_0(t)$, from rest, we find a novel situation. The impulsive voltage appears across R_1, L_1, and C_2 (Fig. 6.15-A*a*). Now $v_{C2} = v_m = 0$ initially, and surely remains so in the charging interval $0 < t < 0+$; there is no possibility of impulsive current to suddenly charge the capacitor. And $i_1 = 0$ initially, and L_1 prevents any sudden change in it, so $v_{R1} = 0$ initially. Hence we have

$$e = v_{R1} + v_{L1} + v_{C2} = 0 + v_{L1} + 0,$$

$$v_{L1} = e = \Lambda_0\,\delta(t) = L_1 \frac{di_1}{dt}, \tag{6.15-7}$$

$$i_1 = \frac{1}{L_1}\int_0^t \Lambda_0\,\delta(x)\,dx = \frac{\Lambda_0}{L_1}u(t),$$

in which δ and u are to be interpreted as in Fig. 6.12-C, and the equations are restricted to the charging interval $0 \leqq t \leqq 0+$. We find that a step occurs in the inductor current i_1, for $i_1(0+) = \Lambda_0/L_1$. This is paradoxical, for we said above that L_1 prevents any sudden change in i_1. Yet there is really no more of a contradiction here than there is in the familiar impulsive charging of an *RC* circuit; in fact this is dual thereto. It is merely unfamiliar. And less realistic, for the *C* element is more like a physical capacitor than the *L* element is like a physical inductor. But in such a model, this is what happens.

Just as an impulsive current through a capacitor causes its voltage to step, so an impulsive voltage applied to an inductor causes its current to step. Subject only to ordinary voltages, an inductor prevents any sudden change in current through it, as does a capacitor prevent any sudden change in voltage across it. But if the voltage applied to an inductor (the current applied to a capacitor) is strong enough, as is an impulse, then a step occurs (Fig. 6.15-A*b*). To be sure, it does not really happen this way in actual networks; our models are oversimplified, especially that of the inductor, and do not express what really happens in $0 < t < 0+$, which is something like the behavior in Fig. 6.12-C. Note the complete duality between the two parts of Fig. 6.15-A*b*, item by item: inductance and capacitance, voltage and current, charge and flux linkage, series connection and parallel connection (see Sec. 5.12).

In Fig. 6.15-A*b*, *X* means that the elements in the dotted path are irrelevant, for the (idealized) current source produces the same current therein regardless of their nature (the idealized voltage source produces the same voltage thereacross regardless of their nature). "Noncapacitive" ("noninductive") means one cannot traverse the dotted path without encountering at least one noncapacitive element (at least one noninductive element); i.e., none of the impulsive current flows in the dotted path (none of the impulsive voltage is used up therein) but all flows through the capacitor (all is across the inductor). For $t > 0+$, of course, we cannot tell what happens without considering these other branches.

To return now to our initial-value calculations: we have found $i_1(0+) = \Lambda_0/L_1 = 0$. Now all this current i_1 flows initially through C_2 (Figs. 6.02-A and 6.15-A*a*) and none through L_3, because L_3 *does* prevent any sudden change in i_3: the presence of C_2 across it insures that the voltage not become infinite. Hence we have

$$i_1(0+) = \frac{\Lambda_0}{L_1} = i_2(0+), \qquad i_3(0+) = 0, \qquad v(0+) = R_3\, i_3(0+) = 0, \quad (6.15\text{-}8)$$

our first initial value. We continue:

$$v_{L3} = v_m - v = L_3 \frac{di_3}{dt}, \qquad v_m(0+) = 0, \qquad (6.15\text{-}9)$$

so that

$$i_3'(0+) = \frac{v_m - v}{L_3} = 0, \qquad v'(0+) = R_3\, i_3'(0+) = 0, \qquad (6.15\text{-}10)$$

our second initial value. Then, since $v'_m = i_2/C_2$,

$$
\begin{aligned}
i_3''(0+) &= \frac{v'_m - v'}{L_3} = \frac{1}{L_3}\left[\frac{i_2(0+)}{C_2} - 0\right] = \frac{\Lambda_0}{L_3 C_2 L_1}, \\
v''(0+) &= R_3\, i_3''(0+) = \frac{R_3 \Lambda_0}{L_1 C_2 L_3},
\end{aligned}
\tag{6.15-11}
$$

our third (and first nonzero) initial value. Note that calculating only $v(0+)$ and $v'(0+)$ gives a very weak check, exactly as does calculating only at infinity in s. And note that because the first two initial values are both zero, it is easy to extend the initial-value check to include $v''(0+)$, which should equal A_2 (Fig. 6.11-A), and does. Note again how similar are the $t = 0+$ and the $s \to \infty$ calculations. The path of impulsive current in $0 < t < 0+$ and the high-frequency (s near infinity) behavior are closely related.

Vivid physical corroboration of these ideas is given by Fig. 6.15-B. There two networks are shown, the transfer function of the first one [part (a)] having three zeros at infinity, that of the second one [part (b)] having seven. Part (c) of the figure is a simultaneous (double-exposure) portrait of the behavior of $|T(j\omega_F)|$ versus ω_F, for both networks. The effect of increasing the number of zeros of $T(s)$ at infinity ought to be to make the "falloff" or "cutoff" faster, and

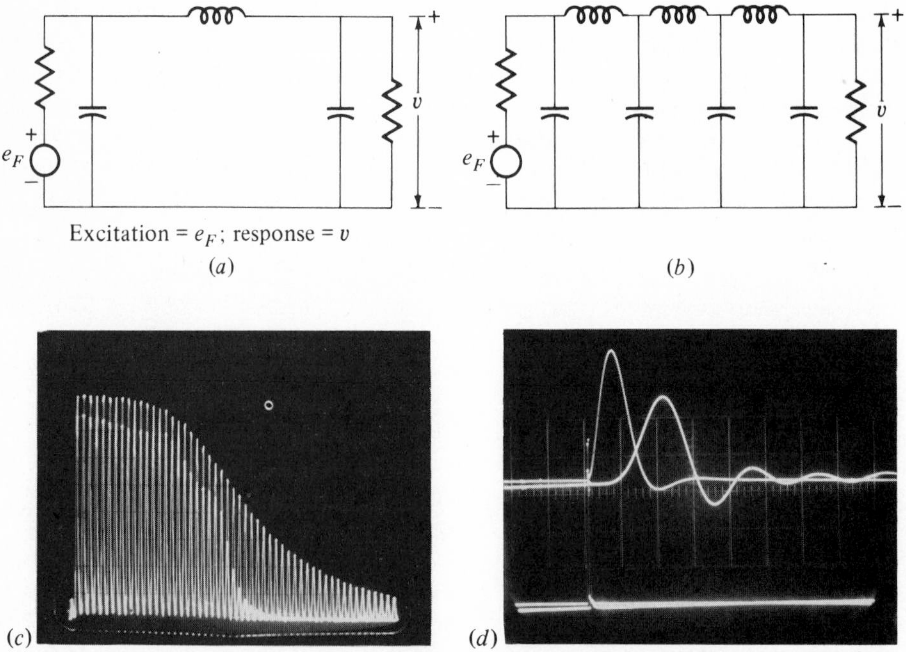

FIG. 6.15-B.

it surely is: there is no need to point out which curve is which! Part (*d*) is a simultaneous portrait of the two impulse responses, and the exciting impulse. The increased number of zero initial values (zero initial derivatives) is strikingly clear; again there is no need to identify the two. We observe that a large number of zero initial values results in a sort of "delay" in that the response is very slow to start; shape distortion also occurs. These two photographs form an object lesson in the tie between the high-frequency behavior of $T(s)$ and the initial behavior of $r_\delta(t)$.

Before we examine other examples in this asymptotic (low-frequency, high-frequency) light, there is here before us the genesis of another useful idea at which we must look.

6.16 THE POINTS OF INFINITE LOSS OF A LADDER NETWORK

Initial values of impulse response we found sometimes easy to obtain by mere physical examination of a network; they are useful in checking its transfer function, already calculated by other means. Asymptotic (high-frequency and low-frequency) transmission is also easily evaluated physically (and is not unrelated to the initial value of impulse response and the final value of step response), and is also useful in checking $T(s)$. We ask now: "Is there something more we can do in the s domain to obtain useful information by physical observation? Are there other values of s, in addition to zero and infinity, for which the transmission is easily evaluated?"

We look again at the transfer function in factored form [see (5.05-3) and (5.13-2), and the discussion of Sec. 5.12]:

$$T(s) = K_0 \frac{(s - z_1)(s - z_2)\cdots}{(s - p_1)(s - p_2)\cdots(s - p_N)}. \tag{6.16-1}$$

The *multiplier* K_0 is immediately determined from the high-frequency asymptotic behavior. The *poles* can*not* be determined by any examination of a small group of elements, for they are the natural frequencies and depend on the communal action of all parts of the network. But the *zeros* are the infinite-loss points (Sec. 5.05) and they *may* be due to the action of a few elements acting alone, resonating or cooperating to prevent transmission for certain values of s.

In the ordinary bridge (Secs. 5.10, 5.13) a relation between four "boxes" determines the balance ($T = 0$) condition. But in *ladder* networks (Sec. 6.01) the zeros of $T(s)$ depend on individual "boxes" and can be found and examined more easily. For a pure ladder network (Fig. 6.16-A) the transfer function can vanish only because a shunt impedance becomes zero, or a series admittance becomes zero: there is no other way. And these are matters of individual rung or side-box behavior, for example, the readily visualized resonances of Fig. 6.01-D. Note that this applies to ladders only: put two networks in parallel, provide

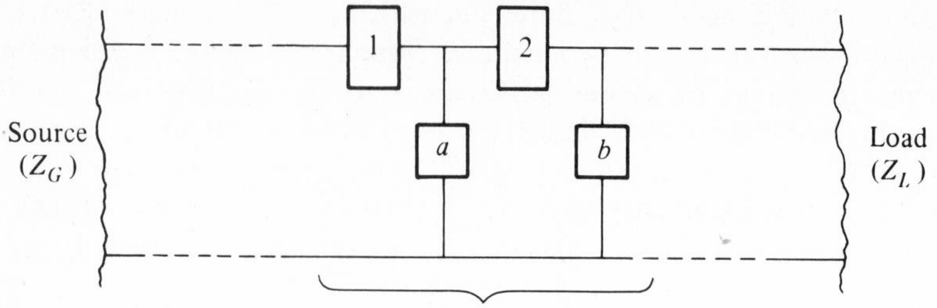

$T = 0$ if (and only if)

$Y_1 = 0$ or $Y_2 = 0$,

or

$Z_a = 0$ or $Z_b = 0$.

NOTE: The ladder network may begin with a series box or a shunt box; it may end in either way. Values of s for which Z_G and Z_L take extreme values (zero, infinity), if they exist, require special consideration, since they place the generator or the load in very special states.

FIG. 6.16-A.

more than one path (Sec. 5.08), and zeros of T can come about by subtle cancellation at values of s that are not at all obvious.

We *can* therefore get more information about *ladder* networks by physical examination in the light of s: the entire numerator of $T(s)$, in fact, can be written by inspection. Consider our last example, the network of Figs. 6.02-A and 6.15-A*a*; look at the ladder arms in turn. To begin with, R_1, the source impedance, and R_3, the load impedance, are independent of s and of no interest here. The first arm of the network proper, L_1 in series connection, resonates or becomes an open circuit only for large s: for it, $Y = 0$ when $s = \infty$. The impedance of the second arm, C_2 in shunt, is zero when $s = \infty$. The third arm, L_3 in series, has zero admittance when $s = \infty$. So we find that T vanishes three times, due to three separate causes, all for s infinite: T has three zeros at infinity.

Now the order of this network is 3, for there are three energy-storing elements in it, and the overall differential equation (Sec. 6.12) is of order 3. Its transfer function has three zeros and three poles [see (5.05-7)]: hence the three zeros we found at infinity are *all* the zeros. We can therefore write merely from inspection of the network:

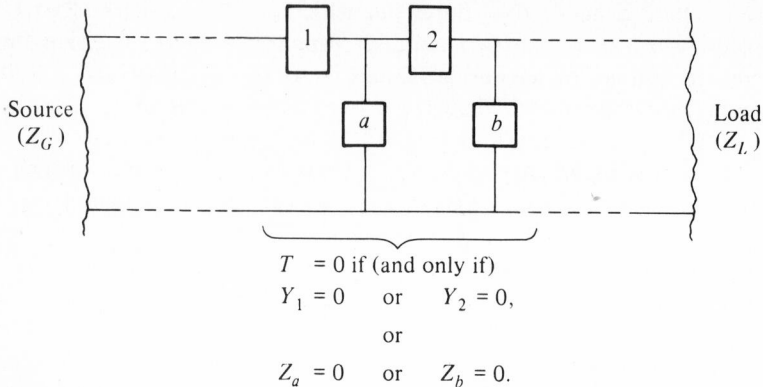

$$\begin{aligned} T(s) &= \frac{A_0 s^2 + A_1 s + A_2}{s^3 + B_1 s^2 + B_2 s + B_3} \\ &= K_0 \frac{(s - z_1)(s - z_2)(s - z_3)}{s^3 + \cdots} \qquad (6.16\text{-}2) \\ &= \frac{\text{constant}}{s^3 + \cdots} = \frac{A_2}{s^3 + B_1 s^2 + B_2 s + B_3}. \end{aligned}$$

Since all three zeros are at infinity, the product $(z_1 z_2 z_3)$, which is A_2/K_0, eclipses all other terms in the numerator, which reduces to the constant A_2. The high-frequency asymptotic behavior gives us $A_2 = 1/(L_1 C_2 L_3 G_3)$ and the low-frequency asymptotic behavior gives us $A_2/B_3 = 1/(1 + R_1 G_3)$, so that only B_1 and B_2 remain unknown after the physical inspection of the network. These we cannot evaluate without determining, somehow or other, the characteristic equation of the network, which involves all five elements: we may calculate the transfer function (Sec. 6.02) or the impedance around some loop [(6.03-3) or (6.03-4)] or the admittance between some pair of nodes [(6.03-6)].

Much of the transfer function of a ladder network can be determined by physical inspection: its zeros (its numerator) by finding the open- or short-circuit s values of each component rung or side piece, its multiplier by examining the high-frequency asymptotic behavior, and then the constant term of the denominator (the product of the negatives of all the natural frequencies) by examining the low-frequency asymptotic behavior.

We can then use these results to obtain an excellent check on the transfer function as computed in some other way. Or we can use them to write down most of the transfer function *ab initio*, and then complete the job by calculating (for example) some loop impedance.

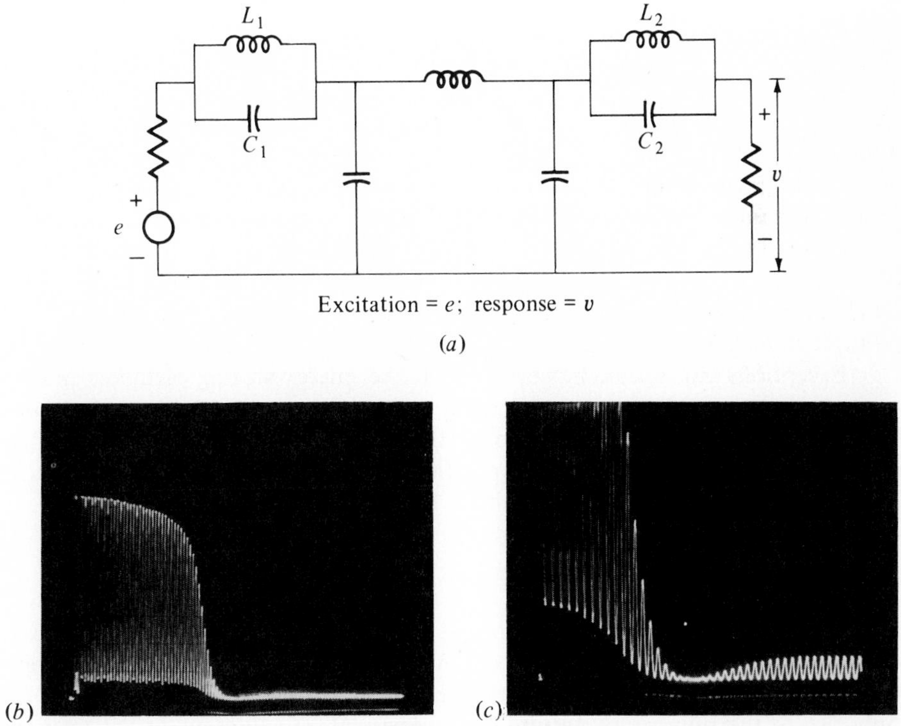

Excitation = e; response = v

(*a*)

(*b*) (*c*)

FIG. 6.16-B.

For the network of Fig. 6.16-B*a*, for example, it is clear that the form of $T(s)$ must be

$$T(s) = \frac{K_0(s^2 + \omega_1{}^2)(s^2 + \omega_2{}^2)}{s^7 + B_1 s^6 + \cdots + B_7}, \tag{6.16-3}$$

the numerator being "obvious" from the two tuned circuits, for which $L_1 C_1 \omega_1{}^2 = 1$ and $L_2 C_2 \omega_2{}^2 = 1$. (There are also three zeros at infinity, due to the other three L and C elements.) Physical corroboration is given by parts (*b*) and (*c*) of the figure, which display the network's transfer function magnitude against frequency, ω_1 and ω_2 coinciding in this case. In part (*b*) the vertical scale is ordinary, and in part (*c*) it is magnified (expanded). The latter shows the "zeros" of $T(s)$ at $j\omega_1$ more clearly. Note that $T(j\omega_1)$ is not really zero, though it *is* very small, because the Q of the tuned circuits is not infinite, and dissipation has its effect in moving the zeros slightly to the left of the ω axis (Sec. 6.19). But our arguments about the points of infinite loss are indeed verified.

Before we illustrate with more examples, the determination of the value of N requires some attention.

6.17 THE ORDER OF A NETWORK

By definition, (3.03-13), the order of a network is the number of its natural frequencies. This number, N, is also the number of poles and the number of zeros in its transfer function, the order of its overall differential equation (Sec. 6.12), the number of constants K_j in its impulse response, and the number of initial conditions which must be known to evaluate that or any other response of the network. It is consequently equal to the number of energy-storing elements in the network in most cases, and extremely easy to determine, as in the example of Sec. 6.16 [(6.16-2)].

Exceptions can occur, however, when the energy-storing elements are so connected that constraints are imposed on the initial conditions. The commonest example of this is presented by three capacitors connected in Π (or Δ) fashion, forming a closed (purely capacitive) loop embedded in a network, as in Fig. 6.17-A*a*. There Kirchhoff's voltage law requires

$$-v_{C1} + v_{C2} + v_{C3} = 0 \tag{6.17-1}$$

at all times. In particular this applies at $t = 0+$, so that only *two* of the three initial-condition voltages can be specified; the third is constrained and determined by (6.17-1). Consequently these three energy-storing elements contribute only *two* to the number of natural frequencies, N. Evaluation of the transfer function, or calculation of some loop impedance (node-pair admittance) will

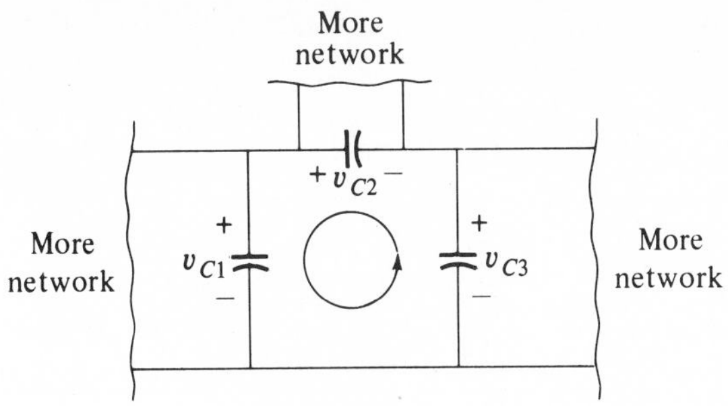

(a) A Π of capacitors

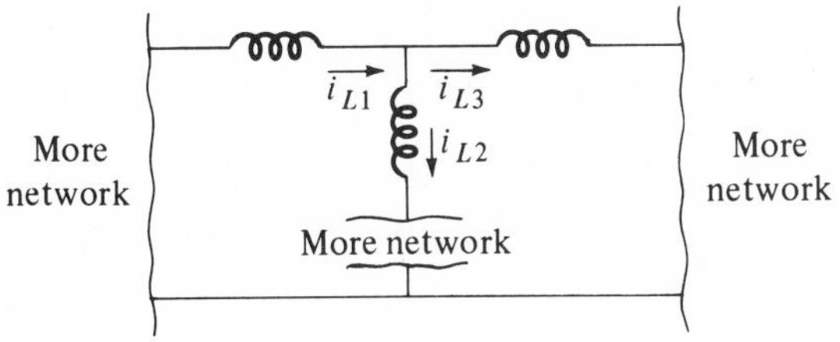

(b) A T of inductors

FIG. 6.17-A.

automatically take care of this; if we determine N by inspection we must count *two* and not *three* for the three capacitors.

Dually, the T(or Y) connection of three inductors (at a node where no other element is connected), as in Fig. 6.17-Ab, imposes the constraint

$$-i_{L1} + i_{L2} + i_{L3} = 0 \tag{6.17-2}$$

according to Kirchhoff's current law. This means that only two of the three initial currents can be specified. Again three energy-storing elements contribute only *two* to the number of natural frequencies, N. Evaluation of the transfer function or loop impedance (node admittance) automatically does this; if we evaluate N by inspection we must count *two* and not *three* for the three inductors. (This T connection of three inductors is useful in the study of transformers and is not uncommon.)

More generally, complicated combinations of such constraints can reduce the value of N appreciably below the number of energy-storing elements present. But these two cases are the only common ones.

We conclude that

> The order of a network is equal to or less than the number of energy- storing elements in the network. (6.17-3)

Commonly the two are equal, but one must be sure that no constraints, such as those of Fig. 6.17-A, exist in the network, before writing N as the number of energy-storing elements.

The common example of this anomaly is illustrated physically in Fig. 6.17-B. Without the capacitor C (or with $C = 0$), all three zeros of the transfer function (of order 3) lie at infinity. The magnitude of $T(j\omega_F)$ falls off with frequency as it should (the slowly falling "curve" of Fig. 6.17-B*b*). With C added, two zeros move from infinity to $\pm j\omega_0$; this is apparent in the second (superposed by double exposure) portrait of Fig. 6.17-B*b*. In both cases $N = 3$ and the impulse response contains three exponential terms and nothing more. But note the differences in the initial values (corresponding to the difference in the number of zeros of the transfer function at infinity):

$$t_\delta(t) = \begin{cases} r_\delta(0) + r'_\delta(0)t + \dfrac{r''_\delta(0)t^2}{2!} + \cdots & \\ 0 \;+\; 0 \;+ \dfrac{r''_\delta(0)t^2}{2!} + \cdots & \text{when } C = 0 \\ (+) + (-)t + \dfrac{r''_\delta(0)t^2}{2!} + \cdots & \text{when } C > 0. \end{cases} \tag{6.17-4}$$

Both responses are shown in Fig. 6.17-B*c*; it is obvious which is which! The physical behavior is just that predicted.

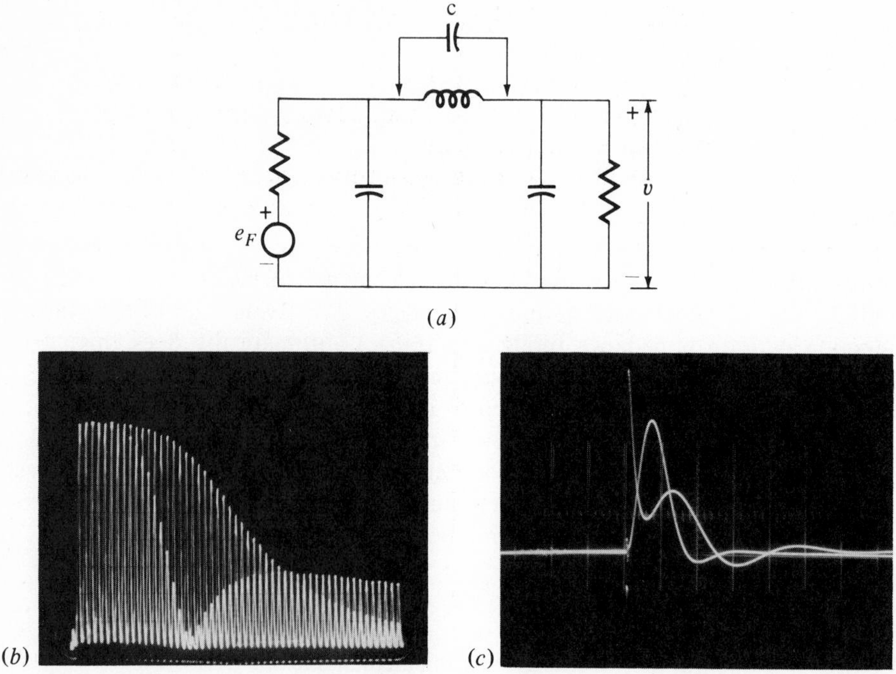

FIG. 6.17-B.

6.18 AN EXAMPLE—ASYMPTOTIC NETWORKS

A few examples will show us how much one can learn from simple asymptotic observations, and how useful the knowledge can be. Let us examine some reasonably complicated networks at or near those values of s for which the calculation of the predominant term in the transfer function is simple because a few elements alone determine the behavior.

Again we choose, for simplicity, a ladder network, that of Fig. 6.18-A*a*. We note that it contains five (idealized) energy-storing elements and that they can be charged initially quite independently of each other. Their five initial states (that is, the three capacitor voltages and two inductor currents) can be specified at will (as by manipulation of previous history); there is no constraint on them; the order of the network is 5. R_1 and R_2 represent source and load (purely resistive) impedances respectively.

Now consider the low-frequency (small-s) behavior. The simplest approach is to clear away the irrelevant, to remove from the network those elements that become relatively unimportant as $s \to 0$ and do not contribute to the leading terms

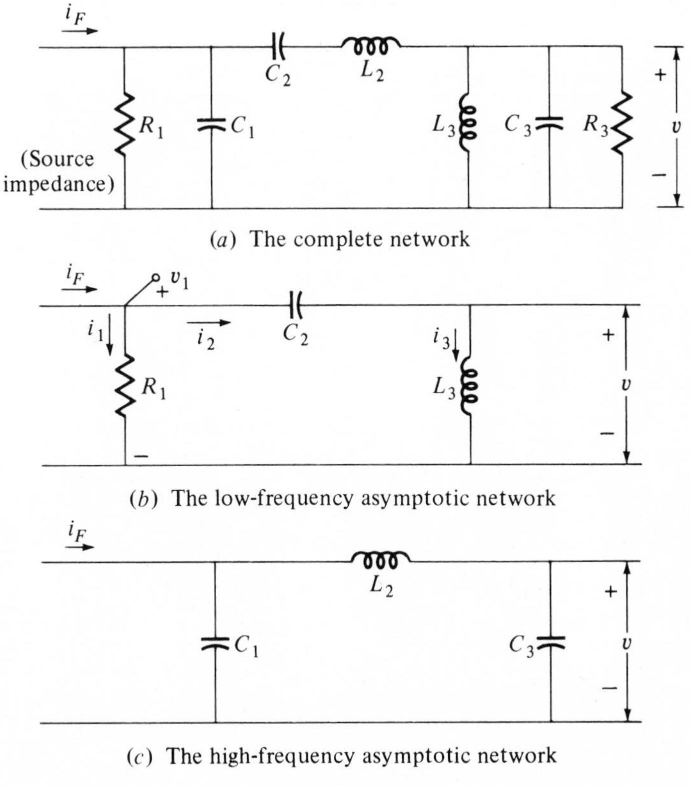

FIG. 6.18-A.

in the series of interest. The (forced component of the) voltage across the combination of R_1 and C_1 in parallel is certainly determined primarily by R_1 for small values of s; the admittance of C_1 is insignificant in comparison. (Formally, $Y = G_1 + C_1 s = G_1$ if we write only the first nonzero term.) The current in C_2 and L_2 is certainly negligibly affected by L_2 when s is small; the impedance $L_2 s$ is insignificant in comparison with the impedance $1/(C_2 s)$. Of the three elements in parallel at the right, L_3 is certainly the most important when s is near the origin; its admittance predominates. Hence the *low-frequency asymptotic network* is that of Fig. 6.18-A*b*. To find the low-frequency transfer function (assuming current excitation) we calculate as we did in Sec. 6.15. In first-approximation terms, assuming exponential (e^{st}) excitation, considering only forced components of currents and voltages, and writing corresponding capital letters for their coefficients of e^{st}, we find, in tabular form:

Item	*Calculation*	
Current in R_1:	$I_1 = \text{excitation current} = I_F$	
Voltage v_1:	$V_1 = R_1 I_F$	
Current i_2:	$I_2 = (C_2 s) V_1$	
Current i_3:	$I_3 = I_2$	(6.18-1)
Voltage v:	$V = (L_3 s) I_3$	
Transfer function:	$T_{LF} = \dfrac{V}{I_F} = (L_3 s)(C_2 s)(R_1) = R_1 C_2 L_3 s^2.$	

The actual transfer function has two zeros at the origin and behaves as (6.18-1) for small values of s. With hindsight we see that this is physically obvious: in the low-frequency asymptotic network there are two elements that block low-frequency transmission; the low-frequency transfer function can be written immediately by inspection [really by mentally going through the steps in (6.18-1)] as the last line in (6.18-1).

Now consider the high-frequency (large-s) behavior. The asymptotic network is that of Fig. 6.18-A*c*, for here R_1, C_2, L_3, and R_3 are of negligible importance. The asymptotic transfer function, constructed as in Sec. 6.15, is

$$T_{HF} = \frac{1}{C_1 s}\,\frac{1}{L_2 s}\,\frac{1}{C_3 s} = \frac{1}{(C_1 L_2 C_3) s^3}. \tag{6.18-2}$$

The actual transfer function has three zeros at infinity, as we can see by inspection: there are three elements that block high-frequency transmission.

We have found five zeros of $T(s)$, and there *are* only five. All the zeros are at the origin or at infinity, there are no others. Therefore

$$
\begin{aligned}
T(s)\Big|_{i_F \text{ to } v} &= \frac{A_0 s^4 + A_1 s^3 + A_2 s^2 + A_3 s + A_4}{s^5 + B_1 s^4 + B_2 s^3 + B_3 s^2 + B_4 s + B_5} \\
&= K_0 \frac{(s-0)(s-0)}{s^5 + \cdots + B_5} \\
&= \frac{A_2 s^2}{s^5 + \cdots + B_5}
\end{aligned} \tag{6.18-3}
$$

in which (from observation of high-frequency behavior) $K_0 = A_2 = 1/(C_1 L_2 C_3)$, and (from observation of low-frequency behavior) $A_2/B_5 = R_1 L_2 C_3$. We note that the dimensions of A_2 and B_5 are correct (dimensionally, T is an impedance), and that specific numerical element values are not needed to obtain these results. We can look on (6.18-3) as a calculation of the transfer function, or use it to check $T(s)$ previously calculated otherwise. There remain undetermined, if we adopt the first point of view, the four constants B_1, B_2, B_3, and B_4. Since these depend on *all* the elements, for they determine the natural frequencies, *no asymptotic calculation can provide their values.* We do know that all five zeros of the characteristic polynomial $(s^5 + \cdots + B_5)$ have negative real parts, because of the resistors in the network. *Where* they lie in the left half plane *does* depend on the numerical values of the elements. Hence the sinusoidal-steady-state $(s = j\omega)$ transmission is unpredictable: there may be two sharp resonances (Fig. 6.18-B*a*) or no resonances at all (Fig. 6.18-B*b*); there may be a single resonance, due to one pole or to two poles that are close to each other or even coincident (Fig. 6.18-B*c*). Correspondingly, the impulse response in t may be highly oscillatory, not oscillatory at all, or take some shape requiring other description.

To obtain a "feel" for the network's properties with minimum effort is one good use of asymptotic (and initial-value and final-value) calculations. Checking a given $T(s)$ is of course another; both applications are worth knowing. We therefore summarize in one place all the information we can obtain from such quick observations, in Fig. 6.18-C.

To complete the analysis of the network it is necessary somehow or other to calculate the characteristic polynomial. And for this it is convenient, if not actually necessary, to have specific element values: there are too many symbols (N is too large) for easy manipulation of the literal symbols as in the *RC* and *RLC* cases. We can then proceed to find the impulse response, in power series or in sum-of-exponentials form (Fig. 6.13-A), the step response, the response to any excitation of interest. We shall find the sinusoidal steady state of great interest and note that it requires only the evaluation of $T(j\omega)$.

We state here without proof, and without explaining the process, that *synthesis* of the network of Fig. 6.18-A, from either of the two prescriptions below, is quite possible, is not unreasonably difficult, and is of course more interesting than analysis of it as a given network. Moreover, other, different network forms can be synthesized from the same prescription(s): synthesis is not unique; analysis is. But competence in analysis is prerequisite to understanding synthesis.

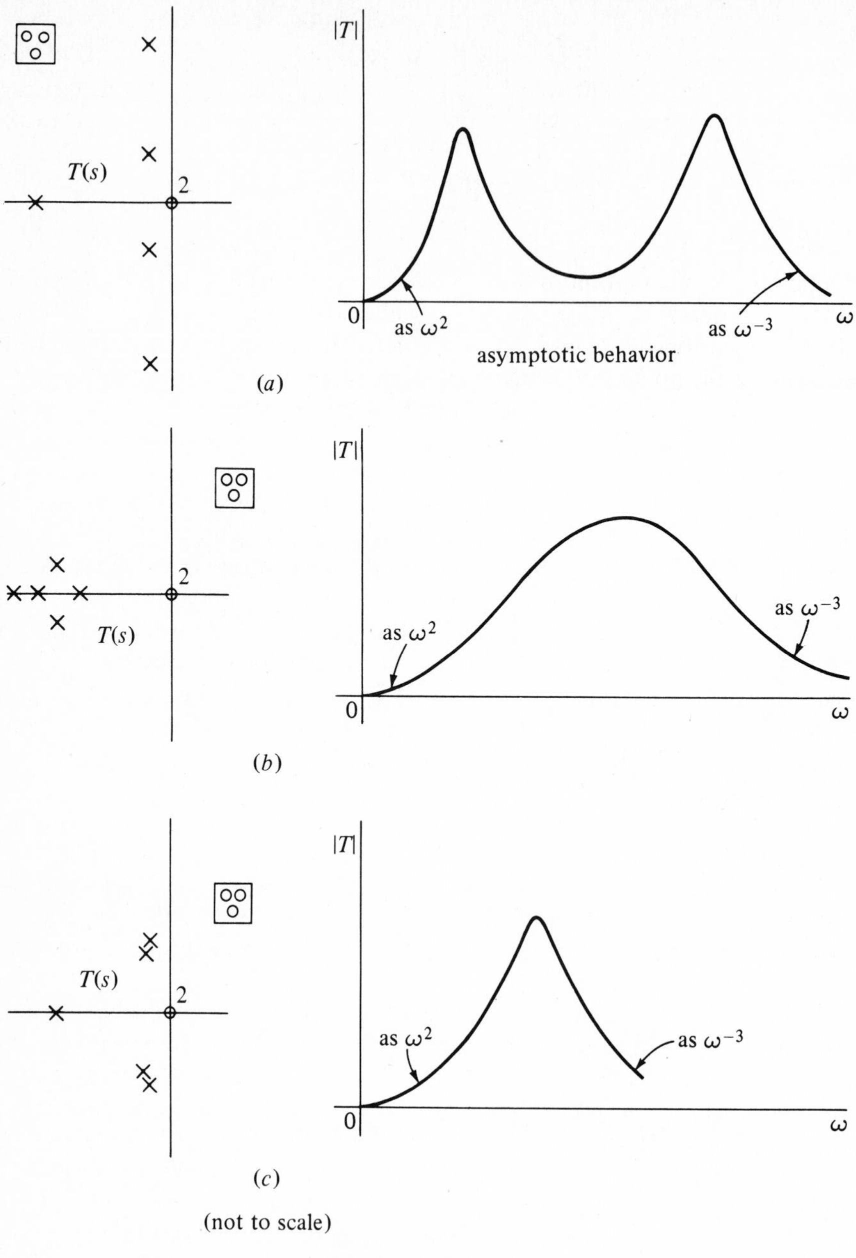

T(s)
2
|T|
0
as ω²
as ω⁻³
ω
asymptotic behavior
(a)
2
T(s)
|T|
as ω²
as ω⁻³
0
ω
(b)
T(s)
2
|T|
as ω²
as ω⁻³
0
ω
(c)
(not to scale)

FIG. 6.18-B.

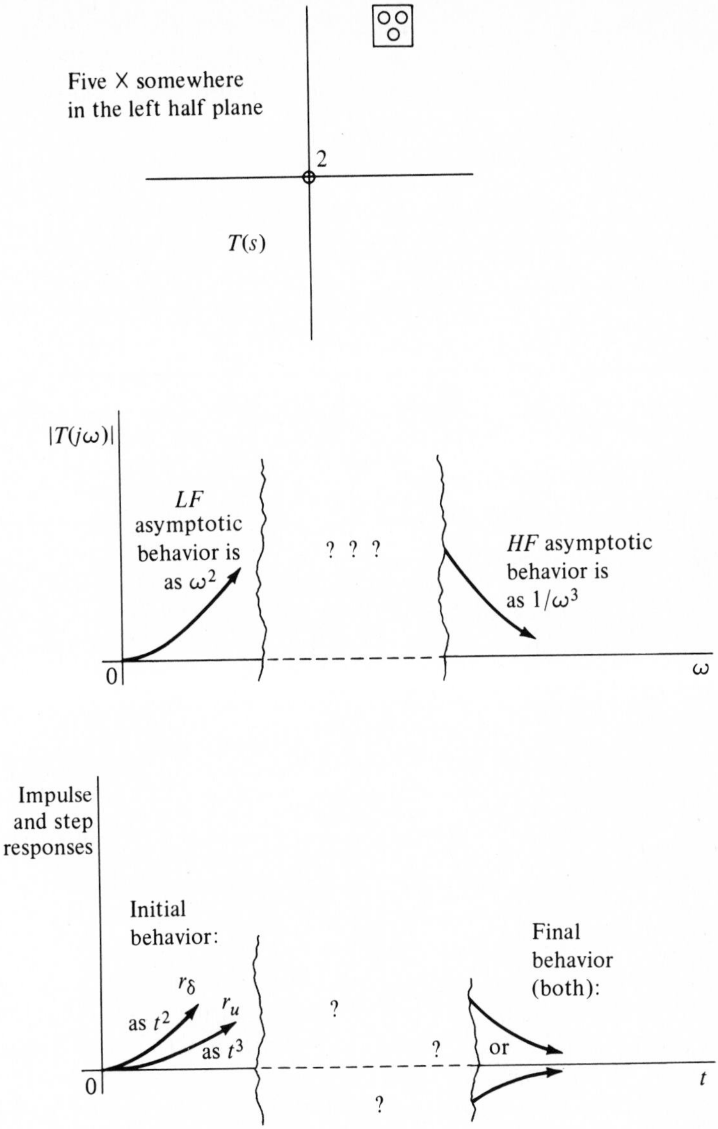

Five X somewhere
in the left half plane
2
T(s)
|T(jω)|
LF
asymptotic
behavior is
as ω²
? ? ?
HF asymptotic
behavior is
as 1/ω³
0
ω
Impulse
and step
responses
Initial
behavior:
r_δ
as t²
r_u
as t³
?
?
?
Final
behavior
(both):
or
0
t

FIG. 6.18–C.

Prescription A: To specify the magnitude of T as a function of ω [this could be a transmission magnitude (filter action) wanted for some system function to be performed by this block]:

$$|T(j\omega)|^2 = \frac{K_0{}^2\omega^6}{\omega^{10} + C_8\omega^8 + C_6\omega^6 + C_4\omega^4 + C_2\omega^2 + C_0}; \qquad (6.18\text{-}4)$$

Prescription B: To specify T as a function of s (this could come, for example, from a desired impulse-response formulation):

$$T(s) = \frac{K_0 s^3}{s^5 + B_1 s^4 + B_2 s^3 + B_3 s^2 + B_4 s + B_5}. \qquad (6.18\text{-}5)$$

The values of the constants are prescribed; the synthesis manipulates them and produces values for the elements of the network. Conditions necessary and sufficient for the synthesis to succeed, that is, to give positive values for the elements in the network of Fig. 6.18-A, are that the numerators have the specific forms given above (other numerator forms can give other network configurations) and that

In A: $$0 \leqq |T(j\omega)|^2 \leqq \frac{R_1 R_3}{4} \quad \text{for all values of } \omega, \qquad (6.18\text{-}6)$$

In B: $|K_0|$ be sufficiently small that (6.18-6) hold, and that all poles of $T(s)$ lie in the left half plane.

Note that in A there are implicit restrictions on the values of the C's to insure that T not be infinite for any value of ω; in B the (equivalent) restrictions are more evident: that the values of the B's be such that the natural frequencies are all in the left half plane. These conditions, essentially restrictions on the attainable magnitude of transmission, are really due to natural limitations on the amount of power that can be delivered by the source (of impedance R_1) to the load R_3.

The network of Fig. 6.18-D*a* could, if properly designed, be a low-pass filter, passing low-frequency signals well and attenuating high-frequency signals, with a transfer function like those of Fig. 6.15-B*c*. Misadjustment of the terminating resistors can radically alter the characteristic, however, and produce one like that of Fig. 6.18-B*a*. Figure 6.18-D*b* shows physically what can happen when the resistors are much larger than they should be. The nature of the impulse response (Fig. 6.18-D*c*) is interesting: it is highly oscillatory (of course—the Q is high) and essentially of one sign. (Figure 6.18-D*d* expands the time scale for a better look.)

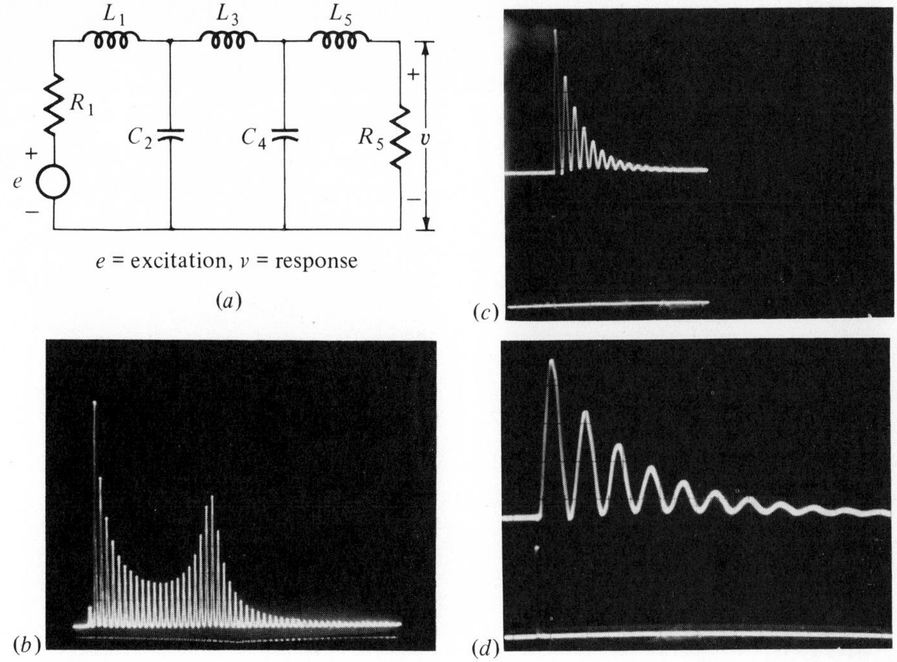

FIG. 6.18-D.

6.19 MORE EXAMPLES; DISSIPATION

In the network of Fig. 6.19-A*a* the capacitor C_2 of Fig. 6.18-A has been differently connected, in parallel with L_2. There are still five energy-storing elements, but now the capacitors form a loop, whose initial-value constraint (Fig. 6.17-A) makes $N = 5 - 1 = 4$. For small values of s the current path is essentially through L_2 and L_3; the low-frequency asymptotic network (Fig. 6.19-A*b*) indicates that the transfer function from i_F to v has one zero at the origin and that

$$T_{LF} = L_3 s. \tag{6.19-1}$$

At high frequencies the current divides between C_1 and the path $C_2 - C_3$; the high-frequency asymptotic network (Fig. 6.19-A*c*) indicates one zero at infinity and

$$T_{HF} = \frac{C_2}{(C_1C_2 + C_2C_3 + C_3C_1)s} \tag{6.19-2}$$

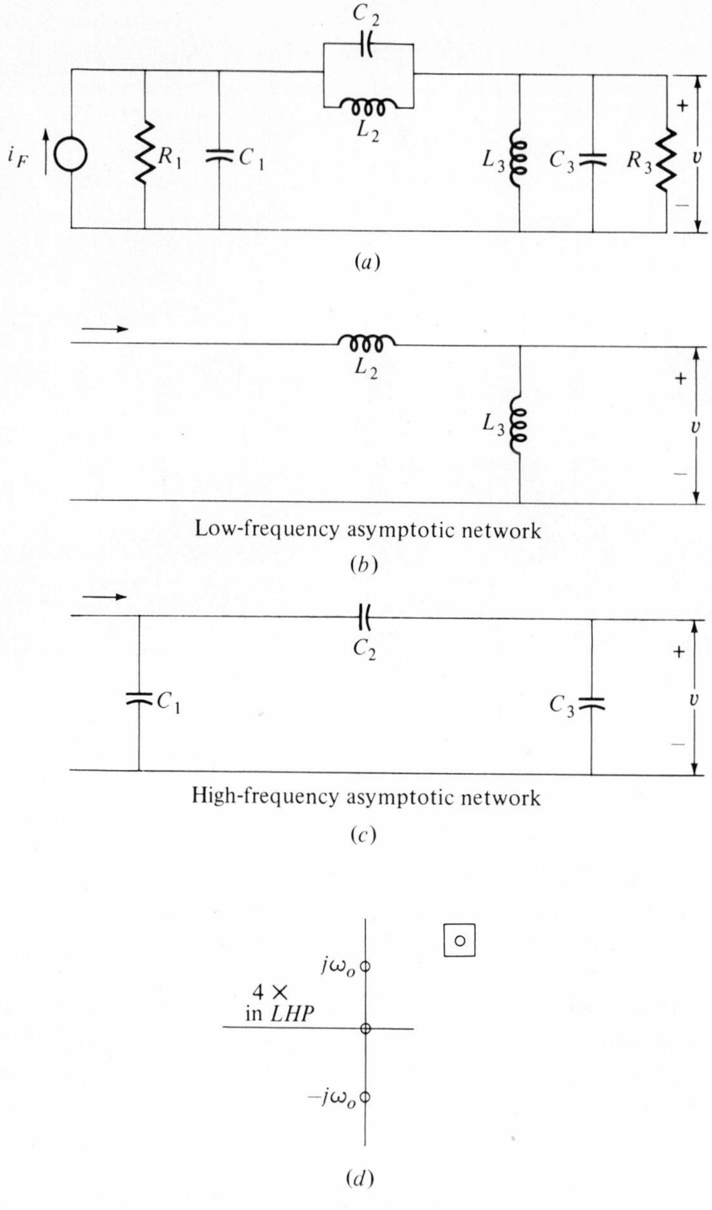

FIG. 6.19-A.

This may be obtained by calculation of T from the asymptotic network, or by calculating first the voltage developed across C_1 and then the fraction produced across C_3 by the voltage divider composed of C_2 and C_3. It is sufficiently complicated to warrant checking, which is easily done by letting each capacitor in turn vanish and checking (6.19-2) against the corresponding physical situation.

We note that there are two other zeros of $T(s)$, for the (infinite-Q) $L_2 - C_2$ tuned circuit completely blocks transmission for $s = \pm j\omega_0 = \pm j1/\sqrt{L_2 C_2}$. Hence

$$T(s) = \frac{K_0(s-0)(s-j\omega_0)(s+j\omega_0)}{s^4 + B_1 s^3 + B_2 s^2 + B_3 s + B_4} = \frac{K_0 s(s^2 + \omega_0^2)}{s^4 + \cdots + B_4}$$
$$= \frac{A_0 s^3 + A_2 s}{s^4 + \cdots + B_4} \tag{6.19-3}$$

in which

$$K_0 = A_0 = \frac{C_2}{C_1 C_2 + C_2 C_3 + C_3 C_1} \tag{6.19-4}$$

from high-frequency behavior, and

$$\frac{K_0 \omega_0^2}{B_4} = \frac{A_2}{B_4} = L_3 \tag{6.19-5}$$

from low-frequency behavior, and

$$\omega_0^2 = \frac{1}{L_2 C_2}. \tag{6.19-6}$$

There remain only three constants in $T(s)$ to be found otherwise; but these have tremendous effect of course on the character of T. Or, we may use the results above to check $T(s)$ calculated otherwise. We note that $r_\delta(0+) = A_0 = K_0$ is not zero because the path of an impulsive charging current is through C_1 in part and $C_2 - C_3$ in part, leaving C_3 initially charged. The final value of r_δ is of course zero, and the final value of r_u is so also, because of the zero of T at the origin. In comparison with Fig. 6.18-A one could say that the action of moving C_2 results in moving one zero from the origin out into the left half plane to cancel a natural frequency, and moving one zero from the origin and one from infinity to $\pm j\omega_0$.

If we further modify the network structure by moving C_3 as in Fig. 6.19-B*a*, we find the order raised to 5 again, the asymptotic networks shown in Fig. 6.19-B, and the transfer function

$$T(s) = \frac{K_0(s^2 + \omega_1^2)(s^2 + \omega_2^2)}{s^5 + B_1 s^4 + B_2 s^3 + B_3 s^2 + B_4 s + B_5}$$
$$= \frac{A_0 s^4 + A_2 s^2 + A_4}{s^5 + \cdots + B_5} \tag{6.19-7}$$

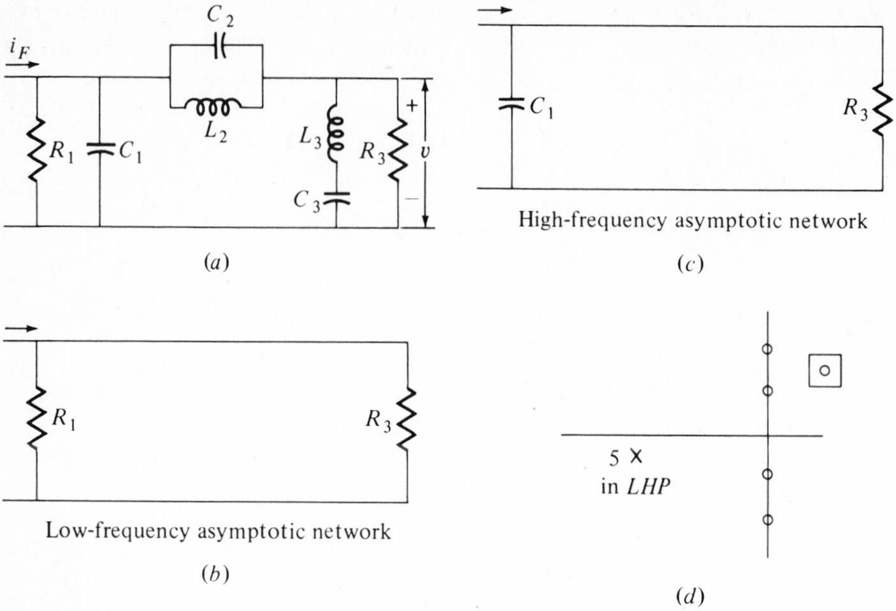

FIG. 6.19-B.

in which

$$K_0 = A_0 = \frac{1}{C_1}, \qquad \frac{K_0\,\omega_1{}^2\omega_2{}^2}{B_5} = \frac{A_4}{B_5} = \frac{1}{G_1 + G_3}, \tag{6.19-8}$$
$$\omega_1{}^2 = \frac{1}{L_2 C_2}, \qquad \omega_2{}^2 = \frac{1}{L_3 C_3}.$$

One might again say that this is obtained from the network of Fig. 6.18-A by moving zeros—two from the origin and two from infinity—to the points $\pm j\omega_1$ and $\pm j\omega_2$.

In these (and previous) models we have ignored the dissipation of energy in inductors and capacitors, and used idealized models. If, to be more realistic, we add some resistance in series with each L (and perhaps in parallel with each C, though this is often less important), we obtain a very good modeling of the inevitable, but often incidental, effect thereof. The effect on the natural-frequency locations, already well within the left half plane because of R_1 and R_2 as a rule, may well be very small. The transfer function zeros at $\pm j\omega_0$, $\pm j\omega_1$, $\pm j\omega_2$ will move somewhat to the left, for the Q's of the individual tuned circuits that produce them become finite (but still large if the element quality is good). But the effect on $r_\delta(t)$ and on $r_u(t)$ may be very small. A plot of $|T(j\omega)|$ will no longer drop to zero at ω_0, ω_1, ω_2, but will have minima of small, nonzero values. Both effects are often of no great practical importance, so long as the element Q's are reasonably high. (See Figs. 6.16-B and 6.17-B.)

Larger amounts of dissipation will move the zeros well out into the left half plane and perhaps to the real axis. There may be applications where this is done deliberately, in order to shape the curve of $|T(j\omega)|$ or those of $r_\delta(t)$ and $r_u(t)$, or for other reasons.

6.20 POINTS OF INFINITE LOSS

We note that if the network is a *ladder* and is *passive*, the zeros of $T(s)$ will always lie in the left half plane, or on the ω axis. For then (Sec. 6.16) the points of infinite loss are due only to zeros and poles of driving-point (two-terminal) immittance functions of passive networks, like those of the combinations of L_2 and C_2, and of L_3 and C_3 above. These are natural frequencies of passive networks (Sec. 5.11), and cannot lie in the right half plane. Hence

> (passive) ladder networks fall in the special "different" category of Sec. 5.09: none of their points of infinite loss are in the right half plane. (6.20-1)

The infinite-loss points of *nonladder* networks are not restricted, nor are those of active networks. The commonest nonladder network is the bridge structure (Secs. 5.10, 5.13). Writing the numerator of its transfer function by inspection is not as easy as it is for ladders, for the zeros are not determined by resonances of individual "boxes." We do know the balance condition for the bridge (Sec. 5.13), to be sure, and hence most of the zeros of $T(s)$. But it is difficult to calculate the characteristic polynomial, to construct the denominator of T [see (6.23-11)]. Only in special cases (zero or infinite source and load impedances, or very special symmetry in the bridge) is this easily done. Generally, the analysis of nonladder networks requires additional tools (Sec. 6.24). This difficulty of analysis is connected with the arbitrariness of the locations of the points of infinite loss and the manner of their generation: they are often caused by the subtle finesse of *cancellation* of signal components, as in a bridge balance.

But in *ladder* networks, where brute-force resonance is used to produce zeros of transmission, the points of infinite loss are clearly visible. Single "separate" elements produce points of infinite loss at dc and at infinity (see, for example, the networks in Secs. 6.07 and 6.09, and Figs. 6.15-B and 6.23-E). Pairs of elements (in tuned circuit form) produce them on the ω axis (see Figs. 6.16-B and 6.17-B). The commonest and most dramatic infinite-loss producer (often used for rejection of signals unwanted) is the latter, the tuned-circuit type. Here is one more example, in Fig. 6.20-A. In (*a*) we see the network, with such infinite-loss producers in two places. (In this example the two resonance frequencies are nominally alike: $L_1C_1 = L_3C_3 = 1/\omega_0{}^2$.) In (*b*) appears its impulse response, showing the expected (step) initial behavior (determined by the number of zeros of $T(s)$ at infinity). And in (*c*) the magnitude of $T(\omega_F)$ versus ω_F

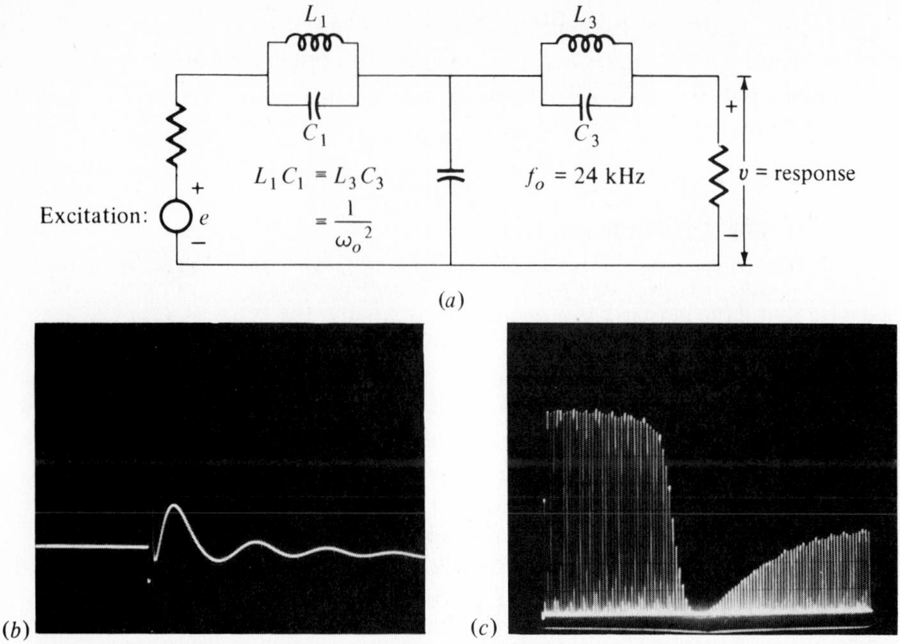

FIG. 6.20-A.

is shown. Note how clearly the (nearly coincident) infinite-loss points appear.

To further study these zeros of $T(s)$, and to physically observe what happens to the forced component of response thereat and nearby, Fig. 6.20-B presents three sinusoidal-steady-state (forced-behavior) response and excitation situations for the same network. The upper wave is in each case the forced component of the response, v_F; the bottom wave is $e(t)$, the sinusoidal excitation. The three frequencies of excitation in Fig. 6.20-B are:

a Just above the nominal resonance,

b At the nominal resonance, (6.20-2)

c Just below the nominal resonance.

Note first the sharp reduction in magnitude of the forced response as ω_F passes through ω_0, just as expected. [The scales (gain settings) are the same in all three photographs.] And this occurs even though the adjustments in the physical network are not very precise, the Q's are not particularly large, and the resonant frequencies are not exactly alike! Then, second, note the dramatic phase shift that v_F undergoes in passing through ω_0, from a reponse lag of about 180° just below resonance, through a phase difference at resonance that is difficult to evaluate, given the very weak response signal, to an appreciable lead (perhaps 30°) just above resonance. That such sudden changes in phase shift should occur is evidenced by a glance at the pole-zero diagram of $T(s)$, using the graphical

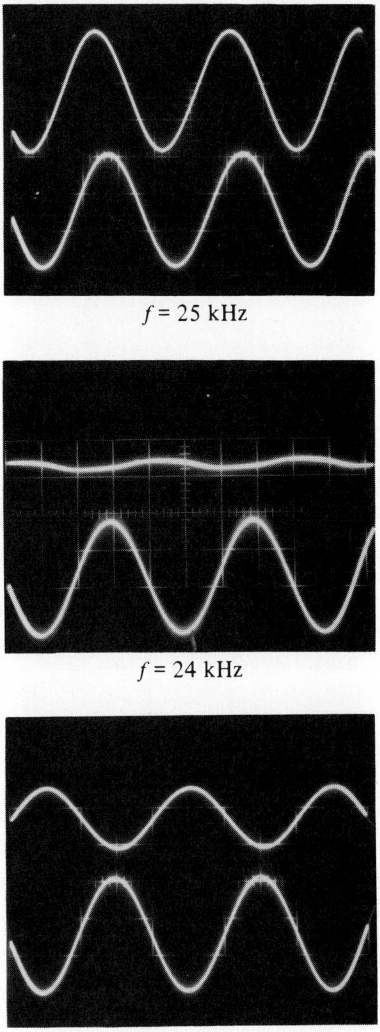

(*a*) $f = 25$ kHz

(*b*) $f = 24$ kHz

(*c*) $f = 23$ kHz

FIG. 6.20-B.

(directed line segment) construction of Sec. 3.17 (see Fig. 6.20-C). It is very clear how the two "sensitive" directed lines, from pointing (almost) straight *down*, contributing 180° lag, move counterclockwise to a pointing *up* position, giving a contribution of 180° lead. [The other factors in $T(j\omega_F)$ change very little in this narrow band; they are essentially constant therein, and T is relatively insensitive to them.] Hence what *should* happen is a sharp 360° change in the phase of v_F relative to e. The phase shift of a simple tuned circuit (Figs. 3.14-D, 3.14-G) is merely "doubled up" here. What actually does happen is *not* this,

$$T(j\omega_F) = H_0 \frac{(\overrightarrow{Z_1P})(\overrightarrow{Z_2P})(\overrightarrow{Z_3P})(\overrightarrow{Z_4P})}{(\overrightarrow{P_1P})(\overrightarrow{P_2P})(\overrightarrow{P_3P})(\overrightarrow{P_4P})(\overrightarrow{P_5P})}$$

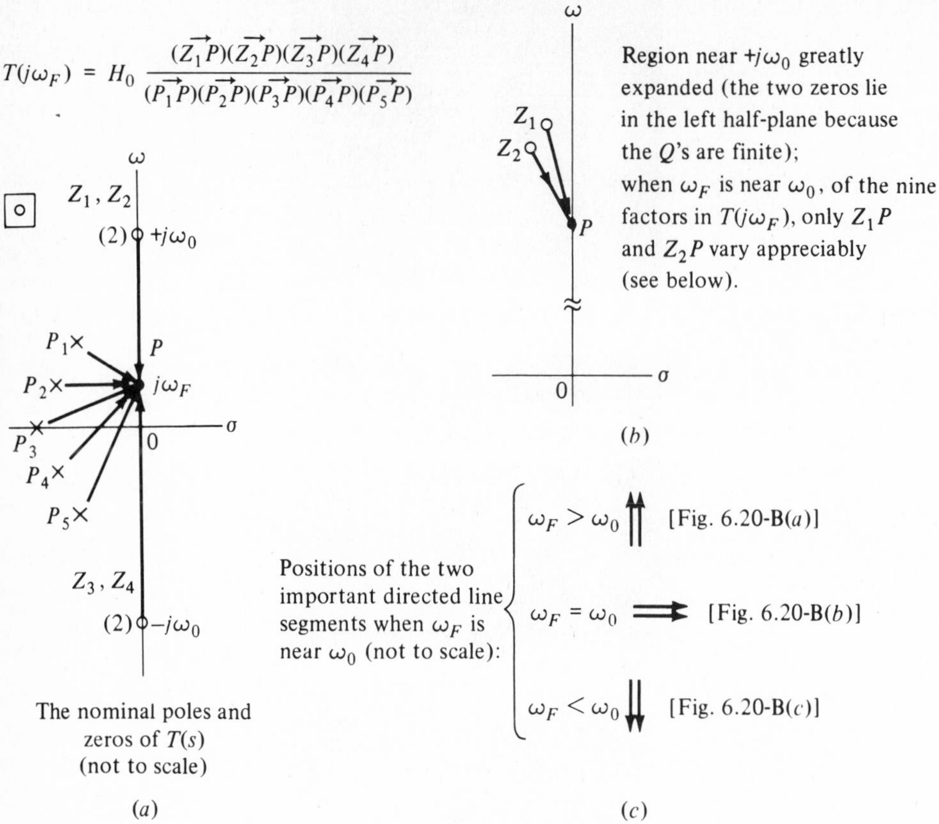

FIG. 6.20-C.

because of the imperfections in the physical network compared to the model. But the photographs do show the two important points:

$$\text{There is virtually no transmission at the resonance frequency, and there is a sharp change in phase at resonance.} \qquad (6.20\text{-}3)$$

"Still" pictures like Fig. 6.20-B are no real substitute for "movies," that is, for actually watching the waves on the oscilloscope as the excitation frequency ω_F is varied slowly (but not *too* slowly) through the resonance: one sees the response wave "walk across the screen" in a very dramatic way!

6.21 THE COMMON-FACTOR ANOMALY

Consider the simple resistance-inductance network of Fig. 6.21-A*a* (which is of interest in the study of transformers). Let us calculate its voltage-to-voltage transfer function by observation. The high-frequency asymptotic network (Fig. 6.21-A*b*) shows one zero at infinity. The low-frequency network is troublesome: Should we simply neglect L_2 and consider L_1 and L_3 to combine in effect into a single inductor (Fig. 6.21-A*c*)? To do so gives us one zero at the origin. Now there are three energy-storing elements and no constraint, so the order is 3. But we have found only two zeros; where is the third one? One senses that there is some sort of degeneracy in the Π connection of three inductors, something that requires investigation. Feeling particularly uneasy about the role of L_2 at low frequencies, let us temporarily dispel all uncertainty by changing the network to remove the degeneracy: we insert a resistor r in series with L_2 (Fig. 6.21-A*d*). The low-frequency asymptotic network is now that of Fig. 6.21-A*e*, with two zeros evident at the origin. The high-frequency behavior is essentially unchanged, so that

$$T(s) = \frac{K_0 s^2}{s^3 + B_1 s^2 + B_2 s + B_3}. \tag{6.21-1}$$

Asymptotic behaviors and conclusions are given in Table 6.21-A. We note that

TABLE 6.21-A Asymptotic Behavior of the Network of Fig. 6.21-A*a*

	From $T(s)$, (6.21-1)	From Physical Behavior of Asymptotic Network	Conclusion
T_{HF}:	$\dfrac{K_0}{s}$	$\dfrac{1}{L_2 s} R_3$	$K_0 = \dfrac{R_3}{L_2}$
T_{LF}:	$\dfrac{K_0 s^2}{B_3}$	$\dfrac{1}{R_1} L_1 s \dfrac{1}{r} L_3 s$	$\dfrac{K_0}{B_3} = \dfrac{L_1 L_3}{R_1 r}$
			$B_3 = r \dfrac{R_1 R_3}{L_1 L_2 L_3}$

all dimensions check. Ordinary complete ladder-network analysis would give the same results and evaluate B_1 and B_2 in addition, but we have no need for them here. We now have

$$T(s) = \frac{(R_3/L_2)s^2}{s^3 + B_1 s^2 + B_2 s + r(R_1 R_3)/(L_1 L_2 L_3)} \tag{6.21-2}$$

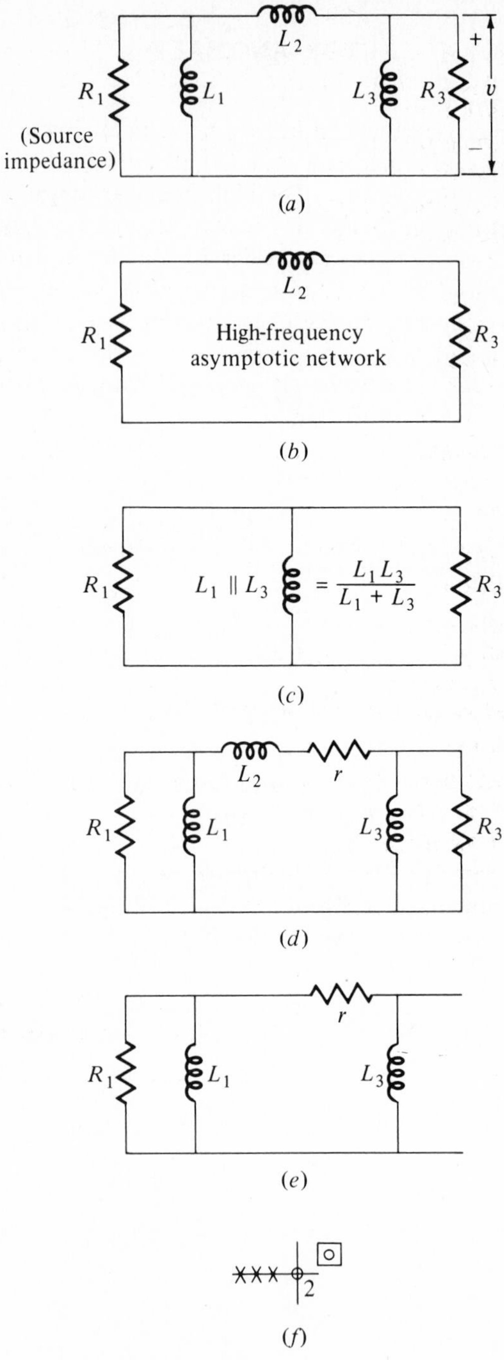
L2
R1
L1
L3
R3
+
v
−
(Source
impedance)
(a)
L2
R1
High-frequency
asymptotic network
R3
(b)
R1
L1 ∥ L3
= L1L3 / (L1 + L3)
R3
(c)
L2
r
R1
L1
L3
R3
(d)
r
R1
L1
L3
(e)
2
(f)

FIG. 6.21-A.

To return to the original, degenerate network, we set $r = 0$ in (6.21-2)—and the paradox is resolved! The order of the network of Fig. 6.21-A*a* is indeed 3; but of the three natural frequencies, one is at the origin. Physically this is evident by inspection: the closed path through the three inductors has zero impedance for $s = 0$. And one should evidently count a zero at dc for L_1 and a second for L_3. The degeneracy of the Π of inductors causes the dc natural frequency to cancel one of the two zeros at the origin.

The transfer function of the network of Fig. 6.21-A*a* is really

$$T(s) = \frac{(R_3/L_2)s^2}{s(s^2 + B_1 s + B_2)} \tag{6.21-3}$$

and is anomalous in having both a zero and a natural frequency at the same point, the origin. If we cancel the common factor $(s - 0)$ from numerator and denominator, we lose the agreement between N and the number of energy-storing elements. We recall our agreement not to cancel common factors, should they appear (Fig. 6.03-A); here is an example of their occurrence, and a good reason for not immediately canceling them without at least noting why they appear. The two factors $(s - 0)$ in numerator and denominator of (6.21-3) are really *surplus* factors; they have no effect on the response, for they cancel in all calculations, for example, of r_δ, of r_u, of $|T(j\omega)|$. Or, one could say that the natural mode e^{0t} *does* appear in the impulse and other responses, but with coefficient *zero*. What really happens in this peculiar network (Fig. 6.21-A*a*) is that of the three natural modes, one exists only internally (direct current flows around the idealized inductor loop) and is not externally evident: it is *unobservable* in the response $v(t)$. Externally it *seems* that there are only two natural frequencies; actually there are three.

We conclude that the number of poles in a transfer function may be less than the number of energy-storing elements for reasons totally different from those of Sec. 6.17. When common factors appear in numerator and denominator they should of course be canceled for simplicity in calculation, but not without noting their existence, to avoid paradoxes and arguments. In the routine calculation of $T(s)$ for the network of Fig. 6.21-A*a*, such cancellations tend to be made automatically, without thinking, and the unobservable natural mode may never be noticed.

The dual situation (Fig. 6.21-B) provides another example of the common-factor effect. Again there is a point of infinite loss at the same place as a

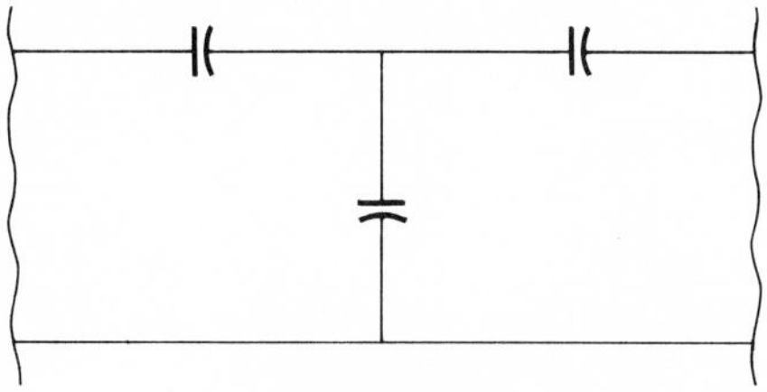

FIG. 6.21-B.

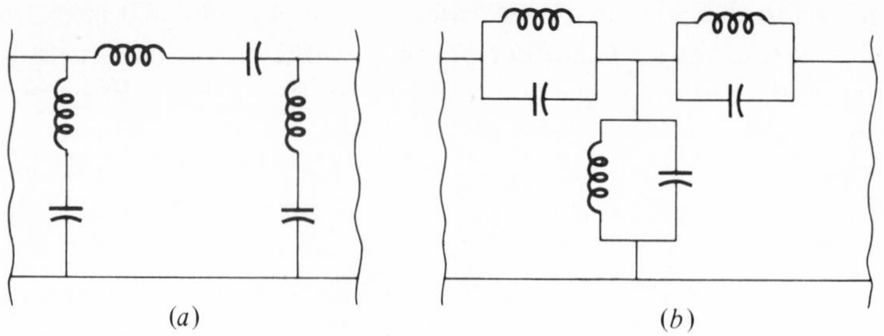

(a) (b)

In (*a*), and in (*b*), the three resonant frequencies are the same.

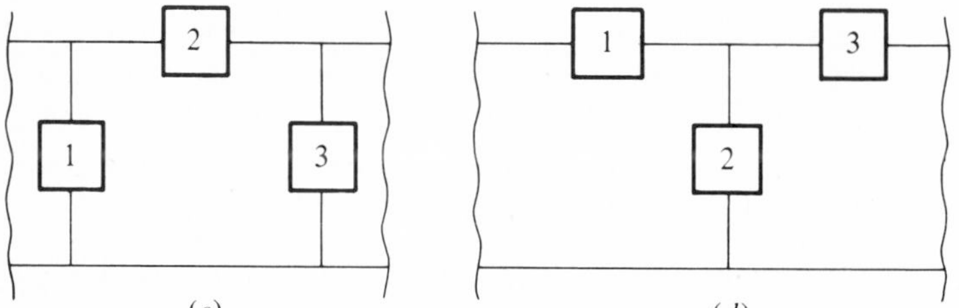

(c) (d)

In (*c*), and in (*d*), Z_1, Z_2, Z_3 differ only by constant multipliers; their poles and zeros are alike.

FIG. 6.21-C.

natural frequency, the origin. A natural mode in the form of constant voltage can exist on the capacitors (as could direct current flow in the inductors in Fig. 6.21-A*a*) regardless of the rest of the network. If these surplus factors are not canceled, then the order equals the number of energy-storing elements. Additional examples can be constructed ad infinitum (Fig. 6.21-C).

One can even regard the order reduction of Sec. 6.17 in this way: in the Π of capacitors there is a natural frequency at infinity and a zero of transmission also at infinity; hence on cancellation, N becomes one less than the number of energy-storing elements. In the T network of inductors the same occurs. But infinity is a very special frequency, and the initial-condition-constraint view of Sec. 6.17 is more reasonable.

6.22 A USE FOR SURPLUS FACTORS: CONSTANT-RESISTANCE NETWORKS

Common factors may be deliberately inserted, wasteful though this may seem, to accomplish certain ends in synthesis. In the elementary bridge network of Fig. 5.10-A, suppose we replace R_2 and R_4 by general function-of-s imittances,

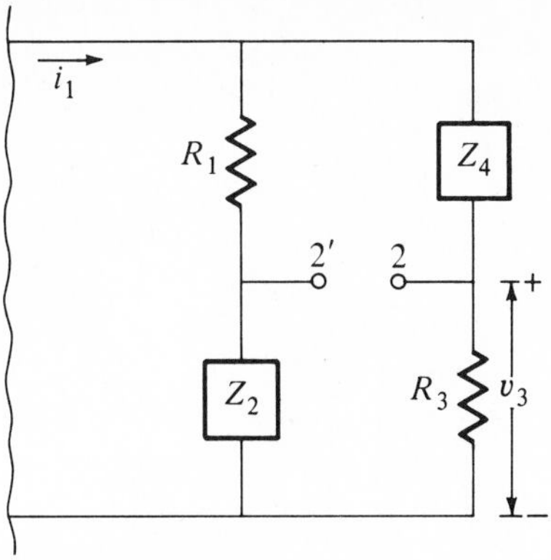

FIG. 6.22-A.

as in Fig. 6.22-A. The transfer function from i_1 (however supplied) to v_3 is easily written, mentally working back from v_3:

$$T = \frac{V_3}{I_1} = \frac{1}{G_3 + (1 + G_3 Z_4)/(R_1 + Z_2)} = \frac{R_3(R_1 + Z_2)}{R_1 + Z_2 + R_3 + Z_4}$$

$$= \frac{R_3(1 + R_1/Z_2)}{Z_4/Z_2 + (R_1 + R_3)/Z_2 + 1}. \tag{6.22-1}$$

The last manipulation, in addition to making the checking of dimensions very easy, leads to an interesting idea, below. Now let us impose some very special requirements on the elements (Fig. 6.22-B):

$$R_1 = R_3 = R_0,$$

$$Z_2(s) = \frac{R_0^{\,2}}{Z_4(s)}, \tag{6.22-2}$$

$$\frac{R_0}{Z_2} = \frac{Z_4}{R_0} = z(s).$$

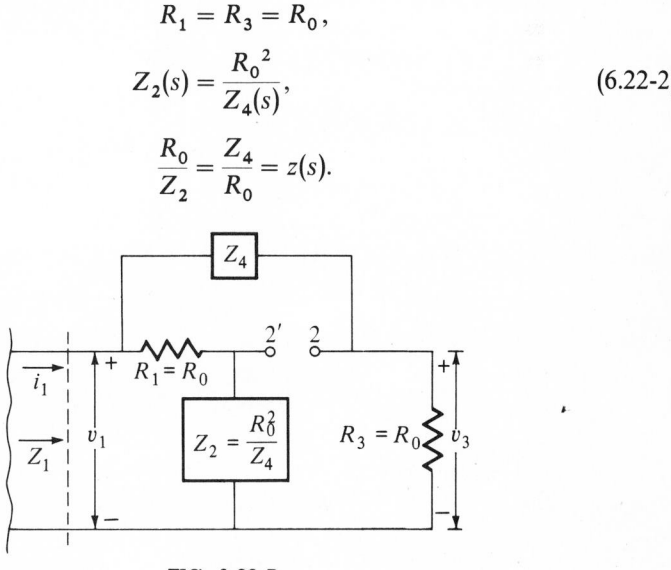

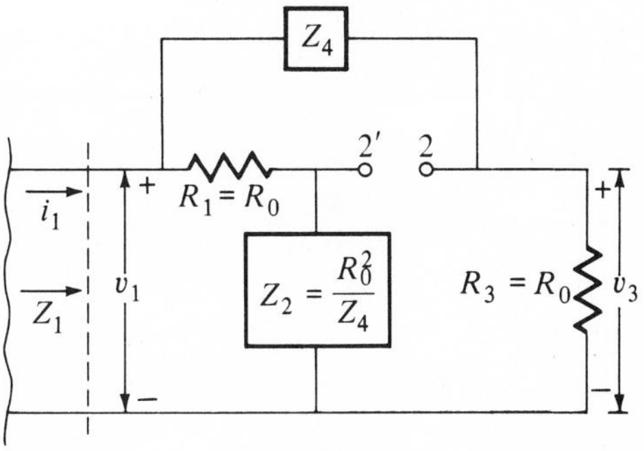

FIG. 6.22-B.

The two resistors are made alike to introduce symmetry, and are called R_0 for convenience. The requirement that $Z_2(s)$ behave as the reciprocal of $Z_4(s)$ seems not impossible to effect, as by using duality (see Secs. 5.12 and 5.11); it can be shown in fact that there are always many ways of realizing an impedance "inverse" to a given one. Finally, we introduce the simple symbol z for the normalized impedance Z_4 or admittance Z_2^{-1}. All these will make for simplicity below. Then

$$\begin{aligned} T(s) &= \frac{R_0(1 + Z_4/R_0)}{Z_4^2/R_0^2 + 2Z_4/R_0 + 1} = R_0 \frac{1 + z}{z^2 + 2z + 1} \\ &= R_0 \frac{1 + z}{(1 + z)^2}. \end{aligned} \tag{6.22-3}$$

After canceling the common factor $[1 + z(s)]$, which can be very complicated so that *many* common factors are really canceled here, we have

$$T(s) = \frac{R_0}{1 + z(s)}. \tag{6.22-4}$$

The apparent reduction in order can be large; many natural modes may be suppressed from external observation.

The utility of this procedure, the reason it is *not* really wasteful, lies in two facts. The first is an unexpected property of the input immittance. Let us calculate the input admittance (Fig. 6.22-B), $Y_1 = 1/Z_1$.

$$\begin{aligned} Y_1 &= \frac{1}{Z_1} = \frac{1}{R_0 + R_0^2/Z_4} + \frac{1}{Z_4 + R_0} \\ &= \frac{1}{R_0 + R_0/z} + \frac{1}{R_0 z + R_0} \\ &= G_0\left(\frac{z}{z + 1} + \frac{1}{z + 1}\right) \\ &= G_0 \frac{z(s) + 1}{z(s) + 1} = G_0. \end{aligned} \tag{6.22-5}$$

The input impedance Z_1 is a constant, R_0, independent of s! But only, of course, because all factors are common and can be canceled. The manipulations and impositions have been aimed at precisely this: making *all* factors in the numerator and the denominator of $Z_1(s)$ common and cancelable. In effect, Z_1 is equivalent to a constant, i.e., a pure resistance, throughout the s plane. In practice, in real networks, this is not exactly possible, of course, but with care it can be very nearly done, at least over the range of values of s of importance.

The utility of this *constant-resistance* property is the following. Suppose

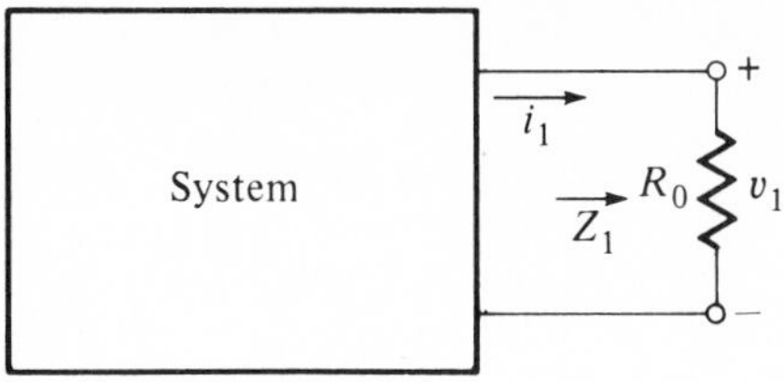

FIG. 6.22-C.

we design the part of the system to the left, up to this point, with a resistor R_0 connected as load (Fig. 6.22-C). The voltage v_1 is developed across R_0 by the system shown. If we now replace R_0 by the constant-resistance network (Fig. 6.22-B), there is no change in v_1, since the impedance presented to the system is the same, $Z_1(s) = R_0$. But the voltage v_3 now differs from v_1 according to the transfer function (6.22-4). So we design $T(s)$, that is, $z(s)$, to effect whatever is necessary in the system at this point. And this is the second advantage, that the design of the network to realize a prescribed $T(s)$ comes down essentially to the design of *one* immittance per

$$z(s) = \frac{R_0}{T(s)} - 1, \tag{6.22-6}$$

provided of course that $z(s)$ be realizable. If difficulty is encountered, the realization of $T(s)$ can be broken into parts by factoring $T(s)$ and realizing each factor in a separate constant-resistance network, all of which can be cascaded (because of the constant-resistance property) without in effect "loading" each other; the overall transfer function (Fig. 6.22-D) is the product of the individual $[1 + z(s)]^{-1}$ (see Sec. 5.07). And there is no reaction back on the system to the left in Fig. 6.22-D in any way different from that in Fig. 6.22-C, again because of the constant-resistance property. This is a very practical and widely used procedure; it is particularly useful for designing networks for insertion in existing systems to correct or change existing transmission properties.

It can be shown that any passive-network-realizable transfer function (a transfer function, all of whose poles are in the left half plane) can be realized in

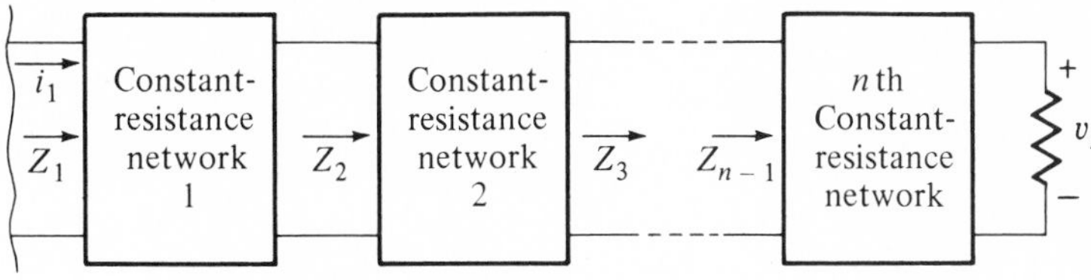

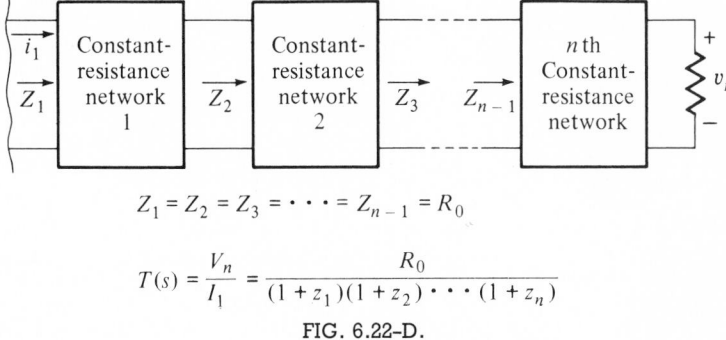

$$Z_1 = Z_2 = Z_3 = \cdots = Z_{n-1} = R_0$$

$$T(s) = \frac{V_n}{I_1} = \frac{R_0}{(1 + z_1)(1 + z_2)\cdots(1 + z_n)}$$

FIG. 6.22-D.

this way (Fig. 6.22-D), though it may require a large number of constant-resistance networks and an attenuation (reduction in signal strength or change in multiplier). (If the transfer function has zeros in the right half plane, a somewhat different sort of bridged-T structure is required also; but these too can be constructed as constant-resistance networks.) Half the energy-storing elements in effect "do no work" as far as transmission is concerned; they provide the cancellations necessary for the constant-resistance property and the elimination of interaction between networks.

Note also that the special element-value requirements imposed above also make the network a balanced bridge, balanced at *all* values of s: the voltage between terminals 2 and 2′ (Fig. 6.22-B) is zero. We can therefore symmetrize by adding another resistor R_0 between these terminals without altering any currents or voltages: the current in it will be zero. The resulting network (Fig. 6.22-E) is more stable with regard to the dependence of the input impedance Z_1 on element variations with aging, adjustment, etc. And the network now operates identically in both directions because of the symmetry. It is known as a *bridged-T constant-resistance* network. It is not a ladder, but is closely related.

Note again that the order is actually N, equal to the number of energy-storing elements in both Z_1 *and* Z_4 (Fig. 6.22-B), and to twice the number of elements in z (Fig. 6.22-E), even though it *seems* from the Z_1 point of view that the order is *zero*, and it *seems* from the $T(s)$ point of view that the order is $N/2$. The number of natural modes observable in the input impedance (by the source) is zero. The number observable in $T(s)$, in transmission, is $N/2$.

The symmetrical lattice networks of Sec. 5.13 give further illustration of the utility of surplus factors. Since only half the elements of the network appear in the transfer function (5.13-14), half the natural modes must be concealed by surplus-factor cancellations in $T(s)$. Here too the input impedance can be made constant-resistance; the lattice offers an additional type of constant-resistance network for use in synthesis.

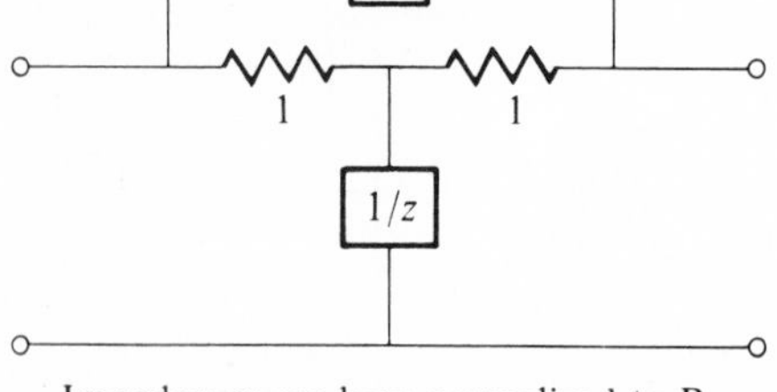

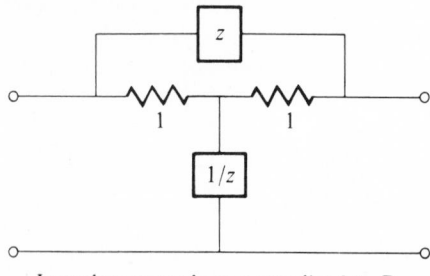

Impedances are here normalized to R_0

FIG. 6.22-E. The bridged-*T* constant-resistance network.

6.23 THE TRANSFER FUNCTION OF THE BRIDGE NETWORK

We have concentrated on ladder networks because their transfer functions are so easily calculated, and because ladder networks are very common. But all of our discussion of transfer functions and responses (of natural frequencies, of initial values, etc.) applies equally well to any network, ladder or not. A particular appeal of ladder networks lies in their ready physical display of many important properties of networks and their close connection with transfer functions, as we have so often seen.

Generally speaking the calculation of a transfer function of a nonladder network is not as simple a task. But there are a few not uncommon networks that are nearly ladder in character and worth attention here. Of these the first is the *bridge* network (Secs. 5.10, 5.13, and 6.22).

The constant-resistance bridged-T network of Fig. 6.22-E is not a ladder network because there is more than one path from input to output (see Secs. 5.09, 5.10, and 5.13). But the resistor connected between terminals 2 and 2′ (Fig. 6.22-B) carries no current because the network is a balanced bridge; it can be disregarded. This gives us therefore a ladder in effect, easy to analyze.

Suppose we now unbalance the bridge and investigate the transfer function calculation. A simple example (Fig. 6.23-A) will indicate what difficulties arise. In Fig. 6.23-A the numbers are the resistor values in ohms; the network is a bridged T but certainly not a constant-resistance one. We begin its analysis by labeling currents and voltages in Fig. 6.23-B. To the current in the 1-Ω resistor (which might be the output point) we give the value unity, following our established ladder analysis procedure. If we attempt to proceed backward (to the left, toward the excitation source e), we encounter the difficulty foreseen in Sec. 6.02 (Fig. 6.02-B): there is more than one path by which current can flow to the output. We do not know how much of the one ampere came by each path. About all we can do is to assign to one path current (that through the 10.8-Ω resistor, say) a symbol x, thus introducing an unknown. This is unfortunate, but necessary. The current i_D in the 2-Ω resistor is then $(1 - x)$.

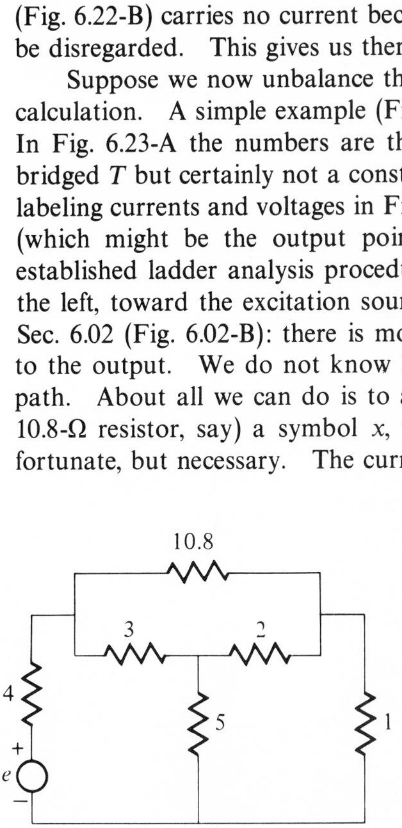

FIG. 6.23-A.

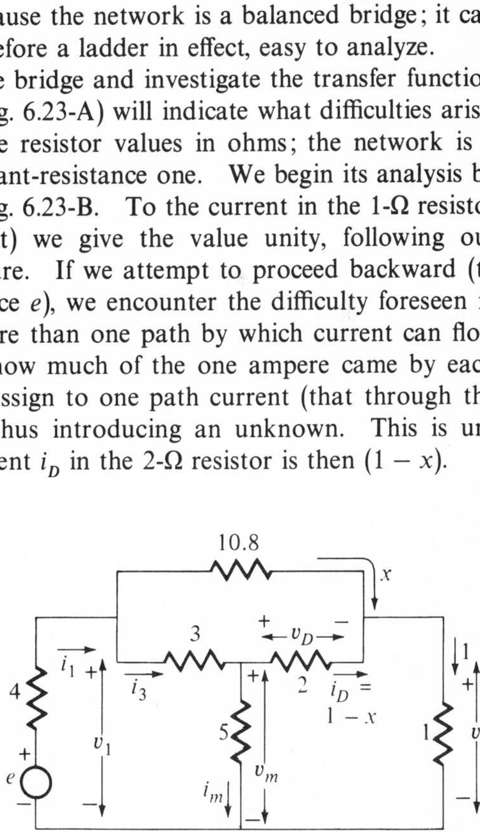

FIG. 6.23-B.

Accepting the nuisance of the unknown x, we continue calculation toward e. In Fig. 6.23-B:

$$
\begin{aligned}
v_m &= v + v_D = 1 + 2(1 - x),\\
i_m &= \frac{1 + 2(1 - x)}{5} = 0.6 - 0.4x,\\
i_3 &= i_m + i_D = 1.6 - 1.4x,\\
v_m &= v_m + 3i_3 = (3 - 2x) + 3(1.6 - 1.4x) = 7.8 - 6.2x.
\end{aligned}
\tag{6.23-1}
$$

At this point we rejoin the other path from input to output. Calculation along that path gives

$$v_1 = 1 + 10.8x. \tag{6.23-2}$$

But this is the same v_1, and therefore

$$1 + 10.8x = 7.8 - 6.2x. \tag{6.23-3}$$

This equation gives us what we need: an equation to determine x, a constraint that expresses the physical connection of the two paths. It is a sort of "lock-up," a remove-the-uncertainty relation. It gives us

$$17.0x = 6.8 \qquad or \qquad x = 0.4. \tag{6.23-4}$$

It is now easy to complete the job: to fill out the values of all the currents and voltages. The results (Fig. 6.23-C) could just as well, with experience, be calculated and written directly on the original diagram.

We can now write down any transfer function we wish (they are all constants, independent of s here, because all elements in the network are resistors). The same reasoning (Sec. 6.02) that *ratios* are the same, regardless

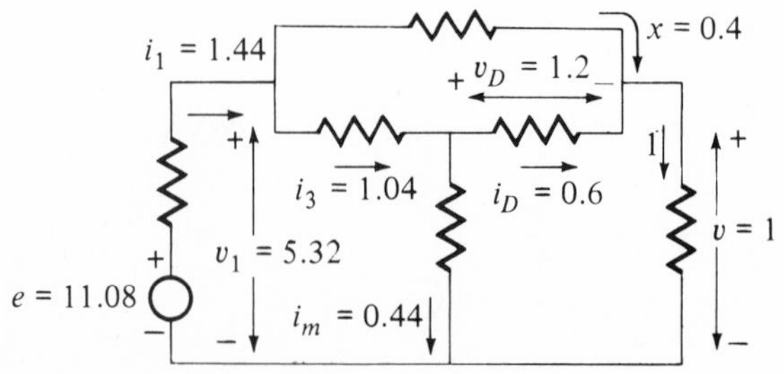

FIG. 6.23-C.

of specific values of currents and voltages, because of linearity, gives us these transfer functions:

$$
\begin{aligned}
\frac{v}{e} &= \frac{1}{11.08} = 0.0903,\\
\frac{v}{i_1} &= \frac{1}{1.44} = 0.6944,\\
\frac{v}{v_1} &= \frac{1}{5.32} = 0.1880,\\
\frac{v_D}{e} &= \frac{1.2}{11.08} = 0.1083,\\
\frac{v_D}{i_1} &= \frac{1.2}{1.44} = 0.8333,\\
\frac{v_D}{v_1} &= \frac{1.2}{5.32} = 0.2256,
\end{aligned}
\tag{6.23-5}
$$

and so on. Note that excitation with a current source (i_1) or a voltage source (of zero internal impedance, v_1), as well as the excitation e shown, are provided for with no extra effort. Transfer functions to v_D have been included because v_D is the output if the network is looked on as a bridge qua bridge (Fig. 6.23-D). The analysis could of course be performed directly on that diagram in the same way.

With this example to guide us, we can now endeavor to calculate the transfer function of a true bridge network (such as those of Sec. 5.13) including a detector impedance. The calculation is perforce not simple, but the result is useful, particularly in dealing with measurement apparatus in bridge form.

We take the general bridge network of Fig. 5.10-C, add a detector arm impedance Z_5, and a source impedance Z_0 (Fig. 6.23-E). Each box now

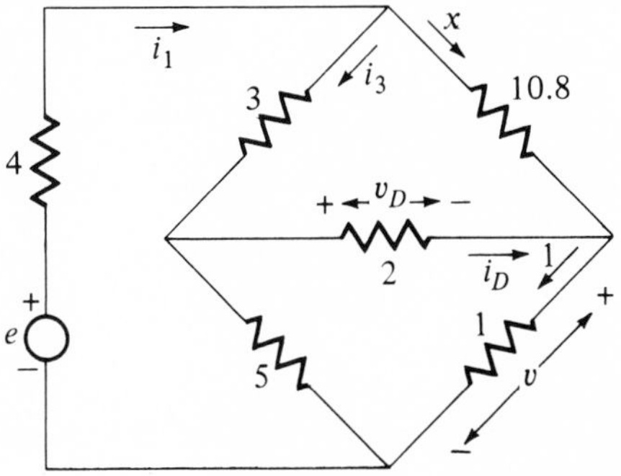

FIG. 6.23-D.

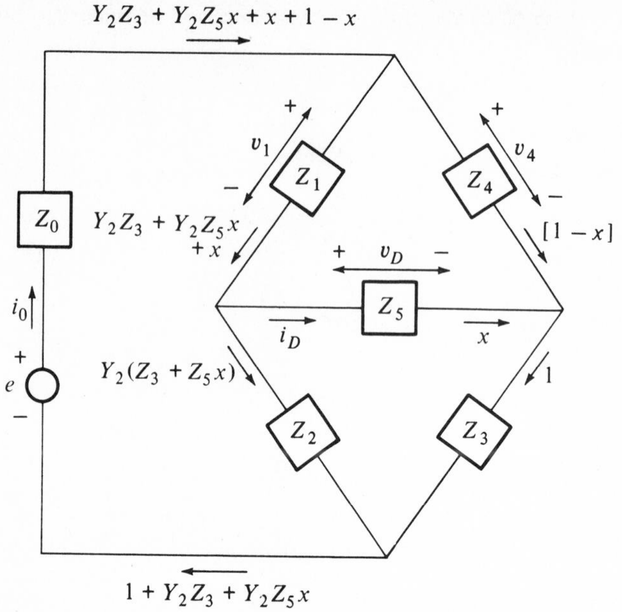

FIG. 6.23-E.

represents an arbitrary collection of R, L, C elements whose driving-point impedance Z is a rational function of s, as in Sec. 5.11. As usual, we assume exponential excitation, in form e^{st}; and it is understood that all voltage and current symbols refer only to the coefficients of the *forced* parts thereof, the natural parts canceling out on their own (Sec. 5.11). We call the current in Z_3 unity and the detector current x (Fig. 6.23-E). These are arbitrary choices, but it seems only reasonable to label the detector current x, for we are very much interested in its value, this being a bridge. We proceed to calculate. The outline below is not necessary, once the idea is clear, for all the calculations are made on the diagram itself; but it is useful for the neophyte. The sequence is:

$$
\begin{aligned}
&\text{Current in } Z_3 && = 1,\\
&\text{Current in } Z_5 && = x,\\
&\text{Current in } Z_4 && = 1 - x,\\
&\text{Voltage across } Z_2 && = Z_3(1) + Z_5 x,\\
&\text{Current in } Z_2 && = Y_2(Z_3 + Z_5 x),\\
&\text{Current in } Z_1 && = Y_2(Z_3 + Z_5 x) + x,\\
&\text{Current } i_0 \text{ at top of diagram} && = [Y_2(Z_3 + Z_5 x) + x] + (1 - x),\\
&\text{Current } i_0 \text{ at bottom of diagram} && = [Y_2(Z_3 + Z_5 x) + 1].
\end{aligned}
\tag{6.23-6}
$$

The last two are of course the same. The "lockup" equation we may now form by equating the voltage rises $(v_D + v_1)$ and (v_4):

$$Z_5 x + Z_1 Y_2 Z_3 + Z_1 Y_2 Z_5 x + Z_1 x = Z_4(1 - x). \tag{6.23-7}$$

Solving for x gives

$$\begin{aligned} x &= \frac{Z_4 - Z_1 Y_2 Z_3}{Z_1 + Z_4 + Z_5 + Z_1 Y_2 Z_5} \\ &= \frac{Y_2(Z_2 Z_4 - Z_1 Z_3)}{Z_1 + Z_4 + Z_5 + Z_1 Y_2 Z_5} \\ &= \frac{Z_2 Z_4 - Z_1 Z_3}{Z_1 Z_2 + Z_2 Z_4 + Z_2 Z_5 + Z_1 Z_5} \end{aligned} \tag{6.23-8}$$

in which the successive manipulations are intended to bring out the bridge balance condition $(x = 0)$ in neater and more familiar, (5.13-16), form.

It remains only to write

$$e = Z_3 + Z_4(1 - x) + Z_0 i_0. \tag{6.23-9}$$

The transfer function from e to v_D, for example, $(Z_5 x/e)$, is now in principle easy to write. The algebra required [substitution of the value of x given by (6.23-8) in the equations and sorting things out] is, however, unpleasantly, unavoidably lengthy. The result is

$$T(s)\Big|_{e \text{ to } v_D} = \frac{v_D}{e} = \frac{Z_5(Z_2 Z_4 - Z_1 Z_3)}{D} \tag{6.23-10}$$

in which

$$\begin{aligned} D = {} & Z_0 Z_1 Z_2 + Z_0 Z_1 Z_3 + Z_0 Z_1 Z_5 + Z_0 Z_2 Z_4 + Z_0 Z_3 Z_4 + Z_0 Z_4 Z_5 \\ & + Z_0 Z_2 Z_5 + Z_0 Z_3 Z_5 + Z_1 Z_2 Z_3 + Z_1 Z_2 Z_4 + Z_1 Z_3 Z_4 + Z_1 Z_3 Z_5 \\ & + Z_1 Z_4 Z_5 + Z_2 Z_3 Z_4 + Z_2 Z_3 Z_5 + Z_2 Z_4 Z_5. \end{aligned} \tag{6.23-11}$$

The natural-frequency equation for the network is of course $D = 0$. The prospect of carrying out the actual, detailed calculation of this, or any other transfer function, and of the corresponding response, for a specific bridge (such as those described in Sec. 5.13) dismays. Since it is seldom called for, we leave it at that, piously noting that we *could* do the job if it were necessary.

We observe (as checks) first that the dimensions are consistent within numerator and denominator, and correct for $T(s)$, and that the numerator exhibits the balance condition correctly.

When Z_5 becomes indefinitely large, we can ignore all terms in (6.23-10) and (6.23-11) that do not contain Z_5; then, on canceling Z_5, we obtain

$$T(s) = \frac{v_D}{e}$$
$$= \frac{Z_2 Z_4 - Z_1 Z_3}{(Z_0 Z_1 + Z_0 Z_2 + Z_0 Z_3 + Z_0 Z_4) + (Z_1 Z_3 + Z_1 Z_4 + Z_2 Z_3 + Z_2 Z_4)}. \tag{6.23-12}$$

This is the transfer function of the bridge when the detector draws no current, the arms being still general and the source having an internal impedance Z_0. Setting $Z_0 = 0$ gives

$$T(s) = \frac{v_D}{e} = \frac{Z_2 Z_4 - Z_1 Z_3}{Z_1 Z_3 + Z_1 Z_4 + Z_2 Z_3 + Z_2 Z_4} \tag{6.23-13}$$

for the general bridge excited by a pure voltage source, the detector arm being of infinite impedance. Finally, to obtain the symmetrical bridge of Fig. 5.13-A, we set

$$Z_1 = Z_3 = Z_a \qquad \textit{and} \qquad Z_2 = Z_4 = Z_b \tag{6.23-14}$$

and obtain

$$T(s) = \frac{Z_b^2 - Z_a^2}{Z_a^2 + 2Z_a Z_b + Z_b^2} = \frac{Z_b - Z_a}{Z_b + Z_a} \tag{6.23-15}$$

which is (5.13-6), another check. Note again the simplification by cancellation of factors which are superfluous in the transfer function (because of the network's symmetry) but not to the network itself.

Once more we recall that all our discussion of transfer functions and responses (initial values, series, asymptotic behavior, ...) applies to our results here, to the bridge (and all networks) as well as to ladder networks.

The bridge network is in a way the next step in network complexity from the ladder. It is widely used, so that we should be familiar with its transfer function—although it is the numerator thereof, with its vivid balance possibility, that is used much more than is the complicated denominator. We have found that we can calculate the transfer function of a bridge, including load and generator impedances, without unreasonable effort. We go no further into more complicated network structures here, for they are uncommon. But a systematic, completely general, and omnipotent analysis method must and can be devised therefor, and we shall briefly describe it here.

6.24 NODES AND LOOPS

To analyze the bridge network in general form we were forced (as foreseen in Fig. 6.02-B) to modify the simple, work-back-from-response-to-excitation method that is so convenient for ladder networks: we introduced an unknown, and then had to solve a supplementary algebraic equation. If the still more complicated situation of Fig. 6.24-A were presented to us, we presumably would have to introduce *two* unknowns, x and y, and solve a *pair* of algebraic, simultaneous equations.

Caught in such algebraic entanglements, it is probably best now to call on well-known, elegant methods of linear algebra. With these one can systematize the analysis and develop a completely general method (or methods) of analysis, usable for *any* network. We shall not discuss the details of this at any length, but it *is* worth doing the analysis of the bridge in such a way. The extension of the ideas so presented, to complete generality, can be carried out elsewhere.

Such systematic formulation of the algebraic equations best follows closely the geometric structure of the network. We accordingly redraw the general bridge of Fig. 6.23-E in purely schematic form, showing only the geometry of its structure, the *graph* of Fig. 6.24-B. We now define (following Kirchhoff and Maxwell—see, for example, GB in Appendix B) two more or less obviously needed terms (of graph theory):

> A *node* is a junction of three or more connections (wires); in Fig. 6.24-C we identify the four nodes by small circles and attach arbitrary counting numbers thereto; (6.24-1)
>
> A *loop* is a closed path through a network; in Fig. 6.24-C we have traced and (arbitrarily) numbered three loops. (6.24-2)

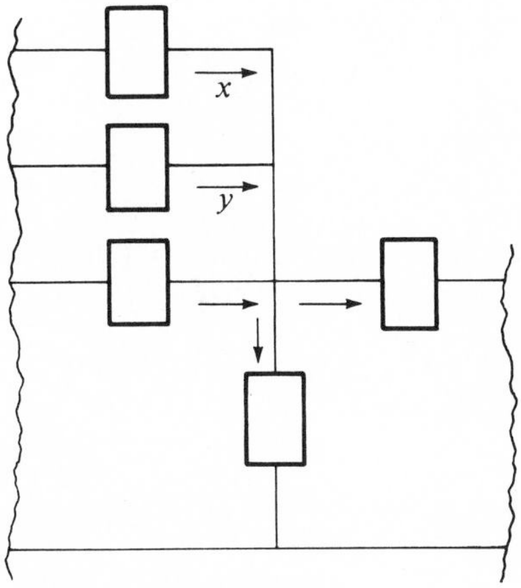

FIG. 6.24-A.

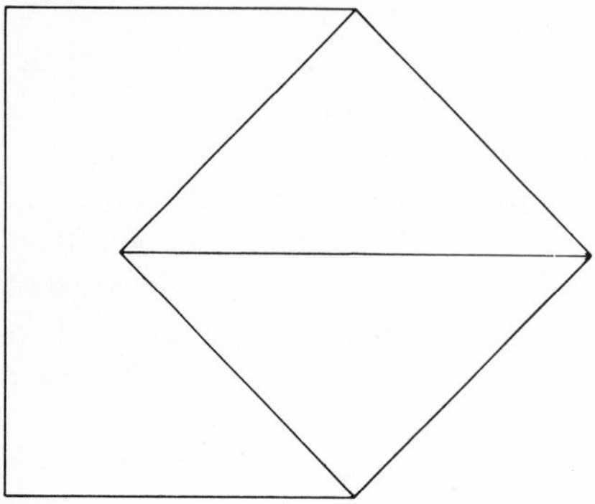

FIG. 6.24-B.

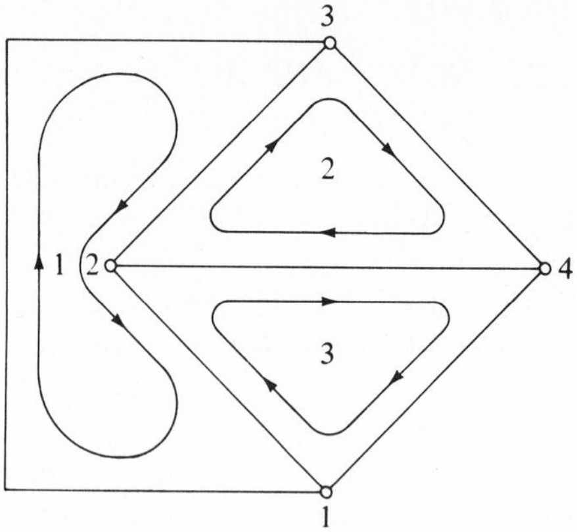

FIG. 6.24-C.

The loops of a network can usually be drawn in various ways. We sense that we need *three* loops here, and use those shown. (In complicated cases where it is not obvious, graph theory has machinery for calculating the number of loops required in an analysis and for selecting them; it also discusses certain arbitrarinesses in node selection.)

We can now apply either or both of Kirchhoff's summation laws (postulates):

At a node the sum of the entering (or leaving) currents is zero; (6.24-3)

Around a loop the sum of the voltage rises (or drops) is zero. (6.24-4)

Either law alone always suffices, though not necessarily with equal facility; sometimes a mixture of the two is convenient. Note that the two laws are *duals* (Sec. 5.12). Neatly and systematically we can now obtain the simultaneous equations to be solved. We use capital letters, for variety and to avoid confusion with previous work; they represent, as usual, only the forced components of currents and voltages, the excitation being $E_0 e^{st}$.

Application of Kirchhoff's current law (6.24-3) will give us simultaneous equations in which node voltages are the unknowns. But "node voltage" is meaningless; a "voltage" must represent the potential *of* some point *with respect to some point.* In defining our node voltages it is convenient to take one of the visible nodes as a common reference point; the choice is arbitrary, but we choose node 4 (Fig. 6.24-D) since then V_2, the voltage of node 2 with respect to node 4, is exactly the unknown of most interest, the detector voltage. Then V_1, V_2, V_3 represent the voltages of nodes 1, 2, 3 with respect to node 4.

In these terms, (6.24-3) applied to node 1 gives

$$Y_2(V_1 - V_2) + Y_3 V_1 + Y_0(V_1 + E_0 - V_3) = 0 \qquad (6.24\text{-}5)$$

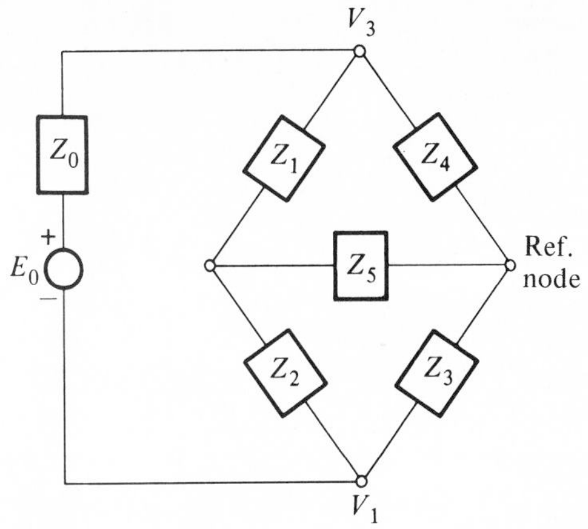

FIG. 6.24-D.

in which each of the three terms represents one of the three currents *leaving* node 1. (Note that the loop currents would be of no use here; we must calculate currents by Ohm's law in terms of node voltages and admittances.) Then at nodes 2 and 3 we find, similarly,

$$Y_2(V_2 - V_1) + Y_1(V_2 - V_3) + Y_5 V_2 = 0 \qquad (6.24\text{-}6)$$

and

$$Y_1(V_3 - V_2) + Y_4 V_3 + Y_0(V_3 - E_0 - V_1) = 0. \qquad (6.24\text{-}7)$$

Here are three equations in three unknowns, ready for simultaneous solution. We first rearrange or symmetrize them, collecting terms and placing knowns on the right side:

$$\left.\begin{aligned}(Y_0 + Y_2 + Y_3)V_1 + (-Y_2)V_2 + (-Y_0)V_3 &= -Y_0 E_0,\\ (-Y_2)V_1 + (Y_1 + Y_2 + Y_5)V_2 + (-Y_1)V_3 &= 0,\\ (-Y_0)V_1 + (-Y_1)V_2 + (Y_0 + Y_1 + Y_4)V_3 &= +Y_0 E_0.\end{aligned}\right\} \qquad (6.24\text{-}8)$$

It is now clear that we could have written (6.24-8) by inspection. The first term, $(Y_0 + Y_2 + Y_3)V_1$, represents the current that flows out of node 1 when all the other voltages are zero. The second term, $(-Y_2)V_2$, represents the current that flows out of node 1 when all voltages except V_2 are zero. The third term, $(-Y_0)V_3$, represents the current that flows out of node 1 when only V_3 is not zero. Their superposition (sum) is the total current that flows out of node 1 under normal conditions, excepting the excitation current. That is in the term $(-Y_0 E_0)$, which represents the (known) current equivalent of the voltage excitation. [Current is an excitation form more fitting to node-voltage-based analysis than voltage, and the mathematics has in effect automatically exchanged the voltage source E_0 of internal impedance Z_0 for an equivalent (Sec. 6.02) current source $I_0 = Y_0 E_0$, with the same internal immittance (Fig. 6.24-E).] This is, with proper sign, of course equal to the sum of the (other) currents leaving node 1, calculated above, which fact the equation states. Note that $(-Y_0 E_0)$ is the current flowing *into* node 1, as excitation, which befits its position on the right side. With hindsight we see that if this source exchange were made first, by us, (6.24-8) would be even easier to write by inspection. (This is but one of various devices and considerations that a complete study of the general method would give.)

The first equation in (6.24-8) superposes the four components, giving the total-current equation (6.24-3) for node 1. The other two equations can similarly be written by inspection of the connections to nodes 2 and 3, respectively.

Before solving (6.24-8) we observe that application of Kirchhoff's voltage law (6.24-4) will give us simultaneous equations in which loop currents are the unknowns. We use the loop currents defined in Fig. 6.24-F and obtain, writing the sums of the voltage drops (following the loop current directions) around each loop in turn,

$$\begin{aligned} Z_0 I_1 + Z_1(I_1 - I_2) + Z_2(I_1 - I_3) - E_0 &= 0, \\ Z_1(I_2 - I_1) + Z_4 I_2 + Z_5(I_2 - I_3) &= 0, \\ Z_2(I_3 - I_1) + Z_5(I_3 - I_2) + Z_3 I_3 &= 0. \end{aligned} \tag{6.24-9}$$

Rewriting in neater (symmetrized) form, as before, gives

$$\left.\begin{aligned} (Z_0 + Z_1 + Z_2)I_1 + (-Z_1)I_2 + (-Z_2)I_3 &= E_0, \\ (-Z_1)I_1 + (Z_1 + Z_4 + Z_5)I_2 + (-Z_5)I_3 &= 0, \\ (-Z_2)I_1 + (-Z_5)I_2 + (Z_2 + Z_3 + Z_5)I_3 &= 0. \end{aligned}\right\} \tag{6.24-10}$$

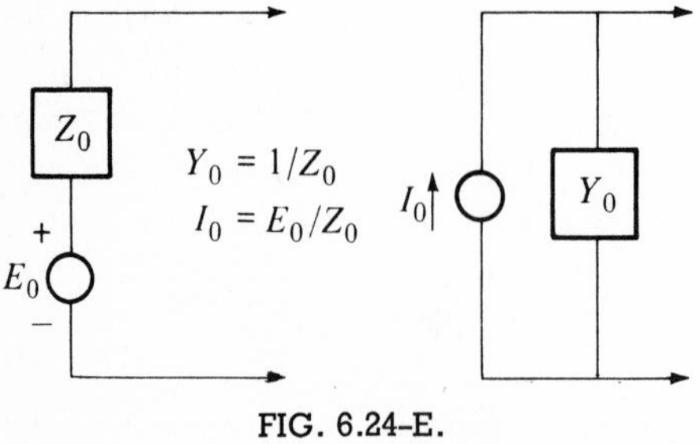

FIG. 6.24-E.

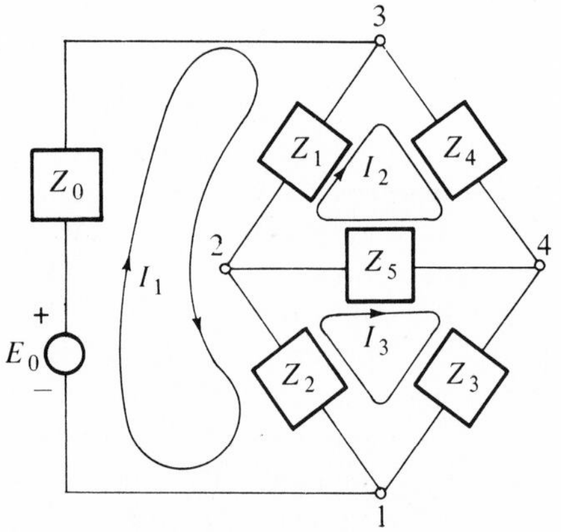

FIG. 6.24-F. The loop basis.

As on the node basis, the symmetrical form (6.24-10) could, with experience, be written by inspection. There is here no source exchange, since a voltage source "fits" the loop basis.

Writing node-based or loop-based equations is not always so simple, but a symmetrical form, like (6.24-8) or (6.24-10), can always be obtained.

With the well-known tools of the algebra of simultaneous linear equations (determinants and Cramer's rule) we can formally write the solutions of (6.24-8) and (6.24-10) immediately. The node voltages are

$$\begin{aligned} V_1 &= Y_0 E_0 \frac{-{}_N\Delta_{11} + {}_N\Delta_{31}}{{}_N\Delta}, \\ V_2 &= Y_0 E_0 \frac{-{}_N\Delta_{12} + {}_N\Delta_{32}}{{}_N\Delta}, \\ V_3 &= Y_0 E_0 \frac{-{}_N\Delta_{13} + {}_N\Delta_{33}}{{}_N\Delta}, \end{aligned} \tag{6.24-11}$$

in which

$${}_N\Delta = \begin{vmatrix} (Y_0 + Y_2 + Y_3) & (-Y_2) & (-Y_0) \\ (-Y_2) & (Y_1 + Y_2 + Y_5) & (-Y_1) \\ (-Y_0) & (-Y_1) & (Y_0 + Y_1 + Y_4) \end{vmatrix} \tag{6.24-12}$$

and the determinants in the numerators are the corresponding cofactors of ${}_N\Delta$. For example,

$${}_N\Delta_{12} = \begin{vmatrix} (-Y_2) & (-Y_1) \\ (-Y_0) & (Y_0 + Y_1 + Y_4) \end{vmatrix} \tag{6.24-13}$$

The subscript N is attached to avoid confusion of these determinants with those of the loop basis (see below).

The loop currents are

$$\begin{aligned} I_1 &= E_0 \frac{{}_L\Delta_{11}}{{}_L\Delta}, \\ I_2 &= E_0 \frac{{}_L\Delta_{12}}{{}_L\Delta}, \\ I_3 &= E_0 \frac{{}_L\Delta_{13}}{{}_L\Delta}, \end{aligned} \tag{6.24-14}$$

in which

$$
{}_L\Delta = \begin{vmatrix} (Z_0 + Z_1 + Z_2) & (-Z_1) & (-Z_2) \\ (-Z_1) & (Z_1 + Z_4 + Z_5) & (-Z_5) \\ (-Z_2) & (-Z_5) & (Z_2 + Z_3 + Z_5) \end{vmatrix} \qquad (6.24\text{-}15)
$$

and ${}_L\Delta_{11}$, ${}_L\Delta_{12}$, and ${}_L\Delta_{13}$ are cofactors thereof.

The expansion of these determinants, however tedious, is straightforward algebra. To do it completely, with specific (rational) functions of s for each immittance is an unpleasant job; fortunately it is seldom demanded in engineering. (The results are, of course, the same as those of Sec. 6.23.) But in principle finding any transfer function is straightforward. For example, from voltage source e to detector voltage v_D (Fig. 6.23-E) we have

$$
T = \frac{v_D}{e} = \frac{V_2}{E_0} = Y_0 \frac{-{}_N\Delta_{12} + {}_N\Delta_{32}}{{}_N\Delta} = Z_5(I_3 - I_2) = Z_5 \frac{{}_L\Delta_{13} - {}_L\Delta_{12}}{{}_L\Delta}. \qquad (6.24\text{-}16)
$$

Also straightforward is a complete analysis, for example finding the impulse response or the response to exponential or other excitation.

The amount of labor involved in analysis of the bridge by each of the three methods—work-back (Sec. 6.23), node equations, loop equations—is about the same. Node basis and loop basis offer a certain elegance but require facility in handling determinants. By no method can the labor be avoided!

The natural-frequency-determining equation can be obtained by each method, as

$$
\begin{aligned}
D &= 0 \qquad \text{from (6.23-11)}, \\
{}_N\Delta &= 0 \qquad \text{from (6.24-11)}, \\
{}_L\Delta &= 0 \qquad \text{from (6.24-14)},
\end{aligned} \qquad (6.24\text{-}17)
$$

all of which lead, of course, to the same characteristic equation, a certain polynomial equals zero.

We note that the two general (node-based and loop-based) formulations of the analysis of the bridge, (6.24-8) and (6.24-10), are strikingly similar. This is related to the fact that the bridge is geometrically its own dual, a property that few networks have. The bridged-T network is of course included; it is simply a redrawn bridge.

It is now interesting to apply such general methods to ladders, to see if they offer better approaches than our familiar work-back (Sec. 6.02) method. The determinants there do have many zero elements, but as one might expect, the work-back method is both simpler and more informative.

The two general methods we have briefly looked at can be used to analyze *any* network. We content ourselves here with this application to the bridge, turning now to another important type of nonladder, yet almost-ladder, network.

6.25 THE TRANSFER FUNCTIONS OF SOME ACTIVE NETWORKS

An *active* element contains a (hidden) controlled energy source; our familiar g_m element (Fig. 6.25-A), introduced originally (Sec. 1.05) as an essential part of an amplifier, is an active element. A ladder network driven (excited) by such an element (Fig. 6.25-B) is still susceptible to our work-back-from-the-output method of analysis, still a ladder. So is a cascade connection of such networks, that of Fig. 6.25-C for example, and its transfer function is equally easy to calculate. Note that the energy delivered in any of these networks to the output (or any other) impedance is infinitely greater than the energy input, which is zero because $i = 0$. There is nothing unreasonable about this, since each controlled current source represents a source of energy. It is one of the important attributes of *active networks*, that is of networks that contain one or more active elements.

Another, hitherto unappreciated attribute of the g_m active element is its *unilateral* character: voltage and current conditions to its right in no way affect conditions to its left (Fig. 6.25-A). The converse is of course *not* true: e_c affects i_c! If, as is common, the amplifier is intended to be a one-way device, this is fine. For amplification in both directions, some more complicated system must be used. But the unilateral property leads to interesting behavior even without such complexity.

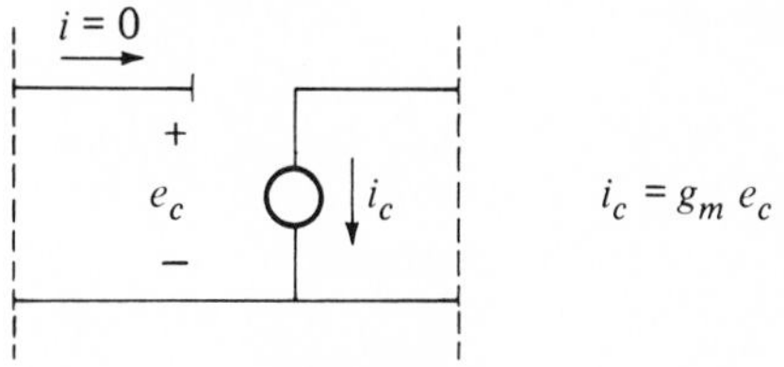

FIG. 6.25-A.

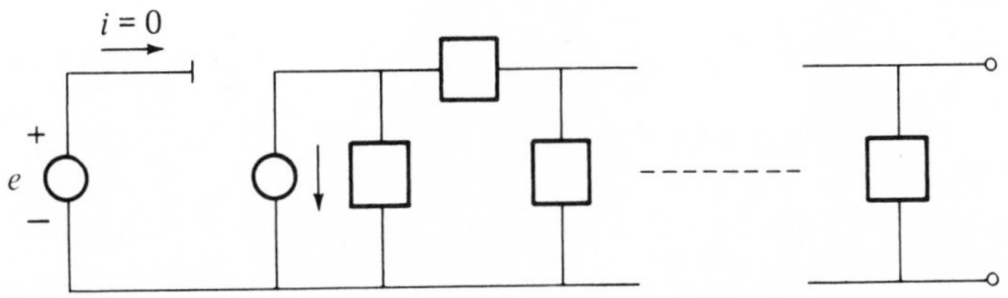

FIG. 6.25-B.

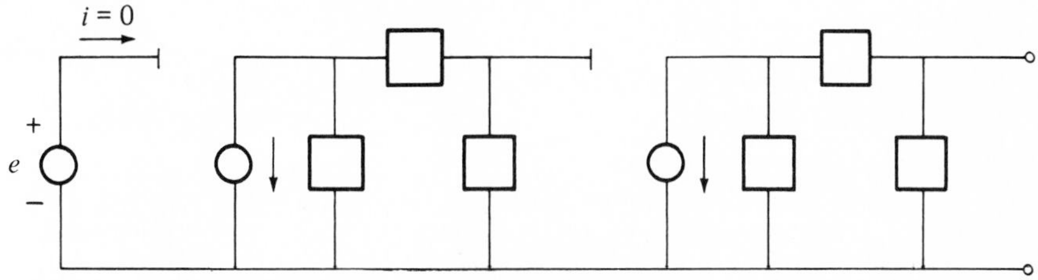

FIG. 6.25-C.

An essential feature of the bridge network's behavior can be vividly if highly schematically represented by the diagram of Fig. 6.25-D. The diagram displays two paths from input to output, emphasizing that the response has two components, the property that makes possible bridge *balance* (reduction of output to zero by subtle cancellation rather than by brute-force or resonance rejection), if only for certain values of s. It is in no sense a diagram of the actual bridge circuit; and it is certainly incorrect even as a purely schematic diagram of the bridge's behavior, for it implies that there is no response at the (usual) excitation point when excitation is applied at the (normal) response (detector) location, which is wrong; the bridge is not at all unilateral.

But the diagram suggests another one, that of Fig. 6.25-E. If this can be implemented it may have interesting properties indeed, because of its closed-loop, signal-feedback nature. And we can surely construct networks with this property by using unilateral elements to implement the unidirectional signal flow implied. A simple specific example is the network of Fig. 6.25-F. It uses one unilateral and active element g_m, and two ordinary (passive) impedances Z_f and Z_L. [The physical apparatus modeled by this one g_m element may well use a number of transistors (or other active devices) and may consist of several stages in tandem; the sign of g_m need not therefore be positive.]

The network seems still essentially a ladder. To find the transfer function

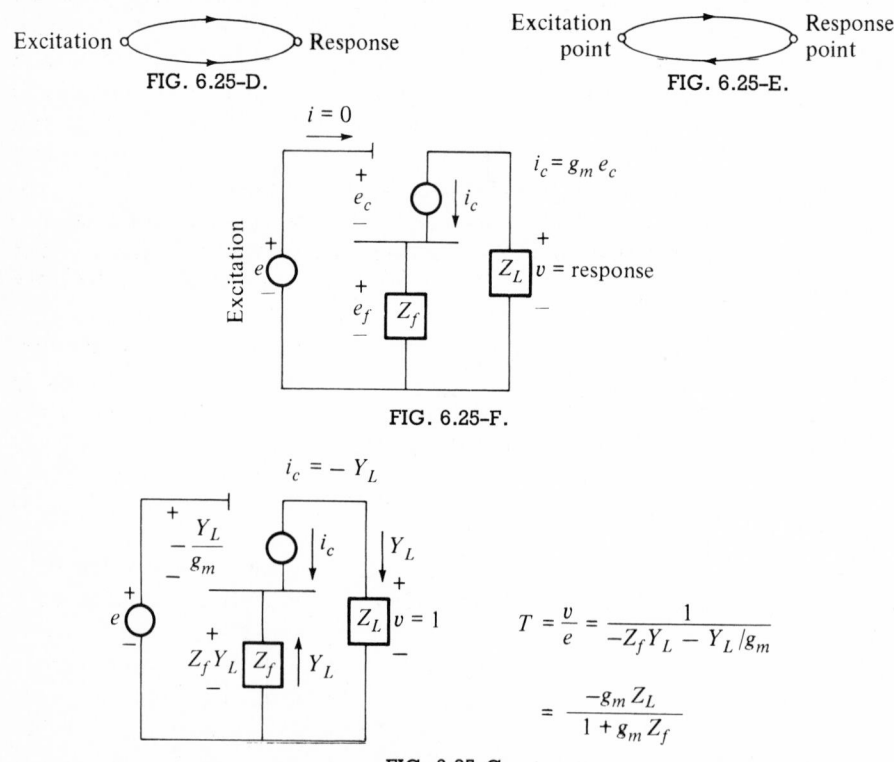

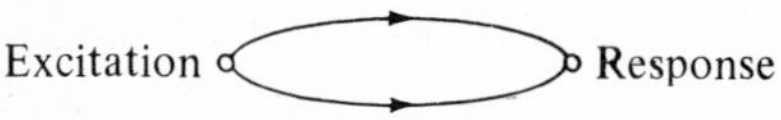

FIG. 6.25-D.

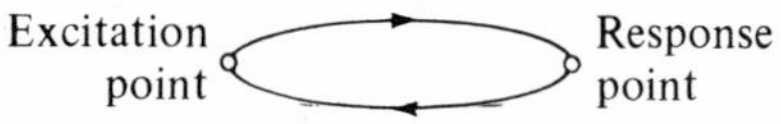

FIG. 6.25-E.

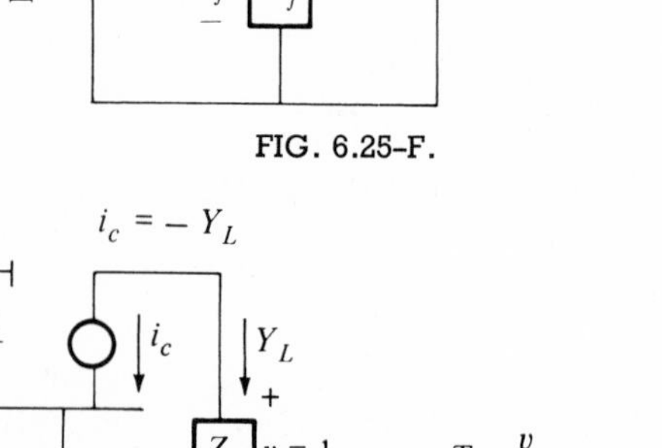

FIG. 6.25-F.

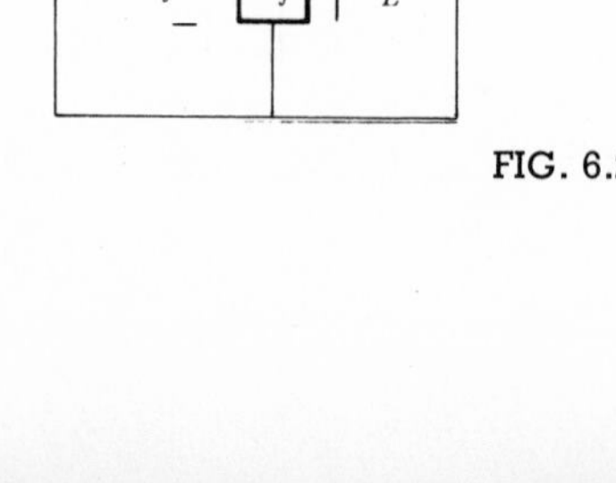

$$T = \frac{v}{e} = \frac{1}{-Z_f Y_L - Y_L/g_m}$$

$$= \frac{-g_m Z_L}{1 + g_m Z_f}$$

FIG. 6.25-G.

we accordingly assume $v = 1$ and try to calculate back to the corresponding value of e (with the usual assumptions as to exponential excitation, etc.). The calculations are simple and can be made directly on the diagram (Fig. 6.25-G); the detailed explanation given in (6.25-1) below is not necessary but may be helpful at first encounter.

$$\begin{aligned}
v &= 1,\\
i_c &= -Y_L = g_m e_c,\\
e_c &= \frac{1}{g_m} i_c = -\frac{Y_L}{g_m},\\
e_f &= Z_f i_c = -Z_f Y_L,\\
e &= e_f + e_c = -Z_f Y_L - \frac{Y_L}{g_m} = -Y_L\left(Z_f + \frac{1}{g_m}\right),\\
T &= \frac{v}{e} = \frac{1}{-Y_L(Z_f + 1/g_m)} = \frac{-g_m Z_L}{1 + g_m Z_f}.
\end{aligned} \tag{6.25-1}$$

The work-back-from-the-output method is here satisfactory: we can calculate the transfer function more or less as though the network were a true ladder. But there is a disturbing element in the analysis: the step of calculating e_c from i_c, according to $e_c = i_c/g_m = -Y_L/g_m$, though mathematically correct, is physically disquieting. The physical action is not that way but the reverse: it is e_c that controls i_c, not i_c that controls e_c. (The actual electronic action is of course not shown in the model, which merely represents, in simplified form, a very complicated electronic phenomenon.) For a ladder network the backward calculation is straightforward, even though one work backward through unilateral elements, because the "flow" is strictly one-way. But here, with feedback around a unilateral element, we become uneasy about it.

This uncomfortable feeling can perhaps be allayed if we start the work-backward calculation at the control point, the heart of the physical action, by writing $e_c = 1$ and then continuing. Then we have $i_c = g_m$, the voltages e_f and v are immediately evident, and the lockup equation closes the feedback loop both mathematically and physically by stating the relation between e_f, e, and e_c. Since e_c was assumed unity we immediately have e and so the transfer function (Fig. 6.25-H).

Why should starting out from e_c rather than from v also make the calculation *simpler*, as it clearly does (see Figs. 6.25-G and 6.25-H)? Fundamentally because we are going *with* Nature, *following* the physical action, and not against Her, not opposing the physical action. We recognize and *use* the cause-effect relation in the controlled source; to do so, because of the feedback, is all-important. The response is fed back to the excitation and affects it (this does not happen in ladders); to recognize and *use* this flow simplifies the analysis.

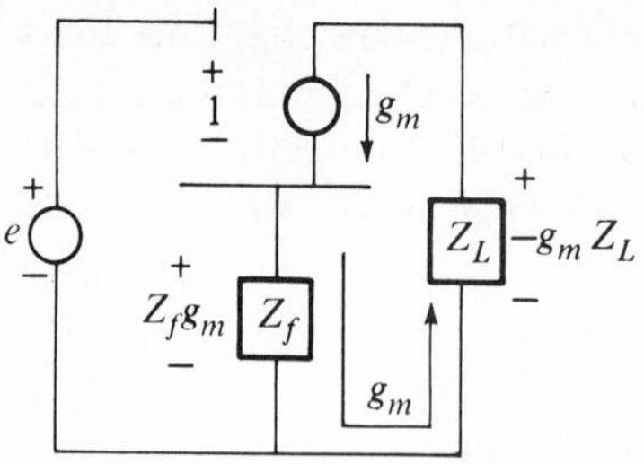

$$T = \frac{v}{e} = \frac{-g_m Z_L}{1 + g_m Z_f}$$

FIG. 6.25-H.

We evidently have here another nonladder network, but one that is again amenable to only slightly modified work-backward analysis.

The concept of signal *flow* becomes important in networks with unilateral elements and feedback. (In ladders the signal flow is so straightforward we do not even think about it.) It is in fact worth developing the idea presented by Fig. 6.25-E, the idea of diagramming the flow of signals; but we must make it precise. Suppose for our present circuit we proceed to draw Fig. 6.25-I*b* as a careful, detailed, and accurate signal-flow diagram. We construct it strictly from the basic equations given in Fig. 6.25-I*a*; these are merely *pictured* by Fig. 6.25-I*b*. The (directed) single lines therein ("branches") and the nodes (small circles) that they join schematically represent the actions described by the equations according to the conventions below.

At each *node* a signal exists, according to its label; this "signal" can be a current or a voltage (or other electrical or mechanical or other variable in

Equation	Physical Action Expressed by Equation
$e_c = e - e_f$	Summation by the series electrical connection
$i_c = g_m e_c$	Behavior of the active, unilateral element
$e_f = Z_f i_c$	The flow of i_c (alone) through Z_f
$v = Z_L i_c$	The flow of i_c (alone) through Z_L

(*a*)

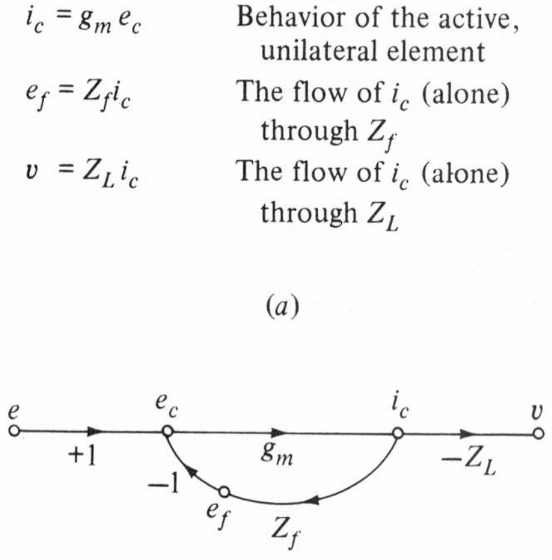

(*b*)

FIG. 6.25-I.

general, since the signal-flow diagram is schematic only and not an actual connection diagram). This signal is made equal, by the node's action, to the *sum* of all incoming signals, signals on branches attached to and arriving at the node; it is sent out by the node on each outgoing branch. A node thus represents a summation action and a transmission.

Each *branch* (line joining two nodes) represents a *unilateral* signal-flow path, passage along which (in the arrow direction, of course) modifies a signal according to the transfer function attached to the branch. [We continue here to make our usual assumptions, discussing only forced components of responses to exponential excitation; but signal-flow graphs can be used to describe actual (total) quantities, differential and integral equations, and other phenomena.]

Note that the (forward) g_m branch is unilateral because of the physical nature of the active element. The (feedback) Z_f branch is *in effect* unilateral because $i = 0$ in Fig. 6.25-F: e cannot transmit forward except through g_m. (When it *can* do so, some of the simplicity of the network's operation departs.)

Further development and discussion of the algebra of manipulation of flow graphs we leave for study elsewhere; they can be rather useful.

Note how clearly the signal-flow diagram displays the inherent closed-loop (feedback) character of the network: contrast Figs. 6.25-F and 6.25-I*b*. Note also the use of nonloop branches for convenience: to set out the excitation e where it can be seen, to attach the correct sign to the feedback signal e_f at the summation point, to display the response v separately and clearly.

If we now recast Fig. 6.25-I*b* in more general, but still single-feedback-loop, form, as suggests itself strongly, we have Fig. 6.25-J, which "fits" many feedback amplifiers and also many automatic control systems. The control systems briefly described in Sec. 1.03 (see Fig. 1.03-B especially) have signal-flow diagrams like that of Fig. 6.25-J, in which

e is the command signal (the desired position of the load),

r is the actual load position,

$$T_{\text{in}} = T_{\text{out}} = 1,$$

$$T_2 = -1, \qquad (6.25\text{-}2)$$

and so

$e_c = e - r =$ error signal, which motivates the apparatus to (try to) reduce the error to zero as desired.

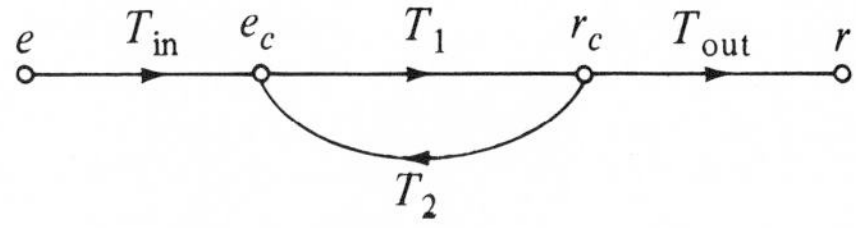

FIG. 6.25-J.

To the extent that the control-system apparatus is linear, such systems are essentially networks, in analog form, and much of our network analysis can be applied thereto. But the subject is very large; we leave it for study elsewhere.

The transfer function from e to r in Fig. 6.25-J is easily derived thereon, almost by inspection:

$$\begin{aligned} e_c &= T_{\text{in}}\, e + T_2 r_c = T_{\text{in}}\, e + T_2(T_1 e_c), \\ T_{\text{in}}\, e &= e_c(1 - T_1 T_2), \\ r &= T_{\text{out}}\, r_c = T_{\text{out}}(T_1 e_c), \end{aligned} \tag{6.25-3}$$

so that

$$T = \frac{r}{e} = \frac{T_{\text{out}}\, T_1}{(1 - T_1 T_2)/T_{\text{in}}} = T_{\text{in}}\, T_{\text{out}} \frac{T_1}{1 - T_1 T_2}. \tag{6.25-4}$$

Or one may write $e_c = 1$ and calculate as in Fig. 6.25-H. Note that there, for the simple feedback amplifier of Fig. 6.25-F [see Fig. 6.25-I],

$$T_{\text{in}} = 1, \qquad T_1 = g_m, \qquad T_2 = -Z_f, \qquad T_{\text{out}} = -Z_L \tag{6.25-5}$$

and so (of course)

$$T = T_{\text{in}}\, T_{\text{out}} \frac{T_1}{1 - T_1 T_2} = -Z_L \frac{g_m}{1 + g_m Z_f} \tag{6.25-6}$$

as in Fig. 6.25-H.

We note that the transfer function without feedback (i.e., with $T_2 = 0$), which is $(T_{\text{in}}\, T_{\text{out}})T_1$, is modified, by feedback, by division by a "feedback factor" $(1 - T_1 T_2)$. This factor, a rational function of s with real coefficients of course (Sec. 5.05), evidently determines the *natural* behavior of the network. For the poles of T, the roots of

$$1 - T_1 T_2 = 0, \tag{6.25-7}$$

are the natural frequencies of the network with feedback.

That this should be so is made physically clearer by the device of Fig. 6.25-K. There we have split the summation-point node into two parts: the signal e_c, here denoted e_R, is still equal to the sum of its two incoming components, but its outward transmission has been interrupted; instead we have attached a separate, artificial, unit ($1e^{st}$) signal generator to feed on forward through the network. (We are concerned here only with the natural behavior of the network, so that the excitation e is irrelevant, as is v, and both are excluded from the diagram.) The signal *returned* at the left of the break, e_R, in response to unit excitation at

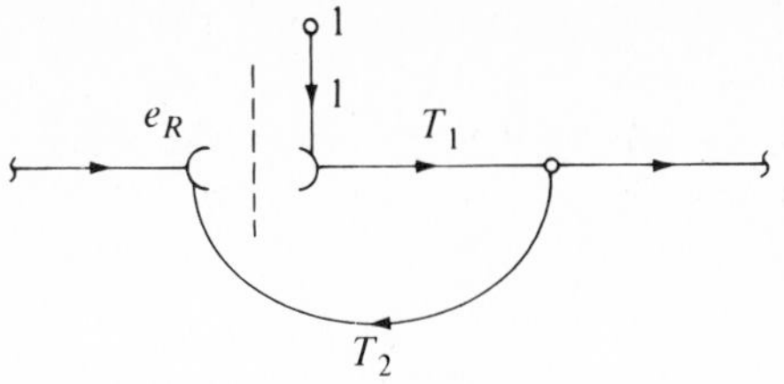

FIG. 6.25-K.

the right, is the *return ratio* for such a break. The *difference* between the signals on the two sides of the break is the

$$\textit{return difference} = 1 - e_R = 1 - T_1 T_2 \qquad (6.25\text{-}8)$$

associated with the break. This (rational) function of s is evidently equal to the feedback factor mentioned above at (6.25-7). It plays an important role in the general theory of feedback (see BO of Appendix B).

That natural behavior corresponds to zero return difference [(6.25-7)] is vividly reinforced now by the following argument. For certain values of s the signal returned to the left of the break in response to the unit excitation $1e^{\alpha t}$ at the right will be exactly equal to that excitation signal; such values of s are the zeros of the return difference. We can then simultaneously remove the generator and reconnect the two halves of the node. The circuit, sensing no difference, will continue to "oscillate" or behave as e^{st}. Since there is now no excitation, e^{st} is a natural mode, or s is a natural frequency. Such values of s, the natural frequencies, are, once more, the roots of the equation (6.25-7), or the equation

$$\text{return difference} = 0. \qquad (6.25\text{-}9)$$

(In complicated systems, return differences may be computed for various elements, at various points; each one set equal to zero, however, will give the same characteristic equation.)

It is a danger of feedback (anticipated in the second example in Sec. 1.03, q.v.) that one or more of the natural frequencies may lie in the right half of the s plane. Then the natural modes contain *increasing* exponentials, as $e^{\alpha t}$ or $e^{\alpha t} \cos \omega_N t$. Such *unstable* behavior is obviously undesirable or even unsafe; practical design problems are associated therewith. In our simple active network (Fig. 6.25-F), the return difference of the active element is $(1 + g_m Z_f)$, which can be written as $g_m(1/g_m + Z_f)$. Since this is proportional to the impedance of a (positive) resistor and the passive impedance Z_f in series, its zeros, the natural frequencies, are in the left half plane and the network is *stable*. More complicated feedback networks are not necessarily stable and must be carefully designed and operated.

The *advantages* of feedback are several, however, and its use is widespread;

its theory is a well-developed, separate art. First of all, if the loop gain, $|T_1T_2|$, is made large compared to unity, for certain important or useful values of s (in certain important frequency bands in the sinusoidal steady state, for example), then

$$T = T_{\text{in}}\,T_{\text{out}}\frac{T_1}{1 - T_1T_2} \cong T_{\text{in}}\,T_{\text{out}}\frac{T_1}{-T_1T_2} \cong (-T_{\text{in}}\,T_{\text{out}})\frac{1}{T_2}. \quad (6.25\text{-}10)$$

This means that the basic amplifier transmission (transfer) function depends chiefly on T_2 and is rather insensitive to changes in T_1. Now the active elements are usually in the forward (T_1) path, and it is they that tend to age, to vary in value with the passage of time. The passive elements that usually form the feedback (T_2) path are relatively stable with time. Feedback, by reducing the sensitivity of the overall transfer function to T_1, therefore *stabilizes* the performance of the amplifier so that it changes little, even though the active (amplifying) elements lose gain. One must pay a price for this of course, in the need for more amplification (gain) in T_1 than would be needed without feedback. For example, 20 dBV (see Appendix D) of amplification in the sinusoidal steady state at a frequency ω_0 can be provided by a simple amplifier with that much gain (Fig. 6.25-L*a*) *or* by a feedback amplifier with a large forward gain and 20 dBV of *loss* (attenuation) in the feedback path, as in Fig. 6.25-L*b*. (We assume no phase shifts at the frequency ω_0.) If, with the passage of time, aging of elements reduces T_1 from $1000\,\underline{/0}$ to $800\,\underline{/0}$ (a drop of 1.9 dBV, from 60 to 58.18 dBV) as in Fig. 6.25-L*c*, then T changes only from 9.90 to 9.88 (a drop from 19.91 to 19.89 dBV). This is a mere 0.2 percent drop in amplification (0.02 dBV) even for a 20 percent (1.9 dBV) drop in the forward-path gain. (The ratio, $20 : 0.2 = 100 : 1$, is essentially the size of the feedback factor or return difference, about 100, to which it may be attributed.)

The improvement in the sensitivity of amplifier gain with respect to active-element variations, evidenced by (6.25-10), has another facet. The functional behavior of the overall transfer function is determined, for values of s for which $|T_1T_2| \gg 1$, chiefly by $T_2(s)$ and very little by $T_1(s)$. Design problems (the variation of amplifier gain with frequency, for example) may thus be moved to the passive elements in T_2, usually simplifying them.

Another important property of feedback is an ability to reduce distortion, which some active elements are prone to introduce, as by having slightly nonlinear characteristics. The impedances presented by an amplifier at its input and its output are also greatly affected by feedback. Feedback evidently offers one more field of intense interest, now ripe for detailed study elsewhere.

The analysis of networks that contain feedback is not necessarily easily made by the work-back method, even when we start at a control point.

As the network's geometry departs more and more from the ladder's, even the signal-flow graph becomes difficult to draw. Eventually one must have recourse to the general loop- or node-based methods of Sec. 6.24, with very careful attention to the active elements.

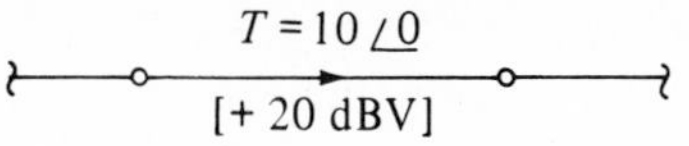

(a)

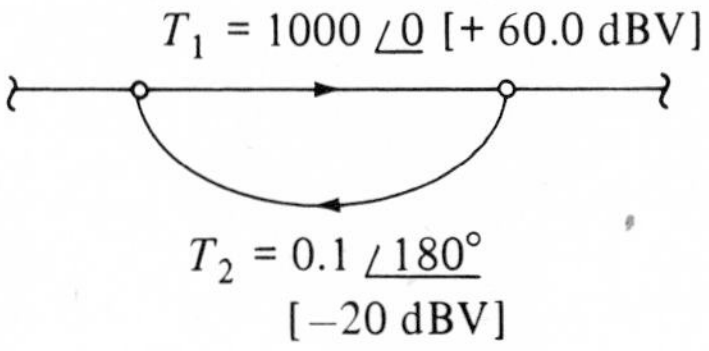

(b)

$$T = \frac{T_1}{1 - T_1 T_2} = \frac{1000}{1-(1000)(-0.1)} = \frac{1000}{1+100} = 9.90 \ [19.91 \text{ dBV}]$$

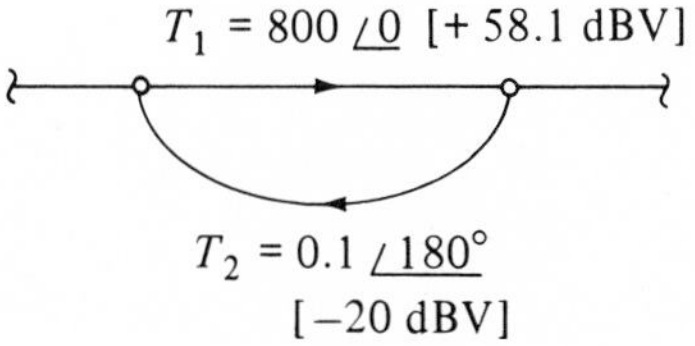

(c)

$$T = \frac{800}{1-(800)(-0.1)} = \frac{800}{81} = 9.88 \quad [+ 19.89 \text{ dBV}]$$

NOTE: All calculations are for $s = j\omega_0$, a particular frequency of interest in the sinusoidal steady state at which all the transfer functions have zero angle (phase shift); the dBV unit is a useful bit of amplifier jargon (see Appendix D).

FIG. 6.25-L.

Let us modify the active element of Fig. 6.25-F by adding a finite output impedance Z_2 (Fig. 6.25-M*a*), making it somewhat more realistic. We proceed to analyze, starting from e_c, $i_c = g_m e_c$. We now have to determine the current in Z_f, perhaps by writing x as we did in Fig. 6.23-B. But this is a much simpler situation: no x is necessary, for the current $g_m e_c$ simply divides between two branches, Z_f and Z_L *and* Z_2. The current-divider relation (5.11-13) gives

$$i_f = \frac{(Z_f + Z_L)^{-1}}{(Z_f + Z_L)^{-1} + (Z_2)^{-1}}\, i_c = \frac{Z_2}{Z_2 + Z_f + Z_L}\, g_m e_c. \tag{6.25-11}$$

Then

$$e_f = Z_f i_f,$$

$$e = e_f + e_c = \left(1 + Z_f g_m \frac{Z_2}{Z_2 + Z_f + Z_L}\right) e_c, \tag{6.25-12}$$

$$T = \frac{v}{e} = \frac{-Z_L i_f}{e_f + e_c} = \frac{-Z_L g_m T_{cf}}{1 + g_m Z_f T_{cf}}.$$

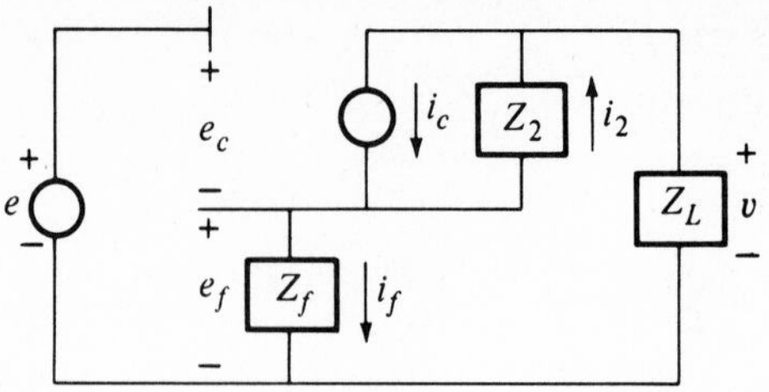

$i_c = g_m e_c$

(*a*)

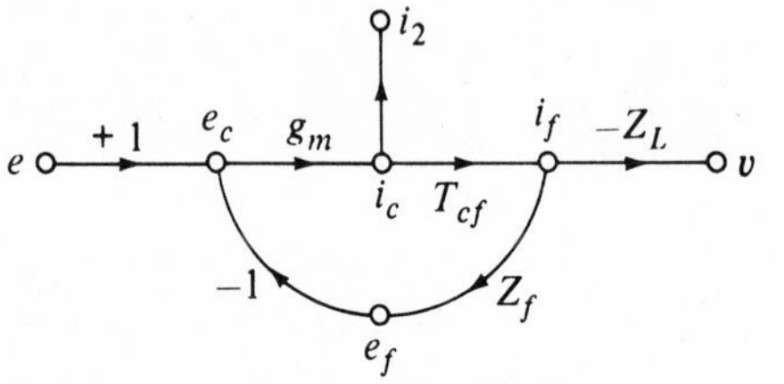

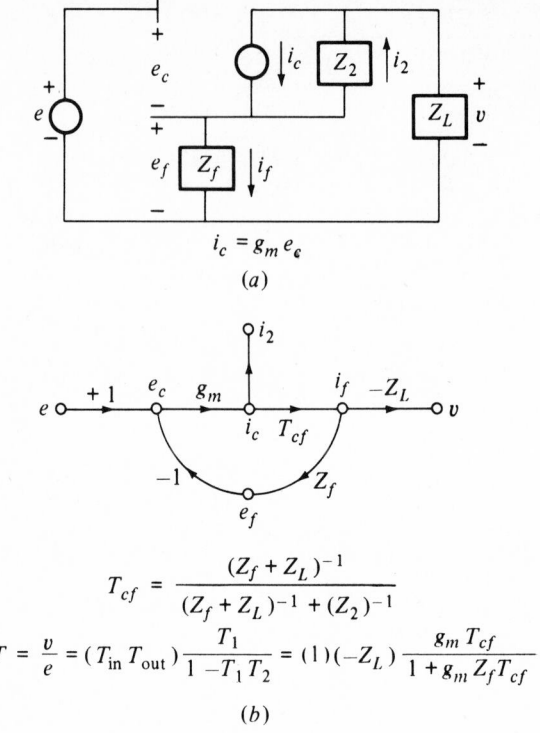

$$T_{cf} = \frac{(Z_f + Z_L)^{-1}}{(Z_f + Z_L)^{-1} + (Z_2)^{-1}}$$

$$T = \frac{v}{e} = (T_{in} T_{out}) \frac{T_1}{1 - T_1 T_2} = (1)(-Z_L) \frac{g_m T_{cf}}{1 + g_m Z_f T_{cf}}$$

(*b*)

FIG. 6.25-M.

The network's signal-flow graph that diagrams these equations and defines T_{cf} (Fig. 6.25-M*b*) shows the same simple, one-loop feedback character. This example has increased only slightly in complexity and presents no real problem in analysis.

Now let us modify the pristine active element of Fig. 6.25-F by adding a finite *input* impedance Z_1, making it more realistic in another way (Fig. 6.25-N*a*). The analysis, starting from e_c, now requires the calculation and superposition of two components for closing the loop to e_c. The controlled current i_c flows through the parallel combination of Z_f and Z_1 to produce the feedback component of e_c, as in Fig. 6.25-N*b*. The excitation *e* acts through a voltage divider to produce the input component (Fig. 6.25-N*c*). [Note that only the current i_c flows in the output (Z_L) loop; *e* has no *direct* effect on the current in Z_L.] The signal-flow diagram of Fig. 6.25-N*d* is again a simple one-loop one, easily drawn once the action is understood. The transfer function, calculated from (6.25-10), is

$$\begin{aligned} T = \frac{v}{e} &= \left[\frac{Z_1}{Z_1 + Z_f}\right]\{-Z_L\} \frac{g_m}{1 - (g_m)[-Z_f Z_1/(Z_1 + Z_f)]} \\ &= \frac{-g_m Z_L[Z_1/(Z_1 + Z_f)]}{1 + g_m[Z_f Z_1/(Z_f + Z_1)]}. \end{aligned} \qquad (6.25\text{-}13)$$

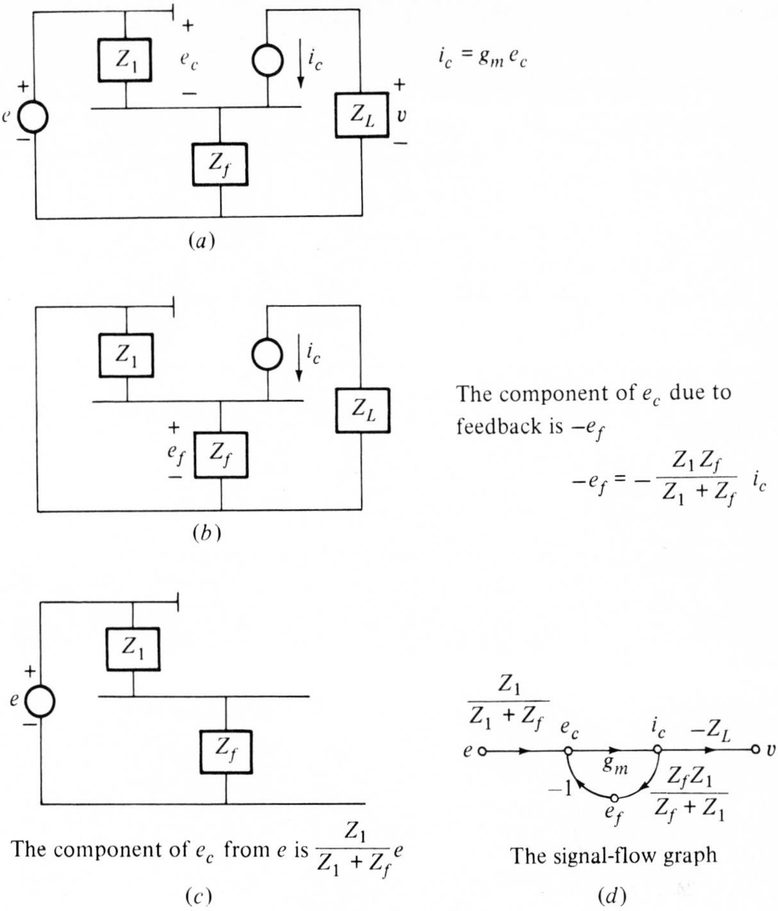

$i_c = g_m e_c$
(a)
The component of e_c due to feedback is $-e_f$
$-e_f = -\frac{Z_1 Z_f}{Z_1 + Z_f} i_c$
(b)
The component of e_c from e is $\frac{Z_1}{Z_1 + Z_f} e$
(c)
The signal-flow graph
(d)

FIG. 6.25-N.

It may equally easily be calculated by working back from $e_c = 1$ (much the same, really).

If we now include in the active-element model *both* Z_1 and Z_2, as in Fig. 6.25-O, a novelty appears: (forward) transmission from e to v can take place not only through the active element ("via g_m") but also along a purely passive-element path ("via Z_f, Z_2, Z_L"). In the previous two cases, infinite impedance in the input loop (Fig. 6.25-M) or in the output loop (Fig. 6.25-N) prevented purely-passive-path transmission and allowed us to calculate as we did. Now, even if the active element is "dead," $g_m = 0$, there is still transmission from e to v. The signal-flow diagram is no longer easy to draw, for the character of the network has changed; it no longer has the very simple one-loop character of Fig. 6.25-J. Nor is working back from $e_c = 1$ now a simple task; the network is far from a ladder! In fact the general loop basis of Sec. 6.24 is now the most straightforward method of analysis. On Fig. 6.25-O we trace the two

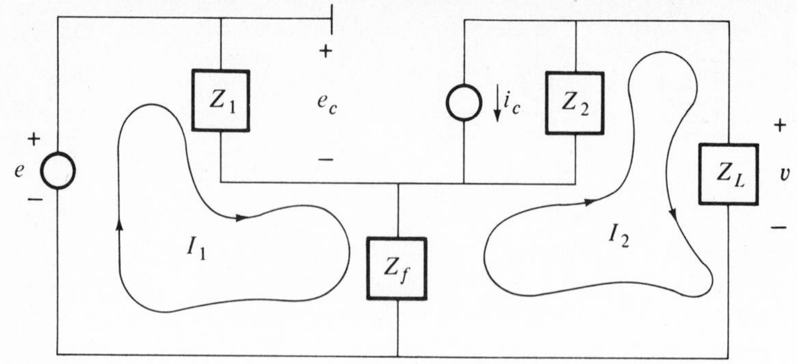

FIG. 6.25-O.

(arbitrary) loops shown. Going clockwise and writing voltage drops with the usual assumptions and $e = E_0 e^{st}$ gives us (note that Z_2 carries *two* currents):

$$\begin{aligned} -E_0 + Z_1 I_1 + Z_f(I_1 - I_2) &= 0, \\ Z_f(I_2 - I_1) + Z_2(I_2 + g_m Z_1 I_1) + I_2 Z_L &= 0, \end{aligned} \tag{6.25-14}$$

or

$$\begin{aligned} (Z_1 + Z_f)I_1 + (-Z_f)I_2 &= E_0, \\ (-Z_f + g_m Z_1 Z_2)I_1 + (Z_f + Z_2 + Z_L)I_2 &= 0. \end{aligned} \tag{6.25-15}$$

Solving gives

$$\begin{aligned} T = \frac{v}{e} = \frac{Z_L I_2}{E_0} &= Z_L \frac{-(-Z_f + g_m Z_1 Z_2)}{\begin{vmatrix} (Z_1 + Z_f) & (-Z_f) \\ (-Z_f + g_m Z_1 Z_2) & (Z_f + Z_2 + Z_L) \end{vmatrix}} \\ &= Z_L \frac{(-g_m Z_2 + Z_f Y_1)/Z_s}{1 + Z_f[(g_m Z_2 - Z_f Y_1)/Z_s + Y_1]} \end{aligned} \tag{6.25-16}$$

in which

$$Z_s = Z_2 + Z_L + Z_f \qquad \text{and} \qquad Y_1 = \frac{1}{Z_1}. \tag{6.25-17}$$

In the second form in (6.25-16) the transfer function has been forced into a form in which the denominator has the familiar return-difference "1 + ( )" form.

Work-back methods, eminently suited to ladder networks and to simple feedback networks, must give way, sooner or later as network complexity increases, to general methods. Fortunately, the former usually suffice.

Here is a simpler active-network example, one that uses a different type of

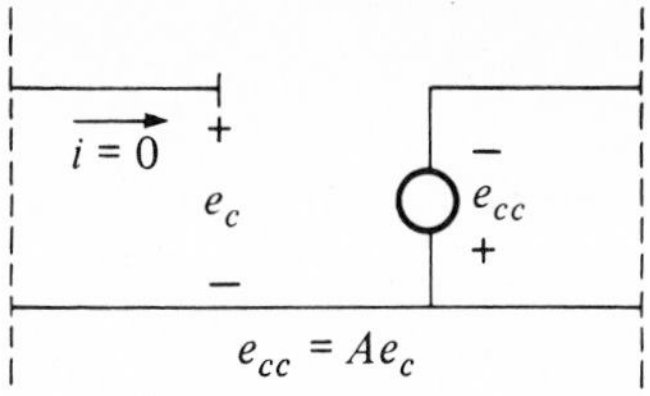

FIG. 6.25-P.

active element (and there are still others). In it the controlling voltage e_c controls a *voltage* source e_{cc} (Fig. 6.25-P). The ratio A is a real (dimensionless) number that can have either sign and can be quite large; it plays a role like that of g_m in Fig. 6.25-A. The physical apparatus modeled by Fig. 6.25-P (often called an *operational amplifier* or simply "opamp") actually contains a number of transistors (as often does that modeled by Fig. 6.25-A); the model is greatly simplified. The device is very useful at low frequencies (values of s) and may well be developed in forms useful at high frequencies.

A simple application of great utility is the network shown in Fig. 6.25-Q. A (voltage) signal source e feeds the (active) opamp augmented by two (passive) impedances Z_0 and Z_f; the output is the voltage v. (The load impedance is irrelevant to the analysis since v is developed across a zero impedance, the source e_{cc}.) The analysis is straightforward. Working from e_c we have

$$
\begin{aligned}
e_{cc} &= Ae_c, \\
v_f &= e_{cc} + e_c, \\
i &= Y_f v_f, \\
e &= e_c + Z_0 i, \\
T &= \frac{v}{e} = \frac{-e_{cc}}{e_c + Z_0 i} = \frac{-Ae_c}{e_c + Z_0 Y_f(A+1)e_c} = \frac{-A}{1 + Z_0 Y_f(A+1)}.
\end{aligned}
\tag{6.25-18}
$$

The signal-flow diagram given in Fig. 6.25-R differs slightly from our previous ones, but is essentially of the same simple, single-feedback-loop character. The output v does seem to come from the middle of the forward portion of the

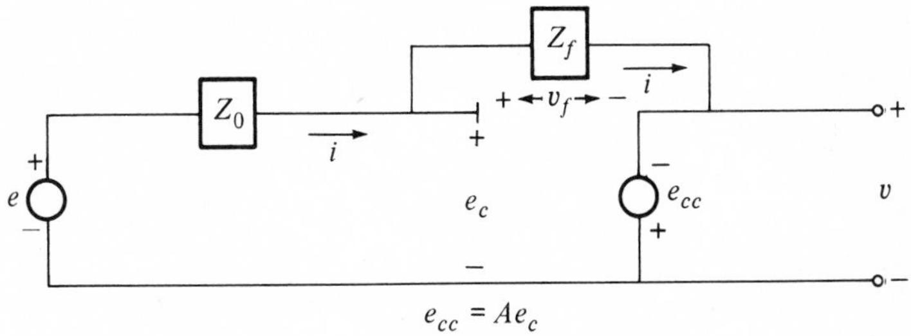

FIG. 6.25-Q.

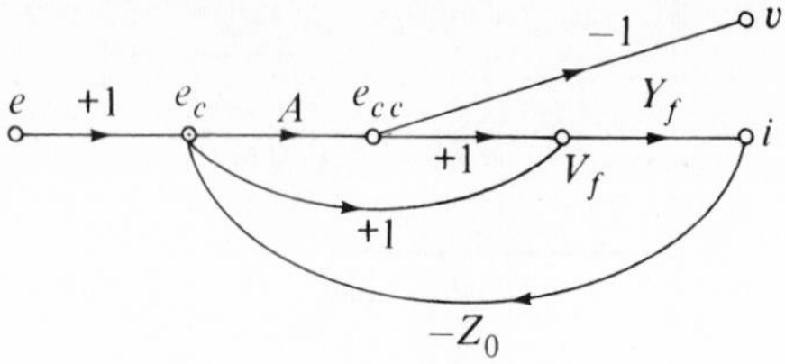

FIG. 6.25-R.

diagram, rather than from the usual place. But v is clearly e_c times A times -1. So we first calculate the transfer function from e to e_c. For that calculation, in which we do not need the connection to v, we combine the two branches from e_c to v_f and obtain Fig. 6.25-S*a*. For the transfer function from e to e_c we can use Fig. 6.25-J, observing that here $T_1 = 1$ (!!) and $T_2 = -(A+1)Y_f Z_0$ (Fig. 6.25-S*b*). Hence from (6.25-4),

$$\frac{e_c}{e} = \frac{1}{1 - (1)[-(A+1)Y_f Z_0]} \qquad (6.25\text{-}19)$$

and

$$T = \frac{v}{e} = -A\frac{e_c}{e} = \frac{-A}{1 + Z_0 Y_f(A+1)}, \qquad (6.25\text{-}20)$$

as before, (6.25-18).

When A is made large, $|A| \gg 1$, then for many practical purposes the transfer function is simply the ratio of Z_f to Z_0:

$$T = \frac{v}{e} = -Y_0 Z_f. \qquad (6.25\text{-}21)$$

Rather useful specific cases are those in Fig. 6.25-T. For Z_0 we use a resistor; then for Z_f a resistor, a capacitor, and the two together, in turn, giving the three transfer functions shown. Another useful property of this network is that since $|A|$ is very large, and e_{cc} is finite, e_c must adjust itself to be practically zero. We can therefore *add* two or more inputs by additional connection

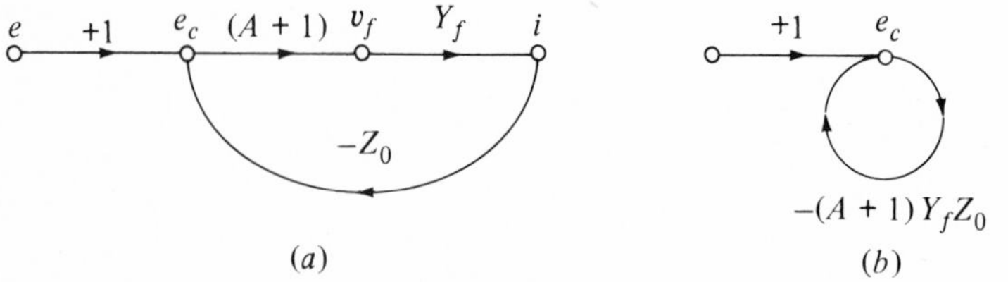

FIG. 6.25-S.

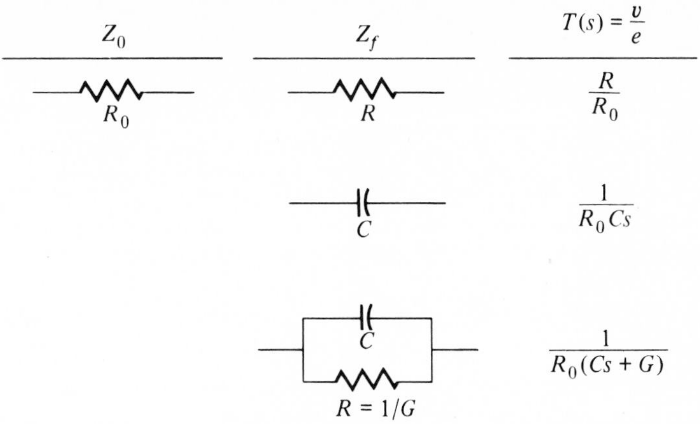

Z_0
Z_f
$T(s) = \frac{v}{e}$
R_0
R
$\frac{R}{R_0}$
C
$\frac{1}{R_0 Cs}$
C
$R = 1/G$
$\frac{1}{R_0(Cs + G)}$

FIG. 6.25-T.

thereof (Fig. 6.25-U). Note that both inputs receive the same "treatment," multiplication by Z_f/R_0, in addition to being added. The signal-flow graph of Fig. 6.25-U*b* is merely a clear statement of this. Note also that such a network is unilateral, and that the transfer function of a cascade connection of several such networks is the product of the individual transfer functions.

The utility of these networks is tremendous. One interesting application is to the realization of a given transfer function. Standing alone, or perhaps a (factored) part of a larger synthesis, suppose the biquadratic ("$N = 2$") transfer function

$$T(s) = \frac{as^2 + bs + c}{s^2 + 2\alpha s + {\omega_0}^2} \tag{6.25-22}$$

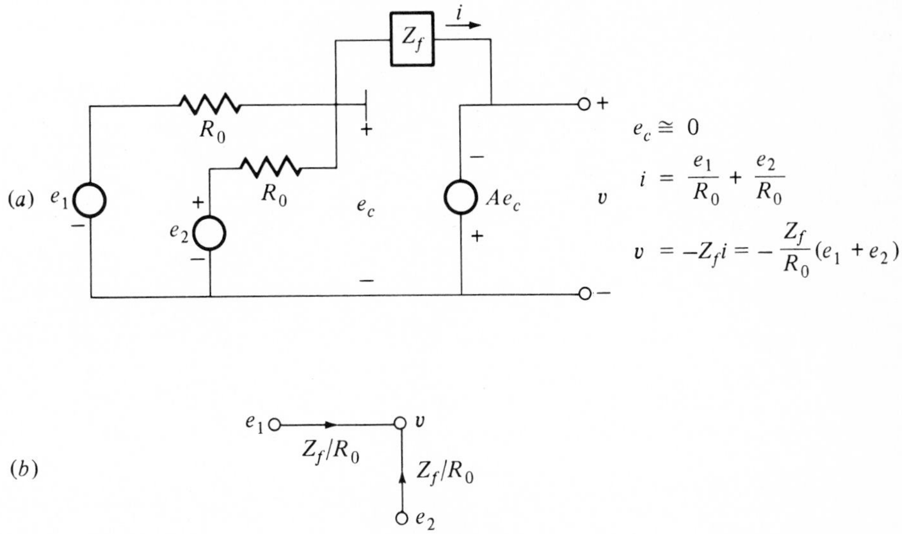

FIG. 6.25-U.

in which a, b, and c can have any (real) values, positive or negative, is to be realized. The poles (natural frequencies) presumably lie in the left half of the s plane, so that α is positive, as well as ω_0^2. But these poles may be "high-Q" ($\alpha \ll \omega_0$), which suggests the use of a resonant circuit, with an inductor. Since inductors tend to be comparatively bulky and heavy, to have external fields, and to be generally undesirable in comparison with resistors, capacitors, and opamps, a realization that uses *no inductors* is highly desirable in practical engineering. And with one each of our three networks (Fig. 6.25-T) in appropriate (feedback) connection, we can accomplish just that. The technique becomes even more interesting as operational amplifiers, resistors, and capacitors in integrated form become smaller and cheaper and give better performance at higher frequencies (s values).

That resistors and capacitors (and active) elements can, in a sense, "replace" inductors is another of the fascinating features of feedback. It seems not unreasonable, since the natural frequencies are the roots of the equation [see Fig. 6.25-J and (6.25-7)]:

$$\text{Loop transmission (transfer) function} = T_1 T_2 = 1. \qquad (6.25\text{-}23)$$

They are no longer simply the poles of a (passive) transfer function, but are determined physically by the (unilateral) transmission around a closed loop or mathematically by equality between numerator and denominator of the product of transfer functions. By using the second and third networks defined by Fig. 6.25-T we can make

$$T_1 T_2 = \frac{\text{constant}}{s(s+\beta)} \qquad (6.25\text{-}24)$$

so that the characteristic equation (with feedback) is

$$s(s+\beta) = s^2 + \beta s = \text{constant} \qquad (6.25\text{-}25)$$

which, it seems, can be made to have the complex roots we wish, without the use of inductance.

To realize (6.25-22) without an inductor, we accordingly construct a feedback loop whose natural frequencies are those required, the zeros of $(s^2 + 2\alpha s + \omega_0^2)$. We need two energy-storing elements (here two capacitors) and several resistors and opamps. Among the various possibilities is that shown in Fig. 6.25-V. This is a pseudo signal-flow diagram as well as a sort of connection diagram of three opamps (the triangle symbol) and associated immittances, each like Fig. 6.25-Q, in tandem. The bottom (ground) line is omitted, but the immittances (taken from Fig. 6.25-T) are specifically shown. The natural-frequency-determining equation, "loop transmission = 1", is

$$\frac{-G_0}{C_1 s + G_1} \frac{-G_0}{C_2 s} \frac{-G_0}{G_3} = 1 \qquad (6.25\text{-}26)$$

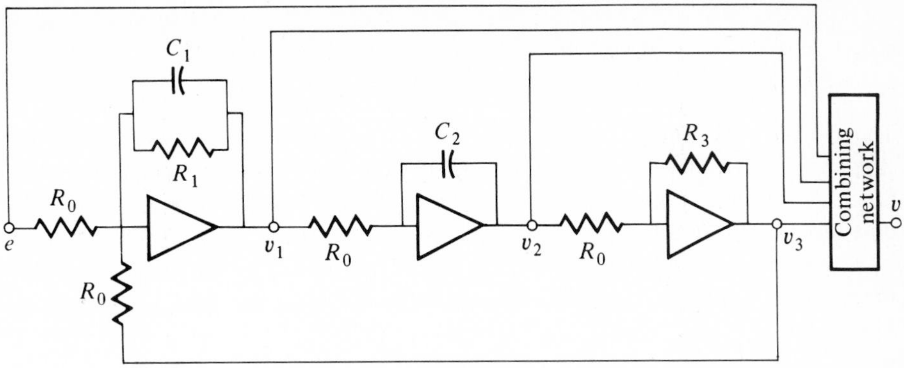

FIG. 6.25-V.

or

$$s^2 + \frac{1}{R_1C_1}s + \frac{R_3}{R_0{}^3C_1C_2} = 0. \tag{6.25-27}$$

Note the need for the third, purely resistive (R_3) stage to obtain the correct sign for the constant term in (6.25-27). Now we need only make $R_1C_1 = 1/(2\alpha)$ and $R_3/(R_0{}^3C_1C_2) = \omega_0{}^2$, which is easy to do, to obtain the desired natural frequencies, i.e., the desired denominator of (6.25-22). As for its numerator, we first rewrite (6.25-22) as

$$T = a + \frac{(b - 2\alpha a)s + (c - \omega_0{}^2 a)}{s^2 + 2\alpha s + \omega_0{}^2}. \tag{6.25-28}$$

Then we note that the transfer function from e to v_1 is

$$T_{e \text{ to } v_1} = \frac{-G_0/(C_1s + G_1)}{1 - (-G_0/C_1s + G_1)(-G_0/C_2s)(-G_0/G_3)} = \frac{(\text{constant})\, s}{s^2 + 2\alpha s + \omega_0{}^2} \tag{6.25-29}$$

and that those from e to v_2 and from e to v_3 are similar, but lack the factor s in the numerator, and have opposite signs. Hence the four leads into the combining network of Fig. 6.25-V carry the necessary parts of (6.25-28), and more. By putting into the combining network appropriate operational amplifiers for scaling, sign changing, and summing, we can realize the transfer function (6.25-22) from e to v, as desired. (There are of course many other possibilities in the details, both within the combining network and in the original feedback loop.)

This last example has, incidentally, presented us with a novelty, a transfer function in which the degree of the numerator is *equal to* (not *less than*) that of the denominator. This requires some attention, and we leave the subject of active networks (a very large one) to discuss it next.

6.26 ONE MORE EXAMPLE: $M = N$

The network of Fig. 6.26-A bears a simple relation to that of Fig. 6.02-A: capacitor has replaced inductor, inductor has replaced capacitor throughout. Since the configuration has not also been changed, this network is *not* dual to that of Fig. 6.02-A. Rather, the relation is one of *frequency inversion*, for s has in a way been replaced by $1/s$. In fact if we make the element values obey the relation

$$L_1 C_1 = L_2 C_2 = L_3 C_3 = \frac{1}{\omega_0^2}, \tag{6.26-1}$$

then replacing s by ω_0^2/s in the transfer function (6.02-5) will give the transfer function of the network of Fig. 6.26-A. The behavior of one network with frequency is the "reverse" of the behavior of the other; the frequency scale has in effect been inverted left to right about the reference or turning point ω_0. What one network does at low frequencies, the other does at high frequencies. This can be a useful design device, but that is not our main concern here.

The transfer function of the network of Fig. 6.26-A, voltage to voltage, has three zeros at the origin (the low-frequency asymptotic network is the whole network) and no zeros at infinity (the high-frequency asymptotic network is simply the two resistors). The order is 3, so that

$$T(s) = \frac{K_0 s^3}{s^3 + B_1 s^2 + B_2 s + B_3} \tag{6.26-2}$$

in which, from the asymptotic behaviors, high-frequency and low-frequency respectively,

$$\begin{aligned} K_0 &= \frac{R_3}{R_1 + R_3}, \\ \frac{K_0 s^3}{B_3} &= (C_1 s)(L_2 s)(C_3 s)(R_3), \\ B_3 &= \frac{1}{C_1 L_2 C_3 (R_1 + R_3)}. \end{aligned} \tag{6.26-3}$$

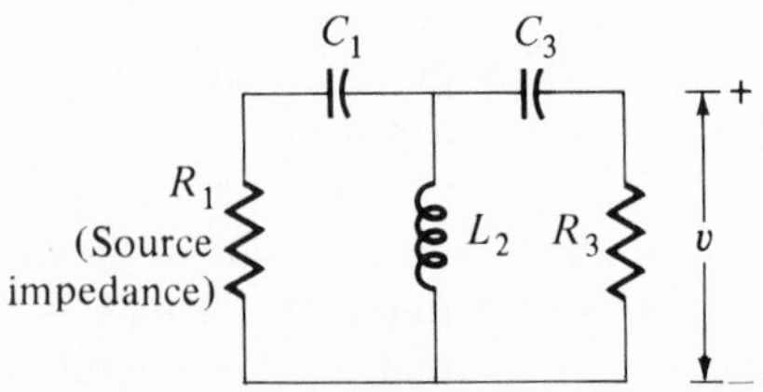

FIG. 6.26-A.

We verify dimensional correctness by inspection. These results can of course also be obtained (with the values of B_1 and B_2 in addition) by straightforward calculation of $T(s)$. We can also obtain them from (6.02-5) with the substitution of $\omega_0{}^2/s$ for s, using (6.26-1).

What is noteworthy in (6.26-2) is that it is a transfer function in which the degree of the numerator is equal to the degree of the denominator [cf. (6.25-22)]. The simple relation between transfer function and impulse response summarized in (5.14-2) and (5.14-3) does *not* hold, for $M = N$ in (5.14-4). Here is an instance of an exceptional answer to the question (5.14-5). It requires investigation.

The outstanding *mathematical* property of (6.26-2) is that this transfer function is not a proper fraction. Its outstanding *physical* property is that it indicates excellent high-frequency transmission: $T(\infty) = K_0 = (1 + R_1 G_3)^{-1}$ [exactly like the dc transmission of (6.02-5)]. This may well be physically unrealistic, for actual networks do not usually transmit high frequencies very well, because of parasitic capacitive and inductive effects. Sometimes, however, parasitic capacitance effects *do* effectively pass high-frequency signals, and that is the nature of this model (Fig. 6.26-A). If we carry out the mathematical division, we obtain

$$\begin{aligned} T(s) &= \frac{K_0 s^3}{s^3 + B_1 s^2 + B_2 s + B_3} \\ &= K_0 + \frac{-K_0 B_1 s^2 - K_0 B_2 s - K_0 B_3}{s^3 + B_1 s^2 + B_2 s + B_3} \\ &= K_0 + T_R(s) \end{aligned} \tag{6.26-4}$$

in which the reduced transfer function $T_R(s)$ has the "standard" proper-fraction form

$$T_R(s) = \frac{A_0 s^2 + A_1 s + A_2}{s^3 + B_1 s^2 + B_2 s + B_3} \tag{6.26-5}$$

with $M = N - 1 = 2$. Schematically we can use the representation of Fig. 6.26-B (like that of Fig. 5.03-B) to analyze the network's behavior. Mathematically, the response in t (and in ω) has two parts: one proportional to the input,

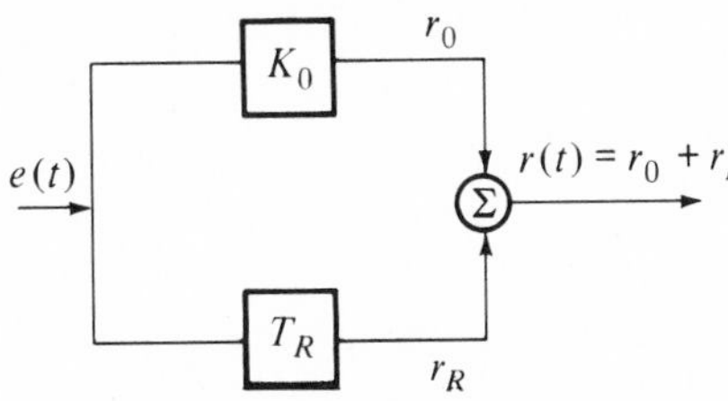

FIG. 6.26-B.

$r_0 = K_0 e$, and one obtained in the usual fashion from T_R, with forced and natural parts.

Physically, the same is true. It is particularly clear for impulse excitation, for the diagram indicates, if $e(t) = \Lambda_0(t)$, that

$$r_\delta(t) = K_0 \Lambda_0 \, \delta(t) + \sum_1^3 K_j e^{p_j t}, \qquad 0 < t \tag{6.26-6}$$

in which the three natural frequencies (the p_j) and the coefficients in the natural part (the K_j) are obtained from T_R in the usual way. The first term indicates an impulse component in $r_\delta(t)$. And examination of the physical behavior of the network confirms this: in $0 < t < 0+$, the impulse excitation finds short circuits in C_1 and C_3 and sees merely the high-frequency asymptotic network composed of R_1 and R_3 alone, so that

$$r_\delta(t) = \frac{R_3}{R_1 + R_3} \Lambda_0 \, \delta(T), \qquad 0 < t < 0+ \tag{6.26-7}$$

exactly as in (6.26-6). The resulting initial values, the conditions *at* $t = 0+$, after the impulse is over, are in principle easy to find. We note that an impulsive voltage exists across R_3, and therefore across L_2, in $0 < t < 0+$; hence i_{L2} steps suddenly, as do v_{C1} and v_{C2}, during the charging interval. The impulsive charging current leaves on each capacitor a charge

$$Q_0 = \int_0^{0+} \frac{\Lambda_0 \, \delta(t)}{R_1 + R_3} \, dt = \frac{\Lambda_0}{R_1 + R_3} \qquad (\text{Wb·turns}/\Omega = \text{V·s}/\Omega = \text{C}). \tag{6.26-8}$$

Hence (Fig. 6.26-C)

$$v_{C1}(0+) = \frac{Q_0}{C_1},$$
$$v_{C3}(0+) = \frac{Q_0}{C_3}, \tag{6.26-9}$$
$$i_{L2}(0+) = \frac{R_3 \Lambda_0}{R_1 + R_3} \frac{1}{L_2}.$$

To find $v(0+)$ from these we may write two simple equations, Kirchhoff's voltage law for the external loop, and Kirchhoff's current law for the bottom node:

$$R_3 i_{R3} + v_{C3} + v_{C1} - R_1 i_{R1} = 0,$$
$$i_{R3} + i_{L2} + i_{R1} = 0. \tag{6.26-10}$$

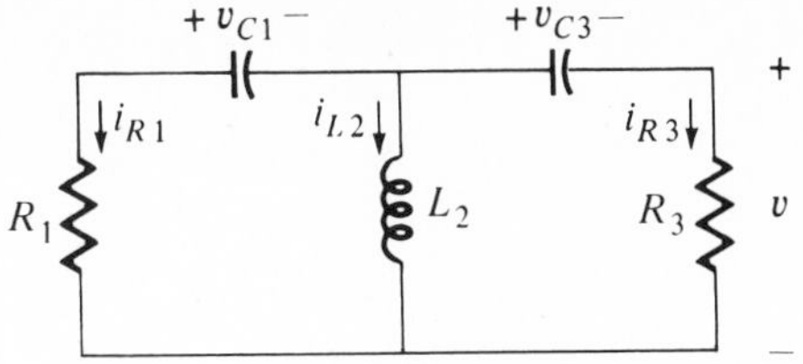

FIG. 6.26-C.

From these, by eliminating i_{R1}, we obtain

$$
\begin{aligned}
(R_3 + R_1)i_{R3} &= -(R_1 i_{L2} + v_{C3} + v_{C1}) \\
&= -\left(\frac{R_1 R_3 \Lambda_0}{L_2(R_1 + R_3)} + \frac{Q_0}{C_3} + \frac{Q_0}{C_1}\right) \qquad (6.26\text{-}11) \\
&= -\frac{\Lambda_0}{R_1 + R_3}\left(\frac{R_1 R_3}{L_2} + \frac{1}{C_1} + \frac{1}{C_2}\right)
\end{aligned}
$$

so that

$$v(0+) = R_3 i_{R3}(0+) = -\frac{\Lambda_0 R_3}{(R_1 + R_3)^2}\left(\frac{R_1 R_3}{L_2} + \frac{1}{C_1} + \frac{1}{C_3}\right). \qquad (6.26\text{-}12)$$

In all these expressions, dimensional checks are easily applied. It is interesting to observe that, because of the good high-frequency transmission (the fact that $M = N$), the impulsive charging current flows "more freely" in the network and its state at $t = 0+$ is more complicated than were conditions in our previous examples: the capacitors carry charge *and* the inductor carries current, and *all* energy-storing elements are charged, at $t = 0+$. The interaction between the loops makes it then necessary to go through calculations like the solution of the simultaneous equations of (6.26-10) to obtain the initial value $v(0+)$.

One could proceed to evaluate $v'(0+)$, but the calculation from the diagram is more involved. Instead, let us try to use the initial-value tableau (Fig. 6.11-A). The expression $(sT)_\infty$ is meaningless. But we recognize from (6.26-4) that the constant term, the nonfractional part of $T(s)$, contributes nothing to $v(t)$ for $t \geqq 0+$; for then the impulsive part of the response is over and gone. Hence we need only consider the reduced, proper-fraction part of T. And $(sT_R)_\infty$ *is* meaningful, and we obtain

$$v(0+) = (sT_R)_\infty = A_0 = -K_0 B_1 \Lambda_0 = -\frac{R_3}{R_1 + R_3}\,\frac{(1/C_1 + 1/C_3 + R_1 R_3/L_2)\Lambda_0}{R_1 + R_3} \qquad (6.26\text{-}13)$$

in agreement with (6.26-12). (It is of course necessary first to calculate B_1 by some other method.) The calculation of $v'(0+)$ is possible, but lengthy and requires B_2 also; we omit it. Physically, however, it is only reasonable that $v'(0+)$ be positive: the amount of energy stored in the network is surely decreasing, and we expect the rate of dissipation of energy also to be decreasing, and $|v|$ to be decreasing.

The behavior of the impulse and step responses is shown in Fig. 6.26-D. In general, in the $M = N$ case, this is the outstanding effect: the good high-frequency transmission "passes" the impulse so that the response contains an

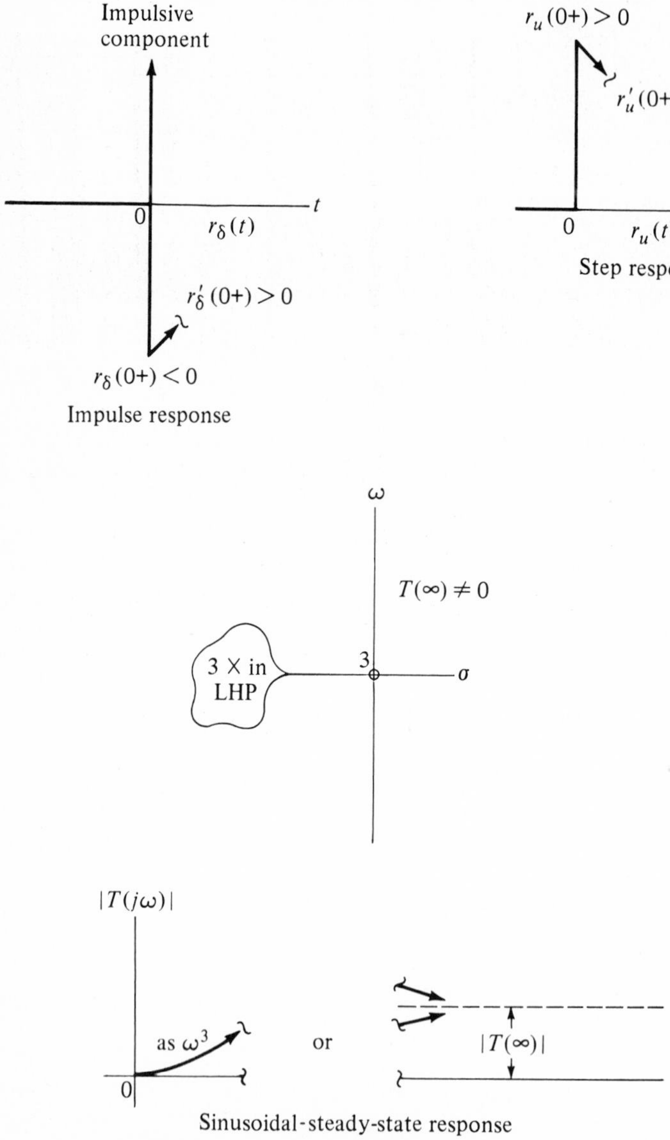

FIG. 6.26-D.

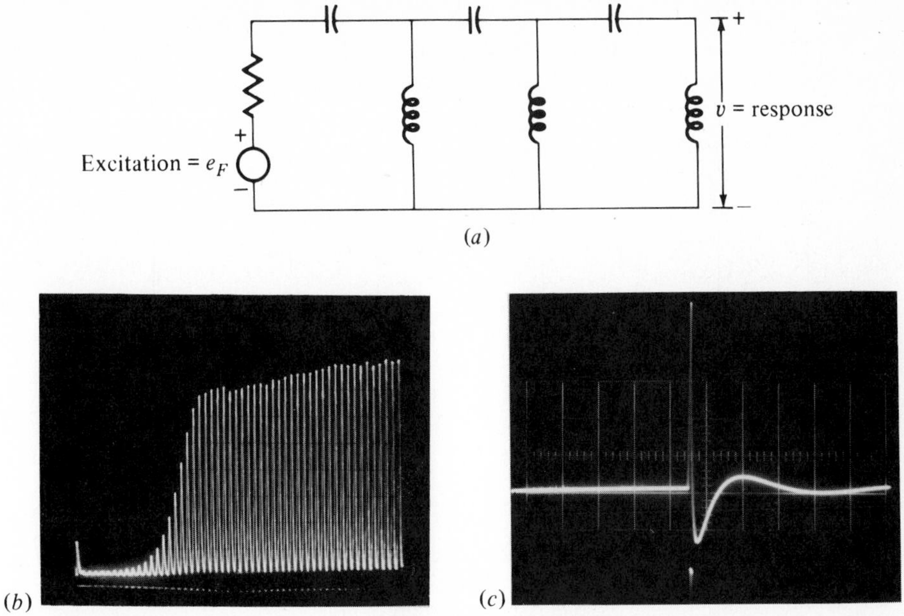

FIG. 6.26-E.

impulsive component. The response then reverses, and thereafter decays; whether it oscillates depends on the locations of the natural frequencies (on the values of the B's).

The theoretical analysis of Fig. 6.26-D is corroborated by the physical action shown in Fig. 6.26-E. The network there is of order 5 (instead of 3) but is essentially the same in character: raising "N" merely emphasizes the high-pass character of the network. $T(s)$ has five zeros at the origin and "levels off" to a constant, nonzero value at infinity; the impulse response has an impulsive component, reverses, and starts its oscillation upward.

The network of Fig. 6.26-F exhibits the same behavior. Here

$$\begin{aligned} T(s) &= \frac{K_0\, s^2(s^2+\omega_0{}^2)}{s^4+B_1 s^3+\cdots} = \frac{K_0\, s^2(s^2+\omega_0{}^2)}{(s^3+B_1 s^2+B_2 s+B_3)s} \\ &= \frac{K_0\, s(s^2+\omega_0{}^2)}{s^3+B_1 s^2+B_2 s+B_3} \qquad (6.26\text{-}14) \\ &= K_0\left[1+\frac{-B_1 s^2+(\omega_0{}^2-B_2)s-B_3}{s^3+B_1 s^2+B_2 s+B_3}\right]. \end{aligned}$$

After canceling the surplus factor s, the order of the transfer function reduces in effect to 3. Step and impulse responses are not unlike those of the previous example; the transfer function has zeros at $s = \pm j\omega_0$, however (due to the L_2C_2 tuned circuit), and only one zero at the origin.

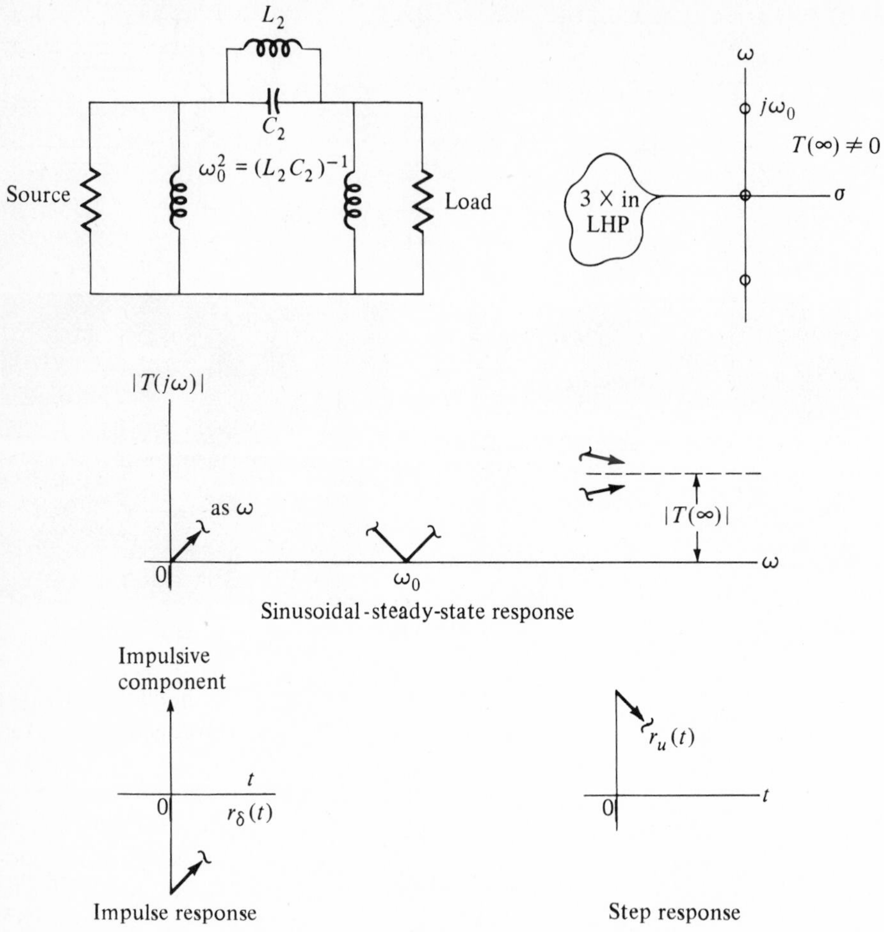

FIG. 6.26-F.

6.27 $M \geqq N$

It is seldom indeed that the degree of the numerator of a transfer function *exceeds* the degree of its denominator. The networks of Fig. 6.27-A*a* and *b* are two examples, the high-frequency behavior of both of which is physically ridiculous. A pole of $T(s)$ is a natural frequency; yet natural frequencies cannot occur at infinity, except on paper, in idealized models. Physically, an inductor behaves according to "$v = L\,di/dt$" or "$Z = Ls$" *only when s is reasonably small;* at high frequencies, parasitic capacitance (and other) effects radically change its behavior, and such network models are there meaningless. They may, with their simplification by omission of parasitic capacitance and other effects important at high frequencies, be adequate and useful for low-frequency, or $t > 0+$, analysis. But the network's actual behavior at high frequencies, or in $0 < t < 0+$ when impulse-excited—these can*not* be obtained from such models. The well-known difficulty of *differentiating* a signal in practice (for example, in analog computation) because of noise problems, stems from the same cause, for noise is effectively high-frequency, or rapidly changing, signals. *Integration* is more practical, for its action is to smooth out rather than to exacerbate "jitterwiggles"; the poles of the transfer function then lie at or near the origin instead of infinity.

Both networks of Fig. 6.27-A*a* and *b* exhibit poles of T at infinity. So also do the two dual networks of Fig. 6.27-A*e* and *f*. In both (*b*) and (*f*) networks the cause can be looked on as another anomaly: one of the three natural frequencies moves to infinity because the two inductors are effectively in series (the

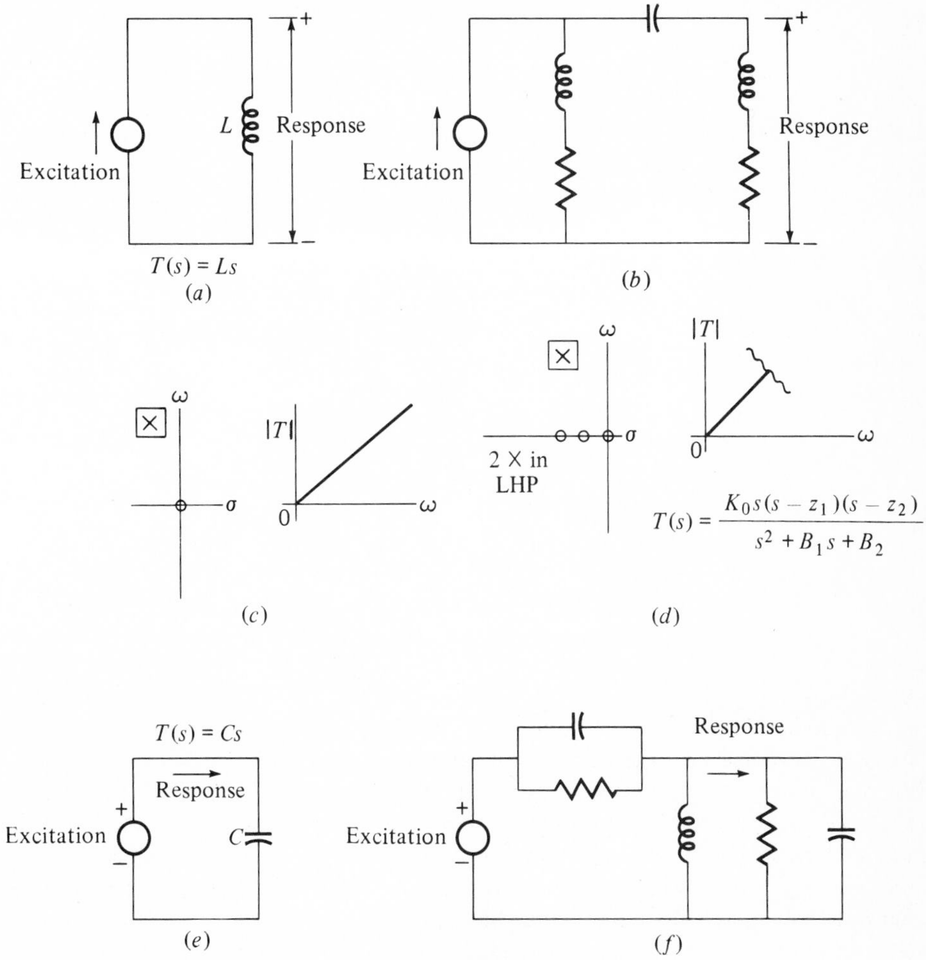

FIG. 6.27-A.

two capacitors are effectively in parallel) and amount to but one energy-storing element. The transfer function, and the diagrams, apply of course to both networks of each pair of duals.

Mathematically, if $M > N$ in (5.05-3) we need only carry out the division until we obtain a reduced, proper-fraction transfer function:

$$\begin{aligned} T(s) &= \frac{K_0 s^M + \cdots}{s^N + B_1 s^{N-1} + \cdots} \\ &= K_0 s^{M-N} + C_{M-N-1} s^{M-N-1} + \cdots + C_1 s + C_0 \qquad (6.27\text{-}1) \\ &+ \frac{A_0 s^{N-1} + \cdots + A_{N-1}}{s^N + B_1 s^{N-1} + \cdots + B_N}. \end{aligned}$$

The polynomial part of this is to be handled by agreement with the model maker; it presumably represents very-high-frequency, or very-small-time $(0 < t < 0+)$ behavior which the model, however, is certainly inadequate to express. A $C_1 s$ term, for example, implies the existence of the *derivative* of an impulse in the impulse response, something a physical inductor cannot produce.

The proper-fraction (reduced) part is to be treated exactly as before: from it come the initial $(t = 0+)$ values, and the power series and sum-of-exponential forms of responses.

6.28 NETWORKS AND TRANSFER FUNCTIONS

This has been a lengthy chapter. It began by defining a category of networks of great practical importance: most of the networks one meets in engineering are of ladder form. We have used them as examples for that reason, and for their simplicity. Discussion of their properties has led us to many ideas about transfer functions that are, however, completely general. The relation between transfer function and overall (input-to-output) differential equation, and the initial-value tableau, are both perfectly general and apply to all networks, not just to ladders. So also do the concepts of low-frequency and high-frequency asymptotic behavior. But the physical delineation of high-frequency and low-frequency asymptotic networks is easiest for ladders.

Ladder networks have a "one-path" nature: the (forced component of the) output signal has but one constituent, and all of it comes from the excitation by one route. It is reduced to zero for certain values of s, the points of infinite loss, which are caused by resonances (open- or short-circuit conditions) in the ladder rungs and sides. Physically, these are therefore quite visible. Since they are zeros or poles of driving-point immittance functions, and hence are natural frequencies of smaller, subsidiary networks, *they cannot lie in the right half plane* if the network is passive. Hence passive ladder networks are in a very special category. Bridges and other networks, in contrast, may have more than one output-signal component, and infinite loss can occur because of cancellation rather than resonance, and hence *anywhere in the s plane*, even if the network is entirely passive.

The one-path nature of ladders makes possible, and easy, the calculation of transfer functions by working back from response to excitation. This is usually *not* possible for nonladder networks: they require something more sophisticated and complicated (the solution of an additional equation, or of two or simultaneous equations). The single path also makes it easy to trace the course of an impulsive excitation through the network, i.e., to delineate the high-frequency asymptotic network. The conditions at $t = 0+$ (the "initial values"), which may be different from the $t = 0$ conditions, are accordingly not difficult to calculate. These provide material for checking a transfer function, calculated in some other way, or for *constructing* $T(s)$ in large part. The single path also makes simple the determination of the low-frequency asymptotic network.

Transfer function denominators are not called *characteristic polynomials* for nothing: they are, implicitly, the all-important natural frequencies, the infinite-gain points. To calculate them from the sum of the impedances around a closed loop, or, dually, from the sum of the admittances across a node pair, is often not difficult. Or, for ladders, one can use the work-backward method of calculating $T(s)$. Actually to evaluate the natural frequencies requires factoring this (characteristic) polynomial. This is but one of a number of numerical chores that *must* be done in network analysis, tedious and often uninteresting though they be. Their profusion and complexity make the digital computer an essential tool.

The numerator of $T(s)$ expresses the infinite-loss points (which in ladders can readily be *seen*). By expanding $T(s)$ in a series of powers of s^{-1} (about infinity, i.e., for high frequencies) one can obtain the initial values of the impulse response, i.e., the initial-value tableau. This is useful for checking, by comparison with the high-frequency asymptotic network. From it we can also obtain the power-series form of $r_\delta(t)$, and of $r_u(t)$, and usually (as we soon shall see) of $r(t)$ in general. Expansion of $T(s)$ in a series of powers of s (for low frequencies, about the origin), is also useful for checking, by comparison with the low-frequency asymptotic network.

The transfer function can do all this because *it is the input-to-output differential equation*, albeit written in a transformed way. Its denominator expresses the conventional left side of the differential equation, operating on the unknown response, and its numerator the operations to be performed on the (known) excitation, the right side. Sometimes $T(s)$ can be reduced, because it is not in lowest terms, by canceling common (surplus) factors from numerator and denominator. The implication is that some natural modes are unobservable from outside, and hence not really a concern of the transfer function. (Constant-resistance networks put this phenomenon to good use.)

The nature of the transfer function, its very existence even, and its remarkable properties, depend on our basic "lulicon" (Sec. 6.12) postulates. We must never forget that. But there are still more properties to study, interesting and useful.

As a guide to action, let us carefully examine the "state of the art" for us at this point, and lay out the next stage of our plan of study. Perhaps it "*would* be just as well if you'd mention what you mean to do next, as I suppose you don't mean to stop here all the rest of your life."

The basic schematic of the network analysis problem (Figs. 1.07-A, 5.01-A) is reproduced in Fig. 6.28-A. It has four elements: the excitation $e(t)$, the initial condition of the network, the network itself, and the response $r(t)$. In the usual analysis problem, the last is to be found from the other three. These other three we have studied, but by no means completely. Let us look at each in turn, to decide what to investigate next.

If the *network's* geometry is simple, specifically, if its form is a ladder, whose components are not unduly complicated, we can readily calculate its transfer function. From this we can in principle obtain the network's impulse response, in power-series or in sum-of-exponential form. The former is straightforward. The latter requires partial-fraction expansion of the transfer function, a process that we have examined only in simple cases. We make it an immediate agenda item to discuss partial-fraction expansion in general terms.

If the network's geometry is *not* simple, a modified or different approach is required to find the transfer function (Secs. 6.23, 6.24). Thereafter, the work is the same.

If the *initial condition* of the network is not rest, we can replace initially charged elements with initially uncharged elements plus step or impulse sources in at least certain cases (see Secs. 2.02, 3.02, 6.06, 6.15). Nonrest initial conditions seem to present nothing really novel. Such substitutions can in fact always be made; we make it part of our plan to study this carefully in Chap. 7 (Sec. 7.22). Then all problems can be reduced to initial-rest analyses.

If the *excitation* is an impulse, the analysis is straightforward. If the excitation is not an impulse, we can convolve it with the impulse response and so obtain the response to any arbitrary excitation: elaborate mathematical function, mere table of numerical values, or curve, whatever it may be. Between (*a*) the impulse, the response to which comes so simply and directly from the transfer function, and (*b*) complicated excitation functions that require lengthy convolution calculations, it seems however that there should be other classes of excitation, intermediate in their analysis demands. An obvious category to examine carefully, since the transfer function concept is based thereon, is the exponential or sum-of-exponentials category. For such excitations the transfer function gives the forced component of the response immediately. Can it also

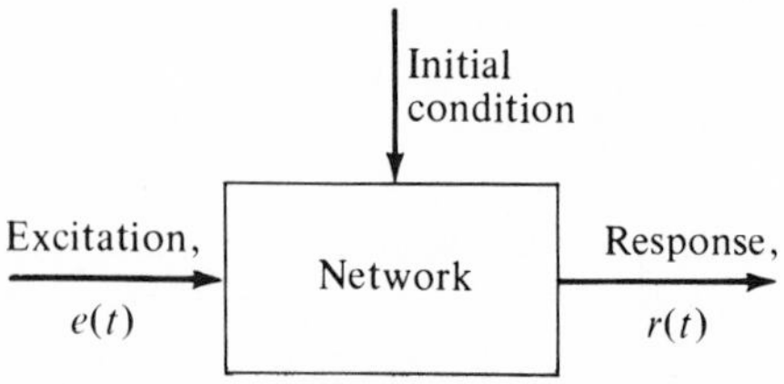

FIG. 6.28-A.

give the *natural* component, without great calculation problems? To answer this is another agenda item.

We have three specific matters for immediate attention. To them we now turn.

6.29 EXERCISES

In all exercises of this section, assume that no energy is stored initially in the network, neglect dissipation of energy in all the L and C elements (except where specifically shown), and take the excitation to be $i_F = Q_0\,\delta(t)$ or $e_F = \Lambda_0\,\delta(t)$ as appropriate, except where otherwise stated. Element values have been normalized and rounded for simplicity.

6-1. (*a*) By inspection of the schematic diagram of Fig. 6-1, using literal element values and only physical reasoning and very simple element-behavior calculations (do not calculate a transfer function or use a differential equation): (1) determine whether there is an impulsive component in the response $r(t)$, and if so, what its strength is; (2) find the initial ($t = 0+$) values of the voltage across and the current in each element, and show them on a circuit diagram; (3) determine which is the first nonzero initial value in the list $r(0+)$, $r'(0+)$, $r''(0+)$, ..., evaluate it, and check its dimensions.

(*b*) Now (using also simple Kirchhoff-law calculations as necessary) evaluate the next initial value in the list of (*a*), 3.

(*c*) Write the response $r(t)$ in power-series form to the extent possible using only the results of (*a*) and (*b*). Sketch $r(t)$ versus t, clearly showing its initial behavior.

(*d*) Repeat (*c*) for the response to *step* excitation.

(*e*) In the transfer function written in literal form

$$T(s) = \frac{A_{N-1-M}\,s^M + A_{N-M}\,s^{M-1} + \cdots + A_{N-1}}{s^N + B_1 s^{N-1} + \cdots + B_N}$$

evaluate M, N, and such coefficients as you can solely on the basis of the results of (*a*) and (*b*).

(*f*) Which of the N zeros of $T(s)$ can be written in terms of the literal element values solely by inspection of the diagram? Which of its N poles?

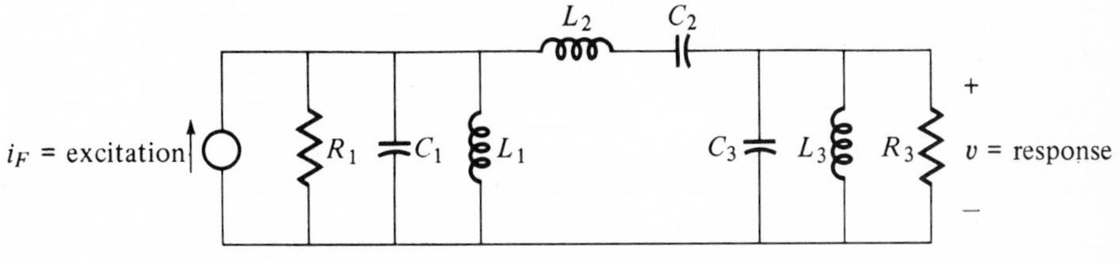

FIG. 6-1.

(*g*) Draw the asymptotic high-frequency network, omitting all elements whose contributions are insignificant for large values of s. Find its transfer function. Which coefficients in $T(s)$, obtained in (*e*), are thereby checked? Which coefficients in addition to those obtained in (*e*) are thereby evaluated?

(*h*) Repeat (*g*) for the low-frequency (small-value-of-s) asymptotic network.

(*i*) Calculate $T(s)$ by analysis, ignoring the work above. For simplicity, take all element values to be unity. Construct the first few lines of the initial-value tableau in the first form of Fig. 6.11A. Which parameters in $T(s)$ as calculated here can now be checked by reference to the results of (*a*) to (*h*)?

6-2. Repeat Exercise 6-1 for the network of Fig. 6-2.

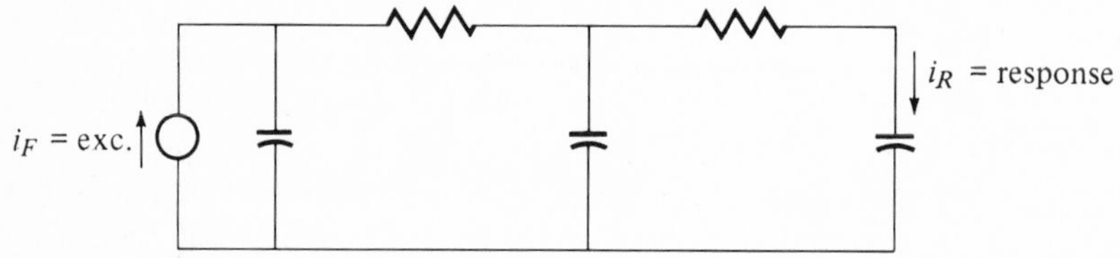

FIG. 6-2.

6-3. Repeat Exercise 6-1 for the two networks of Fig. 6-3. What relation(s) do you observe between the two?

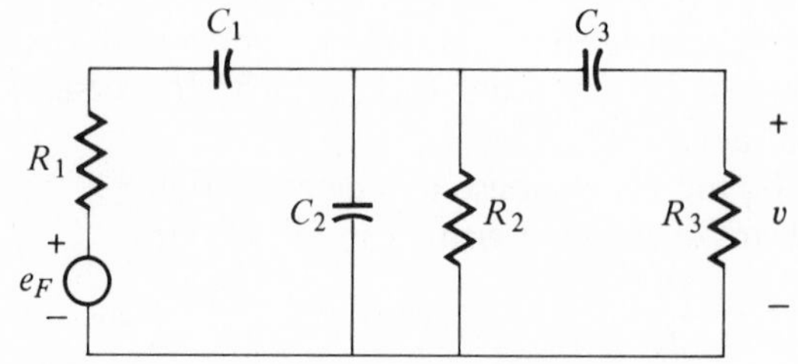

e_F = excitation, v = response

(*a*)

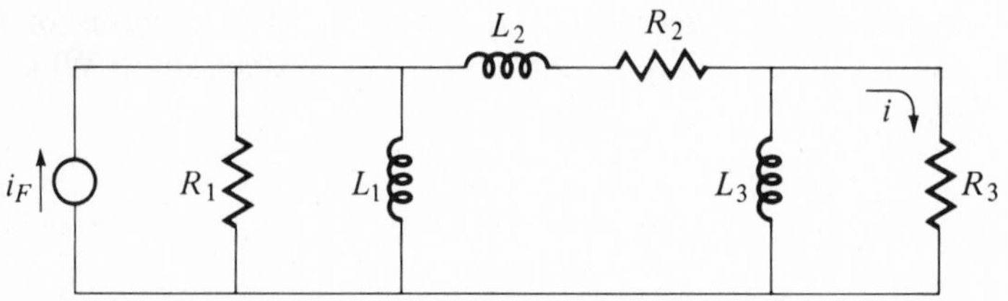

i_F = excitation, i = response

(*b*)

FIG. 6-3.

6-4. Repeat Exercise 6-1 for the network of Fig. 6-4.

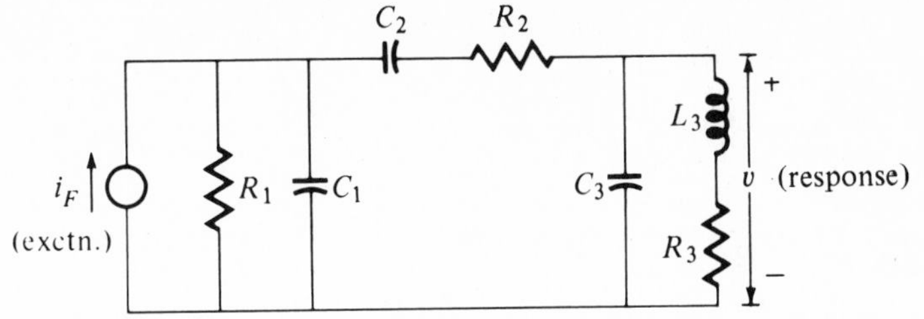

FIG. 6-4.

6-5. Repeat Exercise 6-1 for the network of Fig. 6-5.

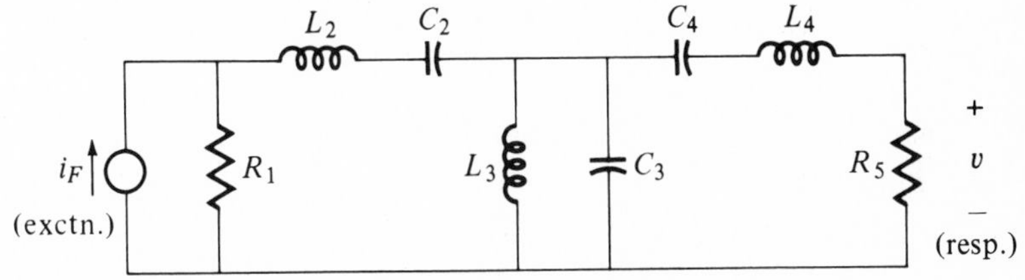

FIG. 6-5.

6-6. Repeat Exercise 6-1 for the network of Fig. 6-6.

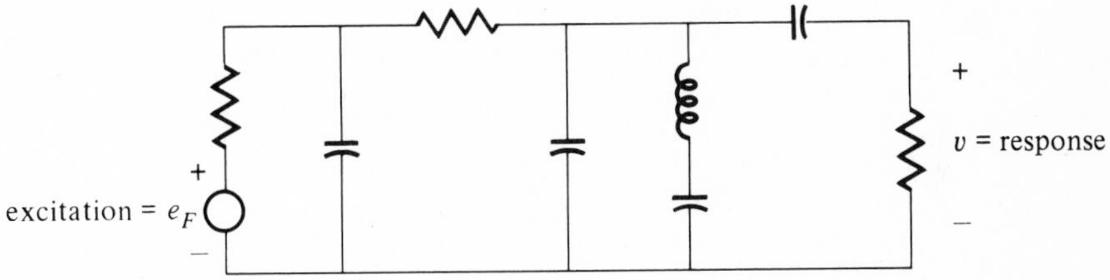

FIG. 6-6.

6-7. Repeat Exercise 6-1 for the network of Fig. 6-7. [In part (*i*), use the element values in Fig. 6-7.]

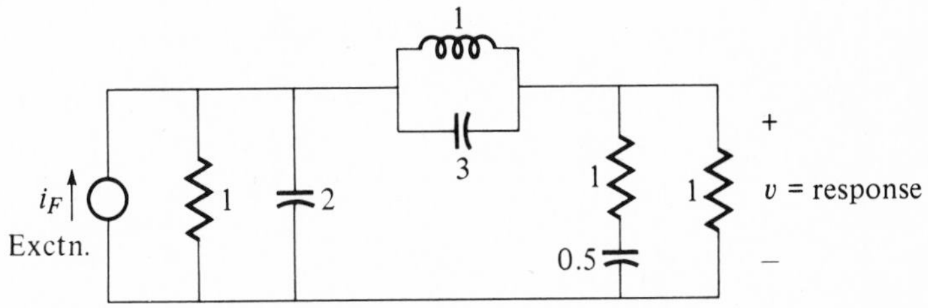

Element values are (normalized) ohms, henrys, farads

FIG. 6-7.

6-8. Repeat Exercise 6-1 for the network of Fig. 6-8. [In part (*i*), use the element values in Fig. 6-8.]

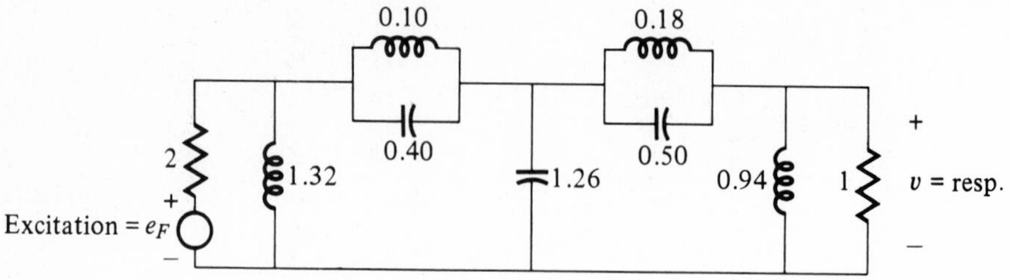

El. vals. are norm. ohms, henrys, farads

FIG. 6-8.

6-9. Repeat Exercise 6-1 for the network of Fig. 6-9. [In part (*i*), use the element values in Fig. 6-9.]

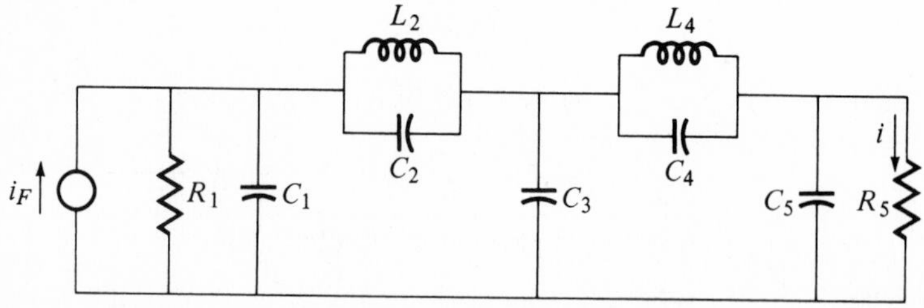

$R_1 = R_5 = 1\ \Omega,\ L_2 = L_4 = 0.5\ \text{H},\ C_1 = C_2 = C_3 = C_4 = C_5 = 1\ \text{F}$

Exc. = i_F Response = i

FIG. 6-9.

6-10. Repeat Exercise 6-1 for the network of Fig. 6-10.

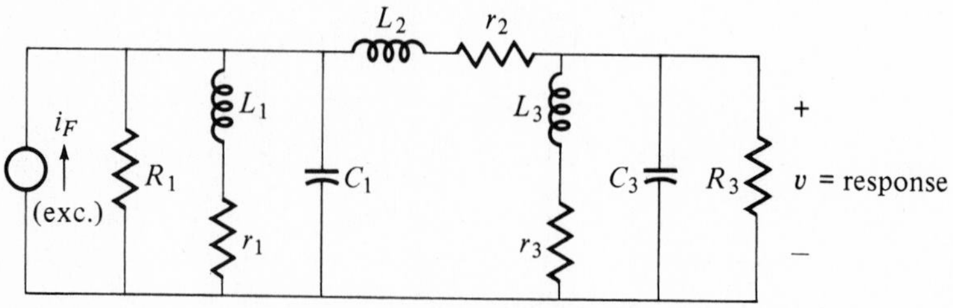

FIG. 6-10.

6-11. Repeat Exercise 6-1 for the network of Fig. 6-11.

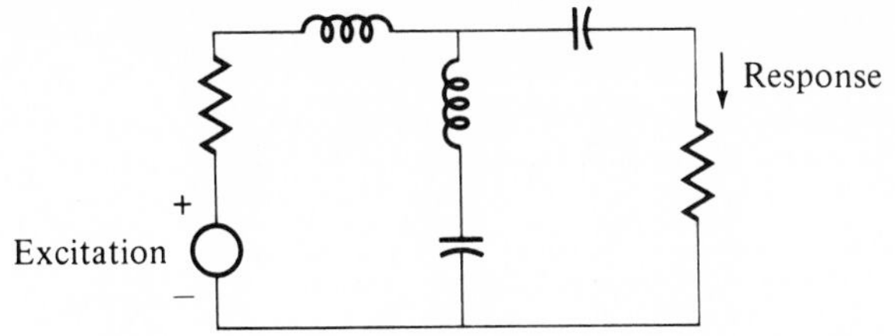

FIG. 6-11.

6-12. Repeat Exercise 6-1 for the network of Fig. 6-12. [In part (*i*), use the element values of Fig. 6-12.]

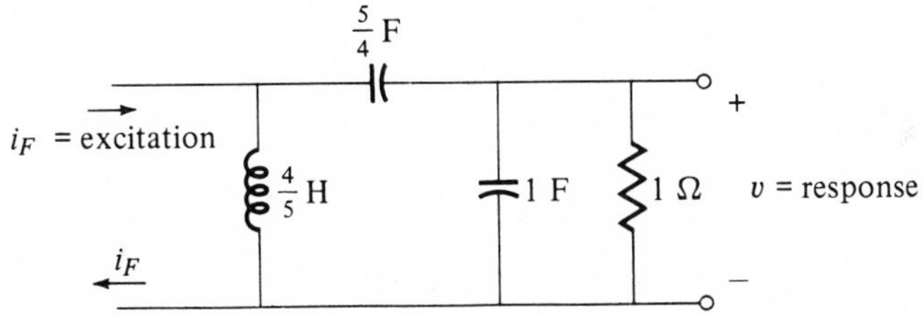

FIG. 6-12.

6-13. Repeat Exercise 6-1 for the network of Fig. 6-13.

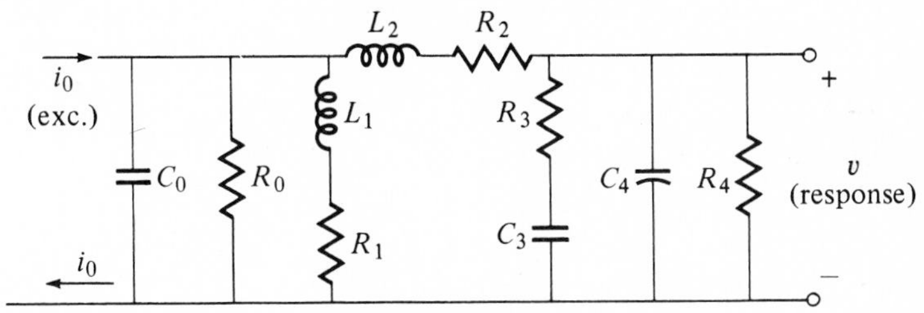

FIG. 6-13.

6-14. Repeat Exercise 6-1 for the network of Fig. 6-14.

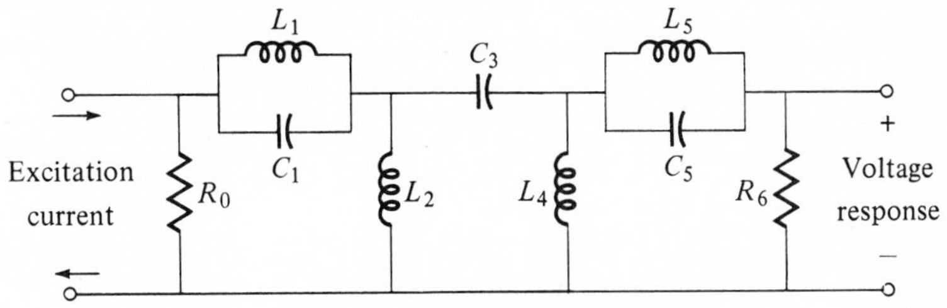

FIG. 6-14.

6-15. Repeat Exercise 6-1 for the network of Fig. 6-15.

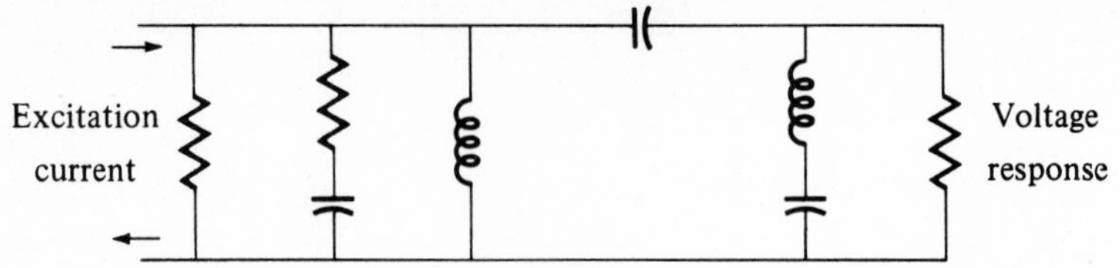

FIG. 6-15.

6-16. Repeat Exercise 6-1 for the network of Fig. 6-16.

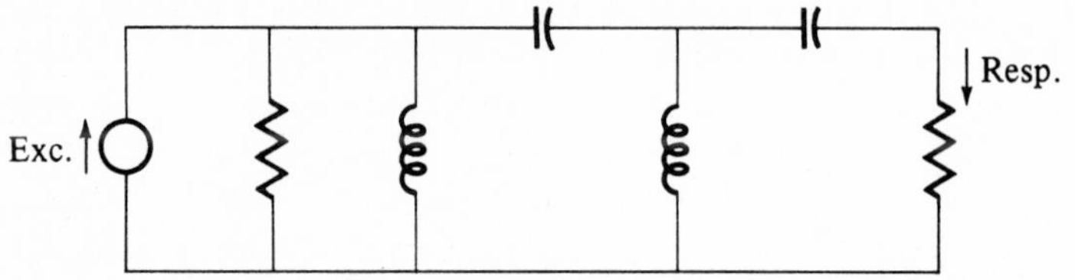

FIG. 6-16.

6-17. Repeat Exercise 6-1 for the network of Fig. 6-17, (*a*) with the switch *closed*, and (*b*) with the switch *open*. Briefly describe the differences between the two analyses.

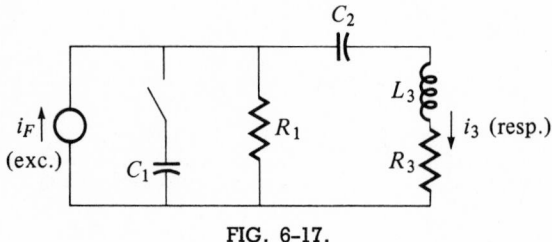

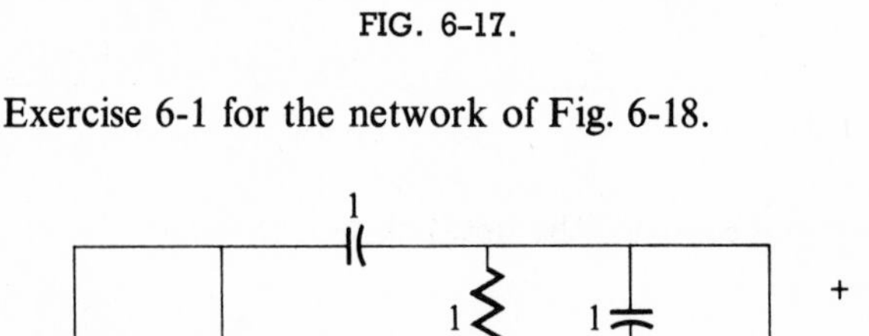

FIG. 6-17.

6-18. Repeat Exercise 6-1 for the network of Fig. 6-18.

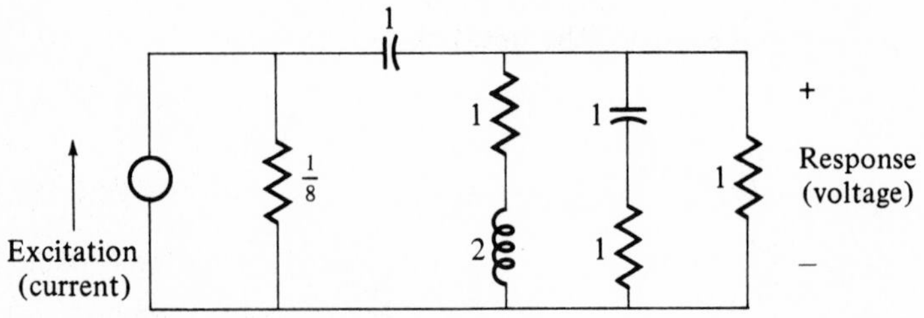

Element values given are: ohms, henrys, farads.

FIG. 6-18.

6-19. Repeat Exercise 6-1 for the network of Fig. 6-19.

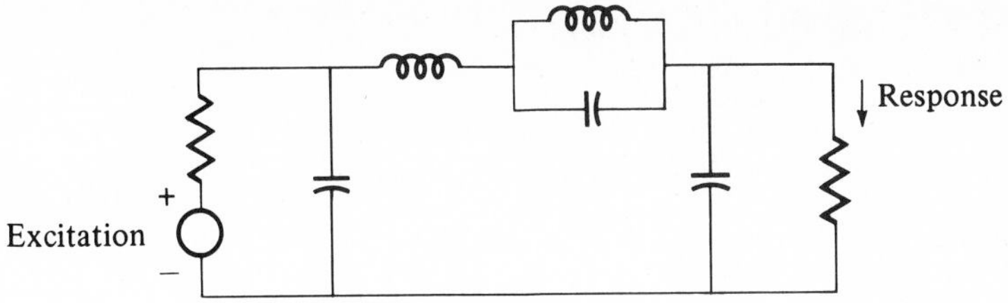

FIG. 6-19.

6-20. Amplifier stages often have resonant circuits in both input and output. Two such stages can be coupled inductively (Z is an inductor in Fig. 6-20) *or* capacitively (Z is a capacitor in Fig. 6-20). (The former also models transformer coupling.) In the sinusoidal steady state, over reasonable frequency bands, the performances of the two are essentially equivalent if properly designed. Compare their performances at very low frequencies (dc) and at very high frequencies (large values of s), using the ideas of Exercise 6-1 with the *minimum* labor necessary. Compare their transmission of steep wavefronts, i.e., the initial-rise response shapes for step excitation. Then do the remaining work of Exercise 6-1 for both networks. [In part (i) choose a resistor value that will make the tuned circuit Qs be 100 at resonance; otherwise use unit element values.]

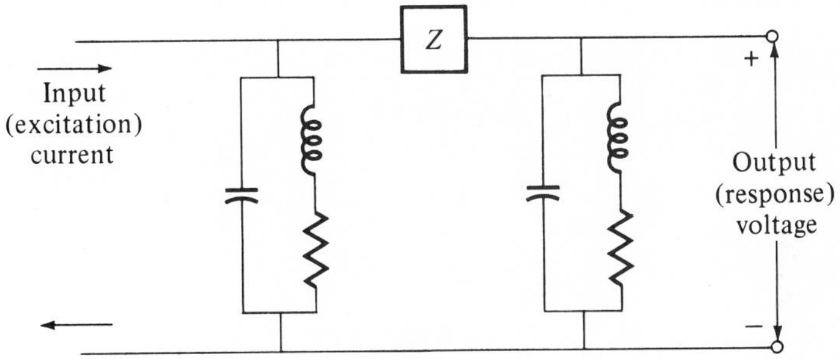

FIG. 6-20.

6-21. Repeat Exercise 6-1 for the (feedback) network of Fig. 6-21. In what way(s) do the results change when R_f is removed? [In part (i) take unit values for L, C, g_m, and make $R_1 = R_f = 10\ \Omega$ and $r = 0.1\ \Omega$.]

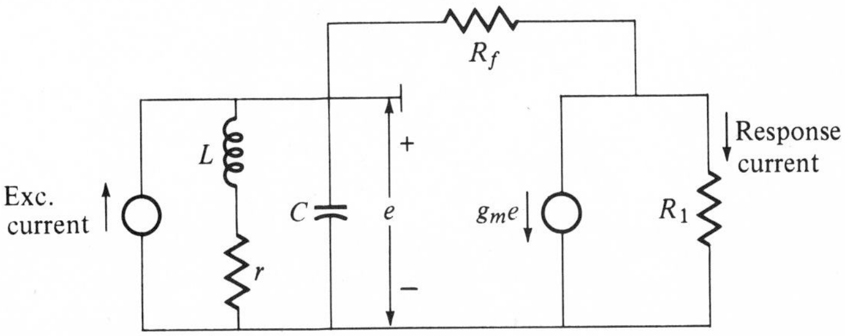

FIG. 6-21.

6-22. Repeat Exercise 6-1 for the network of Fig. 6-22. [Use the element values there shown for part (*i*).]

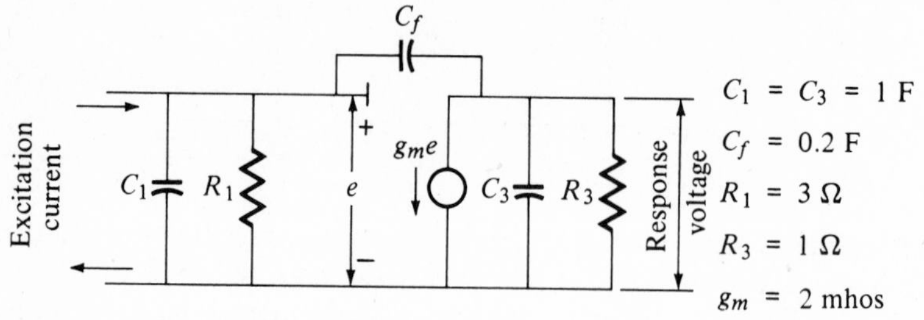

FIG. 6-22.

6-23. Repeat Exercise 6-1 for the network of Fig. 6-23.

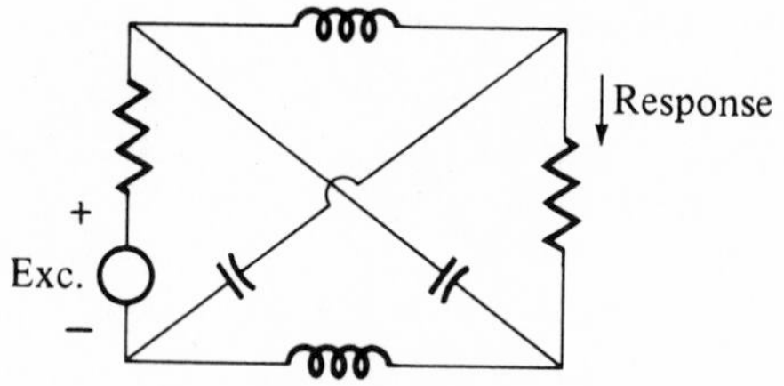

FIG. 6-23.

6-24. Repeat Exercise 6-1 for the network of Fig. 6-24.

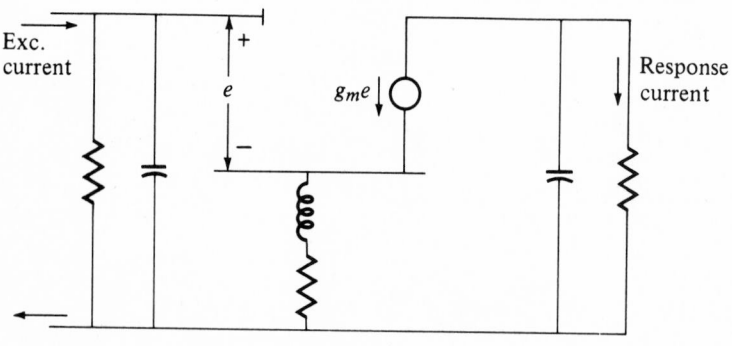

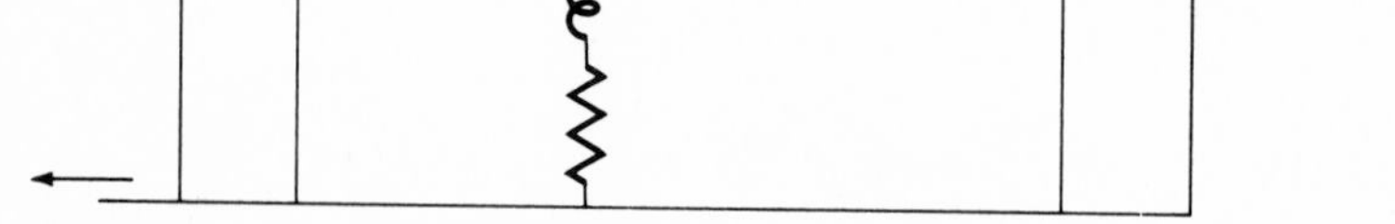

FIG. 6-24.

6-25. Repeat Exercise 6-1 for the network of Fig. 6-25. If the (parasitic) capacitors C_1 and C_3 are ignored, in what way does the initial behavior of the network under impulsive excitation become physically unreasonable? [In part (*i*), use the element values shown]

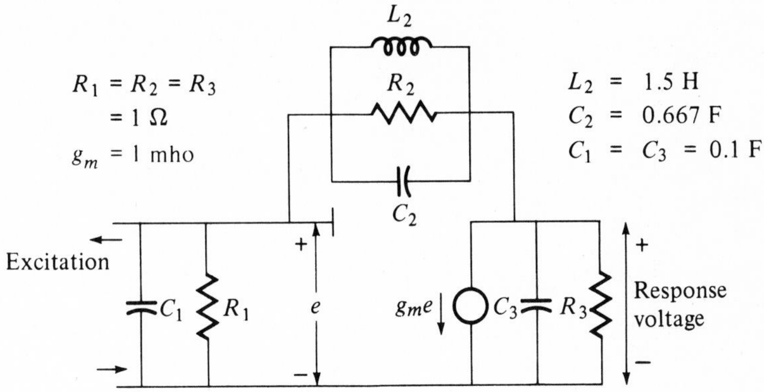

FIG. 6-25.

6-26. Repeat Exercise 6-1 for the network of Fig. 6-26. [In part (*i*), take g_m to be 3 mhos, the other elements to have unit values.]

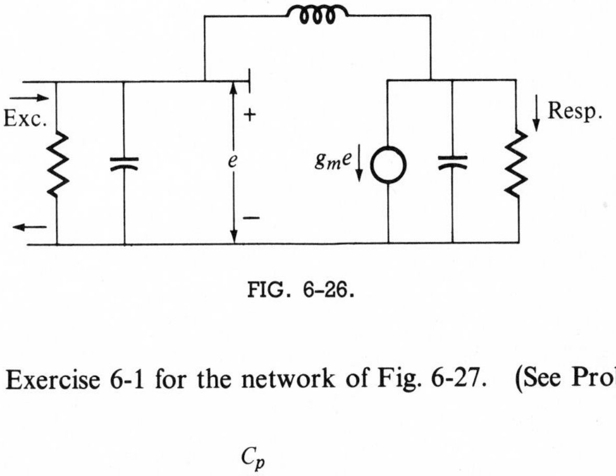

FIG. 6-26.

6-27. Repeat Exercise 6-1 for the network of Fig. 6-27. (See Prob. 6-W.)

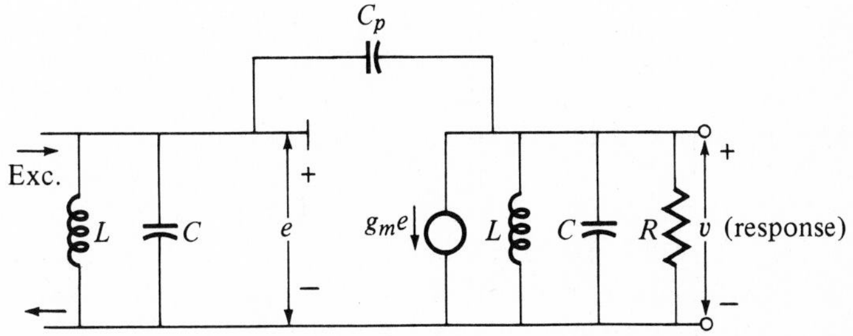

FIG. 6-27.

6-28. Find the transfer function of the (emitter-follower-model) network of Fig. 6-28. What form does it take when Z_1 is removed? (The boxes Z_1 and Z_2 are arbitrary passive impedance combinations.)

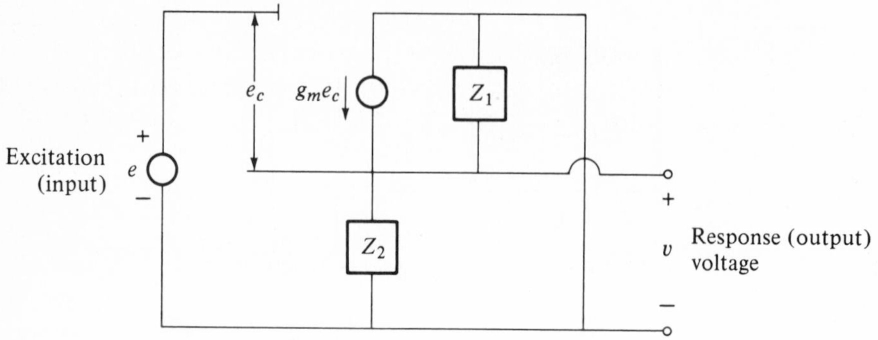

FIG. 6-28.

6-29. Repeat Exercise 6-1 for the network of Fig. 6-29. Compare impulse response and transfer functions for the cases (*a*) $C_2 = 2C_1$, $R_2 = R_1$, and (*b*) $C_2 = C_1$, $R_2 = R_1$.

If, in general, the output waveform must be a replica (perhaps scaled) of the input waveform, what element-value relation do you recommend? [This adjustment is often of practical value in cases where R_1 and C_1 model some connecting apparatus, and C_2 and R_2 model the input to measuring equipment; the waveform of e can thus be accurately transmitted to the measuring apparatus in spite of its (parasitic) R_2 and C_2 elements.]

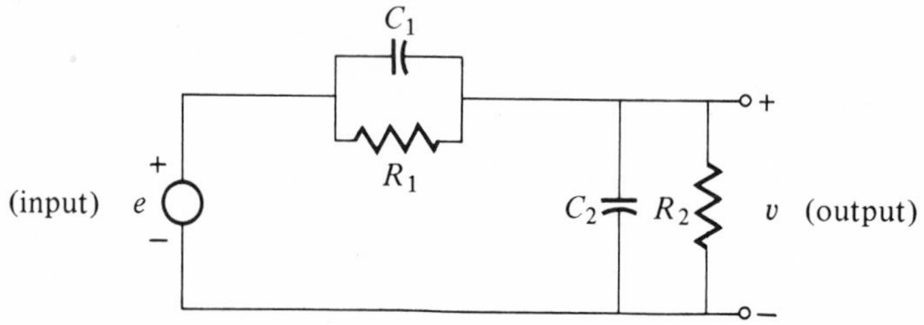

FIG. 6-29.

6-30. Suppose that in Fig. 6.06-B (consider both circuits) the inductor current steps suddenly from zero to a finite value. Find the corresponding voltages and currents of the resistor and the capacitor. What excitation would be necessary to produce them?

6-31. Consider the network of Fig. 6.26-A.

(*a*) Suppose that L_2 is omitted. Show by simple physical reasoning that $v_\delta(0+)$ is negative and $v'_\delta(0+)$ is positive.

(*b*) Suppose L_2 present, but large. Repeat (*a*).

(*c*) Suppose L_2 present but *small*. Discuss the signs of $v_\delta(0+)$ and $v_\delta'(0+)$, making use of the results of Chap. 3 as far as possible.
(*d*) Suppose L_2 present, of arbitrary value. Discuss the signs of $v_\delta(0+)$ and $v_\delta'(0+)$ in general terms, demonstrating all conclusions carefully.

6-32. What is the order of each of the networks shown in Fig. 6-32? Verify your answer (obtained by inspection) by calculation of a transfer function (or simply of the characteristic polynomial) in each case.

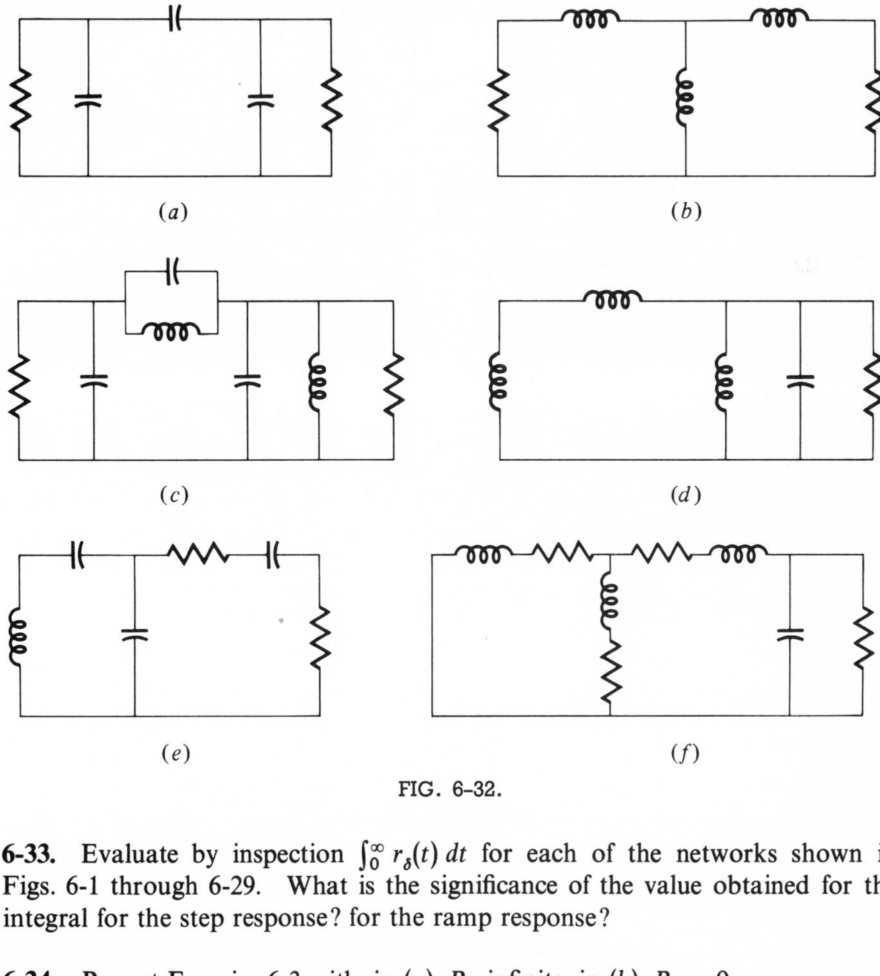

FIG. 6-32.

6-33. Evaluate by inspection $\int_0^\infty r_\delta(t)\,dt$ for each of the networks shown in Figs. 6-1 through 6-29. What is the significance of the value obtained for the integral for the step response? for the ramp response?

6-34. Repeat Exercise 6-3 with, in (*a*), R_2 infinite, in (*b*), $R_2 = 0$.

6-35. One colorful way of explaining asymptotic behavior is exemplified here (see Fig. 6-35): "For very small s (very low frequencies), practically all the exciting current i_F flows through R_1; a little bit of it, however, flows through C_2, and through L_3; hence a little bit of a little bit of response voltage exists as $s \to 0$, and $T(s)$ has two zeros at the origin." Justify this statement, using the MacLaurin expansion of $T(s)$ and other appropriate mathematical apparatus. Do you consider its reasoning (*a*) valid? (*b*) helpful? In similar fashion explain the asymptotic calculations of Sec. 6.15.

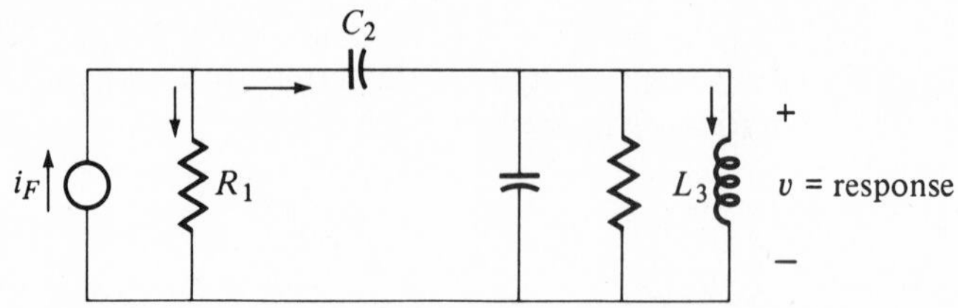

FIG. 6-35.

6-36. In the "twin-T" network of Fig. 6-Q, the capacitors being uncharged, replace the voltage source e by an impulsive current source, $i_F = Q_0\,\delta(t)$. For simplicity, let $R_2 = 2R_1$ and $C_2 = 4C_1$. What are the three initial ($t = 0+$) capacitor voltages? Write the equation whose roots are the natural frequencies of the network. Why is it rather easy to find the values thereof?

6-37. The network shown in Fig. 6-37 is part of a larger network; there is no other connection between the portion at the left (which contains the excitation) and that at the right (in which the response occurs). Assuming no anomalies, discuss the contribution of the portion shown to the transfer function of the complete network as to (*a*) natural frequencies, and (*b*) zeros of transmission. What happens to your answers if all three capacitor values approach infinity? If all three inductor values approach zero? Now replace each series LC combination with a parallel LC combination and repeat the (complete) discussion.

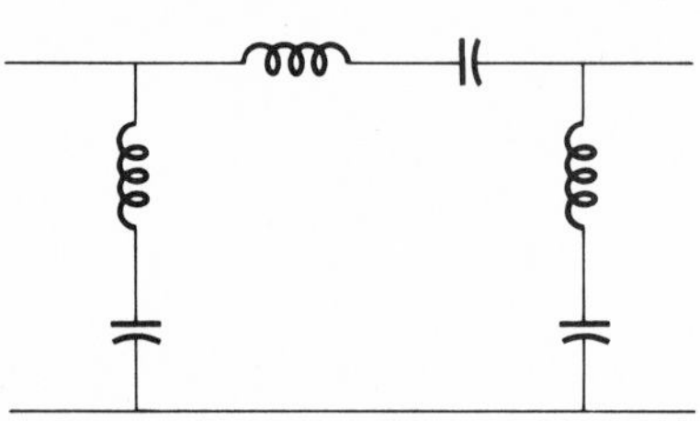

FIG. 6-37.

6-38. (*a*) Repeat Exercise 6-1 for the transmission-line model of Fig. 6-38 (see Prob. 7-Z) in which the two resistors represent excitation and response-measuring device immittances. There are N energy-storing elements, approximately half of them inductors and half capacitors (the number of capacitors is one more than the number of inductors), repetitive in value except for the end capacitors. [In part (*i*), use the element values shown in Fig. 6-38, and a reasonable value for N.]

(*b*) Now replace each inductor L by a capacitor $C' = 0.5$ F and each capacitor C by an inductor $L' = 0.5$ H; make the two "end" inductors of 1 H each. Repeat (*a*) and correlate your results with those of (*a*).

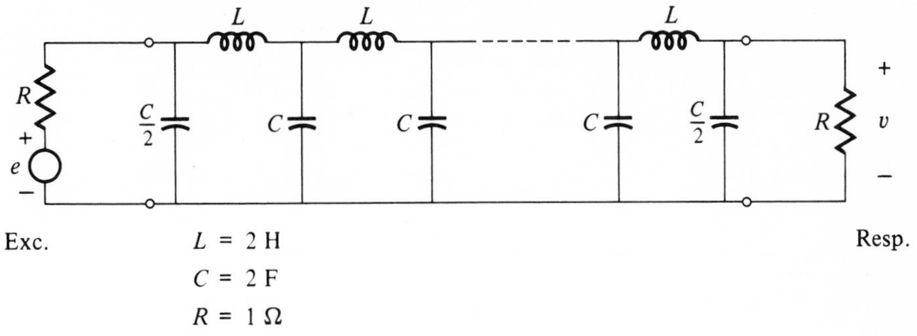

FIG. 6-38.

6.30 PROBLEMS

6-A. Carry out the calculation of the transfer function of the network in Fig. 6.02-A *not* as in the text by assuming the format (6.02-2) for $v(t)$ and working back to $e(t)$ *but* by simply working back in terms of $v(t)$. The first step, to illustrate, is $i_3(t) = G_3\, v(t)$; the second step is $v_m(t) = v(t) + L_3\, G_3\, v'(t)$; the third is $i_2(t) = C_2\, v_m'(t) = \cdots$; etc. In this way find the differential equation that (implicitly) determines $v(t)$ in terms of $e(t)$. Show therefrom that the transfer function is indeed that obtained in Fig. 6.02-A*b*. To what extent is the success of this fashion of analysis dependent on the ladder character of the network?

6-B. Find the transfer function of the network of Fig. 6.02-A by working **forward** from the generator: express the current i_1 in terms of the impedance facing the generator, then divide it into i_2 and i_3. Verify that the transfer function so obtained is the same as that in the figure. Compare the labor for the two methods.

6-C. A rather sly approach to the calculation of ladder-network transfer functions, sometimes useful, is to (*a*) find the characteristic polynomial (as by summing impedances around a loop or the admittances attached to a pair of nodes), (*b*) find the zeros of transmission (the infinite-loss points) by inspection of the arms of the ladder, and (*c*) evaluate the multiplier (scale factor) from the high-frequency or low-frequency asymptotic behavior of the network. For each of the networks of Figs. 6-1, 6-2, and 6-4, so obtain the transfer function, using literal element values. Contrast the labor required with that necessary for the straightforward calculation of $T(s)$ by the work-back or ladder method. Repeat for the network of Fig. 6-9, using the numerical element values.

6-D. A network has the natural frequencies and zeros of transmission shown in Fig. 6-D. The numerical values given are (normalized) radians per second; the sinusoidal-steady-state transmission of the network at unit normalized frequency ($s = j1$) is unity, with zero phase shift. Sketch the impulse response and the step response versus time. Evaluate the two responses at $t = 1$ s (normalized) within 1 percent. Repeat for the network characterized by the same natural frequencies but with three zeros at the origin and one at infinity. What relations do the four responses you have discussed bear, each to the other three?

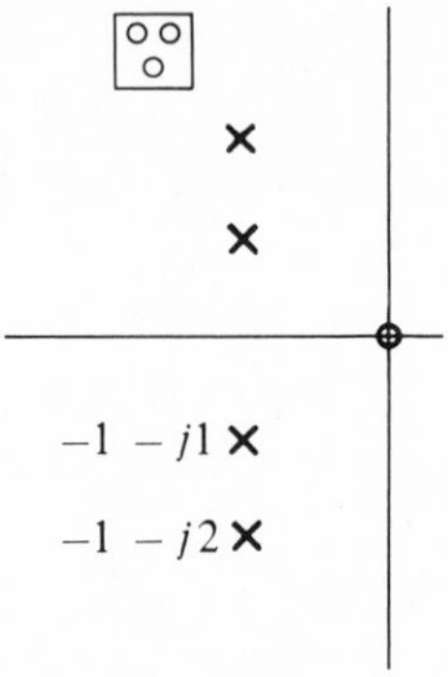

FIG. 6-D.

6-E. Find the transfer function of the network of Fig. 6.12-B, taking all element values to be unity. In which of the seven network formats given in Fig. 6-E (excitation at the left, response at the right in each case) can you find other (equivalent) realizations of the same transfer function? What observation can you make about (*a*) the uniqueness of network *analysis* (response finding), and (*b*) the uniqueness of network *synthesis* (network finding)?

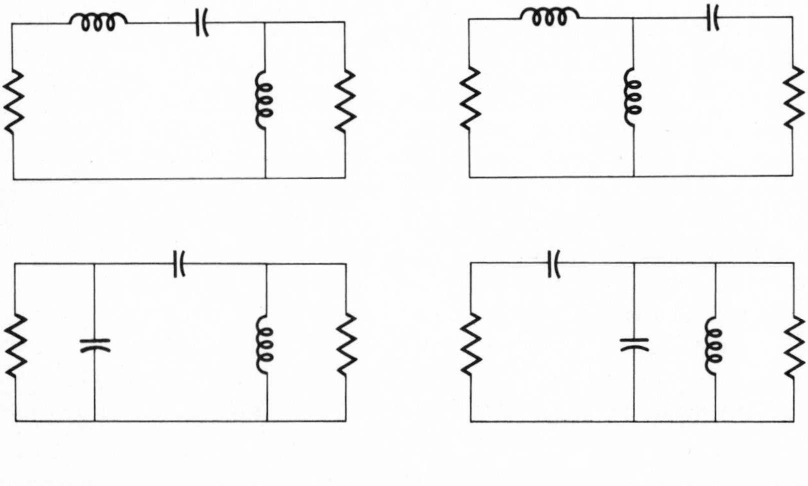

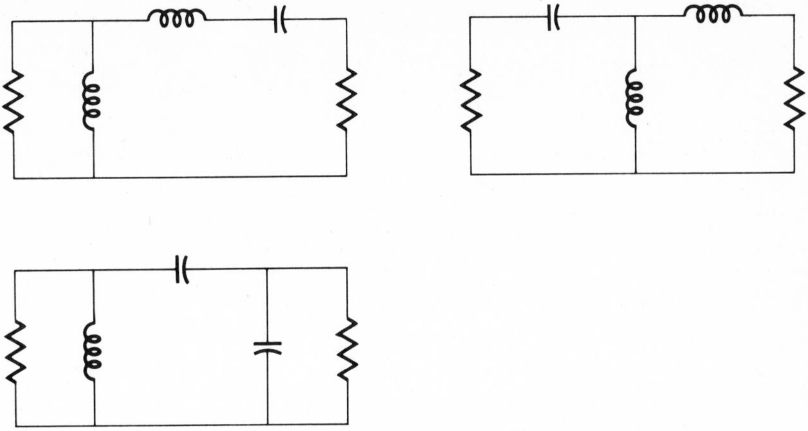

FIG. 6-E.

6-F. The network shown (Fig. 6-F) is initially at rest; $i_F(t) = Q_0\,\delta(t)$. How many natural frequencies does the network have? Where are they in the complex plane? What are the $t = 0+$ conditions in L and in C? What is $i(t)$, $t > 0$? What is $v_x(t)$, $t > 0$?

Repeat when the element values have the special relations $R_1 = R_2 = \sqrt{L/C}$. Which of the natural frequencies are observable in (present in the voltage across) R_2? R_1? Where *are* the natural modes unobservable in R_1 and R_2 actually observable? Explain the anomaly.

In the more general network of Fig. 6.22-B, which has N natural modes, where is each observable?

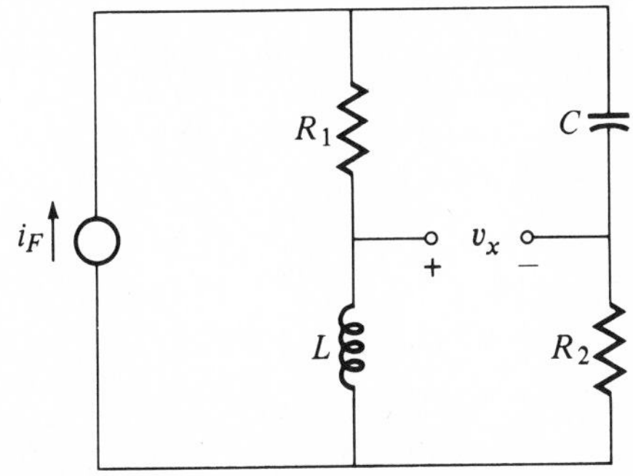

FIG. 6-F.

6-G. (*a*) For each of the eight network situations described below (initially at rest and unit-impulse-excited):

1 Evaluate $v(0+)$ and $v'(0+)$ from the schematic (that is, by physical reasoning);

2 Calculate $T(s)$ and sketch the layout of its poles and zeros;

3 Evaluate $v(0+)$ and $v'(0+)$ from $T(s)$; do your results check?

4 Make the partial-fraction expansion of $T(s)$ and write out the impulse response in sum-of-exponential form;

5 Evaluate $v(0+)$ and $v'(0+)$ from this last result (set $t = 0+$); are the results the same?

6 Sketch the impulse response versus time.

The eight network situations are (see Fig. 6-G):

i Network A with $L = \alpha$, $C = \beta$; in which $\alpha = 2.618 = (3 + \sqrt{5})/2$, $\beta = 0.382 = (3 - \sqrt{5})/2$;
ii Network A with $L = \beta$, $C = \alpha$;
iii Network B with $L = \alpha$, $C = \beta$;
iv Network B with $L = \beta$, $C = \alpha$;
Now replace e_F in series with R_1 by an i_F (equal to $G_1 e_F$) in parallel with R_1;
v repeat (*i*);
vi repeat (*ii*);
vii repeat (*iii*);
viii repeat (*iv*).

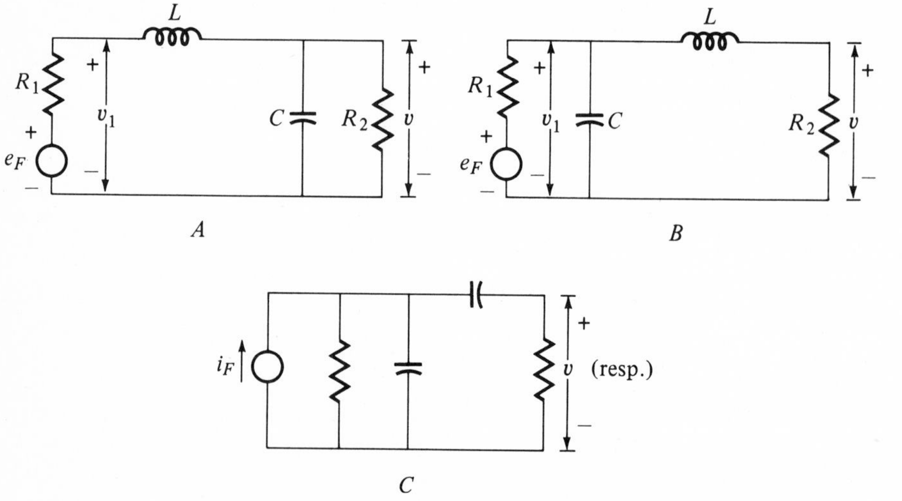

FIG. 6-G.

(*b*) Design a network such that the natural frequencies are unchanged from those of B but the initial values are $v(0+) = +1$, $v'(0+) = 0$. What is the necessary (new) $T(s)$? Sketch the positions of its poles and zeros. Expand it in partial fractions and obtain the (new) impulse response in exponential form. Evaluate $v(0+)$ and $v'(0+)$ therefrom as a check of your work. Sketch it versus time. Can either network format A or B be used, changing only element values, to realize the new transfer function? If so, obtain the new element values of one such realization. If not so, try network format C. If still you have no success, make the response (output) v_1 instead of v and try again. (Synthesis *is* basically trial and error, until you learn enough to know what to do.)

6-H. The impulse response of a network whose details are unknown is found, on measurement, to be something like that sketched in Fig. 6-H*a*. It can be adequately represented by this formula:

$$r_\delta(t) = e^{-t} + e^{-2t} - 2e^{-3t}, \qquad t > 0,$$

in which numbers have been rounded, as well as normalized, for simplicity.

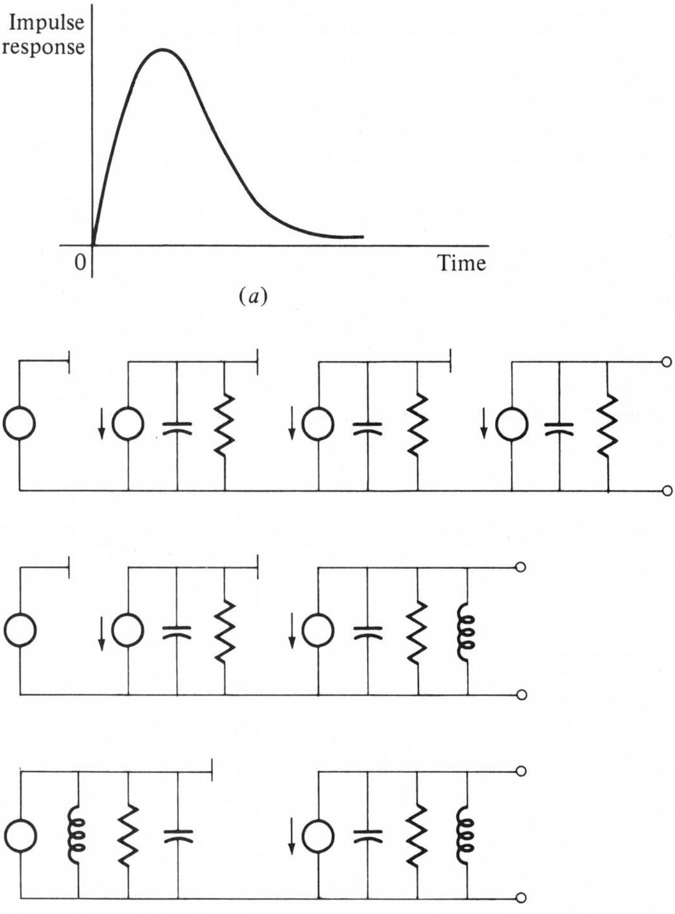

FIG. 6-H.

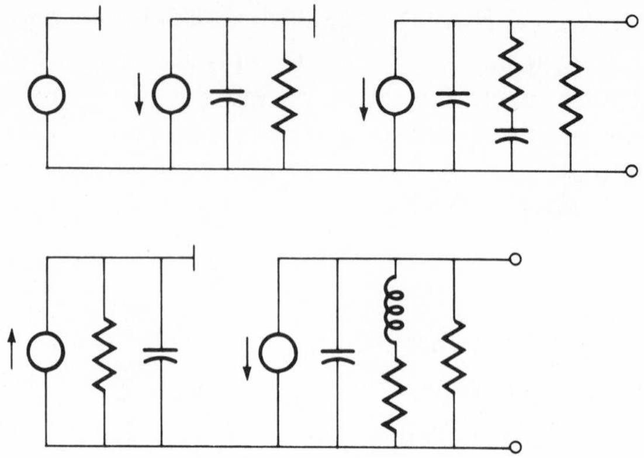

(*b*)

FIG. 6-H. (*Con.*)

(*a*) Let the transfer function of the network be written in the form given in Exercise 6-1(*e*). Determine, by inspection of the data, the values therein of *M*, *N*, A_{N-1-M}. Which zeros of $T(s)$ are *obvious* therefrom? Which poles?

(*b*) Find $T(s)$ in detail. Tabulate and show on a sketch of the complex plane the zeros and poles thereof. What is its asymptotic behavior at very low and at very high frequencies? Which of your last two answers are *obvious* from the impulse response as given? From the integral of the impulse response as calculated mentally?

(*c*) Find and sketch the step response of the network.

(*d*) Which of the five networks shown in Fig. 6-H*b* are possible models for the network, that is, can have exactly the same impulse response? (The excitation, voltage or current, is at the left in each case; the response is the voltage between the two rightmost terminals.) Explain carefully and find the element values in those that can.

6-I. For the network shown in Fig. 6-I*a*, find the form of the transfer function from *e* to *i*. Which of the initial values of the impulse response are zero? Verify your answer in Fig. 6-I*b*, a photograph of the actual impulse response of such a network.

Make as accurate a plot of the *step* response of the network as you can, based on Fig. 6-I*b*. (The time scale there is 20 μs/cm.) From the step-response wave, obtain and plot the response to a rectangular *pulse* of width $T = 70$ μs. Compare your result with the wave of Fig. 6-I*c*, in which the scale is 50 μs/cm (the waves shown are just such an exciting pulse, and the network's response thereto). By altering the pulse-width *T*, can you make your pulse-response wave have the two equal-height "ears" (peaks) of Fig. 6-I*c*? Explain how the width of the pulse that is necessary for such a response is obtained from the impulse response.

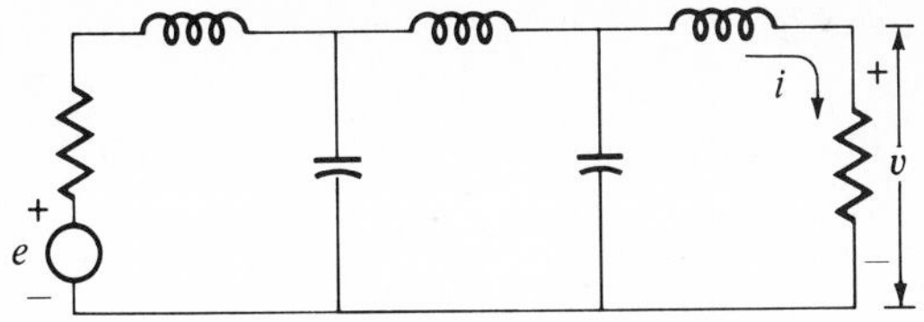

(a)

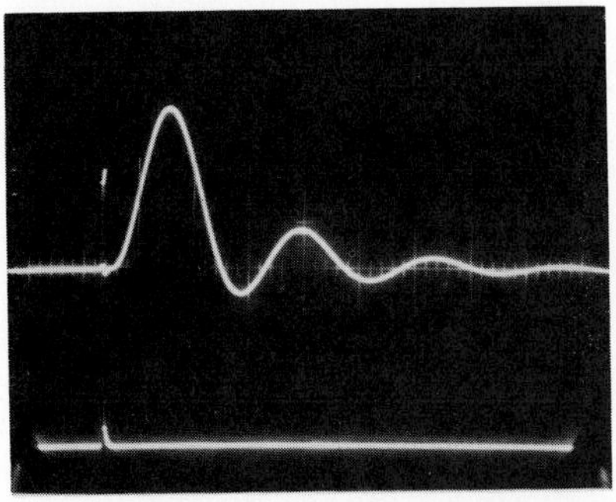

(b)

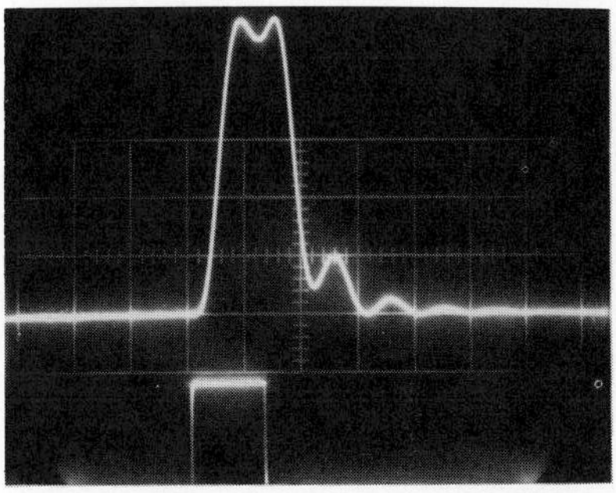

(c)

FIG. 6-I.

6-J. For the network of Fig. 6-J*a* the photograph of Fig. 6-J*b* shows the impulse (narrow-pulse) response at points 1, 2, 3 in turn, with a resistor connected to ground at each observing point in turn. (The photograph is a multiple exposure.) On the assumption that the *L* and *C* elements of the long network that lie beyond the point in question can be ignored in each case, give the *form* of the transfer function from *e* to *v* at 1, at 2, and at 3. In particular, how many initial values of *v* are zero at each point? Explain your reasoning. Why is the assumption above a reasonable one? Verify your answer by correlation with Fig. 6-J*b*. (The initial response at point 1 is large, and off scale in the photograph; the other two, and the exciting impulse, are evident.)

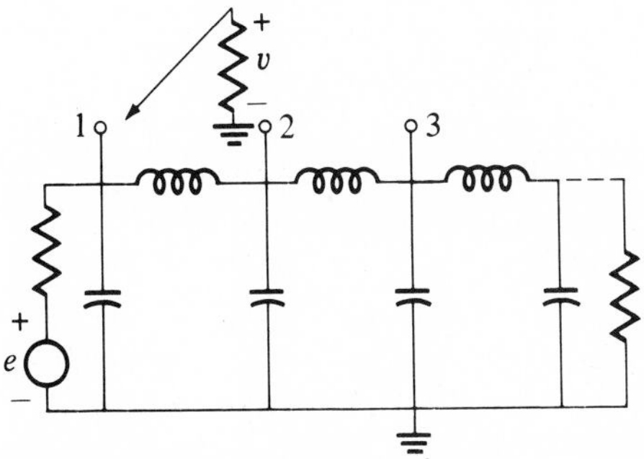

(a)

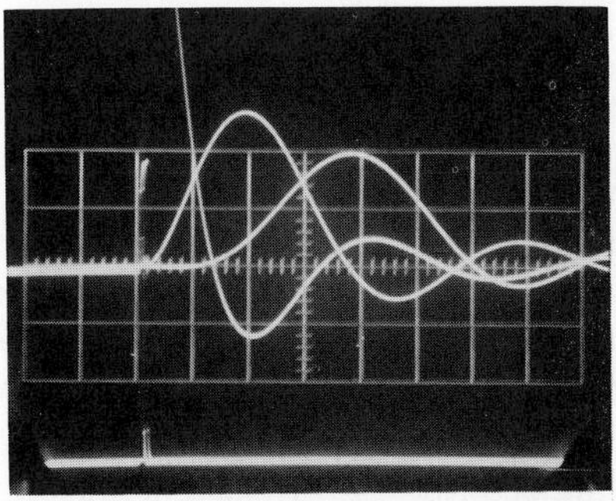

(b)

FIG. 6-J.

6-K. By inspection of Fig. 6-K*a* determine where the zeros of $T(s)$, the transfer function from e to v, lie; show your results on a sketch of the s plane. What cay you say about the locations of the poles of $T(s)$?

Sketch $|T(j\omega_F)|$ versus ω_F. Correlate with Fig. 6-K*b*, a physical view of the same. Constrast Fig. 6-K*b* with Fig. 6-16-B*b* and explain the differences and similarities.

Sketch the impulse response of the network in as much detail as you can with the information given above. Then examine Fig. 6-K*c*, a physical view of the same. What information does Fig. 6-K*c* *add* to your knowledge of the network's transfer function? If you are now told that the minimum point in Fig. 6-K*b* occurs at 25 kHz, can you establish the time scale in Fig. 6-K*c*? Explain.

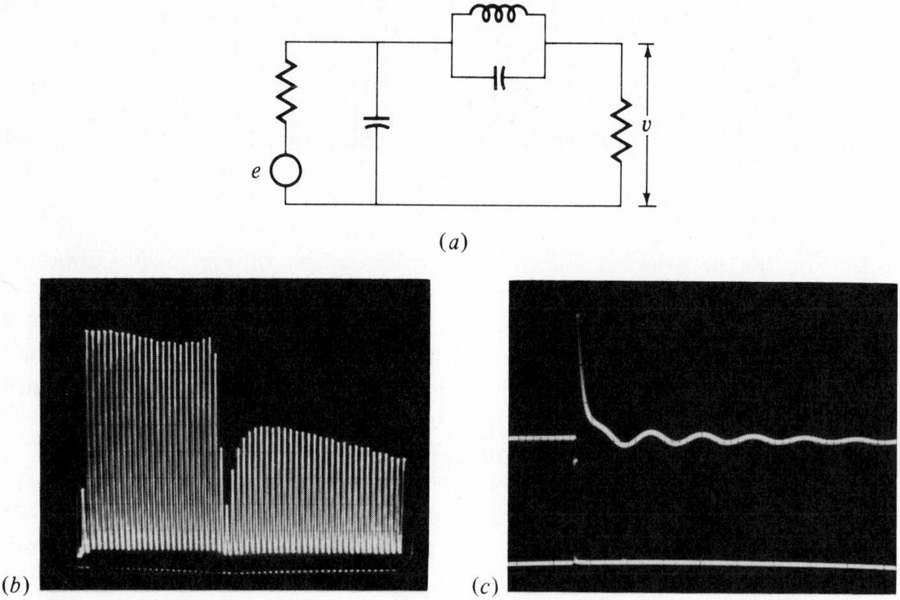

FIG. 6-K.

6-L. The circuit of Fig. 6-L*a* is at rest (its element values, in ohms and farads, are normalized and rounded for simplicity), then impulse-excited: $i_F = Q_0\,\delta(t)$. Trace the path of the impulsive current through the elements of the circuit and determine the three capacitor voltages at $t = 0+$. Then find the two $t = 0+$ currents in the resistors. Why are the three $t = 0+$ *currents* in the capacitors not immediately obvious? Find their values. (If the sum of the voltages around a closed loop is zero, what is true of their derivatives?) State the dual problem and describe the physical actions therein under analogous circumstances. Now add the active element g_m to form the network of Fig. 6-L*b*. For what value of g_m, under the same circumstances, is $i_{C1}(0+) = 0$?

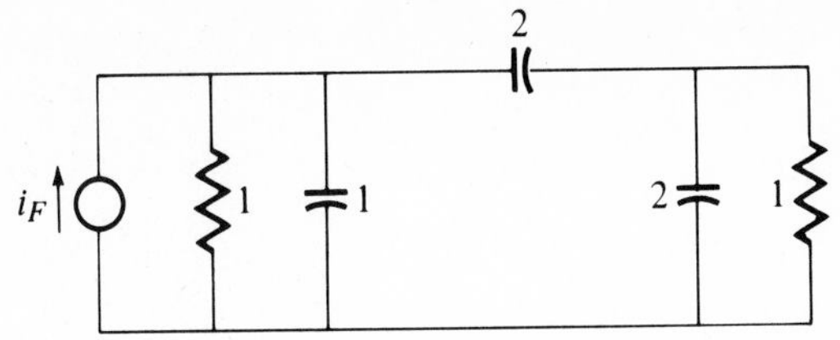

(*a*)

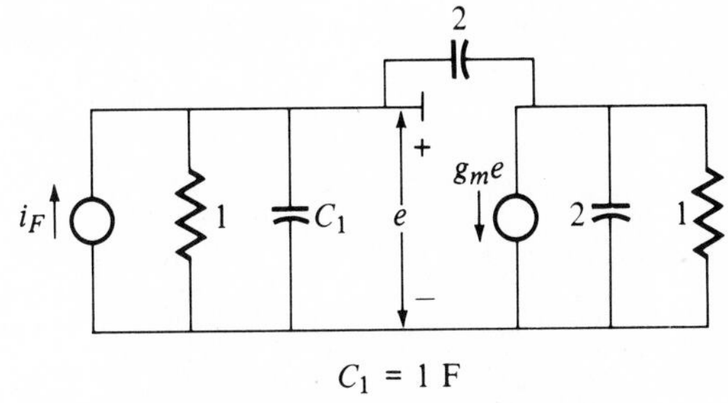

(*b*)

FIG. 6-L.

6-M. The network shown in Fig. 6-M*a* is a very simple one, but useful in certain control systems. Find $T(s)$ (from e to v) in terms of the literal element values given. On an s plane sketch show where the poles and zeros lie, still using literal element values. Explain the action of the infinite-loss-point-producing elements. For what type of excitation wave will the network suppress the excitation wave in the response? If excited from rest by such an excitation wave, of what will the response consist? Illustrate with sketches of excitation and response.

Sketch $|T(j\omega_F)|$ versus ω_F, (*a*) from the pole-zero constellation geometry, (*b*) from the formula found for $T(s)$, and (*c*) from the network schematic, by physical common sense. Why is the shape so simple? Does the physical behavior of such a network, shown in Fig. 6-M*b* corroborate your analysis? Explain.

Sketch $\beta = \overline{T(j\omega_F)}$ versus ω_F, using the same three bases. Locate the maximum of the curve and the value of β there, still in terms of the literal element values. (It is this leading phase shift that is useful in control systems.)

Photographs (*c*), (*d*) and (*e*) (Fig. 6-M) each show a sinusoidal excitation wave and the network's steady-state response thereto; [the three (different) frequencies are (*c*): 0.5 kHz, (*d*): 5.0 kHz, (*e*): 50 kHz]. How can you determine which wave is the excitation and which the response? What, approximately, is the phase shift β in each photograph?

The element values in the network whose behavior is shown in the photographs were: $C_1 = 0.01\ \mu\text{F}$, $R_1 = 2000\ \Omega$, $R_2 = 100\ \Omega$. Plot the β versus ω_F curve to scale, using these values. How well do the phase shifts seen in the photographs agree with the corresponding points on the curve?

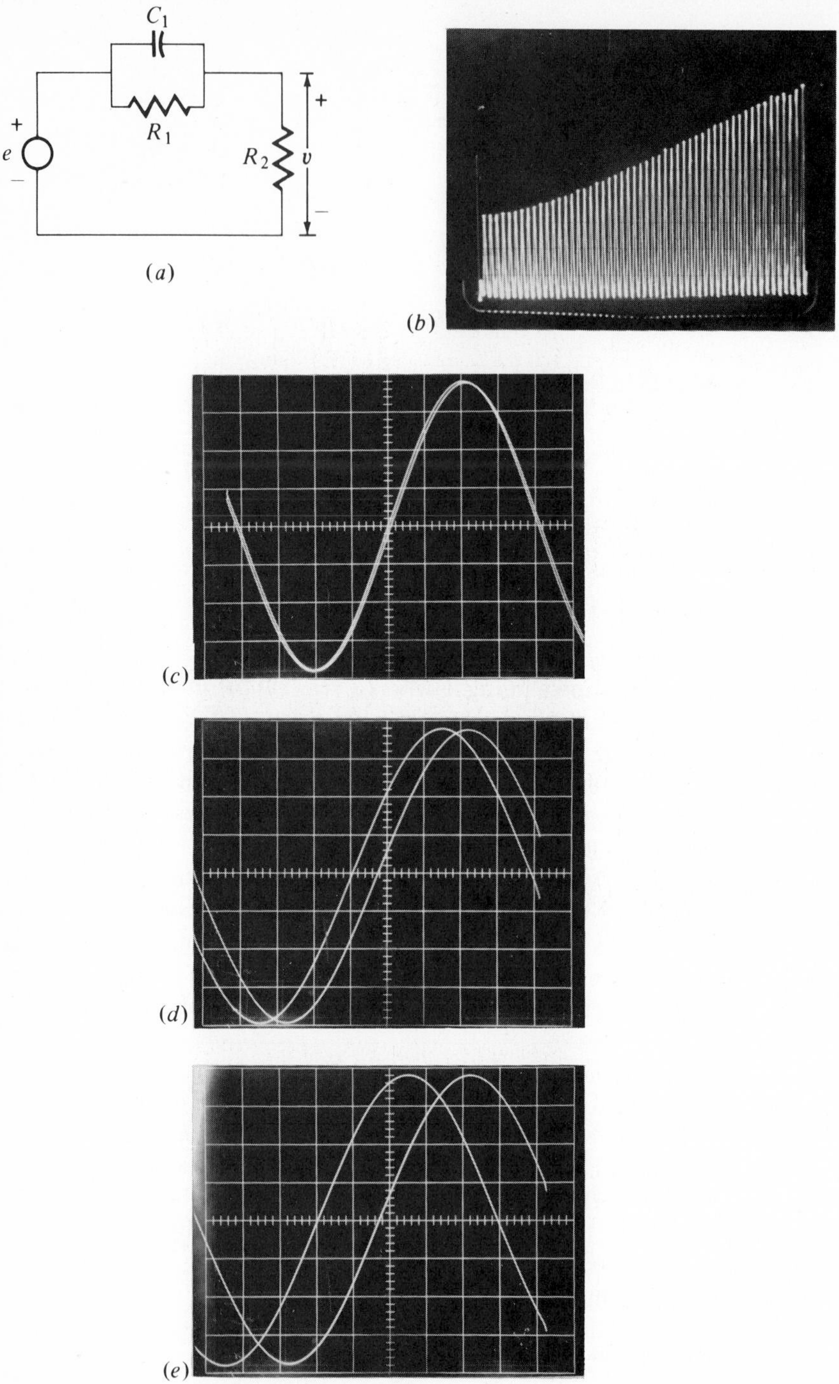

FIG. 6–M.

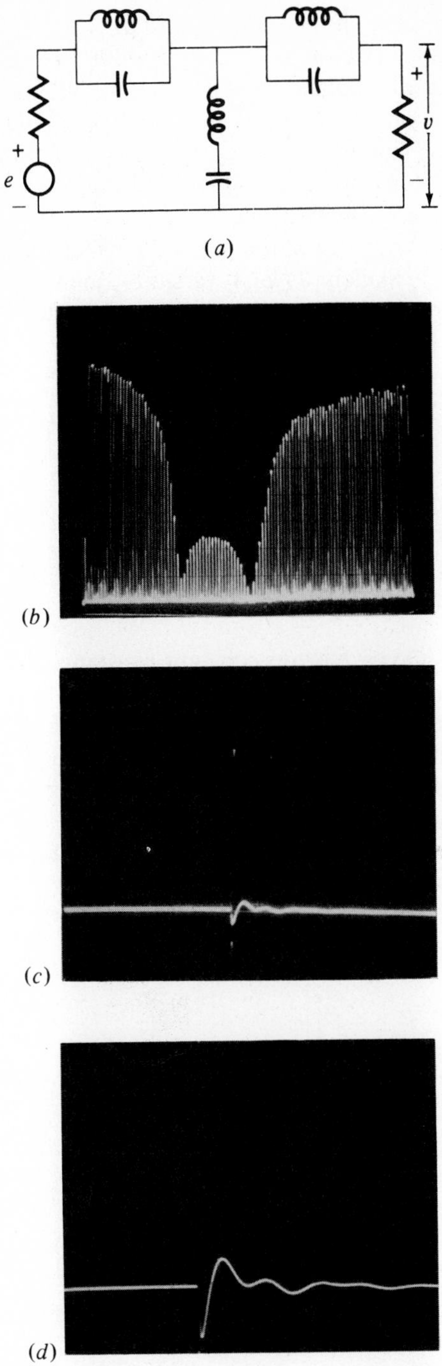

FIG. 6-N.

6-N. (*a*) For the network shown in Fig. 6-N*a* in schematic form, what are the degrees of numerator and denominator of the transfer function $T(s)$ from e to v? On an s plane sketch show, as far as is possible and for general (not special) element values, the poles and zeros of $T(s)$. Set forth and explain the low-frequency and high-frequency asymptotic transfer functions. From both (1) the transfer function and (2) the physical action of the network, obtain and explain the initial behavior of the impulse response of the network. Sketch $r_\delta(t)$ in as much detail as is possible. Sketch $|T(j\omega_F)|$ versus ω_F in as much detail as possible.

(*b*) The photographs (Fig. 6-N*b*,*c*,*d*) show the actual behavior of such a network:

(1) $|T|$ versus ω_F, (2) $r_\delta(t)$, (3) $r_\delta(t)$ with expanded (magnified) vertical scale.

Carefully verify all the information obtained in (*a*), making it more specific where possible.

(*c*) What differences exist in the initial behavior of $r_\delta(t)$ here, and in Fig. 6.20-A*b*? Explain fully.

6-O. Shown in Fig. 6-O is a Norton constant-resistance dividing (or filter-pair) network. Calculate the three driving-point impedances marked, as functions of s. Find the initial ($t = 0+$) voltage and current in each of the nine elements, following impulse excitation from rest. Find the transfer functions from e to v_1 and from e to v_2. Check as many of their coefficients as you can by comparing initial-value tableaux with physically obtained values for $v_1(0+)$, $v_1'(0+)$, ... and $v_2(0+)$, $v_2'(0+)$, ... Explain the two attributes given the network above.

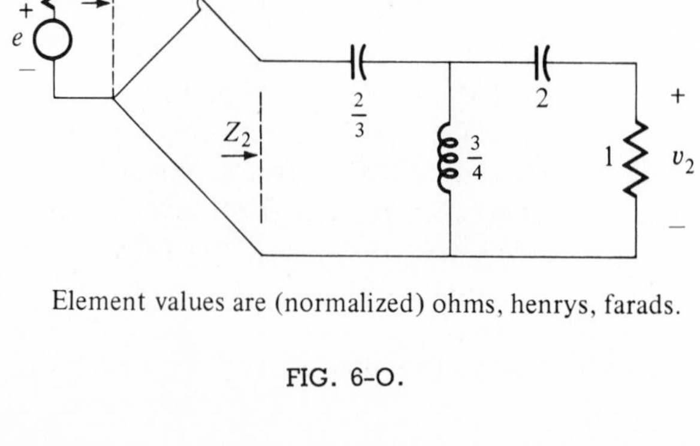

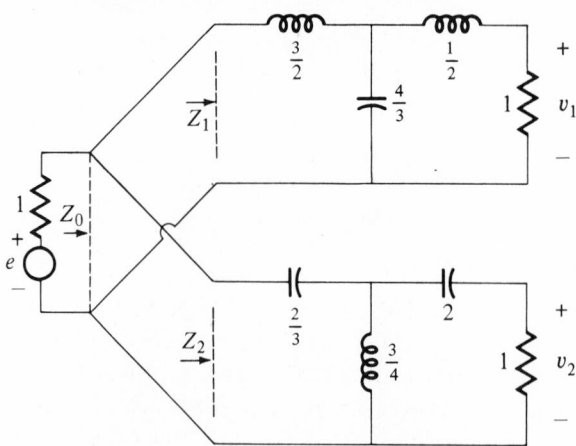

Element values are (normalized) ohms, henrys, farads.

FIG. 6-O.

6-P. In Fig. 6.25-U change the two R_0 symbols to R_1 and R_2 respectively. Find the exact transfer function from e_1 to v and from e_2 to v. What do they become for very large A? From your work here obtain (6.25-18) as a check.

6-Q. The "twin-T" (or "parallel-T") network shown (Fig. 6-Q) acts much as a bridge, but it permits both input (excitation) and output (response-measuring) devices to be grounded. For this reason it can be very useful in measurement, and in selective feedback amplifiers.

(*a*) Explain why in an ordinary bridge signal source and detector cannot *both* be grounded.

(*b*) Calculate the transfer function from e to v. [*Suggestion:* let $v = 1$, denote the current in the right-hand R_2 resistor or the right-hand C_1 capacitor by x, and find a lockup equation that determines x from the fact that $v_1 = v_2$; the notation $a^2 = C_2/(2C_1)$, $b^2 = R_2/(2R_1)$ may be useful.]

(*c*) Show that if $R_2 C_2 = 4R_1 C_1$ the transfer function has a zero ("the bridge balances") at a real frequency, $s = j\omega_0$. What is the value of ω_0?

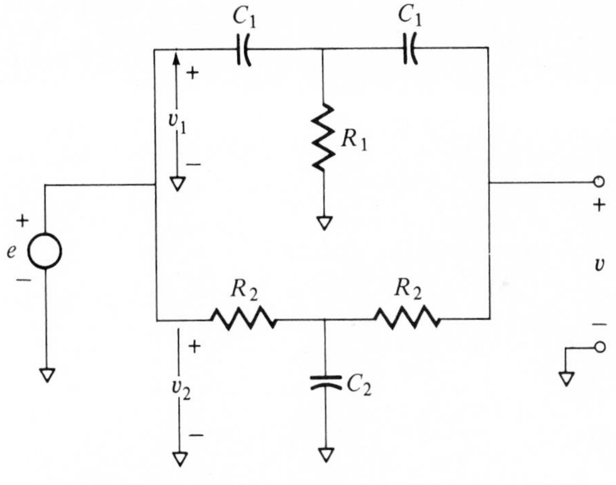

The triangle-marked points are supposed to be connected together (often grounded).

FIG. 6-Q.

6-R. Two networks, each with two accessible ports (terminal pairs), are connected in parallel at their inputs and excited by a common source, e (Fig. 6-R*a*). (The triangle-marked points are to be supposed connected together, perhaps grounded.) Suppose the two transfer functions there defined, $T_1(s)$ and $T_2(s)$, are calculated. Suppose the two networks are now also connected in parallel at their outputs and the new transfer function is denoted T (Fig. 6-R*b*).

(*a*) Why might one well suppose that the two individual transfer functions would add to give the new one, i.e., that $T = T_1 + T_2$?

(*b*) Why, in general, will T *not* necessarily be equal to the sum of T_1 and T_2?

(*c*) Calculate the three transfer functions for each of the two cases of Fig. 6-R*c*. (Take $L = 1$ H and $C = 2$ F for simplicity if you wish.) In which of the two is $T = T_1 + T_2$? Explain. Compare the zeros and poles of the various transfer functions. Is a real-frequency balance ($T = 0$ at a point on the imaginary axis) possible in either case?

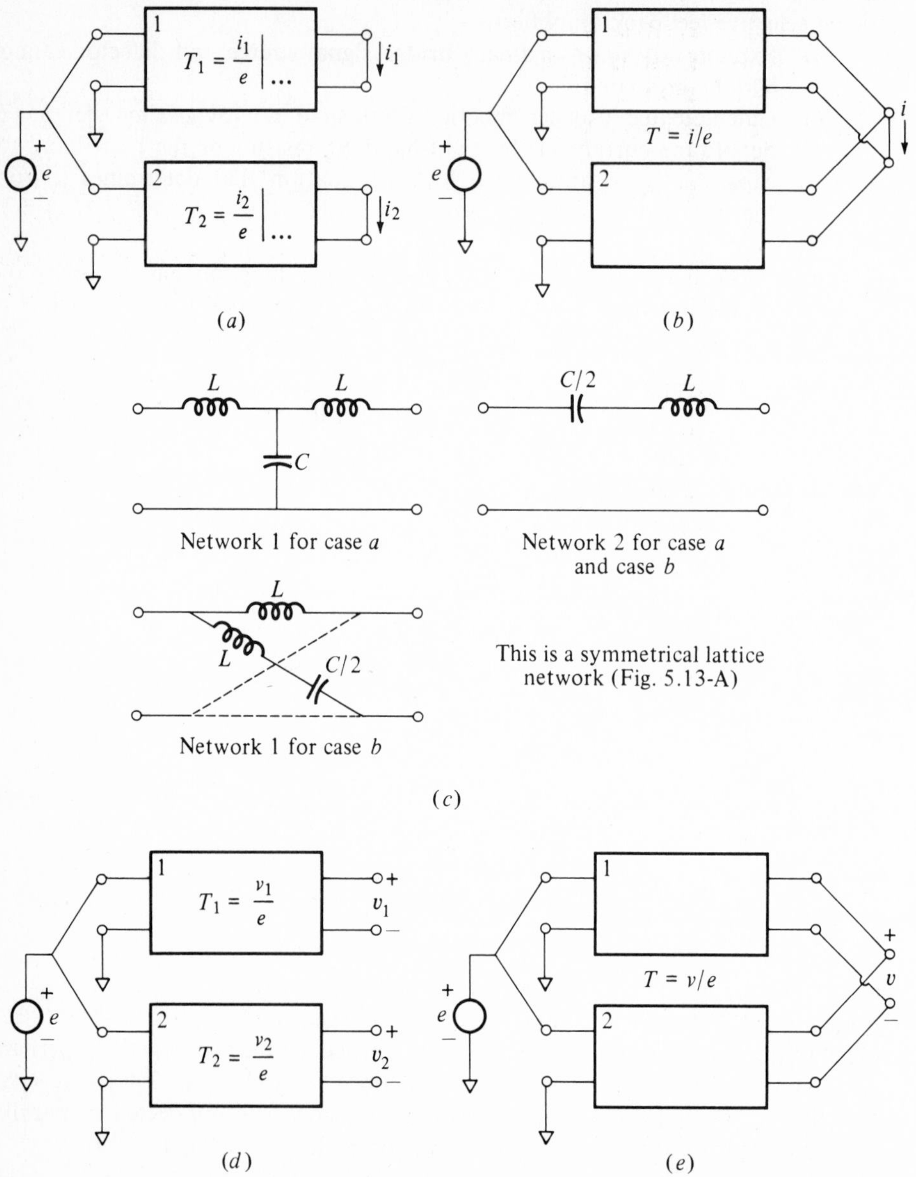

FIG. 6-R.

(*d*) The following test for the additivity of T_1 and T_2 (Fig. 6-R*a*) to give T (Fig. 6-R*b*) is due to Otto Brune. Prove its validity and for illustration apply it to the two cases of (*e*). *The test:* If in Fig. 6-R*a* there is no difference of potential (for all values of s) between the two (short-circuited) output terminal pairs (in which flow i_1 and i_2) then in Fig. 6-R*b*, $T = T_1 + T_2$. (Note that you are asked only to demonstrate the *sufficiency* of the zero potential difference for the transfer function addition to hold; discussion of its necessity is more complicated; in fact "if" can*not* be replaced by "if and only if" in the test statement.)

(*e*) Show, either by example or (better) by more general reasoning, that if T_1 and T_2 are defined with the output terminals *open* (Fig. 6-R*d*) then T_1 and T_2 generally do *not* add to give the transfer function for the combination of the two networks in parallel also at the output (Fig. 6-R*e*). (The calculation of T from T_1 and T_2 is here a more difficult task.)

6-S. In the network of Fig. 6-S element values (ohms, henrys, farads) have been rounded for simplicity. Find the transfer function from e to v_1. Find the transfer function from e to v_2. Find the equation(s) that determine the natural frequencies. Now connect terminal 2′ to terminal 2 (connect the networks in parallel). Why do the natural frequencies change? Find the new natural-frequencies-determining equation(s) (see suggestion in Prob. 6-Q). Find the new transfer function from e to $v_1 = v_2$. Show that a sinusoidal steady-state "balance" condition ($v = 0$) can (cannot) be obtained.

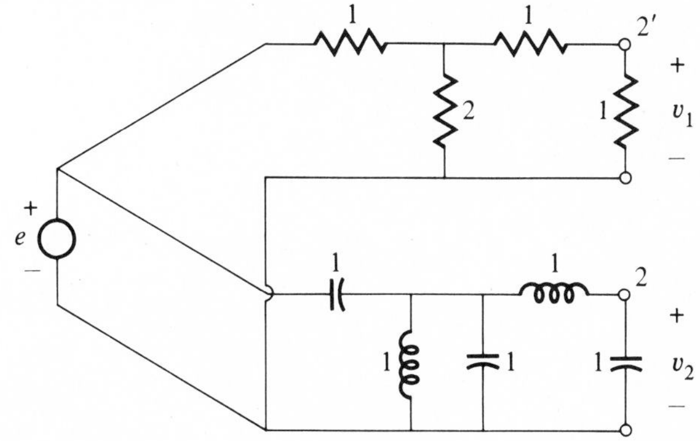

FIG. 6-S.

6-T. A transistor amplifier stage (Fig. 6-T*a*) can be modeled as in Fig. 6-T*b*. Find the input impedance Z_{in} and the current gain (transfer function from i_{in} to i_{out}). What property of the circuit makes the transfer function of a cascade of such amplifier stages simply equal to the product of the individual-stage transfer functions (there is no interaction, in effect)? Does this depend on the particular element values given? (Contributed by Prof. J. F. Gibbons.)

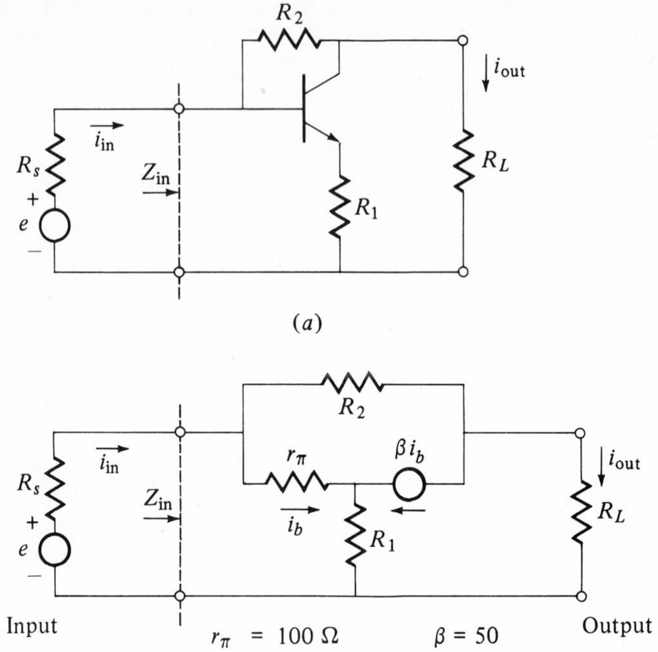

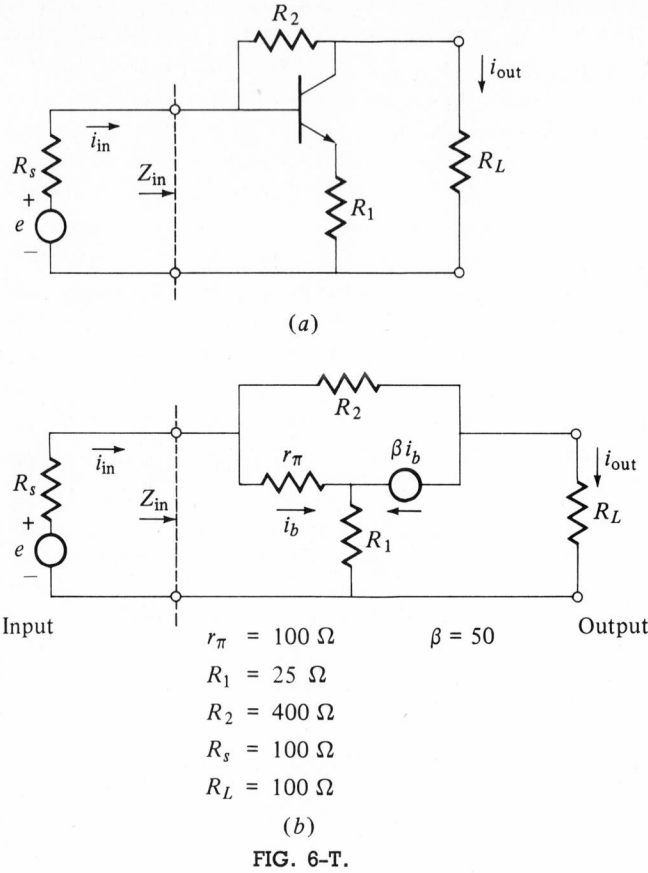

$r_\pi = 100\ \Omega$ $\qquad \beta = 50$

$R_1 = 25\ \Omega$

$R_2 = 400\ \Omega$

$R_s = 100\ \Omega$

$R_L = 100\ \Omega$

(b)

FIG. 6-T.

6-U. Fig. 6-U shows a two-stage amplifier network. The boxes Z contain identical tuned circuits of substantially infinite Q. The resistor R is variable. Find the two transfer functions: from A to B *and* from B to A (from current input to voltage output in each case). In what way do they differ? Explain why the

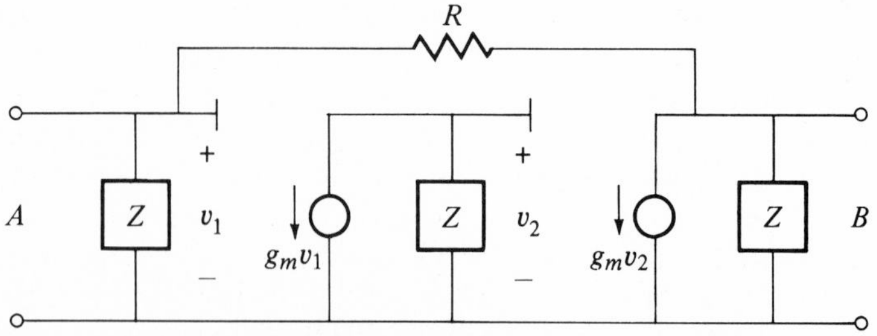

FIG. 6-U.

natural frequencies of the network cannot lie in the right half plane if R is sufficiently large. Is it possible, for some sufficiently small value of R, that one or more of the natural frequencies lie on the ω axis thus making the amplifier into a (sinusoidal) oscillator? If your answer is "yes," for what value of R and what will be the frequency of oscillation? If your answer is "no," explain why oscillation cannot occur. In what way will your answers to these questions change if the tuned circuits have finite Q's?

6-V. For the amplifier shown in Fig. 6-V, define $T(s)$ as the transfer function from input current i_0 to output voltage v, T^0 as $T(s)$ when $g = 0$ (the no-feedback transfer function, analogous to T_1 in Fig. 6.25-J), and T_f as $1/(1+r)$ (the feedback factor analogous to T_2 in Fig. 6.25-J). Calculate $T(s)$ in terms of g, g_m, and the conveniently normalized element values given. Show that if g_m is large and $g = r^{-1}$ is small, and the product gg_m is large, then $T(s)$ is approximately writable in the conventional form [see (6.25-4)], namely, as $T^0/(1 + T^0 T_f)$. For what values of frequency will the approximation fail even though g and g_m retain their (small and large) values? Can you explain why $T(s)$ *cannot* be written exactly in the conventional form $T = T_1/(1 - T_1 T_2)$?

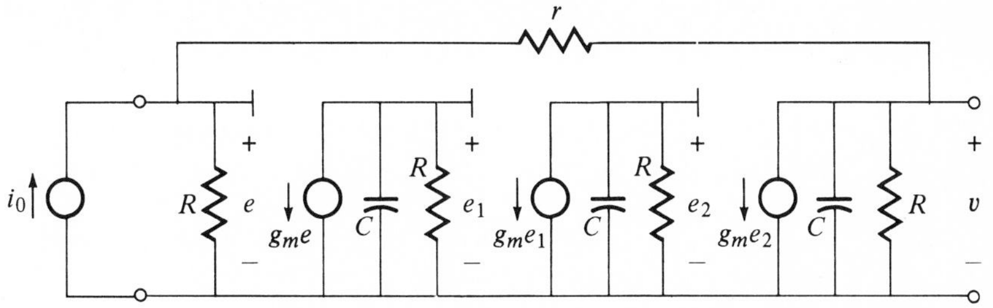

$R = 1\ \Omega,\ C = 1\ \text{F}$

FIG. 6-V.

6-W. Consider the amplifier stage of Fig. 6-27, taking the source impedance to be infinite and R to be a load. C_p is a parasitic capacitance inherent in the active g_m element. Because an energy source is present in this network, one or more natural frequencies *may* lie in the right half of the s plane. Whether this happens, so that the amplifier becomes unstable, may well depend in general on the environment of the network, the value of the load for example. Let $\omega_0 = (LC)^{-1/2}$ and $\omega_f = g_m/C_p$ (which is a sort of figure of merit of the active element and a convenient way to specify C_p). Find the return difference of the g_m element. From it find the equation whose roots are the natural frequencies. In terms of $\beta = C_p/C$ and $k = (g_m R)^{-1}$, taking $\omega_f = \omega_0$, what are the conditions for the network to be a (sinusoidal) oscillator? At what frequency will it oscillate? How do your results corroborate the statements made in the fourth sentence above? (Be sure to find at least one pair of parameter values to verify your answers.)

6-X. Find the input and output impedances (Z_{in} and Z_{out}) of the multiple-feedback amplifier network of Fig. 6-X in terms of the symbols there given. Find also the transfer function from e to V_0. (If interested in the genesis of this circuit, see *Bell System Technical Journal*, pp. 1253 ff, September 1975.)

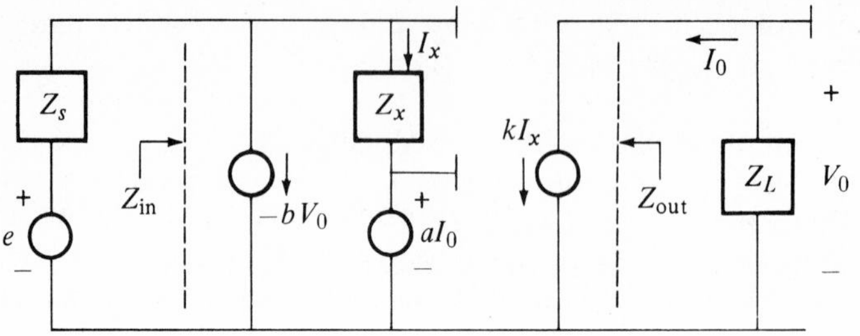

(*d*)

FIG. 6-X.

6-Y. An automatic control system consists of the following parts (Fig. 6-Y): in the forward path a "flat" amplifier of (constant) transfer function K, an amplifier of transfer function $(s^2 + s + 1)^{-1}$, a motor and load whose transfer function is $(s + 1)^{-1}$, and, in the feedback path, a network whose transfer function is $(T_0 s + 1)$. (Numerical values have been scaled in frequency and constants have been amalgamated into K for simplicity.) The feedback signal subtracts at the junction (comparison) point so that K is necessarily positive for the system to be stable for small values of K. Note that the feedback path includes both proportional and derivative (anticipatory) control terms; the former is intended to reduce the error to zero and the latter to allow for large values of K (useful in the control process) without permitting any natural frequencies to enter the right half plane. Show that for $0 < T_0 < 0.5$, the system is stable (all of its natural frequencies have negative real parts) for $0 < K < K_{max}$. What is the value of K_{max}? Show also that for $T_0 > 0.5$ the system is stable for all (positive) values of K (the goal is achieved). What happens if the connection is inadvertently wrongly made at the junction point, that is, if K is negative? (Prob. 5-L may be helpful.)

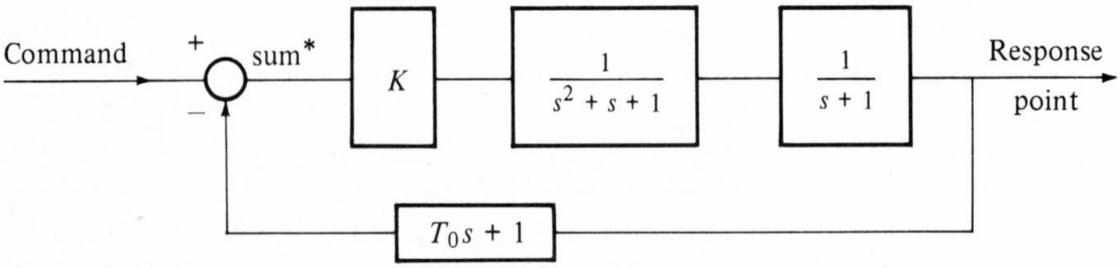

*Algebraic (it is, of course, the *difference* between command and error that is applied to the amplifier)

FIG. 6-Y.

6-Z. When two capacitors at different voltages are suddenly connected together, as by closing the switch X in Fig. 6-Z*a*, a complicated, if very brief, transient phenomenon ensues: the current in the switch is impulsive, there is perhaps an arc, and radiation due to the very rapidly changing current, ...; the model does not tell us. A reasonable and here useful principle is that charge is conserved (see Prob. 2-H) to determine the new capacitor voltage $v(0+)$ at the end of the very brief transient: in effect v_1 and v_2 merely step (from E and 0 respectively) so that $q_{\text{before}} = C_1E + 0 = q_{\text{after}} = C_1v(0+) + C_2\,v(0+)$. Then ensues the (slow) transient return of v to E with time constant $[R(C_1 + C_2)]$.

The dual situation, not as "physically obvious," is not at all academic but very useful in power system analysis, in highly inductive situations where there is much machinery and many transformers (iron), under fault (short-circuit) conditions, for example. Some such situations can sometimes be modelled by the very rapid changing of an inductance, accompanied of course by a (very brief) transient, as in Fig. 6-Z*b*; this is followed by a slow transient return to the

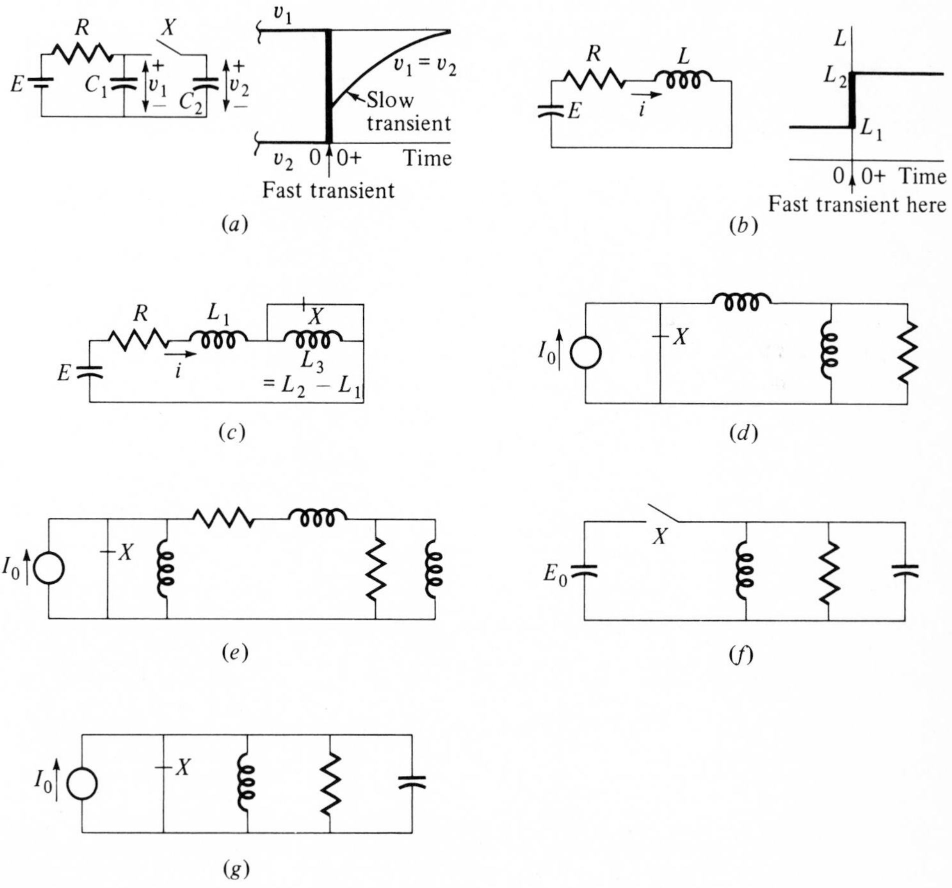

FIG. 6-Z.

original current. Fig. 6-Z*c* is an alternative; there the switch X is opened at $t = 0$, i being then $I_0 = E/R$.

Derive the *principle of constant flux linkages* "Around a closed circuit the net (algebraic) flux linkages cannot change suddenly." [Use the basic form of Faraday's law, voltage is proportional to rate of change of flux linkages (not to $L\,di/dt$), and the fact (assumed) that all noninductive voltage effects on the circuit are in comparison negligible in $0 < t < 0+$.] Apply this principle to the circuit of Fig. 6-Z*c* to determine the current i just *after* the switch X is opened. During the period $0 < t < 0+$ what voltage exists across L_1? Across L_3? Across E and R? What is the energy stored in L_1 and that in L_3 before? After? Explain the change in total stored energy. What happens to i for $t > 0+$? Correlate with the discussion of the "true physical nature" of an impulse in Sec. 2.02.

Finally, draw up a complete list of analogs between the circuits of Fig. 6-Z*a* and Fig. 6-Z*c*, listing *all* items mentioned herein.

For additional practice in resolving such paradoxes, carefully explain all the current and all the voltage phenomena in all the elements in Fig. 6-Z*d* (the switch having long been closed, is suddenly opened), in Fig. 6-Z*e* (the same operation), in Fig. 6-Z*f* (the switch, long open, is suddenly closed), and in Fig. 6-Z*g* (the switch, long closed, is suddenly opened). (In these four circuits, I_0 and E_0 are constants.)

Chapter 7

TRANSFER FUNCTIONS AND NETWORK ANALYSIS

The Seventh Square is all forest—however, one of the Knights will show you the way—and in the Eighth Square we shall be Queens together, and it's all feasting and fun!

—The Red Queen

7.01 INTRODUCTION

From the agenda of Sec. 6.28, we choose first to consider the matter of exponential excitation. We assume an initial condition of rest, and that all N natural frequencies are distinct. [These assumptions are not necessary; the results below, sometimes with appropriate modifications, are valid without them, as we shall see in later sections (Secs. 7.22, 7.23), but the assumptions simplify matters here.] We also assume that $T(\infty) = 0$, that the network does not transmit high-frequency signals, as is commonly true. (Sections 6.26 and 6.27 have shown how to handle those models in which high-frequency transmission is good.)

We start now with the transfer function, so that the detailed construction or form of the actual network under analysis is no longer relevant. Whatever that may be—and there is an infinite number of (equivalent) networks from which a given transfer function could have come—the transfer function has been found (Chap. 6).

Now the response to an *impulse* is usually very easy to obtain, simply by expanding the transfer function in partial fractions (Secs. 5.02, 5.14). True, there are some exceptional cases, but they are ruled out, temporarily, by our assumptions above. Since the impulse response comes so easily from the transfer function, we ask wistfully: "Can other excitations be handled in the same facile way?" The obvious one to investigate first is the exponential type of excitation, that on which the whole transfer-function concept is based.

7.02 THE RESPONSE TO EXPONENTIAL EXCITATION

The response of a network to exponential excitation contains two components. Of these the *forced* component can be evaluated immediately from the transfer function, by definition [(3.10-5)]. The *natural* component is in form $\sum_j K'_j e^{p_j t}$, the p_j being the network's natural frequencies and the K'_j serving to adjust the total response to the network's condition at $t = 0+$. (The prime is used to avoid

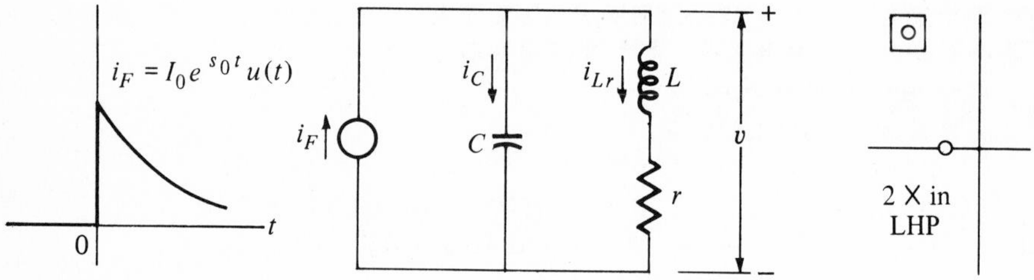

(*a*)

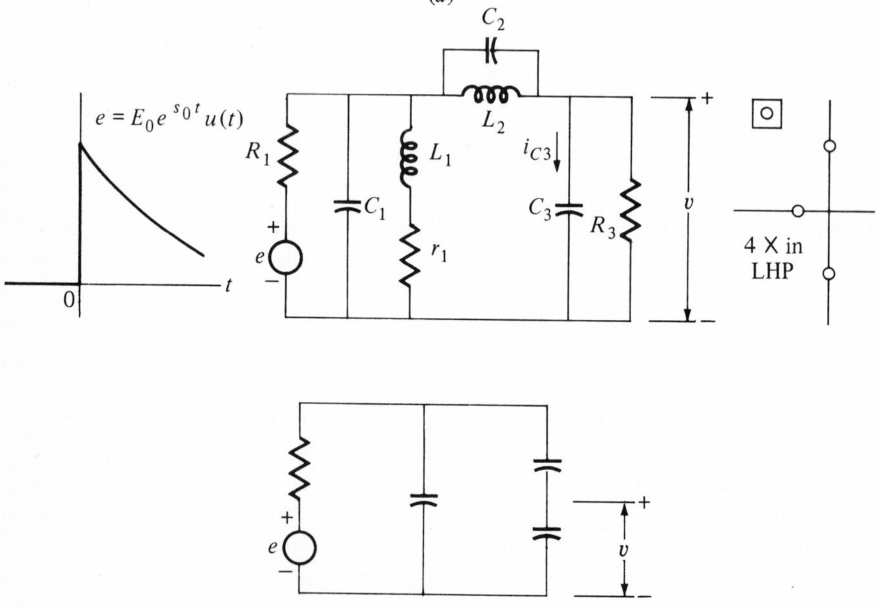

High-frequency asymptotic network

(*b*)

FIG. 7.02-A.

confusion between these coefficients and those of the impulse response.) This condition (at $t = 0+$) is not really complete rest, as it is at $t = 0$. There is no impulsive excitation, and no sudden charging of capacitors, no sudden change in currents or voltages, true. But in the *derivatives*, steps there must be: else there were no response at all! (If all the initial values were zero, the power series terms would all be zero and the response would be simply zero, for all t.)

For example, consider again the RLC circuit (Fig. 7.02-Aa). It is a very simple ladder network; we can write the transfer function from i_F to v by inspection as quickly as we could look it up (Chap. 3):

$$T(s) = \frac{(1/C)(s + r/L)}{s^2 + (r/L)s + 1/LC} \tag{7.02-1}$$

Hence the forced component of the response is

$$v_F(t) = T(s_0)I_0 e^{s_0 t} = \frac{(1/C)(s_0 + r/L)I_0 e^{s_0 t}}{{s_0}^2 + (r/L)s_0 + 1/LC}, \qquad t > 0. \tag{7.02-2}$$

Also by inspection we can immediately find the initial values

$$\begin{aligned} v(0+) &= 0 = v(0), \\ v'(0+) &= \frac{1}{C} i_C(0+) = \frac{1}{C} I_0 \neq v'(0). \end{aligned} \tag{7.02-3}$$

A little thought enables us to calculate the third initial value, $v''(0+)$: At $t = 0+$ we have

$$\begin{aligned} &i_{Lr} = 0, \qquad v_r = 0, \qquad v_L = v = 0, \\ &i'_{Lr} = \frac{1}{L} v_L = 0, \qquad i'_C = i'_F - 0 = s_0 I_0, \\ &v''(0+) = \frac{s_0}{C} I_0 \neq v(0). \end{aligned} \tag{7.02-4}$$

(We could continue and obtain all the initial values thus, but it is a painful process.) We write

$$v_N(t) = \sum_1^2 K'_j e^{p_j t} = v(t) - v_F(t), \qquad t \geqq 0+. \tag{7.02-5}$$

From this come the two simultaneous equations

$$\left.\begin{aligned} v_N(0+) &= K'_1 + K'_2 = 0 - T(s_0)I_0, \\ v'_N(0+) &= K'_1 p_1 + K'_2 p_2 = \frac{1}{C} I_0 - T(s_0) I_0 s_0, \end{aligned}\right\} \tag{7.02-6}$$

which can be solved for the two unknowns K'_1 and K'_2 (see Sec. 3.03). We could even write two more equations:

$$v''_N(0+) = K'_1 p_1{}^2 + K'_2 p_2{}^2 = \frac{s_0}{C} I_0 - T(s_0) I_0 s_0{}^2,$$
$$v'''_N(0+) = K'_1 p_1{}^3 + K_2 p_2{}^3 = v'''(0+) - T(s_0) I_0 s_0{}^3, \qquad (7.02\text{-}7)$$

and solve the set of four for both p's and K's. There is of course no need for that, because we have the p's in the poles of $T(s)$.

For a second example, consider the network of Fig. 7.02-A*b*. By inspection,

$$T(s) = \frac{H_0(s+\alpha)(s^2+\omega_0{}^2)}{s^4 + B_1 s^3 + B_2 s^2 + B_3 s + B_4} \qquad (7.02\text{-}8)$$

in which $\alpha = r_1/L_1$, $\omega_0 = 1/\sqrt{L_2 C_2}$, and the value of the constant multiplier H_0 can be obtained from the high-frequency behavior and then the value of B_4 from the low-frequency behavior. Then, also by inspection,

$$v(0+) = 0 = v(0),$$
$$v'(0+) = \frac{1}{C_3} i_{C3}(0+) = k \frac{1}{C_3} \frac{E_0}{R_1} \neq v'(0), \qquad (7.02\text{-}9)$$

in which k is the fraction (coefficient) obtained by consideration of the voltage divider (Sec. 5.11) of the high-frequency asymptotic network (Fig. 7.02-A*b*). Continuation to the calculation of $v''(0+)$ and so on is possible, but tedious.

To recapitulate: Exponential excitation, which is certainly not impulselike but *is* steplike initially, causes no immediate change in the response value; there is no step (much less impulsive) component in the response. In the *derivatives* of the response, however, there must of course be some changes from the initial conditions (at $t = 0$) to the initial values (at $t = 0+$), else nothing happens, there is no response.

In principle we can obtain the initial values from the schematic diagram by considerations like those in the examples above. And we can thus obtain the power-series-in-t solution. This form does not distinguish forced and natural components; it presents the complete response in a particular, special format. Note that no use is made of the transfer function here at all.

To obtain a closed form of the solution, we can first calculate the transfer function and then use it to obtain the forced component. Then we can in principle use the set of initial values to find the coefficients K'_j in the natural part (and even to find the natural frequencies p_j) by the solution of simultaneous equations. But the labor required, except in very simple cases, is hideous. There should, there *must*, be an easier way!

7.03 THE TRANSFER FUNCTION COMES INTO ITS OWN

In Secs. 5.03 and 5.04 we observed that the transfer function seemed to give the natural part of the response to an exponential excitation, as well, of course, as the forced part. We did not then investigate this, having other things to do first. But the appropriateness of such study now is obvious. Can we really use the transfer function, not just for calculating the forced part, r_F, but also for calculating the natural part, r_N?

Let us now generalize the exponential excitation situation to any network that meets our (temporary) postulates: $M < N$ and all natural frequencies distinct (Fig. 7.03-A). We recapitulate the analysis we have already developed in Secs. 5.03 and 5.04:

$$\begin{aligned} T(s) &= H\frac{s^M + \cdots}{s^N + B_1 s^{N-1} + \cdots + B_N} = \sum_1^N \frac{K_j}{s - p_j} \\ r_{\text{exp}}(t) &= \underbrace{T(s_0)E_0 e^{s_0 t}}_{r_F} + \underbrace{E_0 \sum_1^N K_j' e^{p_j t}}_{r_N}, \qquad t \geqq 0. \end{aligned} \tag{7.03-1}$$

Note that here the coefficients K_j' do not include the excitation coefficient E_0; this notation is simpler and agrees with that of Secs. 5.03 and 5.04.

The natural part of the response serves to adjust the response's initial value(s) to the correct initial value(s): r_F alone as the response is incorrect, for it has nothing to do with the initial condition of the network and is always the

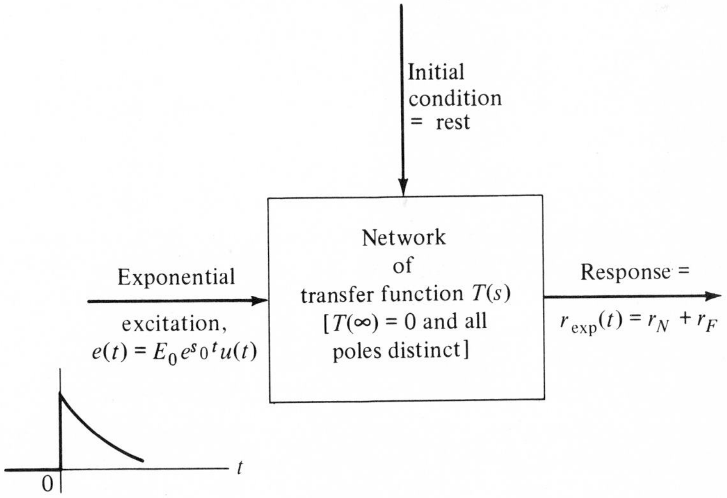

FIG. 7.03-A.

same, regardless of what the initial condition may be. It is the purpose of r_N to "fix r_F up" initially, and then to fade away. A specific requirement on r_N is

$$r_{\mathrm{exp}}(0) = 0 = T(s_0) + \sum_1^N K'_j = \sum_1^N \frac{K_j}{s_0 - p_j} + \sum_1^N K'_j. \tag{7.03-2}$$

The actual relation between the K'_j and the transfer function and excitation parameters is

$$K'_j = \frac{K_j}{p_j - s_0}, \qquad j = 1, 2, 3, \ldots, N, \tag{7.03-3}$$

as might be surmised from (7.03-2); it was formally developed in Sec. 5.04.

We conclude that the answer to the question posed at the beginning of this section is "yes": the transfer function *does* determine $r_N(t)$ as well as r_F. On reflection, this is not surprising, for the transfer function *is* the differential equation, and the differential equation (with the initial condition) is the entire model. The actual manner in which T determines r_N, (7.03-3), can be made much clearer by the following manipulation.

On dividing through (normalizing) by E_0 we have, for $t > 0$,

$$\frac{r_{\mathrm{exp}}(t)}{E_0} = \underbrace{\underbrace{\sum_1^N \frac{K_j}{s_0 - p_j}}_{T(s_0)} e^{s_0 t}}_{r_F} + \underbrace{\sum_1^N \frac{K_j}{p_j - s_0} e^{p_j t}}_{r_N} \tag{7.03-4}$$

$$= K'_0 e^{s_0 t} + K'_1 e^{p_1 t} + K'_2 e^{p_2 t} + \cdots + K'_N e^{p_N t}$$

in which

$$K'_0 = \sum_1^N \frac{K_j}{s_0 - p_j} = T(s_0). \tag{7.03-5}$$

This immediately suggests a new point of view: a substitute analysis problem, an ersatz network situation, equivalent in that the response is the same but different in that the excitation is an impulse, the problem is "reduced to the standard problem," the determination of an impulse response. The actual response $r_{\mathrm{exp}}(t)$ is the same as the impulse response of a different network, one with the same natural frequencies *and one more*, namely, s_0. The transfer function of this ersatz network must be

$$T_{\mathrm{ersatz}}(s) = \frac{K'_0}{s - s_0} + \frac{K'_1}{s - p_1} + \cdots + \frac{K'_N}{s - p_N}$$
$$= \frac{\text{same numerator as in } T}{(s - s_0)(s - p_1)\cdots(s - p_N)} = \frac{T(s)}{s - s_0} = \left[\frac{1}{s - s_0}\right][T(s)], \tag{7.03-6}$$

a very simple relation.

Figure 7.03-B describes this substitution and equivalence, for $t \geqq 0$; in it all initial conditions are rest. Part (a) shows the given analysis problem; part (b) the substitute but overall equivalent problem. The latter is presumably easier to analyze because it is a familiar find-the-impulse-response problem, for which the principal task is the partial-fraction expansion of a transfer function.

The transfer function of two networks connected in tandem, in such a way (without loading) that the individual transfer functions are not thereby altered, is the product thereof (Sec. 5.07). The impulse response of the combination is the convolution of the two individual impulse responses, (4.07-4). These two

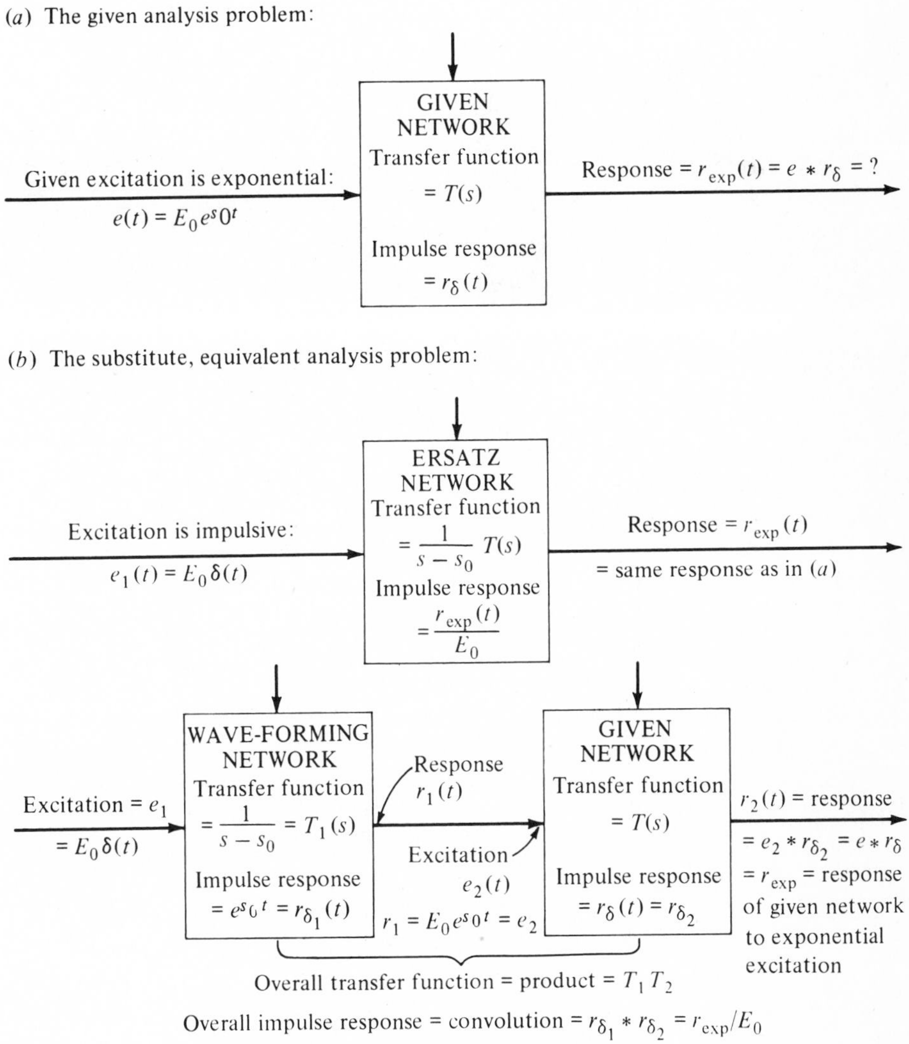

FIG. 7.03-B.

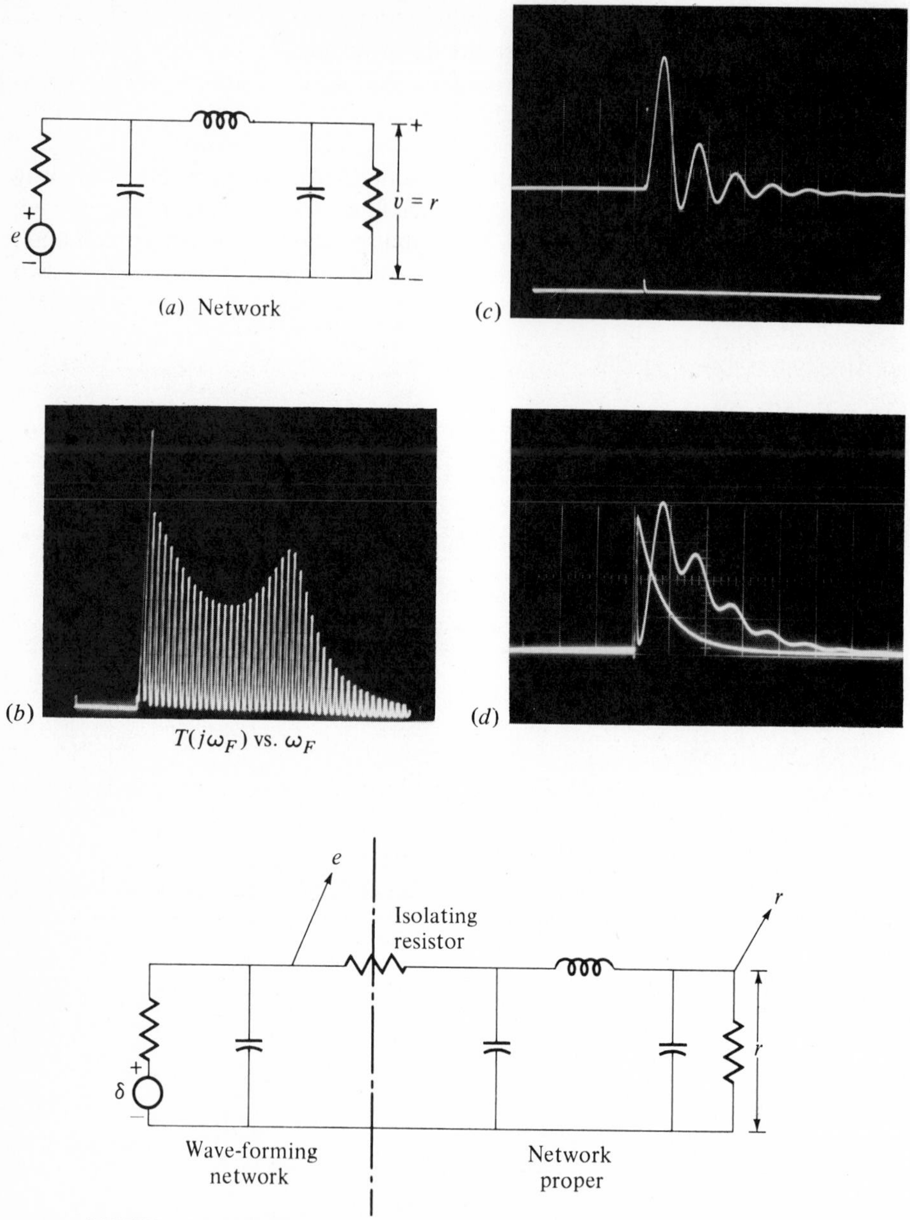

NOTES: The time scales are the same in (*c*) and (*d*): excitation *e* (the simple exponential) and response *r* are superimposed in (*d*). The waveforms seen in (*d*) are excitation *e* and response *r* taken off as shown.

(*e*)

FIG. 7.03-C.

ideas, one from the s domain and one from the t domain, can be considered to be *mates*, another pair of relations between the two domains:

$$T_1(s)T_2(s) \sim r_{\delta 1}(t) * r_{\delta 2}(t). \tag{7.03-7}$$

In the particular case in hand, of an exponentially excited network, the pair is

$$T_{WF}(s)T_{NET}(s) \sim e^{s_0 t} * r_\delta(t), \tag{7.03-8}$$

in which WF refers to the (hypothetical) wave-forming network that converts a unit impulse into the exponential function $\exp(s_0 t)\, u(t)$, and NET to the actual network under analysis, of impulse response $r_\delta(t)$. The last part of Fig. 7.03-B shows this in detail.

To recapitulate: The transfer function of a network determines both forced *and natural* parts of the network's response (from rest) to exponential excitation. The formal details (implied, of course, by convolution, and calculable therefrom) are given specifically in (7.03-2) to (7.03-6) inclusive. A very helpful, vivid physical equivalent is the ersatz network situation of Fig. 7.03-B: the impulse response of the cascade connection of a wave-forming network and the actual network is the same as the actual network's response to exponential excitation, except for the scale factor E_0. The wave-forming network need not actually be synthesized, even on paper, and seldom is; of it all we need is its transfer function, the mate of $e(t)$. But we do look at one example, contrived though it be, to fix ideas. Figure 7.03-C presents a physical realization of Fig. 7.03-B: part (a) shows a network, part (b) its magnitude of $T(j\omega_F)$ "curve," and part (c) its impulse response. In part (d) we see its response to a (superposed) exponential excitation $e(t) = E_0 e^{-\alpha t}$ generated from an impulse by a wave-forming network [see part (e)]. The time constant of the exponential excitation wave is made comparatively long, so that [in (d)] we can actually distinguish r_F (in form like the excitation) and r_N (in form like the network's impulse response of (c)].

The wave-forming network in real-life analysis is only a mental construct. But with its aid, calculation of a network's response to exponential excitation amounts merely to calculation of a different, slightly more complicated network's impulse response.

7.04 WAVE-FORMING NETWORKS

The application of convolution to the exponentially excited network has surprising results. It suggests an excellent tool: to visualize the exponential excitation as produced from an impulse via a wave-forming network of simple transfer function, instead of existing per se. To be sure, this concept is, in retrospect (after the event) physically "obvious" and has no need for the argument of convolution. But the more (varied) points of view we have, the better—and in a way it *was* convolution that first suggested the idea!

Is there any reason to restrict the application of this idea to *exponential* excitation? Could we use it for other excitation wave forms? Again in retrospect we can *see:* in principle it generalizes immediately (Fig. 7.04-A):

> *If* we can find the mate of $e(t)$, that is, $T_{WF}(s)$, the transfer function of the (hypothetical) wave-forming network that converts a unit impulse $\delta(t)$ into $e(t)u(t)$,
>
> *Then* we can imagine such a network placed before the actual network to generate the excitation $e(t)$ therefor. Then we can combine the two networks' transfer functions by multiplication, and so reduce the problem to an impulse-response calculation (Fig. 7.04-A*b*). (7.04-1)

[If the initial condition is not rest, we need only calculate some additional (different) impulse responses, those that provide the response terms that are due to the initial energy storage; this we have glimpsed already, and shall discuss fully in Sec. 7.22.]

If this method of analysis is to be viable, we must be able to find $T_{WF}(s)$ from $e(t)$. True, we have already found many such pairs in the numerous examples we have studied. But they have all been in the category of sums of exponentials on the t side and rational functions on the s side. And we need a *general* process before we can actually generalize. In principle, this is not at all

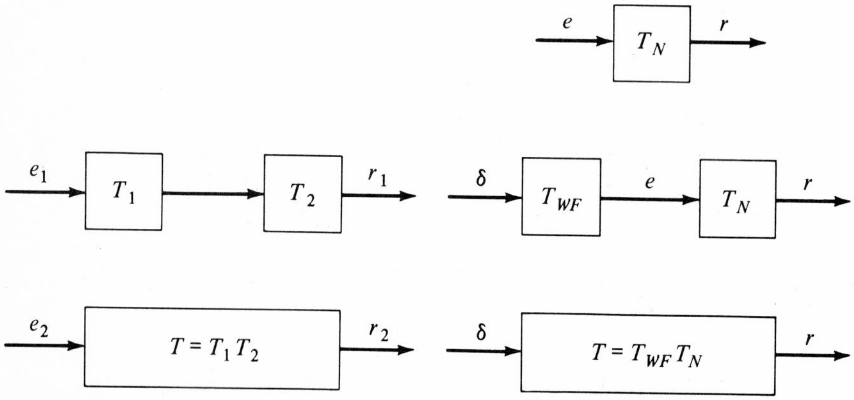

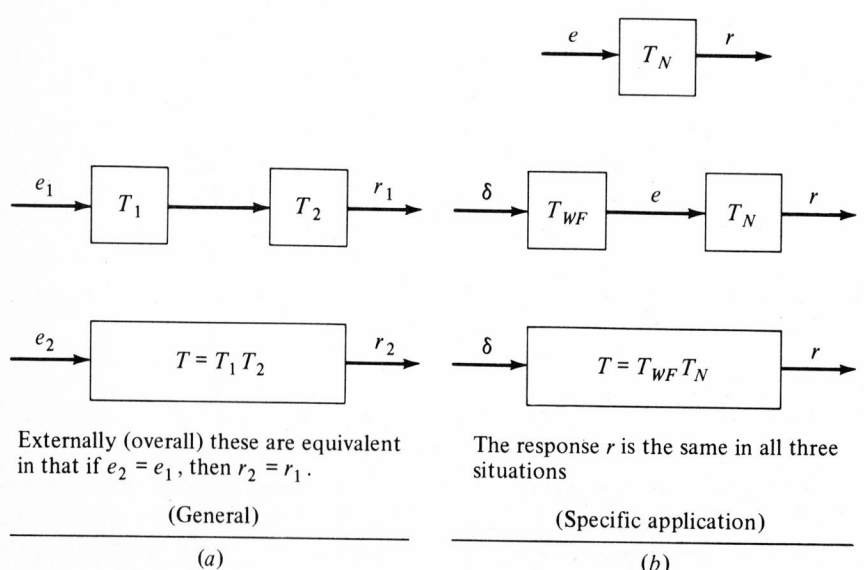

Externally (overall) these are equivalent in that if $e_2 = e_1$, then $r_2 = r_1$.

(General)

(*a*)

The response r is the same in all three situations

(Specific application)

(*b*)

NOTE: all initial conditions are rest; it is assumed that no loading (distortion of transfer function by connection to another network) occurs.

FIG. 7.04-A.

difficult to develop. Practically speaking, however, it is *only* exponential-rational pairs that we need; other pairs are comparatively uncommon and often troublesome.

Suppose the wave-forming network exists, of transfer function $T_{WF}(s)$. This means that the forced part of its response to the exponential excitation $E_0 e^{st}$ is simply that excitation multiplied by $T_{WF}(s)$ [see (3.10-15)]. Its complete response to the exponential excitation, forced part plus natural part, can be obtained by convolution (Chap. 4). Now the wave-forming network, by definition, has an impulse response that is the excitation function $e(t)$ that interests us. Hence the relevant convolution is

$$\begin{aligned}\begin{matrix}\text{Response of wave-forming}\\ \text{network to exponential}\\ \text{excitation } E_0 e^{st}\end{matrix} &= \begin{bmatrix}\text{wave-forming-network}\\ \text{impulse response}\end{bmatrix} * E_0 e^{st} \\ &= \int_0^t e(x) E_0 e^{s(t-x)}\, dx \qquad (7.04\text{-}2) \\ &= E_0 e^{st} \int_0^t e(x) e^{-sx}\, dx \\ &= \text{forced component} + \text{natural component} \\ &= T_{WF}(s) E_0 e^{st} + \text{natural component}\end{aligned}$$

so that

$$T_{WF}(s) = \int_0^t e(x) e^{-sx}\, dx - \frac{e^{-st}}{E_0} \begin{bmatrix}\text{natural component of response}\\ \text{of wave-forming network to}\\ \text{exponential excitation}\end{bmatrix}. \qquad (7.04\text{-}3)$$

If $e(t)$ is exponential (or a sum of exponentials) then of course $T_{WF}(s)$ is rational (Chap. 5) and the natural part of the corresponding impulse response is exponential, so that (7.04-3) becomes

$$T_{WF}(s) = \int_0^t e(x) e^{-sx}\, dx - \frac{1}{E_0} \sum_j K_j e^{(p_j - s)t} \qquad (7.04\text{-}4)$$

in which the p_j are the natural frequencies of the wave-forming network, and K_j are appropriate constants. The real parts of all the p_j are ordinarily negative; but we can imagine circumstances in which the $e(t)$ function of interest increases with time, so that some **RE** (p_j) may be positive. (In such circumstances one expects "the switch to be opened" before too long, of course.)

Suppose now that we consider very large values of time, long after the application of the (exponential) excitation. Mathematically, let $t \to \infty$. The last term in (7.04-3) will presumably vanish, at least if the factor e^{-st} vanishes rapidly enough. We are thoroughly accustomed to s being in general complex: $s = \sigma + j\omega$. Hence the condition is really "at least if $|e^{-st}| = e^{-\sigma t}$ vanishes rapidly enough," i.e., if **RE** (s) is large enough. The natural part of the wave-forming network's response ordinarily *will itself decay eventually to zero;* if it does not, the factor $e^{-\sigma t}$ can generally be made to "take over" and force the product to zero. Those exceptional circumstances in which it is impotent, however large σ, are for us not important.

In the form (7.04-4) the condition is much clearer: it is that **RE** $(s - p_j)$ be positive, taking all p_j into account. In other words

$$\mathbf{RE}\,(s) > \sup \mathbf{RE}\,(p_j) \tag{7.04-5}$$

in which *sup* (least upper bound) picks out the "furthest to the right" of the p_j, that one with the largest real part. If (7.04-5) is fulfilled, then the second term on the right in (7.04-4) vanishes as $t \to \infty$.

One way or another, for the sorts of functions $e(t)$ that we here consider, there exists a lower limit for the real part of s such that for **RE** (s) greater than this limit, this vanishing occurs. We denote the limit σ_a and we so restrict s. Some examples are given in Fig. 7.04-B. In each of the first three α is (real and) positive, and s is not allowed in the crosshatched area. To find σ_a we need merely examine the p_j according to (7.04-5), the X in Fig. 7.04-B, and locate the right-most. These are simply the constants in the exponents, when $e(t)$ is (linearly) exponential. When $e(t)$ is not such, convergence of the integral in (7.04-3) must be tested. The fourth example in Fig. 7.04-B, for instance, requires such a test. Since in $\int_0^\infty e^{\alpha^2 t^2} e^{-st}\,dt$ the exponent of e eventually becomes and remains positive (and increases in value), regardless of how large **RE** (s) is made, there exists no σ_a for this integral (that is, $\sigma_a \to \infty$), and there is no $T_{WF}(s)$ for this excitation. Another interesting example is the excitation function of Sec. 4.08, $e(t) = [1 + (t/T_0)^2]^{-1}$. For it the integral is well known, when $s = 0$, to have the value $T_0\,\pi/2$. For $\sigma > 0$ it certainly "converges even more." But for $\sigma < 0$, investigation shows that it does not converge. Hence for it $\sigma_a = 0$. But the formal evaluation of $T_{WF}(s)$ for it is cumbersome and difficult; it is only the exponential excitations that are readily tractable. And for them, of course, one can construct $T_{WF}(s)$ by inspection (Sec. 5.05).

This is a new concept, that s should be in some way restricted. The present argument confines s to a half plane (as in Fig. 7.04-B), but in our previous work with s we have encountered no such restriction, except for those isolated points that are the poles of transfer functions. (This new restricted area, we note, *includes* the poles of T_{WF}, for they are the p_j to the right of which we must stay.) But we are dealing with the *same s*. The paradox is resolved by examining our work here to find what is novel, compared to our previous work with transfer functions. It is the passage to the limit, letting $t \to \infty$, the waiting for the end of time before calculating T_{WF}, the insistence on seeing *all the future* of the excitation $e(t)$ before using it. But this is surely *not* necessary for analysis up to some particular time; clearly we can find out by analysis (convolution) what occurs up to any time T_1 from a knowledge of $e(t)$ only for $t \leqq T_1$; we do not need to look into the future. (The restrictions on s *are*

$e(t)$	σ_a	$T_{WF}(s)$
$e^{-\alpha t}u(t)$	$-\alpha$	$\dfrac{1}{s+\alpha}$
$e^{+\alpha t}u(t)$	$+\alpha$	$\dfrac{1}{s-\alpha}$
$e^{-\alpha t}\cos(\beta t + 0)$	$-\alpha$	$\dfrac{s\cos\phi + \alpha\cos\phi - \beta\sin\phi}{s^2 + 2\alpha s + \alpha^2 + \beta^2}$
$e^{+(\alpha t)^2}u(t)$	∞	None

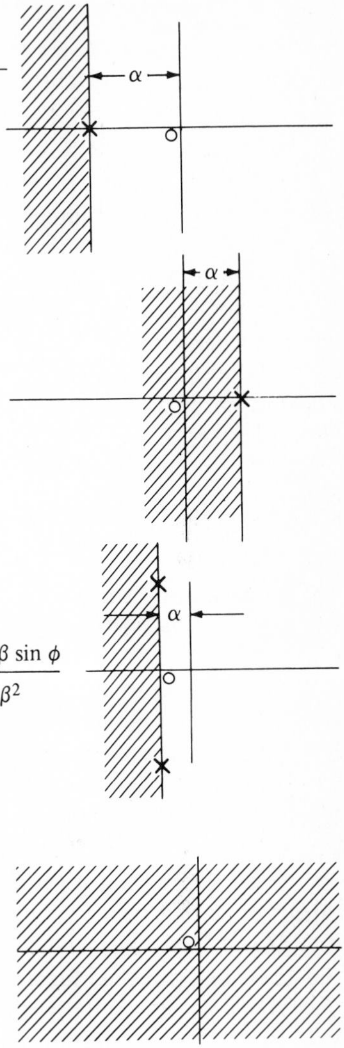

FIG. 7.04-B.

consistent therewith; the paradox is resolved below, at the end of this section.)

Up to now, "$t = \infty$" has never entered our calculations. But to obtain $T_{WF}(s)$ we need to consider all t, even to infinity. We must get rid of r_N so that r_F, with its T_{WF}, stands out; we must arrange things so that r_F eclipses r_N. If the wave-forming network is like our ordinary circuits, we must make e^{st} overcome e^{pt}, and make $|e^{(s-p)t}|$ approach zero; that is, we must obey (7.04-5). If $e(t)$ is "reasonable," so that its generatrix wave-forming network has "reasonable" natural behavior, then making **RE** (s) large enough is feasible, σ_a is finite, and $T_{WF}(s)$ exists.

In general now to find T_{WF}, the wave-forming-network transfer function, we have to carry the convolution integral out to infinity and deal with

$$T_{WF}(s) = \int_0^\infty e(x)e^{-sx}\,dx \tag{7.04-6}$$

which is an improper integral. For such an integral to converge, the integrand must meet certain conditions, and this imposes some restrictions on the parameter s, depending on the large-time behavior of $e(t)$. Analysis thus places a lower limit on **RE** (s), which is the same σ_a. From this comes the formal, technical name given to σ_a, the *abscissa of absolute convergence*. We must add a condition on **RE** (s) to (7.04-6). Since the variable of integration therein disappears on evaluating the limits, we can use for it any letter we choose; we return to the "natural" symbol t and write

$$T_{WF}(s) = \int_0^\infty e(t)e^{-st}\,dt, \qquad \mathbf{RE}\,(s) > \sigma_a. \tag{7.04-7}$$

This formula can be used to evaluate the wave-forming-network transfer function for any $e(t)$, if it exists (if σ_a is finite).

Finally to reconcile all this with our previous freedom for s (the whole plane, except the poles) we can invoke the mathematical doctrine of *analytic continuation*. This will enable us to move from some point where $\sigma > \sigma_a$ to any other point in the plane (the poles of T_{WF} excepted) and so regain our previous freedom of movement for s. [This will be necessary when we apply the wave-forming network concept; we may have to evaluate $T_{WF}(s)$ for some values of s to the left of σ_a.]

We are now in a position, in principle, to find an infinite variety of wave-forming-network transfer functions to add to the already large collection we have established by experiment. But experience will confirm that the exponential functions of t (rational functions of s) are the really useful ones.

We note incidentally that any (finite) excitation wave that eventually vanishes *has* a wave-forming transfer function, since for it σ_a becomes $-\infty$ and there is no restriction on s. An $e^{(\alpha t)^2}$ wave, the third in Fig. 7.04-B, for example, if "truncated" to

$$e(t) = \begin{cases} e^{(\alpha t)^2} & 0 < t < T_1, \\ 0 & T_1 < t, \end{cases} \tag{7.04-8}$$

presents no theoretical problem. For it the integral (7.04-7) converges, regardless of the values of s; $\sigma_a = -\infty$. Actual evaluation of the integral is difficult here, but there is no question of its convergence. This is the wave-forming-network method's answer to the criticism implied above in the statement that to analyze up to $t = T_1$, we do *not* need to know the future of $e(t)$. The paradox is resolved!

7.05 A NEW TRANSFER FUNCTION DEFINITION

The wave-forming-network concept gives us an alternative for defining the transfer function of a network. If we apply (7.04-7) to the hypothetical situation in which we know the impulse response $r_\delta(t)$ (though not necessarily the network), then the corresponding transfer function is given by

$$T(s) = \int_0^\infty r_\delta(t)e^{-st}\,dt, \qquad \mathbf{RE}\,(s) > \sigma_a. \tag{7.05-1}$$

This is basically not new; Sec. 5.05 shows how simple it is to obtain transfer functions from impulse responses that are simply sums of exponentials. And (7.05-1) is of course implied in the convolution integral, from which it was derived.

The merit of (7.05-1) is that in principle the impulse response can now be any (integrable) sort of function. This is not really important for the analysis of networks qua networks, but it can be useful in a sort of dreamland where one sometimes wishes to theorize about unusual, nonexponential functions and "corresponding networks," if only to find out something about *general* network properties, if only to obtain approximations to ideals, if only in an effort to begin a design (synthesis). (See Chap. 8.) There the impulse response need not be exponential in form, even though the definition of transfer function *is* based on the network's response to exponential excitation, which is why the exponential function manifests itself as the *kernel* or *weight function* in the integral of (7.05-1). (The actual evaluation of the integral in such cases may of course be difficult.)

If an impulse response is arbitrarily prescribed, what is the corresponding transfer function? What might be a network realization thereof? If a network's behavior is known only by measurements of its impulse response, what is its transfer function? These are real-life engineering problems, for the solution of which (7.05-1) can be helpful. Such situations can become complicated, and they may well involve lengthy approximation procedures, for it is necessary eventually to warp things into the exponential-rational pair format. We cannot go further into them here, but Sec. 7.06 does present an interesting example.

We now have two definitions of *transfer function*, quite different (but equivalent):

a The ratio of the forced component of the response to an exponential excitation e^{st}, to that excitation [(3.10-15)], (7.05-2)

b The formal integral of the exponentially weighted impulse response, (7.05-1).

The common analysis situation (see Sec. 6.12) is that we have a network in (physical) schematic form, or in (mathematical) differential equation form;

then the calculation of the transfer function does not require any integral at all; the simple procedures of Chaps. 5 and 6 work very well: the impulse response is a sum of exponentials and the transfer function is a rational function of s with real coefficients. Our continuing stress of the importance of the exponential function is not arbitrary; it is established by the experimental fact that all networks we are really interested in have *exponential* impulse responses and *rational* transfer functions.

7.06 A PROBLEM IN SYNTHESIS

To follow the path that determines a network (or at least its transfer function) from a given impulse response is something of a digression at this point, but an interesting one. We shall resume the discussion of analysis qua analysis in Sec. 7.07; here we look briefly into such synthesis problems.

The synthesis of a network from a prescription of its impulse response (or its response to some other excitation) is a tremendous subject. We have already considered one approach thereto in Sec. 6.14; here we consider a somewhat different one.

The integral definition of the transfer function, (7.05-1), makes it possible in principle to find the transfer function that corresponds to any time-domain function taken as an impulse response. For example, the peculiar (piecewise-continuous) wave of Fig. 7.06-A, considered an impulse response, gives us

$$\begin{aligned} T(s) &= \int_0^\infty g(t)e^{-st}\,dt = \int_0^{T_0} E_m \sin\omega_0 t\, e^{-st}\,dt \\ &= \left.\frac{E_m e^{-st}(-s\sin\omega_0 t - \omega_0\cos\omega_0 t)}{s^2+\omega_0^2}\right|_{t=0}^{t=T_0} \qquad (7.06\text{-}1) \\ &= \frac{E_m\omega_0(1+e^{-T_0 s})}{s^2+\omega_0^2}, \qquad \omega_0 T_0 = \pi. \end{aligned}$$

The transfer function which was obtained above is unfortunately not a rational function of s; it contains the nonrational function $e^{-T_0 s}$ (whose real significance will be discussed in Sec. 8.06). Since $g(t)$ is not a sum of exponentials, how could $T(s)$ be rational? (Look at the sharp corner!) Some sort of *approximation* is evidently necessary if $g(t)$ is to be realized as a network impulse response; that can only be done approximately, and we have to find an exponential sum whose behavior is approximately that of $g(t)$.

One obvious approach is to approximate the $T(s)$ of (7.06-1) in the s domain, with a rational function of s. Since we do not yet completely understand the relations between the two domains, and since the prescription is, after all, not in the s domain, we leave this method for study elsewhere.

Another approach is to approximate $g(t)$ itself, in the time domain, with an

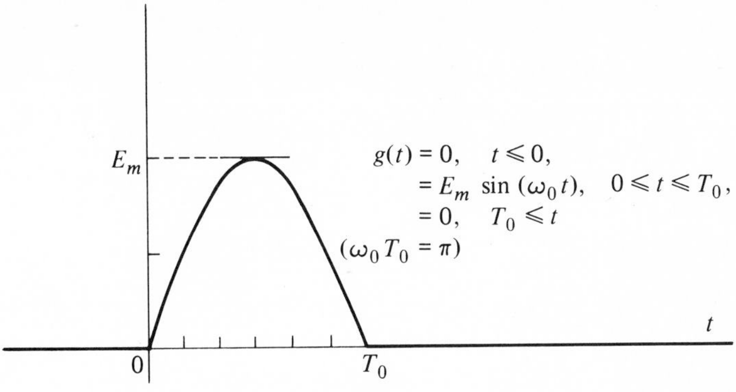

FIG. 7.06-A.

exponential sum. This we shall follow further, since it is closer to the actual prescription, and more relevant to what we are doing in network analysis.

Brief reconsideration of the initial-value matching procedure of Sec. 6.14 shows that it is here of no help whatever. It would give us simply the sinusoid $(E_m \sin \omega_0 t)u(t)$, oscillating for all time after $t = 0$. A procedure based on initial values cannot see beyond $t = 0+$ and can have no knowledge of the change in character of $g(t)$ that is going to occur at $T = T_0$; it is therefore useless here. We need to move out, to include the time range beyond T_0 in our calculations. So, instead of matching $2N$ initial values with an exponential sum $\sum^N Ke^{pt}$, with $2N$ parameters in it, let us try to match at $2N$ *points* spread out in time, say at $t = 0, T, 2T, 3T, \ldots, (2N-1)T$. (The equal spacing is convenient, if not necessary.) This is such a logical extension of the method of Sec. 6.14 that the latter may be helpful here after all, if only in a heuristic way. Perhaps we should not reject it so impetuously.

Suppose the data given as a set of points, as in Fig. 7.06-B. These may

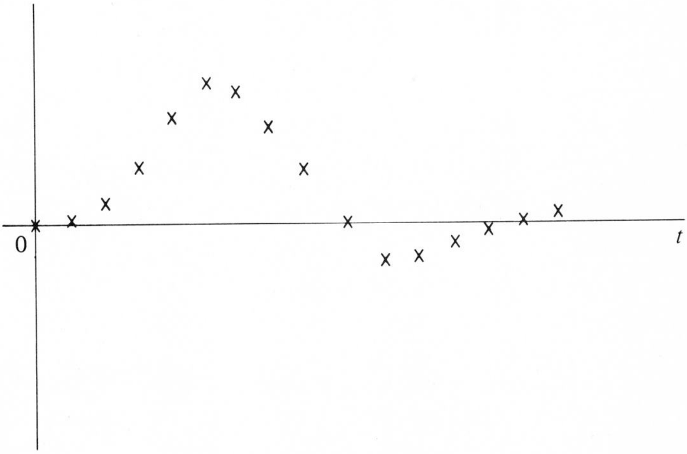

FIG. 7.06-B.

outline the response desired of a network to be designed, *or* they may represent measurements on an extant network whose transfer function we seek; the procedure below is the same. The response outlined by the given points may be the impulse response, the step response, the response to any excitation $e(t)$ whose mate is $E(s)$. We are asked therefrom to find first the transfer function, then the actual network (if the data was measured) or to synthesize a network if the data was a prescription therefor.

It seems not unreasonable to try to extend the method of Sec. 6.14 out in time. This should give a reasonable approximation to the desired behavior, except in unusual cases where we encounter trouble between the match points. Since we are entering uncharted waters we must be especially careful not to take anything for granted and to be alert to the possibility that things may not go as we think they should!

The first part of the problem is essentially, then, a data-fitting or interpolation problem, with an exponential sum:

To determine the $2N$ parameters in the exponential sum

$$f(t) = \sum_{j=1}^{N} K_j e^{p_j t} \tag{7.06-2}$$

so that it passes through the $2N$ data points, i.e., so that

$$\begin{aligned}
f(0) &= \sum K_j &&= \sum K_j \lambda_j^0 &&= f_0, \\
f(T) &= \sum K_j e^{p_j T} &&= \sum K_j \lambda_j &&= f_1, \\
f(2T) &= \sum K_j e^{2p_j T} &&= \sum K_j \lambda_j^2 &&= f_2, \\
f(3T) &= \sum K_j e^{3p_j T} &&= \sum K_j \lambda_j^3 &&= f_3, \\
&\cdots\cdots\cdots\cdots \\
f[(2N-1)t] &= \sum K_j e^{(2N-1)p_j T} &&= \sum K_j \lambda_j^{2N-1} &&= f_{2N-1},
\end{aligned} \tag{7.06-3}$$

in which

$$\lambda_j \equiv e^{p_j T} \tag{7.06-4}$$

and $f_0, f_1, f_2, \ldots, f_{2N-1}$ are the ordinates of the data points.

The equations of (7.06-3) seem at first glance not to resemble the equations of (6.14-1) at all. Yet it seems only reasonable that the process of matching at separate points ought, if we collapse the time scale and make T smaller and smaller, to degenerate into the initial-value-matching process: the various differences ought, in the limit, to become the various initial derivatives. There *must* be a close parallel between the two! Let us first tabulate, in parallel columns, the essence of the two problems (Table 7.06-A).

TABLE 7.06-A

The Initial-Value-Matching Equations of Sec. 6.14 [Equations for Determining the 2N Parameters in the Exponential Sum $r_\delta(t) = \sum K_j e^{p_j t}$]

$r(0+)$	$= \sum K_j$	$= r_0$	(given initial value of r_δ)
$r'(0+)$	$= \sum K_j p_j$	$= r_1$	(given first-derivative initial value)
$r''(0+)$	$= \sum K_j p_j^2$	$= r_2$	(given second-derivative initial value)
....			
$r^{(2N-1)}(0+)$	$= \sum K_j p_j^{(2N-1)}$	$= r_{2N-1}$	(given $2N - 1^{st}$ derivative initial value)

The Equations of the Matching-at-Points Scheme Presently of Interest [Equations for Determining the 2N Parameters in the Exponential Sum $f(t) = \sum K_j e^{p_j t} = \sum K_j (\lambda_j)^{t/T}$]

$f(0)$	$= \sum K_j$	$= \sum K_j \lambda_j^0$	$= f_0$	(given first-point value)
$f(T)$	$= \sum K_j e^{(p_j T)}$	$= \sum K_j \lambda_j$	$= f_1$	(given second-point value)
$f(2T)$	$= \sum K_j e^{(2p_j T)}$	$= \sum K_j \lambda_j^2$	$= f_2$	(given third-point value)
....				
$f[(2N-1)T]$	$= \sum K_j e^{(2N-1)p_j T}$	$= \sum K_j \lambda_j^{(2N-1)}$	$= f_{2N-1}$	(given last-point value)

Table 7.06-A reveals how similar the two problems really are. Prescribed initial values of derivatives become prescribed function values at separated points; unknown natural frequencies p_j are replaced by unknown λ_j. Coefficients K_j remain coefficients K_j. With the (principal) change of searching not for the p_j but for the exponential equivalents, the λ_j, the processes are alike. We can therefore proceed rapidly to write down the steps of the solution process of Sec. 6.14 and then, alongside, in almost perfect parallelism, the steps for our present matching-at-points scheme. Steps (*a*) and (*b*) of Sec. 6.14 (6.14-1) require no further comment. We could continue to exploit the parallel in complete detail, paraphrasing the steps of (6.14-1), to the end. Since the transfer function of Table 7.06-B is here unreal, because of the injection of the λ_j (which are analogous to the p_j of Sec. 6.14, but are not natural frequencies p_j here), we can now proceed more appropriately here in the following manner.

Step (*c*) is of course to solve the N simultaneous equations of Table 7.06-B,

TABLE 7.06-B

Sec. 6.14		Here
The actual transfer function sought is: →	$T(s) = \dfrac{A_0 s^{N-1} + A_1 s^{N-2} + \cdots + A_{N-1}}{s^N + B_1 s^{N-1} + \cdots + B_N}$	← A fictitious transfer function, temporarily useful but to be replaced
Initial-condition-matching equations taken from the initial-value tableau [see (6.14-1c)]:		Matching-at-points equations:
$r_N + r_{N-1}B_1 + \cdots + r_0 B_N = 0$		$f_N + f_{N-1}B_1 + \cdots + f_0 B_N = 0$
$r_{N+1} + r_N B_1 + \cdots + r_1 B_N = 0$		$f_{N+1} + f_N B_1 + \cdots + f_1 B_N = 0$
....		....
$r_{2N-1} + r_{2N-2} B_1 + \cdots + r_{N-1} B_N = 0$		$f_{2N-1} + f_{2N-2} B_1 + \cdots + f_{N-1} B_N = 0$

right half, for B_1, B_2, B_3, ..., B_N. Step (*d*) is now to find the zeros of the characteristic polynomial

$$\lambda^N + B_1\lambda^{N-1} + B_2\lambda^{N-2} + \cdots + B_N. \tag{7.06-5}$$

At this point we take stock [see (6.14-1)*e*, 1]: are the λ values satisfactory? Since by (7.06-4)

$$p_j = \frac{\log_e \lambda_j}{T} \tag{7.06-6}$$

and we wish the p_j to have negative real parts, satisfactory λ values are values whose magnitudes are less than unity: $|\lambda_j| < 1$. If the λ_j, the zeros of the complex characteristic polynomial (generally complex numbers), are satisfactory, we proceed; if they are not, we must revise N, or alter the data.

When the λ_j are satisfactory, we proceed to find the K_j by solving the simultaneous equations of Table 7.06-A (right half). Rewritten and reformed for convenience, these are

$$\begin{aligned}
K_1 + \quad K_2 + \quad K_3 + \cdots + \quad K_N &= f_0, \\
\lambda_1 K_1 + \lambda_2 K_2 + \lambda_3 K_3 + \cdots + \lambda_N K_N &= f_1, \\
\lambda_1^2 K_1 + \lambda_2^2 K_2 + \lambda_3^2 K_3 + \cdots + \lambda_N^2 K_N &= f_2, \qquad (7.06\text{-}7) \\
\cdots\cdots\cdots\cdots\cdots\cdots\cdots\cdots\cdots\cdots\cdots\cdots \\
\lambda_1^{2N-1} K_1 + \lambda_2^{2N-1} K_2 + \lambda_3^{2N-1} K_3 + \cdots + \lambda_N^{2N-1} K_N &= f_{2N-1}.
\end{aligned}$$

We must then [see (6.14-1)*e*, 2] evaluate the behavior of $f(t) = \sum Ke^{pt}$ between the match points and beyond. If this is satisfactory, then $f(t)$ is a suitable $r(t)$ function. Then $T(s) = F(s)/E(s)$ and we can proceed to attempt to synthesize a network [see (6.14-1)*g*]. If the behavior of $f(t)$ is not satisfactory, there is nothing for it but to repeat the process with a different value of N [as in (6.14-1)*f*].

By way of example, let us attempt to apply these ideas to the one-sinusoidal-loop data of Fig. 7.06-A, given as an impulse-response prescription. We must first choose a (starting) value for N. No small value of N will do: $N = 1$ is obviously useless; $N = 2$ can give a crude approximation to $g(t)$ of Fig. 7.06-A in the form of a damped sinusoid, to be sure, but the inevitable second (negative) loop makes the quality of approximation very low. A larger value of N is necessary.

The computations are very lengthy, if straightforward. The equations above state clearly what must be done: solve a set of simultaneous linear algebraic equations, factor a polynomial, solve another set of simultaneous linear algebraic equations, evaluate an exponential sum versus time. Except for trivially small N values, the work must be done by machine (see Sec. 7.17).

We therefore omit the details of the calculations and merely present a few results to show what can be expected.

It rapidly becomes evident that, elegant though the basic idea and the equations be, the approximation must proceed by successive trials. Matching at $2N$ points usually forces the exponential sum through the chosen points, true, but the behavior in between and beyond is often unacceptable, as we soon find out. It then becomes advisable to match, at the $2N$ points, not to the values of $g(t)$, but to values a little bit different, i.e., deliberately to introduce some error at the "match" points in order to improve the overall behavior.

After a number of trials, ordered systematically, each according to the information gained from the previous trial, we arrive at what we shall accept as an end to our problem here: Fig. 7.06-C. This shows the sum of six exponential terms that represent about the best one can do with $N = 6$. [The problem is essentially difficult, because of the complete change in character of $g(t)$ at $t = T_0$.] Note that the time scale here places $T_0 = 6$, for convenience in point matching with six exponentials. With that time scale, and the corresponding s scale, the numerical results are:

$$\left.\begin{aligned} p_1 &= -0.386 + j2.357, & K_1 &= -0.078 + j0.189, \\ p_2 &= -0.890 + j0.305, & K_2 &= -0.272 - j10.419, \\ p_3 &= -0.577 + j1.076, & K_3 &= +0.350 + j2.022. \end{aligned}\right\} \tag{7.06-8}$$

(The three conjugate natural frequencies and their coefficients are omitted.) The period of the sinusoid of which one loop constitutes $g(t)$ is $2 \times 6 = 12$; the corresponding "natural frequency" is $2\pi/12 = 0.52$, and it is interesting to note that the principal component in the approximating exponential sum has an imaginary part of about this magnitude (as would, perhaps, be expected). The others do their best to modify this term to fit $g(t)$.

It is appropriate to display not only the exponential sum itself (Fig. 7.06-C) but also the deviations of $f(t)$ from $g(t)$, what one might call the "error" (Fig. 7.06-D). The deviations, we note, remain within about $\pm 0.06E_m$.

A network realization, to complete the synthesis procedure, can be got in no really simple way, though it can be got in many different ways. We content ourselves with showing that it can be done and shall not here pursue the many alternatives, some much more practical than others. Realization, or synthesis proper, is a lengthy job, in general, a separate discipline, in effect, on which we can spend little time here. The procedure below is not the most elegant or most practical; it is simply the quickest to get, given the tools we already have.

Techniques of approximation of a desired response in the time domain, such as that used here (and that used in Sec. 6.14) lead us to exponential sums for the impulse response. In them can occur two sorts of terms (see Fig. 7.06-E):

a $$\frac{K}{s + \alpha} \tag{7.06-9}$$

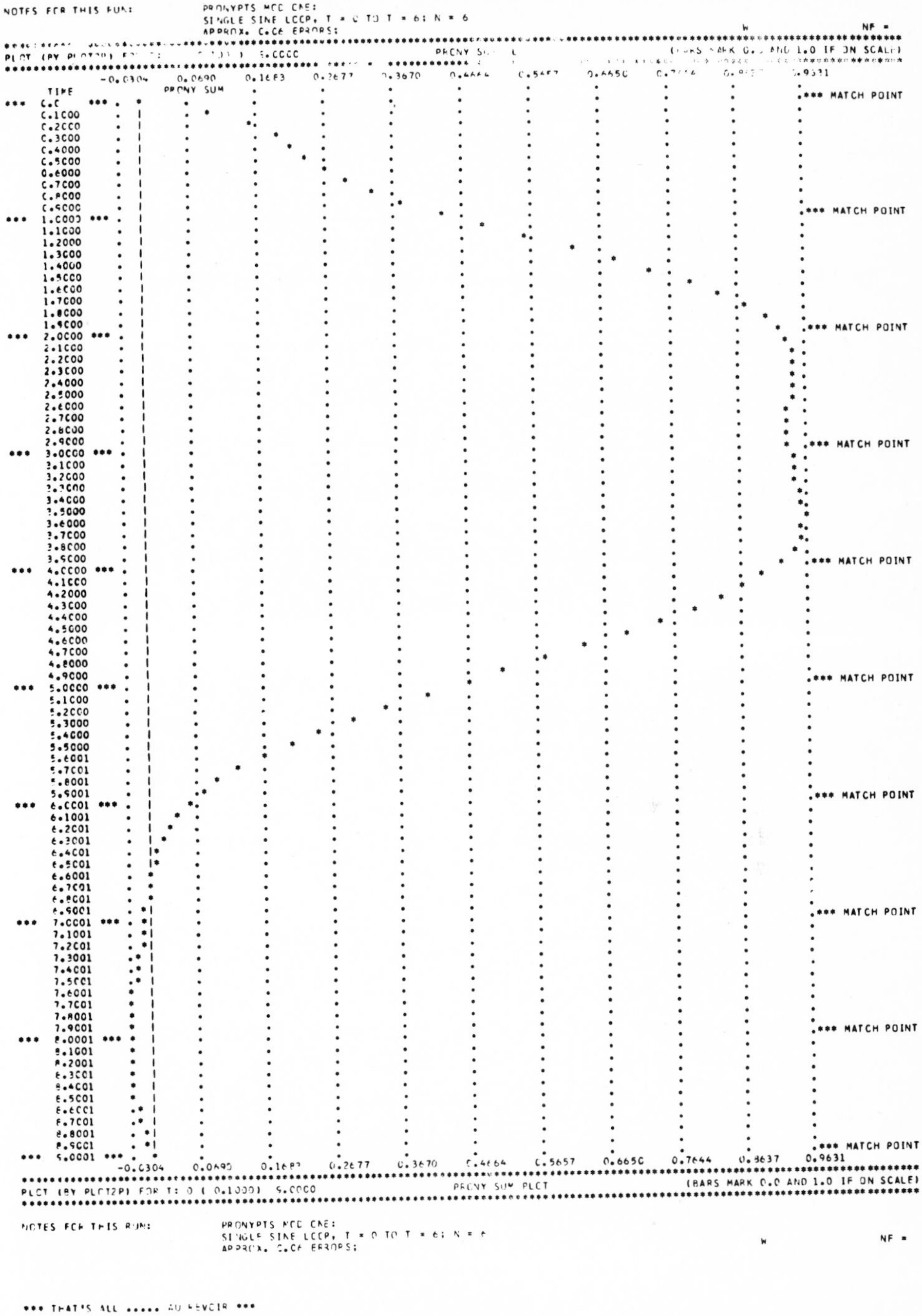
NOTES FOR THIS RUN:
PRONYPTS MOD ONE:
SINGLE SINE LOOP, T = 0 TO T = 6; N = 6
PRONY SUM
TIME
MATCH POINT
-0.0304
0.0690
0.1683
0.2677
0.3670
0.4664
0.5657
0.6650
0.7644
0.8637
0.9631
PLOT (BY PLOT2P) FOR T: 0 (0.1000) 5.0000
PRONY SUM PLOT
(BARS MARK 0.0 AND 1.0 IF ON SCALE)
NF =
*** THAT'S ALL ..... AU REVOIR ***

FIG. 7.06-C.

PROGRAM "PRONYPTS": EVALUATION OF PRONY PARAMETERS AND PRONY SUM THROUGH 2N GIVEN POINTS ******************** NF = 2

NOTES FOR THIS RUN: PRONYPTS MOD ONE FM DISK RUN: W
N = 6, DT = 0.05 NOW W

NF = 2

SUBROUTINE 'PLOT46', BEING CALLED, DULY GIVES:

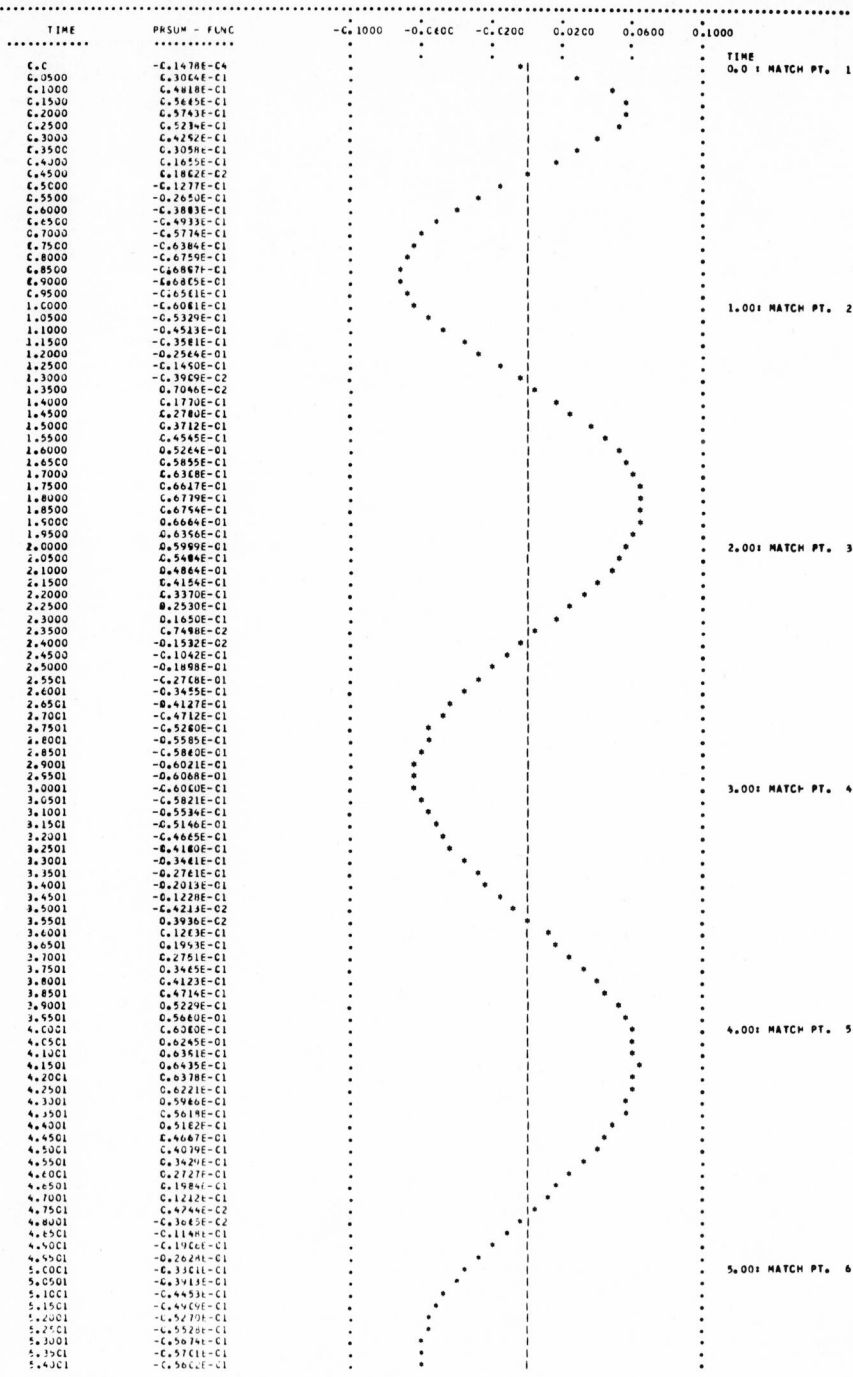

TIME	PRSUM - FUNC
0.0	-0.1478E-04
0.0500	0.3064E-01
0.1000	0.4818E-01
0.1500	0.5665E-01
0.2000	0.5743E-01
0.2500	0.5234E-01
0.3000	0.4292E-01
0.3500	0.3058E-01
0.4000	0.1655E-01
0.4500	0.1862E-02
0.5000	-0.1277E-01
0.5500	-0.2650E-01
0.6000	-0.3803E-01
0.6500	-0.4933E-01
0.7000	-0.5774E-01
0.7500	-0.6384E-01
0.8000	-0.6759E-01
0.8500	-0.6867E-01
0.9000	-0.6805E-01
0.9500	-0.6561E-01
1.0000	-0.6061E-01
1.0500	-0.5329E-01
1.1000	-0.4513E-01
1.1500	-0.3581E-01
1.2000	-0.2564E-01
1.2500	-0.1450E-01
1.3000	-0.3909E-02
1.3500	0.7046E-02
1.4000	0.1770E-01
1.4500	0.2780E-01
1.5000	0.3712E-01
1.5500	0.4545E-01
1.6000	0.5264E-01
1.6500	0.5855E-01
1.7000	0.6308E-01
1.7500	0.6617E-01
1.8000	0.6779E-01
1.8500	0.6794E-01
1.9000	0.6664E-01
1.9500	0.6396E-01
2.0000	0.5999E-01
2.0500	0.5484E-01
2.1000	0.4864E-01
2.1500	0.4154E-01
2.2000	0.3370E-01
2.2500	0.2530E-01
2.3000	0.1650E-01
2.3500	0.7498E-02
2.4000	-0.1532E-02
2.4500	-0.1042E-01
2.5000	-0.1898E-01
2.5501	-0.2708E-01
2.6001	-0.3455E-01
2.6501	-0.4127E-01
2.7001	-0.4712E-01
2.7501	-0.5260E-01
2.8001	-0.5585E-01
2.8501	-0.5840E-01
2.9001	-0.6021E-01
2.9501	-0.6068E-01
3.0001	-0.6000E-01
3.0501	-0.5821E-01
3.1001	-0.5534E-01
3.1501	-0.5146E-01
3.2001	-0.4665E-01
3.2501	-0.4100E-01
3.3001	-0.3461E-01
3.3501	-0.2761E-01
3.4001	-0.2013E-01
3.4501	-0.1228E-01
3.5001	-0.4213E-02
3.5501	0.3936E-02
3.6001	0.1203E-01
3.6501	0.1953E-01
3.7001	0.2751E-01
3.7501	0.3465E-01
3.8001	0.4123E-01
3.8501	0.4714E-01
3.9001	0.5229E-01
3.9501	0.5660E-01
4.0001	0.6000E-01
4.0501	0.6245E-01
4.1001	0.6391E-01
4.1501	0.6435E-01
4.2001	0.6378E-01
4.2501	0.6221E-01
4.3001	0.5966E-01
4.3501	0.5619E-01
4.4001	0.5182E-01
4.4501	0.4667E-01
4.5001	0.4079E-01
4.5501	0.3429E-01
4.6001	0.2727E-01
4.6501	0.1984E-01
4.7001	0.1212E-01
4.7501	0.4744E-02
4.8001	-0.3665E-02
4.8501	-0.1148E-01
4.9001	-0.1906E-01
4.9501	-0.2628E-01
5.0001	-0.3301E-01
5.0501	-0.3913E-01
5.1001	-0.4453E-01
5.1501	-0.4909E-01
5.2001	-0.5270E-01
5.2501	-0.5528E-01
5.3001	-0.5674E-01
5.3501	-0.5701E-01
5.4001	-0.5602E-01

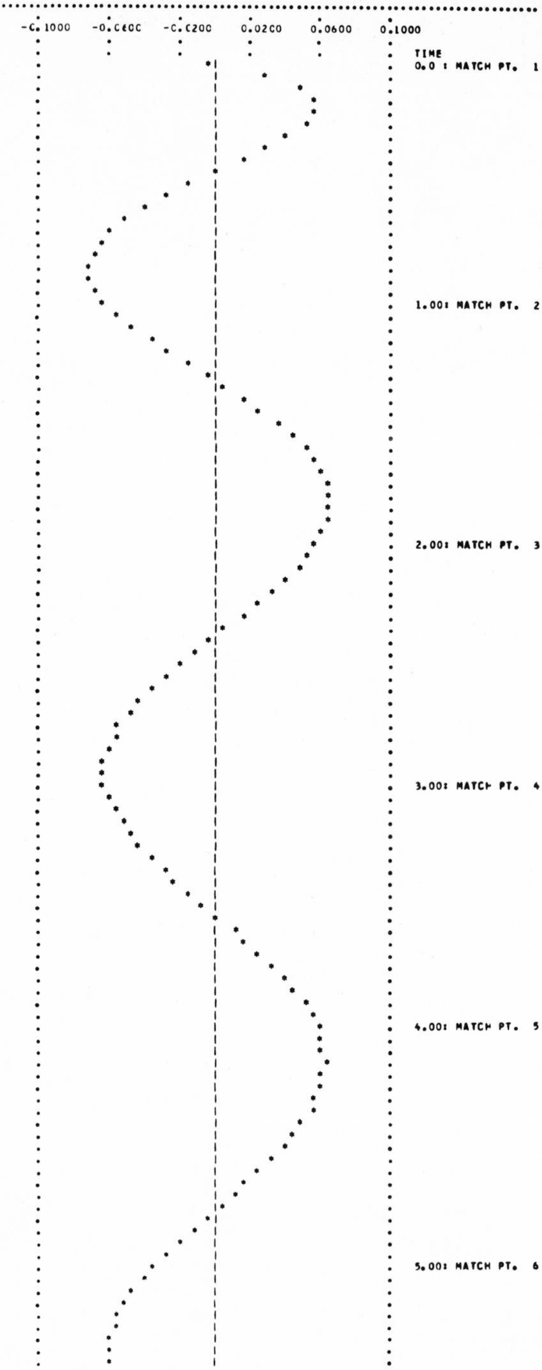

FIG. 7.06-D.

5.4501	-0.5373E-01	
5.5001	-0.5010E-01	
5.5501	-0.4510E-01	
5.6001	-0.3872E-01	
5.6501	-0.3055E-01	
5.7001	-0.2181E-01	
5.7501	-0.1131E-01	
5.8001	0.5059E-03	
5.8501	0.1360E-01	
5.9001	0.2793E-01	
5.9501	0.4344E-01	
6.0001	0.5998E-01	6.00: MATCH PT. 7
6.0501	0.5145E-01	
6.1001	0.4388E-01	
6.1501	0.3722E-01	
6.2001	0.3140E-01	
6.2501	0.2634E-01	
6.3001	0.2159E-01	
6.3501	0.1825E-01	
6.4001	0.1506E-01	
6.4501	0.1235E-01	
6.5001	0.1085E-01	
6.5501	0.8089E-02	
6.6001	0.6409E-02	
6.6501	0.4949E-02	
6.7001	0.3655E-02	
6.7501	0.2477E-02	
6.8001	0.1372E-02	
6.8501	0.3018E-03	
6.9001	-0.7670E-03	
6.9501	-0.1861E-02	
7.0001	-0.3082E-02	7.00: MATCH PT. 8
7.0501	-0.4205E-02	
7.1001	-0.5480E-02	
7.1501	-0.6833E-02	
7.2001	-0.8244E-02	
7.2501	-0.9768E-02	
7.3001	-0.1134E-01	
7.3501	-0.1286E-01	
7.4001	-0.1462E-01	
7.4501	-0.1630E-01	
7.5001	-0.1798E-01	
7.5501	-0.1964E-01	
7.6002	-0.2126E-01	
7.6502	-0.2281E-01	
7.7002	-0.2427E-01	
7.7502	-0.2562E-01	
7.8002	-0.2683E-01	
7.8502	-0.2790E-01	
7.9002	-0.2879E-01	
7.9502	-0.2950E-01	
8.0002	-0.3000E-01	8.00: MATCH PT. 9
8.0502	-0.3029E-01	
8.1002	-0.3036E-01	
8.1502	-0.3020E-01	
8.2002	-0.2981E-01	
8.2502	-0.2918E-01	
8.3002	-0.2833E-01	
8.3502	-0.2726E-01	
8.4002	-0.2597E-01	
8.4502	-0.2449E-01	
8.5002	-0.2281E-01	
8.5502	-0.2096E-01	
8.6002	-0.1896E-01	
8.6502	-0.1682E-01	
8.7002	-0.1457E-01	
8.7502	-0.1223E-01	
8.8002	-0.9824E-02	
8.8502	-0.7372E-02	
8.9002	-0.4900E-02	
8.9502	-0.2432E-02	
9.0002	0.8779E-05	9.00: MATCH PT. 10
9.0502	0.2369E-02	
9.1002	0.4718E-02	
9.1502	0.6945E-02	
9.2002	0.9061E-02	
9.2502	0.1105E-01	
9.3002	0.1289E-01	
9.3502	0.1458E-01	
9.4002	0.1610E-01	
9.4502	0.1744E-01	
9.5002	0.1859E-01	
9.5502	0.1956E-01	
9.6002	0.2033E-01	
9.6502	0.2091E-01	
9.7002	0.2130E-01	
9.7502	0.2151E-01	
9.8002	0.2153E-01	
9.8502	0.2138E-01	
9.9002	0.2107E-01	
9.9502	0.2060E-01	
10.0002	0.2000E-01	10.00: MATCH PT. 11
10.0502	0.1926E-01	
10.1002	0.1842E-01	
10.1502	0.1747E-01	
10.2002	0.1644E-01	
10.2502	0.1535E-01	
10.3002	0.1420E-01	
10.3502	0.1301E-01	
10.4002	0.1180E-01	
10.4502	0.1059E-01	
10.5002	0.9379E-02	
10.5502	0.8190E-02	
10.6002	0.7032E-02	
10.6502	0.5918E-02	
10.7002	0.4857E-02	
10.7502	0.3857E-02	
10.8002	0.2926E-02	
10.8502	0.2071E-02	
10.9002	0.1295E-02	
10.9502	0.6036E-03	
11.0002	-0.2433E-05	11.00: MATCH PT. 12
11.0502	-0.5219E-03	
11.1002	-0.9551E-03	
11.1502	-0.1304E-02	
11.2002	-0.1570E-02	
11.2502	-0.1758E-02	
11.3002	-0.1871E-02	
11.3502	-0.1916E-02	
11.4002	-0.1858E-02	

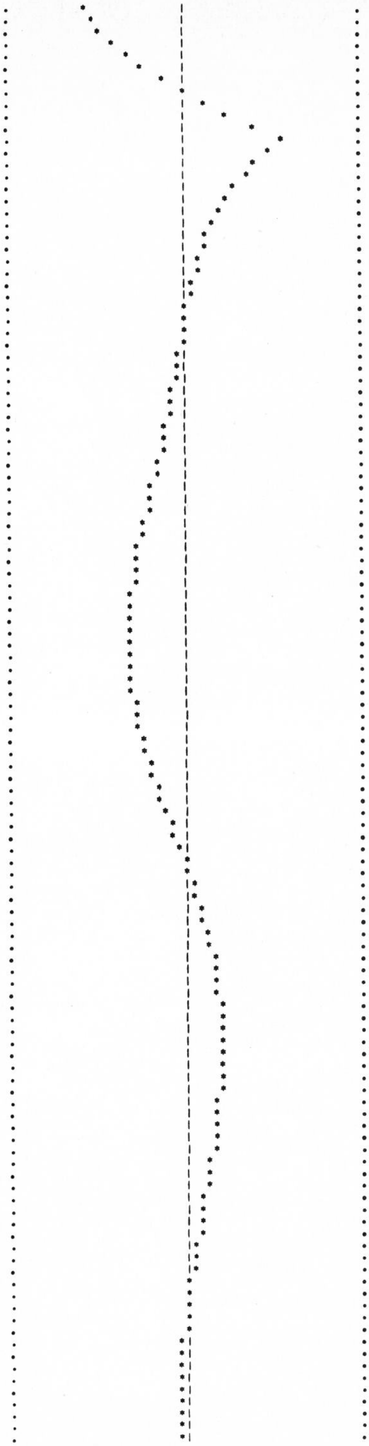

FIG. 7.06-D. (*Con.*)

and

b
$$\frac{K_1}{s-p_1}+\frac{K_2}{s-p_2}. \tag{7.06-10}$$

In the first case α is of course positive; K can have either sign. In the second case, $p_1 = -\alpha + j\omega_N$ (p_1 is complex and lies in the left half of the s plane) and p_2 is the conjugate of p_1; K_2 is the conjugate of K_1, which can have any value.

The first type corresponds to a real natural frequency and is easily realized by the RC network of Chap. 2 and the combining techniques of Chap. 5. It needs no further comment here.

The second type, the complex natural frequency, is the more important and the more troublesome. We can rewrite (7.06-10) thus:

$$\frac{K_1}{s-p_1}+\frac{K_2}{s-p_2}=\frac{K_1}{s-p_1}+*=\frac{K_{1r}+jK_{1i}}{s-(-\alpha+j\omega_N)}+* \tag{7.06-11}$$

in which the star as usual denotes conjugate terms whose effect is important but simple (to double the real part of the first term and to cancel out its imaginary part); they need not be explicitly written. [See Fig. 7.06-E for the (usual) notation in (7.06-11).] It is only this second type that we need in our present example, by (7.06-8). For it the RLC circuit of Chap. 3, used twice and connected by the methods of Sec. 5.08, will suffice. For we can write (see Fig. 7.06-F)

$$\begin{aligned} T(s) &= \frac{(g_{m1}/C_1)(s+\beta_1)-(g_{m2}/C_2)(s+\beta_2)}{s^2+2\alpha s+\omega_0{}^2} \\ &= \frac{(g_{m1}/C_1 - g_{m2}/C_2)+[(g_{m1}/C_1)\beta_1-(g_{m2}/C_2)\beta_2]}{s^2+2\alpha s+\omega_0{}^2} \end{aligned} \tag{7.06-12}$$

in which there are more "disposable" parameters than we really need. We can also manipulate (7.06-11) thus, to obtain a similar form:

$$\begin{aligned} \frac{K_{1r}+jK_{1i}}{(s+\alpha)-j\omega_N}+* &= \frac{2K_{1r}(s+\alpha)-2K_{1i}\omega_N}{(s+\alpha)^2+\omega_N{}^2} \\ &= \frac{(2K1r)s+(2K_{1r}\alpha-2K_{1i}\omega_N)}{s^2+2\alpha s+\omega_0{}^2} \\ &= \frac{as+b}{s^2+2\alpha s+\omega_0{}^2} \end{aligned} \tag{7.06-13}$$

in which $a = 2K_{1r}$, $b = 2K_{1i}\alpha - 2K_{1r}\omega_N$. These two (real) parameters a and b can of course have either sign (or be zero), for they result from the approxima-

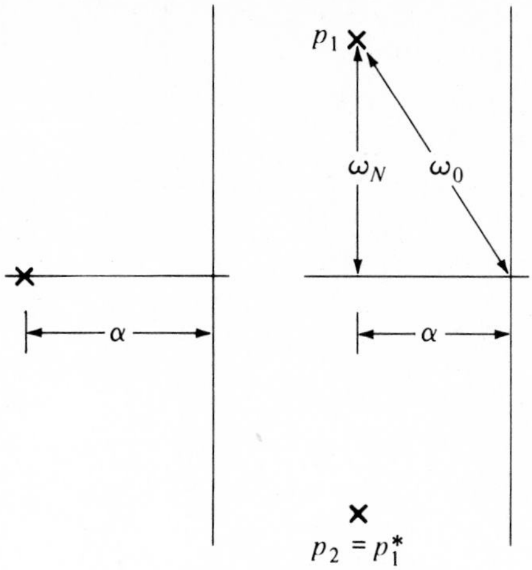

FIG. 7.06-E.

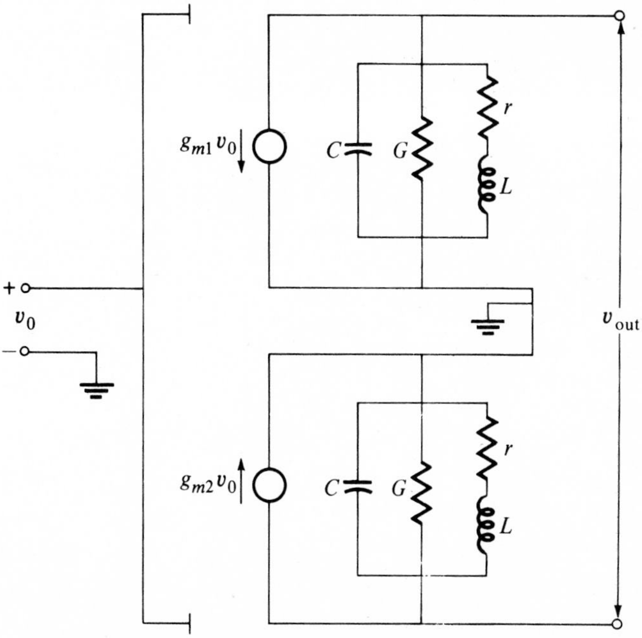

In each RLC circuit: $Z = \dfrac{(1/C)(s+\beta)}{s^2 + 2\alpha s + \omega_0^2}$, $\beta = \dfrac{r}{L}$,

(See Sec. 3.18; see also Fig. 5.08-B.) $2\alpha = \dfrac{r}{L} + \dfrac{G}{C}$,

$$\omega_0^2 = \frac{rG + 1}{LC}$$

FIG. 7.06-F.

tion process. We now need only equate coefficients of the numerators in (7.06-13) and (7.06-12), solve, and obtain the following relations:

$$\left.\begin{aligned} \frac{g_{m1}}{C_1} - \frac{g_{m2}}{C_2} &= a, \\ \beta_1 \frac{g_{m1}}{C_1} - \beta_2 \frac{g_{m2}}{C_2} &= b, \end{aligned}\right\} \tag{7.06-14}$$

and

$$\left.\begin{aligned} \frac{g_{m1}}{C_1} &= \frac{b - a\beta_2}{\beta_1 - \beta_2}, \\ \frac{g_{m2}}{C_2} &= \frac{b - a\beta_1}{\beta_1 - \beta_2}. \end{aligned}\right\} \tag{7.06-15}$$

There are, in effect, three cases to consider. Let us assume that a is positive (if a is not positive, we can make it so, in effect, by mainpulating the summation connections below). Then the sign of b/a determines which of the three cases we have. Table 7.06-C summarizes convenient values of parameters, which lead in

TABLE 7.06-C

$\frac{b}{a}$	β_1	β_2	$\frac{g_{m1}}{C_1}$	$\frac{g_{m2}}{C_2}$
$\frac{b}{a} \leqq 0$	0	2α	$\frac{a}{2\alpha}\left(2\alpha - \frac{b}{a}\right)$	$-\frac{b}{2\alpha}$
$0 \leqq \frac{b}{a} \leqq 2\alpha$	$\frac{b}{a}$	†	a	†
$2\alpha \leqq \frac{b}{a}$	2α	0	$\frac{b}{2\alpha}$	$\frac{a}{2\alpha}\left(\frac{b}{a} - 2\alpha\right)$

† The second *RLC* circuit is here unnecessary (we can set $g_{m2} = 0$).

every case to realizable network results, provided the units are properly interconnected. This can be done quite simply by the methods used in Sec. 5.08. It is of course essential to take care of the various signs in combining the varous terms, by proper additions and subtractions.

Clumsy this synthesis may be, but it makes the point: the exponential sums *can* be network-realized. Our aim here is merely to make that point, not to delve deeply into synthesis (the number of possible network realizations is infinite). It is merely to show that the discussion of partial-fraction expansions, natural frequencies, K's, etc., can be *reversed* to convert analysis into synthesis. (See PR in Appendix B.)

We return to our proper business, network analysis.

7.07 TABLES OF TRANSFER FUNCTIONS (IMPULSE RESPONSES)

It were well to stop here to organize our thoughts, to relate our new-found integral definition of transfer function to our many previously worked examples, to tie up loose ends and to get our intellectual house in order. The ideas of Secs. 7.03 to 7.05, as well as those of Chap. 6 will all be useful in formulating a standard procedure for network analysis. Common sense demands that to be able to use them profitably in the future, we put them into a summary form, as concise as may be consistent with clarity and utility. (Section 7.06 is interesting, useful later and in synthesis, but only ancillary to the present work of network analysis.)

Our experience has many times indicated that "the" network that has a given transfer function is not unique: many different networks can have the same transfer function: for example, the four (simple) networks of Fig. 7.07-A. If we excite the symmetrical lattice network of Fig. 5.13-A with a current source instead of the voltage source there shown, then the transfer function is $(Z_b - Z_a)/2$; one can obviously add *any* impedance to *both* Z_a and Z_b with no effect on $T(s)$ at all.

To any given transfer function, however, there corresponds but one impulse response. Hence, in making our summary, we shall not mention any specific (wave-forming) network realizations of the transfer functions listed, but simply pair impulse response and transfer function. Network realizations can be obtained by synthesis procedures (this is a tremendous subject in itself: see Secs. 6.14 and 7.06); the transfer function of any given network can be obtained by analysis (see Chap. 6).

Also appropriate here is a notation change. The concept of a (nonunique) wave-forming *network*, to convert a unit-impulse excitation into some desired (excitation) wave $e(t)$ is purely a paper fiction, a mental aid; we do not (usually) contemplate actually designing and building such a network; its purpose is purely heuristic. What *is* important is the wave-forming *transfer function*, the $T_{WF}(s)$ of Sec. 7.04, for which the notion of a wave-forming network is but a convenient mental crutch. It is logical then to change *symbols* thus (Fig. 7.07-B):

$$\begin{aligned} r_{\delta_{WFN}}(t) &\to e(t), && \text{the excitation wave function of interest;} \\ T_{WF}(s) &\to E(s), && \text{the corresponding function of } s\text{, the mate of } e(t). \end{aligned} \qquad (7.07\text{-}1)$$

Henceforth we shall denote the mate of a (time) function $f(t)$ by $F(s)$. That is, we shall consistently use lowercase letters for the time function and corresponding uppercase letters for the corresponding s function, its mate, as determined by

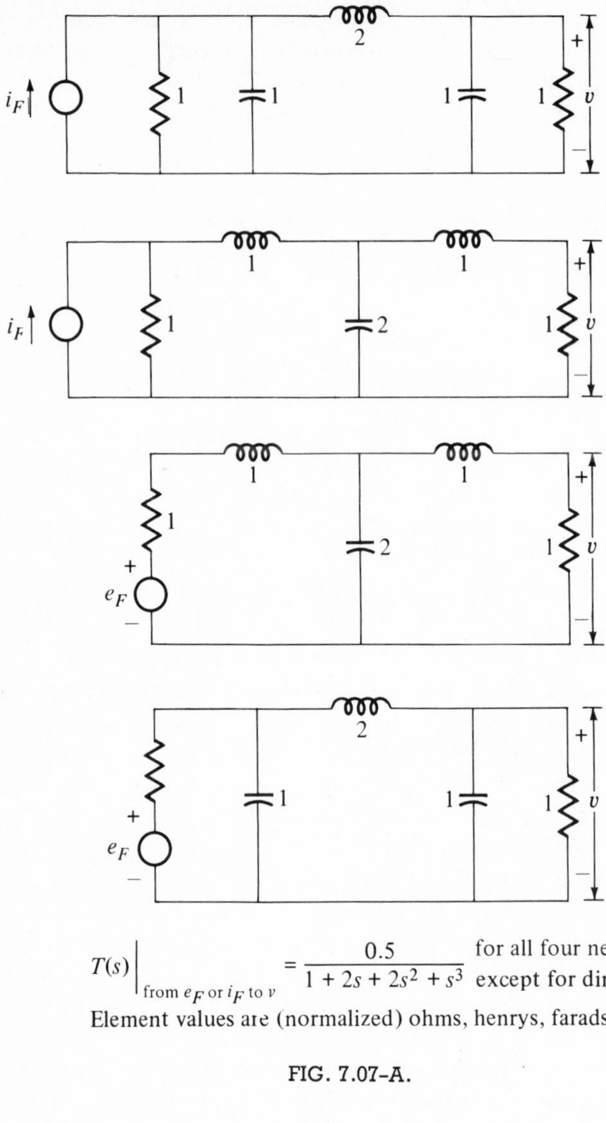

$$T(s)\Big|_{\text{from } e_F \text{ or } i_F \text{ to } v} = \frac{0.5}{1 + 2s + 2s^2 + s^3}$$ for all four networks, except for dimensions

Element values are (normalized) ohms, henrys, farads.

FIG. 7.07-A.

(7.05-1) or by actual analysis of a network that forms the wave $f(t)$ from a unit impulse excitation. The mnemonic value of this system is considerable; examples are

$$\begin{aligned} e(t) &\sim E(s), \\ i(t) &\sim I(s), \\ v(t) &\sim V(s). \end{aligned} \tag{7.07-2}$$

The wave-forming network view

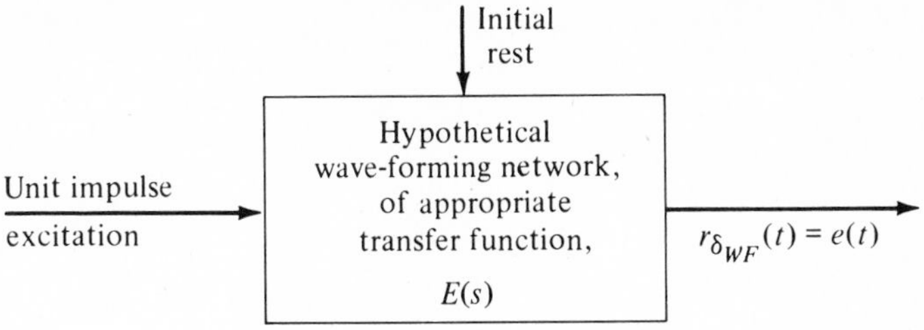

(*a*)

The pair (mathematical transformation or mating) view

$$e(t) \sim E(s) = \int_0^\infty e(t)e^{-st}\,dt, \qquad \text{RE}[s] > \sigma_a.$$

(*b*)

FIG. 7.07-B. Two points of view.

The symbol $\sim$ (Secs. 7.03, 7.04) is a convenient abbreviation of the wave-forming network relation, or of (7.05-1).

This pair (mate) relationship is *very* significant indeed, as we shall find in due course. For example, the mathematical question below arises in one's mind sooner or later and will have to be answered, with due discretion, when we are able:

$$\begin{aligned} &\textit{If:} && F(s) = \int_0^\infty f(t)e^{-st}\,dt, \qquad \mathbf{RE}(s) > \sigma_a, \\ &\textit{Then:} && f(t) = \,? \qquad [\text{in terms of } F(s)]. \end{aligned} \tag{7.07-3}$$

That is, what is the solution of (7.05-1), regarded as an integral equation in the unknown function $f(t)$, $F(s)$ being known?

But our immediate business is far simpler: to organize in tabular form the pairs we know about. We have a bewildering array already, and it is easily extended.

First, let us collect some general relations. Table 7.07-A lists eight such, ranging from pair 1, simply a reminder of notation, to pair 8, convolution in t and multiplication in s, which is very sophisticated. Pairs 6, 7, and 8 are subject to the criticism that we have derived them only for functions in the category of pair 3. But they can now be derived for general application from the fundamental definition (pair 2).

TABLE 7.07-A

Pair Name or Description	t-Domain Member†	s-Domain Member	Reference	Pair Number
General notation scheme	$e(t)$	$E(s)$		1
Mathematical definition	$e(t)$	$E(s) = \int_0^\infty e(t)e^{-st}\,dt, \qquad \mathbf{RE}(s) > \sigma_a$	(7.05-1)	2
The familiar exponential-sum, rational-function pair (with distinct p_j)	$\sum_{j=1}^{N} K_j e^{p_j t}$	$\sum_{j=1}^{N} \frac{K_j}{s - p_j} = \frac{A_0 s^{N-1} + \cdots + A_{N-1}}{s^N + B_1 s^{N-1} + \cdots + B_N}$	Chaps. 5, 6	3
The same plus impulse component	$K_0\,\delta(t) + \sum_{j=1}^{N} K_j e^{p_j t}$	$K_0 + \sum_{j=1}^{N} \frac{K_j}{s - p_j} = \frac{K_0 s^N + \cdots}{s^N + B_1 s^{N-1} + \cdots + B_N}$	Sec. 6.24	4
Series form, displaying initial values‡	$e(0) + e(0) + e''(0)\frac{t^2}{2!} + \cdots$	$\frac{e(0)}{s} + \frac{e'(0)}{s^2} + \frac{e''(0)}{s^3} + \cdots$	Sec. 6.13	5
The integral of $e(t)$	$\int_0^t e(x)\,dx$	$\frac{1}{s}E(s)$	Sec. 6.11	6
The derivative of $e(t)$†	$e'(t)$	$sE(s) - e(0)$	Sec. 6.11	7
Convolution	$e(t) * f(t)$	$E(s)F(s)$	Chap. 4	8

† $u(t)$ is omitted; that is, $t \geq 0$ is understood in the t-domain members;
‡ Initial values, such as $e(0)$, are all to be interpreted as $e(0+)$, etc., in cases where there are differences, due to steps (jumps) at $t = 0$.

The mate of $\int_0^t e(x)\,dx$, for pair 6, is, integration by parts being used to manipulate its definition according to pair 2,

$$\int_0^\infty \left[\int_0^t e(x)\,dx\right] e^{-st}\,dt = \left[\int_0^t e(x)\,dx \frac{e^{-st}}{-s} - \frac{e^{-st}}{-s} e(t)\,dt\right]_0^\infty$$

$$= 0 + \frac{1}{s}\int_0^\infty e(t)e^{-st}\,dt \tag{7.07-4}$$

$$= \frac{1}{s} E(s)$$

on the assumption, of course, that **RE**(s) is sufficiently large, so that the first integral is zero and the second converges, that is **RE**$(s) > \sigma_a$, the value of σ_a being determined by $e(t)$.

As for pair 7, the mate of $e'(t)$ is

$$\int_0^\infty e'(t)e^{-st}\,dt = \int_0^\infty e^{-st}e'(t)\,dt$$

$$= \left[e(t)e^{-st} - \int e(t)(-s)e^{-st}\,dt\right]_0^\infty \tag{7.07-5}$$

$$= 0 - e(0+) + s\int_0^\infty e(t)e^{-st}\,dt$$

$$= sE(s) - e(0+),$$

provided of course that **RE**(s) is sufficiently great [**RE**$(s) > \sigma_a$]. The evaluation of $e(0)$ must be made at $0+$, as indicated, as we well have learned, if there is a sudden change in the value of e at $t = 0$, represented by an impulsive component in $e'(t)$. And it must be understood, of course, that $t \geqq 0+$ in all these pairs, since $u(t)$ is omitted in writing them.

These derivations provide a satisfying check on our previous work, as well as interesting generalizations. Finally we come to pair 8. This is more difficult to generalize. Its form heuristically suggests the following formal manipulations.

$$E(s)F(s) = \int_0^\infty e(x)e^{-sx}\,dx \int_0^\infty f(y)e^{-sy}\,dy$$

$$= \int_0^\infty \int_0^\infty e(x)f(y)e^{-(x+y)}\,dx\,dy \tag{7.07-6}$$

$$= \int_{y=0}^\infty \int_{x=0}^\infty f(t-x)e^{-st}\,dx\,dy$$

in which we have placed $(t - x) = y$. Here t is a parameter; x and y range independently from zero to infinity. If now we perform first the integration on

x, and let t as a variable replace y, we find

$$
\begin{aligned}
E(s)F(s) &= \int_{t=0}^{\infty}\left[\int_{x=0}^{\infty} e(x)f(t-x)\,dx\right]e^{-st}\,dt \\
&= \int_{0}^{\infty}\left[\int_{0}^{t} e(x)f(t-x)\,dx\right]e^{-st}\,dt
\end{aligned}
\tag{7.07-7}
$$

in which of course $f(y) = 0$ for $y < 0$. The result (7.07-7) is formally pair 8, interesting and encouraging, though easily shot full of holes by the guns of rigor.

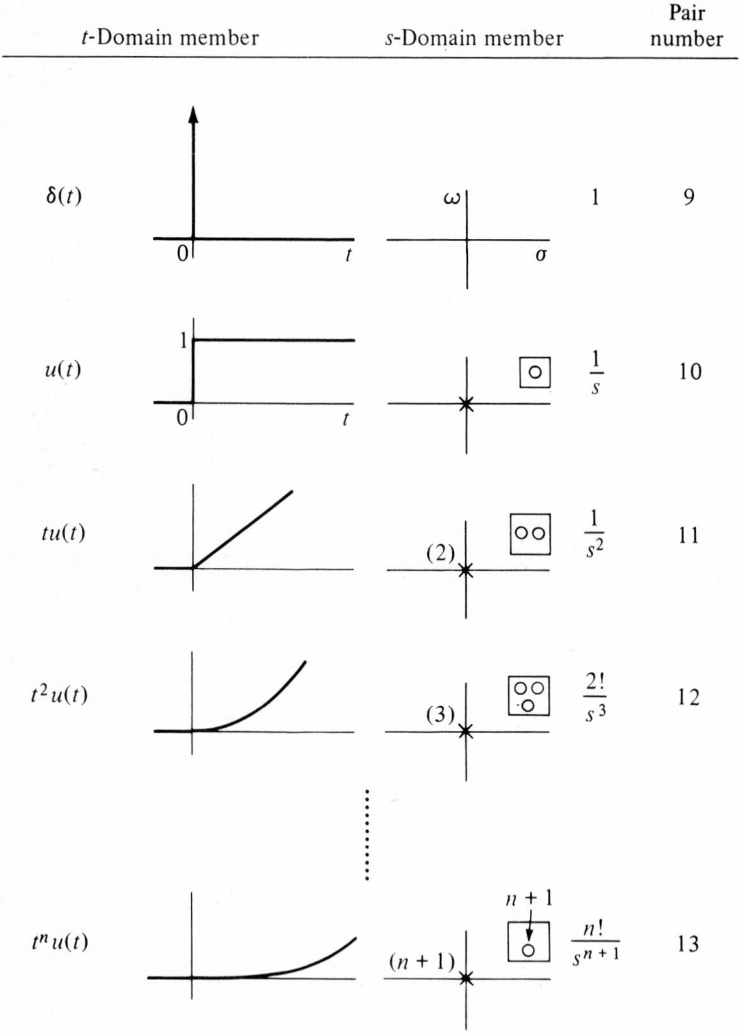

FIG. 7.07-C. Powers of t, poles at the origin pairs.

It can, however, be rigorously demonstrated for all reasonable functions, as we shall see. It is of course valid for functions of the pair-3 type without question, because of our previous work. Extension to those of the type of pair 4 is virtually immediate.

We next collect some groups of specific function pairs. As the first group we list, in Fig. 7.07-C, those pairs that correspond to poles of $E(s)$ at the origin. All were obtained in Sec. 6.11, but are easily checked with the definition, pair 2. It is now possible (because these pairs are specific) and helpful (vividly mnemonic) to characterize them also with diagrams: of t-domain behavior, functional behavior, and s-domain poles and zeros. Note again the importance of these pairs in the power-series form of t-domain expression (Sec. 6.11). They also suggest very clearly what multiple poles mean and lead us to the next, somewhat more general, group.

The second group, in Fig. 7.07-D, corresponds very closely to the first group

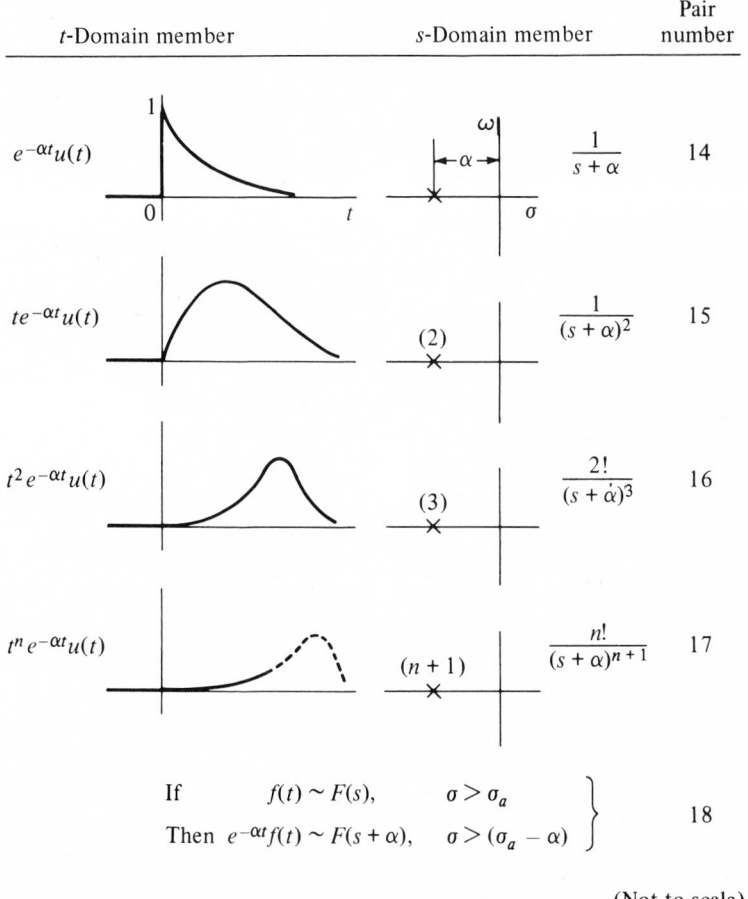

t-Domain member		s-Domain member	Pair number
$e^{-\alpha t}u(t)$	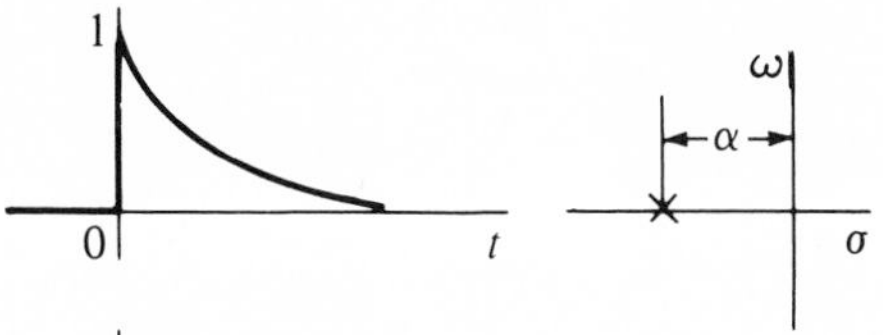	$\dfrac{1}{s+\alpha}$	14
$te^{-\alpha t}u(t)$	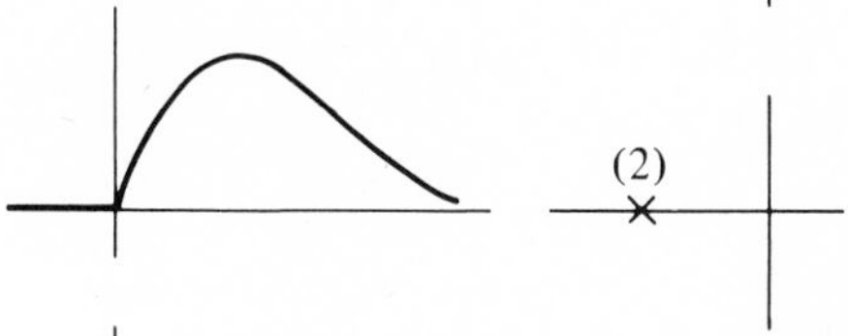	$\dfrac{1}{(s+\alpha)^2}$	15
$t^2e^{-\alpha t}u(t)$		$\dfrac{2!}{(s+\alpha)^3}$	16
$t^n e^{-\alpha t}u(t)$	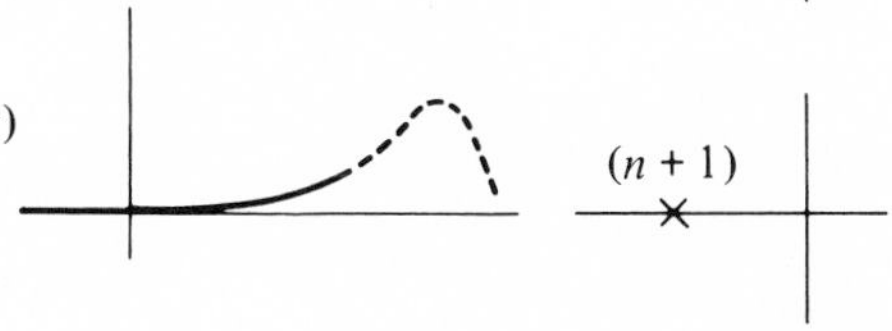	$\dfrac{n!}{(s+\alpha)^{n+1}}$	17

$$\left.\begin{array}{lll} \text{If} & f(t) \sim F(s), & \sigma > \sigma_a \\ \text{Then} & e^{-\alpha t}f(t) \sim F(s+\alpha), & \sigma > (\sigma_a - \alpha) \end{array}\right\} \quad 18$$

(Not to scale)

FIG. 7.07-D.

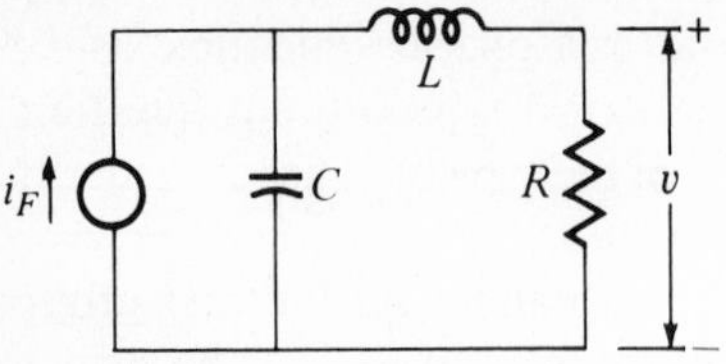

$$R = 2\sqrt{\frac{L}{C}}, \quad K_0 = \frac{R}{LC}$$

$$\alpha = \frac{R}{2L} = \frac{1}{\sqrt{LC}}$$

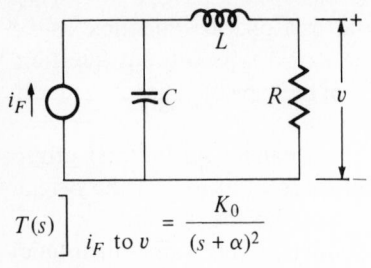

$$T(s)\Big]_{i_F \text{ to } v} = \frac{K_0}{(s+\alpha)^2}$$

$$r_\delta(t) = K_0 t e^{-\alpha t} \quad \text{V/c}$$

(a)

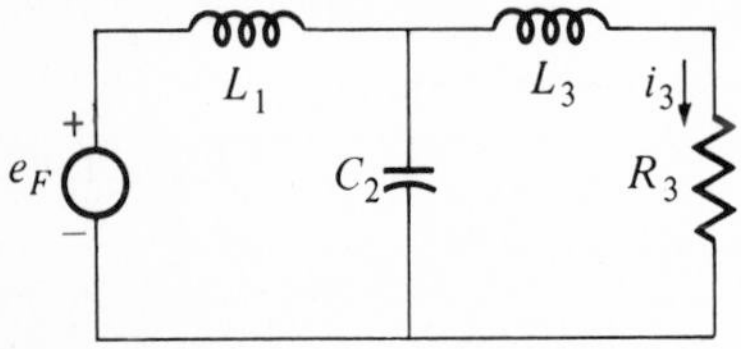

$$L_1 = \frac{8R_3}{3\alpha},$$

$$C_2 = \frac{9}{8R_3\alpha}, \quad K_0 = \frac{1}{L_1 C_2 L_3}$$

$$L_3 = \frac{R_3}{3\alpha},$$

$$r_\delta(t) = K_0 \frac{t^2}{2!} e^{-\alpha t} \quad \text{A/Wb} \cdot \text{turn}$$

$$T(s)\Big]_{e_F \text{ to } i_3} = \frac{K_0}{(s+\alpha)^3}$$

(b)

$$C_1 = \frac{\frac{16}{5}}{R_4\alpha},$$

$$L_2 = \frac{25}{16} \times \frac{R_4}{\alpha},$$

$$C_3 = \frac{\frac{4}{5}}{R_4\alpha},$$

$$L_4 = \frac{1}{4} \times \frac{R_4}{\alpha},$$

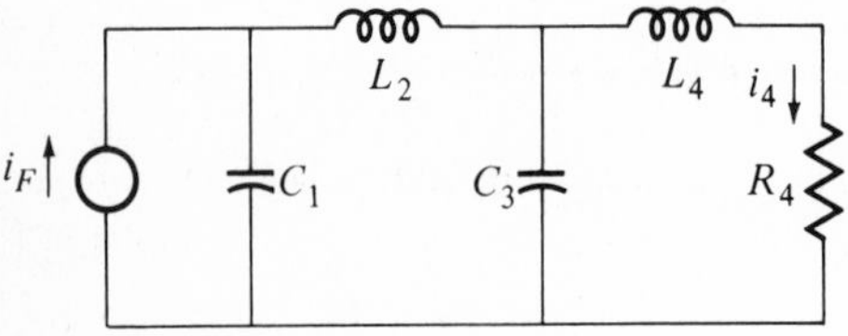

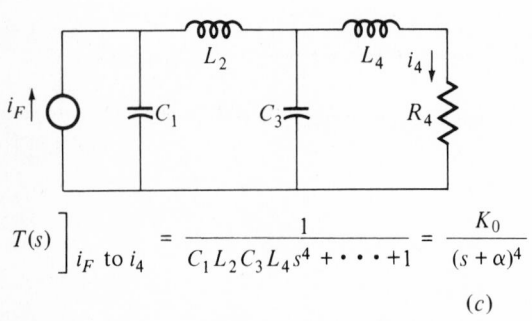

$$T(s)\Big]_{i_F \text{ to } i_4} = \frac{1}{C_1 L_2 C_3 L_4 s^4 + \cdots + 1} = \frac{K_0}{(s+\alpha)^4} \qquad r_\delta(t) = K_0 \frac{t^3}{3!} e^{-\alpha t} \quad \text{A/C}$$

(c)

FIG. 7.07-E.

except for a shift of the (s-domain) poles to the left through a distance α. The corresponding t-domain members experience multiplication by a damping factor, $e^{-\alpha t}$, as our experience with actual networks suggests. Pair 14 was of course the center of discussion in Chap. 2. Pair 15 was found there also, in the discussion of the response of the RC circuit to an excitation identical in form to its own natural behavior (Sec. 2.14, see also Sec. 4.06). Pairs 16 and 17 seem inevitably to follow, by comparison with pairs 14 and 15 (Fig. 7.07-C). They may be verified by using pair 2; and they suggest the theorem that pair 18 is, which is easily demonstrated:

If $f(t)$ and $F(s)$ are mates, σ_a determined by $f(t)$,

Then the mate of $e^{-\alpha t}f(t)$ is (7.07-8)

$$\int_0^\infty e^{-\alpha t}f(t)e^{-st}\,dt = \int_0^\infty f(t)e^{-(s+\alpha)t}\,dt = F(s+\alpha), \qquad \sigma > (\sigma_a - \alpha).$$

Note that the abscissa of absolute convergence is of course also shifted to the left the distance α.

It is of some interest to note that the coincidence of poles in the transfer function of a network does not imply any startling relations among its element values. For example, the networks of Fig. 7.07-E have the transfer functions and impulse responses shown. Slight changes in any element value will separate the natural frequencies and remove the "degeneracy." It is simply that a certain particular set of element values, out of the infinity possible, corresponds to coincidence of all natural frequencies and a particular form of the impulse response (pair 17). Whether this is desirable, or inefficient, or of no consequence at all, depends on the particular application a network is to be designed for.

7.08 REALISM IN INDUCTOR AND CAPACITOR MODELS

The *mathematical* replacement of s by $(s + \alpha)$ suggested itself to us on comparing pairs associated with integrals and power series with those of the *RC* and *RLC* circuits. Pair 18 is the result.

The replacement of s by $(s + \alpha)$ can in effect come about *physically*, too. The idealized inductor element L is not very realistic; a more reasonable model of a real inductor [see Sec. 3.18] consists of the same L in series with a (small) resistor r (Fig. 7.08-A). This accounts for the major lack of realism in L, provided s is small enough that we can neglect parasitic capacitance (which is another nonideal characteristic of real inductors). The impedance of the idealized element is Ls; the impedance of the more realistic model is $L(s + r/L)$. As far as this element is concerned, the change from idealized model to realistic model is accomplished by replacing s by $(s + r/L)$.

Dually the ideal capacitor element C (Fig. 7.08-A) should for realism be associated with an energy-dissipating element, say a small conductance g in parallel. The effect is to replace the admittance Cs by $C(s + g/C)$. Again, the introduction of dissipation into the model replaces s by $(s + \alpha)$ in all appearances of C in immittances and transfer functions, α here being g/C.

In any network with several inductors and capacitors the various (r/L) and (g/C) ratios will of course *differ*. In most practical cases the capacitor dissipation will also be much the smaller: $(g/C) \ll (r/L)$. Hence it is *not* true that if we

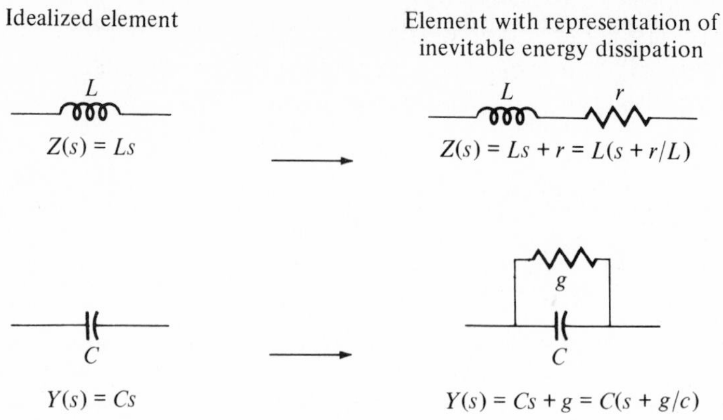

FIG. 7.08-A.

calculate some transfer function $T(s)$ of a network in which all the inductive and capacitive elements are ideal (nondissipative), $T(s + d)$ will then represent the transfer function after dissipation has been added to those elements. If all the dissipation constants (r/L) and (g/C) are small, however, an *average* value may be used without great loss of accuracy. That is, if $T(s)$ is some network transfer function calculated on the basis of ideal, nondissipative L and C elements, then $T(s + d)$ is approximately the transfer function after the addition thereto of small amounts of dissipation for realism, if

$$\begin{aligned} d &= \text{average of all } (r/L) \text{ and } (g/C) \text{ values} \\ &= \frac{\sum_{j=1}^{N_L} (r/L)_j + \sum_{j=1}^{N_C} (g/C)_j}{N} \end{aligned} \tag{7.08-1}$$

in which N_L is the number of inductor elements in the network, N_C is the number of capacitor elements, and $N = N_L + N_C$.

One obvious application of this principle is in synthesis, in situations where the prescription is $T(s)$. One can estimate d in advance, knowing the physical quality of elements available at reasonable cost, then carry out the design work not to realize $T(s)$ but to realize $T(s - d)$, with idealized elements L and C (for which the design process is simpler). This network, when built with real elements, will behave according to $T(s - d)]_{s=s+d} = T(s)$! The *predistortion* required may of course be virtually impossible, for natural frequencies cannot be moved very far to the right if they are already close to the ω axis. This means that the quality of L and C elements needed is extremely high, the resonances involved require very high values of Q. (Practical network types may also place some limitations on the preliminary motion of the *zeros* of T, for quite different reasons into which we do not go here.)

In *analysis*, a saving in work may often be made by carrying out the calculations on the basis of nondissipative, ideal L and C elements. The effect of dissipation in actual elements, if small, can then be estimated by replacing s by $(s + d)$ in the result. Thus pair 18 is useful in both analysis and synthesis.

7.09 TABLES OF TRANSFER FUNCTION PAIRS—II

From poles at the origin (Fig. 7.07-C) we moved to poles on the negative real axis (Fig. 7.07-D). Inevitably we move next to complex poles. Pair 19*a* (Fig. 7.09-A) is formally such a pair; in physical application the imaginary pole must of course be associated with its conjugate, pair 19*b*. Together they give us the sinusoidal pairs 20, 21, and 22. A sinusoidal wave, be it remembered, is essentially an exponential function (with imaginary exponent), but it is *two* such exponentials. Skeptics may easily apply pair 2 to verify these new pairs. Note that all three have the *same poles*, for they are all combinations of "Ke^{pt}" with the same p; they are all sinusoids of the same frequency. Their zeros differ because their K's differ; that is, their initial phases differ, γ of pair 22 being $-\pi/2$ rad in pair 20 and zero in pair 21. The dimensions of the t-domain members here are null; their mates in the s domain have therefore the dimension of time.

Damped-sinusoid pairs follow immediately, on applying pair 18; see pairs 23, 24, 25 (Fig. 7.09-B).

Multiple complex poles must surely correspond to multiplication by powers of t (see pairs 14, 15, 16). Pairs 26 and 27 (Fig. 7.09-C) give two such pairs, which can be verified by using pair 2. The mate of $te^{-\alpha t}\cos(\beta t + \gamma)$ is too complicated, when written in literal terms, to warrant inclusion here.

An alternate route is to use the new pair 28, which introduces us to another way of extending tables: differentiation with respect to parameters. This is a useful device for deriving new pairs from old when the latter contain some parameter that is independent of t and of s. Formally, for example, if we differentiate both members of pair 23 partially, with respect only to the parameter α, we obtain

$$-te^{-\alpha t}\sin\beta t \sim -\frac{2\beta(s+\alpha)}{(s+\alpha)^2+\beta^2} \tag{7.09-1}$$

which is pair 26. Similarly, the operation $\partial/\partial\beta$ applied to pair 23, and the operation $\partial/\partial\alpha$ applied to pair 24, give pair 27. Finally, differentiation of pair 24 with respect to β again gives pair 26. Partial differentiation of pair 25 with respect to each of its three parameters gives corresponding new pairs, which are too complicated to include here.

To *prove* that one may so operate we start from Pair 2 and write

$$\frac{\partial}{\partial p}F(s,p) = \frac{\partial}{\partial p}\int_0^\infty f(t,p)e^{-st}\,dt \tag{7.09-2}$$

in which p is some parameter independent of s and of t. If the functions involved are reasonable [if the integral exists, that is, if $\mathbf{RE}(s)$ is large enough, if $\partial F/\partial p$ exists and is continuous] the sequence of integration and differentiation may be reversed and we have

$$\frac{\partial F}{\partial p} = \int_0^\infty \frac{\partial f}{\partial p} e^{-st}\, dt \tag{7.09-3}$$

t-Domain member [$u(t)$ omitted)]	s-Domain member	Pair number
$e^{j\beta t}$	$\dfrac{1}{s - j\beta}$	19a
$e^{-j\beta t}$	$\dfrac{1}{s + j\beta}$	19b

$\sin \beta t = \dfrac{1}{2j}\left(e^{j\beta t} - e^{-j\beta t}\right)$

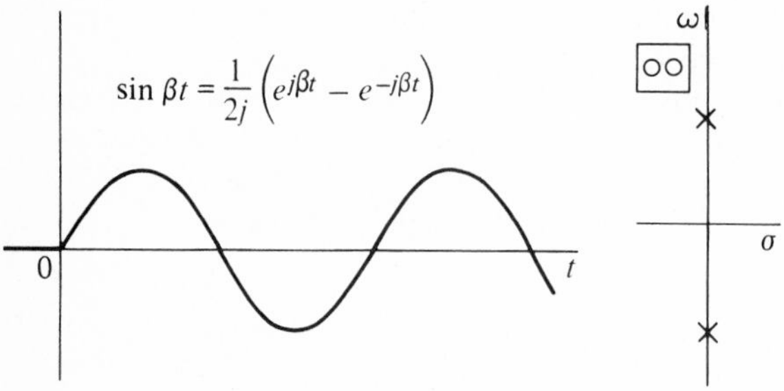

$\dfrac{1}{2j}\left(\dfrac{1}{s - j\beta} - \dfrac{1}{s + j\beta}\right) = \dfrac{\beta}{s^2 + \beta^2}$ 20

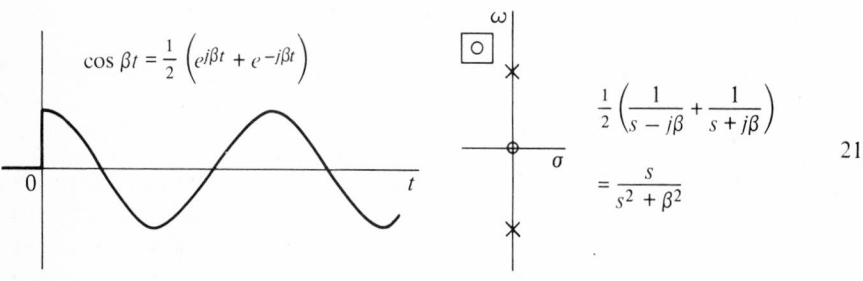

$\cos \beta t = \dfrac{1}{2}\left(e^{j\beta t} + e^{-j\beta t}\right)$

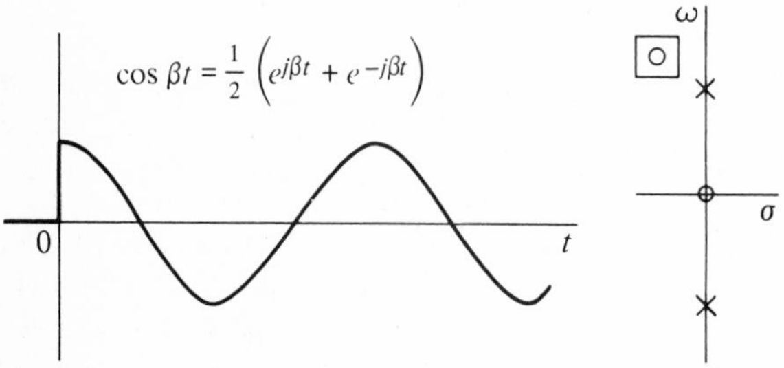

$\dfrac{1}{2}\left(\dfrac{1}{s - j\beta} + \dfrac{1}{s + j\beta}\right) = \dfrac{s}{s^2 + \beta^2}$ 21

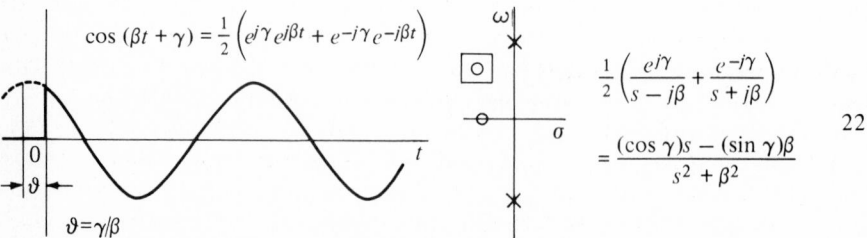

$\cos(\beta t + \gamma) = \dfrac{1}{2}\left(e^{j\gamma}e^{j\beta t} + e^{-j\gamma}e^{-j\beta t}\right)$

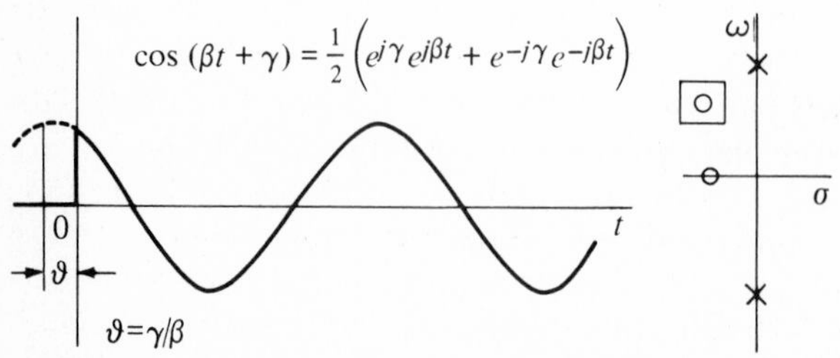

$\dfrac{1}{2}\left(\dfrac{e^{j\gamma}}{s - j\beta} + \dfrac{e^{-j\gamma}}{s + j\beta}\right) = \dfrac{(\cos\gamma)s - (\sin\gamma)\beta}{s^2 + \beta^2}$ 22

FIG. 7.09-A.

t Domain	*s* Domain	Pair number

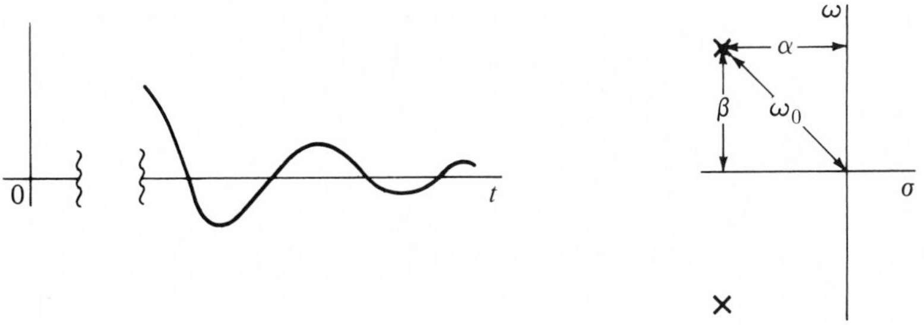

—	$e^{-\alpha t}\sin\beta t \sim \dfrac{\beta}{(s+\alpha)^2+\beta^2} = \dfrac{\beta}{s^2+2\alpha s+(\alpha^2+\beta^2)} = \dfrac{\beta}{s^2+2\alpha s+\omega_0{}^2}$	23
	$e^{-\alpha t}\cos\beta t \sim \dfrac{s+\alpha}{(s+\alpha)^2+\beta^2} = \dfrac{s+\alpha}{s^2+2\alpha s+\omega_0{}^2}$	24
	$e^{-\alpha t}\cos(\beta t+\gamma) \sim \dfrac{(\cos\gamma)(s+\alpha)-(\sin\gamma)\beta}{(s+\alpha)^2+\beta^2} = \dfrac{(\cos\gamma)\,s+(\alpha\cos\gamma-\beta\sin\gamma)}{s^2+2\alpha s+\omega_0{}^2}$	25

FIG. 7.09-B.

which we call pair 28 (Table 7.09-A). (Instances in which this partial differentiation is not valid can be found, but are usually pathological and self-evident.) Correspondingly one may often integrate with respect to a parameter (pair 29).

We conclude this tabulation of pairs with a few special cases of interest and one, the most important pair of all. (This last is in fact the *only* pair one really needs!) Placing $\alpha = 0$ (which we assume, without argument, to be legitimate) in pairs 26 and 27, or differentiation of pairs 20 and 21 with respect to the parameter β, gives the double-pole-on-the-ω-axis pairs 30 and 31 (Fig. 7.09-D). Note, in passing, that these correspond to exciting a nondissipative (infinite-Q) parallel *RLC* circuit by a sinusoid of its own natural frequency, leading of course, to infinite forced response. (Excitation of a finite-Q *RLC* circuit at its own natural frequency corresponds to pairs 26 and 27, in which the forced response reaches a maximum and then decays.) Pair 32 (Fig. 7.09-D) can be derived either by combining pairs 30 and 31, using the trigonometric formula for the cosine of the sum of two angles, or by partial differentiation with respect to β or to γ of

t Domain	s Domain	Pair number

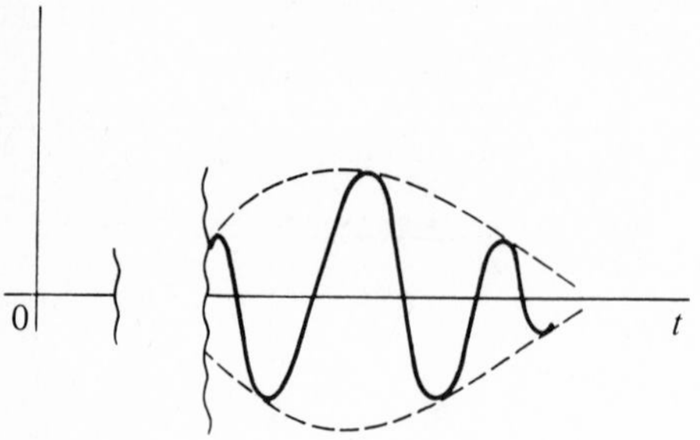

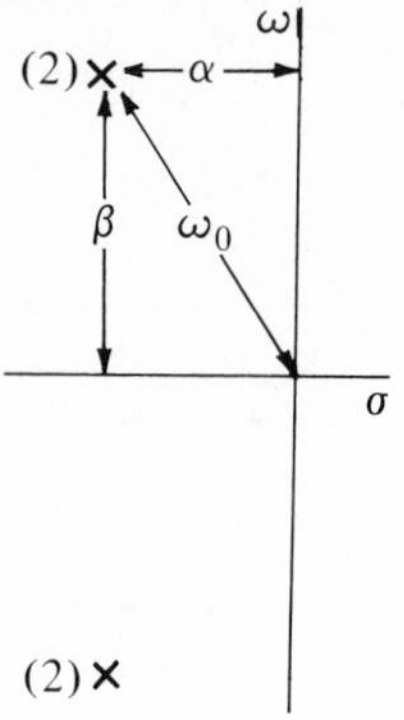

t Domain	s Domain	Pair number
$te^{-\alpha t}\sin\beta t \sim$	$\dfrac{2\beta(s+\alpha)}{[(s+\alpha)^2+\beta^2]^2} = \dfrac{2\beta s + 2\beta\alpha}{(s^2+2\alpha s+\omega_0{}^2)^2}$	26
$te^{-\alpha t}\cos\beta t \sim$	$\dfrac{(s+\alpha)^2-\beta^2)}{[(s+\alpha)^2+\beta^2]^2} = \dfrac{s^2+2\alpha s+(\alpha^2-\beta^2)}{(s^2+2\alpha s+\omega_0{}^2)^2}$	27

FIG. 7.09-C.

pair 22, followed by a shift in γ of $\pi/2$ rad. As for the mate of $t^n \begin{smallmatrix}\sin\\ \cos\end{smallmatrix} (\beta t)$, it is too complicated to write out here. In any event, such pairs are more easily handled by using pair 35.

A little reflection shows that *all* the specific simple-pole ("nontheorem") pairs that we have listed are but special instances and combinations of pair 33

TABLE 7.09-A

If	$f(t, p) \sim F(s, p)$ (p is a parameter independent of t and of s, and the functions are reasonable)	
Then	$\dfrac{\partial f}{\partial p} \sim \dfrac{\partial F}{\partial p}$	pair 28
And	$\int_0^p f(t, x)\,dx \sim \int_0^p F(s, x)\,dx.$	pair 29

t Domain	s Domain	Pair number

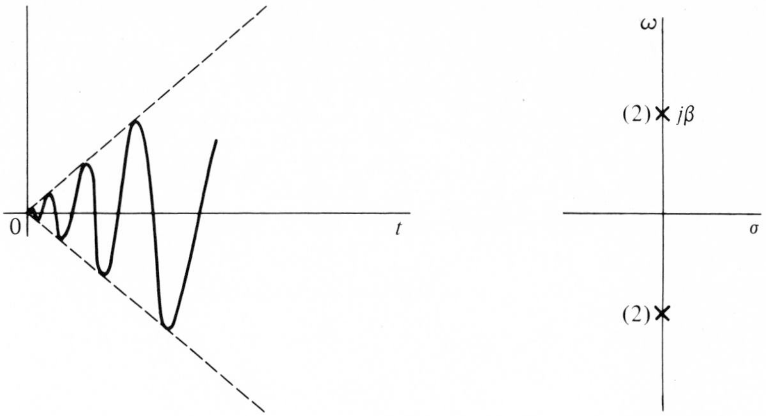

$$t \sin \beta t \sim \frac{2\beta s}{(s^2 + \beta^2)^2} \qquad 30$$

$$t \cos \beta t \sim \frac{(s^2 - \beta^2)}{(s^2 + \beta^2)^2} \qquad 31$$

$$t \cos (\beta t + \gamma) \sim \frac{(\cos \gamma)s^2 - (2\beta \sin \gamma)s - \beta^2 \cos \gamma}{(s^2 + \beta^2)^2} \qquad 32$$

FIG. 7.09-D.

in Table 7.09-B. The various exponential and sinusoidal pairs can be obtained from it by assigning appropriate values to p and combining pairs so derived. The multiple-pole pairs can be obtained therefrom by a limiting process (see Sec. 2.14) or they can be obtained from pair 35, which follows from pair 33 by differentiation with respect to the parameter p. (See pair 34.) Even the impulse function (pair 9) is a (limiting) exponential form, for

$$\lim_{\alpha \to \infty} \alpha e^{-\alpha t} = \delta(t) \tag{7.09-4}$$

and correspondingly, in the s domain,

$$\lim_{\alpha \to \infty} \frac{\alpha}{s + \alpha} = 1 \tag{7.09-5}$$

which is pair 9. The step function (pair 10) and the following powers of t are obviously exponential in character (see pairs 14 to 17).

TABLE 7.09-B

t		s	Pair Number
e^{pt}	$\sim$	$\dfrac{1}{s-p}$	33
te^{pt}		$\dfrac{1}{(s-p)^2}$	34
....			
$t^n e^{pt}$	$\sim$	$\dfrac{n!}{(s-p)^{n+1}}$	35

NOTES: (a) $u(t)$ is omitted, as before;
(b) p is a (constant) parameter, independent of t and of s;
(c) $\sigma_a = \mathbf{RE}\,(p)$.

Pair 33 can therefore truly be called *the fundamental pair*, the foundation stone, of all our pairs. It is the one pair needful! If any verification thereof is necessary, pair 2 will supply it. But the underlying basic reason therefor is the fact that the exponential function is *eigen* to our basic differential equation (Sec. 6.12): the natural modes of our networks are exponential in character, e^{pt}.

Since all transfer functions of interest to us are rational functions of s, the corresponding impulse responses are exponential sums (pairs 3, 4). All important excitation functions appear also to be exponential in character, with mates that are rational functions of s. Consequently the exponential function suffices to describe the responses of all interesting networks to all interesting excitations. Many of the descriptive ("theorem") pairs, such as pairs 6 or 18, in contrast to the specific pairs (33 and those derived therefrom) are not really *necessary* for ordinary network analysis. But they give additional valuable insight and short-cuts. And they will be useful when we discuss the generalization to nonexponential functions made possible by pair 2 (Chap. 8).

7.10 NETWORK ANALYSIS

With statements of the basic problem of network analysis in general terms we are thoroughly familiar (see, for example, Figs. 1.07-A, 5.01-A, 6.25-A, 7.03-AB, 7.07-B). Here, in Fig. 7.10-A, we present still another view of "the network problem." Since there are four entities involved, presumably there are situations in which (any) one of the four is unknown and the other three are to be found.

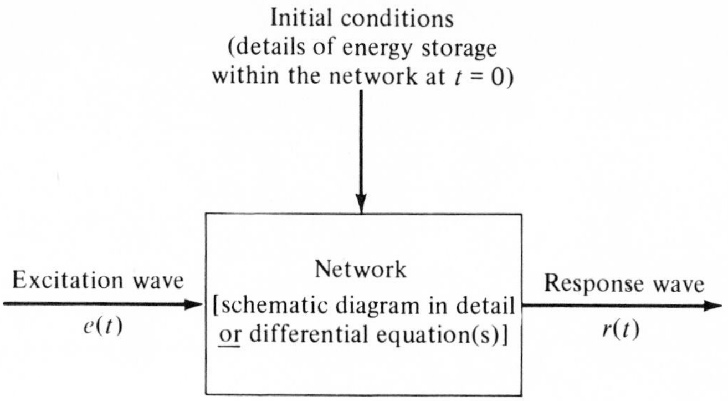

		Network	$e(t)$ excitation	Initial conditions	$r(t)$ response
Problems of network analysis	1	✓	✓	✓	?
	2	✓	✓	?	✓
	3	✓	?	✓	✓
Network synthesis problem	4	?	✓	✓	✓

√ = known (given)
? = unknown (to be found)

FIG. 7.10-A. Network analysis and synthesis.

If the network itself is *known*, the (three) questions are questions of *network analysis*. They have unique answers. Physical common sense dictates that, or at least we assume it does. (The mathematical view, in the theory of differential equations, requires demonstration of the existence and uniqueness of the solution.)

The common, garden variety of network analysis problem is the first one: its study is important as an educational measure, to gain understanding of the operation of networks. The second and third are variants, less commonly encountered, but of essentially the same nature; they arise when an observed response is not what it should be, and one asks whether the trouble is in the initial conditions, or in the excitation. If the trouble is in the network itself, we come to the fourth question.

If the network is *unknown*, the question is one of *synthesis*, and it does not have a unique answer. It then amounts to finding differential equation(s), or network schematics, that transform the given excitation and initial conditions into given responses. This can be done, if at all, in an infinite number of ways, by an infinite number of different networks. The synthesis job is therefore difficult and complicated and we postpone its general consideration (but see Secs. 6.14, 7.06). It is of course the most interesting question; the goal of network-analysis study is to become able to design network(s), usually to be parts of larger systems in which they accomplish specific tasks, such as the separation (filtering out) of one signal from others, or the correction (equalization) of undesirable transmission behavior in some other part of the system, or the shaping of some (excitation) wave into some different (response) form for some purpose of the system.

We have in our study already obtained methods of solution of the analysis problem for many cases, notably ladder and bridge and simple feedback networks that are impulse or step or exponentially excited from rest. These cases actually include nearly all important network analysis problems. To stop to investigate completely all the details of all other conceivable cases would completely obfuscate the matter; it would be a red herring indeed! But Sec. 6.24 has shown how to proceed in even the toughest, most general case.

It is worth our while to carefully codify our present knowledge, the results we have obtained so far. The codification will result in a "mechanization" of the analysis process, a useful tool that we can henceforth use almost automatically. It will also clearly indicate some details that are yet to be completely worked out. So now let us formulate our mechanism for the solution of the analysis problem with care, as far as we are presently able. We shall be surprised how potent it is! We can in the future call on it *routinely* for solving most network analysis problems.

To do a clean job, an honest job, logically requires first that we clearly state our postulates. Restrictions, limitations, speciality of cases must be spelled out with clarity. It is convenient to do this by looking at the three entities (Fig. 7.10-A) in turn.

The *network* (or "circuit") we assume to be "lulicon" (Sec. 6.12) and finite. [One might expect that any discussion of networks with infinite numbers of elements would be silly, but this is not so; such study can be profitable (see Secs. 8.02, 8.03)]. *Physically* the postulate means that the network (model) is composed of a finite number of R, L, C, and g_m elements (Chap. 1). Electromagnetic coupling between inductors can occur, wilfully or parasitically, but it introduces no really new concepts and we postpone its consideration. *Mathematically* the postulate means that we have to deal with a mathematical model that is a finite set of ordinary differential equations that are linear, and whose coefficients are constants. As a consequence of this postulate, the transfer function $T(s)$ exists, and it is a rational function of s with real coefficients, and it can be calculated. True, there is some further assumption here, for it is only for ladder networks (Sec. 6.01) and for some bridges (Sec. 6.23) and active networks (Sec. 6.25) that we have fully demonstrated these things. But it seems entirely reasonable that they be true in general; full discussion would require unreasonable expansion of Sec. 6.24, another red herring.

The *initial condition* of the network we postulate to be *rest* (no energy is stored in any element of the network at $t = 0$). We recall examples (Chaps. 2, 3) in which initially charged elements can be replaced by initially uncharged elements associated with private impulse or step excitations that effectively place the energy thereon all of a sudden, at $t = 0$. This substitution is always possible, and we shall soon demonstrate that nonrest initial conditions require merely the solution of additional, supplementary initial-rest network analysis problems (Sec. 7.22), and the addition of their solutions, to obtain the whole solution. Its discussion here would be one more red herring.

The *excitation* wave is of course arbitrary. But we recognize that a certain class of waves includes almost everything of practical importance and we choose to talk only about that class here. (Later on we can fill in the chinks and generalize.) There is here a dichotomy between (*a*) the important class of excitations, those that are *exponential*, that can be written as sums of exponentials (which includes the impulse, the step, powers of t, and so on) for which the wave-forming transfer function (the "mate" in the s domain) is a *rational* function of s as we have seen in the 30-some pairs we have developed, and (*b*) nonexponential waves. The latter we can presumably convolve with the impulse response, calculated from the transfer function, to obtain the response. But the first category of excitation is the important one: it is the commonly met one, and for it the complicated integration process of convolution can be replaced by simple processes of algebra.

7.11 NETWORK ANALYSIS—THE COMMON CASE

We proceed to the task of "mechanizing" the analysis process in what is by far the commonest case. Figure 7.11-A succinctly defines it and outlines the analysis procedure we have developed (Secs. 7.03, 7.04). The steps are:

a To transform $e(t)$ into $E(s)$, a rational function of s with real coefficients that is its mate (the appropriate wave-forming transfer function);

b To replace the network by its transfer function $T(s)$, another rational function of s with real coefficients, which we may look on as the mate of the network:

c To multiply $E(s)$ by $T(s)$ to form $R(s)$, one more rational function of s with real coefficients; $R(s)$ is the mate of the response $r(t)$;

d To transform back to the t domain by expanding $R(s)$ in partial fractions and using pair 3 or pair 33;

e To apply checks sufficient to be reasonably certain that there are no arithmetical errors, opportunities for which abound!

It remains to comment on the details of each step, some of which do indeed need further explanation.

For step (*a*) the table of pairs usually suffices. Seldom is any interesting

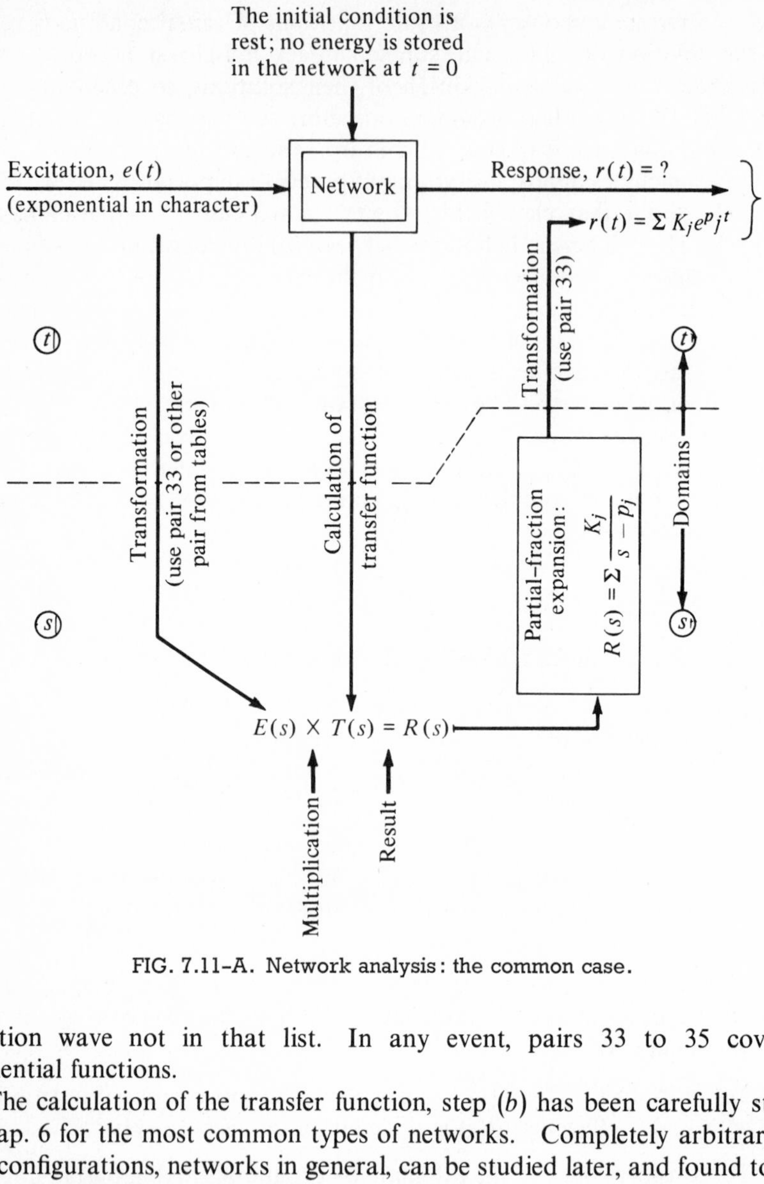

FIG. 7.11-A. Network analysis: the common case.

excitation wave not in that list. In any event, pairs 33 to 35 cover all exponential functions.

The calculation of the transfer function, step (*b*) has been carefully studied in Chap. 6 for the most common types of networks. Completely arbitrary network configurations, networks in general, can be studied later, and found to have transfer functions with exactly the same properties.

Step (*c*) is simple algebra, but it may be advisable merely to indicate it, not actually to perform it (Sec. 7.13).

Step (*d*), though it contains only algebra and arithmetic, does need explanation and illustration by example (Sec. 7.15, 7.17). A discussion of useful checking procedures will accompany them; they constitute step (*e*).

Separation of the response into forced and natural parts is conveniently accomplished in the *s* domain, in step (*d*), before returning to the time domain.

And it is sufficiently important to warrant careful separate examination (Sec. 7.16).

Once we have assimilated the procedure in this, the common case, we shall consider less common cases that can arise (nonrest initial conditions, for example, or coincident poles).

The final step is, of course,

f To examine and discuss the results of the analysis, and to answer as best one can with them, the engineering questions that originally led to posing the network analysis problem.

7.12 PARTIAL-FRACTION EXPANSION

The partial-fraction expansion of $R(s)$ indicated in Fig. 7.11-A, step (*d*) in Sec. 7.11, is in principle a routine algebraic-arithmetic chore. Since it is an essential link in the analysis chain it is extremely important, however dull it may be. Add so its details need discussion and illustration.

In fundamental theorems of algebra and complex-variable analysis we find the basis for the expansion:

a Every polynomial of degree N has N zeros; hence the denominator of $R(s)$ can be factored;

b every rational function that is a proper fraction can be expanded, after factoring its denominator, into a sum of partial fractions, each of which is very simple in form: a constant divided by $(s - p)$, p being one of the zeros of the denominator. (7.12-1)

We continue to assume here, for simplicity, that all the poles of R are separate (distinct); the general case (Sec. 7.23) is the same in essence if not in detail. Proofs of the statements in (7.12-1) can be found in the disciplines indicated. In practice no proof is necessary, for we make the expansions first, and then in effect establish the validity thereof in the course of checking the arithmetic.

Steps (*a*) to (*c*) of the analysis (Fig. 7.11-A) lead us to this rational function of s with real coefficients:

$$E(s)T(s) = R(s) = \frac{d_M s^M + d_{M-1} s^{M-1} + \cdots + d_0}{s^N + b_1 s^{N-1} + b_2 s^{N-2} + \cdots + b_N}. \tag{7.12-2}$$

Lowercase symbols have been used for the constants here to avoid confusion between the coefficients in $R(s)$ and those in $T(s)$ to which we have become accustomed.

We have here *generalized* to include cases where the degree of the numerator of $R(s)$ equals or exceeds that of the denominator: $M \geqq N$. But these are rare indeed in practice, for two reasons:

a usually $T(s)$ has one or more zeros at infinity, because physical systems do not transmit very well at high frequencies. Occasionally the network (model) may be simplified, for convenience, by neglecting the parasitic elements that make $T(\infty) = 0$. In these cases the poor high-frequency transmission is not deliberate; interest is then not in the high-frequency behavior but only in intermediate or low-frequency behavior, and the modeling error is presumably small and tolerable.

b $E(s)$ also is zero at infinity; that is, $e(t)$ has no impulsive component, much less derivatives of impulses! The sole important exception is the impulse itself, so useful as we have seen as index, prototype, and test function (pair 9). Sometimes with the impulse are included some exponentials (pair 4).

If $M \geqq N$, R is not a proper fraction and should immediately be reduced, by ordinary division, to the expanded form

$$R(s) = (c_{M-N}\, s^{M-N} + \cdots + c_2 s^2 + c_1 s + c_0) + \frac{a_0\, s^{N-1} + \cdots + a_{N-1}}{s^N + b_1 s^{N-1} + \cdots + b_N}. \qquad (7.12\text{-}3)$$

The polynomial portion of $R(s)$ in (7.12-3) transforms immediately to

$$c_0\, \delta(t) + c_1\, \delta'(t) + \cdots \qquad (7.12\text{-}4)$$

in $r_\delta(t)$, according to pairs 9 and 7. The peculiar character of the derivative of an impulse requires detailed investigation, either by resorting to limiting processes or by study of the theory of distributions. But the occurrence of $M > N$ is so rare that we pay no further attention to it here; we consider only $M \leqq N$, that is, cases in which R has no pole at infinity.

If R also has no zero at infinity ($M = N$) the division is one term only, $c_0 = d_M = d_N$, and there is an impulsive component in $r(t)$: $c_0\, \delta(t)$. This should not be forgotten, but requires no further comment.

Our concern here is with the principal item, the proper fraction in (7.12-3), to which we now devote attention exclusively. We also return to our familiar pristine notation, writing capitals for the coefficients:

$$R(s) = \frac{A_0\, s^{N-1} + \cdots + A_{N-1}}{s^N + B_1 s^{N-1} + \cdots + B_N}. \qquad (7.12\text{-}5)$$

This is, if you like, the transfer function of the overall network that is the combination of a wave-forming network and the network under study, in cascade, or R is simply the product of E and T (the impulsive part of the response, if there be any, having been computed and set aside). The problem is reduced to that of calculating an impulse response (but see Sec. 7.13).

And this problem consists of two parts:

a The factoring of the denominator,

b The expansion of $R(s)$ in partial fractions. (7.12-6)

The first step requires very little comment here; it has been discussed in Sec. 6.04. Except in the simple cases $N = 1$ and $N = 2$, the factoring is best done by a computer, for which innumerable methods and programs exist (Appendix C). It gives us

$$R(s) = \frac{A_0 s^{N-1} + A_1 s^{N-2} + \cdots + A_{N-1}}{(s - p_1)(s - p_2)\cdots(s - p_N)} = \frac{Q_N(s)}{Q_D(s)}, \tag{7.12-7}$$

in which Q_N and Q_D represent the numerator and denominator polynomials, respectively, a useful notation.

Since, by hypothesis, the N poles of $R(s)$ are distinct, we can for the second step of (7.12-6) write

$$R(s) = \frac{K_1}{s - p_1} + \frac{K_2}{s - p_2} + \cdots + \frac{K_N}{s - p_N}. \tag{7.12-8}$$

(The coefficients K_j in the partial-fraction expansion of $R(s)$ are also known as *residues*, a term relevant to contour integration in the theory of functions of a complex variable, which also makes use of such partial-fraction expansions.) To calculate the K's one can equate the forms in (7.12-7) and (7.12-8), cross multiply by the polynomial Q_D, and then obtain N simultaneous equations in the N unknowns K_j by giving s appropriate specific values. These are linear equations and in principle easily solved. Much more straightforward, however, are the two formulas developed below.

Multiplying through by $(s - p_1)$ in (7.12-8) gives us

$$(s - p_1)R(s) = K_1 + (s - p_1)\left(\frac{K_2}{s - p_2} + \cdots + \frac{K_N}{s - p_N}\right). \tag{7.12-9}$$

If we now let $s = p_1$, a step previously not possible because of the appearance of ∞, a useless value, we find

$$K_1 = [(s - p_1)R(s)]_{s=p_1}. \tag{7.12-10}$$

The expression on the right is not zero by any means, as examination of (7.12-7) will show:

$$(s - p_1)R(s) = \frac{A_0 s^{N-1} + \cdots + A_{N-1}}{(s - p_2)(s - p_3)\cdots(s - p_N)} \tag{7.12-11}$$

which, on setting $s = p_1$, gives neither zero nor infinity, but some finite number which is the value of K_1. The other K_j have similar formulas.

An alternative technique is based on the observation that the derivative (with respect to s) of the denominator of $R(s)$ is, from (7.12-7),

$$\begin{aligned} Q_D'(s) = (s - p_1)\frac{d}{ds}[(s - p_2)(s - p_3)\cdots(s - p_N)] \\ + [(s - p_2)(s - p_3)\cdots(s - p_N)]. \end{aligned} \tag{7.12-12}$$

Since the second term in (7.12-12) is the same as the denominator of the expression in (7.12-11), and the first term vanishes when $s = p_1$, we have

$$K_1 = [(s - p_1)R(s)]_{s=p_1} = \frac{Q_N(p_1)}{Q_D'(p_1)}. \tag{7.12-13}$$

Rewritten to apply to the jth coefficient in the obvious way, (7.12-13) becomes

$$\begin{aligned} K_j = [(s - p_j)R(s)]_{s=p_j} &= \left[\frac{A_0 s^{N-1} + \cdots + A_{N-1}}{(s - p_1)\cdots(s - p_N)}\right]_{\substack{\text{factor } (s - p_j) \text{ omitted,} \\ \text{and } s = p_j}} \\ = \frac{Q_N(p_j)}{Q_D'(p_j)} &= \left[\frac{A_0 s^{N-1} + \cdots + A_{N-1}}{N s^{N-1} + (N-1)B_1 s^{N-2} + \cdots + B_{N-1}}\right]_{s=p_j}. \end{aligned} \tag{7.12-14}$$

In (7.12-14) we have the two formulas previously referred to for the development of (7.12-5) in the form (7.12-8). Both are useful, both are practical, both should be kept at hand.

Actual calculations usually involve complex numbers and easily become complicated. They are perfectly straightforward, however, and easily programmed for machine computation.

7.13 NETWORK ANALYSIS—AN EXAMPLE

As a first illustration of the complete process of analysis of a network, consider the problem posed in Fig. 7.13-A. The network is a simple low-pass filter, designed to pass low-frequency signals quite well, but to impede the passage of

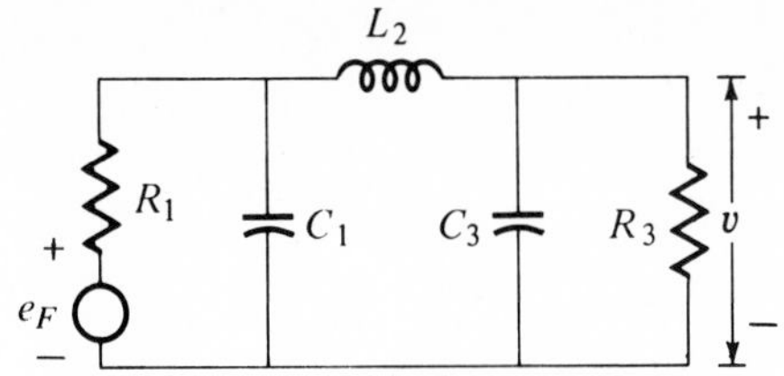

Initial condition: rest

Excitations of interest: impulse, step, exponential and sinusoid

FIG. 7.13-A.

high-frequency signals. The connection of capacitors and inductors shown will certainly do this, the network being in effect transparent at low frequencies and having three zeros at infinity in its transfer function. The actual choice of appropriate element values (the synthesis) is a complicated matter into which we do not here go. Consider here only the analysis of the given network for the various sorts of excitation given. Let us take the various steps of analysis in turn.

To transform the excitation $e_F(t)$ into $E(s)$ is, for the common excitation waves given, a very simple matter: all are found in the tables of pairs, or can mentally be treated with pair 33. We obtain Table 7.13-A.

TABLE 7.13-A

$e_F(t)$	$E(s)$
$\Lambda_0\,\delta(t)$	Λ_0
$E_0\,u(t)$	$\dfrac{E_0}{s}$
$E_0\,e^{-\alpha t}u(t)$	$\dfrac{E_0}{s+\alpha}$
$E_0\cos(\omega_F t+\phi)$	$E_0\dfrac{(\cos\phi)s-(\sin\phi)\omega_F}{s^2+\omega_F{}^2}$

To "transform" the network, that is, to obtain its transfer function, is easy, because of its ladder structure. This calculation has already been made, in effect; it is dual to that of Fig. 6.02-A (see also Fig. 6.07-C) and gives

$$T(s)=\left.\frac{v}{e_F}\right|_{\ldots}=\frac{1}{[(1+R_1G_3)+(L_2G_3+R_1C_1+R_1C_3)s+(L_2C_3+R_1C_1L_2G_3)s^2+(R_1C_1L_2C_3)s^3]}. \qquad (7.13\text{-}1)$$

As a check, we note the dimensions of T (null), of its numerator (null), and of every term in its denominator (each is clearly dimensionless). [See Table 6.05-A.] Asymptotic high- and low-frequency calculations, easily made, provide additional checks: the schematic diagram shows, at high frequencies, $s \to \infty$ (see Sec. 6.15),

$$T(s) \to \frac{1}{R_1 C_1 s L_2 s C_3 s} \tag{7.13-2}$$

and at low frequencies ($s = 0$)

$$T(s) = \frac{R_3}{R_1 + R_3} = \frac{1}{1 + R_1 G_3} \tag{7.13-3}$$

both of which agree, by inspection, with the asymptotic behaviors of (7.13-1). The next step is a very simple one:

$$R(s) = V(s) = E(s)T(s) = \frac{(\text{numerator of } E)(\text{ numerator of } T)}{(\text{denominator of } E)(\text{denominator of } T)} \tag{7.13-4}$$

Here $E(s)$ takes in turn the various forms of Table 7.13-A; $T(s)$ is given by (7.13-1).

Step (d) requires factoring the denominator of $R(s)$, that is, the denominators of E and of T. (There is a little point in multiplying the two together and then refactoring!) We look first at the impulse-response case, the simplest. Here the denominator of E is simply unity and the task is to factor the (cubic) denominator of $T(s)$, that is, to find the three natural frequencies of the network.

In simple "$N = 2$" cases only a quadratic need be factored, the classical formula suffices, and the factoring is almost trivial. But if $N = 3$, as here, or if $N > 3$, factoring by formula is impracticable or impossible (see Sec. 6.04). We must give up the luxury of literal element values, be content with specific numerical element values, and use computing machinery (usually a digital computer).

But before we abandon literal element values, we note that some simplifications can be accomplished without completely giving up the generality of literal element values: scaling (normalization) can be of some help.

7.14 NORMALIZATION

That we can simplify analysis by measuring variables *relatively* rather than *absolutely* we have several times noticed. The response of the RC circuit of Chap. 2 to an impulsive current $Q_0\,\delta(t)$ is, (2.05-6),

$$v(t) = \frac{Q_0}{C}\, e^{-t/(RC)}. \tag{7.14-1}$$

By choosing $V_0 = Q_0/C$ as the unit of voltage rather than the volt, and the time constant $T_0 = RC$ as the unit of time rather than the second, i.e., by changing to *normalized* variables defined by

$$v_n = \frac{v}{V_0}, \qquad t_n = \frac{t}{T_0}, \tag{7.14-2}$$

we change (7.14-1) into

$$v_n(t) = e^{-t_n} \tag{7.14-3}$$

[see Sec. 2.06]. Nonessentials have been removed; the essential character of the response is much clearer without the extra, obfuscating symbols.

In the s domain that circuit's transfer function can similarly be simplified by normalization:

$$T(s) = \frac{1/C}{s + 1/RC} = \frac{R}{1 + RCs} \tag{7.14-4}$$

becomes

$$T_n = \frac{1}{1 + s_n} \tag{7.14-5}$$

in which $T_n = T/R$, and $s_n = RCs$. The circuit has only two parameters; both are "removed" by appropriate scaling of the response $v(t)$ or the transfer function (of s) on the one hand, and time or frequency on the other. It is our basic postulates of linearity and constancy that make this possible.

Note the correspondence between the time and frequency normalizations:

$$t_n = \frac{t}{RC}, \qquad s_n = RCs. \tag{7.14-6}$$

Since the function e^{st} permeates all we do, it is no surprise that s and t undergo inverse operations so that their (dimensionless) product is unchanged.

The RLC circuit (Fig. 3.01-A) has three parameters. For it (Sec. 3.10)

$$T(s) = \frac{(1/C)s}{s^2 + (G/C)s + 1/(LC)} = \frac{(1/C)s}{s^2 + 2\alpha s + {\omega_0}^2}. \tag{7.14-7}$$

Note that in the second form we have not normalized, but have merely introduced new symbols to emphasize the frequency of resonance, and the amount of damping (Sec. 3.04). We may now conveniently normalize the s (frequency) scale thus:

$$s_n = \frac{s}{\omega_0} = \sqrt{LC}s \qquad (7.14\text{-}8)$$

and the response (or transfer function) scale thus:

$$T_n = \frac{T}{\sqrt{L/C}} \quad \text{or} \quad T_n = \frac{T}{R} \qquad (7.14\text{-}9)$$

to obtain

$$T_n = \frac{s_n}{s_n^2 + (1/Q)s_n + 1} \quad \text{or} \quad T_n = \frac{(1/Q)s_n}{s_n^2 + (1/Q)s_n + 1}. \qquad (7.14\text{-}10)$$

Only two parameters are removed by normalizing the (only) two dimensions available. That it is a circuit of more complicated nature is evident in that one parameter remains, taken as Q. Now in the t domain, consistently with (7.14-6) and (7.14-8) we take $t_n = \omega_0 t$ and obtain from

$$v(t) = K_0 e^{-\alpha t} \cos(\omega_N t + \phi) \qquad (7.14\text{-}11)$$

the expression (see Sec. 3.07)

$$v(t) = K_0 e^{-\omega_0 t/(2Q)} \cos(\omega_N t + \phi) = K_0 e^{-t_n/(2Q)} \cos\left(\sqrt{1 - \frac{1}{4Q^2}}\, t_n + \phi\right). \qquad (7.14\text{-}12)$$

Since there are two initial conditions to be inserted, there are two additional constants, and normalization of v can now remove one at best. Normalization cannot of course "remove" all parameters; not all expressions can be made as simple as (7.14-3) and (7.14-5)!

In the present example (Fig. 7.13-A) we see that the first term in the denominator of $T(s)$ in (7.13-1) is already a ratio, and therefore dimensionless; i.e., in normalized form:

$$1 + R_1 G_3 = 1 + \frac{R_1}{R_3}. \qquad (7.14\text{-}13)$$

This should be no real surprise, for "at dc" only the two resistors act, and they merely form a voltage divider. And T itself is dimensionless, being the ratio of two voltages.

These choices of scales (of units) represent useful processes of standardizing or *normalizing*. In the RLC-circuit transfer function, one parameter still remains (most conveniently Q). In the example of Fig. 7.13-A, $T(s)$ of (7.13-1), *more* than one must remain. That is to be expected, for there are five elements in the network, but the paper on which we plot the results of an analysis has only two dimensions. We can scale (normalize) both of these, in fact we *have* to, to make a plot; it is necessary to decide how many units of

a The response to be exhibited, the dependent variable found by the analysis (v, i, $|T|$, ...) (7.14-14)

and of

b The time or the frequency (depending on the domain in which we find ourselves), the independent variable, the prime mover, (7.14-15)

one centimeter of paper shall represent, what scales to use.

It is helpful to anticipate this to some extent, and to normalize from the very beginning of an analysis, or as soon as we can recognize what reference values (units) are convenient or useful. This requires that we shift attention from $T(s)$ or $r(t)$ to something more immediately recognizable—we do not wish to wait until the analysis is complete to choose reference values. (In *synthesis* good reference choices are usually obvious in the very data used, for there we *start* with the response.)

Now we have observed [in (7.14-13)] that normalizing the first fundamental quantity, (7.14-14), may be more or less equivalent to normalizing *impedances* [which we can certainly do *ab initio*]. For another example, see Sec. 6.22, where the transfer function

$$T(s) = \frac{R_0{}^2}{R_0 + Z(s)} \tag{7.14-16}$$

without thinking consciously of normalization at all, was written

$$\frac{T(s)}{R_0} = \frac{1}{1 + z(s)} \tag{7.14-17}$$

The normalization of $Z(s)$ to R_0 scale, i.e., the use of $z(s)$, is "obvious," it screams aloud to be done!

We assume therefore that it may be generally useful to normalize immittances, Z or Y. We combine this of course with normalization of the second fundamental quantity, (7.14-15). Now for suggestions on how to perform these normalizations effectively, let us look at the individual terms in the denominator of $T(s)$ in (7.13-1). The next term is

$$L_2 G_3 s = \frac{L_2}{R_3} s \tag{7.14-18}$$

which suggests measuring L_2 in units of R_3, just as (7.14-13) suggested measuring R_1 in units of R_3. If there is some convenient unit for measuring s also, say ω_0, the term can be written

$$\frac{L_2}{R_3}\omega_0 \frac{s}{\omega_0} = \frac{L_2 \omega_0}{R_3}\frac{s}{\omega_0} = L_{2n} s_n \tag{7.14-19}$$

in which the subscript n denotes a normalized quantity. We note that L_2 is actually normalized by measuring its sinusoidal-steady-state *reactance* at frequency ω_0 in units of R_3: $L_{2n} = (L_2 \omega_0)/R_3$. This is rather neat and simple, and of course dimensionless. We also note that the unit used for s is that same sinusoidal-steady-state frequency, $s_n = s/\omega_0$, also free of dimensions.

The next term, $R_1 C_1 s$, similarly "falls apart" on manipulation consistent with that above. Duality here suggests we use here G instead R; then

$$R_1 C_1 s = \frac{C_1}{G_1} s = \frac{C_1 \omega_0}{G_1}\frac{s}{\omega_0} = C_{1n} s_n \tag{7.14-20}$$

in which the same normalization was used of course for s—we have to be consistent! The C_1 normalization unit is the sinusoidal-steady-state *susceptance* of C_1 at ω_0, in units of G_1. But it is not consistent to use R_1 (or G_1) for normalizing C_1 when R_3 was used for normalizing L_2. Hence we rewrite (7.14-20) as

$$R_1 C_1 s = \frac{C_1 \omega_0}{G_3}\frac{R_1}{R_3}\frac{s}{\omega_0} = \frac{R_1}{R_3} C_{1n} s_n, \tag{7.14-21}$$

in which the normalization of C_1 is consistent with that of L_2. The ratio R_1/R_3 is a parameter that cannot be removed by normalization; see (7.14-13). *Either* R_1 or R_3 could well be used, of course, but consistency is essential.

Continuing to use R_3 and ω_0 as reference units, we have no difficulty in completing the normalization of (7.13-1):

$$\begin{aligned}
R_1C_3s &= \frac{R_1}{R_3}\frac{C_3\omega_0}{G_3}\frac{s}{\omega_0} = \frac{R_1}{R_3}C_{3n}s_n, \\
L_2C_3s^2 &= \frac{L_2\omega_0}{R_3}\frac{C_3\omega_0}{G_3}\left(\frac{s}{\omega_0}\right)^2 = L_{2n}C_{3n}s_n^{\,2}, \\
R_1C_1L_2G_3s^2 &= \frac{R_1}{R_3}\frac{C_1\omega_0}{G_3}\frac{L_2\omega_0}{R_3}\left(\frac{s}{\omega_0}\right)^2 = \frac{R_1}{R_3}C_{1n}L_{2n}s_n^{\,2}, \\
R_1C_1L_2C_3s^3 &= \frac{R_1}{R_3}\frac{C_1\omega_0}{G_3}\frac{L_2\omega_0}{R_3}\frac{C_3\omega_0}{G_3}\left(\frac{s}{\omega_0}\right)^3 = \frac{R_1}{R_3}C_{1n}L_{2n}C_{3n}s_n^{\,3}.
\end{aligned} \tag{7.14-22}$$

Note again that the use of R_3 as the impedance reference unit is arbitrary; R_1 could equally well have been chosen. Their *ratio* is all that is important and it cannot be removed.

On the value of ω_0 we have made no decision. We might use this remaining freedom of choice to "remove" or "unitize" the coefficient of $s_n^{\,3}$. That is, we could require

$$\frac{R_1}{R_3}C_{1n}L_{2n}C_{3n} = 1, \qquad \text{or} \qquad \omega_0^{\,3} = \frac{1}{(R_1/R_3)C_1L_2C_3}. \tag{7.14-23}$$

With this choice of reference frequency, and writing $r = R_1/R_3$, we have

$$T(s) = \frac{1}{(1+r) + (L_2 + rC_1 + rC_3)s + (L_2C_3 + rC_1L_2)s^2 + s^3} \tag{7.14-24}$$

in which all element values are normalized but the subscript n is dropped, as being no longer necessary. Such a choice for ω_0 requires, however, that we first calculate $T(s)$, or at least the s^3 coefficient of (7.14-23), as from the high-frequency asymptote. But we seek to normalize the elements *before* we analyze. A much more convenient method is to choose for ω_0 the frequency that makes the reactance of some element, say L_2, equal to the chosen impedance reference, say R_3. This is easily done at the very beginning of the analysis, and is numerically convenient. Then L_{2n} in (7.14-19) becomes unity. Or one could choose ω_0 so that $C_{1n} = (C_1\omega_0)/G_3$ becomes unity in (7.14-21). Or perhaps the element closest to R_3, C_3, is easiest to use: choose ω_0 so that $C_{3n} = (C_3\omega_0)/G_3 = 1$ and

$$\omega_0 = \frac{G_3}{C_3} = \frac{1}{R_3C_3}. \tag{7.14-25}$$

The choice of ω_0 (the reference value for frequencies) is arbitrary, as is the choice of R_0 (the reference value for impedances). If we follow (17.14-25), then both C_3 and R_3 take unit values upon normalization. The ratio r is now better written as R_1 (normalized value) and we have

$$T(s) = \frac{1}{(1 + R_1) + (L_2 + R_1C_1 + R_1)s + (L_2 + R_1C_1L_2)s^2 + (R_1C_1L_2)s^3} \tag{17.14-26}$$

in which the subscript n has been dropped, but all element values have been normalized to the basis (R_0, ω_0).

It will be numerically very convenient, and not at all confusing, so to normalize element values as soon as possible and then to work entirely with these normalized element values. The aim of normalization is, after all, to simplify. (A disadvantage is that dimensions disappear and dimensional checks are no longer possible; at this stage, however, one is working with numbers and the dimensions are not obvious in any event.) At the end of the analysis, appropriate corrections ("denormalizations") must of course be made.

We note that the original five literal parameters in (7.13-1), the five element values of the network, have been in effect reduced to three: in (7.14-24) the coefficients of s^0, s^1, and s^2 can be considered as the three remaining parameters; in (7.14-26) the parameters left are the normalized element values R_1, C_1, and L_2. Looking at it this way, we could use normalization to make two s-power coefficients unity, *or* to make one coefficient unity and normalize the $|T|$ scale, *or* to make two element values unity. At the same time the other element values, consistently normalized, though not themselves equal to unity of course, will become of the order of unity rather than of 10^{-7} or other unwieldy numbers. All amount to using the scales to make things simpler.

To recapitulate, *normalization* is the act of changing

a The scale of impedances to some convenient reference value of ohms, R_0 (in the discussion above, $R_0 = R_3$ and $R_{3n} = 1$), (7.14-27)

and

b The scale of frequency to some convenient reference value of radians per second, ω_0 [in the discussion above, $\omega_0 = (R_3 C_3)^{-1}$ so that $C_{3n} = 1$]. (7.14-28)

Since we deal so much with the function e^{st} whose argument is dimensionless, (7.14-28) must be equivalent to simultaneously normalizing the time scale so that st remains unchanged: $st = (s/\omega_0)(\omega_0 t) = s_n t_n$, $t_n = \omega_0 t$. The two domains are closely linked, after all: if we normalize one, the other must also be normalized. We shall return to this later.

Then all element values are normalized thus:

$$R_{\text{normalized}} = \frac{R_{\text{actual}}}{R_0},$$

$$L_{\text{normalized}} = \frac{L_{\text{actual}}\,\omega_0}{R_0}, \tag{7.14-29}$$

$$C_{\text{normalized}} = \frac{C_{\text{actual}}\,\omega_0}{G_0}, \qquad R_0 G_0 = 1.$$

And so we shall proceed from now on to choose reference values R_0 and ω_0 at the beginning of analysis, to record them for use in ultimately presenting the final results of the analysis, but to use only normalized element values during the actual analysis. The arithmetic work is simplified considerably by making all element values of the order of unity, and similarly simplifying all frequency values of interest; 10^{-4}, 10^{-11}, 10^6, 10^{13}, and other awkward numbers disappear, as does the troublesome 2π, for ω and f are the same when normalized:

$$\omega_n = \frac{\omega}{\omega_0} = \frac{2\pi f}{2\pi f_0} = \frac{f}{f_0} = f_n. \tag{7.14-30}$$

Since some resistance value is in effect made unity by normalization, and some frequency is also made equal to unity, the expression "to work on a one-ohm, one-radian-per-second basis" may be used. We may also speak of a "per unit" or "percent" basis since the work is carried out with reference to some convenient unit, or basis, of ohms and radians per second. The normalized element values may profitably be thought of and designated as quantities of "normalized ohms," "normalized farads," "normalized henrys," since it is still necessary to distinguish between resistance and conductance, reactance and susceptance, impedance and admittance—and without the usual powers of 10 it is no lônger numerically obvious which is which.

As an example, consider again the network of Fig. 7.13-A, of transfer function (7.13-1), with these element values (rounded off to two significant figures):

$$R_1 = R_2 = 600\ \Omega,$$

$$C_1 = C_3 = 0.066\ \mu\text{F}, \tag{7.14-31}$$

$$L_2 = 48\ \text{mH}.$$

(The symmetry here is not uncommon in real networks.) As impedance reference we (obviously) choose $R_0 = 600\ \Omega$. For frequency reference we might choose the frequency at which the reactance of L_2 is $600\ \Omega$:

$$f_0 = \frac{600}{2\pi(0.048)} = 1.99\ \text{kHz} \tag{7.14-32}$$

or the frequency at which the reactance of either capacitor is 600 Ω:

$$f_0 = \frac{1}{600(2\pi)(0.66 \times 10^{-6})} = 4.01 \text{ kHz}. \tag{7.14-33}$$

Since the uncertainty in element values is about 1 percent (7 parts in 660, or 5 parts in 480), we round these to 2 kHz and 4 kHz. Either will do nicely for f_0; since there are two capacitors and only one inductor, we choose (7.14-33). Then the basis, and the normalized element values, are:

$$\begin{aligned}
&f_0 = 4000 \text{ Hz}, \qquad \omega_0 = 2\pi(4000) \text{ rad/s}, \\
&R_1 = R_2 = 600/600 = 1\ \Omega \text{ (normalized)} \qquad (7.14\text{–}34) \\
&C_1 = C_3 = (0.066 \times 10^{-6})600(2\pi \times 4000) = 0.996 \to 1.00 \text{ F (normalized)}, \\
&L_2 = 48 \times 10^{-3} \times \frac{2\pi \times 4000}{600} = 2.01 \to 2.00 \text{ H (normalized)}.
\end{aligned}$$

The integral (when rounded consistently with the accuracy of the data) element values are not accidental, nor is the example oversimplified. (A simple yet very practical method of filter design gives exactly these values.) Figure 7.14-A*a* shows the normalized analysis problem. Different design methods may of course give different numerical element values. The network of Fig. 7.14-B shows a similar low-pass filter designed in a different way with the same (600 Ω, 4 kHz) normalization basis.

If the signal source is the combination (e_F, $R_1 = 1$) of Fig. 7.14-A*a*, then the transfer function from e_F to v, (7.13-1), is dimensionless and unaffected in value by normalization of the elements. Replacement of the voltage source by the

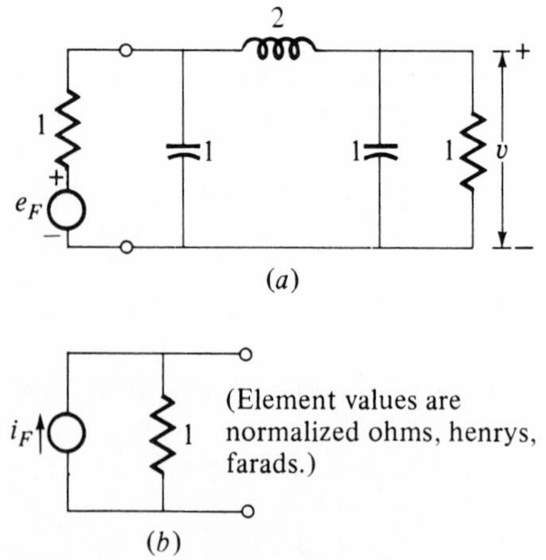

FIG. 7.14-A.

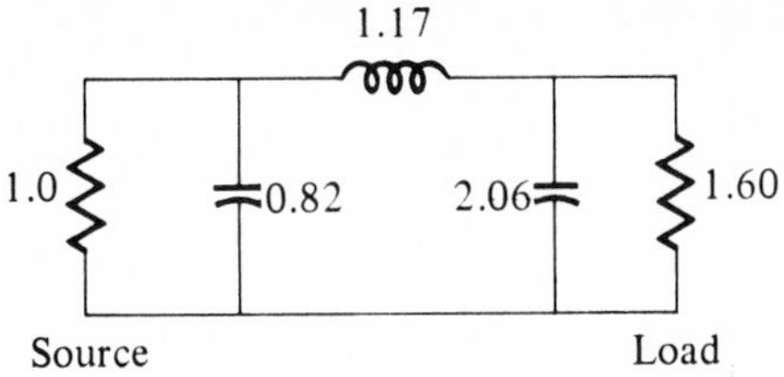

FIG. 7.14-B.

equivalent (Sec. 6.02) current source shown in Fig. 7.14-A*b* introduces dimensions into the transfer function from i_F to v. The only change in $T(s)$, however, is in its (constant) multiplier, by the incorporation of R_1:

$$T_{i_F \text{ to } v} = \left.\frac{v}{i_F}\right|_{\dots} = R_1 \left.\frac{v}{e_F}\right|_{\dots} = R_1 T_{e_F \text{ to } v}. \tag{7.14-35}$$

Normalization of this T, whose dimensions are those of impedance, to the value R_1 gives exactly the same transfer function:

$$\underset{\text{(normalized)}}{T_{i_F \text{ to } v}} = \frac{T_{i_F \text{ to } v}}{R_1} = T_{e_F \text{ to } v}. \tag{7.14-36}$$

Since the natural frequencies of the network are unaltered by the source exchange, and the zeros of transmission are unchanged, there is no *fundamental* change in the transfer function: it is merely a matter of the scale of $T(s)$. Hence in Fig. 7.14-B (and so, often, in the future) there is no real need to show the exact nature (v or i) of the signal (excitation) source: only its impedance is important! The (constant) multiplier is often unimportant. Its specific value can usually be determined by commonsense inspection; in Fig. 7.14-A, for example, the dc ($s = 0$) behavior of either transfer function is obviously 0.5; in Fig. 7.14-B it is 1.6/2.6 = 0.615.

Normalization (the scaling of element values, with corresponding change of scales of t and s) soon becomes automatic, a natural part of network analysis, certainly convenient and usually necessary for numerical comfort. There are no *real* difficulties associated with it; but there *is* a technical problem, usually ignored (and that successfully), that should be considered here. It has to do with dimensions.

In the transformation of the excitation wave $e(t)$ to its mate $E(s)$, a new dimension of time is acquired (Fig. 7.14-C). Since normalization scales time (inversely as it scales frequency s), a scale adjustment should here also be made in E. But, once the mate of the response, $R(s)$, has been found, the reverse transformation, that of $R(s)$ to $r(t)$, *also* adds a dimension, that of frequency (Fig. 7.14-C). Since the corresponding scale adjustment here (in going from s to t) is the inverse or reciprocal of the previous scale adjustment (that in going from t to s), the two adjustments in effect cancel and may be ignored. [Note that the transfer function $T(s)$ itself has either no dimensions or the dimensions

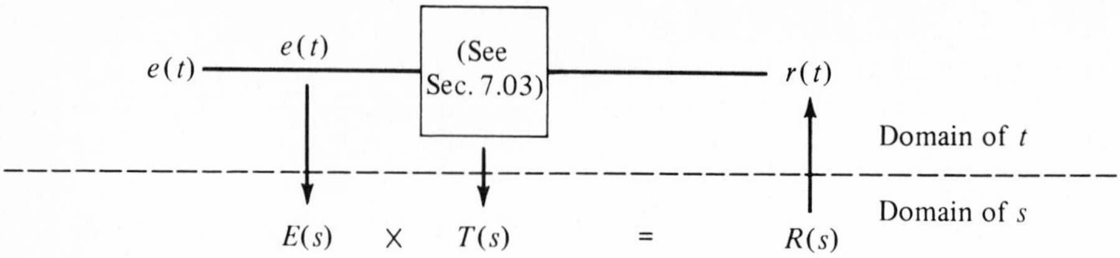

The wave-forming network view | Step a | The formal (integral tranformation) view

$\delta(t) \rightarrow$ [Wave-forming network] $\rightarrow e(t)$

$$E(s) = \int_0^\infty e(t)e^{-st}\,dt$$

Example: $e = E_0 e^{s_0 t}$ (Sec. 7.04)

$e = E_0 e^{s_0 t}$ (Sec. 7.05)

$$T_{WF} = \frac{E_0}{s - s_0} = E(s) \qquad\qquad E = E_0 \frac{1}{s - s_0}$$

In both schemes, note the dimensional transformation $[E] = [e][t]$ and the subsequent relation

$$R = E \times T = \text{rational function of } s = \frac{K_1}{s - p_1} + \ldots$$

Step b

$$r(t) = K_1 e^{p_1 t} + \ldots \qquad\qquad r(t) = \frac{1}{2\pi j}\int R(s)e^{ts}\,ds$$

(Sec. 7.12)

$$= K_1 e^{p_1 t} + \ldots \quad \text{(Sec. 8.08)}$$

In both schemes note the second dimensional transformation

$$[r] = [K_1] = [R][s] = [ET][s] = [e][t][T][s] = [eT]$$

in which the two transformations have canceled, dimensionally.

FIG. 7.14-C.

of Z or of Y; these are appropriate, wanted, cause no difficulty, and have nothing to do with the $s - t$ normalization.]

What this means is that we may use the normalized transfer function, regarded as a function of $s_n = s/\omega_0$, and excitation as a function of $t_n = \omega_0 t$, and response as a function of $t_n = \omega_0 t$, all exactly as though the subscript-n (normalized) variables were the real variables, without difficulty or concern. The only noticeable difference is in the special case of impulse excitation: there confusion about what a *unit* impulse is may arise because of the dimension of time in the multipliers Q_0 (coulombs or ampere-seconds) and Λ_0 (weber-turns or volt-seconds) in the current and voltage excitation functions $Q_0\,\delta(t)$ and $\Lambda_0\,\delta(t)$. But any error here will at the worst be in the form of an incorrect multiplier, an error easily found and corrected by mere commonsense observation and scaling, as exemplified above.

Our network examples from now on will often be given with normalized element values *ab initio*. But normalization and denormalization are important practical (yet not at all troublesome) steps, not to be forgotten.

We return now to our discussion of the numerical analysis procedures of analysis, in particular to step (*d*) of Sec. 7.11, the arithmetic of partial-fraction expansion.

7.15 NETWORK ANALYSIS: AN EXAMPLE, CONTINUED; SIGNAL DISTORTION

We now return to the analysis interrupted at the end of Sec. 7.13 for the introduction of normalization. Let the network of Fig. 7.13-A now have the (normalized) element values of Fig. 7.14-A. The round numbers therein entail only simple arithmetic, excellent for our first example because the trees of the arithmetic will not obscure the forest of the analysis. The transfer function is easily calculated directly from the schematic diagram, or we may take it from (7.13-1) or from (7.14-26):

$$T(s) = \frac{1}{2 + 4s + 4s^2 + 2s^3} = \frac{0.5}{s^3 + 2s^2 + 2s + 1}. \tag{7.15-1}$$

To check, we note that (*a*) as $s \to \infty$, $T(s) \to 1/(2s^3)$, both in (7.15-1) and (physically) on the schematic, and (*b*) $T(s) = 0.5$ from both sources. The second form in (7.15-1), incidentally, is convenient for setting forth the denominator of $R(s)$ ready for factoring.

We recapitulate the steps of analysis (Sec. 7.11): (7.15-2)

- **a** To transform $e(t)$ into $E(s)$ [see Table 7.13-A];
- **b** To "transform the network," i.e., to calculate the transfer function [done, see (7.15-1)];
- **c** To form $R(s) = E(s)T(s)$ (we need only use the results above);
- **d** To expand $R(s)$ in partial fractions, which in each case requires [see (7.12-6)]: (1) factoring the denominator of $R(s)$, and (2) the partial-fraction expansion proper [see (7.12-8)];
- **e** To check the (sometimes lengthy) numerical calculations of (*d*);
- **f** To denormalize, to present the results in the form of tables and plots, and discussion thereof (the "report").

Let us first consider the impulse response, so that $E(s) = \Lambda_0$ (volt-seconds), and

$$R(s) = \frac{0.5\Lambda_0}{s^3 + 2s^2 + 2s + 1} \qquad \text{(V·s).} \tag{7.15-3}$$

Note that if a current source (Fig. 7.14-A*b*) $i_F(t) = Q_0\,\delta(t)$ is considered instead of the voltage source, the only change is to replace Λ_0 by Q_0. We may conveniently let $K_0 = 0.5\Lambda_0$ *or* $0.5Q_0$, as the case may be; it is no longer necessary to distinguish between the two (the distinction is really not at all important in the analysis, though the dimensions do differ: volt-seconds and ampere-seconds, respectively).

Step (*d*), 1, to factor the denominator, is here, because of the simple element values, a simple matter: by trial we find $s = -1$ to be a pole and

$$s^3 + 2s^2 + 2s + 1 = (s + 1)(s^2 + s + 1) = (s - p_1)(s - p_2)(s - p_3) \tag{7.15-4}$$

in which

$$\begin{aligned} p_1 &= -1 + j0, \\ p_2 &= -0.5 + j0.5\sqrt{3}, \\ p_3 &= p_2^*, \end{aligned} \tag{7.15-5}$$

the star denoting the conjugate (see Fig. 7.15-A).

The expansion of $R(s)$, step (*d*), 2, is very simple. All poles are distinct and the degree of the numerator is less than the degree of the denominator. We first note, however, that $R(s)$ may conveniently be normalized by K_0; to carry the latter multiplier along would be a pointless nuisance. We therefore write

$$\begin{aligned} R_n(s) &= \frac{R(s)}{K_0} = \frac{1}{s^3 + 2s^2 + 2s + 1} = \frac{Q_N(s)}{Q_D(s)} \\ &= \frac{1}{(s + 1)(s + 0.5 - j0.5\sqrt{3})(s + 0.5 + j0.5\sqrt{3})} = \frac{1}{(s + 1)(s^2 + s + 1)} \\ &= \frac{K_1}{s - p_1} + \frac{K_2}{s - p_2} + \frac{K_3}{s - p_3} \\ &= \frac{K_1}{s + 1} + \frac{K_2}{s + 0.5 - j0.5\sqrt{3}} + * \qquad \text{(dimensionless)} \end{aligned} \tag{7.15-6}$$

in which we have followed (7.12-8) to indicate the partial-fraction expansion. The star indicates the whole term $[K_3/(s - p_3)]$; it corresponds in a conjugate way to the preceding term and need not be written, because it contains no

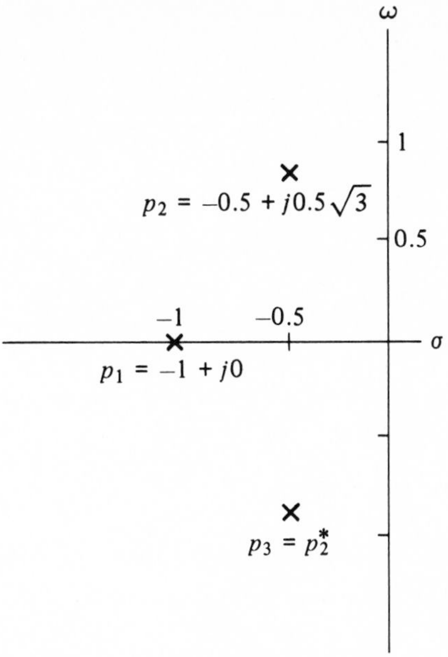

FIG. 7.15-A.

additional information. Using the formulas of (7.12-14) for the coefficients K_j of (7.15-6) gives

$$K_1 = [(s+1)R_n(s)]_{s=-1} = \left[\frac{1}{s^2+s+1}\right]_{s=-1} = 1$$

or

$$K_1 = \frac{Q_N(-1)}{Q_D'(-1)} = \left[\frac{1}{3s^2+4s+2}\right]_{s=-1} = 1. \qquad (7.15\text{-}7)$$

The two formulas of (7.12-14), both used in (7.15-7), give of course the same result; in this simple case there is little to choose between them. For the remaining coefficient we have

$$\begin{aligned} K_2 &= [(s+0.5-j0.5\sqrt{3})R(s)]_{s=p_2} \\ &= \frac{1}{(0.5+j0.5\sqrt{3})(j2\times 0.5\sqrt{3})} = \frac{1}{-1.5+j0.5\sqrt{3}} = \frac{1}{\sqrt{3}\underline{/\ 150^\circ}} \qquad (7.15\text{-}8) \\ &= 0.577\underline{/\ -150^\circ}. \end{aligned}$$

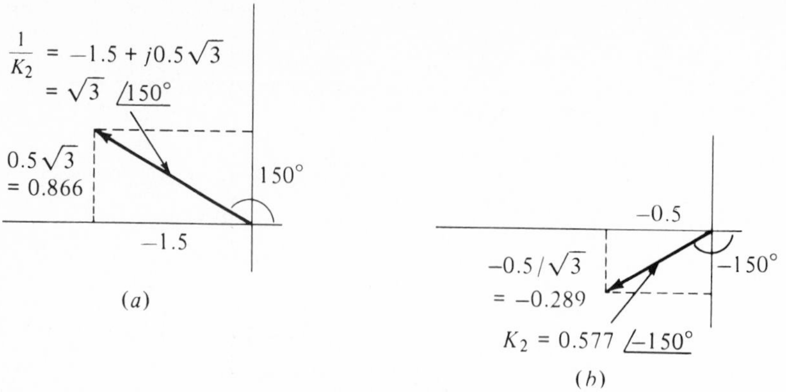

FIG. 7.15-B.

Figure 7.15-B shows both K_2^{-1} and K_2. Alternatively,

$$K_2 = \frac{Q_N(p_2)}{Q'_D(p_2)} = \frac{1}{3(-0.5 + j0.5\sqrt{3})^2 + 4(-0.5 + j0.5\sqrt{3}) + 2} \qquad (7.15\text{-}9)$$
$$= \frac{1}{-1.5 + j0.5\sqrt{3}} = 0.577\angle -150°,$$

the same result. Since $K_3 = K_2^*$, it need not be computed. We note again how, because of the very simple numerical element values, the calculation of the K's is in turn very simple. The first K is of course a real number; Fig. 7.15-B shows the (complex) second K. From these K's we form the impulse response $r_\delta(t)$. Technically we should set $K_0 = 0.5$, since r_δ is by definition the response to a *unit* impulse; it is, however, more convenient to leave the letter K_0 (which is not confusing and does exhibit dimensions), and to write

$$r_\delta(t) = K_0(K_1 e^{p_1 t} + K_2 e^{p_2 t} + *)$$
$$= K_0\{1.00e^{-t} + 2\mathbf{RE}[0.577\angle -150° \times e^{(-0.5 + j0.866)t}]\} \qquad (7.15\text{-}10)$$
$$= K_0[1.00e^{-t} + 1.154e^{-0.5t} \cos(0.866t - 150°)], \qquad t > 0.$$

The argument of the cosine in (7.15-10) is technically gibberish, since $0.866t$ is in radians and 150° in degrees: the units are inconsistent. It is simpler to leave this as it is, however, and to note that calculation of values of $r_\delta(t)$ must of course take account of the inconsistency and convert one or the other.

These arithmetical operations, simple as they are in this first example, need checking. For step (*e*) we use the initial-value tableau (Sec. 6.11) to calculate

$r_\delta(0+)$ and $r'_\delta(0+)$ from $R(s)$, (7.15-3): both are obviously zero. Then from (7.15-10) we find

$$\begin{aligned} r_\delta(0+) &= K_0[1.00 + 1.154 \cos(-150°)] \\ &= K_0(1.00 - 1.154 \times 0.866) \\ &= K_0(1.00 - 1.00) = 0.00 \end{aligned} \tag{7.15–11}$$

and

$$\begin{aligned} r'_\delta(0+) &= K_0\left\{-1.00e^{-t} + 1.154e^{-0.5t}\begin{bmatrix} -0.866 \sin(0.866t - 150°) \\ -0.5 \cos(0.866t - 150°) \end{bmatrix}\right\}_{t=0} \\ &= K_0[-1.00 + 1.154(0.866 \times 0.5 + 0.5 \times 0.866)] \\ &= K_0(-1.00 + 1.00) = 0.00, \end{aligned} \tag{7.15–12}$$

a satisfactory check. Recall again our previous experience (Sec. 6.09) that showed that at least *two* checks are essential: else some of the important calculations are not really verified. Note also that in differentiating the damped sinusoid, the format in the first line of (7.15-12) is both simple and convenient. We conclude from these checks that (7.15-10) is almost certainly correct.

It remains only to present this result in more graphic form, as in Fig. 7.15-C.

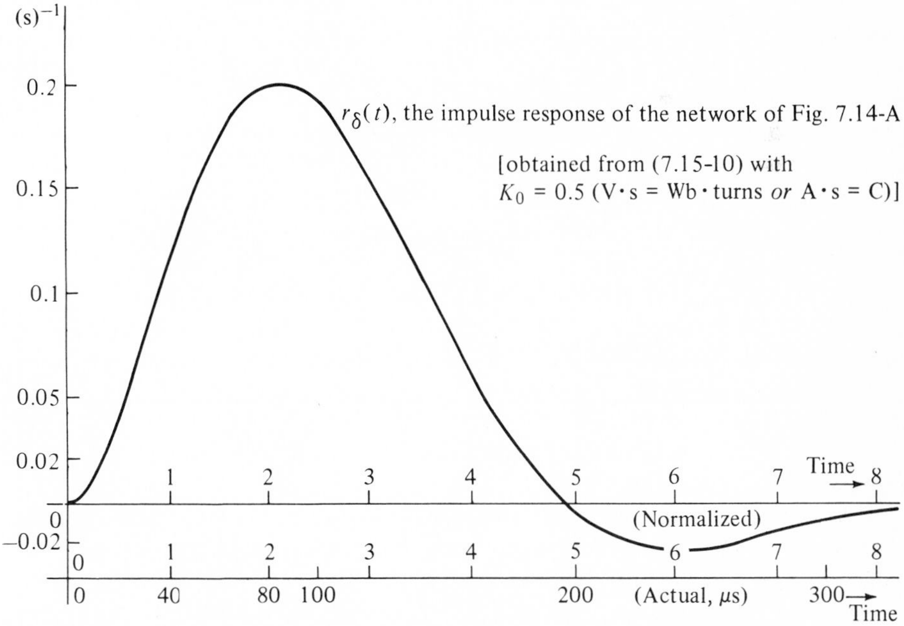

FIG. 7.15-C.

[Calculation of points to determine the plot, from (7.15-10), is entirely straightforward.] We must remember here that t in (7.15-10) and s in (7.15-1) are *normalized* variables. From (7.14-28) and the discussion immediately following it we recall that actual time is equal to normalized time divided by ω_0: $t_{\text{actual}} = t_{\text{normalized}}/\omega_0$. The t in (7.15-10) is normalized and therefore equal to $(\omega_0 t_{\text{actual}})$. But there is no need to rewrite (7.15-10) with actual time, nor any merit in so doing. It is simpler, more illuminating, to plot r_δ versus t from (7.15-10), as in Fig. 7.15-C, and then simply to relabel the time scale with actual values using

$$\begin{aligned} t_{\text{actual}} &= \frac{t_{\text{normalized}}}{\omega_0} = \frac{t_{\text{normalized}}}{2\pi \times 4000} \\ &= 39.8 t_{\text{normalized}} \qquad (\mu s). \end{aligned} \tag{7.15-13}$$

Within the accuracy of the plot, the actual microsecond values of time are simply 40 times the normalized values. Note that this "40" is *not* directly from the first two figures of "4000 Hz" chosen as the reference frequency. If we wished a simple $t_n : t$ ratio, such as 50 μs per unit normalized time, we should take

$$\omega_0 = \frac{t_n}{t} = \frac{1}{50 \times 10^{-6}} = 0.025 \times 10^6 \text{ rad/s}$$

or (7.15-14)

$$f_0 = \frac{\omega_0}{2\pi} = 3.18 \text{ kHz}.$$

instead of 4 kHz. Such a convenient ratio may be desirable of course; and if ω_0 is properly chosen, the *same* time scale may be used for normalized time (without dimensions) *and* actual time in microseconds, or milliseconds, or whatever time unit may be convenient. (A later example will illustrate this, in Sec. 7.19.)

Note that the response, the transfer function being a dimensionless ratio, is practically unaffected by normalization. Its constant multiplier K_0 may be evaluated as 0.5 as in Fig. 7.15-C, or the response may simply be left in terms thereof: it is the *shape* that is most important, and common sense usually suffices to evaluate the multiplier as necessary.

Whatever situation led originally to posing this network-analysis problem can now be reexamined and discussed; then the solution of the network problem will have served its purpose. We observe that the impulse excitation (the input) is distorted by transmission through the network into a finite, wide, rather sloppy pulse whose peak comes about 80 μs after the application of the impulse. The "tail" of the response, the portion after say $t = 200$ μs, is practically negligible, though it does of course, in principle go on forever. Physical corroboration is given in Fig. 6.15-B (for a network very much like this one, and for a more elaborate one of the same general nature), and in Fig. 6.17-B (the $C = 0$ case). Both should be reexamined in the light of our calculations and results here.

We also construct the initial-value tableau (see Sec. 6.10 and Fig. 6.11-A), using (7.15-1). From it (Table 7.15-A) we can write the power-series form of the impulse response,

$$r_\delta(t) = 0 + 0 + \frac{0.5t^2}{2!} - \frac{1.0t^3}{3!} + \frac{1.0t^4}{4!} - \frac{0.5t^5}{5!} + \cdots \qquad (7.15\text{-}15)$$

$$= 0.25t^2 - 0.167t^3 + 0.042t^4 - 0.004t^5 + \cdots.$$

TABLE 7.15-A The Initial-Value Tableau for the Impulse Response that Corresponds to the Transfer Function (7.15-1)

$$\begin{aligned} r &= 0 \\ r' + 2r &= 0 \\ r'' + 2r' + 2r &= 0.5 \\ r''' + 2r'' + 2r' + r &= 0 \\ r^{iv} + 2r''' + 2r'' + r' &= 0 \\ r^{v} + 2r^{iv} + 2r''' + r'' &= 0 \\ &\cdots\cdots \end{aligned}$$

$$\begin{aligned} r(0+) &= & &= 0 \\ r'(0+) &= 0 - 0 & &= 0 \\ r''(0+) &= 0.5 - 0 - 0 & &= 0.5 \\ r'''(0+) &= -2 \times 0.5 & &= -1.0 \\ r^{iv}(0+) &= -2 \times 0.5 - 2(-1) & &= 1.0 \\ r_\delta{}^{v}(0+) &= -0.5 - 2(-1) - 2(1) & &= -0.5 \\ &\cdots\cdots \end{aligned}$$

This is useful of course in checking the above [see (7.15-11) and (7.15-12)]; as a means of calculating $r_\delta(t)$ is is useful only for t up to about unity, though it technically converges for all values of t.

We have worked through this first example in great detail; the numbers are simple and it is easy to follow the calculations. In succeeding examples we shall compress the arithmetical details, but it may be advisable for the reader to carefully work them out for understanding.

Let us now consider the second excitation in Table 7.13-A and calculate the *step* response of the same network. If the excitation is $I_0 u(t)$ or $E_0 u(t)$ (see Fig. 7.14-A) we have for the steps of analysis (Sec. 7.11):

a $E(s) = \dfrac{I_0}{s}$ *or* $\dfrac{E_0}{s}$,

b $T(s) = \dfrac{0.5}{s^3 + 2s^2 + 2s + 1}$ from (7.15-1), (7.15-16)

c $R(s) = \dfrac{K_0}{s(s^3 + 2s^2 + 2s + 1)}$

in which $K_0 = 0.5I_0$ or $0.5E_0$. Note that, since the denominator of $R(s)$ is about to be factored, there is little point in multiplying it out in (7.15-16). We proceed to

d **1** The requisite factoring has already been done; (7.15-5) gives us the poles we need;

2 The partial-fraction expansion proper is

$$R_n(s) = \frac{R(s)}{K_0} = \frac{1}{s(s^3 + 2s^2 + 2s + 1)} = \frac{1}{s^4 + 2s^3 + 2s^2 + s} = \frac{Q_N(s)}{Q_D(s)}$$

$$= \frac{1}{(s)(s+1)(s + 0.5 - j0.5\sqrt{3})(*)} = \frac{1}{s(s+1)(s^2 + s + 1)}$$

$$= \frac{K_F}{s} + \frac{K_1}{s - p_1} + \frac{K_2}{s - p_2} + * \tag{7.15-17}$$

in which the asterisk denotes a term, or factor, of conjugate nature to that preceding and hence unnecessary of explicit expression. The coefficients are easily calculated:

$$K_F = [sR_n(s)]_{s=0} = 1$$

or

$$K_F = \frac{Q_N(0)}{Q'_D(0)} = 1,$$

$$K_1 = [(s+1)R_n(s)]_{s=-1} = -1$$

or

$$K_1 = \frac{Q_N(-1)}{Q'_D(-1)} = \frac{1}{-4+6-4+1} = -1,$$

$$K_2 = [(s + 0.5 - j0.5\sqrt{3})R_n(s)]_{s=p_1}$$

$$= \frac{1}{(-0.5 + j0.5\sqrt{3})(0.5 + j0.5\sqrt{3})(j2 \times 0.5\sqrt{3})}$$

$$= \frac{1}{[(-0.25 - 0.25 \times 3) + j(0.25\sqrt{3} - 0.25\sqrt{3})]j\sqrt{3}} = j\frac{1}{\sqrt{3}}$$

or

$$K_2 = \frac{Q_N(p_2)}{Q'_D(p_2)} = \frac{1}{(4s^3 + 6s^2 + 4s + 1)_{s=p_2}}$$

$$= \frac{1}{\begin{array}{l}4[(0.25 + 0.25 \times 3) + j(0.25\sqrt{3} - 0.25\sqrt{3})] \\ \quad + 6[(0.25 - 0.25 \times 3) - j0.5\sqrt{3}] + 4[-0.5 + j0.5\sqrt{3}] + 1\end{array}}$$

$$= j\frac{1}{\sqrt{3}}. \tag{7.15-18}$$

Then we can immediately write the step response (in which the numerical value of K_o is 0.5 but we continue to write K_0 for clarity and dimensional checking):

$$r_u(t) = K_0[K_F u(t) + K_1 e^{p_1 t} + K_2 e^{p_2 t} + *], \qquad t \geq 0,$$

$$= K_0\left[1 - e^{-t} + 2\,\mathbf{RE}\left(e^{-0.5t}e^{j0.5\sqrt{3}t}\frac{j}{3}\right)\right]$$

$$= K_0\left[1 - e^{-t} + \frac{2}{\sqrt{3}}e^{-0.5t}\cos(0.5\sqrt{3}t + 90^\circ)\right] \tag{7.15-19}$$

$$= K_0\left[1 - e^{-t} - \frac{2}{\sqrt{3}}e^{-0.5t}\sin(0.5\sqrt{3}t)\right].$$

For checking, step (e) of the analysis, we observe first by inspection of $R(s)$ in (7.15-16) that both $r_u(0+)$ and $r'_u(0+)$ should be zero; second, from (7.15-19) we compute

$$r_u(0+) = K_0(1 - 1 + 0) = 0,$$

$$r'_u(0+) = K_0\left[0 + 1 - \frac{2}{\sqrt{3}}\begin{pmatrix}0.5\sqrt{3}t \times 1 \\ -0\end{pmatrix}\right] = 0 \tag{7.15-20}$$

a good check. Note again, in passing, that $r_u(0+)$, because $\sin(0) = 0$, does not really check K_2; two checks (at least) are necessary!

To conclude the analysis, step (f), we construct Fig. 7.15-D. The sharp input step is distorted into a sluggishly rising response. The point of maximum rate of rise (approximately the halfway point) is at $t_n = 2.0+$. This is of course where the peak of the impulse response occurs in Fig. 7.15-C, for that is the derivative of the step response. We could in fact have obtained (7.15-19) by integrating (7.15-10); with $K_0 = 0.5$ we have

$$\begin{aligned} r_u(t) &= \int_0^t r_\delta(x)\,dx \\ &= \left\{ -e^{-x} + 2\,\mathbf{RE}\left[0.577\angle -150^\circ \frac{e^{(-0.5+j0.866)t}}{-0.5+j0.866} \right] \right\}_0^t \\ &= 1 - e^{-t} + 2\mathbf{RE}\left\{ \frac{0.577\angle -150^\circ}{1.0\angle 120^\circ} \left[e^{(-0.5+j0.866)t} - 1 \right] \right\} \qquad (7.15\text{-}21) \\ &= 1 - e^{-t} + 2\,\mathbf{RE}\; 0.577\angle -270^\circ\, e^{(-0.5+j0.866)t} - 1 \\ &= 1 - e^{-t} - 1.154e^{-0.5t}\sin 0.866t \qquad t \geqq 0 \end{aligned}$$

which is the same as (7.15-19). Alternatively, integration of (7.15-15) gives

$$r_u(t) = \frac{0.25t^3}{3} - \frac{0.167t^4}{4} + \frac{0.042t^5}{5} - \frac{0.004t^6}{6} + \cdots. \qquad (7.15\text{-}22)$$

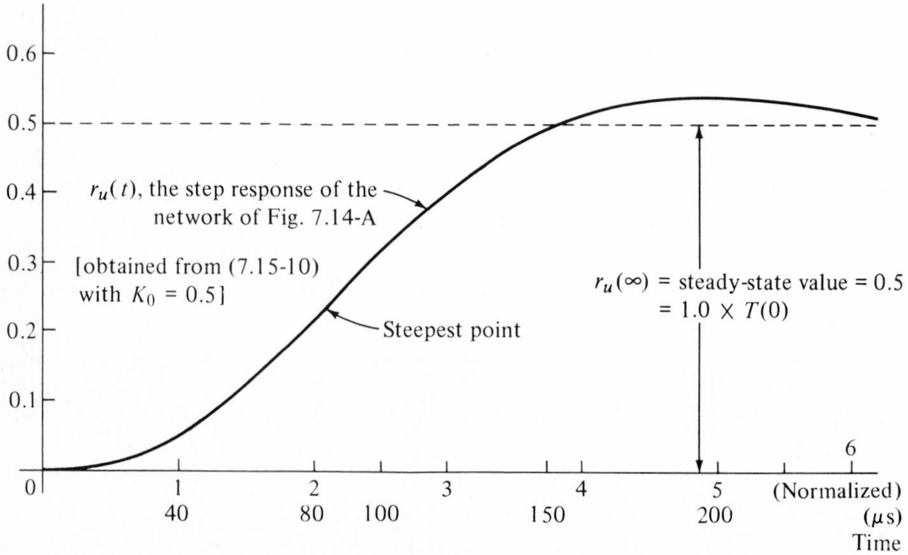

FIG. 7.15-D

TABLE 7.15-B The Initial-Value Tableau for the Step Response that Corresponds to the Transfer Function (7.15-1)

$$\begin{aligned} r &= 0 \\ r' + 2r &= 0 \\ r'' + 2r' + 2r &= 0 \\ r''' + 2r'' + 2r' + r &= 0.5 \\ r^{iv} + 2r''' + 2r'' + r' + 0 &= 0 \\ r^{v} + 2r^{iv} + 2r''' + r'' + 0 &= 0 \\ r^{vi} + 2r^{v} + 2r^{iv} + r''' + + 0 &= 0 \\ \cdots\cdots\cdots\cdots\cdots\cdots \end{aligned}$$

$$\begin{aligned} r_u(0+) &= 0 \\ r_u'(0+) &= 0 \\ r_u''(0+) &= 0 \\ r_u'''(0+) &= 0.5 \\ r_u^{iv}(0+) &= -1 \\ r_u^{v}(0+) &= 2 - 1 = 1 \\ r_u^{vi}(0+) &= -2 + 2 - 0.5 = -0.5 \\ \cdots\cdots\cdots\cdots\cdots\cdots \end{aligned}$$

The initial-value tableau for $r_u(t)$, obtained from (7.15-6), is given in Table 7.15-B. We note that it is (of course) very closely related to that of $r_\delta(t)$, Table 7.15-A, with $K_0 = 0.5$, and that the power-series form of $r_u(t)$ obtained therefrom,

$$\begin{aligned} r_u(t) &= \frac{0.5t^3}{3!} - \frac{t^4}{4!} + \frac{t^5}{5!} - \frac{0.5t^6}{6!} + \cdots \qquad t \geq 0 \\ &= 0.0833t^2 - 0.0417t^4 + 0.0083t^5 - 0.0007t^6 + \cdots \end{aligned} \tag{7.15-23}$$

is the same as (7.15-22). It is useful for values of t up to about 2.0 (because r_u rises more gently than does r_δ).

These (impulse and step) responses are typical for networks of the "low-pass" category, that is, networks with zeros at infinity and poles in the general vicinity of the origin (but none very close to the ω axis, no pronounced resonances). The impulse is distorted into a late-coming, rounded pulse; the step becomes a sluggish rise. Both responses exhibit minor ripples thereafter, around their ultimate (steady-state) values of zero and 0.5 respectively.

Since the transformed excitation $E(s)$ is considerably altered by multiplication by the transfer function of the network [for $T(s)$ is far from constant with s], the output *must* certainly differ from the input: the *distortion* is no surprise. In addition to Figs. 6.15-B and 6.17-B, Fig. 6.18-D shows an interesting physical situation: the network there is like the one we have just analyzed, but the terminating resistors are there purposely given inappropriate values; there result changes in behavior, and particularly distortion of a different sort. It is not easy to predict the nature of the distortion in any detail, but we shall endeavor to develop such ability as we can to predict its general nature. This is an important facet of network analysis: the ability to see globally as well as to make the detailed calculations of analysis.

Now if $T(s)$ is substantially constant for those values of s for which $E(s)$ is large, it should not matter greatly what $T(s)$ is where $E(s)$ is small: the waveshape distortion should then be "small." But if $T(s)$ varies greatly where $E(s)$ is large, then we should expect large distortion. Here is the genesis of an idea which will become more and more important: that there are "important" ranges of s and "less important" ranges of s. Let us examine the two examples in this light:

a For the impulse $E(s) = 1$ and all s are equally important;

b For the step $E(s) = 1/s$ and values of s near zero are the most important.

Of the low-pass network of Fig. 7.13-A, whose natural frequencies are near the origin [see (7.15-5) and Fig. 7.15-A] the transfer function $T(s)$ is more or less constant near the origin: the poles are near but not too near and they "hold T up" for small values of s. But T falls off rapidly for large s, asymptotically as $1/s^3$. Consequently the impulse, for whose mate *all* s are important, is badly distorted (Fig. 7.15-C); but the step, for whose mate those s near the origin are most important and large s are somewhat less important, is less severely distorted (Fig. 7.15-D).

The network's failure to transmit large-s or high-frequency signals well seems to be responsible for the distortion of the sharp, sudden-change features of an input: *all* of the impulse, and the *rise* part of the step. We see again a connection between high-frequency (large-s) network transfer-function behavior and the initial behavior of responses in time (see Secs. 6.06 to 6.11). On the other hand, the network's good transmission of low-frequency (small-s) signals is responsible for the good reproduction of the steady-state part of the step: there is a connection between the behavior of the transfer function at the origin and the ultimate (final) behavior of response in time (see Sec. 6.08).

This suggests that an excitation whose mate's important-s region is essentially the general vicinity of the origin, one for which the high-frequency (large-s) behavior region is much less important, should be less distorted. Such a wave, one with no steep rises, can also be a good illustration of the third type of excitation we are to consider (Table 7.13-A). We consider therefore this combination of two simple exponentials (Fig. 7.15-E):

$$e(t) = K_0(e^{-\alpha t} - e^{-\beta t}) = 3(e^{-0.5t} - e^{-0.6t}). \qquad (7.15\text{-}24)$$

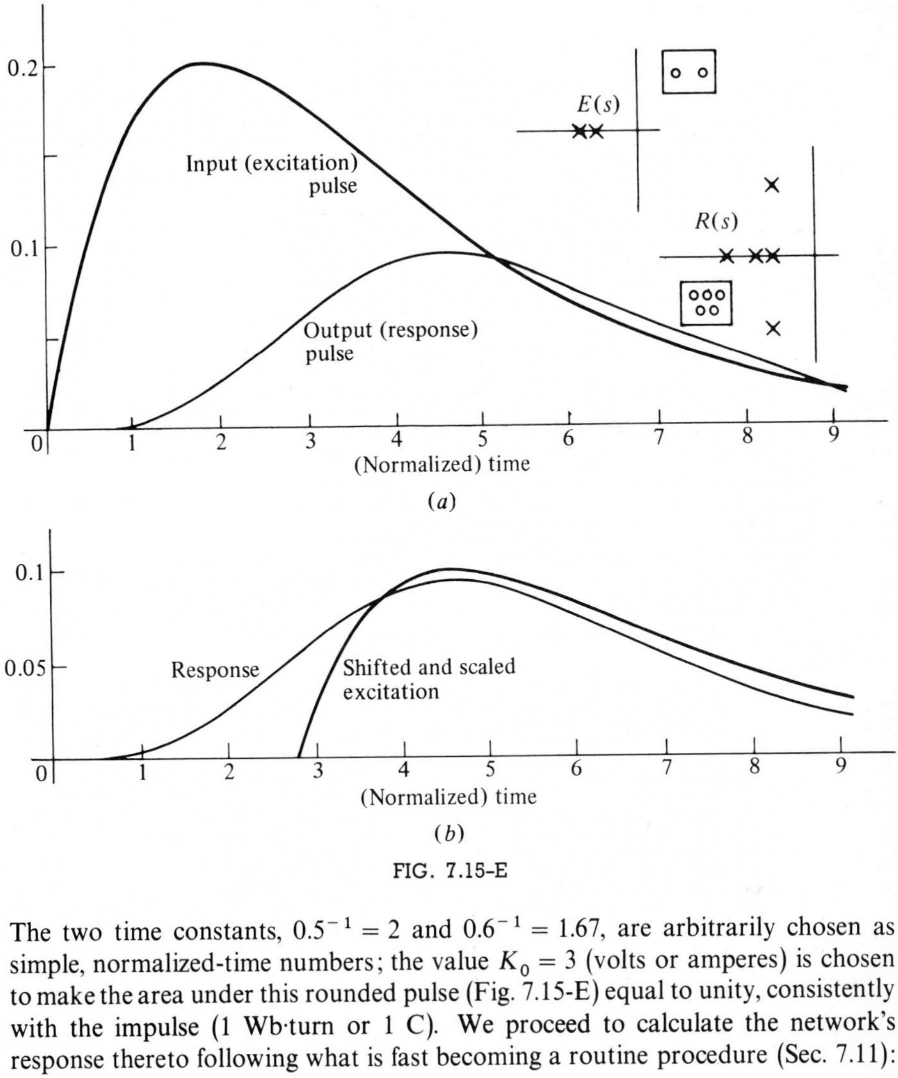

FIG. 7.15-E

The two time constants, $0.5^{-1} = 2$ and $0.6^{-1} = 1.67$, are arbitrarily chosen as simple, normalized-time numbers; the value $K_0 = 3$ (volts or amperes) is chosen to make the area under this rounded pulse (Fig. 7.15-E) equal to unity, consistently with the impulse (1 Wb·turn or 1 C). We proceed to calculate the network's response thereto following what is fast becoming a routine procedure (Sec. 7.11):

a $$E(s) = 3\left(\frac{1}{s+0.5} - \frac{1}{s+0.6}\right) = \frac{0.3}{(s+0.5)(s+0.6)}. \quad (7.15\text{-}25)$$

b $$T(s) = \frac{0.5}{s^3 + 2s^2 + 2s + 1} \quad \text{from (7.15-1).} \quad (7.15\text{-}26)$$

c $$R(s) = E(s)T(s) = \frac{(0.3)(0.5)}{(s^2 + 1.1s + 0.3)(s^3 + 2s^2 + 2s + 1)}$$
$$= \frac{0.15}{s^5 + 3.1s^4 + 4.5s^3 + 3.8s^2 + 1.7s + 0.3} = \frac{Q_N(s)}{Q_D(s)}. \quad (7.15\text{-}27)$$

The multiplied-out denominator form in (7.15-27) is convenient for constructing the initial-value tableau, for checking and for obtaining the power-series form of the response. The factoring operation will of course be performed on the individual parts of the denominator of $R(s)$, avoiding multiplication followed by pointless "unmultiplication." We continue with the partial-fraction expansion.

$$\mathbf{d} \quad R(s) = \frac{0.15}{(s+0.5)(s+0.6)(s-p_1)(s-p_2)(s-p_3)} = \frac{K_a}{s+0.5} + \frac{K_b}{s+0.6} + \frac{K_1}{s+1} - \frac{K_2}{s-p_2} + *. \qquad (7.15\text{-}28)$$

In step (d) the natural frequencies have been taken from (7.15-5) and the poles of $E(s)$ from (7.15-25). The evaluation of the K's begins thus:

$$K_a = [(s+0.5)R(s)]_{s=-0.5} = \frac{0.15}{(0.1)(0.5)(-j0.5\sqrt{3})(+j0.5\sqrt{3})} = 4.00. \qquad (7.15\text{-}29)$$

We note here that the second, differentiate-the-denominator method of computation can be simplified: instead of differentiating the fifth-degree denominator $(s^5 + 3.1s^4 + \cdots + 0.3)$, we need differentiate only the factor $(s^2 + 1.1s + 0.3)$. The reason is worth examining somewhat generally. We have the denominator partially factored in (7.15-27),

$$Q_D(s) = \underbrace{(s^2 + 1.1s + 0.3)}_{Q_{D1}}\underbrace{(s^3 + 2s^2 + 2s + 1)}_{Q_{D2}} \qquad (7.15\text{-}30)$$

and its derivative is

$$Q'_D(s) = Q_{D1}Q'_{D2} + Q'_{D1}Q_{D2}. \qquad (7.15\text{-}31)$$

Since

$$Q_{D1} = (s - s_a)(s - s_b) = (s+0.5)(s+0.6)$$

and

$$Q_{D2} = (s-p_1)(s-p_2)(s-p_3), \qquad (7.15\text{-}32)$$

one of the two terms in (7.15-31) vanishes at each of the poles of $R(s)$. To calculate K_a by the second method we therefore need differentiate only the Q_{D1} factor:

$$K_a = \frac{Q_N(s_a)}{Q'_D(s_a)} = \left.\frac{0.15}{(2s+1.1)(s^3+2s^2+1)}\right|_{s=-0.5} = 4.00. \qquad (7.15\text{-}33)$$

Similar simplification possibilities will often occur.

The remaining K's are

$$K_b = [(s+0.6)R(s)]_{s=-0.6} = \frac{0.15}{(-0.1)(0.4)(-0.1-j0.5\sqrt{3})(-0.1+j0.5\sqrt{3})}$$

$$= \frac{0.15}{(-0.1)(0.4)(0.01+0.75)} = -4.93 \tag{7.15-34}$$

or

$$K_b = \frac{Q_N(s_b)}{Q'_{D1}(s_b)Q_{D2}(s_b)} = \frac{0.15}{(-0.1)(0.304)} = -4.93. \tag{7.15-35}$$

$$K_1 = [(s+1)R(s)]_{s=-1} = \frac{0.15}{(-0.5)(-0.4)(-0.5-j0.5\sqrt{3})(*)} = 0.75$$

or

$$K_1 = \frac{Q_N(-1)}{Q_{D1}(-1)Q'_{D2}(-1)} = \frac{0.15}{(s^2+1.1s+0.3)(3s^2+4s+2)}\bigg|_{s=-1} = \frac{0.15}{(0.2)(1)} = 0.75. \tag{7.15-36}$$

$$K_2 = [(s-p_2)R(s)]_{s=p_2} = \frac{0.15}{(j0.5\sqrt{3})(0.1+j0.5\sqrt{3})(0.5+j0.5\sqrt{3})(j2\times 0.5\sqrt{3})}$$

$$= \frac{0.15\underline{/\ -90^\circ - 83.4^\circ - 60^\circ - 90^\circ}}{(0.5\sqrt{3})(0.01+0.75)^{1/2}(1)^{1/2}(2\times 0.5\sqrt{3})} \tag{7.15-37}$$

$$= 0.1147\underline{/\ 36.6^\circ}$$

or

$$K_2 = \frac{Q_N(p_2)}{Q_{D1}(p_2)Q'_{D2}(p_2)} = \frac{0.15}{(-0.75+j0.05\sqrt{3})(-1.5+j0.5\sqrt{3})}$$

$$= \frac{0.15}{-\sqrt{0.570}\,(-\sqrt{3.00})\underline{/\ (6.57^\circ + 30.0^\circ)}} \tag{7.15-38}$$

$$= 0.1147\underline{/\ 36.6^\circ}.$$

The numerical (and complex-number) calculations have of course been abbreviated above; the student should carefully work each one out, for there is no other way to understand them. For more complicated cases it will be appropriate to use an automatic digital computer; here, for the sake of understanding, it is essential to work out every detail oneself, by hand.

To conclude step (*d*), we write out, from (7.15-28) and the calculated *K*'s, the response

$$\begin{aligned} r(t) &= (4.00e^{-0.5t} - 4.93e^{-0.6t}) \\ &\quad + (0.75e^{-t} + 0.1147\underline{/\ 36.6^\circ}e^{-0.5t}e^{+j0.5\sqrt{3}t} + *) \\ &= [4.00e^{-0.5t} - 4.93e^{-0.6t}] + [0.75e^{-t} + 0.229e^{-0.5t}\cos(0.5\sqrt{3}t + 36.6^\circ)]. \end{aligned} \tag{7.15-39}$$

The brackets in (7.15-39) separate the forced and natural parts of r; we shall discuss this division in more detail in Sec. 7.16.

For checking, step (*e*), and for the power-series form of the response, we construct the initial-value tableau. From (7.15-27) and Fig. 6.11-A we obtain Table 7.15-C and the series

$$\begin{aligned} r(t) &= \frac{0.150t^4}{4!} - \frac{0.465t^5}{5!} + \frac{0.767t^6}{6!} + \cdots \\ &= 0.00625t^4 - 0.00388t^5 + 0.00107t^6 + \cdots \end{aligned} \tag{7.15-40}$$

TABLE 7.15-C The Initial-Value Tableau for (7.15-27), the Response to an Exponential Pulse

$$r = 0$$
$$r' + 3.1r = 0$$
$$r'' + 3.1r' + 4.5r = 0$$
$$r''' + 3.1r'' + 4.5r' + 3.8r = 0$$
$$r^{iv} + 3.1r''' + 4.5r'' + 3.8r' + 1.7r = 0.15$$
$$r^{v} + 3.1r^{iv} + 4.5r''' + 3.8r'' + 1.5r' + 0.3r = 0$$
$$r^{vi} + 3.1r^{v} + 4.5r^{iv} + 3.8r''' + 1.5r'' + 0.3r' = 0$$
$$\cdots\cdots$$

$$\begin{aligned} r(0+) &= 0 \\ r'(0+) &= 0 \\ r''(0+) &= 0 \\ r'''(0+) &= 0 \\ r^{iv}(0+) &= 0.15 \\ r^{v}(0+) &= -3.1 \times 0.15 = -0.465 \\ r^{vi}(0+) &= -3.1 \times (-0.465) - 4.5 \times 0.15 = +0.767 \\ &\cdots\cdots \end{aligned}$$

which is evidently usable, taking only a reasonable number of terms, from $t = 0$ to about $t = 2$. To check our calculations, we calculate, from (7.15-39),

$$r(0+) = 4.00 - 4.93 + 0.75 + 0.23 \cos 36.6^\circ = 4.93 - 4.93 = 0.00 \quad (7.15\text{-}41)$$

to be compared with the initial value (zero) of Table 7.15-C, and

$$\begin{aligned} r'(0+) &= -2.00 + 2.96 - 0.75 - 0.23(-0.5\sqrt{3} \sin 36.6^\circ - 0.5 \cos 36.6^\circ) \\ &= -2.96 + 2.96 = 0.00 \qquad (7.15\text{-}42) \end{aligned}$$

to be compared with the initial value (also zero) of Table 7.15-C. The three-significant-figure (slide-rule precision) checks are satisfactory. [It is not unreasonable that additional checks (of r'', etc.), or more precise ones here, should require more than slide-rule precision.]

Finally, as step (f), we plot the result [the response $v(t)$ or $r(t)$] in Fig. 7.15-Ea. As in the other cases, the response is distorted: it is generally flattened and broadened. But, as we expected because of the zeros of $E(s)$ at infinity (the lesser importance of large values of s), the distortion here can genuinely be considered less severe. In Fig. 7.15-Eb the response is reproduced, and a shifted and scaled version of the excitation superposed. Except for the sluggish initial rise and the "delay," the response evidently does follow the excitation rather well. It appears that the "low-pass, high-stop" character of the network inevitably causes some distortion of the beginning of a waveform. But the distortion seems less for more smoothly beginning waves. The reduction of amplitude from excitation to response, approximately 50 percent here, is not necessarily important. There may be compensating amplification in the system, or, if there is a dimension change from excitation to response (from voltage to current for example), the amplitudes cannot be directly compared.

To the basic purpose of the network in the system in which it operates, these responses may of course be unimportant. If its purpose is to discourage the passage of high-frequency sinusoidal signals, for example, and to encourage the passing of low-frequency sinusoids, these "transient" calculations may be interesting but inconsequential. It depends on a global (large-scale) system analysis. But calculation of the network response to a given excitation is a vital, essential step, and our immediate concern.

Figure 7.15-F shows the response of the same network to an even rounder, smoother pulse, one made up of three exponentials, whose initial slope is zero (whose mate has three zeros at infinity) (Problem 7-K). We can see, (after arbitrarily shifting and scaling the excitation for comparison purposes), how the response here resembles the excitation even more than in previous examples. This is additional evidence that, generally, "the more of $E(s)$ is at low frequencies, and the less at infinity," the better is the (low-pass) network's response. This statement lacks something in formal exactitude, perhaps, but it is very informative!

There remains for study (Table 7.13-A) the case of sinusoidal excitation.

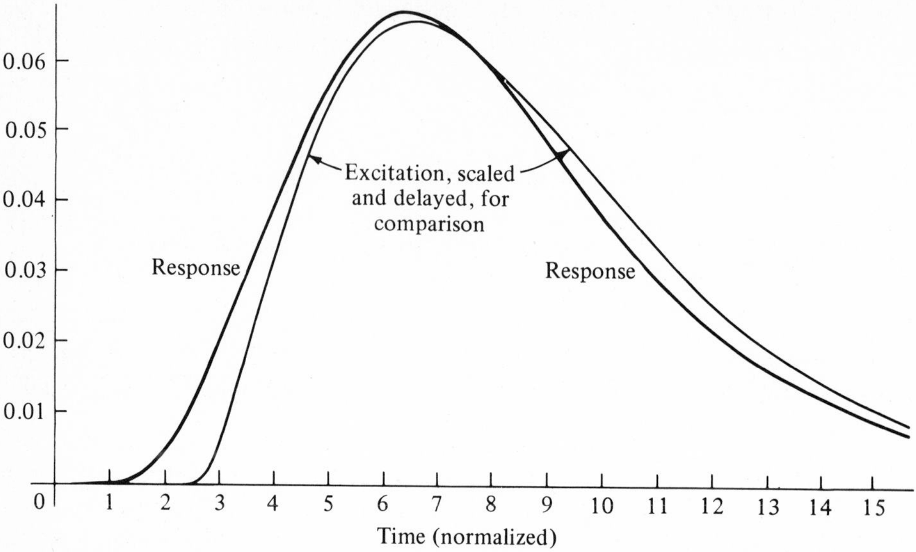

FIG. 7.15-F

We do this rapidly, having a standardized procedure now (Sec. 7.11). If the excitation is

$$e(t) = E_m \cos(\omega_F t + \phi) \tag{7.15-43}$$

the analysis is

$$\begin{aligned}
&\textbf{a} \quad E(s) = E_m \frac{(\cos\phi)s - (\sin\phi)\omega_F}{s^2 + \omega_F^2}, \\
&\textbf{b} \quad T(s) = \frac{0.5}{s^3 + 2s^2 + 2s + 1}, \\
&\textbf{c} \quad R(s) = \frac{0.5E_m[(\cos\phi)s - (\sin\phi)\omega_F]}{(s^2 + \omega_F^2)(s^3 + 2s^2 + 2s + 1)}, \\
&\textbf{d} \quad R(s) = \frac{K_a}{s - s_a} + * + \frac{K_1}{s - p_1} + \frac{K_2}{s - p_2} + *, \\
&\textbf{e} \quad r(t) = [K_a e^{s_a t} + *] + [K_1 e^{p_1 t} + K_2 e^{p_2 t} + *].
\end{aligned} \tag{7.15-44}$$

The evaluation of the K's is straightforward and somewhat tedious; we omit it here. (The value of s_a is of course $+j\omega_F$; the asterisks indicate terms conjugate to those immediately preceding, and unnecessary of expression.) We note that the brackets immediately above have separated the response into its forced and natural components and that the former is of course a sinusoidal wave:

$$r(t) = 2\,\mathbf{RE}\,(K_a e^{j\omega_F t}) + \sum_{k=1}^{k=3} K_k e^{p_k t} = 2|K_a| \cos(\omega_F t + \phi') + r_N(t) \quad (7.15\text{-}45)$$

in which $\phi' = (\phi + \text{angle of } K_a)$ and r_N is the natural part of $r(t)$.

Step (*e*) we omit the details of, except for the power-series form of the response,

$$r(t) = 0 + 0 + 0 + \frac{(\cos\phi)t^3}{3!} + \frac{(-\omega_F \sin\phi - 2\cos\phi)t^4}{4!} + \cdots. \quad (7.15\text{-}46)$$

Presentation of the results encounters a new difficulty: there are several parameters, whose numerical values affect the shape of the response, a complete discussion of which would require a lengthy report.

The forced part of the response is of course a sinusoid of the same frequency as the excitation; its amplitude and phase relation thereto depend on the value of ω_F (through the transfer function), and it continues indefinitely, being "steady-state" in character. Figure 7.15-G shows how the transfer function varies with

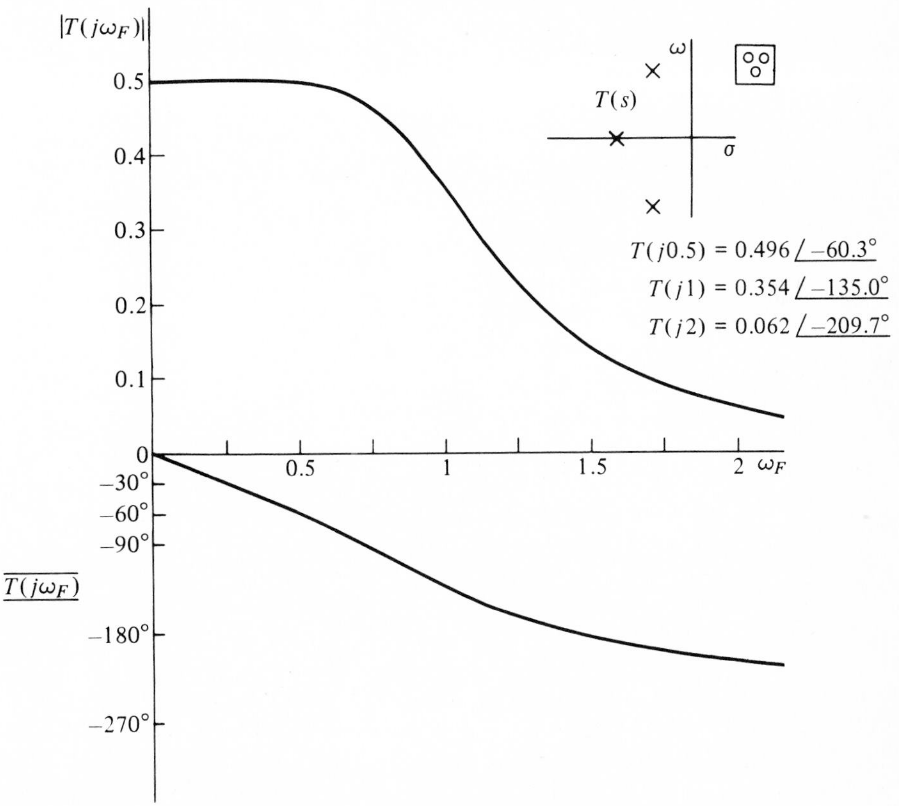

FIG. 7.15-G. Magnitude and angle of $T(j\omega_F)$ versus ω_F.

the excitation frequency in magnitude and angle; the low-pass character of the network is evident.

The natural part consists again of three decaying exponentials (two of them complex), whose amplitudes depend on the value of ω_F, and on the value of the initial phase of the excitation, ϕ. As general indications of the manifold possibilities for response waveshape, we examine the responses for three cases, in all of which E_m has been taken as unity, for simplicity.

Figure 7.15-H shows the responses, from rest, to the three different sinusoidal excitations. The (different) time scales are omitted, and the purely mechanical calculations and printing have been done by a computer, according to a program to be described in Sec. 7.18. [Our interest here is more in the results than in the computational details, the tedious evaluation and plotting of (7.15-44) or (7.15-45).]

For the response shown in Fig. 7.15-H*a* the excitation frequency is rather low ($\omega_F = 0.5$), and the initial phase is zero (the excitation is $\cos \omega_F t$); the response

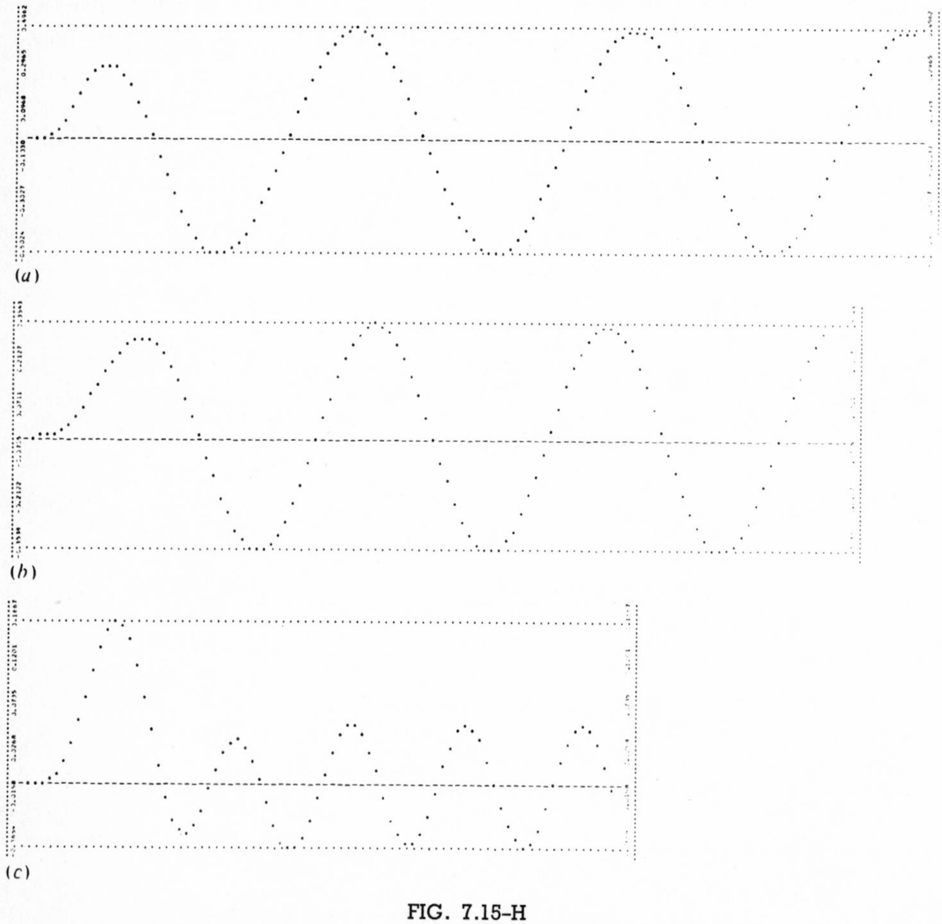

FIG. 7.15-H

shows a transient period of buildup to the steady state, which is a sinusoid of about one-half the amplitude of the excitation, with a phase lag of about 60°. In (*b*), where $\omega_F = 1$ and $\phi = -45°$, there is very little transient behavior, the steady state sets in almost immediately (of amplitude about one-third the excitation's and phase relative to the excitation of $-135°$). In (*c*), where $\omega_F = 2$ and $\phi = -90°$ (the excitation is $\sin \omega_F t$), there is a rather large transient loop in comparison with the steady state, which is only 6 percent of the excitation in size and lags some 210° (or leads about 150°). Infinite variation is possible; these examples suffice for illustration.

What does deserve more attention here, in the case of excitation which is sinusoidal, is the role of the transfer function, and the separate calculation of the forced and natural parts. To these we turn in Sec. 7.16.

7.16 FORCED AND NATURAL PARTS OF THE RESPONSE TO SINUSOIDAL EXCITATION

"*Every excitation can be considered to be a sum of exponentials; nay, every excitation can be considered to be a sum of sinusoids.*" Such a statement is not uncommon; some engineers make wide use of this "fact." There must then be some substance to it, even though we have not proved any such thing. (That the statement is true, subject to proper interpretation of the words, is to be demonstrated elsewhere.)

The voltage and current waveshapes in most electric power systems are practically sinusoidal; a sustained musical tone is a sinusoid together with a few (very important) overtones; communication systems often use pulse trains that may in effect be sums of a few sinusoids, as in Fig. 7.16-A. We can safely assume that *sinusoidal* excitation is at least extremely important, and perhaps something more.

A pulse train made up of several sinusoids:

$$(\cos \omega t)^6 = \frac{1}{32}[10 + 15\cos(2\omega t) + 6\cos(4\omega t) + \cos(6\omega t)]$$

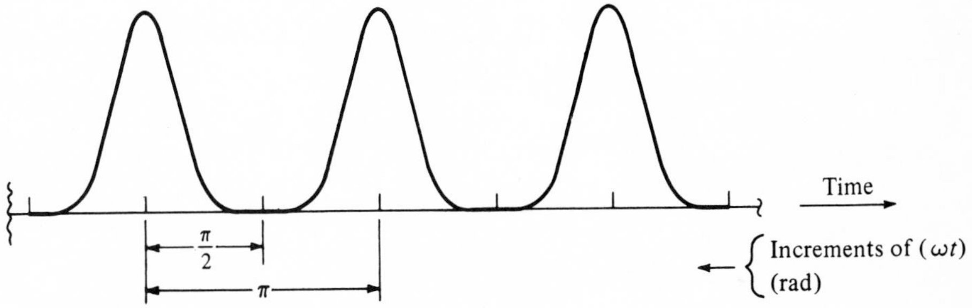

FIG. 7.16–A

The examples of Sec. 7.15 clearly show the following two characteristics of the response to sinusoids:

a $r_F(t)$, the forced component, is a steady (permanent) sinusoid of the same frequency as the excitation wave, usually with a different initial phase; its amplitude, if it has the same dimensions as the excitation, may well be different (if the dimensions differ, one cannot compare amplitudes); (7.16-1)

b $r_N(t)$, the natural component, is a sum of (damped) exponentials, transient (nonpermanent) in character. (7.16-2)

Obvious though these properties should seem by now (given our postulates), they are worth more discussion. In particular the following question arises: "Corresponding to the time-domain decomposition $r = r_F + r_N$, is there a simple decomposition in the s domain, $R = R_F + R_N$?" There must of course be some such decomposition, obtainable (if in no other way) by transforming r_F and r_N individually back to the s domain. Since this process merely retraces a path already trod, it is not necessary: we can look at the "standard" straightforward decomposition in the s domain (as in the examples of Sec. 7.15). But still we ask: "Is there an easier way? Can the work be compressed, made more illuminating, put in more useful form?" One might expect R_F to come simply from E and T: in fact it cannot be greatly different from the simple product $E(s)T(s)$ because a sinusoid is exponential, and the exponential is that function for which the transfer function was invented! One might also expect R_N to be simply related to the impulse response (the natural behavior of the network, par excellence) with some simple correction for the fact that the excitation is sinusoidal and not impulsive. That all the "answers" are "yes," we have in effect already shown. We need simply recapitulate carefully, and codify.

Let us repeat the sinusoidal-excitation analysis in quite general, literal form. It is not difficult and it is very informative. A general sinusoidal excitation is

$$e(t) = E_m \cos(\omega_F t + \phi). \tag{7.16-3}$$

Then (pair 22)

$$E(s) = \frac{E_m[(\cos\phi)s - (\sin\phi)\omega_F]}{s^2 + \omega_F{}^2} \tag{7.16-4}$$

and (see Fig. 7.11-A)

$$R(s) = E(s)T(s)$$

$$= \frac{E_m[(\cos\phi)s - (\sin\phi)\omega_F]}{s^2 + \omega_F{}^2} \times \frac{A_0 s^{N-1} + \cdots + A_{N-1}}{s^N + B_1 s^{N-1} + \cdots + B_N} \tag{7.16-5}$$

in which we have used for $T(s)$ the usual rational form, though we keep $M < N$ for simplicity and we continue to assume that all the natural frequencies are distinct. Then (numerators are not important here—the focus is on the denominators, the poles)

$$R(s) = \frac{\cdots}{(s - j\omega_F)(s + j\omega_F)\,(s - p_1)(s - p_2)\cdots(s - p_N)} \tag{7.16-6}$$

$$= \frac{K_a}{s - j\omega_F} + \frac{K_a^*}{s + j\omega_F} + \frac{K_1'}{s - p_1} + \cdots + \frac{K_N'}{s - p_N},$$

the usual partial-fraction expansion. The decomposition of $R(s)$ into a "forced part" $R_F(s)$, from which comes $r_F(t)$, and a "natural part" $R_N(s)$, from which comes $r_N(t)$, is obvious: the former must consist of the first two terms of (7.16-6), the latter of the N remaining terms. Each of these deserves more investigation.

In the *forced* part the principal quantity is

$$\begin{aligned} K_a &= [(s - j\omega_F)E(s)T(s)]_{s=j\omega_F} \\ &= \left.\frac{E_m[(\cos\phi)j\omega_F - (\sin\phi)\omega_F]T(j\omega_F)}{s + j\omega_F}\right]_{s=j\omega_F} \\ &= \frac{E_m}{2}(\cos\phi + j\sin\phi)T(j\omega_F) \\ &= \frac{E_m}{2}e^{j\phi}T(j\omega_F) \\ &= \tfrac{1}{2}\bar{E}T(j\omega_F), \end{aligned} \tag{7.16-7}$$

$\bar{E}$ being the phasor representation of the excitation, $\bar{E} = E_m\angle\phi$. Hence the "forced part" of $R(s)$ is

$$R_F(s) = \frac{K}{s - j\omega_F} + \frac{K_a^*}{s + j\omega_F} = \frac{1}{2}\frac{\bar{E}T(j\omega_F)}{s - j\omega_F} + * \tag{7.16-8}$$

to which corresponds

$$\begin{aligned} r_F(t) &= K_a e^{j\omega_F t} + * = \tfrac{1}{2}\bar{E}T(j\omega_F)e^{j\omega_F t} + * \\ &= \mathbf{RE}[\bar{E}T(j\omega_F)e^{j\omega_F t}] \\ &= E_m T_F \cos(\omega_F t + \phi + \beta) \end{aligned} \tag{7.16-9}$$

in which the evaluation of the transfer function at the excitation frequency is conveniently symbolized by the resulting magnitude T_F and angle β:

$$T(j\omega_F) = T_F\angle\beta. \tag{7.16-10}$$

If the network is at all complicated, the actual calculation of the values of T_F and β may be laborious, but it is straightforward. The formal expression of $R_F(s)$, the mate of the sinusoid $r_F(t)$, is less useful than the equivalent in phasor terminology:

> The phasor representation of $r_F(t)$ is $2K_a$ or $\bar{E}T(j\omega_F)$; it is equal to the product of (*a*) the phasor representation of the excitation and (*b*) the transfer function evaluated at the frequency of excitation, $T(j\omega_F)$. (7.16-11)

This simple result is in no way a surprise: sinusoids are exponential, and the transfer function was invented to represent the forced response to exponential excitation. It is merely a generalization of the complex-number Ohm's law. (See Secs. 2.15, 3.10, 5.11.)

But its utility is much greater than we can appreciate here, (*a*) because it vastly simplifies routine calculations of sinusoidal response of networks, and (*b*) because of the possibility of expressing *any* excitation in sinusoidal form. We note again, in résumé, [see (7.16-1)] that

> The forced component of the response to a sinusoidal excitation is a sinusoid of the same frequency, of amplitude equal to the excitation amplitude multiplied by the transfer function magnitude, and of angle (initial phase) equal to the excitation angle added to the transfer-function angle. (7.16-12)

We can say loosely that

> "The forced (steady-state) response is equal to the input times the transfer function," (7.16-13)

provided of course that we carefully interpret the terms. This expression is useful because it is so concise; it points to the *basic* action. The accurate expression is (7.16-11) which, rewritten in symbols and using $\bar{R}_F$ for the phasor representation of $r_F(t)$, is

$$\begin{aligned} \bar{R}_F = \bar{E}T(j\omega_F) &= E_m\underline{/\phi}\ T_F\underline{/\beta} \\ &= E_m T_F\underline{/\ (\phi+\beta)}. \end{aligned} \tag{7.16-14}$$

Figure 7.16-B diagrams the process.

Should one be interested only in the forced part of the response, i.e., only in the network's transmission of sinusoids (a very common situation, we shall see, for the reasons above), it appears that the calculation would be simpler if performed in logarithmic terms: multiplication of $\bar{E}$ and $T(j\omega_F)$ would be re-

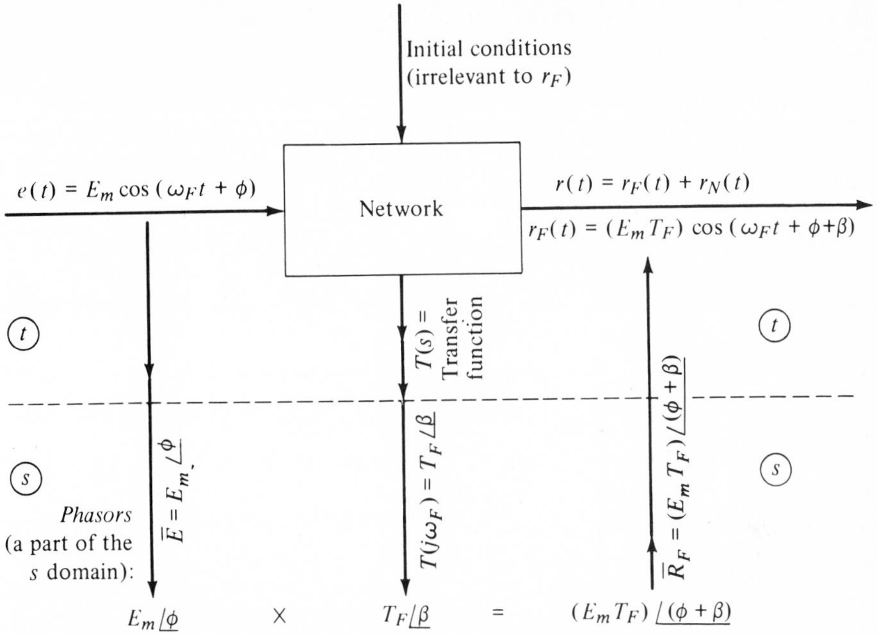

FIG. 7.16-B. Calculation of the forced response of a network to sinusoidal excitation (transmission of a sinusoidal signal through a network).

placed by simple addition of their logarithms. This device is very useful, but it requires careful attention to dimensions and units, discussion of which we defer (see Appendix D).

Note that the calculation of the forced component does not in any way depend on our restrictions imposed on $T(s)$; it is equally valid for any $T(s)$, whatever the relation therein between M and N, whatever the disposition of its poles, etc.

Now let us look at the *natural* part. It is not, of course, obtainable from the excitation by a simple phasor calculation. But it *is* simply calculable from the excitation and the network's impulse response. The principal items in the natural part, from (7.16-6), are the

$$\begin{aligned} K_j' &= [(s - p_j)E(s)T(s)]_{s=p_j} \\ &= E(p_j)[(s - p_j)T(s)]_{s=p_j} \\ &= E(p_j)K_j \end{aligned} \tag{7.16-15}$$

in which K_j (without the prime) is the usual "K_j" in the network's impulse response, found from the partial-fraction expansion of the transfer function,

$$T(s) = \sum_{j=1}^{N} \frac{K_j}{s - p_j}. \tag{7.16-16}$$

From

$$R_N(s) = \sum_{j=1}^{N} \frac{K'_j}{s - p_j} \tag{7.16-17}$$

we obtain

$$\begin{aligned} r_N(t) &= \sum_{j=1}^{N} K'_j e^{p_j t} \\ &= \sum_{j=1}^{N} E(p_j) K_j e^{p_j t}. \end{aligned} \tag{7.16-18}$$

This representation of r_N in terms of the excitation and the network's impulse response is somewhat like the representation of r_N in terms of the excitation and the network's transfer function: (7.16-15) resembles (7.16-7). But in calculating the natural part we have to work *term by term*, we cannot treat it as a whole. This requires that we calculate *each* K'_j individually. (The same is technically true for r_F, but there in effect we have only *one* term; the second term is merely its conjugate.)

The reason lies in *the role of* r_N: its function is to complement r_F to form an expression for the condition of the network at $t = 0+$; r_F alone does not and cannot correctly describe it, except in very unusual cases. It is for r_N to make up the differences between the *actual* $t = 0+$ conditions (determined physically by the network's initial condition of rest) and the initial values expressed by $r_F(0+)$. In mathematical language, it is to match the solution to the boundary conditions. And it is then for r_N to die gracefully away and leave r_F to describe the (steady-state) behavior. This requires that r_N do a number of things, depending on how large N is and on how closely $r_F(t)$ comes to describing, at $t = 0+$, the initial-rest condition of the network. Each term in r_N acts differently, as may be necessary. Independent calculation of r_N is therefore not possible: it is necessary to calculate r_F first, then r_N. The formula (7.16-15) does not contradict this statement; it merely conceals the preliminary calculation of r_F in the factor $E(p_j)$, which clearly depends on the excitation and, together with K_j, somehow implies what r_F is.

For examples we need only turn back to Sec. 7.15. The curves of Fig. 7.15-H*a* and *c* show rather large transient (r_N) components, the curve of Fig. 7.15-H*b* a small one (there the steady state is almost immediately entered). These examples are for differing frequencies of excitation. Figure 7.16-C shows the responses of the same network for three different values of the initial excitation phase ϕ (0, $-20°$, $-90°$, respectively), all for the same excitation frequency, $\omega_F = \sqrt{2}$, and all to the same time scale. We note again the wide range of possibilities, due to the differing initial values of the third derivative of the response, proportional to $\cos \phi$. For a certain initial phase of excitation

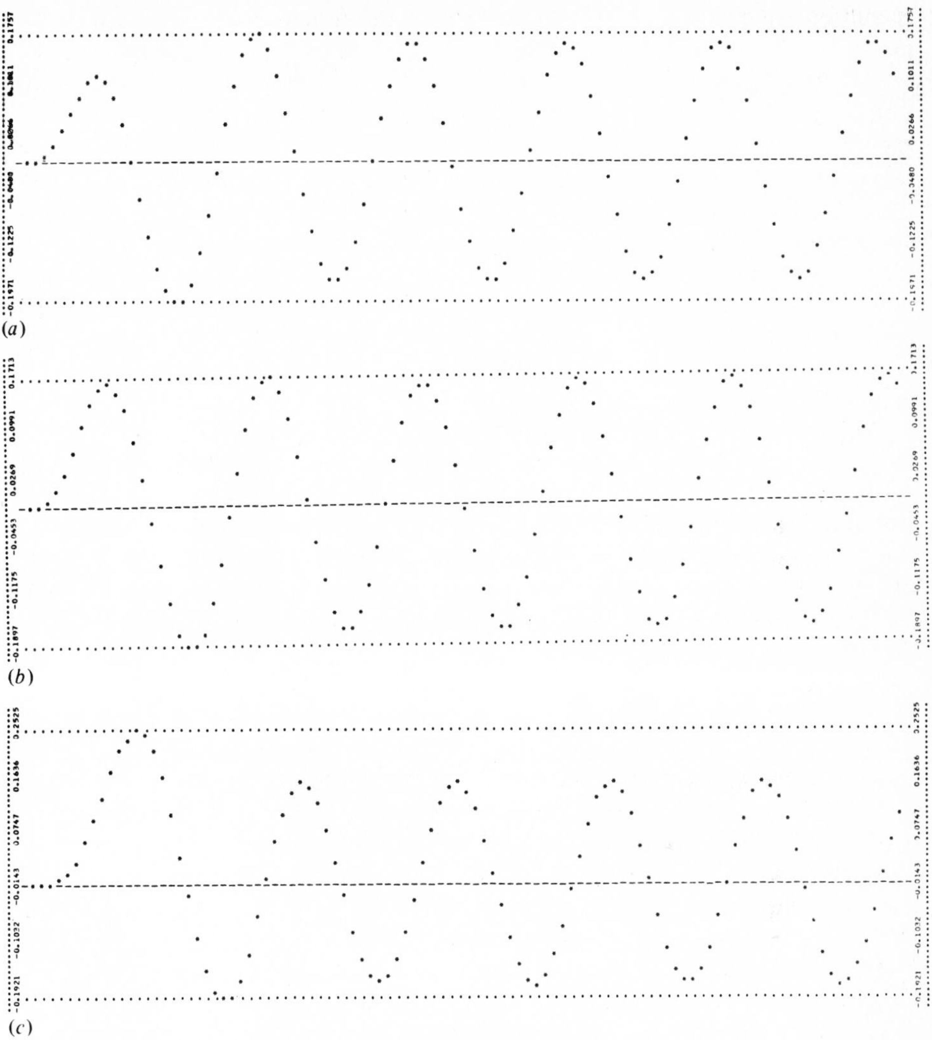

FIG. 7.16-C

($\phi = -20°$), the transient component of the response is very small, though not zero.

A very interesting question is: "Under what circumstances *will* r_N be zero?" For initial conditions of rest, $r_N(t)$ *cannot* be zero: that would mathematically require that $E(s)$ vanish at every natural frequency of the network, so that all $K_j' = 0$ [see (7.16-15)]. Since $E(s)$ has only one finite zero [see (7.16-14)], this can happen only in the extraordinarily simple case of $N = 1$ (for example, in the *RC* network of Chap. 2). In general the natural part of the response cannot vanish because the value of the network's steady-state response to the (sinusoidal) excitation cannot at $t = 0+$ be the same as the initial (rest) condition of the network.

If, instead of an initial condition of rest, certain very special initial conditions obtain in the network, then of course the situation is different. Common sense says that if we adjust the energy storage in each element so that all the initial conditions coincide with those given by $r_F(0+)$, then obviously $r_N(t)$ will be identically zero; there is no need for a transient component. [Under these conditions, $E(s)$ in effect becomes mathematically much more complicated than (7.16-4) and $E(s)$ *does* vanish at *all* the natural frequencies; we shall discuss nonrest initial conditions further in Sec. 7.22.] Such very special initial conditions are technically interesting but practically rare. We note in the examples above (Fig. 7.15-H*b* and Fig. 7.16-C*b*) that the (transient) period during which r_N is appreciable can however be rather short.

Another interesting question is: "Under what circumstances will r_N be as *large* as possible? This is clearly a complicated (though sometimes important) matter, to be worked out in each individual case: it depends of course on the (variable) parameters of the excitation and on the network. We note that r_N can be indeed be comparatively large, depending, for example, (with ω_F fixed) on the value of ϕ (see Fig. 7.16-C) or on the value of ω_F (see Fig. 7.15-H). A natural frequency near the ω axis can, by a sort of resonance, make a component of r_N rather long-lived.

The role of r_F is quite different. The function of r_F is to express the network's *steady-state* response to sinusoidal excitation, the response as it is after the transient has vanished. This is often the "business part" of the response, that which "really counts." As (7.16-1) states, the sinusoid's amplitude and angle are altered by transmission through the network, either deliberately (by the network's design) or unavoidably (by the parasitic system elements that the network models). The transmission effects are different at different frequencies ω_F. And if the excitation contains components at various different frequencies (the excitation is a *sum* of sinusoids), each component will be altered differently, and distortion of the waveshape will occur. (Certain exceptional individual alterations may nevertheless produce no net distortion; see Sec. 8.07.) The pulse train of Fig. 7.16-A provides a simple example. Suppose it is passed through the low-pass network of Fig. 7.14-A, its base frequency being $\omega_F = 0.25$ (normalized to the scale of frequency there used). The resulting steady-state response, calculated simply by finding the network's forced response to each of the components of the excitation and adding them, is shown in Fig. 7.16-D. Comparison with the original pulse (repeated there) makes the point.

The analysis of a network's transmission of sinusoids as a function of ω_F (the calculation of r_F), that is, the study of $T(j\omega_F)$ as a function of ω_F, is both complicated and important. We shall give it considerable attention later.

Two other questions are: "When will r_F be very large? When will r_F be very small?" To the first, a review of Chap. 3 and the subject of *resonance* gives the answer: when a natural frequency of the network is very close to the ω axis, near the point $j\omega_F$. (Complicated networks may well have a number of

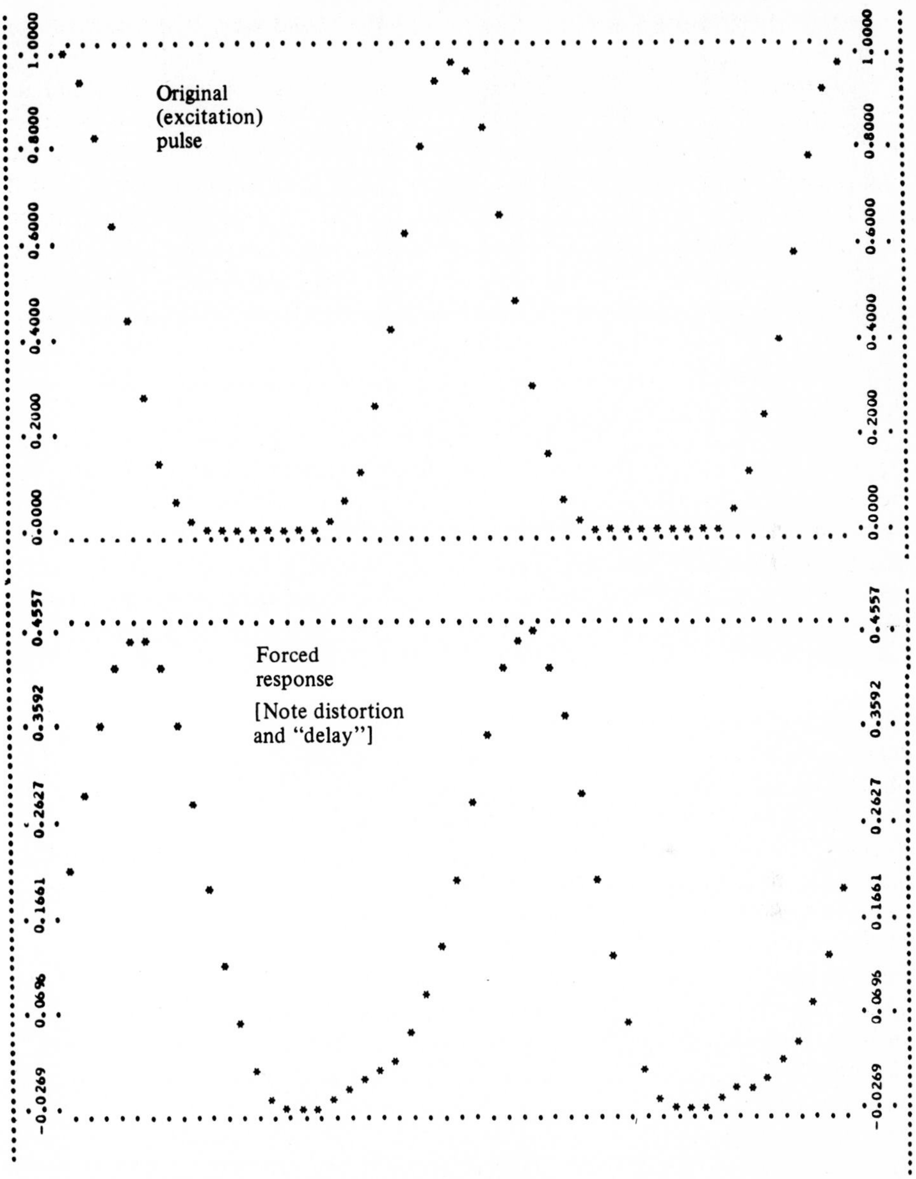

FIG. 7.16-D

resonances; see Fig. 7.16-E, for example.) From this behavior stems the term "point of infinite gain" for a natural frequency.

To the second the answer is clearly: when $T(j\omega_F) = 0$, for then $r_F = 0$, surely the smallest value possible. If a network has a zero of transmission that corresponds to the excitation frequency, then there is no forced response. The term "point of infinite loss" is appropriate therefor. A resonant circuit of infinite

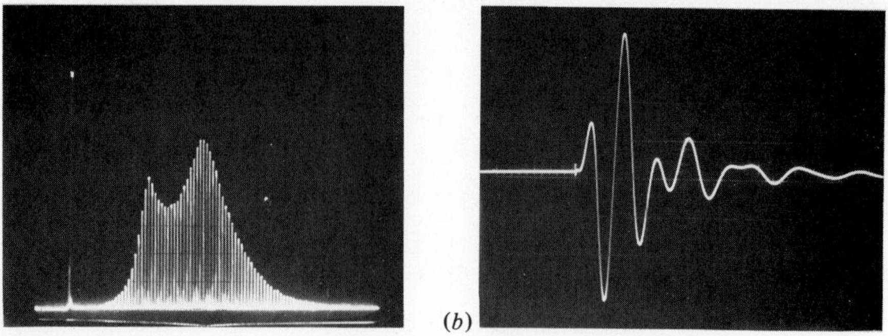

(a) (b)

FIG. 7.16-E. A network with two pronounced resonances (natural frequencies or infinite-gain points near the ω axis): (a) $|T(j\omega_F)|$; (b) impulse response. (See Prob. 7-C.)

Q (without dissipation of energy), or a bridge properly balanced can do this. For example, the first two networks of Fig. 7.16-F will have no steady-state response to sinusoidal excitation of a certain frequency ω_F. Network (*a*) is constructed by resonating the inductor L_2 with a capacitor C_2 such that then $\omega_F = 1/\sqrt{L_2 C_2}$. Network (*b*) is a Wien bridge (Fig. 5.13-C*b*); when the element values obey the condition

$$\frac{R_1}{R_2} = \frac{R_4}{R_3} + \frac{C_3}{C_4} \tag{7.16-19}$$

the bridge is balanced at the frequency $\omega_F = 1/\sqrt{R_3 R_4 C_3 C_4}$. Both rejections are difficult to achieve in practice: the former because nondissipative elements do not exist (at least in passive form) and the latter because a perfect bridge balance is very difficult to maintain. But high-Q circuits can give resonances of the rejection type *near* the ω axis so that r_F is very small (see Fig. 7.16-F*c*). They are very commonly used.

For physical demonstration of the efficacy of such zeros in $T(s)$, see Figs. 6.16-B, 6.17-B, 6.20-A. An additional example, clearly showing two such rejection frequencies in $|T(j\omega_F)|$, is given in Prob. 6-N.

And a bridge *can* be accurately balanced for brief periods of time, as for making measurements.

To recapitulate: We can calculate r_F and r_N for sinusoidal excitation from rest with ease, separately, by using $E(s)$ from $e(t)$, and $T(s)$ from the network. The calculation of r_F is easy because it is simply the forced component of the response to exponential excitation and therefore requires only the evaluation of the transfer function. The calculation of r_N is in principle easy but in practice (if N is large) can involve much numerical work, whether calculated from the (previously obtained) impulse response, or by partial-fraction expansion of $R(s)$, or by brute force, working to make the initial values (at $t = 0+$) correct.

Note once more that the *forced* component of the response can equally well be called the *steady-state* component in all practical cases, because it is permanent

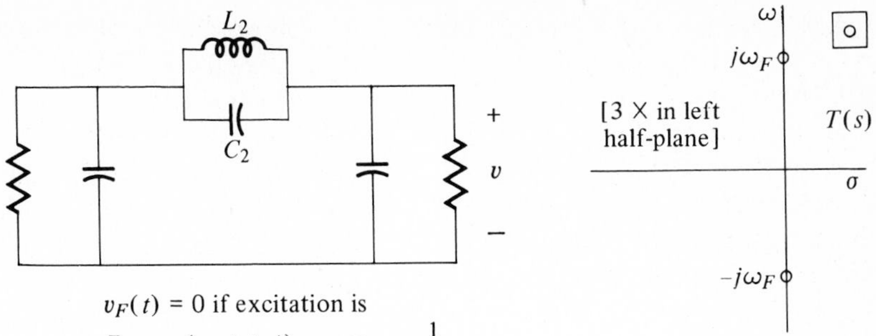

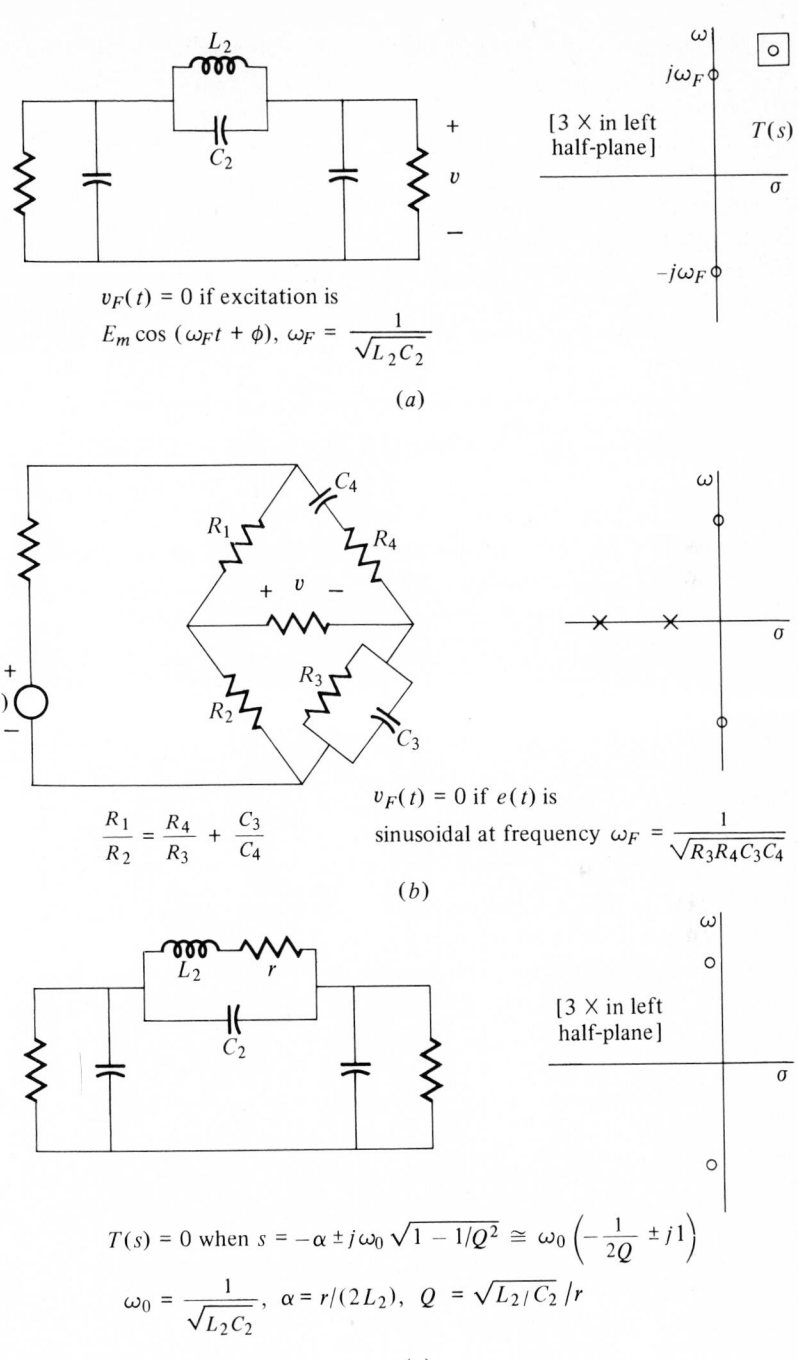

$v_F(t) = 0$ if excitation is

$E_m \cos(\omega_F t + \phi)$, $\omega_F = \dfrac{1}{\sqrt{L_2 C_2}}$

(*a*)

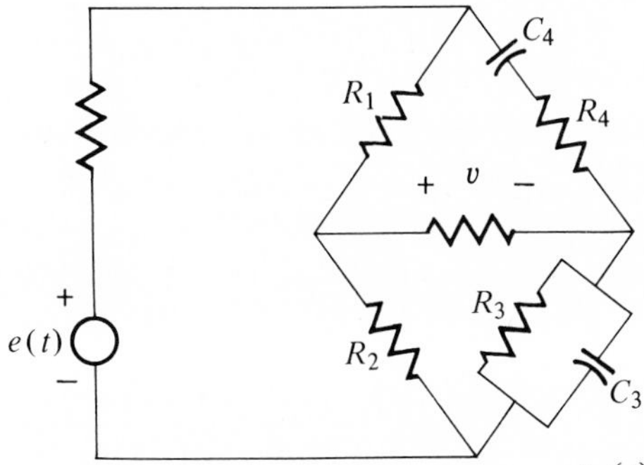

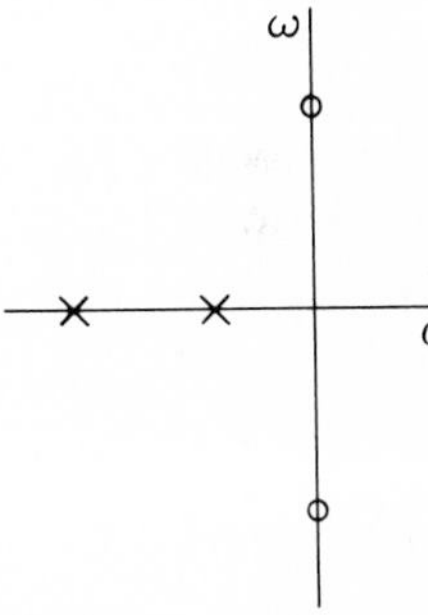

$$\frac{R_1}{R_2} = \frac{R_4}{R_3} + \frac{C_3}{C_4}$$

$v_F(t) = 0$ if $e(t)$ is sinusoidal at frequency $\omega_F = \dfrac{1}{\sqrt{R_3 R_4 C_3 C_4}}$

(*b*)

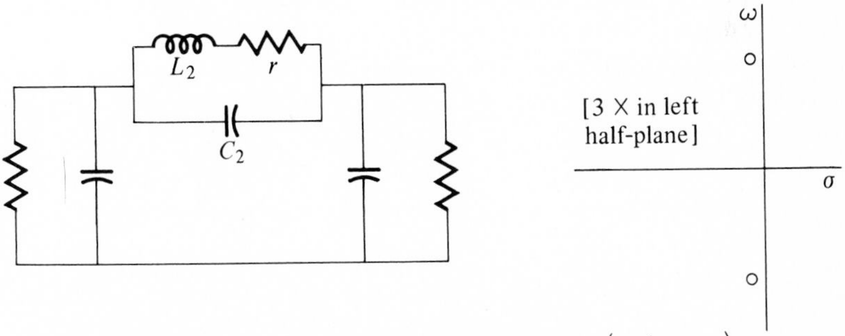

$$T(s) = 0 \text{ when } s = -\alpha \pm j\omega_0\sqrt{1 - 1/Q^2} \cong \omega_0\left(-\frac{1}{2Q} \pm j1\right)$$

$$\omega_0 = \frac{1}{\sqrt{L_2 C_2}}, \quad \alpha = r/(2L_2), \quad Q = \sqrt{L_2/C_2}\,/r$$

(*c*)

FIG. 7.16-F

(not damped) and it is the only part of the response that is so. The *natural* part dies away (because of the resistance present in all realistic models), and it can equally well be called the *transient* component. This usage is very common and ordinarily leads to no confusion. Technically, however, the forced component of a response can be *either* permanent (steady-state) *or* transient: it depends on the character of the excitation. And the natural part can be *either* transient *or* permanent (steady-state): that depends on the presence or absence of resistance.

Finally we repeat the observation that network analysis can be so neatly codified *only* because the network is postulated to be linear and constant, and the excitation is exponential.

7.17 COMPUTATIONAL PROBLEMS OF NETWORK ANALYSIS

Barring certain unusual cases [coincident poles in $R(s)$, for example], the analysis of a given network is now routine. The six-step codification of Sec. 7.11 makes of it a purely mechanical process. The examples of Sec. 7.16 amply illustrate it. And for those interested only in a general overview of network analysis, that may well suffice. But the actual details, in all but the simplest cases, require lengthy numerical calculations that we cannot lightly dismiss: they are not necessarily simple.

Consider, for example, impulse-response analysis of the network of Fig. 7.17-A*a*, designed by a simple technique to be a high-pass filter, with appropriate normalization. The transfer function, easily calculated, is

$$T(s) = \frac{0.5s^5}{s^5 + 4s^4 + 8s^3 + 10s^2 + 8s + 4}. \tag{7.17-1}$$

Asymptotic low-frequency and high-frequency ($s \to 0$, $s \to \infty$) calculations, easily made on both network and transfer function, quickly check this. Its expansion is formally

$$\begin{aligned} T(s) &= 0.5\left(1 + \frac{-4s^4 - 8s^3 - 10s^2 - 8s - 4}{s^5 + 4s^4 + 8s^3 + 10s^2 + 8s + 4}\right) \\ &= 0.5 + \frac{K_1}{s - p_1} + \frac{K_2}{s - p_2} + \frac{K_3}{s - p_3} + \frac{K_4}{s - p_4} + \frac{K_5}{s - p_5} \end{aligned} \tag{7.17-2}$$

in which the p_j are the natural frequencies, the zeros of the polynomial

$$s^5 + 4s^4 + 8s^3 + 10s^2 + 8s + 4. \tag{7.17-3}$$

The transfer function is somewhat unusual in that $M = N$ (see Sec. 6.26), so that the expansion of $T(s)$ requires first the removal of the constant, the one term of the polynomial of (7.12-3) that is here present.

Physically, as the schematic shows, the network passes high-frequency signals very well indeed (T has no zeros at infinity), and therefore the impulse response contains an actual impulse (as in the example of Sec. 6.26). The true behavior of a real high-pass-filter network at high frequencies may not of course be accurately modeled here; but the parasitic effects will probably be much like the capacitors', and not alter the response significantly. The demonstration of Fig. 7.17-B (further discussed below) corroborates this.

Calculation of the impulse response of the network is now reduced to step (d) of Sec. 7.11: transformation of $T(s)$ to the domain of t (since $E = 1$ and so $R = ET = T$). This amounts to partial-fraction expansion of the proper-fraction part of (7.17-2), that is, to (a) the calculation of the p_j, the factoring of (7.17-3), and (b) the calculation of the K_j, as discussed in Sec. 7.12. Both of these, except in simple textbook examples, involve much numerical work.

The factoring of polynomials is a large subject in itself (see Sec. 6.04),

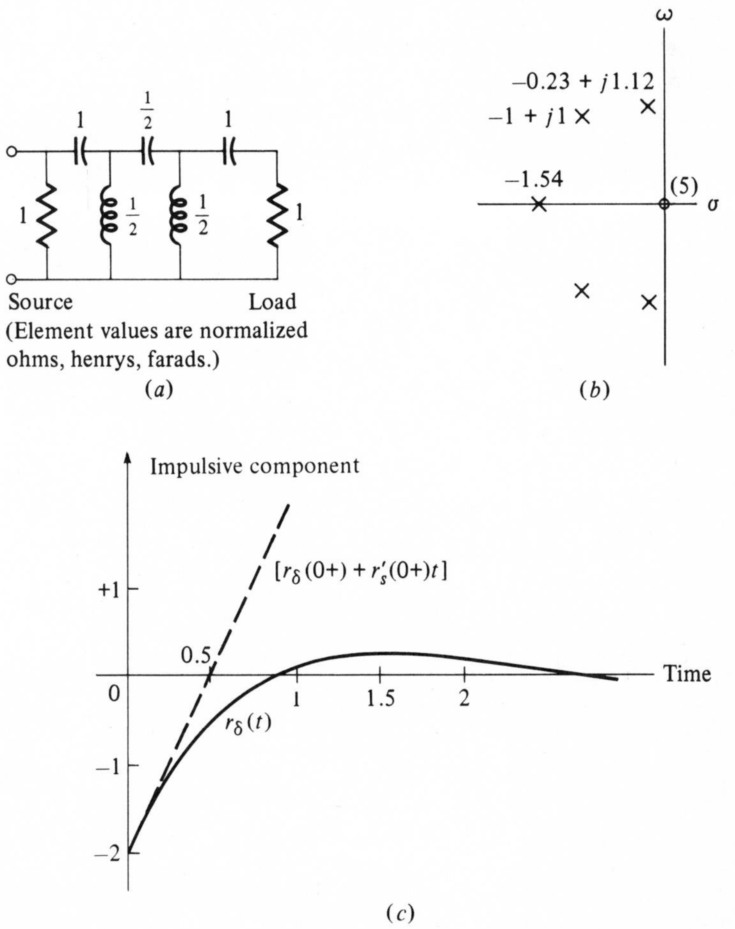

FIG. 7.17-A

```
THIS IS IMPSTEP SEVEN.     NOTES FOR THIS RUN:     MS    SECT. 7.17  FIG. 7.17-A
                                                   HP NET.    M = N = 5
                                                   AUG.  1972  DFT

PARAMETER VALUES READ:     NN =  5, ND =  5, NX = 0, NPLT = 2, NPRD = 1, NPRU = 1, NCODE = 0

FOR THE FOLLOWING CASE THE MULTIPLIER  H =     0.500000      AND THE COEFFICIENTS IN F(S), LOWEST POWER FIRST, ARE:

NUMERATOR:
        0.0            0.0            0.0            0.0            0.0            1.000000

DENOMINATOR:
        4.000000       8.000000       10.000000      8.000000       4.000000       1.000000

THE FOLLOWING ARE THE POLES OF F:

      -1.543688  +J  0.0
      -0.228155  +J -1.115144
      -0.228155  +J  1.115144
      -0.999998  +J -0.999999
      -0.999998  +J  0.999999

THE FOLLOWING ARE THE RESIDUES THEREAT (THE "K'S"):
```

MAGNITUDE	ANGLE (RAD.)	ANGLE (DEG.)	REAL PART	IMAG. PART
1.137445	3.141593	180.000015	-1.137445	-0.000000
0.141354	-5.220115	-299.090576	0.068725	0.123522
0.141354	5.220115	299.090576	0.068725	-0.123522
0.707106	2.356192	134.999863	-0.499998	0.500001
0.707106	-2.356192	-134.999863	-0.499998	-0.500001

```
THE SUM OF THE COMPUTED RESIDUES "K" IS:                                           -0.199999E 01 + J  0.357628E-06

THE SUM OF THE PRODUCTS OF THE COMPUTED RESIDUES AND THE NATURAL FREQUENCIES, "SUM KP", IS:   0.399998E 01 + J -0.476837E-06

SUBROUTINE 'PLOT4Z', BEING CALLED, GIVES (BAR MARKS ZERO ORDINATE):
```

TIME	IMPULSE RESP
0.0	-0.2000E 01
0.2000	-0.1295E 01
0.4000	-0.7593E 00
0.6000	-0.3659E 00
0.8000	-0.8944E-01
1.0000	0.9258E-01
1.2000	0.2001E 00
1.4000	0.2505E 00
1.6000	0.2589E 00
1.8000	0.2378E 00
2.0000	0.1978E 00
2.2000	0.1474E 00
2.4000	0.9324E-01
2.6000	0.4039E-01
2.8000	-0.7470E-02
3.0000	-0.4788E-01
3.2000	-0.7940E-01
3.4000	-0.1014E 00
3.6000	-0.1141E 00
3.8000	-0.1180E 00
4.0000	-0.1141E 00
4.2000	-0.1037E 00
4.4000	-0.8829E-01
4.6000	-0.6937E-01
4.8000	-0.4841E-01
5.0000	-0.2680E-01
5.2000	-0.5820E-02
5.4000	0.1348E-01
5.6000	0.3023E-01
5.8000	0.4382E-01
6.0000	0.5383E-01
6.2000	0.6011E-01
6.4000	0.6269E-01
6.6000	0.6180E-01
6.8000	0.5780E-01
7.0000	0.5119E-01
7.2000	0.4255E-01
7.4000	0.3251E-01
7.6000	0.2170E-01
7.8000	0.1074E-01
8.0000	0.1894E-03
8.2000	-0.9455E-02
8.4000	-0.1779E-01
8.6000	-0.2452E-01
8.8000	-0.2944E-01
9.0000	-0.3247E-01
TIME	IMPULSE RESP

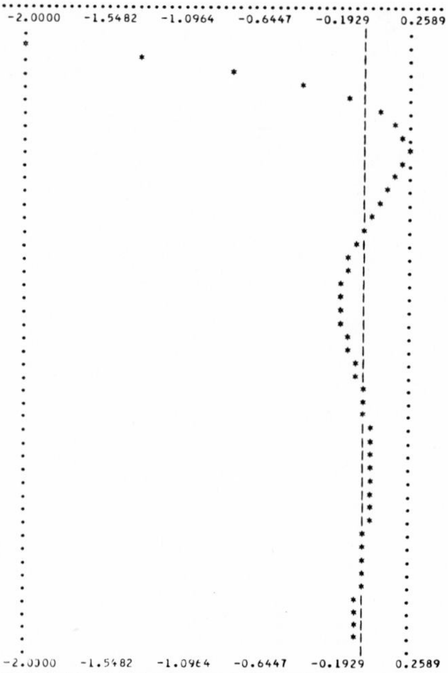

(d)

FIG. 7.17-A. (Con.)

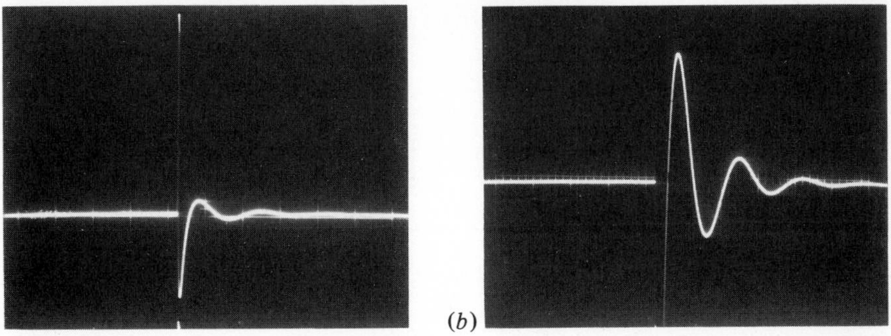

(a) (b)

FIG. 7.17-B

virtually a separate discipline within numerical analysis. Numerical (not literal or formal) it must be, because usable formulas for the zeros exist only for polynomials of low degree. Calculation of zeros must be by successive approximations, calculations that are usually straightforward but lengthy. For them the automatic (programmed) digital computer is admirably suited. Appendix C discusses the process briefly and provides illustrative programs. (The number of zero-finding programs in existence is virtually infinite; every machine library has some.) From this point on we assume the use of such a machine and such a program. Application to (7.17-3) gives the five natural frequencies shown in Fig. 7.17-A*b*. The five zeros of $T(s)$ are all at the origin.

The calculations of the K_j are also lengthy but straightforward (see Sec. 7.12). Very trying of human patience, the numerical work is again easily programmed for quick, accurate performance by a computer. For these particular calculations, however, it may be necessary to write one's own program. This is not a difficult job; Appendix C gives one such program. The results of applying it to our example are shown in Fig. 7.17-A*d*. We now have numerical values for all 10 literal quantities in (7.17-2) and can therefore write

$$r_\delta(t) = 0.5\,\delta(t) + (K_1 e^{p_1 t} + \cdots + K_5 e^{p_5 t})u(t) \tag{7.17-4}$$

or simply

$$r_\delta(t) = K_1 e^{p_1 t} + 2\,\mathbf{RE}\,(K_2 e^{p_2 t}) + 2\,\mathbf{RE}\,(K_4 e^{p_4 t}), \qquad t > 0 \tag{7.17-5}$$

remembering, when using the second form, that there is also an impulsive component that occurs at $t = 0$. Figure 7.17-A*c* shows the first part of the responses, indicating the impulsive component and the first instants of the subsequent behavior (see also Fig. 7.17-B).

Checking the calculations, step (*e*), is not to be ignored, just because a machine did them. Program errors may long lie hidden; data may be entered incorrectly. We therefore separately calculate the first two initial values, $r_\delta(0+)$ and $r'_\delta(0+)$, from the transfer function (initial-value table) and compare them with $\sum K_j$ and $\sum K_j p_j$. The machine (program) can of course be asked to compute the latter, from the natural frequencies and residues that it has found;

this is a simple addition to the program. Now because $M = N$ the initial-value table must be obtained from the coefficients of the proper fraction in (7.17-2) and *not* from those of (7.17-1): the impulsive component must first be removed, since at $t = 0+$ it has vanished. In synthetic form, omitting unnecessary letters and the equality signs (and not forgetting the 0.5 factor) we construct the initial-value table (see Sec. 6.11):

$$\begin{array}{cccc:c} & & & 1 & -2 \\ & & 1 & 4 & -4 \\ & 1 & 4 & 8 & -5 \\ 1 & 4 & 8 & 10 & -4 \\ \hline \end{array} \tag{7.17.6}$$

whence

$$\begin{aligned} r(0+) &= -2, \\ r'(0+) &= -4 - 4(-2) = +4. \end{aligned} \tag{7.17-7}$$

The machine results, from Fig. 7.17-A*d*, are: $\sum K = -2.00 + j0.00$ and $\sum Kp = +4.00 + j0.00$, to five or six significant figures; the check is satisfactory.

These two initial values give the approximate initial response, the straight line Fig. 7.17-A*c*; these two numbers alone are very useful as indicators of how the response starts. (One must not forget the impulsive component, of course, which is *not* given by the initial-value table.)

Finally, step (f) probably requires computation of the actual values of $r_\delta(t)$ over a range of t and presentation of these results in tabular and graphic form. The computer program can easily be extended also to make these computations. In Fig. 7.17-A*d* such a table is combined with a simple, often adequate graphical display of the behavior of $r_\delta(t)$. It was produced quickly and cheaply as part of ordinary computer line-printer output. (Auxiliary plotting equipment can of course be used to make more elegant plots.)

A good network analysis program should, from the coefficients of $T(s)$, produce all these results in one run. Figure 7.17-A*d* shows some of the output of the IMPSTEP program (Appendix C). Construction of the initial-value table and calculation of the values of $r_\delta(0+)$ and of $r'_\delta(0+)$ is purposely *not* asked from the machine, to force users to enter into the analysis themselves.

The physical demonstration of Fig. 7.17-B (the impulse response of just such a network) corroborates the analysis beautifully. The photograph clearly shows an initial impulsive component in the response (note the tip of the exciting impulse directly below the beginning of the response trace). The initial value of the response clearly is negative. Its subsequent behavior is oscillatory, just as calculated (Fig. 7.17-A*d*). In (a) the scale of the response is adjusted so one can see the initial impulsive component (which is large) and the initial value that is negative (also big); the subsequent oscillations are perforce then small. (The tip of the exciting impulse is just visible below the wave.) Part (*b*) expands the response vertically, cutting off the impulse and the negative initial behavior of the response, but making the oscillation larger. The demonstration, in spite of its simplicity and imprecision, gives a vivid verification of the validity of the model and the analysis. Precise element values are not necessary to show these effects; readily obtainable off-the-shelf elements can be used, and readers should perform it for themselves. (See also Fig. 6.23-E.)

7.18 SOME COMPUTER PROGRAMS FOR NETWORK ANALYSIS

Network models are surprisingly accurate as representations of real networks; their accuracy justifies careful analysis. Network analysis is fundamentally the formulation and solution of the corresponding differential (or integrodifferential) equations, those differential equations that are the mathematical form of the model, equivalent to the engineer's schematic-diagram form. These differential equations are very special and simple, and because they are so special and simple, their solution is straightforward. True, it has taken us long to work out actual techniques, because network analysis has so many interesting, relevant, and important facets. In importance therein second only to gaining physical understanding is the transformation of the analysis from the t to the s domain. This changes the sophisticated solution process of the infinitesimal calculus (integration) into routine algebra: network analysis becomes a vast labor of straightforward numerical calculations. For them the programmed digital computer is now the natural tool. It has removed the drudgery of hand computation, with slide rule, logarithm tables, and desk (or pocket) calculators, and makes possible quick, cheap, accurate analysis of even reasonably large, complicated networks. With patience one could (and used to have to) carry out by hand all the calculations that lead to the results of Fig. 7.17-A, however many hours it required. But machine computation, once the programs have been written, requires only one or two seconds. Network analysis is one category of engineering that is largely computer application par excellence!

Since the calculations of natural frequencies and partial-fraction expansions are so quickly performed by computer, we must now ask: "Can we also mechanize the other steps of the analysis?" Let us, from this point of view, review all six steps (Sec. 7.11).

To transform $e(t)$ to $E(s)$, the first step of the analysis, requires nothing more than our tables of pairs in all usual (exponential-excitation) cases. Uncommon, nonexponential excitation functions can be handled by convolution, which we have already seen is an eminently programmable operation (Chap. 4). (Periodic excitation functions have interesting properties of their own; they form a special class of exponentials to be treated elsewhere.)

To "transform the network" into its transfer function, step (*b*), is a different matter. Even ladder-networks analysis by the straightforward process of Chap. 6 can become rather lengthy. But "lengthy, straightforward" is now synonymous with "ideal for a computer to perform." Programming the ladder-network transfer-function calculation (see Sec. 3.12) results, for example, in the program LADANA (Appendix C). It can analyze a ladder network of as many as 16 branches (16 boxes in Fig. 6.01-A), each of which may contain (at most) one element of each type (R, L, C), connected in parallel if the box is a series arm of the ladder, connected in series if it is a shunt arm. The (parallel) dissipative tuned circuit of Fig. 3.18-B (which has four elements) is also admissible as a series branch or "box"; so also is its dual (the four-element series dissipative tuned circuit) as a shunt network branch. The networks of Fig. 6.01-D, for example, and all the ladder networks that we have discussed, are readily analyzed by LADANA. The program is limited in ability, true, but it is powerful enough for us here. When supplied with data, it (*a*) lists the element values given it and draws a crude network schematic (for checking data entry and for later identification of the case), then (*b*) computes the transfer function and the input impedance. Element values are usually normalized, for this makes the size of numbers reasonable and avoids numerical troubles. Hence if $T(s)$ has dimensions, one may have to correct LADANA's results by multiplying by a scale factor, the reference impedance or admittance value. (See Sec. 7.14 for a discussion of this.) The input impedance is of course "off" by exactly such a factor, R_0. We leave such corrections to common sense and do not complicate the program with them. To illustrate the use of LADANA, we apply it first to the network of Fig. 7.17-A, with the result shown in Fig. 7.18-A, to be compared with (7.17-1). A second illustration is given by Fig. 7.18-B. To calculate the transfer function of the 16-element network of Fig. 7.18-C*a* by hand is feasible but dismaying to contemplate, simple though its (normalized) element values are; LADANA computes it in a second, giving us Fig. 7.18-C*b*. Such a transfer function calculating program can of course be extended, and made even more powerful and general. Programming the calculation of the transfer functions of nonladder networks is appreciably more complicated. We content ourselves here with LADANA.

In perspective, we must note that it may sometimes be preferable or even necessary to work in the t domain, with the differential equations, and not to transform to s. This method of analysis is also programmable (Sec. 7.20) but we do not discuss it here.

Step (*c*), the calculation of $R(s) = E(s)T(s)$, requires no great computation; no programming is needed.

Step (*d*), finding the natural frequencies and making the partial-fraction expansion of $R(s)$, we have discussed at length in Sec. 7.17. Its mechanization for impulse and step excitation, cases where $R(s)$ is very simple, gives for example the IMPSTEP program (Appendix C). The same program can be used for any (exponential-form) excitation by using for the transfer function not the network's own transfer function T_{NET}, but the product $T_{WF}T_{NET}$ (Fig. 7.04-A), and asking for the impulse response. The wave-forming network concept that reduces most

```
THIS IS LADANA, A LADDER-NETWORK ANALYSIS PROGRAM

              NOTES:                   MS  SECT.  7.18  FIG. 7.18-A
                                       HP NET. N = 5
                                       AUG. 1972

DATA READ IN:

NUMBER OF BRANCHES =  7,  NUMBER OF ENERGY-STORING ELEMENTS =  5          MWR =  2          MCU = 3          NCON = 0

BRANCH     MSS*     TYPE          R VALUE          L VALUE          C VALUE                              *MSS = 1 : SHUNT
......     ....     ....          ........         ........         ........                                  = 2 :SERIES

  7         1        1            1.000000
  6         2        3                                              1.000000
  5         1        3                             0.500000
  4         2        3                                              0.500000
  3         1        3                             0.500000
  2         2        3                                              1.000000
  1         1        1            1.000000

THAT IS:

 BRANCH NUMBERS:
      7       6       5       4       3       2       1

0---------- C ----------- C ----------- C -------0
      I               I               I               I
      I               I               I               I
      I               I               I               I

      R               L               L               R

      I-              I               I               I
      I               I               I               I
      I               I               I               I
0------------------------------------------------0
      7       6       5       4       3       2       1


THE RESULTS (COEFFICIENTS OF POWERS OF S, IN ASCENDING ORDER, NUMERATORS OVER DENOMINATORS) ARE:

 THE INPUT IMPEDANCE IS:
      1.000000        1.000000        1.500000        1.000000        0.500000        0.125000
...............................................................................................
      1.000000        2.000000        2.500000        2.000000        1.000000        0.250000


 THE CURRENT-TO-VOLTAGE TRANSFER FUNCTION IS :
      0.0             0.0             0.0             0.0             0.0             0.125000
...............................................................................................
      1.000000        2.000000        2.500000        2.000000        1.000000        0.250000

OR (HIGHEST POWER OF S IN NATURAL-FREQUENCY POLYNOMIAL MADE UNITY):

      0.0             0.0             0.0             0.0             0.0             0.500000
...............................................................................................
      4.000000        8.000000       10.000000        8.000000        4.000000        1.000000


THE RESULTS ABOVE WERE PRODUCED BY LADANA.

              NOTES:                   MS  SECT.  7.18  FIG. 7.18-A
                                       HP NET. N = 5
                                       AUG. 1972


THERE SEEMING TO BE NO MORE DATA, I AM, SIR/MADAME (MS.), YOUR MOST OBEDIENT SERVANT, LADANA.
```

FIG. 7.18-A

network analysis problems to impulse-response problems fits the use of IMPSTEP like a glove. The program could readily be expanded to accept the coefficients of both transfer functions (T_{WF} and T_{NET}) separately, and proceed from there. For an example of IMPSTEP'S output, see Fig. 7.17-A; others will follow, beginning in Sec. 7.19.

Step (*e*), checking, in principle requires no further comment. But programs are not infallible: errors can long lie hidden and be inconsequential until some unusual data appear. The entry of data itself can be erroneous. And it is not only educationally important, but important in engineering practice, to keep one's

```
THIS IS LADANA, A LADDER-NETWORK ANALYSIS PROGRAM

          NOTES:          MS - SECT. 7.18
                          EXAMPLE OF USE OF LADANA
                          MISC. NET.

DATA READ IN:

NUMBER OF BRANCHES =  5,  NUMBER OF ENERGY-STORING ELEMENTS =  5        MAP =  2        MCO = 3        NCON = 0

BRANCH    MSS*    TYPE       R VALUE         L VALUE         C VALUE                    *MSS = 1 : SHUNT
......    ....    ....       ........        ........        ........                        = 2 :SERIES

  5        2       3                                         0.940000
  4        1       3                         3.799999
  3        2       5        20.000000        1.370000
  2        2       7        10.000000        0.680000        2.299999
  1        1       1         1.000000

THAT IS:

 BRANCH NUMBERS:
     5       4         3         2         1
                       R         R
0---  C ------------     ----  L --------0
             |         L         C         |
             |                             |
             |                             |
             |                             |
             L                             R
             |                             |
             |                             |
             |                             |
             |                             |
0------------------------------------------0
     5       4         3         2         1


THE RESULTS (COEFFICIENTS OF POWERS OF S, IN ASCENDING ORDER, NUMERATORS OVER DENOMINATORS) ARE:


 THE INPUT IMPEDANCE IS:

      1.000000        5.986499        5.799094        16.020874        6.509497        8.036326
.................................................................................................................
      0.0             0.940000        5.627309        2.093470         7.718063        0.382682



 THE VOLTAGE-TO-VOLTAGE TRANSFER FUNCTION IS :

      0.0             0.0             3.571999        0.487578         5.603240        0.382682
.................................................................................................................
      1.000000        5.986499        5.799094        16.020874        6.509497        8.036326

OR (HIGHEST POWER OF S IN NATURAL-FREQUENCY POLYNOMIAL MADE UNITY):

      0.0             0.0             0.444481        0.060672         0.697239        0.047619
.................................................................................................................
      0.124435        0.744930        0.721610        1.993556         0.810009        1.000000


THE RESULTS ABOVE WERE PRODUCED BY LADANA.

          NOTES:          MS - SECT. 7.18                                                FIG. 7.18-B
                          EXAMPLE OF USE OF LADANA
                          MISC. NET.


THERE SEEMING TO BE NO MORE DATA, I AM, SIR/MADAME (MS.), YOUR MOST OBEDIENT SERVANT, LADANA.
```

FIG. 7.18–B

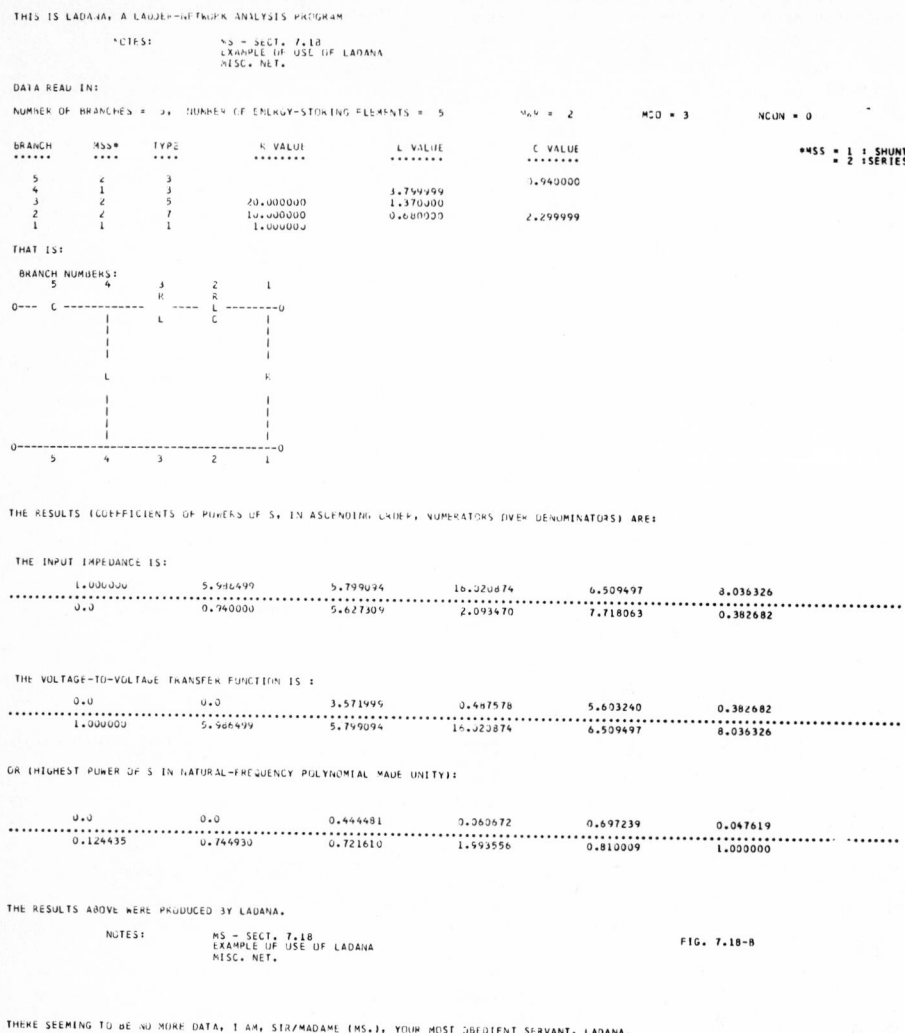

hand in to some extent. We can easily check LADANA's results against the asymptotic low- and high-frequency behavior of the network obtained by inspection of the schematic diagram. For the calculations of Figs. 7.18-A to 7.18-C, these checks are summarily made in Fig. 7.18-D. For checking purposes, IMPSTEP itself should calculate, from the p's and K's it finds, at least the values of $\sum p$ and $\sum Kp$. It is then up to you to calculate what these numbers *should* be, from initial values found from $T(s)$, or from physical examination of the network's initial behavior.

For step (f), discussion of the results of the analysis, tables and plots are useful. Preparation of these is "straightforward and lengthy"—and therefore

THIS IS LADANA, A LADDER-NETWORK ANALYSIS PROGRAM

NOTES: LADANA ILLUSTRATION: N = 14
(14 ELEMENT C-K LPF)
MS FIG. 7.18-C

DATA READ IN:

NUMBER OF BRANCHES = 16, NUMBER OF ENERGY-STORING ELEMENTS = 14   MHR = 2   MCO = 3   NCON = 0

BRANCH	MSS*	TYPE	R VALUE	L VALUE	C VALUE
16	1	1	1.000000		
15	2	2		1.000000	
14	1	2			2.000000
13	2	2		2.000000	
12	1	2			2.000000
11	2	2		2.000000	
10	1	2			2.000000
9	2	2		2.000000	
8	1	2			2.000000
7	2	2		2.000000	
6	1	2			2.000000
5	2	2		2.000000	
4	1	2			2.000000
3	2	2		2.000000	
2	1	2			1.000000
1	1	1	1.000000		

*MSS = 1 : SHUNT
= 2 : SERIES

THAT IS:

```
BRANCH NUMBERS:
 16     15     14     13     12     11     10      9      8      7      6      5      4      3      2      1
0---------- L ---------- L ---------- L ---------- L ---------- L ---------- L ---------- L - -----------0
 |             |             |             |             |             |             |             |      |
 |             |             |             |             |             |             |             |      |
 R             C             C             C             C             C             C             C      R
 |             |             |             |             |             |             |             |      |
 |             |             |             |             |             |             |             |      |
0-------------------------------------------------------------------------------------------------------0
 16     15     14     13     12     11     10      9      8      7      6      5      4      3      2      1
```

THE RESULTS (COEFFICIENTS OF POWERS OF S, IN ASCENDING ORDER, NUMERATORS OVER DENOMINATORS) ARE:

THE INPUT IMPEDANCE IS:

1.000000	13.000000	85.000000	364.000000	1204.000000	2912.000000	6496.000000
9984.000000	16896.000000	16640.000000	22784.000000	13312.000000	15360.000000	4096.000000
4096.000000						
....						
2.000000	26.000000	169.000000	728.000000	2324.000000	5824.000000	11872.000000
19968.000000	28416.000000	33280.000000	34048.000000	26624.000000	19456.000000	8192.000000
4096.000000						

THE CURRENT-TO-VOLTAGE TRANSFER FUNCTION IS :

1.000000	0.0	0.0	0.0	0.0	0.0	0.0
0.0	0.0	0.0	0.0	0.0	0.0	0.0
0.0						
....						
2.000000	26.000000	169.000000	728.000000	2324.000000	5824.000000	11872.000000
19968.000000	28416.000000	33280.000000	34048.000000	26624.000000	19456.000000	8192.000000
4096.000000						

OR (HIGHEST POWER OF S IN NATURAL-FREQUENCY POLYNOMIAL MADE UNITY):

0.000244	0.0	0.0	0.0	0.0	0.0	0.0
0.0	0.0	0.0	0.0	0.0	0.0	0.0
0.0						
....						
0.000488	0.006348	0.041260	0.177734	0.567383	1.421875	2.898438
4.875000	6.937500	8.125000	8.312500	6.500000	4.750000	2.000000
1.000000						

THE RESULTS ABOVE WERE PRODUCED BY LADANA.

NOTES: LADANA ILLUSTRATION: N = 14
(14 ELEMENT C-K LPF)
MS FIG. 7.18-C

THERE SEEMING TO BE NO MORE DATA, I AM, SIR/MADAME (MS.), YOUR MOST OBEDIENT SERVANT, LADANA.

FIG. 7.18-C

Case		From examination of network	From LADANA result
Fig. 7.18-A	HF	$RR/(R+R) = \frac{1}{2}$	$0.5s^5/(1.0s^5) = 0.5$
	LF	$R(Cs)(Ls)\ldots$ $= 1(s)(s/2)^3(s)(1)$ $= s^5/8$	$(0.5s^5)/(4.0) = 0.125s^5$
Fig. 7.18-B	HF	$R/(R+R) = \frac{1}{21}$ $= 0.0476$	$(0.0476s^5)/s^5 = 0.0476$
	LF	$(0.94s)(3.8s)(1.0)$ $= 3.57s^2$	$3.572s^2/1.0 = 3.572s^2$
Fig. 7.18-C	HF	$1(1/s)[1/(2s)]^{12}/s$ $= 1/(4096s^{14})$	$1/(4096s^{14})$
	LF	$\frac{1}{2}$	$\frac{1}{2}$

FIG. 7.18-D. Asymptotic checks of LADANA examples.

easily incorporated into the program. But comment, discussion, the report on the analysis—these are not the machine's job but the engineer's: the engineer must do it.

The case of sinusoidal excitation is important enough to warrant its own program. We accordingly develop RESSINET by incorporating into IMPSTEP the parameters of a sinusoidal excitation $E_m \cos(\omega_F t + \phi)$, in the wave-forming-network manner. RESSINET (Appendix C) provides options for displaying the forced and natural parts of the response separately, as well as the complete response, and various combinations. To illustrate the utility of such a program, consider the calculation of the response of an ordinary *RLC* circuit (Chap. 3) to sinusoidal excitation. The excitation has three parameters (E_m, ω_F, ϕ) and the circuit has three (R, ω_0, Q); but really there are not six of importance but only three: ω_F/ω_0, ϕ, Q. The other three parameters are simply scale factors of no great importance in studying the network's properties and the tuned circuit's response to sinusoidal excitation. But to treat only two values of each requires eight analyses! And even for one set of parameter values, the hand calculation of $r(t)$ from

$$R(s) = \frac{E_m[-(\omega_F \sin\phi) + (\cos\phi)s]s(1/C)}{(s^2 + \omega_F{}^2)(s^2 + 2\alpha s + \omega_0{}^2)} \tag{7.18-1}$$

is too laborious to contemplate with equanimity. RESSINET is very helpful! It computed the curves of Figs. 3.13-A to 3.13-D, and it computes the illustrative responses for the cases of Fig. 7.18-E, each in a fraction of a second.

The curves of Fig. 7.18-E suffice to give a good idea of the nature of RESSINET'S output, and they sample again the variety of responses a tuned circuit can have to sinusoidal excitation. Each of the three parts of Fig. 7.18-E consists of (*a*) the first or "data" page of RESSINET'S output, and (*b*) one of RESSINET'S various pictorial-graphical presentations of the response, this one being the complete response. The latter is considerably reduced in size so that one can clearly see the trend of the response in small compass; the associated table of numerical values is thus illegible here, but is of course full size in the actual computer output. The Q used is 10; the initial conditions are rest. For convenience, $C = 1$ and $E_m = 2$ throughout. In the three parts, $\phi = 0°$, $30°$, and $90°$, respectively, and $\omega_F/\omega_0 = \frac{1}{4}$, 1, and 4, respectively.

We observe that since $M = 2$ and $N = 4$ in (7.18-1), the initial value of the response should be zero in all cases (physically, there is no impulse to charge the capacitor instantaneously). But the initial *slope* of the response, the value of $r'(0+)$, will not be zero unless $\phi = \pm 90°$, for $r'(0+) = E_m \cos \phi$. We observe the RESSINET-computed values of $\sum p$ and of $\sum Kp$, which include of course the terms due to the excitation part of $R(s)$, on the data pages in Fig. 7.18-E. The check is satisfactory.

Most interesting is the transient buildup from rest of the (large) resonant response when driven at a frequency practically equal to the natural frequency ($\omega_F = \omega_0$). If the exciting frequency is appreciably different from the resonant frequency the transient may ride on the back of the steady-state (forced) part, or the steady-state part on the back of the transient (as we have already seen, in Sec. 3.13).

```
THIS IS RESSINET.        NOTES FOR THIS RUN:   MS  SECT. 7.18 - RESSINET EX.          RLC NET., Q = 10
                                               T = S/(S**2+0.1S+1)        EM = 2     PHI = 0                  WF = 1/4
                                               C

PARAMETER VALUES READ:   N =  1,  D =  2, NX = 0, NRPLTS = 1, NCODE = 0

FOR THE FOLLOWING CASE THE TRANSFER FUNCTION COEFFICIENTS, LOWEST POWER FIRST, ARE:

 NUMERATOR:

          0.0              1.000000

 DENOMINATOR:

          1.000000         0.100000          1.000000

 THE FCLLOWING ARE THE POLES OF THIS TRANSFER FUNCTION, T(S):

       -0.05000C  +J -0.998749

       -0.050000  +J  0.998749

THE (SINUSOIDAL) EXCITATION IS EM*COS(WF*T + PHI), IN WHICH:        EM =    2.0000                  WF =    0.2500

                                                                    PHI =    0.0    DEGREES, OR    0.0    RADIANS.

                NOTATION:   E(S) = MATE OF EXCITATION, T(S) = TRANSFER FUNCTION

                            R(S) = E(S)*T(S)

THE FOLLOWING ARE THE RESIDUES OF R(S) THEREAT:

         MAGNITUDE       ANGLE (RAD.)      ANGLE (DEG.)        REAL PART        IMAG. PART
         .........       ............      ............        .........        ..........

          1.067622        -4.705733        -269.618652         -0.007106         1.067598

          1.067622         4.705733         269.618652         -0.007106        -1.067598

THE TRANSFER FUNCTION, EVALUATED AT S = J*WF, IS:  0.2666E 00  AT   88.4724 DEGREES

THE (FORCED) RESPONSE PHASOR IS:  0.5331E 00  AT   88.4724 DEGREES

THE SUM OF THE RESIDUES IS:                                                                 0.1188E-05 + J  0.0

THE SUM OF THE PRODUCTS "KP" IS:                                                            0.2000E 01 + J 0.0

DO THEY CHECK?

THE FOLLOWING PLOTS ARE ASKED FOR:    6
```

FIG. 7.18-Ea

```
THIS IS RESSINET.          NOTES FOR THIS RUN:    MS  SECT. 7.18 - RESSINET EX.          RLC NET., Q = 10
                                                  T = S/(S**2+0.1S+1)        EM = 2      PHI = 0                WF = 1/4
                                                  C
```

PLOT OF (COMPLETE) RESPONSE ALONE:

SUBROUTINE 'PLOT4Z', BEING CALLED, GIVES (BAR MARKS ZERO ORDINATE):

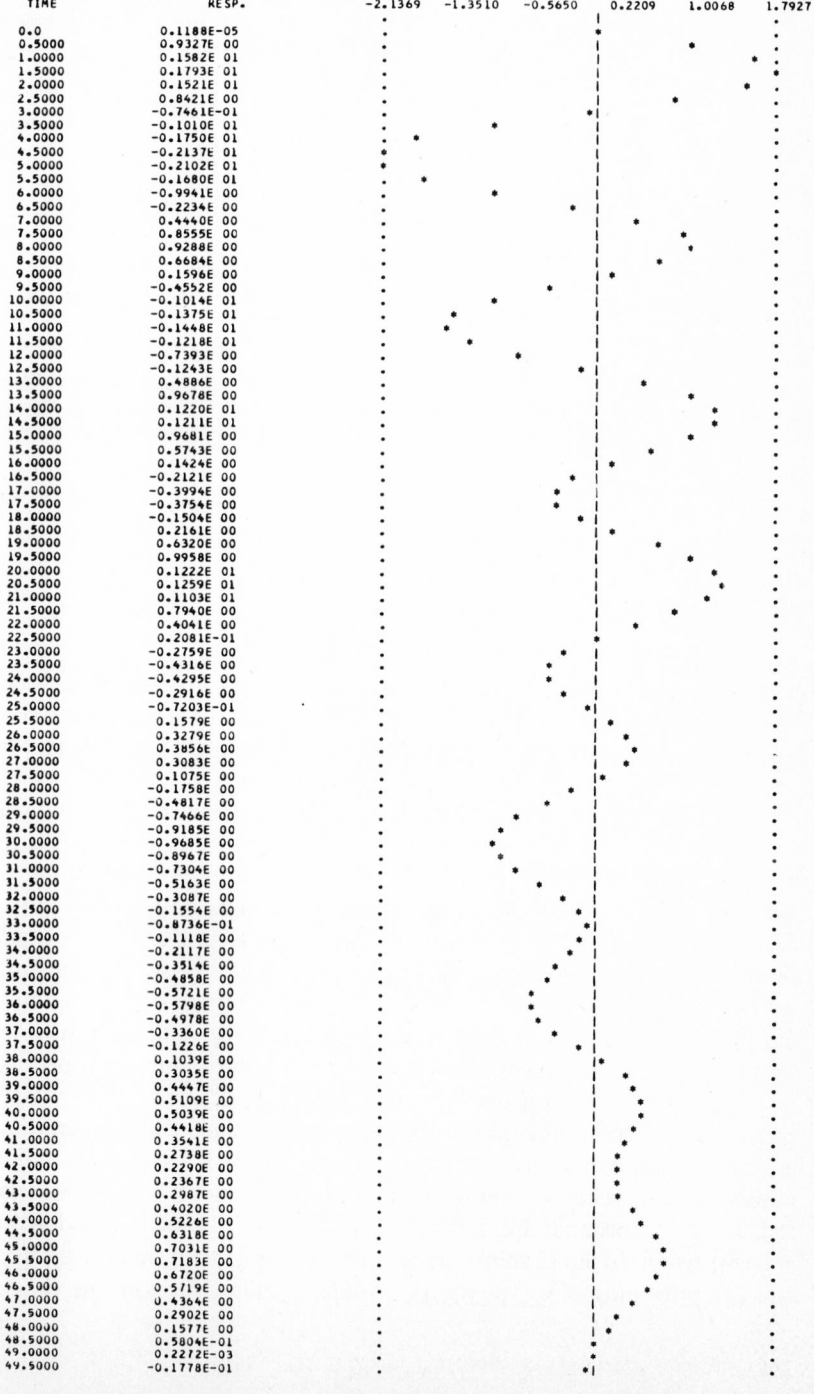

TIME	RESP.
0.0	0.1188E-05
0.5000	0.9327E 00
1.0000	0.1582E 01
1.5000	0.1793E 01
2.0000	0.1521E 01
2.5000	0.8421E 00
3.0000	-0.7461E-01
3.5000	-0.1010E 01
4.0000	-0.1750E 01
4.5000	-0.2137E 01
5.0000	-0.2102E 01
5.5000	-0.1680E 01
6.0000	-0.9941E 00
6.5000	-0.2234E 00
7.0000	0.4440E 00
7.5000	0.8555E 00
8.0000	0.9288E 00
8.5000	0.6684E 00
9.0000	0.1596E 00
9.5000	-0.4552E 00
10.0000	-0.1014E 01
10.5000	-0.1375E 01
11.0000	-0.1448E 01
11.5000	-0.1218E 01
12.0000	-0.7393E 00
12.5000	-0.1243E 00
13.0000	0.4886E 00
13.5000	0.9678E 00
14.0000	0.1220E 01
14.5000	0.1211E 01
15.0000	0.9681E 00
15.5000	0.5743E 00
16.0000	0.1424E 00
16.5000	-0.2121E 00
17.0000	-0.3994E 00
17.5000	-0.3754E 00
18.0000	-0.1504E 00
18.5000	0.2161E 00
19.0000	0.6320E 00
19.5000	0.9958E 00
20.0000	0.1222E 01
20.5000	0.1259E 01
21.0000	0.1103E 01
21.5000	0.7940E 00
22.0000	0.4041E 00
22.5000	0.2081E-01
23.0000	-0.2759E 00
23.5000	-0.4316E 00
24.0000	-0.4295E 00
24.5000	-0.2916E 00
25.0000	-0.7203E-01
25.5000	0.1579E 00
26.0000	0.3279E 00
26.5000	0.3856E 00
27.0000	0.3083E 00
27.5000	0.1075E 00
28.0000	-0.1758E 00
28.5000	-0.4817E 00
29.0000	-0.7466E 00
29.5000	-0.9185E 00
30.0000	-0.9685E 00
30.5000	-0.8967E 00
31.0000	-0.7304E 00
31.5000	-0.5163E 00
32.0000	-0.3087E 00
32.5000	-0.1554E 00
33.0000	-0.8736E-01
33.5000	-0.1118E 00
34.0000	-0.2117E 00
34.5000	-0.3514E 00
35.0000	-0.4858E 00
35.5000	-0.5721E 00
36.0000	-0.5798E 00
36.5000	-0.4978E 00
37.0000	-0.3360E 00
37.5000	-0.1226E 00
38.0000	0.1039E 00
38.5000	0.3035E 00
39.0000	0.4447E 00
39.5000	0.5109E 00
40.0000	0.5039E 00
40.5000	0.4418E 00
41.0000	0.3541E 00
41.5000	0.2738E 00
42.0000	0.2290E 00
42.5000	0.2367E 00
43.0000	0.2987E 00
43.5000	0.4020E 00
44.0000	0.5226E 00
44.5000	0.6318E 00
45.0000	0.7031E 00
45.5000	0.7183E 00
46.0000	0.6720E 00
46.5000	0.5719E 00
47.0000	0.4364E 00
47.5000	0.2902E 00
48.0000	0.1577E 00
48.5000	0.5804E-01
49.0000	0.2272E-03
49.5000	-0.1778E-01

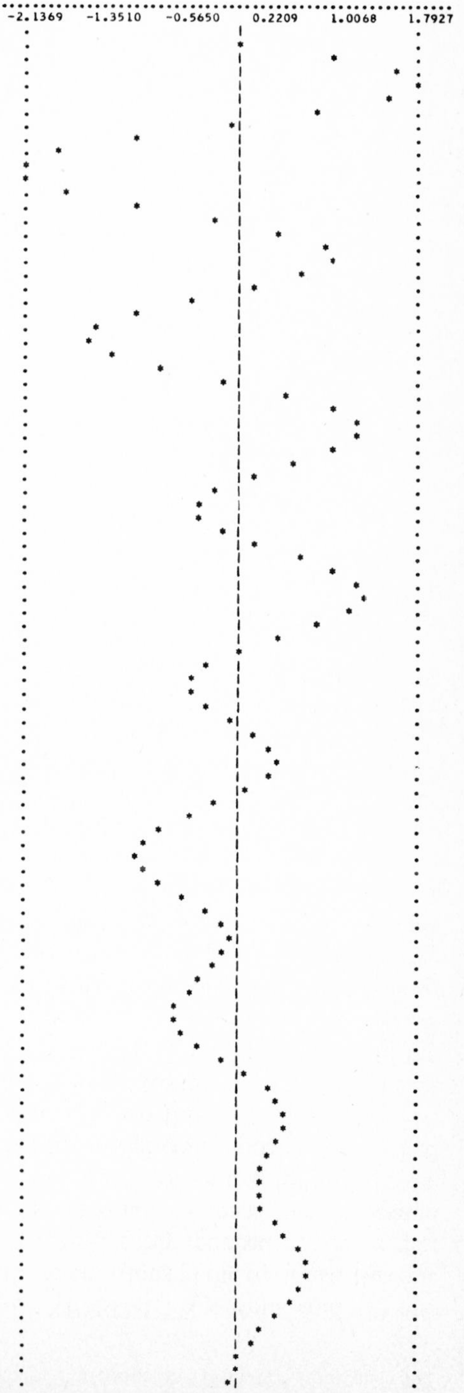

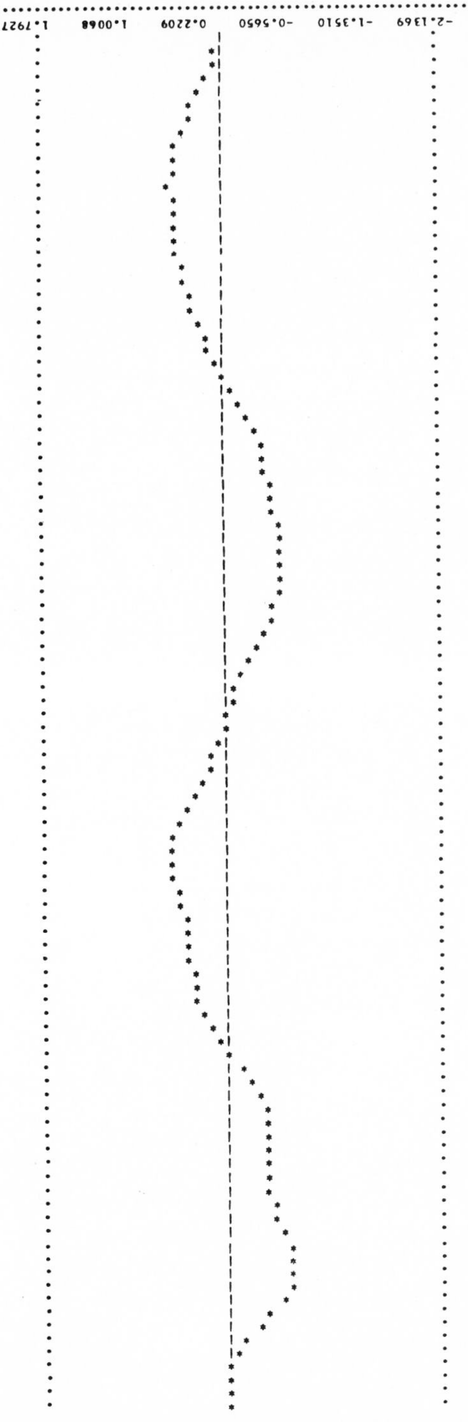

TIME	FORCED RESP.
50.0000	-0.8046E-02
50.5000	0.1098E-01
51.0000	0.1941E-01
51.5000	0.9549E-03
52.0000	-0.5311E-01
52.5000	-0.1420E 00
53.0000	-0.2559E 00
53.5000	-0.3785E 00
54.0000	-0.4911E 00
54.5000	-0.5767E 00
55.0000	-0.6244E 00
55.5000	-0.6312E 00
56.0000	-0.6020E 00
56.5000	-0.5487E 00
57.0000	-0.4861E 00
57.5000	-0.4290E 00
58.0000	-0.3881E 00
58.5000	-0.3680E 00
59.0000	-0.3663E 00
59.5000	-0.3747E 00
60.0000	-0.3814E 00
60.5000	-0.3742E 00
61.0000	-0.3436E 00
61.5000	-0.2851E 00
62.0000	-0.2002E 00
62.5000	-0.9630E-01
63.0000	0.1575E-01
63.5000	0.1236E 00
64.0000	0.2166E 00
64.5000	0.2878E 00
65.0000	0.3350E 00
65.5000	0.3616E 00
66.0000	0.3743E 00
66.5000	0.3819E 00
67.0000	0.3927E 00
67.5000	0.4121E 00
68.0000	0.4419E 00
68.5000	0.4793E 00
69.0000	0.5183E 00
69.5000	0.5508E 00
70.0000	0.5689E 00
70.5000	0.5666E 00
71.0000	0.5412E 00
71.5000	0.4940E 00
72.0000	0.4298E 00
72.5000	0.3556E 00
73.0000	0.2791E 00
73.5000	0.2069E 00
74.0000	0.1432E 00
74.5000	0.8926E-01
75.0000	0.4310E-01
75.5000	0.6853E-03
76.0000	-0.4292E-01
76.5000	-0.9208E-01
77.0000	-0.1493E 00
77.5000	-0.2147E 00
78.0000	-0.2855E 00
78.5000	-0.3569E 00
79.0000	-0.4231E 00
79.5000	-0.4785E 00
80.0000	-0.5189E 00
80.5000	-0.5424E 00
81.0000	-0.5492E 00
81.5000	-0.5421E 00
82.0000	-0.5250E 00
82.5000	-0.5023E 00
83.0000	-0.4774E 00
83.5000	-0.4525E 00
84.0000	-0.4279E 00
84.5000	-0.4021E 00
85.0000	-0.3723E 00
85.5000	-0.3355E 00
86.0000	-0.2892E 00
86.5000	-0.2323E 00
87.0000	-0.1654E 00
87.5000	-0.9063E-01
88.0000	-0.1163E-01
88.5000	0.6741E-01
89.0000	0.1425E 00
89.5000	0.2104E 00
90.0000	0.2695E 00
90.5000	0.3195E 00
91.0000	0.3615E 00
91.5000	0.3971E 00
92.0000	0.4284E 00
92.5000	0.4569E 00
93.0000	0.4829E 00
93.5000	0.5058E 00
94.0000	0.5240E 00
94.5000	0.5352E 00
95.0000	0.5371E 00
95.5000	0.5277E 00
96.0000	0.5062E 00
96.5000	0.4728E 00
97.0000	0.4290E 00
97.5000	0.3769E 00
98.0000	0.3193E 00
98.5000	0.2587E 00
99.0000	0.1972E 00
99.5000	0.1361E 00
100.0000	0.7578E-01

FIG. 7.18–Ea. (*Con.*)

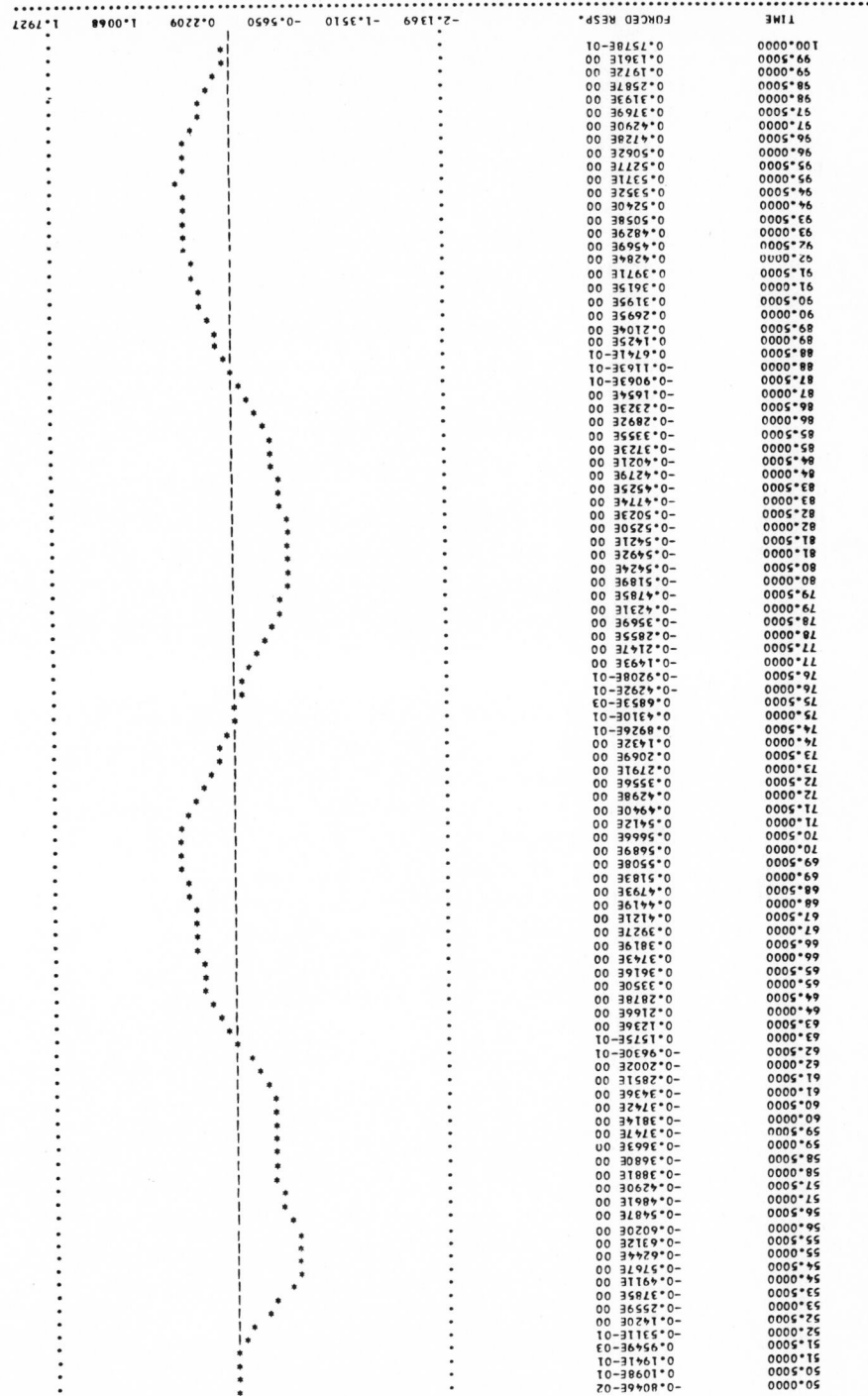

The steady-state response of a network to sinusoidal excitation is in its own right important. To study this we examine the variation of the forced response, or simply of $T(j\omega_F)$ itself, as a function of the frequency of excitation, ω_F. Since this is a complex number, its behavior requires a two-dimensional presentation. One may tabulate the (rectangular-form) real and imaginary parts or the (polar-form) magnitude and angle, and plot them in various ways. Simplest are plots of these parts versus ω_F on an arithmetic scale. Sometimes a logarithmic scale is useful, as for wide-range-of-frequency plots, and to clearly show asymptotic (low- or high-frequency) behavior; logarithmic measure of $|T(j\omega_F)|$ can also be useful (Appendix D). One can suppress ω_F and then plot *both* parts, i.e., the whole of $T(j\omega_F)$ on one diagram, the locus of $T(j\omega_F)$ in (its) complex-number plane. To compute the tables and make the plots versus ω_F (on either arithmetic or logarithmic scale) is obviously a "very straightforward and lengthy" task. Hence it too is eminently programmable. The program FOMEGA (Appendix C) computes these tables and makes various plots versus ω_F on arithmetic or logarithmic scale, on demand. (The line printer's ability to plot complex loci is limited; such plots are better made elsewise.) For illustration of its use, we ask FOMEGA to compute the sinusoidal steady-state behavior of the high-pass filter of Sec. 7.17 [see (7.17-1) and Figs. 7.17-A, 7.18-A]; Fig. 7.18-F shows the resulting magnitude and angle of transfer function plotted on an arithmetic frequency scale. (The table of values and various other plots are omitted.) The program was asked to compute only for $0.2 \leqq \omega_F \leqq 1.8$, since at very low frequencies $|T|$ is practically zero and at very high frequencies it is practically constant. The name "high-pass filter" derives from the sinusoidal-steady-state behavior of $|T|$: high-frequency sinusoidal excitations are "passed" very well, but low-frequency signals are practically rejected. The angle, for programming simplicity, is computed to be 90° (not 450°) at the origin, but its *variation* is correct.

A second example of the use of FOMEGA is the calculation of the steady-state behavior of the large ($N = 14$) low-pass filter of Fig. 7.18-C. Figure 7.18-G shows the results and justifies the name. This calculation is quickly done by machine; by hand it would require many hours of drudgery. For variety, FOMEGA was here asked to use a *logarithmic* scale of frequency, from 0.01 to 4.0, using 41 logarithmically equally spaced points; $|T|$ is also given in logarithmic units (the "dBV" is proportional to the logarithm of $|T|$; see Appendix D). The descent of $|T|$ at $\omega_F = 1$ is so steep in Fig. 7.18-G*a* that we ask for an expansion of the region nearby, which Fig. 7.18-G*b* presents. We note that the flat top of Fig. 7.18-G*a* is not really flat, nor the descent really linear. But the name "low-pass filter" is obviously appropriate. [The program arbitrarily "saturates" at -100 db (where $|T| = 10^{-5}$); the actual characteristic continues down.] Figure 7.18-G*b* also shows the (expanded) angle characteristic.

For (exponential) excitations other than impulse, step, and sinusoid, less common as they are, we may simply form the ersatz transfer function $(T_{WF}\,T_{NFT})$ and ask IMPSTEP for the corresponding impulse response, in true wave-forming network fashion. But these, as well as cases of nonexponential excitation, can also be handled by *convolution*, the programming of which, begun in Chap. 4, now needs more attention. We start with IMPSTEP, and use it to produce the impulse response of the network in question from a given $T(s)$; we then accumulate the products thereof with the excitation that form the integrand (see Sec. 4.08), all of which is "lengthy but straightforward." The CONVOL program (Appendix C) does this, and more.

To it impulse-response information can be supplied not only implicitly in a transfer function (as above), but also as a formula, for example

$$r_\delta(t) = K_0 t^4 e^{-\alpha t}, \tag{7.18-2}$$

from which the program computes the values it needs. This option may be useful if one is interested in the response of a hypothetical, as yet undesigned, network (characterized by that impulse response) to some particular excitation function. The synthesis of the network can well wait until this response is evaluated: it may not be acceptable; the impulse-response prescription may have to be changed. Experiments that use $r_\delta(t)$ instead of the network itself can save work! To facilitate such experiments, CONVOL incorporates a very flexible formula:

$$r_\delta(t) = \frac{Q_N}{Q_D} \exp\,(Q_E) \cos Q_C \tag{7.18-3}$$

in which each Q is a polynomial in t (of degree 7 or less). Only the four

```
THIS IS RESSINET.         NOTES FOR THIS RUN:     MS  SECT. 7.18 - RESSINET EX.          RLC NET., Q = 10
                                                  T = S/(S**2+0.1S+1)       EM = 2.0   PHI = 30 DEG.
                                                  WF = 1.0   CHECK SUM K AND SUM KP                  TEST FROM DISK

PARAMETER VALUES READ:    N =  1,  D =  2, NX = 0, NRPLTS = 1, NCODE = 0

FOR THE FOLLOWING CASE THE TRANSFER FUNCTION COEFFICIENTS, LOWEST POWER FIRST, ARE:

 NUMERATOR:

          0.0                 1.000000

 DENOMINATOR:

          1.000000            0.100000            1.000000

 THE FOLLOWING ARE THE POLES OF THIS TRANSFER FUNCTION, T(S):

      -0.050000  +J -0.998749

      -0.050000  +J  0.998749

THE (SINUSOIDAL) EXCITATION IS EM*COS(WF*T + PHI), IN WHICH:          EM =    2.0000            WF =    1.0000

                                                                      PHI =   30.0000 DEGREES, OR    0.5236 RADIANS.

                 NOTATION:   E(S) = MATE OF EXCITATION, T(S) = TRANSFER FUNCTION

                             R(S) = E(S)*T(S)

THE FOLLOWING ARE THE RESIDUES OF R(S) THEREAT:

         MAGNITUDE        ANGLE (RAD.)       ANGLE (DEG.)         REAL PART          IMAG. PART
         .........        ............       ............         .........          ..........

         10.227001         -3.702444         -212.134460          -8.660250           5.439816

         10.227001          3.702444          212.134460          -8.660250          -5.439816

THE TRANSFER FUNCTION, EVALUATED AT S = J*WF, IS:  0.1000E 02  AT   0.0    DEGREES

THE (FORCED) RESPONSE PHASOR IS:  0.2000E 02  AT   30.0000 DEGREES

THE SUM OF THE RESIDUES IS:                                                             0.0        + J  0.0

THE SUM OF THE PRODUCTS "KP" IS:                                                        0.1732E 01 + J-0.5960E-07

DO THEY CHECK?

THE FOLLOWING PLOTS ARE ASKED FOR:    6
```

FIG. 7.18-E*b*

```
THIS IS RESSINET.        NOTES FOR THIS RUN:   MS  SECT. 7.18 - RESSINET EX.        RLC NET., Q = 10
                                               T = S/(S**2+0.1S+1)      EM = 2     PHI = 30 DEG.
                                               WF = 1.0   CHECK SUM K AND SUM KP              TEST FROM DISK

PLOT OF (COMPLETE) RESPONSE ALONE:

SUBROUTINE 'PLOT4Z', BEING CALLED, GIVES (BAR MARKS ZERO ORDINATE):
```

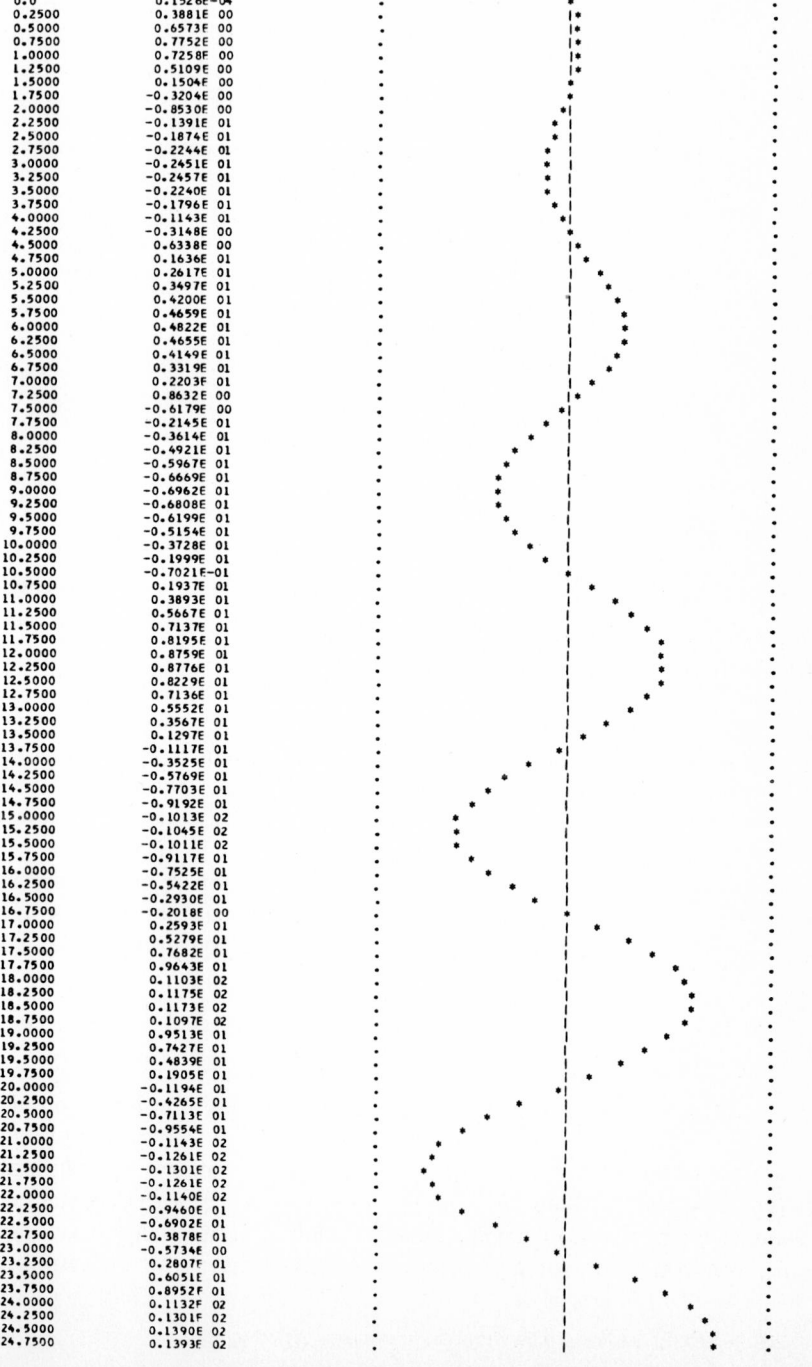

TIME	RESP.
0.0	0.1526E-04
0.2500	0.3881E 00
0.5000	0.6573E 00
0.7500	0.7752E 00
1.0000	0.7258E 00
1.2500	0.5109E 00
1.5000	0.1504E 00
1.7500	-0.3204E 00
2.0000	-0.8530E 00
2.2500	-0.1391E 01
2.5000	-0.1874E 01
2.7500	-0.2244E 01
3.0000	-0.2451E 01
3.2500	-0.2457E 01
3.5000	-0.2240E 01
3.7500	-0.1796E 01
4.0000	-0.1143E 01
4.2500	-0.3148E 00
4.5000	0.6338E 00
4.7500	0.1636E 01
5.0000	0.2617E 01
5.2500	0.3497E 01
5.5000	0.4200E 01
5.7500	0.4659E 01
6.0000	0.4822E 01
6.2500	0.4655E 01
6.5000	0.4149E 01
6.7500	0.3319E 01
7.0000	0.2203E 01
7.2500	0.8632E 00
7.5000	-0.6179E 00
7.7500	-0.2145E 01
8.0000	-0.3614E 01
8.2500	-0.4921E 01
8.5000	-0.5967E 01
8.7500	-0.6669E 01
9.0000	-0.6962E 01
9.2500	-0.6808E 01
9.5000	-0.6199E 01
9.7500	-0.5154E 01
10.0000	-0.3728E 01
10.2500	-0.1999E 01
10.5000	-0.7021E-01
10.7500	0.1937E 01
11.0000	0.3893E 01
11.2500	0.5667E 01
11.5000	0.7137E 01
11.7500	0.8195E 01
12.0000	0.8759E 01
12.2500	0.8776E 01
12.5000	0.8229E 01
12.7500	0.7136E 01
13.0000	0.5552E 01
13.2500	0.3567E 01
13.5000	0.1297E 01
13.7500	-0.1117E 01
14.0000	-0.3525E 01
14.2500	-0.5769E 01
14.5000	-0.7703E 01
14.7500	-0.9192E 01
15.0000	-0.1013E 02
15.2500	-0.1045E 02
15.5000	-0.1011E 02
15.7500	-0.9117E 01
16.0000	-0.7525E 01
16.2500	-0.5422E 01
16.5000	-0.2930E 01
16.7500	-0.2018E 00
17.0000	0.2593E 01
17.2500	0.5279E 01
17.5000	0.7682E 01
17.7500	0.9643E 01
18.0000	0.1103E 02
18.2500	0.1175E 02
18.5000	0.1173E 02
18.7500	0.1097E 02
19.0000	0.9513E 01
19.2500	0.7427E 01
19.5000	0.4839E 01
19.7500	0.1905E 01
20.0000	-0.1194E 01
20.2500	-0.4265E 01
20.5000	-0.7113E 01
20.7500	-0.9554E 01
21.0000	-0.1143E 02
21.2500	-0.1261E 02
21.5000	-0.1301E 02
21.7500	-0.1261E 02
22.0000	-0.1140E 02
22.2500	-0.9460E 01
22.5000	-0.6902E 01
22.7500	-0.3878E 01
23.0000	-0.5734E 00
23.2500	0.2807E 01
23.5000	0.6051E 01
23.7500	0.8952E 01
24.0000	0.1132E 02
24.2500	0.1301E 02
24.5000	0.1390E 02
24.7500	0.1393E 02

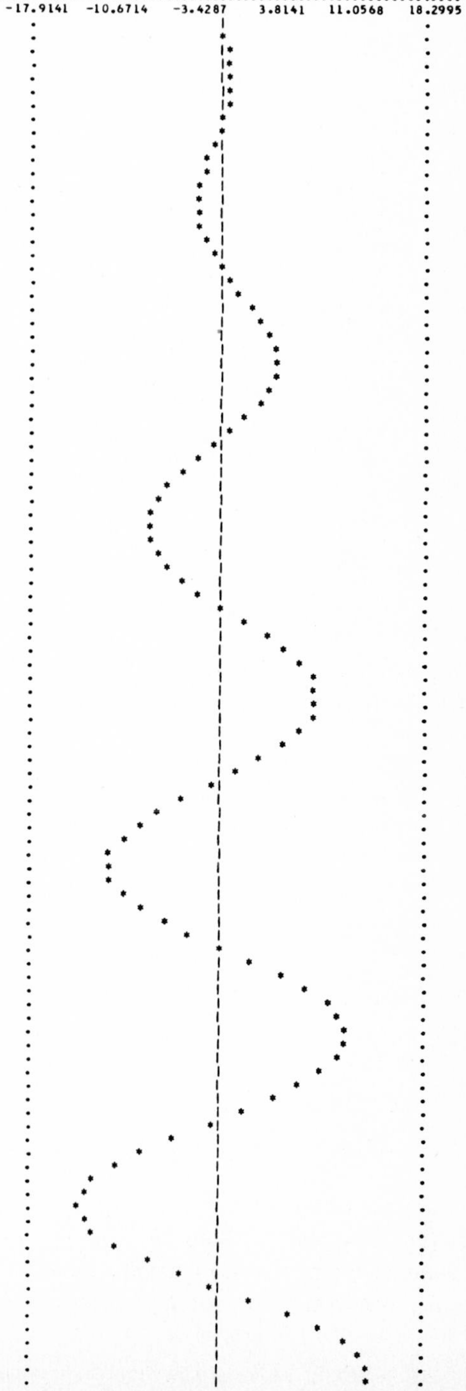

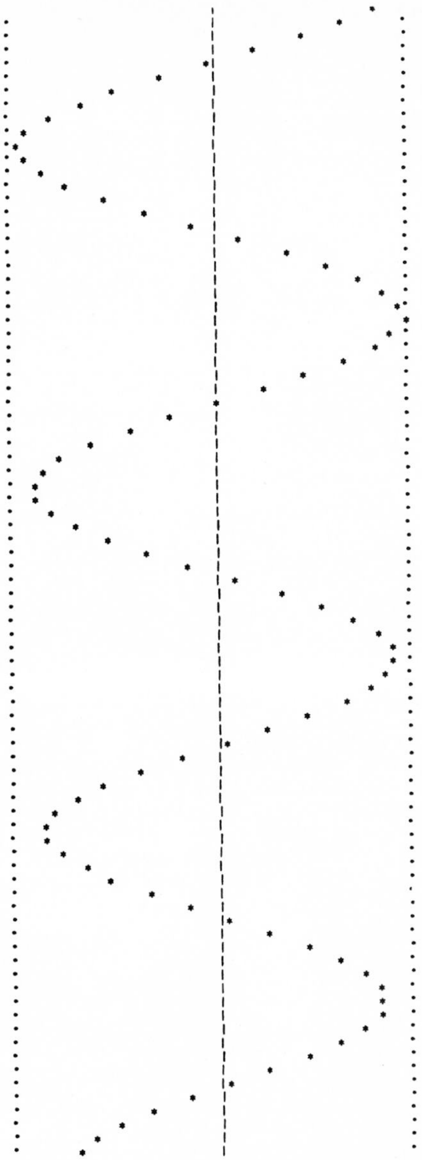

25.0000	0.1309E 02
25.2500	0.1141E 02
25.5000	0.9005E 01
25.7500	0.6010E 01
26.0000	0.2608E 01
26.2500	-0.9903E 00
26.5000	-0.4561E 01
26.7500	-0.7879E 01
27.0000	-0.1073E 02
27.2500	-0.1294E 02
27.5000	-0.1436E 02
27.7500	-0.1489E 02
28.0000	-0.1450E 02
28.2500	-0.1319E 02
28.5000	-0.1105E 02
28.7500	-0.8196E 01
29.0000	-0.4809E 01
29.2500	-0.1094E 01
29.5000	0.2718E 01
29.7500	0.6389E 01
30.0000	0.9689E 01
30.2500	0.1241E 02
30.5000	0.1437E 02
30.7500	0.1545E 02
31.0000	0.1557E 02
31.2500	0.1472E 02
31.5000	0.1294E 02
31.7500	0.1034E 02
32.0000	0.7080E 01
32.2500	0.3355E 01
32.5000	-0.6024E 00
32.7500	-0.4548E 01
33.0000	-0.8234E 01
33.2500	-0.1143E 02
33.5000	-0.1393E 02
33.7500	-0.1557E 02
34.0000	-0.1625E 02
34.2500	-0.1593E 02
34.5000	-0.1460E 02
34.7500	-0.1236E 02
35.0000	-0.9331E 01
35.2500	-0.5705E 01
35.5000	-0.1704E 01
35.7500	0.2425E 01
36.0000	0.6424E 01
36.2500	0.1004E 02
36.5000	0.1305E 02
36.7500	0.1526E 02
37.0000	0.1653E 02
37.2500	0.1678E 02
37.5000	0.1597E 02
37.7500	0.1417E 02
38.0000	0.1148E 02
38.2500	0.8057E 01
38.5000	0.4118E 01
38.7500	-0.9585E-01
39.0000	-0.4322E 01
39.2500	-0.8296E 01
39.5000	-0.1177E 02
39.7500	-0.1452E 02
40.0000	-0.1638E 02
40.2500	-0.1723E 02
40.5000	-0.1701E 02
40.7500	-0.1572E 02
41.0000	-0.1345E 02
41.2500	-0.1033E 02
41.5000	-0.6559E 01
41.7500	-0.2363E 01
42.0000	0.1995E 01
42.2500	0.6245E 01
42.5000	0.1012E 02
42.7500	0.1338E 02
43.0000	0.1581E 02
43.2500	0.1727E 02
43.5000	0.1766E 02
43.7500	0.1695E 02
44.0000	0.1518E 02
44.2500	0.1246E 02
44.5000	0.8951E 01
44.7500	0.4876E 01
45.0000	0.4854E 00
45.2500	-0.3949E 01
45.5000	-0.8150E 01
45.7500	-0.1186E 02
46.0000	-0.1484E 02

FIG. 7.18-E*b*. (*Con.*)

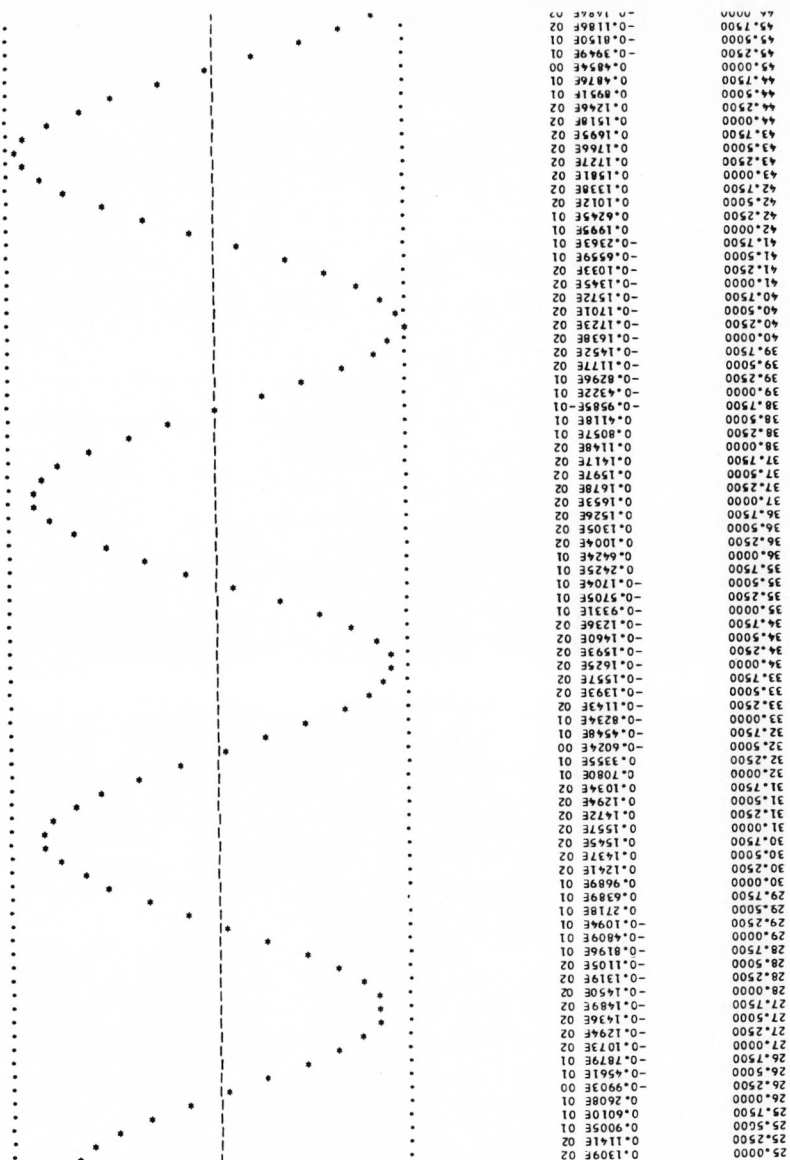

polynomials' coefficients need be supplied as data, the eight C's in

$$Q_N = C_0 + C_1 t + C_2 t^2 + \cdots + C_7 t^7 \tag{7.18-4}$$

and the 24 coefficients of the other three polynomials. This formula includes functions such as (7.18-2), e^{-t^2}, $1/(t + t^4)$, $(3 + 8t^5)e^{-6t^3} \cos(2 + 6t + 13t^2)/(1 + t^2)$; it is comprehensive and meets most needs!

It may be convenient, or necessary, to supply the impulse-response information for convolution simply as a series of values, or points that outline a curve of $r_\delta(t)$. The points may represent measurements of the impulse response of an extant network, easier to obtain than its transfer function; perhaps such measurement is the *only* way to characterize the network. (For some reason the actual response to the excitation of interest cannot itself be measured.) Or, in a sketchier way than by formula, the values may represent some hypothetical network of interest. This option too is provided by CONVOL, which will (linearly) interpolate between the given points to obtain the values it needs. An example is given in Fig. 7.18-H*a*; for others, see Figs. 7.18-J*b* and 7.18-K*a*.

Excitation information can likewise be supplied to CONVOL either formally (by means of the same formula) or as a series of values between which to interpolate. [The ability to accept $e(t)$ in the form of an equivalent (wave-forming) transfer function, i.e., the ability to calculate the impulse response of the cascade connection of two networks by convolution, could easily be added.]

The "read points and interpolate" option also provides for modulating a sinusoidal wave with the piecewise-linear wave: on request, it generates the excitation (or impulse-response) wave

$$e(t) = E_m \cos(\omega_F t + \phi)[f_{PWL}(t)] \tag{7.18-5}$$

in which f_{PWL} is the function defined by the linear interpolation between points process. Such shaped bursts or pulses of sinusoids (see Fig. 7.18-H*b* and *c*), for example) are often useful as test signals in network discussions. [If a more rounded, curved modulation of a sinusoid is wanted, the formula option (7.18-3) can be used.] An additional example is given in Fig. 7.18-L*a*.

We illustrate CONVOL's ability with several examples. First consider a network whose transfer function is

$$T(s) = \frac{K_1}{(s + \alpha)^4}. \tag{7.18-6}$$

(This might be the network of Fig. 7.18-I*a* with proper element values.) Suppose it to be excited by a wave of the same form as its own impulse response (pair 35)

$$e(t) = K_2 t^3 e^{-\alpha t}. \tag{7.18-7}$$

```
THIS IS RESSINET.        NOTES FOR THIS RUN:    MS  CH. 7          SECT. 7.18
                                                Q = 10             WF = 4             PHI = 90 DEG.

PARAMETER VALUES READ:    N =  1,  D =  2, NX = 0, NRPLTS = 1, NCODE = 1

FOR THE FOLLOWING CASE THE TRANSFER FUNCTION COEFFICIENTS, LOWEST POWER FIRST, ARE:

 NUMERATOR:

           0.0                 1.000000

 DENOMINATOR:

           1.000000            0.100000            1.000000

 THE FOLLOWING ARE THE POLES OF THIS TRANSFER FUNCTION, T(S):

       -0.050000  +J -0.998749

       -0.050000  +J  0.998749

THE (SINUSOIDAL) EXCITATION IS EM*COS(WF*T + PHI), IN WHICH:          EM =    2.0000                WF =     4.0000

                                                                      PHI =   90.0000 DEGREES, OR    1.5708 RADIANS.

              NOTATION:  E(S) = MATE OF EXCITATION, T(S) = TRANSFER FUNCTION

                         R(S) = E(S)*T(S)

THE FOLLOWING ARE THE RESIDUES OF R(S) THEREAT:

         MAGNITUDE        ANGLE (RAD.)       ANGLE (DEG.)        REAL PART          IMAG. PART
         .........        ............       ............        .........          ..........

         0.266906         -3.198269          -183.247345         -0.266477          0.015119

         0.266906          3.198269           183.247345         -0.266477         -0.015119

THE TRANSFER FUNCTION, EVALUATED AT S = J*WF, IS:  0.2666E 00   AT  -88.4725 DEGREES

THE (FORCED) RESPONSE PHASOR IS:  0.5331E 00   AT    1.5275 DEGREES

THE SUM OF THE RESIDUES IS:                                                              -0.1788E-06 + J  0.0

THE SUM OF THE PRODUCTS "KP" IS:                                                         -0.3949E-06 + J-0.1048E-07

DO THEY CHECK?

THE FOLLOWING PLOTS ARE ASKED FOR:    6
```

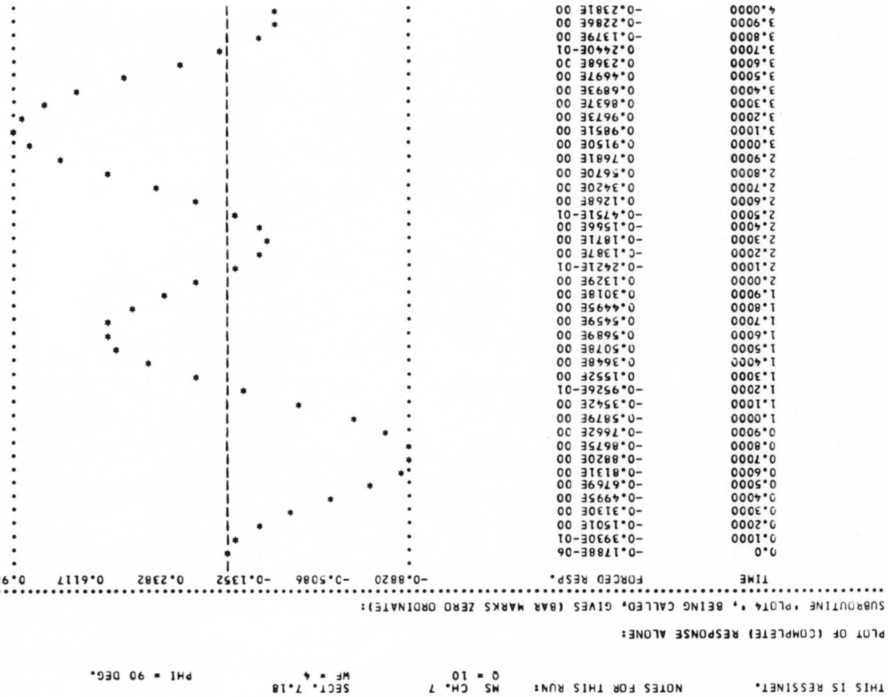

THIS IS RESSINET. NOTES FOR THIS RUN: MS CH. 7 SECT. 7.18 Q = 10 WF = 4 PHI = 90 DEG.
PLOT OF (COMPLETE) RESPONSE ALONE:
SUBROUTINE PLOT4, BEING CALLED, GIVES (BAR MARKS ZERO ORDINATE):
TIME
FORCED RESP.
-0.8820
-0.5086
-0.1352
0.2382
0.6117
0.9851
0.0 -0.1788E-06
0.1000 -0.3930E-01
0.2000 -0.1501E 00
0.3000 -0.3130E 00
0.4000 -0.4995E 00
0.5000 -0.6769E 00
0.6000 -0.8131E 00
0.7000 -0.8820E 00
0.8000 -0.8675E 00
0.9000 -0.7662E 00
1.0000 -0.5879E 00
1.1000 -0.3542E 00
1.2000 -0.9526E-01
1.3000 0.1552E 00
1.4000 0.3644E 00
1.5000 0.5078E 00
1.6000 0.5689E 00
1.7000 0.5459E 00
1.8000 0.4495E 00
1.9000 0.3018E 00
2.0000 0.1329E 00
2.1000 -0.2421E-01
2.2000 -0.1387E 00
2.3000 -0.1871E 00
2.4000 -0.1556E 00
2.5000 -0.4751E-01
2.6000 0.1268E 00
2.7000 0.3420E 00
2.8000 0.5670E 00
2.9000 0.7681E 00
3.0000 0.9150E 00
3.1000 0.9851E 00
3.2000 0.9673E 00
3.3000 0.8637E 00
3.4000 0.6893E 00
3.5000 0.4697E 00
3.6000 0.2368E 00
3.7000 0.2440E-01
3.8000 -0.1379E 00
3.9000 -0.2286E 00
4.0000 -0.2381E 00

FIG. 7.18-E*c*

4.1000	-0.1700E 00	
4.2000	-0.4047E-01	
4.3000	0.1243E 00	
4.4000	0.2924E 00	
4.5000	0.4310E 00	
4.6000	0.5122E 00	
4.7000	0.5168E 00	
4.8000	0.4378E 00	
4.9000	0.2816E 00	
5.0000	0.6670E-01	
5.1000	-0.1787E 00	
5.2000	-0.4215E 00	
5.3000	-0.6285E 00	
5.4000	-0.7721E 00	
5.5000	-0.8340E 00	
5.6000	-0.8086E 00	
5.7000	-0.7033E 00	
5.8000	-0.5379E 00	
5.9000	-0.3411E 00	
6.0000	-0.1458E 00	
6.1000	0.1568E-01	
6.2000	0.1171E 00	
6.3000	0.1421E 00	
6.4000	0.8712E-01	
6.5000	-0.3820E-01	
6.6000	-0.2127E 00	
6.7000	-0.4067E 00	
6.8000	-0.5872E 00	
6.9000	-0.7227E 00	
7.0000	-0.7883E 00	
7.1000	-0.7699E 00	
7.2000	-0.6661E 00	
7.3000	-0.4889E 00	
7.4000	-0.2615E 00	
7.5000	-0.1461E-01	
7.6000	0.2178E 00	
7.7000	0.4044E 00	
7.8000	0.5210E 00	
7.9000	0.5547E 00	
8.0000	0.5053E 00	
8.1000	0.3860E 00	
8.2000	0.2205E 00	
8.3000	0.3999E-01	
8.4000	-0.1225E 00	
8.5000	-0.2370E 00	
8.6000	-0.2814E 00	
8.7000	-0.2451E 00	
8.8000	-0.1305E 00	
8.9000	0.4707E-01	
9.0000	0.2620E 00	
9.1000	0.4821E 00	
9.2000	0.6743E 00	
9.3000	0.8089E 00	
9.4000	0.8653E 00	
9.5000	0.8344E 00	
9.6000	0.7206E 00	
9.7000	0.5408E 00	
9.7999	0.3218E 00	
9.8999	0.9639E-01	
9.9999	-0.1023E 00	
10.0999	-0.2457E 00	
10.1999	-0.3142E 00	
10.2999	-0.3005E 00	
10.3999	-0.2105E 00	
10.4999	-0.6232E-01	
10.5999	0.1164E 00	
10.6999	0.2930E 00	
10.7999	0.4352E 00	
10.8999	0.5160E 00	
10.9999	0.5181E 00	
11.0999	0.4365E 00	
11.1999	0.2797E 00	
11.2999	0.6793E-01	
11.3999	-0.1695E 00	
11.4999	-0.3992E 00	
11.5999	-0.5887E 00	
11.6999	-0.7117E 00	
11.7999	-0.7519E 00	
11.8999	-0.7061E 00	
11.9999	-0.5839E 00	
12.0999	-0.4069E 00	
12.1999	-0.2048E 00	
12.2999	-0.1101E-01	
12.3999	0.1430E 00	
12.4999	0.2323E 00	
12.5999	0.2427E 00	
12.6999	0.1729E 00	
12.7999	0.3442E-01	
12.8999	-0.1496E 00	
12.9999	-0.3487E 00	
13.0999	-0.5295E 00	
13.1999	-0.6614E 00	
13.2999	-0.7209E 00	
13.3999	-0.6958E 00	
13.4999	-0.5870E 00	
13.5999	-0.4084E 00	
13.6999	-0.1847E 00	
13.7999	0.5255E-01	
13.8999	0.2696E 00	
13.9999	0.4361E 00	
14.0999	0.5296E 00	
14.1999	0.5394E 00	
14.2999	0.4676E 00	
14.3999	0.3295E 00	
14.4999	0.1506E 00	
14.5999	-0.3746E-01	
14.6999	-0.2015E 00	
14.7999	-0.3125E 00	
14.8999	-0.3501E 00	
14.9999	-0.3057E 00	

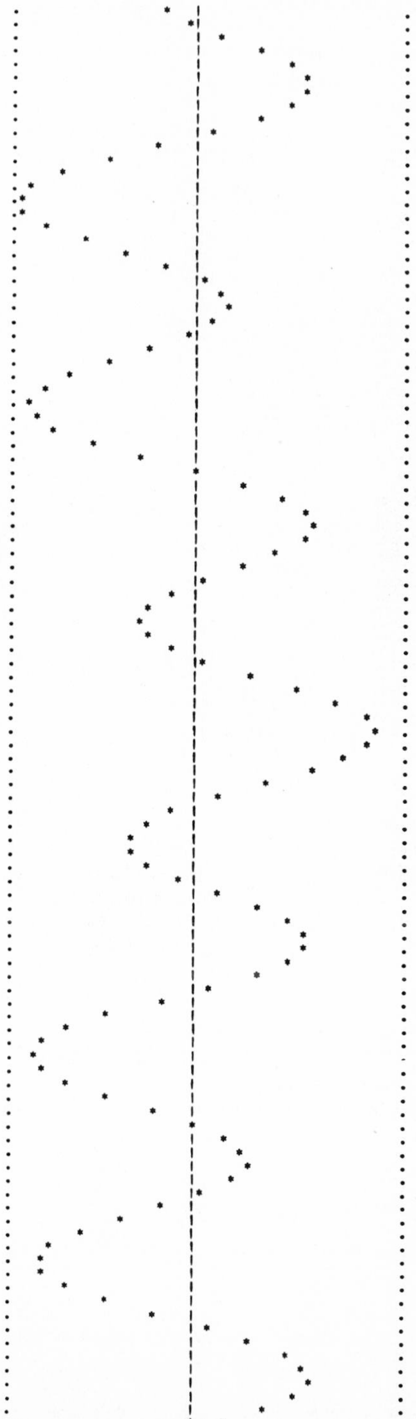

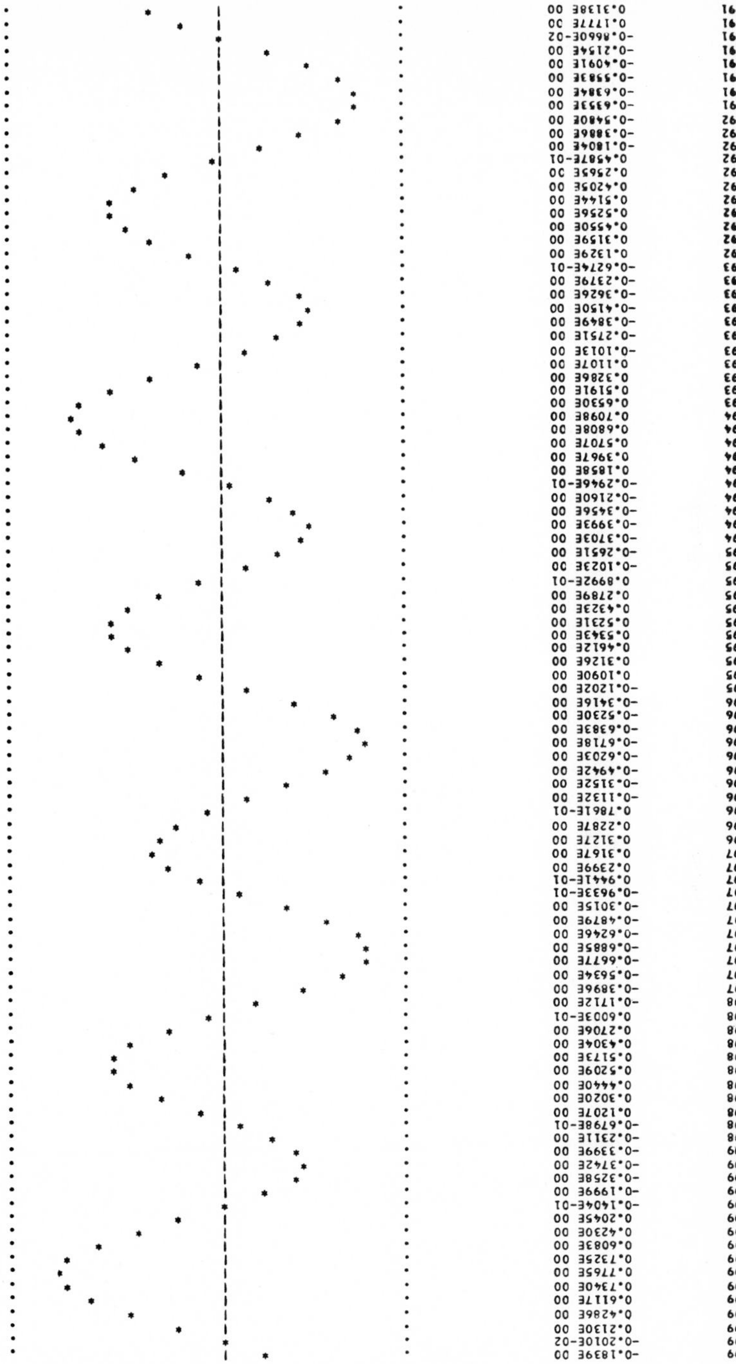

FIG. 7.18-Ec. (*Con.*)

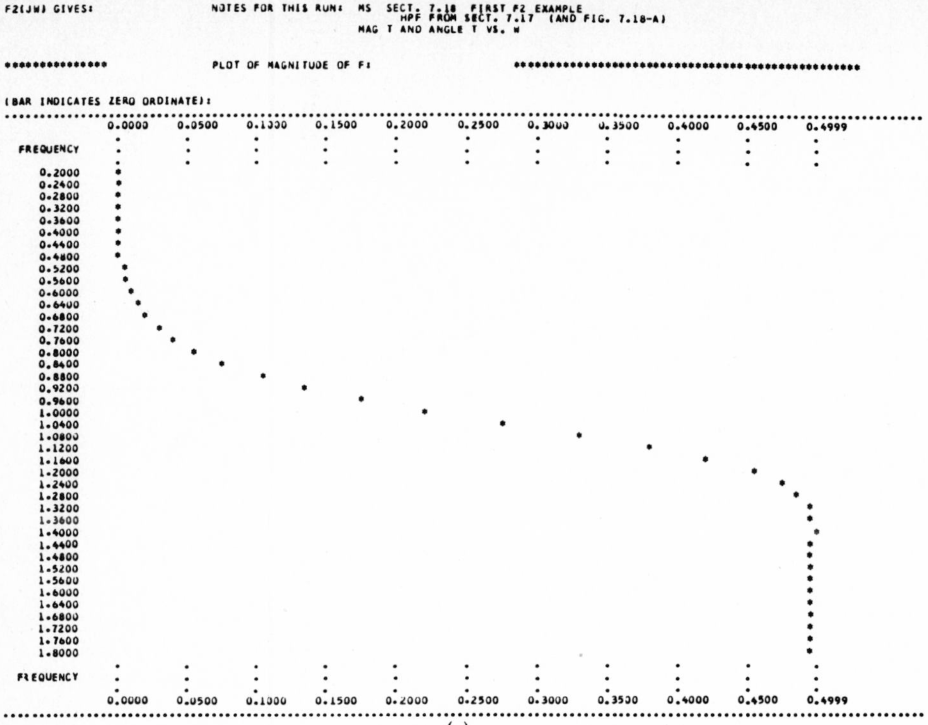
F2(JW) GIVES:
NOTES FOR THIS RUN: MS SECT. 7.18 FIRST F2 EXAMPLE
HPF FROM SECT. 7.17 (AND FIG. 7.18-A)
MAG T AND ANGLE T VS. W
PLOT OF MAGNITUDE OF F:
(BAR INDICATES ZERO ORDINATE):
0.0000 0.0500 0.1000 0.1500 0.2000 0.2500 0.3000 0.3500 0.4000 0.4500 0.4999
FREQUENCY
0.2000 0.2400 0.2800 0.3200 0.3600 0.4000 0.4400 0.4800 0.5200 0.5600 0.6000 0.6400 0.6800 0.7200 0.7600 0.8000 0.8400 0.8800 0.9200 0.9600 1.0000 1.0400 1.0800 1.1200 1.1600 1.2000 1.2400 1.2800 1.3200 1.3600 1.4000 1.4400 1.4800 1.5200 1.5600 1.6000 1.6400 1.6800 1.7200 1.7600 1.8000

(*a*)

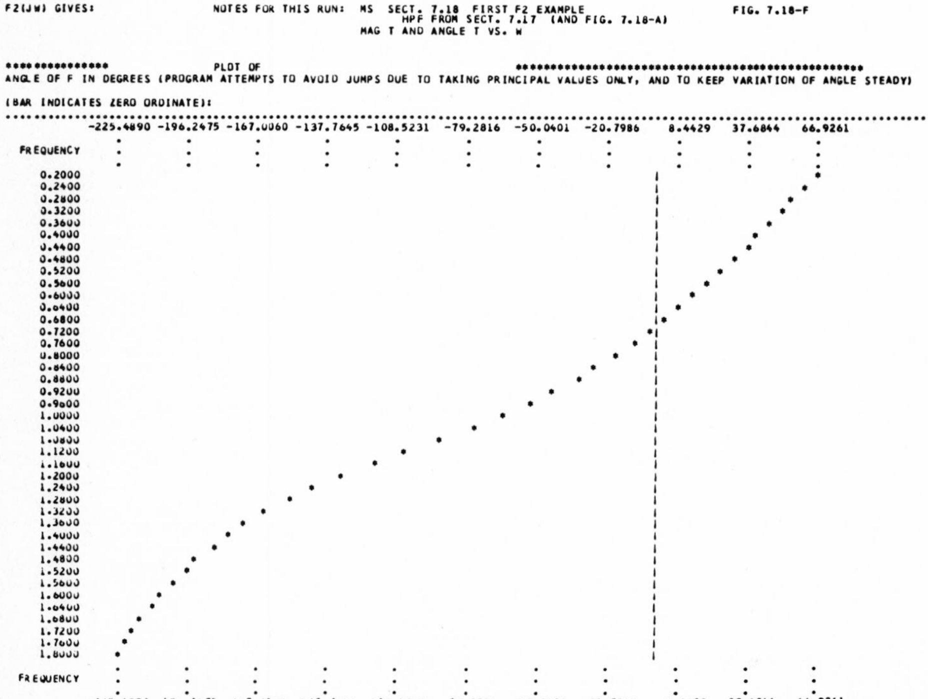
F2(JW) GIVES:
NOTES FOR THIS RUN: MS SECT. 7.18 FIRST F2 EXAMPLE
HPF FROM SECT. 7.17 (AND FIG. 7.18-A)
MAG T AND ANGLE T VS. W
FIG. 7.18-F
PLOT OF
ANGLE OF F IN DEGREES (PROGRAM ATTEMPTS TO AVOID JUMPS DUE TO TAKING PRINCIPAL VALUES ONLY, AND TO KEEP VARIATION OF ANGLE STEADY)
(BAR INDICATES ZERO ORDINATE):
-225.4890 -196.2475 -167.0060 -137.7645 -108.5231 -79.2816 -50.0401 -20.7986 8.4429 37.6844 66.9261
FREQUENCY
0.2000 0.2400 0.2800 0.3200 0.3600 0.4000 0.4400 0.4800 0.5200 0.5600 0.6000 0.6400 0.6800 0.7200 0.7600 0.8000 0.8400 0.8800 0.9200 0.9600 1.0000 1.0400 1.0800 1.1200 1.1600 1.2000 1.2400 1.2800 1.3200 1.3600 1.4000 1.4400 1.4800 1.5200 1.5600 1.6000 1.6400 1.6800 1.7200 1.7600 1.8000

(*b*)

FIG. 7.18–F

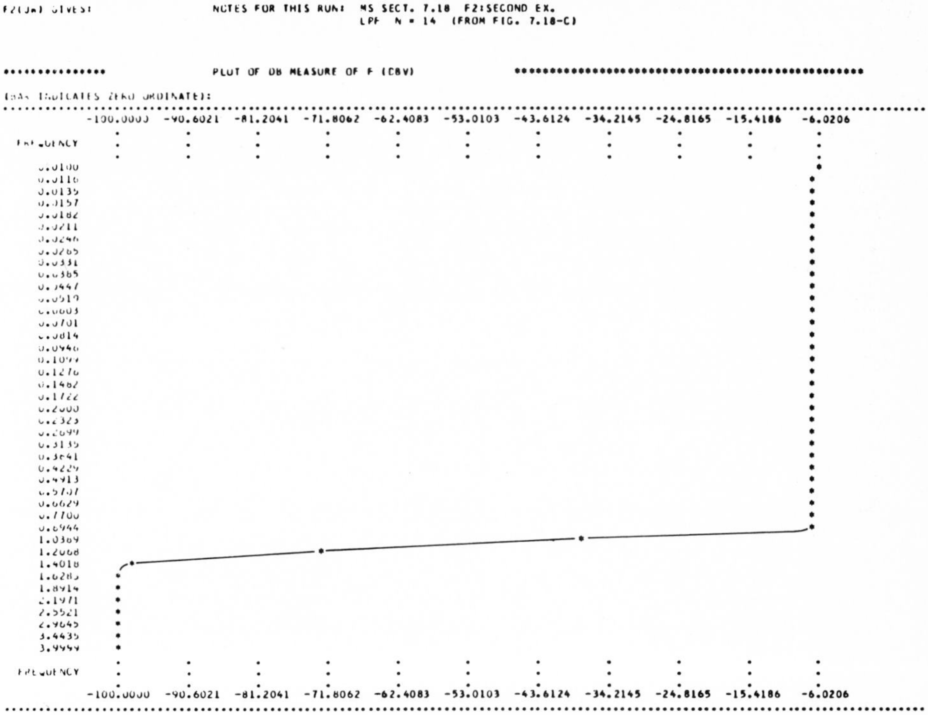
NOTES FOR THIS RUN: MS SECT. 7.18 F2:SECOND EX.
LPF N = 14 (FROM FIG. 7.18-C)
PLOT OF DB MEASURE OF F (CBV)
-100.0000 -90.6021 -81.2041 -71.8062 -62.4083 -53.0103 -43.6124 -34.2145 -24.8165 -15.4186 -6.0206
FREQUENCY

FIG. 7.18-G*a*

F2(JW) GIVES: NOTES FOR THIS RUN: MS SECT. 7.18 F2:SECOND EX.
LPF N = 14 (FROM FIG. 7.18-C)

*************** PLOT OF DB MEASURE OF F (CBV) ***************

(BAR INDICATES ZERO ORDINATE):

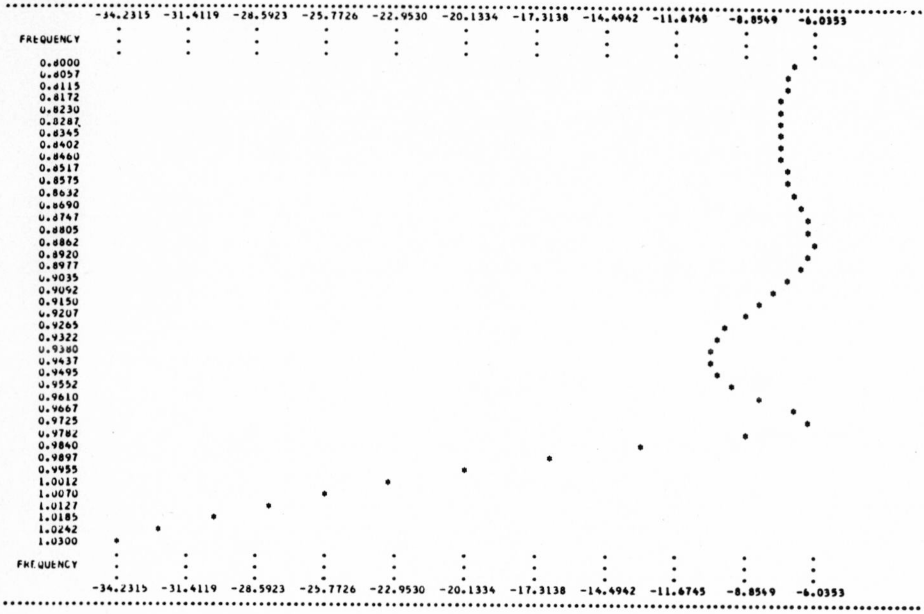

*************** PLOT OF ***************
ANGLE OF F IN DEGREES (PROGRAM ATTEMPTS TO AVOID JUMPS DUE TO TAKING PRINCIPAL VALUES ONLY, AND TO KEEP VARIATION OF ANGLE STEADY)

(BAR INDICATES ZERO ORDINATE):

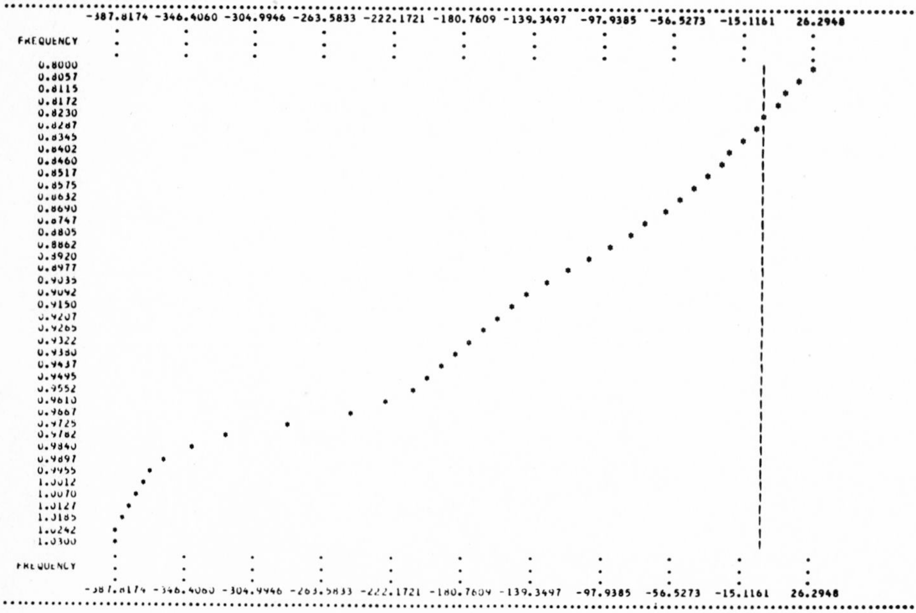

FIG. 7.18-G*b*

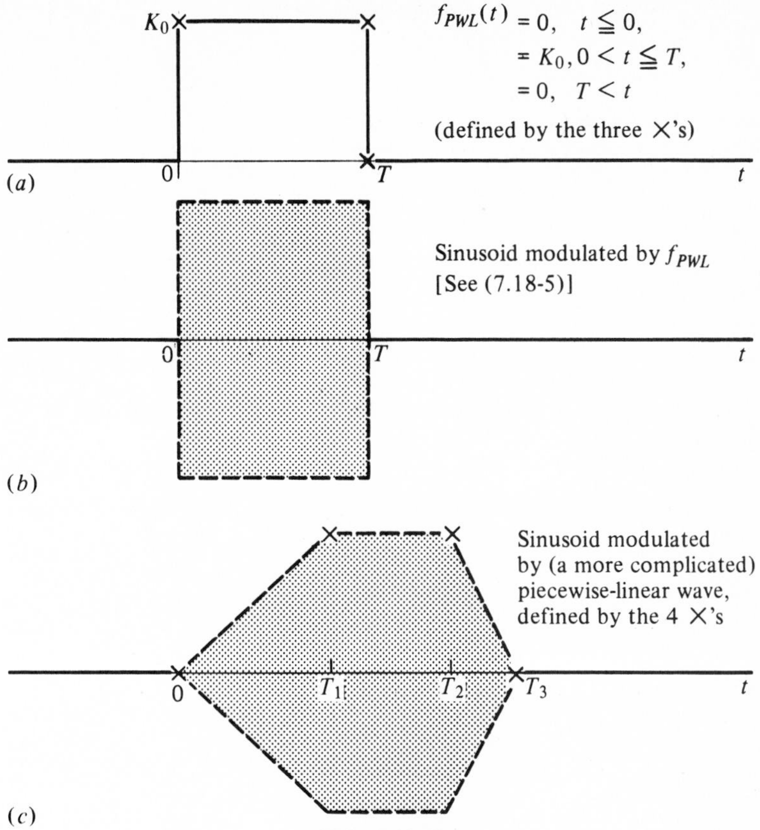

FIG. 7.18-H

The multiple pole in

$$R(s) = E(s)T(s) = \frac{6K_1K_2}{(s+\alpha)^8} \tag{7.18-8}$$

prevents using IMPSTEP (in which the natural frequencies must be separate and distinct). But CONVOL finds no difficulty in evaluating the response; in it we use the formula (7.18-3) for both $r_\delta(t)$ and $e(t)$, setting $6K_1K_2 = 10$ and $\alpha = 1$ for numerical simplicity, and obtain Fig. 7.18-I*b*. The network itself is sluggish, so is the excitation, and the response (Fig. 7.18-I*b*) is of course even more so. (The excitation reaches its peak at $t = 3$, the response only at $t = 7$.) This example, incidentally, gives us a convenient and comforting opportunity to check the operation of CONVOL. For the mate of (7.18-7) can readily be obtained formally, from pair 35:

$$\frac{10}{(s+1)^8} \sim \frac{1}{504} t^7 e^{-t}. \tag{7.18-9}$$

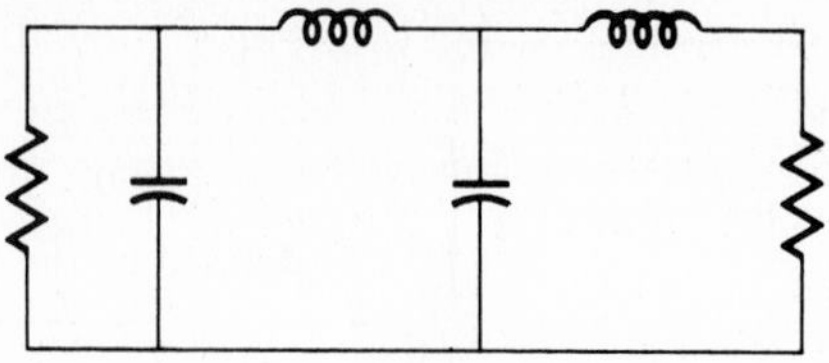

(a)

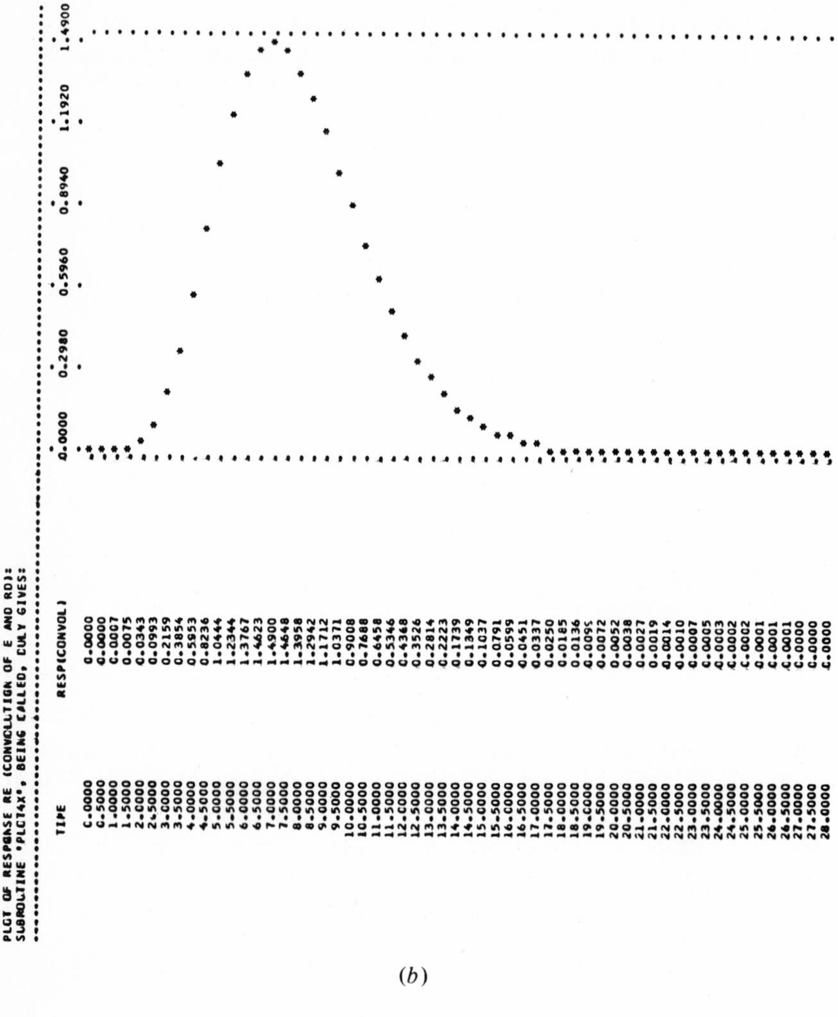
PLOT OF RESPONSE RE (CONVOLUTION OF E AND RD):
SUBROUTINE 'PLOT4X', BEING CALLED, DULY GIVES:
TIME
RESP(CONVOL)
0.0000
0.2980
0.5960
0.8940
1.1920
1.4900
0.0000 0.0000
0.5000 0.0000
1.0000 0.0007
1.5000 0.0075
2.0000 0.0343
2.5000 0.0993
3.0000 0.2159
3.5000 0.3854
4.0000 0.5953
4.5000 0.8236
5.0000 1.0444
5.5000 1.2344
6.0000 1.3767
6.5000 1.4623
7.0000 1.4900
7.5000 1.4648
8.0000 1.3958
8.5000 1.2942
9.0000 1.1712
9.5000 1.0371
10.0000 0.9008
10.5000 0.7688
11.0000 0.6458
11.5000 0.5346
12.0000 0.4368
12.5000 0.3526
13.0000 0.2814
13.5000 0.2223
14.0000 0.1739
14.5000 0.1349
15.0000 0.1037
15.5000 0.0791
16.0000 0.0599
16.5000 0.0451
17.0000 0.0337
17.5000 0.0250
18.0000 0.0185
18.5000 0.0136
19.0000 0.0099
19.5000 0.0072
20.0000 0.0052
20.5000 0.0038
21.0000 0.0027
21.5000 0.0019
22.0000 0.0014
22.5000 0.0010
23.0000 0.0007
23.5000 0.0005
24.0000 0.0003
24.5000 0.0002
25.0000 0.0002
25.5000 0.0001
26.0000 0.0001
26.5000 0.0001
27.0000 0.0000
27.5000 0.0000
28.0000 0.0000

(b)

FIG. 7.18–I

Table 7.18-A (p. 615) lists values of this formal result and those obtained from CONVOL; note the general agreement, to the fourth decimal place. Since CONVOL performs an integration numerically, its results are approximate and of course somewhat in error. But in engineering perspective, three significant figures are ample here; CONVOL acquits itself very well indeed!

For a second illustration of CONVOL's abilities, suppose we are interested in a hypothetical network whose impulse response is the sluggish oscillation sketched (solid curve) in Fig. 7.18-J*a*. We ask, in particular, what would be its response to the positive-negative rectangular-pulse excitation wave of Fig. 7.18-J*b*? We mark six representative points on the impulse-response sketch (the six ×'s in Fig. 7.18-J*a*), and let CONVOL interpolate linearly between them; that is, we assume the dashed lines in Fig. 7.18-J*a* to be an acceptable approximation to $r_\delta(t)$. The excitation, being itself piecewise-linear, is represented accurately by the five points marked × in Fig. 7.18-J*b*. CONVOL, given these six and five points

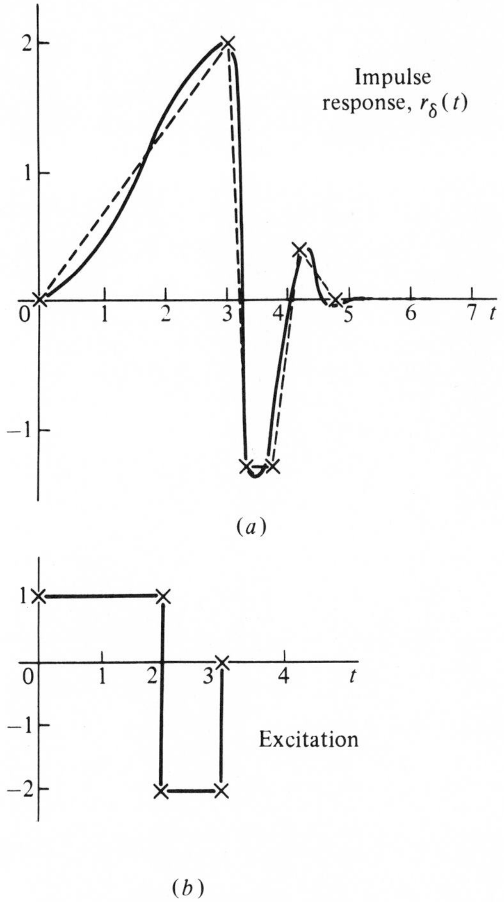

FIG. 7.18-J

PLOT OF RESPONSE RE (CONVOLUTICN OF E AND RO):

SUBROUTINE 'PLOT4Z', BEING CALLED, GIVES (BAR MARKS ZERO ORDINATE):

TIME	RESP(CONVOL)
0.0000	0.0000E 00
0.1000	0.3333E-02
0.2000	0.1333E-01
0.3000	0.3000E-01
0.4000	0.5333E-01
0.5000	0.8333E-01
0.6000	0.1200E 00
0.7000	0.1633E 00
0.8000	0.2133E 00
0.9000	0.2700E 00
1.0000	0.3333E 00
1.1000	0.4033E 00
1.2000	0.4800E 00
1.3000	0.5633E 00
1.4000	0.6533E 00
1.5000	0.7500E 00
1.6000	0.8533E 00
1.7000	0.9633E 00
1.8000	0.1080E 01
1.9000	0.1203E 01
2.0000	0.1333E 01
2.1000	0.1463E 01
2.2000	0.1580E 01
2.3000	0.1683E 01
2.4000	0.1773E 01
2.5000	0.1850E 01
2.6000	0.1913E 01
2.7000	0.1963E 01
2.8000	0.2000E 01
2.9000	0.2023E 01
3.0000	0.2033E 01
3.1000	0.1986E 01
3.2000	0.1842E 01
3.3000	0.1602E 01
3.4000	0.1264E 01
3.5000	0.8756E 00
3.6000	0.4800E 00
3.7000	0.7778E-01
3.8000	-0.3311E 00
3.9000	-0.7167E 00
4.0000	-0.1049E 01
4.1000	-0.1328E 01
4.2000	-0.1553E 01
4.3000	-0.1762E 01
4.4000	-0.1991E 01
4.5000	-0.2240E 01
4.6000	-0.2509E 01
4.7000	-0.2798E 01
4.8000	-0.3107E 01
4.9000	-0.3429E 01
5.0000	-0.3758E 01
5.1000	-0.3997E 01
5.2000	-0.3952E 01
5.3000	-0.3624E 01
5.4000	-0.3013E 01
5.5000	-0.2209E 01
5.6000	-0.1391E 01
5.7000	-0.5600E 00
5.8000	0.2844E 00
5.9000	0.1082E 01
6.0000	0.1713E 01
6.1000	0.2113E 01
6.2000	0.2153E 01
6.3000	0.1907E 01
6.4000	0.1520E 01
6.5000	0.1053E 01
6.6000	0.6267E 00
6.7000	0.2400E 00
6.8000	-0.1067E 00
6.9000	-0.3867E 00
7.0000	-0.5467E 00
7.1000	-0.5867E 00
7.2000	-0.5067E 00
7.3000	-0.3556E 00
7.4000	-0.2311E 00
7.5000	-0.1333E 00
7.6000	-0.6222E-01
7.7000	-0.1778E-01
7.8000	-0.8742E-07
7.9000	0.0000E 00
8.0000	0.0000E 00
TIME	RESP(CONVOL)

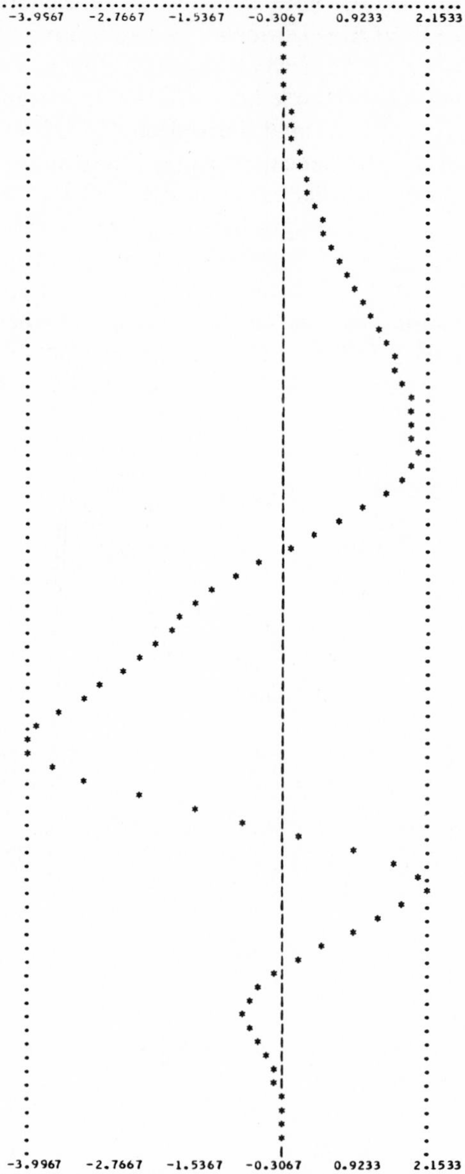

(c)

FIG. 7.18-J. (*Con.*)

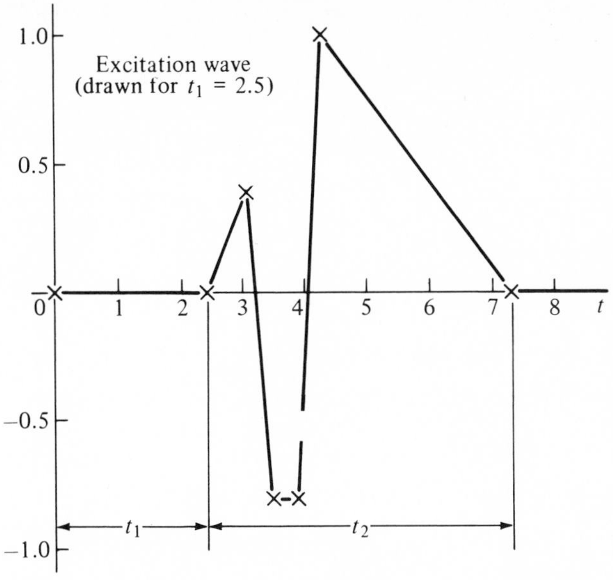

FIG. 7.18-Ka

and asked to use its interpolation ability for both impulse response and excitation, produces the results of Fig. 7.18-J*c*. The response is slowed and generally made more rounded or "curvy," as one would expect. And, since both excitation and impulse response waves are of finite duration, there comes a time ($t = 4.8 + 3.0 = 7.8$) after which the response is zero. For the particular analysis job in hand, one could now proceed to detailed discussion of this response.

It is sobering to consider the number of function evaluations, multiplications, and additions (corresponding to 161 points in the integration) that

TABLE 7.18-A

t	Values of $r(t)$	
	Calculated from (7.18-8) $\left(\frac{t^7 e^{-t}}{504}\right)$	**Obtained from CONVOL (Fig. 7.18-I*b*)**
0	0	0
2	0.0344	0.0343
4	0.5954	0.5953
6	1.3768	1.3767
8	1.3959	1.3958
10	0.9008	0.9008
12	0.4368	0.4368

CONVOL performs in a second or so, and to estimate the time it would take to obtain the data for Fig. 7.18-J*c* by hand!

As a third illustration, consider the same hypothetical network, characterized by the impulse response of Fig. 7.18-J*a*, for which we again use the six-point approximation. As excitation $e(t)$ suppose now, as an experiment, that we use the *same* waveshape, but reverse it in time and shift it to start at a time t_1, as in Fig. 7.18-K*a*. In the convolution operation $e(t)$ will be again reversed in time (see Fig. 4.07-A) so that the two factors in the convolution integrand become identical, except for a shift in time. As convolution proceeds, and the waves "slide over each other," there will come a time when the two exactly coincide; at this instant the magnitude of the integral (the response) will take its maximum value. At other instants, when the waves do not coincide, there will usually be a reduction in area; the response values will be smaller. The output of CONVOL (Fig. 7.18-K*b*) shows a marked peak at the coincidence time $t = 7.3$. This is of course equal to $(t_1 + t_2)$ or the sum of the "delay" t_1 and the pulse duration t_2. Now if t_1 is not known, but is to be measured by observing the response, this particular sort of *matching* of excitation-signal shape to impulse-response shape can be helpful because it produces so clearly defined a maximum. For example, in air-traffic control using radio-pulse ranging (radar), the excitation $e(t)$ may be the echo returned from an airplane in response to a signal emitted at $t = 0$; the network is then the receiver. The echo may be weak, and confused by noise, so that the correlation and the marked peaking of the response because of the matching can be useful: since t_2 is known, t_1 is easily determined from the time of the peak value of response $(t_1 + t_2)$. Compare the response of the same network to an arbitrary, unmatched pulse (Fig.

```
CONVOLUTION PROGRAM 'CONVOL'    .....*****

        GENERAL NOTES FOR THIS RUN:   MS SECT. 7.18 - CONVOL ILL. #3
                                      PWL EXPRESSIONS FOR BOTH E AND RD
                                      EXCITATION MATCHED TO IMPULSE RESPONSE

 CODES READ IN:     NPLOT =  1,    NEXC = -1,    NRD = -1,  NWR = 0      DATA READ: DT =  0.050, T2 = 12.000, NRT = 241,  I.E.
 UTILIZES THESE VALUES OF THE TIME (NRT = 241):    0.0     (  0.050)  12.000     **********  NRT MUST BE ODD; VERIFY THIS HERE********
 AND COMPUTES RESPONSE AT THESE VALUES OF TIME:    0.0     (  0.100)  12.000

PROGRAM  READ VALUES OF EXCITATION FROM INPUT (CARD) DATA AND INTERPOLATED:

           RDANDI DATA: NCODE = 0, WF =    0.0   PHI =    0.0   DEGREES

           AND POINTS:            TIME                FUNCTION

                               0.0                 0.0
                               2.500000            0.0
                               3.099999            0.800000
                               3.500000           -1.599999
                               3.900000           -1.599999
                               4.299999            2.000000
                               7.299999            0.0

PROGRAM READ VALUES OF IMPULSE RESPONSE FROM INPUT   (CARD) DATA AND INTERPOLATED:

           RDANDI DATA: NCODE = 0, WF =    0.0   PHI =    0.0   DEGREES

           AND POINTS:            TIME                FUNCTION

                               0.0                 0.0
                               3.000000            2.000000
                               3.400000           -1.599999
                               3.799999           -1.599999
                               4.200000            0.800000
                               4.799999            0.0

PLOT OF RESPONSE RE (CONVOLUTION OF E AND RD):

SUBROUTINE 'PLOT4 ', BEING CALLED, GIVES (BAR MARKS ZERO ORDINATE):
...........................................................................................................
           TIME                 RESP(CONVOL)             -0.1767    1.0298    2.2364    3.4429    4.6494    5.8559
                                                            .I                                                 .
            0.0                 0.0                         .*                                                 .
            0.1000              0.0                         .*                                                 .
            0.2000              0.0                         .*                                                 .
```

FIG. 7.18-K*b*

TIME	RESP(CONVOL)
0.3000	0.0
0.4000	0.0
0.5000	0.0
0.6000	0.0
0.7000	0.0
0.8000	0.0
0.9000	0.0
1.0000	0.0
1.1000	0.0
1.2000	0.0
1.3000	0.0
1.4000	0.0
1.5000	0.0
1.6000	0.0
1.7000	0.0
1.8000	0.0
1.9000	0.0
2.0000	0.0
2.1000	0.0
2.2000	0.0
2.3000	0.0
2.4000	0.0
2.5000	0.0
2.6000	0.1481E-03
2.7000	0.1185E-02
2.8000	0.4000E-02
2.9000	0.9481E-02
3.0000	0.1852E-01
3.1000	0.3200E-01
3.2000	0.5000E-01
3.3000	0.6933E-01
3.4000	0.8600E-01
3.5000	0.9600E-01
3.6000	0.9600E-01
3.7000	0.8533E-01
3.8000	0.6400E-01
3.9000	0.3200E-01
4.0000	-0.9666E-02
4.1000	-0.5600E-01
4.2000	-0.1010E 00
4.3000	-0.1387E 00
4.4000	-0.1641E 00
4.5000	-0.1766E 00
4.6000	-0.1767E 00
4.7000	-0.1647E 00
4.8000	-0.1413E 00
4.9000	-0.1067E 00
5.0000	-0.6141E-01
5.1000	-0.5924E-02
5.2000	0.5933E-01
5.3000	0.1339E 00
5.4000	0.2174E 00
5.5000	0.3093E 00
5.6000	0.4071E 00
5.7000	0.4996E 00
5.8000	0.5733E 00
5.9000	0.6151E 00
6.0000	0.6136E 00
6.1000	0.5653E 00
6.2000	0.4809E 00
6.3000	0.4181E 00
6.4000	0.4477E 00
6.5000	0.6459E 00
6.6000	0.1070E 01
6.7000	0.1693E 01
6.8000	0.2469E 01
6.9000	0.3341E 01
7.0000	0.4241E 01
7.1000	0.5050E 01
7.2000	0.5632E 01
7.3000	0.5856E 01
7.4000	0.5632E 01
7.5000	0.5050E 01
7.6000	0.4241E 01
7.7000	0.3341E 01
7.8000	0.2469E 01
7.9000	0.1693E 01
8.0000	0.1070E 01
8.1000	0.6459E 00
8.2000	0.4477E 00
8.3000	0.4181E 00
8.4000	0.4809E 00
8.5000	0.5653E 00
8.6000	0.6136E 00
8.7000	0.6151E 00
8.8000	0.5733E 00
8.9000	0.4996E 00
9.0000	0.4071E 00
9.1000	0.3093E 00
9.2000	0.2174E 00
9.3000	0.1339E 00
9.4000	0.5934E-01
9.5000	-0.5925E-02
9.6000	-0.6141E-01
9.7000	-0.1067E 00
9.8000	-0.1413E 00
9.9000	-0.1647E 00
10.0000	-0.1767E 00
10.1000	-0.1766E 00
10.2000	-0.1641E 00
10.3000	-0.1387E 00
10.4000	-0.1010E 00
10.5000	-0.5600E-01
10.6000	-0.9667E-02
10.7000	0.3200E-01
10.8000	0.6400E-01
10.9000	0.8533E-01
11.0000	0.9600E-01
11.1000	0.9600E-01
11.2000	0.8600E-01
11.3000	0.6933E-01
11.4000	0.5000E-01
11.5000	0.3200E-01
11.6000	0.1852E-01
11.7000	0.9482E-02
11.8000	0.4000E-02
11.9000	0.1135E-02
12.0000	0.1482E-03

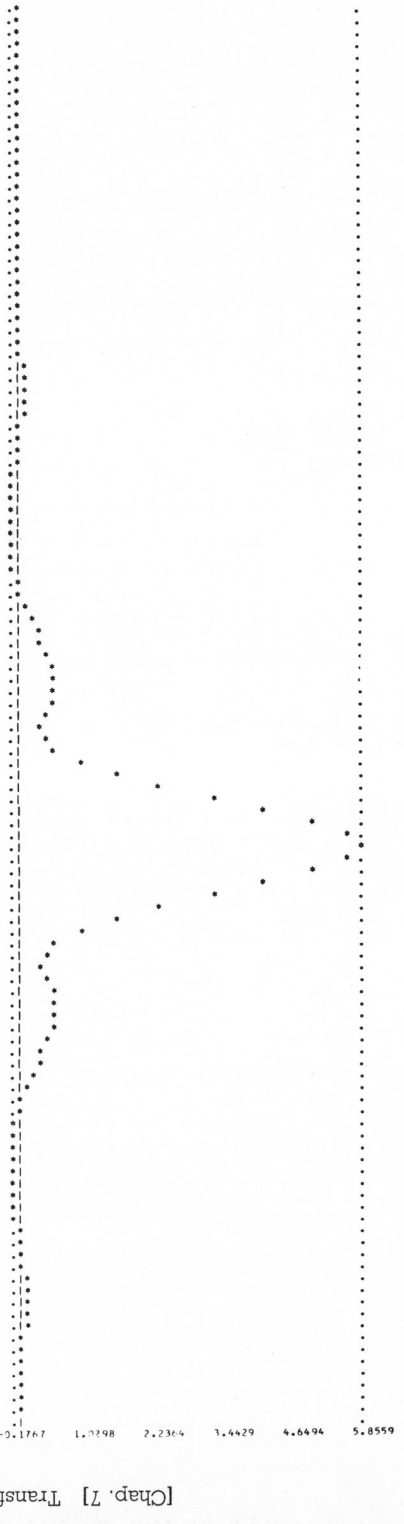

FIG. 7.18-Kb. (Con.)

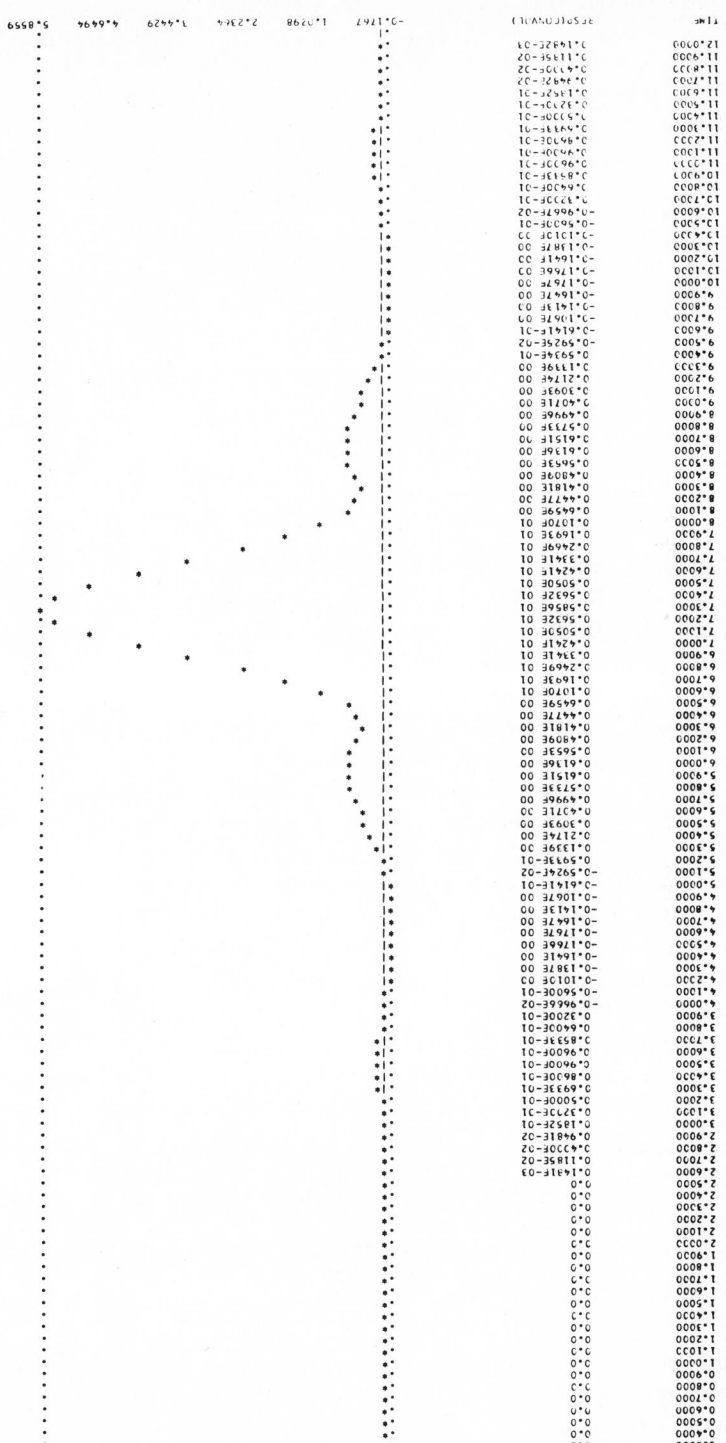

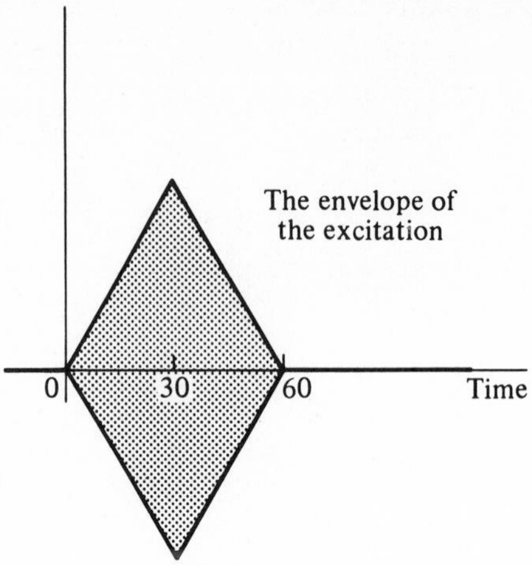

FIG. 7.18-La

7.18-J*c*) with this response to an excitation pulse that is matched to the network's impulse response (Fig. 7.18-K*b*). Convolution here provides not only an analysis method but also an idea for system design, we note. (The detection process can be improved by choice of the particular (common) pulse shape used; consult the literature on "matched filters.")

As a final (fourth) example of CONVOL's versatility, suppose we use it to find the response of the high-pass network of Sec. 7.17 (Fig. 7.18-A) to a triangular-pulse modulated sinusoid, a sort of wave packet or "r-f burst" (Fig. 7.18-L*a*). We give the network's transfer-function data and this excitation to CONVOL. The result, the network's output, is shown in Fig. 7.18-L*b*. We note some distortion, and some retardation of the signal, the discussion of which is the next step in the analysis; into that we do not go. We do observe once more how simple it is to use CONVOL for such lengthy calculations; to do them by hand, though of course perfectly possible *in principle*, is not practical.

Out of the infinity of useful network-analysis programs that might be written, we now have five:

LADANA	(for transfer-function computation),	
IMPSTEP	(for impulse and step response),	
RESSINET	(for sinusoidal response),	(7.18-10)
FOMEGA	(for steady-state behavior),	
CONVOL	(for convolution).	

We could embroider these, as by providing more options (flexibility) and greater accuracy. We could combine two or more into one program: we could, for example, develop a program which, given the schematic of a network, would

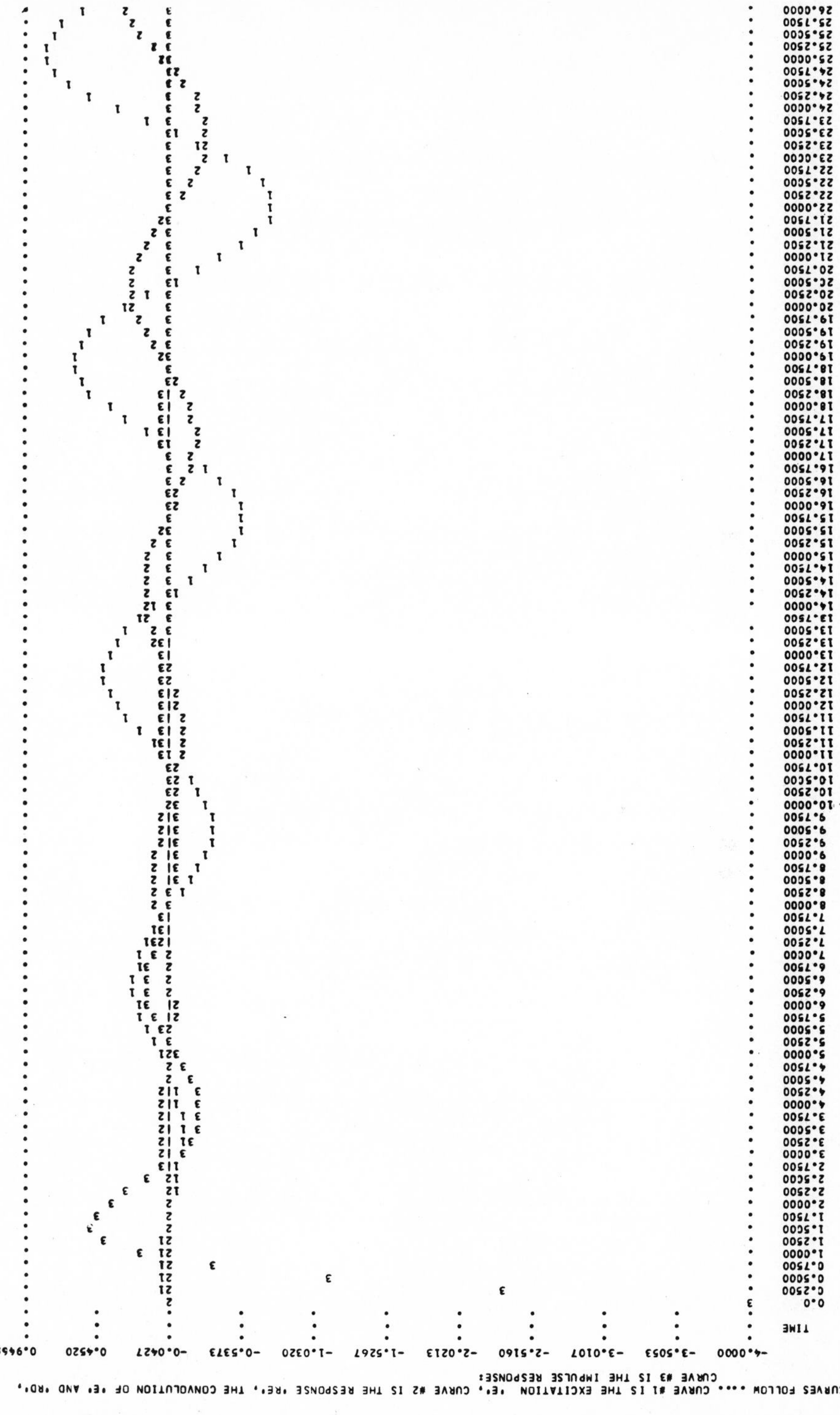
3 CURVES FOLLOW .... CURVE #1 IS THE EXCITATION 'E', CURVE #2 IS THE RESPONSE 'RE', THE CONVOLUTION OF 'E' AND 'RD',
CURVE #3 IS THE IMPULSE RESPONSE:
TIME

FIG. 7.18-Lb

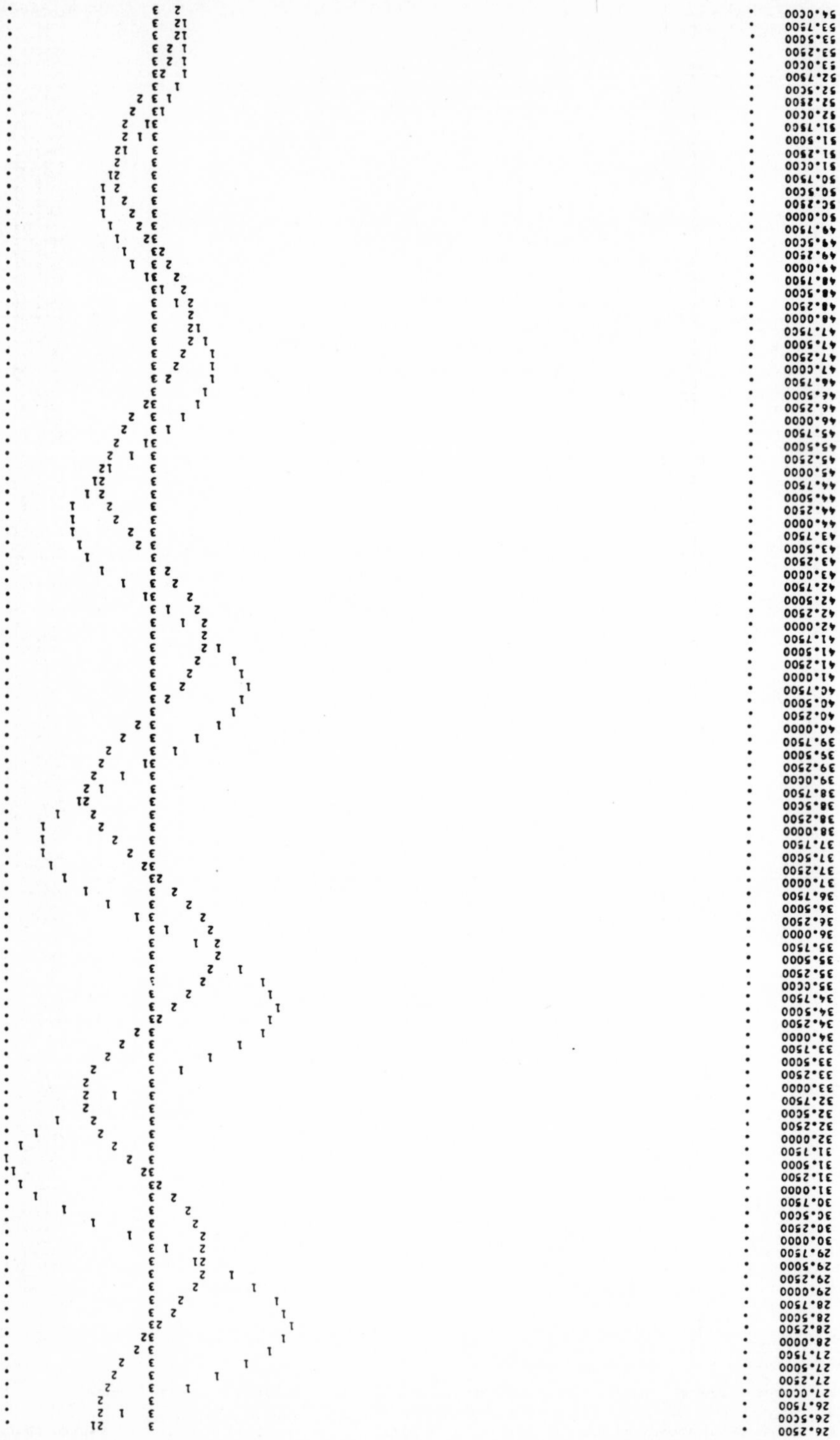
54.0000
53.7500
53.5000
53.2500
53.0000
52.7500
52.5000
52.2500
52.0000
51.7500
51.5000
51.2500
51.0000
50.7500
50.5000
50.2500
50.0000
49.7500
49.5000
49.2500
49.0000
48.7500
48.5000
48.2500
48.0000
47.7500
47.5000
47.2500
47.0000
46.7500
46.5000
46.2500
46.0000
45.7500
45.5000
45.2500
45.0000
44.7500
44.5000
44.2500
44.0000
43.7500
43.5000
43.2500
43.0000
42.7500
42.5000
42.2500
42.0000
41.7500
41.5000
41.2500
41.0000
40.7500
40.5000
40.2500
40.0000
39.7500
39.5000
39.2500
39.0000
38.7500
38.5000
38.2500
38.0000
37.7500
37.5000
37.2500
37.0000
36.7500
36.5000
36.2500
36.0000
35.7500
35.5000
35.2500
35.0000
34.7500
34.5000
34.2500
34.0000
33.7500
33.5000
33.2500
33.0000
32.7500
32.5000
32.2500
32.0000
31.7500
31.5000
31.2500
31.0000
30.7500
30.5000
30.2500
30.0000
29.7500
29.5000
29.2500
29.0000
28.7500
28.5000
28.2500
28.0000
27.7500
27.5000
27.2500
27.0000
26.7500
26.5000
26.2500

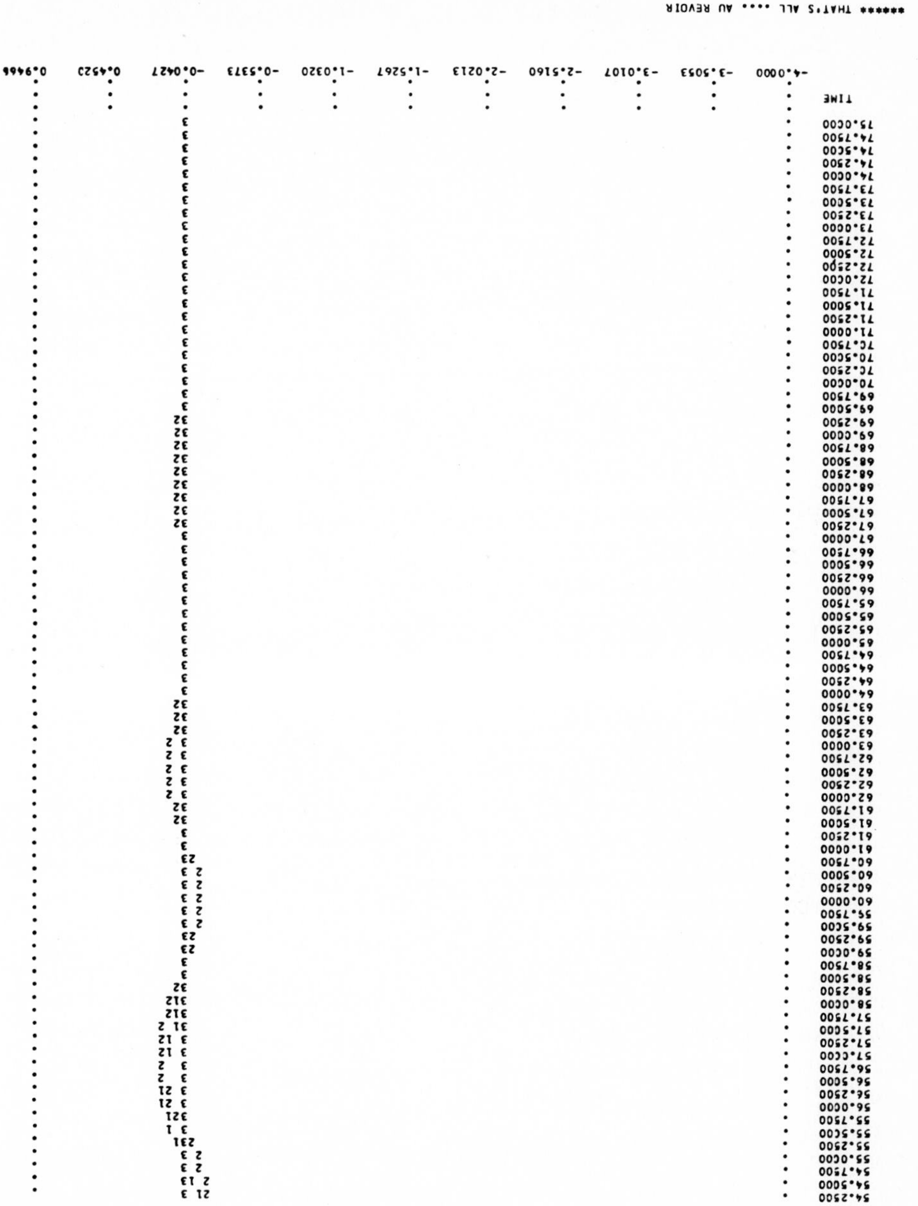

FIG. 7.18-L*b*. (*Con.*)

compute all the responses mentioned. But this would destroy the educational value of taking results from one program, examining them, thinking about them, and then feeding them into the next program. We leave the composite, large-scale programs for the professional, whose work may be far more sophisticated.

There remain two lacunae in our programs (as in our formal analysis ability): we do not know what to do if:

a The initial condition of the network is not rest,

b Poles coincide in $R(s)$. (7.18-11)

If energy-storing elements initially contain energy, there is no great difficulty: certain additional components occur in the response, and they can be computed by the same methods (Sec. 7.22). Nor do coincident poles in $R(s)$ present new problems; we shall discuss them in Sec. 7.23. But we first (Sec. 7.19) use these programs on a somewhat larger scale to learn more about network properties, as well as to further illustrate their use in network analysis.

7.19 USING THE NETWORK-ANALYSIS PROGRAMS; DELAY, ASYMPTOTIC BEHAVIOR

We have now mechanized most of the chores of network analysis and can perform many types of analysis with ease and even pleasure. We can now easily perform very instructive experiments; cases where large amounts of computation are necessary no longer trouble us. So, before filling the gaps in our analysis ability, let us apply the programs both for practice in their use and to learn more about properties of networks.

Consider first the network of Fig. 7.19-A*a*, obviously of low-pass character, generally speaking, but of unknown antecedents. Its element values are real (not normalized), so we first normalize them, to make the numbers computationally convenient: the powers of 10, the "milli" and "micro" values, are awkward. Let us this time base the normalization on a convenient time scale (see Sec. 7.15) to make denormalization of time values in the results of the analysis effectively unnecessary; that is, we aim to make the normalized time scale differ from the actual time scale merely by a factor of 10 or 100 or 1000 or some such convenient number. We can then easily read the (normalized) time scale of the results directly in actual time units. Since the element values are fractions of millihenrys and microfarads, the interesting time region of impulse and step response is probably a few microseconds (10 or 100 at the most); the normalized time unit should correspond then to a microsecond or perhaps a millisecond. We also want the magnitude of the normalized element values to be of the order of unity. An obviously convenient choice for R_0, the impedance reference level (Sec. 7.14), is 72 Ω. We note that $L_1 = 0.160$ mH and its

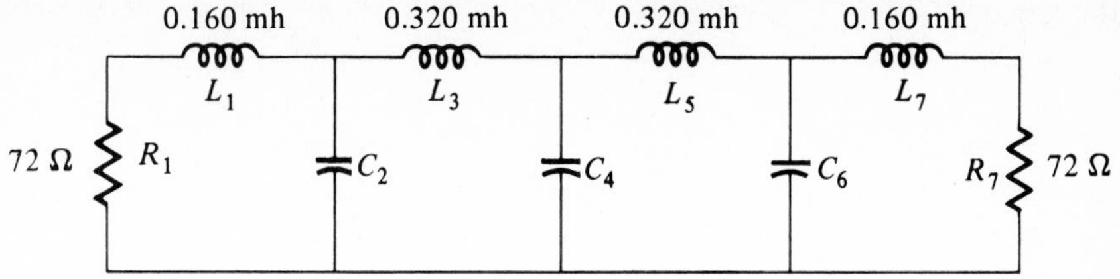

$C_2 = C_4 = C_6 = 0.0640\ \mu F$

FIG. 7.19-A*a*

```
HIS IS LADANA, A LADDER-NETWORK ANALYSIS PROGRAM

        NOTES:        MS  SECT. 7.19              EXAMPLE OF USE OF LADANA              FIG. 7.19-A b
                      NORMALIZATION: ONE MICROSECOND ACTUAL CORRESPONDS TO ONE NOMALIZED SECOND
                                                  (N = 7   C-K   LPF) FIG. 7.19-A

DATA READ IN:

NUMBER OF BRANCHES =  9,  NUMBER OF ENERGY-STORING ELEMENTS =  7          MWR =  2          MCO = 3          NCON = 0

BRANCH     MSS*      TYPE          R VALUE          L VALUE          C VALUE                        *MSS = 1 : SHUNT
......     ....      ....          ........         ........         ........                            = 2 :SERIES

   9         1         1           1.000000
   8         2         2                            2.220000
   7         1         2                                             4.600000
   6         2         2                            4.440000
   5         1         2                                             4.600000
   4         2         2                            4.440000
   3         1         2                                             4.600000
   2         2         2                            2.220000
   1         1         1           1.000000

THAT IS:

 BRANCH NUMBERS:
      9        8        7        6        5        4        3        2        1

0----------- L ------------ L ------------ L ------------ L --------0
      |                 |                 |                 |                 |
      |                 |                 |                 |                 |
      |                 |                 |                 |                 |
      |                 |                 |                 |                 |

      R                 C                 C                 C                 R

      |                 |                 |                 |                 |
      |                 |                 |                 |                 |
      |                 |                 |                 |                 |
      |                 |                 |                 |                 |
0-------------------------------------------------------------------0
      9        8        7        6        5        4        3        2        1

THE RESULTS (COEFFICIENTS OF POWERS OF S, IN ASCENDING ORDER, NUMERATORS OVER DENOMINATORS) ARE:


 THE INPUT IMPEDANCE IS:

        1.000000       13.320000       91.907990      430.741943     1251.418213     3704.200439     4259.828125
     9456.816406
.......................................................................................................................
        2.000000       27.119995      183.815979      806.543213     2502.837158     5623.039063     8519.656250
     9456.816406



 THE CURRENT-TO-VOLTAGE TRANSFER FUNCTION IS :

        1.000000        0.0             0.0             0.0             0.0             0.0             0.0
        0.0
.......................................................................................................................
        2.000000       27.119995      183.815979      806.543213     2502.837158     5623.039063     8519.656250
     9456.816406

OR (HIGHEST POWER OF S IN NATURAL-FREQUENCY POLYNOMIAL MADE UNITY):


        0.000106        0.0             0.0             0.0             0.0             0.0             0.0
        0.0
.......................................................................................................................
        0.000211        0.002868        0.019437        0.085287        0.264660        0.594602        0.900901
        1.000000



THE RESULTS ABOVE WERE PRODUCED BY LADANA.

        NOTES:        MS  SECT. 7.19              EXAMPLE OF USE OF LADANA              FIG. 7.19-A
                      NORMALIZATION: ONE MICROSECOND ACTUAL CORRESPONDS TO ONE NOMALIZED SECOND
                                                  (N = 7   C-K   LPF) FIG. 7.19-A


THERE SEEMING TO BE NO MORE DATA, I AM, SIR/MADAME (MS.), YOUR MOST OBEDIENT SERVANT, LADANA.
```

FIG. 7.19-A*b*

reactance, $L_1\omega$, is 72 Ω at $\omega = 72/(0.16 \times 10^{-3}) = 0.45 \times 10^6$ rad/s. This suggests a value of 0.5×10^6, more or less, for the frequency reference value ω_0 (Secs. 7.14, 7.15). But $t_n = \omega_0 t$ and we want $t_n = 1$ to correspond to one millisecond or one microsecond. Choosing the latter, we set $\omega_0 = 10^6$ rad/s. [Note in passing that $f_0 = \omega_0/(2\pi) = 156$ kHz; for easy *frequency* (in Hertz) scale denormalization we have made a poor choice of ω_0—but one cannot eat one's cake and have it too!]

Normalization of the element values now proceeds straightforwardly (see Sec. 7.14):

$$
\begin{aligned}
&\omega_0 = 10^6 \text{ rad/s}, \qquad R_0 = 72\ \Omega, \\
&L_{1n} = \frac{L_1\omega_0}{R_0} = \frac{(0.16 \times 10^{-3})(10^6)}{72} = 2.22 \text{ H (normalized)} = L_{7n}, \\
&L_{3n} = L_{5n} = 4.44 \text{ H (normalized)}, \\
&C_{2n} = C_2\,\omega_0\,R_0 = (0.0640 \times 10^{-6})(10^6)(72) = 4.60 \text{ F (normalized)} = C_{4n} = C_{6n}.
\end{aligned}
\tag{7.19-1}
$$

```
THIS IS IMPSTEP SEVEN.    NOTES FOR THIS RUN:    MS  SECT. 7.19                          NORMALIZED N=7 LPF
                                                 IMPULSE AND STEP RESPONSES
                                                 1 MILLISEC ACTUAL CORR. 1 UNIT NORMALIZED TIME

PARAMETER VALUES READ:   NN =  0, ND =  7, NX = 0, NPLT = 2, NPRD = 1, NPRU = 1, NCODE = 0

FOR THE FOLLOWING CASE THE MULTIPLIER  H =    1.000000     AND THE COEFFICIENTS IN F(S), LOWEST POWER FIRST, ARE:

NUMERATOR:

        1.000000

DENOMINATOR:

        2.000000        27.119995        183.815994        806.542969        2502.834961        5623.031250
     8519.636719      9456.804688

THE FOLLOWING ARE THE POLES OF F:

     -0.229319  +J  0.0

     -0.193940  +J -0.148179

     -0.193940  +J  0.148179

     -0.031284  +J -0.399991

     -0.031284  +J  0.399991

     -0.110567  +J -0.289744

     -0.110567  +J  0.289744

THE FOLLOWING ARE THE RESIDUES THEREAT (THE "K'S"):

       MAGNITUDE        ANGLE (RAD.)       ANGLE (DEG.)         REAL PART       IMAG. PART
       .........        ............       ............         .........       ..........

       0.233249          0.0                0.0                 0.233249        0.0

       0.186568          2.278032         130.521683           -0.121220        0.141822

       0.186568         -2.278032        -130.521683           -0.121220       -0.141822

       0.018326          1.082005          61.994339            0.008605        0.016180

       0.018326         -1.C82005         -61.994339            0.008605       -0.016180

       0.084395         -1.618368         -92.725662           -0.004013       -0.084300

       0.084395          1.618368          92.725662           -0.004013        0.084300

THE SUM OF THE COMPUTED RESIDUES "K" IS:                                             -0.684708E-05 + J  0.0

THE SUM OF THE PRODUCTS OF THE COMPUTED RESIDUES AND THE NATURAL FREQUENCIES, "SUM KP", IS:    0.219792E-05 + J -0.349246E-08

THE FIRST NONZERO IV, CALCULATED FROM THE GIVEN TRANSFORM, IS:    0.105744E-03    (THIS IS THE INITIAL VALUE CF THE   6-TH
DERIVATIVE OF RD)
```

FIG. 7.19-Ac

```
THIS IS IMPSTEP SEVEN.    NOTES FOR THIS RUN:    MS  SECT. 7.19                          NORMALIZED N=7 LPF
                                                 IMPULSE AND STEP RESPONSES
                                                 1 MILLISEC ACTUAL CORR. 1 UNIT NORMALIZED TIME

SUBROUTINE 'PLOT4Z', BEING CALLED, GIVES (BAR MARKS ZERO ORDINATE):
```

TIME	IMPULSE RESP
0.0	-0.6903E-05
1.0000	-0.4672E-05
2.0000	0.4053E-05
3.0000	0.6795E-04
4.0000	0.3328E-03
5.0000	0.1071E-02
6.0000	0.2660E-02
7.0000	0.5514E-02
8.0000	0.9971E-02
9.0000	0.1618E-01
10.0000	0.2402E-01
11.0000	0.3303E-01
12.0000	0.4245E-01
13.0000	0.5130E-01
14.0000	0.5848E-01
15.0000	0.6300E-01
16.0000	0.6404E-01
17.0000	0.6121E-01
18.0000	0.5453E-01
19.0000	0.4449E-01
20.0000	0.3202E-01
21.0000	0.1836E-01
22.0000	0.4860E-02
23.0000	-0.7135E-02
24.0000	-0.1650E-01
25.0000	-0.2248E-01
26.0000	-0.2474E-01
27.0000	-0.2343E-01
28.0000	-0.1913E-01
29.0000	-0.1274E-01
30.0000	-0.5370E-02
31.0000	0.1862E-02
32.0000	0.7952E-02
33.0000	0.1215E-01
34.0000	0.1405E-01
35.0000	0.1362E-01
36.0000	0.1114E-01
37.0000	0.7205E-02
38.0000	0.2549E-02
39.0000	-0.2045E-02
40.0000	-0.5871E-02
41.0000	-0.8402E-02
42.0000	-0.9354E-02
43.0000	-0.8710E-02
44.0000	-0.6703E-02
45.0000	-0.3760E-02
46.0000	-0.4208E-03
47.0000	0.2751E-02
48.0000	0.5259E-02
49.0000	0.6747E-02
50.0000	0.7046E-02
51.0000	0.6182E-02
52.0000	0.4368E-02
53.0000	0.1950E-02
54.0000	-0.6525E-03
55.0000	-0.3014E-02
56.0000	-0.4776E-02
57.0000	-0.5692E-02
58.0000	-0.5667E-02
59.0000	-0.4755E-02
60.0000	-0.3148E-02
61.0000	-0.1134E-02
62.0000	0.9500E-03
63.0000	0.2777E-02
64.0000	0.4079E-02
65.0000	0.4683E-02
66.0000	0.4535E-02
67.0000	0.3699E-02
68.0000	C.2344E-02
69.0000	0.7074E-03
70.0000	-0.9437E-03
71.0000	-0.2355E-02
72.0000	-0.3323E-02
73.0000	-0.3727E-02
74.0000	-0.3537E-02
75.0000	-0.2817E-02

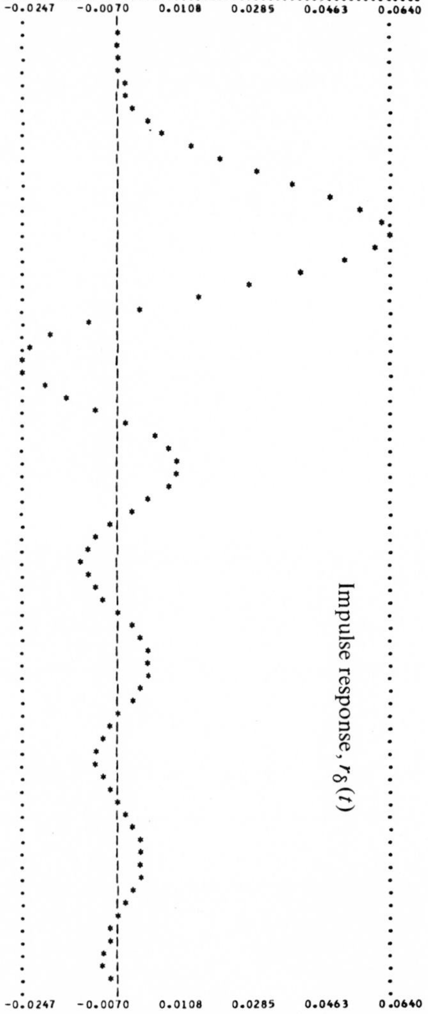

FIG. 7.19-Ac. (Con.)

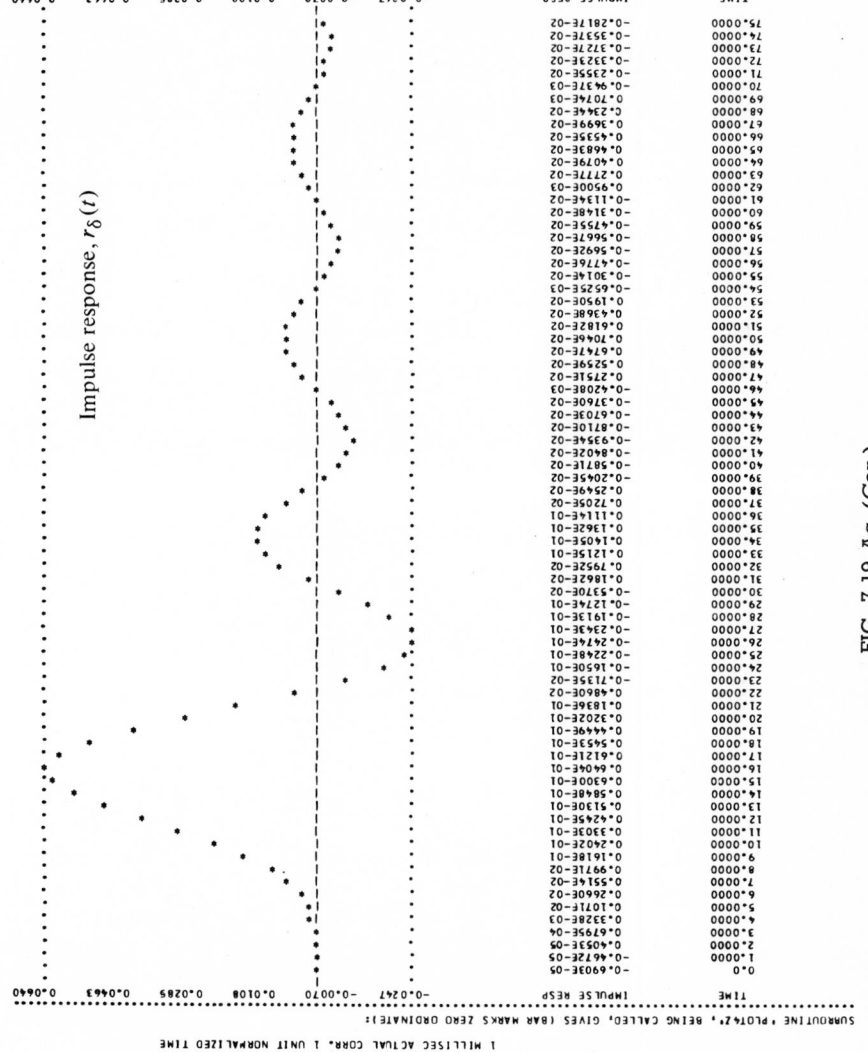

THIS IS IMPSTEP SEVEN.   NOTES FOR THIS RUN: VS SECT. 7.15   NORMALIZED N=7 LPF
IMPULSE AND STEP RESPONSES
1 SECOND OF ACTUAL CORR. 1 UNIT NORMALIZED TIME

SUBROUTINE "PLIT4Z", BEING CALLED, GIVES (BAR MARKS ZERO ORDINATE):

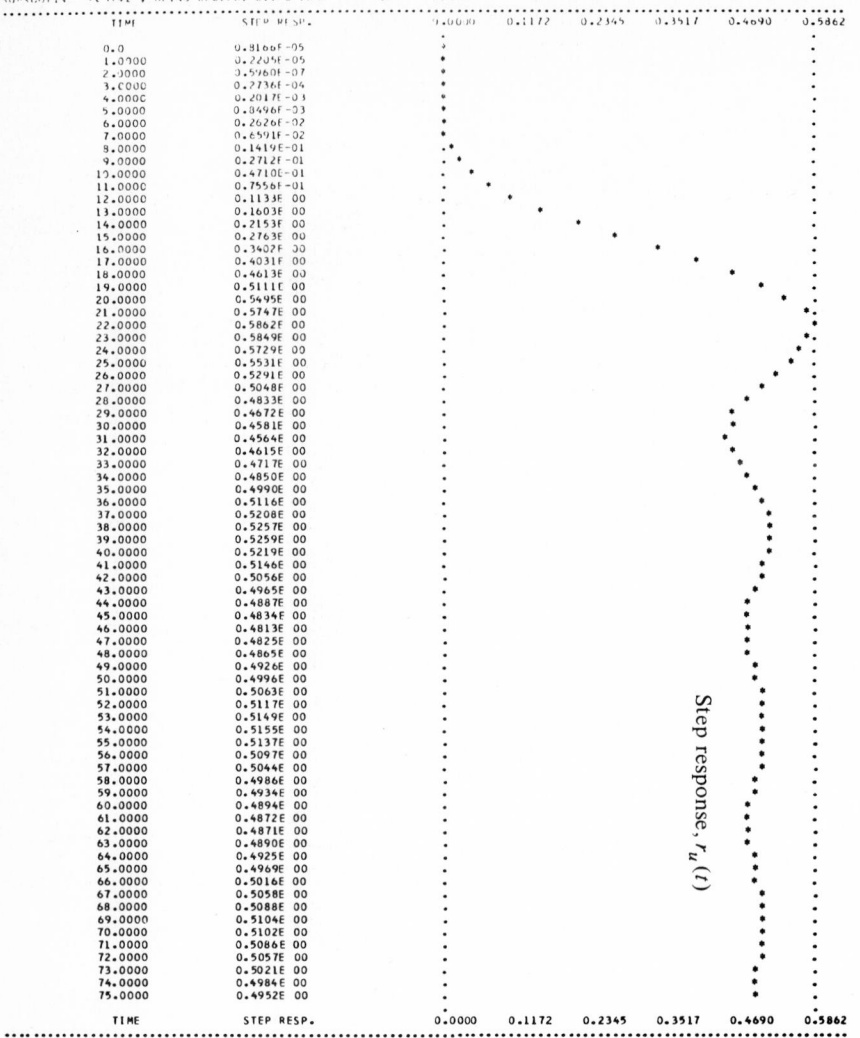

TIME	STEP RESP.
0.0	0.8166E-05
1.0000	0.2205E-05
2.0000	0.5960E-07
3.0000	0.2736E-04
4.0000	0.2017E-03
5.0000	0.8496E-03
6.0000	0.2626E-02
7.0000	0.6591E-02
8.0000	0.1419E-01
9.0000	0.2712E-01
10.0000	0.4710E-01
11.0000	0.7556E-01
12.0000	0.1133E 00
13.0000	0.1603E 00
14.0000	0.2153E 00
15.0000	0.2763E 00
16.0000	0.3402E 00
17.0000	0.4031E 00
18.0000	0.4613E 00
19.0000	0.5111E 00
20.0000	0.5495E 00
21.0000	0.5747E 00
22.0000	0.5862E 00
23.0000	0.5849E 00
24.0000	0.5729E 00
25.0000	0.5531E 00
26.0000	0.5291E 00
27.0000	0.5048E 00
28.0000	0.4833E 00
29.0000	0.4672E 00
30.0000	0.4581E 00
31.0000	0.4564E 00
32.0000	0.4615E 00
33.0000	0.4717E 00
34.0000	0.4850E 00
35.0000	0.4990E 00
36.0000	0.5116E 00
37.0000	0.5208E 00
38.0000	0.5257E 00
39.0000	0.5259E 00
40.0000	0.5219E 00
41.0000	0.5146E 00
42.0000	0.5056E 00
43.0000	0.4965E 00
44.0000	0.4887E 00
45.0000	0.4834E 00
46.0000	0.4813E 00
47.0000	0.4825E 00
48.0000	0.4865E 00
49.0000	0.4926E 00
50.0000	0.4996E 00
51.0000	0.5063E 00
52.0000	0.5117E 00
53.0000	0.5149E 00
54.0000	0.5155E 00
55.0000	0.5137E 00
56.0000	0.5097E 00
57.0000	0.5044E 00
58.0000	0.4986E 00
59.0000	0.4934E 00
60.0000	0.4894E 00
61.0000	0.4872E 00
62.0000	0.4871E 00
63.0000	0.4890E 00
64.0000	0.4925E 00
65.0000	0.4969E 00
66.0000	0.5016E 00
67.0000	0.5058E 00
68.0000	0.5088E 00
69.0000	0.5104E 00
70.0000	0.5102E 00
71.0000	0.5086E 00
72.0000	0.5057E 00
73.0000	0.5021E 00
74.0000	0.4984E 00
75.0000	0.4952E 00

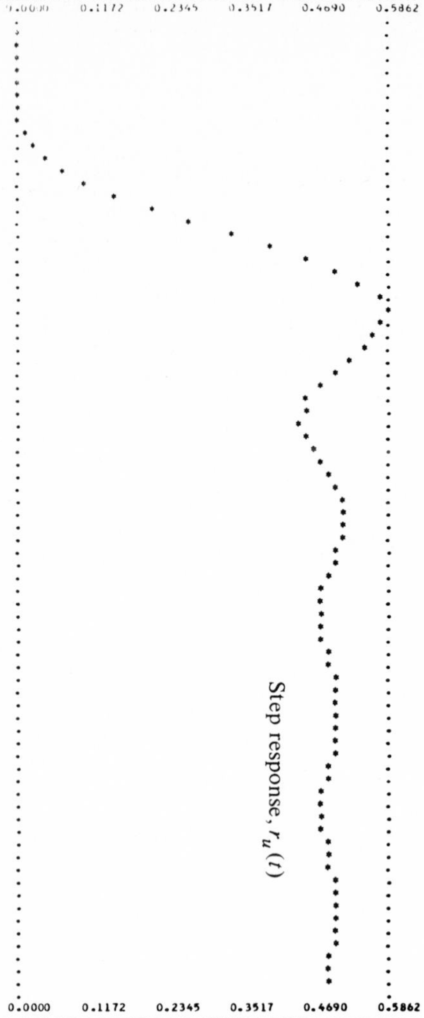

FIG. 7.19-Ac. (*Con.*)

************** A MESSAGE FROM IMPSTEP7: DID YOU GET ENOUGH POINTS, IS THE "GRANULARITY" FINE ENOUGH?
OR DID YOU, PERCHANCE, OVERWORK ME AND GET TOO MANY?
TAKE A GOOD LOOK BEFORE YOU LEAVE. **************************************

ANOTHER MESSAGE: HERE ARE THE NATURAL FREQUENCIES ONCE MORE, TOGETHER WITH THE RECIPROCALS OF THE NEGATIVES OF THEIR REAL PARTS AND THE PERIODS OF THE OSCILLATIONS:

-0.229319	+ J 0.0	4.360730	999599.937500
-0.193940	+ J -0.148179	5.156237	6.748591
-0.193940	+ J 0.148179	5.156237	6.748591
-0.031284	+ J -0.399991	31.965363	2.500057
-0.031284	+ J 0.399991	31.965363	2.500057
-0.110567	+ J -0.289744	9.044268	3.451316
-0.110567	+ J 0.289744	9.044268	3.451316

YOU CHOSE TIME POINTS SPACED 1.000; DID YOU CHOOSE WELL ?

ARE YOU HAPPY ?

THE END.

*** THIS IS F2 (JW) *** NOTES FOR THIS RUN: MS SECT. 7.19 NORMALIZED N=7 LPF FIG. 7.19-Ad
1 MILLISEC ACTUAL CORR. 1 UNIT NORMALIZED TIME
SINUSOIDAL STEADY STATE; FOMEGA.

THE VALUE OF THE MULTIPLIER H IS 1.00000, THE DEGREE OF THE NUMERATOR IS 0, THE DEGREE OF THE DENOMINATOR IS 7 AND THE COEFFICIENTS IN F(S), NUMERATOR ABOVE DENOMINATOR, LOWEST (ZEROTH) POWERS FIRST, ARE:

1.0000

2.0000 27.1200 183.8160 806.5430 2502.8350 5623.0313 8519.6367 9456.8047

NB:PROPER NORMALIZATION IS ASSUMED; IF H IS TOO SMALL, YOU MAY GET USELESS RESULTS.

*IN ANGLE OF F, PROGRAM ATTEMPTS TO AVOID JUMPS DUE TO TAKING PRINCIPAL VALUES ONLY, AND TO KEEP VARIATION OF ANGLE STEADY
*THE DBV VALUE "SATURATES" AT -100 FOR CALCULATION CONVENIENCE

PLOT OF MAGNITUDE OF F:

(BAR INDICATES ZERO ORDINATE):

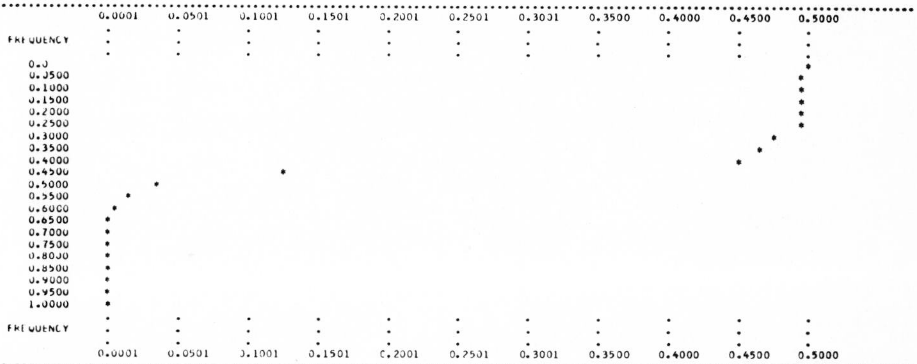

PLOT OF ANGLE OF F IN DEGREES (PROGRAM ATTEMPTS TO AVOID JUMPS DUE TO TAKING PRINCIPAL VALUES ONLY, AND TO KEEP VARIATION OF ANGLE STEADY)

(BAR INDICATES ZERO ORDINATE):

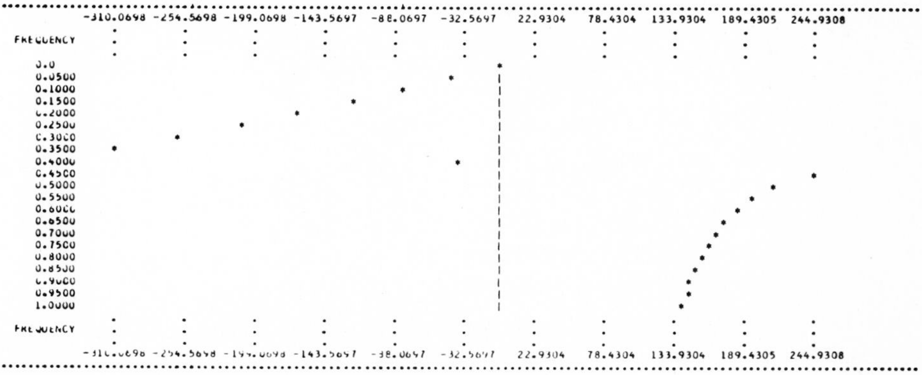

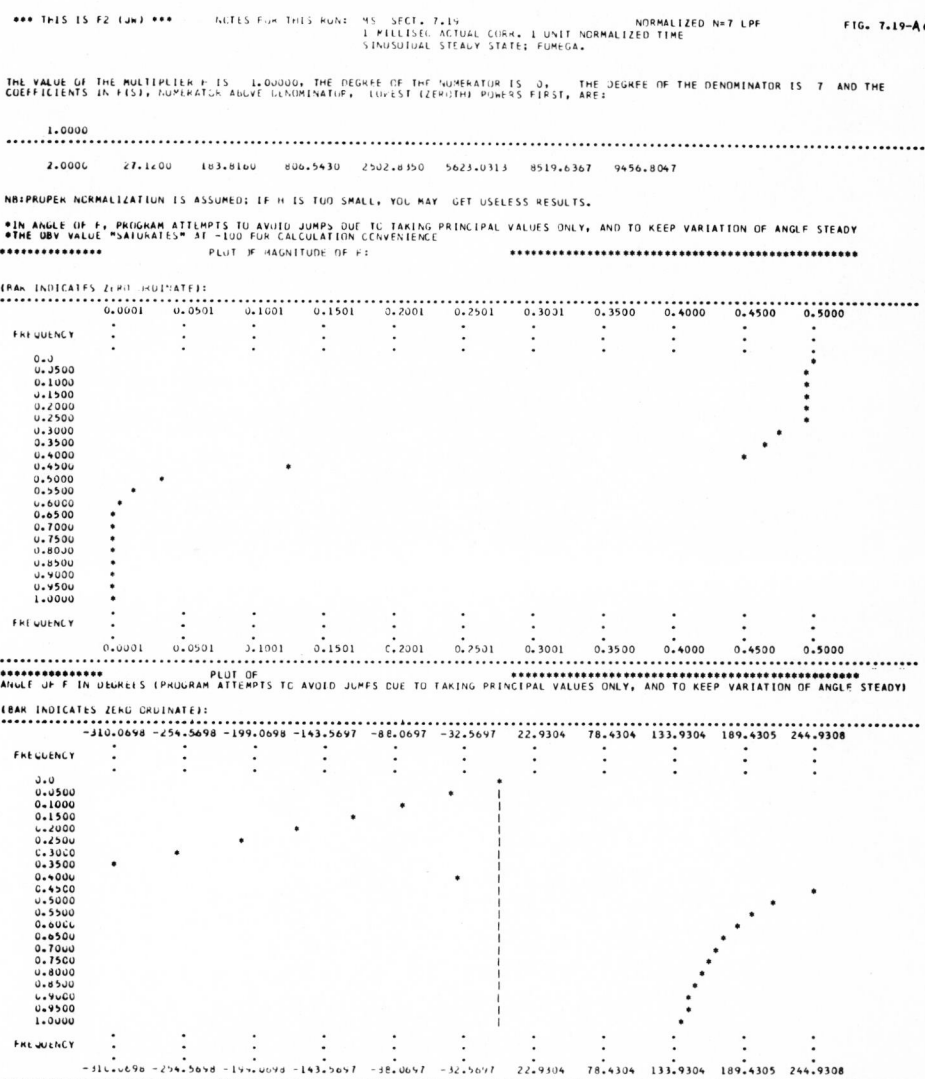

FIG. 7.19-Ad

F2(JW) GIVES: NOTES FOR THIS RUN: MS SECT. 7.19 NORMALIZED N=7 LPF FIG. 7.19-Ae
1 MILLISEC ACTUAL CORR. 1 UNIT NORMALIZED TIME
SINUSOIDAL STEADY STATE; FOMEGA.

PLOT OF DB MEASURE OF F (DBV)

(BAR INDICATES ZERO ORDINATE):

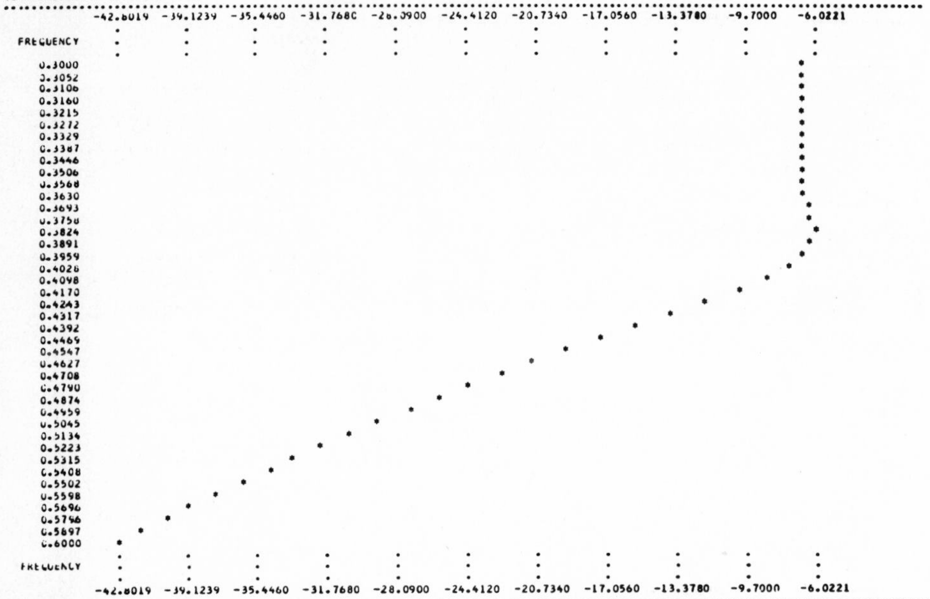

PLOT OF
ANGLE OF F IN DEGREES (PROGRAM ATTEMPTS TO AVOID JUMPS DUE TO TAKING PRINCIPAL VALUES ONLY, AND TO KEEP VARIATION OF ANGLE STEADY)

(BAR INDICATES ZERO ORDINATE):

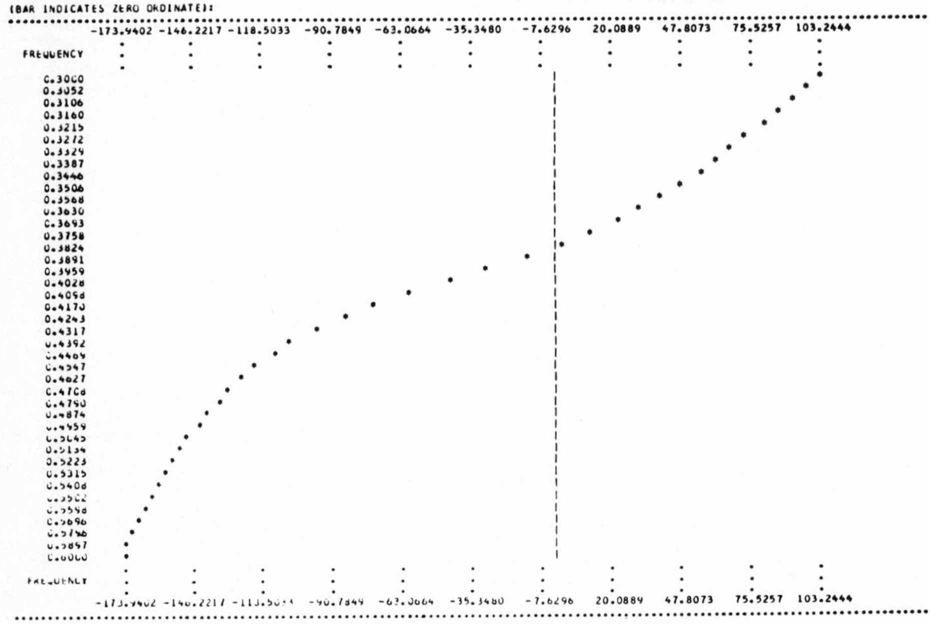

FIG. 7.19-Ae

F2(JW) GIVES: NOTES FOR THIS RUN: MS SECT. 7.19   NORMALIZED N=7 LPF   FIG. 7.19-Af
1 MILLISEC ACTUAL ... 1 UNIT NORMALIZED TIME
SINUSOIDAL STEADY STATE; FOMEGA.

*************** PLOT OF MAGNITUDE OF F: **

(BAR INDICATES ZERO ORDINATE):

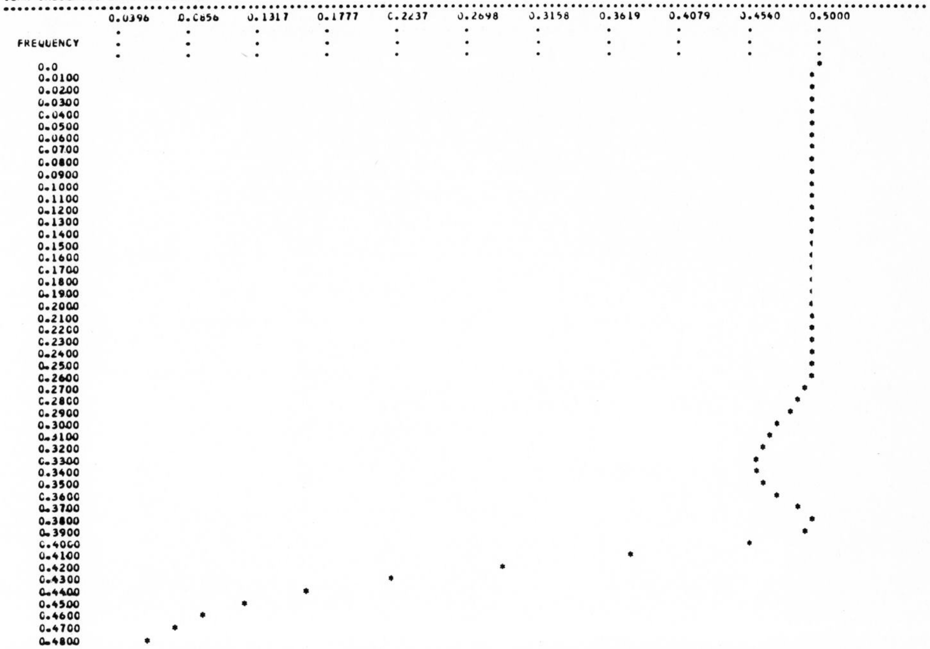

*************** PLOT OF **
ANGLE OF F IN DEGREES (PROGRAM ATTEMPTS TO AVOID JUMPS DUE TO TAKING PRINCIPAL VALUES ONLY, AND TO KEEP VARIATION OF ANGLE STEADY)

(BAR INDICATES ZERO ORDINATE):

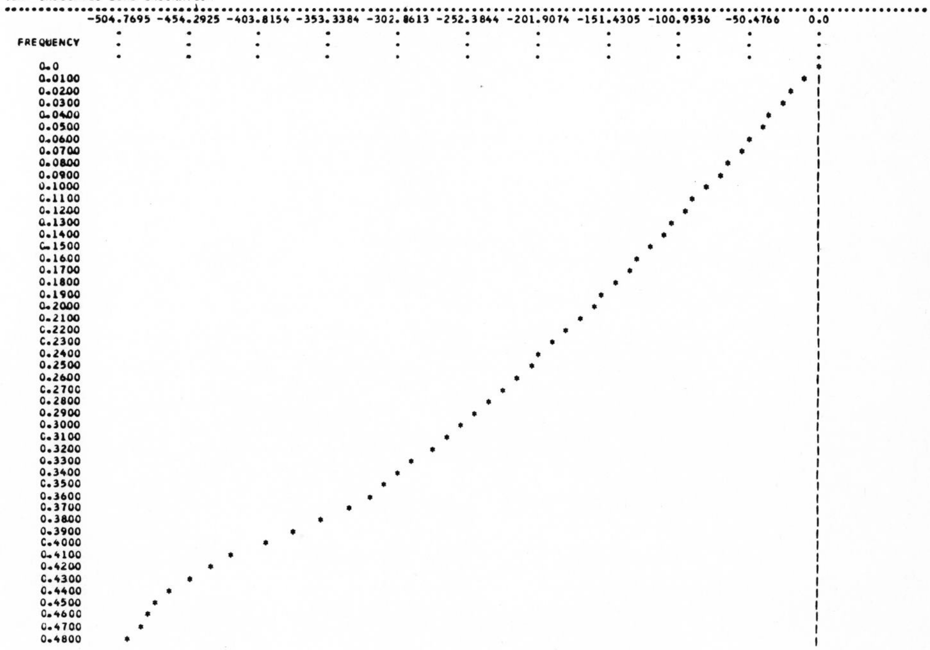

FIG. 7.19-Af

THIS IS F2(JW) WITH THESE FEATURES: JP = 2, NLCG = 0, NWR = 1

RESULTS (VALUES OF F(JW)) FOR OMEGA = 0.0 TO 0.5000, TAKING 51 POINTS (EQUALLY SPACED ON ARITH. FREQ. SCALE):

POINT NO.	OMEGA	MAGNITUDE	ANGLE (DEGREES)*	ANGL. INCR.	REAL PART	IMAGINARY PART	DB MEASURE(DBV)*
1	0.0	0.5000	0.0	0.0	0.5000	0.0	-6.0206
2	0.0100	0.5000	-7.7700	-7.7700	0.4954	-0.0676	-6.0206
3	0.0200	0.5000	-15.5441	-7.7741	0.4817	-0.1340	-6.0207
4	0.0300	0.5000	-23.3263	-7.7823	0.4591	-0.1980	-6.0209
5	0.0400	0.5000	-31.1208	-7.7945	0.4280	-0.2584	-6.0211
6	0.0500	0.4999	-38.9317	-7.8109	0.3889	-0.3142	-6.0216
7	0.0600	0.4999	-46.7629	-7.8312	0.3424	-0.3642	-6.0223
8	0.0700	0.4998	-54.6183	-7.8555	0.2894	-0.4075	-6.0233
9	0.0800	0.4998	-62.5019	-7.8836	0.2308	-0.4433	-6.0246
10	0.0900	0.4997	-70.4174	-7.9155	0.1675	-0.4708	-6.0264
11	0.1000	0.4995	-78.3684	-7.9509	0.1007	-0.4893	-6.0286
12	0.1100	0.4994	-86.3586	-7.9902	0.0317	-0.4984	-6.0312
13	0.1200	0.4992	-94.3521	-8.0334	-0.0382	-0.4978	-6.0341
14	0.1300	0.4991	-102.4732	-8.0811	-0.1078	-0.4873	-6.0369
15	0.1400	0.4989	-110.6073	-8.1341	-0.1756	-0.4670	-6.0395
16	0.1500	0.4988	-118.8014	-8.1942	-0.2403	-0.4371	-6.0412
17	0.1600	0.4988	-127.0639	-8.2625	-0.3006	-0.3980	-6.0419
18	0.1700	0.4988	-135.4059	-8.3420	-0.3552	-0.3502	-6.0409
19	0.1800	0.4990	-143.8406	-8.4347	-0.4029	-0.2944	-6.0382
20	0.1900	0.4992	-152.3837	-8.5430	-0.4424	-0.2314	-6.0339
21	0.2000	0.4995	-161.0527	-8.6691	-0.4725	-0.1622	-6.0285
22	0.2100	0.4998	-169.8668	-8.8141	-0.4920	-0.0879	-6.0234
23	0.2200	0.5000	-178.8438	-8.9770	-0.4999	-0.0101	-6.0207
24	0.2300	0.4999	-187.9996	-9.1558	-0.4950	0.0696	-6.0232
25	0.2400	0.4992	-197.3439	-9.3443	-0.4765	0.1488	-6.0345
26	0.2500	0.4978	-206.8786	-9.5347	-0.4440	0.2251	-6.0588
27	0.2600	0.4955	-216.5943	-9.7158	-0.3978	0.2954	-6.0998
28	0.2700	0.4920	-226.4708	-9.8765	-0.3389	0.3567	-6.1603
29	0.2800	0.4875	-236.4782	-10.0073	-0.2692	0.4064	-6.2409
30	0.2900	0.4820	-246.5823	-10.1041	-0.1916	0.4423	-6.3390
31	0.3000	0.4760	-256.7554	-10.1731	-0.1091	0.4634	-6.4476
32	0.3100	0.4701	-266.9888	-10.2334	-0.0247	0.4695	-6.5554
33	0.3200	0.4652	-277.3093	-10.3206	0.0592	0.4615	-6.6465
34	0.3300	0.4623	-287.7998	-10.4905	0.1413	0.4402	-6.7014
35	0.3400	0.4624	-298.6250	-10.8252	0.2215	0.4059	-6.6989
36	0.3500	0.4667	-310.0698	-11.4448	0.3004	0.3571	-6.6195
37	0.3600	0.4756	-322.5898	-12.5200	0.3778	0.2889	-6.4549
38	0.3700	0.4882	-336.8689	-14.2791	0.4490	0.1913	-6.2281
39	0.3800	0.4990	-353.7854	-16.9165	0.4960	0.0540	-6.0387
40	0.3900	0.4939	-373.9785	-20.1931	0.4793	-0.1193	-6.1274
41	0.4000	0.4542	-396.6787	-22.7002	0.3643	-0.2713	-6.8544
42	0.4100	0.3794	-419.0469	-22.3682	0.1952	-0.3254	-8.4173
43	0.4200	0.2947	-438.2490	-19.2021	0.0600	-0.2886	-10.6110
44	0.4300	0.2221	-453.5183	-15.2693	-0.0136	-0.2217	-13.0684
45	0.4400	0.1671	-465.4973	-11.9790	-0.0446	-0.1610	-15.5416
46	0.4500	0.1270	-475.0691	-9.5718	-0.0538	-0.1151	-17.9227
47	0.4600	0.0980	-482.9294	-7.8[illegible]4	-0.0532	-0.0922	[illegible]

FIG. 7.19–Af. (Con.)

From LADANA, given our data, we obtain the transfer function (Fig. 7.19-A*b*)

$$T(s) = \frac{1.0}{2.0 + 27.12s + \cdots + 8519.7s^6 + 9456.6s^7} \tag{7.19-2}$$

without difficulty. (The operation is now routine, but one might pause a moment to estimate how long it would take to carry it out by hand.)

Suppose we now inquire of IMPSTEP what the impulse and step responses of the network are. Figure 7.19-A*c* tells us. We note that the natural frequencies ($-0.23 \pm j0$, $-0.19 \pm j0.15$, etc.) are clustered in the vicinity of the origin as one expects from low-pass devices. The impulse response is a spread-out and rounded "pulse" that "occurs" at $t = 16$ μs (remember that normalized time values can be read as microsecond actual-time values), followed by decaying oscillations. The contributions of the seven natural frequencies interact in such a way, difficult to describe in numerical detail, that this very distorted and retarded "impulse" emerges. We observe that $r_\delta(0+)$ should be zero since in (7.19-2) $M = 0$ and $N = 7$; according to IMPSTEP $\sum K = -0.000007$ (Fig. 7.19-A*c*) which *is* zero for all practical purposes (the "7" is certainly in the nonsignificant or "noise" region of the single-precision machine computations). Similarly $r_\delta'(0+)$ should be zero and we see that $\sum Kp = (0.2 - j0.0003) \times 10^{-5}$ (Fig. 7.19-A*c*), again zero for practical purposes. Also equal to zero are the initial values of the next four derivatives of $r_\delta(t)$. But $\mathrm{r}^{\mathrm{vi}}(0+)$ should equal $1/9457 = 1.06 \times 10^{-4}$ according to (7.19-2); this is confirmed by IMPSTEP (Fig. 7.19-A*c*). Mathematically the impulse response begins *very* slowly indeed; physically, the inductors and capacitors greatly delay the reaction to the excitation. The fact that eventually a semblance of an impulse does emerge is due to certain relations between the elements that could be traced to the networks's crude resemblance to a transmission line (Prob. 7-Z). The step response, the integral of $r_\delta(t)$, rises slowly (most steeply at $t = 16$ μs), overshoots its final steady-state value of 0.5, and then oscillates thereabout.

These two responses are more or less characteristic of low-pass networks that depart not too far from being transmission-line models [see discussion following (7.15-23)]. But it is interesting to pursue further the ideas raised in Sec. 7.15 about ranges of s important for preserving the signal's shape, and less important ranges of s. So we call on FOMEGA to give us the sinusoidal-steady-state ($s = j\omega$) characteristics of the network. The results are shown in Fig. 7.19-A*d*. The low-pass character is verified, the cutoff of transmission around $\omega = 0.4$ seeming very steep indeed. But the frequency increment used, $\omega = 0.05$, is evidently unsatisfactory, so we ask FOMEGA for more detail near $\omega = 0.4$ (on a logarithmic frequency scale for variety) and obtain Fig. 7.19-A*e*; the behavior from zero up to about 0.5 is more clearly shown in Fig. 7.19-A*f*.

A certain correlation can be found, as intimated in Sec. 7.15. The impulse is grossly distorted in passing through the network, because its mate or transform is a constant and has "components" equally distributed over all values of s;

at the higher frequencies it is violently manhandled. Nevertheless a recognizable, if very rounded pulse eventually does emerge. The *step* is less severely distorted since its mate is $1/s$ and for it "high frequencies" are somewhat less important: $r_u(t)$ is, in fact, a not unrespectable step.

We have so far paid no attention to the other part of $T(j\omega)$, its *angle*. Surely it too has some effect on signal transmission! Since for sinusoidal waves it is concerned with phase shift and hence with the time scale, it may be to some extent responsible for the delay in occurrence of the (rounded) pulse. To test the validity of this idea, consider temporarily the response to a very simple, single-sinusoid signal in the steady state. Take for the excitation

$$e(t) = E_0 \cos(\omega_F t + \phi), \tag{7.19-3}$$

and evaluate the transfer function at the corresponding point [as in (7.16-10)]:

$$T(j\omega_F) = T_F \underline{/ \beta}. \tag{7.19-4}$$

Then

$$\begin{aligned} r_F(t) &= E_0 T_F \cos(\omega_F t + \phi + \beta) \\ &= E_0 T_F \cos\left[\omega_F\left(t + \frac{\beta}{\omega_F}\right) + \phi\right] \end{aligned} \tag{7.19-5}$$

is the resulting forced response. This sinusoid has the same frequency (of course) but its amplitude is different and it is shifted ahead in time an amount

$$t_{\text{shift}} = \frac{\beta}{\omega_F} \qquad \text{(s)} \tag{7.19-6}$$

If $\omega_F = 0.2$, for example, then (from Fig. 7.19-Af) $T_F = 0.5$ and $\beta = -161°$ or -2.81 rad. Hence

$$t_{\text{shift}} = \frac{\beta}{\omega_F} = \frac{-2.81}{0.2} = -14\ \mu\text{s}. \tag{7.19-7}$$

(Note that β must be in appropriate units, consistent with those of ω_F.)

Since the emerging signals (Fig. 7.19-Ac) are distorted, one cannot speak in any precise sense of the network's *delay* in responding. Nevertheless a consensus would probably admit the use of "delay" to describe the time of occurrence of the peak of the (rounded) pulse, or of the steepest (or perhaps half-way up) point on the rounded step, the value of about 16 μs. Is it mere accident that this value is so close to that of (7.19-7)?

We have some additional, very informative data at hand in the pulse-train discussion of Sec. 7.16. That wave is the sum of several sinusoids (Fig. 7.16-A).

Look again at Fig. 7.16-D, the forced (steady-state) response thereto of the network of Fig. 7.14-A. There is a "delay" clearly enough, even if the shape *is* distorted. This delay is about 5 plotted points, one period being 25 points. Since the basic frequency is $2\omega_F = (2)(0.25) = 0.5$, the corresponding period is $2\pi/0.5$. Hence the observed "delay" is $(\frac{5}{25})(2\pi/0.5) = 2.51$ s (normalized). Now from (7.15-1) we calculate the angle of the network's transfer function at the frequencies of interest, those of the components of the pulse train. The results (Table 7.19-A) indicate that the various sinusoidal components are individually

TABLE 7.19-A

Frequency, ω	Angle of $T(j\omega)$	t_{shift}, Normalized
0	0	0
$2\omega_F = 0.5$	$-60.4°$	-2.11
$4\omega_F = 1.0$	$-135.0°$	-2.36
$6\omega_F = 1.5$	$-186.1°$	-2.16

time-shifted (delayed) varying amounts; but all the shifts are of the same order of magnitude, about 2.2 s (normalized). We calculated the actual signal "delay" to be 2.51. Now 2.5 and 2.2 differ only by some 10–15 percent. The evidence seems to indicate that

If β/ω is more or less constant, about the same at all the frequencies of signal components, then $t_{\text{shift}} = \beta/\omega$ for the signal itself, more or less. (7.19-8)

In the present example (Fig. 7.19-A*c*) we do not have a simple sum of sinusoids: neither $\delta(t)$ nor $u(t)$ is such. If a similar argument is to hold water at all, we must consider the phase shift (angle of T) at *all* frequencies, or at least at all frequencies where $|T|$ is appreciable, where the transmission is not negligible. Here we do have in the ω domain a transmission region with no great variation in $|T|$, for $0 < \omega < 0.4$, say. There β/ω is also more or less constant or, better, $\beta(\omega)$ is more or less *linear:*

$$\beta(\omega) = \beta_0 + T_1\omega + T_2\omega^2 + T_3\omega^3 + \cdots \qquad (7.19\text{-}9)$$

in which $\beta_0 = 0$. The increments of β, taken from Fig. 7.19-Af for a fixed increment of ω (Table 7.19-B), show that at low frequencies $T_1 \cong -14\ \mu s$ [see (7.19-7)] and T_2, T_3, ... are negligible. So long as $\beta_0 = 0$ in (7.19-9), and the higher powers of ω (that is, T_2, T_3, ...) are negligible, then $T_1 = \beta/\omega$ and there is no distinction between β/ω and $d\beta/d\omega$, the slope of $\beta(\omega)$. [For the present we do not consider what happens when $|\beta/\omega|$ (the "phase delay") and $|d\beta/d\omega|$ (the "group delay") are *not* the same.]

TABLE 7.19-B

Frequency, ω	Phase Increment, Degrees	t_{shift}, μs
0.01	−7.8	−14
0.1	−8.0	−14
0.2	−8.7	−15
0.3	−10.2	−18

The correlation we have found in this example reinforces (7.19-8) as an experimentally obtained rule of thumb (which can however be placed on a more solid theoretical foundation as we shall later see, in Sec. 8.06). For the present we simply recapitulate our experimental results:

> A low-pass network (in which $|T|$ is more or less constant at low frequencies, $\beta_0 = 0$, and $d\beta/d\omega$ is more or less constant at a value T_1 at low frequencies) can be expected to distort signals that pass through it, but the distortion may be small enough to admit of defining a "delay" for the signals; it is more or less equal to the value of $|d\beta/d\omega|$ or of $|\beta/\omega|$ at low frequencies, where transmission is good. [For an extension see (7.19-24).] (7.19-10)

For a second instructive illustration of the use of our network analysis programs, consider the network of Fig. 7.19-B*a*. It certainly is not a low-pass network, for the transfer function has three zeros at the origin (due to L_1, C_2, and L_3). The other three zeros are at infinity (due to C_1, L_2, and C_3). The network is probably, then, of "band-pass" character. We call on LADANA and on FOMEGA (Fig. 7.19-B*b* and *c*) and find confirmation: over a narrow band of frequencies ($0.96 < \omega < 1.04$, say), $|T|$ is approximately constant. Outside this band it falls off rapidly, so the adjective *band-pass* is indeed applicable and we adopt it. We note with interest that the "out-band" parts, the "skirts" of the curve, seem to be straight lines. Now the asymptotic behavior of the transfer function (from Fig. 7.19-B*b*)

$$T(s) = \frac{0.005s^3}{1 + 0.2s + 3.02s^2 + 0.401s^3 + 3.02s^4 + 0.2s^5 + s^6} \tag{7.19-11}$$

is, at low frequencies,

$$\begin{aligned} T_{LF}(s) &= 0.005s^3, \\ |T_{LF}(j\omega)| &= 0.005\omega^3, \end{aligned} \tag{7.19-12}$$

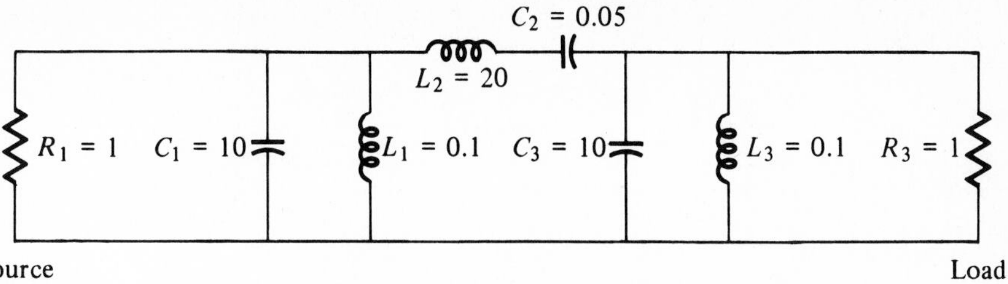

(Normalized ohms, henrys, farads)

FIG. 7.19–B*a*

```
THIS IS LADANA, A LADDER-NETWORK ANALYSIS PROGRAM

              NOTES:            MS SECT. 7.19  BPF  LADANA     FIG. 7.19-
                                (FM LP PROTOTYPE BY TRFMTN)
                                C

DATA READ IN:

NUMBER OF BRANCHES =  8,  NUMBER OF ENERGY-STORING ELEMENTS =  6          MWR =  2          MCO = 3          NCON = 0

BRANCH     MSS*     TYPE          R VALUE          L VALUE          C VALUE                              *MSS = 1 : SHUNT
......     ....     ....          ........         ........         ........                                  = 2 :SERIES

  8         1        1           1.000000
  7         1        3                             0.100000
  6         1        2                                              10.000000
  5         2        2                            20.000000
  4         2        3                                               0.050000
  3         1        3                             0.100000
  2         1        2                                              10.000000
  1         1        1           1.000000

THAT IS:

 BRANCH NUMBERS:
    8       7       6       5       4       3       2       1

0-------------------------  L ----  C ----------------------0
    I       I       I                       I       I       I
    I       I       I                       I       I       I
    I       I       I                       I       I       I
    R       L       C                       L       C       R
    I       I       I                       I       I       I
    I       I       I                       I       I       I
    I       I       I                       I       I       I
0-----------------------------------------------------------0
    8       7       6       5       4       3       2       1

THE RESULTS (COEFFICIENTS OF POWERS OF S, IN ASCENDING ORDER, NUMERATORS OVER DENOMINATORS) ARE:

 THE INPUT IMPEDANCE IS:

        0.0            0.100000          0.010000          0.200500          0.010000          0.100000          0.0
.......................................................................................................................
        1.000000       0.200000          3.019997          0.400999          3.019993          0.200000          0.999999

 THE CURRENT-TO-VOLTAGE TRANSFER FUNCTION IS :

        0.0            0.0               0.0               0.000500          0.0               0.0               0.0
.......................................................................................................................
        1.000000       0.200000          3.019997          0.400999          3.019993          0.200000          0.999999

OR (HIGHEST POWER OF S IN NATURAL-FREQUENCY POLYNOMIAL MADE UNITY):

        0.0            0.0               0.0               0.000500          0.0               0.0               0.0
.......................................................................................................................
        1.000000       0.200000          3.019999          0.401000          3.019995          0.200000          1.000000

THE RESULTS ABOVE WERE PRODUCED BY LADANA.

              NOTES:            MS SECT. 7.19  BPF  LADANA     FIG. 7.19-
                                (FM LP PROTOTYPE BY TRFMTN)
                                C

THERE SEEMING TO BE NO MORE DATA, I AM, SIR/MADAME (MS.), YOUR MOST OBEDIENT SERVANT, LADANA.
```

FIG. 7.19–B*b*

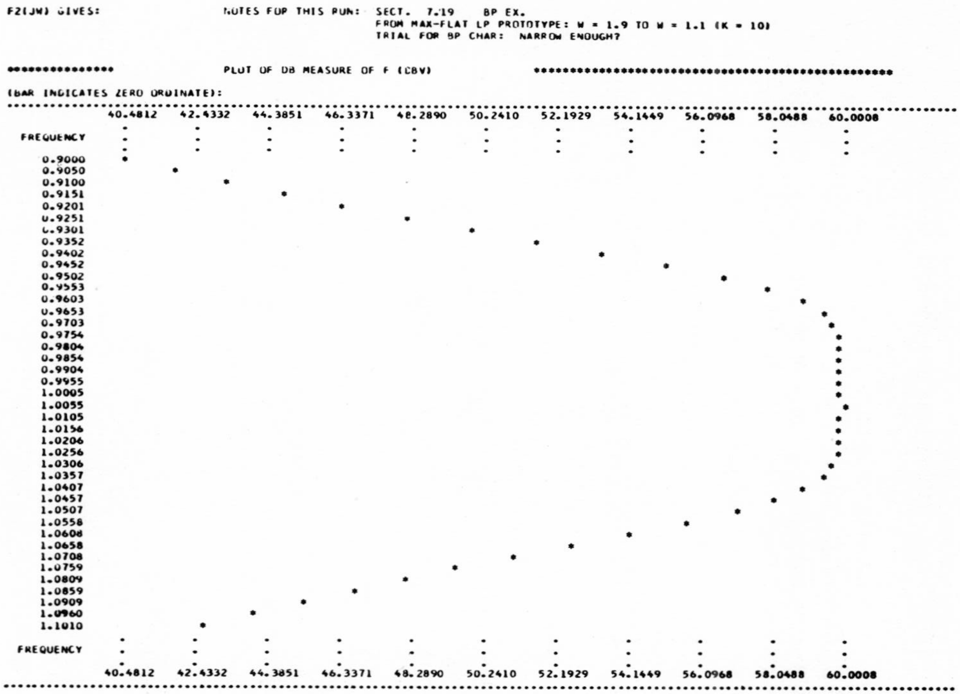
F2(JW) GIVES:
NOTES FOR THIS RUN: SECT. 7.19 BP EX.
FROM MAX-FLAT LP PROTOTYPE: W = 1.9 TO W = 1.1 (K = 10)
TRIAL FOR BP CHAR: NARROW ENOUGH?
PLOT OF DB MEASURE OF F (CBV)
(BAR INDICATES ZERO ORDINATE):
40.4812 42.4332 44.3851 46.3371 48.2890 50.2410 52.1929 54.1449 56.0968 58.0488 60.0008
FREQUENCY
0.9000
0.9050
0.9100
0.9151
0.9201
0.9251
0.9301
0.9352
0.9402
0.9452
0.9502
0.9553
0.9603
0.9653
0.9703
0.9754
0.9804
0.9854
0.9904
0.9955
1.0005
1.0055
1.0105
1.0156
1.0206
1.0256
1.0306
1.0357
1.0407
1.0457
1.0507
1.0558
1.0608
1.0658
1.0708
1.0759
1.0809
1.0859
1.0909
1.0960
1.1010
FREQUENCY
40.4812 42.4332 44.3851 46.3371 48.2890 50.2410 52.1929 54.1449 56.0968 58.0488 60.0008

FIG. 7.19-B*c*

and, at high frequencies,

$$T_{HF}(s) = \frac{0.005}{s^3},$$
$$|T_{HF}(j\omega)| = \frac{0.005}{\omega^3} \tag{7.19-13}$$

which are cubic and inversely cubic, respectively. These are certainly not linear! But for Fig. 7.19-B*b* the machine was asked to compute the *logarithmic* measure of $|T|$ and to space points equally on a logarithmic frequency scale. Such logarithmic measures (Appendix D) are often convenient for various reasons. One such reason is the simple asymptotically *linear* behavior of $|T|$ curves suggested by Fig. 7.19-B*b* and easily confirmed by analysis. Let us define

$$y = \log\left|\frac{T(j\omega)}{T_0}\right|,$$
$$x = \log\frac{\omega}{\omega_0}, \tag{7.19-14}$$

in which the base used for the logarithms is arbitrary; T_0 and ω_0 are any convenient references (the argument of a logarithm must be dimensionless). Then, asymptotically, on the assumption that T in (7.19-11) has been normalized and is free of dimensions, i.e., that T_0 may be set equal to unity,

At low frequencies:

$$\begin{aligned} y &= \log 0.005\omega^3 = 3\log\frac{\omega}{\omega_L} \\ &= 3\log\frac{\omega}{\omega_0} + 3\log\frac{\omega_0}{\omega_L} \\ &= 3x + \text{constant}, \end{aligned} \tag{7.19-15}$$

At high frequencies:

$$\begin{aligned} y &= \log\frac{0.005}{\omega^3} = 3\log\frac{\omega_H}{\omega} \\ &= -3\log\frac{\omega}{\omega_0} - 3\log\frac{\omega_0}{\omega_H} \\ &= -3x + \text{constant} \end{aligned} \tag{7.19-16}$$

in which ω_L and ω_H are additional convenient reference frequencies, those at which y takes the value zero. If we plot y against x, that is, $|T|$ versus ω, using logarithmic scales for both, the low-frequency asymptote is clearly a straight line of positive slope of three units; the high-frequency asymptote is clearly a straight line of negative slope of three units. The suggestion of Fig. 7.19-B*b* has become an established fact. We observe that to draw the asymptotes

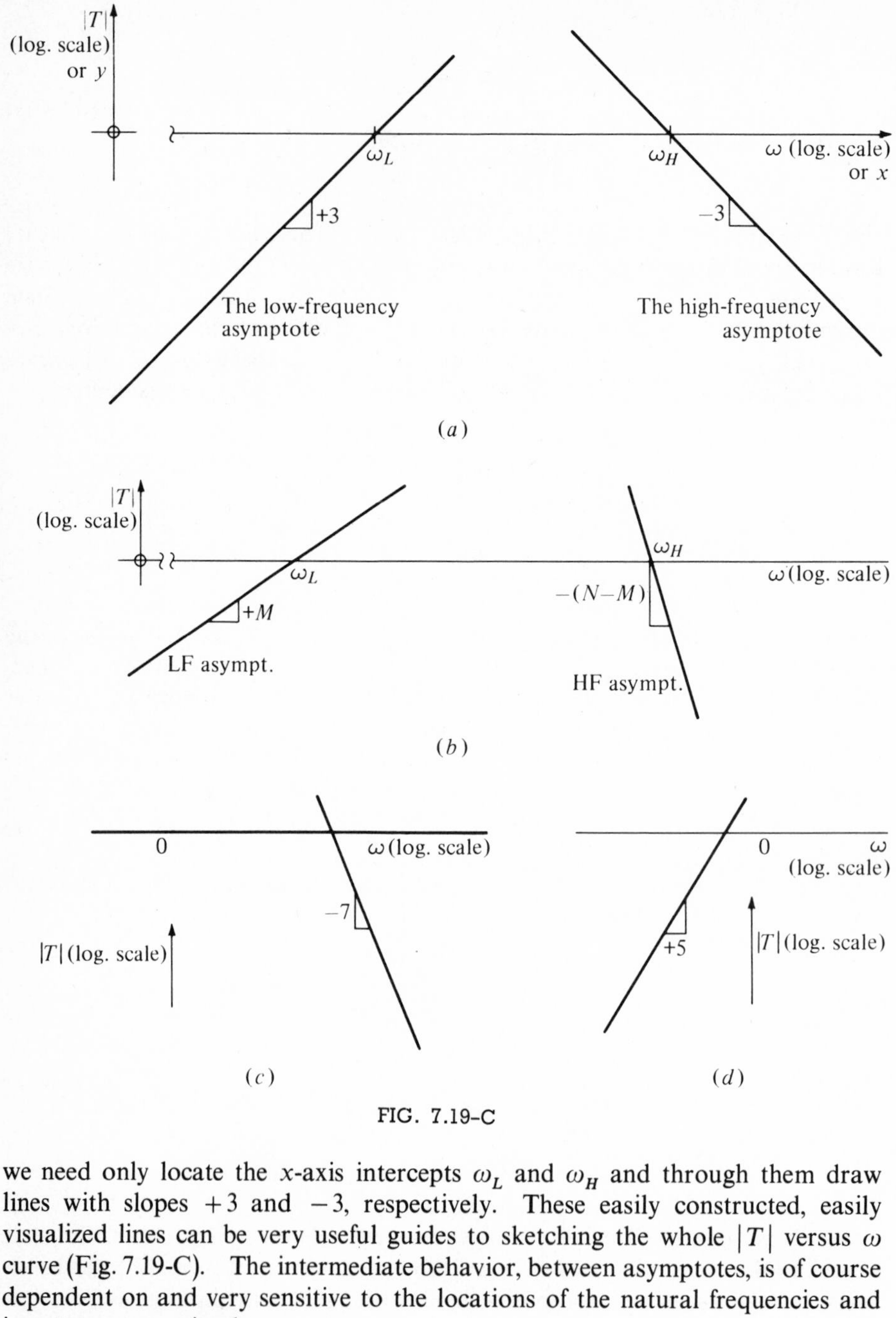

FIG. 7.19-C

we need only locate the x-axis intercepts ω_L and ω_H and through them draw lines with slopes $+3$ and -3, respectively. These easily constructed, easily visualized lines can be very useful guides to sketching the whole $|T|$ versus ω curve (Fig. 7.19-C). The intermediate behavior, between asymptotes, is of course dependent on and very sensitive to the locations of the natural frequencies and is not necessary simple.

In terms somewhat more general as to form of $T(s)$ but more specific as to base of logarithms used, consider

$$T(s) = \frac{Ks^M}{s^N + B_1 s^{N-1} + \cdots + B_N} \tag{7.19-17}$$

a transfer function with M zeros at the origin and $(N - M)$ zeros at infinity, normalized and without dimensions. We alter (7.19-14) to

$$y = 20 \log_{10} |T(j\omega)|, \tag{7.19-18}$$

which is the decibel measure of T (see Appendix D), and obtain for the asymptotes

$$\begin{aligned} y_{LF} &= 20 \log \frac{K\omega^M}{B_N} = 20M \log \frac{\omega}{\omega_L}, \\ \omega_L &= \sqrt[M]{B_N/K} \end{aligned} \tag{7.19-19}$$

and

$$\begin{aligned} y_{HF} &= 20 \log K\omega^{M-N} = -20(N - M) \log \frac{\omega}{\omega_H}, \\ \omega_H &= \frac{1}{\sqrt[N-M]{K}}. \end{aligned} \tag{7.19-20}$$

The low-frequency asymptote is linear with slope $+M$ units [Fig. 7.19-C*b*)], the number of zeros at the origin; the high-frequency asymptote is linear with slope $-(N - M)$ units, $(N - M)$ being the number of zeros at infinity. The intermediate part of the curve cannot even be sketched without knowledge of the poles of $T(s)$: the possibilities there are infinite in number, from smooth, broad sweeps to violent oscillations.

The *unit of slope* mentioned above is

$$20 \log 2 = 6.02 \text{ dBV per octave} \tag{7.19-21}$$

or

$$20 \log 10 = 20 \text{ dBV per decade.} \tag{7.19-22}$$

The term *octave* refers to a 2 : 1 frequency interval, the term *decade* to a 10 : 1 frequency interval. (The former word stems from musical usage in which a 2 : 1 frequency or pitch interval is divided to contain eight notes.) Any rational function, logarithmically plotted, will exhibit such asymptotically linear behavior at both low and high frequencies. The slopes will be integral multiples of this basic unit (positive, zero, or negative). For examples we can draw from networks already discussed (see Table 7.19-C).

TABLE 7.19-C

Network (Example)	General Character of Network	N	M	Asymptotic Slopes Low-Frequency	Asymptotic Slopes High-Frequency	See Fig.
Fig. 7.19-B	Band-pass	6	3	+3	−3	7.19-C*a*
Fig. 7.19-A	Low-pass	7	0	0	−7	7.19-C*c*
Fig. 7.18-A	High-pass	5	5	+5	0	7.19-C*d*
Fig. 7.18-B	Band-pass	5	2	+2	−3	
Fig. 7.18-C	Low-pass	14	0	0	−14	
Fig. 7.18-I	Low-pass	4	0	0	−4	
Chap. 3 (tuned circuit)	Band-pass	2	1	+1	−1	7.19-D

This observation about asymptotic low- and high-frequency behavior is obviously valid for all rational transfer functions, to which our interest is confined. Such drawings were originally applied in engineering by Hendrick W. Bode (Appendix B) and are sometimes called *Bode plots* or *Bode diagrams.* They can be extremely useful because the two asymptotes can be sketched immediately after a glance at the transfer function or simply at the network schematic,

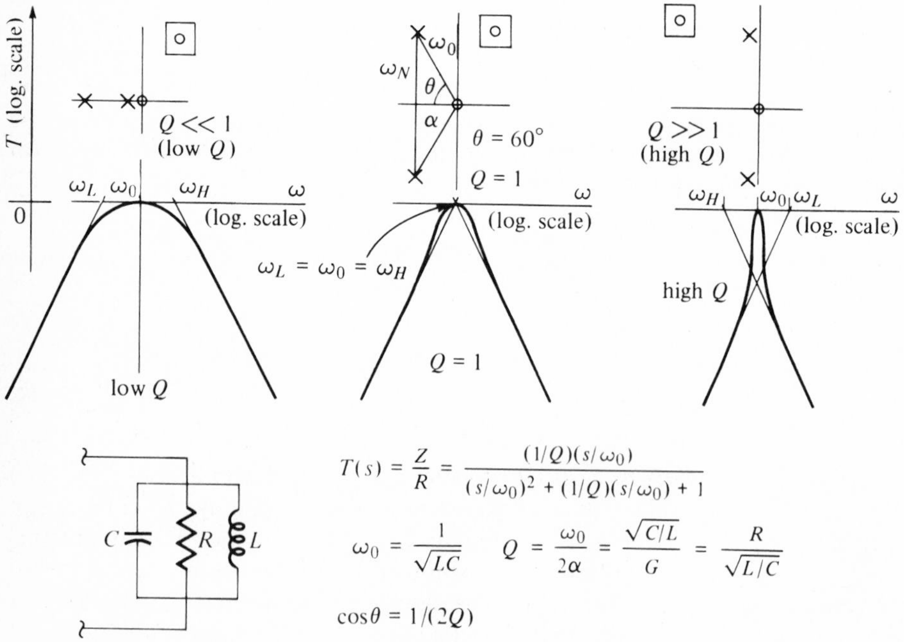

FIG. 7.19-D

where N and M are usually obvious. The central part of the curve can*not* be so sketched immediately (unless one has other knowledge of the network's intermediate behavior, either from its design or because of some simplifying characteristic). But only this central part requires calculation, and its general shape is often evident from the location of the natural frequencies or other knowledge—but by no means always. To illustrate the variety possible in even a simple circuit, consider the ordinary tuned circuit (Chap. 3) in this light (Fig. 7.19-D). Though asymptotically the same, at intermediate frequencies the curves vary considerably in character!

We should not leave this example without also examining at least some aspects of its time-domain behavior. Given the transfer function (7.19-11) IMPSTEP produces the impulse response of Fig. 7.19-E*a*. We observe a "pulse" of oscillation, followed by a second highly damped pulse (and presumably by others later). The numerical values of the natural frequencies (Fig. 7.19-E) show that the natural behavior consists of three lightly damped sinusoids of nearly equal frequencies. The observed phenomenon of *beats* (the mixing of frequencies to give a slowly modulated sinusoidal wave) is thus to be expected (Prob. 7-P).

```
THIS IS IMPSTEP SEVEN.    NOTES FOR THIS RUN:    MS  SECT. 7.19   BP  EXAMPLE

PARAMETER VALUES READ:   NN =  3, ND =  6, NX = 0, NPLT = 2, NPRD = 1, NPRU = 0, NCODE = 0

FOR THE FOLLOWING CASE THE MULTIPLIER  H =     0.100000      AND THE COEFFICIENTS IN F(S), LOWEST POWER FIRST, ARE:

NUMERATOR:
        0.0            0.0            0.0            1.000000
DENOMINATOR:
        1.000000       0.200000       3.020000       0.401000       3.020000       0.200000
        1.000000

THE FOLLOWING ARE THE POLES OF F:
      -0.023906  +J -0.957332
      -0.023906  +J  0.957332
      -0.050009  +J -0.998634
      -0.050009  +J  0.998664
      -0.026101  +J -1.043967
      -0.026101  +J  1.043967
THE FOLLOWING ARE THE RESIDUES THEREAT (THE "K'S"):
        MAGNITUDE      ANGLE (RAD.)    ANGLE (DEG.)      REAL PART      IMAG. PART
        .........      ............    ............      .........      ..........

        2.763163        3.639635       208.547241       -2.427244      -1.320477
        2.763183       -3.639635      -208.547241       -2.427244       1.320477
        5.005014       -0.047368        -2.713989        4.999400      -0.236989
        5.005014        0.047368         2.713989        4.999400       0.236989
        3.006412        2.595942       148.736542       -2.569350       1.560248
        3.006412       -2.595942      -148.736542       -2.569850      -1.560248

THE SUM OF THE COMPUTED RESIDUES "K" IS:                                        0.461197E-02 + J  0.0
THE SUM OF THE PRODUCTS OF THE COMPUTED RESIDUES AND THE NATURAL FREQUENCIES, "SUM KP", IS:   0.626564E-02 + J -0.953674E-06
THE FIRST NONZERO IV, CALCULATED FROM THE GIVEN TRANSFORM, IS:   0.100000E 00    (THIS IS THE INITIAL VALUE OF THE   2-TH
DERIVATIVE OF RO)
IT WERE WELL TO CHECK THESE: DO "SUM OF K'S" AND "SUM OF "KP'S" , ETC. CHECK? *****
```

FIG. 7.19-E*a*.

THIS IS IMPSTEP SEVEN. NOTES FOR THIS RUN: MS SECT. 7.19 BP EXAMPLE
DERIVED FROM 1 + W**6 LP PROTOTYPE BAND NARROWED
ARE POLES OK? BEATS IN RO?

SUBROUTINE 'PLOT42', BEING CALLED, GIVES (EAR MARKS ZERO ORDINATE):

TIME	IMPULSE RESP
0.0	0.4610E-02
0.5000	0.1837E-01
1.0000	0.4346E-01
1.5000	0.5836E-01
2.0000	0.4251E-01
2.5000	-0.1190E-01
3.0000	-0.9310E-01
3.5000	-0.1712E 00
4.0000	-0.2080E 00
4.5000	-0.1721E 00
5.0000	-0.5409E-01
5.5000	0.1253E 00
6.0000	0.3163E 00
6.5000	0.4531E 00
7.0000	0.4745E 00
7.5000	0.3465E 00
8.0000	0.7874E-01
8.5000	-0.2713E 00
9.0000	-0.6092E 00
9.5000	-0.8286E 00
10.0000	-0.8426E 00
10.5000	-0.6141E 00
11.0000	-0.1735E 00
11.5000	0.3816E 00
12.0000	0.9071E 00
12.5000	0.1249E 01
13.0000	0.1290E 01
13.5000	0.9810E 00
14.0000	0.3703E 00
14.5000	-0.4061E 00
15.0000	-0.1155E 01
15.5000	-0.1663E 01
16.0000	-0.1778E 01
16.5000	-0.1432E 01
17.0000	-0.6780E 00
17.5000	0.3145E 00
18.0000	0.1301E 01
18.5000	0.2020E 01
19.0000	0.2263E 01
19.5000	0.1935E 01
20.0000	0.1086E 01
20.5000	-0.9643E-01
21.0000	-0.1323E 01
21.5000	-0.2273E 01
22.0000	-0.2703E 01
22.5000	-0.2453E 01
23.0000	-0.1567E 01
23.5000	-0.2412E 00
24.0000	0.1208E 01
24.5000	0.2438E 01
25.0000	0.3046E 01
25.5000	0.2942E 01
26.0000	0.2038E 01
26.5000	0.6709E 00
27.0000	-0.9522E 00
27.5000	-0.2394E 01
28.0000	-0.3277E 01
28.5000	-0.3363E 01
29.0000	-0.2608E 01
29.5000	-0.1179E 01
30.0000	0.5795E 00
30.5000	0.2234E 01
31.0000	0.3367E 01
31.5000	0.3663E 01
32.0000	0.3087E 01
32.5000	0.1711E 01
33.0000	-0.1145E 00
33.5000	-0.1940E 01
34.0000	-0.3310E 01
34.5000	-0.3877E 01
35.0000	-0.3489E 01
35.5000	-0.2232E 01
36.0000	-0.4086E 00
36.5000	0.1533E 01
37.0000	0.3111E 01
37.5000	0.3932E 01
38.0000	0.3787E 01
38.5000	0.2706E 01
39.0000	0.9524E 00
39.5000	-0.1042E 01
40.0000	-0.2785E 01
40.5000	-0.3840E 01
41.0000	-0.3961E 01
41.5000	-0.3101E 01
42.0000	-0.1480E 01
42.5000	0.5028E 00
43.0000	0.2357E 01
43.5000	0.3626E 01
44.0000	0.4001E 01
44.5000	0.3393E 01
45.0000	0.1956E 01
45.5000	0.4981E-01
46.0000	-0.1855E 01
46.5000	-0.3291E 01
47.0000	-0.3910E 01
47.5000	-0.3566E 01
48.0000	-0.2355E 01
48.5000	-0.5811E 00
49.0000	0.1314E 01
49.5000	0.2664E 01
50.0000	0.3696E 01
50.5000	0.3615E 01
51.0000	0.2654E 01
51.5000	0.1061E 01
52.0000	-0.7652E 00
52.5000	-0.2374E 01
53.0000	-0.3376E 01
53.5000	-0.3541E 01
54.0000	-0.2642E 01
54.5000	-0.1464E 01
55.0000	0.2413E 00
55.5000	0.1852E 01
56.0000	0.2979E 01
56.5000	0.3357E 01
57.0000	0.2913E 01
57.5000	0.1773E 01

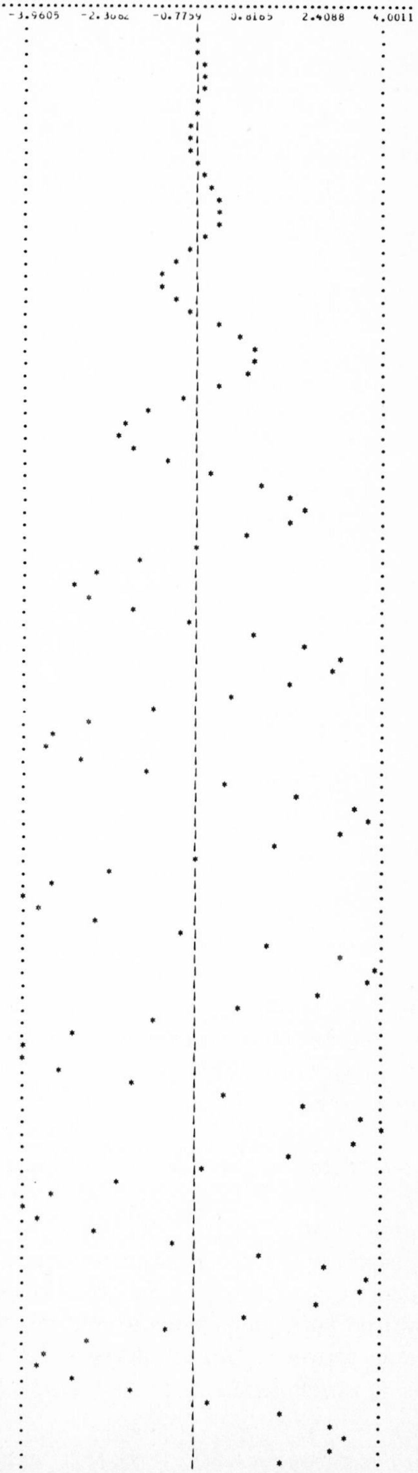

58.0000	0.2314E 00
58.5000	-0.1328E 01
59.0000	-0.2525E 01
59.5000	-0.3080E 01
60.0000	-0.2874E 01
60.5000	-0.1979E 01
61.0000	-0.6315E 00
61.5000	0.8292E 00
62.0000	0.2040E 01
62.5000	0.2731E 01
63.0000	0.2735E 01
63.5000	0.2080E 01
64.0000	0.9445E 00
64.5000	-0.3798E 00
65.0000	-0.1567E 01
65.5000	-0.2334E 01
66.0000	-0.2512E 01
66.5000	-0.2080E 01
67.0000	-0.1163E 01
67.5000	-0.2264E-02
68.0000	0.1113E 01
68.5000	0.1910E 01
69.0000	0.2220E 01
69.5000	0.1990E 01
70.0000	0.1286E 01
70.5000	0.3050E 00
71.0000	-0.7051E 00
71.5000	-0.1500E 01
72.0000	-0.1899E 01
72.5000	-0.1825E 01
73.0000	-0.1319E 01
73.5000	-0.5227E 00
74.0000	0.3579E 00
74.5000	0.1107E 01
75.0000	0.1553E 01
75.5000	0.1605E 01
76.0000	0.1272E 01
76.5000	0.6555E 00
77.0000	-0.8138E-01
77.5000	-0.7555E 00
78.0000	-0.1210E 01
78.5000	-0.1349E 01
79.0000	-0.1160E 01
79.5000	-0.7089E 00
80.0000	-0.1201E 00
80.5000	0.4572E 00
81.0000	0.8804E 00
81.5000	0.1076E 01
82.0000	0.9990E 00
82.5000	0.6927E 00
83.0000	0.2474E 00
83.5000	-0.2205E 00
84.0000	-0.5967E 00
84.5000	-0.8054E 00
85.0000	-0.8064E 00
85.5000	-0.6196E 00
86.0000	-0.3061E 00
86.5000	0.4862E-01
87.0000	0.3572E 00
87.5000	0.5521E 00
88.0000	0.5995E 00
88.5000	0.5044E 00
89.0000	0.3049E 00
89.5000	0.5984E-01
90.0000	-0.1684E 00
90.5000	-0.3291E 00
91.0000	-0.3943E 00
91.5000	-0.3624E 00
92.0000	-0.2556E 00
92.5000	-0.1100E 00
93.0000	0.3441E-01
93.5000	0.1452E 00
94.0000	0.2040E 00
94.5000	0.2086E 00
95.0000	0.1709E 00
95.5000	0.1101E 00
96.0000	0.4636E-01
96.5000	-0.5530E-02
97.0000	-0.3919E-01
97.5000	-0.5644E-01
98.0000	-0.6424E-01
98.5000	-0.7025E-01
99.0000	-0.7877E-01
99.5000	-0.8845E-01
100.0000	-0.9274E-01
100.5000	-0.8277E-01
101.0000	-0.5160E-01
101.5000	0.1785E-02
102.0000	0.7013E-01
102.5000	0.1385E 00
103.0000	0.1879E 00
103.5000	0.2003E 00
104.0000	0.1652E 00
104.5000	0.8383E-01
105.0000	-0.2940E-01
105.5000	-0.1492E 00
106.0000	-0.2453E 00
106.5000	-0.2901E 00
107.0000	-0.2671E 00
107.5000	-0.1757E 00
108.0000	-0.3364E-01
108.5000	0.1270E 00
109.0000	0.2669E 00
109.5000	0.3494E 00
110.0000	0.3501E 00
110.5000	0.2644E 00
111.0000	0.1092E 00
111.5000	-0.7982E-01
112.0000	-0.2570E 00
112.5000	-0.3776E 00
113.0000	-0.4096E 00
113.5000	-0.3419E 00
114.0000	-0.1832E 00
114.5000	0.1594E-01
115.0000	0.2212E 00
115.5000	0.3769E 00
116.0000	0.4434E 00
116.5000	0.4023E 00
117.0000	0.2626E 00
117.5000	0.5030E-01
118.0000	-0.1664E 00
118.5000	-0.3539E 00
119.0000	-0.4516E 00
119.5000	-0.4426E 00

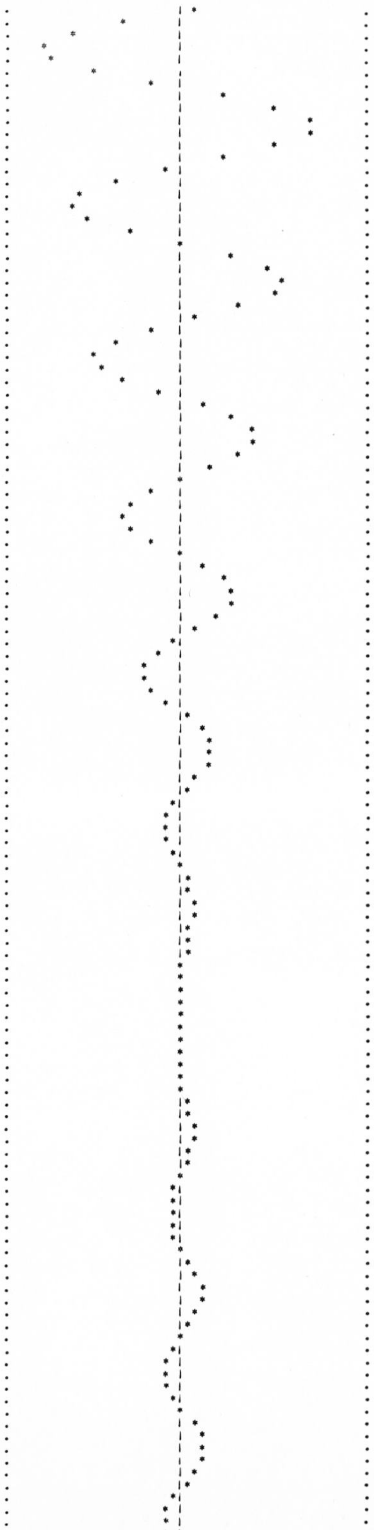

FIG. 7.19-Ea. (*Con.*)

TIME	IMPULSE RESP
120.0000	-0.3261E 00
120.5000	-0.1293E 00
121.0000	0.9971E-01
121.5000	0.3049E 00
122.0000	0.4561E 00
122.5000	0.4611E 00
123.0000	0.3741E 00
123.5000	0.1964E 00
124.0000	-0.2819E-01
124.5000	-0.2447E 00
125.0000	-0.4005E 00
125.5000	-0.4550E 00
126.0000	-0.4040E 00
126.5000	-0.2527E 00
127.0000	-0.4135E-01
127.5000	0.1703E 00
128.0000	0.3494E 00
128.5000	0.4335E 00
129.0000	0.4143E 00
129.5000	0.2944E 00
130.0000	0.1051E 00
130.5000	-0.1062E 00
131.0000	-0.2873E 00
131.5000	-0.3965E 00
132.0000	-0.4073E 00
132.5000	-0.3198E 00
133.0000	-0.1574E 00
133.5000	0.3900E-01
134.0000	0.2212E 00
134.5000	0.3455E 00
135.0000	0.3835E 00
135.5000	0.3285E 00
136.0000	0.1962E 00
136.5000	0.2076E-01
137.0000	-0.1543E 00
137.5000	-0.2869E 00
138.0000	-0.3465E 00
138.5000	-0.3212E 00
139.0000	-0.2201E 00
139.5000	-0.6989E-01
140.0000	0.9152E-01
140.5000	0.2251E 00
141.0000	0.2998E 00
141.5000	0.3002E 00
142.0000	0.2290E 00
142.5000	0.1064E 00
143.0000	-0.3640E-01
143.5000	-0.1645E 00
144.0000	-0.2474E 00
144.5000	-0.2661E 00
145.0000	-0.2242E 00
145.5000	-0.1293E 00
146.0000	-0.8568E-02
146.5000	0.1081E 00
147.0000	0.1932E 00
147.5000	0.2262E 00
148.0000	0.2075E 00
148.5000	0.1391E 00
149.0000	0.4192E-01
149.5000	-0.5939E-01
150.0000	-0.1407E 00
150.5000	-0.1840E 00
151.0000	-0.1815E 00
151.5000	-0.1368E 00
152.0000	-0.6321E-01
152.5000	0.2008E-01
153.0000	0.9279E-01
153.5000	0.1387E 00
154.0000	0.1492E 00
154.5000	0.1243E 00
155.0000	0.7286E-01
155.5000	0.8832E-02
156.0000	-0.5135E-01
156.5000	-0.9545E-01
157.0000	-0.1134E 00
157.5000	-0.1041E 00
158.0000	-0.7209E-01
158.5000	-0.2708E-01
159.0000	0.1950E-01
159.5000	0.5661E-01
160.0000	0.7725E-01
160.5000	0.7657E-01
161.0000	0.6257E-01
161.5000	0.3525E-01
162.0000	0.3385E-02
162.5000	-0.2404E-01
163.0000	-0.4304E-01
163.5000	-0.5047E-01
164.0000	-0.4672E-01
164.5000	-0.3454E-01
165.0000	-0.1795E-01
165.5000	-0.1065E-02
166.0000	0.1292E-01
166.5000	0.2224E-01
167.0000	0.2650E-01
167.5000	0.2657E-01
168.0000	0.2344E-01
168.5000	0.1825E-01
169.0000	0.1164E-01
169.5000	0.4063E-02
170.0000	-0.4435E-02

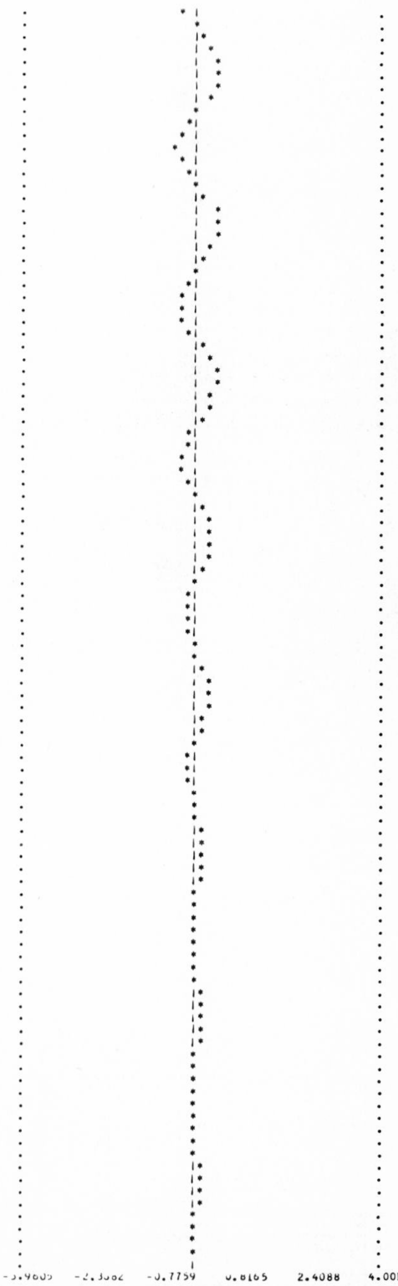

FIG. 7.19-Ea. (Con.)

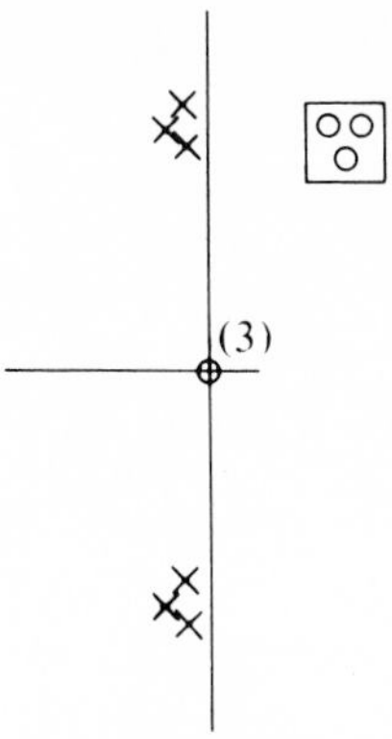

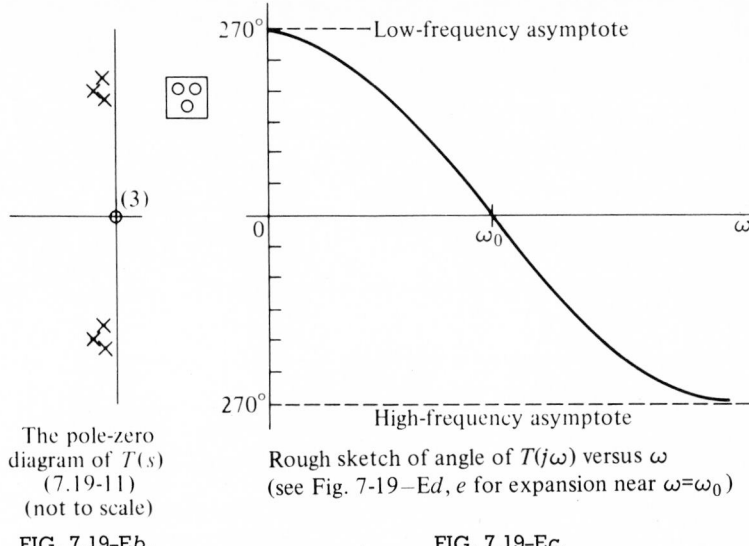

The pole-zero diagram of $T(s)$ (7.19-11) (not to scale)

FIG. 7.19-E*b*.

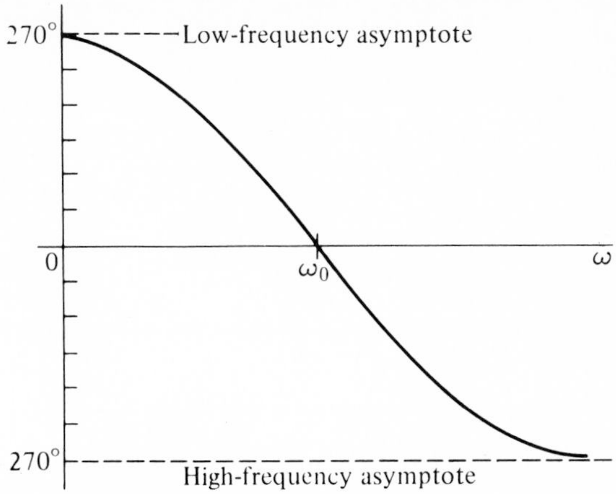

Rough sketch of angle of $T(j\omega)$ versus ω (see Fig. 7-19–E*d*, *e* for expansion near $\omega=\omega_0$)

FIG. 7.19-E*c*.

The exercise of a little imagination shows us the impulse response as a distorted and delayed "pulse" in the form of an envelope for oscillations at a "frequency" that approximates the three "ω_N" (which are nearly equal). This pulse seems to be centered at about 44 s (normalized time). Curious, we ask once more about a connection between this "delay" and the sinusoidal-steady-state characteristic, the angle $\beta(\omega)$ of (7.19-4). A rough sketch of $\beta(\omega)$, based on the pole-zero diagram of Fig. 7.19-E*b*, shows (Fig. 7.19-E*c*) that β/ω is here virtually meaningless, since it is infinite at $\omega = 0$ and zero at $\omega = 1$. But, at least at frequencies where $|T|$ is appreciable, β *is* approximately linear and $d\beta/d\omega$ may be meaningful [see (7.19-9) and (7.19-10) and associated discussion]. We call on FOMEGA to complement the $|T|$ versus ω curve of Fig. 7.19-B*c* with a curve for $\beta(\omega)$ for the region of good transmission, using (7.19-11). In Fig. 7.19-E*d* and *e* we see the results: first over $0.95 < \omega < 1.05$ and then (expanded) over $0.995 < \omega < 1.005$. We observe that near the pass-band center ω_0 ($\omega_0 = 1$)

$$\beta(\omega) = \beta_0 + \left(\frac{d\beta}{d\omega}\right)_0 (\omega - \omega_0) + \left(\frac{d^2\beta}{d\omega^2}\right)_0 \frac{(\omega - \omega_0)^2}{2!} + \cdots, \qquad (7.19\text{-}23)$$

in which $\beta_0 = 0$ and the terms beyond the linear one are small. Hence near ω_0, $\beta(\omega) \cong -T_D(\omega - \omega_0)$ in which $T_D = |(d\beta/d\omega_0)|$ is about $[(270°)(\pi/180)/0.1] =$ 47 s (normalized) if we calculate from Fig. 7.19-E*d*, and about $[(22.9°)(\pi/180)/0.01] = 40$ s (normalized) if we use Fig. 7.19-E*e*. [The computer's uncertainty—is $\beta(\omega_0)$ zero or 360° or 720° or ...?—need not concern us, since we are now interested only in the derivative of $\beta(\omega)$.] The average of these two numbers is 44 s (normalized), to be compared with the location of the "pulse center" in Fig. 7.19-E*a*(!).

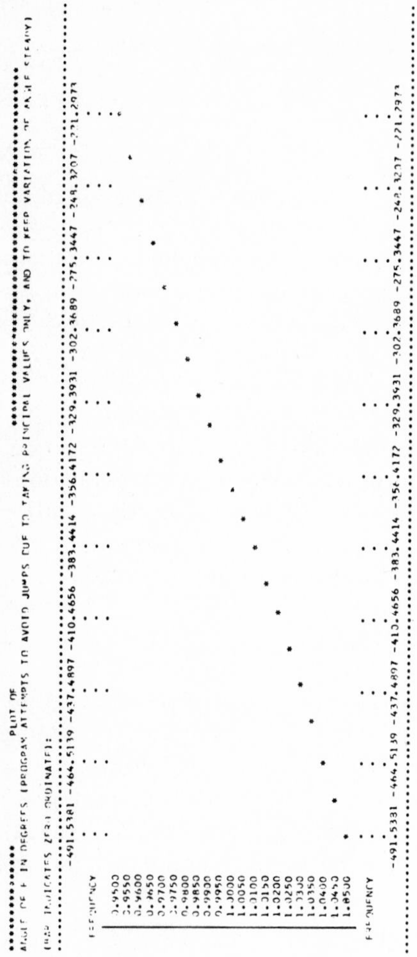

FIG. 7.19-E*d*

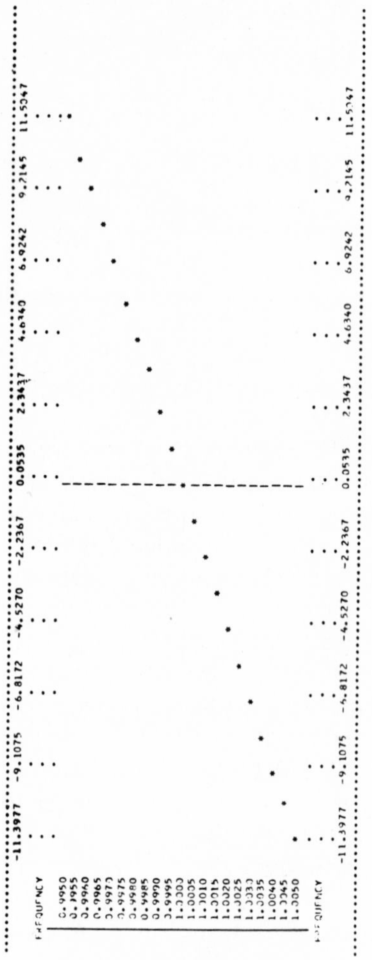

FIG. 7.19-Ee

Our conjecture of (7.19-10) evidently should be modified (β_0 being zero) to focus entirely on $d\beta/d\omega$ (rather than β/ω) and that in any frequency range in which $|T|$ is appreciable. It may give a good measure of the time delay of signals as they pass through the network, if the signals are such that the distortion is small enough to admit of defining a "delay." Note that we have not contradicted (7.19-10), but have made a more general statement. The question of distortion is now more troublesome, since, in the vernacular, the input pulse is "dc" and the output pulse is "ac." It were well then to calculate one more response: the response to a signal whose "components lie in the pass band," an "ac pulse" or an "ac step," for example. We accordingly apply RESSINET to obtain Fig. 7.19-F, the network's response to a "step of ac," $e(t) = [E_m \cos(\omega_F t)u(t)]$, $\omega_F = 1$ (the band-center frequency). The mate of this is $[(E_m s)/(s^2 + \omega_F{}^2)]$ from pair 21; and its "components lie mostly in the pass band" in the sense of the discussion of the latter part of Sec. 7.15. The response consists of a slow buildup to the sinusoidal steady state as the natural (transient) part decays. To define a "delay" is obviously difficult—but we could say, on some not entirely unreasonable basis, that the halfway point [see the discussion following (7.19-7)] occurs in Fig. 7.19-F at about $t = 44$ s (normalized) (!).

We shall not belabor the point any longer here. To recapitulate, and revise (7.19-10) in accord with the results of our experiments:

> If the concept of "signal delay" is meaningful, if a delay can be said to exist, the signal distortion being not too great, that delay seems to be measurable in the ω domain by $T_D = |d\beta/d\omega|$ over the range of frequencies (low-pass or band-pass) where $|T|$ is appreciable and $\beta(\omega)$ is more or less linear; the "delay" is, of course, only approximate, not an exact figure. (7.19-24)

We note again, with a feeling of satisfaction, that our computer programs make very useful "laboratory equipment" for experiments. Our results, (7.19-24) in particular, are purely empirical to be sure—but their evidence is strong. Theoretical confirmation can be attempted and discussed later (Sec. 8.07).

Physical corroboration, refreshing and reassuring, is given by Fig. 7.19-G. The network described there schematically [in (*a*)] has the definitely low-pass if not very sharp cutoff $|T|$ characteristic shown in (*b*). The impulse response (*c*) is a rather distorted but still recognizable pulse followed by oscillations during the settling-down-again period. We may consider the location in time of the pulse peak to be at one square or 1 cm after the excitation. This is at $t = 40$ μs since the scale is 40 μs/cm. Calculations from measurements of angle (phase shift) $\beta(\omega)$ [Prob. 7-B] give a value for $\beta'(0)$ of about 42 or 43 μs, a phenomenal and satisfying agreement! The step response, in (*d*), is just what one would expect for the integral of the impulse response.

A less spectacular, less corroborative, but interesting and relevant example is

THIS IS RESSINET.   NOTES FOR THIS RUN: MS SECT. 7.19
B-P NETWORK, SINUSOIDAL INPUT
RESSINET USED   W1 = W2 = 1 COSINE WAVE

PLOT OF (COMPLETE) RESPONSE ALONE:

SUBROUTINE 'PLOT4Z', BEING CALLED, GIVES (BAR MARKS ZERO ORDINATE):

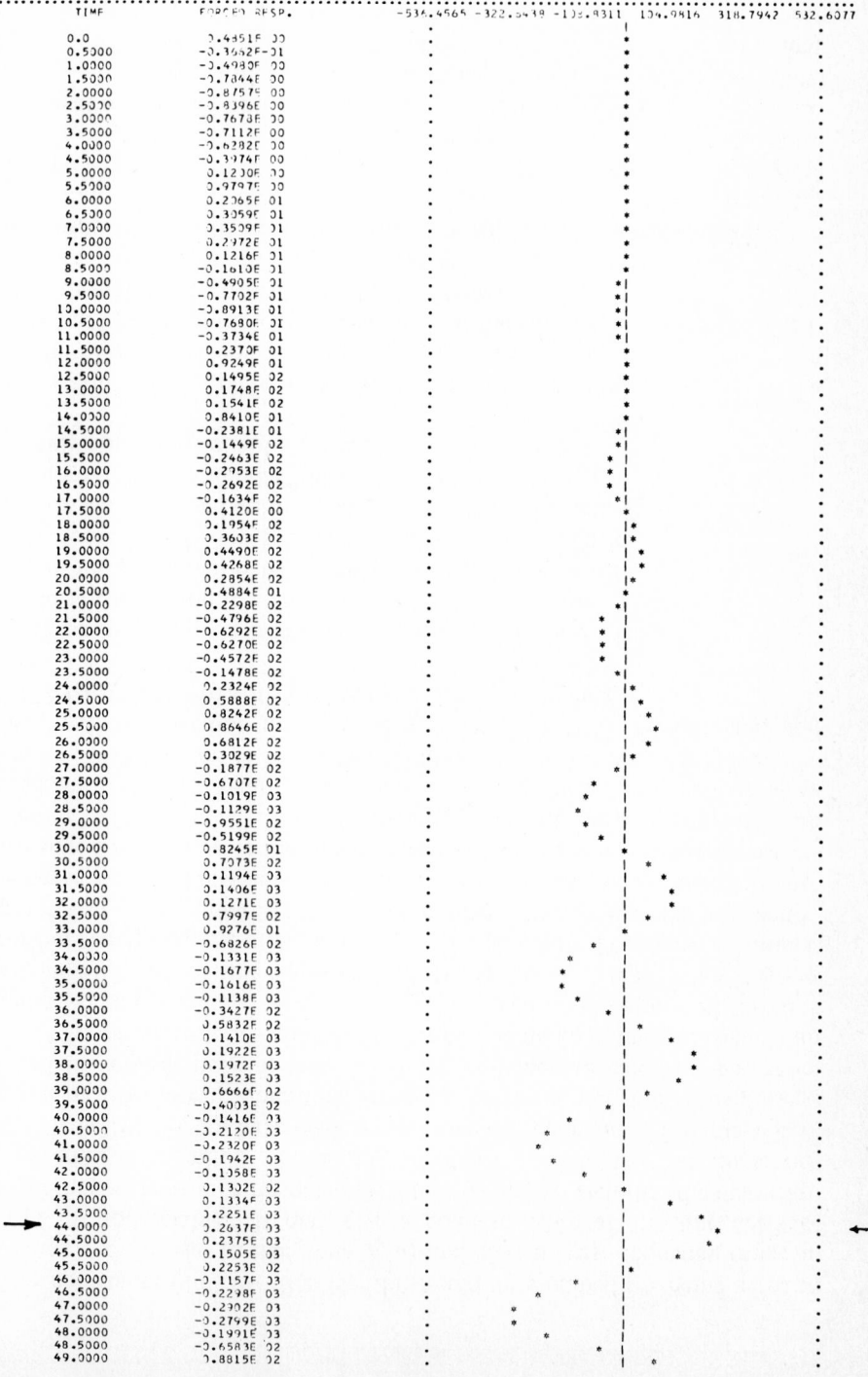

TIME	FORCED RESP.
0.0	0.4851E 00
0.5000	-0.3662E-01
1.0000	-0.4980E 00
1.5000	-0.7844E 00
2.0000	-0.8757E 00
2.5000	-0.8396E 00
3.0000	-0.7678E 00
3.5000	-0.7112E 00
4.0000	-0.6282E 00
4.5000	-0.3974E 00
5.0000	0.1200E 00
5.5000	0.9797E 00
6.0000	0.2065E 01
6.5000	0.3059E 01
7.0000	0.3509E 01
7.5000	0.2972E 01
8.0000	0.1216E 01
8.5000	-0.1610E 01
9.0000	-0.4905E 01
9.5000	-0.7702E 01
10.0000	-0.8913E 01
10.5000	-0.7680E 01
11.0000	-0.3734E 01
11.5000	0.2370E 01
12.0000	0.9249E 01
12.5000	0.1495E 02
13.0000	0.1748E 02
13.5000	0.1541E 02
14.0000	0.8410E 01
14.5000	-0.2381E 01
15.0000	-0.1449E 02
15.5000	-0.2463E 02
16.0000	-0.2953E 02
16.5000	-0.2692E 02
17.0000	-0.1634E 02
17.5000	0.4120E 00
18.0000	0.1954E 02
18.5000	0.3603E 02
19.0000	0.4490E 02
19.5000	0.4268E 02
20.0000	0.2854E 02
20.5000	0.4884E 01
21.0000	-0.2298E 02
21.5000	-0.4796E 02
22.0000	-0.6292E 02
22.5000	-0.6270E 02
23.0000	-0.4572E 02
23.5000	-0.1478E 02
24.0000	0.2324E 02
24.5000	0.5888E 02
25.0000	0.8242E 02
25.5000	0.8646E 02
26.0000	0.6812E 02
26.5000	0.3029E 02
27.0000	-0.1877E 02
27.5000	-0.6707E 02
28.0000	-0.1019E 03
28.5000	-0.1129E 03
29.0000	-0.9551E 02
29.5000	-0.5199E 02
30.0000	0.8245E 01
30.5000	0.7073E 02
31.0000	0.1194E 03
31.5000	0.1406E 03
32.0000	0.1271E 03
32.5000	0.7997E 02
33.0000	0.9276E 01
33.5000	-0.6826E 02
34.0000	-0.1331E 03
34.5000	-0.1677E 03
35.0000	-0.1616E 03
35.5000	-0.1138E 03
36.0000	-0.3427E 02
36.5000	0.5832E 02
37.0000	0.1410E 03
37.5000	0.1922E 03
38.0000	0.1972E 03
38.5000	0.1523E 03
39.0000	0.6666E 02
39.5000	-0.4003E 02
40.0000	-0.1416E 03
40.5000	-0.2120E 03
41.0000	-0.2320E 03
41.5000	-0.1942E 03
42.0000	-0.1058E 03
42.5000	0.1302E 02
43.0000	0.1334E 03
43.5000	0.2251E 03
44.0000	0.2637E 03
44.5000	0.2375E 03
45.0000	0.1505E 03
45.5000	0.2253E 02
46.0000	-0.1157E 03
46.5000	-0.2298E 03
47.0000	-0.2902E 03
47.5000	-0.2799E 03
48.0000	-0.1991E 03
48.5000	-0.6583E 02
49.0000	0.8815E 02

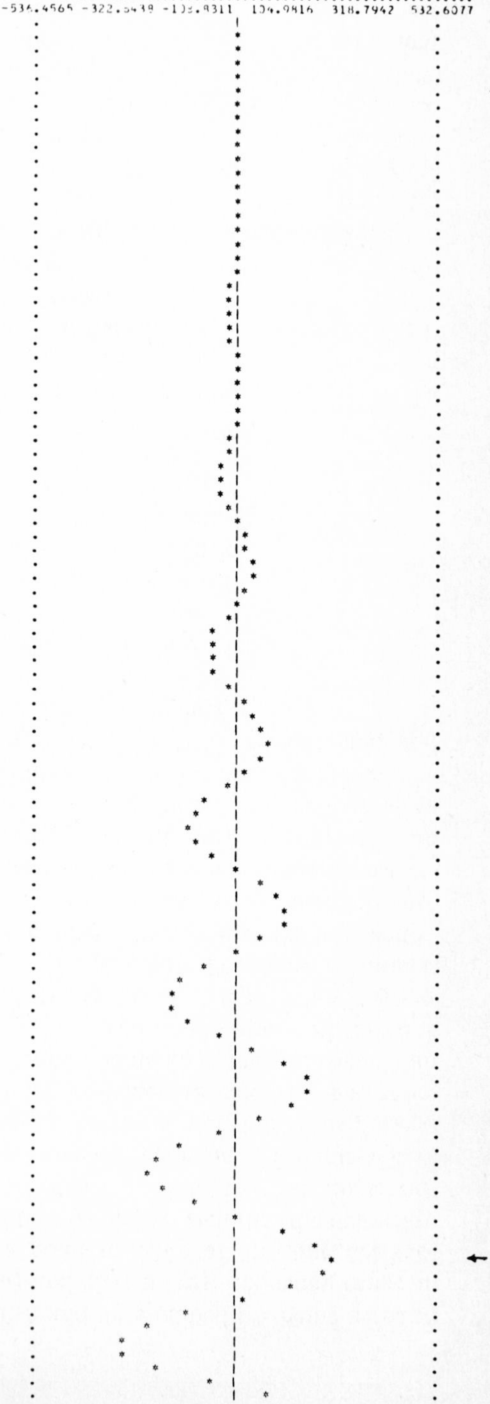

TIME	FORCED RESP.
49.5000	0.2249E 03
50.0000	0.3075E 03
50.5000	0.3193E 03
51.0000	0.2496E 03
51.5000	0.1156E 03
52.0000	-0.5096E 02
52.5000	-0.2093E 03
53.0000	-0.3197E 03
53.5000	-0.3533E 03
54.0000	-0.2997E 03
54.5000	-0.1701E 03
55.0000	0.4033E 01
55.5000	0.1830E 03
56.0000	0.3196E 03
56.5000	0.3799E 03
57.0000	0.3470E 03
57.5000	0.2272E 03
58.0000	0.4846E 02
58.5000	-0.1461E 03
59.0000	-0.3083E 03
~~59.5000~~	~~-0.3472E 03~~
60.0000	-0.3893E 03
60.5000	-0.2847E 03
61.0000	-0.1076E 03
61.5000	0.9935E 02
62.0000	0.2853E 03
62.5000	0.4038E 03
63.0000	0.4243E 03
63.5000	0.3401E 03
64.0000	0.1704E 03
64.5000	-0.4416E 02
65.0000	-0.2510E 03
65.5000	-0.3988E 03
66.0000	-0.4502E 03
66.5000	-0.3910E 03
67.0000	-0.2344E 03
67.5000	-0.1780E 02
68.0000	0.2060E 03
68.5000	0.3818E 03
69.0000	0.4654E 03
69.5000	0.4352E 03
70.0000	0.2973E 03
70.5000	0.8448E 02
71.0000	-0.1515E 03
71.5000	-0.3527E 03
72.0000	-0.4690E 03
72.5000	-0.4709E 03
73.0000	-0.3567E 03
73.5000	-0.1536E 03
74.0000	0.8920E 02
74.5000	0.3122E 03
75.0000	0.4603E 03
75.5000	0.4963E 03
76.0000	0.4104E 03
76.5000	0.2229E 03
77.0000	-0.2094E 02
77.5000	-0.2614E 03
78.0000	-0.4392E 03
78.5000	-0.5103E 03
79.0000	-0.4563E 03
79.5000	-0.2898E 03
80.0000	-0.5110E 02
80.5000	0.2016E 03
81.0000	0.4062E 03
81.5000	0.5121E 03
82.0000	0.4928E 03
82.5000	0.3523E 03
83.0000	0.1247E 03
83.5000	-0.1347E 03
84.0000	-0.3621E 03
84.5000	-0.5016E 03
85.0000	-0.5185E 03
85.5000	-0.4083E 03
86.0000	-0.1974E 03
86.5000	0.6257E 02
87.0000	0.3081E 03
87.5000	0.4788E 03
88.0000	0.5326E 03
88.5000	0.4559E 03
89.0000	0.2673E 03
89.5000	0.1259E 02
90.0000	-0.2457E 03
90.5000	-0.4444E 03
91.0000	-0.5345E 03
91.5000	-0.4938E 03
92.0000	-0.3320E 03
92.5000	-0.8864E 02
93.0000	0.1768E 03
93.5000	0.3993E 03
94.0000	0.5242E 03
94.5000	0.5209E 03
95.0000	0.3900E 03
95.5000	0.1635E 03
96.0000	-0.1033E 03
96.5000	-0.3448E 03
97.0000	-0.5021E 03
97.5000	-0.5365E 03
98.0000	-0.4395E 03
98.5000	-0.2350E 03
99.0000	0.2704E 02
99.5000	0.2824E 03
100.0000	0.4687E 03

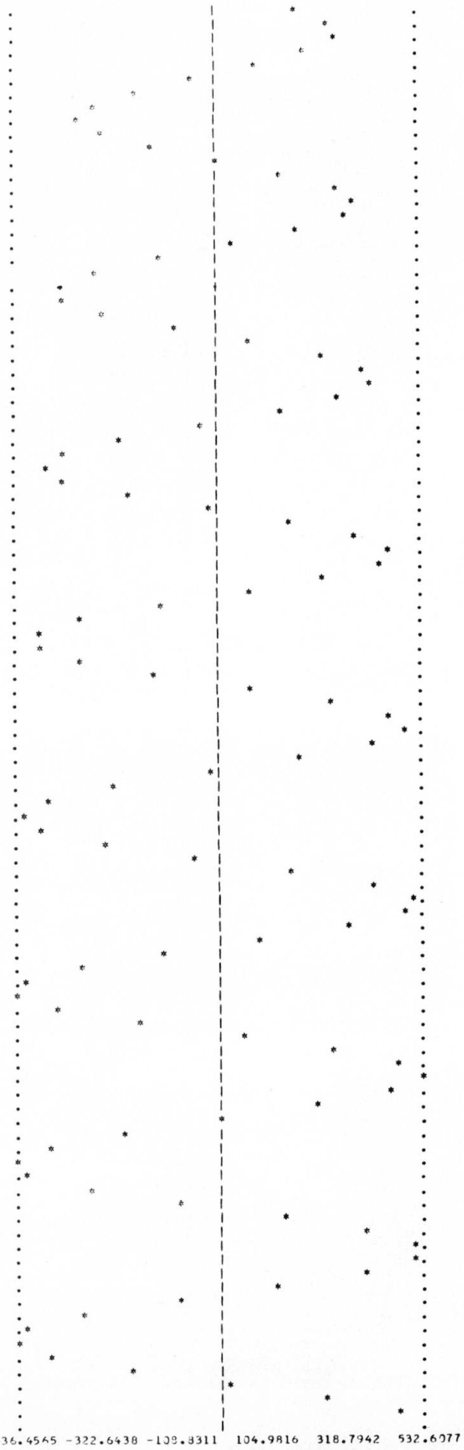

FIG. 7.19-F.

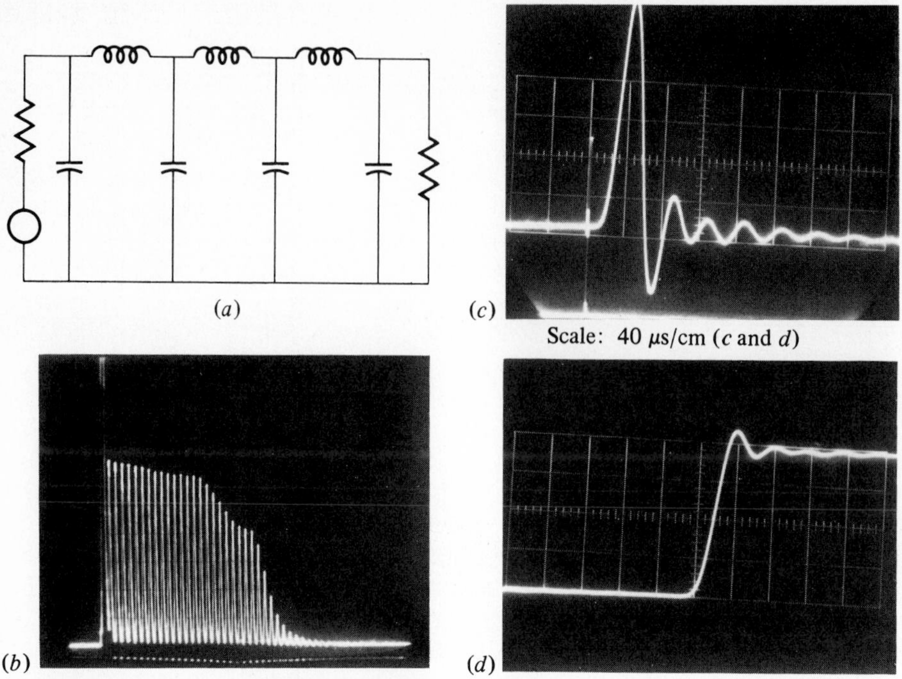

(*a*) (*c*) Scale: 40 μs/cm (*c* and *d*)

(*b*) (*d*)

Zero-frequency (dc) marker at left; bars are 1 kHz apart.

FIG. 7.19-G

given in Fig. 7.19-H. For the band-pass network whose $|T|$ versus ω characteristic is shown there in *a*, $\beta'(\omega_0) = 100\ \mu$s (Prob. 7-B). The impulse response [part (*b*)] is now much less of a pulse, hardly a pulse at all, since the low frequencies are suppressed by the network. But if we look on $r_\delta(t)$ as a sort of converted, "ac" pulse of one or two oscillations, we might say it occurs at 2 cm, that is, 80 μs after the excitation. It is stretching things a little, perhaps, to say

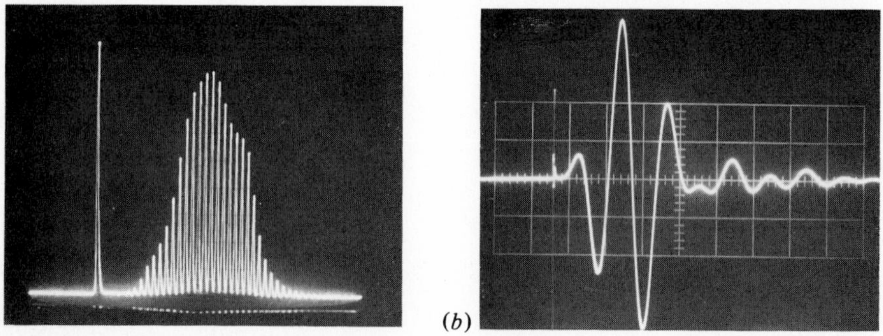

(*a*) (*b*)

Dc (zero-frequency) marker at left; bars are 1 kHz apart.

Scale is 40 μs/cm (division)

FIG. 7.19-H

there is corroboration of a sort here also; one should excite the network with an "ac pulse."

As a final illustration here of the use of the network-analysis programs, for one more experiment intended also to add to our knowledge of networks, consider a network for which $|T|$ is constant for *all* ω. Such networks do indeed exist [see (5.13-12) and (5.13-13) and associated discussion] and are properly denominated *all-pass*. The response of such a network should be very interesting, because we cannot blame the inconstancy of $|T|$ for distortion; the common failure of a network to transmit some "frequency components" of the signal, as evidenced by previous experiments, does not here occur. There remains only the other part of $T(j\omega)$, its angle, to cause distortion. If this angle were truly linear as a function of frequency, for all ω, we should expect perfect transmission, except for a delay: no distortion of signal shape, merely retardation. If the angle is approximately linear, what distortion occurs? To find out, we construct a $T(s)$ of all-pass character by placing the natural frequencies and the zeros of $T(s)$ as in Fig. 7.19-I*a*. Their relative left-right symmetry makes $|T(j\omega)|$ constant as ω varies; their equal vertical spacing should make $\beta(\omega)$ more or less linear as a function of ω, at least for low frequencies: as the angle contribution of one pole peters out, the next one should take over and keep the angle going. To verify this design idea, we call on FOMEGA, with the results shown in Fig. 7.19-I*b*. The phase $\beta(\omega)$ is indeed nearly linear for low frequencies; its slope is $7.54° (\pi/180)/0.05 = 2.6$ s (normalized). At high frequencies (Fig. 7.19-I*a*) β is far from linear: it is nearly constant at 1440°. Since the transmission *is* appreciable at high frequencies in an all-pass structure we therefore expect an impulse input to be distorted, a step to be somewhat less distorted since its mate, $1/s$, is small at high frequencies, and a

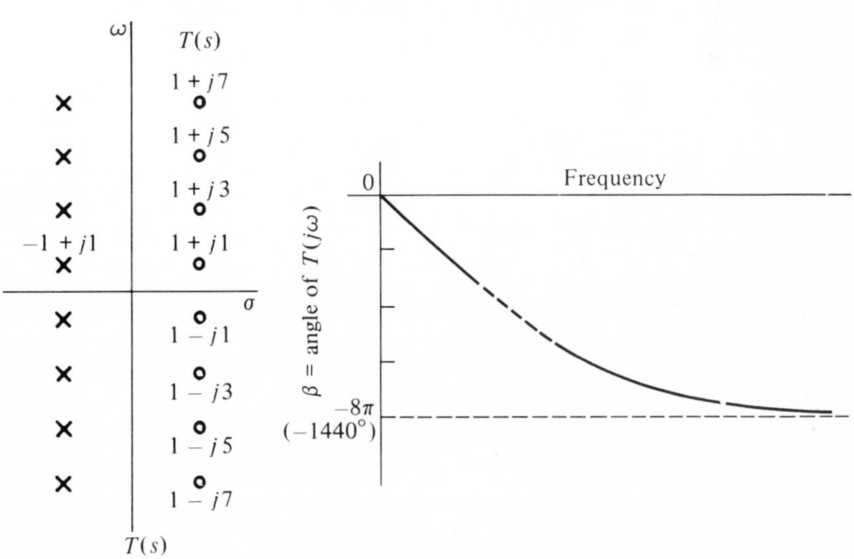

FIG. 7.19–Ia

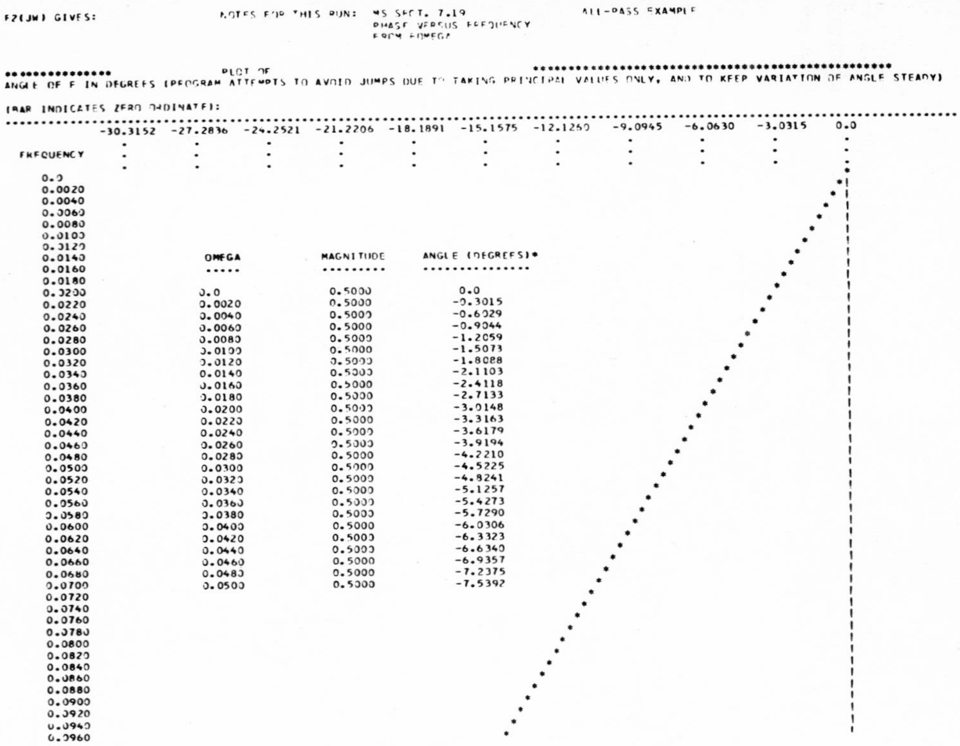

OMEGA	MAGNITUDE	ANGLE (DEGREES)*
0.0	0.5000	0.0
0.0020	0.5000	-0.3015
0.0040	0.5000	-0.6029
0.0060	0.5000	-0.9044
0.0080	0.5000	-1.2059
0.0100	0.5000	-1.5073
0.0120	0.5000	-1.8088
0.0140	0.5000	-2.1103
0.0160	0.5000	-2.4118
0.0180	0.5000	-2.7133
0.0200	0.5000	-3.0148
0.0220	0.5000	-3.3163
0.0240	0.5000	-3.6179
0.0260	0.5000	-3.9194
0.0280	0.5000	-4.2210
0.0300	0.5000	-4.5225
0.0320	0.5000	-4.8241
0.0340	0.5000	-5.1257
0.0360	0.5000	-5.4273
0.0380	0.5000	-5.7290
0.0400	0.5000	-6.0306
0.0420	0.5000	-6.3323
0.0440	0.5000	-6.6340
0.0460	0.5000	-6.9357
0.0480	0.5000	-7.2375
0.0500	0.5000	-7.5392

FIG. 7.19-I*b*

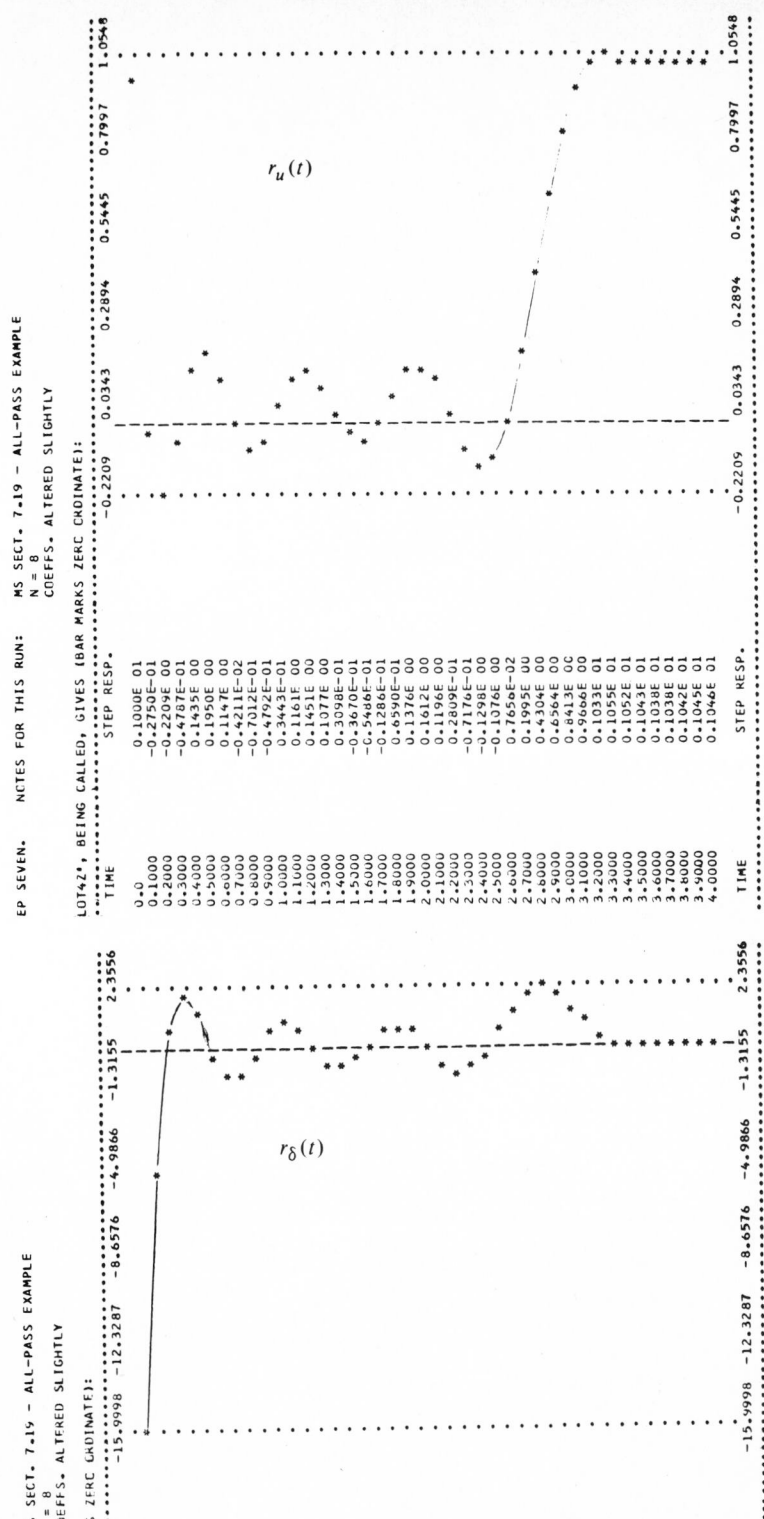

S SECT. 7.19 - ALL-PASS EXAMPLE
= 8
UEFFS. ALTERED SLIGHTLY

S ZERC CRDINATE):

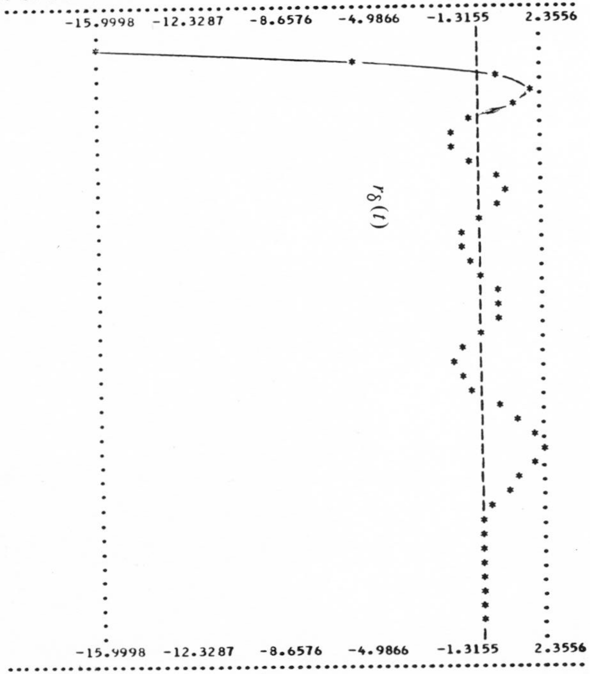

EP SEVEN. NCTES FOR THIS RUN: MS SECT. 7.19 - ALL-PASS EXAMPLE
N = 8
COEFFS. ALTERED SLIGHTLY

LOT4Z', BEING CALLED, GIVES (BAR MARKS ZERC CRDINATE):

TIME	STEP RESP.
0.0	0.1000E 01
0.1000	-0.2750E-01
0.2000	-0.2209E 00
0.3000	-0.4787E-01
0.4000	0.1435E 00
0.5000	0.1950E 00
0.6000	0.1147E 00
0.7000	-0.4211E-02
0.8000	-0.7012E-01
0.9000	-0.4792E-01
1.0000	0.3443E-01
1.1000	0.1161E 00
1.2000	0.1451E 00
1.3000	0.1077E 00
1.4000	0.3098E-01
1.5000	-0.3670E-01
1.6000	-0.5486E-01
1.7000	-0.1286E-01
1.8000	0.6590E-01
1.9000	0.1376E 00
2.0000	0.1612E 00
2.1000	0.1196E 00
2.2000	0.2809E-01
2.3000	-0.7176E-01
2.4000	-0.1298E 00
2.5000	-0.1076E 00
2.6000	0.7656E-02
2.7000	0.1995E 00
2.8000	0.4304E 00
2.9000	0.6564E 00
3.0000	0.8413E 0C
3.1000	0.9666E 00
3.2000	0.1033E 01
3.3000	0.1055E 01
3.4000	0.1052E 01
3.5000	0.1043E 01
3.6000	0.1038E 01
3.7000	0.1038E 01
3.8000	0.1042E 01
3.9000	0.1045E 01
4.0000	0.1046E 01
TIME	STEP RESP.

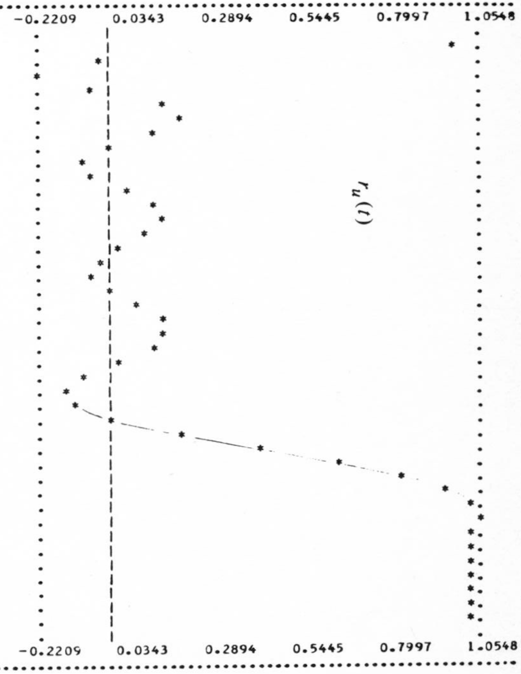

FIG. 7.19-Ic

PLOT OF RESPONSE RF (CONVOLUTION OF E AND RC):

SUBROUTINE 'PLOT4 ', BEING CALLED, GIVES (BAR MARKS ZERO ORDINATE):

TIME	RESP(CONVCL)
0.0	0.0
0.1000	0.2285E 00
0.2000	-0.1679E-01
0.3000	-0.1211E 00
0.4000	-0.3323E-01
0.5000	0.9750E-01
0.6000	0.1448E 00
0.7000	0.8688E-01
0.8000	-0.1474E-01
0.9000	-0.8121E-01
1.0000	-0.7218E-01
1.1000	-0.3824E-02
1.2000	0.7131E-01
1.3000	0.1022E 00
1.4000	0.7022E-01
1.5000	-0.2981E-02
1.6000	-0.7165E-01
1.7000	-0.9404E-01
1.8000	-0.5612E-01
1.9000	0.2134E-01
2.0000	0.9455E-01
2.1000	0.1202E 00
2.2000	0.7803E-01
2.3000	-0.1765E-01
2.4000	-0.1232E 00
2.5000	-0.1837E 00
2.6000	-0.1551E 00
2.7000	-0.2165E-01
2.8000	0.1988E 00
2.9000	0.4621E 00
3.0000	0.7137E 00
3.1000	0.9065E 00
3.2000	0.1013E 01
3.3000 ←	0.1031E 01
3.4000	0.9759E 00
3.5000	0.8752E 00
3.6000	0.7560E 00
3.7000	0.6383E 00
3.8000	0.5326E 00
3.9000	0.4416E 00
4.0000	0.3632E 00
4.1000	0.2951E 00
4.2000	0.2359E 00
4.3000	0.1859E 00
4.4000	0.1459E 00
4.5000	0.1156E 00
4.6000	0.9315E-01
4.7000	0.7589E-01
4.8000	0.6094E-01
4.9000	0.4662E-01
5.0000	0.3287E-01
5.1000	0.2094E-01
5.2000	0.1252E-01
5.3000	0.8616E-02

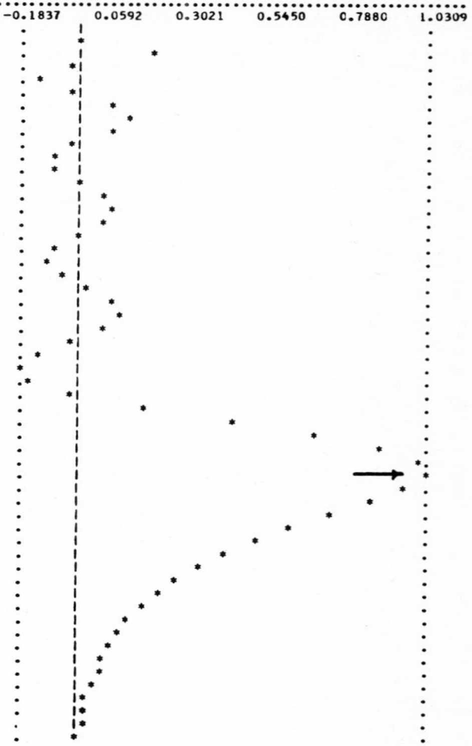

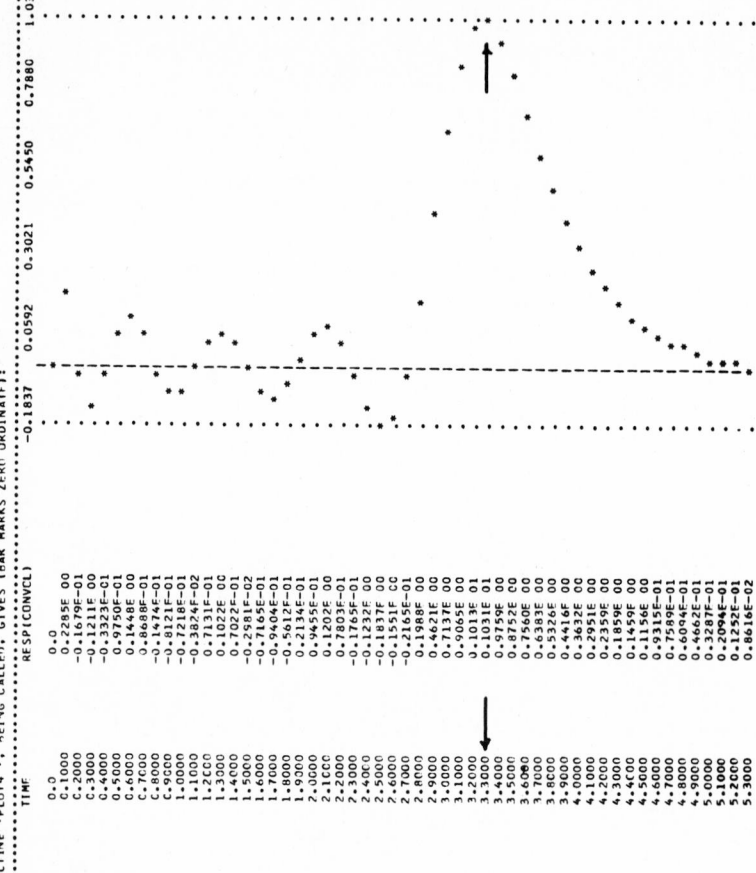

FIG. 7.19-I*d*. Response of all-pass network of Fig. 7.19-I*a* to "slug pulse" ($9te^{-3t}$) of unit area.

"low-frequency" signal such as $te^{-\alpha t}$, whose mate is $(s + \alpha)^{-2}$, to be much less distorted if $\alpha \ll 7$, say (Fig. 7.19-I*a*). We call on IMPSTEP for the first two responses (Fig. 7.19-I*c*) and on CONVOL for the third, taking $\alpha = 3$ and adjusting the multiplier for unit pulse area (Fig. 7.19-I*d*).

In the impulse response (which contains also an impulsive component not shown) we observe severe distortion as expected: it is practically impossible to define a "delay" because the output is distorted beyond recognition. The step response is more respectable; a delay of about 2.8 s (normalized) is definable without much argument. This figure (2.8) should be contrasted with the 2.6 s (normalized) low-frequency phase slope calculated above. The response to the low-frequency "slug pulse" is even less distorted: a delay of about 3.0 s (normalized) is readily definable (the peak of the excitation wave occurs at about $t = 0.3$, that of the response at about $t = 3.3$).

We conclude again that (7.19-24) is at least empirically valid here too: so long as the magnitude of $T(j\omega)$ is approximately constant at "important frequencies" and the angle thereof is nearly linear thereat (β_0 being zero or a multiple of π rad), the response is a recognizable, if somewhat distorted, version of the input wave, delayed in time.

We have found our programs (which can of course be improved) to be genuinely useful not only as tools of analysis qua analysis, but as instruments of education. Our examples in this section have taught us a great deal about distortion and delay, not to mention time and element-value normalization, and the linear nature of the low-frequency and high-frequency asymptotic behavior of the magnitude of transfer functions on logarithmic scales.

7.20 ANOTHER APPROACH

Computers can be used for network analysis in other, quite different ways. Programs can be written to provide direct "conversational" interaction with the computer by means, for example, of typewriters. Data about the network schematic is typed in; the appropriate differential equations (numerous because usually set up loop by loop or node by node, but each of low order) are then formulated within the computer. The analysis proceeds by high-speed step-by-step (finite-difference) integration of these equations. Operations are entirely in the t domain and sufficiently rapid to give the impression of instantaneous analysis, reported back immediately via the typewriter or a cathode-ray tube, or other "mouthpiece."

Such programs exist in various forms and are continually developing.† Each one requires considerable discussion to understand and to use and to make the actual installation on the computer (the programs are large); hence we make no attempt to discuss such interactive programs here. They do afford a different, often useful approach to network analysis. [If *nonlinear* elements are

† See, for example, *IEEE Spectrum*, pp. 14 ff., June, 1971.

present in a network, such programs are extremely useful; their utility is not affected by the nonlinearities (but our s-based analysis then becomes almost powerless).]

For synthesis such analysis programs can be very helpful because they are so rapid; synthesis is in effect carried out by making a succession of intelligent trials (analyses). The direct coupling of measuring instruments to computers is also practical and can be extremely useful in making final adjustments of actual networks.

7.21 THE STATE OF THE ART

We pause to consider where we are: what we have accomplished and where next to go. Network analysis, even to the limited extent that we have studied it, is rich in alternatives. To diagram the methods we now have available is to create a bewildering maze, Fig. 7.21-A.

The *differential-equation* route from excitation to response is the basic one, for our basic (lumped, linear, constant) postulates state simply that networks are physical systems modelable by differential equations of very special sort:

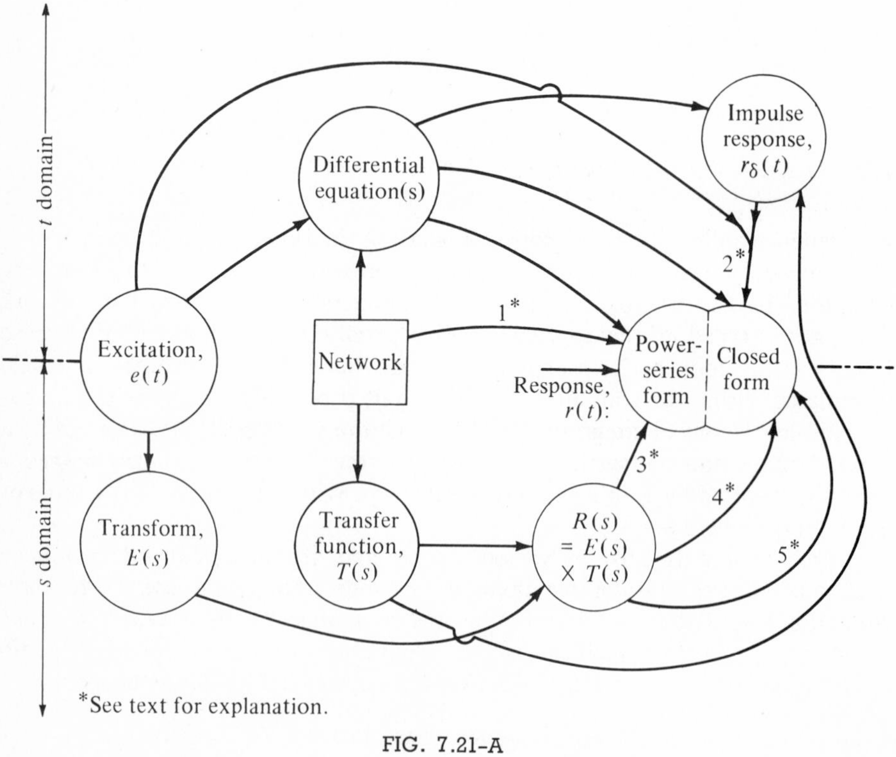

FIG. 7.21-A

ordinary, linear, constant-coefficient. Because these equations are so special they can be solved, and that with little effort. Our analysis processes are equivalent to so doing; they are merely variations, schemes that reduce the labor of differential-equation solving to a minimum, and in so doing often conceal the true nature thereof. We see in Fig. 7.21-A not one but a number of such devices, ways of calculating the response of a network to a given excitation. (And of course one could be asked, in analysis, to calculate "backward," to find the excitation from a given response: then $R = ET$ becomes $E = R/T$, for example.)

Most of the paths in Fig. 7.21-A are familiar; but some need distinguishing identification:

1 Represents the calculation of initial values by physical reasoning from the network schematic,

2 Is convolution: $r = e * r_\delta$,

3 Represents expansion of $R(s)$ about infinity, i.e., developing $R(s)$ in a series of powers of $(1/s)$,

4 Is the partial-fraction expansion of $R(s)$, the "standard" path, the high road that leads to a sum of exponentials,

5 Represents another path from $R(s)$ to $r(t)$, not yet discussed, and useful in difficult or exotic cases; it solves the integral equation (7.07-3) by a method to be discussed in Chap. 8.

Most of the paths are in themselves multiple; they have variations. We may use formal calculation, or numerical calculation (often with the computer's help), or make actual measurements of the network's behavior. Reference back to similar diagrams from which this one has developed, such as Figs. 6.13-A and 7.11-A, would be good here. See also the six-step breakdown of what analysis is (Sec. 7.11).

Still our knowledge is not complete. Let us consider it now for each of these four elements of analysis in turn: excitation, network, initial conditions, response.

Excitation: We can handle any given excitation wave, formal or numerical (table). We may of course have to resort to purely numerical methods, but we are not limited in the sorts of excitation waves we can treat.

Network: If the network is given in schematic form and is a ladder, we can easily calculate its transfer function(s). Many nonladders (bridges, simple feedback networks) we can also handle easily. The completely general case, networks of arbitrary complexity (configuration), we have also briefly discussed (Sec. 6.27). If the network is described by its impulse response, we can of course find its response to any excitation, whether $r_\delta(t)$ be formula or table of values, by convolution.

Initial conditions: We have generally assumed an initial network state of complete rest: no energy is stored in the network "at $t = 0$." If one or more elements do initially contain energy, these in effect merely apply additional excitations, the responses to which are to be superposed on the response to the principal excitation. We shall clarify this in Sec. 7.22.

Response: The various methods of response calculation found in Fig. 7.21-A have all been explained in detail, almost completely. But we have not completely examined the partial-fraction expansion of $R(s)$, path 4, for *multiple* poles can occur; this is a detail we shall discuss in Sec. 7.23. And path 5, yet to be discussed (Chap. 8), may require recourse to elaborate transform tables, or detailed study of contour integration in the theory of functions of a complex variable.

There remain also questions that can arise when N is large, when the network is extensive and complicated. There may then be difficulties with numerical accuracy that require double-precision calculations, or a change of variable, or an entirely different approach to the transformation back from the s domain to the t domain. Discussion of all of these we postpone. We turn now to the lacunae in the matters of initial conditions and multiple poles.

7.22 NONREST INITIAL CONDITIONS

In Sec. 7.21 we remarked a lacuna in our knowledge: what to do if the network under analysis is not initially quiescent. For simplicity's sake we have generally postulated an initial state of complete *rest:* that no energy be stored in any energy-storing element at $t = 0$. When this is not true, additional work is required. But it is merely *supplemental* work, merely the addition to the response of certain other components that are due to those initial energies, and are separately calculable. The energies act merely as additional excitations, whose effects are superposable because of linearity.

We recall [(2.02-9) and Secs. 3.02 and 3.08—see also Sec. 6.02] that an initial voltage V_0 on the capacitor of an RC or an RLC circuit can be replaced by an

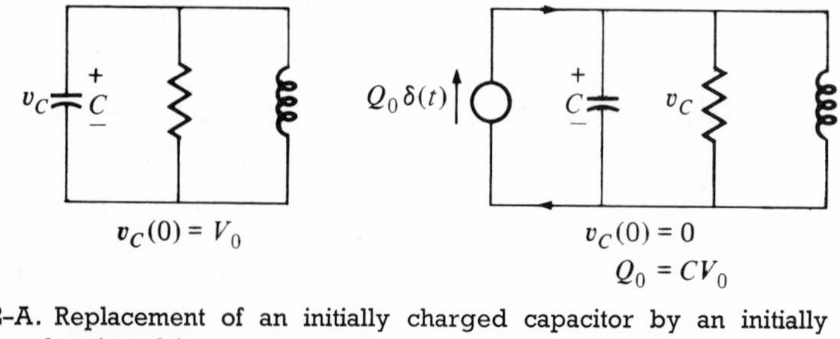

FIG. 7.22-A. Replacement of an initially charged capacitor by an initially uncharged capacitor and an impulsive current generator.

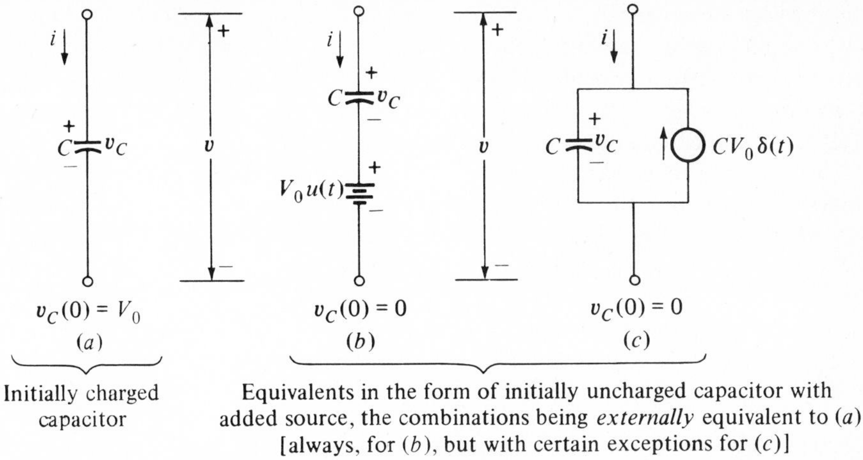

FIG. 7.22-B. The exchange of initially charged capacitors for initially uncharged capacitors with additional sources ($t > 0$).

impulsive current generator, $Q_0\,\delta(t)$, in parallel with the same capacitor initially *un*charged, the charge Q_0 being equal to CV_0, as in Fig. 7.22-A. Such an exchange can evidently be made for any initially charged capacitor C in any network (Fig. 7.22-B*a*, *c*) if attached to the terminal nodes of the capacitor there is no additional (parallel, purely capacitive) path for the impulsive charging current, no other path by means of which C can be robbed of part of its initial charge Q_0. In other words, no purely capacitive closed loop through C can exist, if the replacement is to be valid.

The proviso that the replacement impulsive current pass entirely through C is essential, else some of Q_0 will go elsewhere, as in Fig. 7.22-C. If any of the three capacitors (say C_2) is initially charged, and we replace this charge by an impulsive current generator connected in parallel with that capacitor, some of the impulsive current will flow through the other two capacitors and not all of its charge will be deposited on the appropriate capacitor (C_2). But in such a situation, one capacitor cannot alone be initially charged, for the sum of the three capacitor voltages must always be zero: at least two capacitor voltages must initially be nonzero (unless all three are zero). At least *two* equivalent impulse current generators must therefore be used; simultaneous consideration

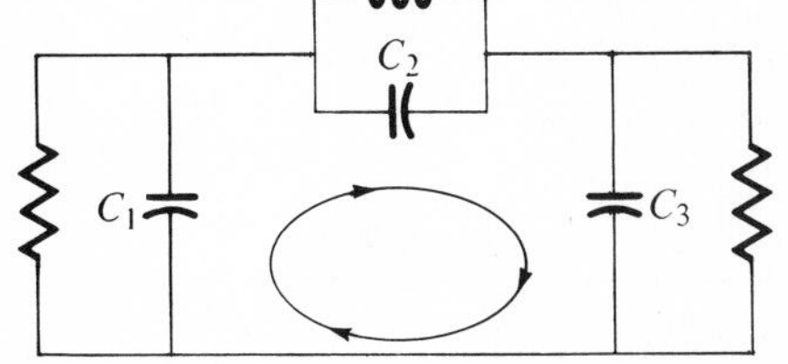

FIG. 7.22-C. Network with a purely capacitive loop.

will determine appropriate values for them. The matter is essentially the same as the reduction in network order caused by the purely capacitive path (see Sec. 6.17, in particular Fig. 6.17-A). We leave such complications for later discussion; they are here merely red herrings across our path. The "capacitively isolated" capacitor is the commoner case.

We observe now that another type of replacement for the initially charged capacitor is possible: the series combination of the initially *un*charged capacitor and a step voltage generator (a battery switched in at $t = 0$) of Fig. 7.22-B*b*. The equivalence is easily demonstrated by writing

$$v(t) = \frac{1}{C}\int_0^t i(x)\,dx + V_0, \qquad t > 0 \tag{7.22-1}$$

which clearly holds for all three of the situations in Fig. 7.22-B, with the previously mentioned qualification for (*c*). There is no possibility in (*b*) of the capacitor's being robbed by other capacitors of its rightful initial voltage. Hence the (*b*) replacement is *always* possible.

An initial charge on a capacitor amounts therefore to an additional excitation applied to the network initially at rest. All we need do to take account of it is to calculate the contribution of this equivalent source (as by calculating and using the appropriate transfer function) and add that in to the response due to the principal excitation. Note again that the equivalences of Fig. 7.22-B are *external* only; the voltage across the capacitor in (*b*) is *not* the same as the voltage across the capacitor in (*a*), but the external (between the two terminals) voltages *are* identical.

An initially charged inductor represents the dual situation; Fig. 7.22-D applies. We construct it from Fig. 7.22-B by blindly applying the rules of duality (Sec. 5.12). Then, as a precaution, we carefully verify the result by observing that all three branches obey

$$i(t) = \frac{1}{L}\int_0^t v(x)\,dx + I_0, \qquad t > 0. \tag{7.22-2}$$

[See also (3.08-8).] Again there is a proviso, for (*c*): such an exchange (Fig. 7.22-D*a*, *c*) can evidently be made for any initially charged inductor L if on the loop(s) that close the circuit for L and the ersatz voltage source there is no additional (in series, inductive) element to consume part of the impulsive charging voltage, no other element by means of which L can be robbed of part of its initial flux linkages $\Lambda_0 = LI_0$. In other words, no purely inductive node attached to L can exist.

The proviso that the replacement impulsive voltage appear entirely across L is essential, else some of Λ_0 will go elsewhere, as in Fig. 7.22-E. The situation is completely dual and needs no further comment.

We can, in sum, always replace initially charged inductor and capacitor

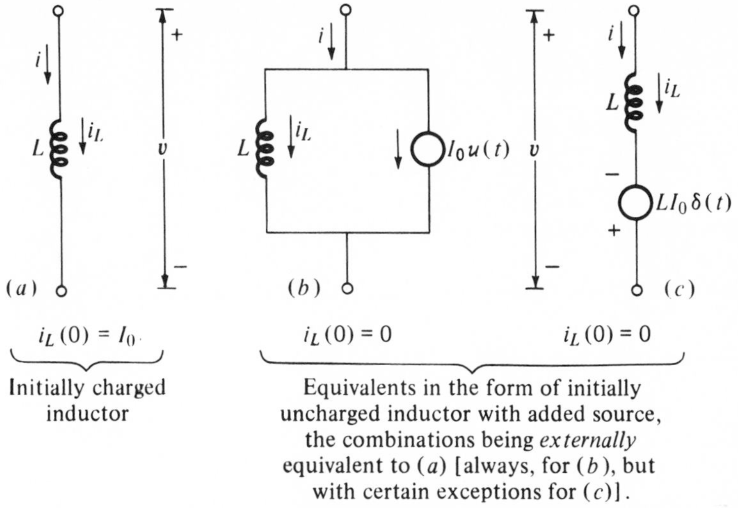

FIG. 7.22-D

elements by sources equivalent to the initial charges together with the same uncharged elements. We can always use the (*b*) equivalents as they are, and often the (*c*) ones. And the latter can always be used if we investigate and take account of the additional sources necessary for the other ("robbing") elements. The equivalences are *external* to the terminal pairs shown in Figs. 7.22-B and 7.22-D, and external *only*. An appropriate transfer function must of course be found for each equivalent excitation.

To illustrate the handling of nonrest initial conditions, consider the network of Fig. 7.22-F. In addition to the nominal excitation i_F, there may be excitation due to the four possible initial energy stores, expressed by V_2, V_3, I_4, and V_5 (initial voltages and current). Let us calculate the corresponding transfer functions.

We choose first to obtain the (common) denominator thereof, since all transfer functions of a given network have the same denominator, except possibly for a constant multiplier. [The natural frequencies belong to the network as a whole, they do not depend on any particular path from any particular excitation point to any particular response point; they are the zeros of every transfer function denominator (except in cases where common factors may have been

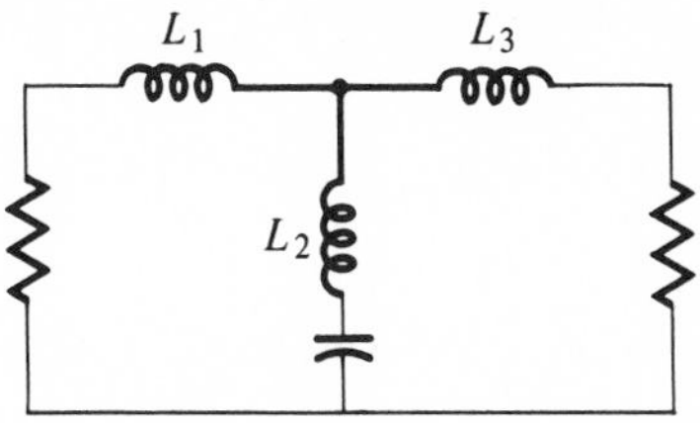

FIG. 7.22-E. Network with a purely inductive node.

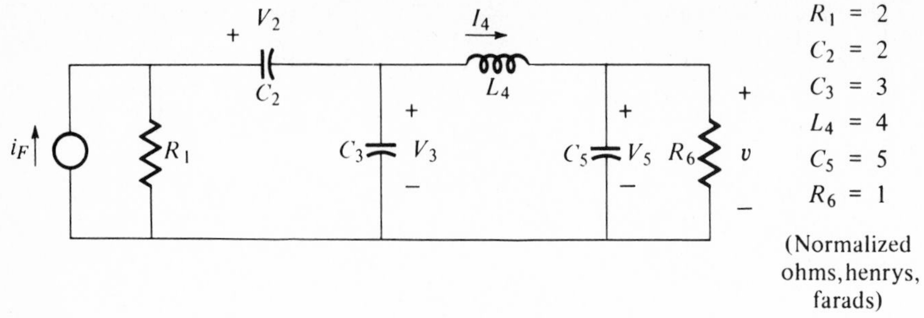

FIG. 7.22-F

canceled from numerator and denominator, as in Secs. 6.21 and 6.22).] To find them we choose to sum the impedances around a loop (Sec. 6.03), as

$$\frac{1}{C_5 s + G_6} + L_4 s + \frac{1}{C_3 s + 1/(1/(C_2 s) + R_1)} = 0 \tag{7.22-3}$$

from which comes

$$D(s) = R_1 C_2 C_3 C_5 L_4 s^4 + (\quad)s^3 + (\quad)s^2 + (\quad)s + G_6 = 0. \tag{7.22-4}$$

We take this polynomial $D(s)$ as the common denominator of all the relevant transfer functions. By inspection of the (redrawn, for visual convenience, in Fig. 7.22-G) networks, using for each energy-storing element's initial charge the (*c*) equivalent excitations (ersatz sources) of Figs. 7.22-B and -C, as is here possible in every case, we obtain the zeros of the transfer functions and write:

$$T_{i_F \text{ to } v} = K_1 \frac{s}{D}, \tag{7.22-5}$$

$$T_{i_{C2} \text{ to } v} = K_2 \frac{1}{D}, \tag{7.22-6}$$

$$T_{i_{C3} \text{ to } v} = K_3 \frac{s+\alpha}{D}, \qquad \alpha = \frac{1}{R_1 C_2} \tag{7.22-7}$$

$$T_{e_{L4} \text{ to } v} = K_4 \frac{s(s+\beta)}{D}, \qquad \beta = \frac{C_2 + C_3}{R_1 C_2 C_3} \tag{7.22-8}$$

$$T_{i_{C5} \text{ to } v} = K_5 \frac{as^3 + bs^2 + cs + d}{D}, \tag{7.22-9}$$

$$a = R_1 C_2 C_3 L_4, \qquad b = L_4(C_2 + C_3), \qquad c = R_1 C_2, \qquad d = 1.$$

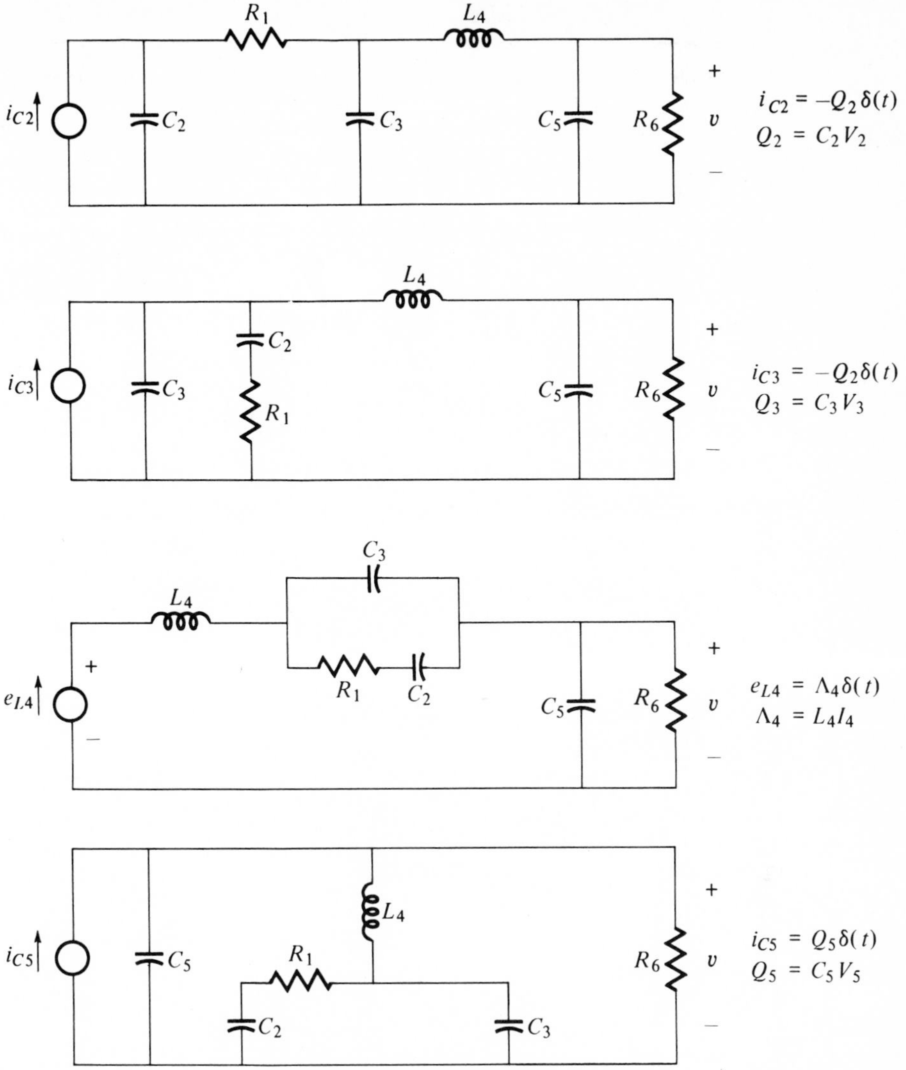

FIG. 7.22-G. All elements are initially *un*charged.

In the last transfer function, the numerator polynomial comes from the (shunt) impedance

$$L_4 s + \frac{1}{C_3 s + 1/[R_1 + 1/(C_2 s)]}, \tag{7.22-10}$$

set equal to zero. Examination of asymptotic low-frequency and high-frequency behavior of networks and of transfer functions, and comparison of results, give us the values of the K's (Table 7.22-A), and each of them in two ways so that

TABLE 7.22-A

Transfer function	High-Frequency ($s \to \infty$) Asymptotic Behavior of $T(s)$	High-Frequency ($s \to \infty$) Asymptotic Behavior of Network	Low-Frequency ($s \to 0$) Asymptotic Behavior of $T(s)$	Low-Frequency ($s \to 0$) Asymptotic Behavior of Network
i_F to v	$\frac{K_1}{Bs^3}$	$\frac{1}{C_3 sL_4 sC_5 s}$ $\therefore K_1 = \frac{B}{C_3 L_4 C_5} = R_1 C_2$	$\frac{K_1 s}{G_6}$	$R_1 C_2 sR_6$ $\therefore K_1 = G_6 R_1 C_2 R_6 = R_1 C_2$
i_{C2} to v	$\frac{K_2}{Bs^4}$	$\frac{1}{C_2 sR_1 C_3 sL_4 sC_5 s}$ $\therefore K_2 = 1$	$\frac{K_2}{G_6}$	R_6 $\therefore K_2 = 1$
i_{C3} to v	$\frac{K_3}{Bs^3}$	$\frac{(1)(1)}{C_3 sL_4 sC_5 s}$ $\therefore K_3 = R_1 C_2$	$\frac{K_3 \alpha}{G_6}$	R_6 $\therefore K_3 = \frac{1}{\alpha} = R_1 C_2$
e_{L4} to v	$\frac{K_4 s^2}{Bs^4}$	$\frac{1}{L_4 sC_5 s}$ $\therefore K_4 = R_1 C_2 C_3$	$\frac{K_4 s\beta}{G_6}$	$(C_2 + C_3)sR_6$ $\therefore K_4 = \frac{C_2 + C_3}{C_2 + C_3} R_1 C_2 C_3 = R_1 C_2 C_3$
i_{C5} to v	$\frac{K_5 as^3}{Bs^4}$	$\frac{1}{C_5 s}$ $\therefore K_5 = \frac{R_1 C_2 C_3 L_4 C_5}{C_5 R_1 C_2 C_3 L_4} = 1$	$\frac{K_5 d}{G_6}$	R_6 $\therefore K_5 = \frac{1}{d} = 1$

NOTES: $B = R_1 C_2 C_3 L_4 C_5$ from leading coefficient of $D(s)$, (7.22-4).
$a = R_1 C_2 C_3 L_4$ from (7.22-9).
$d = 1$ from (7.22-9).

we have a check on each (because the denominator D has already been computed by independent means). The dimensions should in each case be carefully checked also.

The complete response of the network is, in the s domain,

$$V(s) = T_{i_F \text{ to } v} I_F(s) + T_{i_{C2} \text{ to } v} Q_2 + T_{i_{C3} \text{ to } v} Q_3 + T_{e_{L4} \text{ to } v} \Lambda_4 + T_{i_{C5} \text{ to } v} Q_5 \tag{7.22-11}$$

$$= \frac{K_1 s I_F + K_2 Q_2 + K_3(s+\alpha)Q_3 + K_4 s(s+\beta)\Lambda_4 + K_5(as^3 + bs^2 + cs + d)Q_5}{D}.$$

This form of $V(s)$ shows clearly how each initial energy store acts as a source, exactly as does the nominal excitation $i_F(t)$, except for poverty of waveform. The transfer functions for each differ among themselves in their zeros and multipliers (because the sources are located at different points in the network) but not in their denominators. Calculation of the actual time-domain response is now straightforward. If for simplicity's sake we introduce the (normalized) numerical element values of Fig. 7.22-F, then (7.22-4) becomes

$$D(s) = 240s^4 + 148s^3 + 52s^2 + 14s + 1 \tag{7.22-12}$$

and the K's take these values:

$$K_1 = 4, \qquad K_2 = 1, \qquad K_3 = 4, \qquad K_4 = 12, \qquad K_5 = 1. \tag{7.22-13}$$

In the s domain the complete response is

$$V(s) = \frac{(Q_2 + Q_3 + Q_5) + (4I_F + 4Q_3 + 5\Lambda_4 + 4Q_5)s + (12\Lambda_4 + 20Q_5)s^2 + (48Q_5)s^3}{1 + 14s + 52s^2 + 148s^3 + 240s^4} \tag{7.22-14}$$

in which I_F is the s-domain mate of $i_F(t)$, and the constants Q_2, Q_3, Λ_4, and Q_5 represent the initial conditions in the four energy-storing elements (Fig. 7.22-G). Various combinations of initial conditions can give various responses in time, to be added, of course, to the response to $i_F(t)$, the "true excitation," as from rest. For example Fig. 7.22-H shows the nature of the four responses due to the four possible initial conditions acting individually; Fig. 7.22-I shows what comes from several combinations thereof. The number of possibilities is of course tremendous. In both figures the (normalized) time range is from zero to 50; the points are 1 s (normalized) apart.

As an example of the effect of nonrest initial conditions when the excitation is sinusoidal, Fig. 7.22-J shows the response (R) to such excitation (E) with $\omega_F = 0.3$

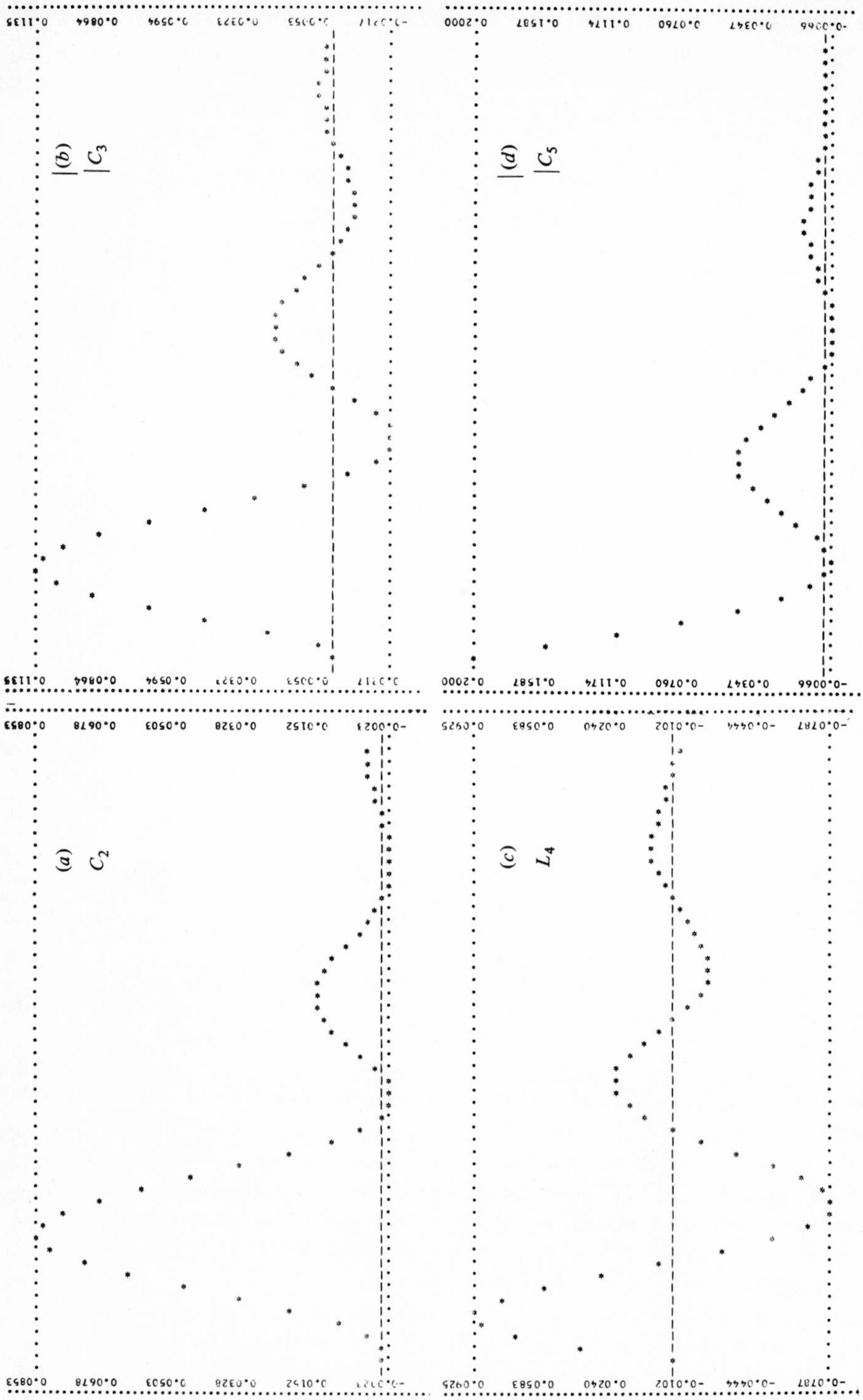

FIG. 7.22–H. Responses due to initial energy in one element alone, as indicated.

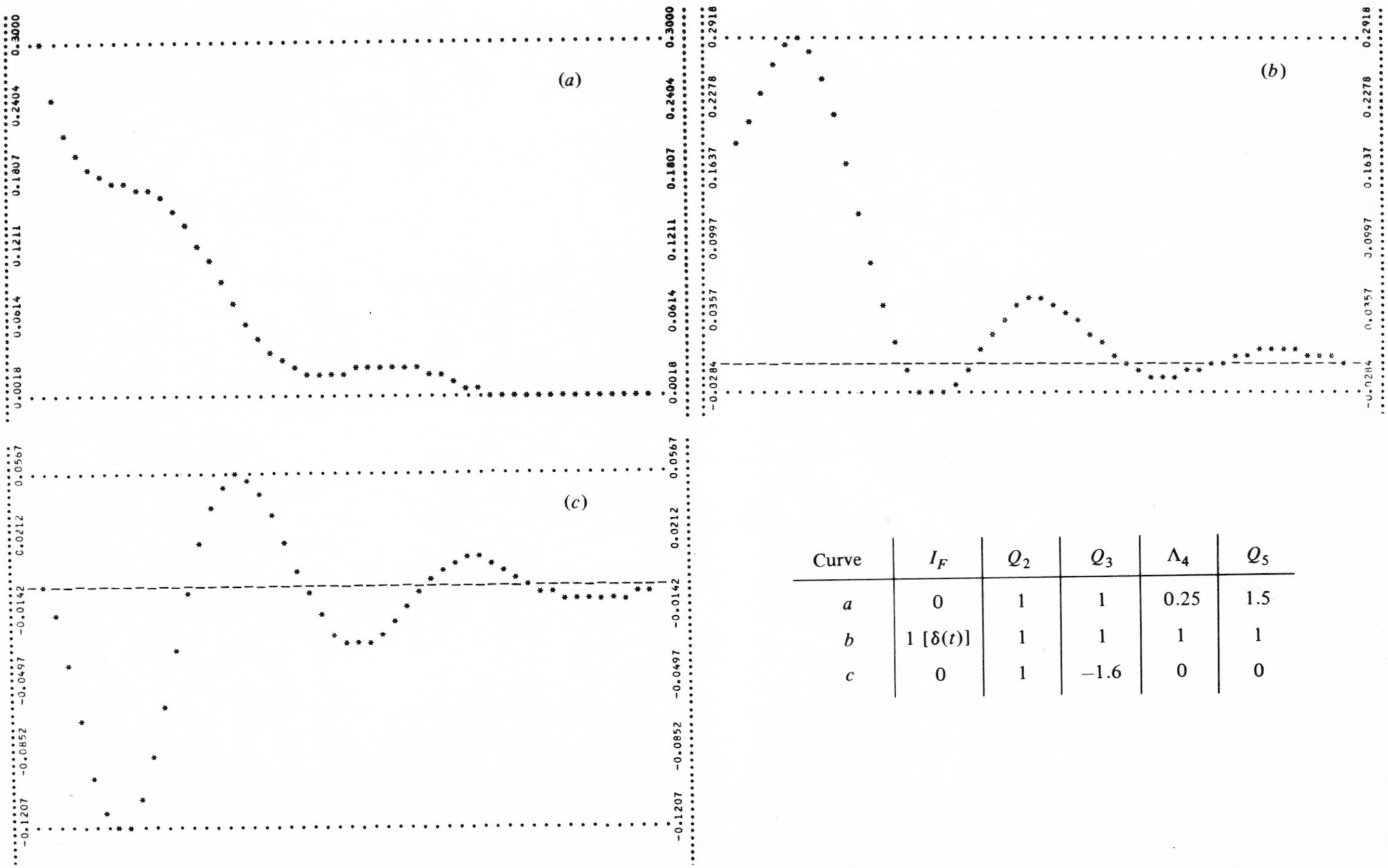

Curve	I_F	Q_2	Q_3	Λ_4	Q_5
a	0	1	1	0.25	1.5
b	1 $[\delta(t)]$	1	1	1	1
c	0	1	−1.6	0	0

FIG. 7.22–I. Responses due to excitation and to initial energy in various elements as indicated.

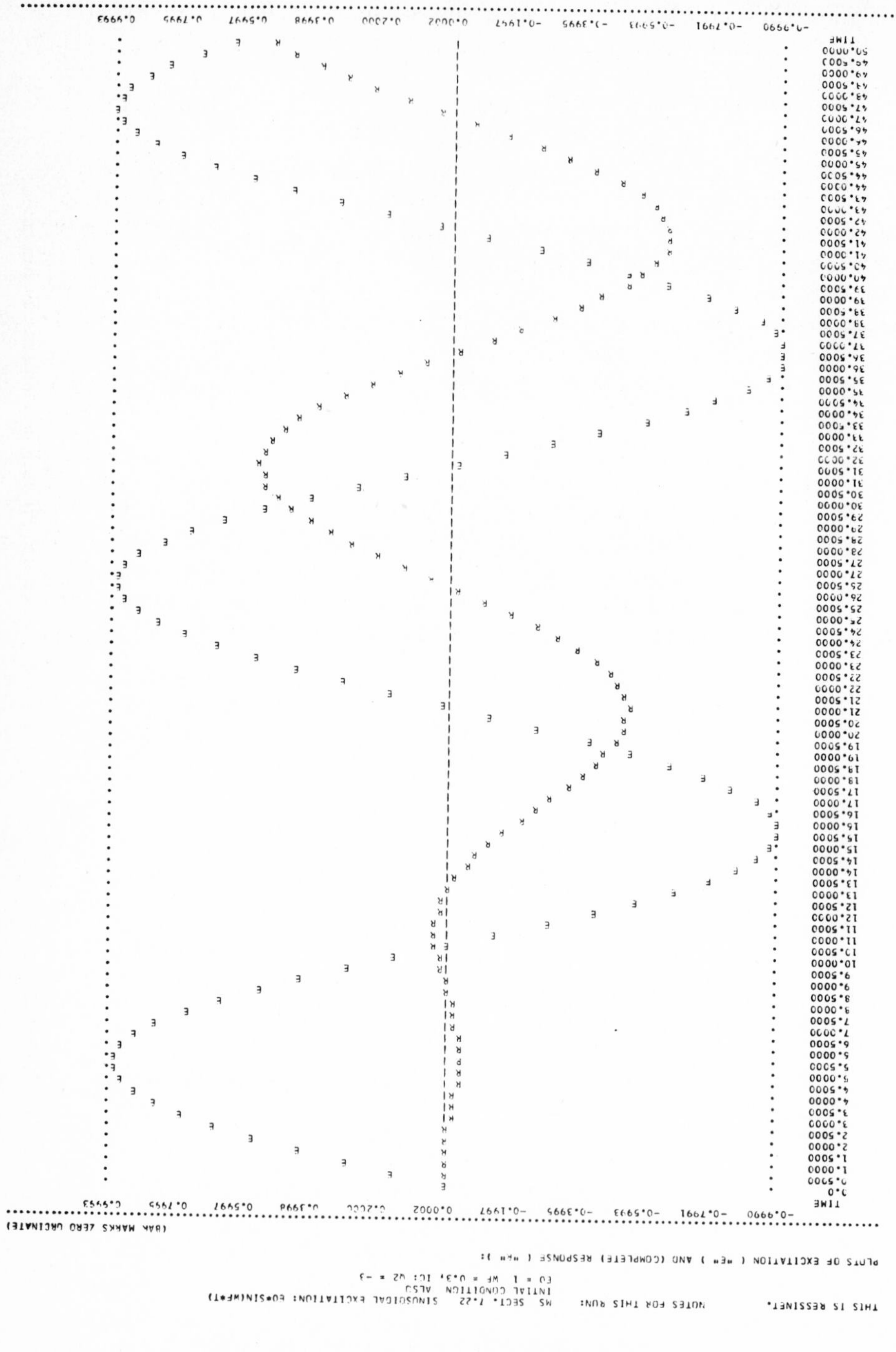
TIME
-0.9990 -0.7991 -0.5993 -0.3995 -0.1997 0.0002 0.2000 0.3998 0.5997 0.7995 0.9993
(BAR MARKS ZERO ORDINATE)
PLOTS OF EXCITATION ("E") AND (COMPLETE) RESPONSE ("R"):
THIS IS RESSINET. NOTES FOR THIS RUN: MS SECT. 7.22 SINUSOIDAL EXCITATION: E0*SIN(WF*T)
INITIAL CONDITION ALSO
E0 = 1 WF = 0.3, IC: U2 = -3

FIG. 7.22-J

and initial condition $V_{C2} = -3$. The response is now initially very sluggish but eventually, of course, reaches a sinusoidal steady state (with 83° lag and 0.7 relative amplitude).

An interesting application of the response due to initial-condition excitation is the pulse-forming network of Fig. 7.22-K. It is designed to "key" an oscillator (as in a radar or nuclear particle accelerator system) represented by the load resistor R_L. At first the switch *Sw* is open, and C_1 charges to a steady-state voltage, E_0. When ready for the pulse, the switch *Sw* is closed, current flows through R_L due to the charge previously established on C_1, that is, to that initial condition; Fig. 7.22-L shows that a fairly respectable, more-or-less rectangular pulse is then generated at R_L. The design can of course be modified and greatly improved; this is merely an example.

The foregoing discussion and examples have amply demonstrated the fact that energy initially stored in a network presents no new analysis problem: it merely adds to the labor of analysis by requiring the calculation of additional (additive) terms. These terms may well alter the shape of the response tremendously, at least in the transient state. Certain special initial conditions may even reduce the natural (transient) part of the response to zero.

Nonrest initial conditions, in summary, simply represent additional excitations whose contributions die out in any reasonable (dissipative) network; they do not affect the ultimate steady state if there is one. (See the discussion in Sec. 7.16.)

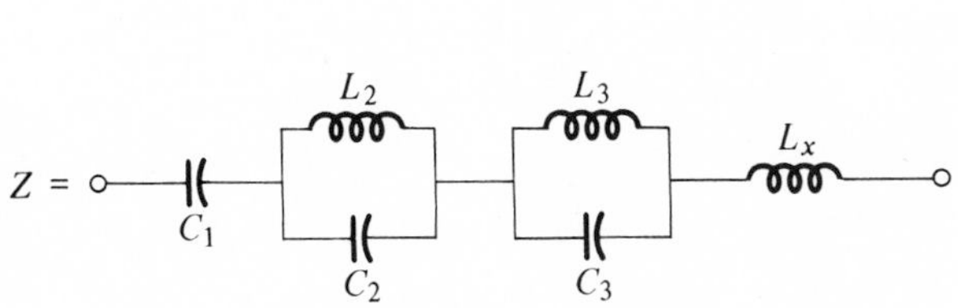

$C_1 = 3, \; C_2 = C_3 = 1.5$

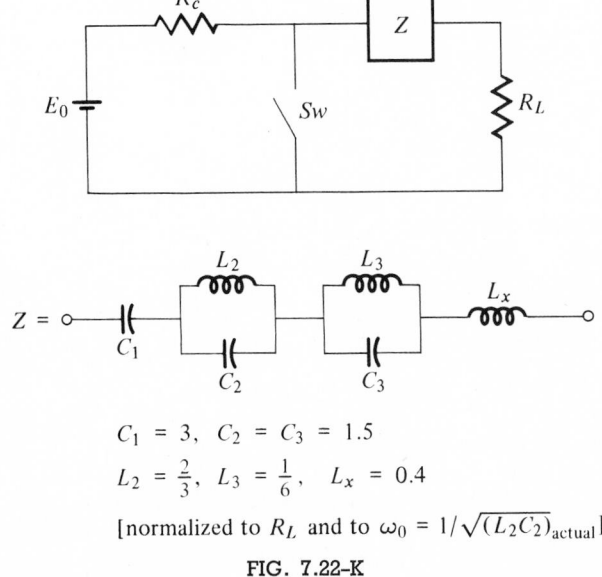

$L_2 = \frac{2}{3}, \; L_3 = \frac{1}{6}, \; L_x = 0.4$

[normalized to R_L and to $\omega_0 = 1/\sqrt{(L_2C_2)_{\text{actual}}}$]

FIG. 7.22-K

THIS IS IMPSTEP SEVEN. NOTES FOR THIS RUN: MS SECT. 7.22 PFN EXAMPLE
N = 6, RO = 1, K = 1/3, X = 0.4
T = PI FIRST TRY:

SUBROUTINE 'PLOT4Z', BEING CALLED, GIVES (BAR MARKS ZERO ORDINATE):

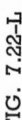

TIME	STEP RESP.
0.0	-0.3219E-05
0.1000	0.2197E 00
0.2000	0.3827E 00
0.3000	0.4958E 00
0.4000	0.5664E 00
0.5000	0.6020E 00
0.6000	0.6103E 00
0.7000	0.5985E 00
0.8000	0.5733E 00
0.9000	0.5407E 00
1.0000	0.5056E 00
1.1000	0.4723E 00
1.2000	0.4438E 00
1.3000	0.4221E 00
1.4000	0.4085E 00
1.5000	0.4032E 00
1.6000	0.4058E 00
1.7000	0.4153E 00
1.8000	0.4304E 00
1.9000	0.4494E 00
2.0000	0.4705E 00
2.1000	0.4918E 00
2.2000	0.5119E 00
2.3000	0.5292E 00
2.4000	0.5427E 00
2.5000	0.5515E 00
2.6000	0.5553E 00
2.7000	0.5541E 00
2.8000	0.5482E 00
2.9000	0.5381E 00
3.0000	0.5247E 00
3.1000	0.5090E 00
3.2000	0.4922E 00
3.3000	0.4755E 00
3.4000	0.4599E 00
3.5000	0.4464E 00
3.6000	0.4359E 00
3.7000	0.4291E 00
3.8000	0.4263E 00
3.9000	0.4277E 00
4.0000	0.4333E 00
4.1000	0.4427E 00
4.2000	0.4553E 00
4.3000	0.4705E 00
4.4000	0.4874E 00
4.5000	0.5049E 00
4.6000	0.5221E 00
4.7000	0.5379E 00
4.8000	0.5514E 00
4.9000	0.5617E 00
5.0000	0.5679E 00
5.1000	0.5696E 00
5.2000	0.5662E 00
5.3000	0.5576E 00
5.4000	0.5436E 00
5.5000	0.5245E 00
5.6000	0.5005E 00
5.7000	0.4722E 00
5.8000	0.4401E 00
5.9000	0.4049E 00
6.0000	0.3675E 00
6.1000	0.3286E 00
6.2000	0.2892E 00
6.3000	0.2500E 00
6.4000	0.2117E 00
6.5000	0.1752E 00
6.6000	0.1409E 00
6.7000	0.1094E 00
6.8000	0.8104E-01
6.9000	0.5605E-01
7.0000	0.3457E-01
7.1000	0.1662E-01
7.2000	0.2115E-02
7.3000	-0.9136E-02
7.4000	-0.1739E-01
7.5000	-0.2295E-01
7.6000	-0.2619E-01
7.7000	-0.2749E-01
7.8000	-0.2722E-01
7.9000	-0.2578E-01
8.0000	-0.2353E-01
8.1000	-0.2078E-01
8.2000	-0.1783E-01
8.3000	-0.1493E-01
8.4000	-0.1220E-01
8.5000	-0.9856E-02
8.6000	-0.7957E-02
8.7000	-0.6548E-02
8.8000	-0.5634E-02
8.9000	-0.5186E-02
9.0000	-0.5149E-02
9.1000	-0.5450E-02
9.2000	-0.6006E-02
9.3000	-0.6729E-02
9.4000	-0.7534E-02
9.5000	-0.8345E-02
9.6000	-0.9096E-02

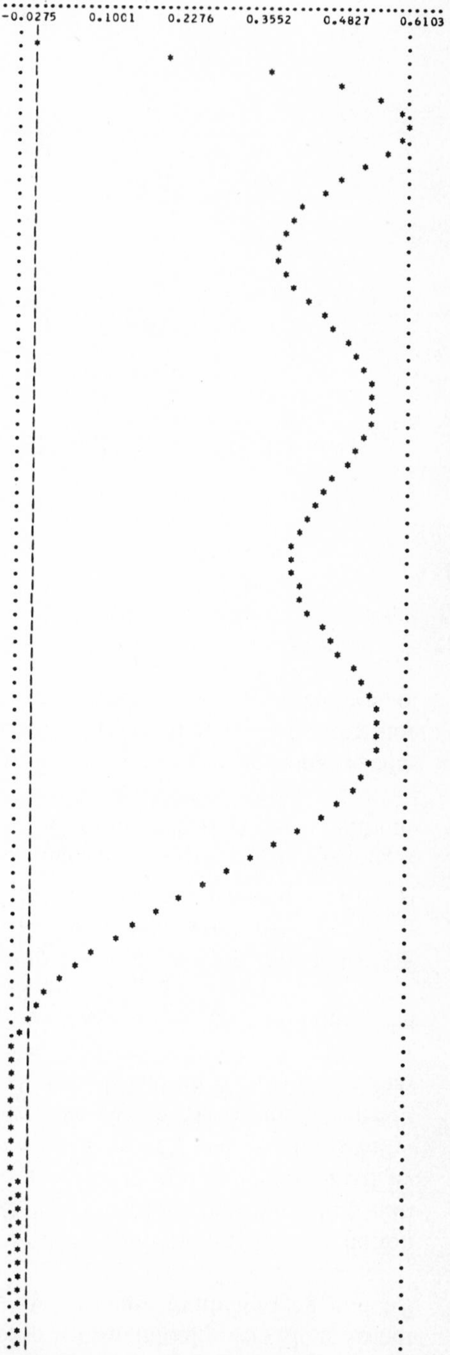

FIG. 7.22-L

7.23 MULTIPLE (COINCIDENT) POLES

In Sec. 7.21 we remarked on another lacuna in our knowledge: what to do if some property of the network under analysis, or of the excitation applied, or of both, makes two (or more) poles of the response transform R coincide. For simplicity's sake we have generally postulated that all the poles of $R(s)$ be *simple* (separate and distinct, one from another). When this is not true, how does the analysis differ?

Coincidence of two or more poles in $R(s) = E(s)T(s)$ represents nothing extraordinary, no radically different situation. Surely there can be no great difference between (*a*) the response when two poles are distinct but very close together (a case we can handle) and (*b*) the response when the two poles actually coincide! The coincidence of poles can merely mean that:

> Some of the elements in the network have values such that two or more of the network natural frequencies are the same: $T(s)$ has multiple poles (see Fig. 7.23-A), (7.23-1)

or

> The excitation's waveform has certain shape peculiarities: perhaps it contains $te^{-\alpha t}$ instead of the sum of two different exponentials (see Fig. 7.23-B); these lead to multiple poles in $E(s)$, (7.23-2)

or

> The network is being excited at one of its natural frequencies (E and T have common poles). (7.23-3)

The third possibility arouses more interest than the other two: it seems more dramatic because it suggests resonant response, of large amplitude. (But the other two possibilities are mathematically the same as this one, and therefore

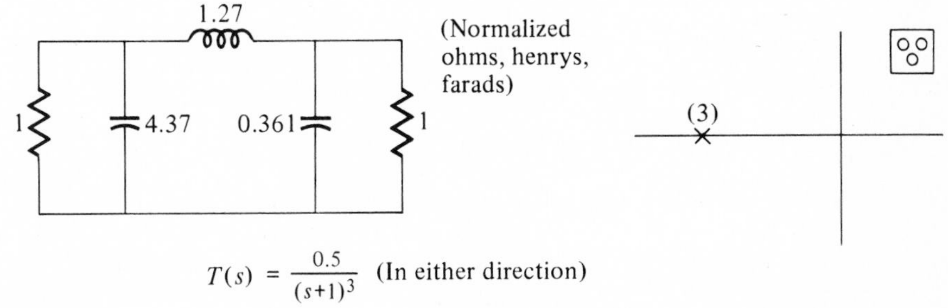

FIG. 7.23-A

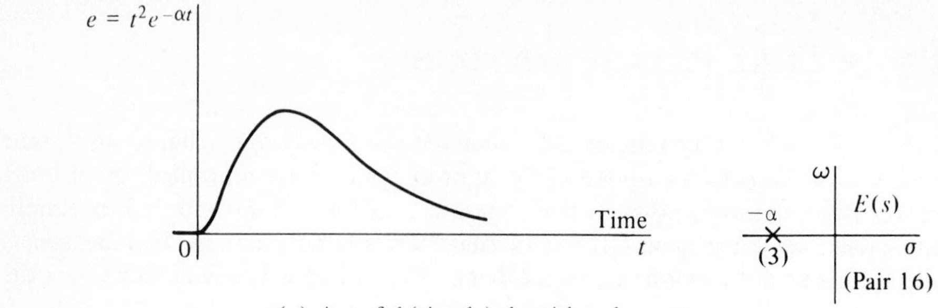

(*a*) A useful (simple) sluggish-pulse wave.

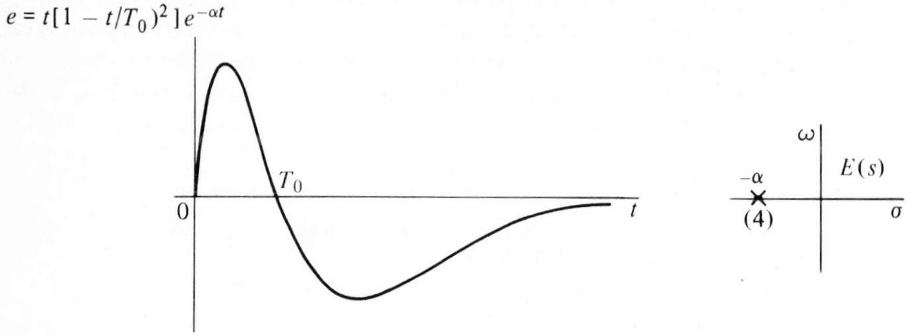

(*b*) A one-oscillation variation.

FIG. 7.23-B. Some excitations whose transforms leave multiple poles.

must be equally dramatic.) But if the natural frequency is in the left half plane as it usually is, the most that can occur is that some of the e^{pt} terms in the response are replaced by terms like $t^n e^{pt}$, and some sort of short-lived pulse occurs: witness the result of excitation of the *RC* circuit by its own natural-mode wave, $e^{-\alpha t}$ (Sec. 2.14). The response must die out, however large it may temporarily become. True, if we consider an *RLC* (tuned) circuit (Fig. 7.23-C) whose Q is very high, then excitation at the natural frequency may temporarily give a very large response. [And excitation at the nearby sinusoidal mode $s = j\omega_0$, that is, excitation by $\sin \omega_0 t$, will certainly give a large (the normal resonant) steady-state response.] But that is the most than can occur. In any event, infrequent though it be, it can happen that $R(s)$ have "multipoles." Hence we must study the phenomenon enough to find out how to handle it in analysis.

To get some feeling for the necessary modifications in our now-standard procedure, let us attempt to perform the analysis summarized in Fig. 7.23-C. We recall that the analysis of the *RLC* circuit excited at its own natural frequency was not made in Chap. 3 because it seemed too complicated. It can of course be performed by convolution; a curve such as that of Fig. 7.23-C is therefore available as a guide. But surely analysis by the standard partial-fraction expansion method can also be made, if with some modification. We have (Fig. 7.23-C)

$$R(s) = E(s)T(s) = \frac{\dfrac{I_0 \omega_N}{C} s}{(s^2 + 2\alpha s + \omega_0^{\,2})^2}. \qquad (7.23\text{-}4)$$

Since there are four poles, the partial-fraction expansion of $R(s)$ surely has four terms. Since there are only two pole locations, we have available, in the usual partial-fraction expansions of Sec. 7.12, only two types of terms: $1/(s - p_1)$ and $1/(s - p_2)$. But we expect terms like te^{pt} in $r(t)$ and therefore terms like $1/(s - p_1)^2$ and $1/(s - p_2)^2$ should appear in the expansion of $R(s)$: it should include *both* sorts of terms.

A simpler case worth investigating first is that of one "multipole":

$$F(s) = \frac{a_0 + a_1 s}{(s - p_1)^2}. \qquad (7.23\text{-}5)$$

The numerator is of as high a degree as it can be with $M < N$. Since the numerator is not a constant, however, none of our pairs will transform (7.23-5). If written

$$F(s) = \frac{a_0}{(s - p_1)^2} + \frac{a_1 s}{(s - p_1)^2}, \qquad (7.23\text{-}6)$$

however, pair 34 can handle the first term. We focus therefore on the second term [which is much like (7.23-4)]. The important point here is that the numerator is not a constant and therefore the term is not yet expanded in partial fractions (it is not yet in *simple* elements). What *is* the mate of $[s/(s - p_1)^2]$? An obvious way to find the answer is to consider first the perturbed situation of Fig. 7.23-D*b*, for the function

$$F_1(s) = \frac{s}{(s - p_1)(s - p_1 - \delta)} \qquad (7.23\text{-}7)$$

(in which δ is small, complex, but otherwise arbitrary) *is* tractable in the standard way. We need only transform it and then take the limit as $\delta \to 0$. We proceed in the standard way:

$$F_1(s) = \frac{K_1}{s - p_1} + \frac{K_2}{s - p_1 - \delta}, \qquad (7.23\text{-}8)$$

$$K_1 = (s - p_1)F_1(s)\Big|_{s = p_1} = -\frac{p_1}{\delta},$$
$$K_2 = (s - p_1 - \delta)F_1(s)\Big|_{s = p_1 + \delta} = 1 + \frac{p_1}{\delta} = \frac{p_1}{\delta}\left(1 + \frac{\delta}{p_1}\right). \qquad (7.23\text{-}9)$$

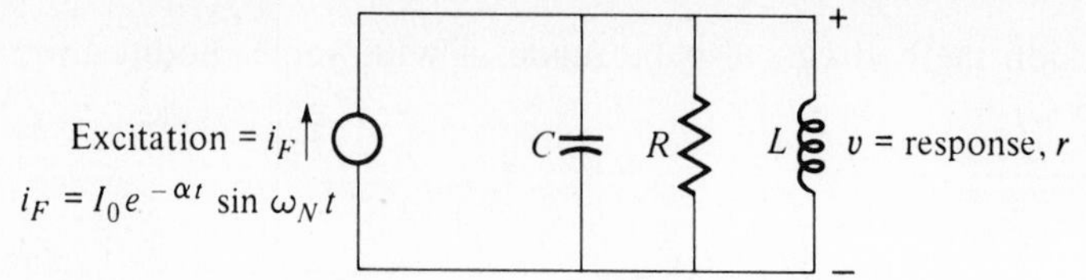

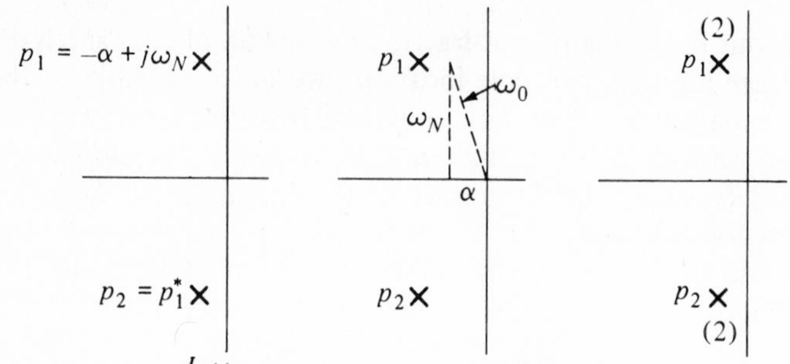

$$I_F(s) = \frac{I_0 \omega_N}{(s+\alpha)^2 + \omega_N^2} \qquad T(s) = \frac{(1/C)s}{s^2 + 2\alpha s + \omega_0^2} \qquad R(s) = I_F(s)\mathrm{T}(s)$$

[Pair 23] [(3.17-1)]

THE FOLLOWING 5 PAGES IS OUTPUT FROM PLOTTING ROUTINE "FGRAPH"

TIME	RESPONSE
0.0000	0.0000
0.2000	0.0193
0.4000	0.0738
0.6000	0.1564
0.8000	0.2578
1.0000	0.3677
1.2000	0.4752
1.4000	0.5694
1.6000	0.6403
1.8000	0.6793
2.0000	0.6800
2.2000	0.6382
2.4000	0.5523
2.6000	0.4238
2.8000	0.2567
3.0000	0.0575
3.2000	-0.1651
3.4000	-0.4007
3.6000	-0.6377
3.8000	-0.8638
4.0000	-1.0671
4.2000	-1.2362
4.4000	-1.3612
4.6000	-1.4340
4.8000	-1.4487
5.0000	-1.4023
5.2000	-1.2944
5.4000	-1.1279
5.6000	-0.9080
5.8000	-0.6430
6.0000	-0.3431
6.2000	-0.0206
6.4000	0.3111
6.6000	0.6379
6.8000	0.9458
7.0000	1.2212
7.2000	1.4518
7.4000	1.6273
7.6000	1.7394
7.8000	1.7826
8.0000	1.7544
8.2000	1.6551
8.4000	1.4882
8.6000	1.2601
8.8000	0.9798
9.0000	0.6585
9.2000	0.3092
9.4000	-0.0538
9.6000	-0.4158
9.8000	-0.7620
10.0000	-1.0781
10.2000	-1.3514
10.4000	-1.5708
10.6000	-1.7274
10.8000	-1.8150
11.0000	-1.8301
11.2000	-1.7726
11.4000	-1.6448
11.6000	-1.4524
11.8000	-1.2034
12.0000	-0.9081
12.2000	-0.5787
12.4000	-0.2287
12.6000	0.1278
12.8000	0.4764
13.0000	0.9032

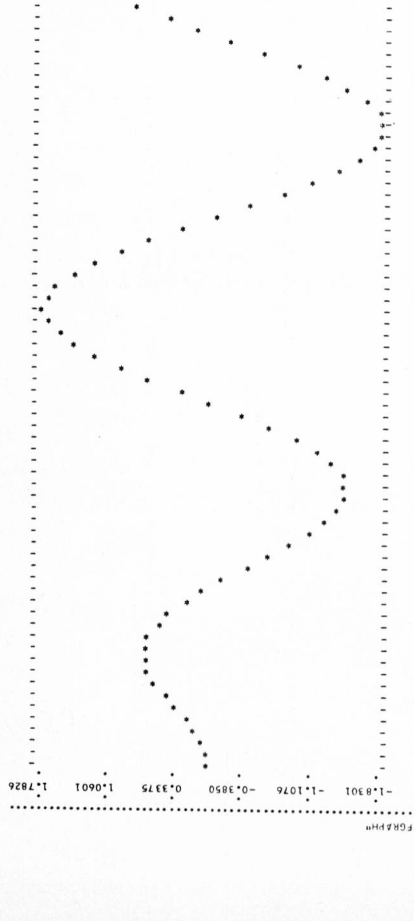

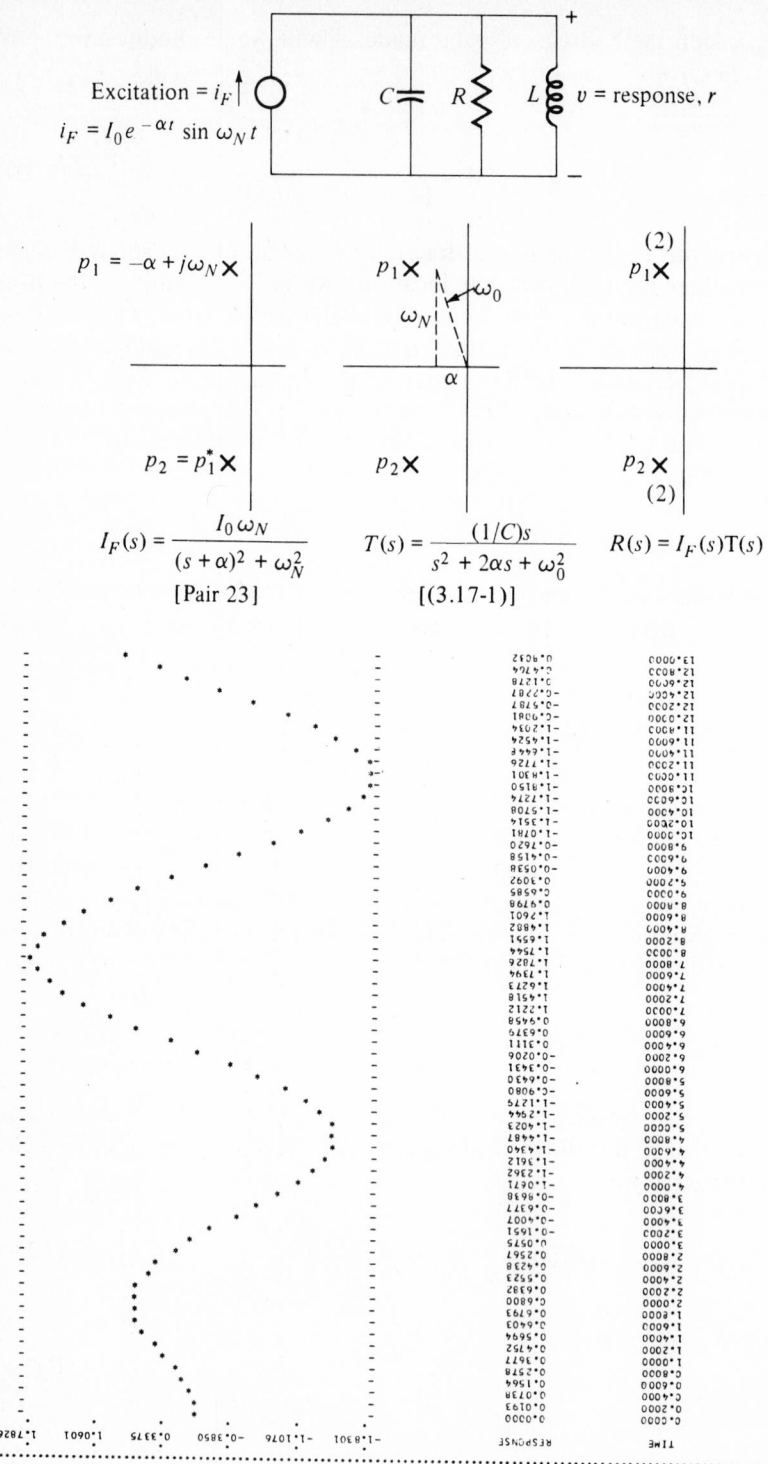

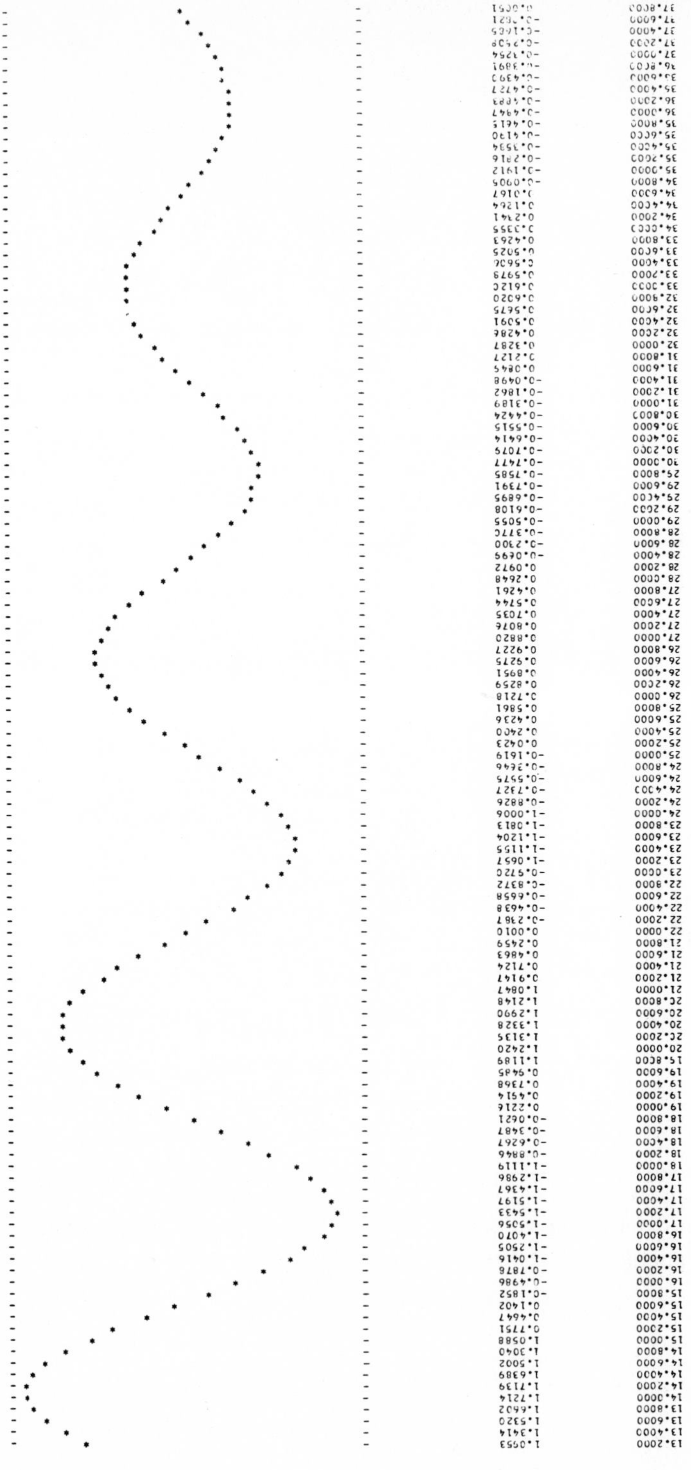

FIG. 7.23-C

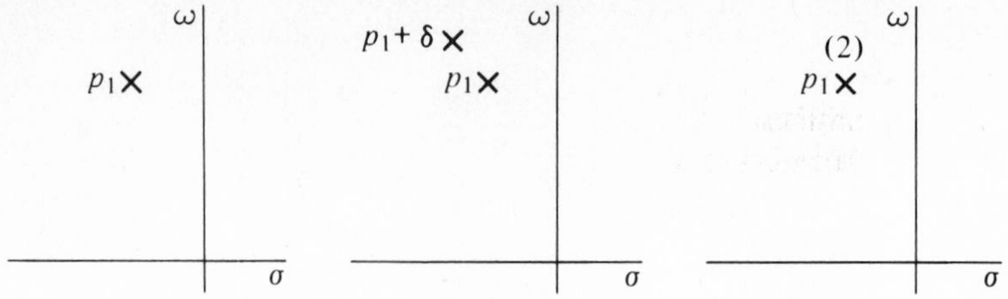

(*a*) A single pole: p_1.

(*b*) Two nearly coincident poles: p_1 and $(p_1 + \delta)$.

(*c*) Two coincident poles: p_1 and p_1.

FIG. 7.23-D

Hence the mate of $F_1(s)$ is

$$
\begin{aligned}
f_1(t) &= \frac{p_1}{\delta}\left[\left(1+\frac{\delta}{p_1}\right)e^{(p_1+\delta)t} - e^{p_1 t}\right] \\
&= \frac{p_1}{\delta}\, e^{p_1 t}\left[\left(1+\frac{\delta}{p_1}\right)e^{\delta t} - 1\right] \\
&= \frac{p_1}{\delta}\, e^{p_1 t}\left[\left(1+\frac{\delta}{p_1}\right)\left(1+\delta t+\frac{\delta^2 t^2}{2}+\cdots\right) - 1\right] \qquad (7.23\text{-}10) \\
&= \frac{p_1}{\delta}\, e^{p_1 t}\left(\frac{\delta}{p_1}+\delta t+\frac{\delta^2 t}{p_1}+\cdots\right) \\
&= e^{p_1 t}(1 + p_1 t + \delta t + \cdots),
\end{aligned}
$$

and in the limit as $\delta \to 0$, $f_1(t)$ becomes simply $(1 + p_1 t)e^{p_1 t}$. We have then a new pair:

$$\frac{s}{(s-p_1)^2} \sim (1 + p_1 t)e^{p_1 t} \qquad (7.23\text{-}11)$$

[Here comparison with (2.14-11) and (2.14-12) is interesting: in effect we used (7.23-10) in calculating the response of the *RC* network to excitation at its own natural frequency.] But the time-domain member of (7.23-11) itself transforms (pairs 33, 34) to

$$\frac{1}{s-p_1} + \frac{p_1}{(s-p_1)^2} \qquad (7.23\text{-}12)$$

which suggests again that the s-domain member of (7.23-11) is not properly broken down or expanded, is not in true partial-fraction form, with *simple* elements. And now the light dawns! The s-domain member [of (7.23-11) and of (7.23-5) and of similar functions] should first (before transforming) be expanded thus:

$$F(s) = \frac{a_0 + a_1 s}{(s - p_1)^2} = \frac{K_{12}}{(s - p_1)^2} + \frac{K_{11}}{s - p_1}. \tag{7.23-13}$$

How should such an expansion be made? We take a leaf from the procedure for simple poles: first to multiply through by the trouble-making term, which here is $(s - p_1)^2$, to form

$$G(s) = (s - p_1)^2 F(s) = a_0 + a_1 s = (a_0 + a_1 p_1) + a_1(s - p_1) = K_{12} + K_{11}(s - p_1) \tag{7.23-14}$$

which is really the Taylor-series expansion of $G(s)$ about $s = p_1$. Thus clearly

$$K_{12} = G(p_1) = [(s - p_1)^2 F(s)]_{s=p_1} \tag{7.23-15}$$

and

$$K_{11} = G'(p_1) = \left[\frac{d}{ds}(s - p_1)^2 F(s)\right]_{s=p_1} \tag{7.23-16}$$

and the mystery is solved.

Generalization is simple, now that the principle is out. For one multiple pole of order m the procedure is to form from

$$F(s) = \frac{Q_N(s)}{(s - p_1)^m (s - p_2)(s - p_3)\ldots} \tag{7.23-17}$$

the related function

$$G(s) = (s - p_1)^m F(s) = \frac{Q_N(s)}{(s - p_2)(s - p_3)\ldots} \tag{7.23-18}$$

and to expand $G(s)$ in Taylor series about the point $s = p_1$:

$$G(s) = G(p_1) + G'(p_1)(s - p_1) + \frac{G''(p_1)(s - p_1)^2}{2!} + \cdots. \tag{7.23-19}$$

Then

$$F(s) = \frac{G(s)}{(s-p_1)^m} = \frac{G(p_1)}{(s-p_1)^m} + \frac{G'(p_1)}{(s-p_1)^{m-1}} + \frac{G''(p_1)/2!}{(s-p_1)^{m-2}} + \cdots$$
$$+ \frac{G^{(m-1)}(p_1)/(m-1)!}{s-p_1} + (\cdots)$$
$$= \underbrace{\frac{K_{1m}}{(s-p_1)^m} + \frac{K_{1,m-1}}{(s-p_1)^{m-1}} + \cdots + \frac{K_{11}}{(s-p_1)}}_{F_1(s)} + F_2(s) \tag{7.23-20}$$
$$= F_1(s) + F_2(s)$$

in which

$$K_{1,j} = \frac{G^{(m-j)}(p_1)}{(m-j)!}, \qquad j = m, m-1, \ldots, 1. \tag{7.23-21}$$

The function $F_1(s)$ contains the m terms of the partial-fraction expansion of $F(s)$ that correspond to the m poles at p_1; the reduced function $F_2(s) = F - F_1$ contains the remaining poles of $F(s)$. We now calculate the K's associated with F_2 in the usual way, and finally obtain the expansion

$$F(s) = F_1 + F_2 = \underbrace{\cdots}_{F_1(s)} + \underbrace{\frac{K_2}{s-p_2} + \frac{K_3}{s-p_3} + \cdots}_{F_2(s)} \tag{7.23-22}$$

in which

$$K_2 = [(s-p_2)F(s)]_{s=p_2} \tag{7.23-23}$$

and so on.

Our final generalization is to the function with several multiple poles

$$F(s) = \frac{Q_N(s)}{(s-p_1)^{m_1}(s-p_2)^{m_2}\cdots}$$
$$= \frac{K_{1,m_1}}{(s-p_1)^{m_1}} + \frac{K_{1,m_1-1}}{(s-p_1)^{m_1-1}} + \cdots + \frac{K_{11}}{s-p_1} \tag{7.23-24}$$
$$+ \frac{K_{2,m_2}}{(s-p_2)^{m_2}} + \cdots + \cdots$$

in which

$$K_{k,j} = \frac{G^{(m_k - j)}(p_k)}{(m_k - j)!}, \qquad k = 1, 2, \ldots, \quad j = m_k, m_k - 1, \ldots, 1. \tag{7.23-25}$$

These two relations, (7.23-24) and (7.23-25), together with pair 35, indicate how to perform the most general partial-fraction expansion and to return to the time domain. Virtually useless as they stand, because of their hideous complexity, we nevertheless include them for completeness. In any actual problem one would work out the K's by first multiplying through by each multipole factor, and then calculating the corresponding Taylor series as in (7.23-17), (7.23-18), ... above.

To illustrate, we return to the analysis of the tuned circuit excited at its own natural frequency, to (7.23-4). We have

$$R(s) = \frac{I_0 \omega_N}{C} \frac{s}{(s - p_1)^2 (s - p_2)^2} \tag{7.23-26}$$

in which $p_1 = -\alpha + j\omega_N$ and p_2 is the conjugate of p_1 (Fig. 7.23-C). The calculation (setting aside the coefficient $I_0 \omega_N / C$) is simply

$$\begin{aligned} G(s) = (s - p_1)^2 R(s) = \frac{s}{(s - p_2)^2} &= G(p_1) + G'(p_1)(s - p_1) + F_2(s) \\ &= K_{12} + K_{11}(s - p_1) + F_2(s) \end{aligned} \tag{7.23-27}$$

in which

$$K_{12} = G(p_1) = \frac{p_1}{(p_1 - p_2)^2} = \frac{-\alpha + j\omega_N}{(j2\omega_N)^2} = \frac{\alpha - j\omega_N}{4\omega_N^{\ 2}}, \tag{7.23-28}$$

$$\begin{aligned} K_{11} = G'(p_1) &= \left. \frac{(s - p_2)^2 - 2s(s - p_2)}{(s - p_2)^4} \right|_{s = p_1} \\ &= \left. \frac{-s - p_2}{(s - p_2)^3} \right|_{s = p_1} = \frac{2\alpha}{(j2\omega_N)^3} = \frac{j\alpha}{4\omega_N^{\ 3}}. \end{aligned} \tag{7.23-29}$$

The terms that correspond to the second (conjugate) pole, p_2, need not be calculated, since they are respectively conjugate to those just calculated. Hence (still without the multiplier)

$$\begin{aligned} R(s) &= \frac{s}{(s - p_1)^2 (s - p_2)^2} = \frac{G(s)}{(s - p_1)^2} \\ &= \frac{K_{12}}{(s - p_1)^2} + \frac{K_{11}}{(s - p_1)} + * \end{aligned} \tag{7.23-30}$$

in which the asterisk indicates the terms that correspond to the second pole [the F_2 part of (7.23-27) above] which need not be calculated: because they are conjugate to those already written, they merely double the real part and cancel the imaginary part thereof. Therefore, without the multiplier (pairs 33, 34)

$$\begin{aligned} r(t) &= K_{12} t e^{p_1 t} + K_{11} e^{p_1 t} + * \\ &= 2\,\mathbf{RE}\left(\frac{\alpha - j\omega_N}{4\omega_N^{\,2}}\right) t e^{(-\alpha + j\omega_N)t} + \frac{j\alpha}{4\omega_N^{\,3}} e^{(-\alpha + j\omega_N)t} \\ &= \frac{\omega_0 t}{2\omega_N^{\,2}} e^{-\alpha t} \cos(\omega_N t - 90^\circ + \phi) + \frac{\alpha}{2\omega_N^{\,3}} e^{-\alpha t} \cos(\omega_N t + 90^\circ) \\ &= \frac{\omega_0 t}{2\omega_N^{\,2}} e^{-\alpha t} \sin(\omega_N t + \phi) - \frac{\alpha}{2\omega_N^{\,3}} e^{-\alpha t} \sin \omega_N t. \end{aligned} \tag{7.23-31}$$

in which $\phi = \sin^{-1}(\alpha/\omega_0)$, the small angle by which the angle of p_1 exceeds 90° (Fig. 7.23-C).

Finally, on restoring the constant multiplier, we obtain

$$r(t) = \frac{I_0}{2C\omega_N} e^{-\alpha t}\left[\omega_0 t \sin(\omega_N t + \phi) - \frac{\alpha}{\omega_N} \sin \omega_N t\right]. \tag{7.23-32}$$

This is the response plotted in Fig. 7.23-C, for $Q = 5$ ($\alpha = 0.1$), on the scale $\omega_0 = 1$. As checks, we should calculate from (7.23-32) the initial values of $r(t)$ and of $r'(t)$, both of which should be zero, since the denominator of (7.23-4) is higher in degree than the numerator by 3. It is not difficult also to calculate $r''(0)$, the first nonzero term of the power series for $r(t)$, as an additional check. By Taylor-series expansion of $\sin(\omega_N t + \phi)$ about $t = 0$, and of $e^{-\alpha t}$ and of $\sin \omega_N t$, we obtain from (7.23-32)

$$\begin{aligned} r(t) &= \frac{I_0}{2C\omega_N} e^{-\alpha t}\left\{\omega_0 t[(\sin\phi) + (\cos\phi)\omega_N t + \cdots] - \frac{\alpha}{\omega_N}(\omega_N t + \cdots)\right\} \\ &= \frac{I_0}{2C\omega_N} e^{-\alpha t}\left[\omega_0 t\left(\frac{\alpha}{\omega_0} + \frac{\omega_N}{\omega_0}\omega_N t + \cdots\right) - \alpha t\right] \\ &= \frac{I_0}{2C\omega_N}(1 - \alpha t + \cdots)(\omega_N^{\,2} t^2 + \cdots) \\ &= \frac{I_0}{2C}\omega_N t^2 + \cdots. \end{aligned} \tag{7.23-33}$$

Hence $r(0) = 0$, $r'(0) = 0$, and $r''(0) = I_0 \omega_N / C$. From (7.23-4) we have, for large s,

$$R(s) = \frac{I_0 \omega_N}{C}\left(\frac{1}{s^3} + \cdots\right) \tag{7.23-34}$$

whence

$$r(t) = \frac{I_0 \omega_N}{C}\left(\frac{t^2}{2!} + \cdots\right) \qquad (7.23\text{-}35)$$

so that in (7.23-32) the first three initial values of $r(t)$ are correct. This practically verifies the calculation, though in theory we need eight checks.

It is the factor t in the response (7.23-32) that indicates the double pole(s); not simply e^{pt} but also te^{pt} is the response that corresponds to a double pole. A third-order complex pole would give

$$(K_{11} + K_{12}t + K_{1e}t^2)e^{pt} \qquad (7.23\text{-}36)$$

and so on. The polynomial multipliers simply represent coincident poles; they are no exception to the principle that responses of networks are usually sums of exponentials and nothing else.

Multiple poles, to recapitulate, are uncommon. Coincidence of natural frequencies in a network represents a very special condition, as does coincidence of poles in $E(s)$; so also does coincidence between network natural frequencies and poles of $E(s)$, that is, excitation at a network's natural frequency. For completeness's sake we have developed the necessary extensions of the basic partial-fraction expansion [see (7.23-24), (7.23-25), and pair 35]. But it is a rare occurrence to need them.

7.24 THE STATE OF THE ART

Let us return to Sec. 7.21 where we examined the state of the art, our knowledge of network analysis. We found certain lacunae, which Secs. 7.22 and 7.23 have filled (the treatment of arbitrary initial conditions and of multiple poles). There remains the problem of the general use of the integral formulation of the impulse-response transfer function relation (path 5 of Sec. 7.21).

The completely general network-transfer-function-calculation problem we have only briefly touched on, true, and we have used ladder networks for our examples. Note again, however, that once $T(s)$ is found, for whatever network, by whatever means, with whatever labor, all that we have learned about transfer functions still applies. Nothing in our definition, use, or general discussion of $T(s)$ depends on its belonging to or being calculated from any particular class of network analysis. We found certain lacunae, which Secs. 7.22 and 7.23 have network's configuration. We have generally used ladders as examples only because (*a*) their transfer functions are easy to calculate, (*b*) their physical behavior is easy to understand, and (*c*) they and simple feedback structures constitute practically all the networks met in real apparatus.

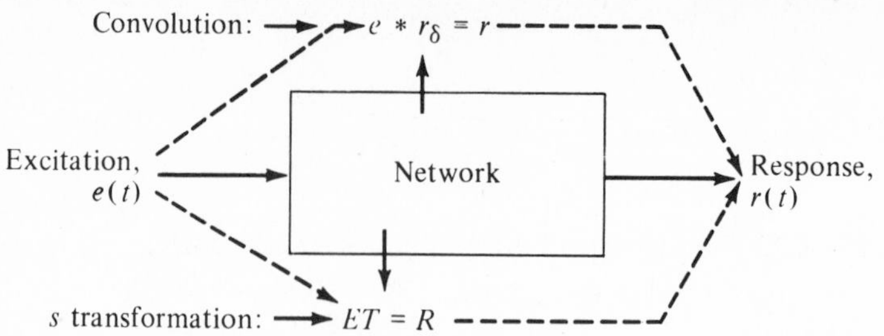

FIG. 7.24-A

Another problem that we leave for study elsewhere is analysis when N is very large. A transfer function with 20 poles can in principle be expanded in partial fractions as well as one with 2 poles, but in practice numerical difficulties arise. They can be circumvented, but we set aside that discussion also.

The methods summarized in Fig. 7.21-A enable us to analyze all ordinary networks. When the excitation is a sum of exponentials (which includes impulses and steps, the commonest of all) we can transfer $e(t)$ immediately to the s domain; when $e(t)$ is not exponential, we can resort to convolution, provided the network's impulse response (or transfer function) is available. We note once more that the former, the usual transforming to and working in the s domain, is equivalent to the latter, convolution in the t domain. But working in the s domain (algebra) is much the simpler process. Hence we prefer the lower route in Fig. 7.24-A.

There are many corollaries, and many developments important in their own right that we have to leave for study elsewhere, such as the detailed study of the sinusoidal steady state and of the properties of networks of various sorts. But one "corollary" at which we have promised to look further does deserve attention here. We turn now (in Chap. 8) to a more detailed, albeit brief, study of path 5 of Sec. 7.21.

7.25 EXERCISES

7-1. The pole-zero pattern of a network's transfer function is shown in Fig. 7-1*a* (in which the numbers have been simplified). What part of the transfer function is unknown? Find formal expressions for the network's response from rest to (*a*) $\delta(t)$, (*b*) $\cos t\, u(t)$, (*c*) $e^{-0.5t}u(t)$. As a check, in each case calculate the initial ($t = 0+$) values of the response and of its first derivative both from your result and from the transfer function. How many initial values would it be necessary to calculate for a *complete* check on your work?

Repeat for the pattern of Fig. 7-1*b* and that of Fig. 7-1*c*.

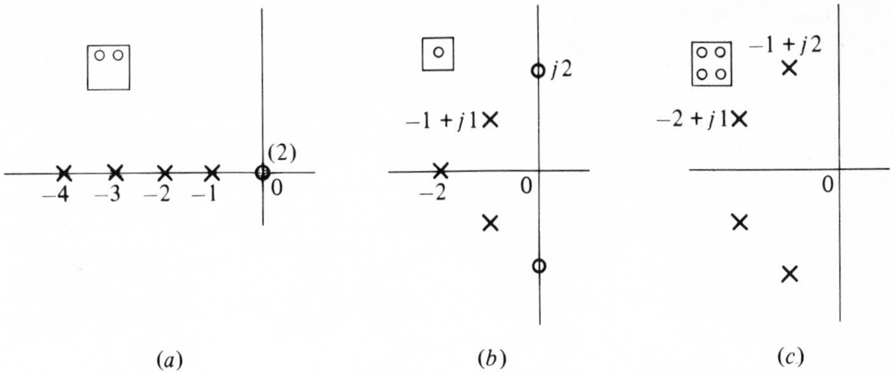

FIG. 7-1

7-2. The circuit of Fig. 7-2 was designed to counteract the inherent tendency of a system to transmit better, in the sinusoidal steady state, at low frequencies that at high frequencies. The excitation is the voltage $e(t)$, the response is the voltage $v(t)$. Explain briefly in physical terms why the transmission is greater at high than at low frequencies. Obtain the transfer function and show its poles and zeros on a diagram. Sketch $T(j\omega)$ in both magnitude and angle versus ω. Sketch the locus of $T(j\omega)$ in its (complex) plane. Why might it be called a "lead network"? (It is sometimes useful in compensating instability tendencies in control systems.)

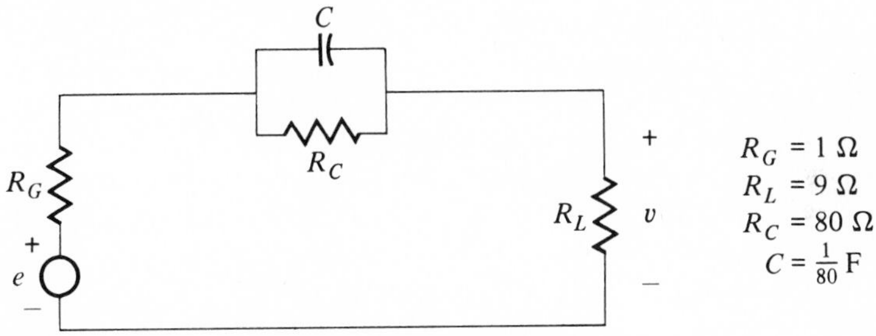

FIG. 7-2

7-3. Calculate (and sketch) the response of the network of Fig. 7-3 (from rest) to each of these excitation functions i_F: (*a*) an impulse, (*b*) a step, (*c*) the exponential e^{-t}, (*d*) the sinusoid $\sin t$, (*e*) the sinusoid $\cos t$.

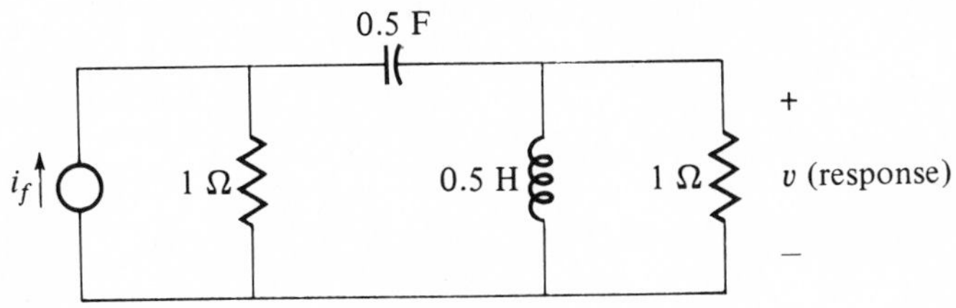

FIG. 7-3

7-4. For the simple network of Fig. 7-4, excited sinusoidally at frequency ω_F, find the forced (steady-state) component of response. Sketch it, and the excitation wave, together, for $\omega_F =$ (*a*) 1, (*b*) 2. Is it possible to apply the excitation to the network without stored energy at such a point on the excitation wave that there is no transient (natural) component? If the initial condition is not rest, will your reply change? Explain.

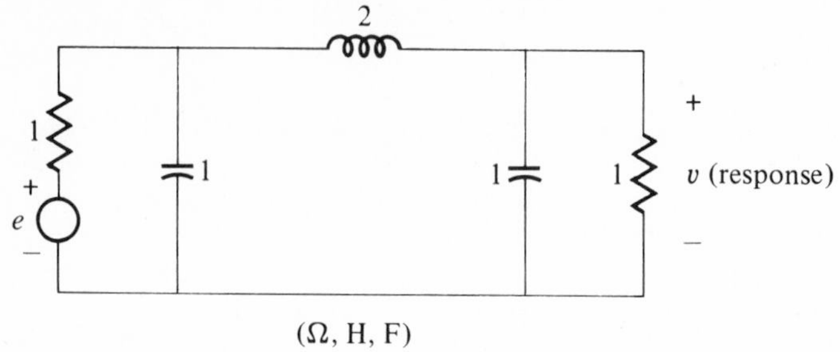

FIG. 7-4

7-5. A (passive) network with N natural frequencies is excited from rest by the sinusoidal wave $E_m \cos(\omega_F t + \phi)$. Show that, if one of the natural frequencies is real, then by proper adjustment of the excitation initial phase and amplitude, the corresponding natural mode may be removed from the response. What adjustment will also remove a second natural mode? What roles do the network's infinite-loss points play in your calculations?

Suppose the excitation includes (additively) a second sinusoidal component, $E_{m2} \cos(2\omega_F t + \phi_2)$. How many natural modes can in principle now be removed from the response? What equations must be solved to determine the necessary adjustments of excitation parameters?

7-6. Calculate and sketch the voltage v (Fig. 7-6) for each of the following cases. Check each step of your calculations carefully.

Initial Condition	**Excitation** e
Rest	Impulse
Rest	Step
Rest	Ramp
Rest	Sin $2t$
Rest	Cos $2t$
$v_C(0) = 1$, $i_L(0) = 0$	None
$v_C(0) = 0$, $i_L(0) = 1$	None

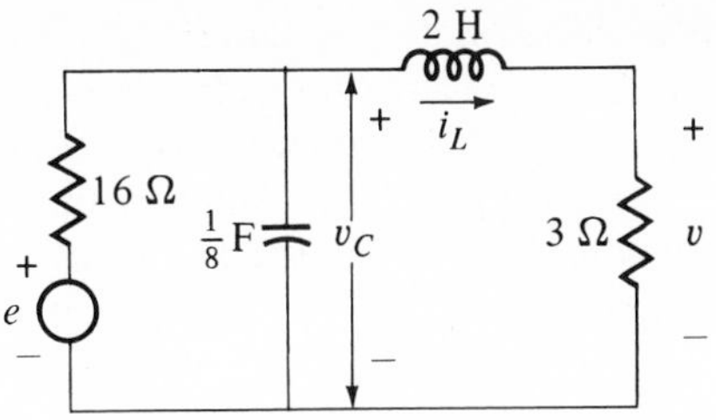

FIG. 7-6

7-7. Construct a table with five columns, headed "excitation," "mate of excitation," "transfer function," "mate of response," "response." For each example of Sec. 7.15, fill in the columns with the physical nature (current, voltage, ...) and the *dimensions* of each quantity. Consider both current and voltage excitation in each possible case.

7-8. Verify that all four networks of Fig. 7.07-A do have the same transfer function, that given. Explain the relations between the four in terms of duality and source exchanges.

7-9. For the data of (7.14-31), what frequency base (reference) will make equal to unity the coefficient of s^3 in the denominator of $T(s)$, as in (7.14-24)? With this choice for f_0 and $R_0 = 600\ \Omega$, what are the normalized element values? Is the analysis now simpler than that based on (7.14-34)? Explain.

7-10. If in Fig. 7.15-C the ratio of actual time to normalized time is to be exactly 40 μs : 1, what should be the value of f_0?

7-11. Find the impulse response and the step response of two networks whose transfer functions are $(21s^2 + 189)/(s^3 + 6s^2 + 11s + 6)$ and $(2s + 1)/(s^3 + 5s^2 + 16s + 30)$. What property (perhaps artificial) makes these calculations quite simple?

Using the series form (7.15-15), how many terms are needed to obtain the value of the impulse response correct to three significant figures for (normalized) $t = 1.0$? 2.0?

7-12. The natural frequencies of the network of Fig. 7-12 are -1 and -2 in normalized units. Sketch its impulse response, $r_\delta(t)$. Now write down the

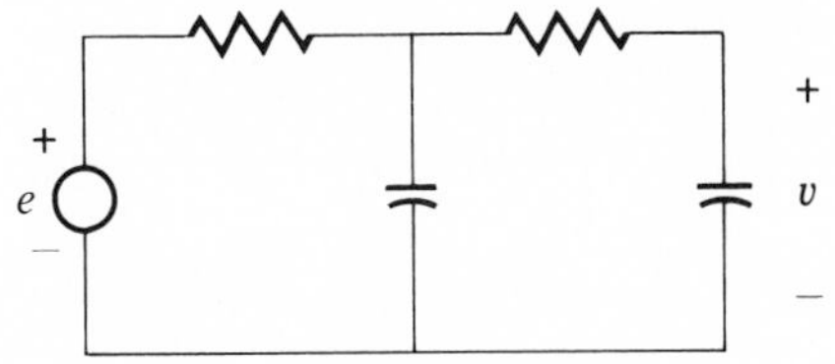

FIG. 7-12

formula therefor, making all calculations mentally. Then write down the step- and ramp-response formulas. What is the sinusoidal-steady-state transmission from e to v in amplitude ratio? In phase shift? (Again calculate mentally, taking ω_F as first 1, then 2 normalized units; accept the limited accuracy obtained.) Now write down, using previous results and such additional (mental) calculations as necessary, the response from rest to excitation $E_m \cos \omega_F t$, taking $\omega_F = 1$, then 2, and the response to $E_m \sin \omega_F t$.

7-13. Obtain the time-domain mate of each of the functions given below. In each case check your work using four initial values, each calculated in two ways. Is your check complete?

(*a*) $\dfrac{12s^2 + 14s + 3}{2s^3 + 3s^2 + s}$ (*b*) $\dfrac{s + 2}{s^2 + 4s + 40}$ (*c*) $\dfrac{1}{s^3 + 2s^2 + 2s + 1}$

7-14. It is commonly said that the steady-state response to sinusoidal excitation is got by altering the excitation sinusoidal wave in magnitude and initial phase by operating on it with the transfer function. Explain carefully (*a*) why this statement is essentially correct, but (*b*) is also open to misconstruction. Illustrate with examples.

7-15. Prove that the *magnitude* of $T(j\omega)$ is an *even* function of ω and that the *angle* of $T(j\omega)$ is an *odd* function of ω. Upon what postulates do these facts rest?

7-16. Show that the impulse response of a network whose transfer function is $T(s)$ can be expressed thus:

$$r_\delta(t) = \frac{1}{\pi}\int_0^\infty |T(j\omega)| \cos(\omega t + \beta)\, d\omega$$

in which β is the angle of $T(j\omega)$. Upon which of our assumptions does this depend? Given plots based on sinusoidal-steady-state measurements of both magnitude and angle of $T(j\omega)$, the response of the network to any excitation can in principle be computed. Outline a procedure for doing so and indicate where you think the principal difficulties would be found.

7-17. Find and sketch the impulse response of the network of Fig. 7-17. For simplicity, take all element values to be unity.

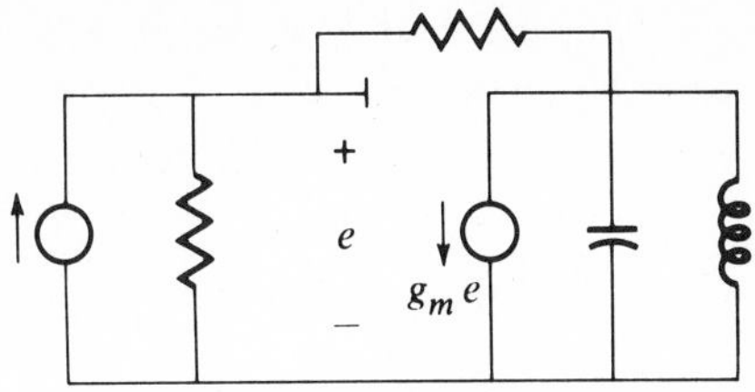

FIG. 7-17

7-18. Find and sketch the impulse response of the simple bridge network shown in Fig. 7-18. The response is the voltage v. [See (6.23-10).]

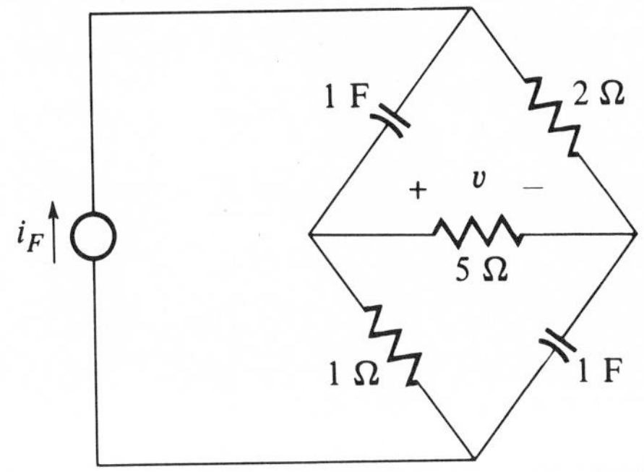

FIG. 7-18

7-19. Let the element values of the network of Fig. 7-19 be of unit value for simplicity. The load Z_L is infinite. The source impedance Z_G is zero. Find the transfer function, show its zeros and poles, sketch $|T(j\omega)|$ versus ω, and also the angle of $T(j\omega)$. Repeat for Z_G infinite. In what ways do the two transfer functions differ? In what ways are they the same? Explain fully.

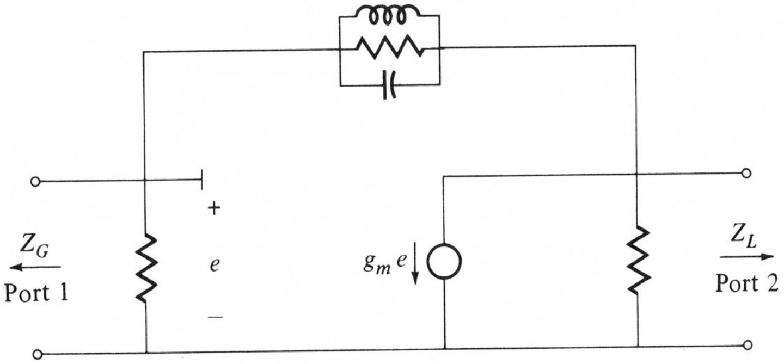

FIG. 7-19

7-20. The network of Fig. 7-20 is excited by injecting an impulse of current as shown. (There is no stored energy.) Find and sketch both voltages (v_R and v_L)

(*a*) with switch *X* open, (*b*) with switch *X* closed at and after the time of excitation. Do the minimum work required for sketches that are roughly correct, not necessarily precise.

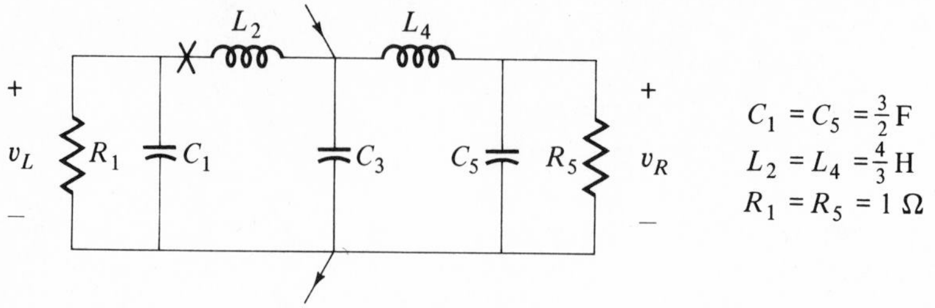

FIG. 7-20

7-21. Explain how, in general, the parameters of the impulse response of a network in the form $\sum Ke^{pt}$ determine the *zeros* of its transfer function. Illustrate by finding the zeros when the impulse response has each of the following simplified forms (1) as given, (2) after reversing the sign of one of the two terms.

$$(a)\ \ e^{-t} - e^{-2t}, \qquad (b)\ \ e^{-t} + e^{-2t}\cos t,$$
$$(c)\ \ 2e^{-t} - e^{-2t}, \qquad (d)\ \ 2e^{-t} - \cos t.$$

7-22. The responses of three networks, from rest, to very short pulses of unit "effect" (area) are found, by experiment, to be representable by the form $r_\delta(t) = (K_1 e^{-t} + K_2 e^{-2t} + K_3 e^{-3t})u(t)$. The values of the *K*'s are given below. Which of the networks can be modeled exactly (except perhaps for a constant multiplier in transfer function or impulse response) by a cascade of noninteracting *RC* and *RLC* networks? Find, for each network, the forced component of the response from rest to (*a*) sinusoidal excitation of frequency $\omega_F = 1$ and unit amplitude, (*b*) unit step excitation, (*c*) exponential excitation of increasing size, $E_0 e^{0.5t}$, (*d*) exponential excitation of decreasing size, $E_0 e^{-0.5t}$. Why is it much more difficult to find the *natural* parts of the responses?

Network	K_1	K_2	K_3
A	+1	−5	+5
B	−1	+4	−3
C	+3	−30	+35

7-23. An interesting theorem from network synthesis states that *if* $f(t) = \sum K_j/(s - p_j)$, in which all the p_j have negative real parts, are simple (noncoincident), and if complex occur in conjugate pairs with conjugate *K*'s, *then* $f(t)$ can be realized as the impulse response of a (passive) network that uses only *RC* and *RLC* circuits. Prove the theorem, making use of a bridge network whose "detector" arm has infinite impedance and whose excitation is by a current (infinite-impedance) source. (Such a realization is not necessarily efficient; it may be wasteful and otherwise impractical, and there may well exist better realization techniques and networks.)

7-24. In the pair

$$\frac{(as + b)}{(s + \alpha)^2 + (\beta)^2} \sim e^{-\alpha t} \cos(\beta t + \phi),$$

how does the initial phase ϕ depend on a? or b? What is ϕ for $a = 0$? For $b = 0$? Sketch ϕ versus b for $a = 1$ (include all values possible, positive and negative).

7-25. Show that the transfer function of the network of Fig. 7-25 has the same form as that of the parallel *RLC* circuit (Chap. 3), $(1/C)s/(s^2 + 2\alpha s + \omega_0^2)$. What is the maximum possible value of its Q, defined as $\omega_0/(2\alpha)$? To what part of the plane are its natural frequencies therefore restricted? Find the impulse and step responses when $R_1 = 5000\ \Omega$, $C_2 = 0.01\ \mu\text{F}$, $R_3 = 1000\ \Omega$, $C_3 = 20$ pF. If possible, use approximate methods to simplify the work, *explaining why* this is possible. (An accuracy of 2 percent is ample.)

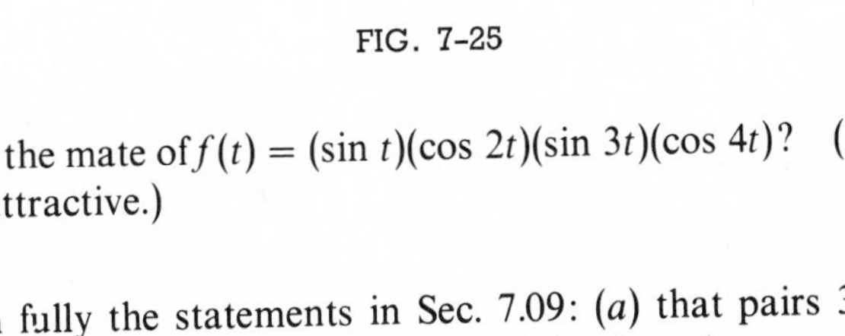

FIG. 7-25

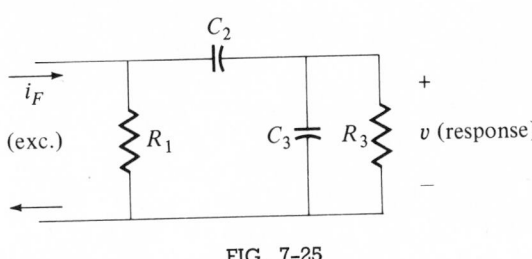

7-26. What is the mate of $f(t) = (\sin t)(\cos 2t)(\sin 3t)(\cos 4t)$? (The exponential form may be attractive.)

7-27. Explain fully the statements in Sec. 7.09: (*a*) that pairs 30 and 31 "correspond to exciting a nondissipative ... to infinite forced response," and (*b*) "excitation of a finite-*Q* *RLC* circuit ... then decays."

7-28. Obtain the mate of each of the time functions given below. Present your result in the form of the quotient of two multiplied-out polynomials. In each case how many initial values must be computed from the time-domain function and from your result, to obtain a complete check on your work? How many do you judge will give a satisfactory check? Why? Perform such a check.

(*a*) $3 \sin 3t - 4 \sin 4t$, (*b*) $e^{-2t} + 2 \sin t$,

(*c*) $3e^{-t} \cos 3t + 2e^{-t/2} \cos 2t$, (*d*) $e^{-t} \cos 2t + e^{-2t} \cos t + e^{-3t}$,

(*e*) $e^{-t} \cos 2t + e^{-2t} \sin t + e^{-3t}$, (*f*) $e^{-t} - e^{-2t} + e^{-3t} - e^{-4t}$,

(*g*) $e^{-t}(1 + 0.1 \sin 0.2t) \cos 10$.

(Recall, from trigonometry, the expression for the product of two sinusoids in a here more useful form, *or* simply use exponentials.)

7-29. The response of a network, initially at rest, to the excitation $E_0 e^{s_0 t} u(t)$ is $r(t)$. Suppose the network now to be excited by the derivative of that excitation, $[E_0 s_0 e^{s_0 t} u(t) + E_0 \delta(t)] T_0$. (The constant T_0 rectifies the dimensions.) Find the response thereto in terms of $r(t)$ and $r'(t)$, giving a full explanation of difficulties encountered and how you overcome them.

7-30. Verify, using pair 2, all the pairs of Fig. 7.07-C. Show also how mathematical induction can be used to demonstrate pair 13, given pair 10. Explain the relation of these pairs to pair 5.

7-31. Show that pair 8 is valid for functions of the pair-4 type, using only pair 3 and (simple) evaluation of a convolution integral.

7-32. Pair 9 can be derived (and the impulse function can be defined) by limiting processes in countless ways. Equations (7.09-4) and (7.09-5) give one example. Show that each of the following pairs gives another derivation. Illustrate with sketches and carry out the limiting process, first verifying the area from zero to infinity in the time domain, and the pair itself.

(*a*) $$\frac{u(t) - u(t - T_0)}{T_0} \sim \frac{1 - e^{-T_0 s}}{T_0 s} \qquad T_0 \to 0,$$

(*b*) $$\alpha e^{-\alpha t} u(t) \sim \frac{\alpha}{s + \alpha} \qquad \alpha \to \infty,$$

(*c*) $$\frac{1}{T_0}\left[\frac{t}{T_0} e^{-t/T_0} u(t)\right] \sim \frac{1}{(1 + T_0 s)^2} \qquad T_0 \to 0,$$

(*d*) $$\frac{2t}{T_0^2}[u(t) - u(t - T_0)] \sim \frac{2}{T_0^2 s^2}[1 - (1 + T_0 s)e^{-T_0 s}].$$

7-33. Use the initial-value tableau to calculate from the s-domain members of pairs 20 to 30 inclusive the first three nonzero initial values of the t-domain member; check the pairs by direct calculation in the t domain.

7-34. Use pair 18 to derive pairs 23, 24, and 25 from pair 22.

7-35. Derive pairs 30 and 31 by setting α equal to zero in pairs 26 and 27. Is justification of the validity of this step necessary? Derive pairs 30, 31, and 32 by partial differentiation with respect to a parameter in pairs 20, 21, 22.

7-36. Show that the mate of $(t^n \sin \beta t)$ is

$$\frac{n!\left[\dfrac{n+1}{1}s^n\beta - \dfrac{(n+1)n(n-1)}{(1)(2)(3)}s^{n-2}\beta^3 + \dfrac{(n+1)n(n-1)(n-2)(n-3)}{(1)(2)(3)(4)(5)}s^{n-4}\beta^5 + \cdots\right]}{(s^2+\beta^2)^{n+1}}$$

Find similar formulas for the mates of $(t^n \cos \beta T)$ and $[t^n \cos (\beta t + \gamma)]$. (*Hint:* pair 35 and the binomial theorem may be useful.)

7-37. Of a network we know only that its (normalized) transfer function is

$$T(s) = K\frac{s^3+1}{s^5+5s^4+9s^3+9s^2+3s+1}.$$

It is excited (from rest) by the sinusoid $E_m \sin \omega_F t$, $\omega_F = 1$.

(*a*) Calculate the first two nonzero initial values of the response and sketch its beginning versus t.
(*b*) Calculate the steady-state response and express it both as a function of time and as a phasor; sketch it, along with the excitation, in both forms.
(*c*) Sketch the complete response versus time, from $t = 0$ to the steady state, as best you can with only the results of (*a*) and (*b*).

7-38. For the network analysis problems below (see Fig. 7-38), carefully examine every path in Fig. 7.21-A. Explain why each one is usable or not usable. Where there is a choice, compare the labor involved in each with that of the others. The initial condition is rest; the excitations to be considered are: (*a*) e^{st}, (*b*) $\delta(t)$, (*c*) $e^{-(t/T_0)^2}$, (*d*) $J_0(\alpha t)$, (*e*) $\cos \omega_F t$.

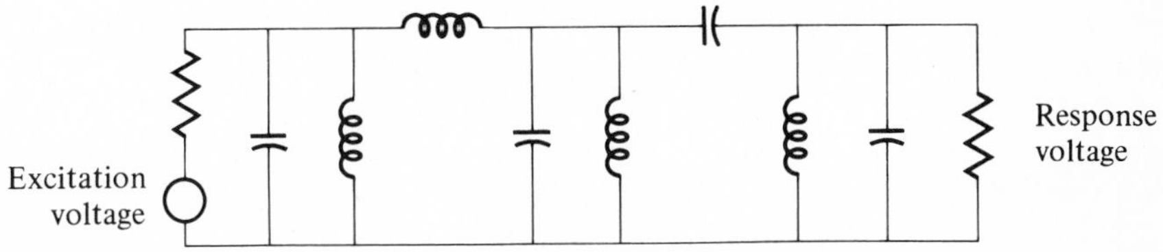

FIG. 7-38

7-39. In what ways do the curves of $|T(j\omega)|$ versus ω differ for the two networks of Fig. 7-39? The schematic diagram is the *same* for both, with element values such that the low-frequency and high-frequency asymptotes are the same for the two networks, but their natural frequencies lie as shown. Explain with sketches.

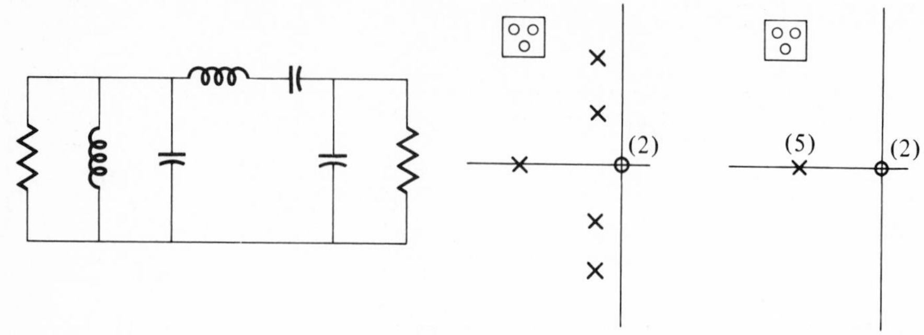

FIG. 7-39

7-40. In the network of Fig. 7.13-A let the (normalized) natural frequencies be $(-1+j0)$, $(-1 \pm j1)$, and let $T(0) = 0.5$. Show that the two resistor values must be equal; let their values be unity. Show now that the remaining three element values must be roots of certain algebraic equations; find their values. Do you encounter more than one solution? Can you determine whether these three element values are rational or irrational numbers? (In network synthesis there exists a much more straightforward method for determining these element values; a certain amount of numerical calculation cannot however be avoided.)

7-41. The two functions shown in Fig. 7-41 are, both of them, eminently reasonable excitation waves. Why is analysis of a network's response to (*a*) simpler than that of its response to (*b*)? Explain carefully why only one of the two cases can be readily handled by using the transfer function, the mate of the excitation, partial-fraction expansion, and either a displaced replica of that response or an initial-condition-no-excitation response. What method(s) have we that can handle the other? Repeat for the two excitations $e^{-\alpha^2 t^2}u(t)$ and $e^{-\alpha t}u(t)$.

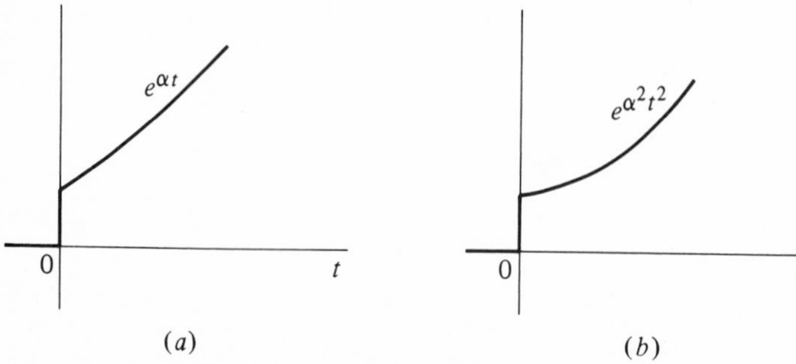

FIG. 7-41

7-42. Suppose the response of a network (from rest) to the excitation $e_1(t)$ is $r_1(t)$. Both $r_1(t)$ and $e_1(t)$ are known formally but you have no other information. Describe (working as far as possible in the s domain) how to find the response of the same network from rest to a different excitation $e_2(t)$. Illustrate by finding the response r_2 to the excitation e_2 in each of the following cases ["$u(t)$" is omitted].

Case	$e_1(t)$	$r_1(t)$	$e_2(t)$
1a	$e^{-t} + e^{-2t}$	e^{-2t}	e^{-3t}
1b	"	"	$\delta(t)$
1c	"	"	$\cos 4t$
2a	$\sin 5t$	$e^{-t} + e^{-2t} + 0.2 \sin 5t$	$\cos 5t$
2b	"	"	$\sin 2t$
2c	"	"	$u(t)$
3a	$e^{-\alpha t}$	$Ae^{-\alpha t} + Be^{-\beta t}$	$\delta(t)$
3b	"	"	$e^{-\beta t} \cos \beta t$
3c	"	"	$e^{-\alpha t}[u(t) - u(t - 1/\alpha)]$

Why can you not find the network's responses from nonrest initial conditions? Can your procedure be used if you have only *numerical* (as opposed to formal) data on e_1 and r_1? (See Prob. 4-W.)

7-43. If the calculation of $r_2(t)$ in Exercise 7-42 is to be performed by *convolving* $r_1(t)$ with some function $f(t)$, what must $f(t)$ be in each case?

7-44. The calculations of partial-fraction expansion must essentially be performed according to Sec. 7.12. But certain variations have their appeal. Complex numbers can be completely avoided, once the denominator of $R(s)$ has been factored, by solving simultaneous linear algebraic equations in real variables. The two terms in the expansion that correspond to two (conjugate) complex poles can be combined into the form $(as + b)/(s^2 + 2\alpha s + \omega_0{}^2)$ and the two corresponding unknowns, a and b, can be found by establishing and solving simultaneous equations by setting s equal to some convenient real values (such as 0, +1, −1, ..., ∞); other "K's" in the expansion may be found at the same time by enlarging the number of equations to be solved. Work out the method for

$$R(s) = \frac{A_0 s^2 + A_1 s + A_2}{(s^2 + 2\alpha s + \omega_0{}^2)(s + \beta)} = \frac{K_1}{s + \beta} + \frac{as + b}{s^2 + 2\alpha s + \omega_0{}^2}$$

or

$$A_0 s^2 + A_1 s + A_2 = K_1(s^2 + 2\alpha s + \omega_0{}^2) + (s + \beta)(as + b).$$

Then illustrate specifically for $R(s) = (3s^2 + 2s + 4)/[(s + 1)(s^2 + s + 10)]$.

An alternative finds complex K's with only moderate use of complex numbers thus: by using the initial-value tableau, find the values of the right-hand sides in the simultaneous equations below:

$$K_1 + K_2 + K_3 = (\ldots),$$
$$p_1 K_1 + p_2 K_2 + p_3 K_3 = (\ldots),$$
$$p_1{}^2 K_1 + p_2{}^2 K_2 + p_3{}^2 K_3 = (\ldots).$$

Solve these for the K's. Compare this method of obtaining the values of the K's with that above and those of Sec. 7.12. To what extent do they differ? What are the advantages of each? Illustrate by placing the expansions of the example above, made by each method, side by side.

7.26 PROBLEMS

7-A. Analyze the network of Fig. 5.14-A, taking the end inductors to have unit inductance, the other inductors to have inductance 2, the capacitors to have capacitance 2, the source resistor to be 4 and the load resistor to be 10 (all values are, of course, normalized ohms, henrys, farads). In this way verify Fig. 4.09-A.

7-B. (*a*) Here are sinusoidal-steady-state phase-shift measurements made when the photographs of Fig. 7.19-G were taken:

Frequency kHz	**Phase, Degrees Lag**
6.0	90
11.8	180
16.5	270
21	360
24	450

From these estimate the value of $\beta'(0)$. Does Fig. 7.19-G*c* confirm (7.19-24) or not? Explain.

(*b*) When the photographs of Fig. 7.19-H were taken, rough phase-shift measurements gave the following as the frequencies of 0, 90, 180, 270, and 360° phase lag, respectively: 12.8, 15.7, 18.2, 20.6, 23.7 (kHz).

Evaluate $\beta'(\omega_0)$ approximately, ω_0 being more or less at the center of the "pass" frequency band. Can you correlate this value with Fig. 7.19-H*b* in any way? (Consider "delay" and oscillation frequencies.) Explain.

7-C. Refer to Fig. 7.16-E for the actual transfer function and impulse response of a network whose schematic is shown here (Fig. 7-C).

How many points of infinite loss are there? Where are they?

How many points of infinite gain (natural frequencies) are there? Sketch a reasonable set of locations therefor, consistent with Fig. 7.16-E*a*. Then sketch several other, different, possible sets of locations and the corresponding $|T(j\omega_F)|$ versus ω_F curves. Explain each.

Correlate Fig. 7.16-E*b* with Fig. 7.16-E*a* as well as you can, given the scale information below. Can you distinguish the two principal natural frequencies in (*b*)? Can you distinguish the other natural frequencies? Explain. [*Data* for Fig. 7.16-E: (*a*) scale is 500 Hz per line; the marker at the left is zero frequency; (*b*) scale is 50 μs/cm (per square).]

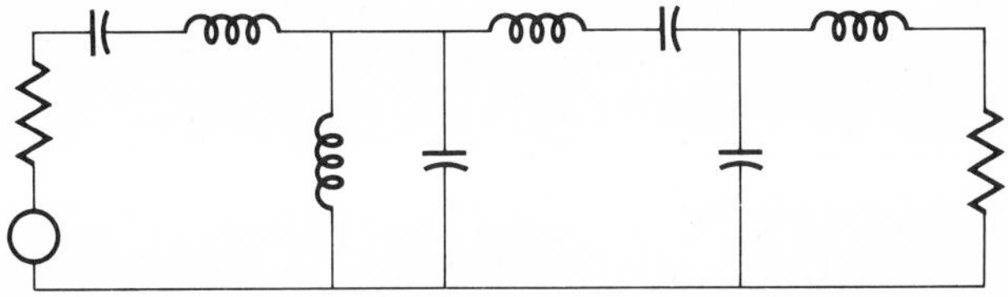

FIG. 7-C

7-D. The impulse-response wave shown in Fig. 7-D is vertically magnified to show the nature of the oscillations (the initial spike is quite large in comparison). Check the network's behavior at infinity against the initial impulse-response behavior. Do they agree? Explain. Check the final behavior of $r_\delta(t)$ in the same way. Is the area under the impulse-response curve zero, approximately? Should it be? Explain, and sketch the step-response curve.

The scales of the photographs are: 20 μs/cm (large division) and 1.25 kHz per bar. The exciting impulse identifies the time origin; the frequency origin is the leftmost (shorter) bar. What is the band of elimination of frequencies? What, approximately, are the resonant frequencies of the tuned circuits? How do they produce the band-elimination effect? Is there any correlation between the oscillations of the impulse response and the $|T|$ outline? Should there be?

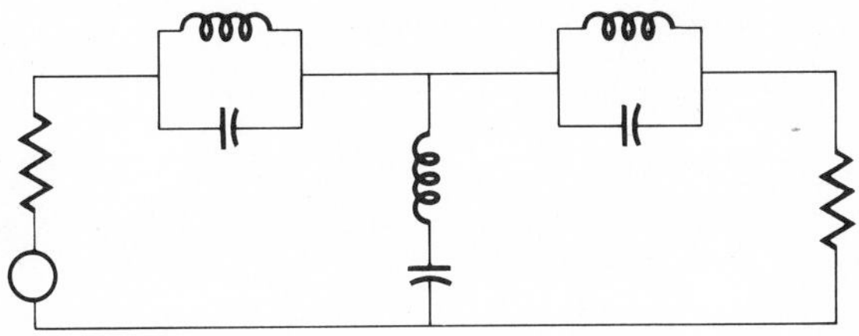

FIG. 7-D

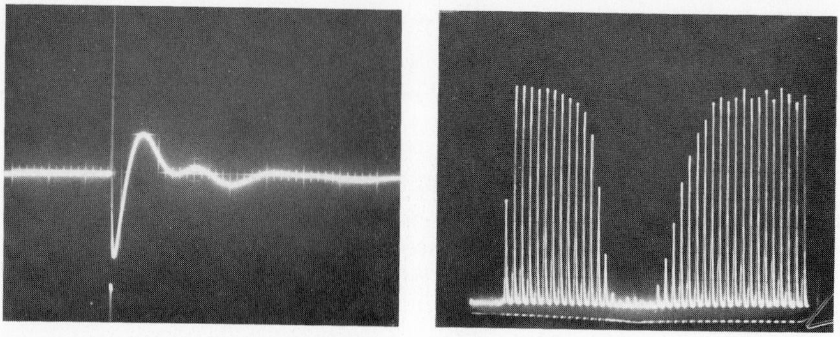

FIG. 7-D. (*Con.*)

7-E. The network shown in Fig. 7-E is excited sinusoidally. The two photographs show the excitation e and the response v waves for two cases: (*a*) $C' = 0.01\ \mu F$, (*b*) $C' = 100$ pF. The frequency of excitation, 25 kHz, is the resonant frequency (where v is a maximum) for case (*b*).

In each case, which wave is the excitation and which is the response? The scales are the same in the two photographs: 4 μs/cm (division) in time and arbitrary but unchanged in amplitude; the excitation e is not changed.

Explain fully why the reduction of C' from a practically infinite value [in (*a*)] to a moderate value [in (*b*), q.v.] *advances* the response essentially 90°. Use physical and s-plane arguments.

Discuss the change in location of the resonance between the two cases.

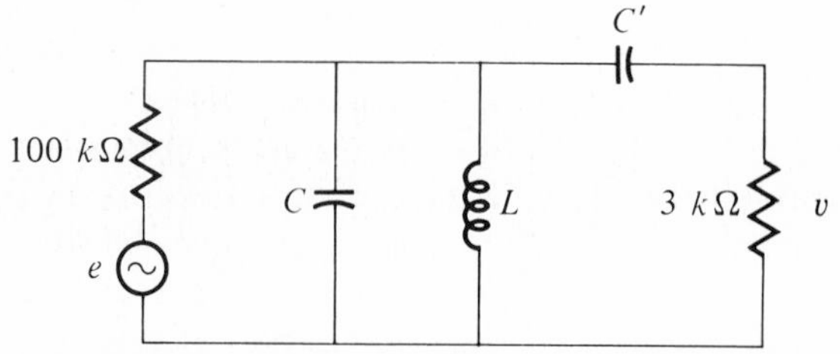

$C = 0.016\ \mu F$

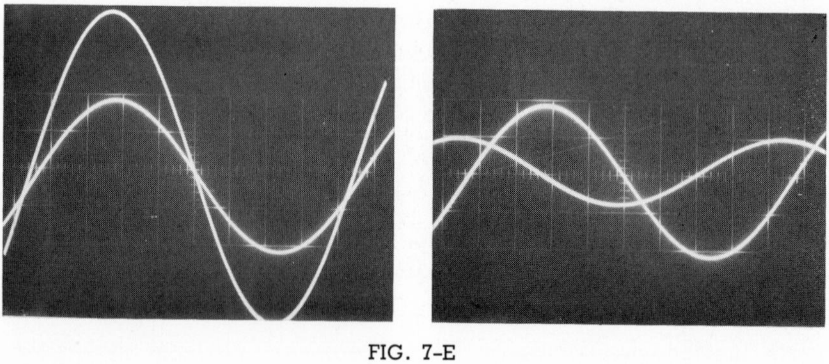

FIG. 7-E

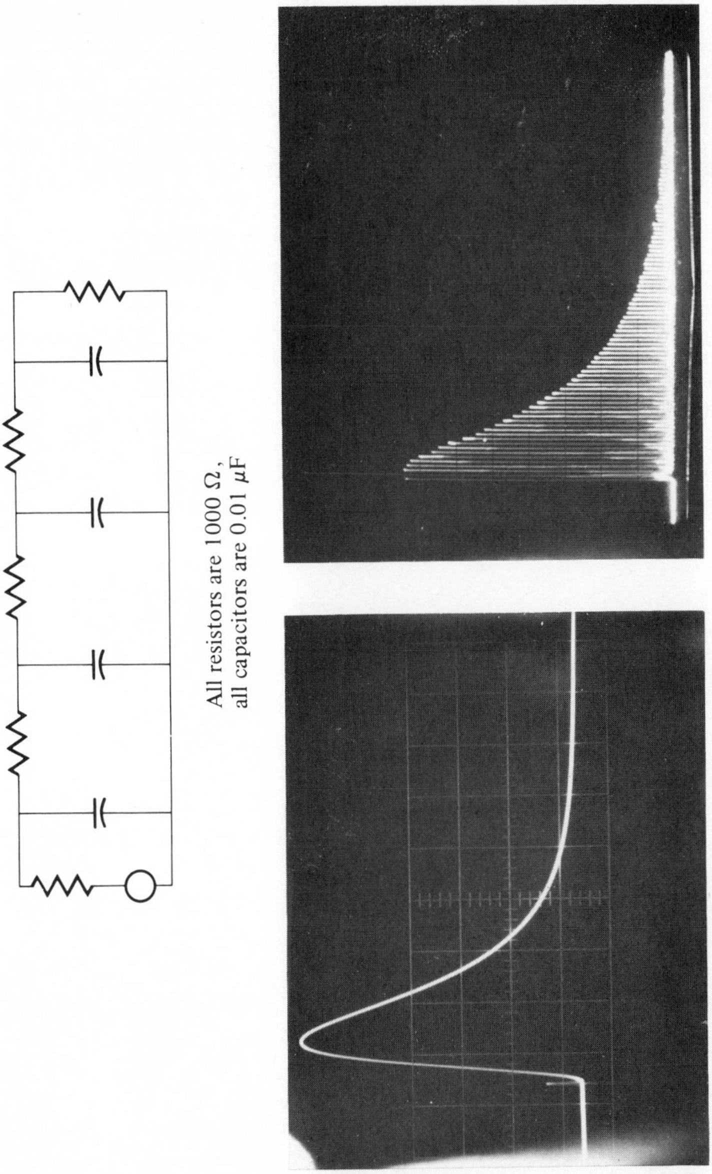
All resistors are 1000 Ω,
all capacitors are 0.01 μF

FIG. 7-F

7-F. The network of Fig. 7-F (p. 697) has no inductance; its behavior in both domains is therefore comparatively simple, in no way dramatic. The scales are 25 μs/cm (division) and 500 Hz per bar; the time origin is marked by the exciting impulse "spike," the frequency origin by a zero-frequency marker "spike."

Could these be photographs of the behavior of a simple, one-R one-C circuit? Explain your answer carefully, using the "time constant" and the "half-power frequency" of the photographs.

Using the element values given on the schematic, compute both curves and compare your results with the photographs. Discuss discrepancies and agreements.

Point out, in the computed results, the relative importances of the several natural frequencies in the impulse response, at various portions of the time scale. Repeat, in an analogous way, in the frequency domain.

7-G. The scales in the photographs (Fig. 7-G) are 40 μs/cm and 1 kHz per bar (zero frequency is marked by the large bar at the left, zero time by the spike).

How many zeros does $T(s)$ have at infinity? Does the initial behavior of $r_\delta(t)$ conform thereto? Explain. What is the principal oscillation frequency in

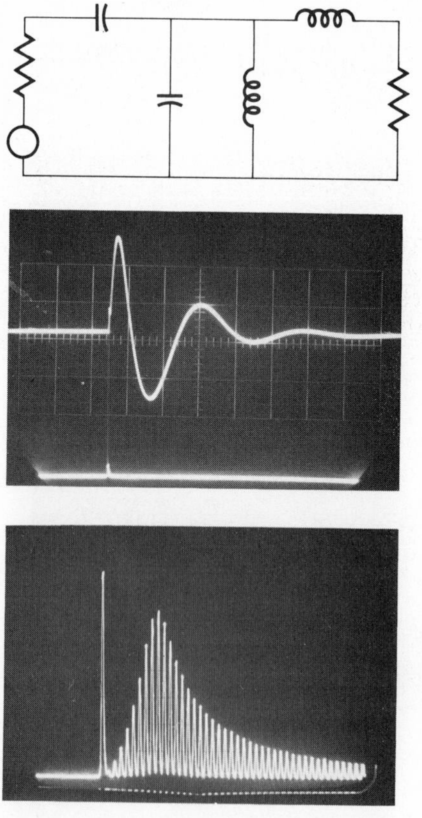

Impulse response, scale is 40 μs/cm (large div.)

$|T(j\omega_F)|$,
Scale is 1 kHz per bar, zero frequency is at large bar at left

FIG. 7-G

$r_\delta(t)$? What are the principal natural frequencies, roughly? Does the $|T|$ outline confirm your answers?

Assume reasonable values for the elements (say, for example, that the resistors are equal and the reactances of the L and C elements are equal to the resistor values at the principal oscillation frequency). Compute both $|T|$ versus ω_F and $r_\delta(t)$. How well do your results agree with the photographs? Can you improve them?

7-H. The two photographs in Fig. 7-H exhibit $|T(j\omega_F)|$ versus ω_F, and $r_\delta(t)$ versus t for a band-pass network. Estimate the number of natural frequencies and their locations. From the initial behavior of $r_\delta(t)$ estimate the number of zeros of $T(s)$ at infinity. Explain your reasoning. Do all your answers, thus far, cohere?

The scales are 40 μs/cm (division) in time, and the excitation impulse is seen at the left; zero frequency is marked by the isolated bar at the far left, and the bar separation is 1 kHz. What relation can you find between the "oscillation frequency" of $r_\delta(t)$ and the $|T|$ plot? Explain.

Either by physical construction and measurement or by computer calculations try to reproduce these photographs by a "real" network.

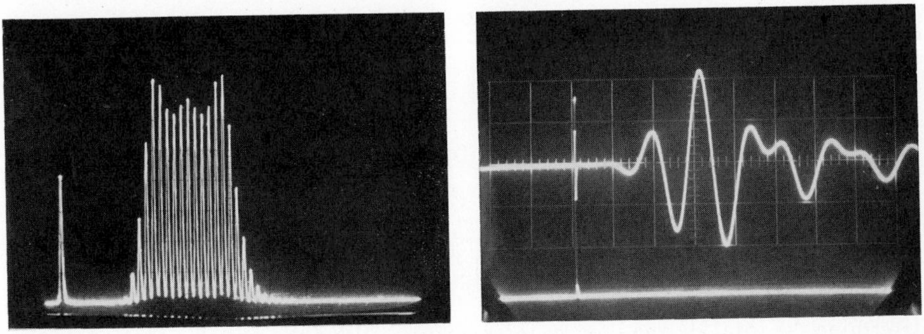

FIG. 7-H

7-I. Consider a network whose points of infinite loss (zeros of transmission) lie far enough from the origin that it may properly be called *low-pass*. In its transfer function

$$T(s) = \frac{A_0 s^{N-1} + A_1 s + \cdots + A_{N-2} s + A_{N-1}}{s^N + B_1 s^{N-1} + \cdots + B_{N-1} s + B_N}$$

why are A_{N-1} and B_N not equal to zero? Show, by considering the behavior of $T(s)$ near the origin, that the "delay" [see (7.19-10)] experienced in passing through the network by signals whose important frequency ranges are "low," is

"Delay" = the sum of the reciprocals of the points of infinite loss
– the sum of the reciprocals of the natural frequencies.

[*Note:* (5.05-4) may be useful here.] Verify this in the example of Fig. 7.19-A. Construct other examples and verify the statement thereby.

Show also that this "delay" may be expressed as $\int_0^\infty t r_\delta(t)\,dt$ (known as one of the *moments* of the impulse response). [*Suggestion:* find the values of $T(0)$ and of $T'(0)$ from (7.05-1).]

7-J. Network analysis is not always: given e, find r. Sometimes it is: given r, find e, as, for instance, in the calibration or correction of distortion in instruments (measurements). For example, a ballistic galvanometer is an instrument for measuring the total charge delivered to it by a (brief) current pulse: the swing of the indicating needle is supposed to be proportional to the charge in coulombs. The mechanical parts of such an instrument, with some simplification, can be modeled by an electrical tuned circuit to which the current pulse is applied. Take its parameters as $Q = 1$, $R = 1000\ \Omega$, resonant frequency $1/(2\pi\sqrt{LC}) =$ 0.5 Hz. Ignore inductor dissipation. The input current pulse is not square, but neither is it very remarkable in shape: it is merely widened out. [The analogs, incidentally, are: voltage of model *to* angular velocity of indicator (needle), inductor flux linkages of model *to* angular position of the indicator.]

Work here entirely in terms of the electric analog (model) for simplicity. A pulse of somewhat sloppy but recognizably pulselike shape produces (from rest) a voltage wave that rises, falls, sighs a little, and decays, with these characteristics, carefully measured:

$$v(0) = 0, \qquad v'(0) = 0, \qquad v''(0) = +0.8\ \text{V/s}^2, \qquad v(\infty) = 0.$$

Careful analysis of the output voltage wave shows that it consists of a damped exponential (due to the instrument, or the tuned circuit) and two exponentials, of time constants 1 and 2 seconds. Plot the input current pulse waveform and determine the total charge therein in coulombs. Plot the angular motion of the galvanometer indicator versus time and obtain its calibration constant (number of coulombs per maximum degree swing of indicator) in terms of the modeling scale factor K (degrees per weber-turn).

7-K. Calculate the coefficients K for the response of Fig. 7.15-F in the form $\sum_1^6 K_j e^{p_j t}$. Obtain also the first four nonzero terms of the power-series form of the response. Compare the utility of the two forms for calculating the values of the response for $t_n = 0.5, 1.0, 1.5, 2.0, 2.5, 3.0$. Why is the exponential form not necessarily the best to use for small values of time? To what precision must each K be calculated to obtain three significant figures correctly in each of the six values of the response?

The (rounded-pulse) excitation is $C_1 e^{-0.5t} + C_2 e^{-0.6t} + C_3 e^{-0.7t}$ in which the coefficients C are chosen (*a*) to make the initial values of pulse and first derivative equal to zero and (*b*) to make the total area under the pulse equal to unity. The network transfer function is (7.15-1).

7-L. (*a*) Consider the network of Fig. 7.13-A, with the (real) element values of (7.14-31). Let the transfer function be $T = K/(s^3 + B_1 s^2 + B_2 s + B_3)$, from voltage to voltage, or from current to voltage ad libitum. Let its partial-fraction expansion be

$$T(s) = \frac{K_1}{s - p_1} + \frac{K_2}{s - p_2} + \frac{K_3}{s - p_3}.$$

Show that the three K's and the three p's are of the same order of magnitude. What are their dimensions? Let the excitation be exponential, $E_0 e^{s_0 t} u(t)$, and the initial condition be rest. In the response thereto, $r(t) = K_0 e^{s_0 t} + K'_1 e^{p_1 t} + K'_2 e^{p_2 t} + K'_3 e^{p_3 t}$, show that the four K's have the same dimensions as does E_0, and the same order of magnitude as E_0.

(*b*) Now starting afresh from the network of Fig. 7.14-A*a* with the normalized element values of (7.14-34), repeat the analysis of (*a*), answering the same questions but using normalized time and frequency variables t_n and s_n. Show that the four K's are identical to those found in (*a*).

(*c*) Place the essentials of the two analyses in parallel columns and compare them.

(*d*) Repeat (*c*) with this change: let the excitation be impulsive, $Q_0\,\delta(t)$ or $\Lambda_0\,\delta(t)$. Where, as you go through the analyses, do you find dimensional differences from the work of (*c*)? Why do the final K's differ in the two columns? What difference is there between *unit* impulse response (that for $Q_0 = 1$ or $\Lambda_0 = 1$) in the two columns? Explain carefully.

(*e*) Repeat (*d*) with excitation $Q_0\,\alpha e^{-\alpha t}$ or $\Lambda_0\,\alpha e^{-\alpha t}$, sluggish pulses of unit area. [Use the results of (*a*), (*b*), and (*c*).] In the light of this work, explain the anomaly found in (*d*).

7-M. The natural frequencies of a network are:

$$-9.94987 + j0.0, \qquad -0.00366 \pm j0.92401, \qquad -0.02140 \pm j0.3830;$$

all of the zeros of the transfer function lie at infinity. What is the slope of the high-frequency asymptote of $|T(j\omega)|$ in logarithmic measure when plotted on a logarithmic frequency scale? What is the frequency where the high-frequency asymptote of $|T(j\omega)|$ intersects its low-frequency asymptote? Sketch $|T(j\omega)|$ versus ω explaining departures from the asymptotes. Now let the network be excited, from rest, by a step function. In the response expression $\sum_1^5 K_j e^{p_j t} + K_0$, what is the value of K_0? The usual, straightforward (hand) evaluation of the other five K's is tedious. Show that, for cases like this one, they can be evaluated approximately but rapidly, and with very little error, from values of the transfer function *on the ω axis*, that is, from values of $T(j\omega)$. Evaluate the five K's by hand, within 2 percent. Sketch the behavior of the response versus time. Now repeat the analysis by machine and compare results.

7-N. A theorist proposes three different configurations (Fig. 7-N) for the five natural frequencies of a network, all of whose zeros of transmission are to lie at infinity. Carefully compare the three propositions in each of these respects: (*a*) in the frequency domain, the behavior, in magnitude and in angle, of $T(j\omega)$; (*b*) in the time domain, the form of the impulse and step responses. What do you suppose the theorist had in mind in each case?

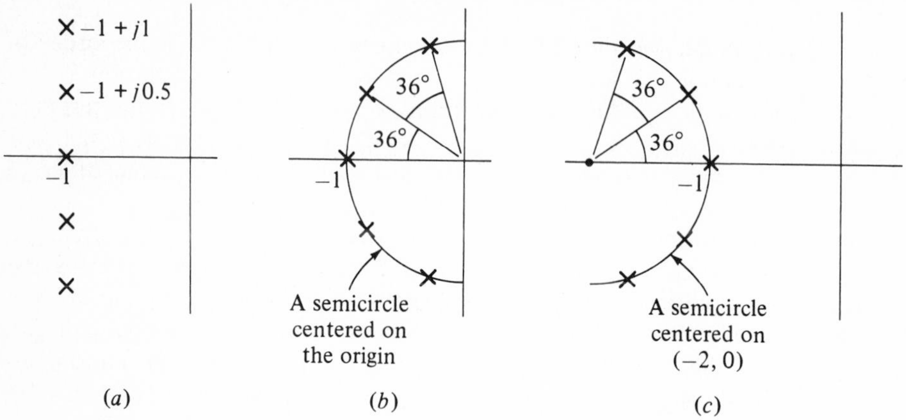

FIG. 7-N

7-O. Figure 7-O shows an amplifier composed of three identical stages of the "shunt-peaked" type. For simplicity, let the stages be identical, with $R = 1\ \Omega$, $L = 0.4$ H, $C = 1$ F, $g_m = 10$ mhos; let the other three unmarked resistors have unit value also.

(*a*) Plot the gain (magnitude of transfer function, voltage to voltage) for a single amplifier stage (1) with $L = 0$, (2) with $L = 0.4$ H. Explain why it is aptly denoted "shunt-peaked."

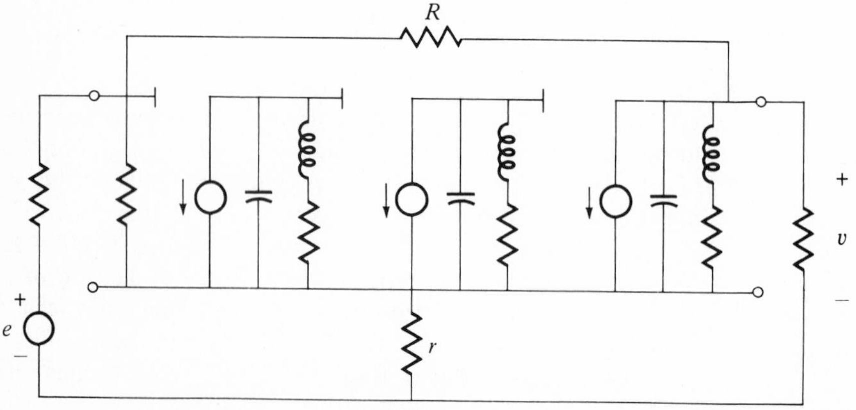

FIG. 7-O

(*b*) Carefully prepare plots of the magnitude of the transfer function from e to v for $s = j\omega$, versus ω, for three cases: (1) $r = 0$ and $G = 0$ (no feedback), (2) $r = 0$, $G = 0.01$ (shunt feedback), (3) $r = 0.01$, $G = 0$ (series feedback). In (2) and (3) use the 0.01 (mho or ohm) value given only if the amplifier is then stable; if it is unstable, reduce G or r until it becomes stable. With the aid of your plots, discuss generally the effects of the two kinds of feedback, at least for this example.

7-P. Consider the impulse response of Fig. 7.19-E, which displays the phenomenon of *beats*.

(*a*) Neglecting the damping, show how two (or three) sinusoids of nearly equal frequencies can alternately reinforce and cancel so that "pulses" of oscillation occur in their sum. (Trigonometric formulas for expanding the sine and cosine of the sum of two angles may be helpful.) Illustrate with sketches as well as formal results. Explain the term "beat."

(*b*) Show how the presence of (light) damping need not remove the beat phenomenon, though it may become attenuated.

(*c*) Verify that your analysis is relevent to Fig. 7.19-E.

(*d*) Suppose a network to have two neighboring high-Q resonances, the natural frequencies being $p_1 = -0.01 \pm j1.01$ and $p_2 = -0.01 \pm j0.99$. The transfer function has two zeros at the origin and two at infinity. In the impulse response point out and explain all interesting features. Repeat for the step response. Repeat for the response to sinusoids of frequency 0.99, 1.00, 1.01.

7-Q. Refer to Fig. 7.19-I*a*. Find the response of the all-pass network to the sluggish pulses (*a*) Kte^{-t}, (*b*) Kte^{-2t}, (*c*) Kte^{-4t}, (*d*) Kte^{-5t}, and compare with the other parts of Fig. 7.19-I. Point out and explain the interesting features.

7-R. Apply a sinusoidal signal of frequency $w_F = 1$ to the band-pass network of Fig. 7.19-B. Can you find, in the response, an indication of a "delay" (of about 44 normalized seconds)? Explain.

7-S. Consider the high-pass network of Figs. 7.18-A and 7.18-F. Apply first a "sinusoidal step" $E_m(\cos \omega_F t)u(t)$, and then a "wave packet" or "sinusoidal pulse" $E_m(\cos \omega_F t)[u(t) - u(t - T)]$. Discuss the responses as to distortion, and "delay" if the word can be made meaningful. Choose informative and reasonable parameter values.

7-T. For each of the "*RC* filter" networks shown in Fig. 7-T determine the frequency at which the sinusoidal-steady-state response is shifted 90° in phase from the excitation. Plot $|T(j\omega)|$ in dBV versus ω for each network. Compare with the corresponding curve of Fig. 7.19-A. Both sorts of networks may be called *low-pass filters;* point out and explain, in terms of network configurations, the differences between those of Fig. 7-T and that of Fig. 7.19-A. Now study and compare the distortion produced by the network of Fig. 7.19-A and that produced by network (*c*) of Fig. 7-T; use these signals: (*a*) the impulse (Fig. 7.19-A*c*), (*b*) the step (Fig. 7.19-A*c*), (*c*) the somewhat rounded pulse of (7.15-24), (*d*) a very smooth pulse like that of Fig. 7.15-F. Explain the differences in terms of $|E(j\omega)|$ and $|T(j\omega)|$ as functions of ω.

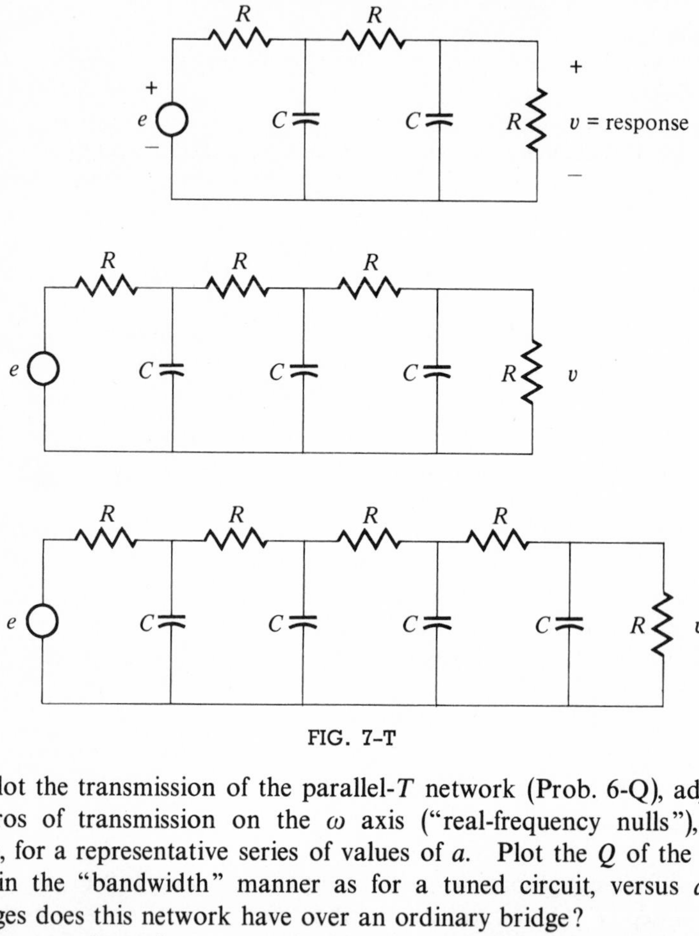

FIG. 7-T

7-U. Plot the transmission of the parallel-*T* network (Prob. 6-Q), adjusted to have zeros of transmission on the ω axis ("real-frequency nulls"), $|T(j\omega)|$ versus ω, for a representative series of values of a. Plot the Q of the network, defined in the "bandwidth" manner as for a tuned circuit, versus a. What advantages does this network have over an ordinary bridge?

Suppose such a twin-*T* network placed in the feedback path of an amplifier: let T_2 in Fig. 6.25-*J* be the parallel-*T*-network transfer function used above. Let the amplifier's forward transmission T_1 be "flat" or constant. What will be the general nature of the amplifier's transfer function? Illustrate with plots of the magnitude thereof for $s = j\omega$, versus ω, for reasonable values of a and T_1.

7-V. The two-port network shown in Fig. 7-V is a high-pass device: all of its zeros of transmission lie at the origin ("at dc"). The input impedance of the terminated network is

$$Z(s) = \frac{4s^5 + 8s^4 + 8s^3 + 6s^2 + 2s + 1}{4s^5 + 8s^4 + 8s^3 + 4s^2 + 2s},$$

normalized to: $R_0 = R_L = R_S = 75\ \Omega$ and $\omega_0 = 10^6$ rad/s. The input signal $e(t)$ is a sloppy pulse made up of two exponentials, $e^{-\alpha t}$ and $e^{-\beta t}$, such that $e(0) = 0$, $e'(0) = +1$, $e''(0) = -1$, $e'''(0) = +\frac{13}{16}$. The same time scale is of course used here also: t is in microseconds.

(*a*) Obtain a plot of the voltage v in response (from rest) to this excitation. Compare it with the signal applied (1) as to waveshape and (2) as to time of occurrence or "delay."

(*b*) Use the facts above and your knowledge of low-frequency asymptotic network behavior to (1) lay out a suitable schematic diagram for the contents of the network and (2) to find the values of its elements. Then (3) calculate its transfer function and check with that used in (*a*).

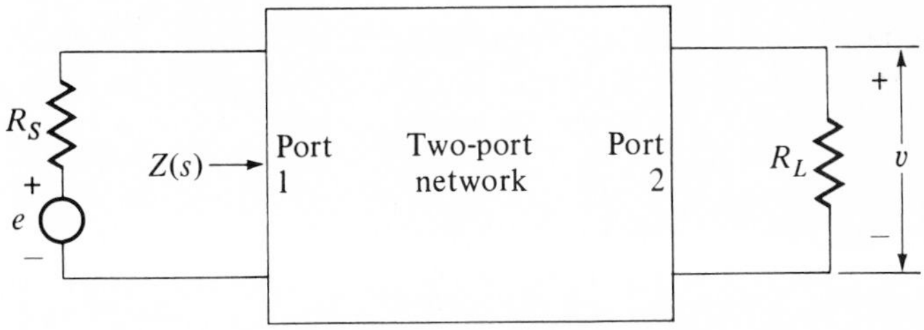

FIG. 7-V

7-W. It is alleged, on the basis of discussion of the mate of the s-domain function whose normalized magnitude, squared, when $s = j\omega$, is $e^{-\omega^2}$, a discussion into which we do not go, that its time-domain mate should be a very smooth function of time

(*a*) Explain why it is difficult, to say the least, to find a network whose transfer function $T(s)$ is such that $|T(j\omega)|^2 = e^{-\omega^2}$. Consider a series of networks with *approximately* such a transfer function, so designed that

$$|T_N(j\omega)|^2 = \frac{1}{1 + \omega^2 + \omega^4/2 + \omega^6/6 + \omega^8/24 + \cdots + \omega^{2N}/N!}.$$

(*b*) Why might such a network transfer function well be considered an approximation of the desired (unrealizable) one? Why is it more likely to be realizable?

(*c*) Find corresponding transfer functions for $N = 1, 2, 3, 4, 5$, in turn. Since $|T(j\omega)|^2 = T(s)T(-s)|_{s=j\omega}$, factoring the denominator of each approximant to $|T|^2$, having first replaced ω^2 by $-s^2$, will lead to the natural frequencies and their negatives; from these the denominator of $T_N(s)$ can be computed. Digest, verify, and explain this statement carefully. Then plot the step response in each case and explain whether, in your judgment, the goal is achieved. [Contrast the step responses with, for example, the step response of a network whose transfer function is $1/(s^2 + 0.2s + 1)$, in which there is pronounced "overshoot."]

7-X. Figure 7-X shows the schematic diagram of a low-pass filter, with source (R_1) and load (R_5) terminating resistors. Below are given two sets of normalized element values therefor (henrys and farads), based on $R_1 = R_5 = 1\ \Omega$ and infinite-Q inductors and capacitors. One set (A) was obtained by a quick but inelegant "conventional" filter theory; the other (B) comes from a very sophisticated design procedure. Prepare, for comparison, detailed plots for each design, of the transfer function at "real frequencies" (for $s = j\omega$) in the following forms: (*a*) magnitude of $T(j\omega)$ versus ω, (*b*) angle of $T(j\omega)$ versus ω, (*c*) locus of $T(j\omega)$ in its own complex plane, ω being the (suppressed) parameter. Point out and discuss the differences in performance. Now suppose each inductor has a Q of 50 at (normalized) $\omega = 1$. Find and display the effect thereof, still assuming infinite capacitor Q.

Element	A	B
C_1	0.856	2.190
C_2	0.312	0.151
C_3	1.522	2.777
C_4	0.836	0.410
C_5	0.666	1.975
L_2	0.856	0.952
L_4	0.666	0.815

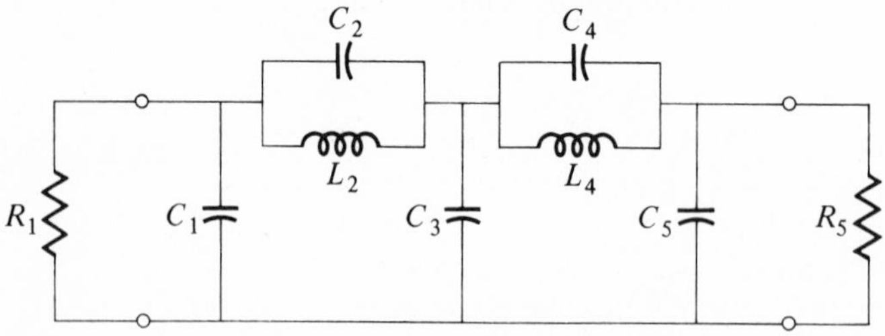

FIG. 7-X

7-Y. Some of the advantages of feedback require that the loop gain, $|T_1 T_2|$, be large for important values of s [see (6.25-10) and the subsequent discussion]. At the same time, of course, it is necessary that the natural frequencies all lie in the left half plane. The two aims frequently conflict, of course. The following simple illustration will give some idea of the difficulty.

Consider a forward path that consists of three identical $g_m RC$ stages, so that in Fig. 6.25-J, $T_1(s) = [-g_m/(Cs + G)]^3$ or, in normalized form, $T_1 = -K/(s + 1)^3$. Let the feedback path be unity: $T_2 = +1$. Note that, for stability, the feedback signal must be *added* (not subtracted) at the junction point. Determine the maximum (positive) value of K for which the amplifier is stable, in three ways as set forth below.

(*a*) The simplicity of the network makes possible immediate determination of the natural frequencies as K varies: sketch their loci (motions) as K increases from zero, and determine the maximum allowable value of K. Note that the change of frequency scale (of frequency reference, $\omega_0 = 1/(RC)$ as $K = (g_m R)^3$ varies is of no consequence since no frequency values here used depend on it.

(*b*) Apply the Hurwitz test of Prob. 5-L.

(*c*) Use the *Nyquist diagram*, a new device which we here briefly introduce.) Sketch the locus of the loop transmission function, $-T_1(j\omega)T_2(j\omega)$, in its complex plane, omitting the negative sign in T_1 as indicated, and letting ω vary from zero to infinity. Do this for a series of increasing values of K, as in (*a*). Note that the curve is a simple one and that instability occurs when the curve has expanded to pass through the critical point $(-1, 0)$. [This point could equally well be at $(+1, 0)$; it depends on whether the minus sign is included or omitted in the calculations; its omission (as here) is somewhat more common.] Describe how the Nyquist diagram, so easily sketched in this example, here graphically determines the maximum allowable value of K. (In general, the Nyquist diagram is more complicated and requires much more careful treatment.)

We now ask: "Can the amplifier be modified to permit an increase of K beyond that value, so that $|T_1 T_2|$ will be larger, at least at low frequencies?" [This is very desirable; see Sec. 6.25.] Sketch $|T_1(j\omega)T_2(j\omega)|$ versus ω, and its angle also. (Here, and in the curves that follow, a logarithmic frequency scale, and logarithmic units for $|T_1 T_2|$, will improve the presentation. Why?)

One might begin, prompted by these curves and the Nyquist diagram, by introducing into the loop transfer function additional factors, factors that will attenuate $|T_1 T_2|$ at low (but not too low) frequencies, hoping thereby to change the Nyquist diagram by pulling it away from the critical point, to the right, and hence to permit a larger value of K. Indicate by sketches what the hope is. Now try incorporating into the loop transfer function the factor that corresponds to multiplying its magnitude by $(1 + \omega^4)^{-1/2}$, namely $(s^2 + \sqrt{2}s + 1)^{-1}$. First sketch the magnitude, angle, and Nyquist curves that you expect and then plot the actual new curves. What is the effect of the new factor on each of the three? What is the new maximum value of K? [The fact that a more rapid cutoff defeats its purpose by introducing more (lagging) phase shift at low frequencies was one of Bode's early discoveries in his famous work on feedback amplifiers.]

Now attempt to improve matters by retaining the faster cutoff at low frequencies but returning to the original cutoff rate before it is too late: incorporate still one more factor, $(s^2 + 0.5s + 1)$, into the loop transfer function. (Note that this restores the original high-frequency asymptote, determined by C and g_m, which is usually fixed by the devices used, anyway.) What is its effect on each of the three presentations? What is now the maximum usable value of K?

Experiment further with loop transfer modifying functions of the form $(s^2 + As + 1)/(s^2 + Bs + 1)$, to see if you can improve the maximum usable value of K. The theoretical maximum depends on more complicated considerations, including of course the frequency *band* (width) over which large loop gain is wanted; into these we do not go; see BO in Appendix B. Nor do we discuss the implementation (realization) of the correcting factors, once decided on.

7-Z. A "smooth" (uniform) transmission line can be thought of as an infinite succession of infinitesimal series inductors and shunt capacitors, distributed uniformly along the line. To thoroughly analyze such a line (a distributed system) requires partial differential equations. But an often adequate model, in lumped form, can be constructed by concentrating the distributed inductance and capacitance as (for example) in the "π model" of Fig. 7-Z. Here L is made equal to half the total distributed inductance of the length of line to be modeled, and C to half the total capacitance. (Certain symmetries and simplifications arise if this "halving" is performed.)

(*a*) For a coaxial cable with the parameters given, find the value of L and C for a single-π model of 1.5 km of transmission line.

(*b*) Suppose this simple model to be excited by a voltage source of impedance R and terminated at the far end in a load resistor, also R. Find the transfer function (voltage to voltage) in terms of L, C, and R (literal elements). Now insert the numerical values and find the numerical coefficients, the A's and B's in the transfer function. If you were to insert such coefficient values into a computer program such as IMPSTEP, what difficulties might well be expected? These difficulties can be circumvented by normalization. Adopt these scalings (normalizations):

$$L_n = \frac{L\omega_0}{R_0}, \qquad C_n = C\omega_0 R_0, \qquad s_n = \frac{s}{\omega_0}, \qquad t_n = \omega_0 t.$$

(*c*) Now return to the transfer function computed in (*b*) above. Show that by proper choice of the two scale factors ω_0 (rad/s) and R_0 (Ω) the transfer function can be altered to the much simpler (and not unfamiliar) form

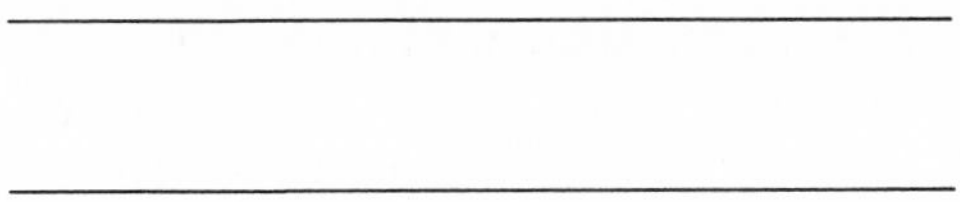

Smooth transmission line (two wires)

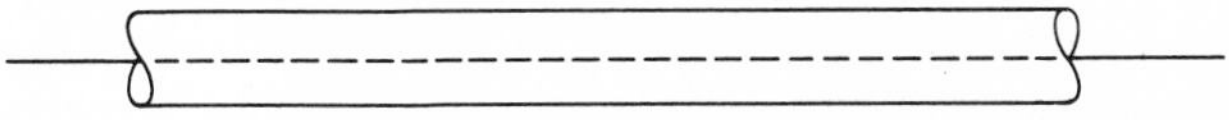

Smooth transmission line (coaxial)

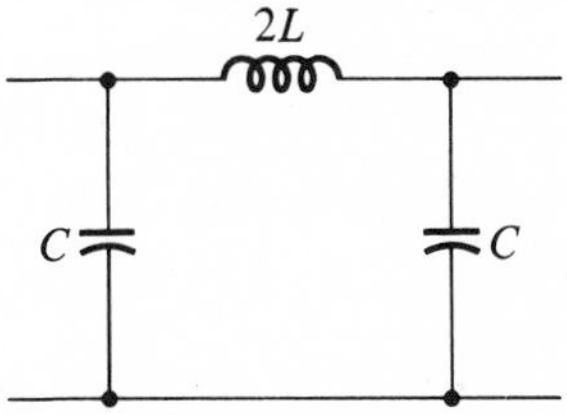

π model of line section

For coaxial line, with a = external radius of inner conductor,
b = internal radius of outer conductor,

the inductance $= 0.461 \log_{10} (b/a)\ \mu\text{H/m}$,

the capacitance $= 24.16/[\log_{10} (b/a)]$ pF/m;

take $a = 1$ mm and $b = 4$ mm.

FIG. 7-Z

$$T(s) = \frac{0.5}{s^3 + 2s^2 + 2s + 1} \quad \text{in which } s \text{ is normalized.}$$

What are the scale factors, and what are the normalized element values of the single-π model and of the terminations? (*Note:* the choice of reference of normalizing values ω_0 and R_0 is entirely arbitrary; the values you have used do, however, seem to "fit nicely" in this problem, no?)

(*d*) Find and plot the impulse response of the three-element π model (with its two resistive terminations) versus t_n. Then relabel the time axis in actual time units, explaining why it is not necessary to replot the curve.

(*e*) Now construct a better model for the 1.5 km of coaxial cable by first dividing it in half and then modeling each half as above (with one π model), then joining the two networks to form a single (5 element) model. If the same normalizing (scaling) factors ω_0 and R_0 are used, what are the normalized element values of this model? What (simple) change in scaling will now make each inductor equal to 2 H (normalized) and each end capacitor equal to 1 F (normalized)? What then is the value of the internal C?

(*f*) Find the impulse response of this (five-element) model, with the same two resistor terminations, and correct its scale to real time units.

(*g*) Now compare the two impulse responses of the same line, differently modeled. What changes do you see? What would you expect if you were now to subdivide each half of the line and repeat the process?

(*h*) Do that.

(*i*) Suppose a certain reasonable (short) length of transmission line is representable well enough by a single π model as above. Now consider a very long line, modeled by a very long succession of such (identical) π models, each corresponding to the agreed-on short line length. Call the elements involved $2L$ and C in each individual π model, as above. On the assumption that the line is very long indeed, so that the impedance looking into the line model after one π section is exactly the same as that looking in at the beginning of the model, find, as a function of L, C and s, the input impedance of the long line model, $Z(s)$. This is a transfer function, from i_F (source) to v (response, at the input). Is it a rational function of s with real coefficients? Explain.

The low-pass property of all the line models is evident in the series inductors and shunt capacitors, which give rise to many zeros of transmission at infinity. You are now to explain the low-pass property in a different way, by considering the input impedance of the long-line model, $Z(s)$.

Find and plot the resistance (real) and reactance (imaginary) parts of this impedance versus frequency, for $s = j\omega$. Suppose the model to be driven by a sinusoidal voltage source of internal resistive impedance $R = \sqrt{L/C}$. Plot the steady-state average power taken by the network as a function of frequency. Explain, solely in terms of this plot, as you see it, the applicability of the terms "low-pass," "cutoff frequency," "characteristic impedance," and "filter" to the line model. Note that these properties are consequences of the lumping (modeling) and are *not* properties of the line itself. But they are interesting and played a very important role in the actual genesis of filter theory.

Chapter **8**

THE LAPLACE TRANSFORMATION

"Curiouser and curiouser!" cried Alice....
"Now I'm opening out like the largest telescope that ever was!"

8.01 INTRODUCTION

We are now in a position to solve any ordinary network-analysis problem. Since we can handle every common, ordinary situation, we might rest on our laurels and be content. But the spirit of curiosity is seldom content. In what we've done there are still many ideas undeveloped, many loose ends left hanging. Some of them are well worth attention. Some have tremendous educational value. Some have practical value as indicators of where to go next with profit, as guides to useful action. Some delineate limits or bounds in the form of useful theorems we should know and respect in practice. Some would stimulate thought and lead eventually to new tools. One such tool in particular is the *Laplace transformation*, of which we have heard rumors.

A *specific* loose end, one troublesome thought in the "solution of any ordinary network analysis problem" is this: "What if N, the order of the network, is large, very large?" In principle, we can proceed as usual. In practice we shall surely encounter computational problems of numerical precision. These can be solved, true, by methods we leave for consideration elsewhere. But an interesting thought about "limits, bounds, theorems" immediately presents itself: What if N becomes indefinitely large? Is there anything of interest in such a hypothetical situation? Educational it may well be. And it may lead to useful practical approximations for large-N cases. It might even yield *simple*, useful guides. (Infinite series, after all, do often sum to simple forms.)

Let us perform an experiment with an infinite network and see what comes of it.

8.02 AN INFINITE NETWORK

Consider the analysis problem of Fig. 8.02-A. Such networks exist in transmission-line theory as models of ideal lines by long, even infinite successions of identical small π networks (see Prob. 7-Z); their analysis may therefore be useful. Or we may, simply out of curiosity, ask what happens when a network is very long, even infinite, taking the element values to repeat numerically for simplicity's sake.

In the usual way we transform the excitation $i_F(t)$ to $I(s)$. We assume

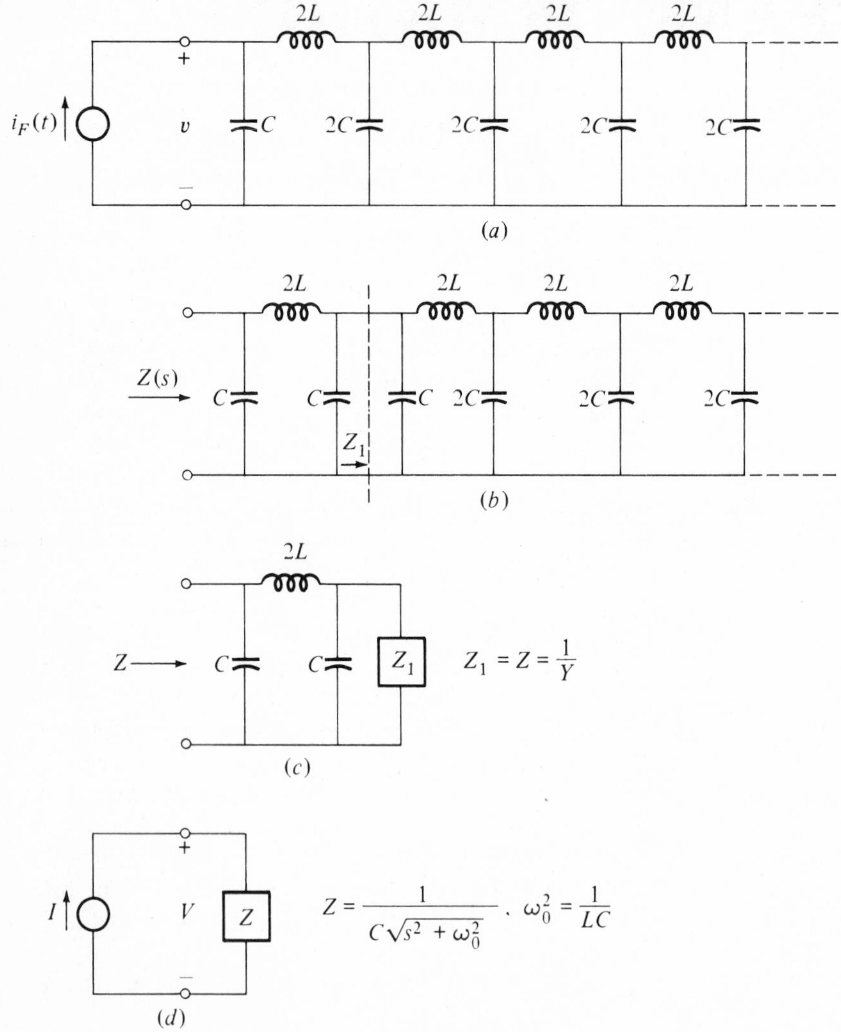

FIG. 8.02-A

initial rest. We consider as response the voltage at the input, $v(t)$, or $V(s)$ when transformed. The appropriate transfer function is then

$$\frac{V}{I} = Z(s) = \cfrac{1}{Cs + \cfrac{1}{2Ls + \cfrac{1}{2Cs + \cfrac{1}{2Ls + \cfrac{1}{2Cs + \cdots}}}}} \tag{8.02-1}$$

The repetitive character of such an endless (continued) fraction suggests that it may be amenable to simple analysis. We assume that it converges and attempt to find a simple form. After dividing the second capacitor in two for symmetry's sake we observe that the impedance "one step down the line," Z_1 in Fig. 8.02-A*b*, must, because the network is infinite, be indistinguishable from the input impedance Z. Hence $Z_1 = Z$ and we may redraw the network, as far as the source and response voltage are concerned, as in Fig. 8.02-A*c*. Then we observe that

$$Y(s) = \frac{1}{Z(s)} = Cs + \frac{1}{2Ls + 1/(Cs + Y)} \tag{8.02-2}$$

an equation easily solved for Y, or for Z. We find

$$Y(s) = Cs + \frac{Cs + Y}{2LCs^2 + 2LsY + 1}, \tag{8.02-3}$$

then

$$2LCs^2Y + 2LsY^2 + Y = 2LC^2s^3 + 2LCs^2Y + 2Cs + Y \tag{8.02-4}$$

and

$$Y^2 = \frac{2LC^2s^3 + 2Cs}{2Ls} = C^2s^2 + \frac{C}{L} = \frac{C}{L}(LCs^2 + 1). \tag{8.02-5}$$

On setting

$$\omega_0{}^2 = \frac{1}{LC} \tag{8.02-6}$$

we obtain

$$Y(s) = C\sqrt{s^2 + \omega_0{}^2} = \frac{1}{Z(s)}. \tag{8.02-7}$$

This transfer function is not rational and is not in our tables. It is irrational because the number of elements is not finite, something new to us. But the analysis proceeds formally, at least, in the usual s-domain fashion (Fig. 8.02-A*d*):

$$V(s) = I(s)Z(s) = \frac{I(s)}{C\sqrt{s^2 + \omega_0{}^2}}. \tag{8.02-8}$$

We have no pair to help us back to the t domain. Nor should we expect to have one in our collection, for this network is not in the lumped-linear-constant-finite category. But we do have one tool that might be useful: the series expansion of $F(s)$ in powers of s^{-1} developed in Sec. 6.13. Let us attempt to apply it here. (Note carefully at each step that *dimensionally* all is well.)

For simplicity let the excitation be an impulse, $i_F(t) = Q_0\,\delta(t)$, so that $I(s) = Q_0$. We now have

$$\begin{aligned} V(s) &= \frac{Q_0/C}{\sqrt{s^2 + \omega_0{}^2}} = \frac{Q_0/C}{s\sqrt{1 + (\omega_0/s)^2}} \\ &= \frac{Q_0/C}{s}\left[1 - \frac{1}{2}\left(\frac{\omega_0}{s}\right)^2 + \frac{3}{8}\left(\frac{\omega_0}{s}\right)^4 - \frac{5}{16}\left(\frac{\omega_0}{s}\right)^6 + \cdots\right] \\ &= \frac{Q_0}{C}\left(\frac{1}{s} - \frac{\omega_0{}^2}{2}\frac{1}{s^3} + \frac{3}{8}\omega_0{}^4\frac{1}{s^5} - \frac{5}{16}\omega_0{}^6\frac{1}{s^7} + \cdots\right), \end{aligned} \tag{8.02-9}$$

the expansion being made by the binomial theorem or by ordinary Maclaurin series expansion in $(1/s)$. Then from pair 35 we obtain

$$\begin{aligned} v(t) &= \frac{Q_0}{C}\left[1 - \frac{1}{2}\frac{\omega_0{}^2t^2}{2!} + \frac{3}{8}\frac{\omega_0{}^4t^4}{4!} - \frac{5}{16}\frac{\omega_0{}^6t^6}{6!} + \cdots\right] \qquad t > 0 \\ &= \frac{Q_0}{C}\left[1 - \tfrac{1}{4}(\omega_0 t)^2 + \tfrac{1}{64}(\omega_0 t)^4 - \frac{1}{2304}(\omega_0 t)^6 + \cdots\right] \end{aligned} \tag{8.02-10}$$

which seems to be a well-behaved series, and from which we can plot, at least for small values of t, the behavior of $v(t)$. This is done in Fig. 8.02-B, which therefore shows the response of the infinite network, at its input, to an impulsive current excitation. The function defined by this series (8.02-10) [the constant (Q_0/C) apart] has in fact been thoroughly studied and is well known. It is the *Bessel function* of the first kind of order zero, denoted $J_0(x)$. The integral

$$\int_0^\infty J_0(\omega_0 t)e^{-st}\,dt = \frac{1}{\sqrt{s^2 + \omega_0{}^2}}, \qquad \mathbf{RE}(s) > 0 \tag{8.02-11}$$

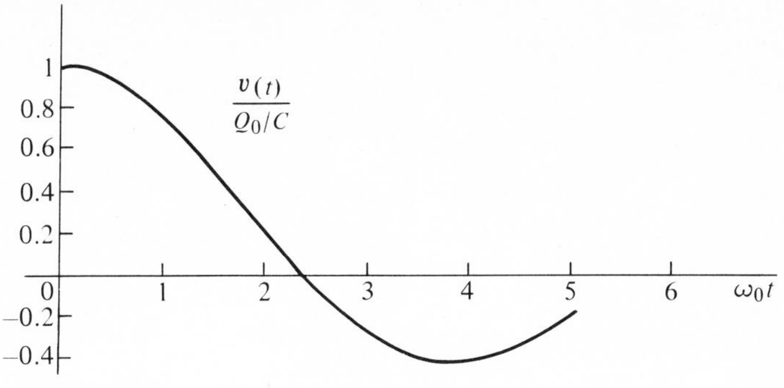

FIG. 8.02-B

is derived in works on Bessel functions. We can therefore add another pair to our catalog:

$$\frac{1}{\sqrt{s^2 + \omega_0^{\ 2}}} \sim J_0(\omega_0 t). \qquad (8.02\text{-}12)$$

From tables of the Bessel functions the curve of Fig. 8.02-B can be extended indefinitely. We observe that it seems not radically different from the ordinary (damped-sinusoidal) response of a two-natural-frequency tuned circuit. But mathematically it is essentially different, not being exponential in character; physically also it must differ, being the response of an infinite network.

It is both logically important, and not at all difficult to check this calculation physically. Suppose again that $i_F(t)$, the excitation in Fig. 8.02-A, is the impulse $Q_0\,\delta(t)$. By familiar physical reasoning (Fig. 8.02-C), that the inductor cannot immediately respond, we find that the impulsive current flows entirely through the first capacitor and the voltage across the second capacitor does not immediately change. Hence we have

$$v(0+) = \frac{Q_0}{C}, \qquad i_2(0+) = 0, \qquad v_1(0+) = 0. \qquad (8.02\text{-}13)$$

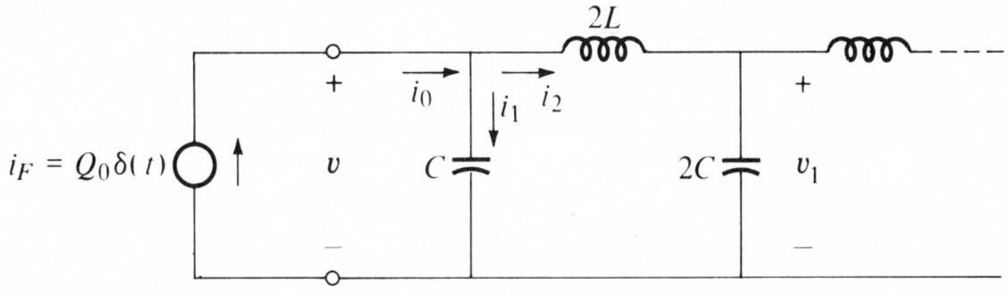

FIG. 8.02-C

Then we observe that

$$i_0(0+) = 0, \qquad i_1(0+) = -i_2(0+) = 0. \tag{8.02-14}$$

Now

$$v'(0+) = \frac{1}{C} i_1(0+) = 0. \tag{8.02-15}$$

Then from

$$C\frac{dv}{dt} = i_1 = -i_2, \tag{8.02-16}$$

we obtain

$$C\frac{d^2v}{dt^2} = i_1' = -i_2'. \tag{8.02-17}$$

We can evaluate $i_2'(0+)$ by first observing that

$$\begin{aligned} v_1 + (2L)i_2' &= v, \\ i_2'(0+) &= \frac{v(0+) - v_1(0+)}{2L} = \frac{Q_0/C - 0}{2L} = \frac{Q_0}{C}\frac{1}{2L} \end{aligned} \tag{8.02-18}$$

so that

$$v''(0+) = \left(\frac{1}{C}\right)[-i_2'(0+)] = -\frac{Q_0}{C}\frac{1}{2LC} = -\frac{Q_0}{C}\frac{{\omega_0}^2}{2} \tag{8.02-19}$$

in which, as before [(8.02-6)], ${\omega_0}^2 = 1/(LC)$. We have now

$$\begin{aligned} v(t) &= v(0+) + v'(0+)t + \frac{v''(0+)t^2}{2} + \cdots \\ &= \frac{Q_0}{C}\left[1 + 0 - \frac{1}{2}\frac{(\omega_0 t)^2}{2} + \cdots\right] \end{aligned} \tag{8.02-20}$$

in which, as before, ${\omega_0}^2 = 1/(LC)$. The first three terms of (8.02-10) (the second one being zero) are thus verified by physical reasoning; we could continue indefinitely. The pair (8.02-12) is thus placed on as solid a footing as can be established for such a theoretical physical situation.

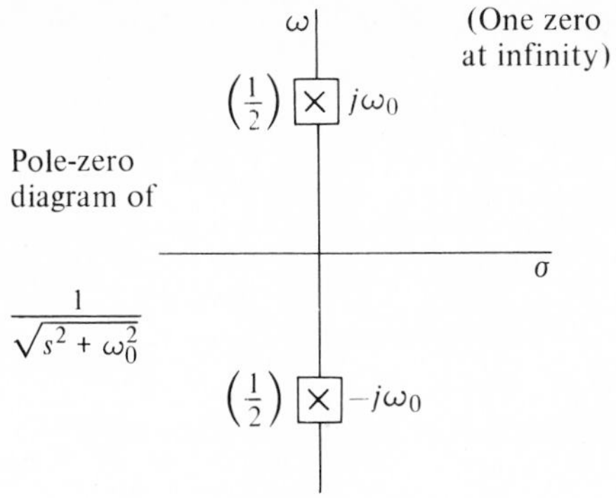

FIG. 8.02-D

To complete the parallel with "real" pairs we observe that a pole-zero diagram can even be constructed. We rewrite the transfer function (8.02-7) thus:

$$Z(s) = \frac{1}{C\sqrt{s^2 + \omega_0{}^2}} = \frac{1}{C\sqrt{(s - j\omega_0)(s + j\omega_0)}} = \frac{1}{C(s - j\omega_0)^{1/2}(s + j\omega_0)^{1/2}} \tag{8.02-21}$$

so that we can place "half poles" at $\pm j\omega_0$, and a zero at infinity (Fig. 8.02-D) to make such a diagram. Such *branch points* as they are called in mathematical analysis, can be thought of as "fractional" poles (or zeros) but require much more study for real understanding. We content ourselves with what we have so far developed.

8.03 ANOTHER INFINITE NETWORK

Sufficiently different to be interesting in its own right, yet sufficiently similar to be equally easy to analyze is the semi-infinite network of Fig. 8.03-A. Analysis in the fashion of Sec. 8.02 gives us, by physical reasoning

$$v(t) = \frac{Q_0}{C}\left[1 - \tfrac{1}{2}(\omega_0 t) + \tfrac{3}{16}(\omega_0 t)^2 - \tfrac{5}{96}(\omega_0 t)^3 + \cdots\right], \qquad t > 0, \tag{8.03-1}$$

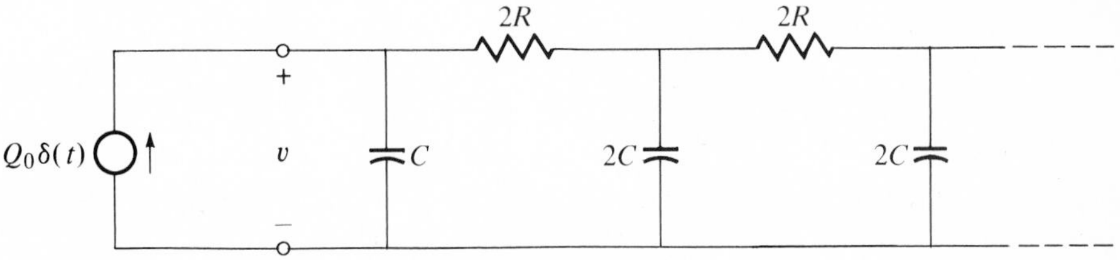

FIG. 8.03-A

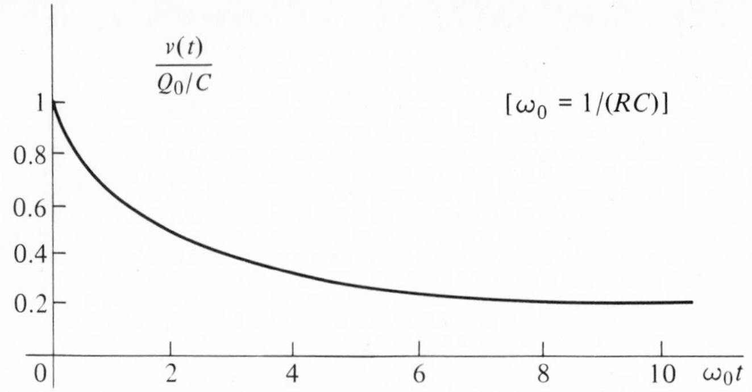

FIG. 8.03-B

and by formal, s-domain reasoning,

$$V(s) = Q_0\,Z(s) = \frac{Q_0}{\sqrt{(Cs)^2 + GCs}} = \frac{Q_0}{Cs\sqrt{1 + \omega_0/s}}, \qquad \omega_0 = \frac{1}{RC},$$

$$= \frac{Q_0}{Cs}\left(1 - \frac{\omega_0}{2s} + \frac{3}{8}\frac{\omega_0^{\,2}}{s^2} - \frac{5}{16}\frac{\omega_0^{\,3}}{s^3} + \cdots\right) \tag{8.03-2}$$

$$= \frac{Q_0}{C}\left(\frac{1}{s} - \frac{\omega_0}{2}\frac{1}{s^2} + \frac{3}{8}\omega_0^{\,2}\frac{1}{s^3} - \frac{5}{16}\omega_0^{\,3}\frac{1}{s^4} + \cdots\right).$$

From the series of (8.03-2) follows immediately the series of (8.03-1), on applying pair 35, as one would, by now, naturally expect. The time-domain series (8.03-1) has been thoroughly studied; it is again known as a Bessel function, but of a different kind, and modified by exponential damping:

$$v(t) = \frac{Q_0}{C}\,e^{-(1/2)\omega_0 t}\,I_0\!\left(\frac{\omega_0 t}{2}\right). \tag{8.03-3}$$

The symbol I_0 represents the Bessel function of the first kind with imaginary argument; $I_0(z) = J_0(jz)$ as may be verified by using the series of Sec. 8.02. The corresponding Laplace integral and many interesting properties of the function are discussed in Bessel-function literature.

The behavior of the response is shown in Fig. 8.03-B. Note the sluggish, nonoscillating character, typical of RC networks. (See also Fig. 8.04-A.)

8.04 OTHER INTERESTING SQUARE ROOTS

A related type of pair (discussed at length in Oliver Heaviside's work with transmission-line models) arises from the network of Fig. 8.03-A when the elements become infinitesimal. There let

$$R = r\,\Delta x,$$
$$C = c\,\Delta x, \tag{8.04-1}$$

in which r and c represent amounts of resistance and capacitance per unit distance along a transmission line, and Δx is a small distance there along. Then

$$Z(s) = \frac{Q_0}{\sqrt{(c\,\Delta xs)^2 + (c/r)s}}. \tag{8.04-2}$$

In the limit as Δx tends to zero, $Z(s)$ becomes

$$Z(s) = \frac{Q_0\sqrt{r/c}}{\sqrt{s}} \tag{8.04-3}$$

but the corresponding limit in the time domain (as ω_0 becomes infinite) is difficult to evaluate. So also is physical analysis of the situation; for which partial differential equations are necessary. (The *lumped* postulate of Sec. 4.02 no longer holds, although the "network" is linear and constant.) We content ourselves here with formal evaluation of the mate of $1/\sqrt{s}$. It can be found in tables, of course, but it is instructive first to attempt its evaluation ourselves. We recall pair 35,

$$\frac{1}{s^n} \sim \frac{t^{n-1}}{(n-1)!} \tag{8.04-4}$$

in which, of course, n is a positive integer. It suggests, that if it were legitimate to place $n = \frac{1}{2}$, the mate of $1/\sqrt{s}$ would be proportional to $1/\sqrt{t}$. [Even the formalism $(-\frac{1}{2})!$ has meaning to those who have studied gamma functions.] To pursue this idea, since direct Laplace transformation is certainly easier than inverse transformation (it is easier to find the mate of $1/\sqrt{t}$ than that of $1/\sqrt{s}$), let us try to calculate

$$\int_0^\infty \frac{1}{\sqrt{t}} e^{-st}\,dt. \tag{8.04-5}$$

On substituting x^2 for st we find

$$\int_0^\infty \frac{1}{\sqrt{t}} e^{-st}\,dt = \int_0^\infty \frac{\sqrt{s}}{x} e^{-x^2} \frac{2x}{s}\,dx = \frac{2}{\sqrt{s}} \int_0^\infty e^{-x^2}\,dx = \frac{2}{\sqrt{s}} \frac{\sqrt{\pi}}{2}, \tag{8.04-6}$$

the value of the infinite integral of e^{-x^2} being well known. Hence we have the pair

$$\frac{1}{\sqrt{s}} \sim \frac{1}{\sqrt{\pi t}} \tag{8.04-7}$$

precisely as anticipated.

It is not difficult to verify this peculiar, $1/\sqrt{}$ behavior in s and in t, at least approximately. Figure 8.04-A*a* shows the impulse response of the network of Fig. 8.03-A truncated (made finite) after three capacitors and three resistors. The attenuation in the network makes the input impedance virtually the same as that of the network of Fig. 8.03-A, so that Fig. 8.04-A*a* is practically a demonstration of that of Fig. 8.03-B. The infinite-network transfer function is, from (8.03-2),

$$T = \frac{V}{I} = Z = \frac{\sqrt{R/C}}{\sqrt{s}\sqrt{1 + RCs}}, \tag{8.04-8}$$

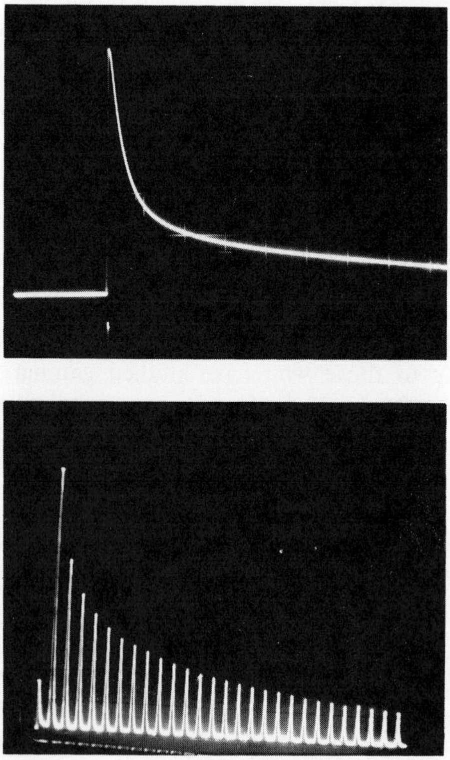

(*a*) $RC = 5\ \mu s$
One square = 20 μs

(*b*) Bar spacing is 2000 Hz

FIG. 8.04-A

which is essentially $1/\sqrt{s}$ at low frequencies. The initial behavior of the impulse response is of course a step, corresponding to the high-frequency $1/s$ nature of the transfer function, as the photograph shows. But for positive t, corresponding in a way to low frequencies where T behaves as $1/\sqrt{s}$, $r_\delta(t)$ does resemble $1/\sqrt{t}$: note the sudden change of heart, from steep (exponential-like) decay to the much slower decay representative of $1/\sqrt{t}$; there is almost a "corner" in the wave. The behavior of $|T(j\omega_F)|$ (Fig. 8.04-A*b*) shows a corresponding change: it is as $1/\sqrt{s}$ at low frequencies, then as $1/s$ at high frequencies, as the exercise of a little imagination will show. This "verification" of the pair (8.04-7) is somewhat whimsical, to be sure, but it is nevertheless interesting and very informative.

Related pairs, obtainable by integration in time (pair 6) and by other means, are

$$\frac{1}{s^{3/2}} \sim \frac{2}{\sqrt{\pi}}\sqrt{t}, \tag{8.04-9}$$

$$\frac{1}{s^{3/4}} \sim \frac{0.816\ldots}{t^{1/4}}, \tag{8.04-10}$$

and many others (see, for example, the Campbell and Foster table, Appendix B).

8.05 A FEW MORE NOVEL PAIRS

The infinitely long ladder network of Sec. 8.02 introduced us to irrational functions of s and to a novel time function, the Bessel function J_0. Associated therewith are many more of similar sort, for example the pair

$$\frac{1}{\sqrt{s^2+\omega_0{}^2}\,(s+\sqrt{s^2+\omega_0{}^2})^n} \sim \frac{1}{\omega_0{}^n} J_n(\omega_0 t), \qquad n = 0, 1, 2, 3, \ldots, \tag{8.05-1}$$

and the pair

$$\frac{1}{(s+\sqrt{s^2+\omega_0{}^2})^n} \sim \frac{n}{\omega_0{}^n}\frac{J_n(\omega_0 t)}{t}, \qquad n = 0, 1, 2, 3, \ldots. \tag{8.05-2}$$

The literature of Bessel functions is rich in other, related pairs.

Another interesting pair, fairly simple, but of different character for it is transcendental on the s side, can be got by integration with respect to the parameter β (see pair 29) of pair 21,

$$\frac{s}{s^2+\beta^2} = \frac{1/s}{1+\beta^2/s^2} \sim \cos \beta t. \tag{8.05-3}$$

The result is

$$\tan^{-1}\frac{\beta}{s} \sim \frac{\sin \beta t}{t}. \tag{8.05-4}$$

Still another transcendental (in s) pair, not unrelated to (8.05-4), can be reached by a "backward" application of the series in powers of $1/s$. Consider the difference of two ordinary exponential functions and their power series:

$$\begin{aligned} e^{-\alpha t} - e^{-\beta t} &= \left(1 - \alpha t + \frac{\alpha^2 t^2}{2!} - \frac{\alpha^3 t^3}{3!} + \cdots\right) \\ &\quad - \left(1 - \beta t + \frac{\beta^2 t^2}{2!} - \frac{\beta^3 t^3}{3!} + \cdots\right) \\ &= (\beta - \alpha)t - \frac{(\beta^2 - \alpha^2)t^2}{2!} + \frac{(\beta^3 - \alpha^3)t^3}{3!} - \cdots. \end{aligned} \tag{8.05-5}$$

From this we can obtain

$$\frac{e^{-\alpha t} - e^{-\beta t}}{t} = (\beta - \alpha) - \frac{(\beta^2 - \alpha^2)t}{2!} + \frac{(\beta^3 - \alpha^3)t^2}{3!} - \cdots \tag{8.05-6}$$

whose mate is

$$\frac{\beta - \alpha}{s} - \frac{\beta^2 - \alpha^2}{2s^2} + \frac{\beta^3 - \alpha^3}{3s^3} - \cdots. \tag{8.05-7}$$

The familiar series

$$\log_e (1 + x) = x - \tfrac{1}{2}x^2 + \tfrac{1}{3}x^3 - \cdots \tag{8.05-8}$$

suggests rewriting (8.05-7) in the form

$$\begin{aligned} &\left(\frac{\beta}{s} - \frac{\beta^2}{2s^2} + \frac{\beta^3}{3s^3} - \cdots\right) - \left(\frac{\alpha}{s} - \frac{\alpha^2}{2s^2} + \frac{\alpha^3}{3s^3} + \cdots\right) \\ &= \log\frac{1 + \beta/s}{1 + \alpha/s} = \log\frac{s + \beta}{s + \alpha} \end{aligned} \tag{8.05-9}$$

(the logarithms being, of course, *natural* logarithms). Putting these results together gives us the pair

$$\frac{e^{-\alpha t} - e^{-\beta t}}{t} \sim \log\frac{s + \beta}{s + \alpha}. \tag{8.05-10}$$

Questions anent the rigor of some of these developments we leave for expert analysis elsewhere; but all the pairs given here are indeed valid and may be found in advanced references, together with the appropriate values of σ_a.

8.06 DELAY

From ordinary networks' rational functions of s with real coefficients we have come to irrational and even to transcendental functions of s. The irrational functions seem to have to do with (hypothetical) networks that are infinite in extent. They can be useful as guides to thought and action, but are not strictly physically realizable; but infinitesimal elements, infinite in number, *are* physically meaningful, in transmission-line theory (into which we do not go).

The transcendental functions of Sec. 8.05 seem to have little to do with reality, even with infinite networks. But there is one transcendental function of considerable physical significance. It is (of course!) the *exponential* function, this time in the s domain, and it merits discussion here because it is very important and does have physical significance. It relates to the transfer function that corresponds to the retarding or delaying of a signal in time, without distortion. Let us see exactly what such a transfer function is.

Suppose the "box" of Fig. 8.06-A to contain whatever physical apparatus is necessary to delay the excitation wave $e(t)$ by D (seconds). There seems to be nothing physically unreasonable about the existence of such a device, provided of course that D is positive and not negative (that the response occurs *after* the excitation and not before). We require that there be no distortion of the waveshape. (Amplification or attenuation, as expressed by multiplication by a constant, is not to be considered distortion.) That is, we require the response $r(t)$ to be a (scaled) replica of the excitation, occurring D seconds later in time:

$$r(t) = ke(t - D), \tag{8.06-1}$$

as shown in Fig. 8.06-B. Note that the delayed signal output response *must* be zero for $0 < t < D$ (Fig. 8.06-B*b*) because we analyze the situation from $t = 0$ onward, starting from rest (Fig. 8.06-B*a*). The device can reasonably emit a delayed replica of $e(t)$ but therein cannot supply any signal in $0 < t < D$. In the response (8.06-1) k is a constant, positive or negative, large or small, representing amplification or attenuation, and perhaps polarity reversal. D is the delay (in seconds), a positive number.

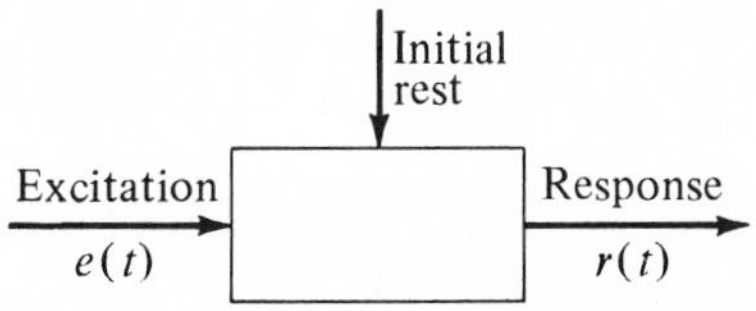

FIG. 8.06-A

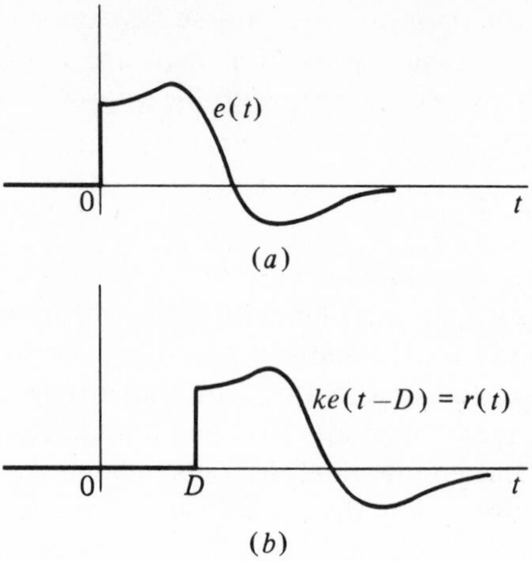

FIG. 8.06-B

The transfer function of such a device is easily derived. The output's mate is

$$\begin{aligned} R(s) &= \int_0^\infty r(t)e^{-st}\,dt = \int_D^\infty e(t-D)e^{-st}\,dt \\ &= \int_0^\infty e(x)e^{-s(x+D)}\,dx \qquad [x = t - D] \\ &= e^{-Ds}\int_0^\infty e(x)e^{-sx}\,dx \\ &= e^{-Ds}E(s). \end{aligned} \tag{8.06-2}$$

[The two meanings of e in (8.06-2) should cause no confusion if one recognizes that the symbol *does* have two distinct meanings.] Since $E(s)$ is the mate of the excitation $e(t)$, the transfer function of a (hypothetical) ideal delay device is simply e^{-Ds}, D being the amount of delay in seconds. The exponential function, ubiquitous in the time domain, also has s-domain significance, corresponding to a shift in the other domain. (See pair 18 in which the shift is in the s domain, and the exponential appears in the t domain.)

Such a device cannot be a finite, lumped-linear-constant-element network, for its transfer function is transcendental and not rational in s. Ideal transmission lines do, however, have such transfer functions, for they contain infinite numbers of infinitesimal elements (are described by partial differential equations). Hence physical realizations of ideal-delay transfer functions do exist, though not as networks.

Many are the examples of analysis in which we have already seen evidence that pointed toward this conclusion, that the ideal *delaying* transfer function is e^{-Ds}. Here, in (8.06-2) we have formal proof and explicit statement. Necessary and sufficient for (distortionless) delay is a transcendental transfer function e^{-Ds}. Equivalent is the *delay pair:*

$$\begin{aligned} &\textit{If:} && f(t)u(t) \sim F(s), \\ &\textit{Then:} && f(t-D)u(t-D) \sim e^{-Ds}F(s). \end{aligned} \qquad (8.06\text{-}3)$$

Time-domain shifting (delay) is accomplished in the s domain by multiplication by e^{-Ds}. [The $u(t)$ is incorporated in (8.06-3) to make absolutely clear the need for the gap in $0 < t < D$ in the delayed function (see Fig. 8.06-Bb).]

The development of (8.06-2) and (8.06-3) required the use of the integral formulation of the relation between the t and s domains. Our ordinary analysis method, finding the forced component of response to exponential (e^{st}) excitation and hence the transfer function, can*not* obtain these results. (The reason is, of course, that the "network" is not a network in the sense we use it, is not a finite collection of lumped elements.) But the pair (8.06-3), ideal and nonnetworklike though it is, is practically very important as a theoretical *limit*, an indication of what can be done at the most, an ideal at which we can aim but never reach. We have seen it approached many times; in fact (7.19-24) is really an empirical derivation of the same thing.

We conclude that the integral (Laplace) method is useful in addition to the mere e^{st} method. We may not ordinarily use it in actual network analysis, but it's good to have, it's informative and useful. We shall discuss it further in Sec. 8.07.

8.07 DISTORTIONLESS TRANSMISSION

The relation (8.06-2) is extremely important for another reason: it gives us (theoretical) conditions necessary and sufficient for *distortionless transmission*, usually a desirable system characteristic. We define "distortionless transmission" as transmission such that in the t domain the output is a replica of the input. Scaling (multiplication by a positive or negative constant of any size, the result of amplification or attenuation) we do not consider distortion. Nor is a *delay* in time to be considered distortion. Hence (Sec. 8.06) a distortionless transmission transfer function is (ke^{-Ds}), in which the constant k represents the scaling, and D the delay.

Such a characteristic is an end to be sought in any transmission system. Real ones do not have it, for all components of a system (amplifiers, cables, other apparatus) distort signals as they pass. But it is a criterion, an ultimate, a guide to what the system ought to be. It is useful in studying and describing a system as characterized by measurements or calculations made, for it tells us what needs to be done in the way of compensation for undesirable distortion. A common use of networks is in fact to correct distortion in an extant system (which for some reason cannot be altered except by adding to it such "equalizer" networks).

The distortionless transmission condition (8.06-2) is frequently more useful when expressed (equivalently) in ω-domain terms. Setting $s = j\omega$ gives us

$$T(j\omega) = (ke^{-Ds})_{s=j\omega} = ke^{-jD\omega} = k\underline{/-D\omega} \qquad (8.07\text{-}1)$$

or (see Fig. 8.07-A)

$$\begin{aligned} |T(j\omega)| &= k = \text{constant}, \\ \underline{/T(j\omega)} &= -D\omega = \text{linear, with slope } -D, \text{ function} \\ &\qquad \text{for which } d\beta/d\omega = -D = \text{constant.} \end{aligned} \qquad (8.07\text{-}2)$$

The design of corrective networks (equalizers) must aim to

a Straighten out (level off) the amplitude as a function of ω,
b Straighten out (linearize) the angle (phase shift) as a function of ω. (8.07-3)

Such a task is impossible of fulfillment by a (finite) lumped, linear, constant network of the sort we are studying, for it would require transcendental transfer functions. Rational transfer functions can at best *approximate* this ideal behavior (but often with fairly good results).

Our many previous examples (Secs. 7.15, 7.16, 7.19, for example) corroborate the following statements:

a Constant amplitude and linear phase are needed, yet
b Networks (systems) with only *approximately* those characteristics can still give fairly good transmission. (8.07-4)

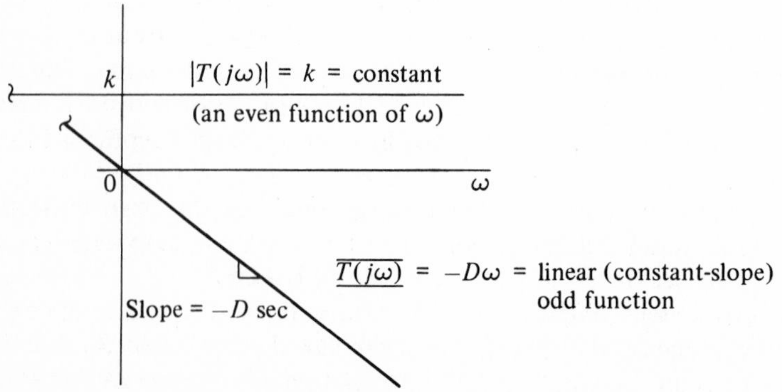

FIG. 8.07-A. The ideal, distortionless transfer function in the ω domain.

The approximation, we learn from these examples as well as from mere common sense, should be good over the important bands of frequency (important from the signal's point of view). Elsewhere the characteristics are practically less important, though theoretically it is essential to have ideal behavior at *all* frequencies. If the output is, though not a replica of the input, yet recognizable as an only reasonably distorted version of the input, the transmission is adequate. Standards differ, according to the job: a system transmission that is "excellent" for 1000 speech channels may be completely unacceptable for television signals, even though the important frequency ranges be about the same.

We here look briefly at two more specific examples. The first is a low-pass filter (see Sec. 7.15 and Fig. 7.19-A) whose "cutoff" is fairly gentle (smooth, rounded); there is, however, a clear transition from "pass" to "stop" frequency regions in the sinusoidal-steady-state characteristic, as Fig. 8.07-B*a* shows. (Details of the network itself, of less immediate importance, can be found in Prob. 8-H.) From Fig. 8.07-B*b* we see that the slope of the phase (angle) characteristic at low frequencies, where the phase is more or less linear and the transmission is more or less constant, is -3.3 s (normalized). The "delay" to be expected for "low-frequency" signals is thus about 3.3 s. In the time domain, the impulse response (Fig. 8.07-B*c*) is recognizable as a "distorted impulse" that occurs about 3.7 s after the exciting impulse. The 10 percent difference between these two numbers (3.3 and 3.7) is what one might expect for such an inexact, though useful correlation. We note that (8.07-4) is borne out well enough: the signal is *not* of course really "low-frequency," but enough of it does lie in the "good-transmission" range that (8.07-4) has some meaning.

The second example is also a low-pass filter (Prob. 8-I) but one designed for much more rapid transition from "pass" to "stop," as its $|T(j\omega)|$ characteristic shows (Fig. 8.07-C*a*). The frequency scale is adjusted (normalized) so that the "beginning" of the "cutoff" occurs at approximately the same frequency, $\omega_n = 0.7$, as it does in Fig. 8.07-B*a*. This is of course arbitrary (it could be placed, for example, so that the points where $|T| = 0.5$, or some other value, coincide) but this seems a "sensible" basis for comparison. With this frequency scaling we find a low-frequency phase slope (Fig. 8.07-C*b*) whose magnitude is 4.1 s (normalized). The "delay" to be expected for low-frequency signals is thus appreciably larger for this second example. The impulse response (Fig. 8.07-C*c*) is still recognizable and the principal peak occurs some 5.1 s after the excitation, agreeing roughly with the 4.1 figure. The impulse response is much more oscillatory here than it was in the first example. (Note that Fig. 8.07-C*c* shows the response over more than twice as long a time as does Fig. 8.07-B*c*; compare also the positive and negative peak values of the two.) And its initial value, though small, is here not zero because this network has only one zero of transmission at infinity.

It appears from this experiment that increasing the rapidity of cutoff, starting from a given frequency, tends to increase the magnitude of the low-frequency phase slope, and hence the "delay." We further note an increase in the distortion in the output pulse (oscillation or "ringing"). The steeper cutoff and perhaps the

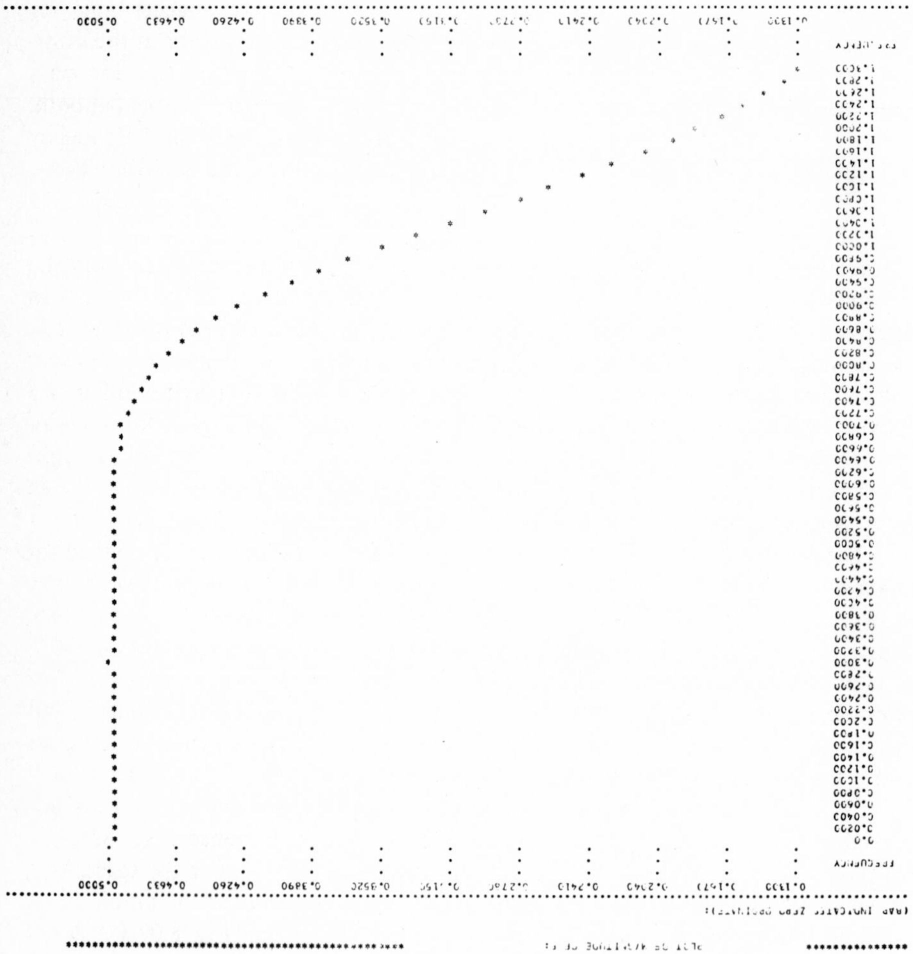

FIG. 8.07-Ba

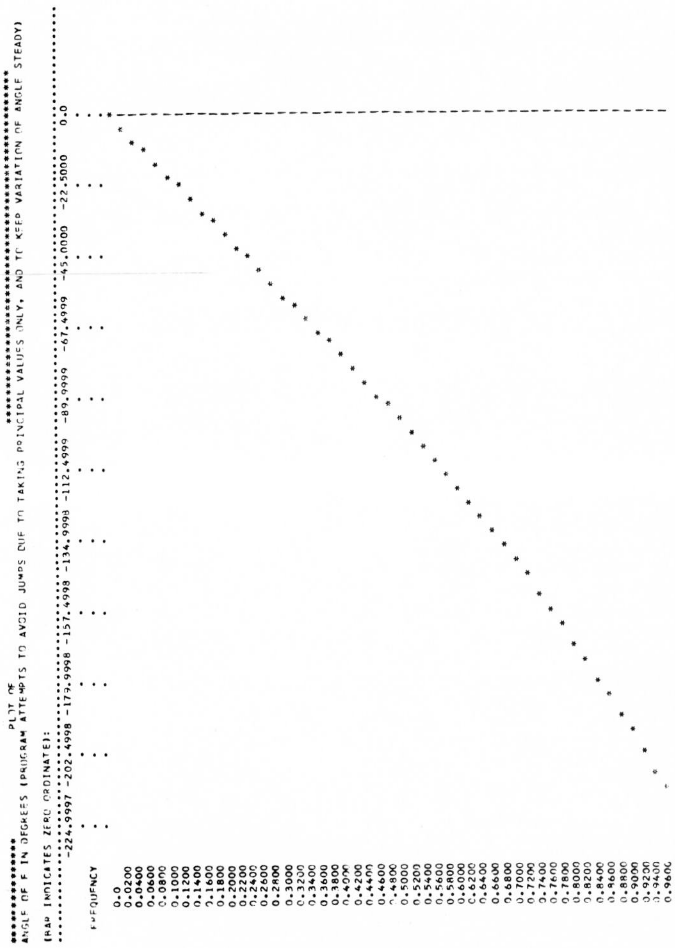
PLOT OF
ANGLE OF F IN DEGREES (PROGRAM ATTEMPTS TO AVOID JUMPS DUE TO TAKING PRINCIPAL VALUES ONLY, AND TO KEEP VARIATION OF ANGLE STEADY)
(BAR INDICATES ZERO ORDINATE):
-224.9997 -202.4998 -179.9998 -157.4998 -134.9998 -112.4999 -89.9999 -67.4999 -45.0000 -22.5000 0.0
FREQUENCY
0.0
0.0200
0.0400
0.0600
0.0800
0.1000
0.1200
0.1400
0.1600
0.1800
0.2000
0.2200
0.2400
0.2600
0.2800
0.3000
0.3200
0.3400
0.3600
0.3800
0.4000
0.4200
0.4400
0.4600
0.4800
0.5000
0.5200
0.5400
0.5600
0.5800
0.6000
0.6200
0.6400
0.6600
0.6800
0.7000
0.7200
0.7400
0.7600
0.7800
0.8000
0.8200
0.8400
0.8600
0.8800
0.9000
0.9200
0.9400
0.9600

FIG. 8.07-Bb

```
THIS IS IMPSTEP SEVEN.     NOTES FOR THIS RUN:     MAX FLAT N = 5  LPF
                                                   DATA FROM KUO
                                                   IMPSTEP7

SUBROUTINE 'PLCT4Z', BEING CALLED, GIVES (BAR MARKS ZERO ORDINATE):
```

TIME	IMPULSE RESP
0.0	-0.7868E-05
0.2500	0.6402E-04
0.5000	0.9291E-03
0.7500	0.3970E-02
1.0000	0.1050E-01
1.2500	0.2133E-01
1.5000	0.3659E-01
1.7500	0.5573E-01
2.0000	0.7765E-01
2.2500	0.1009E 00
2.5000	0.1237E 00
2.7500	0.1445E 00
3.0000	0.1619E 00
3.2500	0.1746E 00
3.5000	0.1818E 00
3.7500	0.1831E 00
4.0000	0.1785E 00
4.2500	0.1685E 00
4.5000	0.1537E 00
4.7500	0.1350E 00
5.0000	0.1136E 00
5.2500	0.9054E-01
5.5000	0.6710E-01
5.7500	0.4431E-01
6.0000	0.2312E-01
6.2500	0.4286E-02
6.5000	-0.1161E-01
6.7500	-0.2423E-01
7.0000	-0.3341E-01
7.2500	-0.3921E-01
7.5000	-0.4184E-01
7.7500	-0.4163E-01
8.0000	-0.3905E-01
8.2500	-0.3460E-01
8.5000	-0.2882E-01
8.7500	-0.2223E-01
9.0000	-0.1534E-01
9.2500	-0.8589E-02
9.5000	-0.2338E-02
9.7500	0.3124E-02
10.0000	0.7600E-02
10.2500	0.1098E-01
10.5000	0.1322E-01
10.7500	0.1437E-01
11.0000	0.1451E-01
11.2500	0.1379E-01
11.5000	0.1237E-01
11.7500	0.1044E-01
12.0000	0.8181E-02
12.2500	0.5771E-02
12.5000	0.3373E-02
12.7500	0.1124E-02
13.0000	-0.8679E-03
13.2500	-0.2524E-02
13.5000	-0.3797E-02
13.7500	-0.4669E-02
14.0000	-0.5146E-02
14.2500	-0.5258E-02
14.5000	-0.5050E-02
14.7500	-0.4580E-02
15.0000	-0.3913E-02

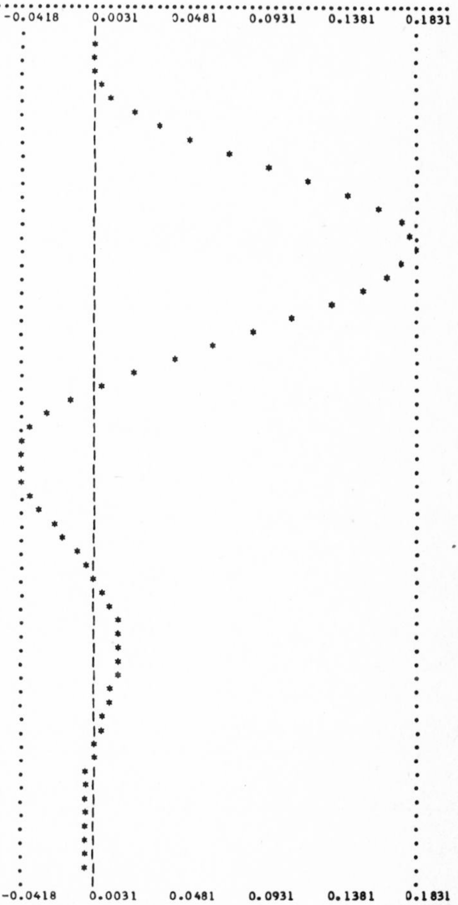

FIG. 8.07-Bc

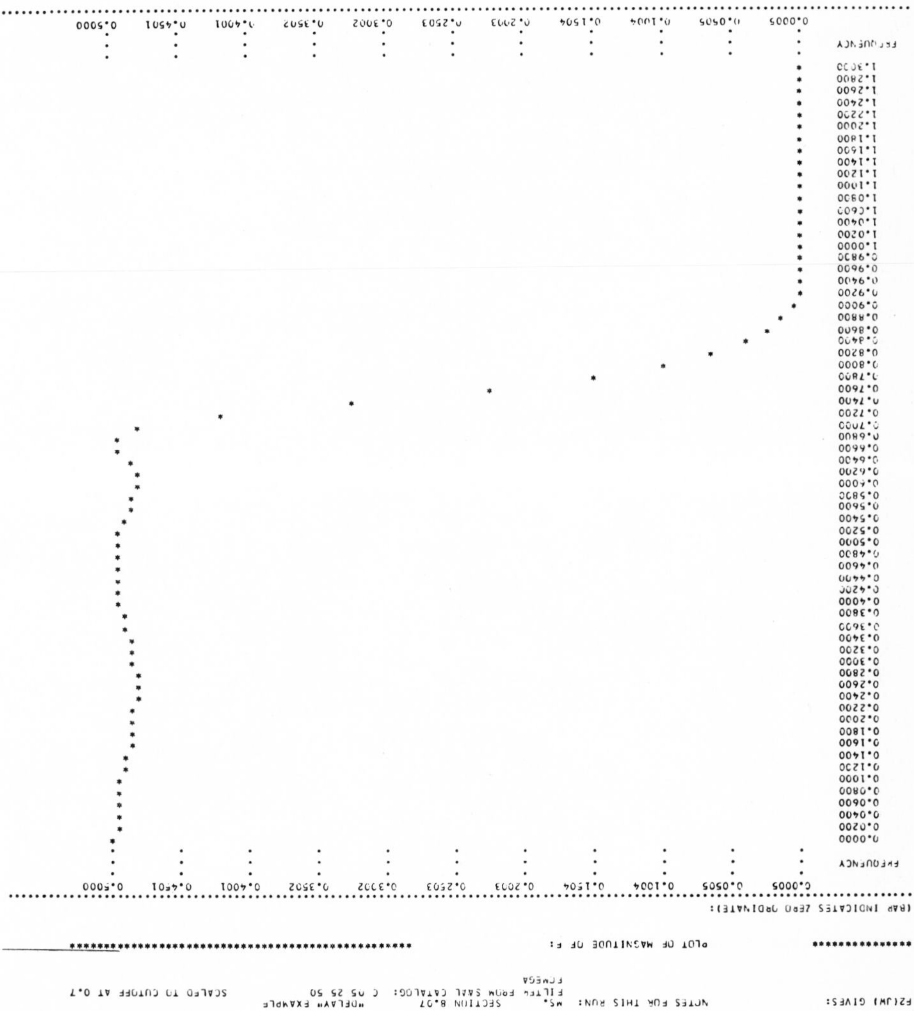
F2(JW) GIVES:
NOTES FOR THIS RUN: MS. SECTION 8.07 WDELAY EXAMPLE
FILTER FROM SAAL CATALOG: C 05 25 50 SCALED TO CUTOFF AT 0.7
OMEGA
PLOT OF MAGNITUDE OF F:
(BAR INDICATES ZERO ORDINATE):
FREQUENCY

FIG. 8.07-Ca

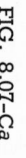

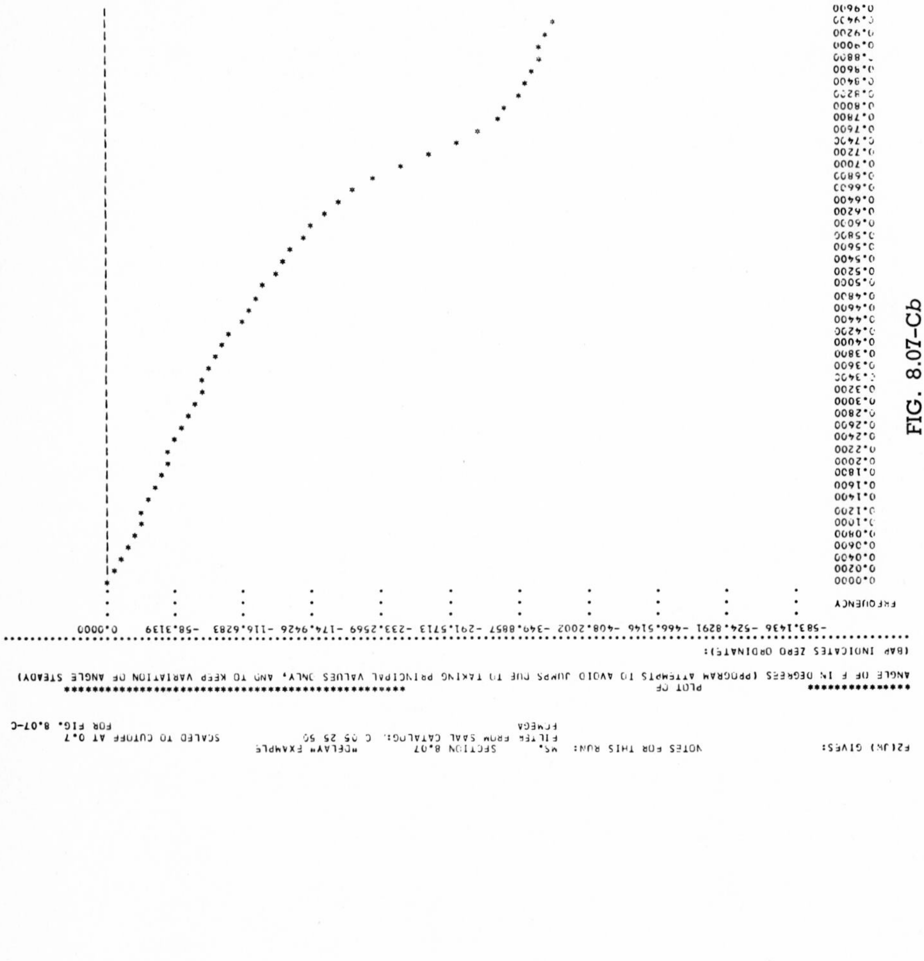
F2(JW) GIVES:
NOTES FOR THIS RUN: MS. SECTION 8.07 "DELAY" EXAMPLE
FILTER FROM SAAL CATALOG: C 05 25 50
FOMEGA
SCALED TO CUTOFF AT 0.7
FOR FIG. 8.07-C
PLOT OF
ANGLE OF F IN DEGREES (PROGRAM ATTEMPTS TO AVOID JUMPS DUE TO TAKING PRINCIPAL VALUES ONLY, AND TO KEEP VARIATION OF ANGLE STEADY)
(BAR INDICATES ZERO ORDINATE):
-583.1436 -524.8291 -466.5146 -408.2002 -349.8857 -291.5713 -233.2569 -174.9426 -116.6283 -58.3139 0.0000
FREQUENCY
0.0000
0.0200
0.0400
0.0600
0.0800
0.1000
0.1200
0.1400
0.1600
0.1800
0.2000
0.2200
0.2400
0.2600
0.2800
0.3000
0.3200
0.3400
0.3600
0.3800
0.4000
0.4200
0.4400
0.4600
0.4800
0.5000
0.5200
0.5400
0.5600
0.5800
0.6000
0.6200
0.6400
0.6600
0.6800
0.7000
0.7200
0.7400
0.7600
0.7800
0.8000
0.8200
0.8400
0.8600
0.8800
0.9000
0.9200
0.9400
0.9600

FIG. 8.07-Cb

```
THIS IS IMPSTEP SEVEN.    NOTES FOR THIS RUN:    MS.      SECTION 8.07        "DELAY" EXAMPLE
                                                 FILTER FROM SAAL CATALOG:  C 05 25 50        SCALED TO CUTOFF AT 6.7
                                                 IMPSTEP7

SUBROUTINE 'PLOT4Z', BEING CALLED, GIVES (BAR MARKS ZERO CRDINATE):
```

TIME	IMPULSE RESP
0.0000	0.2039E-01
0.5000	0.1641E-01
1.0000	0.2101E-01
1.5000	0.3094E-01
2.0000	0.4382E-01
2.5000	0.5793E-01
3.0000	0.7198E-01
3.5000	0.8492E-01
4.0000	0.9582E-01
4.5000	0.1038E 00
5.0000	0.1081E 00
5.5000	0.1082E 00
6.0000	0.1036E 00
6.5000	0.9432E-01
7.0000	0.8069E-01
7.5000	0.6346E-01
8.0000	0.4376E-01
8.5000	0.2299E-01
9.0000	0.2706E-02
9.5000	-0.1553E-01
10.0000	-0.3034E-01
10.5000	-0.4067E-01
11.0000	-0.4588E-01
11.5000	-0.4586E-01
12.0000	-0.4100E-01
12.5000	-0.3218E-01
13.0000	-0.2060E-01
13.5000	-0.7720E-02
14.0000	0.4979E-02
14.5000	0.1611E-01
15.0000	0.2456E-01
15.5000	0.2957E-01
16.0000	0.3084E-01
16.5000	0.2850E-01
17.0000	0.2309E-01
17.5000	0.1548E-01
18.0000	0.6722E-02
18.5000	-0.2056E-02
19.0000	-0.9805E-02
19.5000	-0.1567E-01
20.0000	-0.1909E-01
20.5000	-0.1983E-01
21.0000	-0.1800E-01
21.5000	-0.1401E-01
22.0000	-0.8517E-02
22.5000	-0.2319E-02
23.0000	0.3755E-02
23.5000	0.8946E-02
24.0000	0.1265E-01
24.5000	0.1450E-01
25.0000	0.1437E-01
25.5000	0.1239E-01
26.0000	0.8939E-02
26.5000	0.4533E-02
27.0000	-0.2093E-03
27.5000	-0.4665E-02
28.0000	-0.8282E-02
28.5000	-0.1065E-01
29.0000	-0.1153E-01
29.5000	-0.1090E-01
30.0000	-0.8908E-02
30.5000	-0.5891E-02
31.0000	-0.2280E-02
31.5000	0.1435E-02
32.0000	0.4780E-02
32.5000	0.7352E-02
33.0000	0.8868E-02
33.5000	0.9190E-02
34.0000	0.8340E-02
34.5000	0.6486E-02
35.0000	0.3913E-02
35.5000	0.9818E-03
36.0000	-0.1922E-02
36.5000	-0.4434E-02
37.0000	-0.6262E-02
37.5000	-0.7211E-02
38.0000	-0.7208E-02
38.5000	-0.6304E-02
39.0000	-0.4660E-02
39.5000	-0.2520E-02
40.0000	-0.1772E-03

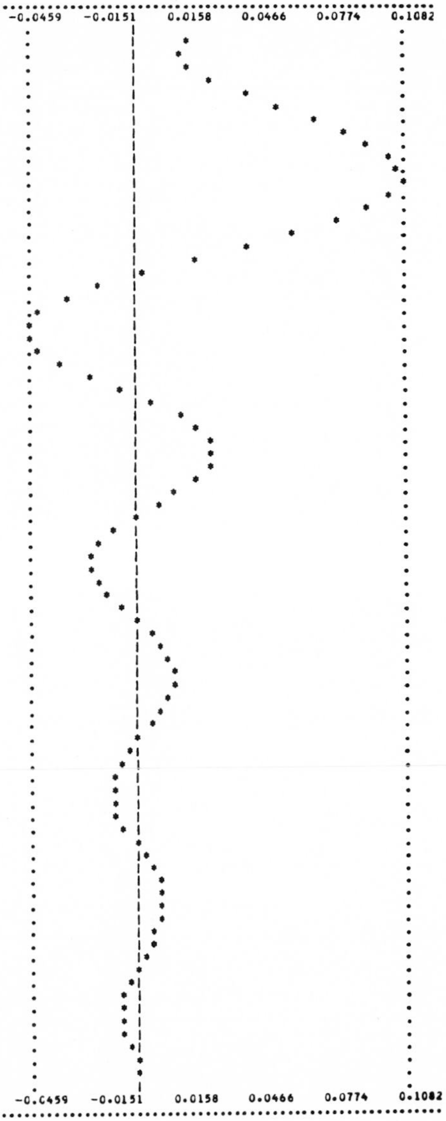

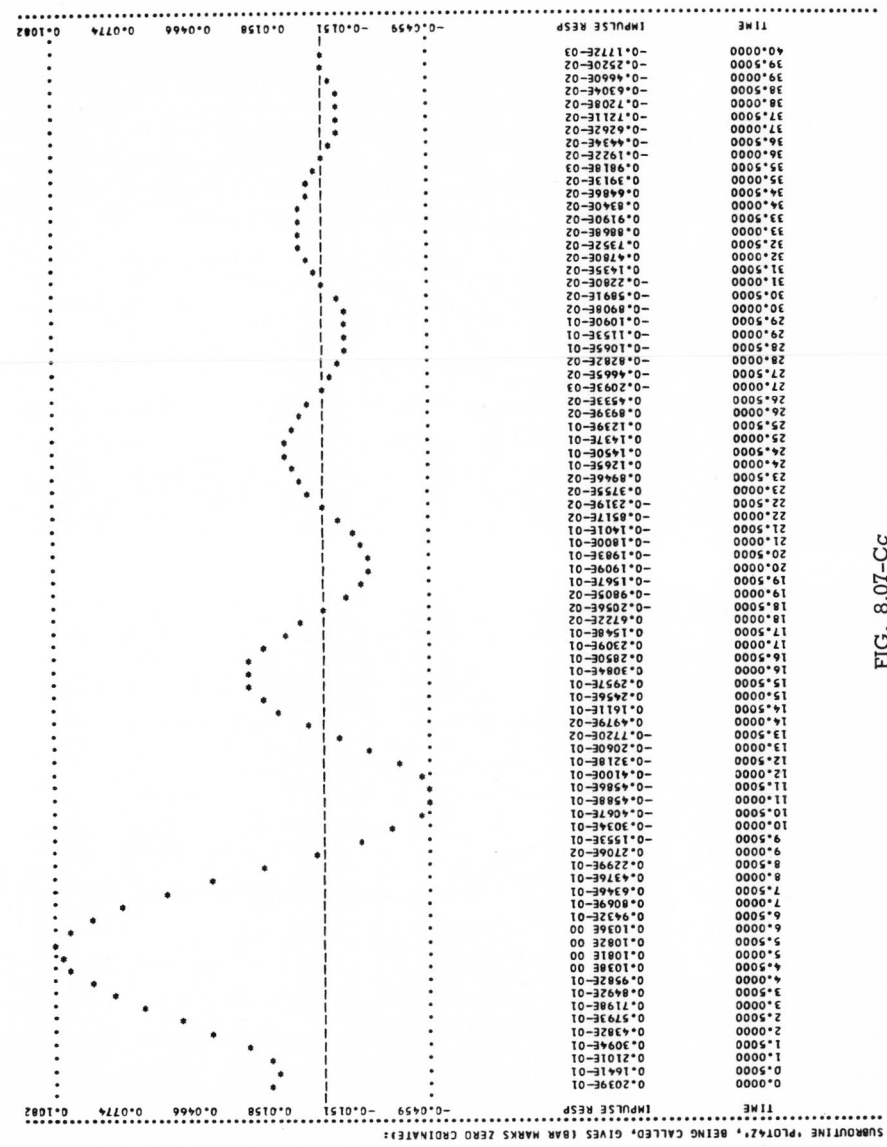

FIG. 8.07-Cc

"bump" that appears in the phase characteristic are likely causes thereof; the somewhat less constant pass-band characteristic may also have its effect. That all these things are intimately related can indeed be shown in various ways. The relations are not simple, however, and the whole ω range from zero to infinity must technically be taken into account. Relations between magnitude and angle of $T(j\omega)$ do exist, and their study is fascinating, but we cannot go into them here.

These (and previous) examples begin to give us some feeling for what good transmission is. A complicated subject, it is not merely a matter of flat (constant) magnitude and linear phase over a certain frequency band (since no signal is strictly limited as to frequency content), though that may well be one's major concern.

One example of the value of such understanding is in the compensation of systems already built, for good transmission. Another is in the design of networks whose purpose is simply to *delay* signals (to gain time for some other operation on the signal, or to simulate propagation of some sort). Again we must be content here with mere mention of these subjects.

The delay pair (8.06-3) and its ω-domain formulation (8.07-1) or (8.07-2), nonrational though the s function be, is evidently sufficiently important that it should be a part of our repertoire.

8.08 THE LAPLACE TRANSFORMATION

Our theoretical experiments with extensions of network analysis to physically unreal yet interesting (and *almost* real) cases have been adventurous but successful. The results are informative too, unreal though the "networks" be.

The experiments were of course carefully chosen for those ends. But they do imply the existence of a larger subject in mathematics, of which our ordinary circuit analysis is but one segment. "Ordinary circuit analysis," as we have developed it, comes down essentially to (see Fig. 8.08-A, Secs. 7.24, 7.21, etc.):

a **1** Transformation of $e(t)$ to $E(s)$,

2 Calculation of $T(s)$,

3 Calculation of $R(s) = E(s)T(s)$, (8.08-1)

b Transformation of $R(s)$ to $r(t)$, the response itself.

The transformation of $e(t)$ to $E(s)$ we can accomplish by direct integration [see (7.04-7)]:

$$E(s) = \int_0^\infty e(t)e^{-st}\,dt, \qquad \mathbf{RE}(s) > \sigma_a. \tag{8.08-2}$$

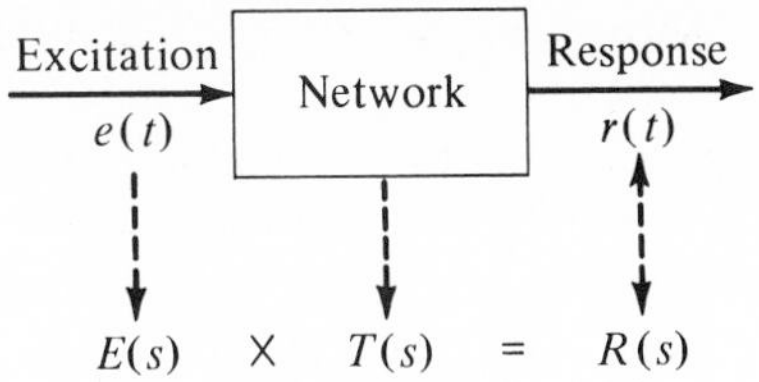

FIG. 8.08-A

But the transformation of $R(s)$ to $r(t)$ we have accomplished not directly, as by integration, but by examining the results of many such e-to-E transformations, our collection of pairs, in which we have always found what we need. It is necessary only to assume uniqueness to be able thus to "work backward"; and because our $R(s)$ have always been *rational* functions of s, the pair collection has sufficed.

In cases like those of the previous sections of this chapter, we do *not* find $R(s)$ in our collection of results (pairs). We have resorted to tricks (use of series, differentiation and integration with respect to parameters) which though valid are still not direct and honest approaches. There ought to be a parallel to (8.08-2)—a straightforward, direct method of transforming back from s to t. By this we could mathematically accomplish the *inverse* transformation, that of $R(s)$ to $r(t)$, in straightforward fashion, regardless of the character of $R(s)$. (Some restrictions on R may be necessary, but certainly not the very severe one that it be rational.)

Such an inverse transformation does exist, being the solution of (8.08-2) regarded as an integral equation in which $e(t)$ is unknown and $E(s)$ is known (see Sec. 7.07). Elegant though it is, its derivation, and even mere demonstration of its validity, are not simple. We therefore simply state it:

$$r(t) \ (=) \ \frac{1}{2\pi j} \int_{\Gamma} R(s) e^{ts} \, ds. \qquad (8.08\text{-}3)$$

The integral is to be evaluated in the (complex) s plane along some contour Γ whose nature is somewhat restricted by the character of $R(s)$, that is, by σ_a, but which essentially runs from "minus infinity" to "plus infinity." For some values of t the value of the integral may not be the appropriate value for $r(t)$ (for example, at a step); the symbol $(=)$ (read "equals almost everywhere") calls attention to this. Evidently some knowledge of the theory of functions of a complex variable is necessary to understand all this. We omit such discussion,

a Because it would be far too lengthy, and (8.08-4)

b Because it is completely unnecessary for our purposes.

For us $T(s)$ is always rational and $E(s)$ usually is, so that $R(s)$ is usually rational; for that case we are fully equipped. For unusual excitations one can always fall back on convolution [obtaining $r_\delta(t)$ from $T(s)$ if necessary], or the use of power series.

The two transformations (8.08-2) and (8.08-3) are known respectively, in mathematics, as the (unilateral) *direct* and *inverse* operations of the *Laplace transformation.* Their history is long; its roots even antedate Laplace's work (1779). Their literature is enormous. It consists not only of theory (developments, proofs, etc.), but it also contains collections of pairs—extensive tables constructed by experiment, by exploration of the mathematical literature of integration, by application of the transformation formulas themselves. One extremely important such table, worth knowing about, is that of Campbell and Foster (Appendix B), for example. The student interested in pursuing the theory and history of the Laplace transformation might well start with Gardner and Barnes (Appendix B), particularly its appendixes.

The direct Laplace transformation also exists in a *bilateral* form:

$$E(s) = \int_{-\infty}^{\infty} e(t)e^{-st}\,dt. \tag{8.08-5}$$

The use of (8.08-5) requires knowing $e(t)$ all the way back in history. It has its uses, but the unilateral form (8.08-2) is the one appropriate for most network-analysis problems, which usually start at some particular time, with given initial conditions. The inverse of (8.08-5) is formally the same as (8.08-3), but both (8.08-5) and its inverse are governed by *two* restrictions, upper and lower, on **RE** (s). We say no more about it here.

8.09 THE IDEAL FILTER; CUTOFF RATE, DELAY, RISE TIME

The Laplace transformation is unnecessary for ordinary network analysis. But certain extraordinary network-analysis problems, unreal though they be, have engineering utility as limiting or ideal situations. For them the Laplace transformation can be very useful. *Delay* presents one such problem; the infinite networks of the earlier sections of this chapter provide other examples; here is one more.

Consider a (hypothetical) device with the characteristics of Fig. 8.09-A. In the ω domain its sudden cutting off of transmission at $\omega = \omega_c$ suggests the name "ideal low-pass filter." We endow it with a linear phase characteristic also.

The transfer function is obviously not a rational function of s; in fact we do not know what $T(s)$ is, or even how to go about looking for it, or whether it exists. But if the σ_a associated with it is negative, as we now assume, then we can calculate the (hypothetical) impulse response from the characteristics of Fig. 8.09-A by using (8.08-3) with $\sigma = 0$, $s = j\omega$:

$$r_\delta(t)(=)\frac{1}{2\pi j}\int_{-\infty}^{\infty} E_0\,T(j\omega)e^{jt\omega}j\,d\omega. \tag{8.09-1}$$

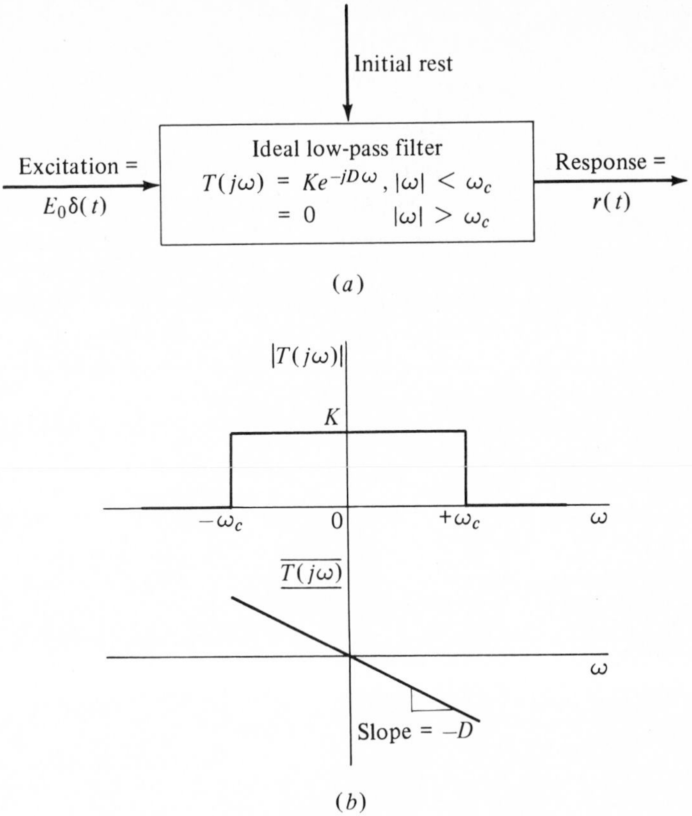

FIG. 8.09-A

The integral is in this case not difficult to evaluate because the integrand is zero except in the range $-\omega_c < \omega < +\omega_c$, and because the integrand has odd and even symmetries that cause the imaginary part to vanish (of course!—the response *must* be real). The real part then "doubles up" and we have only a simple integral from $\omega = 0$ to $\omega = \omega_c$:

$$
\begin{aligned}
r_\delta(t) &= \frac{E_0}{2\pi}\int_{-\omega_c}^{+\omega_c} Ke^{-jD\omega}e^{jt\omega}\,d\omega \\
&= \frac{E_0 K}{2\pi}\int_{-\omega_c}^{+\omega_c} [\cos{(t-D)\omega} + j\sin{(t-D)\omega}]\,d\omega \\
&= \frac{E_0 K}{2\pi}\,2\int_{0}^{+\omega_c} \cos{(t-D)\omega}\,d\omega \\
&= \frac{E_0 K}{\pi}\,\frac{\sin{[\omega_c(t-D)]}}{t-D}.
\end{aligned}
\tag{8.09-2}
$$

Most interesting about this response (Fig. 8.09-B*a*), which seems superficially to resemble those of the examples of Sec. 8.07 (Figs. 8.07-B*c*, 8.07-C*c*), is that it extends indefinitely back in time: D (seconds) before the peak is *not* its beginning; there is no proviso "$t > 0$" in (8.09-2). The device, according to this analysis, responds before it is excited. (The term *noncausal* is sometimes used to describe this unreal property.) This is not paradoxical, for we have no assurance that the device postulated has any physical reality. Nor is the Laplace-transformation calculation in error. The only reasonable conclusion is that the device (transfer function) so blithely assumed is not physically realizable, for its response is absurd. It is therefore impossible to present physical corroboration, any demonstration of such a network. It is interesting, nevertheless, to review previous low-pass-filter examples and to note how their responses *do* more or less (in a satisfying way) corroborate the discussion here. See Fig. 6.15-B, for example, and Figs. 6.17-B, 7.15-C, 7.19-A, 8.07-C, and more. Ideal low-pass filters they certainly do not show, but the impulse responses do have appearances more or less that of Fig. 8.09-B*a*, except that they definitely start at $t = 0$.

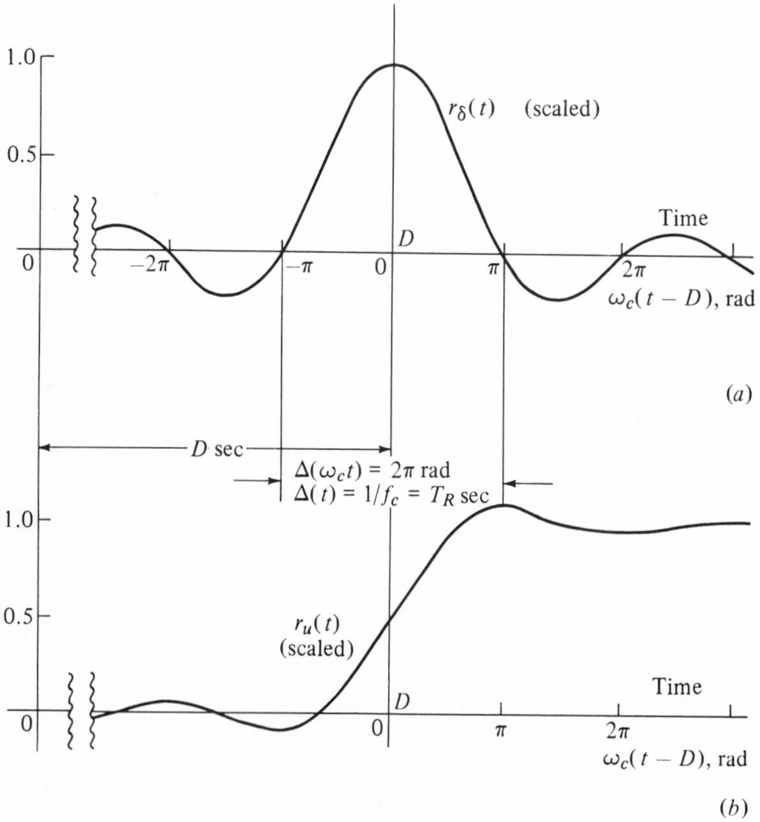

FIG. 8.09-B

The impossible properties of the transfer function of Fig. 8.09-A are:

a Sudden, infinitely steep cutoff,

b Exactly constant $|T|$ over a band of frequencies, (8.09-3)

c Exactly linear phase over a band of frequencies.

One might say, to be sure, that the temporal difficulty disappears and that these properties become possible when the "delay" D is infinite. And this is simply the logical, limiting conclusion of the rule (Sec. 8.07): "The sharper the cutoff, the longer the delay." Infinitely sharp cutoff means infinite delay, the response signal never appears. (Analysis in terms of the relations between magnitude and angle of transfer function leads to the same conclusion.) But this is all preposterous.

Also interesting, however, is the (hypothetical) step response of this "ideal" network, which integrating the impulse response gives (Fig. 8.09-B*b*). The distortion here is principally the finite steepness of its rise. That it resembles the step responses of the two real examples of Sec. 8.07 is clear, without plotting them, because of the simple, integral relation between step and impulse responses.

We observe that a convenient (though arbitrary) definition of the *rise time* of the step response is the time between the two zero crossings that define the principal lobe of the impulse response. For this is, one might say, the principal part of the step response. It is (Fig. 8.09-B)

$$T_R = \frac{2\pi}{\omega_c} = \frac{1}{f_c} = \frac{1}{\text{bandwidth}}. \tag{8.09-4}$$

The implied definition of a *bandwidth* for the device of Fig. 8.09-A*a*, as f_c Hz (or ω_c rad/s) is natural and hardly arbitrary. Very interesting is the relation between bandwidth and rise time that (8.09-4) states. Let us see if it holds, at least approximately, in real cases.

In Fig. 8.07-B*a*, a bandwidth is difficult to define because the cutoff is slow. The ω values (*a*) 0.7 (where attenuation begins), (*b*) 1.0 (where it is well under way), and (*c*) 1.3 (where the signal is no longer appreciable) are all "reasonable." The corresponding rise times, calculated from (8.09-4), are 9.0, 6.3, and 4.8 s. The actual rise time, measured by the principal lobe of the impulse response (Fig. 8.07-B*c*), is about $(6.3 - 0.7) = 5.6$ s. The "middle-attenuation bandwidth" 1.0 gives a fairly good check ($\omega_c = 1.0$ leads to $T_R = 6.3$). As a rough rule of thumb, (8.09-4) seems to have some validity.

In Fig. 8.07-C*a* the bandwidth is much easier to define because the cutoff is sharp. It is surely 0.75 closely, so that $T_R = 8.4$. From Fig. 8.07-C*c* we obtain a rise time of $(9.0 - 0.5) = 8.5$ s. In this second case, where there is little room for argument about the value of the bandwidth, the agreement with (8.09-4) is excellent. In the first case, where the value of the bandwidth is debatable, the applicability of (8.09-4) is equally uncertain. But we have in (8.09-4) at least a rule of thumb:

$$\text{Step rise time in seconds} = \frac{1}{\text{bandwidth in hertz}}. \tag{8.09-5}$$

The accuracy of both the first rule of thumb (Sec. 8.07),

$$\text{Delay} = \text{magnitude of low-frequency phase slope}, \tag{8.09-6}$$

and this second rule (8.09-5) depends on circumstances; both can be useful but are not to be taken as absolutes. (One can even formulate a third rule, one that describes distortion oscillation magnitudes, from Fig. 8.09-B.)

The ideal low-pass filter, intended primarily to illustrate the utility of Laplace-transformation analysis, seems also to have engineering utility as a limiting-case guide. Ideal from the ω-domain view, it is certainly not ideal from the t-domain view. Some rounded cutoff shape in ω is probably more desirable. What compromise is best is a complicated question we leave unanswered. But the rules of thumb developed above are useful.

Laplace-transformation analysis can similarly be applied to many other hypothetical but interesting models. In principle we can use it to find the response of any imagined device: $R(s)$ need not be rational, the integration can often be carried out easily, and the results, however absurd, can yet be useful if taken in perspective.

Once more we note that the Laplace (integral) transformation has practical utility in the discussion of unreal yet interesting (as limiting or ideal) situations.

8.10 A REAL NETWORK

It is appropriate now, as well as interesting and very instructive, to examine the operation of the Laplace transformation on a real network-analysis problem, one which deals with a rational function of s. Is the work as simple as in the ideal-network case of Sec. 8.09? How does it compare with our standard (partial-fraction expansion) technique?

For simplicity consider the RC circuit of Chap. 2, impulse-excited. We have

$$R(s) = Q_0 \frac{1/C}{s + \alpha} \tag{8.10-1}$$

and, by (8.08-1),

$$r_\delta(t) \ (=) \ \frac{1}{2\pi j} \int_\Gamma \frac{Q_0/C}{s + \alpha} e^{ts} \, ds. \tag{8.10-2}$$

Since σ_a is negative ($\sigma_a = -\alpha$), we can set $\sigma = 0$ and obtain a real integral in ω:

$$r_\delta(t) \;(=)\; \frac{1}{2\pi}\int_{-\infty}^{\infty} \frac{Q_0/C}{\alpha + j\omega}\, e^{jt\omega}\, d\omega. \tag{8.10-3}$$

The evaluation is straightforward; we encounter and use the same symmetries as in Sec. 8.09. Using the abbreviation $V_0 = Q_0/C$ we obtain:

$$\begin{aligned} r_\delta(t) \;(=)\; &\frac{V_0}{2\pi}\int_{-\infty}^{\infty} \frac{\cos t\omega + j \sin t\omega}{\alpha + j\omega}\, d\omega \\ = &\frac{V_0}{2\pi}\int_{-\infty}^{\infty} \frac{(\alpha \cos t\omega + \omega \sin t\omega) + j(\alpha \sin t\omega - \omega \cos t\omega)}{\alpha^2 + \omega^2}\, d\omega \\ = &\,2\frac{V_0}{2\pi}\int_{0}^{\infty} \frac{\alpha \cos t\omega + \omega \sin t\omega}{\alpha^2 + \omega^2}\, d\omega. \end{aligned} \tag{8.10-4}$$

Hence

$$\begin{aligned} r_\delta(t) \;(=)\; &\frac{V_0}{\pi}\left(\int_0^\infty \frac{\alpha \cos t\omega}{\alpha^2 + \omega^2}\, d\omega + \int_0^\infty \frac{\omega \sin t\omega}{\alpha^2 + \omega^2}\, d\omega\right) \\ = &\frac{V_0}{\pi}\left(\frac{\pi}{2} e^{-\alpha t} + \frac{\pi}{2} e^{-\alpha t}\right), \qquad t > 0 \\ = &\frac{V_0}{\pi}\left(\frac{\pi}{2} e^{-\alpha t} - 0\right), \qquad t = 0 \\ = &\frac{V_0}{\pi}\left(\frac{\pi}{2} e^{-\alpha t} - \frac{\pi}{2} e^{-\alpha t}\right) = 0, \qquad t < 0. \end{aligned} \tag{8.10-5}$$

In this calculation we have used two well-known definite integral values, obtainable from any good table of integrals:

$$\int_0^\infty \frac{\cos mx}{1 + x^2}\, dx = \frac{\pi}{2} e^{-m}, \qquad m > 0, \tag{8.10-6}$$

and

$$\int_0^\infty \frac{x \sin mx}{1 + x^2}\, dx = \frac{\pi}{2} e^{-m}, \qquad m > 0. \tag{8.10-7}$$

The conclusion is

$$\begin{aligned} r_\delta(t) &= V_0 e^{-\alpha t}, \qquad t > 0, \\ &= 0, \qquad t < 0, \\ &= \tfrac{1}{2} V_0, \qquad t = 0. \end{aligned} \tag{8.10-8}$$

This is, of course, correct. The step at $t = 0$, on account of which $r_\delta(0)$ is not really defined, is the reason for the symbol $(=)$, which states just that. The inverse Laplace transformation gives the average value, $r_\delta(0) = \frac{1}{2}[r_\delta(0-) + r_\delta(0+)]$, a most reasonable compromise. Our normal formulation

$$r_\delta(t) = V_0 e^{-\alpha t} u(t) \tag{8.10-9}$$

begs the question but removes the parentheses. In any physical case, of course, discussion of the exact value of $r_\delta(0)$ is pointless (see Fig. 2.02-B).

We observe that the labor involved in Laplace-transformation analysis in even this extremely simple RC case is much greater than that of our ordinary analysis. It is only fair to say that this sort of integral evaluation can be circumvented by using contour integration from the theory of functions of a complex variable; and it leads, in fact, to exactly our standard partial-fraction, table-of-pairs technique. But that technique can, as we well know, be developed without the Laplace transformation.

8.11 CONCLUSION

To the question: "What do we have in the Laplace transformation?" we can now reply "a very elegant, powerful mathematical tool, with many applications." It can be used, for example, in the straightforward solution of differential equations qua differential equations, particularly of the ordinary, linear, constant-coefficient type which, in the mathematical view, network analysis is. The differential equations transform into rational functions of s, then to be treated exactly as we have been doing with our $T(s)$, $E(s)$, and $R(s)$ functions. The parallel is closer than that: the two are the same thing, viewed slightly differently.

The use of the Laplace transformation for what we call nonrational network functions affords one example of its additional capabilities; there are many more. But for ordinary network analysis the Laplace transformation is unnecessary (Chaps. 1 to 7), though it gives an alternate approach to (derivation of) the same analysis technique. In ordinary, everyday, bread-and-butter network analysis, the actual work is the same, whichever is used.

We accordingly leave the Laplace transformation for deeper study elsewhere.

8.12 EXERCISES

8-1. In terms of $F(s)$, the mate of $f(t)$, find the mates of (*a*) $tf(t)$, (*b*) $f(t)/t$. For what sort(s) of circuit analysis problems might these pairs be useful?

8-2. Let $f(t)$ and $F(s)$ be mates. Show that, *properly interpreted*, the transform of $f'(t)$ is $[sF(s)]$. Explain carefully why the transform of $f'(t)$ is usually given, however, as $[sF(s) - f(0)]$; see Fig. 6.09-D and restate its "theorem" in Laplace-transform language.

If a network model transmits well at high frequencies (its transfer function numerator and denominator are of the same degree), application of the initial-value theorem (tableau) encounters difficulties. What are they? How do they relate to the matter of the previous paragraph?

8-3. A theorem of Laplace-transformation theory is

If: $$f(t) \sim F(s),$$

Then: $$\omega_0 f(\omega_0 t) \sim F\left(\frac{s}{\omega_0}\right),$$

in which ω_0 is real and positive, a reference frequency for scaling. This is a formal statement of the frequency and time normalization we have often practiced [see (7.14-28)]. First prove the theorem. Then verify the statement above, and explain the role of the factor ω_0 in $[\omega_0 f(\omega_0 t)]$. In actual network analysis, using normalization by ω_0, how is the factor handled? Illustrate with examples from Chap. 7 or elsewhere.

We are also familiar with impedance-level scaling [see (7.14-27)]. Is there, or should there be, a corresponding Laplace-transformation theorem? Explain.

8-4. Let the impulse response of a network be $r_\delta(t) = \sum_1^N K_j e^{p_j t}$. Let the excitation applied to the network be time-limited: $e(t) = 0$, $T < t$. For $t > T$, is the response thereto $r(t) = \sum_1^N K_j e^{p_j t}$? Explain your answer fully, being careful to clarify any symbol ambiguities. Will the response waveshape for $t > T$ closely resemble the impulse-response waveshape? Illustrate.

In general, will the natural part of a network response, r_N in $r = r_N + r_F$, have the same waveshape as the network's impulse response? Will it have the same sort of formal representation, $\sum Ke^{pt}$? Illustrate the possibilities.

8-5. Why is it difficult to find the response of a network to the simple excitation $Kt/(1 + t/T_0)$, whose waveshape is very much like that of $(1 - e^{-\alpha t})$, the response to which is very easily obtained? How would you go about finding the actual difference between the two responses?

8-6. Compare the distorting oscillations of Fig. 8.09-B*a* with those of the examples of Sec. 8.07. Can you devise a rule to predict, approximately, the size of oscillations in the impulse response of a (real) low-pass filter?

8-7. Find the maximum value of the r_δ curve in Fig. 8.09-B in terms of E_0, K, and ω_c; explain the dimensions. Repeat for the r_u curve and correlate with your previous answers.

8-8. Consider a single *RC* circuit (Chap. II). To apply the two rules of Sec. 8.09 (about delay and rise time) is here difficult indeed. Explain why, and formulate definitions that will force the rules to have some validity here, explaining their arbitrariness. Repeat for a cascade of two (noninteracting) *RC* circuits, of transfer function proportional to $(s + \alpha)^{-2}$. Repeat for cascades of 3, 4, and 5 *RC* circuits.

8-9. Go over the examples of Chap. 7, applying the two rules of thumb (Exercise 8-8) where it is reasonable to do so. Discuss and explain their accuracy and utility in each case.

8-10. The arguments of Sec. 1.10 and elsewhere, that impulse excitation and *short*-pulse excitation give indistinguishable responses, can be strengthened mathematically by using the delay pair of Sec. 8.06. Let $T(s)$ be the transfer function of a network, and $r_\delta(t)$ its impulse response. Consider the network's response $r(t)$ to the square-pulse excitation $E_0[u(t) - u(t - T_0)]$. Show that the mate of $r(t)$ is expressible as

$$R(s) = E_0\,T_0\left(1 - \frac{T_0\,s}{2!} + \frac{T_0{}^2 s^2}{3!} + \cdots\right)T(s)$$

so that the response can be written

$$r(t) = E_0\,T_0[r_\delta(t) + k_1 r_\delta'(t) + k_2\,r_\delta''(t) + \cdots].$$

After properly adjusting the pulse strength $(E_0\,T_0)$, evaluate the difference between impulse and square-pulse response and explain the conditions [on the value of T_0 and on the nature of $T(s)$] under which the argument of Sec. 1.10 is corroborated. When is it *not* corroborated?

8-11. Evaluate the general or nth term of (8.02-10). Show that the ratio of the term in t^{2m} to that preceding it approaches $t^2/(4m^2)$. Discuss the range of convergence of the series in t.

8-12. Carefully verify both (8.03-1) and (8.03-2), using methods suggested in Sec. 8.02.

8-13. Carefully verify the three formulas and the plot given in Sec. 8.03. (The methods of Sec. 8.02 suffice.) Look up in tables and thus corroborate the pair

$$\frac{1}{\sqrt{(s/\omega_0)^2 + s/\omega_0}} \sim e^{-\omega_0 t/2} I_0\left(\frac{\omega_0\,t}{2}\right).$$

8-14. Show that the integrals that express the mates (Laplace transforms) of $t^{-1/2}$ and of $t^{-1/4}$ converge at the lower ($t = 0$) limit as well as at the upper limit. Then verify, by calculation or by examination of at least two different tables, the pairs given in Sec. 8.04. Does $u(t)/t$ have a mate?

8-15. Find $G(s)$, the mate of $g(t)$ of Fig. 7.06-A, by using (8.06-3) and thus verify (7.06-1). Where are the poles of $G(s)$? Explain carefully. In what way is this related to (7.04-8)? Where are the zeros of $G(s)$? Is the usual rule "number of zeros = number of poles" obeyed? [See (5.05-7).]

8.13 PROBLEMS

8-A. Consider the excitation wave defined by (7.04-8). Prove that for it $\sigma_a \to -\infty$ so that its Laplace transform $E(s)$ is meaningful throughout the s plane. Why is calculation of $E(s)$ difficult? Is $E(s)$ obtainable from (extensive) Laplace-transform tables? What is the most practical way of obtaining the response of a given network to such an excitation? Illustrate by obtaining the response from rest of the networks of Sec. 8.07 to this excitation, taking $\alpha = 1$ and $T_1 = 10$ (normalized units).

8-B. A so-called delay network has poles and zeros of its transfer function at $s = \pm 1 + j0,\ \pm 1 \pm j1,\ \pm 1 \pm j2,\ \pm 1 \pm j3$, with the symmetry of (5.13-12). Evaluate its performance as such by obtaining and criticizing its response to various signals of pulselike nature.

8-C. The network used to obtain Fig. 8.04-A was the finite RC network shown in Fig. 8-C. The excitation e was a 0.5-μs pulse, repeated 2000 times per second; the bar separation in the $|T(j\omega)|$ figure is 2 kHz; the small bar at the left is "dc." The time scale in the $r_\delta(t)$ picture is 20 μs/cm (one square represents

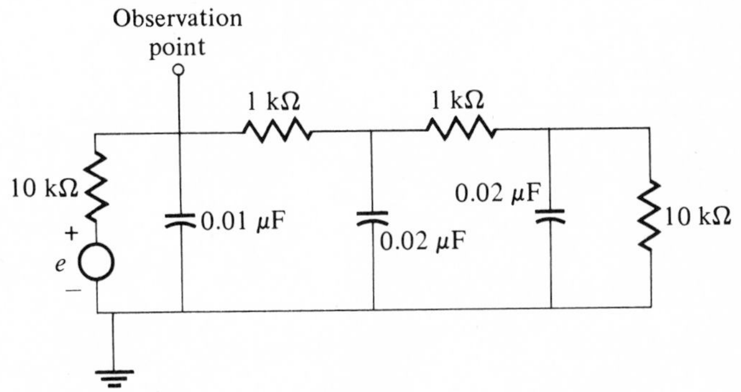

FIG. 8-C

20 μs). Obtain the network's transfer function and its impulse response. Carefully compare them with the theoretical $\omega^{-1/2}$ and $t^{-1/2}$ responses of Sec. 8.04. Are the comments of Sec. 8.04 about Fig. 8.04-A justified?

8-D. On the assumption that

$$Z(s) = \frac{\sqrt{1 + s^2}}{1 + (1 - m^2)s^2}$$

is a viable transfer function in spite of its irrational character (see Fig. 8-D), find the network's (voltage) response to impulse and step current excitation in power series form through the fifth power of t. Check, by physical reasoning, at least the first three terms. Plot the responses as far as you can. (Take $m = 0.6$ if you wish.) Verify formally that $Z(s)$ is indeed the driving-point impedance of the infinite network of Fig. 8-D. Obtain its actual impulse response from a Laplace-transformation table, plot it, and compare with your plot.

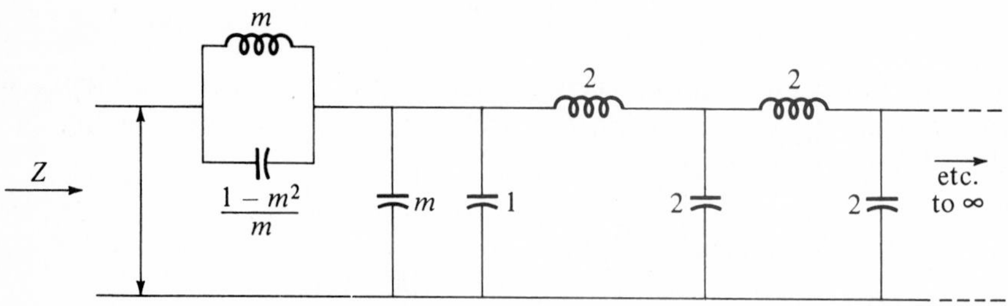

FIG. 8-D

8-E. Select several low-pass networks from the examples in Chap. 7. Define the "rise time" of the step responses as the time elapsing between the instants when $r_u = 0.05$ and $r_u = 0.95$, the final value being $r_u = 1.0$. (This might be called "the 5 to 90 percent rise time.") Define the bandwidth as the frequency at which $|T(j\omega)|$ has dropped to half its dc value. What are the values of the various [(rise time)(bandwidth)] products? Compare with (8.09-5) and explain the differences.

What are the values for a rise time defined by "10 to 90 percent" rise?

Point out the arbitrary definitions in the three versions above. Which do you prefer? Can you formulate a better one?

Apply your rule to the simple RC circuit of Chap. 2.

8-F. Consider an hypothetical "ideal" band-pass filter whose transfer function has the form shown in Fig. 8-F. Calculate and plot the impulse response. Compare your result with (8.09-2). Discuss the concept of "delay" therein and therefor. Show that your result can in fact be written in almost that form, the single sinusoid being replaced by the product of two sinusoids, of frequencies ω_0 and ω_c. Carefully point out similarities and differences. In what sense can

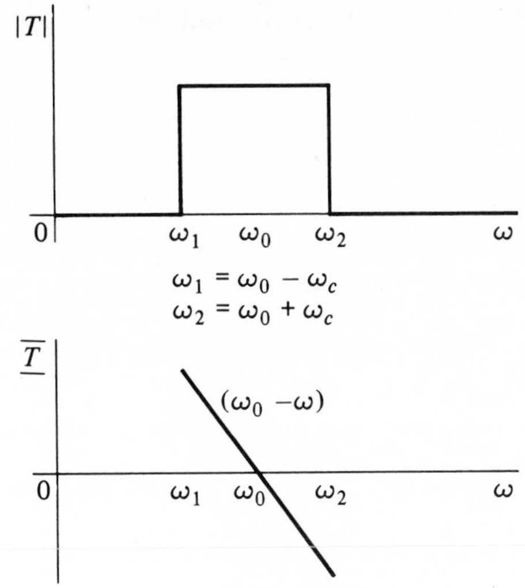

FIG. 8-F

the response of the "ideal" band-pass filter be considered to be a "modulation" of that of the "ideal" low-pass filter? Find its response to a brief sinusoidal pulse $e(t) = E_0 \cos \omega_0 t[u(t) - u(t - T_0)]$. Take $\omega_0 = \sqrt{\omega_1 \omega_2}$ and $\omega_0 T_0 = 10\pi$. Discuss the concepts of "delay" and "rise time" in this situation. Repeat for other interesting values of T_0 and ω_0.

8-G. Postulate a device with linear phase shift and the magnitude-of-transfer function characteristic of Fig. 8-G*a*. Thoroughly discuss its impulse and step responses. Repeat for the other three characteristics given in Fig. 8-G. Are any of the "devices" noncausal?

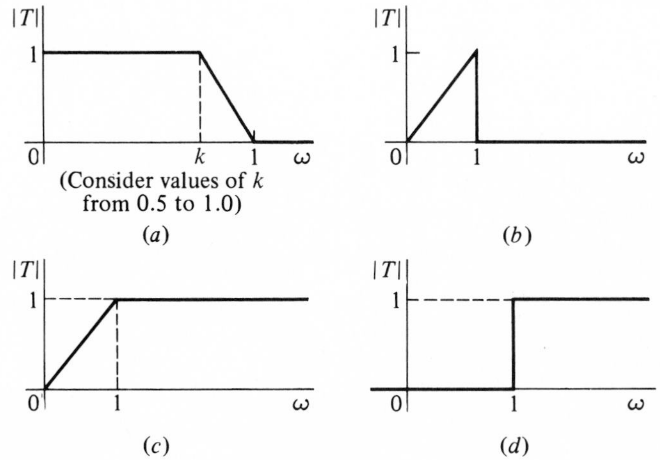

FIG. 8-G

8-H. The network of Fig. 8.07-B has the transfer function

$$T(s) = \frac{0.5}{1.000 + 3.836s + 5.236s^2 + 5.236s^3 + 3.836s^4 + 1.000s^5}.$$

Verify this in detail. Construct also a pole-zero diagram and explain how the sinusoidal-steady-state characteristics are consistent therewith. What seems to be the rule used in laying them out? Find also the response to the more "low-frequency" excitation $te^{-\alpha t}u(t)$, for $\alpha = 1$ and for $\alpha = 0.5$. Discuss the distortion and "delay," and compare with those of Fig. 8.07-B*c*.

8-I. The network of Fig. 8.07-C has the transfer function

$$T(s) = \frac{1 + 1.660918s^2 + 0.604115s^4}{2 + 8.314996s + 18.3666251s^2 + 33.30288s^3 + 27.969579s^4 + 29.628803s^5}.$$

Follow the directions of Prob. 8-H for this network. If the element values are *not* scaled so that cutoff begins at $\omega_n = 0.7$, what are the low-frequency phase slope and the "impulse delay"?

Chapter **9**

CONCLUSION

Begin at the beginning and go on till you come to the end: Then stop.

—The King of Hearts

In Chap. 1 we began, not at the real beginning, for that had been long ago. And we have gone on at length, at great length, for many, many pages. Have we come to the end? No, far from it!

We can *analyze* any ordinary circuit or network, yes. (Only certain unusual circumstances will baffle us now, but with our present background we can easily look into their analysis when the need arises.)

Do we understand network properties? Their skillful use? To some extent, yes. But there is clearly much more to study: signal transmission theory and practice is a large subject. The imaginary axis where $\sigma = 0$ and $s = j\omega$ is far more important and useful a place than we yet appreciate: *there*, for instance we can make measurements at leisure, the state being steady and not transient; *there* is the domain of *Fourier*, a name to conjure with!

Can we *synthesize* networks for desired, prescribed behavior? Only in very simple cases, and that only by a sort of backwards analysis. Network synthesis is a tremendous, fascinating field, of prime engineering importance.

Very useful though it is, we have not discussed mutual induction, or transformers. Nor have we discussed what can be done (with or without mutual inductance) by R, C, and g_m elements used in great profusion, in ever increasingly practical miniaturized physical formats.

We're ready for these things—yes. Our preparation in network analysis is excellent now. But no more can we do here; whether we have come to the end or no, we must halt. The book is long enough.

Most important to recognize is that we have learned enough to understand how little we know and to see dimly how much more there is yet to study in "circuits and networks."

... to make an end is to make a beginning.
The end is where we start from.

—T. S. Eliot

APPENDIX A

DEMONSTRATIONS

Actually *seeing*, with one's own eyes, the demonstrations that the photographs in the book display is an extremely important part of studying network analysis. Obvious though the point may seem, it needs iteration and reiteration! The mathematics of analysis is beautiful, necessary, important, basic. But it is merely logic applied to the paper models of networks set up for the purpose; the *reality* is appreciated only by seeing, or, better, designing, building, and testing the networks.

Physical corroboration is essential to the study of networks. It is logically necessary to check our analyses; educationally it is imperative so to do. Physical viewing is essential to getting any *feeling* for networks' behavior. Hence the photographs in the text.

But looking at the photographs is not enough. Every student ideally, every teacher certainly, should actually perform the demonstrations for himself. It is easy to do; the apparatus is simple and available in every reasonably well-equipped laboratory. There is seldom anything critical about the values of elements or the settings of apparatus; none of the demonstrations (except for certain balance situations) are oversensitive or difficult to perform. And many more ideas will occur as you do them. Try them out! Experiment!

This appendix A gives the essential data for the demonstration results photographed in the text. The signal generating and observing apparatus needed is standard, the networks can be constructed from off-the-shelf elements or adjustable laboratory "boxes": they need not be precise to make their points. We tabulate here in concise form the data for the photographs in the book (made by simple Polaroid cameras). Good luck!

There follow: Fig. A-A, which gives the basic connection diagram, Table A-A, which gives details on important settings, element values, and connections; and Fig. A-B, which describes the networks (circuits) used in the various demonstrations (photographs) of the text.

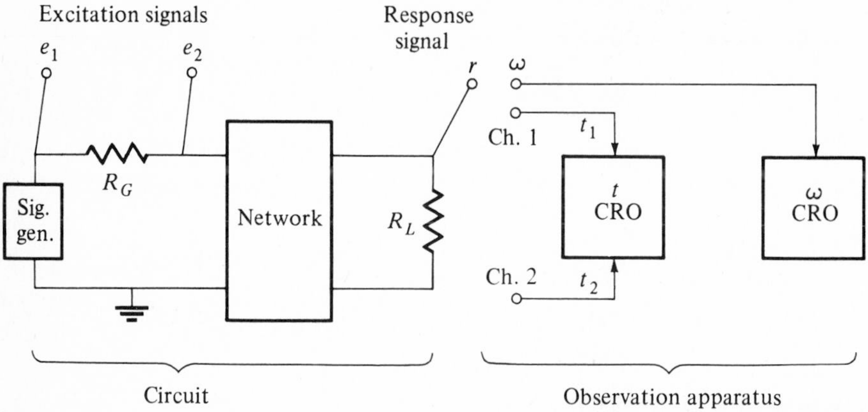

NOTES:

(1) Connection code:

A: e_1 to t_2, r to t_1
B: e_2 to t_2, r to t_1
C: r to ω

(2) Frequency scale in ω (on ω CRO) is fixed by the fact that bar separation = PG rep. rate (Hz, pulses per second).

(3) t domain CRO is ordinary, two-channel, cathode-ray oscilloscope; ω domain CRO is (low-frequency) spectrum analyzer with storage capacity (see Fig. 3.14-H discussion).

(4) *Signal generators*: PG = pulse generator with adjustable rep. rate and pulse width.

SQ = square-wave generator, symmetrical (equal on-off periods) setting, adjustable frequency.

SIN = sinusoid generator, adjustable frequency.

FIG. A-A. Basic demonstration arrangement.

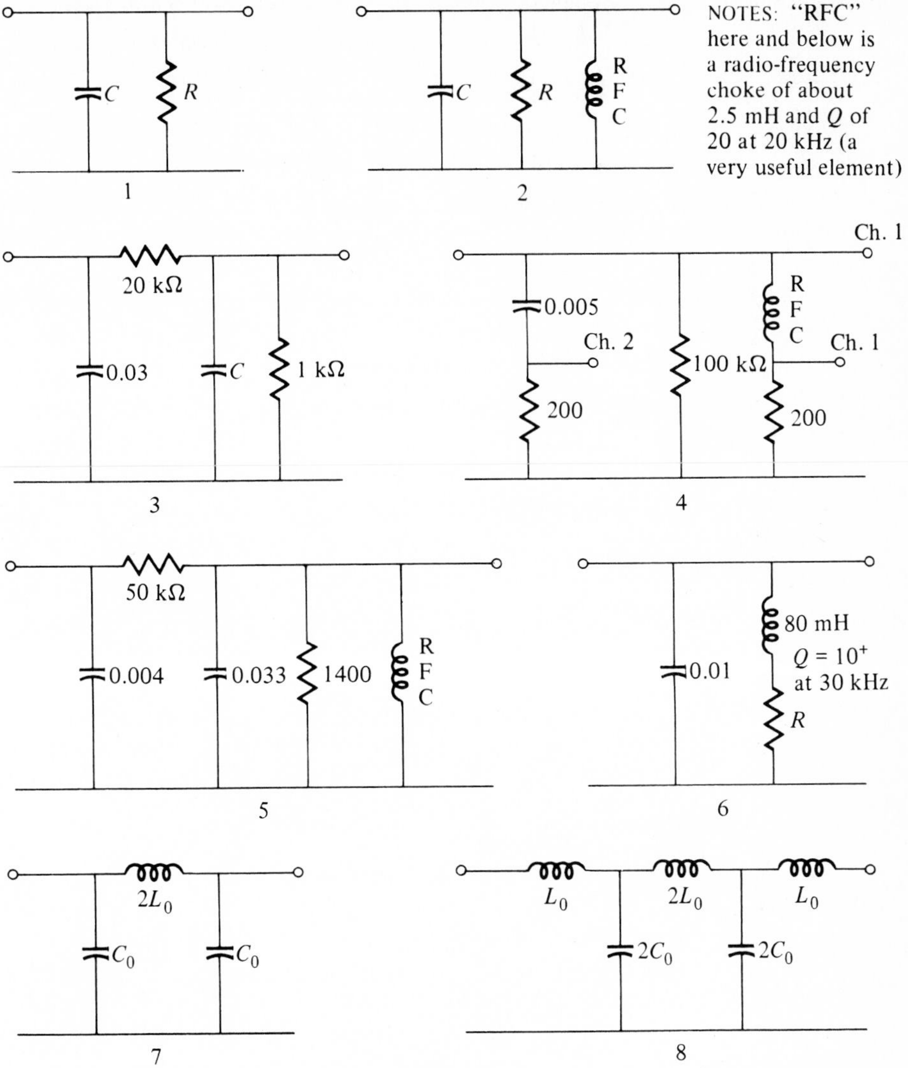

C
R
1
C
R
RFC
2
NOTES: "RFC" here and below is a radio-frequency choke of about 2.5 mH and Q of 20 at 20 kHz (a very useful element)
20 kΩ
0.03
C
1 kΩ
3
Ch. 1
0.005
Ch. 2
100 kΩ
RFC
Ch. 1
200
200
4
50 kΩ
0.004
0.033
1400
RFC
5
80 mH
Q = 10⁺
at 30 kHz
0.01
R
6
2L0
C0
C0
7
L0
2L0
L0
2C0
2C0
8

FIG. A-B

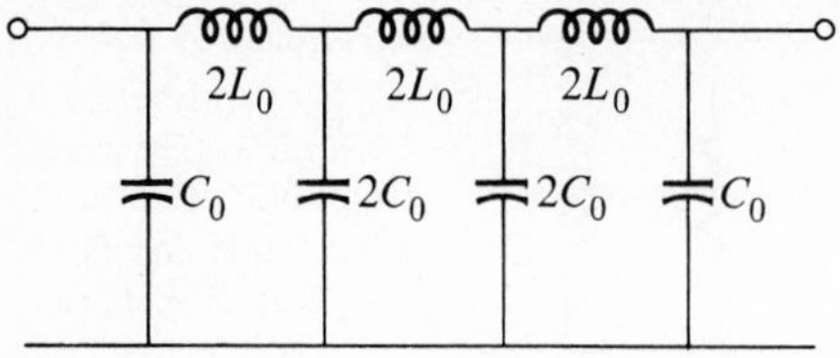

9

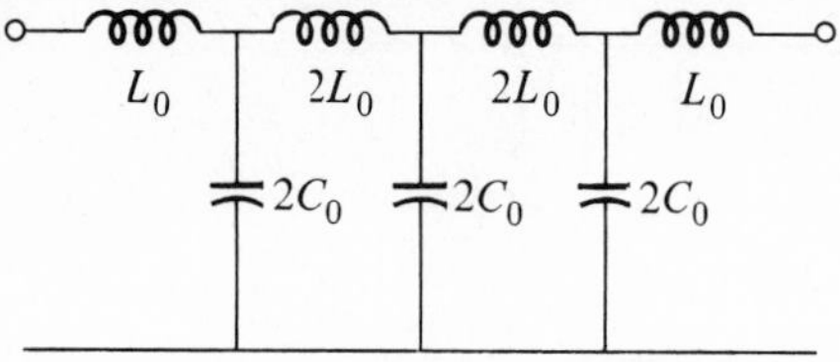

10

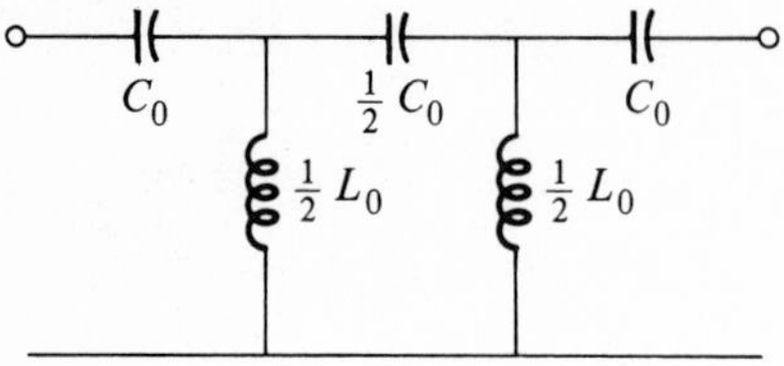

11

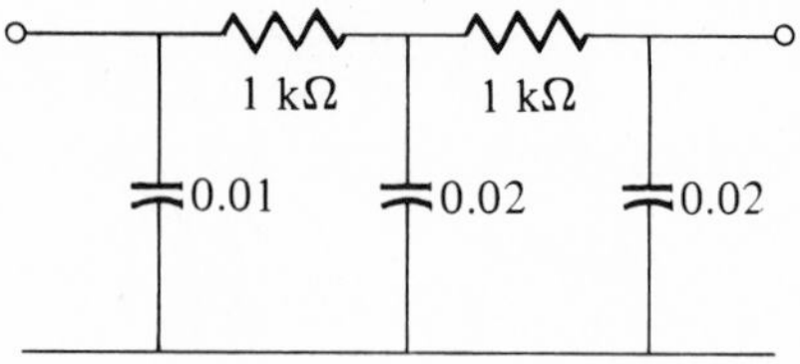

12

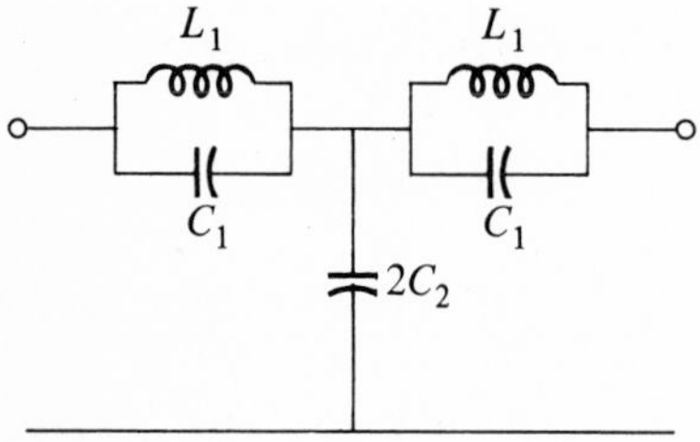

13

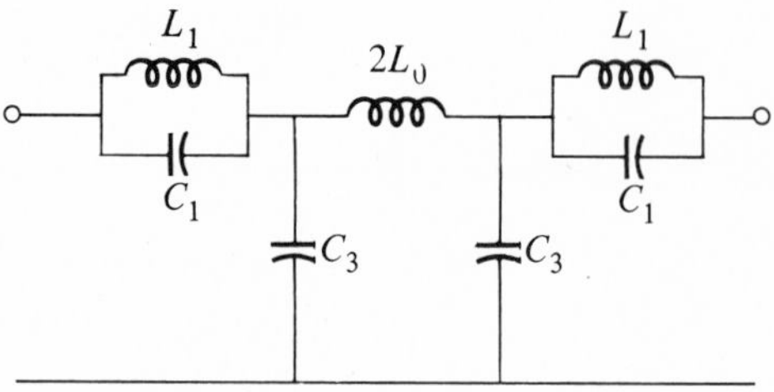

14

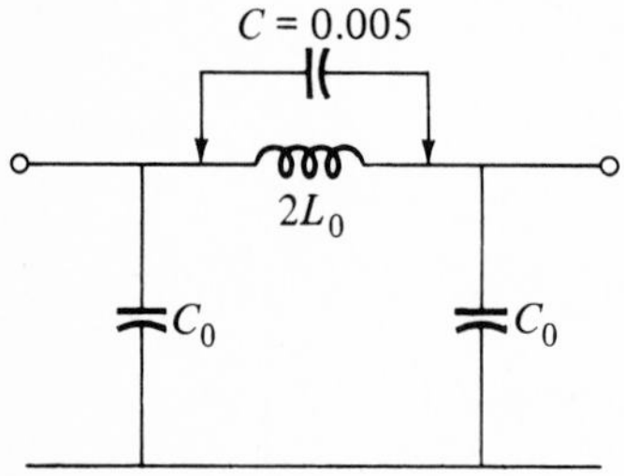

15

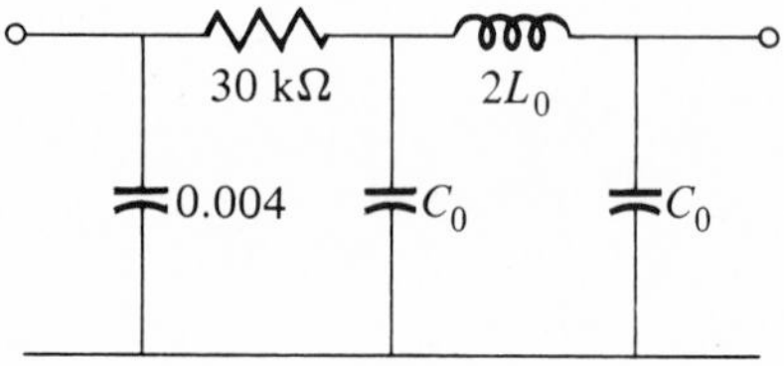

16

FIG. A-B. (*Con.*)

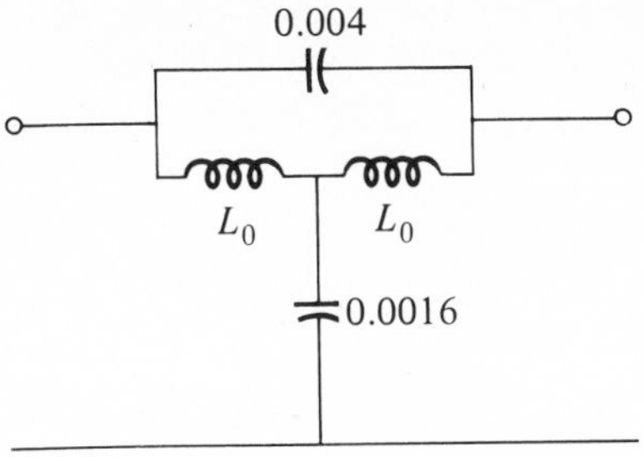

17

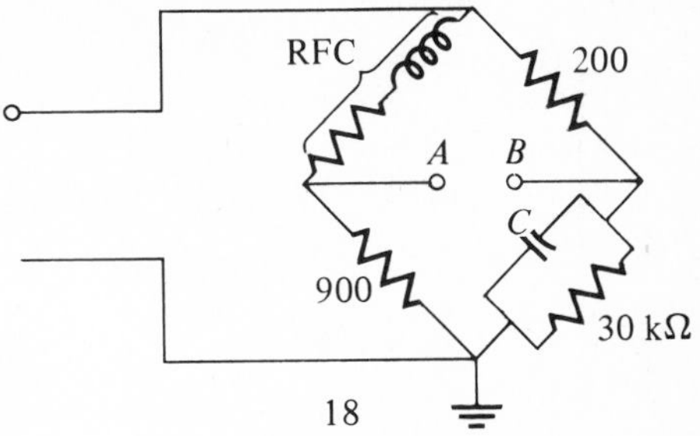

18

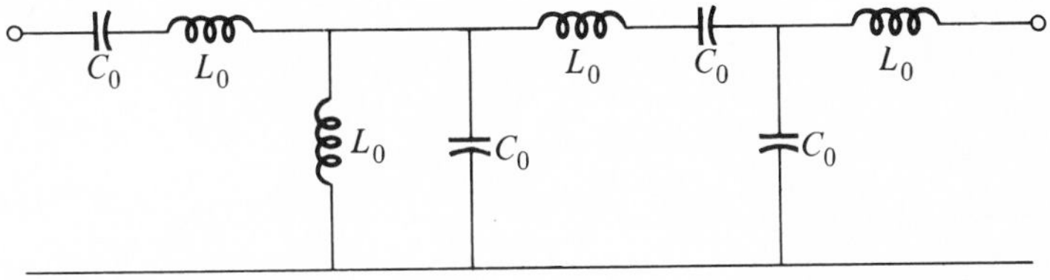

19

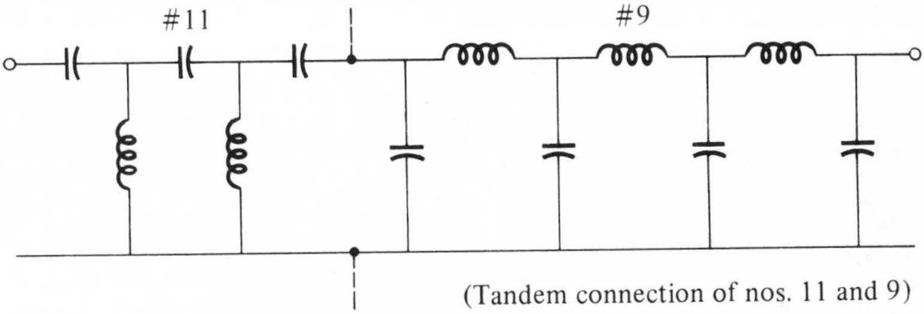

20

NOTES:

R values are in ohms, C values are in microfarads, kΩ = 1000 ohms; L_0 = 8 mH (1000 ohms reactance at 20 kHz), C_0 = 0.008 μF (same comment); L_1 = 5 mH (m = 0.6), C_1 = 0.009 μF, C_2 = 0.005 μF, C_3 = 0.013 μF.

FIG. A-B. (*Con.*)

TABLE A–A

Figure Number (See Text)	Signal Generator: Type Used (Fig. A-A)	Signal Generator Settings: Frequency Hz or pps	Signal Generator Settings: Pulse Width, μs	Termination Values, Ω: R_G	Termination Values, Ω: R_L	Network Used (Fig. A-B)	Connections (Fig. A-A)	t *CRO* Sweep Speed, $\mu s/cm$	Comments (Element Values Are in Ω and μF, $K = 1000$)
1.08-C (1)	SQ	1600	—	none	none	none	e_1 to t	50 and 5	double exposure
1.08-C (2)	SQ	4200	—	none	none	none	e_1 to t	20 and 4	double exposure
1.10-C	PG	1100	1	$30K$	—	none	A	20	$C = 0.002$, $R = 3K$
1.10-C	PG	1100	2	$30K$	—	none	A	20	same
1.10-C	PG	1100	4	$30K$	—	none	A	20	same
2.05-A	PG	500	2	$10K$	—	1	A	40	$R = 10K$, $C = 0.001$ (*a*), 0.002 (*b*), $C = 0.004$ (*c*)
2.07-B	PG	500	1	$10K$	—	1	A	—	$C = 0.01$, $R = 1$ (*a*), 4 (*b*), 10 (*c*)
2.09-D	SQ	1000	—	$30K$	—	1	A	20	$R = 5K$, $C = 0$ (*a*), 0.001 (*b*), 0.003 (*c*), 0.005 (*d*)
2.11-C	PG	1000	150	$10K$	—	1	A	40	$R = 10K$, $C = 0.008$ (*a*), 0.004 (*b*), 0.002 (*c*), 0.001 (*d*)
2.12-D	SQ	1000	—	$10K$	—	1	A	200	$R = 2K$, $C = 0$ (*a*), 0.02 (*b*), 0.04 (*c*), 0.08 (*d*)
2.12-E	PG	1000	200	$10K$	—	1	A	—	$R = 2K$, $C = 0.08$
2.14-D	PG	800	1	$1K$	—	3	B	40	$C = 0.01$ (*a*), 0.03 (*b*), 0.05 (*c*), 0.10 (*d*)
2.15-E	SIN	100 (*a*) 1000 (*b*) 10,000 (*c*)	—	$10K$	—	1	A	2000 (*a*) 200 (*b*) 20 (*c*)	$R = 2K$, $C = 0.08$
2.15-G	SIN	0.5	—	$100K$	—	1*	A	0.5 sec/cm	*short C initially; open to start transient; $R = 10^6$, $C = 3$
3.02-A	PG	3000	2	$10K$	—	4	see network Fig. A-B	10	(*a*) top ch. 1 connected to t_1; (*b*) bottom ch. 1 connected to t_1
3.07-B	PG	2000	3/4	$30K$	—	2	A	100 (*a*) 20 (*b*)	$R = 100K$ (open) $C = 0.003$

3.07-C	PG	2000	3/4	30K	—	2	A	20	$R=1K$ (*a*), $3K$ (*b*), $10K$ (*c*)
3.08-D	SQ	900	—	30K	—	2	A	(*a*) 100, (*b*) 20	$R=100K$ (open), $C=0.003$
3.08-E	SQ	900	—	30K	—	2	A	20	$C=0.003$, $R=10K$ (*a*), $3K$ (*b*), $1K$ (*c*), 500 (*d*)
3.09-B (*a, b, c, d, e, f, h*)	PG	1700	(*a*) 4 approx. (*b*) 8 ,, (*c*) 12 ,, (*d*) 17 ,, (*e*) 24 ,, (*f*) 36 ,, (*h*) 16 ,,	30K	—	2	A	20	$R=11K$, $C=0.003$
3.09-B(*g*)	SQ	4200	—	30K	—	2	A	20	$R=11K$, $C=0.003$
3.10-A	SQ	700	—	30K	—	5	B	(*a*) 100, (*b*) 20	$C=0.033$, $R=1400$
3.14-G	SIN	(*a*) 17.7K (*b*) 18.2K (*c*) 18.6K	—	30K	—	2	A	10	$C=0.03$, $R=100K$ (open)
3.14-H	PG	500	2.5	30K	—	2	C	—	$C=0.03$, $R=100K$ (*a*), $3K$ (*b*), $1K$ (*c*)
3.14-I	PG	500	2.5	30K	—	1	C	—	$C=0.03$, $R=1K$
3.18-D	PG	1000	2	60K	—	6	C	—	(*a*) $R=0$, (*b*) $R=200$, (*c*) is double exposure of the two, (*a*) and (*b*)
4.09-A	PG	1000	1^-	4K	10K	10	r to t_1	45	—
5.13-B	PG	1000	1	1K	600	17	(*g*) C (*h*) A	(*g*) — (*h*) 40	
5.13-D	SIN	20,000	—	10K	—	18	e_1 to t_2 and (A—B), difference, to t_1 (Fig. A-B)	50	$C=0.003$ (*a*), 0.013 (*b*), 0.023 (*c*)
6.15-B	PG	1000	1	1K	1K	7 and 9*	A and C	20	*double exposure in t, stored in ω
6.16-B	PG	(*b*) 1000 (*c*) 500	(*b*) 1 (*c*) 2	1K	1K	14	C	—	(*b*) normal, (*c*) expanded ω scale
6.17-B	PG	1000	1	1K	1K	15	A and C	20	double exposure in t, stored in ω; C connected and disconnected
6.18-D	PG	1000	2	10K	60K	8	(*b*) C (*c*) A (*d*) A	(*b*) — (*c*) 100 (*d*) 20	
6.20-A	PG	500	2	1K	300	13	A and C	25	

TABLE A-A. (*Con.*)

Figure Number (See Text)	Signal Generator: Type Used (Fig. A-A)	Signal Generator Settings: Frequency Hz or pps	Signal Generator Settings: Pulse Width, μs	Termination Values, Ω: R_G	Termination Values, Ω: R_L	Network Used (Fig. A-B)	Connections (Fig. A-A)	*t CRO* Sweep Speed, $\mu s/cm$	Comments (Element Values Are in Ω and μF, $K = 1000$)
6.20-B	SIN	(*a*) 25*K* (*b*) 24*K* (*c*) 23*K*	—	1*K*	300	13	A	10	
6.23-E	PG	1*K*	1	1*K*	1*K*	11	A and C	10	
7.03-C	PG	1000	1	10*K*	3*K*	(*b*) 7 (*c*) 7 (*d*) 16	(*b*) C (*c*) A (*d*) B	(*b*) 40 (*c*) — (*d*) 40	
7.16-E	PG	500	1	0	2*K*	19	(*a*) C (*b*) A	(*a*) — (*b*) 50	
7.17-B	PG	1000	2	1*K*	4*K*	11	A	40	(*a*) normal, (*b*) expand vertical scale 10×
7.19-G	(*b*) PG (*c*) PG (*d*) SQ	1000	(*b*) 1 (*c*) 1 (*d*) —	1*K*	1*K*	9	(*b*) C (*c*) A (*d*) A	(*b*) — (*c*) 40 (*d*) 40	
7.19-H	PG	1000	1	1*K*	3400	20	(*a*) C (*b*) A	(*a*) — (*b*) 40	
8.04-A	PG	2*K*	1/2	10*K*	10*K*	12	(*a*) e_1 to t_1, e_2 to t_2 (*b*) C	(*a*) 20 (*b*) —	

APPENDIX B

READINGS

This is *not* a bibliography of network analysis; that would run to hundreds of pages. But interested students should feel now both the need and the ability to continue their studies. One objective of this book is simply to give them the ability to read more rigorous works. The few suggestions below provide a good start, opening out into an infinity of other works.

Students might well begin with some of the classic works of Euler, Faraday, Maxwell, Kirchhoff, Heaviside, and Carson. Or with some of these specific suggestions from modern works:

GB: GARDNER, M. F., and J. L. BARNES, "Transients in Linear Systems" (New York: Wiley). A very readable, rigorous treatment of the Laplace transformation, with an excellent bibliography, particularly of network history.

CF: CAMPBELL, G. A., and R. M. FOSTER, "Fourier Integrals for Practical Applications" (New York: Van Nostrand). One of the best tables of Laplace transforms, with a readable introduction to the subject from quite a different point of view.

BU: BUSH, V., "Operational Calculus" (New York: Wiley). Excellent continuation and interpretation of Heaviside's work; a very interesting alternative, if out-of-fashion, point of view.

GU: GUILLEMIN, E. A., A series of books by a great teacher ("Introductory Circuit Theory" and many others), well worth attention.

BO: BODE, H. W., "Network Analysis and Feedback Amplifier Design" (New York: Van Nostrand; reprinted 1975, Huntington, N.Y.: Krieger). A modern classic, essential to anyone who really wants to understand network theory; more advanced in many ways, but profitable reading.

DI: DIRAC, P. A. M., see *Proceedings of the Royal Society, London*, series A, vol. 113, for the original introduction of the impulse function, worth looking up.

PR: TUTTLE, D. F., "On Fluids, Networks and Engineering Education," in "Aspects of Network and System Theory" (New York: Holt). Gives more details on the initial-value and point-matching techniques of Secs. 6.14 and 7.06.

Finally, of course, is the contemporary literature, both books and journals: ever expanding, often trivial, always of some interest, occasionally significant. Good luck!

APPENDIX C

COMPUTATION

Engineering has long since advanced to the use of accurate models in the analysis of which very precise computation is often appropriate and required. In this respect, network analysis has been in the forefront. Detailed, complicated, lengthy computation is the essence thereof.

Engineering perspective is, however, *extremely* important. The advent of the digital computer places temptations in the engineer's path: it is easy to lose one's head and succumb to the computer's wiles.

Sometimes a one-significant-figure *mental computation* is sufficient and quickly obtained: more elaborate computation would be downright ridiculous! In such common operations as the solution of a right triangle, finding a square root, or the value of a trigonometric function, and the like, the first one or two terms of a power series often suffice, when some quantity is comparatively small, to give an adequate answer, rapidly. Examples are given in Fig. C-A and at many points in the text (Secs. 3.06, 3.07, 3.14, 6.09, for example). No engineer should be unskilled in this sort of computation! [The asymptotic calculations of analysis, in both time and frequency domains (Chaps. 3, 6, 7) are more elaborate examples of such simplifying calculations.]

Hand computation with a slide rule or a small calculator is the next procedure. Quick, convenient, often adequate, this requires no comment.

$$\frac{1}{1+x} = 1 - x + x^2 + \cdots = 1 - x \text{ for all practical purposes if } x \text{ is small.}$$

Example: $(9.8)^{-1} = \dfrac{1}{10 - 0.2} = \dfrac{1}{10(1 - 0.02)} = 0.102$ (exact value is 0.102041).

If $b \ll a$, $c = \sqrt{a^2 + b^2}$

$$= a\sqrt{1 + b^2/a^2}$$

$$= a\left[1 + \tfrac{1}{2}(a^2/b^2) + \cdots\right]$$

$$= a\left[1 + \tfrac{1}{2}(a^2/b^2)\right] \text{ practically.}$$

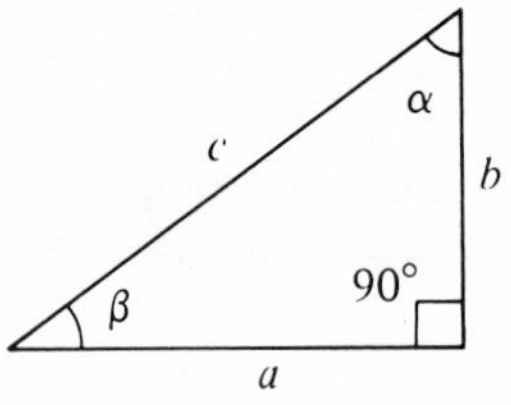

Example:

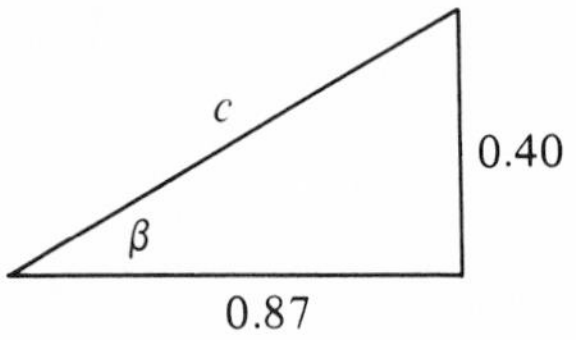

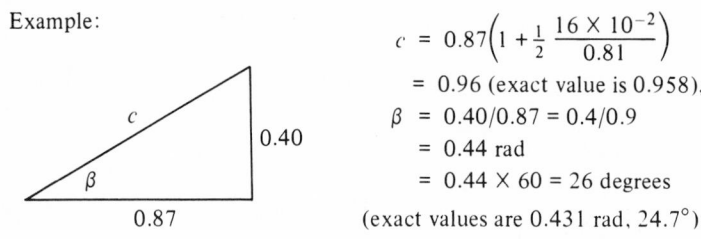

$$c = 0.87\left(1 + \tfrac{1}{2}\,\frac{16 \times 10^{-2}}{0.81}\right)$$

$= 0.96$ (exact value is 0.958),

$\beta = 0.40/0.87 = 0.4/0.9$

$= 0.44$ rad

$= 0.44 \times 60 = 26$ degrees

(exact values are 0.431 rad, 24.7°).

Innumerable similar, legitimate shortcuts occur in engineering. Watch for them!

FIG. C-A. Some mental calculation aids.

But the processes of network analysis more often require lengthy and complicated calculations for which the large automatic (programmed) digital computer is admirably suited. Network analysis and the digital computer seem actually "made for each other," as so many times the text suggests. Here, in Appendix C, we need only codify such remarks and provide the computer programs so often mentioned (especially in Sec. 7.18).

The writing of computer programs is an art: individual programmers will devise their own variations, their own styles, making some difference between their work and that of others, even if the backbone, the basis, is the same. No two persons will write "the" program for a given task in the same way; the possibilities are essentially infinite. Every student should ideally write his or her own programs. Practically, of course, this is impossible: the time demands are too great, and the work has already been done, not once but many times. Students should, nevertheless, endeavor to write, "debug," and use at least one program of this sort "on their own." Here we provide the programs mentioned in the text, in one man's version, sufficient for the network-analysis demands of this book. But they are by no means the best programs possible. They are written for educational purposes and can often be made more efficient, at the cost perhaps of disguising educational points to be made. They can be made more accurate, by using devices of numerical analysis. They can be combined into larger programs that do network analysis in more integrated (fewer steps) fashion. Such refinements are not the province of this book.

The language used herein is the *lingua franca* FORTRAN. Each program starts with a "masthead," a series of comments which is intended to explain the purpose of the program and to give all necessary instructions for its use, once the program itself is in the machine. Teachers and students should write their own variations, implement their own ideas. Comments, criticisms, improvements, suggestions, ... will be welcomed. Copies of the programs in card-deck or computer-listing form can be obtained at cost. For both purposes write to Prof. D. F. Tuttle, Electrical Engineering Department, Stanford University, Stanford, Calif. 94305.

First, certain chores of frequent occurrence are: (*a*) factoring of polynomials ("root finding"), (*b*) evaluation of polynomials for specific values of s, and (*c*) the plotting of results.

Finding natural frequencies (polynomial factoring) is briefly discussed in Sec. 6.04 (see also Secs. 7.17 to 7.19). The two subroutines there mentioned, POLRTT and DALBER, are listed below; variations of the former occur within certain of the large programs described below.

Polynomial evaluation can be accomplished by straightforward multiplication, or by using the remainder theorem and synthetic division. The subroutine EVLPOL uses the latter method, and can also be found in the large programs to follow. The remainder theorem is simply the fact that, for example, the value of the polynomial $Q(s) = a_0 + a_1 s + \cdots + a_n s^n$ for $s = j\omega$ is the value thereat of the remainder $R(s)$ in the division:

$$\frac{Q(s)}{s^2 + \omega^2} = Q_1(s) + \frac{R(s)}{s^2 + \omega^2}, \qquad Q(j\omega) = R(j\omega).$$

The subroutine first computes $R(s)$, a simple binomial, and then evaluates it.

Plotting of results directly on the same printer with the numerical output is quick, simple, and cheap. Various plotting subroutines are in the large programs below. (This is a field particularly rich in opportunities for innovation.) Peripheral equipment for producing more elaborate plots, at a price, is usually available also.

Here are listings of the five large programs described in Sec. 7.18 [cf. (7.18-10), p. 618]:

These are followed by illustrative (sample) data-card sets for each. Finally, the two factoring subroutines POLRTT and DALBER are listed.

LADANA (for calculating transfer functions of ladder networks)

```
C       LADDER-NETWORK ANALYSIS PROGRAM "LADANA"
C                                                                     LADANA PAGE 1
C       CALCULATES TRANSFER FUNCTIONS OF LADDER NETWORKS WHOSE BRANCHES
C                    HAVE AT MOST 3 ELEMENTS EACH ( IN SERIES IF
C                    A SHUNT BRANCH, IN PARALLEL IF A SERIES BRANCH);
C                    THE DISSIPATIVE TUNED CIRCUIT IS ALSO ALLOWED (TYPE 8)
C                    (MAXIMUM NUMBER OF BRANCHES = 16)
C       NOTATION:
C
C           NBR = NUMBER OF BRANCHES
C                 BRANCH = SHUNT-CONNECTED RLC COMBINATION (1,2 OR 3 ELEMENTS
C                          IN SERIES)  OR SERIES-CONNECTED RLC COMBINATION
C                          ( 1, 2 OR 3 ELEMENTS IN PARALLEL ) OR DISS. T.C.
C              NB: COUNT FROM LOAD TO SOURCE, I.E.
C              *** BRANCH NUMBER ONE IS AT THE LOAD END. ***
C
C           N = NUMBER OF ENERGY-STORING ELEMENTS
C           NCON = NUMBER OF CONSTRAINTS THAT IN EFFECT REDUCE NUMBER OF NATURAL
C                  FREQUENCIES ( PURE C LOOPS, PURE L NODES, BRUNE NETWORK SECTIONS
C                  ETC. ) SO THAT ACTUAL NUMBER OF NATURAL FREQUENCIES = N - NCON
C           MSS = 1 INDICATES SHUNT BRANCH
C               = 2 INDICATES SERIES BRANCH
C           MTYPE =  CODE FOR BRANCH TYPE:
C                 =  1 : RESISTOR ALONE
C                 =  2 : CAPACITOR ALONE IF SHUNT, INDUCTOR ALONE IF SERIES
C                 =  3 : INDUCTOR ALONE IF SHUNT,CAPACITOR ALONE IF SERIES
C                 =  4 : RESISTOR AND INDUCTOR IN SERIES IF SHUNT, RESISTOR
C                        AND CAPACITOR IN PARALLEL IF SERIES
C                 =  5 : RESISTOR AND CAPACITOR IN SERIES IF SHUNT, RESISTOR
C                        AND INDUCTOR IN PARALLEL IF SERIES
C                 =  6 : INDUCTOR AND CAPACITOR IN SERIES IF SHUNT, IN PARALLEL IF
C                        SERIES
C                 =  7 : RESISTOR, INDUCTOR AND CAPACITOR IN SERIES IF SHUNT,
C                        IN PARALLEL IF SERIES
C                 =  8 : DISSIPATIVE TUNED CIRCUIT: L, R, R2, C:          ---------
C                                                                                 I
C                           --L--R---                                             L
C                           I       I                                             I
C                   --------C--------                    OR                       R
C                           I       I                                             I
C                           ---R2----                                       -------
C                                                                           I     I
C                                                                           C     R2
C                   -----------------                                       I     I
C                                                                           -------
C                                                                              I
C           MWR = 1 : WRITE OUT INTERMEDIATE DETAILS (VERY LENGTHY)            I
C               = 2 :  WRITE RESULTS ONLY                                      I
C           MCO = 1 : MAKE COEFFICIENT OF LEADING POWER OF S IN CHARACTERISTIC
C                     POLYNOMIAL EQUAL TO UNITY
C               = 2 : LEAVE COEFFICIENT AS IT COMES
C               = 3 : WRITE OUT BOTH FORMS
C
C *** DATA CARDS REQUIRED:
C
C           (1) TITLE - NOTE (3 CARDS)
C           (2) NBR, N, MWR, MCO, NCON (FORMAT 5I10)
C               MAX. VALUE NBR = 16                      MAX. VALUE N = 32
C               ( NCON USUALLY IS ZERO; MAY BE OMITTED)
C           (3) MSS, MTYPE, BRR, BRL, BRC
C                     BRR = VALUE OF RESISTOR ELEMENT (OHMS)
C                     BRL = VALUE OF INDUCTOR ELEMENT (HENRIES)
C                     BRC = VALUE OF CAPACITOR ELEMENT (FARADS)
C               ( ONE SUCH CARD FOR EACH BRANCH, IN ORDER, FORMAT 2I10,3F20.6)
C               ( INCLUDE ARBITRARY  DUMMY VALUES FOR NONEXISTENT ELEMENTS IN
C               ( BRANCHES.)
C               FOR TYPE 8 BRANCH, PUT VALUE OF R2 ON A SECOND CARD (F20.6)
C ***********NB: START FROM LOAD END REPEAT START FROM LOAD END AND MOVE
C                STEADILY TO GENERATOR END. *****
C           (4) ADDITIONAL COMPLETE SETS OF DATA CARDS MAY BE PLACED HERE
C
    1 FORMAT (8I10)
    2 FORMAT (2I10, 3F20.6)
    3 FORMAT (1H0, 15X, 'NOTES:', 10X, 20A4, 2(/, 32X, 20A4))
    4 FORMAT (1H0, 'DATA READ IN:')
    5 FORMAT(1H0, 'CONDITIONS AT THE RIGHT OF  BRANCH ',     I3, ':',10X
     1, 70('*'),/)
    6 FORMAT (1H0,'(COMMON) DENOMINATOR:')
    7 FORMAT (1H , 'VOLTAGE NUMERATOR:')
    8 FORMAT (1H0, 'BRANCH', 6X, 'MSS*',5X, 'TYPE', 12X, 'R VALUE', 13X,
     1 'L VALUE', 13X, 'C VALUE', 12X, 'R2 VALUE')
    9 FORMAT (1H0, 'CURRENT NUMERATOR:')
   10 FORMAT (1H0, 'CONDITIONS AT THE LEFT OF ELEMENT(S)', I3, ': ',/)
   11 FORMAT (1H0,  6F20.6)
```

```
   12 FORMAT (1H ,I4, I10, I10, 3F20.6, / )
   13 FORMAT (/////,' MSS(J) =',  I2, ' DATA IS BAD.')
   14 FORMAT (1H1, 'THIS IS LADANA, A LADDER-NETWORK ANALYSIS PROGRAM')
   15 FORMAT (///)
   16 FORMAT (1HO,   4X, 'INPUT IMPEDANCE:', /)
   17 FORMAT (1HO, 12X,  6F20.6 )
   18 FORMAT (1H ,  132('.') )
   19 FORMAT (1HO, 4X, 'CURRENT-TO-VOLTAGE TRANSFER FUNCTION:', /)
   20 FORMAT (1HO, 'NUMBER OF BRANCHES = ', I2,',  NUMBER OF ENERGY-STOR
     1ING ELEMENTS = ', I2, 10X, ' MWR = ', I2, 10X, 'MCO =', I2, 10X, '
     2NCON =', I2,/)
   21 FORMAT (1HO, 4X, 'VOLTAGE-TO-VOLTAGE TRANSFER FUNCTION:', /)
   22 FORMAT (1HO, 12X, 6F20.6)
   23 FORMAT (1H ,  F18.6,  6F19.6)
   24 FORMAT (20A4)
   25 FORMAT (///,' OR (HIGHEST POWER OF S IN NATURAL-FREQUENCY POLYNOMI
     1AL MADE UNITY):',///)
   26 FORMAT (1H+,116X, '*MSS = 1 : SHUNT',/,' ......', 6X, '....', 5X,
     1 '....', 11X,               4(8('.'), 12X), 5X, '= 2 :SERIES', /)
   27 FORMAT (1H1, 'THE RESULTS (COEFFICIENTS OF POWERS OF S, IN ASCENDING
     1NG ORDER, NUMERATORS OVER DENOMINATORS), AS COMPUTED BY LADANA, FO
     2LLOW:', //)
   28 FORMAT (1HO,'CTE = CN(N1) = CN(',I2,') = ',E12.3,' IS TOO SMALL TO
     1 DIVIDE BY.')
   29 FORMAT (1HO,'CTE = VN(N1) = CN(',I2,') =', E12.3,' IS TOO SMALL TO
     1 DIVIDE BY.')
   30 FORMAT (1H , 16X, 'THERE ARE ONLY', I3, ' NATURAL FREQUENCIES BECA
     1USE NCON =', I2,'; YOU MAY FIND SUPERFLUOUS ZERO COEFFICIENTS IN
     2THE RESULTS,',/, 17X, 'WHICH ASSUME', I3, ' NATL. FREQS.')
   31 FORMAT (1H , 33A4)
   32 FORMAT (1HO, 'THAT IS:')
   33 FORMAT (1H , I7, 15I8)
   34 FORMAT (1HO,' BRANCH NUMBERS:', /)
   35 FORMAT (///  , ' THERE SEEMING TO BE NO MORE DATA, I AM, SIR/MADAM
     1E (MS.), YOUR MOST OBEDIENT SERVANT, LADANA.', //// )
   36 FORMAT (/////,' THE RESULTS ABOVE WERE PRODUCED BY LADANA.', //)
   37 FORMAT (1H )
   38 FORMAT (4F20.6)
   39 FORMAT (1H )
   41 FORMAT (1H , I4, 2I10, F20.6 )
   42 FORMAT (1H , I4, 2I10, F60.6 )
   43 FORMAT (1H , I4, 2I10, F40.6 )
   44 FORMAT (1H , I4, 2I10, 4F20.6 )
   45 FORMAT (1H , I4, 2I10, F20.6, F40.6 )
   46 FORMAT (1H , I4, 2I10, F40.6, F 20.6 )
C
      REAL LSYM, LRSYM
      DIMENSION D(34), CN(34), VN(34), MSS(16), MTYPE(16), BRR(16),BPL(
     116), BRC(16), VNT(34), CNT(34), DT(34), TITLE(60), DIAG(14,33)
      DIMENSION NRBR(16), BRR2(16)
C *** DIAGRAM SYMBOLS:
      DATA BLANK, RSYM, LSYM, CSYM/'    ', ' R ', ' L ', ' C '/
      DATA BEGIND, DASHS, BARD/'0---', '----', '  | '/
      DATA ENDD /'0   '/
      DATA LRSYM, CRSYM, R2SYM/' L R', 'C R2', ' R 2'/
C *** INITIALIZE:
C         CN = NUMERATOR OF CURRENT COEFFICIENT ARRAY
C         VN = NUMERATOR OF VOLTAGE COEFFICIENT ARRAY
C         D = DENOMINATOR COEFFICIENT ARRAY
   99 DO 100 I = 2, 34
      D(I) = 0.0
      CN(I) = 0.0
  100 VN(I) = 0.0
      D (1) = 1.0
      CN(1) = 0.0
      VN(1) = 1.0
      DO 51 I = 1, 14
      DO 51 J = 1,33
   51 DIAG(I,J) = BLANK
      DIAG(2,1) = BEGIND
      DIAG(14,1) = BEGIND
C *** READ DATA AND WR[illegible] IT OUT:                                    ********
      READ (5,24,END = [illegible]) TITLE
      WRITE (6,14)
      WRITE (6,3) TITLE
      WRITE (6,4)
      READ (5,1) NBR,N, MWR, MCO, NCON
      NBR2 = NBR + NBR
      NBR3 = NBR2 + 1
      WRITE (6,20) NBR, N, MWR, MCO, NCON
      DO 55 I = 2, NBR2
      DIAG(2,I) = DASHS
   55 DIAG(14,I) = DASHS
      DIAG(2, NBR3) = ENDD
```

```
      DIAG (14, NBR3) = ENDD
      NN = N-NCON
      IF (NCON .NE. 0)  WRITE (6,30) NN, NCON, N
      N1 = N + 1
      N2 = N + 2
      DO 150  I = 1,NBR
      READ (5,2)  MSS(I),  MTYPE(I),  BRR(I),  BRL(I),  BRC(I)
  150 IF (MTYPE(I) .EQ. 8) READ (5, 38) BRR2(I)
C *** PREPARE TABLE AND DIAGRAM:                                     *************
      WRITE (6,8)
      WRITE (6,26)
      DO 90 I = 1, NBR
      J = NBR-I + 1
      ITO = MTYPE(J)
      JJ = I + I
      IF (MSS(J)-2)  60, 70, 9009
C *** MSS = 1, SHUNT BRANCH:                                           SHUNT BR
   60 DO 59 L = 3, 5
      LL = L + 8
      DIAG(LL, JJ) = BARD
   59 DIAG(L ,JJ) = BARD
      GO TO (61, 62, 63, 64, 65, 66, 67, 68),ITO
   61 DIAG(8,JJ) = RSYM
      DIAG(6,JJ) = BARD
      DIAG(10,JJ) = BARD
      WRITE (6,41) J, MSS(J), MTYPE(J), BRR(J)
      GO TO 90
   62 DIAG(8,JJ) = CSYM
      DIAG(6,JJ) = BARD
      DIAG(10,JJ) = BARD
      WRITE (6,42) J, MSS(J), MTYPE(J), BRC(J)
      GO TO 90
   63 DIAG(8,JJ) = LSYM
      DIAG(6,JJ) = BARD
      DIAG(10,JJ) = BARD
      WRITE (6,43) J, MSS(J), MTYPE(J), BRL(J)
      GO TO 90
   64 DIAG(7,JJ) = RSYM
      DIAG(9,JJ) = LSYM
      WRITE (6,44) J, MSS(J), MTYPE(J), BRR(J), BRL(J)
      GO TO 90
   65 DIAG(7,JJ) = RSYM
      DIAG(9,JJ) = CSYM
      WRITE (6,45) J, MSS(J), MTYPE(J), BRR(J), BRC(J)
      GO TO 90
   66 DIAG(7,JJ) = LSYM
      DIAG(9,JJ) = CSYM
      WRITE (6,46) J, MSS(J), MTYPE(J), BRL(J), BRC(J)
      GO TO 90
   67 DIAG(7,JJ) = RSYM
      DIAG(8,JJ) = LSYM
      DIAG(9,JJ) = CSYM
      WRITE (6,44) J, MSS(J), MTYPE(J), BRR(J), BRL(J), BRC(J)
      GO TO 90
   68 DIAG (7, JJ) = LSYM
      DIAG (8, JJ) = RSYM
      DIAG (9, JJ) = CRSYM
      WRITE (6, 44) J, MSS(J), MTYPE(J), BRR(J), BRL(J), BRC(J), BRR2(J)
      GO TO 90
C *** MSS = 2, SERIES BRANCH:                                          SERIESBR
   70 GO TO ( 71, 72, 73, 74, 75, 76, 77, 78), ITO
   71 DIAG(2,JJ) = RSYM
      WRITE (6,41) J, MSS(J), MTYPE(J), BRR(J)
      GO TO 90
   72 DIAG(2,JJ) = LSYM
      WRITE (6,43) J, MSS(J), MTYPE(J), BRL(J)
      GO TO 90
   73 DIAG(2,JJ) = CSYM
      WRITE (6,42) J, MSS(J), MTYPE(J), BRC(J)
      GO TO 90
   74 DIAG(1,JJ) = RSYM
      DIAG(2, JJ) = BLANK
      DIAG(3,JJ) = CSYM
      WRITE (6,45) J, MSS(J), MTYPE(J), BRR(J), BRC(J)
      GO TO 90
   75 DIAG(1,JJ) = RSYM
      DIAG(2, JJ) = BLANK
      DIAG(3,JJ) = LSYM
      WRITE (6,44) J, MSS(J), MTYPE(J), BRR(J), BRL(J)
      GO TO 90
   76 DIAG(1,JJ) = LSYM
      DIAG(2, JJ) = BLANK
      DIAG(3,JJ) = CSYM
      WRITE (6,46) J, MSS(J), MTYPE(J), BRL(J), BRC(J)
```

```
      GO TO 90
   77 DIAG(1,JJ) = RSYM
      DIAG(2,JJ) = LSYM
      DIAG(3,JJ) = CSYM
      WRITE (6,44) J, MSS(J), MTYPE(J), BRR(J), BRL(J), BRC(J)
      GO TO 90
   78 DIAG (1, JJ) = LRSYM
      DIAG (3, JJ) = R2SYM
      DIAG (2, JJ) = CSYM
      WRITE (6, 44) J, MSS(J), MTYPE(J), BRR(J), BRL(J), BRC(J), BRR2(J)
   90 CONTINUE
      WRITE (6,32)
      WRITE (6,34)
C  ** DRAW DIAGRAM:                                              DRAWDIAG
      DO 177 I = 1,NBR
  177 NRBR(I) = NBR-I + 1
      WRITE (6,33) (NRBR(I), I = 1,NBR)
      WRITE (6, 39)
      DO 175 I = 1,14
  175 WRITE (6,31) (DIAG(I,J), J = 1,33)
      WRITE (6,33) (NRBR(I), I = 1,NBR)
C END TABLE AND DIAGRAM.
C *** LADDER-NETWORK (MESH ) CALCULATION:                        ********
      DO 800 I = 1,NBR                                              BEGIN
      IF (MWR .EQ. 2)  GO TC 200                                     MAIN
      WRITE (6,15)                                                   LOOP
      WRITE (6,5) I                                                ( MESH
      WRITE (6,7)                                                  (CALC.
      WRITE (6,17) (VN(J),  J = 1,N2)
      WRITE (6,6)
      WRITE (6,17) (D(J), J = 1, N2)
      WRITE (6,9)
      WRITE (6,17)(CN(J), J = 1, N2)
  200 CONTINUE
      DO 301 J = 1,N1
      DT(J) = D(J)
      CNT(J) = CN(J)
  301 VNT(J) = VN(J)
      MMMT = MTYPE(I)
      IF (MSS(I) .EQ. 2)  GO TO 500
C *** SHUNT BRANCH  (MSS = 1) ******* SHUNT BRANCH              SHUNT CASE
      GO TO ( 310, 320, 330, 340, 350, 360, 370, 380 ), MMMT     ********
  310 CONTINUE
C     RESISTOR ALONE: D AND VN REMAIN AS IS
C     D REMAINS AS IS
C     VN REMAINS AS IS
      G = 1.0/BRR(I)
      DO 312 J = 1, N1
  312 CN(J) = CNT(J) + G * VNT(J)
      GO TO 600
  320 CONTINUE
C     CAPACITOR ALONE: D, VN AND CN(1) REMAIN AS IS
C     D REMAINS AS IS
C     VN REMAINS AS IS
C     CN(1) REMAINS AS IS
      C = BRC(I)
      DO 322 J = 2, N1
  322 CN(J) = CNT(J) + C * VNT(J-1)
      GO TO 600
  330 CONTINUE
C     INDUCTOR ALONE: ALL 3 ARRAYS CHANGE
C     ALL THREE ARRAYS CHANGE:
      XL = BRL(I)
      DO 332  J = 2, N1
      D(J) = XL*DT(J-1)
      CN(J) = XL*CNT(J-1) + VNT(J)
  332 VN(J) = XL*VNT(J-1)
      D(1) = 0.0
      VN(1) = 0.0
      CN(1) =  VNT(1)
      GO TO 600
  340 CONTINUE
C     RESISTOR AND INDUCTOR IN SERIES (IN SHUNT)
C     ALL THREE ARRAYS CHANGE;
      XL = BRL(I)
      R = BRR(I)
      DO 342  J = 2,  N1
      D(J) = R * DT(J) + XL * DT(J - 1)
      CN(J) = R * CNT(J) + XL * CNT(J-1) + VNT(J)
  342 VN(J) = R * VNT(J) + XL * VNT(J-1)
      D(1) = R * DT(1)
      CN(1) = R * CNT(1) + VNT(1)
      VN(1) = R * VNT(1)
      GO TO 600
```

```
  350 CONTINUE
C      RESISTOR  AND CAPACITCR IN SERIES ( IN SHUNT)
C      ALL THREE CHANGE:
       R = BRR(I)
       C = BRC(I)
       RC = R * C
       DO 352  J = 2, N1
       D(J) = DT(J) + RC * DT(J-1)
       CN(J) = CNT(J) + RC * CNT(J-1) + C * VNT(J-1)
  352  VN(J) = VNT(J) + RC * VNT(J-1)
       GO TO 600
  360 CONTINUE
C      INDUCTOR AND CAPACITOR IN SERIES( IN SHUNT )
C      ALL CHANGE:
       XL = BRL(I)
       C = BRC(I)
       CL = C*XL
       DO 362  J = 3, N1
       D(J) = DT(J) + CL*DT(J-2)
       CN(J) = CNT(J) + CL*CNT(J-2) + C*VNT(J-1)
  362  VN(J) = VNT(J) + CL*VNT(J-2)
       CN(2) = C*VNT(1) + CNT(2)
       GO TO 600
   70 CONTINUE
C      INDUCTOR, RESISTOR AND CAPACITOR IN SERIES (IN SHUNT )
C      ALL THREE ARRAYS CHANGE:
       XL = BRL(I)
       C = BRC(I)
       R = BRR(I)
       RCX = R*C
       XLC = XL*C
       D(2) = DT(2) + RCX*DT(1)
       CN(2) = CNT(2) + RCX*CNT(1) + C*VNT(1)
       VN(2) = VNT(2) + RCX*VNT(1)
       DO 372 J=3, N1
       D(J) = DT(J) + RCX*DT(J-1) + XLC*DT(J-2)
       CN(J) = CNT(J) + RCX*CNT(J-1) + XLC*CNT(J-2) + C*VNT(J-1)
  372  VN(J) = VNT(J) + RCX*VNT(J-1) + XLC*VNT(J-2)
       GO TO 600
  380 CONTINUE
C      DISSIPATIVE TUNED CIRCUIT IN SHUNT:
C      ALL 3 ARRAYS CHANGE:
       XL = BRL(I)
       C = BRC(I)
       R = BRR(I)
       R2 = BRR2(I)
       G2 = 1.0/BRR2(I)
       RG = 1 + R*G2
       RC = R*C + G2**L
       XLC = XL*C
       D(1) = RG*DT(1)
       D(2) = RG*DT(2) + RC*DT(1)
       VN(1) = RG*VNT(1)
       VN(2) = RG*VNT(2) + RC*VNT(1)
       CN(1) = G2*VNT(1) + RG*CNT(1)
       CN(2) = C*VNT(1) + G2*VNT(2) + RG*CNT(2) + RC*CNT(1)
       DO 382 J = 3, N1
       D(J) = RG*DT(J) + RC*DT(J-1) + XLC*DT(J-2)
       CN(J) = G2*VNT(J) + C*VNT(J-1) + RG*CNT(J) + RC*CNT(J-1) + XLC*CNT
      1(J-2)
  382  VN(J) = RG*VNT(J) + RC*VNT(J-1) + XLC*VNT(J-2)
       GO TO 600
C ***  SERIES BRANCH(MSS = 2) *******                                SERIES CASE
  500  GO TO (510, 520, 530, 540, 550, 560, 570, 580), MMMT             ********
  510  CONTINUE
C      RESISTOR ALONE: D AND CN REMAIN AS IS
C      D REMAINS AS IS
C      CN REMAINS AS IS
       R = BRR(I)
       DO 512  J = 1, N1
  512  VN(J) = VNT(J) + R * CNT(J)
       GO TO 600
  520  CONTINUE
C      INDUCTOR ALONE : D, CN AND VN(1) REMAIN AS IS
C      D REMAINS AS IS
C      CN REMAINS AS IS
C      VN(1) REMAINS AS IS
       XL = BRL(I)
       DO 522 J = 2,N1
  522  VN(J) = VNT(J) + XL * CNT(J-1)
       GO TO 600
  530  CONTINUE
C      CAPACITOR ALONE: ALL 3 ARRAYS CHANGE
C      ALL THREE ARRAYS CHANGE:
```

```
      C = BRC(I)
      DO 532  J = 2,N1
      D(J) = C * DT(J-1)
      VN(J) = C*VNT(J-1) + CNT(J)
  532 CN(J) = C * CNT (J - 1)
      D(1) = 0.0
      CN(1) = 0.0
      VN(1) = CNT(1)
      GO TO 600
  540 CONTINUE
C     RESISTOR AND CAPACITOR IN PARALLEL ( IN SERIES )
C     ALL THREE ARRAYS CHANGE:
      C = BRC(I)
      G = 1.0 / BRR(I)
      DO 542  J = 2,  N1
      D(J) = G * DT(J) + C * DT(J - 1)
      VN(J) = G * VNT(J) + C * VNT(J-1) + CNT(J)
  542 CN(J) = G * CNT(J) + C * CNT(J-1)
      D(1) = G * DT(1)
      VN(1) = G * VNT(1) + CNT(1)
      CN(1) = G * CNT(1)
      GO TO 600
  550 CONTINUE
C     RESISTOR AND INDUCTOR IN PARALLEL ( IN SERIES )
C     ALL THREE ARRAYS CHANGE:
      G = 1 / BRR(I)
      XL = BRL(I)
      GL = G*XL
      DO 552  J = 2,N1
      D(J) = DT(J) + GL*DT(J-1)
      VN(J) = VNT(J) + GL*VNT(J-1) + XL*CNT(J-1)
  552 CN(J) = CNT(J) + GL*CNT(J-1)
      GO TO 600
  560 CONTINUE
C     INDUCTOR AND CAPACITOR IN PARALLEL ( IN SERIES )
C     ALL CHANGE:
      C = BRC(I)
      XL = BRL(I)
      CL = C*XL
      DO 562 J = 3,N1
      D(J) = DT(J) + CL*DT(J-2)
      VN(J) = VNT(J) + CL*VNT(J-2) + XL*CNT(J-1)
  562 CN(J) = CNT(J) + CL*CNT(J-2)
      VN(2) = XL*CNT(1) + VNT(2)
      GO TO 600
  570 CONTINUE
C     INDUCTOR, CAPACITOR AND RESISTOR IN PARALLEL ( IN SERIES )
C     ALL THREE ARRAYS CHANGE:
      C = BRC(I)
      XL = BRL(I)
      G = 1.0/BRR(I)
      GLX = G*XL
      XLC = XL*C
      D(2) = DT(2) + GLX*DT(1)
      VN(2) = VNT(2) + GLX*VNT(1) + XL*CNT(1)
      CN(2) = CNT(2) + GLX*CNT(1)
      DO 572 J = 3,N1
      D(J) = DT(J) + GLX*DT(J-1) + XLC*DT(J-2)
      VN(J) = VNT(J) + GLX*VNT(J-1) + XLC*VNT(J-2) + XL*CNT(J-1)
    5 CN(J) = CNT(J) + GLX*CNT(J-1) + XLC*CNT(J-2)
      GO TO 600
  580 CONTINUE
C     DISSIPATIVE PARALLEL TUNED CIRCUIT (IN SERIES):
C     ALL THREE ARRAYS CHANGE:
      C = BRC(I)
      XL = BRL(I)
      R = BRR(I)
      R2 = BRR2(I)
      G2 = 1.0/R2
      GLX = G2*XL + R*C
      XLC = XL*C
      RG1 = R*G2 + 1.0
      D(1) = RG1*DT(1)
      D(2) = RG1*DT(2) + GLX*DT(1)
      CN(1) = RG1*CNT(1)
      CN(2) = GLX*CNT(1) + RG1*CNT(2)
      VN(1) = RG1*VNT(1) + R*CNT(1)
      VN(2) = GLX*VNT(1) + RG1*VNT(2) + XL*CNT(1) + R*CNT(2)
      DO 582 J = 3, N1
      D(J) = RG1*DT(J) + GLX*DT(J-1) + XLC*DT(J-2)
      CN(J) = RG1*CNT(J) + GLX*CNT(J-1) + XLC*CNT(J-2)
      VN(J) = RG1*VNT(J) + GLX*VNT(J-1) + XLC*VNT(J-2) + R*CNT(J) + XL*C
     1NT(J-1)
  582 CONTINUE
```

```
  600 CONTINUE
      IF ( MWR .EQ. 2 )  GO TO 800
  602 WRITE (6,10) I
      WRITE (6,7)
      WRITE (6,17) (VN(J), J = 1,N2)
      WRITE (6,6)
      WRITE (6,17) (D(J), J = 1, N2)
      WRITE (6,9)
      WRITE (6,17) (CN(J), J = 1, N2)
      WRITE (6,15)
  800 CONTINUE                                                    *********
C ******* END MAIN LOOP (CALC.)                                   END LOOP
C *** NOW WRITE OUT RESULTS:
      WRITE (6, 27)
      WRITE (6,16)
      WRITE (6,23) (VN(J), J = 1,N1)
      WRITE (6,18)
      WRITE (6,23) (CN(J), J = 1,N1)
        IF (MSS(NBR) .EQ. 2) GO TO 1000
      WRITE (6,15)
      WRITE (6,19)
      IF (MCO .EQ. 1)  GO TO 960
      WRITE (6,23) (D(J), J = 1,N1)
      WRITE (6,18)
      WRITE (6,23) (CN(J), J = 1,N1)
      IF (MCO .EQ. 3)  GO TO 955
      GO TO 1700
  955 WRITE (6,25)
  960 CTE = CN(N1)
      IF ( ABS(CTE) .LE. 1.0E-06 )  GO TO 979
      DO 970  J = 1, N1
      CN(J) = CN(J)/CTE
  970 D(J) = D(J)/CTE
      WRITE (6,23) (D(J), J = 1,N1)
      WRITE (6,18)
      WRITE (6,23) (CN(J), J = 1,N1)
      GO TO 1700
  979 WRITE (6,28) N1, CTE
      GO TO 1700
 1000 WRITE (6,15)
      WRITE (6,21)
      IF ( MCO .EQ. 1 )  GO TO 1060
      WRITE (6,23) (D(J), J = 1,N1)
      WRITE (6,18)
      WRITE (6,23)  (VN(J), J = 1,N1 )
      IF (MCO .EQ. 3)  GO TO 1055
      GO TO 1700
 1055 WRITE (6,25)
 1060 CTE = VN(N1)
      IF ( ABS(CTE) .LE. 1.0E-06 )  GO TO 1079
      DO 1070  J = 1,N1
      D(J) = D(J)/CTE
 1070 VN(J) = VN(J)/CTE
      WRITE (6,23)  (D(J), J = 1,N1)
      WRITE (6,18)
      WRITE (6,23) (VN(J), J = 1,N1)
      GO TO 1700
 1079 WRITE (6,29) N1, CTE
 1700 WRITE (6,36)
      WRITE (6,3) TITLE
      GO TO 99
C RETURN TO 99 FOR MORE DATA (CASES); IF FINISHED:
 9000 WRITE (6,35)
      STOP
 9009 WRITE(6,13) MSS(J)
      STOP
      END
```

.... illustrative (sample) data-card set for LADANA:

```
LADANA CALC
MS FIG.6.18-A, G = 1 + K(W**2-1)**4(W**2 + 1)/W**4
NO. 1
          6          5          0         3         0
          1          1     0.718355              0.0                  0.0
          1          3      0.0                       0.435791            0.0
          1          2       0.0                 0.0              2.413603
          2          2      0.0                1.541553            0.0
          2          3       0.0               0.0                0.972437
          1          2       0.0                 0.0               0.634212
```

IMPSTEP (for calculating impulse and step responses from transfer functions)

```
C    *******     THIS IS IMPSTEP EIGHT      *****      *****
C                                                       VERSION: JAN 1974
C
C THIS PROGRAM EVALUATES THE INVERSE TRANSFORM OF A RATIONAL FUNCTION:
C
C               H (AN S**N + ....... + A1 S + A0)
C       F(S) = ......................................
C               BD S**D + ... + B1 S + B0
C
C   OR, IF YOU PREFER, THE IMPULSE RESPONSE CORRESPONDING TO THAT AS A
C       TRANSFER FUNCTION. IT ALSO EVALUATES THE MATE OF F(S)/S. IN OTHER WORDS,
C       BOTH IMPULSE AND STEP RESPONSE OF A NETWORK WHOSE TRANSFER FUNCTION
C       IS F(S).  IT THEN PLOTS THEM PER INSTRUCTIONS (MAXIMUM NUMBER OF POINTS
C       SET INTO DIMENSIONS BELOW IS 500). *** IT IS ASSUMED THAT N .LE. D AND
C       ALL POLES (NATURAL FREQUENCIES) ARE SIMPLE AND THE DENOMINATOR IS GIVEN
C       IN POLYNOMIAL FORM; MAXIMUM VALUE OF D IS 20 IN DIMENSIONS OF PROGRAM.
C
C ***** DATA CARDS REQUIRED:
C
C *** WATCH THE FORMATS *** WATCH THE FORMATS *** WATCH THE FORMATS ***
C
C      (1) TITLE-REFERENCE-COMMENT-WHATEVER NOTES YOU WANT PRINTED (3 CARDS)....
C
C      (2) H    (MULTIPLIER) (FORMAT F20.6)
C
C      (3)     NN(=N), ND(=D), NX, NPLT, NPRD, NPRU, NCODE (FORMAT 7I10)
C
C              MAXIMUM VALUE OF N, AND OF D, IS 20
C              NX = 1 GIVES TABULATED RESULTS AS WELL AS PLOTS
C              NX = 0 GIVES PLOTS ONLY, NO TABLE OF VALUES
C              NPLT = 0 GIVES NO PLOTS AT ALL
C              NPLT = 1 GIVES COMBINED PLOT (BOTH RESPONSES ON SAME PAGE) ONLY
C              NPLT = 2 GIVES SEPARATE PLOTS ONLY, ACCORDING TO VALUES OF NPRD
C                     AND NPRU:    NPRD=1 GIVES IMPULSE-RESPONSE PLOT,
C                                       0 GIVES NO IMP-R. PLOT;
C                                  NPRU=1 GIVES STEP-RESPONSE PLOT; =0, NO.
C              NPLT = 3 GIVES BOTH OF THE ABOVE (EQUIVALENT TO NPLT=1 AND NPLT=2)
C              NCODE = 0   NORMAL USAGE
C                    = 1 SUPPRESS ALL OUTPUT EXCEPT ESSENTIAL DATA AND PLOTS
C
C      (4) A0, A1, A2 ..... AN (NUMERATOR COEFFICIENTS)    (FORMAT 4F20.6)
C
C      (5) B0, B1, ..... BD (DENOMINATOR COEFFICIENTS) (FORMAT 4F20.6)
C
C      (6) T1, DT, T2 ..... T1, DT, T2 ..... ETC (TIME INTERVALS AND ENDS
C          FOR COMPUTATION OF INVERSE TRANSFORM)   (FORMAT  3F20.6)
C          ****** NB: AFTER THE LAST SUCH GROUP IN EACH CASE PUT A CARD
C          WITH A NEGATIVE VALUE OF T1 AND TWO DUMMY NUMBERS
C          ...THIS STARTS THE NEXT CASE, OR TERMINATES THE CALCULATIONS IF THERE
C          ARE NO MORE CARDS.   EACH TIME INTERVAL MAY HAVE UP TO 500 POINTS.
C
C **** (7) ADDITIONAL COMPLETE SETS (1)-(6) MAY BE PLACED HERE
C
      REAL IMG, MAG, MC, MN, MS, K(21, 2), IMK
      DIMENSION A(21), B(21), C(21), BB(21), ARR(1503), SR(21), SI(21),
     1 SMAG(21), TITLE(60), PLT1(1002), PLT2(1002), TTT(3), RDL(3), RUL(3)
     23)
      DATA TTT/'    ', '  TI', 'ME  '/, RDL/'IMPL', 'LSE ','RESP'/,
     1 RUL/'  ST', 'EP R', 'ESP.'/
      CTE = 57.295780                                                   0
    1 FORMAT (8I10)                                                     0
    2 FORMAT (4F20.6)                                                   0
    3 FORMAT (F20.6, 6I10)
    4 FORMAT (1H0, 'PARAMETER VALUES READ:    NN =', I3, ', ND =', I3,
     1 ', NX =', I2, ', NPLT =', I2 , ', NPRD =', I2, ', NPRU =', I2,
     2 ', NCODE =', I2, / )
    5 FORMAT (1H1, /// )
    6 FORMAT ( 20A4 )
    7 FORMAT (1H+, 25X, 'T1 =', F6.2,', DT =', F6.3, ', T2 =', F6.2,
     1 ', NRPTS =', I4, ', NX =', I3, / )
    9 FORMAT ( 1H+,  26X,  'NOTES FOR THIS RUN:',   4X, 20A4,
     1 2(/, 50X, 20A4) )
   10 FORMAT (1H0, /, ' THE FOLLOWING ARE THE POLES OF F: ' )
   11 FORMAT (1H0, 'DENOMINATOR:', / )                                  0
   12 FORMAT (1H0, 'NUMERATOR:' ,/)                                     C
   13 FORMAT (1H0, ///, ' *****    NO TABLE OF VALUES WAS ASKED FOR *****
     1 ', 15X, 'PLOTS FOLLOW')
   14 FORMAT (1H0, 'THE FIRST NONZERO IV, CALCULATED FROM THE GIVEN TRAN
     1SFORM, IS:',  E16.6, 5X, '(THIS IS THE INITIAL VALUE OF THE ', I3,
     2'-TH')
   15 FORMAT ( // ,' THE SUM OF THE COMPUTED RESIDUES "K" IS:', 50X, E18
     1.6, ' + J ',  E13.6)
   16 FORMAT (1H0,  'THE SUM OF THE PRODUCTS OF THE COMPUTED RESIDUES AN
     1D THE NATURAL FREQUENCIES, "SUM KP", IS:', E17.6, ' + J', E14.6)
```

```
   17 FORMAT (///, 15('*'),   5X,  'A MESSAGE FRCM IMPSTEP8:', 5X,
     1 'DID YCU GET ENOUGH PCINTS,  IS THE "GRANULARITY" FINE ENCUGH?',
     2 /, 49X, 'OR DID YOU, PERCHANCE,  CVERWCRK ME AND GET TOC MANY?',
     3 /, 49X, 'TAKE A GOOD LOOK BEFORE YCU LEAVE. *******', 42('*') )
   18 FCRMAT (1HO, 'IT WERE WELL TC CHECK THESE: CC "SUM CF K''S" AND "S
     1UM CF "KF''S"CHECK?    , 30('*'))
   19 FORMAT (1HO,  'ANCTHER MESSAGE: HERE ARE THE NATURAL FREQUENCIES O
     1NCE MORE, TOGETHER WITH THE RECIPROCALS OF THE NECATIVES OF THEIR
     2REAL PARTS', /, '   (THE "TIME CCNSTANTS"), ANC THE PERICCS OF THE
     3CSCILLATING PARTS:', /)
   2C FCRMAT (1HO,   'THE FCLLOWING ARE THE RESICUES THEREAT (THE "K''S"
     1):', //, 10X, 'MAGNITUDE',    7X, 'ANGLE (RAD.)',   6X, 'ANGLE (DEG.
     2)',    8X,   'REAL PART',   9X, 'IMAG. PART',/, 10X, 9('.'), 7X, 12('
     3.'), 6X, 12('.'), 8X, 9('.'), 9X, 10('.'))
   21 FCRMAT (1H+, 85X, 'ARE YOU HAPPY?  DO YOU LIKE ME?   /S/  IMPSTEP8
     1', /)
   22 FORMAT (1HO, F16.6,'  + J  ',  3F20.6)
   23 FORMAT (1HO, 'YCU CHOSE TIME POINTS SPACED', F7.3,  ': DID YCU CHOO
     1SE WELL? ')
   24 FCRMAT(1H1,       'TIME DATA READ IN: ')
   25 FORMAT (1H1, 50('*'), //, ' DC YCU REALLY WANT T2 LESS THAN T1? ........
     1... YOU SAY T1 =', F6.2, ' ANC T2 =', F6.2, //,  ' IMPSTEP8 THINKS
     2 NCT.', 5X, 2C('*'))
   26 FCRMAT (1H1, 50('*'), //, ' YCU ASK FCR', I5, ' POINTS.  TCC MANY.
     1 SCRRY.')
   27 FCRMAT (1HO,'FROM THE SMALLNESS OF AT LEAST CNE RESIDUE ("K") IT S
     1EEMS THAT THE DATA MAY NOT BE WELL NCRMALIZEC. RESULTS MAY BE DISA
     2PPCINTING.')
   28 FORMAT (1HO, 'THE END.')
   29 FCRMAT (1H                     ,'DERIVATIVE CF RC)')
   3C FCRMAT (1HO, 2HCR   )
   31 FORMAT (1HO, 'THIS IS  NUMBER', I3, ';  THAT RESIDUE IS ', E16.6,
     1 2CX, 50('*'))
   32 FORMAT (1H , 76X, 'IT HAS BEEN CHANGEC TC', E16.6,'  BY YCUR LEAVE
     1')
   33 FORMAT (1H , F1C.4, I5, 2E20.6)
   34 FORMAT (1HO, 'AT T =', F10.4, ', SET TERM DUE TO ', I2, '-TH NATUR
     1AL FREQUENCY EQUAL TC ZERO BECAUSE T*(REAL PART OF NF) =', E15.4,/
     2'  WHICH WOULC CAUSE UNDERFLCW. OK?',18X,'NB: THIS MESSAGE APPEARS
     3 CNLY CNCE; THERE MAY BE MORE SUCH ZERO-INGS HEREAFTER.')
   40 FORMAT (1HO, 53HINVERSE TRANSFORMATICN CF F(S) AND CF F(S)/S GIVES
     1:        )
   5C FORMAT (1HO, 49HA CCEFFICIENT IS INFINITE ******* TRCUBLE........)
   60 FORMAT (1HO, 5X, 4HTIME, 7X, 16HIMPULSE RESPONSE, 5X, 13HSTEP RESP
     1CNSE,  8X, 'IMPULSE RESP.',    10X,  'STEP RESP.',
     2       /, 16X, 19H(INVERSE TRANSFORM),    24X, 'IN E FCRMAT',  11X,
     3 'IN E FCRMAT ', / )
   61 FCRMAT (1H+, 5X, 4('.'), 6X, 19('.'), 4X, 13('.'), 8X, 13('.'), 9X
     1, 11('.'))
   70 FCRMAT (F10.4, F2C.6, F18.6, 2E22.4)
   80 FORMAT (1HO, 28X, I2, F16.6, ' + J', F11.6)
   90 FCRMAT (1HO, I2, F16.6, 4F18.6)
  100 FCRMAT (F10.4, F18.4, 22X, E22.4 )
  110 FORMAT (1HO,   'FCR THE FCLLCWING CASE THE MULTIPLIER  H = ', F12.
     16, 6X, 'AND THE COEFFICIENTS IN F(S), LOWEST POWER FIRST, ARE:',/)
  120 FCRMAT ( 3(' ',  6F20.6, / ) )
  130 FORMAT (1HO, 89X, 11HDENOMINATOR)
  140 FCRMAT (1HO, 58HTHE IMPULSE RESPONSE ALSC CCNTAINS AN IMPULSE OF
     1STRENGTH      , F7.4, 3X, 22HAT T=C, NCT SHOWN HERE  , / )
  150 FCRMAT (1H , 86HTHIS PROGRAM CANNCT CALCULATE THE STEP RESPONSE BE
     1CAUSE F(S) HAS A POLE AT THE CRIGIN   , / )
  160 FCRMAT (1HO, 80HTHE RESPONSE CONTAINS SINGULARITY FUNCTIONS WHICH
     1THIS PROGRAM CANNCT CALCULATE      , / )
  212 FCRMAT (1H1, 'THIS IS IMPSTEP EIGHT.')
***** END CF FCRMATS *****
  200 REAC (5,6,END=710) TITLE
      READ (5, 2) H
      REAC (5, 1)      NN, ND,  NX, NPLT, NPRO, NPRU, NCCDE
      NNN=NN+1
      NDD=ND+1
      NCCC = ND - 1
C *** ZERC ALL COEFFICIENTS BEFCRE READ ANY IN, SC IV CHECK WILL WORK:
      CO 202 I=1,NDD
      A(I) = 0.0
  202 B(I) = 0.0
      SUMKRE = 0.0
      SUMKIM = 0.0
      SUMKPR = 0.0
      SUMKPI = 0.0
      REAC (5,2) ( A(J), J=1,NNN )
C A(1) = CCEFFICIENT CF ZERCTH POWER TERM
C A(NNN) = COEFFICIENT CF NNTH POWER TERM
      READ (5,2) ( B(J), J=1,NDD)
      WRITE (6, 212)
```

```
      WRITE (6,9) TITLE
      WRITE (6, 4) NN, ND, NX, NPLT, NPRD, NPRU, NCODE
      WRITE (6,110) H                                                     0
      WRITE (6,12)                                                        0
      WRITE (6,120) ( A(J), J=1,NNN )                                     0
      WRITE (6,11)                                                        0
      WRITE (6,120) ( B(J), J=1,NDD)                                      0
      CALL POLRTT(NDD, B, C, SR, SI)
      DO 215 J=2,NDD                                                      0
  215 BB(J-1) = B(J)*(J-1)
      WRITE (6,10)
      DO 216 J=1,ND
  216 WRITE (6, 80) J, SR(J), SI(J)
      WRITE (6,20)
      DO 399 J=1,ND
C CALCULATE RESIDUES:
      CALL EVLPOL( NN, A, SR(J), SI(J), UN, VN, MN, AN)
      CALL EVLPOL (NDDD, BB, SR(J), SI(J), UC, VC, MC, AC)
      IF (MC .EQ. 0) WRITE (6,50)
      IF (MC .EQ. 0) RETURN
      K(J,1) = H*MN/MC
      ANAC = AN - AC
      K(J,2) = ANAC/CTE
      PTRE = K(J,1)*COS(K(J,2))
      PTIM = K(J,1)*SIN(K(J,2))
      WRITE (6, 90) J, K(J,1), K(J, 2), ANAC, PTRE, PTIM
      IF (SI(J) .EQ. 0) K(J,1)=H*UN/UC
      REK = K(J,1)*COS(K(J,2))
      IF (SI(J) .EQ. 0) REK = K(J,1)
      IMK = K(J,1)*SIN(K(J,2))
      SUMKRE = SUMKRE + REK
      SUMKIM = SUMKIM + IMK
      SUMKPR = SUMKPR + REK*SR(J) - IMK*SI(J)
      SUMKPI = SUMKPI + IMK*SR(J) + REK*SI(J)
      SMAG(J)=SQRT(SR(J)**2+SI(J)**2)
  399 CONTINUE
C *** CHECK SIZE OF NUMBERS(NORMALIZATION):
      DO 405 J = 1,ND
      TEST = ABS(K(J,1))
      TX = 0.001
      IF (TEST .GE. TX)  GO TO 405
      WRITE (6, 27)
      WRITE (6, 31) J, K(J, 1)
      TX2 = 0.0001
      IF (TEST .GE.TX2)  GO TO 405
      K(J, 1) = 0.0
      WRITE (6, 32) K(J, 1)
  405 CONTINUE
C *** CALCULATE AND WRITE SUM OF RESIDUES, AND SUM OF (RESIDUES*N.F'S) FOR
C      CHECKING:
      WRITE (6, 15) SUMKRE, SUMKIM
      WRITE (6, 16) SUMKPR, SUMKPI
C *** CALCULATE FIRST NONZERO IV  PER  H*A(NNN)/B(NDD):
      FIRIV = H*A(NNN)/B(NDD)
      NDM = ND-NN - 1
      WRITE (6, 14) FIRIV, NDM
      WRITE (6,29)
      WRITE (6, 18)
      BBD = B(1)
      IF(BBD .EQ. 0) FO=909090.0
      IF (BBD .NE.0) FO=H*A(1)/BBD
C ABOVE PRINTS APPROPRIATE ERROR MESSAGE  IF F HAS A POLE AT ORIGIN
  500 READ (5,2)                    T1, DT, T2
C *** LOOK FOR NEXT CASE(JOB)IF T1 IS NEGATIVE:
      IF (T1 .LT. 0.0)  GO TO 200
  501 CONTINUE
C *** CHECK ON T2 VIS-A-VIS T1:
      IF (T2 .LE. T1) GO TO 9001
      NRPTS  = (T2 - T1)/DT  + 1
C *** AND ON NRPTS:
      IF (NRPTS .GT. 501) GO TO 9002
      NUNDER = 0
      DTT = DT
      IF(NCODE .EQ. 1) GO TO 502
      WRITE (6,24)
      WRITE (6, 7) T1, DT, T2, NRPTS, NX
      IF (FO .EQ. 909090.0) WRITE (6,150)
      IF (NX .EQ. 0) WRITE (6,  13)
      IF (NX .EQ. 0)  GO TO 502
      WRITE (6,40)
      WRITE (6,60)
      WRITE (6, 61)
  502 CONTINUE
      T=T1
```

```
      NPT = 0
C CALCULATE RESPONSE(  ) AT T= : 510-600
  510 RESP =0
      SRESP=F0
      DO 600 J=1,ND
      MS=SMAG(J)
      XX1 = SR(J)*T
C PROTECT AGAINST UNDERFLOW FROM VERY LARGE NEGATIVE EXPONENT:
      IF (XX1 .LT. -100.0)  GO TO 525
C WHAT ABOUT OVERFLOW?
      XX2 = EXP(XX1)
      XPT = K(J, 1)*XX2
      GO TO 527
  525 XPT = 0.0
      IF (NUNDER .EQ. 0)  WRITE (6, 34) T, J, XX1
      NUNDER = 1
  527 APT = SI(J)*T
      IF (SI(J))550, 540, 530
  530 TERM=2*XPT*COS(APT+K(J,2))
      AS=ATAN2(SI(J), SR(J) )
      STERM = 2*XPT*COS(APT+K(J,2)-AS     )/MS
      GO TO 560
  540 TERM=XPT
      IF (SR(J) .EQ. 0) STERM=909090.0
      IF (SR(J) .NE. 0) STERM = XPT/SR(J)
      GO TO 560
  550 TERM=0
      STERM=0
  560 CONTINUE
      RESP=RESP+TERM
      SRESP = SRESP + STERM
  600 CONTINUE
      NPT = NPT + 1
      NPT2 = NPT + NRPTS
      NPT3 = NPT2 + NRPTS
      ARR(NPT) = T
      ARR(NPT2) = RESP
C ARR SERVES PLOT22 (BOTH RESPONSES);
C PLT1 SERVES PLOT4Z (IMPULSE RESPONSE);
C PLT2 SERVES PLOT 4Z (STEP RESPONSE);
      PLT1(NPT) = T
      PLT2(NPT) = T
      PLT1(NPT2) = RESP
      IF (F0 .EQ. 909090.0 ) ARR(NPT3) = 0.0
      IF (F0 .EQ. 909090.0 ) PLT2(NPT2) = 0.0
      IF (F0 .NE. 909090.0 ) ARR(NPT3) = SRESP
      IF (F0 .NE. 909090.0 ) PLT2(NPT2) = SRESP
      IF (NCODE .EQ. 1) GO TO 695
      IF (NX) 690, 695, 690
  690 IF (F0 .EQ. 909090.0) WRITE (6,100) T, RESP, RESP
      IF (F0 .NE. 909090.0) WRITE (6, 70) T, RESP, SRESP, RESP, SRESP
  695 CONTINUE
      IF (NPT .EQ. NRPTS) GO TO 698
      T = T  + DT
      GO TO 510
C *** END CALCULATION OF POINTS IN TIME: NOW TO
C     MISCELLANEOUS COMMENTS AND PLOTS:
  698 CONTINUE
      IF (NN .EQ. ND) STR = H*A(NNN)/B(NDD)
      IF (NCODE .EQ. 1) GO TO 699
      IF (NN .EQ. ND) WRITE (6,140) STR
      IF (NN .GT. ND) WRITE (6, 160)
  699 IF (NPLT-1) 900, 811, 821
  811 CONTINUE
      WRITE (6, 212)
      WRITE (6, 9) TITLE
C: *** SKIP PLOT22 IF THERE IS A POLE AT THE ORIGIN:
      IF (F0 .EQ. 909090.0) WRITE (6,150)
      IF (F0 .EQ. 909090.0) GO TO 821
      CALL PLT22N (ARR, NRPTS, NRPTS)
      GO TO 900
  821 CONTINUE
      WRITE (6, 212)
      WRITE (6, 9) TITLE
      IF (NPRD.EQ. 1) CALL PLOT4Z (NRPTS, PLT1, TTT, RDL)
C *** IF THERE IS A POLE AT THE ORIGIN, SKIP PLOT4Z OF STEP RESPONSE:
      IF (F0 .EQ. 909090.0) WRITE (6,150)
      IF (F0 .EQ. 909090.0) GO TO 500
      WRITE (6, 212)
      WRITE (6, 9) TITLE
      IF (NPRU .EQ. 1) CALL PLOT4Z (NRPTS, PLT2, TTT, RDL)
      IF (NPLT .EQ. 3) GO TO 811
      GO TO 900
C ***
```

```
  710 WRITE (6, 28)
      STOP
C
  900 CONTINUE
C *** AT END OF COMPUTATION: WRITE GRANULATORY MESSAGES:
      WRITE (6,17)
      WRITE (6,19)
      DO 5000 J = 1,ND
      XXX = 999999.999
      YYY = 999999.999
      IF ( SR(J) .NE. 0.0 )  XXX = -1.0/SR(J)
      IF ( SI(J) .NE. 0.0 )  YYY = 1/ABS  (SI (J) )
      IF ( SI(J) .NE. 0.0 )  YYY = 6.28318/ABS(SI(J))
 5000 WRITE (6,22) SR(J), SI(J), XXX, YYY
      WRITE (6,23) DTT
      WRITE (6,21)
      GO TO 500
   01 WRITE (6, 25) T1, T2
      GO TO 500
 9002 WRITE (6, 26) NRPTS
      GO TO 500
      END
      SUBROUTINE POLRTT (M1, ACOF, COF, ZR, ZI)
C SIMPLE POLYNOMIAL ZERO FINDER, DERIVED FROM IBM SSP "POLRT": USES NEWTON'S
C METHOD, WITHOUT ZERO IMPROVEMENT (NO "IFIT" REFINEMENT ON ORIGNAL POLYNOMIAL)
C HAS SEVERAL (10) SCATTERED STARTING POINTS, TRIED AS NECESSARY (LIMIT: 25
C   ERATIONS ON EACH).IF COMPONENT(S) OF DERIVATIVE IS TOO SMALL, RESTARTS AT
C NEXT TRIAL POINT TO AVOID UNDERFLOW OR OVERFLOW ON DIVISION BY NEARZERO. GIVES
C ERROR MESSAGE IF NO CONVERGENCE.  NOTATION:  M1 = 1 + DEGREE OF POLYNOMIAL
C              ACOF = ARRAY OF COEFFICIENTS,   COF = WORKING (SCRATCH-PAD) ARRAY
C              ZR (+ J) ZI ARE ARRAYS INTO WHICH GO ZEROS FOUND.
      DIMENSION ACOF(M1), COF(M1), ZR(M1), ZI(M1), XA(10), YA(10)
      DATA XA/ 4*-0.5, 2*0.2, 4*1.0/
      DATA YA/ 0.28, 0.47, 1.03, 0.79, 0.10, 0.80, 0.0, 0.5, 2.0, 4.0/
      N = M1 - 1
      NX=N
      NXX=M1
      N2=1
C *** PUT COEFFICIENTS INTO WORKING ARRAY (IN REVERSE ORDER):
      DO 40 L=1, M1
      ZR(L) = 0.999999E11
      ZI(L) = 0.999999E11
      MT= M1-L+1
   40 COF(MT)=ACOF(L)
C *** PICK APPROXIMATE ZERO:
   45 IN=0
   50 CONTINUE
      IN=IN+1
      IF (IN-10) 51, 51, 95
   51 X = XA(IN)
      Y = YA(IN)
      ICT=0
C    * EVALUATE POLYNOMIAL AND DERIVATIVE:
   60 UX=0.0
      UY=0.0
      V =0.0
      YT=0.0
      XT=1.0
      U=COF(N+1)
      IF(U) 65,130,65
   65 DO 70 I=1,N
      L =N-I+1
      XT2=X*XT-Y*YT
      YT2=X*YT+Y*XT
      U=U+COF(L )*XT2
      V=V+COF(L )*YT2
      FI=I
      UX=UX+FI*XT*COF(L )
      UY=UY-FI*YT*COF(L )
C *** IF UX OR UY IS TOO SMALL, GO TO 50 AND TRY NEW START:
      T2 = 1.0E-08
      TUX = ABS(UX)
      TUY = ABS(UY)
      IF (TUX .LT. T2 .AND. TUX .NE. 0.0)  GO TO 50
      IF (TUY .LT. T2 .AND. TUY .NE. 0.0)  GO TO 50
      XT=XT2
   70 YT=YT2
      SUMSQ=UX*UX+UY*UY
C *** ELSE CONTINUE TO ITERATE ON THIS ONE; CORRECTION:
      DX = (V*UY - U*UX)/SUMSQ
      DY = -(U*UY + V*UX)/SUMSQ
      X = X +DX
      Y = Y + DY
      IF (ABS(DY)+ ABS(DX)-1.0E-05) 100,80,80
```

```
C *** INCREMENT LARGE, GO 'ROUND AGAIN: 80
   80 ICT=ICT+1
      IF (ICT-25) 60, 50, 50
C *** INCREMENT SMALL, SATISFIED, GO OUT: 100
  100 IF ( ABS(Y)-1.E-4* ABS(X)) 135,125,125
  125 ALPHA=X+X
      SUMSQ=X*X+Y*Y
      N=N-2
      GO TO 140
  130 X=0.0
      NX=NX-1
      NXX=NXX-1
  135 Y=0.0
      SUMSQ=0.0
      ALPHA=X
      N=N-1
  140 COF(2)=COF(2)+ALPHA*COF(1)
      NSX   =N
      IF(NSX    .LT. 2) NSX   =2
      DO 150 L=2,NSX
  150 COF(L+1)=COF(L+1)+ALPHA*COF(L)-SUMSQ*COF(L-1)
  155 ZR(N2) = X
      ZI(N2) = Y
      N2=N2+1
      IF(SUMSQ) 160,165,160
  160 Y=-Y
      SUMSQ=0.0
      GO TO 155
  165 IF (N) 200, 200, 45
  200 RETURN
   95 WRITE (6, 1) N2, IN, ICT, (I, ZR(I), ZI(I), I = 1, M1)
    1 FORMAT (1H0,    '********* ERROR RETURN FROM SUBROUTINE POLRTT (DOES
     1 NOT CONVERGE):  N2 =', I3,    ', IN =',  I2,  ', ICT = ', I3, 10X,
     2 27('*'),/,   10X,  'ZEROS FOUND,   IF ANY:', 21(I10, F12.6, '  + J'
     3, F14.6, /, 31X ))
      RETURN
      END
            SUBROUTINE PLT22N (A, N, NL)
C     THIS IS PLOT22 MODIFIED TO PRINT ZERO AXIS IF IT'S ON SCALE
      DIMENSION IOUT(101), YPR(11), ICAR(3), A(1), B(1)
      DATA BLANK, ICAR(1), ICAR(2), ICAR(3) /1H , 1HI, 1HS, 1HI/
      DATA IDOT /1H./
    1 FORMAT (1H , 132 ('.'))
    2 FORMAT (1H , F11.4,  5X, 101A1)                                    C
    3 FORMAT (1H0, 'PLOTS OF IMPULSE RESPONSE ( "I" ) AND STEP RESPONSE(
     1 "S" ):', 45X, '(BAR MARKS ZERO ORDINATE)')
    4 FORMAT (1H , 6X, 'TIME')
    8 FORMAT (1H , 10X,          11F10.4)
C *** CALCULATE SCALE OF THE "Y'S"                                        0
      M1 = N+1                                                           0
      M2 = 3*N                                                           0
      YMIN = A(M1)                                                       0
      YMAX = YMIN                                                        0
      M3 = N+2                                                           0
      DO 40 I=M3,M2                                                      0
      IF (A(I)-YMIN) 28,26,26                                            0
   26 IF(A(I)-YMAX) 40,40,30                                             0
   28 YMIN = A(I)                                                        0
      GO TO 40                                                           0
   30 YMAX = A(I)                                                        0
   40 CONTINUE                                                           0
      YSCAL = (YMAX-YMIN)/100.0                                          C
      JPZ = -YMIN/YSCAL + 1
      YPR(1) = YMIN                                                      0
      DO 42 J=1,9                                                        0
   42 YPR(J+1) = YPR(J) + YSCAL*10.0                                     0
      YPR(11) = YMAX                                                     0
      WRITE (6,3)
      WRITE (6,1)
C *** PRINT SCALE OF THE Y'S                                             0
      WRITE (6, 8) (YPR(I), I=1,11)                                      C
      WRITE (6,4)
C *** DETERMINE VALUE OF "X" TO PRINT                                    0
      L = 1                                                              0
   45 XPR = A(L)
C *** DETERMINE "Y" CHARACTERS TO PRINT                                  C
      DO 55 IY = 1,101
   55 IOUT(IY) = BLANK                                                   C
      IOUT(1) = IDOT
      IOUT(101) = IDOT
      IF (JPZ .GE. 1 .AND. JPZ .LE. 101) IOUT(JPZ) = ICAR(3)
      DO 60 J=1,2                                                        0
      LL = L+ J*N                                                        0
      JP = (A(LL)-YMIN)/YSCAL + 1                                        0
```

```
      IOUT(JP) = ICAR(J)
   60 CONTINUE
C *** PRINT LINE
      WRITE (6, 2) XPR, (IOUT(IZ), IZ=1,101)
      L = L+1
      IF (L-NL) 45, 45, 86
C *** PRINT SCALE OF THE "Y'S"
   86 WRITE (6,4)
      WRITE (6, 8) (YPR(I), I=1,11)
      WRITE (6,1)
      RETURN
      END
      SUBROUTINE EVLPCL( M,C,SRE,SIM, PCLYRE, PCLYIM, POLMAG,POLANG)
C *** MODIFIED FOR ANGLE VALUE WHEN S=0 AND POLY = 0 + J 0; SEE BELOW  MR71 DFT
C THIS IS "EVALPCLY" OF ALGOL TRANSLATED INTO A FORTRAN IV SUBROUTINE AS
C "EVLPCL", WITH THE NOTATION:  M = DEGREE OF PCLYNOMIAL
C        M = DEGREE OF PCLYNOMIAL
C S = SRE + J SIM    IS THE VALUE OF S AT WHICH TO EVALUATE THE PCLYNOM.
C C(1) = CONSTANT TERM, C(M+1) = COEFFICIENT OF MTH POWER TERM IN PCL.
C RESULT OF EVALUATION IS(PCLYRE + J PCLYIM)= PCLMAG AT ANGLE POLANG
      DIMENSION C(1), C1(51)
      MM = M+1
       MMM=M-1
      DO 5 I=1,MM
    5 C1(I) = C(I)
      IF (MMM) 40, 30, 10
   10 D = -2*SRE
       E = SRE**2 + SIM**2
       DO 20 III=1,MMM
       II=M+2-III
      C1(II-1) = C1(II-1) - D*C1(II)
   20 C1(II-2) = C1(II-2) - E*C1(II)
   30 PCLYRE = C1(2)*SRE + C1(1)
      PCLYIM  =C1(2)*SIM
   35  PCLMAG = SQRT(PCLYRE**2 + POLYIM**2)
   36 IF (PCLYRE .EQ. 0 .AND. PCLYIM .EQ. 0) GO TO 37
       PCLANG = ATAN2(PCLYIM,POLYRE)*57.295780
       RETURN
   37 POLANG = 909090.9
C      THE FF. CDS ARE NOT NEC. IN I7, NC?
C      IF (SRE .EQ. C .AND. SIM .EQ. C) GO TO 38
C
      RETURN
C  38 IF ( C1(1) .EQ. 0 .AND. C1(2) .NE. C) PCLANG  = SIGN (90.0, C1(2))
C     I.E. IF POLYNOMIAL = 0 AT S = 0, SET ANGLE = (SIGN OF COEFF. OF S)*90:
C       POLANGL = 90.0*C1(2)/ABS(C1(2))
C     RETURN
   40 POLYRE = C1(1)
       PCLYIM = 0
      GO TO 35
       END
      SUBROUTINE PLCT4Z (N, A, X, Y)
C
C  ... USES ONLY HALF THE WIDTH OF THE PAPER FOR THE PLOT, BUT PRINTS VALUES OF
C       BOTH "X" AND "Y"  WITH LEGEND AT LEFT. PRINTS ZERO AXIS IF ON SCALE.
C
      REAL ICUT
      DIMENSION IOUT( 51), YPR( 6), A(1), X(3), Y(3)
      DATA BLANK, ASTX, BAR, DOT /1H , 1H*, 1H|, 1H./
    1 FORMAT (1H , 115('.') )
    2 FORMAT (1H ,         F20.4, E22.4, 18X, 51A1  )
    3 FORMAT (1H , 60X, 51A1)
    4 FORMAT (1HO,'SUBROUTINE ''PLCT4Z'', BEING CALLED, GIVES (BAR MARKS
     1 ZERO ORDINATE):')
    8 FORMAT (1H ,  9X, 3A4, 10X, 3A4, 12X, 6F10.4)
      NL = N
      M1 = N+1
      M2 = 2*N
      YMIN = A(M1)
      YMAX = YMIN
      M3 = N+2
      DO 40 I=M3,M2
       IF (A(I)-YMIN) 28,26,26
   26 IF (A(I)-YMAX) 40,40,30
   28 YMIN = A(I)
      GO TO 40
   30 YMAX = A(I)
   40 CONTINUE
      YSCAL = (YMAX-YMIN)/50.0
      JP2 = -YMIN/YSCAL + 1
      YPR(1) = YMIN
      DO 42 J=1,4
   42 YPR(J+1) = YPR(J) + YSCAL*10.0
      YPR( 6) = YMAX
```

```
      WRITE (6, 4)
      WRITE (6, 1)                                                        ALIGNER*
      WRITE (6,8) X, Y, (YPR(I), I=1,6)
      DO 43 IY = 2,50
   43 IOUT(IY) = BLANK
      IOUT(1) = DOT
      IOUT(51) = DOT
      IF  (JPZ .GE. 1 .AND. JPZ .LE. 51) IOUT(JPZ) = BAR
      WRITE (6,3) (IOUT(IZ), IZ = 1,51)
      L = 1                                                               5C.
   45 XPR = A(L)
      LL = L + N
      JP = (A(LL)-YMIN)/YSCAL + 1                                         6C.
      IOUT(JP) = ASTX
   60 CONTINUE                                                            64.
      WRITE (6,2) XPR, A(LL), (IOUT(IZ), IZ=1,51)
      IOUT(JP) = BLANK
      IOUT(1) = DOT
      IOUT(51) = DOT
      IF (JPZ .GE. 1 .AND. JPZ .LE.51) IOUT(JPZ) = BAR
      L = L+1                                                             68.
      IF (L-NL) 45,45,86
   86 CONTINUE
      WRITE (6,3) (IOUT(IZ), IZ = 1,51)
      WRITE (6,8) X, Y, (YPR(I), I=1,6)
      WRITE (6, 1)                                                        ALIGNER*
      RETURN
      END
```

.... **illustrative (sample) data-card set for IMPSTEP:**

```
FIRST NOTE CARD (ANY NOTES, SUCH AS DATE, PURPOSE, ETC.)
SECOND ONE
THIRD ONE ( THE THREE WILL BE PRINTED VERBATIM)
     1.0
          3          7          0          2          1          1          0
     1.0                     0.0                     3.6                     1.0
     1.0                     6.3                      12.0                 8.3
     7.8                     9.0                     3.1                     1.0
     0.0                     0.1                     5.0
     0.0                    0.05                     5.0
    -1.0                     0.0                     0.0
```

RESSINET (for calculating responses to sinusoidal excitations)

```
C *** RESSINET ***
C RESSINET CALCULATES THE RESPONSE OF A NETWORK WHOSE TRANSFER FUNCTION IS:
C   T(S) = (AN S**N + ... + +A1 S + A0)/(BD S**D + ... + B0)
C TO THE SINUSOIDAL EXCITATION  EM*COS(WF*T + PHI)
C       IT TABULATES AND PLOTS PARTS PER INSTRUCTIONS (MAXIMUM NUMBER OF POINTS
C       SET INTO DIMENSIONS BELOW IS 500). *** IT IS ASSUMED THAT N .LE. D AND
C       ALL POLES (NATURAL FREQUENCIES) ARE SIMPLE AND THE DENOMINATOR IS GIVEN
C       IN POLYNOMIAL FORM; MAXIMUM VALUE OF D IS 20 IN DIMENSIONS OF PROGRAM.
C ***** DATA CARDS:      (ONE SET FOR EACH CASE TO BE "DONE")
C      (1) TITLE-REFERENCE-COMMENT-WHATEVER NOTES YOU WANT PRINTED (3 CARDS)....
C      (2) EM, WF, PHI (IN DEGREES) (FORMAT 3F20.6)
C      (3) N, D, NX, NRPLTS, NCODE (FORMAT 5I10)
C             NX = 1 GIVES TABULATED RESULTS AS WELL AS PLOTS
C             NX = 0 GIVES PLOTS ONLY, NO TABLE OF VALUES
C             NRPLTS = TOTAL NUMBER OF PLOTS WANTED (0 .LE. NRPLTS .LE. 7)
C             NCODE = 0   NORMAL USAGE
C                   = 1 SUPPRESS ALL OUTPUT EXCEPT ESSENTIAL DATA AND PLOTS
C      (4) N1, N2, N3, ETC. (FORMAT 7I10)
C              THESE ARE THE CODE NUMBERS FOR THE PLOTS WANTED; SEE CODE BELOW:
C             OMIT CARD (4) IF NRPLTS = 0; CARD (4) SHOULD CONTAIN "NRPLTS"
C             INTEGERS, TOTAL
C             PLOT CODES:
C                   0 : NO PLOTS                          E  = EXCITATION
C                   1 : E AND RF                          R  = (TOTAL) RESPONSE
C                   2 : E AND R                           RN = NATURAL COMPONENT OF R
C                   3 : RN AND RF                         RF = FORCED COMPONENT OF R
C                   4 : R AND RF
C                   5 : RN
C                   6 : R
C                   7 : R AND RN
C      (4) A0, A1, A2 ..... AN (NUMERATOR COEFFICIENTS)  (FORMAT 4F20.6)
C      (5) B0, B1, ..... BD (DENOMINATOR COEFFICIENTS) (FORMAT 4F20.6)
C      (6) T1, DT, T2 ..... T1, DT, T2 ..... ETC (TIME INTERVALS AND ENDS
C          FOR COMPUTATION OF INVERSE TRANSFORM)  (FORMAT  3F20.6)
C         ******* NB: AFTER THE LAST SUCH GROUP IN EACH CASE PUT A CARD
C          WITH "T1" .GT. 1000000.0 AND ANY TWO NUMBERS IN THE NEXT TWO FIELDS..
C          ...THIS STARTS THE NEXT CASE, OR TERMINATES THE CALCULATIONS IF THERE
C          ARE NO MORE CARDS.    EACH TIME INTERVAL MAY HAVE UP TO 500 POINTS.
C      EACH CASE (SET OF DATA PER ABOVE 6 PARTS) IS SEPARATELY TREATED.
       REAL IMQ, MAG, MC, MN, MS, K(21, 2), IMK
C THIS PROGRAM WAS DEVELOPED BY DAVID F. TUTTLE, FOR USE IN EE101-102-103
C AND ASSOCIATED TEXTS
      DIMENSION A(21), B(21), C(21), BB(21), ARR(1503), SR(21), SI(21),
     1 SMAG(21), TITLE(60), PLT1(1002),             TTT(3), RDL(3), RUL(
     23), AE(2), BE(3), EST(501), RFST(501), RNST(501), MPLT(8), ICAR(2)
      DATA ICARE, ICARR, ICARN, ICARF/'E', 'R', 'N', 'F'/
      DATA TTT/'    ', '  TI', 'ME  '/, RDL/'NATL', '.PT.','RESP'/,
     1RUL/'FORC', 'ED P', 'ESP.'/
       CTE = 57.295780
     1 FORMAT (8I10)                                                        0
     2 FORMAT (4F20.6)                                                      0
     3 FORMAT (F20.6, 6I10)
     4 FORMAT (1H0, 'PARAMETER VALUES READ:     N =', I3, ',   D =', I3,
      1 ', NX =', I2, ', NRPLTS =', I2, ', NCODE =',I2, /)
     5 FORMAT (1H1, /// )
     6 FORMAT ( 20A4 )
     7 FORMAT (1H+, 25X, 'T1 =', F6.2,', DT =', F6.3, ', T2 =', F6.2,
      1 ', NRPTS =', I4, ', NX =', I3, / )
     8 FORMAT (1H0, 'THE (SINUSOIDAL) EXCITATION IS EM*COS(WF*T + PHI), I
      1N WHICH:',10X,' EM =', F10.4, 18X, 'WF =',F10.4,//,72X,'PHI =', F1
      20.4, ' DEGREES, OR', F10.4,' RADIANS.')
     9 FORMAT ( 1H+,  26X,  'NOTES FOR THIS RUN:',   4X, 20A4,
      1 2(/, 50X, 20A4) )
    10 FORMAT (1H0, /,'  THE FOLLOWING ARE THE POLES OF THIS TRANSFER FUN
      1CTION, T(S):')
    11 FORMAT (1H0,' DENOMINATOR:', / )
    12 FORMAT (1H0,' NUMERATOR:' ,/)
    13 FORMAT (1H0, ///, ' *****   NO TABLE OF VALUES WAS ASKED FOR *****
      1')
    14 FORMAT(1H0, 'THE SUM OF THE RESIDUES IS:',75X, E11.4, ' + J ', E11
      1.4,/)
    15 FORMAT (1H , 'THE SUM OF THE PRODUCTS "KP" IS:' 70X, E11.4, ' + J'
      1, E11.4,/)
    16 FORMAT (1H0, 8X, 'THE PERIOD OF EXCITATION IS', F10.6)
    17 FORMAT (///, 15('*'),   5X,  'A MESSAGE FROM RESSINET:', 4X,
      1 'DID YOU GET ENOUGH POINTS,  IS THE "GRANULARITY" FINE ENOUGH?',
      2 /, 44X, 'OR DID YOU, PERCHANCE,  OVERWORK ME AND GET TOO MANY?',
      3 /, 44X, 'TAKE A GOOD LOOK BEFORE YOU LEAVE. *******', 30('*') )
    18 FORMAT (1H , 'DO THEY CHECK?')
    19 FORMAT (1H0,  'ANOTHER MESSAGE: HERE ARE THE NATURAL FREQUENCIES O
      1NCE MORE, TOGETHER WITH THE RECIPROCALS OF THE NEGATIVES OF THEIR
      2REAL PARTS', /, ' AND THE PERIODS OF THE OSCILLATIONS:',/)
    20 FORMAT (1H0,   'THE FOLLOWING ARE THE RESIDUES OF R(S) THEREAT:',
      1     //, 10X, 'MAGNITUDE',   7X, 'ANGLE (RAD.)',  6X, 'ANGLE (DEG.
```

```
     2)',   8X, 'REAL PART',  9X, 'IMAG. PART',/, 10X, 9('.'), 7X, 12('
     3.'), 6X, 1('.'), 8X, 9('.'), 9X, 10('.'))
  21 FORMAT (1H+, 65X, 'ARE YOU HAPPY?',/)
  22 FORMAT (1H0, F16.6,'  + J ', F10.6, 2F20.6)
  23 FORMAT (1H0, 'YOU CHOSE TIME POINTS SPACED', F7.3, '; DID YOU CHOO
     1SE WELL ?', /)
  24 FORMAT(1H1,        'TIME DATA READ IN: ')
  25 FORMAT (1H1, 50('*'), //, ' DO YOU REALLY WANT T2 LESS THAN T1? ........
     1... YOU SAY T1 =', F6.2, ' AND T2 =', F6.2, //,  ' RESSINET THINKS
     2 NOT.', 5X, 20('*'))
  26 FORMAT (1H1, 50('*'), //, ' YOU ASK FOR', I5, ' POINTS.  TOO MANY.
     1 SORRY.')
  27 FORMAT (1H0,'FROM THE SMALLNESS OF AT LEAST ONE RESIDUE ("K") IT S
     1EEMS THAT THE DATA MAY NOT BE WELL NORMALIZED. RESULTS MAY BE DISA
     2PPOINTING.')
  28 FORMAT (1H0, 'THE END.')
  30 FORMAT (1H0, 2H0R  )
  31 FORMAT (1H0, 'THE TRANSFER FUNCTION, EVALUATED AT S = J*WF, IS:',
     1E12.4, 2X, 'AT',F10.4, ' DEGREES')
  32 FORMAT (1H0, 'THE (FORCED) RESPONSE PHASOR IS:', E12.4,2X,'AT', F1
     10.4, ' DEGREES')
  33 FORMAT(1H0, 18X,              'NOTATION:    E(S) = MATE OF EXCITATIO
     1N, T(S) = TRANSFER FUNCTION', //, 31X, 'R(S) = E(S)*T(S)')
  34 FORMAT(1H0,'A COEFFICIENT IS INFINITE... ERGO TROUBLE: J=',I3,', M
     1C =',E14.6,', EDM =',E14.6)
  35 FORMAT(1H0, 'THE FOLLOWING PLOTS ARE ASKED FOR:', 8I5)
  40 FORMAT (1H0,    'INVERSE TRANSFORMATION OF  R(S) GIVES:')
  60 FORMAT (1H0, 4X, 'TIME', 12X, 'EXCITATION',  6X, 40('.'), ' RESPON
     1SE ', 45('.'))
  61 FORMAT(1H , 37X, 'FORCED COMP.', 8X, 'NATURAL COMP.',11X, 'SUM TOT
     1AL')
  62 FORMAT(1H+, 102X, 'RN IN', 16X, 'RF IN',/, 102X, 'E FORMAT', 13X,
     1'E FORMAT',/)
  70 FORMAT(F10.4, F20.6, F18.6, 2F21.6, 2E21.4)
  71 FORMAT(1H0, 'PLOTS OF EXCITATION ( "E" ) AND FORCED RESPONSE ( "F"
     1 ):')
  72 FORMAT(1H0, 'PLOTS OF EXCITATION ( "E" ) AND (COMPLETE) RESPONSE (
     1 "R" ):')
  73 FORMAT(1H0,'PLOTS OF NATURAL ( "N" ) AND FORCED ( "F" ) COMPONENTS
     1 OF RESPONSE:')
  74 FORMAT(1H0,'PLOTS OF (COMPLETE) RESPONSE ( "R" ) AND FORCED COMPON
     1ENT OF RESPONSE ( "F" ):')
  75 FORMAT(1H0, 'PLOT OF NATURAL (TRANSIENT) COMPONENT OF RESPONSE:')
  76 FORMAT(1H0, 'PLOT OF (COMPLETE) RESPONSE ALONE:')
  77 FORMAT(1H0, 'PLOT OF NATURAL ( "N" ) PART OF RESPONSE AND COMPLETE
     1 RESPONSE ( "R" ):')
  80 FORMAT (1H0, F16.6, 2X, 2H+J, F10.6)
  90 FORMAT (1H0, 5F18.6)
 100 FORMAT (F10.4, F18.4, 22X, E22.4 )
 110 FORMAT (1H0,    'FOR THE FOLLOWING CASE THE TRANSFER FUNCTION COEFF
     1ICIENTS, LOWEST POWER FIRST, ARE:',/)
 120 FORMAT ( 3(' ',  6F20.6, / ) )
 130 FORMAT (1H0, 89X, 11HDENOMINATOR)
 212 FORMAT (1H1, 'THIS IS RESSINET.')
C ***** END OF FORMATS *****
C
 200 READ (5,6,END=9009) TITLE
     READ (5, 2) EM, WF, PHI
     PHIR = PHI/CTE
     READ (5, 1)      NN, ND,  NX, NRPLTS, NCODE
     NNN=NN+1
     NDD=ND+1
     NDDD = ND - 1
     IF (NRPLTS .EQ. 0) GO TO 201
     READ(5,1) (MPLT(I), I=1, NRPLTS)
 201 CONTINUE
     READ (5,2) ( A(J), J=1,NNN )
C A(1) = COEFFICIENT OF ZEROTH POWER TERM
C A(NNN) = COEFFICIENT OF NNTH POWER TERM
     READ (5,2) ( B(J), J=1,NDD)
     WRITE (6, 212)
     WRITE (6,9) TITLE
     WRITE(6,4) NN, ND, NX, NRPLTS, NCODE
     WRITE (6,110)
     WRITE (6,12)
     WRITE (6,120) ( A(J), J=1,NNN )
     WRITE (6,11)
     WRITE (6,120) ( B(J), J=1,NDD)
C EVALUATE TRANSFER FUNCTION:
     AE(1) = -WF*SIN(PHIR)
     AE(2) = COS(PHIR)
     BE(1) = WF*WF
     BE(2) = 0.0
     BE(3) = 1.0
```

```
      CALL EVLPOL (NN, A, 0.0, WF, UN, VN, MN, AN)
      CALL EVLPOL (ND, B, 0.0, WF, UC, VC, MC, AC)
      TFMAG = MN/MC
      TFANG = AN-AC
      TFANR = TFANG/CTE
C FIND NATURAL FREQUENCIES:
      CALL POLRTD (B, C, NDD, SR, SI)
      DO 215 J=2,NDD                                                    0
  215 BB(J-1) = B(J)*(J-1)                                              0
      WRITE (6,10)                                                      0
      DO 216 J=1,ND                                                     0
  216 WRITE (6,80) SR(J), SI(J)                                         0
C
      WRITE (6,8) EM, WF, PHI, PHIR
      WRITE(6,33)
      WRITE (6,20)                                                      0
      SUMKR = 0.0
      SUMKI = 0.0
      SUMKPR = 0.0
      SUMKPI = 0.0
C CALCULATE RESIDUES:
      DO 399 J=1,ND
      CALL EVLPOL( NN, A, SR(J), SI(J), UN, VN, MN, AN)                 0
      CALL EVLPOL (NDDD, BB, SR(J), SI(J), UC, VC, MC, AC)              0
C
      CALL EVLPOL(1,AE,SR(J), SI(J), UE1, VE1, ENM, ENA)
      CALL EVLPOL(2, BE, SR(J), SI(J), UE2, VE2, EDM, EDA)
      CT1 = EDM*MC
      IF (CT1 .EQ. 0) GO TO 9003
      K(J,1) = EM*ENM*MN/CT1
      ANAC = AN - AC + ENA - EDA
      K(J,2) = ANAC/CTE
      PTRE = K(J,1)*COS(K(J,2))
      PTIM = K(J,1)*SIN(K(J,2))
      WRITE (6,90) K(J,1), K(J,2), ANAC, PTRE, PTIM
      IF (SI(J) .EQ. 0) K(J,1)=EM*UN*UE1/(UC*UE2)
      IF (SI(J) .EQ. 0) PTRE= K(J,1)
      SUMKR = SUMKR + PTRE
      SUMKI = SUMKI + PTIM
      SUMKPR = SUMKPR + PTRE*SR(J)-PTIM*SI(J)
      SUMKPI = SUMKPI + PTRE*SI(J) + PTIM*SR(J)
  399 CONTINUE                                                          0
C *** CHECK SIZE OF NUMBERS(NORMALIZATION):
      DO 405 J = 1,ND
      TEST = ABS(K(J,1))
      IF (TEST .LE. 0.001)  GO TO 407
  405 CONTINUE
      GO TO 419
  407 WRITE (6,27)
C
  419 CONTINUE
C CALCULATE FORCED PART OF RESPONSE:
      RFMAG = EM*TFMAG
      RFANG = PHI + TFANG
      RFANGR = RFANG/CTE
      WRITE (6, 31) TFMAG, TFANG
      WRITE (6,32) RFMAG, RFANG
C
C *** CHECK INITIAL VALUES:
      SUMKR = SUMKR +       RFMAG*COS(RFANGR)
      SUMKPR = SUMKPR-WF*RFMAG*SIN(RFANGR)
C
      WRITE(6,14) SUMKR, SUMKI
      WRITE (6,15) SUMKPR, SUMKPI
      WRITE (6,18)
      IF (NRPLTS .EQ.0) GO TO 451
      WRITE(6,35) (MPLT(I), I=1, NRPLTS)
  451 CONTINUE
  500 READ (5,2)                   T1, DT, T2                           0
C *** AT END OF COMPUTATION: WRITE GRANULATORY MESSAGES:
      IF (T1 .LT. 1000000.0) GO TO 501
C *** LOOK FOR NEXT CASE(JOB):
      GO TO 200
  501 CONTINUE
      IF (T2 .LE. T1) GO TO 9001
      T3 = T2 + 0.0001
      NRPTS  = (T3 - T1)/DT  + 1
      IF (NRPTS .GT. 501) GO TO 9002
      DTT = DT
      IF(NCODE .EQ. 1) GO TO 502
      WRITE (6,24)
      WRITE (6, 7) T1, DT, T2, NRPTS, NX
      IF (NX .EQ. 0) WRITE (6,  13)
      IF (NX. EQ. 0)  GO TO 502
```

```
      WRITE (6,40)
      WRITE (6,60)
      WRITE(6,61)
      WRITE(6,62)
C
  502 CONTINUE
      T=T1
      NPT = 0
C CALCULATE RESPONSE(S) AT T= : 510-600 AND 600 - 698
C
  510 RN = 0
      WFT = WF*T
      RF = RFMAG*COS(WFT + PFANGR)
      E = EM * COS(WFT + PHIR)
C
      DO 600 J=1,ND
      XPT = K(J,1)*EXP(SR(J)*T)
      APT = SI(J)*T
      IF (SI(J))550, 540, 530
  530 TERM=2*XPT*COS(APT+K(J,2))
      GO TO 560
  540 TERM=XPT
      GO TO 560
  550 TERM=0
  560 CONTINUE
      RN = RN + TERM
  600 CONTINUE
C STORE FOR PLOTS LATER:
      R = RF + RN
      NPT = NPT + 1
      EST(NPT) = E
      RFST(NPT) = RF
      RNST(NPT) = RN
C FILL FIRST SECTION OF ARRAYS ARR AND PLT1 WITH TIME VALUES:
      ARR(NPT) = T
      PLT1(NPT) = T
      IF (NCODE .EQ. 1) GO TO 695
      IF (NX .EQ. 0) GO TO 695
      WRITE(6,70) T, E, RF, RN, R, RN, R
  695 CONTINUE
      IF (NPT .EQ. NRPTS) GO TO 698
      T = T  + DT
      GO TO 510
  698 CONTINUE
      IPLT = 0
  700 CONTINUE
      IF (IPLT .EQ. NRPLTS) GO TO 900
C ELSE DRAW PLOTS;
      WRITE (6, 212)
      WRITE (6, 9) TITLE
      IPLT = IPLT + 1
      L = MPLT(IPLT)
      GO TO (710, 720, 730, 740, 750, 760, 770), L
  710 CONTINUE
      DO 715 I = 1, NRPTS
      NPT2 = I + NRPTS
      NPT3 = NPT2 + NRPTS
      ARR(NPT2) = RFST(I)
  715 ARR(NPT3) = EST(I)
      ICAR(2) = ICARE
      ICAR(1) = ICARF
      WRITE(6,71)
      CALL  PLT22N(ARR,NRPTS, NRPTS, ICAR)
      GO TO 700
  720 CONTINUE
      DO 725 I = 1, NRPTS
      NPT2 = I + NRPTS
      NPT3 = NPT2 + NRPTS
      ARR(NPT2) = RFST(I) + RNST(I)
  725 ARR(NPT3) = EST(I)
      ICAR(2) = ICARE
      ICAR(1) = ICARR
      WRITE(6,72)
      CALL PLT22N(ARR, NRPTS, NRPTS, ICAR)
      GO TO 700
  730 CONTINUE
      DO 735 I = 1, NRPTS
      NPT2 = I + NRPTS
      NPT3 = NPT2 + NRPTS
      ARR(NPT2) = RNST(I)
  735 ARR(NPT3) = RFST(I)
      ICAR(1) = ICARN
      ICAR(2) = ICARF
      WRITE(6,73)
```

```
      CALL PLT22N(ARR, NRPTS, NRPTS  ICAR)
      GO TO 700
  740 CONTINUE
      DO 745 I = 1, NRPTS
      NPT2 = I + NRPTS
      NPT3 = NPT2 + NRPTS
      ARR(NPT2) = RNST(I) + RFST(I)
  745 ARR(NPT3) = RFST(I)
      ICAR(1) = ICARR
      ICAR(2) = ICARF
      WRITE(6,74)
      CALL PLT22N(ARR, NRPTS, NRPTS, ICAR)
      GO TO 700
  750 CONTINUE
      DO 755 I = 1, NRPTS
      NPT2 = I + NRPTS
  755 PLT1(NPT2) = RNST(I)
      WRITE(6,75)
      CALL PLOT4Z(NRPTS, PLT1, TTT, RDL)
      GO TO 700
  760 CONTINUE
      DO 765 I = 1, NRPTS
      NPT2 = I + NRPTS
  765 PLT1(NPT2) = RFST(I) + RNST(I)
      WRITE(6,76)
      CALL PLOT4Z(NRPTS, PLT1, TTT, RUL)
      GO TO 700
  770 CONTINUE
      DO 775 I = 1, NRPTS
      NPT2 = I + NRPTS
      NPT3 = NPT2 + NRPTS
      ARR(NPT2) = RFST(I) + RNST(I)
  775 ARR(NPT3) = RNST(I)
      ICAR(1) = ICARR
      ICAR(2) = ICARN
      WRITE(6,77)
      CALL PLT22N(ARR, NRPTS, NRPTS, ICAR)
      GO TO 700
C
C ***
  900 WRITE (6, 212)
      WRITE (6, 9) TITLE
      WRITE (6,17)
      WRITE (6,19)
      DO 5000 J = 1,ND
      XXX = 999999.999
      YYY = 999999.999
      IF ( SR(J) .NE. 0.0 )  XXX = -1.0/SR(J)
      IF ( SI(J) .NE. 0.0 )  YYY = 1/ABS  (SI (J) )
 5000 WRITE (6,22) SR(J), SI(J), XXX, YYY
      PER = 6.28318/WF
      WRITE (6,16) PER
      WRITE (6,23) DTT
      WRITE (6,21)
      GO TO 500
 9001 WRITE (6, 25) T1, T2
      GO TO 500
 9002 WRITE (6, 26) NRPTS
      GO TO 500
 9003 WRITE(6,34) J, MC,EDM
      STOP
 9009 WRITE(6,28)
      STOP
      END
      SUBROUTINE POLRTD (XCOF, COF, M1, ROOTR, ROOTI)
      DIMENSION XCOF(M1),  COF(M1),  ROOTR(M1), ROOTI(M1)
      IFIT=0
      N = M1 - 1
      NX=N
      NXX=M1
      N2=1
      DO 40 L=1, M1
      MT= M1-L+1
   40 COF(MT)=XCOF(L)
   45 XO=.005
      YO=0.01
      IN=0
   50 X=XO
      XO=-10.0*YO
      YO=-10.0*X
      IN=IN+1
      X=XO
      Y=YO
      GO TO 59
```

```
 55 IFIT=1
    XPR=X
    YPR=Y
 59 ICT=0
 60 UX=0.0
    UY=0.0
    V =0.0
    YT=0.0
    XT=1.0
    U=COF(N+1)
    IF(U) 65,130,65
 65 DO 70 I=1,N
    L =N-I+1
    XT2=X*XT-Y*YT
    YT2=X*YT+Y*XT
    U=U+COF(L )*XT2
    V=V+COF(L )*YT2
    FI=I
    UX=UX+FI*XT*COF(L )
    UY=UY-FI*YT*COF(L )
    XT=XT2
 70 YT=YT2
    SUMSQ=UX*UX+UY*UY
    IF (SUMSQ) 75, 110, 75
 75 DX = (V*UY - U*UX)/SUMSQ
    DY = -(U*UY + V*UX)/SUMSQ
    X = X +DX
    Y = Y + DY
 78 IF (ABS(DY)+ ABS(DX)-1.0E-05) 100,80,80
 80 ICT=ICT+1
    IF(ICT-500) 60,85,85
 85 IF(IFIT)100,90,100
 90 IF(IN-5) 50,95,95
 95 WRITE (6, 1) N2, IN, ICT, IFIT
  1 FORMAT (1H0,    '********* ERROR RETURN FROM SUBROUTINE POLRTD (DOES
   1 NOT CONVERGE):  N2 =', I3,    ', IN =',   I2,   ', ICT = ', I3,
   2  ', IFIT =', I2, ' **********', // )
    RETURN
100 DO 105 L=1,NXX
    MT= M1-L+1
    TEMP=XCOF(MT)
    XCOF(MT)=COF(L)
105 COF(L)=TEMP
    ITEMP=N
    N=NX
    NX=ITEMP
    IF(IFIT) 120,55,120
110 IF(IFIT) 115,50,115
115 X=XPR
    Y=YPR
120 IFIT=0
    IF ( ABS(Y)-1.E-4* ABS(X)) 135,125,125
125 ALPHA=X+X
    SUMSQ=X*X+Y*Y
    N=N-2
    GO TO 140
130 X=0.0
    NX=NX-1
    NXX=NXX-1
135 Y=0.0
    SUMSQ=0.0
    ALPHA=X
    N=N-1
140 COF(2)=COF(2)+ALPHA*COF(1)
    NSX   =N
    IF(NSX    .LT. 2) NSX   =2
145 DO 150 L=2,NSX
150 COF(L+1)=COF(L+1)+ALPHA*COF(L)-SUMSQ*COF(L-1)
155 ROOTI(N2)=Y
    ROOTR(N2)=X
    N2=N2+1
    IF(SUMSQ) 160,165,160
160 Y=-Y
    SUMSQ=0.0
    GO TO 155
165 IF (N) 200, 200, 45
200 RETURN
    END
          SUBROUTINE PLT22N (A, N, NL, ICAR)
C   THIS IS PLOT22 MODIFIED TO PRINT ZERO AXIS IF IT'S ON SCALE
      DIMENSION IOUT(101), YPR(11), ICAR(2), A(1), B(1)
      DATA BLANK, IDOT, IBAR/' ','.','|'/
    1 FORMAT (1H , 132 ('.'))
    2 FORMAT (1H , F11.4,   5X, 101A1)
```

```
    3 FORMAT(1H0, 107X, '(BAR MARKS ZERO ORDINATE)')
    4 FORMAT (1H , 6X, 'TIME')
    8 FORMAT (1H , 10X,          11F10.4)
C *** CALCULATE SCALE OF THE "Y'S"
      M1 = N+1
      M2 = 3*N
      YMIN = A(M1)
      YMAX = YMIN
      M3 = N+2
      DO 40 I=M3,M2
      IF (A(I)-YMIN) 28,26,26
   26 IF(A(I)-YMAX) 40,40,30
   28 YMIN = A(I)
      GO TO 40
   30 YMAX = A(I)
   40 CONTINUE
      YSCAL = (YMAX-YMIN)/100.0
      JPZ = -YMIN/YSCAL + 1
      YPR(1) = YMIN
      DO 42 J=1,9
   42 YPR(J+1) = YPR(J) + YSCAL*10.0
      YPR(11) = YMAX
      WRITE (6,3)
      WRITE (6,1)
C *** PRINT SCALE OF THE Y'S
      WRITE (6, 8) (YPR(I), I=1,11)
      WRITE (6,4)
C *** DETERMINE VALUE OF "X" TO PRINT
      L = 1
   45 XPR = A(L)
C *** DETERMINE "Y" CHARACTERS TO PRINT
      DO 55 IY = 1,101
   55 IOUT(IY) = BLANK
      IOUT(1) = IDOT
      IOUT(101) = IDOT
      IF (JPZ .GE. 1 .AND. JPZ .LE. 101) IOUT(JPZ) = IBAR
      DO 60 J=1,2
      LL = L+ J*N
      JP = (A(LL)-YMIN)/YSCAL + 1
      IOUT(JP) = ICAR(J)
   60 CONTINUE
C *** PRINT LINE
      WRITE (6, 2) XPR, (IOUT(IZ), IZ=1,101)
      L = L+1
      IF (L-NL) 45, 45, 86
C *** PRINT SCALE OF THE "Y'S"
   86 WRITE (6,4)
      WRITE (6, 8) (YPR(I), I=1,11)
      WRITE (6,1)
      RETURN
      END
      SUBROUTINE EVLPOL( M,C,SRE,SIM, POLYRE, POLYIM, POLMAG,POLANG)
C *** MODIFIED FOR ANGLE VALUE WHEN S=0 AND POLY = 0 + J 0; SEE BELOW  MR71 DFT
C THIS IS "EVALPOLY" OF ALGOL TRANSLATED INTO A FORTRAN IV SUBROUTINE AS
C "EVLPOL", WITH THE NOTATION:  M = DEGREE OF POLYNOMIAL
C        M = DEGREE OF POLYNOMIAL
C S = SRE + J SIM   IS THE VALUE OF S AT WHICH TO EVALUATE THE POLYNOM.
C C(1) = CONSTANT TERM, C(M+1) = COEFFICIENT OF MTH POWER TERM IN POL.
C RESULT OF EVALUATION IS(POLYRE + J POLYIM)= POLMAG AT ANGLE POLANG
      DIMENSION C(1), C1(51)
      MM = M+1
       MMM=M-1
      DO 5 I=1,MM
    5 C1(I) = C(I)
      IF (MMM) 40, 30, 10
   10 D = -2*SRE
       E = SRE**2 + SIM**2
       DO 20 III=1,MMM
       II=M+2-III
      C1(II-1) = C1(II-1) - D*C1(II)
   20 C1(II-2) = C1(II-2) - E*C1(II)
   30 POLYRE = C1(2)*SRE + C1(1)
      POLYIM  =C1(2)*SIM
   35  POLMAG = SQRT(POLYRE**2 + POLYIM**2)
   36 IF (POLYRE .EQ. 0 .AND. POLYIM .EQ. 0) GO TO 37
       POLANG = ATAN2(POLYIM,POLYRE)*57.295780
       RETURN
   37 POLANG = 909090.9
C      THE FF. CDS ARE NOT NEC. IN I7, NO?
C      IF (SRE .EQ. 0 .AND. SIM .EQ. 0) GO TO 38
C
       RETURN
C  38 IF ( C1(1) .EQ. 0 .AND. C1(2) .NE. 0) POLANG  = SIGN (90.0, C1(2))
C      I.E. IF POLYNOMIAL = 0 AT S = 0, SET ANGLE = (SIGN OF COEFF. OF S)*90:
```

```
C        POLANGL = 90.0*C1(2)/ABS(C1(2))
C       RETURN
   40 POLYRE = C1(1)
        POLYIM = 0
       GO TO 35
        END
       SUBROUTINE PLOT4Z (N,A, X, Y)
C  ... USES ONLY HALF THE WIDTH OF THE PAPER FOR THE PLOT, BUT PRINTS VALUES OF
C        BOTH "X" AND "Y"  WITH LEGEND AT LEFT. PRINTS ZERO AXIS IF ON SCALE.
      REAL IOUT
      DIMENSION IOUT( 51), YPR( 6), A(1), X(3), Y(3)
      DATA BLANK, ASTX, BAR, DOT /1H , 1H*, 1H|, 1H./
    1 FORMAT (1H , 115('.') )
    2 FORMAT (1H ,       F20.4, E22.4, 18X, 51A1   )
    3 FORMAT (1H , 60X, 51A1)
    4 FORMAT (1H0,'SUBROUTINE ''PLOT4 '', BEING CALLED, GIVES (BAR MARKS
     1 ZERO ORDINATE):')
    8 FORMAT (1H ,  9X, 3A4, 10X, 3A4, 12X, 6F10.4)
      NL = N
      M1 = N+1                                                          21.
      M2 = 2*N                                                          27.
      YMIN = A(M1)
      YMAX = YMIN                                                       29.
      M3 = N+2                                                          30.
      DO 40 I=M3,M2                                                     31.
       IF (A(I)-YMIN) 28,26,26                                          32.
   26 IF (A(I)-YMAX) 40,40,30                                           33.
   28 YMIN = A(I)                                                       34.
      GO TO 40
   30 YMAX = A(I)                                                       36.
   40 CONTINUE                                                          37.
      YSCAL = (YMAX-YMIN)/50.0                                          38.
      JPZ = -YMIN/YSCAL + 1
      YPR(1) = YMIN
      DO 42 J=1,4                                                       40.
   42 YPR(J+1) = YPR(J) + YSCAL*10.0
      YPR( 6) = YMAX                                                    42.
      WRITE (6, 4)
      WRITE (6, 1)
      WRITE (6,8) X, Y, (YPR(I), I=1,6)                            ALIGNER*
      DO 43 IY = 2,50
   43 IOUT(IY) = BLANK
      IOUT(1) = DOT
      IOUT(51) = DOT
      IF  (JPZ .GE. 1 .AND. JPZ .LE. 51) IOUT(JPZ) = BAR
      WRITE (6,3) (IOUT(IZ), IZ = 1,51)
      L = 1
   45 XPR = A(L)                                                        50.
      LL = L + N
      JP = (A(LL)-YMIN)/YSCAL + 1
      IOUT(JP) = ASTX                                                   60.
   60 CONTINUE
      WRITE (6,2) XPR, A(LL), (IOUT(IZ), IZ=1,51)                       64.
      IOUT(JP) = BLANK
      IOUT(1) = DOT
      IOUT(51) = DOT
      IF (JPZ .GE. 1 .AND. JPZ .LE.51) IOUT(JPZ) = BAR
      L = L+1
      IF (L-NL) 45,45,86                                                68.
   86 CONTINUE
      WRITE (6,3) (IOUT(IZ), IZ = 1,51)
      WRITE (6,8) X, Y, (YPR(I), I=1,6)
      WRITE (6, 1)
      RETURN                                                       ALIGNER*
      END
```

.... **illustrative (sample) data-card set for RESSINET:**

```
SECT. 7.19       C-K LONG LADDER (FIG. 7.18-C),   NOMINAL CUTTOFF = 1.0
WF = 0.5               PHI = -90 DEG. (SINE)
COMPLETE RESPONSE   -    NOTE LONG "DELAY"
      1.0                 0.5                      -90.0
        0          14          0          1          0
        6
    0.000244
   0.000488281       0.00634766           0.0412598          0.177734
   0.567383            1.421875           2.898438          4.875
     6.9375            8.125               8.3125           6.5
     4.75               2.0                1.0              0.0
    0.0                 0.5               50.0
     0.0               0.5               100.0
    0.0                 1.0              100.0
  2000000.0            0.0                  0.0
```

FOMEGA (for calculating transfer functions for $s = j\omega$)

```
C    FCMEGA           FOMEGA
C
C      IT TABULATES VALUES OF A RATICNAL FUNCTION,
C
C      F(S) = H (AN S**N + ........ + AO) / (BO S**C + ........ +BO)
C
C      FCR  S = 0 + JW,   W TAKING VALUES FRCM W1 TC W2, EQUALLY SPACED
C      CN LOGARITHMIC OR ON ARITHMETIC (FREQUENCY) SCALE
C
C
C NOTATICN:
C
C         H   =  MULTIPLIER IN F(S)
C         NN = DEGREE OF NUMERATOR OF F(S)   ("N" ABOVE) (MAX.VALUE = 50)
C         ND = DEGREE CF DENOMINATCR OF F(S)  ("C" ABOVE) (MAX.VALUE = 50)
C
C  DATA CARCS REQUIREC (ONE SET FOR EACH CASE):
C
C      (1) THREE CARCS (EACH FORMAT 20A4)  WITH LITERAL INFORMATICN TC
C          BE PRINTED AS AIDE-MEMCIRE
C      (2) H, NN (= N ABOVE), ND (=D) (FCRMAT F2C.6, 2I1C)
C      (3) AO, A1, A2, ....... AN     (4F20.6)
C      (4) BC, B1, B2, ....... BD (   "   )
C      (5) W1,W2,NRPTS,JP,NWR,NLOG      (FCRMAT 2F2C.6, 4I1C)
C
C         NRPTS = NUMBER OF POINTS WANTED, EQUALLY SPACED CN LOGARITHMIC
C                    SCALE, WHEN THAT IS USED, ECUALLY SPACEC ON
C                    ARITHMETIC FREQUENCY SCALE IF THAT IS ASKED FOR
C                    MAX. NRPTS  = 101
C
C         JP = CODE FOR PLOT:     JP = 0   ...   NO PLOT
C                                      1 PLCTS MAGNITUDE OF F(JW)
C                                      2 PLOTS ANGLE OF F(JW)
C                                      3 PLCTS DB MEASURE OF MAGNITUDE OF F
C                                      4 PLCTS REAL AND IMAGINARY PARTS OF F
C                                      5 PLOTS DE MEASURE AND ANGLE OF F
C                                      6 PLCTS DE MEASURE,ANGLE,ANC ANGLE INCREMENT
C                   (DB MEASURE = 20*LOG|F| I.E. "DBV")
C                   NB: IF A POWER RATIC IS READ IN FOR "F", DB VALUES
C                       WILL BE TOO LARGE BY A FACTOR OF TWO
C
C         NWR = CODE FOR DETAILED TABLE CF NUMERICAL RESULTS:
C                   NWR = 0 ... NO TABLE,          NWR = 1 ... PRINT TABLE
C
C         NLCG = 0 FCR ARITHMETIC FREQUENCY SCALE.
C                1 FCR LCGARITHMIC FREQUENCY SCALE
C
C         THIS CARD (5) MAYBE REPEATED, FOR ADDITICNAL RANGES OF FREQUENCY,
C         (TABLE), PLOTS ....... EACH GRCUP CF 6 NUMBERS CN A SEPARATE CARC
C
C  ******* NB: IF THERE IS A FOLLCWING CASE, THE LAST CARC IN THIS GROUP (5)
C          MUST HAVE W1 .GT. 1,0C0,CC0.0 ANC ARBITRARY NUMBERS IN THE OTHER FIVE
C          FIELDS, TC MCVE TO NEXT CASE - ELSE PRCGRAM STOPS.
C
   1 FORMAT (1H1,  'THIS IS FOMEGA WITH THESE FEATURES:     JP =',I2,
    1 ', NLOG =',   I3,', NWR = ',I2   )
   2 FCRMAT (1H+,  30X,  'NCTES FCR THIS RUN:  ',20A4,/,  2(52X, 20A4,
    1 / ) )
   3 FCRMAT (1H ,    'COEFFICIENTS IN F(S), NUMERATOR ABOVE DENOMINATOR,
    1  LOWEST (ZERCTH) PCWERS FIRST, ARE:',/)
   4 FORMAT (1HO,    1X, 'POINT NO.',   4X,   'OMEGA',   11X,  'MAGNITUDE',
    1  5X, 'ANGLE (DEGREES)*', 5X, 'ANGL. INCR.',  7X, 'REAL PART', 7X,
    2 'IMAGINARY PART',  3X, 'DB MEASURE(DBV)*', /,1X, 10('.'),   4X,
    3 5('.'),  11X, 9('.'), 5X, 15('.'), 6X, 11('.'), 7X, 9('.'), 7X,
    4 14('.'), 3X, 16('.'),/ )
   5  FCRMAT (1HO, 57HSIREN....A VALUE OF CMEGA USED IS A POLE OF THE
    1FUNCTION      , / )
   6  FCRMAT (F20.6, 2I10)
   7   FORMAT (4F20.6)
   8 FCRMAT( 20A4 )
   9 FORMAT ( 1HO,  2(10F12.4, /, 1X) )
  10 FORMAT (1HO,    /, ' RESULTS ( VALUES CF F(JW) ) FCR CMEGA =',
    1 F8.4, '    TO', F10.4,  ', TAKING',  I4,  '   POINTS', 2X,
    2'(EQUALLY SPACED ON')
  11 FCRMAT ( I6, F14.4, 6F18.4)
  12 FORMAT (1HO, ///,' THE END (NC MORE DATA, NC MORE CASES). *******
    1         *******   '   , /// )
  13 FORMAT (1HC, 'THE VALLE OF THE MULTIFLIER H IS', F10.5,', THE DEGR
    1EE OF THE NUMERATCR IS', I3, ',     THE DEGREE OF THE DENOMINATOR I
    2S', I3,'  AND THE' )
  14 FCRMAT (1H ,   132('.') )
  15 FCRMAT (1H1,' FOMEGA GIVES:')
  16 FORMAT (8I10)
  17 FCRMAT ( 2F20.6, 4I10 )
```

```
   18 FORMAT (1H0,  15('*'), 15X, 'FLCT CF ', 35X, 50('*'))
   19 FCRMAT (1H+,  38X,  'DB MEASURE OF F (DBV)')
   20 FORMAT (1H0, '*IN ANGLE OF F, PROGRAM ATTEMPTS TO AVOID JUMPS DUE
     1TO TAKING PRINCIPAL VALUES ONLY, ANC TO KEEP VARIATICN OF ANGLE ST
     2EADY',/,' *THE DBV VALUE "SATURATES" AT -100 FOR CALCULATION CONVE
     3NIENCE; ALSO AT +100' )
   21 FORMAT (1H+,  38X, 'ANGLE INCREMENTS (DEGREES) ' )
   22 FORMAT ( 1H+  , 103X, 'ARITH. FREQ. SCALE):'  )
   23 FCRMAT (1H+   , 103X, 'LOG. FREQ. SCALE):'   )
   24 FORMAT (1H0, 'NOTE: THE VALUE OF W1 HAS BEEN CHANGED FRCM ZERC ( O
     1R A NEGATIVE VALUE ) AS GIVEN, TC + 0.01 ',/,' BECAUSE A LOGARITHM
     2IC FREQUENCY SCALE WAS ASKED FOR *** CK?',/)
   25 FORMAT (1H0, 25X, '***** NO TABULATION OF VALUES, NOT ASKED FOR')
   26 FORMAT (1H+, 38X,  'MAGNITUDE OF F:', / )
   27 FCRMAT (1H+, 38X,  'REAL ANC IMAGINARY PARTS OF F:', / )
   28 FORMAT (1H1,' *** THIS IS F2 (JW) ***')
   29 FCRMAT (1H ,      'ANGLE OF F IN DEGREES (PROGRAM ATTEMPTS TC AVOID
     1D JUMPS DUE TO TAKING PRINCIPAL VALUES ONLY, AND TO KEEP VARIATIONS
     2 OF ANGLE',               ' STEADY)' )
   30 FORMAT (1H0,'****** THE VALUE CF H READ IN WAS NEGATIVE;    FOR CON
     1VENIENCE F2 TAKES THE LIBERTY OF CHANGING ITS SIGN TC +. IT APOLOG
     2IZES FOR',// ' ANY ANNOYANCE YOU MAY FEEL.',/)
   31 FORMAT (//  ,' NB:PROPER NORMALIZATICN IS ASSUMED; IF H IS TOO SMA
     1LL, YOU MAY  GET USELESS RESULTS.')
   32 FCRMAT (1H0, 20('*'),' H =', E11.4, ' SEEMS RATHER SMALL TO ME', 3
     1X, '(....FOMEGA)',  5X, 'MIGHT IT NOT BE WELL TO READJUST IT ',//,
     2' AND SC AVOID SMALL NUMBERS?' )
C
      REAL MN, MD
      DIMENSION A(51), B(51), TITLE(6C), PLT(404),
     1 PLT1(101), PLT2(101), PLT3(101), PLT4(101), PLT5(101), PLT6(101)
C
      CTE = 57.295780
  100 READ (5,8)  TITLE
      READ (5,6          ) H, NN, ND
      WRITE (6,28)
      WRITE (6,2) TITLE
      NNN = NN+1
      NDD = ND + 1
      IF (H .GT. 0.0)  GO TO 140
      H = -H
      WRITE (6,30)
  140 CONTINUE
      WRITE (6,13) H, NN, ND
      WRITE (6,3)
      READ (5,7) ( A(I), I=1, NNN)
      WRITE (6,9) ( A(I), I=1,NNN)
      WRITE (6,14)
      READ (5,7) ( B(I), I=1,NDD)
      WRITE (6,9) ( B(I), I=1,NDD)
      IF ( H .LT. 0.001  )  WRITE (6,32) H
      WRITE (6,31)
      WRITE (6,20)
C
C     READ DESIRED FREQUENCY RANGE AND
C     BEGIN CALCULATIONS THEREFOR
C
  150 READ (5,17,END=900)  W1,      W2, NRPTS,JP, NWR, NLOG
      IF (W1 .GT. 1000000. ) GO TO 100
      NY = 1
      WRITE (6,1) JP, NLOG, NWR
      WRITE (6,10) W1, W2, NRPTS
      IF ( NLOG.EQ. 0 ) WRITE ( 6,22 )
      IF ( NLOG .EQ. 1 ) WRITE (6,23 )
      IF (NWR .EQ. 0) WRITE (6, 25)
      IF ( NLOG .EQ. 1 )  WRITE (6,24)
      IF (NWR .EQ. 1)  WRITE (6,4)
      IF (NLOG) 158, 158, 152
C     *** THIS IS THE PATH: NLOG = 1, LOG. FREQ. SCALE
  152 CONTINUE
      IF (W1 .GT. 0.0) GO TO 155
      W1 = 0.01
  155 WO = W1
      X1 = ALOG(W1/WO)
      X2 = ALOG(W2/WO)
      DX = (X2 - X1)/(NRPTS-1)
      X = X1
  158 W = W1
      DW = (W2-W1)/(NRPTS-1)
C *** CALCULATE POINTS:
      J = 0
      FA1 = 0.0
C      SEE NOTE ABOUT REMOVING JUMPS BELOW (AT 200+)
      TEST = 0.0
```

```
      XXX = 75.0
      YYY = -75.0
      ADDR = 0.0
      DELTFA  = 0.0
C
  160 CALL EVLPOL (NN, A, 0.0, W, XN, YN, MN, AN)
      CALL EVLPOL (ND, B, 0.0, W, XD, YD, MD, AD)
      J = J+1
      IF (MD .EQ. 0) GO TO 200
      FM = H*MN/MD
 *** TO AVOID TROUBLE, LET FM "SATURATE" AT -100 OR +100 DBV:
      IF (FM .LT. 0.00001) FM = 0.00001
      IF (FM .GT. 100000.0) FM = 100000.0
      GO TO 210
  200 WRITE (6,5)
      FM = 909090.9
C ***** NOTE: STATEMENTS 210-215  ARE INTENDED TO REMOVE JUMPS IN ANGLE DUE TO
C        CALCULATED ANGLES REMAINING IN "PRINCIPAL-VALUE" OR "PRINCIPAL-CYCLE"
C        CIRCLE:
  210 FA = AN - AD + ADDR
      FDB = 20.0*ALOG10(FM)
      IF ( J .EQ. 1 ) GO TO 215
      TEST = FA - FA1
      IF (TEST .GT. XXX) ADDR = ADDR - 360.0
      IF (TEST .LT. YYY) ADDR = ADDR + 360.0
      FA = AN - AD + ADDR
      DELTFA = FA - FA1
  215 FA1 = FA
C
      FAR = FA/CTE
      FRE = FM*COS(FAR)
      FIM = FM*SIN(FAR)
      IF (NWR) 630,630,620
  620 WRITE (6, 11) J, W, FM, FA, DELTFA, FRE, FIM, FDB
  630 IF (JP) 650, 650, 640
C *** STORE DATA FOR PLOTS LATER:
  640 PLT(J) = W
      PLT1(J) = FM
      PLT2(J) = FA
      PLT3(J) = FRE
      PLT4(J) = FIM
      PLT5(J) = FDB
      PLT6 (J) = DELTFA
  650 CONTINUE
      IF (NLOG) 680, 680, 660
  660 X = X + DX
      W = W0*EXP(X)
      GO TO 682
  680 W = W + DW
  682 IF (W .LE. W2) GO TO 160
      IF (NWR .GT. 0) WRITE (6, 20)
  685 IF (JP) 150, 150, 698
C
  698 CONTINUE
      JJJ = J
      DO 750 I=1,JJJ
      JJ2 = I + JJJ
      JJ3 = JJ2 + JJJ
      IF (JP-2) 701, 702, 703
C *******      JP = 1:
  701 PLT(JJ2) = PLT1(I)
      GO TO 750
C *******      JP = 2:
  702 PLT(JJ2) = PLT2(I)
      GO TO 750
C *******      JP = 3, 4, 5, .... :
  703 IF (JP-4) 711, 713, 715
C *******      JP = 3 :
  711 PLT(JJ2) = PLT5(I)
      GO TO 750
C  ******      JP = 4 :
  713 CONTINUE
C THIS ONE IS FOR REAL AND IMAGINARY PARTS
      NY = 2
      PLT(JJ2) = PLT3(I)
      PLT(JJ3) = PLT4(I)
      GO TO 750
C *******      JP = 5, 6, 7, ..... :
  715 CONTINUE
      PLT(JJ2) = PLT5(I)
  750 CONTINUE
      WRITE (6, 15)
      WRITE (6, 2) TITLE
      WRITE (6,18)
```

```
C       WRITE TITLE PER VALUE OF JP:
        IF (JP .EQ. 1) WRITE (6, 26)
        IF (JP .EQ. 2)  WRITE (6, 29)
        IF (JP .EQ. 3) WRITE (6, 19)
        IF (JP .EQ. 4) WRITE (6, 27)
        IF (JP .GE. 5) WRITE (6, 19)
        CALL PLOTF2 (JJJ, PLT, NY)
        IF (JP .LT.5) GO TO 150
C JP .GE. 5:
        DO 860 I=1,JJJ
        JJ2 = I + JJJ
  860 PLT(JJ2) = PLT2(I)
        WRITE (6, 15)
        WRITE (6, 2) TITLE
        WRITE (6,18 )
        WRITE (6, 29)
        CALL PLOTF2(JJJ, PLT, NY)
        IF (JP .EQ. 5) GO TO 150
C JP = 6:
        WRITE (6, 15)
        WRITE (6, 2) TITLE
        WRITE (6, 18)
        WRITE (6, 21)
        DO 880 I=1,JJJ
        JJ2 = I + JJJ
  880 PLT(JJ2) = PLT6(I)
        CALL PLOTF2 ( JJJ, PLT, NY)
        WRITE (6,20)
        GO TO 150
  900 WRITE (6, 12)
         STOP
        END
        SUBROUTINE PLOTF2 (N,A,NY)
C  DERIVED FROM PLOT4S  ... PRINTS THE N PAIRS OF VALUES OF X AND Y AS GIVEN (NO
C       SORTING OR SPACING) ... OR OF X, Y1 AND Y2 (TWO   DEP. VARIABLES)
C
C  NOTATION... N =  NUMBER OF POINTS (X,Y) OR (X,Y1,Y2) IN DATA ARRAY A
C              A = ARRAY OF DATA,
C                                      NY = 1: ARRAY A IS N VALUES OF X FOLLOWED BY N
C                                              VALUES OF Y
C                                      NY = 2: ARRAY A IS N VALUES OF X FOLLOWED BY N
C                                              VALUES OF Y1 FOLLOWED BY N VALUES OF Y2
C                                          DIMENSION OF ARRAY A IS 2N OR 3N
      DIMENSION IOUT(101), YPR(11), A(1)
    1 FORMAT (1H0,      '(BAR INDICATES ZERO ORDINATE):')
    2 FORMAT (1H , F11.4, 5X, 101A1)
    3 FORMAT (1H )
    4 FORMAT (1H , 16X, 11('.', 9X))
    5 FORMAT (1H , 132 ('.'))
    7 FORMAT (1H ,'  FREQUENCY', 5X, 11('.', 9X))
    8 FORMAT (1H ,11X, 11F10.4)
      DATA IBLANK, IASTX,IIM, IRE/1H ,1H*,1HI, 1HR/
      DATA IBAR / 1H|/
      NL = N
C *** CALCULATE SCALE OF "Y":
      M1 = N+1
      M2 = 2*N
      IF (NY .EQ. 2) M2 = 3*N
      YMIN = A(M1)
      YMAX = YMIN
      M3 = N+2
      DO 40 I=M3,M2
       IF (A(I)-YMIN) 28,26,26
   26 IF (A(I)-YMAX) 40,40,30
   28 YMIN = A(I)
      GO TO 40
   30 YMAX = A(I)
   40 CONTINUE
      YSCAL = (YMAX-YMIN)/100.0
      JPZ = -YMIN / YSCAL + 1
      YPR(1) = YMIN
      DO 42 J=1,9
   42 YPR(J+1) = YPR(J) + YSCAL*10.0
      YPR(11) = YMAX
C *** PRINT SCALE OF "Y":
      WRITE (6,1)
      WRITE (6,5)
      WRITE (6,8) (YPR(I), I=1, 11)
      WRITE (6,4)
      WRITE (6,7)
      WRITE (6,4)
C *** ZERO THE ARRAY IOUT:
   50 DO 55 IY = 1,101
   55 IOUT(IY) = IBLANK
```

```
      IF (JPZ .GE. 1  .AND.  JPZ .LE.  101)  IOUT(JPZ) = IBAR
      IYCH    IASTX
      IF (    .EQ. 2) IYCH = IRE
C *** PRIN. POINTS
      JPZ = 101
      DC 100 L=1,N
      LL = L + N
      JP = (A(LL)-YMIN)/YSCAL + 1                                       6C.
      IOUT( JP) = IYCH
      IF (NY .EQ. 1) GO TO 6C
        LLL = LL + N
      JP2 = (A(LLL)-YMIN)/YSCAL + 1
      IOUT(JP2) = IIM
   6C CONTINUE                                                          64.
      WRITE (6,2) A(L),(IOUT(IZ), IZ=1,101)
      IOUT(JP2) = IBLANK
      IOUT(JP) = IBLANK
      IF (JPZ .GE. 1  .AND.  JPZ .LE.  101)  IOUT(JPZ) = IBAR
  100 CCNTINUE
C *** PRINT SCALE OF "Y" AGAIN:
   8C WRITE (6,4)
      WRITE (6,7)
      WRITE (6,4)
      WRITE (6,8) (YPR(I), I=1,11)
      WRITE (6,5)
      RETURN
      END
      SUBROUTINE EVLPOL( M,C,SRE,SIM, PCLYRE, PCLYIM, PCLMAG,PCLANG)
C *** MODIFIED FOR ANGLE VALUE WHEN S=0 AND POLY = C + J C: SEE BELOW  MR71 DFT
C THIS IS "EVALPOLY" OF ALGOL TRANSLATED INTO A FORTRAN IV SUBROUTINE AS
C "EVLPCL", WITH THE NOTATION:  M = DEGREE OF POLYNOMIAL
C S = SRE + J SIM   IS THE VALUE OF S AT WHICH TO EVALUATE THE POLYNOM.
C C(1) = CONSTANT TERM, C(M+1) = COEFFICIENT OF MTH POWER TERM IN POL.
C RESULT OF EVALUATION IS(POLYRE + J POLYIM)= PCLMAG AT ANGLE PCLANG
      DIMENSION C(1), C1(51)
      MM = M+1
       MMM=M-1
      DC 5 I=1,MM
    5 C1(I) = C(I)
      IF (MMM) 40, 30, 10
   1C D = -2*SRE
       E = SRE**2 + SIM**2
       DC 2C III=1,MMM
       II=M+2-III
      C1(II-1) = C1(II-1) - D*C1(II)
   20 C1(II-2) = C1(II-2) - E*C1(II)
   30 PCLYRE = C1(2)*SRE + C1(1)
      POLYIM =C1(2)*SIM
   35  PCLMAG = SQRT(POLYRE**2 + POLYIM**2)
   36 IF (POLYRE .EQ. C .AND. PCLYIM .EQ. 0) GC TC 37
       PCLANG = ATAN2(POLYIM,POLYRE)*57.295780
       RETURN
   37 PCLANG = 909090.9
      IF (SRE .EQ. 0.0 .AND. SIM .EQ. 0.0)  GO TO 38
      RETURN
   38 CC 39 I = 2, MM
      J = I
      IF (C1(I) .NE. C.C)  GO TO 5C
   39 CCNTINUE
   50 POLANG  = (J-1)*9C.0*C1(J)/ABS(C1(J))
      RETURN
   4C POLYRE = C1(1)
       PCLYIM = 0
      GC TO 35
       END
```

.... illustrative (sample) data-card set for FOMEGA:

```
EE266 FOMEGA EXAMPLE                                               (NOTE CARD ONE)
POWER RATIO = 1 + W**6/(W**2-1)**2    INSERTED FOR "F":               NOTE S=W/J
POWER RATIO (WATCH DOUBLE DB)                                     (AND S**2 = -W**2)

    1.0                    6          4
    1.0                   0.0                   2.0                 0.0
   1.0                    0.0                   1.0
     1.0                      0.C                 2.0                0.0
    1.0
    0.0                    1.0                  51         1         0         0
   0.5                    1.1                   61         3         0         0
```

CONVOL [for convolution (Chap. 4)]

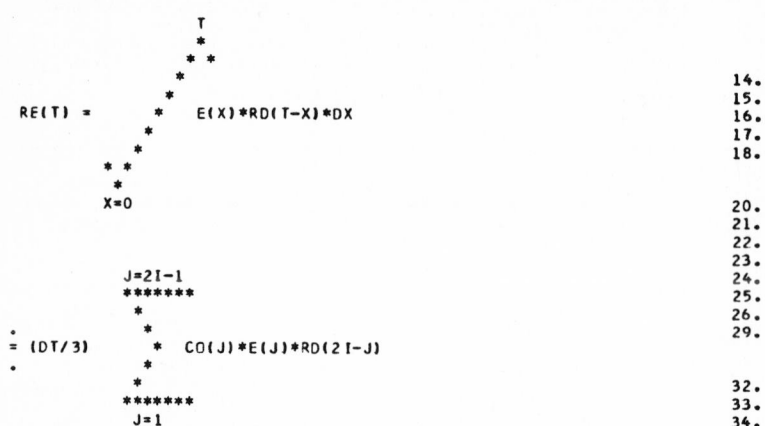

```
C             *****      CONVCL       *****       (CONVOLUTION)
C
C  HIS IS "CONVOL", A CONVOLUTICN PROGRAM FOR EE101,.....
C
C     CALCULATES RESPONSE  RE(T)  TO EXCITATION  E(T)  BY CONVOLUTION OF
C     EXCITATION WITH INDICIAL (IMPULSE) RESPONSE, RD(T). TABULATES;
C     OPTIONALLY, IT ALSO CALLS    PLOTTING SUBROUTINES AND SKETCHES CURVES OF
C     RD(T), OF E(T), AND OF RE(T), PER INSTRUCTIONS. IT THEN REPEATS THE WHOLE
C     FOR SEVERAL CASES.
C
C
C
C                          T
C                          *
C                         * *
C                        *                                           14.
C                       *                                            15.
C     RE(T) =          *    E(X)*RD(T-X)*DX                          16.
C                     *                                              17.
C                    *                                               18.
C                  * *
C                   *
C                  X=0                                               20.
C                                                                    21.
C                                                                    22.
C                                                                    23.
C                  J=2I-1                                            24.
C                  *******                                           25.
C                   *                                                26.
C   .                *                                               29.
C   = (DT/3)          *   CO(J)*E(J)*RD(2I-J)
C   .                *
C                   *                                                32.
C                  *******                                           33.
C                   J=1                                              34.
C
C
C     EXCITATION, E, IS CALCULATED FROM A FORMULA (SUBROUTINE) OR INTERPCLATED
C     BETWEEN POINTS  READ FROM DATA CARDS.
C
C     IMPULSE RESPONSE, RD, IS CALCULATED FROM A GIVEN TRANSFER FUNCTION, OR BY
C     A FORMULA (SUBROUTINE), OR INTERPOLATED BETWEEN POINTS  READ FROM DATA
C     CARDS, AS DESIRED.
C
C     IF RD IS CALCULATED FROM A TRANSFER FUNCTION, USE
C
C                 A(1) + A(2)S + A(3)S**2 + ..... + A(M+1)S**M
C        T(S) = C ..............................................., M .LE. N
C                 B(1) + B(2)S + B(3)S**2 + ..... + B(N+1)S**N
C
C     THEN CONVOL CALCULATES ITS POLES (WHICH SHOULD ALL BE SEPARATE AND
C     DISTINCT), CALCULATES THE RESIDUES, CALCULATES THE VALUES OF RD(T), THE
C     THE INVERSE TRANSFORM OF T(S), CONVOLVES RD(T), WITH THE EXCITATION
C     FUNCTION E(T)   (SEE ABOVE)  AND THUS OBTAINS THE FUNCTION RE(T).
C
C     FORMULA (SUBROUTINE FORMLA) NOTES:
C
C                     THE FORMULA USED IS
C
C           F(T) = ( QN/QD )*EXP(QE)*COS(QC)
C
C                  QN = CN(1)  + CN(2)*T  +.....+ CN(NN+1)*T**NN
C                  QD = CD(1)  + CD(2)*T  +----  +CD(ND+1)*T**ND
C                  QE = CE(1)  + CE(2)*T  +----.+ CE(NE+1)*T*NNE
C                  QC = CC(1)  + CC(2)*T  +.....+ CC(NC+1)*T**NC
C
C           FOR WHICH THE DATA CARDS NEEDED ARE:
C
C                  (1) NN, ND, NE, NC (FORMAT4I10)
C                  (2) CN(I), I=1,2, ... ,(NN+1)  (4 PER CARD, FORMAT 4F20.6)
C                  (3) CD(I), I=1,2, ... ,(ND+1)
C                  (4) CE(I), I=1,2, ... ,+(NE+1)
C                  (5) CC(IE, I=1,2, ... ,+(NC+1)
C
C     RANDI (PIECEWISE-LINEAR READ-AND-INTERPOLATE SUBROUTINUE ) NOTES:
C
C       THIS SUBROUTINE READS FUNCTION VALUES AND INTERPOLATES LINEARLY BETWEEN
C       THEM: ASSUMES FUNCTION CONSTANT BEYOND LAST POINT GIVEN, AT LAST VALUE
C       GIVEN. AT A DISCONTINUITY, USES "LEFT" VALUE. MAX. NO. POINTS: 51
C       THE RESULTING PIECEWISE LINEAR FUNCTION MAY ALSO MODULATE THE SINUSOID
C       COS(WF*T+PHI) (SET NCODE = 1); ELSE NCODE = 0.
C
C      DATA CARDS FOR RDANDI:
C      (1) NCODE, WF, PHI (FORMAT I10, 2F20.6)
```

```
C        (2): THEN
C                         VALUE OF T, VALUE OF FUNCTION, ONE PAIR ON EACH
C      CARD, FORMAT 2F20.6; FIRST POINT SHOULD BE THAT FOR T=0; AFTER LAST
C      DATA CARD PLACE A CARD WITH A NEGATIVE VALUE OF T AND ARBITRARY VALUE
C      OF FUNCTION, TO TERMINATE READING.
C
C      NOTATION:
C
C                T = 2(I-1)DT = 0, 2DT, 4DT, ........ (NRT-1)DT
C                  = VALUES OF TIME AT WHICH RESPONSE RE(T) IS CALCULATED.
C                J = 1,2,3, ........ (2I-1)  (COUNTER IN INTEGRATION)
C                I = 1,2,3, ........ NRT32=(NRT+1)/2    (COUNTER FOR RESPONSE CALC
C                                                         TIME VALUES)
C                NRT MAX = 601; NRT32 MAX = 301
C              ICO =    1,4,2,4,2,4, ........ 4,2,4,1   (SIMPSON'S RULE WEIGHTS)
C               E(J)  AND  RD(J)  ARE FIRST EVALUATED FOR J=1,2,3, ..... , NRT
C                     THAT IS FOR T = (J-1)DT = 0,DT,2DT,3DT, .... , (NRT-1)DT
C              THEN RE(T) IS EVALUATED FOR T = 0,2DT, 4DT,......(NRT + 1)DT
C
C         NRD = CODE FOR CALCULATION OF IMPULSE RESPONSE:
C                 NRD = NEGATIVE INTEGER: READ VALUES (CARDS) AND INTERPOLATE
C                        ZERO: CALCULATE RD FROM T(S)
C                        POSITIVE INTEGER: CALCULATE RD FROM SUBROUTINE FORMLA
C         NEXC = CODE FOR EXCITATION CALCULATION METHOD:
C                 NEXC = NEGATIVE INTEGER: READ VALUES (CARDS) AND INTERPOLATE
C                      = POSITIVE INTEGER: CALCULATE E FROM SUBROUTINE FORMLA
C         NPLOT : CODE FOR DETERMINING WHAT PLOTS TO PRINT:
C              = 0 NO PLOTS PRINTED
C              = 1 PLOTS RESPONSE RE ONLY (WITH TABLE)
C              = 2 PLOTS E, RD, RE ON ONE PLOT, IN ADDITION
C              = 3 PLOTS EXCITATION (ALONE) ALSO
C              = 4 PLOTS IMPULSE RESPONSE (ALONE) ALSO
C         NWR =0: NO TABLES PRINTED
C              1: TABLES PRINTED
C             =2: ALSO PRINTS SEPARATE DETAILED TABLES OF EXC. AND IMP.
C                 RESP.
C         TITLE ... ARRAY OF TITLE TO BE PRINTED ON THE RESULTS (TABLE
C                 AND CURVE) AS AN AIDE-MEMOIRE AS TO WHAT EXCITATION WAS USED,
C                 ETC.    (3 CARDS)
C         T1 = INITIAL VALUE OF TIME (= 0 IN THIS PROGRAM)
C         DT = INCREMENT OF TIME
C         T2 = FINAL VALUE OF TIME
C         NRT = TOTAL NUMBER OF POINTS OF TIME TO BE USED = (T2-T1)/DT + 1
C              ..... MUST BE ODD INTEGER ** MUST * MUST BE ODD INTEGER *********
C         E ... ARRAY OF VALUES OF THE EXCITATION FUNCTION
C         RE ... ARRAY OF RESPONSE (CONVOL.)
C         RD ... ARRAY OF VALUES OF IMPULSE RESPONSE
C         A ... ARRAY OF COEFFICIENTS OF NUMERATOR OF T(S)     (DEGREE = M)
C         B ... ARRAY OF COEFFICIENTS OF DENOMINATOR OF T(S)   (DEGREE = N)
C         C = CONSTANT MULTIPLIER OF T(S)
C
C *** DATA CARDS REQUIRED:
C
C      (1) TITLE (20A4, 20A4, 20A4))    (3 CARDS)
C      (2) NEXC, NRD, NPLOT, NWR (FORMAT 4I10)
C      (3)   DT, T2, NRT   (2F20.6, I10)    (NRT MAX = 601)
C      NB:     NRT32 = (NRT+1)/2  =  NUMBER OF TIME POINTS IN OUTPUT (RESPONSE)
C              MAX. VALUE OF NRT32 = 301 (MAX NRT = 601)
C                            *** NB:  NRT MUST BE ODD  ********
C
C *** IF NRD = 0,  THEN (TRANSFER FUNCTION DATA):
C      ( ) C, M, N,                 (F20.6,2I10)    (MAX. M, N = 10)
C      ( ) A(I)    (I=1,M+1)     4F20.6 FORMAT)
C      ( ) B(I)    (I=1,N+1)     (FORMAT 4F20.6)
C
C *** THEN EXCITATION'S RDANDI OR FORMLA DATA:
C
C      IF NEXC .LT. 0 CARDS WITH DATA FOR RDANDI (SEE RDANDI NOTE):
C      IF NEXC .GT. 0 THEN CARDS WITH DATA FOR FORMLA: ( SEE FORMLA NOTE)
C
C *** THEN, IF NRD .NE. 0, IMPULSE RESPONSE'S RDANDI OR FORMLA DATA:
C      IF NRD  .LT. 0 CARDS WITH DATA FOR RDANDI (SEE RDANDI NOTE):
C      IF NRD .GT. 0 THEN CARDS WITH DATA FOR FORMLA: ( SEE FORMLA NOTE )
C
C  **** REPEAT THE ENSEMBLE, CASE BY CASE ... WHEN DATA ENDS, PROGRAM ENDS.
C
      REAL KM(11), KAR(11), KAD(11)
      DIMENSION A(11), B(11), PORE(11), POIM(11), WK(11), TITLE(60),
     1      E(601), RD(601), TETFN(1204),BI(11), RE(601),
     2  ARR(602), XT(3), YE(3), YD(3), YR(3)
      DATA XT, YE, YD, YR/'    ', 'TIME', '    ', ' EXC', 'ITAT', 'ION ',
     1, ' IMP', '. RE', 'SP. ', 'RESP', '(CON', 'VOL)'/
    1 FORMAT (20A4)
    2 FORMAT (F20.6, 6I10)
```

```
    3 FORMAT (4F20.6, 2E24.6)
    4 FORMAT (4F20.6)
    5 FORMAT (1H ,       'T(S) = ',  F8.4, '----', 20('-----')    )
    6 FORMAT (1H ,  16X,  7F15.4, /)
    7 FORMAT (1H , ' UTILIZES THESE VALUES OF THE TIME (NRT =',              1
     1 I4,'):', F8.3, '  (', F7.3,')', F8.3,  5X, 10('*'), '  NRT MUST B
     2E ODD; VERIFY THIS HERE', 8('*'))
    8 FORMAT (1H0, 'CALCULATION OF THE POLES OF T(S) BY THE SUB-ROUTINE
     1"POLRTD" GIVES THESE POLES:', / )
    9 FORMAT (1H0, F20.6, '     + J', F12.6)
   10 FORMAT (1H0, 'THE RESIDUES OF T(S) ARE (MODULUS, ANGLE IN  ',     '
     1DEGREES, IF COMPLEX....IF NOT, VALUE THEN ZERO)', // )                24.
   11 FORMAT (20A4)                                                          25.
   12 FORMAT (1H0, 'CONVOLUTION PROGRAM ''CONVOL''    .....***** ',/)
   13 FORMAT (1H ,' CODES READ IN:     NPLOT =', I3, ',    NEXC =', I3,
     1 ',    NRD =', I3,',   NWR =', I2)
   14 FORMAT (1H0,  'THERE IS BUT ONE POLE, SO POLRT WAS NOT CALLED; THE
     1 POLE IS:', / )
   15 FORMAT (1H1,'PLOT OF EXCITATION, E:')
   16 FORMAT (8I10)
   17 FORMAT ((2F20.6, / ))
   18 FORMAT (1H0, 10X, 'NRD =', I2, ', M =', I3, ', N =', I3, '; THE TR
     1ANSFER FUNCTION USED FOR CALCULATING THE IMPULSE RESPONSE IS, ',
     2 /, 35X, ' (TERMS WRITTEN IN ASCENDING ORDER: 0, 1, 2, ...):', //)
   19 FORMAT (1H , 10X,  'GENERAL NOTES FOR THIS RUN:', 3X, 3(20A4,/,40X
     1))
   20 FORMAT (1H1, 'PLOT OF RESPONSE RE (CONVOLUTION OF E AND RD):')
   21 FORMAT (1H1, '3 CURVES FOLLOW .... CURVE #1 IS THE EXCITATION  ''E
     1'', CURVE #2 IS THE RESPONSE ''RE'', THE CONVOLUTION OF ''E'' AND
     2''RD'',',/, 21X, 'CURVE #3 IS THE IMPULSE RESPONSE:')
   22 FORMAT (1H1, 'PLOT OF IMPULSE RESPONSE, RD:')
   23 FORMAT (1H0,            '*** THE VALUES OF THE IMPULSE RESPONSE ARE
     1CALCULATED FROM FORMLA, PER:')
   24 FORMAT (1H0, ///,    ' ****** THAT''S ALL .... AU REVOIR', /// )
   25 FORMAT (2F20.6, I10)
   26 FORMAT (1H0, 'DATA HAS NEXC = 0, WHICH IS NOT PROVIDED FOR IN PROGRAM
     1RAM. STOP HERE. CHANGE DATA AND TRY AGAIN. $$$$$$$$$$$$$$$$$$')
   27 FORMAT (1H1, 'RESULTS RE(T) OF THE CONVOLUTION OF RD(T) AND ',    '
     1 E(T):', / )
   28 FORMAT (1H0, 13X,  'TIME', 13X, 'EXCITATION', 13X, 'RD', 18X,        8.
     1 'RE', 15X, 'EXCIT. (E FMT.)',   11X, 'RE (E FMT.)', /, 13X, '....
     2..', 12X, '..........',  11X,  8('.'), 12X, 8('.'), 11X,15('.'),
     3 10X,  12('.'), / )
   29 FORMAT (1H ,' AND COMPUTES RESPONSE AT THESE VALUES OF TIME:',
     1F8.3, '  (', F7.3, ')', F8.3)
   30 FORMAT (1H+, 70X, 'DATA READ: DT =', F7.3, ', T2 =', F7.3, ', NRT
     1=', I4, ',  I.E.')
   31 FORMAT (1H0, 'IMPULSE RESPONSE:',/)
   32 FORMAT (1H0,        '*** THE VALUES OF THE EXCITATION ARE CALCULATED
     1 FROM FORMLA, PER:')
   33 FORMAT (4F20.6)
   34 FORMAT (1H0, 'PROGRAM  READ VALUES OF EXCITATION FROM INPUT (CARD)
     1 DATA AND INTERPOLATED:')
   35 FORMAT (1H0, 'PROGRAM READ VALUES OF IMPULSE RESPONSE FROM INPUT
     1 (CARD) DATA AND INTERPOLATED:')
   36 FORMAT (1H0, 16X, 'TIME', 16X, 'EXCITATION',/)
   37 FORMAT( 1H0, 16X, 'TIME', 12X, 'IMPULSE RESPONSE',/)
   38 FORMAT( 1H0, 'EXCITATION:',/)
C *** END OF FORMATS ***                                                      15.
   50 READ (5,1, END=900) TITLE
      WRITE (6,12)                                                           20.
      WRITE (6, 19) TITLE
      READ (5, 16) NEXC, NRD, NPLOT, NWR
      WRITE (6, 13)        NPLOT, NEXC, NRD, NWR
      READ (5,25) DT, T2, NRT
C COMPUTE NRT? VERIFY IS ODD? MAKE ODD?
      WRITE (6,30) DT, T2, NRT
      T1 = 0.0
      WRITE (6, 7) NRT, T1, DT, T2
      DT2 = 2*DT
      WRITE (6,29) T1, DT2, T2
      IF (NRD) 204,  75, 204
C CASE NRD = 0 (TRANSFER FUNCTION USED FOR RD):
   75 READ (5,2) C, M, N
      M1 = M+1
      N1 = N+1
      N11 = N-1                                                              24.
      READ (5,4) (A(I), I=1, M1)                                             25.
      READ (5,4) (B(I), I=1, N1)                                             26.
      WRITE (6,31)                                                           27.
      WRITE (6, 18) NRD, M, N
      WRITE (6,6) (A(I), I=1, M1)
      WRITE (6,5) C
      WRITE (6,6) (B(I), I=1, N1)
```

```
      CCC = C*A(M1)/B(N1)
      IF (N .EQ. 1) GO TO 101
      CALL POLRTD (B, WK, N1, PORE, POIM)
      GO TO 112
  101 PORE(1) = -B(1)/B(2)
      POIM(1) = 0.0
      WRITE (6, 14)
      GO TO 110
  112 WRITE (6, 8)
  110 WRITE (6, 9) (PORE(I), POIM(I), I=1, N)
      DO 120 I=1, N
  120 BI(I) = B(I+1)*I
      DO 200 JZ = 1,N
      CALL EVLPOL (M,A,    PORE(JZ),POIM(JZ),QNRE,QNIM,QNMG,QNAGD)
      CALL EVLPOL(N11,BI,    PORE(JZ),POIM(JZ),QDRE,QDIM,QDMG,QDAGD)
      IF(POIM(JZ))140,135,140
  135 KM(JZ) =C*QNRE/QDRE
      KAR(JZ) = 0.0
      KAD(JZ) = 0.0
      GO TO 200                                                        10.
  140 KM(JZ) = C*QNMG/QDMG
      KAD(JZ) = QNAGD - QDAGD
      KAR(JZ) = KAD(JZ)/57.29578
  200 CONTINUE                                                         15.
      WRITE (6,10)                                                     16.
      WRITE (6,17) (KM(I), KAD(I), I=1,N)
  204 CONTINUE
      WRITE (6,38)
C ***** BEGIN CALCULATION OF EXCITATION VALUES *******        *******        *****
      IF (NEXC) 340, 350, 325
  325 CONTINUE
C SUBROUTINE FORMULA OPTION FOR EXCITATION:
      WRITE (6,32)
      CALL FORMLA (NRT, DT, E)
      GO TO 370
  340 CONTINUE                                                         35.
      WRITE(6,34)
C *** **** READ CARDS WITH EXCITATION VALUES AND INTERPOLATE OPTION:
      CALL RDANDI( NRT, DT, E )
      IF (NWR .LE.1)   GO TO 370
      WRITE (6,36)
      DO 345 I = 1,NRT
      T = (I-1)*DT
  345 WRITE (6,6) T, E(I)
      GO TO 370
  350 CONTINUE
C    THIS IS NEXC=0    THIS CODE IS NOT USED.  WRITE ERROR MESSAGE AND STOP:
      WRITE (6, 26)
      STOP
C ***** END CALCULATION EXCITATION VALUES **************               ********
  370 CONTINUE
C  ***  BEGIN CALCULATE IMPULSE RESPONSE *************                 ********
      IF (NRD) 380, 400, 390
  380 CONTINUE
C     READ CARDS AND INTERPOLATE OPTION ON RD CALCULATION:
      WRITE(6,35)
      CALL RDANDI ( NRT, DT, RD )
      IF (NWR .LE. 1)  GO TO 501
      WRITE (6,37)
      DO 385 I = 1,NRT
      T = (I-1)*DT
  385 WRITE (6,6) T, RD(I)
      GO TO 501
C SUBROUTINE FORMULA OPTION FOR RD:
  390 CONTINUE
      WRITE (6, 23)
      CALL FORMLA (NRT, DT, RD)
      GO TO 501
C *** USE-TRANSFER-FUNCTION OPTION FOR RD:
  400 DO 500  LT=1,NRT
      TEMPS = (LT-1)*DT
      SUM = 0.0                                                        38.
      DO 450 LL=1,N                                                    39.
      XPT = KM(LL)*EXP(PORE(LL)*TEMPS)                                 40.
      IF(POIM(LL))430,410,420
  410 TERM = XPT
      GO TO 440                                                        43.
  420 TERM = 2*XPT*COS(POIM(LL)*TEMPS + KAR(LL))
      GO TO 440
  430 TERM = 0.0
  440 SUM = SUM + TERM                                                 51.
  450 CONTINUE                                                         52.
  500 RD(LT) = SUM
  501 CONTINUE
```

```
C  ***     END CALCULATE IMPULSE RESPONSE *****                              ********
C *** BEGIN CONVOLUTION:
      NRT32 = (NRT+1)/2                                                        1.
      NRT322 = 2*NRT32
      IF (NWR .EQ. 0)  GO TO 600
      WRITE (6,27)                                                             2.
      WRITE (6,28)                                                             3.
  600 NRT22 = NRT + 2                                                          4.
      T = 0.0
      TETFN(1) = 0.0
      ARR(1) = 0.0
      ADD = 0.0
      IF (NRD .NE. 0)  GO TO 601
      IF                    (M .EQ. N)  ADD = CCC * E(1)
  601 RE(1) = ADD
      TETFN(NRT22) = ADD
      IF ( NWR .EQ. 1 )
     1WRITE (6,3) T, E(1), RD(1), TETFN(NRT22), E(1), TETFN(NRT22)
C *** THE CONVOLUTION PROPERLY SPEAKING BEGINS HERE ***                      ********
      DO 650 I=2, NRT32
      II = 2*(I-1)
      T = II*DT
C *** STORE T FOR PLOTS:
      TETFN(I) = T
      ARR(I) = T
      I2 = I+I
      I1 = I2-1                                                                14.
      ICO = 1                                                                  16.
      SUM = 0
      DO 640 J=1,II
      IJ = I2 - J
C NEED INSURANCE AGAINST UNDERFLOW HERE? (FORMLA PROVIDES SOME)
      TERM = ICO*E(J)*RD(IJ)
      IF (ICO-2) 604, 604, 602                                                 23.
  602 ICO = 2                                                                  24.
      GO TO 606                                                                25.
  604 ICO = 4
  606 CONTINUE
  640 SUM = SUM + TERM
      SUM = SUM + E(I1)*RD(1)
      SUM = SUM * DT/3.0
  *** SUM = RESPONSE = RE(T) HERE
      IF (NRD .NE. 0) GO TO 641
C     (THIS ALLOWS FOR IMPULSE IN RD IF M = N)
      IF                    (M .EQ. N) SUM = SUM + CCC * E(I1)
  641 I1NRT = I + NRT322
C STORE RE VALUE FOR PLOTTING:
      TETFN(I1NRT) = SUM
      RE(I) = SUM
      IF (NWR .EQ. 0)  GO TC 650
      WRITE (6,3) T, E(I1), RD(I1), TETFN(I1NRT), E(I1), TETFN(I1NRT)
  o50 CONTINUE                                                                 33.
C *** THE CONVOLUTION PROPERLY SPEAKING ENDS HERE ***                          34.
C
C *** HERE BEGINS STORAGE FOR PLCTTING:                                      ********
C        J = 1,3,5,7,9, ..... (NRT)                                            36.
C        NRT32 = (NRT+1)/2
C                                                                              37.
C   STORAGE FOR PLOTTING CONTAINS (IN ARRAY TETFN(JJ)  ):
C        (JJ)       (A) JJ=1,2,3,......,NRT32=(NRT+1)/2)  VALUES OF THE TIME
C                              (2JJ-1)DT,   THAT IS  0,2DT,4DT, .... (NRT-1)DT
C        (JJ2)      (B)  JJ = (NRT32+1), (NRT32+2), .... (2NRT32) VALUES CF
C                                 THE EXCITATION E,  FOR THE SAME VALUES OF TIME
C        (JJ3)      (C) JJ = (2NRT32+1), (2NRT32+2), ..... (3NRT32)  VALUES
C                                 OF THE RESPONSE RE,  AT THE SAME VALUES OF TIME
C        (JJ4)
C
      DO 700 JJ=1,NRT32
      J = 2*JJ - 1
C     TETFN (JJ) = (J-1)*DT  ALREADY DONE; SEE 6600+12
C     ARR(JJ) = SAME
C
      JJ2 = JJ + NRT32
      JJ3 = JJ2 + NRT32
      JJ4 = JJ3 + NRT32
      ARR(JJ2) = TETFN(JJ3)
      TETFN(JJ2) = E(J)                                                        40.
C *** TETFN(JJ3) = RE(J), ALREACY DONE ABOVE AT 640 + 5.
      TETFN(JJ4) = RD(J)
  700 CONTINUE                                                                 41.
      IF (NPLOT) 800, 800, 750                                                 42.
  750 CONTINUE
      WRITE (6,20)
      CALL PLOT4X(NRT32, ARR, XT, YR)
```

```
      IF (NPLOT .EQ. 1)  GO TO 800
      WRITE (6,21)
      CALL PLT23C(TETFN, NRT32)
      IF (NPLOT .EQ. 2)  GO TO 800
      WRITE (6,15)
      CALL PLOT4X(NRT32, TETFN, XT, YE)
      IF (NPLOT .EQ. 3)  GO TO 800
      WRITE (6,22)
      DO 777 JJ=1, NRT32
      J = 2*JJ-1
      JJ2 = JJ+NRT32
      ARR(JJ2) = RD(J)
  777 CONTINUE
      CALL PLOT4X (NRT32, ARR, XT, YD)
  800 CONTINUE                                                          46.
      GO TO 50                                                          47.
  900 WRITE (6,24)
      STOP
      END                                                               51.
      SUBROUTINE EVLPOL( M,C,SRE,SIM, POLYRE, POLYIM, POLMAG,POLANG)
      DIMENSION C(1), C1(51)
      MM = M+1
       MMM=M-1
      DO 5 I=1,MM
    5 C1(I) = C(I)
      IF (MMM) 40, 30, 10
   10 D = -2*SRE
       E = SRE**2 + SIM**2
       DO 20 III=1,MMM
       II=M+2-III
      C1(II-1) = C1(II-1) - D*C1(II)
   20 C1(II-2) = C1(II-2) - E*C1(II)
   30 POLYRE = C1(2)*SRE + C1(1)
      POLYIM  =C1(2)*SIM
   35  POLMAG = SQRT(POLYRE**2 + POLYIM**2)
   36 IF (POLYRE .EQ. 0 .AND. POLYIM .EQ. 0) GO TO 37
       POLANG = ATAN2(POLYIM,POLYRE)*57.295780
       RETURN
   37 POLANG = 909090.9
      WRITE (6,1) M, (C(I), I = 1,4)
    1 FORMAT (1H0, 'A MESSAGE FROM EVLPOL: EVALUATION OF POLYNOMIAL OF
     1 DEGREE', I2, ' AND COEFFICIENTS:',/, 4E15.4,' ETC. GIVES 0 + J 0.
     2 WHAT ANGLE SHOULD I GIVE YOU? HELP.')
      RETURN
   40 POLYRE = C1(1)
       POLYIM = 0
      GO TO 35
       END
      SUBROUTINE POLRTD (XCOF, COF, M1, ROOTR, ROOTI)
      DIMENSION XCOF(M1),  COF(M1),  ROOTR(M1), ROOTI(M1)
      IFIT=0
      N = M1 - 1
      NX=N
      NXX=M1
      N2=1
      DO 40 L=1, M1
      MT= M1-L+1
   40 COF(MT)=XCOF(L)
   45 XO=.005
      YO=0.01
      IN=0
   50 X=XO
      XO=-10.0*YO
      YO=-10.0*X
      IN=IN+1
      X=XO
      Y=YO
      GO TO 59
   55 IFIT=1
      XPR=X
      YPR=Y
   59 ICT=0
   60 UX=0.0
      UY=0.0
      V =0.0
      YT=0.0
      XT=1.0
      U=COF(N+1)
      IF(U) 65,130,65
   65 DO 70 I=1,N
      L =N-I+1
      XT2=X*XT-Y*YT
      YT2=X*YT+Y*XT
      U=U+COF(L )*XT2
```

```
      V=V+COF(L )*YT2
      FI=I
      UX=UX+FI*XT*COF(L )
      UY=UY-FI*YT*COF(L )
      XT=XT2
   70 YT=YT2
      SUMSQ=UX*UX+UY*UY
      IF (SUMSQ) 75, 110, 75
   75 DX = (V*UY - U*UX)/SUMSQ
      DY = -(U*UY + V*UX)/SUMSQ
      X = X +DX
      Y = Y + DY
   78 IF (ABS(DY)+ ABS(DX)-1.0E-05) 100,80,80
   80 ICT=ICT+1
      IF(ICT-500) 60,85,85
   85 IF(IFIT)100,90,100
   90 IF(IN-5) 50,95,95
   95 WRITE (6, 1) N2, IN, ICT, IFIT
    1 FORMAT (1H0,   '********* ERROR RETURN FROM SUBROUTINE POLRTD (DOES
     1 NOT CONVERGE):  N2 =', I3,   ', IN =',  I2,  ', ICT = ', I3,
     2  ', IFIT =', I2, ' **********', // )
      RETURN
  100 DO 105 L=1,NXX
      MT= M1-L+1
      TEMP=XCOF(MT)
      XCOF(MT)=COF(L)
  105 COF(L)=TEMP
      ITEMP=N
      N=NX
      NX=ITEMP
      IF(IFIT) 120,55,120
  110 IF(IFIT) 115,50,115
  115 X=XPR
      Y=YPR
  120 IFIT=0
      IF ( ABS(Y)-1.E-4* ABS(X)) 135,125,125
  125 ALPHA=X+X
      SUMSQ=X*X+Y*Y
      N=N-2
      GO TO 140
  130 X=0.0
      NX=NX-1
      NXX=NXX-1
  135 Y=0.0
      SUMSQ=0.0
      ALPHA=X
      N=N-1
  140 COF(2)=COF(2)+ALPHA*CCF(1)
      NSX   =N
      IF(NSX    .LT. 2) NSX   =2
  145 DO 150 L=2,NSX
  150 COF(L+1)=COF(L+1)+ALPHA*CCF(L)-SUMSQ*COF(L-1)
  155 ROOTI(N2)=Y
      ROOTR(N2)=X
      N2=N2+1
      IF(SUMSQ) 160,165,160
  160 Y=-Y
      SUMSQ=0.0
      GO TO 155
  165 IF (N) 200, 200, 45
  200 RETURN
      END
      SUBROUTINE PLT23C (A, N)
      INTEGER BAR, DOT, BLANK
      DIMENSION A(1), YPR(11), IOUT(101), ICAR(3)
      DATA BLANK, BAR, DOT, ICAR/' ', '|', '.', '1', '2', '3'/
    1 FORMAT (1H , 7X, 'TIME', 4X, 11('.', 9X))
    2 FORMAT (1H , F11.4,  4X, 101A1)
    3 FORMAT (1H )                                                      24.
    4 FORMAT (16X, 11('.', 9X))
    8 FORMAT (1H ,10X, 11F10.4)
C *** CALCULATE SCALE OF THE "Y'S"                                       4.
      M1 = N+1                                                          5.
      M2 = 4*N
      M3 = N+2                                                          9.
      YMIN = A(M1)                                                      7.
      YMAX = YMIN                                                       8.
      DO 40 I=M3,M2                                                     10.
      IF (A(I) .LT. YMIN) YMIN = A(I)
      IF (A(I) .GT. YMAX) YMAX = A(I)
   40 CONTINUE
      YSCAL = (YMAX-YMIN)/100.0                                         17.
      YPR(1) = YMIN                                                     18.
      DO 42 J=1,9
```

```
 42 YPR(J+1) = YPR(J) + YSCAL*10.0
    YPR(11) = YMAX                                                        21.
    JPZ = -YMIN/YSCAL + 1
    WRITE (6,3)                                                           23.
    WRITE (6, 8)YPR
    WRITE (6,4)
    WRITE (6,1)
    WRITE (6,4)
    L = 1                                                                 28.
 45 XPR = A(L)
    DO 55 IY = 1,101
 55 IOUT(IY) = BLANK                                                      35.
    IOUT(1) = DOT
    IOUT(101) = DOT
    IF (JPZ .GT. 0 .AND. JPZ .LT. 102)  IOUT(JPZ) = BAR
    DO 60 J=1,3
    LL = L+ J*N                                                           37.
    JP = (A(LL)-YMIN)/YSCAL + 1                                           38.
    IOUT(JP) = ICAR(J)                                                    39.
 60 CONTINUE                                                              40.
    WRITE (6, 2) XPR, IOUT
    L = L+1                                                               43.
    IF (L .EQ. M1)  GO TO 86
    GO TO 45                                                              49.
 86 WRITE (6,4)
    WRITE (6,1)
    WRITE (6, 4)
    WRITE (6, 8)YPR
    RETURN                                                                56.
    END                                                                   57.
      SUBROUTINE RDANDI ( NRT,DT, ARR)
    DIMENSION ARR(1), FF(51), TT(51)
  1 FORMAT (4F20.6)
  3 FORMAT (1H0, 13X, 'RDANDI DATA: NCODE =', I2, ', WF =', F8.3, ' PH
   1I =', F8.3, ' DEGREES ')
  2 FORMAT (I10, 2F20.6)
  4 FORMAT (1H , 20X, 2F20.6)
  5 FORMAT (1H0, 13X, 'AND POINTS:', 10X, 'TIME', 14X, 'FUNCTION',/)
    READ (5,2) NCODE, WF, PHIDEG
    WRITE (6,3) NCODE, WF, PHIDEG
C READ DATA:
    I = 0
    WRITE (6,5)
 50 READ (5,1) T, F
    IF (T .LT. 0.0)  GO TO 60
    I = I + 1
    FF(I) = F
    TT(I) = T
    WRITE (6,4) T, F
    GO TO 50
 60 N = I
C NOW FORM TABLE BY INTERPOLATION:
    DO 500 J = 1, NRT
    T = (J-1)*DT
    DO 100 I = 1,N
    TEST = T-TT(I)
    IF (TEST) 300, 200, 100
100 CONTINUE
    FUNC = FF(N)
    GO TO 499
200 FUNC = FF(I)
    GO TO 499
300 FUNC = FF(I-1) + (T-TT(I-1))*(FF(I)-FF(I-1))/(TT(I)-TT(I-1))
499 ARR(J) = FUNC
    IF (NCODE .EQ.0)  GO TO 500
    ANGL = WF*T + PHIDEG/57.29578
    ARR(J) = FUNC*COS(ANGL)
500 CONTINUE
    RETURN
    END
    SUBROUTINE FORMLA (NRT, DT, ARR)
  1 FORMAT (8I10)
  2 FORMAT (4F20.6)
  3 FORMAT (1H0, 7X, 'FORMLA CALLED; NN =', I2, ', ND =', I2, ', NE ='
   1,I2, ', NC =', I2, '  POLYNOMIAL COEFFICIENTS:')
  4 FORMAT (1H0, 8F16.6)
    DIMENSION ARR(1), CN(8), CD(8), CE(8), CC(8)
    DO 100 I = 1,8
    CN(I) = 0.0
    CD(I) = 0.0
    CE(I) = 0.0
100 CC(I) = 0.0
    READ (5,1) NN, ND, NE, NC
    NN1 = NN + 1
```

```
      ND1 = ND + 1
      NE1 = NE + 1
      NC1 = NC + 1
      READ (5,2) (CN(I), I = 1, NN1)
      READ (5,2) (CD(I), I = 1,ND1)
      READ (5,2) (CE(I), I = 1, NE1)
      READ (5,2) (CC(I), I = 1, NC1)
      WRITE (6,3) NN, ND, NE, NC
      WRITE (6,4) CN, CD, CE, CC
      DO 300 I = 1, NRT
      T = (I-1)*DT
      PN = CN(7)
      PD = CD(7)
      PE = CE(7)
      PC = CE(7)
      DO 200 J = 1,6
      L = 7-J
      PN = CN(L) + T*PN
      PD = CD(L) + T*PD
      PE = CE(L) + T*PE
  200 PC = CC(L) + T*PC
      ARR(I) = PN * EXP(PE)*COS(PC)/PD
      IF (ABS(ARR(I)) .LT. 1.0E-10)  ARR(I) = 0.0
  300 CONTINUE
      RETURN
      END
      SUBROUTINE PLOT4X (N,A,X,Y)
C ... USES ONLY HALF THE WIDTH OF THE PAPER FOR THE PLOT, BUT PRINTS VALUES OF
C       BOTH "X" AND "Y"  WITH LEGEND AT LEFT. PRINTS ZERO AXIS IF ON SCALE.
      REAL IOUT
      DIMENSION IOUT( 51), YPR( 6), A(1), X(3), Y(3)
      DATA BLANK, ASTX, BAR, DOT /1H , 1H*, 1H|, 1H./
    1 FORMAT (1H , 115('.') )
    2 FORMAT (1H ,       F20.4, E22.4, 18X, 51A1   )
    3 FORMAT (1H , 60X, 51A1)
    4 FORMAT (1H0,'SUBROUTINE ''PLOT4 '', BEING CALLED, GIVES (BAR MARKS
     1 ZERO ORDINATE):')
    8 FORMAT (1H ,  9X, 3A4, 10X, 3A4, 12X, 6F10.4)
      NL = N                                                            21.
      M1 = N+1                                                          27.
      M2 = 2*N
      YMIN = A(M1)                                                      29.
      YMAX = YMIN                                                       30.
      M3 = N+2                                                          31.
      DO 40 I=M3,M2                                                     32.
       IF (A(I)-YMIN) 28,26,26                                          33.
   26 IF (A(I)-YMAX) 40,40,30                                           34.
   28 YMIN = A(I)
      GO TO 40                                                          36.
   30 YMAX = A(I)                                                       37.
   40 CONTINUE                                                          38.
      YSCAL = (YMAX-YMIN)/50.0
      JPZ = -YMIN/YSCAL + 1
      YPR(1) = YMIN                                                     40.
      DO 42 J=1,4
   42 YPR(J+1) = YPR(J) + YSCAL*10.0                                    42.
      YPR( 6) = YMAX
      WRITE (6, 4)
      WRITE (6, 1)                                                 ALIGNER*
      WRITE (6,8) X, Y, (YPR(I), I=1,6)
      DO 43 IY = 2,50
   43 IOUT(IY) = BLANK
      IOUT(1) = DOT
      IOUT(51) = DOT
      IF  (JPZ .GE. 1 .AND. JPZ .LE. 51) IOUT(JPZ) = BAR
      WRITE (6,3) (IOUT(IZ), IZ = 1,51)
      L = 1                                                             50.
   45 XPR = A(L)
      LL = L + N
      JP = (A(LL)-YMIN)/YSCAL + 1                                       60.
      IOUT( JP) = ASTX
   60 CONTINUE
      WRITE (6,2) XPR, A(LL), (IOUT(IZ), IZ=1,51)
      IOUT( JP) = BLANK
      IOUT(1) = DOT
      IOUT(51) = DOT
      IF (JPZ .GE. 1 .AND. JPZ .LE.51) IOUT(JPZ) = BAR
      L = L+1                                                           68.
      IF (L-NL) 45,45,86
   86 CONTINUE
      WRITE (6,3) (IOUT(IZ), IZ = 1,51)
      WRITE (6,8) X, Y, (YPR(I), I=1,6)
      WRITE (6, 1)                                                 ALIGNER*
```

```
RETURN
END
```

.... illustrative (sample) data-card set for CØNVØL:

```
CONVOL TEST                                                         (FIRST NOTE CARD)
ILLUSTRATIVE EXAMPLE FOR EE101                                      (SECOND "     ")
BOTH E AND RD DATA IN FORM OF POINTS                                (3RD        )
     -1          -1           1          0
   0.05             12.0                        241
        0·             0.0                                  0.0
    0.0               1.0
   3.0              0.0
    3.0                 -1.0
    4.0             -0.5
     5.0             0.0
  -3.6              0.0
      0                0.0                 0.0
    0.0                0.0
     1.0             0.8
    2.0              0.8
    2.0             0.0
   -2.0            0.0
```

```
      SUBROUTINE POLRTT (M1, ACOF, COF, ZR, ZI)
C THIS PROGRAM WAS DEVELOPED BY DAVID F. TUTTLE, FOR USE IN EE101-102-103
C AND ASSOCIATED TEXTS
C SIMPLE POLYNOMIAL ZERO FINDER, DERIVED FROM IBM SSP "POLRT": USES NEWTON'S
C METHOD, WITHOUT ZERO IMPROVEMENT (NO "IFIT" REFINEMENT ON ORIGNAL POLYNOMIAL)
C HAS SEVERAL (10) SCATTERED STARTING POINTS, TRIED AS NECESSARY (LIMIT: 25
C ITERATIONS ON EACH).IF COMPONENT(S) OF DERIVATIVE IS TOO SMALL, RESTARTS AT
C NEXT TRIAL POINT TO AVOID UNDERFLOW OR OVERFLOW ON DIVISION BY NEARZERO. GIVES
C ERROR MESSAGE IF NO CONVERGENCE.  NOTATION:  M1 = 1 + DEGREE OF POLYNOMIAL
C              ACOF = ARRAY OF COEFFICIENTS,   COF = WORKING (SCRATCH-PAD) ARRAY
C              ZR (+ J) ZI ARE ARRAYS INTO WHICH GO ZEROS FOUND.
      DIMENSION ACOF(M1), COF(M1), ZR(M1), ZI(M1), XA(10), YA(10)
      DATA XA/ 4*-0.5, 2*0.2, 4*1.0/
      DATA YA/ 0.28, 0.47, 1.03, 0.79, 0.10, 0.80, 0.0, 0.5, 2.0, 4.0/
      N = M1 - 1
      NX=N
      NXX=M1
      N2=1
C *** PUT COEFFICIENTS INTO WORKING ARRAY (IN REVERSE ORDER):
      DO 40 L=1, M1
      ZR(L) = 0.999999E11
      ZI(L) = 0.999999E11
      MT= M1-L+1
   40 COF(MT)=ACOF(L)
C *** PICK APPROXIMATE ZERO:
   45 IN=0
   50 CONTINUE
      IN=IN+1
      IF (IN-10) 51, 51, 95
   51 X = XA(IN)
      Y = YA(IN)
      ICT=0
C *** EVALUATE POLYNOMIAL AND DERIVATIVE:
   60 UX=0.0
      UY=0.0
      V =0.0
      YT=0.0
      XT=1.0
      U=COF(N+1)
      IF(U) 65,130,65
   65 DO 70 I=1,N
      L =N-I+1
      XT2=X*XT-Y*YT
      YT2=X*YT+Y*XT
      U=U+COF(L )*XT2
      V=V+COF(L )*YT2
      FI=I
      UX=UX+FI*XT*COF(L )
      UY=UY-FI*YT*COF(L )
C *** IF UX OR UY IS TOO SMALL, GO TO 50 AND TRY NEW START:
      T2 = 1.0E-08
      TUX = ABS(UX)
      TUY = ABS(UY)
      IF (TUX .LT. T2 .AND. TUX .NE. 0.0)  GO TO 50
      IF (TUY .LT. T2 .AND. TUY .NE. 0.0)  GO TO 50
      XT=XT2
   70 YT=YT2
      SUMSQ=UX*UX+UY*UY
C *** ELSE CONTINUE TO ITERATE ON THIS ONE; CORRECTION:
      DX = (V*UY - U*UX)/SUMSQ
      DY = -(U*UY + V*UX)/SUMSQ
      X = X +DX
      Y = Y + DY
      IF (ABS(DY)+ ABS(DX)-1.0E-05) 100,80,80
C *** INCREMENT LARGE, GO 'ROUND AGAIN: 80
   80 ICT=ICT+1
      IF (ICT-25) 60, 50, 50
C *** INCREMENT SMALL, SATISFIED, GO OUT: 100
  100 IF ( ABS(Y)-1.E-4* ABS(X) ) 135,125,125
  125 ALPHA=X+X
      SUMSQ=X*X+Y*Y
      N=N-2
      GO TO 140
  130 X=0.0
      NX=NX-1
      NXX=NXX-1
  135 Y=0.0
      SUMSQ=0.0
      ALPHA=X
      N=N-1
  140 COF(2)=COF(2)+ALPHA*COF(1)
      NSX   =N
      IF(NSX   .LT. 2) NSX   =2
      DO 150 L=2,NSX
```

```
150 CCF(L+1)=COF(L+1)+ALPHA*CCF(L)-SUMSQ*COF(L-1)
155 ZR(N2) = X
    ZI(N2) = Y
    N2=N2+1
    IF(SUMSQ) 160,165,160
160 Y=-Y
    SUMSQ=0.0
    GO TO 155
165 IF (N) 200, 200, 45
200 RETURN
 95 WRITE (6, 1) N2, IN, ICT, (I, ZR(I), ZI(I), I = 1, M1)
  1 FORMAT (1HO,    '******** ERROR RETURN FROM SUBROUTINE POLRTT (DOES
   1 NOT CONVERGE):  N2 =', I3,    ', IN =',  I2,  ', ICT = ', I3, 10X,
   2 27('*'),/,  10X,  'ZEROS FOUND,  IF ANY:', 21(I10, F12.6, '  + J'
   3, F14.6, /, 31X ))
    RETURN
    END
```

```
      SUBROUTINE DALBLR (A,     M1, ROOTR, ROOTI, C, PTST)                JULY 74
C THIS PROGRAM WAS DEVELOPED BY DAVID F. TUTTLE, FOR USE IN EE266 AND ASS. TEXTS
C
C   SOURCE: TOM HATLEY TO OFT (1968)... MODIFIED BY INCORPORATING SUBROUTINES
C        DPNORM, DPCOMP, EVAL INTO SUBROUTINE FNDRUT AND IN VARIOUS OTHERWAYS
C
C    THIS SUBROUTINE USES THE METHOD OF D'ALEMBERT FOR FINDING ZERO(S) OF THE
C       POLYNOMIAL "A",  THEN DIVIDES OUT THE ZERO FACTOR, AND CONTINUES UNTIL
C       ALL ZEROS HAVE BEEN FOUND. IT IS NOT PARTICULARLY FAST, BUT HANDLES
C       PRACTICALLY ALL POLYNOMIALS
C
C   NOTATION:      A IS A DOUBLE-PRECISION ARRAY CONTAINING THE POLYNOMIAL'S
C                         COEFFICIENTS (THE CONSTANT TERM IS IN A(1) )
C                  M = INTEGER = DEGREE OF POLYNOMIAL;      M1 = M + 1
C                  ROOTR  AND ROOTI ARRAYS CONTAIN, ON RETURN, THE ZEROS (REAL
C                         AND IMAGINARY PARTS, RESPECTIVELY).
C                  C = WORKING ARRAY
C                  PTST = TEST VALUE FOR DECIDING WHETHER ZERO HAS BEEN REACHED
C                         ETC.,    1.0E-10 IS A REASONABLE VALUE
C                  FNDRUT IS SUBROUTINE FOR LOCATING ONE ZERO
C
C *** REVISED JULY, 1974
C
      DOUBLE PRECISION C(M1),   ROOTR(M1), ROOTI(M1), A(M1), RR, RI, TEST
     1, P, Q, B, DIS, DSQRT, DABS, T2, EPS, EPS1
      INTEGER ABRTN
C
C EPS GOES TO FNDRUT
C  2 IS  TEST ON NORM OF POLYNOMIAL AT ALLEGED ZERO
C EPS1 IS  TEST FOR "PURE REAL" OR "PURE IMAGINARY"
C
      EPS = PTST
      EPS1 = PTST
      T2 = PTST
      J = 1
      N = M1 - 1
      IF (N .EQ. 0)  GO TO 92
      N1 = M1
      DO 190 I = 1, N
      ROOTR(I) = 9.99999D10
  190 ROOTI(I) = 9.99999D10
C *** REVERSE SEQUENCE OF COEFFICIENTS:
      DO 200 K=1,N1
      J1 = N1 - K + 1
  200 C(K) = A(J1)
      TEST = DABS(C(1))
      IF (TEST .LE. 1.0E-06)  GO TO 94
C *** HANDLE CASES N = 2, 1, 0..... :
      IF (N-1) 92, 80, 222
  222 IF (N .EQ. 2)  GO TO 70
C *** REMOVE ZEROS AT THE ORIGIN, IF ANY:
  225 IF (C(N1) .NE. 0.0 ) GO TO 5
      IF (N .EQ. 0)  GO TO 93
      ROOTR(J) = 0.0
      ROOTI(J) = 0.0
      J = J + 1
      N = N-1
      N1 = N+1
      GO TO 225
    5 CONTINUE
C *** HANDLE CASES N = 2, 1, 0..... :
      IF (N .EQ. 0)  RETURN
      IF (N-1) 92, 80, 230
  230 IF (N. EQ. 2)  GO TO 70
C *** NOW START ON NON-ZERO ZEROS:
      CALL FNDRUT (N1,C, EPS, RR, RI, ABRTN )
C *** IF FNDRUT DID NOT CONVERGE, RETURN:
      IF (ABRTN .NE. 0) RETURN
C *** TEST NORM (SIZE) OF ZERO ALLEGEDLY FOUND; IF IT IS TOO SMALL
C        (NEARLY ZERO), RETURN VIA 90:
      TEST = DSQRT (RR*RR + RI*RI)
      IF (TEST .LE. T2)  GO TO 90
C *** CLEAN UP NEARLY REAL AND NEARLY IMAGINARY ZEROS:
      IF (DABS(RR/TEST) .LT. EPS1) RR=0.0
      IF (DABS(RI/TEST) .LT. EPS1) RI=0.0
      IF (RI .NE. 0.0) GO TO 50
C *** LINEAR FACTOR; REMOVE ONE (REAL) ZERO:
      N = N-1
      N1 = N+1
      ROOTR(J) = RR
      ROOTI(J) = 0.0
      J = J+1
      DO 30 I=2,N1
```

```
   30 C(I) = C(I) + RR*C(I-1)
   31 IF (N .EQ. 2) GO TO 70
      IF (N .EQ. 1) GO TO 80
C *** ELSE CONTINUE TO NEXT ZERO:
      GO TO 5
  *** COMPLEX FACTOR; REMOVE TWO (CONJUGATE, COMPLEX) ZEROS:
   50 N = N-2
      N1 = N+1
      ROOTR(J) = RR
      ROOTR(J+1) = RR
      ROOTI(J) = RI
      ROOTI(J+1) = -RI
      J = J+2
      P = -2.0*RR
      Q = RR*RR + RI*RI
      C(2) = C(2) - P*C(1)
      IF (N .EQ. 1) GO TO 80
      DO 60 I=3,N1
   60 C(I) = C(I) - P*C(I-1) - Q*C(I-2)
      GO TO 31
C *** LAST QUADRATIC FACTOR:
   70 TEST = DABS(C(1))
      IF (TEST .LE. 1.0E-06)  GO TO 94
      P = C(2)/(2.0*C(1))
      Q = C(3)/C(1)
      DIS = P*P - Q
      IF (DIS .GE. 0.0) GO TO 72
      DIS = DSQRT(-DIS)
      ROOTR(J) = - P
      ROOTI(J) = DIS
      ROOTR(J+1) = -P
      ROOTI(J+1) = -DIS
      RETURN
   72 DIS = DSQRT(DIS)
      ROOTR(J) = -P + DIS
      ROOTI(J) = 0.0
      ROOTR(J+1) = -P - DIS
      ROOTI(J+1) = 0.0
      RETURN
  *** LAST LINEAR FACTOR
   80 TEST = DABS(C(1))
      IF (TEST .LE. 1.0E-06)  GO TO 94
      ROOTR(J) = -C(2)/C(1)
      ROOTI(J) = 0.0
      RETURN
   90 WRITE (6, 1) TEST
    1 FORMAT (1H0, 'THE NORM AT A NON-ZERO ZERO IS', E14.6,' WHICH SEEMS
     1 UNREASONABLE. INSTRUCTIONS, PLEASE. /S/ DALBER.')
      RETURN
   92 WRITE (6, 2) N
    2 FORMAT (1H0, 'N =',I2,'; ARE YOU SERIOUS? A MISTAKE? /S/ DALBER.')
      RETURN
   93 WRITE (6, 3) N, C(N1)
    3 FORMAT (1H0, 'DALBER FINDS N =', I2,' AND COEFF. =', E14.6,'; SOME
     1THING IS WRONG.')
      RETURN
   94 WRITE (6, 4) N, C(1)
    4 FORMAT (1H0, 'DALBER FINDS DEGREE N = ',I2,' AND COEFFICIENT OF N-
     1TH POWER OF S IS', E14.6,'; I GIVE UP. /S/ DALBER.')
      RETURN
      END
      SUBROUTINE FNDRUT (M1,C, EPS, ROOTR, ROOTI, ABRTN)
C *** THIS  SUBROUTINE FINDS ONE ZERO OF A POLYNOMIAL USING TECHNIQUE BASED ON
C       D'ALEMBERT'S LEMMA
C EPF IS TEST ON NORM AT 102+ (BOWL TEST); ALSO AT 101+, AND AT 103+
C       EPF = 1.0E-10 IS REASONABLE
C EPD IS TEST ON SIZE OF DELR AND DELI AT 116+, 117+, 118+, 120+
C EPD SHOULD BE APPRECIABLY SMALLER THAN EPF
      DOUBLE PRECISION GESR, GESI, POINTR(4), POINTI(4), RR, RI, ROOTR,
     1 ROOTI, DELR, DELI, C(M1), EPS, FN(5), A, TEST, DABS, TR, TI, YR,
     2 YI, DSQRT, EPF, EPD
      INTEGER ABRTN
      EPF = EPS
C *** NEED DELTA TEST APPRECIABLY SMALLER THAN NORM TEST TO GET ZEROS:
      EPD = EPF/100.0
      ABRTN = 0
C *** CHANGE MADE TO AVOID EVEN-POLYNOMIAL TROUBLES
C       (BOWL TRAPPING AT ORIGIN)  (GESR WAS 0.0):
      GESR = -1.0
      GESI = 0.0
      I3 = 0
      GO TO 2050
```

```
C *** COME BACK TO 2000 (FROM 102+) IF TRAPPED IN BOWL: MAKE NEW GUESS:
 2000 CONTINUE
C *** I3 INCREMENTS WHEN GUESS IS CHANGED TO GET OUT OF BOWL
      I3 = I3 + 1
C *** IF BOWL TRAP HAS OPERATED 3 TIMES, GIVE UP: RETURN (VIA 131):
      IF (I3 .GE. 3)  GO TO 131
      GESR = GESR-1.0
      GESI = GESI + 1.0
C 050 CONTINUE
C *** GUESS IS CENTER POINT OF SQUARE WITH VERTICES POINT(I), I = 1, 2, 3, 4
C *** GUESS = GESR + J GESI  IS CENTER  POINT OF SQUARE: VERTICES ARE AT
C
C *** START CYCLE:
C *** EVALUATE POLYNOMIAL AT GUESS:
      YR = C(1)
      YI = 0.0
      DO 52 J=2,M1
      TR = YR*GESR          - YI*GESI
      TI = YI*GESR          + YR*GESI
      YR = TR + C(J)
   52 YI = TI
      FN(1) = DSQRT (YR*YR + YI*YI)
C *** END EVALUATE AT GUESS
C *** SET STEP SIZE:
      DELR = 0.250
      DELI = 0.250
      I1 = 0
      I2 = 0
C *** I1 INCREMENTS WHEN CENTER POINT IS MOVED, STEP SIZE REMAINS CONSTANT
C *** I2 INCREMENTS WHEN STEP SIZE IS DECREASED, CENTER POINT REMAINS SAME.
   10 CONTINUE
C *** QUERY: HAVE THERE BEEN MORE THAN 7 ATTEMPTS WITH  THE SAME STEP SIZE?
C         IF YES, GO TO 30; IF NO, KEEP TRYING ( GO TO 20 )
      IF (I1 .GE. 7) GO TO 30
   20 CONTINUE
      I1 = I1 + 1
C *** IF 200 CENTER POINTS HAVE BEEN TRIED, RETURN (VIA 131); ELSE GO TO 50
C         AND KEEP TRYING
      IF (I1 - 200) 50, 131, 131
   30 CONTINUE
C *** NOW ASK: HAS STEP SIZE EVER BEEN DECREASED?
      IF (I2) 40, 40, 20
  *** IF YES, KEEP TRYING (GO TO 20)
  *** IF NO: INCREMENT STEP SIZE AND START AGAIN:
   40 I1 = 1
      DELR = 8.0*DELR
      DELI = 8.0*DELI
C *** ESTABLISH NEW POINTS (VERTICES):
   50 POINTR(1) = GESR + DELR
      POINTI(1) = GESI
      POINTR(2) = GESR - DELR
      POINTI(2) = GESI
      POINTR(3) = GESR
      POINTI(3) = GESI + DELI
      POINTR(4) = GESR
      POINTI(4) = GESI - DELI
C  *** EVALUATE POLYNOMIAL AT POINT(I), I = 1, 2, 3, 4 (THE VERTICES):
      DO 60 I=2,5
C     N1 = N+1
      YR = C(1)
      YI = 0.0
      DO 51 J=2, M1
      TR = YR*POINTR(I-1) - YI*POINTI(I-1)
      TI = YI*POINTR(I-1) + YR*POINTI(I-1)
      YR = TR + C(J)
   51 YI = TI
      FN(I) = DSQRT (YR*YR + YI*YI)
   60 CONTINUE
  *** QUERY: HAS STEP SIZE BEEN DECREASED MORE THAN TEN TIMES?
      IF (I2 - 10) 70, 70, 101
C     IF YES, GO TO 101; ELSE, KEEP ON, GO TO 70:
   70 CONTINUE
C *** ARE VALUES OF POLYNOMIAL ALL REASONABLE ( .LE. 10.0)?
C         IF YES,  GO TO 101, IF NO,  GO TO 81
      DO 80 I = 1, 5
      IF (FN(I) - 10.0) 80, 80, 81
   80 CONTINUE
      GO TO 101
   81 CONTINUE
C *** VALUES OF POLYNOMIAL ARE NOT ALL REASONABLY SMALL;
C     ASK: ARE THEY ALL CLOSE?
C     IF YES: GO TO 100+:  NO: GO TO 101
      DO 100 I=1,5
      DO 90 J=I,5
```

```
      IF (DABS(FN(1) - FN(J)) .GT. 1.0D-03) GO TO 101
   90 CONTINUE
  100 CONTINUE
C *** QUERY:  HAS STEP SIZE BEEN DECREASED? (IS I2 .GT. 0?)
C           IF NO, INCREMENT STEP SIZE (GO TO 40); IF YES, KEEP GOING:
      IF (I2 .EQ. 0) GO TO 40
  101 CONTINUE
C *** QUERY: IS GUESS A ZERO?  IF NO,  GO TO 103; IF YES,  GO ON (TO 102):
      IF (FN(1) .GT. EPF)  GO TO 103
  102 ROOTR = GESR
      ROOTI = GESI
C *** TEST IF TRAPPED IN BOWL (LOCAL MINIMUM, NOT ZERO);
C     IF SO, GO TO 2000 AND MAKE NEW START; IF NO, RETURN (HAVE ZERO)
      IF (FN(1) .GT. EPF)  GO TO 2000
      RETURN
  103 CONTINUE
C *** GUESS IS NOT A ZERO; QUERY: IS ANY VERTEX A ZERO?
C *** IF SO, GO TO 111 AND RETURN, ELSE GO TO 112
      DO 110 I=2,5
      J = I
      IF (FN(I) .LE. EPF)  GO TO 111
  110 CONTINUE
      GO TO 112
C *** YES, VERTEX (J-1) IS A ZERO:
  111 ROOTR = POINTR(J-1)
      ROOTI = POINTI(J-1)
      J1 = J-1
      RETURN
  112 CONTINUE
C *** NO;  NO VERTEX IS A ZERO.   COMPARE NORMS AT CENTER AND VERTICES:
C            IF CENTER NORM .LE. ALL VERTEX NORMS, GO TO 116;
C            IF CENTER NORM .GT. A VERTEX NORM, GO TO 123
      DO 115 I=2,5
      IF (FN(1) .GT. FN(I)) GO TO 123
  115 CONTINUE
  116 CONTINUE
C *** ZERO LIES WITHIN PRESENT SQUARE.  DECREASE STEP SIZE:
      DELR = DELR/2.0
      DELI = DELI/2.0
C *** IS STEP SIZE VERY SMALL?  TEST REAL, THEN IMAG. (116-120); THEN:
C                           IF YES, GO TO 102, AND (IF NOT IN BOWL) RETURN WITH Z
C                           IF NO,  GO TO 121 AND ON
      A = GESR
      IF (A .NE. 0.0) GO TO 117
      IF (DELR .LE. EPD) GO TO 118
      GO TO 121
  117 CONTINUE
      IF (DABS(DELR/A) .GT. EPD)  GO TO 121
  118 A = GESI
      IF (A .NE. 0.0) GO TO 120
      IF (DELR .LE. EPD) GO TO 102
      GO TO 121
  120 CONTINUE
      IF (DABS(DELR/A) .LE. EPD) GO TO 102
  121 CONTINUE
C *** QUERY: HAVE THERE BEEN TOO MANY ITERATIONS? IF YES, RETURN VIA 131;
C             ELSE CONTINUE:
      IF (I1 .GT. 200) GO TO 131
  122 I2 = I2 + 1
      GO TO 50
C *** (FROM 112) ZERO LIES OUTSIDE PRESENT SQUARE;
C     FIND VERTEX WITH MINIMUM NORM, ESTABLISH IT AS CENTER OF NEW SQUARE:
  123 TEST = FN(2)
      J = 1
      DO 130 I=3,5
      IF (TEST .LE. FN(I)) GO TO 130
  124 TEST = FN(I)
      J = I-1
  130 CONTINUE
      GESR = POINTR(J)
      GESI = POINTI(J)
      FN(1) = FN(J+1)
C *** NOW GO TO 10 AND CONTINUE:
      GO TO 10
C *** DOES NOT CONVERGE; SET FLAG AND RETURN:
  131 ABRTN = 1
C *** WRITE ERROR MESSAGE:
      WRITE (6, 132) I1, I2, I3
  132 FORMAT (1H0, 'SUBROUTINE DALBERT DOES NOT CONVERGE. RETURN WITH I1
     1 =', I3, ', I2 =', I3, ', I3 =', I3)
      ROOTR = GESR
      ROOTI = GESI
      RETURN
      END
```

APPENDIX D

LOGARITHMIC MEASURE

The utility of logarithmic (as opposed to arithmetic) measure is discussed in the text at various points. We here draw those discussions together and add certain elements of precision.

The utility of logarithmic measure is manifold:

a It can contract large, cumbersome regions of a scale into tractable ranges (see, for example, Sec. 3.14);

b It can make certain circuit symmetries more evident (Sec. 3.14);

c It can make certain asymptotic behaviors more evident; this is discussed in detail in Sec. 7.19; see also Secs. 3.14, 6.15, 6.18;

d It provides computational convenience in calculations with cascaded networks whose transfer functions multiply, for their logarithmic measures simply add (Secs. 5.07, 7.16, 7.18; Fig. 6.25-L); this is extremely convenient in discussing the overall behavior of systems composed of a number of parts connected in tandem.

Logarithmic measure has still other advantages, but these suffice for here. We add, for precision, certain standard definitions of logarithmic-measure "units."

The *decibel* is used to measure power ratios according to the definition

$$10^{x/10} = \frac{P_1}{P_2},$$

in which P_1 and P_2 are the two amounts of power to be compared and x is the decibel measure of their ratio:

$$x = \log_{10} \frac{P_1}{P_2} \qquad \text{(dB)}.$$

(The *bel* is inconveniently large and seldom used; the *deci*bel is more convenient.) It is obviously simpler, for example, when comparing 100 W and 1 mW to use 5 dB instead of the actual ratio, 100,000. Note that x is positive when $P_1 > P_2$ and negative when $P_1 < P_2$. Some convenient numerical values are given in Table D-1.

Note that when the two amounts of power being compared are developed in the *same* load impedance, we can also compare the voltages and the currents

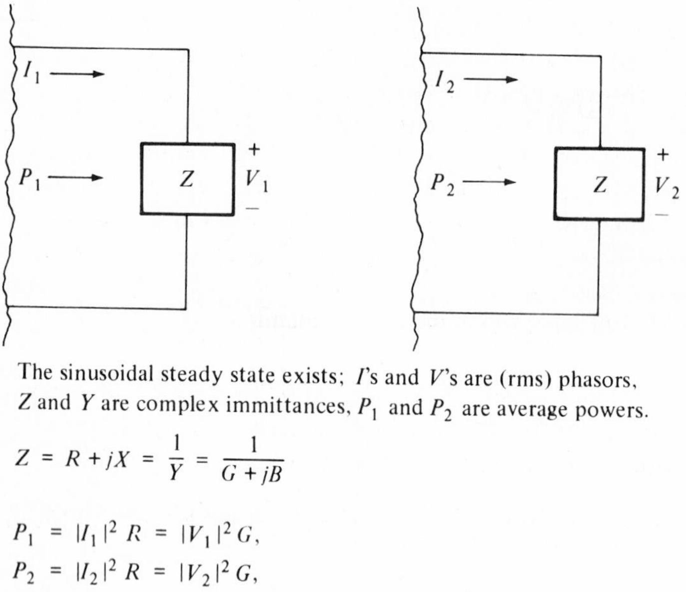

The sinusoidal steady state exists; I's and V's are (rms) phasors, Z and Y are complex immittances, P_1 and P_2 are average powers.

$$Z = R + jX = \frac{1}{Y} = \frac{1}{G + jB}$$

$$P_1 = |I_1|^2 R = |V_1|^2 G,$$
$$P_2 = |I_2|^2 R = |V_2|^2 G,$$

$$\frac{P_1}{P_2} = \left|\frac{I_1}{I_2}\right|^2 = \left|\frac{V_1}{V_2}\right|^2 = 10^{x/10}$$

$$x = 10 \log_{10} (P_1/P_2) = 20 \log_{10} \left|\frac{I_1}{I_2}\right| = 20 \log_{10} \left|\frac{V_1}{V_2}\right| \text{ dB}.$$

FIG. D-A

TABLE D-1.

Power Ratio	Decibel Measure	Power Ratio	Decibel Measure	Power Ratio	Decibel Measure
1	0	1	0	1.00	0
2	+3.0	$\frac{1}{2}$	−3.0	1.26	1
3	4.8	$\frac{1}{3}$	−4.8	1.59	2
4	6.0	$\frac{1}{4}$	−6.0	2.00	3
8	9.0	$\frac{1}{8}$	−9.0	2.51	4
10	10.0	$\frac{1}{10}$	−10.0	3.16	5
20	13.0			3.98	6
30	14.8			5.01	7
40	16.0			6.31	8
80	19.0			7.94	9
100	20.0			10.00	10
1000	30.0				
10,000	40.0				

in decibels (Fig. D-A). In general, when P_1 and P_2 refer to *different* impedances, the measures of current and voltage ratios in Fig. D-A are not correct; an additional term, such as $10 \log (R_1/R_2)$, appears in the expression for x. The magnitude of a transfer function is often not a power ratio or even the square root of a power ratio. It is nevertheless very convenient to measure $|T(j\omega)|$ in "decibels." For exactness, we define a "voltage decibel" to be used when the ratio under discussion is a voltage ratio, a current ratio, the magnitude of a transfer function, or the like:

$$10^{x/20} = \left|\frac{T(j\omega)}{T_{\text{ref}}}\right|, \qquad x = 20 \log \left|\frac{T(j\omega)}{T_{\text{ref}}}\right| \qquad \text{(dBV)},$$

in which T_{ref} is some convenient reference number (perhaps unity in value) added to remove the dimensions of T so that the logarithm can properly be taken. The 20 is used instead of 10 to make the agreement of the dBV and the dB as close as possible, since $|T|^2$ is often proportional to power. By further extensions, one can similarly measure almost anything in "decibels," even dollar amounts, or the GNP (!).

Any table of common logarithms is a table of decibel measure.

We mention here, for completeness, these other logarithmic units that need no further discussion (see Secs. 3.04, 3.06, 3.07): the *neper*, the *radian*, the *degree*.

There is no truly logarithmic unit for the measure of *frequency*, but the terms *octave* and *decade* are in common use [see (7.19-21) and (7.19-22)].

For exact definitions, see IEEE Standard 100–1972 ["IEEE Standard Dictionary of Electrical and Electronic Terms" (New York): Wiley, 1972)], where these and various other logarithmic units (dBm, dBa, etc.) are defined.

INDEX

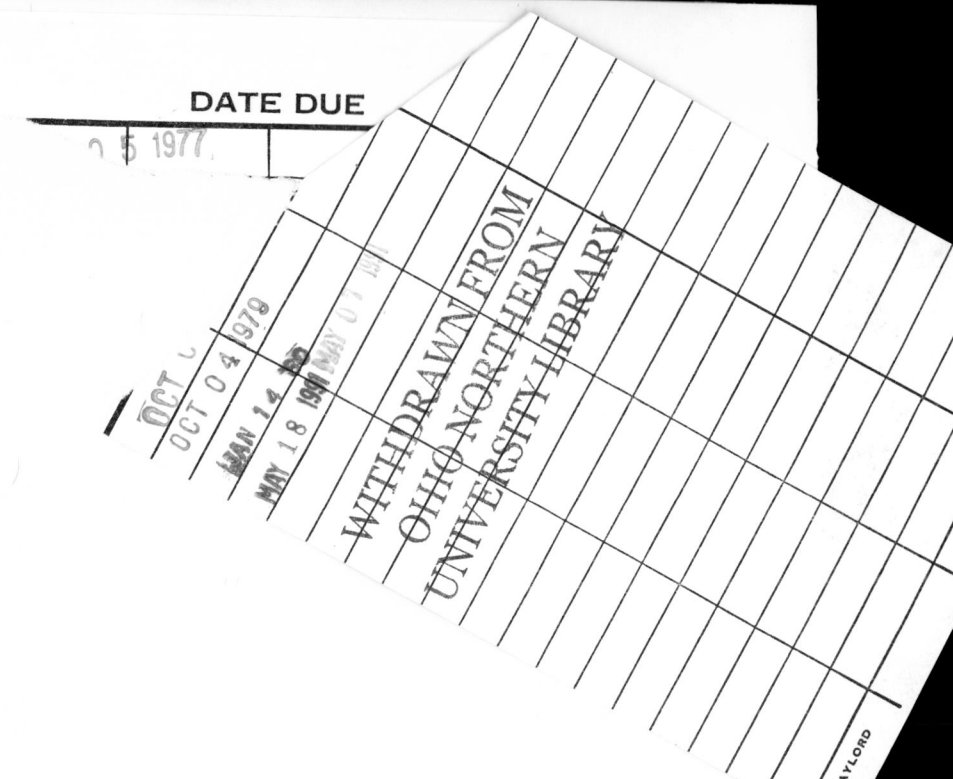